THE ENCYCLOPEDIA OF CIVIL AIRCRAFT

THE ENCYCLOPEDIA OF CIVIL AIRCRAFT

General Editor
DAVID DONALD

Thunder Bay
P·R·E·S·S

This edition published in the United States by
Thunder Bay Press
5880 Oberlin Drive, Suite 400
San Diego, CA 92121-4794
http://www.advmkt.com

Copyright © 1999 Orbis Publishing Ltd.
Copyright © 1999 Aerospace Publishing

Copyright under International, Pan American, and Universal Copyright Conventions. All rights reserved. No part of this publication may be reproduced, stored in a retrieval system, or transmitted, in any form, by any means, electronic, mechanical, photocopying, recording or otherwise, without prior written permission of the copyright holder. Brief passages (not to exceed 1,000 words) may be quoted for reviews.

ISBN 1-57145-183-8

Library of Congress Cataloging-in-Publication Data available upon request

Conceived and produced by Amber Books Ltd.
Bradley's Close, 74-77 White Lion Street,
London N1 9PF

Editorial consultant: Chris Chant

Printed in Italy

1 2 3 4 5 99 00 01 02 03

This revised and updated edition is based on material previously published by Orbis Publishing Ltd. in 1993 as part of the reference set *Airplane*.

CONTENTS

Introduction 7	**Chapter 5** 88	**Chapter 9** 182
Chapter 1 10	ATR 42 & 72 89	Boeing 747 183
Aérospatiale Caravalle 11	Avro/Hawker Siddeley	Boeing 747-400 190
Ascendant Airbus A300 18	HS 748 96	**Chapter 10** 200
Airbus A310 26	**Chapter 6** 104	Boeing 757 201
Chapter 2 34	Ayres Thrush 105	Boeing 767 209
Airbus A320 35	BAC One Eleven 111	Boeing 777 216
Airbus A330/340 42	Beech King Airs 118	Boeing C-97 224
Antonov An-2 Colt 47	**Chapter 7** 124	**Chapter 11** 232
Chapter 3 52	Beech Model 18 125	Bristol Britannia 233
Antonov An-12 Cub 53	Boeing 247 131	Bristol Freighter 239
Antonov An-24 61	Boeing 307 Stratoliner 135	**Chapter 12** 246
Chapter 4 66	Boeing 314 Clipper 140	British Aerospace Ae125 247
Antonov An-124 &	**Chapter 8** 147	British Aerospace Ae146 252
An-225 67	Boeing 707 148	**Chapter 13** 260
Antonov Twin	Boeing 720 155	Business Jets 261
AN-24/26/30/32 73	Boeing 727 159	Canadair CL-44 265
Armstrong Whitworth	Boeing 737 163	Canadair CL-215 271
Argosy 81	Boeing 737	Canadair CL-600
	300/400/500 series 168	Challenger 277
	Boeing 737	
	600/700/800 series 176	

CONTENTS

Chapter 14 — 282
- Cessna Citation — 283
- Cessna Family — 288
- Concorde — 296

Chapter 15 — 304
- Conroy/Aérospatiale Guppy — 305
- Convair Twins CV 240/640 series — 309
- Convair 880/990 — 314

Chapter 16 — 322
- Curtiss Condor II — 323
- Dash 7 — 328
- Dash 8 Canadian Commuter — 334

Chapter 17 — 340
- Dassault Falcons — 341
- De Havilland Canada Beaver — 345
- De Havilland Canada DHC-3 Otter — 352
- De Havilland Canada DHC-6 Otter — 360
- De Havilland Comet — 366

Chapter 18 — 374
- De Havilland Dove & Heron — 375
- De Havilland Dragon — 379
- De Havilland's Moths — 384

Chapter 19 — 390
- Dornier Do X — 391
- Dornier Flying boat — 397
- Douglas DC2 — 402

Chapter 20 — 408
- Douglas DC-4 — 409
- Douglas DC6/7 — 414
- Douglas DC8 — 423
- Embraer Bradeirante — 432
- Fokker F.VII — 441

Chapter 21 — 446
- Fokker F.27 Friendship — 447
- Fokker F28 Fellowship — 453
- Ford Tri-Motor — 458
- Gulfstream — 466

Chapter 22 — 472
- Hamilton Metalplane — 473
- Handley Page Herald — 479
- Handley Page HP-42 — 485
- Hawker Siddeley Trident — 489
- Heinkel He 70 — 493

Chapter 23 — 500
- Islanders and Trislanders — 501
- Ilyushin 11-12 & 11-14 — 507
- Ilyushin Il-18 — 510

Chapter 24 — 516
- Ilyushin Il-62 — 517
- Ilyushin Il-76 — 522
- Ilyushin Il-86 & 96 — 530

Chapter 25 — 536
- The Jet Commander, Westwind, Astra Story — 537
- Jetstream — 541

Chapter 26 — 546
- Junkers F13 — 547
- Junkers G38 — 553
- Learjet — 559

Chapter 27 — 564
- Let L-410/610 — 565
- Lockheed Constellation — 570
- Lockheed L-188 Electra — 577
- Lockheed 'Singles': the Plywood Bullets — 582
- Lockheed Tristar — 590
- Lockheed Twins — 596

Chapter 28 — 604
- Martin 2-0-2 & 4-0-4 — 605
- McDonnell Douglas DC9 & MD-80 — 609
- McDonnell Douglas DC10 — 614

Chapter 29 — 622
- McDonnell Douglas MD11 — 623
- McDonnell Douglas MD-80 & MD-90 — 628
- Metro II — 635

Chapter 30 — 642
- The Microlight Revolution — 643
- Mils Civilian Helos — 648
- Mitsubishi MU-2 — 655
- NAMC YS11 — 660

Chapter 31 — 668
- Piper Cherokee — 669
- Piper Navajo & Cheyenne — 674
- Pitts Special — 681
- RFB Fantrainer — 685

Chapter 32 — 692
- Reno-Formula One — 693
- Saab 340 & 200 — 698
- Sabreliner — 704
- Shorts 330 — 709
- Shorts 360 — 715
- Shorts Civil Flying Boats — 720

Chapter 33 — 728
- Tupolev Tu-104 — 729
- Tupolev Tu-114 'Rossiya' — 733
- Tupolev Tu-134 — 741
- Tupolev Tu-144 'Charger' — 745

Chapter 34 — 752
- Tupolev Tu-154 — 753
- Vickers Vanguard — 758
- Vickers Viking — 765
- Vickers Viscount — 770
- Vickers VC10 — 776

Chronology — 782

Specifications — 789

Index — 804

INTRODUCTION

Above: The development of large carrier aircraft has opened up the four corners of the world, more than the early pioneers of flight could ever have imagined.

Although it can be claimed, truthfully, that the first air passengers were carried as early as 1783, this was by balloon – never a very effective means of transport, as it is forever at the mercy of the winds. Propulsion and a rudder had to be added to convert an elongated balloon into the airship, which only started to become practical in the latter part of the 19th century. Graf von Zeppelin, a former general in the German Army and a brilliant engineer, built the first useful airships in 1900. By 1908, his large but flimsy airships were flying for 12 hours, covering 100 miles (160 km). At the end of 1909, Delag had been established to manufacture and operate airships of increasing size and capability. Several airship passenger services were located in Central Europe by 1912. Operations resumed after World War I, but were still fraught with problems and dangers. They ended in 1937 after the great Hindenburg caught fire and was destroyed, with considerable loss of life, as it tried to land in the USA after a transatlantic flight.

Passenger services

Riding on the back of the Wright Brothers' success at Kitty Hawk, the first regular passenger service, using a small Benoist flying boat, was operating across Tampa Bay, Florida in the summer of 1914. Unfortunately, the service was suspended during World War I. Immediately after the end of the war, several countries began using converted bombers as mail carriers and staff transports. During this period aircraft manufacturers began converting warplanes for civil transport. The four most important of these planes were originally bombers: three British types – the de Havilland D.H.4, the Handley Page O/400 and the Vickers Vimy; and one French type, the Breguet Bre.14. The advantages of these aircraft – apart from their availability – were that they could carry an adequate payload and were moderately reliable. These conversions effectively made possible the first true air transport services of 1919–20.

These and other aircraft, however, were not in themselves enough to form the foundations of a fully-fledged air transport industry. They were unreliable, often uncomfortable and cold, and extremely expensive. The situation began to improve when manufacturers started to build the first purpose-built airliners. Here, sales were hindered by prohibitive purchase costs. A D.H.16, for example (with cabin accommodation for four passengers), could cost about four times as much as a war-surplus D.H.4A, which could be converted to carry two passengers. Similarly, the early conversions of Handley Page bombers could carry six passengers, while the W/8 and W/8B, designed as airliners, carried 12 passengers but cost more than twice the price of a surplus bomber. One of the few true airliners of the period to be built and sold in useful quantities over an extended time was a French machine, the Farman Goliath, which carried 12 passengers and was moderately cheap to buy and operate as it had two low-powered engines.

Developments after 1914

Members of the Alliance who had emerged victorious at the end of World War I singularly failed to create sure foundations for their fledgling air transport industries. Germany was altogether more successful, largely through the efforts of Dr Hugo Junkers, a pioneering exponent of the low-wing cantilever monoplane layout and all-metal construction. The first successful product of this design concept was the F.13 four-passenger transport that made its maiden flight in June 1919. It was subsequently manufactured in what were, for the time, prodigious numbers. The F.13 was slow, but it was reliable under many operating conditions and in many climates, and served for more than 30 years alongside larger and more powerful variants, produced by Junkers in a steady stream.

Another country to secure an early foothold in the air transport business was The Netherlands, largely through the output of the Fokker company. Anthony Fokker escaped from Germany in 1919 with six train loads of of aircraft materials and equipment, and quickly re-established himself in The Netherlands. He produced the F.II high-wing transport, which was capable of carrying four passengers and which had a fabric-covered, steel-tube fuselage and a plywood-covered, cantilever wing of wooden construction. Its basic concept was retained in a series of steadily improving aircraft that included the five-passenger F.III of 1921 and the six-passenger F.VII of 1923. In 1925, Fokker began manufacturing the eight-passenger F.VII/3m, a development of the F.VII, with three-engined powerplant. More than 150 were built.

In the USA, early air transport was concerned primarily with the delivery of mail, by single-engined biplanes. In the latter part of the 1920s, with better aircraft now being built and changes in American air transport legislation in place, the situation changed. In September 1928, the Ford Tri-Motor, effectively an F.VII/3m recast with an all-metal airframe and capable of carrying up to 14 passengers, triumphantly entered into service. Other technical leaps forward were the 18-passenger Boeing 80A twin-engined biplane and the Curtiss Condor, whose later developments included a supercharged powerplant and retractable landing gear.

By this time, the UK was no longer at the cutting edge of the aeronautical industry, largely as a result of the dictates of Imperial Airways. However, aircraft of note that were built in small numbers during this period, with three or four radial engines, included the D.H.66, the Armstrong Whitworth Argosy, the Handley Page H.P.42 landplanes and the Short Calcutta and Kent flying boats. Though these British aircraft were the largest and most comfortably appointed airliners of their day, they were all slow.

Materials advancement

The first product of a new and developing technology was stressed-skin or semi-monocoque construction. This technique involved the use of light alloys such as Dural, which allowed the skinning to bear a substantial part of the airframe's loads, which in turn made possible the design of a less-substantial primary structure under the skinning. Lockheed, a small American company whose early wood-based creations included the Vega of 1927 and the Orion (the latter, incidentally, switching to the low-set cantilever wing and retractable main landing gear units of 1931), was a pioneer of monocoque airliners.

Larger American companies, later followed by Lockheed, went straight to stressed-skin, light-alloy construction. The most important of these larger companies was Boeing, whose Monomail of 1930 could accommodate up to eight passengers as well as carry a substantial airmail load. The Monomail was an aerodynamically 'clean' cantilever monoplane of light-alloy construction with retractable, main-landing gear units. In 1933, Boeing capitalised on the success of the Monomail by introducing the first 'modern airliner' in the shape of the 247. The 247 was powered by two nicely cowled radial engines driving variable-pitch propellers. It could also accommodate 10 passengers and a stewardess, and had retractable main landing gear units, and rubber de-icing boots along the leading edges of the wings and tail unit.

Sadly, the 247 was quickly to prove too small to cope with increased passenger demands, and as a result was rapidly overtaken by the larger and more sophisticated airliners of its rivals. American domination of the airliner market continued, with little opposition, until recently. The largest and fastest of its rivals was Douglas's 'modern airliner', the DC-1, with its clean lines (especially in the fuselage and engine cowlings), a wing fitted with trailing-edge flaps for better take-off and landing performance, and a larger cabin that benefited from not having the wing spars running through it. The 1933 DC-1 was a prototype and paved the way for the DC-2 (with 14-seat accommodation) and then the classic DC-3, which boasted a number of refinements, including more powerful engines and a wider fuselage providing 21-seat accommodation. The 220 DC-2s that were built were followed by 430 DC-3s, which were still selling very well at the outbreak of World War II. They were then developed for military use as the C-47, R4D and Dakota series, of which more than 10,000 were delivered. These craft were to prove vital to the Allied war effort, and the flooding of the civil market with surplus aircraft after the end of the war in 1945 was largely responsible for the creation of the modern air transport industry as a low-cost, worldwide phenomenon.

The only two serious rivals to Douglas in the late 1930s were Lockheed in the USA and Junkers in Germany. Lockheed concentrated on smaller, twin-engined aircraft that were, nevertheless, considerably faster than the DC-3. These included the eight-passenger Model 10 Electra of 1934, which was followed, in quick succession, by the consistently improved Models 12, 14 and 18, of which more than 1,000 were delivered. The Model 14 and Model 18 were notable for the introduction of the Fowler flap, which increased the area of the wing when extended backward before being lowered to reduce landing speed. Another type of landing aid, by Junkers, was the double-wing flap, which was a separate unit hinged to the trailing edge below and behind the wing proper. This was a key feature of the Ju 52, the most significant European transport type of the 1930s. The original Ju 52 single-engined type was developed into the definitive Ju 52/3m three-engined transport; more than 4,800 were completed, most of them for military use.

Trans-oceanic service

The late 1930s were also notable for the emergence of a number of four-engined airliners. The most important of these, in the short term, were the Martin M-130, Short 'C' class and Sikorsky S-42 long-range flying boats, which pioneered trans-oceanic services. Of greater long-term importance was the Douglas DC-4E landplane, which made its first appearance in 1938. Although a failure, it paved the way for the successful Boeing 307 Stratoliner, the Douglas DC-4 and the Lockheed Constellation transports, which were originally built in modest numbers for military service and then appeared in civil service from 1945 onwards. They in turn were largely responsible for the emergence of the long-range air transport market, which was dominated by the DC-4, DC-6 and DC-7 from Douglas, and the Constellation, Super Constellation and Starliner from Lockheed. These airliners, with their high-powered and reliable radial engines, could carry up to 100 passengers and, with the advent of cabin pressurisation, could cruise at high altitude at greater speeds and less expense.

Pressurisation was also a feature of the Convair CV-240, the best high-speed transport developed for short-range services. The CV-240 was then developed into the more capable CV-340 and CV-440, which also had two powerful radial engines. A number of these aircraft, under the designations CV-580, CV-600 and CV-640, were then converted to turboprop power. These were comparable to the piston-engined types Martin 2-0-2 and 4-0-4, which were produced in smaller numbers.

Below: The Lockheed L12 Electra Junior had accommodation for six passengers and a crew of two. Its performance was enhanced by having two Pratt and Whitney Wasp radial piston engines.

After the war, the Brabazon Committee in the UK published its report on the the future of the airline industry. Two stipulations in the report were to have a significant impact on the development of civilian air transport. The Brabazon IIB requirement called for a short-range type with turboprop power. The result was the Vickers Armstrong Viscount of 1948 with four Rolls-Royce Dart engines. The Viscount was the world's first commercial aircraft with turbine power, and ushered in the era of genuinely quiet and smooth passenger flight. The technical success of the Viscount was reflected in a number of variants that maintained high performance with more powerful versions of the Dart engine, and raised the payload from 32 to 75 passengers. Another requirement relating to shorter-range routes resulted in the de Havilland D.H.106 Comet airliner – the world's first commercial aircraft with turbojet power in the form of four de Havilland Ghosts. The Comet was fast and could carry up to 36 passengers, in exceptional comfort, over a range of up to 2,000 miles (3,220 km) at altitudes well above even the most adverse weather conditions. It entered service in 1952 but was then beset by fatigue problems that resulted in the loss of two aircraft. It was 1958 before a new generation of safer, larger and more capable Comet airliners began to appear.

The four-year gap in the Comet's career was to prove disastrous, for in the interim the Americans, having realised the importance of turbine power, pressed ahead with the development and marketing of a second generation of advanced aerodynamic airliners with better engines. These were the Boeing 707, Convair CV-880 and Douglas DC-8, all of them swept-wing aircraft with four turbojet engines that were gradually replaced by more capable turbofan units offering greater power in concert with reduced noise and lower fuel consumption.

The Bristol Britannia, intended for medium-range services on British Empire routes, was the result of a misguided requirement of 1947. The Britannia was an efficient aeroplane but its development was protracted and, as a result, it had already been rendered technically obsolete by the new American long-range transports by the time it entered service. The 707, 720 and DC-8 underwent extensive development in the course of their lives so that they could carry greater loads over longer distances.

France, at this time, developed the dual-engine, turbine-powered Sud-Est Caravelle, a short-range type that first flew in 1952. Uniquely, its engines were installed on the sides of the rear fuselage in podded nacelles specifically positioned to leave the wing uncluttered. The Caravelle was moderately successful, and a few are still in service. Unfortunately, the same cannot be said of two rival designs with propulsion by four turboprops: the Lockheed Electra and Vickers Armstrong Vanguard, from the USA and UK respectively. Although a few Electras are still extant as freighters, the Vanguard has disappeared completely – even as a cargo-carrier.

The USSR also moved into the field of turbine propulsion for civil aircraft, and soon produced a number of interesting types, some of which performed excellently. Some were derived from the nation's pioneering turbine-powered bombers, but none could match the ever-lower operating costs of their Western counterparts, and therefore achieved little commercial success, except in the closed market of the USSR and its immediate allies.

Commercial competition

The advent of turbojet and turbofan propulsion was marked by the increasing stratification of airliners into the short-, medium- and long-range types. Long-range types were typified by the American transports mentioned above and one British offering, the Vickers VC10, which was only produced in small numbers. By far the most successful of the medium-range types was the Boeing 727 with three engines grouped on the sides of, and in the rear of, the fuselage. This appeared after a conceptually similar British type, the Hawker Siddeley Trident. In the short-range market, the Europeans were dominant early on, with the twin-turboprop Fokker F.27 Friendship and Hawker Siddeley 748, and the twin-turbofan BAC One-Eleven and Fokker F.28 Fellowship.

Once again, these were easily outclassed by the American Boeing 737 and Douglas DC-9, each with a pair of turbofan engines under its wing. Douglas delivered 976 DC-9s before the type was further developed, following the company's merger with McDonnell, as the MD-80 and MD-90 series. Following McDonnell Douglas's later merger with Boeing, a further variant, the Boeing 717, was developed and is still on the market. Most successful of all, however, is the Boeing 737, which is still in production.

In the 1950s and 1960s, considerable thought was also given to the development of supersonic transports for long-range services. This bore little fruit save for the Anglo-French type, the Aérospatiale/British Aerospace Concorde, and one Soviet type, the Tupolev Tu-144. Only 13 Concordes were ever delivered but all are still in service. The Tu-144, on the other hand, suffered major problems and was withdrawn before entering full service.

Wide-body aircraft

The next step up the technical ladder for commercial aircraft was the introduction of much increased seating capacity. This saw the development of larger 'wide-body' aircraft that were only made possible by the concurrent development of a new generation of very powerful turbofan engines. The first of these was the Boeing 747, which entered service in 1970 with a fuselage 20 ft (6.1 m) wide, and which could accommodate up to 500 passengers or the equivalent weight in freight. There have since been several variants, with substantial payloads, of the all-passenger, all-freight, passenger and freight, and alternative passenger and freight types of the 747 for different ranges.

Two smaller-capacity, 'wide-body' aircraft intended to fill market niches not occupied by the 747 were the Lockheed TriStar and McDonnell Douglas DC-10. Each boasted a three-turbofan powerplant, one installed at the tail and the other two under the wing.

Europe was becoming more and more concerned by the domination of the civil air transport market by the American 'big three' manufacturers, with the lion's share going to Boeing. The result was the creation of the Airbus consortium, whose first product was the A300B, a twin-turbofan type for the short-range market. Airbus initially found it hard to penetrate the market but started to make considerable inroads as it moved towards a full product line of 'wide-body' aircraft covering the full spectrum of capacities and ranges. All are powered by two turbofan engines except for the four-turbine A340.

In recent years, the main priorities for airliner manufacturers have been greater economy of operation, higher levels of reliability, reduced manning requirements and greater flexibility of service. Achieving results in all these areas has been made possible – at least in part – by the adoption of improved types of turbofan engine, the use of more advanced materials in the airframe, and the introduction of sophisticated computer-controlled avionics to carry out tasks previously performed by additional personnel on the flight deck.

Below: The Hawker Siddeley HS748 was a short/medium range feeder airline. It was valued for its simplicity and reliability by less-developed nations, who were the main customers.

Chapter 1

Aérospatiale Caravelle

Ascendant Airbus A300

Airbus A310

Aérospatiale Caravelle

When war-ravaged Europe returned to peace in 1945, the aviation manufacturing companies turned to commercial aircraft. In France the major type to appear was the attractive Caravelle, an airliner that provided work for over 20 years.

Once World War II was over, the French aviation industry revived itself and came forward with a profusion of new projects which covered every area of military, air carrier and light aviation. The French government was very supportive and sponsored many prototypes which, in fact, never reached production. However, this support did stimulate a great surge of research and development, including a number of projects for turbine-engined airliners. Designs started to appear at the start of the 1950s from Hurel Dubois, Dassault, SNCASE and Breguet, based upon a tentative specification drawn up by the Sécretariat Générale de l'Aviation Commerciale et Civile (SGACC), the government body responsible for civil aviation. It was felt that there would be a need (and a good market) for a medium-sized airliner with 55/65-passenger capacity and a 2000-km (1,243-mile) range with a 7-tonne payload. The specification demanded that the aircraft be turbine powered since it would be expected to achieve sector block speeds of over 600 km/h (373 mph).

Some of the submissions to the SGACC in March 1952 were quite unrealistic. The Hurel Dubois proposal was for a jet version of the high-aspect-ratio wing design which later became the HD-31. It was to have been powered by a pair of SNECMA Atar jet engines mounted on the struts which supported the enormous wing, but the drag from the HD.45 structure might have made it difficult for this aircraft to meet the SGACC performance requirements. Both Dassault and Breguet proposed turboprop designs and SNCASE (Société Nationale de Construction Aéronautique du Sud-Est) came out with its X-200 family, and it was a member of this group of SNCASE designs which was selected as the most promising for further development. The X-210 was a streamlined aircraft with three SNECMA Atar engines arranged in a group in the rear fuselage. At this time, the Atar was able to develop some 2722 kg (6,000 lb) of thrust and was at a fairly early stage in its development history. It became evident fairly shortly after selection of the X-210 that SNCASE would be on safer ground if they used the 4082-kg (9,000-lb) thrust Rolls-Royce Avon, which was at a more mature stage, and this led to the SGACC approving a revision to the X-210 project. The higher power output of the Avon meant that only two engines would be required, and SNCASE was therefore able to eliminate the centre engine in the definitive SE 210 design.

The design which the SGACC approved in July 1952 led to a production contract from the Sécretariat d'Etat de l'Air, signed on 3 January 1953. The SE 210 Caravelle, as the type was named, had a mildly swept wing and a graceful swept tail with the tailplane fitted one-third of the way up the fin. Of course, the main recognition feature was the pair of Avon engines in pods mounted on the rear fuselage behind the wing, and in this respect SNCASE set a fashion which was to be followed by numerous other civil aircraft types. The podded engines had the great advantage of ease of access for maintenance and isolation from the rest of the airframe from noise or fire. The Caravelle fuselage was well streamlined, and the nose section

The two prototype Caravelles, F-WHHH and F-WHHI, are seen here on a test flight in 1956. The second aircraft was assigned to the vital task of sales demonstrations once the certification test period was over, and in the spring of 1957 it went on a marathon tour of South America, followed by the United States and Canada.

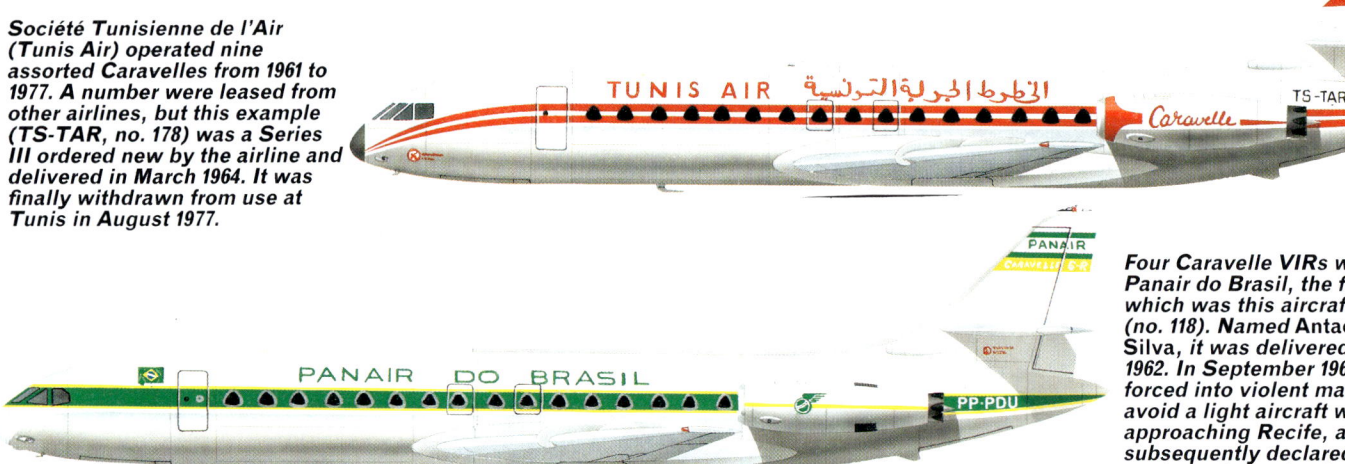

Société Tunisienne de l'Air (Tunis Air) operated nine assorted Caravelles from 1961 to 1977. A number were leased from other airlines, but this example (TS-TAR, no. 178) was a Series III ordered new by the airline and delivered in March 1964. It was finally withdrawn from use at Tunis in August 1977.

Four Caravelle VIRs were sold to Panair do Brasil, the first of which was this aircraft, PP-PDU (no. 118). Named Antao Leme da Silva, it was delivered in July 1962. In September 1963 it was forced into violent manoeuvres to avoid a light aircraft while approaching Recife, and was subsequently declared unairworthy and written off.

was identical to the design used by de Havilland for the Comet 1 airliner. In fact, SNCASE actually speeded up the completion of the first two prototypes by purchasing standard Comet nose and cabin sections from de Havilland and grafting these on to the Toulouse-built main fuselages. Caravelles which went into service with customers had a very similar cockpit layout to that of the British aircraft.

The company built four development airframes. Two were used for flight testing, one was installed in the CEAT water tank for pressure fatigue testing and the remaining example was subjected to normal airframe static stress tests. The first flying prototype (F-WHHH c/n 01) was rolled out at the Toulouse factory in mid-April 1955 and made its first flight on 27 May 1955 in the hands of Pierre Nadot, the SNCASE senior test pilot. In deference to the national airline (and the expected launch customer) F-WHHH was painted with Air France titles and the famous sea-horse logo of the airline. In fact, the formal order for 12 Caravelles was placed by Air France on 3 February 1956, and this provided the manufacturer with a startup quantity for the production line and the promise of a further order of 12 aircraft at a later stage.

While SNCASE started to make preparations for full-scale output of the Caravelle, the business of certification was well in hand. Over 400 hours had been flown by the first prototype by the time the basic certificate of airworthiness was granted on 2 April 1956. This was only the first hurdle, however, because the testing had to take the Caravelle forward to its full transportation category certificate, and the all-important United States Approved Type Certificate. The first of these was finally awarded on 2 April 1958 with the American approval following on 8 April. In carrying out the flight testing necessary for these approvals, SNCASE had been fortunate in being able to take advantage of the performance improvements achieved by Rolls-Royce with the Avon, and both F-WHHH and the second prototype (F-WHHI), which joined the test programme on 6 May 1956, were fitted with the 4536-kg (10,000-lb) thrust Avon RA.26 engines.

Worldwide sales tours

With two flying examples now available, SNCASE embarked on a comprehensive series of sales tours with the Caravelle. In late 1956 F-WHHI, fully furnished in the Air France cabin layout, flew to Amsterdam, Rome, Stockholm, Brussels, Lisbon and Helsinki to give demonstrations to European national airlines. With the setting up of the production line at an advanced stage it was clear that more customers for the type were an urgent necessity. F-WHHI was also sent on a long South American tour, followed by demonstrations in the United States which touched Dallas, Kansas City, Denver, Washington, Atlanta, San Francisco and Seattle, to name but a few. This did not yield any immediate orders, but there was encouraging interest from TWA and United Airlines. Fortunately, the European tour started to bear fruit and Scandinavian Airlines System (SAS) placed an order for six Caravelles, with 19 on option, at the end of June 1957.

The first four Caravelles from the production line (two for Air France and two for SAS) were delivered during March and April 1959. SAS had been using F-WHHI (c/n 02) for crew training and was able to catch the headlines with the first revenue service flown from Copenhagen to Beirut on 26 April using the first of its new aircraft. The initial series of Caravelles was designated Series I and they were broadly similar to the prototypes except that they were powered by Avon RA.29/1 Mk 522 engines, had the fuselage lengthened by 1.5 m (4.9 ft) and were fitted with an extended dorsal fin which housed radio antennae. The reaction of airline customers was very positive because the Caravelle was the first short/medium-haul

Caravelle no. 63 was originally intended as the prototype Model VII, but was finally completed as the first Caravelle 10A. Entitled Caravelle Horizon, it featured a modified wing with greater root chord, a raised cabin window line, a fairing at the rear of the tailplane and double slotted flaps. F-WJAO was scrapped in 1969.

Originally delivered as a Series III to Scandinavian Airlines System, this Caravelle (no. 186) was later leased to Panair do Brasil, who operated it until July 1969. It then went to SOGERMA for evaluation of the SNECMA M53 engine and later became the testbed for the CFM56 turbofan, making its first test flight on 17 March 1977.

*Finnair was a major Caravelle user, with an initial fleet of three Caravelle IIIs. In December 1962 Finnair became launch customer for the Series 10B3 and took delivery of 10 of this new model, with first revenue services starting in August 1964. The machine illustrated is the fourth aircraft delivered, **OH-LSD** (no. 187).*

*Named Teva, this Caravelle 11R is no. 264 and was originally delivered to Iberia as **EC-BRY**. It was sold to the Armeé de l'Air in October 1976 fitted with the forward cargo door shown here. It is in service with ETOM82 based at Papeete in French Polynesia, and until recently operated to CEP, France's Pacific nuclear test range.*

pure jet airliner to appear on European airways, and it set new standards of speed, comfort and quietness.

Eventually, Air France received 10 Caravelle Is, SAS received six and two were delivered to Air Algérie. The other early customer for the Caravelle I was the Brazilian airline VARIG, which had placed an order for two in October 1957. The first of these was flown out to Rio de Janeiro in September 1959, and it was placed in service on the Buenos Aires to New York route operating via Montevideo, Sao Paulo, Rio de Janeiro, Belem, Port of Spain and Nassau. This route was later taken over by Boeing 707s which were more suited to the distance involved, and the Caravelles then flew VARIG's domestic services for a number of years. The second VARIG aircraft (PP-VJD) was destroyed by fire after a landing accident at Brasilia in the summer of 1961 and its sister ship was then sold to Sud-Aviation, which leased it to Air Vietnam, Middle East Airlines and, finally, Royal Air Maroc.

The advent of the Avon Mk 522A resulted in the production line changing over to the Caravelle IA, 12 of which were built and delivered to Air France, SAS, Finnair, Air Algérie and Royal Air Maroc. The first significant change to the type came, however, when the Avon RA.29/3 Mk 527 became available. This engine had a power rating of 5171kg (11,400lb), which brought a substantial improvement in the performance of the Caravelle. Sud-Aviation (as SNCASE had become on its merger with Sud-Ouest in March 1957) used part of the Avon engine's designation to identify the new model, and for this reason the aircraft was known as the Caravelle III. The development prototype (F-WJAQ, c/n 19) made its first flight on 30 December 1959, and this sparked off orders from a crop of new airline customers. It also prompted the operators of Caravelle I and IA machines to have them upgraded to Caravelle III standards by the substitution of new engines and enlarged nacelles. A completely new batch of this series was set in hand for Air France, which eventually received a total of 43 Caravelles for service on its European route network until the last was withdrawn in favour of Boeing 727s and Boeing 737s in 1981. Other users of the Caravelle III included Swissair, Alitalia, SAS and Royal Air Maroc, and a total of 79 Caravelles was built from scratch as Series III machines.

American turbofans

Sud-Aviation had always had its sights set on the American market and in July 1960 Caravelle number 42 (a standard Series III aircraft) was delivered to General Electric in the United States. After an extended publicity tour, sponsored by Sud and by Douglas Aircraft with whom a sales and technical support agreement had been signed earlier in that year, the aircraft was re-engined with a pair of General

*The prototype Caravelle VIR was first flown as **F-WJAP** on 6 February 1961 and bore the livery of United Airlines, the North American launch customer. In the event, the aircraft flew extensively on demonstrations in the USA but was not delivered to United. It passed to Cruzeiro in Brazil as **PP-CJC**, and was withdrawn in 1975.*

The tenth Caravelle had a colourful history. It was delivered to VARIG in August 1959 as PP-VJC and was later converted to a Model III. It returned to Sud in May 1964 and was leased to Air Vietnam, who used it until 1968. It then spent short periods with Air Maroc and MEA before entering service with the République Centrafricaine.

Electric CJ805-23C turbofans. This was the intended configuration of the proposed Caravelle VII, which would take advantage of the 7303-kg (16,100-lb) rating of the General Electric engine to meet the performance requirements of TWA in particular. In fact, TWA did place an order for 20 Caravelle VIIs, but this was subsequently cancelled when Douglas decided to withdraw from its arrangement with Sud and launch its own DC-9 twin-jet airliner.

As it turned out, Sud-Aviation did not proceed with the General Electric project because the uprated Avon RA.29/6 Mk 532R became available and the much improved Caravelle VI was introduced. Not only were the engines changed, but Sud took the opportunity to enlarge the cockpit area, which resulted in a slightly bulged roof contour and side windows of greater area. At last the hoped-for order from the USA was placed, in this instance by United Airlines. This was for 20 Caravelle VI aircraft configured for 64 first-class passengers. The first United Caravelle (N1001U c/n 86) was named *Ville de Toulouse*, and was delivered to the airline in June 1961. United's aircraft were actually designated Caravelle VIR because they used the Avon Mk 533 engine which was equipped with reverse thrust, and other examples carried the designation Caravelle VIN to identify the Avon Mk 531 which was fitted with special noise-suppression equipment. The United Caravelle VIRs were flown on routes covering the mid-west and western sectors of the carrier's network, and they were finally withdrawn from service and placed in storage at Denver's Stapleton Airport in 1970. They were gradually sold off, several going into service with Transavia in the Netherlands and with the Copenhagen-based Sterling Airways for inclusive tour work.

The Caravelle VIR and Caravelle VIN brought new customers to Sud-Aviation and, for the first time, real penetration of the South American market was achieved. Panair do Brazil received four and these joined three aircraft in service with Cruzeiro do Sul when Panair's organisation was liquidated in February 1965. Three Caravelle VIRs went to LAN-Chile, and throughout the 1960s the Argentine national carrier, Aerolineas Argentinas, used three of the Caravelle VIN variant on domestic routes and subsequently passed them to the Fuerza Aérea Argentina for military transport operations. A batch of nine Caravelle VINs went into service with Indian Airlines to replace Vickers Viscounts on long-haul domestic services, and at least five of these met with accidents while operating with the airline.

Airframe modifications

The style of designation now changed with the introduction of the Caravelle 10 series. The abortive General Electric project had stimulated Sud-Aviation to think of new powerplants for the Caravelle. The Caravelle 10A Horizon was therefore launched and this model had a stretched airframe, altered wing leading-edge design, higher-set cabin windows and a 'bullet' fairing behind the tailplane. It was powered by two CJ805 engines and Sud's prototype, F-WJAO (c/n 63) made its first flight at Toulouse on 31 August 1962. As it turned out, the main production version was the Caravelle 10B (otherwise known as the 'Super B' or 'Super Caravelle'), which used the Pratt & Whitney JT8D-1 rather than General Electric's engine.

Finnair was the principal operator of the Super B with a fleet of eight, but Sud also sold a number of Caravelle 10B1.R aircraft which carried the Caravelle VIN airframe to the JT8D powerplant. Five of these had a somewhat complex history because they were ordered originally by the Spanish airline Aviaco and then cancelled, with the result that they were eventually sold to various airlines including Sterling, LTU and Iberia. Other examples were delivered to the Jordanian carrier, ALIA, and to the French company JTA, which used them on routes in the Pacific operating from Noumea.

The final variants of the Caravelle were the Caravelle 11R and Caravelle 12. The Caravelle 11R was a passenger/cargo derivative of the Caravelle 10R with a large cargo door and a 1-m (3.28-ft) extension to the fuselage. Only six were built, and these were sold to Air Afrique, Air Congo and Trans Europe. The Caravelle 12 was a stretched model with two fuselage plugs, behind and in front of the wing, and it was produced to the requirements of Sterling Airways which had a need for a dozen examples. The Pratt & Whitney JT8D-9 engine rated at 6804 kg (15,000 lb) was used in the Caravelle 12, and a number of detailed changes were also made to the landing gear and flaps. In fact, Sterling took delivery of only seven aircraft and the remaining Caravelle 12s were leased to Air Inter. The last of these aircraft was aircraft number 280, and its first flight on 8 March 1973 brought the Caravelle production line to a close. Remarkably, there was no successor.

Today, the Caravelle still gives faithful service to a few airlines although it has been retired from the fleets of the main national carriers. It will be a sad day when the last Caravelle is finally grounded, because this first successful European short-haul jet airliner has earned for itself a very special place in history.

Aérospatiale Super Caravelle cutaway drawing key

1 Radome
2 Weather radar scanner
3 Front pressure bulkhead
4 Windscreen panels
5 Instrument panel shroud
6 Windscreen wipers
7 Rudder pedals
8 Landing/taxiing lamp
9 Cockpit floor level
10 Control column
11 Pilot's seat
12 Co-pilot's seat
13 Overhead systems switch panel
14 Flight engineer's position
15 Electrical system fuse panels
16 Crew baggage locker
17 Cockpit section joint frame
18 Radio and electrical equipment racks, port and starboard
19 Cockpit doorway
20 Galley
21 Service door/emergency exit
22 Cabin attendants' folding seats
23 Wardrobe
24 Nose undercarriage mounting frame
25 Oxygen cylinder
26 Nose undercarriage leg strut
27 Twin nosewheels
28 Nosewheel leg door
29 Hydraulic retraction jack
30 Entry lobby
31 Forward 'up-and-over' door
32 Door surround structure
33 Cabin bulkhead
34 Fuselage skin panelling
35 Frame and stringer fuselage construction
36 First-class four-abreast seating
37 Overhead luggage racks
38 Curtained window panel
39 Cabin floor panels
40 Seat mounting rails
41 Forward underfloor cargo hold
42 Ventral cargo hold door
43 Removable cabin partition
44 Floor beam construction
45 Wing root fillet
46 Wing centre section carry through
47 Skin panel centreline joint
48 Optional centre section fuel tank, capacity 660 Imp gal (3000 litres)
49 Front spar/fuselage attachment main frame
50 Starboard emergency exit windows
51 VHF aerial
52 Starboard wing inboard integral fuel tank
53 Inboard wing fence
54 Fuel pumps
55 Outboard wing fence
56 Outer wing fuel tank, total system capacity 4,180 Imp gal (19000 litres)
57 Starboard navigation light
58 Wing tip fairing
59 Starboard aileron
60 Aileron balance weights
61 Aileron hydraulic jack
62 Starboard double slotted flap, down position
63 Flap screw jacks
64 Starboard spoilers, open
65 Airbrake, upper and lower surfaces, open
66 Spoiler hydraulic jack
67 Flap drive and interconnection shafts

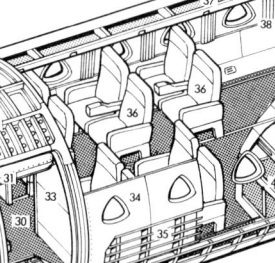

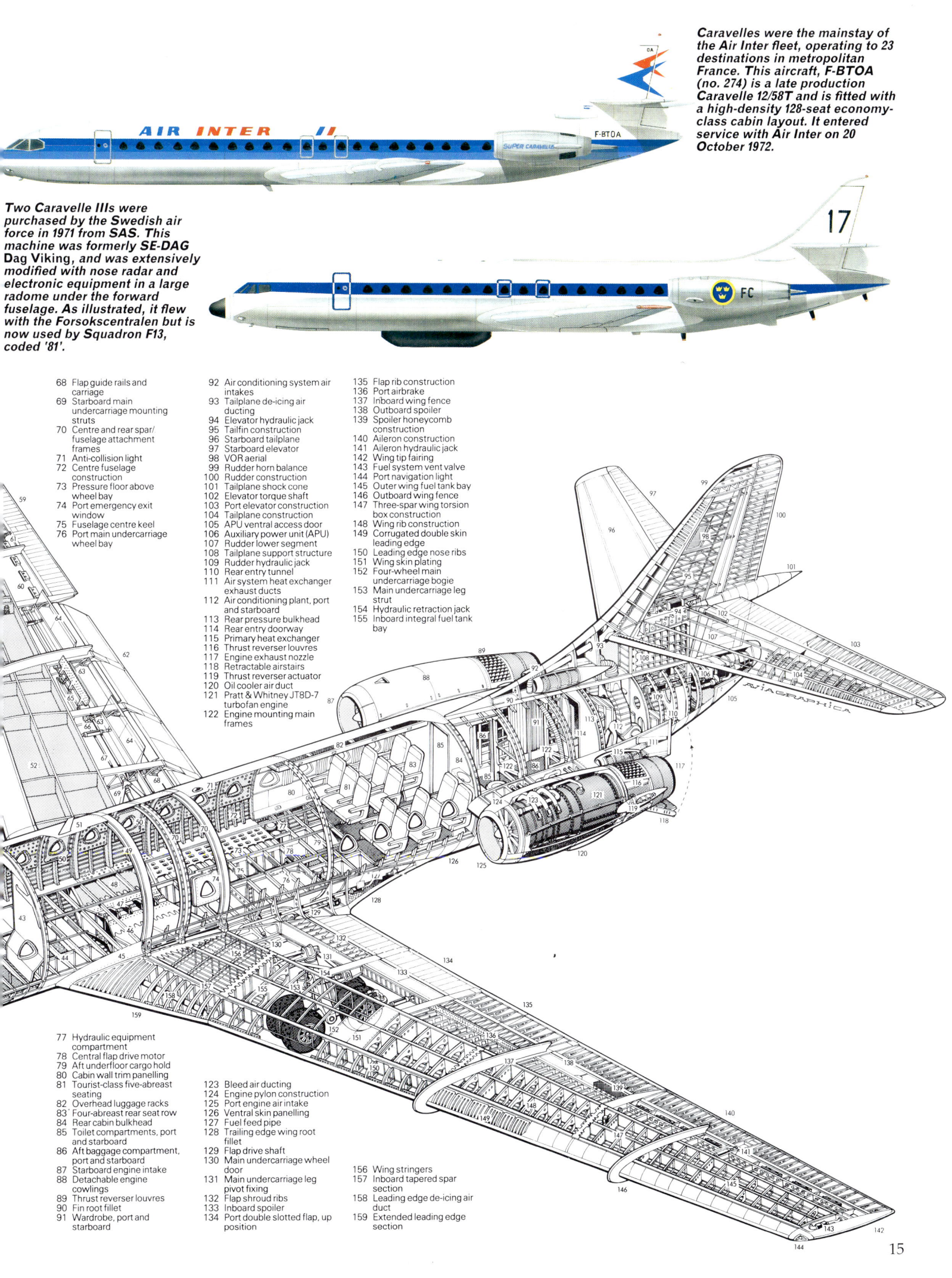

Caravelles were the mainstay of the Air Inter fleet, operating to 23 destinations in metropolitan France. This aircraft, F-BTOA (no. 274) is a late production Caravelle 12/58T and is fitted with a high-density 128-seat economy-class cabin layout. It entered service with Air Inter on 20 October 1972.

Two Caravelle IIIs were purchased by the Swedish air force in 1971 from SAS. This machine was formerly SE-DAG Dag Viking, and was extensively modified with nose radar and electronic equipment in a large radome under the forward fuselage. As illustrated, it flew with the Forsokscentralen but is now used by Squadron F13, coded '81'.

68 Flap guide rails and carriage
69 Starboard main undercarriage mounting struts
70 Centre and rear spar/fuselage attachment frames
71 Anti-collision light
72 Centre fuselage construction
73 Pressure floor above wheel bay
74 Port emergency exit window
75 Fuselage centre keel
76 Port main undercarriage wheel bay
77 Hydraulic equipment compartment
78 Central flap drive motor
79 Aft underfloor cargo hold
80 Cabin wall trim panelling
81 Tourist-class five-abreast seating
82 Overhead luggage racks
83 Four-abreast rear seat row
84 Rear cabin bulkhead
85 Toilet compartments, port and starboard
86 Aft baggage compartment, port and starboard
87 Starboard engine intake
88 Detachable engine cowlings
89 Thrust reverser louvres
90 Fin root fillet
91 Wardrobe, port and starboard
92 Air conditioning system air intakes
93 Tailplane de-icing air ducting
94 Elevator hydraulic jack
95 Tailfin construction
96 Starboard tailplane
97 Starboard elevator
98 VOR aerial
99 Rudder horn balance
100 Rudder construction
101 Tailplane shock cone
102 Elevator torque shaft
103 Port elevator construction
104 Tailplane construction
105 APU ventral access door
106 Auxiliary power unit (APU)
107 Rudder lower segment
108 Tailplane support structure
109 Rudder hydraulic jack
110 Rear entry tunnel
111 Air system heat exchanger exhaust ducts
112 Air conditioning plant, port and starboard
113 Rear pressure bulkhead
114 Rear entry doorway
115 Primary heat exchanger
116 Thrust reverser louvres
117 Engine exhaust nozzle
118 Retractable airstairs
119 Thrust reverser actuator
120 Oil cooler air duct
121 Pratt & Whitney JT8D-7 turbofan engine
122 Engine mounting main frames
123 Bleed air ducting
124 Engine pylon construction
125 Port engine air intake
126 Ventral skin panelling
127 Fuel feed pipe
128 Trailing edge wing root fillet
129 Flap drive shaft
130 Main undercarriage wheel door
131 Main undercarriage leg pivot fixing
132 Flap shroud ribs
133 Inboard spoiler
134 Port double slotted flap, up position
135 Flap rib construction
136 Port airbrake
137 Inboard wing fence
138 Outboard spoiler
139 Spoiler honeycomb construction
140 Aileron construction
141 Aileron hydraulic jack
142 Wing tip fairing
143 Fuel system vent valve
144 Port navigation light
145 Outer wing fuel tank bay
146 Outboard wing fence
147 Three-spar wing torsion box construction
148 Wing rib construction
149 Corrugated double skin leading edge
150 Leading edge nose ribs
151 Wing skin plating
152 Four-wheel main undercarriage bogie
153 Main undercarriage leg strut
154 Hydraulic retraction jack
155 Inboard integral fuel tank bay
156 Wing stringers
157 Inboard tapered spar section
158 Leading edge de-icing air duct
159 Extended leading edge section

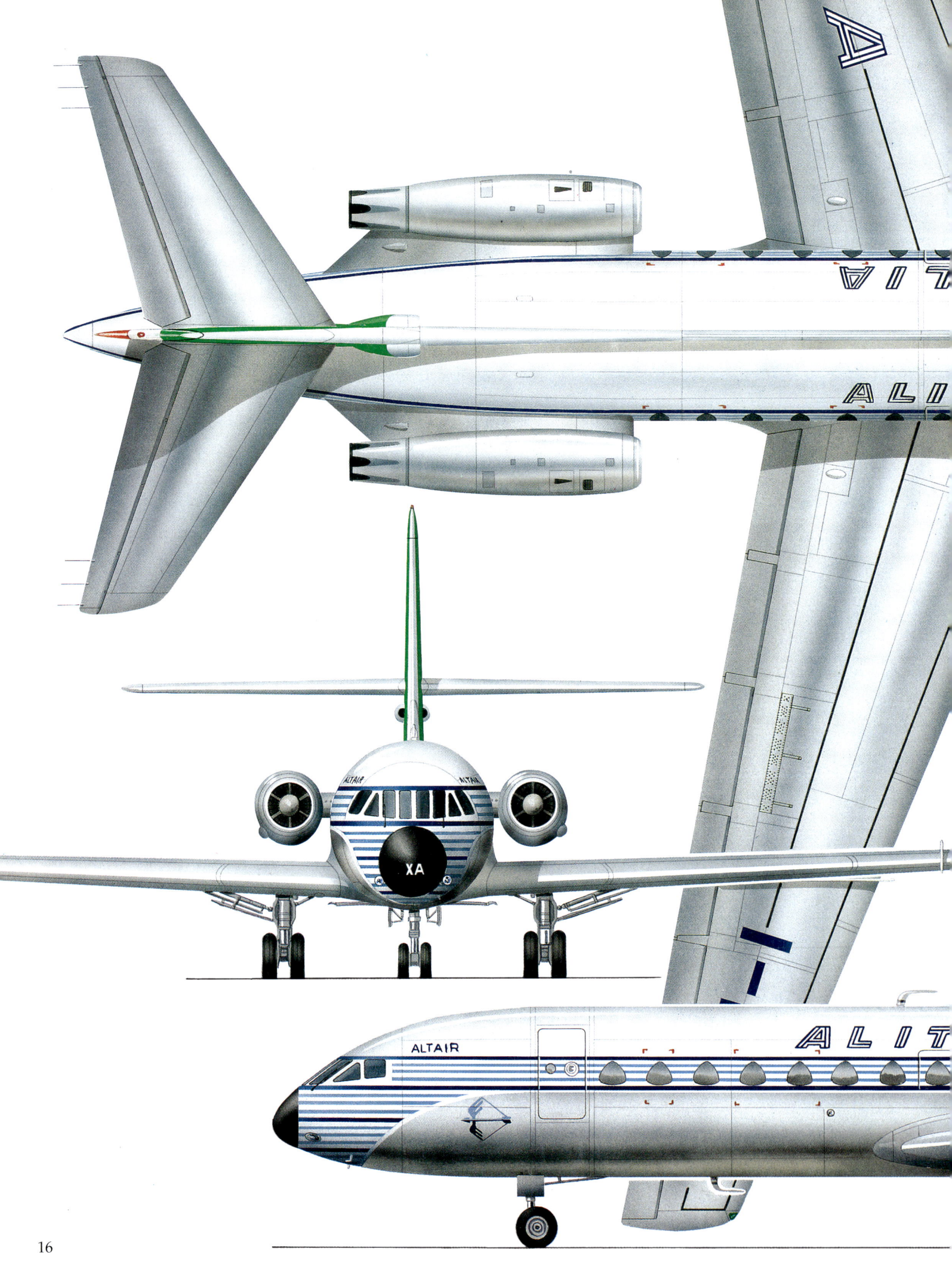

The Italian airline Alitalia was one of the largest operators of the Caravelle, with some 21 aircraft on strength during 1965. This aircraft, I-DAXA, was a Series III which received the fleet name Altair and first entered service in April 1960. It later became a Series VIN with Avon RA.29/6 Mk 531 engines, and was used by Alitalia's charter subsidiary, SAM, in the late 1960s. I-DAXA was finally sold to the Ecuadorian operator SAETA as HC-BAD, and flew until it was withdrawn from use at Quito.

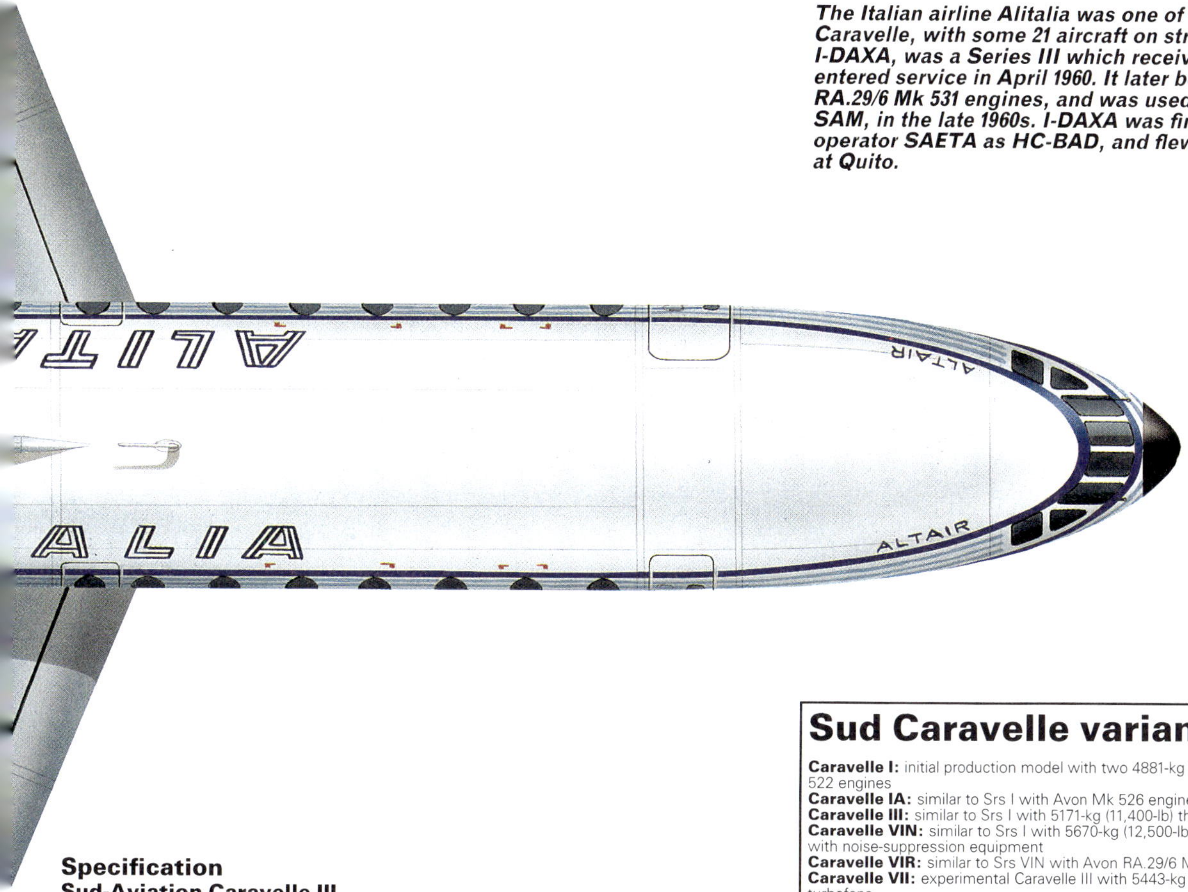

Sud Caravelle variants

Caravelle I: initial production model with two 4881-kg (10,760-lb) thrust Rolls-Royce Avon RA.29/1 Mk 522 engines
Caravelle IA: similar to Srs I with Avon Mk 526 engines
Caravelle III: similar to Srs I with 5171-kg (11,400-lb) thrust RA.29/3 engines
Caravelle VIN: similar to Srs I with 5670-kg (12,500-lb) thrust Avon RA.29/6 Mk 531 engines fitted with noise-suppression equipment
Caravelle VIR: similar to Srs VIN with Avon RA.29/6 Mk 532R engines with thrust reversal
Caravelle VII: experimental Caravelle III with 5443-kg (12,000-lb) thrust General Electric CJ805-23C turbofans
Caravelle 10A: Caravelle Horizon with modified wing, windows and flaps and with tailplane fairing; powered by CJ805-23C turbofans
Caravelle 10B1.N: Srs VII with 6350-kg (14,000-lb) thrust Pratt & Whitney JT8D-1 engines
Caravelle 10B1.R: Srs 10B1.N with JT8D-1(R) or JT8D-7(R) engines with thrust reversal
Caravelle 10B3: Srs 10B1.R with 1-m (3.28-ft) fuselage extension; some fitted with 6577-kg (14,500-lb) thrust JT8D-9(R) engines; also known as **Horizon B** or **Super B**
Caravelle 11R: Srs 10B1.R with 0.93-m (3.05-ft) fuselage extension and convertible passenger/freight interior, cargo door, strengthened landing gear
Caravelle 12: Srs 10B3 with 3.23-m (10.6-ft) fuselage extension and strengthened landing gear

Specification
Sud-Aviation Caravelle III
Type: medium-haul commercial transport
Powerplant: two 5171-kg (11,400-lb) thrust Rolls-Royce Avon RA.29/3 Mk 527 turbojets
Performance: maximum cruising speed at 7620 m (25,000 ft) at a gross weight of 41000 kg (90,390 lb) 805 km/h (503 mph); service ceiling 12000 m (39,370 ft); maximum range with full fuel, maximum payload and normal reserves 1845 km (1,153 miles); take-off balanced field length at maximum take-off weight ISA at sea level 1830 m (6,000 ft); landing distance at maximum landing weight 1800 m (5,900 ft)
Weights: empty 24185 kg (53,318 lb); maximum take-off 46000 kg (101,411 lb); maximum payload 8400 kg (18,520 lb)
Dimensions: span 34.30 m (112 ft 6.4 in); length 32.01 m (105 ft 0.25 in); height 8.72 m (28 ft 7.3 in); wing area 146.7 m² (1,579.12 sq ft).

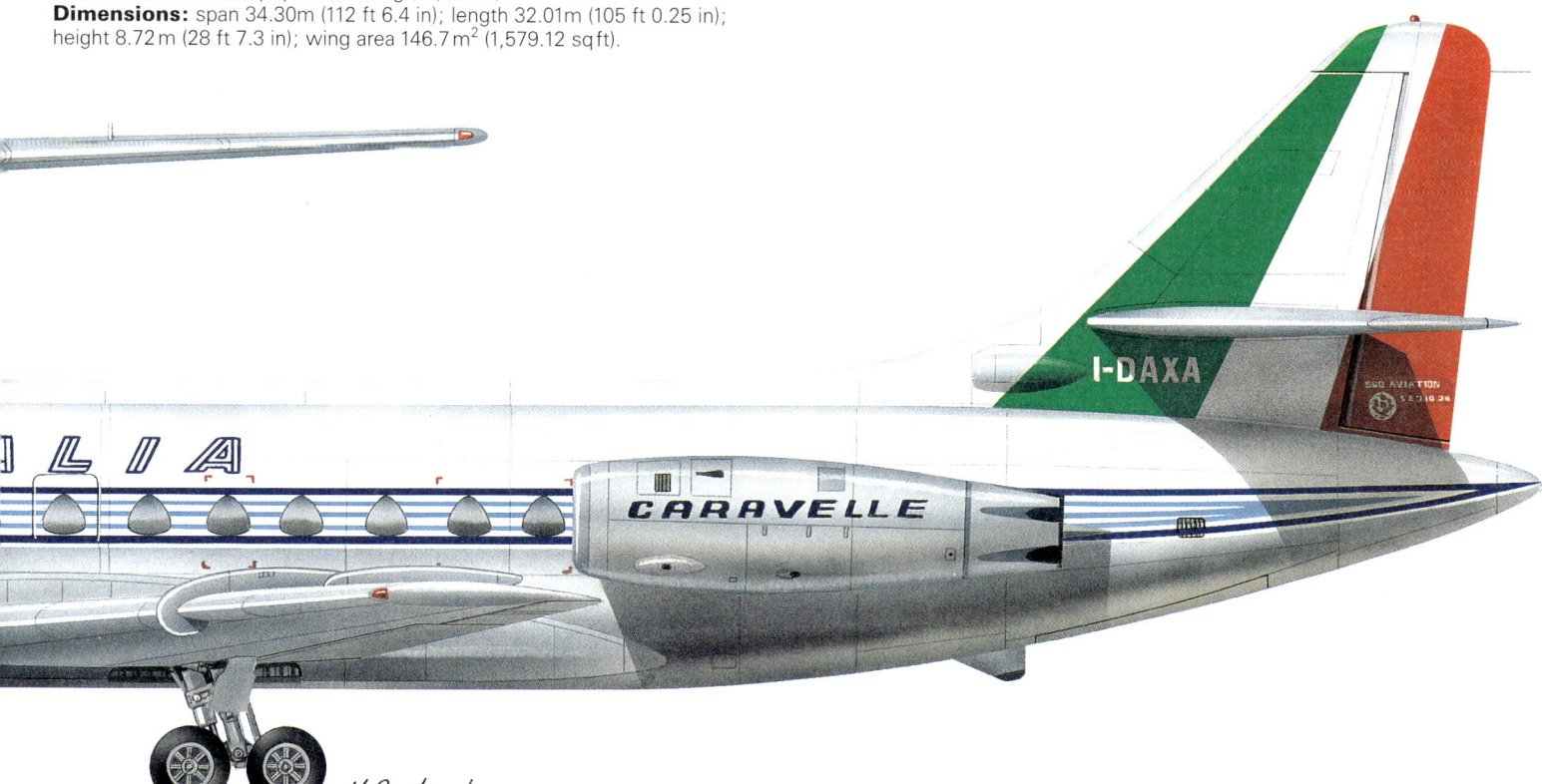

Ascendant Airbus

The first wide-body twin-engined commercial aircraft to enter service, bringing new standards of quietness and fuel economy to regional routes around the world, the A300 family is outstanding proof of Europe's technological, industrial and financial capabilities, making Airbus Industrie the principal competitor to the mighty Boeing concern in the world airliner market.

In 1981, Airbus Industrie won more than a half-share (55 per cent) of the world market for twin-aisle transport aircraft, while Boeing won only a third (36 per cent) and Douglas and Lockheed shared the remainder. Lockheed subsequently announced its intention to terminate manufacture of the TriStar, while Douglas kept the DC-10 production line open entirely on the basis of orders for the military KC-10 Extender tanker/transport derivative.

To have emerged thus, as a major force in the wide-body airliner market, and Boeing's principal competitor in the overall commercial transport field, Airbus Industrie had clearly achieved a meteoric rise since the A300 first flew in 1972. This new form of corporate entity, described in French law as a 'Groupement d'Intérêt Economique' achieved its present position by developing a radically new type of airliner, geared to high-density short/medium-length routes, and bringing major advances in economy of operation, reliability and safety. What follows is the story of how a number of European companies combined their technological and industrial capabilities successfully to challenge the virtual monopoly of commercial transports enjoyed by US manufacturers throughout most of the post-war years.

Studies of wide-body airliners began on both sides of the Atlantic in the 1960s, with American manufacturers concentrating on larger, longer-range aircraft such as the four-engined Boeing 747, which flew in 1969, and the three-engined DC-10 and TriStar, which both

Line-up of A300s being prepared for delivery at the flight test centre at Toulouse in the spring of 1982. The Garuda aircraft is the first of nine A300B4-200s ordered, and Eastern Air Lines, after operating four leased A300B2-400s, ordered 25 more, putting the Airbus on the American map. Six JT9D-powered B4s were delivered to Iberia between February 1981 and February 1982.

flew in 1970. European manufacturers were meanwhile projecting smaller, short-range aircraft, and in 1965 French and West German manufacturers combined to establish a study group (Studiengruppe Airbus). At that time British European Airways was in the market for a 200-seater to replace the Vickers Vanguard and, in view of the impact of rising fuel prices on traffic growth in the 1970s, this would have been a very practical size of aircraft. However, the other major European airlines felt at that time that a capacity of 250-300 seats would be more economical in terms of seat-mile costs, and that such a size was essential if aircraft movements at the main airports were to be kept in check.

Competitive mix

An element of competition was introduced by the industrial groupings that developed. Sud-Aviation worked with Dassault on the Galion, while Hawker Siddeley Aviation combined efforts with Breguet and Nord-Aviation on the HBN-100, the latter eventually forming the basis for the A300. The principal contractors for the three countries were to have been HSA, Sud-Aviation and Arbeitsgemeinschaft Airbus, which last became Deutsche Airbus GmbH in 1967. The two engines were to have been Rolls-Royce RB.178-51s, later superseded by RB.207s. However, at that time the TriStar appeared to be moving ahead much more quickly and to have an assured future, so Rolls-Royce dropped development of the RB.207 to concentrate on the RB.211 for the Lockheed aircraft. (The engine was later found to require so much financial investment that Rolls-Royce had to be taken over by the British government.)

The UK role in what was to prove the most successful European collaborative air transport venture of all time then suffered a further blow when the British government declined to join in funding development of the A300 because of lack of orders. Nevertheless, in May 1969 the French and German governments decided to go ahead, and construction of the first prototype A300B1 began in October of that year. The UK's only major part in the programme was that HSA acted as a subcontractor to Airbus Industrie, designing, developing and manufacturing the wings (although the leading and trailing edges were built by VFW-Fokker, later Fokker BV). This arrangement subsequently provided a great deal of 'tinbashing' work for the UK, but the country's failure to seize the lead in the Airbus programme gave to France key areas such as the front fuselage, which was of far greater significance to equipment suppliers. The Airbus thus began on the basis of French and German funding, with Fokker-VFW of the Netherlands and CASA of Spain joining the industrial grouping later, in December 1970 and December 1971.

The aircraft that emerged was somewhat smaller than the 300-seater envisaged when the designation A300 had been chosen, hence the name A300B for the (approximate) 250-passenger redesign. The reason for this reduction in size was partly that potential customers had begun saying that the original concept would have had

The purchase in 1980 of five A300B4-200s by Trans Australia Airlines signalled a change in that country's two domestic airline policy of identical equipment for TAA and Ansett, the latter choosing the Boeing 767. A300s, then the only wide-bodied airliners on Australian domestic routes, linked Brisbane, Sydney, Melbourne and Perth.

Airbus Industrie A300-600R cutaway drawing key

1 Radome
2 Weather radar scanner
3 Scanner mounting and tracking mechanism
4 VOR localiser aerial
5 Front pressure bulkhead
6 Windscreen panels
7 Windscreen wipers
8 Instrument panel shroud
9 Control column
10 Rudder pedals
11 Cockpit floor level
12 ILS aerial
13 Pitot heads
14 Access ladder to lower deck
15 Captain's seat
16 Centre control pedestal
17 Direct vision opening side window panel
18 First officer's seat
19 Overhead systems switch panel
20 Maintenance side panel
21 Observer's seat
22 Folding fourth seat
23 Cockpit bulkhead
24 Air conditioning ducting
25 Crew wardrobe/locker
26 Nose undercarriage wheel bay
27 Hydraulic retraction jack
28 Taxiing lamp
29 Twin nosewheels, forward retracting
30 Hydraulic steering jacks
31 Nosewheel leg doors
32 Nose undercarriage pivot fixing
33 Forward toilet compartment
34 Wash hand basin
35 Galley
36 Starboard entry/service door
37 Door mounted escape chute
38 Cabin attendant's folding seat
39 Curtained cabin divider
40 Forward main entry door
41 Door latch
42 Door surround structure
43 Underfloor avionics equipment racks
44 Runway turn-off light
45 Fuselage lower lobe frame and stringer construction
46 Floor beam construction
47 Cabin window panels
48 Forward freight hold door
49 Cabin wall trim panelling
50 VHF communications aerial
51 Overhead stowage bins
52 Curtained cabin divider
53 First-class passenger seating, 26 seats
54 Underfloor air system ducting
55 Door mounted escape chute
56 Main cabin entry door
57 Overhead stowage bins
58 Central galley unit
59 Starboard General Electric CF6-80C2 engine nacelle
60 Pratt & Whitney JT9D-7R4H1 or PW4156 alternative engine installation
61 Common nacelle pylon beam
62 Pylon attachment links
63 Pylon tail fairing
64 Starboard wing engine pylon
65 Tourist-class passenger cabin seating, 241 seats (267 seats total in mixed-class layout)
66 Air system distribution ducting
67 Conditioned air delivery ducting
68 LD3 baggage container, 12 in forward hold
69 Water tank
70 Slat drive shaft motor and gearbox
71 Wing spar centre-section carry-through
72 Ventral air conditioning packs (two)
73 Wing centre box fuel tank
74 Three-spar wing centre-section construction
75 Centre-section floor beams
76 Front spar attachment main frame

One of eight A300B4-100s operated by Olympic Airways. Now entirely owned by the Greek government, Olympic was originally founded in 1957 by Aristotle Onassis, who bought TAE Greek National Airlines and thus gained a monopoly of internal routes and national designation for overseas services.

Airbus Industrie A300 variants

(Note: the following list deals only with the principal variants; the model number (A300) is followed by a function letter, B indicating basic passenger version, C convertible, and F freighter; the airframe designation number as set out below has its last two digits modified to show the type of powerplant, numbers 01 to 19 for General Electric engines, 20 to 39 for Pratt & Whitney engines, and 40 to 59 reserved for Rolls-Royce; it is followed by two digits indicating variations in design weights

A300B1: designation allocated to the two prototype aircraft, of which only one is now flying (a/c SN2)
A300B2-100: initial production short-haul version with 'simple wing' leading edge, as used by Air France to introduce the series into regular service in May 1974
A300B2-200: similar to A300B2-100 series, but with wing-root leading-edge Krüger flaps (as developed for the A300B4 series) and with the A300B4's wheels and brakes, to give the aircraft the capability to operate from hot/high airfields (e.g. by South African Airways); examples of this series include the **A300B2-201** with CF6-50C and the **A300B2-220** with JT9D-59A engines
A300B2-300: similar to A300B2-200, but with increased zero-fuel and landing weights for improved payload and multi-stop flexibility; first operator was SAS
A300B2-600: based on A300B2-200, but stretched to provide two extra seat rows and space for two extra LD3 containers; it also has slightly uprated engines, the rear fuselage and tailplane of the A310, more composite materials, and slightly more range
A300B4-100: basic medium-haul version, with additional fuel tank in the centre wing box, slightly stronger structure, uprated engines, wheels and brakes, and Krüger flaps at the wing leading-edge root; first operator was Germanair; this series includes the **A300B4-101** with CF6-50C and the **A300B4-120** with JT9D-59A engines
A300B4-200: similar to A300B4-100, but with increased take-off weight by virtue of strengthened wings, fuselage and landing gear, allowing a higher payload or a range up to 5795 km (3,600 miles) for transcontinental operations
A300B4-600: stretched version derived from A300B4-200, with modifications as for A300B2-600; first ordered by Saudi Arabian Airlines
A300C4: convertible freighter version with large upperdeck cargo door on port side of aircraft, a reinforced cabin floor, smoke detection system in the main cabin, and interior trim adaptable to the freighter role; first operated by Hapag-Lloyd
A300F4: all-freighter version similar to A300C4, but with all passenger provisions removed and cabin windows blanked over; first ordered by Korean Air
A300-600: current designation of A300B4-200
A300-600C: convertible version of the A300-600 with forward upperdeck cargo door and same engine options. In service with Kuwait Airways and Abu Dhabi Private Flight
A300-600ER: original designation of current A300-600R
A300-600F: freighter version of A300-600 with forward upperdeck cargo door but the same engine options. Built without passenger systems or windows. None ordered
A300-600R: current designation of A300B2-600

77 Fuselage centre-section construction
78 Starboard wing inboard main fuel tank; standard fuel capacity 13,628 Imp gal (62000 l)
79 Outer wing skin panel joint strap
80 Fuel system piping
81 Pressure refuelling connections
82 Refuelling valves
83 Fuel feed tank and pumps
84 Fuel tank dividing ribs
85 Leading-edge slat drive shaft
86 Three-segment leading-edge slats, open
87 Wing fence
88 Slat screw jacks
89 Outer wing panel integral fuel tank
90 Fuel vent tank
91 Starboard navigation light (green)
92 Wing tip fairing
93 Starboard wing-tip fence
94 Tail navigation and strobe lights (white)
95 Static dischargers
96 Fixed portion of trailing edge
97 One-piece single-slotted Fowler-type flap, down position
98 Flap guide rails
99 Fuel jettison pipe
100 Outboard roll-control spoilers/lift dumpers (two)
101 Inboard airbrakes/lift dumpers (three)
102 Spoilers/airbrake hydraulic jacks
103 Flap screw jacks
104 Flap drive shaft
105 Starboard all-speed aileron
106 Aileron triplex hydraulic actuators
107 Wing root spoilers/lift dumpers (two)
108 Inboard flap segment
109 Cabin air system recirculation fan
110 Pressure floor above wheel bay
111 Rear spar attachment main frame
112 Starboard main undercarriage retracted position
113 Undercarriage door jack
114 Equipment bay walkway
115 Undercarriage bay pressure bulkhead
116 Flap drive motor and gearbox
117 Hydraulic reservoir, triplex section
118 Eight-abreast tourist-class passenger seating
119 Starboard Type 1 emergency exit door
120 Upper fuselage frame and stringer construction
121 Rear underfloor freight hold door
122 Freight/cargo compartment dividing bulkhead
123 Cabin wall insulating blankets
124 Cargo hold door
125 Cabin floor panelling
126 Seat mounting rails
127 Rear cabin air recirculation fan
128 ADF aerials
129 Fuselage skin panelling
130 Ceiling trim/lighting panels
131 Central overhead stowage bins
132 Rear galley
133 Fin root fairing
134 Fin spar attachment joints
135 Three-spar fin torsion box construction
136 Starboard trimming tailplane
137 Tailplane trim fuel tank; additional capacity 1,342 Imp gal (6100 l)
138 Starboard elevator
139 Glassfibre-reinforced fin leading-edge
140 Fin rib construction
141 Fin tip fairing
142 Static dischargers
143 Carbon fibre rudder skin panelling
144 Honeycomb core construction
145 Rudder triplex hydraulic actuators
146 APU equipment bay
147 Garrett GTCP331-250 auxiliary power unit (APU)
148 Tailcone fairing
149 APU exhaust duct
150 Port elevator construction
151 Elevator triplex hydraulic actuators
152 Static dischargers
153 Port tailplane rib construction
154 Leading-edge nose ribs
155 Port tailplane integral fuel tank
156 Tailplane pivot fixing
157 Moving tailplane sealing plate
158 Tailplane centre-section carry-through
159 Tailplane trim screw jack
160 Fin support structure
161 Rear pressure bulkhead
162 Rear toilet compartments (four)
163 Cabin attendant's folding seat
164 Rear entry door
165 Rear cabin seven-abreast passenger seating
166 Cabin side-wall frames
167 Underfloor bulk cargo hold, 610 cu ft (17.3 m³)
168 Cabin window panels
169 LD3 baggage containers, 10 in rear hold
170 Port Type 1 emergency exit door
171 Lower fuselage skin panelling
172 Wing root trailing-edge fillet
173 Port inboard single-slotted flap
174 Wing root spoilers/lift dumpers
175 Flap guide rail
176 Spoiler hydraulic jacks
177 Auxiliary spar
178 Main undercarriage side struts
179 Retractable ventral landing lamp, port and starboard
180 Hydraulic retraction jack
181 Main undercarriage pivot fixing
182 Inboard flap track mechanism
183 Aileron triplex hydraulic actuators
184 Port all-speed aileron construction
185 Port airbrakes/lift dumpers
186 Flap down position
187 Flap guide rails
188 Fuel jettison pipe
189 Flap track fairings
190 Roll control spoilers/lift dumpers
191 Fixed portion of trailing edge
192 Trailing-edge composite construction
193 Static dischargers
194 Tail navigation and strobe lights (white)
195 Port wing-tip fence
196 Wing tip fairing
197 Port navigation light (red)
198 Rear spar
199 Outer wing panel rib construction
200 Front spar
201 Port leading-edge slat segments
202 Slat screw jacks
203 Slat guide rails
204 Wing leading-edge de-icing air pipes
205 Telescopic de-icing air delivery ducts
206 Port wing integral fuel tank
207 Outer wing panel skin joint strap
208 Port main wheel undercarriage four-wheel bogie
209 Main undercarriage leg strut
210 Nacelle pylon attachment joint
211 Engine pylon construction
212 Exhaust nozzle plug fairing
213 Core engine, hot stream, exhaust nozzle
214 Engine turbine section
215 Fan air, cold stream, exhaust duct
216 Reverser cascade, closed
217 Engine bleed air ducting
218 General Electric CF6-80C2-A1 turbofan engine
219 Engine fan blades
220 Noise attenuating intake lining
221 Intake cowling nose ring
222 Detachable engine cowling panels
223 Bleed air system pre-cooler
224 Inboard leading-edge slat
225 Bleed air delivery ducting
226 Inner wing panel three-spar construction
227 Inboard integral fuel tank
228 Inboard wing ribs
229 Wing root skin joint strap
230 Krüger flap actuator
231 Wing root Krüger flap, extended

The first of three General Electric-powered A300B4-200s for Air Afrique was delivered in May 1981. Constituted in 1961 by agreement between 11 independent African states that were formerly French colonies, Air Afrique operates services linking 22 African countries with France, Italy, Switzerland and the USA.

One of two A300-B4-100s flown by Philippine Airlines, which once operated 12 A300 series airliners. One of the older airlines in the Far East, PAL was formed in 1941, but soon had to terminate operations due to the war. Services were resumed in February 1946.

excessively high seat/kilometre costs, and partly that the high thrust of the RB.207 was no longer available. The A300B was designed so that, in principle, it could accept the General Electric CF6, the Pratt & Whitney JT9D, or the RB.211, although all the initial aircraft used the CF6 in a standard McDonnell Douglas wingpod. Some later A300s have the JT9D engine, but the RB.211 turbofan failed to find a launch customer in an A300 application.

Further co-operation

Airbus Industrie was formally constituted in December 1970 with Sud-Aviation and Deutsche Airbus as the principal industrial partners. As further countries joined in the programme, the work-split became more complex. Sud-Aviation (later part of Aérospatiale) secured the all-important forward fuselage, lower centre fuselage and engine pylons. MBB (later Daimler-Benz Aerospace Airbus) was given the rear fuselage, vertical tail, and upper centre fuselage, plus the flaps guides. VFW-Fokker (at the time a subsidiary of MBB) built the cylindrical fuselage section ahead of the wing. Fokker (as mentioned earlier) made the wing leading and trailing edges, while HSA (now part of British Aerospace) was given responsibility for the torsion box. CASA made the horizontal tail and front fuselage door. Messier of France produced the landing gear. Both SNECMA in France and MTU in Germany made parts for the CF6 engine, which for the Airbus series was assembled by SNECMA. Major components of the A300 were flown between factories in Airbus Industrie's Super Guppies and Belugas for assembly.

Behind the three-crew flight deck, the fuselage of 5.64 m (18 ft 6 in) diameter provided six-abreast seating in first class, seven in business, eight in economy, or nine in charter, while the underfloor hold accommodated up to 20 LD3 containers in pairs abreast. All current models of the A300B (the 600R) take 220-345 seats, depending on cabin configuration, although 269 is typical of all-tourist (eight-abreast) seating, and have a two-crew flight deck.

The wing is moderately swept, with a very advanced aerofoil section that permits a high cruise speed without significant compressibility effects, i.e. drag rise due to shock-induced separations. It is fitted with leading-edge flaps and two-slot Fowler trailing-edge flaps. Each wing has two ailerons, the inboard one functioning at all speeds and drooping in sympathy with the flaps, and the outboard one working only at low speeds. Four spoilers on each side act as airbrakes, and three as lift-dumpers. The flying controls are powered by triple hydraulic systems, with manual reversion. The landing-gear has a twin-wheel nose unit and four-wheel bogies on each main leg.

Success in America

Reverting to the development history of the A300B, the first of two B1s (F-WUAB, later F-OCAZ) made its maiden flight on 28 October 1972, followed by the second (F-WUAC, later OO-TEF) on 5 February 1973. The first of the initial production series (F-WUAD, the first A300B2 short-haul version) took to the air on 28 June 1973 and the second A300B2 (F-WUAA) on 20 November 1973. The A300B2 was awarded Franco-German certification on 15 March 1973, and FAA certification on 30 May 1974. The A300B2 entered service with Air France on the Paris–London run on 23 May 1974.

Thus ended the first phase of the struggle to develop a major European commercial transport aircraft that could be sold successfully all over the world. However, some of the most difficult days for Airbus Industrie were yet to come. The critics who had claimed that the aircraft would be a technical flop now changed their argument to say that it would never be sold in economical numbers. In the mid-1970s sales were certainly slow to materialise (only one aircraft was sold in 1976). One of the keys to the success of Airbus was the launching of the A300B4 for medium-haul routes, a heavier version which offered a range of 4825 km (3,000 miles) and later 5800 km (3,600 miles) by comparison with the 2970-km (1,845-mile) range of the A300B2, all these figures relating to 269 passenger loads. Four of the A300B4s were taken by Eastern Air Lines on a six-month lease, beginning late in 1977. The following April, delighted with the results obtained with the A300B4s, Eastern bought them and ordered 25 more with options on a further nine. The acceptance of the aircraft by a major US operator was a turning point in sales of the A300B series. Later Airbus succeeded in winning over other American airlines including Continental and Pan American and orders started pouring in from the rest of the world, including certain areas (such as the Far East) that were enjoying a substantial growth in traffic. By 1978 Airbus (with 23 per cent of the market) was second only to Boeing in wide-body sales, and within three years had pushed Boeing out of the lead.

Efficiency, reliability and sales

As sales boomed, development continued on variants of the A300B theme. The A300C4 was produced as a convertible freighter version, which entered service with Hapag-Lloyd of Germany in 1980. An A300B equipped with Pratt & Whitney JT9D engines flew in May 1979, and in the following January the A300B2 with JT9D-59As entered service with Scandinavian Airlines. Garuda Indonesian Airways and Iberia also use the Airbus with JT9Ds. It may

Egyptair ordered eight A300B4-200s in 1978-79 and while awaiting delivery leased the Series B4-100 SU-AZY from Bavaria-Germanair, from May 1977 to January 1978, by which time the German airline had merged into Hapag-Lloyd. Formed in 1932 as Misr Airwork (later Misrair) Egyptair became United Arab Airlines in 1969, and adopted its present name in 1971.

The Japanese internal airline Toa Domestic (renamed Japan Air System in April 1988) ordered eight A300B2s in 1979 including JA8466 seen here. TDA's management liked the Airbus 'house colours' so much that they adopted it as the airline's new livery. TDA subsequently bought five used A300s and later seven A300-600Rs delivered from February 1991 to JAS, which now flies 36 A300 series airliners.

be added that Garuda was the first airline to take delivery of the A300 equipped with the two-crew 'forward-facing crew cockpit' (FFCC), as developed for the later A310, the delivery taking place in January 1982. Two months later, on 4 March, the A300B celebrated its millionth accident-free flying hour, a remarkable achievement by any standard. Since its introduction into service, the A300B had averaged a technical despatch reliability of 98.5 per cent, higher than that of any other wide-body. Airbus Industrie is also rightly proud of the fact that US Civil Aeronautics Board figures showed that the A300B's direct maintenance costs in the USA were the lowest for all wide-bodies then in service.

Singapore Airlines ordered 12 A300B4-200s in 1978-79 but only took delivery of eight, cancelling the others in favour of A310s. The A300s were delivered between December 1980 and September 1983 but were phased out of service from December 1984 onwards.

Larger and larger again

Although chronologically out of step, since it comes after the short-fuselage A310, it is convenient to complete the current story of A300B development with mention of the A300-600, which Airbus decided to fund in December 1980. The 'Dash-600' is a slightly stretched version, with two extra seat rows, typically seating 285 in all-tourist class, and providing space for two extra LD3 containers under the floor. It has uprated CF6-80C or JT9D-7R4H engines, the more sharply tapered rear fuselage and the smaller horizontal tail developed for the A310, considerably more components made of composite materials, and slightly more range. The Dash 600 first flew on 8 July 1983, and the type entered service with Saudia in April 1984, following certification with Pratt & Whitney engines in March. Certification with General Electric engines followed in September 1985.

Formerly the Shanghai region of the Civil Aviation Administration of China (CAAC), China Eastern operates on internal services and to neighbouring countries. Among its fleet are 10 A300-605Rs.

The -600R extended-range version first flew on 9 December 1987 and features the tailplane trim fuel tank developed for the A310 and a higher maximum take-off weight. The A300-600 has received wide acceptance and fleets of both versions have subsequently been ordered by 15 airlines, five leasing companies and the Abu Dhabi Royal Flight. Best customer for the Dash 600 (and launch airline for the -600R) has been American Airlines which ordered four aircraft during 1990, bringing its fleet to 35. Other major customers include Korean Air with 21, Thai International with 16, and Lufthansa and Saudia with 11 each.

In November 1990 Airbus announced that it was studying the 1995-98 replacement of the four Super Guppies used to transport major components between participating factories to Toulouse for final assembly, with a special version of the A300-600R, featuring an enlarged unpressurised upper fuselage with a new cargo door in the nose, and able to carry a payload of between 45 and 50 tonnes, or double the existing figure. Airbus announced the programme for this A300-600ST Super Transporter (renamed the Beluga) in 1990. Construction by SATIC, an Airbus subsidiary, led to a first flight in September 1994, and four aircraft had been delivered by 1998.

In order primarily to share in the A310 programme, British Aerospace became a full partner in Airbus Industrie in January 1979. In the combined A300/310 programme, Aérospatiale (France) has a 37.9 per cent share, as does Daimler-Benz Aerospace Airbus (previously Deutsche Airbus and originally MBB) of Germany. British Aerospace has 20 per cent, and CASA (Spain) has 4.2 per cent. Fokker of the Netherlands and Belairbus of Belgium, the latter involved only in the A310 programme, are both Airbus associates.

By the end of June 1998, Airbus had secured 488 firm orders for the A300 from customers in all parts of the world. In 1998 the joint A300 and A310 production line was running at about six aircraft per year after peaking at 4.5 aircraft per month in 1990, and A300 deliveries amounted to 470 aircraft, the backlog suggesting production into the early 2000s. There is now a considerable market for older aircraft to be converted as A300-600F pure freighters. The consortium claims that an Airbus-built airliner takes off somewhere in the world every 20 seconds, and with the targeted 35 per cent of the jet airliner market captured, the multi-national Airbus Industrie programme can justly be described as one of the great success stories of post-war commercial aviation.

Based at Paris-Orly, Air Liberté is a French charter carrier that flies to the Mediterranean and North America. Pride of the fleet are a pair of A300-622Rs, configured for 345 passengers in one class.

Airbus A310

While the first Airbus, the A300, introduced the new marque to an appreciative world, and the succeeding A320 introduced the first truly modern airliner, the A310 in the middle has been somewhat overshadowed. However, this aircraft has achieved superb sales, largely on account of its excellent economy and range performance.

It is ironic that at the beginning of the story of Airbus Industrie the project staff planned nine versions of the basic A300B, whereas the only one actually to be built was a tenth version that came later: this A300B10 eventually matured as the A310 airliner that is today recognised as the most fuel-efficient medium/long-haul transport in the world, despite the keen competition of the 767. Compared with the American aircraft, the European jetliner has a smaller wing and a bigger and more capacious cabin.

Originally the Airbus partners planned a huge twin designated A300, with a fuselage diameter of 6.4m. Airlines thought it too big, and it was cut back to a diameter of 5.64m, resulting in the A300B. The number of seats (typical mixed-class) was reduced from 300 to around 250. But throughout the 1970s, though the A300B gradually won orders – which after ten years became a torrent – some airlines still regarded even this aircraft as too big. Two, Lufthansa and Swissair, specifically asked Airbus Industrie to study a smaller version.

The trouble with a smaller version is that it calls for massive development expenditure (which has to be recovered from sales) in order to produce an aircraft that is probably more expensive to buy and carries fewer people. Of course, the simplest answer is just to put fewer seats in the original, bigger, aircraft. This might be popular with passengers, but is bound to result in sharply increased seat-kilometre costs. Another alternative is just to make the fuselage shorter, but then the wing, engines and landing gear are all unnecessarily large and heavy. Ideally, the smaller aircraft should be an almost fresh design, but this involves astronomic development costs, which – because the process takes many years – are compounded by inflation, which was severe in the late 1970s.

New wing

Throughout the mid-1970s Airbus Industrie was in severe political difficulties, half the management wanting to tie up a collaboration deal with Boeing. British Aerospace wanted to join as a full partner but was thwarted by the implacable opposition of British Airways and Rolls-Royce. Despite all this, in late 1977 the brilliant aerodynamics team of British Aerospace at Hatfield – which a decade earlier had designed the wing of the A300B – began scheming a totally new wing for the proposed 'all-new' A300B10. But BAe was not a member of

The first of three A310s destined for Kenya Airways demonstrates the steep climb-out from the manufacturer's airfield in Toulouse. These are also A310-304s, the extended-range version of the basic aircraft with drag-reducing winglet fences.

Pan Am was one of the US majors that embraced Airbus products, ordering significant batches of the A300 and A310. Its A310 fleet once stood at 19, made up of seven Series 222s (illustrated) and 12 Series 324s, the latter featuring Pratt & Whitney PW4152 engines. Some served on the airline's European network.

the consortium, so other new wing designs were started by MBB, VFW-Fokker and Aérospatiale. Meanwhile, studies were also made of a wing that retained the existing A300B inter-spar box structure and added new leading and trailing edges.

The public first saw the A310 in the form of a model at the Hanover airshow in April 1978. It looked very much like an A300B but the wing was smaller and the fuselage was 12 frames shorter. The engines were expected to be lower-thrust versions of those used in the A300B. Soon afterwards, on 9 June 1978, Lufthansa and Swissair defined a joint final specification, and a month later these important airlines announced their intentions to place launch orders. This enabled Airbus Industrie to swing into full engineering development. The only real argument still concerned the wing, but a solution was in sight when, in January 1979, British Aerospace returned to Airbus as a full partner. BAe accordingly handled the main responsibility for the wing, but 33 German and French engineers came to Hatfield to integrate features of the Bremen-based wing with that designed by BAe.

In January 1979 the Executive Vice-President and General Manager of Airbus Industrie, Roger Béteille, said "British Aerospace is doing a fantastic job pulling the whole thing together. The wing is now back on schedule..." And Production Director Felix Kracht added "The discussions between the different wing schools have been very fruitful, and the final wing will be significantly better than the best originally achieved." It differed in significant ways from the wing of the A300B. Seen from the front, while most of it looked very thin, the portion inboard of the engines grew rapidly thicker, with the undersurface curving sharply down towards the very bottom of the fuselage. Fowler flaps were fitted outboard, while the inboard flaps were called 'vaned Fowlers' by the Bremen team and 'double-

The high-density Asian routes operated by Air India are flown by A310-304s, 11 of which were once on strength. The large internal carrier, Indian Airlines, operates a substantial fleet of A300s and A310s.

One of the most distinctive schemes applied to airliners anywhere is that of the Royal Jordanian Airline (previously Alia). It currently adorns four A310-300 and two A310-200 aircraft.

slotted' by Hatfield.

As finally agreed, the A310 fuselage had a redesigned rear end with the rear pressure bulkhead nearer the tail. So while the fuselage itself is shorter than that of the A300B by 13 frames (not 12), the cabin is only 11 frames shorter. This new, refined, rear fuselage was later adopted for the big A300-600 family. Other changes include a smaller horizontal tail, revised engine pylons able to accept either General Electric or Pratt & Whitney engines without modification, and landing gears modified to match the A310's reduced weights. Of course, the A310 has, from the start, been equipped with the pioneering Airbus FFCC (forward-facing crew cockpit) with multifunction cathode-ray tube displays and digital avionics.

Workshare changes

In organising the A310 manufacturing programme there were a few changes compared with that for the A300B. Aérospatiale builds the nose, including the cockpit, the lower centre section including the wing box under the floor, the engine pylons and airbrakes, and carries out assembly and flight test at Toulouse-Blagnac. MBB makes almost the entire fuselage, as well as the vertical tail, flaps and spoilers, and final assembly of the wings before despatch to Toulouse. British Aerospace at Chester makes the wing inter-spar boxes, which are by far the strongest and heaviest parts of the whole aircraft. CASA's plants in Spain make the horizontal tail, forward passenger doors and landing-gear doors. In the Netherlands Fokker makes the all-speed ailerons between the flaps, wingtips, main-leg

Airbus Industrie A310 Cutaway Drawing Key

1. Radome
2. Weather radar scanner
3. Radar scanner mounting
4. VOR localiser aerial
5. Front pressure bulkhead
6. Windscreen panels
7. Windscreen wipers
8. Instrument panel shroud
9. Control column
10. Rudder pedals
11. Cockpit floor level
12. ILS aerial
13. Pitot tubes
14. Access ladder to lower deck
15. Captain's seat
16. Centre control pedestal
17. Opening side window panels
18. First Officer's seat
19. Overhead systems control panel
20. Maintenance side panel
21. Observer's seat
22. Folding fourth seat
23. Cockpit bulkhead
24. Air conditioning ducting
25. Crew wardrobe/locker
26. Nose undercarriage wheel bay
27. Hydraulic retraction jack
28. Taxiing lamp
29. Steering jacks
30. Nosewheel doors
31. Forward toilet
32. Wash hand basin
33. Galley
34. Starboard entry/service door
35. Door mounted escape chute
36. Cabin attendant's folding seat
37. Hand baggage locker/wardrobe

Another striking scheme is that of Emirates Airlines, which has 10 A310-304s. This view graphically shows the large flap guides, wingtip fences and full-span leading edge slats that help to give the A310 its impressive take-off performance.

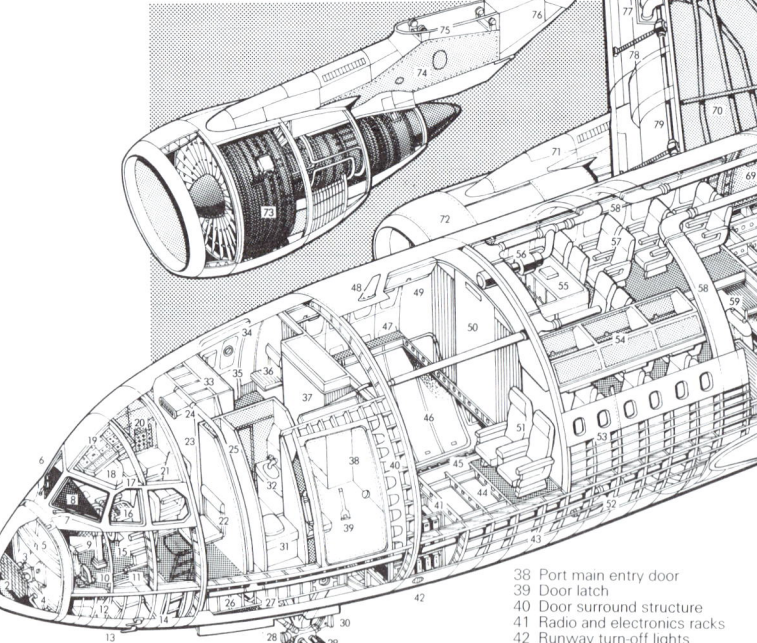

38. Port main entry door
39. Door latch
40. Door surround structure
41. Radio and electronics racks
42. Runway turn-off lights
43. Fuselage frame and stringer construction
44. Floor beam construction
45. Forward freight hold

doors and flap-track fairings. A specially formed Belgian consortium, Belairbus, makes the full-span slats and the forward wing/body fairings. Prime contractor for the landing gears is MHB (Messier-Hispano-Bugatti).

Continued production

Aérospatiale did not set up any separate assembly line for the A310 but slotted the smaller aircraft in among the A300Bs and A300-600s. It did not even start a new numbering sequence, so that the first two (for example) were aircraft Nos 162 and 172 on the line. There was no prototype, though the first four differed from standard in various ways. For example, the first two had an emergency crew escape route through the nose-gear bay, and another had provision for explosively blowing out one of the main cargo doors to demonstrate compliance with the harsh FAA certification requirements.

Airbus had planned a short-haul A310-100, but this was never built and the first aircraft were A310-200s, which became the basic passenger version, with internal fuel capacity of 55,000 litres. The first off the line was uniquely painted in the liveries of the launch customers: from the left side it looked like a Lufthansa aircraft (except for having Pratt & Whitney logos on the engine pods, the German airline having selected GE engines) while from the right side there was no way to tell it was not a Swissair aircraft (apart from the experimental French registration F-WZLH). The first flight took place on 3 April 1982. From the very beginning it was clear that the new Airbus was a winner, but nobody had suspected how far it would perform beyond prediction.

Virtually all the unexpected performance came from the beautiful new wing, which is absolutely 'clean' and has no flaws (such as gaps in the slats at the inboard end or above the pylon strut), and an outstanding aerofoil profile with a flat-top section and average thickness/chord ratio of almost 12 per cent. Perhaps the greatest surprise was that the buffet boundary (the combination of speed and altitude at which aerodynamic buffet will be encountered even in level flight) exceeded expectation by very nearly 10 per cent, which meant either that the A310 could cruise 900m higher for any given weight or else carry 11 tonnes more payload at a typical cruise height. The drag measures were so much better than prediction that optimum cruise Mach number was raised from the planned 0.78 to 0.805.

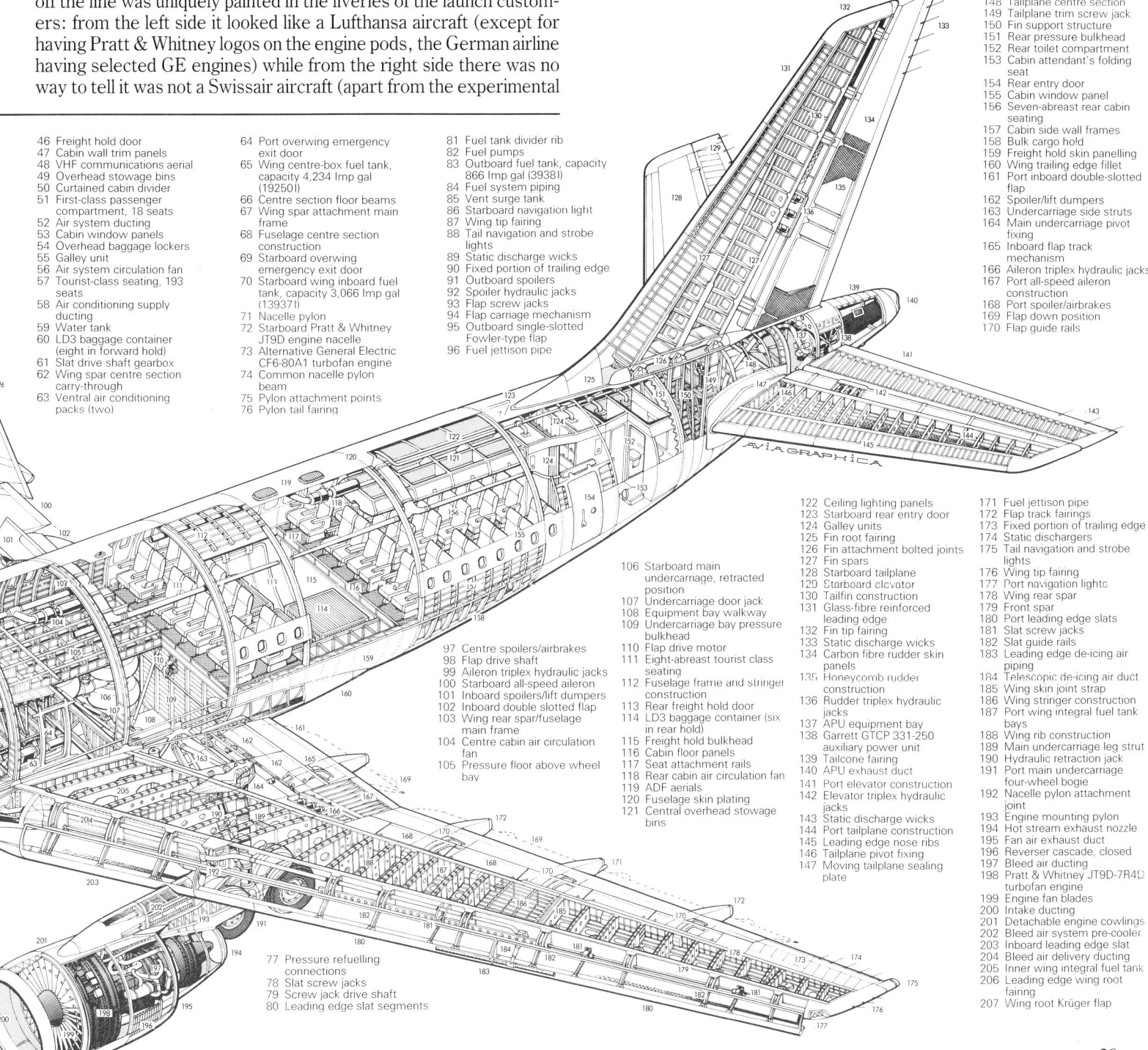

46 Freight hold door
47 Cabin wall trim panels
48 VHF communications aerial
49 Overhead stowage bins
50 Curtained cabin divider
51 First-class passenger compartment, 18 seats
52 Air system ducting
53 Cabin window panels
54 Overhead baggage lockers
55 Galley unit
56 Air system circulation fan
57 Tourist-class seating, 193 seats
58 Air conditioning supply ducting
59 Water tank
60 LD3 baggage container (eight in forward hold)
61 Slat drive shaft gearbox
62 Wing spar centre section carry-through
63 Ventral air conditioning packs (two)
64 Port overwing emergency exit door
65 Wing centre-box fuel tank, capacity 4,234 Imp gal (19250l)
66 Centre section floor beams
67 Wing spar attachment main frame
68 Fuselage centre section construction
69 Starboard overwing emergency exit door
70 Starboard wing inboard fuel tank, capacity 3,066 Imp gal (13937l)
71 Nacelle pylon
72 Starboard Pratt & Whitney JT9D engine nacelle
73 Alternative General Electric CF6-80A1 turbofan engine
74 Common nacelle pylon beam
75 Pylon attachment points
76 Pylon tail fairing
77 Pressure refuelling connections
78 Slat screw jacks
79 Screw jack drive shaft
80 Leading edge slat segments
81 Fuel tank divider rib
82 Fuel pumps
83 Outboard fuel tank, capacity 866 Imp gal (3938l)
84 Fuel system piping
85 Vent surge tank
86 Starboard navigation light
87 Wing tip fairing
88 Tail navigation and strobe lights
89 Static discharge wicks
90 Fixed portion of trailing edge
91 Outboard spoilers
92 Spoiler hydraulic jacks
93 Flap screw jacks
94 Flap carriage mechanism
95 Outboard single-slotted Fowler-type flap
96 Fuel jettison pipe
97 Centre spoilers/airbrakes
98 Flap drive shaft
99 Aileron triplex hydraulic jacks
100 Starboard all-speed aileron
101 Inboard spoilers/lift dumpers
102 Inboard double slotted flap
103 Wing rear spar/fuselage main frame
104 Centre cabin air circulation fan
105 Pressure floor above wheel bay
106 Starboard main undercarriage, retracted position
107 Undercarriage door jack
108 Equipment bay walkway
109 Undercarriage bay pressure bulkhead
110 Flap drive motor
111 Eight-abreast tourist class seating
112 Fuselage frame and stringer construction
113 Rear freight hold door
114 LD3 baggage container (six in rear hold)
115 Freight hold bulkhead
116 Cabin floor panels
117 Seat attachment rails
118 Rear cabin air circulation fan
119 ADF aerials
120 Fuselage skin plating
121 Central overhead stowage bins
122 Ceiling lighting panels
123 Starboard rear entry door
124 Galley units
125 Fin root fairing
126 Fin attachment bolted joints
127 Fin spars
128 Starboard tailplane
129 Starboard elevator
130 Tailfin construction
131 Glass-fibre reinforced leading edge
132 Fin tip fairing
133 Static discharge wicks
134 Carbon fibre rudder skin panels
135 Honeycomb rudder construction
136 Rudder triplex hydraulic jacks
137 APU equipment bay
138 Garrett GTCP 331-250 auxiliary power unit
139 Tailcone fairing
140 APU exhaust duct
141 Port elevator construction
142 Elevator triplex hydraulic jacks
143 Static discharge wicks
144 Port tailplane construction
145 Leading edge nose ribs
146 Tailplane pivot fixing
147 Moving tailplane sealing plate
148 Tailplane centre section
149 Tailplane trim screw jack
150 Fin support structure
151 Rear pressure bulkhead
152 Rear toilet compartment
153 Cabin attendant's folding seat
154 Rear entry door
155 Cabin window panel
156 Seven-abreast rear cabin seating
157 Cabin side wall frames
158 Bulk cargo hold
159 Freight hold skin panelling
160 Wing trailing edge fillet
161 Port inboard double-slotted flap
162 Spoiler/lift dumpers
163 Undercarriage side struts
164 Main undercarriage pivot fixing
165 Inboard flap track mechanism
166 Aileron triplex hydraulic jacks
167 Port all-speed aileron construction
168 Port spoiler/airbrakes
169 Flap down position
170 Flap guide rails
171 Fuel jettison pipe
172 Flap track fairings
173 Fixed portion of trailing edge
174 Static dischargers
175 Tail navigation and strobe lights
176 Wing tip fairing
177 Port navigation lights
178 Wing rear spar
179 Front spar
180 Port leading edge slats
181 Slat screw jacks
182 Slat guide rails
183 Leading edge de-icing air piping
184 Telescopic de-icing air duct
185 Wing skin joint strap
186 Wing stringer construction
187 Port wing integral fuel tank bays
188 Wing rib construction
189 Main undercarriage leg strut
190 Hydraulic retraction jack
191 Port main undercarriage four-wheel bogie
192 Nacelle pylon attachment joint
193 Engine mounting pylon
194 Hot stream exhaust nozzle
195 Fan air exhaust duct
196 Reverser cascade, closed
197 Bleed air ducting
198 Pratt & Whitney JT9D-7R4E turbofan engine
199 Engine fan blades
200 Intake ducting
201 Detachable engine cowlings
202 Bleed air system pre-cooler
203 Inboard leading edge slat
204 Bleed air delivery ducting
205 Inner wing integral fuel tank
206 Leading edge wing root fairing
207 Wing root Krüger flap

Along with Lufthansa, Swissair was the launch customer for the A310 – the aircraft above being the prototype in Swiss colours. Five of the original Series 221 are in service, and four of the Series 322; Swissair again being the launch customer for the extended range version.

The No. 2 aircraft (F-WZLI, No 172) flew on 13 May 1982. It soon made a route-proving trip to the Far East. Up to this time an army of rivals, led by Boeing, had poured scorn on the A310, pointing out how poor the high-altitude performance would be, penalised by so small a wing. On this trip the No. 2 aircraft silenced such critics for ever. It carried a full load, equivalent to 218 passengers and baggage, 4818 km from Toulouse to Kuwait. Then it bucked an 85 km/h headwind all the way to Singapore – a distance of 7415 km, again with a full load. But it was the sector from Kuala Lumpur to Bangkok that surprised the critics. Again with a full payload, F-WZLI took off and climbed straight up to 13,100 m! Throughout the 26,000 km trip the Mach 0.8 cruise returned an average fuel burn of 2.7 Imp gal per nautical mile (6.67 litres/km). This was 6.5 per cent below the best prediction.

Powerplant choices

It should be stated that, while the first two aircraft were powered by Pratt & Whitney JT9D-7R4D1 engines, each rated at 213.5 kN (48,000 lb), an alternative launch engine was the JT9D-7R4E1 or the somewhat lighter General Electric CF6-80C2A2, both rated at 222.4 kN (50,000 lb). Since 1982 some customers have specified the PW4152, rated at 231.2 kN (52,000 lb), the biggest operator of this version being Pan Am. In 1980 several impending orders for the A310 from airlines in the Middle East appeared to offer an opening to Rolls-Royce, which by this time was beginning to realise that it had been mistaken not to 'get aboard the Airbus', but no A310 has been sold with an RB.211 engine.

In the course of A310-200 production various improvements have been introduced, the only visible one being the addition of winglets. Airbus Industrie calls these 'wingtip fences', and they are larger than those of the A300-600, and of a quite different shape. They were first fitted to a Thai Airways aircraft, delivered on 7 May 1986, and can be retrofitted. Another version, first delivered to Martinair of the Netherlands in November 1984, is the A310-200C convertible. This has an upper-deck cargo door with a width of 3.58 m, and a convertible interior which can accept 16 standard cargo pallets or any combination of cargo and passengers. The Dash-200F is a dedicated all-cargo version, with a weight-limited payload of 39.4 tonnes.

The impressive margin of actual over predicted performance naturally opened the door to new versions of the A310 to operate at heavier weights. First, the basic A310-200 was upgraded from the original maximum take off weight of 132 tonnes to 138.6 tonnes; then customers were given the option of a second upgraded weight of 142 tonnes. But perhaps more important was the development of a new longer-ranged version – the A310-300. This introduced an Airbus idea – the tailplane trim tank. Most conventional aircraft, with positive static ability, have the centre of gravity ahead of the centre of lift of the wings. Thus, the aircraft has a natural tendency to go into a steep dive or outside loop. It is prevented by the tailplane or elevators continuously being angled to push down on the tail. This fights against the lift of the wing and increases drag. In the A310-300 the tailplane is full of fuel, and this exerts the required download with no drag. As a byproduct it makes possible a fuel capacity of 61,100 litres. If necessary a further 7,200 litres can be accommodated in the aft cargo hold, but without this the A310-300 still has remarkable range, such as 8300 km with full payload. With the cargo-bay tanks fitted the full-payload (243-seat configuration) goes up to 9175 km. And this was originally planned as a smaller version of a short-haul aircraft!

The A310-300 first flew on 8 July 1985, powered by JT9D-7R4E1 engines. On 6 September 1985 the first -300 flew with CF6-80C2 engines, Air India being the first customer for this version. The normal MTO (maximum take off) weight is 150 tonnes, but customers

The A310-304s of Wardair had even greater range than the standard A310s, being provided with an additional centre tank in part of the cargo hold. So configured, the aircraft regularly crossed the Atlantic on services to the UK. Some 12 such aircraft were in service.

China's continuing demand for Western technology to improve the level of its aviation scene has resulted in the acquisition of five A310s for CAAC (Civil Aviation Administration of China). This is one of the three A310-222s: a pair of A310-304s also serve.

have the option of 153 or 157 tonnes. The first customer for the upgraded version with underfloor fuel was Wardair of Canada. This variant was certificated in November 1987. By this time the A310 had been upgraded in other ways, all of them decreasing empty weight or in some other way improving efficiency, though they were not obvious externally.

Plastic parts

Physically the biggest change came in spring 1985 when MBB introduced a completely new fin structure, made almost entirely from CFRP (carbon fibre reinforced plastics). The largest primary structure in the world in bulk production in CFRP, it saved 115 kg and had only one-twentieth as many parts as the original metal fin. The new rudder is partly of CFRP and partly of carbon fibre. There are major parts of glass, carbon and aramid fibre (Kevlar) in the horizontal tail. In 1989 the elevators were switched from metal to CFRP construction. Even bigger weight-savings were made possible by the switch, from mid-1986, to carbon brakes. The discs are actually made of RCC (reinforced carbon/carbon), and compared with the original steel brakes they last much longer and save the weight of about four or five passengers and their baggage.

Altogether the A310 has proved a smash hit, and as big a success story as its bigger ancestor. There is no doubt that its efficiency is so very high that it will continue in production even when – as seems likely – the A300-600 family are eventually replaced on the production line by the even bigger A330. With its capacity of 180 up to about 280 passengers, combined with intercontinental range, the A310 fills a very important niche in Airbus Industrie's product spectrum. It is also a very good-looking aircraft, and some of the most handsome are those of Deutsche Lufthansa in its new 1989 livery.

By the end of June 1998, A310 orders amounted to 261 aircraft, of which 255 had been delivered, and Airbus hopes for increased sales as early A310, Boeing 757 and Boeing 767 aircraft are retired.

Receiving its first aircraft on 29 March 1983, Lufthansa was actually the first airline to put the aircraft into service, this occurring on 12 April. The fleet has built-up to nine A310-203s, with the first of at least three A310-304s delivered recently.

Air Portugal ordered five A310-304s to serve its middle-distance, high-density routes, filling the gap between the Boeing 733 and the Lockheed TriStar 500.

Airbus A310-204
Thai Airways International

Chapter 2

Airbus 320

•

Airbus A330/340

•

Antonov AN-2 'Colt'

AIRBUS A320

After successes with its large A300 and A310 airliners, Airbus was hoping that the smaller A320 would equal the sales of these reliable carriers. A delay in development allowed the A320 to become easily the most advanced airliner currently in service, and sales have raced past the 1,700 mark for the A320 and its derivatives.

The attractive A320 has taken its place among the massed 737 ranks on short-haul routes in Europe, in the hands of Air France, Air Inter and British Airways. Here the three initial operators are seen in company with a development aircraft.

Airbus Industrie has from the start been identified in the minds of the public and of its customers with fat wide-body, or twin-aisle, aircraft. But for the fullest commercial success a manufacturer (of almost anything) needs to be able to offer a broad spectrum of products. From as early as 1970 the Airbus management studied possible single-aisle (SA) aircraft, much smaller than the A300B and A310 and more in the class of the 737 or DC-9. But there were so many possibilities nothing could be launched.

Dassault was building the Mercure (and losing all its investment, with a production run of ten). BAC wanted to build the One-Eleven 700/800 and a wide-body Three-Eleven, and was also a partner in the rival Europlane consortium. Hawker Siddeley wanted to build a QTOL (quiet take-off and landing) in partnership with Dornier and VFW-Fokker. Later came a whole spate of national or international projects, notably including the European JET family and the Dutch Fokker F29. But in 1979 British Aerospace became a member of Airbus Industrie, and the JET family became the Airbus SA-1 and SA-2, single-aisle projects with different body lengths to seat from 130 to 180.

Even then there was one final hurdle. In 1980 it looked as if such aircraft should be designed for propfan propulsion, offering reduced fuel consumption, without any significant loss in cruising speed. Airbus carried out a prolonged study of propfans, in partnership with propfan pioneers Hamilton Standard and Pratt & Whitney. The outcome was a decision to stick to advanced turbofans. The result was the A320, announced in February 1981 and given the go-ahead at the Paris airshow four months later. But the go-ahead was more in spirit than in practice. For one thing, no decision had been made between a range of candidate engines in the thrust range 9072 to 12250 kg (20,000 to 27,000 lb), some of which existed only on paper. But far more serious was the fact that, running true to form, the British government failed to reach any agreement with British Aerospace regarding support and launch costs. Not until 1 March 1984 could BAe at last announce an agreement, in which the early part of the programme was to be assisted by a loan of £250 million, repayable by BAe, to help the company with massive tooling costs in building the wings.

Design refinement

The 'three lost years' were not in fact lost, because throughout this period both the design of the A320 and its exceedingly advanced systems were refined and improved. The main externally obvious development was to make all aircraft the same length, and though this was fixed at only 37.7 m (123 ft 3 in), not much longer than the 'short' Dash-100, the internal cabin length was almost the same as that of the previous 'long' Dash-200. The wing passed through several evolutionary stages with six different spans, finally settling at 33.91 m (111 ft 3 in). It was long and slender, with an aspect ratio (wing length divided by mean chord) of over 9, making for aerodynamic efficiency significantly higher than the wings of the 737 or MD-80. Of course, another advantage over those older-technology rivals is that the fuselage diameter is greater; the internal cabin width is 3.7 m (12 ft 1 in), compared with 3.25 m (10 ft 8 in) for the 737 and 757 and 3.07 m (10 ft 1 in) for the MD-80 series. This enabled Airbus to make the central aisle much wider than in any previous single-aisle aircraft whilst still working with seat suppliers to develop the widest [1 m 57 cm (5 ft 2 in)] and most comfortable triple-seat unit in the world. These seats have exceptional room underneath for extra baggage, but this is seldom needed because the overhead bins offer "greater capacity per passenger than in any previous airliner of this type, at over 2.1 cubic feet."

No current airliner can match the A320 in terms of high technology, in both its aerodynamics and onboard equipment. Airbus has already developed a stretched A321 version capable of carrying up to 185 passengers over normal sectors.

To look at an A320 it hardly seems revolutionary. An aerodynamicist might comment on the almost perfect external form, and on the unique way the wings have only two sections of slotted flap on each side running aft and down on tracks to form an unbroken lifting surface from root to aileron. Outboard ailerons were retained, partly in order to form a major part of the LAF (load alleviation function). There are five spoilers above each wing, and the LAF uses the two outermost plus the ailerons. Whenever an A320 flies through turbulence (at over 200 knots and with flaps retracted) accelerometers in the fuselage sense vertical accelerations and, via the universal electrical FBW (fly-by-wire) signalling, power the ailerons and spoilers to damp out and virtually eliminate disturbance. The Australians somehow got the idea this was so that the wing could be designed weaker, the LAF meaning it would be subject to less aerodynamic forces (!!); in fact it is to reduce fatigue and give passengers a smooth ride. The four outermost spoilers control the A320 in roll, and three innermost are speed brakes and all five are lift dumpers after landing. Along the leading edge are five sections of slat.

In 1982-83 Airbus took yet a further long look at propfans, notably with a single-rotation pusher project called SBP-8, but it scored as many minus points as pluses. In the end CFM International came up with an engine specially tailored to the new airliner, and Airbus went ahead with this. The CFM56-5-A1 has improved aerodynamics throughout, with active clearance control (automatically keeping clearances very tight as the engine expands and contracts) and full-authority digital control, with a take-off thrust of 11340 kg (25,000 lb). The nacelle is made by Rohr, and Hispano Suiza (SNECMA) makes the unusual reverser which opens four giant petal doors round the pod to direct the whole airflow forwards.

When the A320 was launched, this was the only engine available. Subsequently IAE (International Aero Engines, a powerful five-nation consortium including Rolls-Royce and Pratt & Whitney) offered a basically newer and even more advanced engine, the V2500. Because it was timed later than the FCFM engine the V2500 tended to lose out in orders; indeed this problem actually led to Lufthansa switching from the V2500 to CFM. By late 1988 it was clear that the V2500 was the quietest in its class, and from the late 1980s it offered the keenest of propulsion competition. It is cowled in a completely different nacelle, with bucket type reverser, made by Rohr and Shorts. Thrust is again 11,340 kg (25,000 lb) in the initial V2500-A1 version.

State-of-the-art equipment

These are some of the visible features. Invisible, but probably the most significant part of the whole aircraft, is the amazing avionics system. As Airbus was not forced to modify the A320 from anything else, but started with the proverbial clean sheet of paper, the entire avionics are 'state of the art' and more advanced than anything in any other airliner of the period. The differences are to be found in every aspect of the avionics, from overall architecture to the appearance of the cockpit. Of course, everything is digital, and there are

Air France took delivery of its first A320-100 on 28 March 1988 to become the first operator. The publicised crash at an air show at Mulhouse cast a shadow over both airline and aircraft, but official reports have vindicated both.

numerous computers throughout the aircraft.

The fact that the flight controls are fly-by-wire has already been mentioned, but not the fact that the various flight-management and guidance systems (which are very complicated) make the A320 the safest aircraft ever built. (Even they cannot save you if you try hard enough; the first A320 to run into trouble was simply flown too low and hit trees!) The computer memories know the safe flight envelope and prevent any part of it being exceeded. One simple way of demonstrating this is to fly level and then haul hard back into a steep climb or loop. If the aircraft is flying at very high indicated airspeed, for example at full power low down, the computers reduce the input to a safe demand that would not overstress the wings; having reached a steep angle of climb, the computers progressively reduce the angle to maintain safe flying speed. If the aircraft begins at low speed the throttles will be opened fully in order to stabilise the maximum steep climb possible, never quite reaching the point of stalling. If the air is rough, the safe margin away from the stall is precisely increased. If severe windshear is encountered the A320 automatically pulls up the nose and increases power to maintain exactly the desired flight path. Not even a violent downburst encountered on the approach to a runway can disturb the serene control of the situation constantly exerted by the A320 avionics systems. No pilot could respond in the same way (after all, by the time a pilot has recognised windshear or a downburst he may be in mortal danger).

Acquiring its A320s by default, British Airways has found its Airbus aircraft efficient and reliable in operation, despite problems in setting up a maintenance infrastructure. Originally flying from Gatwick, the A320s are now based at Heathrow.

Sidestick control

Probably the most visible aspect of the avionics is the flight deck. At first sight there seems to be no way to fly the aircraft, because the traditional control yoke is missing! In its place is a small SSC (sidestick controller), as in the F-16, which responds electronically to the

The A320 has followed modern fighter jets such as the F-16 in dispensing with the conventional control column. A sidestick relays pilot input commands to the fly-by-wire system.

'State-of-the-art' is the only way to describe the A320's cockpit, featuring multi-display screens and few dials. A central fault-finding system displays impending faults to the crew.

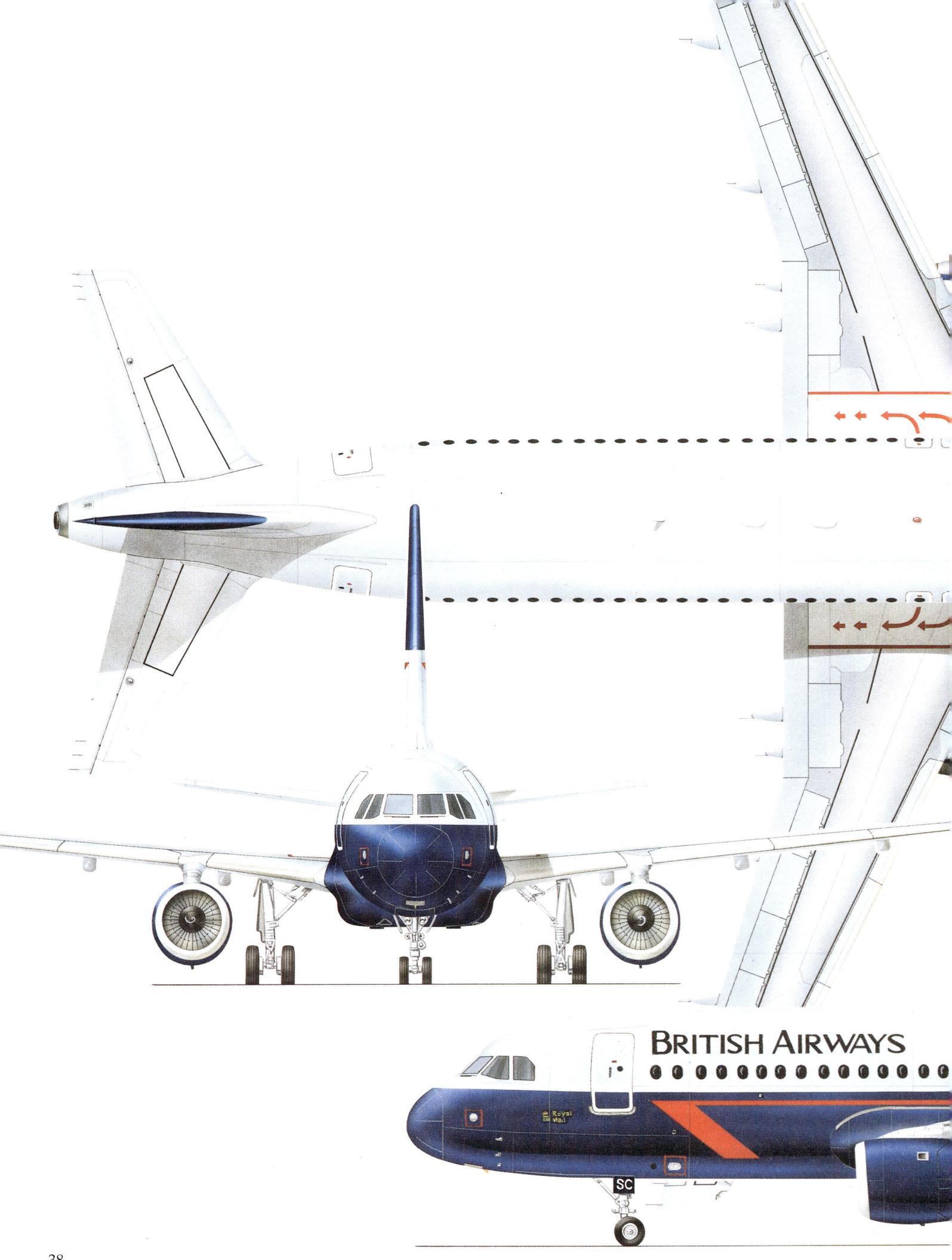

Specification: Airbus A320-110
Accommodation: flight crew of two; up to 179 passengers in high density single-class arrangement; 12 four-abreast in first class and 138 six-abreast in economy class; 84 in business and 68 in economy class (typical)
Powerplant: two CFM International CFM56-5-A1 turbofans of 104.5 kN thrust each
Dimensions: wing span 33.9 m; length 37.57 m; height 11.8 m; wing area 122.4 m^2; tailplane area 31.0 m^2
Weights: empty 38201 kg; 68000 kg; maximum payload 18801 kg
Performance: (at max T/O weight) cruise speed approx Mach 0.8; take-off distance 1707 m; landing distance 1540 m; range with reserves approx 5000 km

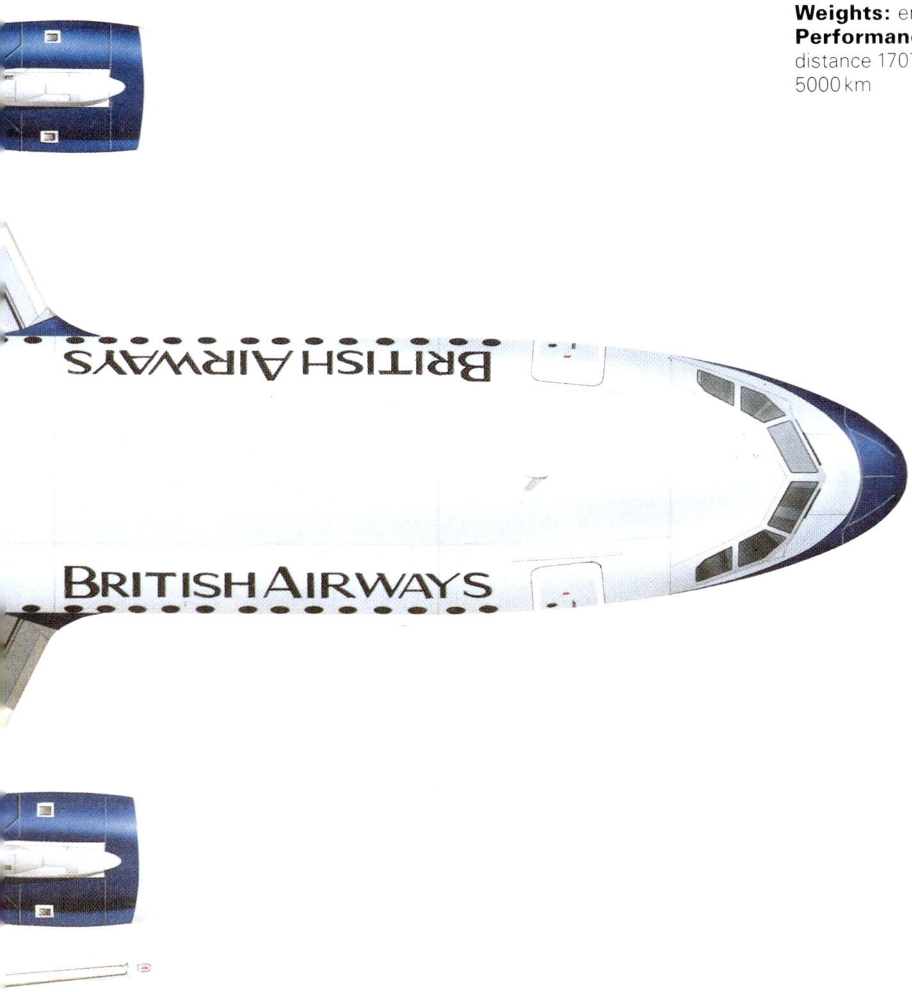

The UK national carrier British Airways has always favoured Boeing products, and despite British Aerospace playing a major role in Airbus design and production, had chosen US products in preference for several years. What was once Britain's second airline, British Caledonian, had ordered the A320 for its short-haul routes to Europe, but the two airlines merged before delivery, meaning that the Airbus aircraft started their careers in British Airways colours. Having somewhat embarrassingly become an Airbus operator, BA is now well-pleased with its A320 fleet, these posting superb reliability rates.

The first non-European operator was Ansett, the Australian domestic carrier. Here 13 A320-200s have joined Boeing 727s, 737s and 767s in connecting the major Australian cities. Note the wingtip fences, denoting this as the current Series 200 production version.

Before its integration into Air France as Air France Europe, Air Inter operated, as its name suggested, exclusively on domestic services within France. Airbus A300s formed the bulk of the fleet, but these were complemented by A320s for lower-density sectors.

Indian Airlines had a fleet of 10 A300s, and its 30 A320-200 aircraft fill the payload/range gap between the A300 and its 21 Boeing 737s. The airline operates domestic services only.

force the pilot exerts on it (complex laws have been followed to enable two pilots to transfer control, and to allow one pilot to override another, should this ever become necessary). Absence of the yoke makes room for a pull-out table in front of each pilot, and also gives a perfect view of the instruments. But again there is a surprise, because there seem to be no instruments. Instead there are just six multifunction colour displays, two facing each pilot and two on the centreline. These displays are bigger than any seen previously, at 18.4 cm (7¼ in) square. They comprise a PFD (primary flight display) and an ND (navigation display) for each pilot, while the two in the middle are the ECAM (electronic centralised aircraft monitor). Each display can be made to tell the pilot almost anything, driven by three management computers. When one counts all the displays and instruments the total comes to 12, compared with 42 in the 737-300 and 43 in the MD-80, yet the A320 pilot can call up a far greater amount of information.

Computer control

On the central console are the usual throttles (which are not connected to the engines but to the Fadecs (full-authority digital engine controls) and other controls. There are also new items: a compact and totally integrated RMP (radio management panel) on each side of the engine controls, giving instant and faultless control of every radio and navaid in the aircraft, and two MCDUs. The latter, the multi-function controller display units, are the human interfaces with another feature older aircraft do not have, a CFDS (centralised fault display system). Previous BITE (built-in test equipment) developed haphazardly, but in the A320 the CFDS is a perfect system from nose to tail and from wingtip to wingtip. It notices and records every error or fault, or even a fault about to happen. Anything really significant is told to the pilots via an ECAM. Details can be radioed ahead to the engineering base at the destination. Ground crew can enter the cockpit, touch a key on an MCDU and immediately get a long paper print-out of every detail of system function, and especially of any malfunction. Full details are given of any work to be done or tests needed, and no further forms have to be filled in or documentation raised.

With so much 'Year-2000 stuff' on board it is no wonder that this ordinary-looking aircraft is also made of high-tech materials. For example the big belly fairing is made of Kevlar aramid fibre, many parts of the wing secondary structure is special titanium alloy 'blown' to shape by SPF/DB (superplastic forming and diffusion bonding) and the entire tail is of carbon fibre composite except for the fin leading edge which is glassfibre.

Cyprus Airways was the first operator of the IAE V2500-powered A320-200. The delivery of the five aircraft allowed the carrier to retire its BAC1-11s and Boeing 707s to become an all-Airbus operator.

The first A320 was ceremonially rolled out at Toulouse on 14 February 1987 and flew on 22 February. Certification was achieved on 26 February 1988, and scheduled services began with Air France and British Airways (which took over British Caledonian, which placed the order) in April 1988. Very soon after these operators got into business, Air Inter began replacing Super Caravelles and Mercures with the A320. These three airlines are the only ones to have bought the A320-100, at a gross weight of 68 tonnes (149,915 lb), total production of this model being 21. It was quickly replaced on the assembly line by the A320-200, which accounts for the remainder of the 1,017 orders placed for the A320 by June 1998, a time by which 650 of the aircraft had been delivered.

Range extension

Except for the A320-100s all aircraft have a large centre-section fuel tank, which increases fuel capacity from 15588 litres (3,429 Imp gal) to 23430 litres (5,154 Imp gal), and makes a very large difference to the range. For example, range with 150 passengers and baggage is about 3700 km (2,000 nautical miles) for the Dash-100s, but more than 5500 km (3,000 nautical miles) for all others. A visible difference is that, except for the Dash-100s, all A320s have winglets, or as Airbus calls them, 'wingtip fences', which further enhance aerodynamic efficiency and save a significant amount of fuel. No more of the Dash-100 type will be produced, because obviously the emphasis is on ever greater capability.

In early 1989 aircraft started coming off the line with the new V2500 engine. This engine first flew in a Boeing 720 in Canada, and powered an Airbus Industrie test A320 from 28 July 1988. Two things were at once apparent. Visually, the V2500 pod looks different, with a smooth unbroken curve from the inlet to the single nozzle at the rear. In contrast, the CFM56 pod has a large forward section which discharges fan air followed by a slim aft fairing over the core. Another startling feature was the dramatically reduced noise level with the IAE engine. All A320s are quiet, but the V2500 had been measured to make the A320 with these engines, "the quietest twin-jet by a significant margin."

The first customer to put the V2500 into service was Cyprus Airways, which began receiving a fleet of eight in 1989. Soon afterwards V2500-powered aircraft were on the way to Indian Airlines, which had an initial order for 19. These were the first A320s fitted with four-wheel bogie main landing gears. Originally it was expected that customers would be offered as an option a two-wheel main landing gear with enlarged tyres inflated to reduced pressure, giving a bigger footprint area and thus enabling the A320 to operate from unimproved surfaces. Eventually it was found possible to do even better with bogie main gears, retracting into a modified bay. This landing gear was first flown on a test aircraft (Air France livery) on 19 November 1988.

Future developments

Airbus studied a number of derived versions of the A320 from 1985, and to date there have been three main developments. The first, which was launched in 1989, is the A321-100 with accommodation for 185 passengers in a lengthened fuselage. This model entered service with Lufthansa in March 1994. The A321-200 is an extended-range version with increased fuel capacity and a maximum take-off weight increased from 83,000 kg (182,980 lb) to 89,000 kg (196,210 lb) with an option at 93,000 kg (205,025 lb), and this variant was introduced in 1997. Orders for the A321 totalled 237 by June 1998, with 99 of these delivered. Launched in June 1993 and entering service with Swissair in May 1996, the second developed model is the A319-100, with a shortened fuselage for the carriage of up to 124 passengers. The type has a maximum take-off weight of 64,000 kg (141,095 lb) but is also offered in three higher-weight versions. Orders for the A319 totalled 477 by June 1998, by which time 90 had been delivered. The latest development up to 1998 has been the A318, originally schemed as the A319M5, with a five-frame reduction in the fuselage length to provide accommodation for 107 passengers. Airbus decided to go ahead with this model in 1998; all these models are offered with uprated V2500 and CFM56 turbofans.

Northwest Airlines is one of the world's largest Airbus operators, its fleet including 70 A320-200s (50 delivered and 20 on order) in mid-1998 as well as 50 and 16 examples of the A319 and A330 respectively.

A330/340

The A330/A340 is probably the most important Airbus programme of the moment. Important in so far as the future success of the manufacturer rests squarely on its capable shoulders, for these are the aircraft which the company hopes will wrest the lead in long-range aviation from Boeing's hands once and for all.

When Airbus Industrie was formally constituted in 1970 its multinational management recognised the fact that, to succeed in the long term, they would have to develop a range of products. At the start they had only one, the A300B, a TA (twin aisle) jet with a fuselage diameter of 5.6 m (18 ft 6 in), seating around 250 passengers, and two large turbofan engines. AI quickly schemed nine variations called A300B1 to B9, but was unable to launch any. Ironically, the first variant actually to go ahead was the tenth, the B10, which in 1978 became the smaller A310. But the need for more variants never went away, and year after year AI refined them and looked for a launch opening.

By 1977 work was concentrated on the B9 and B11. The B9 was to be an A300B with a stretched body and the most powerful engines available, to carry heavy loads over short/medium sectors. The B11 was to have the short body of the A310 riding on a bigger longer-span wing carrying four engines in the 98-kN (22,000-lb) class. It was to replace the 707 and DC-8 on long-haul routes where traffic did not justify use of 747s. AI kept refining these projects, in 1980 restyling them TA9 and TA11 (from twin-aisle, to differentiate them from the SA studies which led to the A320).

By 1983 AI had realised that the two TA projects could, in effect, be twin- and four-engined versions of the same aircraft, and that these could use major portions of the fuselage, tail, landing gear and systems which were common (or almost common) to the A300/310 family. This, combined with the promise of suitable engines, opened the way to a truly brilliant extension of what was fast becoming a globally successful family of jetlines. As far as the engines were concerned there was no problem: the four-engined long-hauler could use uprated versions of the CFM56, while the big twin would have a choice of Pratt & Whitney, General Electric or Rolls-Royce engines in the 267-kN (60,000-lb) class.

The greatest single breakthrough was the gradual realisation that both aircraft could use virtually the same wing. It was partly the wish to use a building-block approach, with many parts common to all the TA aircraft, that prompted British Aerospace, the principal wing designer, to study this possibility. Even though, during the period 1980-87, both the TA projects naturally grew in weight and capability, it was always obvious that the short-bodied ultra-long-haul TA11 would be considerably heavier, because of its greater fuel capacity. Fuel would fill the entire integral-tank wing box from the centreline to the tip on each side. This, together with the presence of the outboard engines, would greatly relieve the bending moment on the wing. In contrast, the big twin would have less outboard fuel and no outboard engines, and bending moment would be further increased by the longer body with more payload, thus cancelling out the advantage of the lower total weight.

On 27 January 1986 the AI supervisory board announced: "We are now in a position to finalise the detailed technical definition of the TA9, which is now designated the A330, and the TA11, henceforth called the A340. . . . This will be the last major investment to be made by the partners in . . . offering a complete aircraft family from 150 to 400 seats, optimised for stage-lengths from 300 nm (550 km) to 7,000 nm (13,000 km)." A few days later Bernard Ziegler, Vice-president Engineering and former chief test pilot, announced AI's careful study of four new technologies: propfans, active controls, all-electric systems and variable camber. The problem was: how does one judge when a new idea is mature enough to be adopted in a vehicle to do a tough non-stop job for the world's airlines, with complete reliability?

Propfans are a half-way house between the turbofan and the turboprop. Some are essentially turboprops with a large number of thin scimitar-like blades. Others are just turbofans with a bigger than

The pace of the A340's test flying programme was dramatic, and in less than one year six aircraft took to the air. The example in the foreground is the A340-200 model: 4.25 m (14 ft) shorter than the initial A340-300 machine alongside it and powered by CFM56-5C turbofans, this was the first A340 version to enter airline service.

The A340 and A330 are essentially built around the same wing. Fitted with high-lift devices and full-span spoilers (for improved braking), the wing is the key to the aircraft's great range.

Airbus Skylink's trusty fleet of five Aerospacelines Guppys has been entrusted with the job of carrying finished aircraft sections from as far apart as Bremen, Filton and Madrid to the Toulouse assembly line.

normal fan, possibly driven slower than normal by a reduction gear but enclosed in a fan duct in the usual way. They promised 'jet speed with less noise and turboprop fuel economy', and around 1980 looked very attractive. Boeing even tried to launch an all-new propfan transport, the 7J7. Sadly, there was no propfan in prospect powerful enough for the A330, but for the A340 a new five-nation consortium, IAE (International Aero Engines), offered a seemingly fabulous new engine, the SuperFan, in 1986. It was basically IAE's V2500 turbofan with the existing 160-cm (63-in) fan replaced by a fan of 280-cm (110-in) diameter, with variable-pitch blades and driven by a reduction gear at one-third of the LP (low-pressure) turbine speed. Rated in the 133-kN (30,000-lb) class, it was just what the A340 wanted. A bonus was its ability to go into reverse pitch, like a propeller. AI took this idea on board with alacrity, and on this basis the first airline commitments to the new Airbus family were announced in March 1987.

AI had said it would go ahead if it got 40 commitments from five customers, but by March 1987 it already had 104 signed up for nine airlines. A few days later one of the most respected airlines, Northwest, signed up for 20 Superfan A340s, plus options on 10 A330s. The future looked bright.

Maintaining the balance

Of the other new technologies, two were judged too much of a risk to incorporate. Today, electrical machines can be made several times more powerful than previously, for any given weight and bulk, because of the use of rare-earth magnetic materials such as SmCo (samarium cobalt). Using such motive power throughout the aircraft looks very attractive from many points of view. It would save weight and complication, eliminate hydraulics (always a source of leaks and maintenance tasks) and probably improve safety, but AI did not think all-electric aircraft, and especially all-electric flight controls, were ready for the airlines (this is quite a different matter from FBW, fly-by-wire, electrical signalling to the flight controls, which AI took for granted for both the new aircraft). Likewise, active controls, such as are needed on today's totally unstable new fighters, were judged not yet quite ready. By the year 2000 the unstable jetliner may be acceptable, but only after a few million hours with active controls on fighters.

The fourth new technology, variable camber, appeared to offer almost 'something for nothing'. Ordinary wings have essentially fixed camber (curvature), and so fly at their best L/D (lift drag ratio) at only one speed and altitude. At any other height or speed the wing is less efficient, but (reasoned AI) there is no reason why the wing should not redesign itself in flight to have different shapes. Already 'swing wing' aircraft have different degrees of sweepback, so why not a wing that can adjust its aerofoil profile? The change did not have to be large; indeed it could be provided merely by mounting the giant Fowler flaps on carriages running on two railed tracks. Thus, at take-off the flaps would be run far out behind the wing to increase area and moved down to give a low-drag high-lift setting. As the aircraft climbed to altitude the flaps would be run slowly back into their stowed position, until at some point in the flight, after burning off up to 70 tonnes of fuel, the flaps would be completely housed flush with the wing. For landing they would be extended out again, finally rotating downwards to the desired angle to give high lift and high drag for landing.

At this point the A330/340 programme underwent some major hiccups. One was that AI decided to drop the idea of variable camber, even though this promised significant gains with hardly any penalty. The reason was basically a wish for simplicity and to throw out any idea that might pose even the slightest development risk, though as the flaps and their mechanism altered only in minor details it seemed a pity. A much bigger blow was the announcement by IAE on 7 April 1987 that it was not going ahead with the SuperFan! There was again nothing wrong with the idea, simply a feeling by IAE that they might not be able to deliver on time. This caused major reverberations, Lufthansa saying, "We cannot be sure we can be confident any more on what the consortium is promising" (the airline even switched to the rival CFM engine for its big fleet of A320s).

CFM, in fact, was the only alternative for the A340, producing the Dash-5C2 version of the CFM56 engine with major advances, including a new four-stage LP compressor, active-clearance (i.e. tight

The A340's primary function is to carry passengers over ultra-long range routes without the costs incurred by operating the same routes with bigger aircraft (such as the Boeing 747-400), which previously were all that was available. With 295 passengers on board the -300 series Airbus can travel 6,650 nautical miles, and with the same passenger load and a further 38400 lb (17.4 tonnes) of cargo it can still cover a very impressive 5,100 nm. By June 1998 orders, for both the -200 and -300 variants, stood at 230, of which 140 had been delivered.

fit with almost no leak) HP compressor, new five-stage LP turbine and integrated mixer nozzle. Thus, the new engine fits in a full-length nacelle, complete with Hispano-Suiza reverser with four pivoted blocker doors. The -5C2 was certificated in December 1991 at 139 kN (31,200 lb) thrust. CFM has never stopped their aggressive development, and today is able to offer the -5C3 rated at 145 kN (32,500 lb) for A340s delivered in late 1993 and the CFM56-5C4 rated at 151 kN (34,000 lb) for early 1995.

These uprated engines have enabled AI to offer even greater payload/range performance on the A340, just as more powerful giant engines are progressively upgrading the performance of the A330. It had always been recognised that, though the A340 was the biggest advance in that it offered range capability well beyond any previous AI product – indeed, greater than any previous airliner in history – the A330 would sell in greater numbers. In terms of fuel burn per passenger-mile the A330 promised to be the most efficient vehicle ever created, and thanks to continued engine development this big twin swiftly grew to have no mean range capability itself. It was launched with engines in the 267-kN (60,000-lb) class, with a range with full payload of some 324 passengers in the 7700-km (4,800-mile) class. As now in production it has GE or P&W engines in the 302-kN (68,000-lb) class, or RR Trents of 316 kN (71,100 lb), enabling it to fly over 850 km (5,500 miles) with 335 passengers.

Originally AI found it hard to pin down the correct lengths of fuselage for the projected A330 and A340 variants, but today all versions have the same overall length of 63.5 m (208 ft 10 in), except for a special ultra-long-range model, the A340-200. This is identical to the standard A340-300 in fuel capacity, gross weight, engines and everything else, except that it has a length reduced to 59.39 m (194 ft 10 in). This reduces passenger capacity from 295 in three-class layout to only 262, though provided there are enough emergency exits the original number could be carried. The difference in length is hard to detect with such a big aircraft, but the basic effect is to increase range from 12510 km (7,770 miles) to 13990 km (8,695 miles).

Meeting thrust demands

Even the A340-200 is 12 frames (16 m/20 ft) longer than the A300-600, and the two new aircraft are easily the biggest, heaviest and most powerful aircraft ever put into production in Western Europe. The wing, developed at several BAe sites (mainly Hatfield and Bristol, is a masterpiece, with a fully supercritical aerofoil profile and larger winglets (AI calls them wingtip fences) than on the A300 or 310. The giant wings are made at Chester, with parts contributed by Textron Aerostructures in Nashville. Though the span is 60.3 m (197 ft 10 in) AI never considered making the wingtips fold, as on the Boeing 777. Virtually the only difference between the 330 and 340 wing is that the latter is reinforced in the region of the outer pylon and that slat sections 4 and 5 are modified to clear the outboard pylon strut.

A large proportion of the entire airframe is made of composite materials, including CFRP (carbon-fibre reinforced plastics), GFRP (glass) and AFRP (aramid). Major composite parts include almost the entire wing apart from the main tank box, flap tracks and slats, the giant wing/body fairing and almost the entire tail unit apart from the centre section of the tailplane inside the fuselage. As on earlier Airbuses, the tailplanes form integral tanks both to increase capacity and provide perfect longitudinal trim without aerodynamic drag.

For the first time on an Airbus the prime contractor for the landing gear is Britain's Dowty Group. The main gears are the largest and strongest ever produced in Europe. The rocking bogie units have unexpectedly resulted in 'kiss' landings undetectable by those on board. The heavier A340 has an additional twin-wheel main gear on the centreline, not found on the A330.

As before, almost the whole fuselage will be made by DASA's Deutsche Airbus division. The available capacity is impressive. Above the floor as many as 440 passengers can be carried, but such a configuration is unlikely on aircraft able to fly 18 hour sectors, so a typical two-class cabin layout in the initial A330-300 provides for 335 passengers. The A330-300 Combi is offered with a large cargo door aft of the wing, a typical load with this version being 201 passengers and six 2.44 × 3.1-m (96 × 125-in) cargo pallets, or 231 passengers and four pallets. With take-off weight in-

The use of six aircraft in the A340's flight test programme permitted the rapid accumulation of flight hours, and the aircraft flew such non-stop sectors as Toulouse-Singapore and Fairbanks (Alaska)-Toulouse. All the aircraft carried comprehensive data-recording instruments in the cabin, along with water ballast instead of passengers.

Launch customers for the A340, Air France and Lufthansa were soon joined by operators of all sizes from United Airlines and All Nippon Airways to Austrian Airlines and TAP Air Portugal. Many of these are already Airbus operators and see the A340 and A330 as a logical progression in their re-equipment plans.

creased from the regular 253 tonnes (558,870 lb) to 267 tonnes (588,600 lb), this load would be flown over 9660 km (6,000 miles). Under the floor there is room for no fewer than 32 standard LD3 containers, or 11 standard 2.24 × 3.17 m (88 × 125 in) pallets. Right at the back is a tapering bay for 19.68 m³ (695 cu ft) of bulk cargo. This exceeds the capacity of the MD-11 (which in many respects is the nearest competitor, though it has shorter range).

Among the final changes thought of before everything was locked up in concrete and metal jigs were two options for the cabin accommodation. Airbus had already pioneered flexible sybsystems linking the overhead passenger services for each seat, so that moving the seat rows became a relatively minor problem. For the A330/340 a comprehensive communications system was designed, with a low-rate satellite antenna above the fuselage (a high-rate system is an option) and a communications centre in the cabin. This centre is a true innovation, to be used by passengers and cabin crew alike. It includes telephones, fax, telex, and computer facilities, for such passenger-related tasks as onward booking of air flights, hire cars and hotels, and such cabin crew duties as updating galley

The first A330 is seen here almost complete and undergoing vibration trials on the Toulouse test rig. By June 1998 Airbus had received orders for 238 aircraft, of which 71 had been delivered.

status, bar stocks, passenger manifests, duty-free sales and other clerical tasks. The other new idea was to use a central underfloor area for off-duty crew rest, with seats, seven full-length bunk beds each with intercom and emergency oxygen, changing room, inflight entertainment/TV, refrigerator and other facilities.

By 1991 it could be said that the new family was in production, with final assembly in the gigantic new 'Clement Ader' hall, one of the largest single buildings in Europe, built at Toulouse in 1989-91. The first six aircraft were A340s. Nos 1, 2, 3 and 5 are Dash-300s, while Nos 4 and 6 are short-body Dash-200s. The first aircraft was the subject of a ceremonial 'roll-in' (the gleaming white monster being towed into the hall from outside) in the presence of almost 6,000 guests on 14 October 1991.

Into the great wide-open

This aircraft, with French test registration F-WWAI, made a flawless 4 h 47 min first flight on 25 October 1991, and was quite soon joined by the rest of the six-strong certification fleet. No flight-test programme has ever gone better. The problems have been trivial and almost all performance results have been better than predicted. Fuel burn and drag are several percentage points lower than estimated, stalls are immaculate, and there is no doubt that the biggest Airbuses are going to be built in large numbers. A340 No. 4, the first with a representative passenger cabin, began flying on 15 June 1992, and the first A330 was to fly in October 1992. All flight trials with the GE CF6-80E1, the A330 launch engine, were completed on an A300 at Mojave, California, in May 1992, and the PW4168 and Trent 770 series are on schedule. TWA specified the uprated Trent 772 to enable their A330s to carry full payload from the US West Coast to Europe.

The first A330 flew in 1992, and the first A340 and A330 flew in 1993 and 1994 respectively. By 1998 orders for the two types stood at 230 and 238. A330 variants include the baseline A330-300 for 335–440 passengers, and the A330-200 with its fuselage shortened by 10 frames for 253–405 passengers, over a longer range. Current A340 models are the A340-300 for 375–440 passengers, and the A330-200 with a shortened fuselage for 263–303 passengers.

Antonov An-2 'Colt'

Today the biplane seems as much a prehistoric anachronism as the steam locomotive. How strange, then, that one of these outdated species should have been built in huge quantities since the end of World War II, and still be in production and widespread service today.

When pictures of the Antonov An-2 reached the West in 1947, typically Soviet in being so heavily retouched as to be almost drawings, some people probably laughed openly: this so-called 'new prototype' was a biplane. It is easy to jump to ill-founded conclusions, but this really did seem to indicate a certain backwardness on the part of the Soviet designers. Even allowing for a harsh operating environment, the choice of a biplane configuration was difficult to comprehend.

The An-2 is therefore the outstanding example of how a successful aircraft may, in the Soviet context, be quite different from its counterparts elsewhere. In a capitalist society the pressure of competition results in product-improvement on a bigger scale and at a faster rate than anything so far demonstrated by Soviet industry (except in the matter of armaments) but at the start of design it eliminates anything that seems less than the best. Over the years Western aircraft companies have produced many STOL (short take-off and landing) aircraft, but they have automatically started with a monoplane wing, and achieved STOL capability not by increasing its area but by adding liberal high-lift devices. In the Soviet Union unbreakable simplicity counts for much more than appearance, and the front-line troops in World War II never scorned the Polikarpov Po-2 or newly designed Shcherbakov Shche-2 because they cruised only at 130 km/h (81 mph). As the Duke of Edinburgh said in the 1950s, "You don't have to be supersonic if your only competitor is a bullock cart."

Another factor that can easily distort one's opinion of the An-2 is size. It is natural to group single-engine biplanes together, thinking of them as being basically trainers or club aircraft. The An-2, however, is entered via a door which leads to a cabin very like that of the Douglas DC-3, though shorter, leading at the front to a real airliner flight deck. With 1,000 very fully used horsepower, the result is an aircraft of extraordinary capability and versatility. One might almost claim that '20,000 customers cannot be wrong.' Moreover, while the West has agonised for decades about the problem of the 'DC-3 replacement', the Soviets found little difficulty in building the Yakovlev Yak-40 to do that job. Their difficulty was thus to find an An-2 replacement, and after much thought their decision was overtaken by the collapse of the Soviet Union, and the An-2 therefore soldiers on.

Altogether Antonov's hefty biplane, universally loved and called *Annushka*, has succeeded in beating the amazing career of the Po-2. This much smaller biplane was admittedly built in even greater numbers, but it established the single-engine biplane as a good formula for an amazing number of contrasting missions, and over a long period of time. *Annushka* is hardly a trainer, and is not known to have been used for front-line attack missions, as was the 74.6-kW (100-hp) Po-2, but it makes up for this by doing many things beyond the Po-2's capability.

Oleg K. Antonov, like several other OKB (experimental construction bureau) leaders, began his career by building a glider for the 1924 national meet at Koktebel. He became by far the most important designer of Soviet gliders, but in World War II he had to work as first deputy to Yakovlev at Factory No. 153, Novosibirsk. In 1945 he was permitted to begin recruiting for his own OKB once more, opening it at Novosibirsk on 31 May 1946. He then moved to Kiev, where he then built up one of the top OKBs with impressive facilities now used for large and small transports. The first task assigned to the new bureau, on its formation, was to create a replacement for the Po-2 to meet the varied demands of the Ministry of Agriculture and Forestry.

Just before the war Antonov had built a version of the Fieseler Fi

Ungainly it may be as it sits on the ground, but Antonov's massive single-engine biplane has proved its critics wrong over the years through its sterling service in a multitude of roles. These examples wear Nepalese civil registration, an example of the many foreign operators who have found the An-2 an invaluable asset.

A standard An-2TD of DOSAAF, whose initials appear on the fuselage. This aircraft was delivered before 1956 and operated from Moscow Tushino. The Annushka is still used by DOSAAF and its successors for pilot and navigation training, initial air experience flying and, especially, parachute practice.

156 Storch as the OKA-38. This seemed a good starting point, and it was slightly enlarged to match the 522-kW (700-hp) ASh-21 engine, but its overall dimensions were greater than a comparable biplane. Prolonged discussion and research eventually led, in about July 1946, to a win for the biplane layout, which in parametric studies consistently showed the lower structural weight. It never occurred to Antonov that anyone might consider a biplane outdated; the calculations simply showed it to be the better choice. Authority was accordingly given for two prototypes of what was known as the SKh-1 (rural economy 1), the first having the 566.7-kW (760-hp) ASh-21 and the second the 745.7-kW (1,000-hp) ASh-62IR. The first machine was flown by N. P. Volodin on 31 August 1947.

Consideration was given to using welded steel tube and fabric fuselage construction, for cheapness and ease of local repair, but various factors, including structure weight and the fact that 'blacksmith' tube repairs might not be airworthy, resulted in light-alloy stressed-skin construction from the firewall to the tail. The wings are simple light-alloy two-spar structures, skinned with aluminium back to the front spar and fabric elsewhere, except for ribbed aluminium access panels above the six fuel cells which fill the space between the spars of the upper wing out as far as the ailerons. All control surfaces, and the tailplane, are fabric covered. So too are the four slotted flaps, which are lowered electrically to 17° for take-off; for landings they are ignored, but in really tight situations can be depressed to 45°, the ailerons (upper wing only) drooping in sympathy by 20°. The upper wings also have full-span metal slats, which open automatically at high angles of attack. The interplane struts are of the I-type, made of pressed light alloy, and the wings are braced by double sets of steel wires.

The engine is mounted on steel tubes ahead of the firewall, with the carburettor air inlet above (with filter), oil cooler below and exhaust stack on the right. In early aircraft the favoured propeller was the V-509A type with four scimitar blades of 3.6-m (11 ft 9.7-in) diameter; today the usual type is one of the AV-2 series with four broader but shorter and straighter blades with diameter of 3.35 m (10 ft 11.9 in). The fixed landing gear has a castoring tailwheel and long-stroke main oleos (ultra-long in agricultural versions). An engine-driven compressor charges a bottle in the rear fuselage which among other things pressurises the main oleos according to terrain. In most aircraft there is a cargo door on the left side which hinges upwards: originally 1.53 by 1.45 m (60.24 by 57.1 in), this is now usually 1.65 by 1.67 m (65 by 65.75 in). Inset in this is the inward-opening passenger door. The basic cabin width and height are respectively 1.6 by 1.8 m (63 by 70.9 in), giving room for 2 + 1 seating (typically four rows or 12 seats) in An-2P (*Passazhirski*) configuration. Soviet aircraft have a tough aluminium floor for cargo and, invariably, no interior trim. A bulkhead at the front, often with twin spring-loaded central doors, gives access to the flight deck. This is dominated by the fantastic area of glazing, the side windows overhanging the side of the fuselage by 0.3 m (1 ft) to give undistorted vision through flat panels straight downwards or to the rear. By modern standards the instruments and controls have always been about 20 years behind the times, though they have changed with different production models. The main feature is that modern variants can be flown solo, whereas until 1964 an engineer had to be carried as a matter of policy, the pilot not normally being permitted to touch any engine controls! The centre and left windscreens have electric heating, while the right side uses hot air from an exhaust muff which also heats the cabin. As in British wartime aircraft, the pneumatic brakes are controlled from the pilot's 'spectacles' wheel, differential action only being applied via the pedals.

Design modifications

Flight development was generally very favourable, and at an early stage it was agreed that the 745.7-kW (1,000-hp) engine was the better choice because it increased payload from 1300 to 2140kg (2,866 to 4,718 lb), with virtually unchanged flight performance. The relatively narrow chord of the biplane wings did raise problems in travel of the centre of gravity, and in the first 60 production aircraft a full-load take-off generally needed the yoke to be held at the forward stop to avoid uncontrollable pitch-up. This initial version, available from October 1948, was designated An-2T (*Transportnyi*) which at first accounted for nearly half the output. On the 61st aircraft a larger horizontal tail made full-load operation safer, but tail heaviness has always been a problem and precludes carrying anything in the capacious rear fuselage except the battery and other small items. For the record, take-offs are made tail-down, the An-2 flying off at 90 km/h (56 mph) at full load after a run of 170 m (558 ft) on grass, or less on a runway. Subsequent climb, however, is slow. Apart from low-geared ailerons, and need to hold off bank firmly in a sustained turn, the An-2 is simple to fly and the only pilot action commonly needed is to work the electric elevator trim continuously during parachute training.

The numerous known variants are listed separately. As the original customer, the agriculture ministry had first call on production aircraft, this version being the An-2SKh. At first this had an eight-blade fan driving the pump for the fluid contained in a 1400-litre (308-Imp gal) copper tank in the cabin and distributed via a spraybar system hung on six lower-wing attachments. Solids were spread

Several examples may have been built of the An-2ZA high-altitude air-sampling version. The ASh-62IR engine was blown by a TK-19 turbocharger to maintain 634 kW (850 hp) to 9500 m (31,170 ft) and the meteorological observer occupied a heated cabin at the tail, communicating via radio with the rest of the crew.

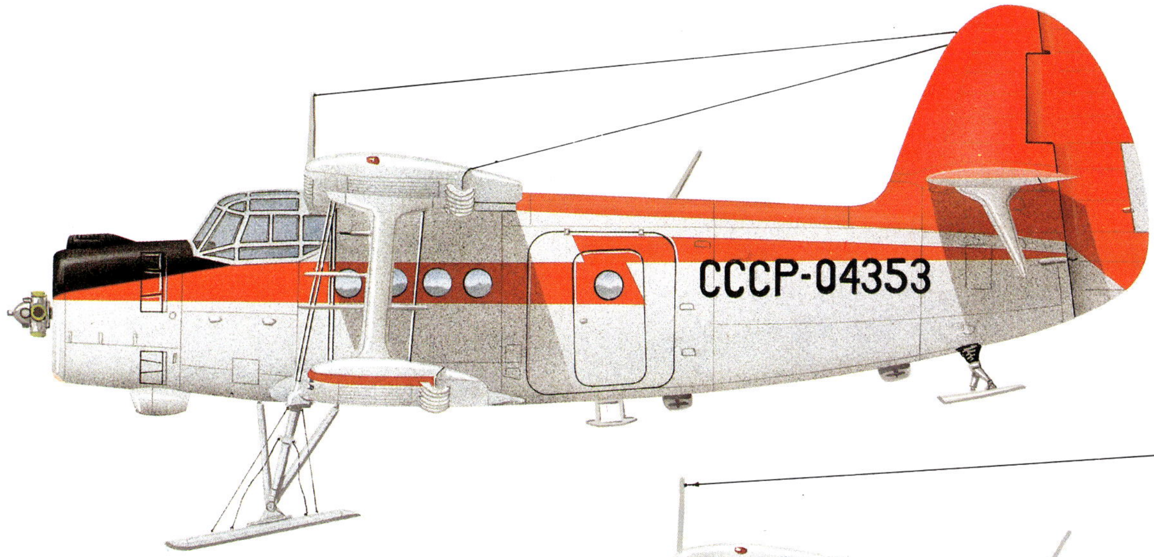

This attractive An-2 is a civil aircraft which probably operates with Aeroflot's Northern or Arctic Directorate, though Aeroflot titling is absent. Skis can quickly be fitted to all An-2 variants, though conversion to floats, as in the An-2V version, is not possible outside a factory. The bright colours are to show up against snow.

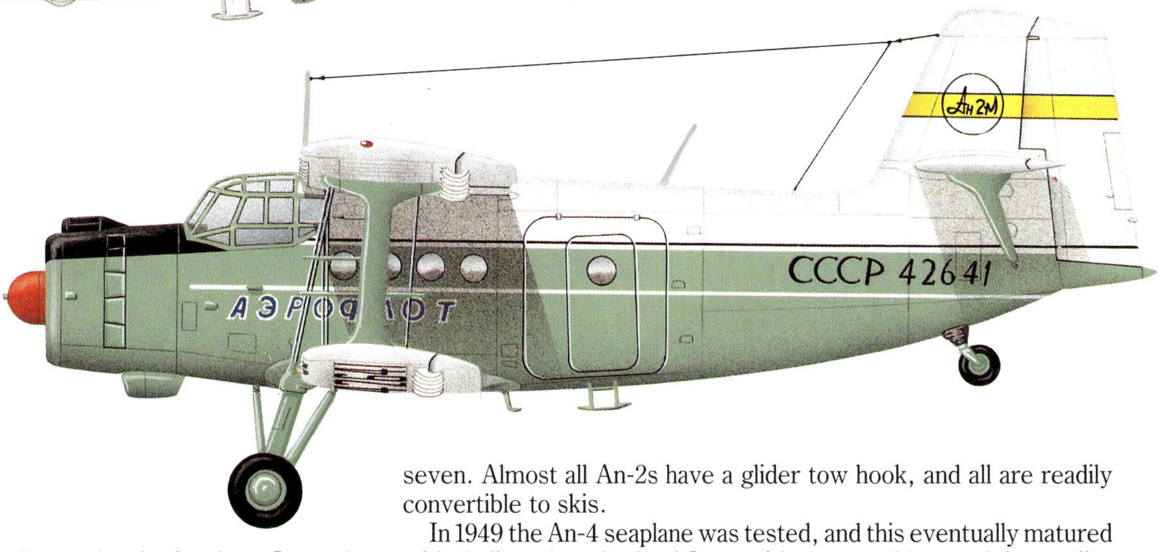

The An-2M was developed not in Poland but by Antonov's bureau at Kiev, to improve operating economics. It can be flown by a pilot without an engineer, and also has a glass-fibre hopper of increased capacity (1960 litres/431 gal). The main visible changes are to the engine cowling and vertical tail, but there are numerous others.

from a 300-mm (11.8-in) diameter duct under the fuselage. Later the fan was replaced by the VD-10 with four variable-pitch blades. The first An-2SKh was evaluated by G. I. Lysenko on a farm near Kiev in June 1948. Operating at 155 km/h (96 mph) with a swath width of 60-100 m (197-328 ft) it completed the established (Po-2) monthly work quota in three days. In 1948 5.2 million acres (2.1 million hectares) were treated by aircraft, but by 1963 the An-2 was doing 26 million acres (64.4 million hectares).

The other basic mass-production Soviet versions were the utility An-2T used by the armed forces and Aeroflot, the An-2P 14-passenger model for Aeroflot, the An-2TP for Aeroflot scheduled routes with a soundproofed and lined cabin usually with 12 seats, the military/civil An-2S (*Sanitarnyi*) with mounts for three stretchers on each side of a central aisle and an attendant seat, and the An-2TD used in vast numbers by DOSAAF (the colossal 'voluntary society for support of the navy, army and air force') which organises all sporting aviation and uses the An-2TD for parachuting with static lines, removable cargo door and folding bench seats along the sides, usually four on the left and six on the right but alternatively five and seven. Almost all An-2s have a glider tow hook, and all are readily convertible to skis.

In 1949 the An-4 seaplane was tested, and this eventually matured with shallow-draught dural floats with water rudders and the smaller V514-D9 propeller which led to today's AV-2 series. It went into limited production in 1954 as the An-2V, which has a 550-kg (1,213-lb) winch and special seals for such items as the landing and taxi lights in the lower leading edge. The An-2L (*Leso-okhraneniye*, or forest protection) carried glass flasks of fire-extinguishing fluid. It is less common than the An-2PP (*Protivo pozharnyi,* or protection against fire) seaplane water bomber which takes aboard 1260 litres (277 Imp gal) of water in five seconds' taxiing.

Soviet prototypes were numerous. One, without designation, tested soft-field landing gear (inspired by that of Bonmartini in Italy) which added front and rear wheels to each mainwheel, giving three in a row on each side, and a large ski to the tailwheel. The An-6 was fitted with a TK-19 turbocharger to maintain 633.8 kW (850 hp) to 9500 m (31,168 ft), far above the normal ceiling; with other changes it was intended for use in mountainous regions. A variant of this model was the An-2ZA (*Zondirovanyi atmosferii,* or air sampling), with a heated compartment for a scientific observer faired into the fin. Even stranger was the An-2F (*Fedya*) or An-2K (shortened from NRK,

The An-2F Fedya, also designated An-2NRK (night artillery correction), was a major rebuild for tactical observation. The gun was usually a 12.7-mm (0.5-in) UBT, but the big NS-23 was also tested. Though the twin-finned tail appears to have remained attached, Fedya was not built in quantity.

A recent oddball was this LALA-1 used by Poland's WSK-PZL-Mielec (where most An-2s have been built) to support development of the M-15, the planned successor in the agricultural role. LALA was fitted with the M-15's engine, the AI-25 turbofan, in the rear of the truncated fuselage, and with its vertical hopper fairings.

nochnoi razvedchik korrectirovshcik, or night reconnaissance and correction) for observation and artillery fire control. It had a totally glazed rear cabin surmounted by a hand-aimed gun, tapering to a slim boom carrying a twin-finned tail.

In the late 1950s about 100 An-2Ps, often with rectangular cabin windows, were produced at East Germany's Dresden plant. In 1959 general aviation was assigned, almost across the board, to Poland. Manufacture in the Soviet Union ceased at about 5,450 airframes in August 1960, since when the only production in that country has been a small batch of An-2Ms embodying 290 major changes to improve life and efficiency. The most obvious are a redesigned vertical tail and engine installation, but other features include a larger glass-fibre chemical tank of 1960-litre (431-Imp gal) capacity, a 37.3-kW (50-hp) power take-off from the engine to a pump which can double dispensing rates, a better spraying or dusting installation (which can quickly be removed) and extensive bonding and welding of metal airframe parts.

Licence production begins

In 1959 future production was assigned to the PZL factory at Mielec, Poland, where the first Polish An-2 Antek flew on 23 October 1960. Following Soviet tests of a preliminary batch of 10, the An-2 went into full production at Mielec in versions identified by different suffix letters (see list). Some of these have been developed in Poland, and in any case the Mielec engineers have introduced progressive improvements, the airframe overhaul period on the An-2R (agricultural version) being raised from 900 hours in 1961 to 1,500 hours in 1970 and 2,000 hours in 1973. Production at Mielec reached 5,000 in February 1973 and 11,950 in 1995. Soviet An-2s were exported to all Warsaw Pact countries and to Afghanistan, Cuba, Greece, India, Mali and Nepal. PZL-Mielec has exported over 10,000 to the Soviet Union, and to Bulgaria, Czechoslovakia, Egypt, France, East Germany, Hungary, North Korea, Mongolia, the Netherlands, Romania, Sudan, Tunisia and Yugoslavia.

In 1957 a licence was granted to China, where the first Yunshuji-5 (Y-5) was completed in December of that year. Production was centred at Nanchang, using the HS-5 (Chinese designation for ASh-62IR) produced at Quzhou. Y-5 production was later switched to Harbin, and since 1981 has been centred at Shijiazhuang, Hebei province. Total Y-5 production, and local Chinese variants, are not known, but the total is about 1,000.

Some of the Y-5s have PT6A turboprops, and this leads to the problem of replacing the An-2. On Aeroflot scheduled passenger routes, where reasonable airstrips are available, the replacement is the twin-turboprop Antonov An-28, seating up to 20 and cruising much faster and more smoothly than the old biplane (the An-28 is produced by PZL-Mielec). The agricultural market was assigned to Mielec's own design team under an agreement of 1 March 1971, as a result of which the prototype M-15 was flown on 20 May 1973. This was a biplane of An-2 weight but with long-span wings and powered by an AI-25 turbofan. It had many extraordinary features, and five were subjected to prolonged testing in the Soviet Union before being rejected in 1981. By this time Antonov had already completed preliminary flight trials with the first An-3 (SSSR-30576), which was virtually a standard An-2 with a 1081.3-kW (1,450-ehp) Glushenkov TVD-10 turboprop. A pre-production batch followed, with a chemical tank reportedly of 2200 litres (484 Imp gal) and a TVD-20 turboprop in a longer nose. Further trials were undertaken in 1982-83 after the cancellation of the PZL M-15 but no production or conversions of existing aircraft ensued. In 1988 another powerplant, the 1140 kW (1,528 ehp) Glushenkov TVD-1500, was being developed for the An-3, leading to renewed speculation concerning a conversion programme or renewed construction efforts.

Not including this possibility the total of An-2s so far cannot be much less than 20,000, which is not bad for an aircraft thought to be obsolete at the start!

Antonov An-2 cutaway drawing key

1. Starboard navigation light
2. Wing tip fairing
3. Wing panel fabric covering
4. Starboard leading edge slat (open position)
5. Slat guide rails
6. Interplane strut attachments
7. Aileron external hinges
8. Starboard fabric-covered aileron
9. Aileron control rod
10. Fuel tank access panel
11. Starboard upper single-slotted flap
12. Flap control rod
13. Starboard fuel tanks (total fuel capacity 264 Imp gal/1200 litres)
14. Fuel filler caps
15. Lower wing trailing edge flaps
16. Bracing wire tie rod
17. Diagonal wire bracing
18. Starboard interplane strut
19. Fabric-covered lower wing panel
20. Starboard landing light
21. Aluminium leading edge skinning
22. Engine cowling panels
23. Cowling nose ring
24. AV-2 four-bladed variable-pitch propeller (diameter 11 ft 0¾ in/3.37 m)
25. Spinner
26. Propeller hub pitch change mechanism
27. Propeller reduction gearbox
28. Carburettor air intake
29. Air intake filters
30. Oil filler cap
31. Oil tank
32. Exhaust collector ring: exhaust duct on starboard side
33. Shvetsov ASh-62IR nine-cylinder radial engine
34. Oil cooler intake
35. Starboard mainwheel
36. Oil radiator
37. Engine cooling air flaps
38. Accessory equipment compartment
39. Engine bearer struts
40. Fireproof bulkhead
41. Rudder pedals
42. Underfloor control runs
43. Control system access panel
44. Cockpit floor level
45. Ventilating air scoops
46. Side console panel
47. Instrument panel
48. Control column
49. Windscreen wipers
50. Stand-by compass
51. Instrument panel shroud
52. Windscreen panels
53. Cockpit roof glazing
54. Co-pilot's seat
55. Downward vision windows
56. Safety harness
57. Pilot's seat
58. Cockpit bulkhead
59. Main cabin doorway
60. Wing spar carry-through
61. Aerial lead-in
62. Radio aerial mast
63. Gull-wing root fillet
64. Flush D/F aerial panels
65. Passenger overhead baggage racks
66. Wing spar attachment joints
67. Upper wing root rib
68. Three-abreast passenger seating (maximum 12 passengers)
69. Cabin window panels
70. Port upper wing fuel tank bays
71. Flap shroud ribs
72. Baggage stowage areas
73. Cabin wall trim panels
74. Fuselage skin panelling

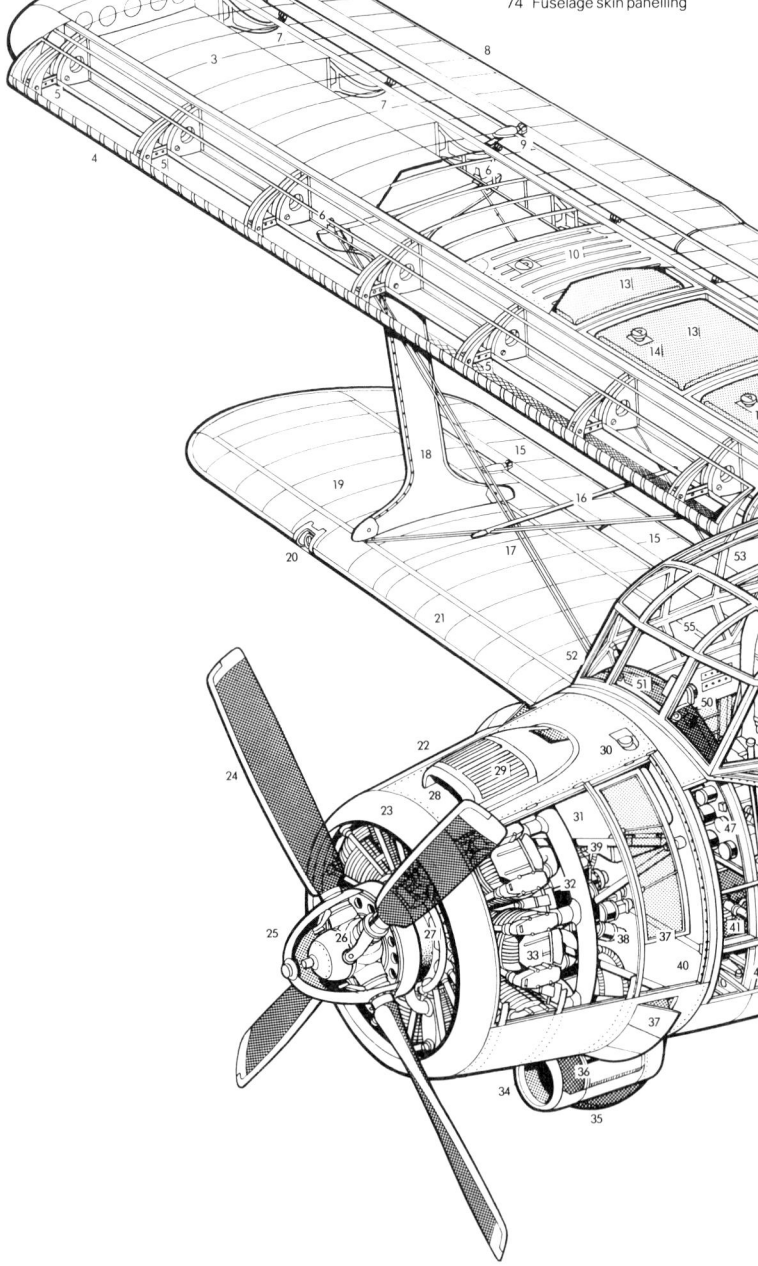

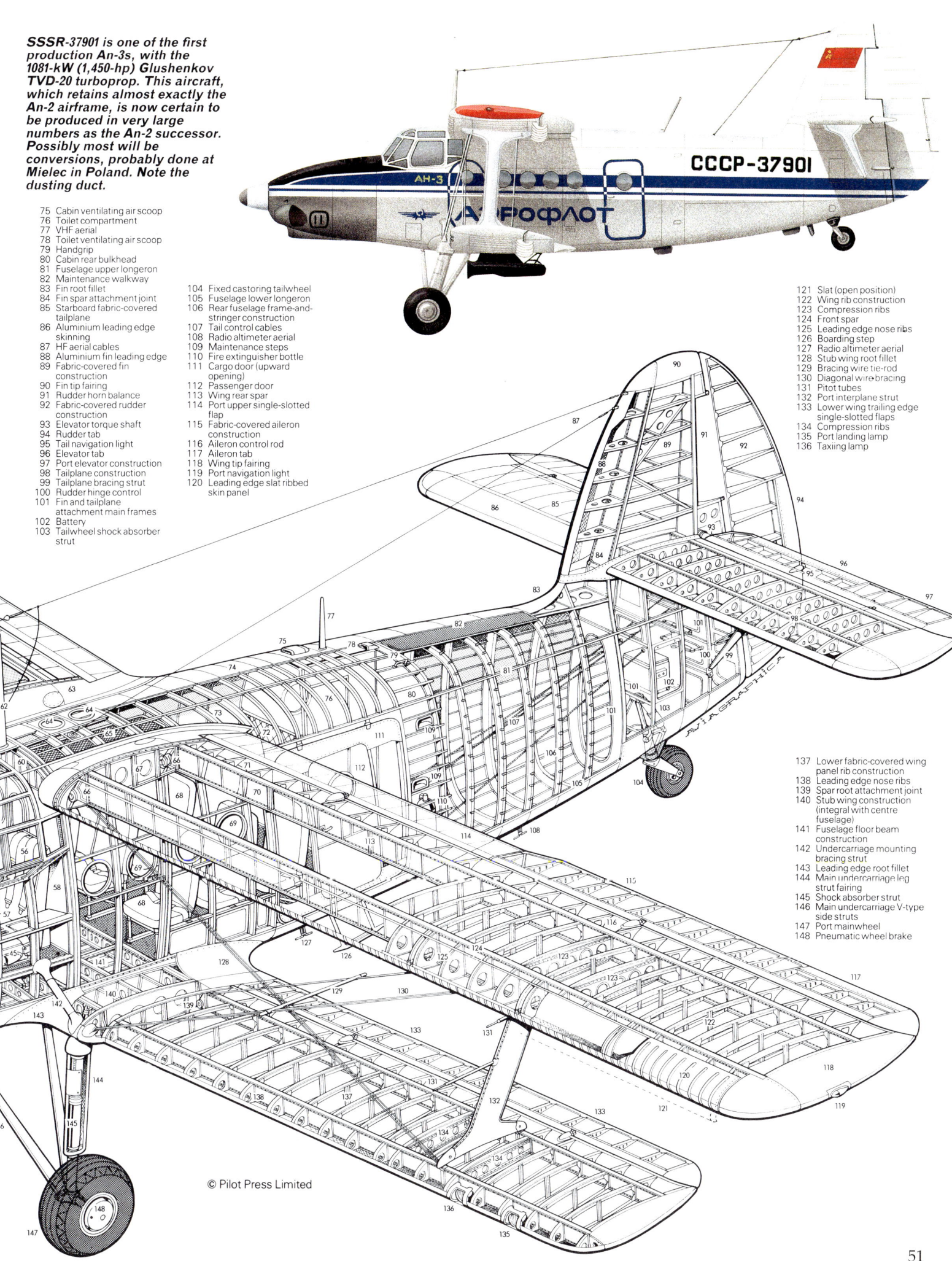

SSSR-37901 *is one of the first production An-3s, with the 1081-kW (1,450-hp) Glushenkov TVD-20 turboprop. This aircraft, which retains almost exactly the An-2 airframe, is now certain to be produced in very large numbers as the An-2 successor. Possibly most will be conversions, probably done at Mielec in Poland. Note the dusting duct.*

75 Cabin ventilating air scoop
76 Toilet compartment
77 VHF aerial
78 Toilet ventilating air scoop
79 Handgrip
80 Cabin rear bulkhead
81 Fuselage upper longeron
82 Maintenance walkway
83 Fin root fillet
84 Fin spar attachment joint
85 Starboard fabric-covered tailplane
86 Aluminium leading edge skinning
87 HF aerial cables
88 Aluminium fin leading edge
89 Fabric-covered fin construction
90 Fin tip fairing
91 Rudder horn balance
92 Fabric-covered rudder construction
93 Elevator torque shaft
94 Rudder tab
95 Tail navigation light
96 Elevator tab
97 Port elevator construction
98 Tailplane construction
99 Tailplane bracing strut
100 Rudder hinge control
101 Fin and tailplane attachment main frames
102 Battery
103 Tailwheel shock absorber strut
104 Fixed castoring tailwheel
105 Fuselage lower longeron
106 Rear fuselage frame-and-stringer construction
107 Tail control cables
108 Radio altimeter aerial
109 Maintenance steps
110 Fire extinguisher bottle
111 Cargo door (upward opening)
112 Passenger door
113 Wing rear spar
114 Port upper single-slotted flap
115 Fabric-covered aileron construction
116 Aileron control rod
117 Aileron tab
118 Wing tip fairing
119 Port navigation light
120 Leading edge slat ribbed skin panel
121 Slat (open position)
122 Wing rib construction
123 Compression ribs
124 Front spar
125 Leading edge nose ribs
126 Boarding step
127 Radio altimeter aerial
128 Stub wing root fillet
129 Bracing wire tie-rod
130 Diagonal wire bracing
131 Pitot tubes
132 Port interplane strut
133 Lower wing trailing edge single-slotted flaps
134 Compression ribs
135 Port landing lamp
136 Taxiing lamp
137 Lower fabric-covered wing panel rib construction
138 Leading edge nose ribs
139 Spar root attachment joint
140 Stub wing construction (integral with centre fuselage)
141 Fuselage floor beam construction
142 Undercarriage mounting bracing strut
143 Leading edge root fillet
144 Main undercarriage leg strut fairing
145 Shock absorber strut
146 Main undercarriage V-type side struts
147 Port mainwheel
148 Pneumatic wheel brake

© Pilot Press Limited

Chapter 3

Antonov AN-12 'Cub'

•

Antonov AN-24

Antonov An-12 'Cub'

The Antonov has been called the Soviet Union's Hercules. While the sturdy 'Cub' may not be quite in the same class as the illustrious C-130, it has been a dependable and long-lived aircraft.

An ASU-85 air-portable self-propelled anti-tank gun is disgorged from the cavernous hold of an An-12BP's 'Cub' of Military Transport Aviation. A separate ramp is carried for the loading and unloading of vehicles.

Oleg K. Antonov, most approachable of all Soviet designers, loved to show off the products of his big design bureau at Kiev. Previously known only for sailplanes and small aircraft, he sprang a major surprise in 1956 with the An-8, a big military cargo transport powered by two 3803-kW (5,100-shp) turboprops. In 1999 a few of these machines remained in service. In 1957 Antonov followed with a slightly bigger civil transport powered by four 2983-kW (4,000-shp) turboprops. Built for Aeroflot, the civil airline, this An-10 was given the NATO name 'Cat'.

It took some time to get a satisfactory production version, but eventually about 200 of these big pressurised machines went into service carrying up to 110 passengers and sometimes cargo. In terms of numbers, however, the biggest market was clearly for a military transport in the class of the Lockheed C-130. This aircraft, the An-12BP, called 'Cub' by NATO, appeared in 1959. When production ended in 1973 well over 900 had been delivered. Most went to the V-TA, the Soviet military air transport force. At peak strength these aircraft could simultaneously airlift more than two army divisions totalling over 14,000 men and all their equipment and vehicles, anywhere within a radius of 1200 km (745 miles).

In many respects the An-12 is broadly similar to the C-130, but it has a smaller wing, lighter structure, lower gross weight and engines of less power. Its usefulness is considerably curtailed by two rather strange design features, both of which stemmed from extreme difficulty in designing a good rear cargo door: it is basically unpressurised, and the rear door has no integral ramp. The only pressurised regions are the cockpit and a small cabin immediately to the rear which normally seats 14. The main cargo hold is unpressurised, though it is provided with heating. As for the rear doors, there are three of these, the two biggest being hinged along each side and arranged to open upwards and inwards. The rear section, as in many military cargo aircraft, is hinged at the rear and opens upwards to increase clearance under the tail. If it is necessary to load or unload vehicles, a separate ramp has to be carried. Normally stowed in the fuselage, this has to be physically manhandled by a large team, or slung to the overhead loading hoist, and finally pinned into place on the rear face of the open loading aperture under the floor. In combat this can be a major disadvantage. One recalls the siege of Khe Sanh, in Vietnam, where C-130s got in the habit of landing under direct hostile fire and loading and unloading every kind of commodity, before slamming the powered rear doors shut and taking off again with a total elapsed time of two minutes. With the An-12BP it takes fully two minutes to get the ramp into place so that unloading can begin. This problem does not affect air dropping, however, for which the three hinged sections of door can be hydraulically opened in flight.

One rather unexpected feature of the An-12BP is that it has a tail gun turret. Installed for self-defence, to discourage attacking fighters, this carries twin NR-23 cannon of 23-mm calibre. The turret is manned by a tail gunner who aims the guns via an optical sight system, without assistance from radar.

Flight deck

The main flight deck reflects the large specialist crews long favoured by the USSR. Two pilots sit side-by-side with dual controls, confronted by an immense array of traditional electromechanical dial instruments. In typical Soviet style air can be blown at them by two rubber-bladed fans hung from above the front windscreens, and the glass panes themselves have de-icing. To the rear on the left is the 'radist', the operator of the radio and radar,

Antonov An-12s have been the backbone of the Soviet transport fleet since the 1960s, serving with both Aeroflot and the V-TA. Despite the introduction of the Il-76, they will serve for many years yet.

The Egyptian Air Force flew the An-12 until quite recently, when US equipment, in the shape of the Lockheed Hercules, supplanted it in the final aftermath of the Soviet-Egyptian split in the 1970s.

facing forward. On the right is the engineer, again facing outward to manage giant systems panels covering the starboard wall. In the glazed nose is the navigator, who assists in lining up the aircraft for supply-dropping runs. At least in the early period of service his task involved direct map-reading by studying the features of the terrain coming up ahead. In the chin position is the main navigator and surveillance radar, which in late An-12BPs has been updated to an I-band set with the name of 'Toadstool', the radome being visibly larger. Though the rear crewman helps manage the set, the display screen is in the navigator's compartment in the nose.

Conventional structure

The basic aircraft is quite unconventional. The finely streamlined circular section fuselage was inherited from the pressurised An-10, but the internal cargo/troop hold is of rectangular section with a maximum width of 3.5 m (11 ft 5.8 in) and height of 2.6 m (8 ft 6.3 in). This cross-section is bigger than that of the C-130, especially in width. As a cargo aircraft the steel and titanium floor is stressed to support loads of 1500 kg/m^2 (307 lb/sq ft), while for loading and unloading there is an electric travelling gantry crane with a rating of 2300 kg (5,070 lb) able to reach anywhere inside the cabin. In the troop role it is possible to add seats bolted to the floor cargo attachments for a total of 90 passengers. In the paratroop role 60 can be carried, on tip-up seats which are normally carried around the walls, even in all-cargo operation. The paratroops hitch up to a static line along the interior and all can leave in 20 seconds.

The airframe has a flush-riveted stressed skin, leading edges being electro-thermally de-iced. The engines are completely underslung, and have hot bleed air anti-icing. They turn four-bladed AV-68 or AV-68B reversing propellers with electric de-icing and a diameter of 4.5 m (14 ft 9 in). Fuel to a maximum of 18100 litres (3,981 Imp gal) is housed in 22 flexible bag tanks filling the main spar box almost from tip to tip. The landing gear comprises twin steerable nosewheels and two main four-wheel bogies with anti-skid brakes. In common with most Soviet aircraft the tyre pressure is low enough for repeated operations from unpaved airstrips, being only 5.6-6.7 kg/cm^2 (80-95 lb/sq in). For operation from icy and snow-covered airstrips it is possible to replace the wheels with light alloy skis with electric heating (to avoid sticking to the ice) and integral brakes.

Unpressurised

Vehicles which can be transported include the PT-76 amphibious tank, ASU-57 and ASU-85 self-propelled artillery, ZSU-23-4 flak vehicle and all the former Soviet and Warsaw Pact range of ICVs, APCs and scout cars. The maximum cargo payload is 20000 kg (44,092 lb). This can be carried 3600 km (2,236 miles) in high-altitude cruise, but troops have to be conveyed at much lower levels because of the lack of pressurisation: here the fuel burn is greater and range less. Mention should also be made of the An-12B civil transport, which has been widely sold both to the Soviet Aeroflot and to many foreign airlines and air forces. This differs mainly in having the rear turret removed and the turret shape retained by a fairing. In a few cases the turret is retained except for removal of the guns and ammunition.

From the late 1960s An-12s have been used for various tests and trial purposes, one early conversion (probably one of many for engine testing) being used by Egypt to test the stillborn Brandner E-300 turbojet. In the Indo-Pakistan war several regular squadron aircraft (16 of which were supplied to India) were converted into rather haphazard bombers by being fitted with up to 16000 kg (35,273 lb) of conventional HE bombs (mainly thousand-pounders) carried on pallets which were manhandled out of the open rear doors when approximately over the target.

Flying testbeds

There is evidence that a number of An-12BPs have been used to test-fly many items of equipment, including sensors for reconnaissance and for atmospheric research and meteorological forecasting. At least one has been used for engine testing in the USSR, though details are lacking. Others have been used for developing ASW systems, including a MAD in a long tubular fairing projecting behind the tail and with quite different surveillance radar. Another type packages some of the special avionics and sensors into large blisters ahead of the main landing-gear pods, ahead of the rear doors and ahead of the tail turret, and with a giant radome under the tail turret which must severely restrict the allowable ground angle on take-off and landing. Another testbed, thought to be associated with ASW research, has a very large cylindrical projection extending ahead of the nose, a long ogival tapered extension aft of the tail turret, small chin radar and multiple blade antennas under the forward fuselage and outer wings. The most remarkable trials conversion yet seen, in 1986, has a gigantic box under the rear fuselage with a flat undersurface which is apparently subdivided into five dielectric rectangular panels of different sizes; this aircraft also has a long ogival tailcone extension, various small blade antennas and a long fence (possibly an antenna) along the top of the rear fuselage.

Non-transport versions

All these trials aircraft have, or had, civil registration. In contrast, there are several conversions of former transport An-12BPs which are fully operational military types, often encountered over international waters or used, in one case, operating from Egypt with that country's air force. The designations of these re-builds are unknown, so they are generally known by their NATO reporting names.

'Cub-A' is an Elint conversion. Most of the modification in this version concerns equipment, visible changes being restricted to blade antennas projecting from the fuselage around the rear part of the flight deck and from the pressurised forward cabin area.

The 'Cub' has provided the Indian Air Force with its main heavy-airlift capability since the early 1960s, earning a reputation for rugged reliability in all climates and conditions.

Antonov An-12 in service

Algeria
About six of the eight 'Cubs' delivered remain in service, wearing semi-civil registrations 7T-WAA to 7T-WAH.

Bangladesh
One An-12 may be in service.

Antonov An-12 'Cub' of the Algerian air force.

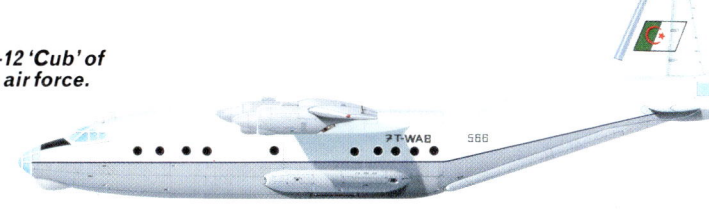

Algeria operates about half a dozen An-12s in the transport role, and these aircraft are sometimes seen in Europe now that Algeria is loosening its ties with the Soviet Union. The 'Cubs' are augmented by a fleet of US-built Lockheed Hercules.

Specification: An-12BP

Wings
Span	38.00 m	(124 ft 8 in)
Gross area	121.7 m²	(1,310.0 sq ft)

Fuselage and tail unit
Accommodation: flightcrew of five, plus pressurized accommodation for 14, and unpressurized accommodation for 90 troops or 60 paratroops

Length	33.10 m	(108 ft 7.1 in)
Cabin volume	97.2 m³	(3,432.6 cu ft)
Height	10.53 m	(34 ft 6.6 in)
Tailplane span	12.20 m	(40 ft 0.3 in)

Landing gear
Hydraulically retractable tricycle landing gear with twin-wheel nose unit and four-wheel bogie main units

Wheel track	5.42 m	(17 ft 9.4 in)
Wheelbase	10.82 m	(35 ft 6 in)

Weights
Empty	28000 kg	(61,728 lb)
Empty equipped	32000 kg	(70,547 lb)
Normal take-off	55100 kg	(121,473 lb)
Maximum take-off	61000 kg	(134,482 lb)
Payload	20000 kg	(44,092 lb)

Powerplant
Four Ivchenko AI-20K turboprops each driving an AV-68 or AV-68B four-blade propeller

Rating, each	2983 ekW	(4,000 eshp)

Antonov An-12 'Cub' recognition features

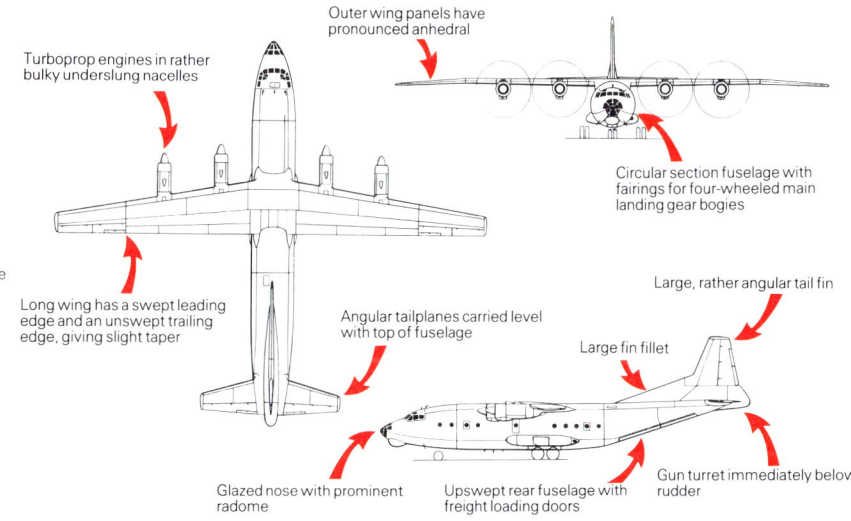

- Turboprop engines in rather bulky underslung nacelles
- Outer wing panels have pronounced anhedral
- Circular section fuselage with fairings for four-wheeled main landing gear bogies
- Long wing has a swept leading edge and an unswept trailing edge, giving slight taper
- Angular tailplanes carried level with top of fuselage
- Large, rather angular tail fin
- Large fin fillet
- Glazed nose with prominent radome
- Upswept rear fuselage with freight loading doors
- Gun turret immediately below rudder

Performance

Maximum cruising speed at 25,000 ft (7620 m)	361 kts; 670 km/h	(416 mph)
Minimum flying speed at sea level	88 kts; 163 km/h	(101 mph)
Service ceiling	33,465 ft	(10200 m)
Take-off run, concrete	700 m	(2,290 ft)
Landing run, concrete	500 m	(1,646 ft)
Range, with maximum fuel	5700 km	(3,542 miles)
Range, with maximum payload	3600 km	(2,237 miles)

Service ceiling
- An-124 'Condor' 32,810 ft
- Il-76T 'Candid' 39,370 ft
- C-141B StarLifter 38,000 ft E
- C-5A Galaxy 35,750 ft
- An-12BP 'Cub' 33,500 ft
- C-130H Hercules 33,000 ft
- An-22 'Cock' 32,800 ft E

Maximum payload
- An-124 'Condor' 150000 kg E
- C-5A Galaxy 118388 kg
- An-22 'Cock' 80000 kg
- C-141B StarLifter 41222 kg
- Il-76T 'Candid' 40000 kg E
- An-12BP 'Cub' 20000 kg
- C-130H Hercules 19356 kg

Speed at low altitude
- C-5A Galaxy 460 kts E
- C-141B StarLifter 432+
- An-124 'Condor' 432 kts E
- Il-76T 'Candid' 430 kts E
- An-12BP 'Cub' 419 kts
- An-22 'Cock' 399 kts
- C-130H Hercules 325 kts

Range (maximum payload)
- C-5A Galaxy 5526 km
- Il-76T 'Candid' 5000 km
- An-22 'Cock' 5000 km
- C-141B StarLifter 4725 km
- An-124 'Condor' 4500 km E
- C-130H Hercules 3791 km
- An-12BP 'Cub' 3600 km

Take-off run at sea level
- An-12BP 'Cub' 700 m
- Il-76T 'Candid' 850 m
- C-130H Hercules 1090 m
- An-22 'Cock' 1300 m
- C-141B StarLifter
- An-124 'Condor' 2440 m E
- C-5A Galaxy 2530 m

Czechoslovakia
Two An-12s operate with the 1st Transport Regiment based at Mosnov.

China
The An-12 is in production in China as the Y-8. The aircraft is produced as a standard military transport/freighter and as an inflight-refuelling tanker. A specialised maritime patrol version with a Litton APS-504 search radar is also under development.

One of about 17 maritime versions of the Shaanxi Y-8, which are in service with the Chinese People's Liberation Navy.

Egypt
About five of the 24 An-12s delivered remain in service. Elint 'Cub-Bs' operated in Egyptian markings during the 1970s were Soviet aircraft, and were flown by Soviet crews. Civil type registrations (e.g. SU-AOJ, SU-ARA) are worn.

An Antonov An-12 of the Egyptian air force.

Ethiopia
Fifteen An-12s were delivered, and about nine remain in service.

Guinea Republic
At least two An-12s are in service.

India
A total of 41 An-12s was delivered, eight in 1961 being followed by further batches of eight and 25 ordered in 1962 and 1963 respectively. Indian An-12s operated as bombers, and as maritime reconnaissance, SAR and airborne defence control post platforms, until replaced by Il-76 jet transports.

One of No. 24 Squadron's An-12 'Cubs'.

Iraq
Ten An-12s were originally delivered, but few, if any, remain in service. Some were used as inflight-refuelling tankers for Dassault-Breguet Mirage F1EQs during long-range attacks on Iranian oil installations. Some of these aircraft occasionally wear Iraqi Airways colours.

Jordan
Three An-12s were operated by the Jordanian Air Force until 1985.

Madagascar
At least one Antonov An-12 has been delivered to the Armée de l'Air Malgache, and serves alongside a mix of Soviet and Western transport types.

Poland
Between 12 and 20 Antonov An-12s are in service, a handful of them configured for VIP transport duties, but most serving as cargo transports.

A VIP configured An-12 of the Polish air force.

Syria
Six transport-configured 'Cubs' are in service with the Syrian air force.

USSR
Although the An-12 is being replaced by the Ilyushin Il-76, large numbers remain in front-line service. Aeroflot and military 'Cubs' have seen extensive use in Afghanistan, and several have been lost. About 260 are estimated to be in service with the Soviet air force, with increasing numbers being converted to the ECM and Elint roles. About 25 equip the navy.

Yugoslavia
Twelve 'Cubs' served with a heavy transport squadron at Beograd (Belgrade) until retired in November 1990.

This Soviet air force 'Cub-B' Elint-gathering platform was intercepted over the Baltic by a Swedish fighter aircraft.

Antonov An-12 variants
An-12 'Cub': normal designation of civil and demilitarised versions, notably with the turret or guns removed
An-12BP: standard V-TA assault transport, basically unpressurised
An-12 'Cub-A': first Elint rebuild, mainly distinguished by blade antenna
An-12 'Cub-B': major Elint conversion with mainly ventral blisters and blade antenna; operated by AV-MF
An-12 'Cub-C': major active ECM rebuild, with ventral canoe jammer installations, dispensers and extra antenna; operated at one time by Egyptian Air Force as well as USSR
An-12 'Cub-D': further active ECM version; details not yet available
Shaanxi Y-8: Chinese-built version with many minor changes
Shaanxi Y-8X: Chinese maritime version

The transport version of the Shaanxi Y-8 is basically a Chinese copy of the An-12, differing only in having an extended transparent nosecone similar to that fitted to Chinese-built 'Badgers'. The maritime reconnaissance variant has a distinctive chin radome.

Antonov An-12BP 'Cub-A' cutaway drawing key
1. Nose compartment glazing
2. Optically flat lower viewing panel
3. Nose radome
4. Weather and navigational radar scanner
5. Chart table
6. Navigator's station
7. Nose compartment entry hatch
8. 'Odd Rods' IFF aerials
9. Windscreen panels
10. Windscreen wipers
11. Instrument panel shroud
12. Pilot's instrument panel
13. Control column
14. Rudder pedals
15. Boarding ladder
16. Door mounted retractable taxiing lamp
17. Blade antennas
18. Crew entry door/escape hatch, open
19. Avionics equipment racks
20. Flight deck floor level
21. Pilot's seat
22. Cockpit eyebrow windows
23. Co-pilot's seat
24. Overhead systems switch panel
25. Aerial lead-in
26. Cockpit roof escape hatch
27. Flight engineer's instrument panels
28. Engineer's swivelling seat
29. Flight deck doorway
30. Cockpit pressure bulkhead
31. Radio operator's station
32. Nose landing gear pivot fixing
33. Pitot head
34. Nosewheel hydraulic steering control unit
35. Twin nosewheels, aft retracting
36. Blade antenna
37. Cargo deck floor level
38. Ventral access hatch
39. Cabin window panels
40. Port side emergency exit window hatch
41. Paratroop seating, 100-troops maximum
42. Central 'back-to-back' seat rows, removable
43. Cabin wall removable troop seats
44. Cabin wall insulating and trim panelling
45. Starboard side emergency exit window hatch
46. D/F loop aerials
47. Fuselage frame and stringer construction
48. Cargo deck floor beams
49. Underfloor bulk stowage compartment
50. Main cargo loading deck
51. Crew/passenger entry door
52. Wing spar/fuselage attachment main frame
53. Engine floodlight
54. Wing root fillet fairings
55. Front spar centre-section carry-through
56. Wing panel bolted root joints
57. Centre section ribs
58. Inboard bag-type fuel tanks (three), total fuel capacity 13,901 litres (3058 Imp gal), 18,100 litres (3981 Imp gal) with overload tanks
59. Starboard inner engine nacelle
60. Hinged engine cowling panels
61. Ventral oil cooling intake
62. Propeller spinners
63. AV-68 four-bladed fully feathering and reversible, variable pitch propellers

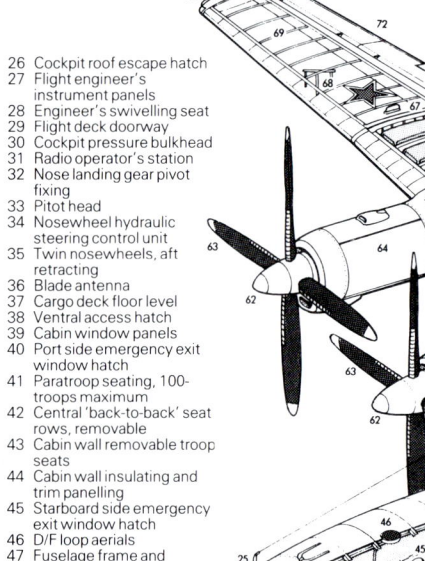

64 Starboard outer engine nacelle
65 Intermediate bag-type fuel tanks (five)
66 Outboard bag-type fuel tanks (three)
67 Outer wing panel joint rib
68 Ventral navigational antennas
69 Anhedral outer wing panel
70 Starboard navigation lights
71 Static dischargers
72 Starboard two-segment aileron
73 Aileron trim tab
74 'Cub-B' electronic intelligence variant (Elint)
75 Variant with maritime surveillance radar
76 'Cub-C' Electronic countermeasures variant (ECM)
77 ASW version with magnetic anomaly detection equipment (MAD)
78 Starboard double-slotted flap, down position
79 Flap guide rails
80 Wing root trailing edge fillet
81 ADF sense aerial, port and starboard
82 Starboard emergency exit window hatch
83 Overhead travelling cargo handling crane
84 Rear cabin roof escape hatches
85 Starboard ramp door, open
86 Fin root fillet
87 Tailfin support structure
88 Two-spar torsion box tailfin construction
89 Starboard tailplane
90 Starboard elevator
91 Fin leading edge thermal de-icing
92 HF aerial cables
93 Short wave ground-control communications antennas
94 Anti-collision light
95 Static dischargers
96 Rudder
97 Rudder trim tabs
98 Tail navigation light
99 Tail warning radar antenna
100 Rear gunner's station
101 Gun turret, two x 23-mm NR-23 cannon
102 Elevator tab
103 Port elevator
104 Static dischargers
105 Tailplane leading edge thermal de-icing
106 Two-spar torsion box tailplane construction
107 Ventral radar altimeter antenna
108 Tailplane centre-section carry-through
109 Ventral tail gunner's access door/escape hatch
110 Ramp door hydraulic jack
111 Rear ramp door, raised position
112 Cargo crane travelling rail
113 Port cargo ramp door, open
114 Flush communications aerials
115 Detachable vehicle loading ramps
116 Rear cargo loading deck
117 Vehicle loading guide rails
118 Wing root trailing edge fillet
119 Port emergency exit window hatch
120 Port double-slotted flap
121 Flap vane
122 Flap guide rails
123 Aileron tab
124 Port two-segment aileron
125 Static dischargers
126 Leading edge de-icing air exit louvres
127 Port navigation lights
128 Leading edge corrugated inner skin de-icing air duct
129 Port anhedral outer wing panel
130 Outer wing panel bolted joint rib
131 Port outboard fuel tanks
132 Engine exhaust nozzle
133 Port outer engine nacelle
134 Ivchenko AI-20K turboprop engine 2982 ekW (4000 eshp)
135 Engine bearer struts
136 Accessory equipment gearbox
137 Ventral oil cooler
138 Propeller hub pitch change mechanism
139 Propeller blade root electrical de-icing
140 Engine cowling annular air intake
141 Compressor intake
142 Engine driven generator
143 Generator cooling air duct
144 Port wing intermediate fuel tanks
145 Airborne auxiliary power unit (APU)
146 Main engine mounting wing ribs
147 Port inboard fuel tanks
148 Front spar
149 Detachable leading edge panels (engine control systems access)
150 Port inboard engine nacelle
151 Main landing gear pivot fixing
152 Hydraulic retraction jack
153 Mainwheel leg door
154 Four-wheel main landing gear bogie
155 Main landing gear sponson fairing
156 Air conditioning plant
157 Port AV-68 propellers
158 Retractable landing lamp, port and starboard
159 Air conditioning system cooling air ram intake

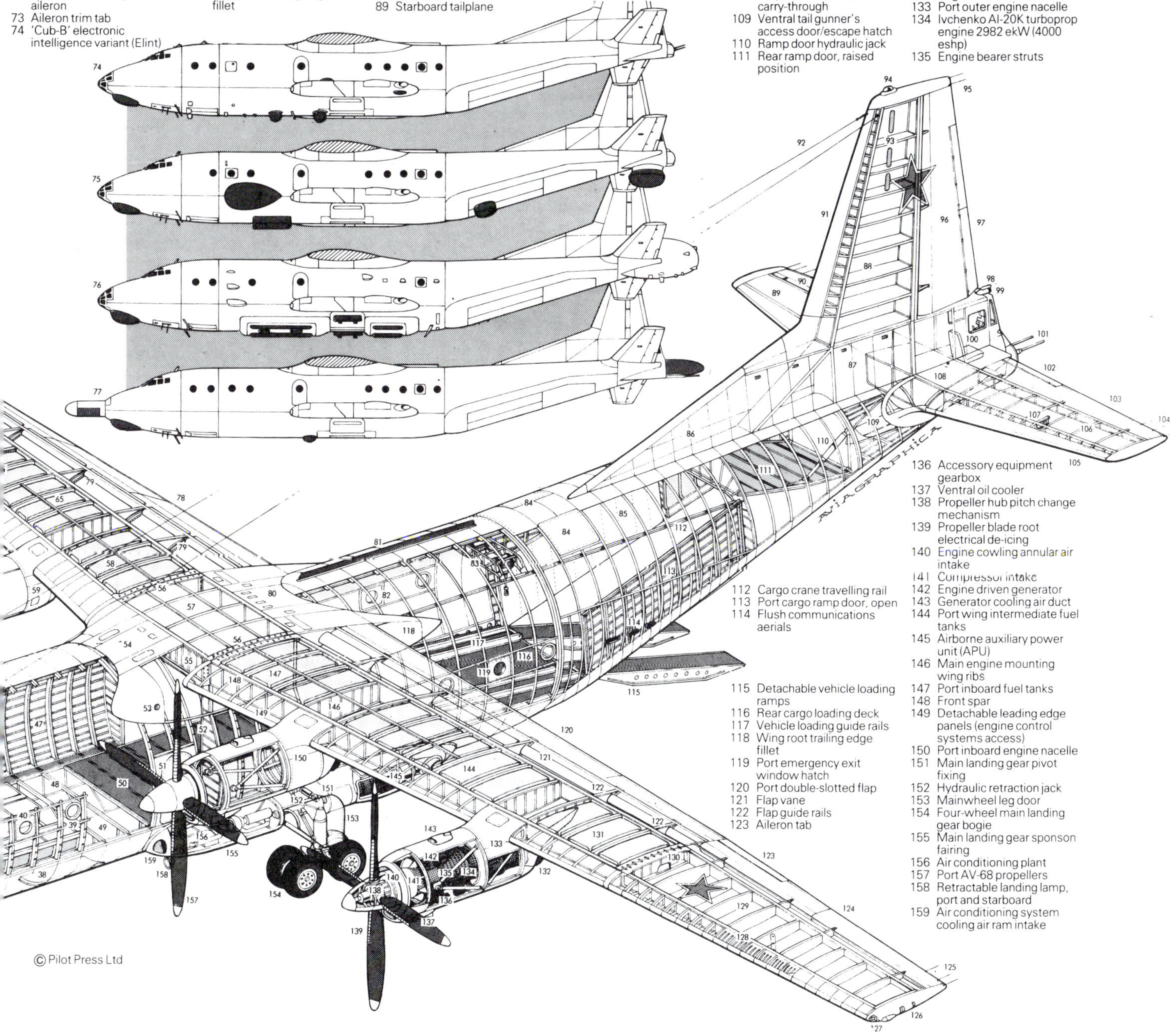

57

Antonov An-12BP 'Cub-B'
Soviet naval aviation
Aeroflot

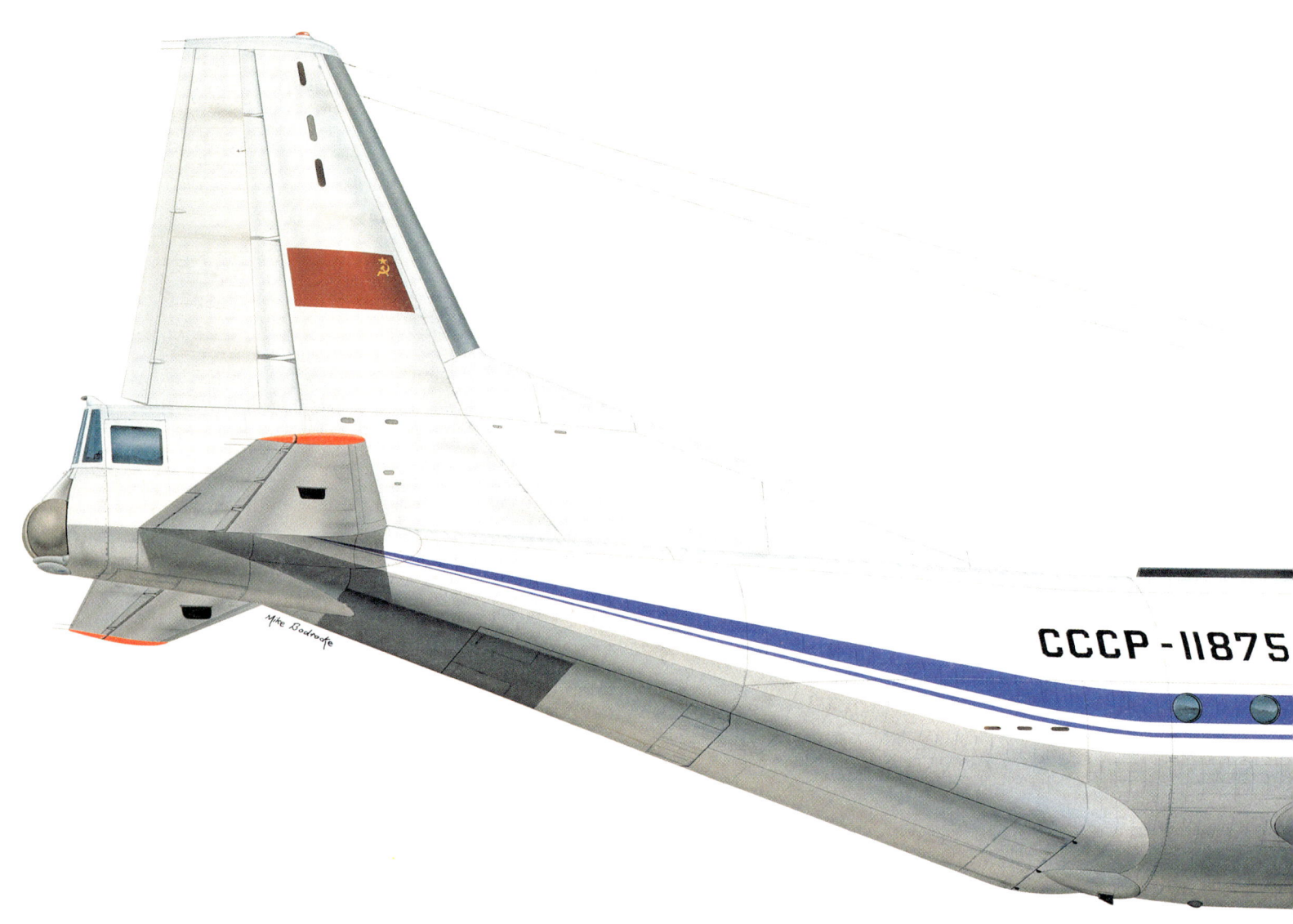

An An-12 'Cub-D' ECM platform of the Soviet Air Force, intercepted over the Baltic by a Swedish interceptor. These aircraft specialise in active ECM but may have a secondary Elint role.

'Cub-B' is an Elint conversion operated over sea areas by the AV-MF (naval aviation). This version retains the tail turret and original surveillance radar, but uses the underfloor area for approximately nine new antennas which include two hemispherical radomes. The main cabin is probably occupied by passive receivers, analysers and recording equipment,

The Zvolenský Stíhaci Letecký (Czech and Slovakia Air Force) became a 'Cub' operator fairly late in life for a Warsaw Pact air force and at one time had a pair of An-12s on strength. Both aircraft were operated by the 1st Transport Regiment based at Mosnov, near the Polish border.

and special communications radios. There are additional blades and blisters above the fuselage, as well as the fence along the top of the rear fuselage. It is thought that about 10 of this version are in service.

Active jammers

'Cub-C' is an active ECM conversion carrying several tonnes of generating, distribution and control equipment in the cabin, plus pallet-mounted jammers for at least five and probably 10 different wavebands faired into the underside of the fuselage. The forward part of the rear cargo door installation is bulged, there are dispensers for chaff and other payloads, and the tail turret is replaced by a swollen rear fuselage of almost circular cross-section tapering sharply to a rather obtuse rear fairing well aft of the tail. This tailcone may be related to that seen replacing the turret on some Tupolev 'Bear' aircraft and on the Tu-126 'Moss' AWACS platform. There are various other changes in 'Cub-C', including groups of blisters around the forward fuselage in line with the small pressurised cabin compartment. The surveillance radar is the bulged 'Toadstool'.

'Cub-D' is yet another ECM conversion. Details are lacking, but the US Department of Defense states that it carries equipment which differs considerably from that of 'Cub-C', including active jammers or decoy systems. The number of 'Cub-C' and 'Cub-D' conversions in service is estimated at 40.

New variants

New variants of these trusty former transports are likely to appear as they are replaced in the V-TA by the Ilyushin Il-76, Antonov An-124 and similar much bigger and more capable transports. In 1987 it was estimated that the number of unchanged An-12BP transports remaining in V-TA service had been reduced to about 260, probably with as many more either in store or serving in civil roles and forming an immediately available reserve force. Foreign customers include Algeria, China, Czechoslovakia, Ethiopia, India, Iraq, Madagascar, Poland and Yugoslavia.

Today the An-12 continues in production, and as an intensive development programme, but in China! Following a major detail redesign, the first Shaanxi Yunshuji-8 (Y-8) made its first flight in 1974, and since 1981 more than 50 have been delivered including a static-test specimen. A fully-pressurised 100-passenger version has been developed, and a maritime-surveillance version with an H-6 (Tu-16 'Badger') nose, Litton APS-504 surveillance radar, inertial and Omega navigation systems and mainly Collins radio. All Y-8 versions have numerous detail structural and systems changes, improved protection against icing and AI-20 K modified engines produced as the Wojiang-6. Flight Refuelling Ltd also completed a study for a Y-8 modified as a flight-refuelling tanker with three hose-drum units.

Glossary
APC Armoured Personnel Carrier
ASW Anti-Submarine Warfare
AWACS Airborne Warning And Control System
ECM Electronic CounterMeasurers
Elint Electronic intelligence
ICV Infantry Combat Vehicle
MAD Magnetic Anomaly Detector

A Chinese-built Shaanxi Y-8 of the People's Liberation Army Air Force is loaded at a Chinese air base. The Chinese have developed a dedicated radar-equipped maritime reconnaissance variant of the An-12 'Cub'.

Antonov An-24 family

In the Soviet system, if something was tough and simple it was probably made in thousands, and this applied to aircraft as much as anything. Thus Antonov's An-24 and its derivatives, though on paper rather uncompetitive, have sold in greater numbers than all the foreign rivals combined.

After World War II the 'DC-3 replacement' question bothered the USSR as much as it did the rest of the world, though the DC-3s on Aeroflot routes were called Lisunov Li-2s. S. V. Ilyushin provided the first-generation replacements in the form of Il-12s and Il-14s, but these were not a lot better than the Li-2s, and the latter continued to operate in large numbers. By 1955 the GUGVF, the chief administration of the civil air fleet, had begun to study requirements for a replacement for all these piston-engine aircraft, which at that time carried more than 90 per cent of Soviet air traffic. For longer-range routes the Antonov An-10 and Ilyushin Il-18 were eventually ordered, and rather unexpectedly the transport conversion of the Tupolev Tu-16 bomber also appeared as the Tu-104. Settling the requirement for the true mass market for the short-haul routes was more difficult. At last it was issued in December 1957.

The main reason for the delay had been uncertainty over the type of engine. Soviet conservatism if anything exceeded that of Western air carriers, and there was great reluctance to abandon the established piston engine. It must be remembered that in the mid-1950s the Il-14 was only just entering service in Comecon countries; Czechoslovakia, for example, built up its fleet from zero in 1956 to 25 in 1958. Thus piston engines were by no means regarded as in any way obsolescent. Lack of competition meant that there was no pressure to be 'first', or 'fastest', or even 'most comfortable'. On the other hand, reliability was very important, as was the ability to be maintained by tens of thousands of ground engineers who were distributed across the Soviet Union to the far Pacific, and who were certainly not PhDs!

The 1957 specification did not demand a turboprop, but it was significant that a special engine, the AI-24, had in 1955 been ordered from the A. G. Ivchenko bureau. A massive single-shaft engine, it was scaled down from the established AI-20 to reduce power from 2983 kW (4,000 hp) to 1864 kW (2,500 hp). It was deliberately made conservative in design so that it would withstand the often brutish treatment to which Soviet hardware was sometimes subjected and still run without trouble. Again, pressurisation was not demanded, but the OKB (experimental aircraft construction bureau) of Oleg K. Antonov, which was charged with designing the new transport, received permission to construct a water tank to carry out fatigue testing on pressurised fuselages, and though this was initially for the big An-10 Ukraina it would serve just as well for a pressurised local-service transport. All Antonov was specifically told was '32 to 40 seats, and it must operate from small unpaved airports'.

Antonov later said that the two-year delay in the order, from 1955 to 1957, resulted in a totally different aircraft. From the start he had decided on a high-wing monoplane (having looked carefully at multi-engined biplanes, which were not as archaic as they sounded), in order to put the cabin floor near the ground and the engines and propellers well above it, to avoid slush, stones and other material. In 1955 he had agreed with Handley Page's Herald and expected to use four piston engines, probably 522-kW (700-hp) ASh-21s. With great reluctance he again followed the thinking at Handley Page and changed to two of the new turboprops. From the operating point of view this was considered retrograde: there were half as many

Six different types of multi-blade propeller were tested in the Soviet Union from 1979, two of them on An-24 aircraft. The Soviets did not like the term 'propfan' for their subsonic straight-blade propellors which merely put more thrust through a given diameter, with less noise. This An-24RV is instrumented internally but is otherwise standard.

Typical of the civil operators of the An-24 is the Polish state airline LOT (Polskie Linie Lotnicze), which ordered a total of 14 An-24Vs to replace the piston-engined Ilyushin Il-14 on its domestic and short-haul international routes. This was the first aircraft delivered, making its first service flight between Warsaw and Wroclaw on 22 March 1966.

engines, reliability was probably going to be poorer (at least initially) and cost would be slightly higher. What tipped the scales was that in the USSR almost everything has a military angle. It was considered highly desirable, first, to get the gigantic GVF (civil air fleet) manpower trained in the rudiments of gas turbines, and familiar with them on a daily basis, and, second, to get high-octane petrol replaced by standard T-1 or TS-1 turbine fuel, so that warplanes could if necessary refuel at even the smallest civil airport.

Thus, to Oleg Konstantovich's delight, the new transport was designed in 1958 as a truly modern machine, with a big, almost circular fuselage of 0.3-kg/cm^2 (4.27-$lb/sq in$) pressure differential and two of Ivchenko's new AI-24 turboprop engines. By this time the possibility of substantial exports was obvious. Whereas few Soviet aircraft had ever been sold abroad (though plenty of Il-14s had been donated to heads of state), the new turboprop appeared to have a good chance of big sales among the growing airlines of the Third World. By the time it was designed the front-runner among Western rivals, the Fokker F.27, was already in service. Antonov never dreamed of trying to beat this in any numerical way (his duty was to meet Soviet needs), but thanks to the Soviet economic system it was obvious Fokker could be handsomely undercut on price.

Far more than with the big An-10, Antonov pulled out all the stops to make the An-24, as it was designated, totally modern and efficient. Though almost the whole primary structure was made of aluminium alloy, it included very extensive use of precision forgings, large machined planks with integral stiffening, and, above all, welding and metal bonding. The latter was a copy (without licence) of Ciba-ARL's patented method, and it was used for the entire pressure cabin and for 67 per cent of the skin overall. Throughout the airframe there were over 120,000 spot welds, most performed by precision machines. Design life of the primary structure was 30,000 hours, possibly the first time such a life had been considered in Soviet aircraft design.

Very curiously, the wing was made quite small in relation to the fuselage size and aircraft gross weight, so that the wing loading accepted was $289.7 kg/m^2$ ($59.35 lb/sq ft$) compared with (for example) $236.7 kg/m^2$ ($48.5 lb/sq ft$) for the British Herald 200 at maximum weight. Reasonable field length was achieved by powerful slotted area-increasing (almost Fowler-type) flaps. The wing was made in five sections, the four outermost having anhedral. The fuselage was truly circular at the front, and then abruptly adopted a section made up of three large circular arcs meeting at the dorsal centre-line and at a chine low down on each side (making a sharp kink at the level of the floor). This resulted in the floor having excellent width and the cockpit being amazingly broad, but with the side windows so angled that there is not much vision downwards (except on the ground through the open direct-vision panel). Another result was that there was no depth available for cargo or baggage under the floor, all such loads having to go above the floor in place of seats (in early An-24s, on each side aft of the flight deck, in line with the propellers).

All flight controls were made manual, with servo and trim (and, on the rudder, spring) tabs being made of glassfibre. High-pressure bleed air from the engines was used both to heat the leading edges for de-icing purposes and also to drive an air-cycle machine, with heat exchanger, in each nacelle to provide cabin air. The fuel system was served by flexible cells in the centre section (four or eight, depending on range requirement) and integral tanks outboard of the engines. The AV-72 four-blade propellers, of 3.9-m (127.95-in) diameter and with electrothermal de-icing, were designed specially for the task. The twin-wheel landing gears were arranged to have tyres inflated at variable pressures down to $3.5 kg/cm^2$ ($50 lb/sq in$) according to airstrip softness. Like the flaps, steering and brakes, the units were retracted forwards hydraulically, system pressure being only $154 kg/cm^2$ ($2,200 lb/sq in$). Most electrical power was DC, but raw AC was used for heating the windscreens, inlets, pitot head and propellers.

Trials and modifications

Thanks to the new engine, which initially gave $1902 kW$ ($2,550$ ehp), Antonov was able to exceed the specification and provide a cabin big enough for 50, even with the typically Soviet four-place flight deck. At first, however, Antonov hardly mentioned this possibility, and the first brochures did not go beyond a 40-seat 'all-tourist' version, as required. The prototype (L1959) was flown at Kiev by Yuri Kurlin and G. Lysenko on 20 December 1959. There were no major problems, but on the second aircraft (L1960) vertical tail area was increased by enlarging the dorsal fin and adding a ventral fin, and the engine nacelles were extending further behind the wings. A TG-16 APU (auxiliary power unit) was installed in the rear of the right nacelle, mainly to provide ground power. Five development aircraft were completed in 1961, these introducing RO-3 weather

Different in appearance from L-1959, the prototype, this An-24 was the first to be exported. It went to Lebanese Air Transport (later Lebanese Overseas) but was sold to Egypt as SU-AOM. It joined a fleet of 10 An-24Vs, but was never brought up to An-24V standard. Misrair (United Arab and today Egyptair) withdrew the Antonovs by 1975.

In the late 1960s three An-24s of early vintage were supplied to the air force of the Congo Republic, a former French colony in Africa. Ten years later they were replaced by five new An-24RVs, one of which is pictured. Utilisation with most third-world customers is extremely low, and this is especially true of air forces.

Generally similar to the earlier An-24T, the An-26 'Curl' is distinguished by a 'beaver tail' rear fuselage incorporating a loading ramp which can be slid forward for direct loading, or air-dropping operations. Serving with many air arms and a few quasi-civil operators in the transport role, the An-26 can be quickly adapted for passenger, paratroop or ambulance duties.

SSSR 30022 was the prototype An-30 photographic and survey aircraft, which triggered off modest sales which probably covered the cost of development. It uses the same basic airframe as the An-24RV apart from the new forward fuselage. At around 6095 m (20,000 ft) the photostrip width can be 14.4 km (9 miles), and sufficient oxygen is carried for eight-hour missions at up to 8230 m (27,000 ft).

radar in a longer and more pointed nose.

First delivery, to Aeroflot's Ukrainian Directorate, took place in April 1962. Flight testing was announced as complete in September, by which time regular cargo services were being flown in the Ukraine. Passenger service opened on the Kiev-Kherson route in October 1962, usually with 32 seats but from spring 1963 with nine instead of eight passenger windows on each side and 40 seats. A 44-seat version began flying from Moscow in September 1963. All these initial versions were designated An-24V Series I; NATO allotted the name 'Coke'.

By 1965 the fact that shorter field length was needed was obvious, especially in hot/high conditions. The An-24V Series II replaced the Srs I in production, with 2103-kW (2,820-ehp) AI-24T engines, with water injection, and with the inboard flaps extended in chord and slightly in span. Later in 1967 a further boost was added in the form of a Tumansky RU-19-300 turbojet/APU in the rear of the right nacelle, fed by an inlet on the nacelle inner side (opposite the main-engine jet-pipe). This could give 900-kg (1,985-lb) extra thrust, but was normally used to give about 218-kg (480-lb) thrust and also provide all electric power, thus putting more power into the propellers. Jet-equipped aircraft were designated An-24RV. A third 1967 prototype was the An-24TV (later An-24T) with the passenger door replaced by a broader rear fuselage with twin canted ventral fins and an upward-hingeing cargo door. Items could be loaded by a 1500-kg (3,307-lb) electric hoist running the length of the 15.68 m (51 ft 5 in) reinforced floor. There were fewer windows, folding seats along the sides and a crew of five.

Antonov was proud of 'his next trick', which he used on several types of aircraft. He designed a clever cargo ramp door, which, instead of hingeing up from the rear, hinged down from the front, to form a vehicle ramp. What was new was that the door could be un-pinned from its hinges and driven forward, suspended from a tracked arm on each side, until it was under the fuselage entirely ahead of the door sill. This facilitated direct loading from trucks, and also air dropping. In 1970 this aircraft was redesignated An-26 (NATO 'Curl'), and as well as being fully pressurised (which the An-24T was not) it

Yugoslavia's An-26 transports were delivered camouflaged, unlike most of those acquired by operators of twin-turboprop Antonovs. This example on climb-out shows the flaps at take-off setting, and clearly reveals the fairings along each side of the fuselage for the railed tracks which can carry the unshipped cargo door forward for direct loading from trucks.

Cuba, one of 30 air force users of the An-26, refuelled its first batch of five at London's Gatwick Airport, where this example was photographed in 1978. Cuba's aircraft wore Aeroflot livery. A further 15 passed through later, and these An-26s have subsequently worked hard on both military and civil transport duties.

was restressed for operation at weight increased from 21000 kg (46,296 lb) to 24000 kg (52,911 lb), has two additional fuel cells, a belly skinned with Bimetal (bonded ply of aluminium faced with hard titanium to resist gravel and other impacts), a 4500-kg (9,921-lb) cargo handling system flush with the floor and, usually, a bulged window on the left of the rear flight deck for contact navigation and precision drop guidance. Not obvious from the side is the upswept rear fuselage, this being hidden by deep glassfibre ventral fins on each side. After 1981 production at Kiev and Ulan-Ude centred on the An-26B, with stowable Rollgangs which, when positioned in the floor, enable two handlers to unload and load three standard pallets (total 5500 kg/12,125 lb) in as many minutes. There are many dedicated medevac, executive, paratroop and even fire-fighting versions. Over 1,000 were built before production switched to the An-32.

'Clank' and 'Cline'

A variant so different it was given a new designation is the An-30 (NATO 'Clank'), first flown in 1974. This was incorrectly said to be 'the first specialised aerial survey aircraft produced in the Soviet Union', but it may be the best. The fuselage is redesigned to accommodate a giant glazed nose for the navigator and a darkroom in the main cabin, access between the two being achieved by bodily raising the flight deck. The navigator has special precision aids for accurate positioning of the aircraft (these are progressively being updated, for example with satellite guidance), while the photographic staff have film stores, consoles for controlling the cameras and for processing film. If required magnetometers, bolometers or microwave radiometers can be carried for many kinds of geophysical, meteorological or prospecting duties.

Last of the production derived versions, the An-32 (NATO 'Cline') is a full-blooded attempt to achieve a dramatic improvement in the payload that can be carried in extremely adverse hot/high conditions. Oddly, the wing is still of the original small size, but it has leading-edge slats and triple-slotted flaps. The tailplane is redesigned, with slightly more span and chord and full-span fixed inverted slats. Much more powerful AI-20 engines are fitted, in the form of the 3670 kW (5,042 ehp) AI-20D Series 5 unit for improved take-off, service ceiling and payload capabilities under hot/high conditions. The engines are mounted extraordinarily high, the lower nacelles still housing the main gears. Curiously, despite the massive onslaught on STOL capability, take-off run is longer than for the original An-24 (760 m/2,495 ft compared with 500 m/1,640 ft), but payload is increased from 5500 kg (12,125 lb) to 6700 kg (14,770 lb). A large order from the Indian Air Force enabled the An-322 to go into production, the IAF name being Sutlej (a Punjabi river). Other, much smaller, orders have been received, and the An-32 has been bought by the home air force.

Soviet production of the An-24 ended in 1978, with about 1,100 delivered. However, the type has been in production in China as the Xian Y-7 since 1983. The first Y-7 flew on 1 February 1984, and since then an updated version has appeared – the Y7-100. This has redesigned cockpit and cabin, and drag-reducing winglets fitted. A Y7-200 version is also available with a large measure of Western technology in its powerplant and avionics, the former based on the 2051 kW (2,750 shp) Pratt & Whitney Canada PW127 turboprop.

Antonov An-24V Series II cutaway drawing key

1. Radome
2. Weather radar scanner
3. Scanner tracking mechanism
4. Radome hinges
5. ILS glideslope aerial
6. VOR localizer aerial
7. Radar transmitters and receivers
8. Forward pressure bulkhead
9. Nose undercarriage wheel bay
10. Rudder pedals
11. Instrument panel shroud
12. Radar display
13. Curved windscreen panels
14. Windscreen wipers
15. Cockpit eyebrow windows
16. Overhead systems switch panel
17. Co-pilot/navigator/radio operator's seat
18. Instrument panel
19. Control column
20. Cockpit floor level
21. Nose undercarriage pivot fixing
22. Twin steerable nosewheels, forward retracting
23. Lower electrical equipment bay, port and starboard
24. Underfloor control runs
25. Space provision for radio operator
26. Side console panel
27. Pilot's seat
28. Opening (direct vision) side window panel
29. Space provision for flight engineer
30. Circuit breaker panels
31. Aerial lead-in
32. Cockpit roof escape hatch, interchangeable with jettisonable astrodome observation hatch
33. Cockpit doorway
34. Control linkages
35. Cockpit rear bulkhead
36. Radio and electronics equipment racks
37. Baggage compartment
38. Baggage loading shelving
39. Starboard side 'up-and-over' baggage door
40. Crew wardrobe
41. Curtained cabin doorway
42. Passenger cabin front bulkhead
43. Fuselage skin doubler in line with propellers
44. Four-abreast passenger seating, 50-seat all tourist class layout

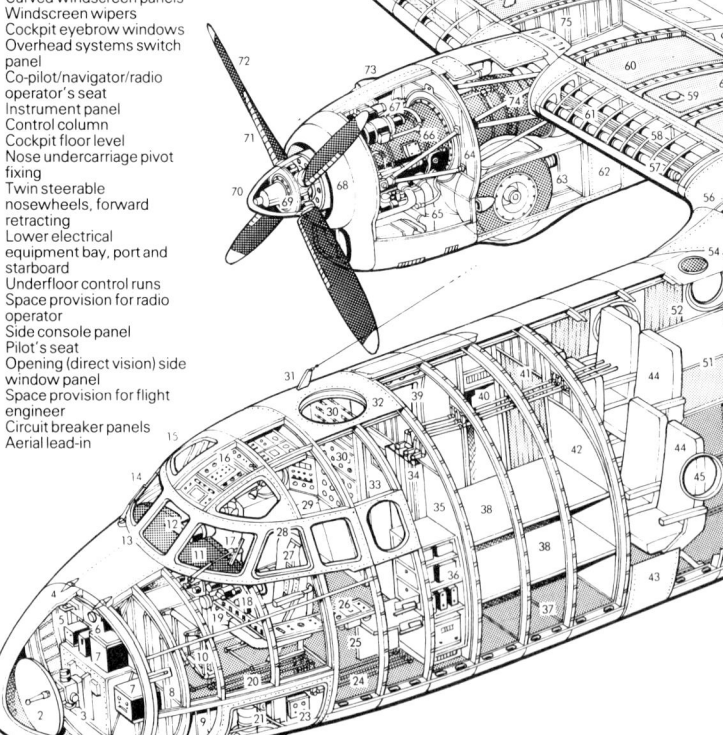

SSSR-83966 was the prototype An-32, and the only member of this family until 1982. It originally flew with 3128-kW (4,195-ehp) AI-20M engines, but was subsequently fitted with the 3863-kW (5,180-ehp) AI-20DM, giving even more outstanding 'hot and high' performance. The Sutlej transports of the Indian Air Force have the standard powerplant, but are non-standard in having Indian avionics.

Developed from the An-26, the An-32/AI-20M 'Cline' short/medium-range transport is easily distinguishable by the much enlarged ventral fin and the overwing location of the engines in much deeper nacelles. Tip-up seats along each cabin wall allow carriage of 39 passengers or 30 paratroops, and low-pressure tyres permit operation from unpaved strips.

45 Cabin window panels
46 Passenger cabin floor panelling
47 VHF aerial
48 Seat mounting rails
49 Emergency exit window hatch
50 Floor beam construction
51 Cabin wall trim panelling
52 Curtained window panels
53 Centre fuselage frame and stringer construction
54 D/F loop aerial
55 Air supply ducting
56 Wing root fillet
57 Leading-edge de-icing air duct
58 Cabin air supply duct
59 Fuel filler cap
60 Inboard bag-type fuel tanks
61 Leading-edge engine control runs
62 Starboard nacelle
63 Starboard main undercarriage, stowed position
64 Fireproof bulkhead
65 Air conditioning system, hot air supply
66 Ivchenko AI-24A turboshaft engine
67 Engine auxiliary equipment
68 Hot-air de-iced intake lip
69 Propeller hub pitch change mechanism
70 Spinner
71 Propeller blade root electric de-icing
72 AV-72 four-bladed, constant speed propeller
73 Engine cowling panels
74 Exhaust duct, exhausts on outboard side of nacelle
75 Wing panel joint rib
76 Fuel vent
77 Fuel filler cap
78 Outer wing panel integral fuel tank; total system capacity 1,220-Imp gal (5550 litres)
79 Leading edge de-icing air duct
80 Retractable landing/taxiing lamp
81 Outer wing panel joint rib
82 Anhedral outer wing panel
83 Starboard navigation light
84 Wing tip fairing
85 Starboard two-segment aileron
86 Aileron tabs
87 Outboard double-slotted Fowler-type flap, down position
88 Flap guide rails and screw jacks
89 Nacelle tail fairing
90 TG-16 turbine starter/generator, starboard side only
91 Inboard double-slotted flap segment, down position
92 Flap guide rails
93 Flap screw jacks
94 Optional long-range fuel tanks (four), capacity 228 Imp gal (1037 litres)
95 Central flap drive electric motor
96 Wing/fuselage attachment main rib
97 Wing attachment joints
98 Control access panels
99 Wing root trailing edge fillet
100 Cabin roof lighting panels
101 Overhead light luggage racks
102 Detachable ceiling panels, systems access
103 Cabin warm air ducting
104 Galley/buffet unit
105 Cabin attendant's folding seat
106 Toilet compartment
107 Coat rails
108 Tailplane de-icing air duct
109 Fin root fillet construction
110 HF notch aerial
111 Starboard tailplane
112 Starboard elevator
113 Fin leading-edge de-icing
114 Fin rib and stringer construction
115 HF aerial cable
116 De-icing air exit louvres
117 Static discharger
118 Rudder construction
119 Rudder tabs
120 Tail navigation light
121 Elevator tab
122 Port elevator rib construction
123 Static discharger
124 Tailplane leading-edge de-icing
125 Tailplane rib construction
126 Elevator hinge control
127 Radar altimeters
128 Rudder torque shaft
129 Ventral fin
130 Fin tailplane construction
131 Tailcone construction
132 Tailplane control rods
133 Rear pressure bulkhead
134 Emergency flare chutes, port and starboard
135 Tailcone access door
136 Rear baggage/wardrobe compartment
137 Sliding main entry door, open
138 Folding airstairs
139 Entry doorway
140 Passenger cabin rear bulkhead
141 Cabin fresh air supply duct
142 Cot, port and starboard, infant accommodation
143 Rear cabin passenger seating
144 Port inboard double-slotted Fowler-type flap
145 Flap screw jacks
146 Engine mounting main ribs
147 Control access panels
148 Nacelle tail fairing construction
149 Port outer double-slotted flap
150 Flap shroud ribs
151 Flap rib construction
152 Rear spar
153 Aileron tabs
154 Port two-segment aileron construction
155 Wing tip fairing
156 De-icing air outlet louvres
157 Port navigation light
158 Outer wing panel rib construction
159 Aileron segment interconnection
160 Leading-edge corrugated inner skin panel, de-icing air ducts
161 Front spar
162 Outer wing panel joint rib
163 Port wing integral fuel tank bay
164 Retractable landing/taxiing lamp
165 Wing stringers
166 Wing skin panelling
167 Hydraulic reservoir
168 Main undercarriage pivot fixing
169 Hydraulic retraction jack
170 Port engine exhaust pipe
171 Mainwheel leg doors
172 Main undercarriage leg strut
173 Twin mainwheels, forward retracting
174 Main undercarriage front strut
175 Mainwheel doors, closed after cycling of undercarriage leg
176 Mainwheel bay
177 Engine bearer struts
178 Inboard leading-edge de-icing air ducting
179 Inner wing panel fuel tank bays
180 Wing attachment fuselage main frames
181 Port engine cowling panels
182 Fireproof bulkhead
183 Main engine mounting ring frame
184 Forward engine mounting struts
185 Cabin air system cold air and pressurizing supply
186 Oil cooler
187 Engine annular air intake
188 Propeller spinner
189 Oil cooler and air system intake
190 Intake lip hot air de-icing

Chapter 4

Antonov AN-124 & AN-225

Antonov Twin AN-24/26/30/32

Armstrong Whitworth Argosy

Antonov An-124 and An-225

When it first appeared in the West at the 1985 Paris air show, there was great debate as to whether the An-124 was actually bigger than the Galaxy or not. Such chauvinism on the part of those anxious to see the West remain the holders of the 'world's biggest aircraft' title could not hide the fact that the An-124 was a superb design of immense potential. When the awesome An-225 flew in 1988 it became clear that the leadership in large aircraft design lay firmly in Antonov's hands.

Though Lockheed-Georgia proudly proclaims to be THE AIRLIFT CAPITAL OF THE WORLD, the Ukrainian city of Kiev has actually produced more different basic types of heavy airlift aircraft which, when added together, can lift more than all the Giants from Georgia. Most have been turboprops, but when in 1975 a successor was required in the heaviest category, to augment and eventually replace the An-22 Antei, there was never much doubt it would be a jet.

Engines take longer to develop than aircraft, and this programme was clearly going to need a large HBPR (high bypass ratio) turbofan in the class of the JT9D, CF6 and RB 211. The Soviet Union was very aware that this was a gap in its spectrum of engines, and a gap that was beginning to hurt the future development of civil passenger aircraft as well. Consideration was given to importing foreign (probably Rolls-Royce) engines, but this would have been totally contrary to the basic policy of self-sufficiency in all aerospace and defence capabilities. Accordingly, the ZMKB (Zaporozhye engine construction bureau), headed by General Designer Vladimir Lotarev, was asked to produce the missing engine. Designated D-18T, it benefited by being started a full ten years later than (for example) the TF39 engine of the C-5 Galaxy. Compared with that engine the D-18T came out with a smaller fan diameter, a length of 538.4 cm (212 in) compared with 688.3 cm (271 in) and lower installed pod drag, despite having a greater airflow, higher efficiencies and takeoff thrust 23400 kg (51,590 lb) compared with 18600 kg (41,000 lb). The D-18T first ran in 1981, and was designed into a very neat nacelle complete with reverser.

Having a firm engine programme is half the battle won for the aircraft designer. The other half is having a firm requirement from the customer. The Antei successor was clearly going to be a vehicle of enormous importance for many years – probably until at least 2020 – and it also had to meet the demands of the VTA, the military air-transport force, and Aeroflot, the operating arm of GVK, the civil-aviation organisation. Literally hundreds of requirements required discussion and horsetrading. Some of the civil loads were bulkier than any likely military load, whereas none of the civil loads were individually as heavy as a T-80 main battle tank. The VTA obviously needed to be able to drop heavy loads by parachute,

Wearing its Paris show number from the previous year on its nose, the prototype An-225 (having been reregistered) taxis out at the 1990 Farnborough show. This is the aircraft modified to serve as a 'Buran' transporter, but it had left its charge at home for this display.

Antonov sensibly adopted a straight-through loading method for the An-124 with a hinged nose and rear ramp. However, the An-225 is not fitted with a cargo ramp due to the different role envisaged for it.

Six Zaporozhye/Lotarev D-18T turbofans provide the power for the Mriya. The wing is essentially that of the An-124 attached to a new centre-section, with extensive use of carbon-fibre panelling along the top surface.

something hardly ever done by Aeroflot. The VTA also wanted to extract even the largest and heaviest loads by low-level cable schemes, while Aeroflot wanted compatibility with foreign civil airports. It took three years to talk everything through. Meanwhile, design of the An-124 went ahead, and the design progressed with exceptional speed and sureness. The design objectives were higher than those of the C-5 Galaxy, yet the newer technology enabled some remarkable results to be achieved. For example, the wing was given sweepback varying from 32° to 35°, which would tend to make it much heavier than the 25° Lockheed wing and also greatly lengthen the takeoff and landing runs. Nevertheless, despite the fact that its area was 6280 m^2 (6,760 sq ft) compared with 5759 (6,200), it came out at almost precisely the same structure weight. Of course the bigger and more sharply swept wing allowed the Soviet aircraft to be heavier and faster.

Cross-section of the cargo hold, determined

This, the second prototype An-124, was named Ruslan. The name has stuck with the subsequent aircraft far more satisfactorily than the NATO reporting code of 'Condor'.

by the customers' needs, was fixed at 6.4 m (21 ft) wide and over 4.4 m (14 ft 5 in) high, both considerably bigger than for the C-5. The lower deck was given full-section access at both ends, with an upward-hinged 'crocodile' nose revealing triple-hinged front ramps and a conventional pair of rear doors and upper and lower ramps. The landing gear comprised two steerable nose gears, one on each side of the fuselage, and ten twin-wheel main units in rows of five along each side of the fuselage, the two front units on each side being steerable. This arrangement conferred the required 'flotation' capability of being able to use unpaved airstrips, for example in Siberia or in military front-line operations. Like most large Soviet aircraft the An-124 was designed for inflight adjustment (up or down) of all tyre pressures. A new feature was that, by retracting the nose gears and adjusting pressure in the main oleos the whole aircraft could be made to beg nose-up or kneel nose-down to ease loading or unloading of especially massive items.

Under pressure

The requirement for full-width nose access could have been met with a flight deck in the front of the crocodile jaw nose, but a much better answer was to put it at the upper level. After some research it was considered best to pressurise the main cargo hold only to 2.44 kPa (3.55 lb/sq in), the full 5.3 kPa (7.8 lb/sq in) differential being reserved for the complete upper deck. This upper deck begins with the six-seat flight deck, with accommodation at the rear for a relief crew, needed because of the ability of the aircraft to fly for over 30 hours. Aft of the wing was arranged a passenger compartment, normally seating 88.

The design wing group achieved complete success both structurally (suffering none of the Galaxy's prolonged troubles which eventually necessitated re-winging) and aerodynamically. The entire leading edge was fitted with powered slats (called leading-edge flaps), while each trailing edge was provided with six tracks for three huge sections of Fowler flap. Outboard were placed conventional ailerons, backed up by eight sections of upper-surface spoiler and four slightly larger sections of airbrake. The leading edge was arranged to be de-iced by hot bleed air. Antonov's design team, led by Piotr Balabuyev, looked carefully at a T-tail but found the lowest weight and drag, with good control and stability in all conditions, was with a conventional tailplane attached to the fuselage. The tailplane was made fixed, and like the fin was equipped with

The An-225 was developed specifically with the Soviet space program in mind and was the first aircraft to be flown at a gross weight of 1,000,000 lb. Any future sales will now be due solely to Western interest.

The An-124 was soon used to generate hard currency for Antonov as aircraft were leased out to Western Cargo airlines. Britain's Air Foyle was the first such operator to obtain a Ruslan and six are currently in the fleet.

the unusual deicing method using giant electrical impulses. All flight controls were made fully powered, signalled by a quadruply redundant FBW (fly by wire) system, with a fifth mechanical linkage driving the surface power units directly.

Hi-tech airlift

Even in the materials used the An-124 made the maximum use of new technology. No less than 15000 m^2 (16,150 sq ft) of the exterior skin was designed in carbon and glassfibre composites, the total being 5500 kg (12,125 lb) in weight. These parts include the huge wing/body fairings and the panels along the sides which cover the re-entrant part where the upper and lower lobes of the fuselage are joined. The main cargo floor was made of high-strength and impact-resistant titanium, and special forged links were inserted between this floor and the massive forged frames encircling the fuselage so that no stress problems should result from a temperature difference of 70 °C. The great size of the wing, combined with sealing the main torsion box to form integral tanks, enabled fuel capacity to be set at 230000 kg (507,063 lb) (compared with 150820 kg (332,500 lb) for the Galaxy). The maximum payload was also established at a record level: 150000 kg (330,693 lb), compared with 118390 kg (261,000 lb) for the US aircraft. Cargo handling devices included two electric travelling cranes, together able to cover the entire cargo floor, each with a rating of 20000 kg (44,090 lb).

From the start there was no doubt the 124 was a winner, and a large manufacturing programme was organised. Each wing, left or right, is made at Tashkent, Uzbekistan, and flown to the assembly hall at Kiev bolted on top of an An-22. Many other factories contribute

The An-225 is a fully fly-by-wire design, unusual for an aircraft of its class but hardly surprising when one considers the sheer size of its control surfaces. Each wing is fitted with eight air brakes.

major parts. The first to fly, SSSR-680125, opened the flight test programme on 26 December 1982, the test crew being led by the Antonov KB's chief pilot Vladimir I. Terski. Hardly anything needed to be modified, and even the detailed software of the flight-control system needed only minor changes. The second aircraft, SSSR-82002, bore the name Ruslan, the giant folk-hero of Soviet literature and music. It took part in the 1985 Paris airshow. By this time trials were involving load-carrying flights all over the Soviet Union, and from the end of 1985 it could be said the An-124 was in service with Aeroflot. One load, in January 1986, was a 154-tonne Euclid dump truck for the Yakut diamond mines. In May 1987 Terski and two crews flew 20150 km (12,521 miles) round the Soviet Union in 25 hrs 30 mins, landing with a 120000 kg (26,450 lb) fuel reserve.

By 1991 at least 23 An-124s were in service, and the Antonov bureau had formed a special company to sell cargo space all over the world. With the slogan "The 124 can carry anything anywhere" they are achieving increasing success, especially when the load cannot be carried by anything else. Even when it can, the sheer capability and efficiency of the 124 has won major contracts, including that between Montreal and Toulouse carrying all Canadair's A330/340 fuselage parts and to Chester carrying A330/340 leading edges. Among 124 loads are 50 normal cars or 70 compact ones, 100000 kg (22,045 lb) of fresh meat (never an easy load) or 176 head of bloodstock. Production of this important aircraft is continuing.

In 1988 the Antonov OKB announced that it was building a single aircraft even larger than the An-124. Three years earlier it had been assigned the task of designing an aircraft able to carry, as a piggy-back load, such enormous items as the Buran manned space orbiter, Energiya space launch vehicle and various gigantic structures needed by the burgeoning oil, gas, petrochemical and construction industry, especially in Siberia. A preliminary investigation showed that it would be a fairly straightforward exercise to use the An-124 as a basis and stretch it in ways involving the minimum technical risk. The main changes agreed were: to add a new wing centre section with two additional engines; add plugs to lengthen the fuselage ahead of and behind the wing; fit a new twin-finned tail to a modified rear fuselage; add

extra units to the main landing gears; and add secondary dorsal structures to carry the various payloads. At first a daunting challenge, the problem did not seem to pose any high-risk areas when it was broken down and studied piecemeal. On the other hand the new aircraft, by far the largest in the modern world, did seem to pose some basic piloting difficulty when on the ground, and one study addressed the problem of matching the aircraft to existing aprons, taxiways and runways.

Strong shoulders

The work went ahead under Deputy General Constructor Anatoli Bulanenko. The extra centre section is the strongest structure in the history of aviation. It is horizontal, whereas the left and right wings attached to it slope downwards with anhedral as in the An-124. The trailing edge of this centre section is straight, unlike the original wings which sweep back from the roots. On it is attached a single section of slotted Fowler flap, which travels out on two tracks. The leading edge, however, is fixed. Along the top of the wing the spoilers/airbrakes have been redesigned to cover the entire span, the centre section having four airbrakes and the main wing four airbrakes inboard and eight spoilers outboard, a total of 32 powered surfaces for the whole aircraft. The new tail was given a fin and two-part (upper and lower) rudder mounted at right angles on each end of an almost untapered horizontal tail with dihedral; thus, the fins and rudders slope inwards. Each tailplane is fixed to a broad but flat rear fuselage and carries three sections of elevator. As in the

When displaying at air shows, the An-225 has proved itself to be a spectacular performer with startling turns and low-speed handling. At Paris it flew just such a routine while carrying the Buran shuttle.

An-124, all control surfaces are fully powered and controlled by a quadruple fly-by-wire system.

Much thought was given to the length of fuselage. The intention was that the new aircraft should carry outsize payloads on top of the fuselage, not inside the hold. In theory stability and control would have been adequate with the original fuselage, which of course would also have been lighter. The longer fuselage improves the aerodynamics, puts the tail further behind the payloads (some of which

71

The An-124 still owes much to 'old-world' aircraft design. It retains a distinctly non-user friendly cockpit with banks of dials and gauges. In Aeroflot service it was quite normal to have a flight deck crew of seven.

were found in model testing to cause very turbulent wakes) and enabled the number of main landing gears on each side to be increased from five to seven. Testing showed that manoeuvrability on the ground would be increased by removing the steering from the forward main units and fitting it to the rear four units on each side. Thus the pilot can power-steer no fewer than 20 of the 32 wheels.

Space flight

Though it was thought that internal cargo would seldom be carried, the front end was made pure An-124. At the rear, however, a new and simplified rear fuselage was designed, longer than before and with no cargo door or ramp. On top the fuselage structure was stressed to carry the various heavy loads, which impart weight, drag and inertial loads (as, for example, during applied roll manoeuvres). It was decided to equip the first aircraft to carry Buran spacecraft, and it was accordingly given two large axial beams, raised above the fuselage at the roots of the wings to serve as the main load pylons, together with multiple attachments in line with the leading edge, aft of the wing and in line with the leading edge of the tailplane to anchor the Buran bracing struts.

The Antonov bureau assigned the designation An-225. Bulanenko said "We felt this aeroplane deserved a name, but most classical mythology had already been used. As we are based in the Ukraine we picked a Ukrainian word, Mriya, meaning dream". The first An-225, SSSR-480182, was ceremonially rolled out from the Kiev assembly hall at the end of November 1988. It was the first time such an event had taken place in public in the Soviet Union, and reported around the world. The occasion was reminiscent of the launch of a giant ship, the crowd of workers and visitors looking like ants by comparison. Pre-flight preparations were quite brief, and the first flight took place on 21 December.

So successful were the flight trials that the Ministry of Aviation Industry notified the GIFAS organisation in Paris that the world's biggest aircraft would take part in the forthcoming airshow in June 1989. On the first day of the show the Mriya, re-registered as SSSR-82060 and bearing airshow number 387, took off from Kiev and was then escorted by Armée de l'Air Mirages from the French frontier to Le Bourget. On arrival, it had everyone standing spellbound. It was dull and drizzling, with a low cloudbase, but pilot Alexander Galunenko electrified the whole show by the brief show he put on before landing. With the six D-18T engines at quite high power he orbited the airfield in a haze of vapour, white vortices writhing from the wingtips like a fighter, the bank angle never less than 45°, and with the 62000 kg (136,700 lb) Buran riding on top without even an aerodynamic fairing over its blunt base. Galunenko said afterwards "You can see, the 225 handles like a fighter. You do not need to use two hands."

Holding court at Paris

During that show the Mikoyan bureau invited the entire Paris press corps to a conference to meet test pilot Kvotchur and hear the story of his amazing last-second escape from a doomed MiG-29. Of course, a vast crowd attended, but there was room for everyone: the conference was held in one corner of the hold of the An-225. On the walls were big diagrams showing some of the alternative payloads, and Bulanenko said he hoped further aircraft would be funded beyond the existing Nos 1 and 2. Details were also given of a single flight by No. 1, on 22 March 1989, which broke no fewer than 106 world and class records. Taking off from Kiev at 508350 kg (1,120,370 lb), the Mriya carried a payload of 156300 kg (344,576 lb) (more than 90720 kg (200,000 lb) below the maximum) round a 2000 km (1,242 mile) circuit at an average speed of 813 km/h (505.24 mph), reaching a cruise height of 12340 m (40,485 ft) in the process. Maximum fuel capacity is 300000 kg (660,000 lb), "but", said Bulanenko, "we've never had to put in more than 200 so far."

Only one An-225 exists, but the Antonov bureau, now a proudly Ukranian company, have plans to build at least one more in the near future.

Antonov's Twins: An-24/26/30/32

Dependable, versatile and above all simple to run, Antonov's family of short-range transports, beginning with the An-24 and continuing to the An-32, was one of the Soviet Union's most successful exports. The aircraft are still a common sight around the world.

On 20 December 1959 the prototype An-24 made its maiden flight. It looked like a copy of the Western Fokker F27 Friendship and Handley Page Dart-Herald, which were already in production. In fact it had a smaller and much more heavily loaded wing than either of the Western transports, and bigger engines. This might have been expected to make it much faster, but in fact it had a pedestrian flight performance, and not only needed a longer airstrip but also burned more fuel than the Western machines. A 1959 observer might even have dismissed the An-24 as uncompetitive. He would certainly have been unlikely to guess that more An-24s and derived aircraft would be sold than of all similar Western types combined!

The An-24 was designed as a 40-seat passenger aircraft to replace the Lisunov Li-2 (Soviet-built Douglas DC-3) for the vast Soviet airline Aeroflot. The need was for a vehicle that could operate reliably from scorching desert to Arctic wastes, while being maintained by

Wearing the old-style red star insignia of the Magyar Légierö (Hungarian air force), an An-26 of 'Szolnok' regiment approaches the runway on short, short finals. The newly-reconstituted Hungarian air force now carries a revised national marking consisting of a red, white and green arrowhead on the tailfin.

ground staff with a generally low level of skill or experience (especially of what might be termed 'modern' aircraft). Thus, the An-24 almost appeared with piston engines and an unpressurised fuselage, and Oleg K. Antonov even considered making it a biplane with fixed landing gear. That it finally flew as a modern pressurised monoplane with Ivchenko AI-24 turboprop engines has probably played a major part in keeping derivatives in production right up to the present day.

An-24 service with Aeroflot began in the winter of 1962-63, at first with 32 seats but later with 40 and finally 44. From the start the An-24 proved popular. A basic design feature was to make the fuselage of a curved triangular section so that the floor was placed right at the bottom. This made the floor wider than normal but eliminated the carriage of cargo under the floor, so all cargo and baggage had to be stowed above the floor in large compartments at the front and rear which would otherwise have been used for seating. In almost every version there are pairs of seats on each side of a central aisle, so that a 44-seater has 11 rows, and in most aircraft a folding airstair is hinged at the main door aft on the left side. It was expected that the nose would be made in the form of a glazed compartment for the navigator, but in

the end a radar was fitted instead and the flight deck arranged for a pilot on the left, co-pilot/navigator on the right and engineer behind them facing large panels up the right and rear walls.

Flight controls were made manual, and fair effort was needed. Full de-icing was required, and (though not much air can be bled from a 1491-kW/2,000-hp turboprop) the entire leading edges of the wings and tail are fitted with pipes for hot bleed air. This is expelled through rows of holes and runs back along chordwise corrugations in a thin inner skin to heat the entire leading-edge region before finally escaping through louvres at the tips. To enable the highly loaded wing to give good lift at low speeds it was fitted with powerful hydraulically-driven flaps of the double-slotted type, each section of flap having a small auxiliary van ahead of it. An unusual feature, necessitated by the absence of facilities at many Aeroflot airfields, is a TG-16 gas-turbine at the rear of the right nacelle. This can be started from the aircraft's own batteries, and provides electric power for starting the main engines and also for running various services on the ground. Fuel is housed in bag tanks in the centre-section of the wing and in integral (sealed structure) tanks farther outboard.

One of the most obvious differences between the An-26 and An-24 is the former's bulged observation window on the port side. This houses the cargo/paradropping 'bombsight'.

The main feature of the An-26 is the rear loading ramp. It is seen here on this East German machine, slung underneath the fuselage, allowing vehicles to drive right up to the cabin. In another configuration, the door acts as a drive-on ramp for the loading of small vehicles.

Specification: Antonov An-26B 'Curl'

Wings
Span	29.20 m	(95 ft 9.6in)
Area	74.98 m²	(807.1 sq ft)

Fuselage and tail unit
Accommodation	flight-deck crew of five, plus up to 40 passengers	
Length overall	23.80 m	(78 ft 1 in)
Width of fuselage (external)	2.90 m	(9 ft 6.2 in)
Tailplane span	9.973 m	(32 ft 8.6 in)
Height overall	8.575 m	(28 ft 1.6 in)

Landing gear
Retractable tricycle landing gear with twin wheels on all units
Wheel track	7.90 m	(25 ft 11 in)
Wheelbase	7.651 m	(25 ft 1.25 in)

Weights
Empty (typical)	15020 kg	(33,113 lb)
Maximum take-off	24000 kg	(52,911 lb)
Maximum payload	5500 kg	(12,125 lb)

Powerplant
Two Ivchenko AI-24VT turboprops
Rating, each	2103 ekW	(2,820 ehp)

An-26 recognition features

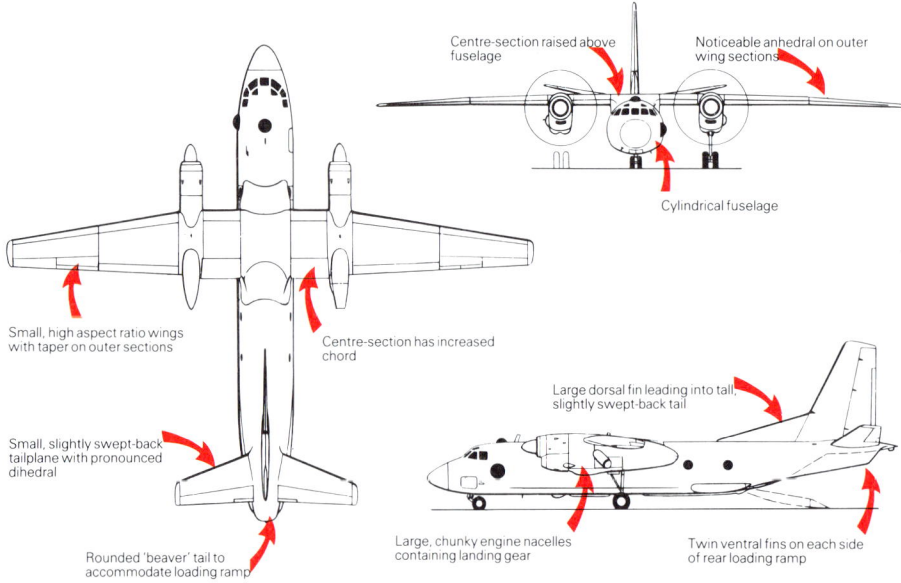

Antonov An-24V Series II cutaway drawing key

1. Radome
2. Weather radar scanner
3. Scanner tracking mechanism
4. Radome hinges
5. ILS glideslope aerial
6. VOR localizer aerial
7. Radar transmitters and receivers
8. Forward pressure bulkhead
9. Nose undercarriage wheel bay
10. Rudder pedals
11. Instrument panel shroud
12. Radar display
13. Curved windscreen panels
14. Windscreen wipers
15. Cockpit eyebrow windows
16. Overhead systems switch panel
17. Co-pilot/navigator/radio operator's seat
18. Instrument panel
19. Control column
20. Cockpit floor level
21. Nose undercarriage pivot fixing
22. Twin steerable nosewheels, forward retracting
23. Lower electrical equipment bay, port and starboard
24. Underfloor control runs
25. Space provision for radio operator
26. Side console panel
27. Pilot's seat
28. Opening (direct vision) side window panel
29. Space provision for flight engineer
30. Circuit breaker panels
31. Aerial lead-in
32. Cockpit roof escape hatch, interchangeable with jettisonable astrodome observation hatch
33. Cockpit doorway
34. Control linkages
35. Cockpit rear bulkhead
36. Radio and electronics equipment racks
37. Baggage compartment
38. Baggage loading shelving
39. Starboard side 'up-and-over' baggage door
40. Crew wardrobe
41. Curtained cabin doorway
42. Passenger cabin front bulkhead

Antonov An-24/26/30/32 'Coke' variants

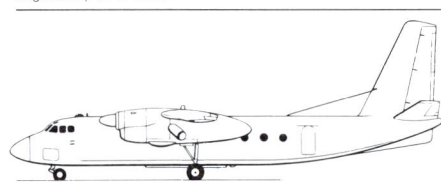

An-24: original passenger transport, 1875-ekW (2,515-ehp) AI-24 engines, up to 44 seats

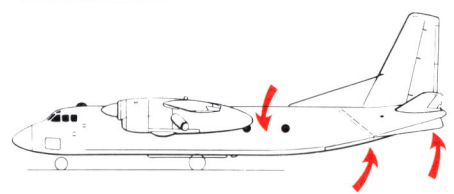

An-24V Srs II: AI-24A engines with power maintained by water injection to high ambient temperatures, up to 52 seats, greater take-off weight
An-24P: special version for forest firefighting

An-24T: version with rear cargo ramp door hinged up at rear (but unable to load vehicles)
An-24RV: An-24V Srs II with booster turbojet in right nacelle
An-24RT: An-24T with booster turbojet

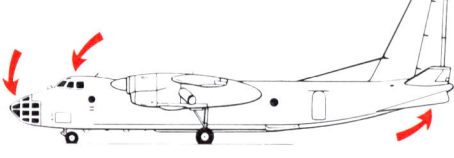

An-26: standard multi-role transport with 2103-ekW (2,820-ehp) AI-24VT engines and dual hinged/swinging rear cargo ramp door; booster jet fitted
An-26B: version specially equipped for handling palletized cargo

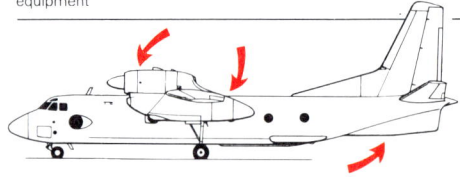

An-30: specialized photographic and surveying aircraft based on An-24RV but with redesigned forward fuselage and special equipment

An-32: STOL version with 3863-ekW (5,180-ehp) AI-20DM engines driving larger propellers; fitted with slats, triple-slotted flaps and slatted tailplane

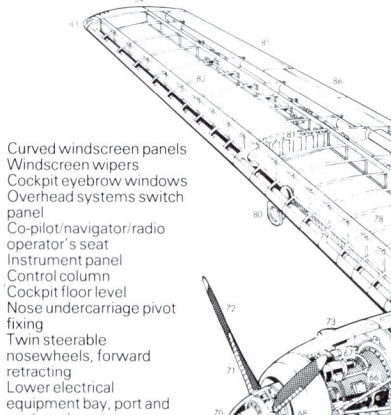
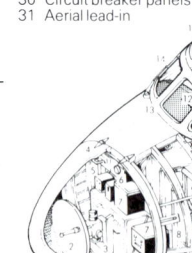
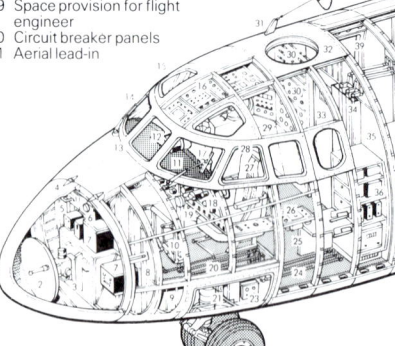

Performance (at 23000 kg/50,705 lb)

Maximum cruising speed at medium heights	237 kts	440 km/h (273 mph)
Landing speed	102 kts	190 km/h (118 mph)
Maximum rate of climb at sea level	1,575 ft	(480 m) per minute
Service ceiling	24,600 ft	(7500 m)
Range with maximum payload and no reserves	1100 km	(683 miles)
Take-off run at sea level, on concrete	780 m	(2,559 ft)

Maximum payload
- Lockheed C-130H 19356 kg
- Antonov An-72 10000 kg
- Aeritalia G222 9000 kg
- DHC-5D Buffalo 8165 kg
- BAe Andover C.Mk 1 6936 kg
- Fokker F27 Mk 400M 6438 kg
- Antonov An-26 5500 kg

Maximum rate of climb per minute
- Lockheed C-130H 1,900 ft
- DHC-5D Buffalo 1,820 ft
- Aeritalia G222 1,705 ft
- Fokker F27 Mk 400M 1,620 ft
- Antonov An-26 1,575 ft
- BAe Andover C.Mk 1 1,180 ft
- Antonov An-72 not quoted

Cruising speed
- Antonov An-72 at optimum altitude 388 kts
- Lockheed C-130H at optimum altitude 300 kts
- Fokker F27 Mk 400M at 20,000 ft 259 kts
- Aeritalia G222 at 19,685 ft 237 kts
- Antonov An-26 at 19,685 ft 237 kts
- BAe Andover C.Mk 1 at 15,000 ft 230 kts
- DHC-5D Buffalo at 10,000 ft 227 kts

Range, maximum payload
- Lockheed C-130H 3791 km
- Fokker F27 Mk 400M 2213 km
- Aeritalia G222 1371 km
- DHC-5D Buffalo 1112 km
- Antonov An-26 1100 km
- Antonov An-72 1000 km
- BAe Andover C.Mk 1 454 km

Take-off run
- BAe Andover C.Mk 1 1,260 ft
- Antonov An-72 1,540 ft
- Aeritalia G222 2,170 ft
- DHC-5D Buffalo 2,300 ft
- Fokker F27 Mk 400M 2,310 ft (field length, no TO distance quoted)
- Antonov An-26 2,560 ft
- Lockheed C-130H 3,580 ft

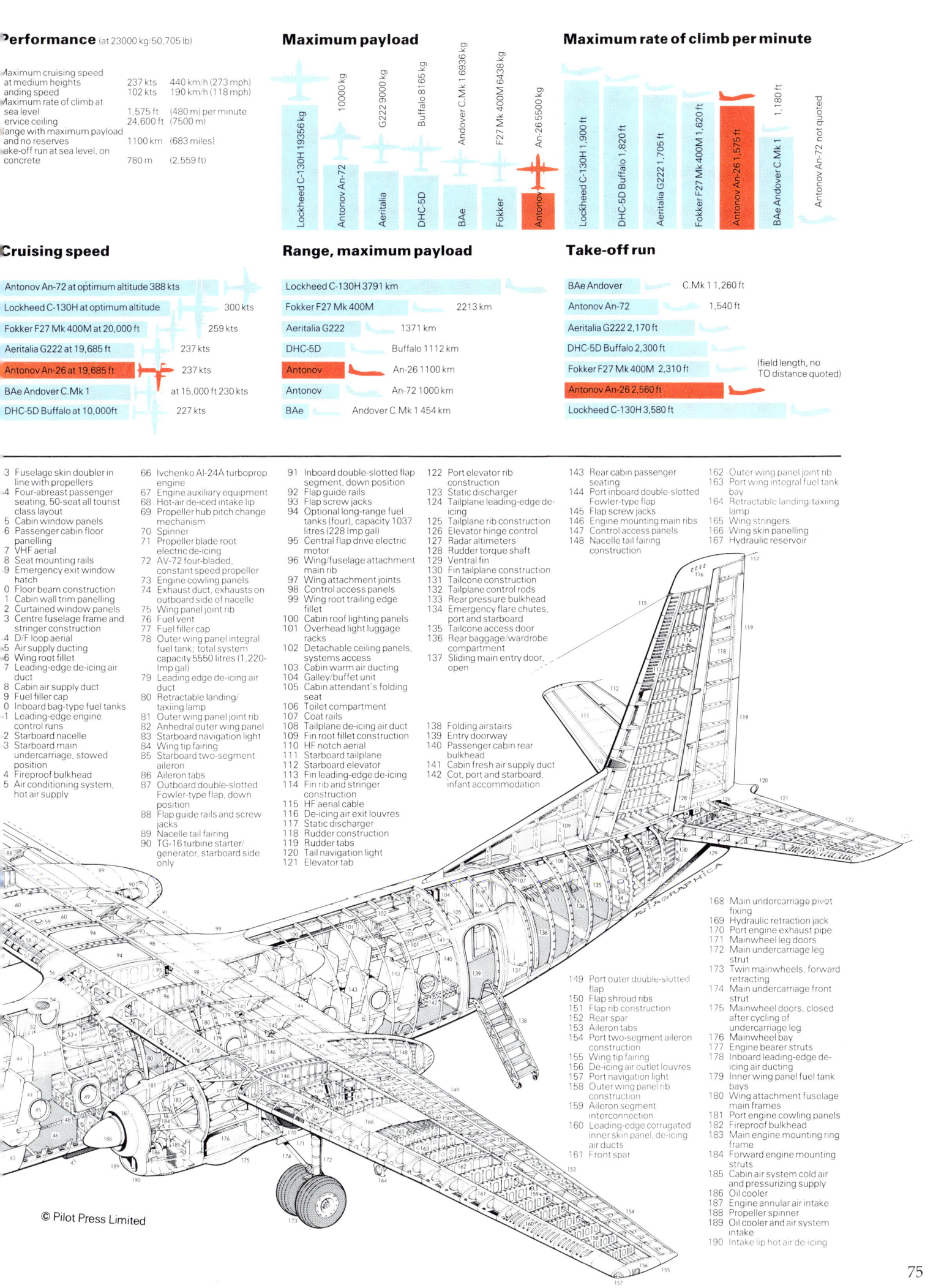

3 Fuselage skin doubler in line with propellers
4 Four-abreast passenger seating, 50-seat all tourist class layout
5 Cabin window panels
6 Passenger cabin floor panelling
7 VHF aerial
8 Seat mounting rails
9 Emergency exit window hatch
10 Floor beam construction
11 Cabin wall trim panelling
12 Curtained window panels
13 Centre fuselage frame and stringer construction
14 D/F loop aerial
15 Air supply ducting
16 Wing root fillet
17 Leading-edge de-icing air duct
18 Cabin air supply duct
19 Fuel filler cap
20 Inboard bag-type fuel tanks
21 Leading-edge engine control runs
22 Starboard nacelle
23 Starboard main undercarriage, stowed position
24 Fireproof bulkhead
25 Air conditioning system, hot air supply
66 Ivchenko AI-24A turboprop engine
67 Engine auxiliary equipment
68 Hot-air de-iced intake lip
69 Propeller hub pitch change mechanism
70 Spinner
71 Propeller blade root electric de-icing
72 AV-72 four-bladed, constant speed propeller
73 Engine cowling panels
74 Exhaust duct, exhausts on outboard side of nacelle
75 Wing panel joint rib
76 Fuel vent
77 Fuel filler cap
78 Outer wing panel integral fuel tank; total system capacity 5550 litres (1,220-Imp gal)
79 Leading edge de-icing air duct
80 Retractable landing/taxiing lamp
81 Outer wing panel joint rib
82 Anhedral outer wing panel
83 Starboard navigation light
84 Wing tip fairing
85 Starboard two-segment aileron
86 Aileron tabs
87 Outboard double-slotted Fowler-type flap, down position
88 Flap guide rails and screw jacks
89 Nacelle tail fairing
90 TG-16 turbine starter/generator, starboard side only
91 Inboard double-slotted flap segment, down position
92 Flap guide rails
93 Flap screw jacks
94 Optional long-range fuel tanks (four), capacity 1037 litres (228 Imp gal)
95 Central flap drive electric motor
96 Wing/fuselage attachment main rib
97 Wing attachment joints
98 Control access panels
99 Wing root trailing edge fillet
100 Cabin roof lighting panels
101 Overhead light luggage racks
102 Detachable ceiling panels, systems access
103 Cabin warm air ducting
104 Galley/buffet unit
105 Cabin attendant's folding seat
106 Toilet compartment
107 Coat rails
108 Tailplane de-icing air duct
109 Fin root fillet construction
110 HF notch aerial
111 Starboard tailplane
112 Starboard elevator
113 Fin leading-edge de-icing
114 Fin rib and stringer construction
115 HF aerial cable
116 De-icing air exit louvres
117 Static discharger
118 Rudder construction
119 Rudder tabs
120 Tail navigation light
121 Elevator tab
122 Port elevator rib construction
123 Static discharger
124 Tailplane leading-edge de-icing
125 Tailplane rib construction
126 Elevator hinge control
127 Radar altimeters
128 Rudder torque shaft
129 Ventral fin
130 Fin tailplane construction
131 Tailcone construction
132 Tailplane control rods
133 Rear pressure bulkhead
134 Emergency flare chutes, port and starboard
135 Tailcone access door
136 Rear baggage/wardrobe compartment
137 Sliding main entry door, open
138 Folding airstairs
139 Entry doorway
140 Passenger cabin rear bulkhead
141 Cabin fresh air supply duct
142 Cot, port and starboard, infant accommodation
143 Rear cabin passenger seating
144 Port inboard double-slotted Fowler-type flap
145 Flap screw jacks
146 Engine mounting main ribs
147 Control access panels
148 Nacelle tail fairing construction
149 Port outer double-slotted flap
150 Flap shroud ribs
151 Flap rib construction
152 Rear spar
153 Aileron tabs
154 Port two-segment aileron construction
155 Wing tip fairing
156 De-icing air outlet louvres
157 Port navigation light
158 Outer wing panel rib construction
159 Aileron segment interconnection
160 Leading-edge corrugated inner skin panel, de-icing air ducts
161 Front spar
162 Outer wing panel joint rib
163 Port wing integral fuel tank bay
164 Retractable landing/taxiing lamp
165 Wing stringers
166 Wing skin panelling
167 Hydraulic reservoir
168 Main undercarriage pivot fixing
169 Hydraulic retraction jack
170 Port engine exhaust pipe
171 Mainwheel leg doors
172 Main undercarriage leg strut
173 Twin mainwheels, forward retracting
174 Main undercarriage front strut
175 Mainwheel doors, closed after cycling of undercarriage leg
176 Mainwheel bay
177 Engine bearer struts
178 Inboard leading-edge de-icing air ducting
179 Inner wing panel fuel tank bays
180 Wing attachment fuselage main frames
181 Port engine cowling panels
182 Fireproof bulkhead
183 Main engine mounting ring frame
184 Forward engine mounting struts
185 Cabin air system cold air and pressurizing supply
186 Oil cooler
187 Engine annular air intake
188 Propeller spinner
189 Oil cooler and air system intake
190 Intake lip hot air de-icing

© Pilot Press Limited

The main feature of the An-26 is the rear loading ramp, which allows cargo and vehicles to drive on and off. The ramp forms a seal, allowing the fuselage to be pressurised, and can also be slung round under the fuselage for paradropping and loading directly from trucks.

Production soon switched to the An-24V Series II, in which engines fitted with water injection maintain power in hot conditions, to enable the maximum weight to be increased. This made possible a welcome increase in range with full payload of 5500 kg (12,125 lb) to about 550 km (340 miles). Versions appeared with interiors stripped out for cargo, with a seat for a cargo handler. These led to the An-24T, a specialised cargo version. The entire rear fuselage was redesigned, the door on the left being removed and a full-width ventral door being added, hinged upwards at the rear. The ventral fin had to be replaced by twin inclined ventral fins at the rear, one on each side of the tailcone. When the door of this version is opened the entire interior is exposed, and small vehicles can drive on board if ramps are available. Alternatively, cargo can be loaded with the aid of an electric hoist of 1500-kg (3,307-lb) capacity, which runs the length of the cabin along a rail in the ceiling. There is also a floor conveyor with a capacity of 4500 kg (9,921 lb), flush with the fixed floor on each side, and this can be winched along electrically or manually. There are fewer windows, and the floor is reinforced to bear heavy local loads.

The interior of the An-24T can be arranged with just one or two loadmaster seats, or folding seats (in twin, triple or quadruple units) can be attached round the walls, facing inwards, for 38 equipped troops or 30 paratroops. Alternatively, pillars can be inserted to carry 24 stretchers (litters), with a seat for an attendant. An OPB-1R sight can be fitted, resembling a simple optical bombsight, for precision dropping of cargo or paratroops.

The An-24T was first seen in 1967, and in the same year the An-24RV appeared, with the auxiliary generator in the right nacelle replaced by an Ru-19-300 turbojet. This drives an electric generator to provide ground power for all

Principal recipient of all Antonov's designs was the former Soviet Union, and its Transport Aviation arm was the largest user of the An-26. 'Curls' operated not only as straightforward transports, as shown here, but as dedicated Elint aircraft festooned with aerials.

purposes (including main-engine starting) and also provides auxiliary thrust on take-off, enabling maximum weight to be increased by 800 kg (1,764 lb) at sea level, and by 2000 kg (4,409 lb) at high tropical temperatures. The An-24RT is the An-24T cargo version fitted with the booster turbojet. Brochures describe provision for booster take-off rockets on the cargo versions, but these have not been seen in action. Another specialised version is the An-24P for dropping crews of firefighters over forest fires.

Y-7 production version

All An-24 versions are called 'Coke' by NATO, and over 1,100 were built. The type is no longer made in the USSR, but in February 1984 the first full production Y-7 was announced as having flown at Xian in the People's Republic of China. This country imported 40 An-24s from the USSR, and survivors are expected to be rebuilt to the same standard as the production Y-7, which is claimed to be considerably improved over the An-24.

Nine pre-series Y-7s were built, and one of these was sent to Hong Kong Aircraft Engineering Company (HAECO) in 1985 for completion as the Y7-100, with advanced equipment supplied by 17 American, French and British companies. Fitted with winglets, the Series 100 is now in low-rate production at Xian, which is taking each aircraft to the finished and equipped state. More than 115 had been sold by May 1993. The Y7-200A first flew in 1993 with WJ-5A engines (Chinese-built AI-24As) replaced by Pratt & Whitney Canada PW124Bs, driving Western propellers. The further improved Y7-200B has a lower empty weight despite a fuselage stretched by 0.74 m (2 ft 6 in), and the first production examples are scheduled for delivery later in 1992 following a first flight on 27 November 1990. The Series 200B retains WJ-5E turboprop engines but has new three-bladed propellers plus improved avionics and operating characteristics.

An-26

First revealed at the 1969 Paris air show and designated 'Curl' by NATO, the An-26 is the most numerous of all versions. It is basically the An-24RT but has a further redesigned rear fuselage to increase convenience. O. K. Antonov himself devised the new loading arrangement, which can be identified by the broad underside of the rear fuselage with a much larger ventral fin extending from the bottom of the fuselage to the tail on each side. Between these deep sidewall fins the fuselage sweeps sharply up under the tailplane. This area is sealed by a special power-driven cargo door, which in flight can seal the pressurised fuselage or be opened by hingeing downwards for airdropping. Alternatively, the door can be disconnected from the main hinges and swung right round under electric power on rails along each side (covered by bulged blister fairings) to lie under the rear fuselage. This facilitates loading directly from trucks, using the roof-rail electric hoist which is increased in capacity to 2000 kg (4,5000 lb).

First seen in 1969, the An-26 soon proved to have a tremendous worldwide market. Though more than 200 are used by Aeroflot, and small numbers by airlines in other countries, the vast majority of An-26 customers have been air forces. Early in production the standard An-26 was further improved with more powerful AI-24T engines (from 1980 upgraded to VT standard) with various fault-protection systems and 3.90-m (12.8-ft) four-bladed propellers, enabling weights to be increased. Another improvement was to skin the underside of the fuselage with abrasion-resistant 'bimetal' comprising a sandwich of hard titanium on a backing of aluminium alloy to stand up better to prolonged operations from unpaved airstrips. (Yet the rear-fuselage strakes are of soft glass-fibre.) A customer option, fitted to almost all An-24s, is a bulged observation blister on the left side of the flight deck, next to the crew toilet. The blister is used in conjunction

with the optical sight for accurate air dropping of cargo or troops.

Since 1981 the standard production model has been the An-26B 'Curl-A', specially equipped to carry palletised cargo. Roll-gangs (panels fitted with rollers) in the floor can be swung up against the sides of the cabin when not required. When they are in use two men can unload the maximum cargo load of three pallets, each 2.43 m (8 ft) long, 1.46 m (9.5 in) wide and with a combined weight of 5500 kg (12,125 lb), and then reload three pallets, all in 30 minutes. As in other versions the interior can be converted in about 30 minutes to troop transport or as a casevac ambulance.

Most recent variant of the An-26 is the 'Curl B' Sigint version for the Soviet air force. The type has also been adopted by China, who produce the Y7H-500 variant (originally designated Y14-100) of the basic An-26, powered by Chinese-built WJ-5A-1 engines driving four-bladed propellers. The initial Y7H-500 flew for the first time on 8 December 1988. By the time An-26 production ceased in favour of the An-32 in 1985, 1,410 had been built.

An-30

Only small numbers were built of the An-30 specialised photographic and survey aircraft, codenamed 'Clank' by NATO. First flown in 1974, the most obvious external differences from the An-26 from which it it derived is the glazed nose housing the navigator, whose role is to guide the An-30 on exact survey runs for mapping and cartographic purposes. To provide access to the new nose compartment, the entire cockpit was raised, radically changing the appearance of the aircraft. Equipment includes advanced precision radio navigation aids and a battery of cameras in the cabin, which can be augmented or replaced by other sensors such as magnetometers for mineral prospecting. Standard An-30s are used by the Afghan, Bulgarian, Romanian and Russian air forces, and can be converted into transports. A special version, designated the An-30M 'Sky Cleaner', is used by Aeroflot for seeding clouds with chemicals in order to promote rainfall. It is equipped with pods on each side of the lower fuselage housing 384 meterological cartridges fired into clouds to artificially induce rainfall.

An-32

Last of the twin-turboprops of the family, the An-32 is basically a slightly-altered An-26 fitted with extra high-lift devices and much more powerful engines. The most remarkable thing about it is that, despite the complicated high-lift devices and 85 per cent more power, it offers no obvious advantage over the An-26 and even has the same take-off run!

The An-32, disclosed in 1977, is fitted with 3760 kW (5,042 ehp) AI-20D Series 5 engines similar to those of the Antonov An-12 'Cub' and Ilyushin Il-18 'Coot' and related versions. The nacelles look amazingly deep because, while the lower parts still have to accommodate the landing gears, the upper parts have been raised to enable the propellers (of diameter increased to 4.70 m/15 ft 5 in) to clear the fuselage. On the prototype the jetpipes were very short, stopping above the wing, but on the production An-32 (called 'Cline' by NATO) they extend to a point near the tips of the nacelles aft of the trailing edge. There is no need for a booster turbojet, the right nacelle merely housing a TG-16M gas-turbine electric generator. The wings are similar in appearance but are fitted with full-span leading-edge slats and triple-slotted flaps, while the tailplane is fitted with a full-span fixed inverted slat along the leading edge. The rear-fuselage ventral strakes are enlarged, to counter the greater side area of the big propellers well forward. The rest of the An-32 is basically like an An-26.

There appears to have been no immediate requirement for this overpowered 'hot-and-high' transport, and in 1979 Antonov said production would depend entirely on sufficient orders being forthcoming. Since then the type has entered service with the air forces of the Soviet Union, Afghanistan, Bangladesh, Mongolia and Peru, all customers who appreciate the type's suitability for high-altitude operations over difficult terrain. Other 'Clines' have been delivered to civil operators such as Aeroflot and the Nicaraguan airline Aeronica. However, the real trigger for the An-32 to begin rolling off the production line was an Indian air force requirement for a replacement for its elderly Fairchild C-119 Boxcars, and 95 An-32s

Poland's fleet of An-26s is operated by the 13th PWL (Pulk Lotnictwa Transportowego – Air Transport Regiment) based at Kraków/Balice in the south of the country. They form the bulk of the air force's transport capability and are based alongside a handful of Il-14s. This aircraft carries a little 'regional' nose-art in the firey shape of a Pomeranian dragon.

were ordered in late 1979, to allow the ancient twin-boom transports to fade from the scene. Of the initial order, 45 were to be built by the Kanpur factory of Hindustan Aeronautics Ltd (HAL). However, these plans foundered and, after much delay, the type entered service with the air force in July 1984 devoid of any local assembly. Indian An-32s, locally named 'Sutlej' after a Punjabi river, are fitted with much improved avionics and navigation gear as per India's special requirements. In 1987 a further batch of 25 was ordered to bring the total in service with five squadrons, the paratroop training school, and the Aviation Research and Analysis wing, to 123 aircraft.

The first thing to strike an onlooker about the An-32 is the size of its enormous Ivchenki AI-40DM engines, perched high above the fuselage. Aeroflot operated the aircraft mostly in its Far Eastern and Arctic divisions, where its improved performance was most valuable.

Antonov An-26 'Curl'
Transportfliegerstaffel 24
East German Air Force

Iain Wyllie

Antonov An-24/26/30/32 in service

Afghanistan
By 1987 as many as 40 An-24/26s had been delivered to the Soviet-backed air force. Several An-30s and An-32s were also in service but the war and subsequent Soviet withdrawal have left an uncertain number of survivors.

Angola
At least 30 An-26s are believed to have been delivered to the air force, but several of these have quasi-military status flying for the government or the national airline TAAG. Some Angolan An-26s have been modified by Antonov to carry bomb racks on the side of the fuselage, beneath the wing. These aircraft have seen action against UNITA rebels and others.

Bangladesh
One An-24 was delivered to the air force in the mid-1970s, followed more recently by three An-26s for operation under civil markings. At least one An-26 returned to the USSR in 1989 and appears to have been replaced by at least two An-32s.

Bulgaria
A single transport regiment is based at Sofia/Vrjdebna, which has at least seven An-24s, nine An-26s and a further two An-30s on strength.

Cape Verde
Two An-26s were delivered in 1982 and another is thought to have entered service since then.

China
Before its split with the Soviet Union, China received 40 examples of both An-24s and An-26s. The survivors are being supplemented by indigenously-built XAC Y-7s (in both An-24 and An-026 versions), which fly with the air force and the pseudo-military national airline CAAC.

CIS
The main V-TA (transport aviation) arm of the air force (in whatever form it survives today) operated an estimated 400 An-26s with smaller numbers of An-24 passenger aircraft. AV-MF (naval aviation) had its own force of some 350 transports, of which roughly 120 were An-24/An-26s. Elements of the Soviet air force also operated small numbers of An-30s and An-32s.

Congo
The largest transport element at the air force's disposal is five An-24s delivered in the mid-1970s, plus one An-26.

Cuba
Once one of the Soviet Union's 'most favoured nations', Cuba received up to 30 An-26s between 1975 and 1985, plus a further small number of An-24s. The number in military service is now uncertain, as several aircraft were passed on to Cubana and the air force is suffering a severe financial crisis. Two An-26s are also believed to be in service.

Czech and Slovakia
The newly reorganised Czech and Slovakian air force retains at least 11 An-24s and six An-26s at its main transport base at Mosnov.

Ethiopia
Ethiopia was a client state of the Soviet Union for many years, but with the withdrawal of its benefactor's assistance and the continuing bitter civil war the status of its small number of An-26s is unknown.

Germany
With the re-unification of the two Germanies, the 12 An-26s of the former East German air force Transportfliegergeschwader (TFG) 24, based at Dresden, became part of the Luftwaffe which has retained the Antonovs in service for the time being.

Guinea Republic
The air force was reported to have received four ex-Air Guinea An-24s, but the status of these aircraft is unknown.

Hungary
A unit of at least two An-24s and 12 An-26s has been retained by the air force since its Warsaw Pact days. The aircraft are flown by 'Vitez Hari Laszlo' and 'Szolnok' regiments, based at Tököl.

India
A large number of An-32s currently equip Nos 12, 19, 33, 33, 43, and 48 Squadrons in addition to the parachute school and several test centres. A total of 123 aircraft, known locally as 'Sutlej', has been delivered since 1979.

Iraq
Iraq received a squadron's worth of An-24s, some of which were operated in civil markings, plus at least two An-26s, before the war with Iran. Since that long conflict and Iraq's subsequent invasion of and expulsion from Kuwait, any additional aircraft that may have been obtained are unlikely to remain serviceable.

Laos
A single transport squadron flies four types, including a possible seven An-24s and three An-26s.

Libya
Unable to obtain Western aircraft, Libya took delivery of at least 13 An-26s from 1983 onwards.

Madagascar
Four An-26s are believed to be still in service.

An-26 'Curl' of the Afghan air force.

An-24 'Coke', used for staff transport by Czech and Slovakia.

Romania is one of the few military users of the An-30 'Clank'.

An-26 'Curl' in what used to be standard Soviet colours.

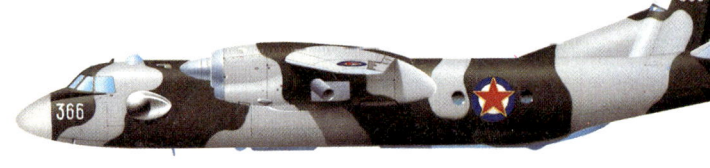

Yugoslavia previously had 30 An-26 'Curl' aircraft in service.

Mali
The air force transport squadron operates a pair of both An-24s and An-26s.

Mongolia
Only single examples of the An-26 and An-32 are thought to still be in service with the air force and airline, along with six or seven survivors of 18 An-24s originally received.

Mozambique
From 1978 onwards at least 11 An-26s were delivered to the air force at Maputo. Since then several have been destroyed. The aircraft have also been fitted with fuselage bomb racks similar to those operated by Angola.

Nicaragua
Despite denials, a pair of An-26s was delivered to the Sandanista regime in 1982.

North Korea
This air force obtained some 30 An-24s since the 1960s and is possibly also an An-26 operator.

Pakistan
Pakistan became an unlikely An-26 user when it obtained two aircraft whose crews had defected from the Afghan air force. Both of these are now believed to be out of service due to spares shortages.

Peru
Based at Lima's Callao air force base, 8 Grupo, comprised of Nos 841, 842 and 843 Squadrons, operated 16 An-26s until 1987/88 when they were exchanged for a similar number of more suitable An-32s.

Poland
The Polish air force's main transport base is at Kracöw/Balice where the 13th Transport Regiment operates a dozen An-26s. All its An-24s have been passed to LOT, the national airline, along with some An-26s. Other An-26s have been transferred to Air Batory, a new civil operator.

Romania
The majority of Romania's transport assets are based at Otopeni, where at least six An-24s, three An-26s and four An-30s can be found.

Somalia
Very few Antonovs, perhaps only as many as three An-24s and a single An-26, have survived the loss of Somalia's backing by the Soviet Union and the current civil war.

Sudan
Five An-24s were operated by the air force from Khartoum, but all are thought to be in storage.

Syria
Syria's two transport squadrons are believed to have a total of three An-24s and five An-26s in service.

Vietnam
This sizeable air force had three transport regiments which counted among its strength some 10 An-24s and 50 An-26s.

Yemen
With the amalgamation of the former Arab (South) and People's (North) Republics of Yemen into a single nation, the North's small number of An-24s and An-26s now serve with the newly established unified air force.

Yugoslavia
Yugoslavia no longer exists as a single recognisable nation and the Yugoslav air force, as an entity, is now merely a historical footnote. In its Warsaw Pact days Yugoslavia had two squadrons of 30 An-26s. The fact that these units were based at Zagreb and Belgrade, now in two separate states, means the status of any aircraft is uncertain in the extreme.

Zambia
Between 1981 and 1982 four An-26s were delivered as part of a Soviet air package.

Armstrong Whitworth Argosy

That the Argosy was destined to become one of aviation's 'also-rans' was not the fault of its creators who, at the time of its conception, could rightly see a large market for a dedicated freighter and short-range high-density passenger transport. Subsequent unforeseen developments and political decisions robbed the Argosy of its success, rather than any design or initial marketing faults.

Design of the Argosy began in late 1955, initially to answer the RAF's OR323, although this was subsequently cancelled. Undeterred, Armstrong Whitworth continued work on the design on a privately-funded basis, as the civil potential of the type was considerable. The aircraft emerged as bearing all the hallmarks of other contemporary freighters: a twin-boom layout provided easy access to both ends of the cargo hold, with a swing door in the nose and clamshell doors in the rear. The high-set wing put the cabin floor as near to the ground as possible while still providing clearance for the propellers. The flight deck was set high above the cabin so that maximum internal capacity was maintained, and was accessed via a ladder from the cabin floor. The normal crew was two, although a third place was provided.

Early surveys dictated a measure of proven technology, so the wing was based on that of the Shackleton Mk 3, and the engines chosen were the trusted Rolls-Royce RDa.7/2 Dart 526 turboprops. In fact, Armstrong Whitworth used Viscount nacelles, identical back to the firewall. The baseline version, designated A.W.650 and initially dubbed 'Freightercoach', could be configured for freight, seats or a mixture of both, with freight forward and 36 passengers in six-abreast seating in the rear.

At an early stage in the design, the aircraft that was to be renamed the Argosy was proposed as a family of aircraft to suit various requirements. A military specification freighter was designated A.W.660, while A.W.670 covered a variant with a much larger double-deck unpressurised fuselage which was intended as a car ferry for six vehicles and 30 passengers. A variant of this was A.W.671, a short-range 'air bus' with accommodation for 130 on its two decks. Then at some point, Armstrong

British European Airways ordered the Argosy Mk 220 (designated Mk 222 for the airline) to replace its interim Mk 100s. The first of these more-capable aircraft was delivered in 1965.

Above: Argosy 100 G-APRN was displayed at the 1959 SBAC show, at Farnborough, in basic Riddle Airways colours. The previous year it had made the type's public debut at Paris.

Below: G-AOZZ was the first Argosy and, with Eric Franklin at the controls, it took to the air for the first time at Bitteswell on 8 January 1959. It was later delivered to BEA.

Whitworth envisaged swapping the four Darts with two of the much more-powerful Tyne engines, offering far greater efficiency and load-carrying capability. A.W.651 and A.W.661 covered Tyne-engined versions of the civil and military freighters, although in the event no Argosy was ever to feature the larger engine. Later, a VTOL Argosy was proposed, powered by two Tynes inboard, and featuring outboard pods each containing 20 lift-jets. These would have been arranged in two banks fore and aft to provide control in the hover.

Straight into production

Confidence in the design was such that Armstrong Whitworth (part of the Hawker Siddeley Group) began the construction of 10 aircraft without orders, plus two static test articles. The main factory was at Baginton, Coventry, but assembly and flight tests were to be undertaken at Bitteswell, Leicestershire. There was no prototype as such, all 10 aircraft being built to production standards and modified as necessary as flight trials progressed.

G-AOZZ was the first to be completed, and was rolled out after assembly at Bitteswell on 21 December 1958, just 28 months after the initiation of the programme. After a series of taxi trials, the aircraft lifted off for its maiden flight at noon, 8 January 1959, piloted by Eric Franklin (chief test pilot), and William Else (assistant chief test pilot), with Roy Hadley and Kenneth Oldfield as engineers. The 62-minute flight was highly successful, showing no design flaws.

American orders

Early 1959 was a good time for Armstrong Whitworth: the Air Staff had decided to adopt the A.W.660 for the RAF, flight trials were proceeding apace, and the first civil customer had been attracted. US operator Riddle Airlines placed an order for four (later raised to seven), impressed by the confidence of the manufacturer and the speed with which it had transformed a concept into reality. The second A.W.650 (G-APRL) flew on 14 March, the third (G-APRM) on 26 April, and the fourth (G-APRN) on 13 May, in time to appear at the Paris air show, wearing Riddle colours. It flew in September at Farnborough, in company with no. 5 (G-APVH). Argosy no. 6 (G-APWW) flew on 23 September, but it was to be another year before the remaining four were flown.

Early trials were aimed at general handling, radio and autopilot (no. 1), engine and propeller trials plus tropical testing at Kkartoum and Nairobi (no. 2), and performance (no. 3). The only major fault lay in some vibration in the tailplane and rear fuselage, but this was easily cured by the addition of a row of vortex-generating fins behind the flight deck. In late 1960 the fourth aircraft undertook an unproductive sales tour of Europe before joining the test fleet which was working hard on attaining

Above: Registered to Sir W. G. Armstrong Whitworth Aircraft Limited on 12 February 1957, the prototype G-AOZZ undertook a specific portion of the flight test programme, charged with handling and centre-of-gravity trials. Delivered to BEA in 1961, it was one of several aircraft to serve later with Universal Airlines.

Left: Now an almost forgotten name (having changed its identity to Airlift International in November 1963), Riddle Airlines ordered four Argosies in 1959, then lost the USAF contract it had hoped to employ them on. This Logair contract was then renewed in 1960 and Riddle reconfirmed and increased their order. Sadly, the airline had its CAB approval suspended, due to losses, in 1962 and the Argosies were returned to the manufacturer.

Above: Inflight-refuelling trials were conducted using a Argosy C.Mk 1 fitted with a bolt-on probe. The disparity between its speed and the RAF's jet tankers proved too much to overcome.

type approval from the CAA and FAA, a goal achieved on 5 December 1960.

A week later, the first aircraft for Riddle (no. 9, N6501R) was delivered to its new owner via Prestwick. The remaining six followed across the Atlantic, delayed somewhat by a 26 January grounding following the discovery of cracks. After a swift rectification of the problem, deliveries continued, G-APRL/N6507R completing the order on 17 August 1961.

The Logair routes

In Riddle service, the Argosy was pushed quickly into service on the Logair contract, which the airline had won on 17 June 1960. Logair was a MATS requirement for supplemental freight carriage between US air bases. There were three routes, all starting and ending at Tinker AFB, Oklahoma, and each taking in a number of (mainly SAC) bases in a circuit. Each route was flown daily, eventually requiring the services of five Argosies. The remaining pair was used on scheduled freight runs. In June 1962, Riddle's Logair contract was ended, and the airline had to return its Argosies to Armstrong Whitworth. However, the Logair contracts were assigned to other companies, and five Argosies were bought by Capitol Airlines and two for lease to Zantop.

Above right: Armstrong Whitworth revised their original A.W.66 proposal to include the swinging nose after examining customer requirements for a quick turn-around.

The rounded rear empennage of civilian Argosies differed substantially from those completed for the RAF, which were fitted with air-operable doors for dropping heavy loads.

Specification
A.W.650 Argosy 100
Powerplant: four Rolls-Royce RDa.7/2 526 turboprops, 1567 kW (2,100 shp) each
Wing span: 35.05 m (115 ft)
Length: 26.44 m (86 ft 9 in)
Height: 8.23 m (27 ft)
Wing area: 135.45 m² (1,458 sq ft)
Empty weight: 22000 kg (48,500 lb)
Maximum take-off weight: 39917 kg (88,000 lb)
Payload: 12700 kg (28,000 lb)
Cruising speed: 476 km/h (296 mph)
Maximum range: 4345 km (2,700 miles)

Specification
A.W.650 Argosy 222
Powerplant: four Rolls-Royce Dart 532/l turboprops, 1664 kW (2,230 shp) each
Wing span: 35.05 m (115 ft)
Length: 26.44 m (86 ft 9 in)
Height: 8.91 m (29 ft 3 in)
Wing area: 135.45 m² (1,458 sq ft)
Empty weight: 22136 kg (48,800 lb)
Maximum take-off weight: 42185 kg (93,000 lb)
Payload: 14061 kg (31,000 lb)
Cruising speed: 462 km/h (287 mph)
Range: maximum 3510 km (2180 miles), with maximum payload 1667 km (1035 miles)

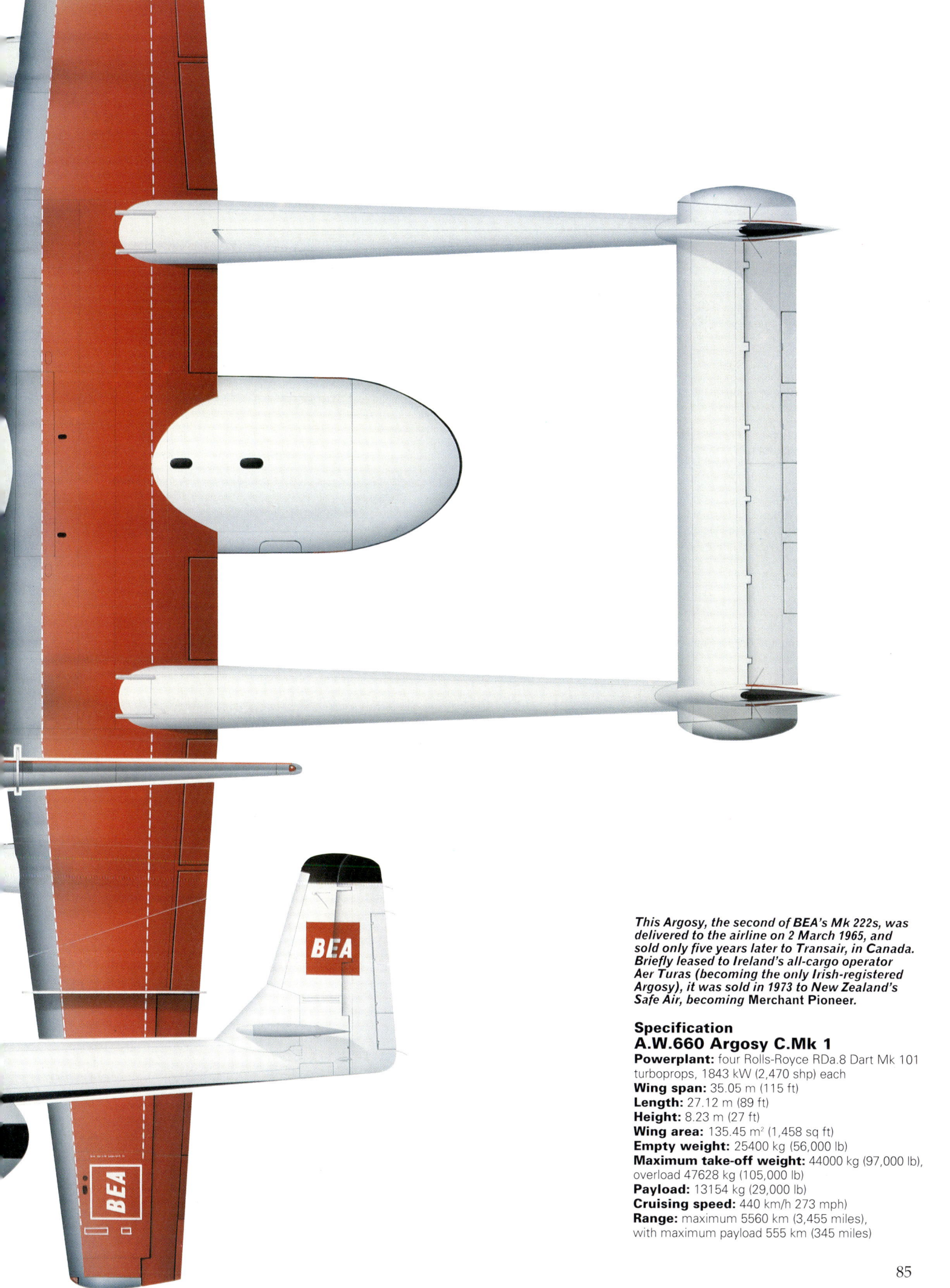

This Argosy, the second of BEA's Mk 222s, was delivered to the airline on 2 March 1965, and sold only five years later to Transair, in Canada. Briefly leased to Ireland's all-cargo operator Aer Turas (becoming the only Irish-registered Argosy), it was sold in 1973 to New Zealand's Safe Air, becoming **Merchant Pioneer.**

Specification
A.W.660 Argosy C.Mk 1
Powerplant: four Rolls-Royce RDa.8 Dart Mk 101 turboprops, 1843 kW (2,470 shp) each
Wing span: 35.05 m (115 ft)
Length: 27.12 m (89 ft)
Height: 8.23 m (27 ft)
Wing area: 135.45 m² (1,458 sq ft)
Empty weight: 25400 kg (56,000 lb)
Maximum take-off weight: 44000 kg (97,000 lb), overload 47628 kg (105,000 lb)
Payload: 13154 kg (29,000 lb)
Cruising speed: 440 km/h 273 mph)
Range: maximum 5560 km (3,455 miles), with maximum payload 555 km (345 miles)

In RAF service the 'wheelbarrow' was a popular aircraft, and the fleet was heavily tasked with maintaining the service's Far East commitments.

Zantop later acquired the other Argosies, although one crashed while attempting an emergency landing on a stretch of highway in Ohio. The crew might have succeeded if it had not been for the intervention of a bridge across the road! The company changed its name to Universal and operated Logair flights from Wright-Patterson, in addition to commercial work. Two ex-BEA aircraft were later added to the fleet.

American sales still left Armstrong Whitworth with three Argosies. British European Airways bought these on 27 April 1961, after specifying a Rolamat cargo floor, lashing points and metal lining in the cargo hold. In so doing, BEA became the first European civil operator to acquire a dedicated freighter. G-APRN was the first to fly in its new Argosy 102 configuration, on 21 September 1961. G-APRM and G-AOZZ followed later in the year, allowing BEA to fly its first service on 2 January 1962, between Heathrow and Frankfurt/Düsseldorf. Soon the three aircraft had assumed all the European freight operations, including those from Manchester, replacing BEA's DC-3s, Viscounts and chartered Yorks.

Military freighter

Having sold its initial production run of civil A.W.650s, Armstrong Whitworth next concentrated on the A.W.660 military freighter. The Royal Air Force, in fact, was to become the largest and best-known operator of the type. Built to military specifications, the most obvious feature of the A.W.660 was the beaver tail, which could be opened in flight to allow the air-dropping of large items or paratroops. G-APRL tested the door, first flying as such on 28 July 1960. The 56 production Argosy C.Mk 1s (XN814-821, XN847-858, XP408-413, XP437-450, XR105-109, XR133-143) had numerous other differences, including military systems and a closed front door incorporating a glazed panel for accurate para-dropping, plus a weather radar mounted in a small radome. XN814 was the first A.W.660, first flying on 4 March 1961 and later delivered to the A&AEE at Boscombe Down. Service aircraft were delivered to an un-numbered OCU at Benson on 21 November, although the training commitment was later shifted to Thorney Island and 242 OCU.

Operational transport squadrons to fly the Argosy were No. 114 (Benson, from 1 February 1962), No. 105 (Benson/Khormaksar/Muharraq, from 21 February 1962), No. 267 (Benson, from 23 November 1962), No. 215 (Benson/Singapore, from 1 July 1963) and No. 70 (Akrotiri, from 9 October 1967). The Benson-based aircraft flew regularly on supply flights to Germany, while the overseas units supported Britain's waning colonial interests. The aircraft was popular in service, and proved highly effective when compared with the Hastings and Beverley. There were three major accidents, one belly landing at Benson, one low-level accident in Libya when an aircraft struck a water tower (the only accident to be fatal) and a short-landing into Khormaksar harbour. Remarkably, the latter aircraft was salvaged from the water, dried out, overhauled and promptly returned to flying service.

Withdrawal from overseas bases in the late 1960s saw the gradual return of Argosies to Benson, and the type began to be withdrawn from its original role. Benson's No. 114 Sqn retired its aircraft in October 1971, and No. 70 flew the last transport Argosy flight in February 1975. However, in November 1967, No. 115 Squadron at Watton had received Argosy E.Mk 1s for the landing and navigation aid calibration role, with equipment fitted internally. The squadron moved to Cottesmore, and then to Brize Norton, from where the last operational Argosy sortie was mounted on 24 January 1978.

Many of the retiring Argosies had been assembled in 1972 at Kemble, where 14 were to have been converted to T.Mk 2 standard to take over the navigator and air electronics officer training role with 6 FTS at Finningley. Two conversions were completed, identified by a large nose radome, of which only one, XP411, entered service. The T.Mk 2 programme was cancelled in 1975 and the sole service example became a ground maintenance trainer. The RAF Argosies were dispersed to ground schools or fire dumps, and many were sold to civilian customers for scrap or continued use. The last was sold on 10 February 1976. Argosies also served with trials units, including the Empire Test Pilot School (which crashed XR105 on 27 April 1976) and the A&AEE, which flew the last military Argosy flight on 1 October 1984, which ended with XN817's undercarriage collapsing on the runway at West Freugh.

Second generation

As the RAF was receiving its brand new transports, BEA was starting on its all-freight operations. Unexpectedly, the Argosy 102s were found to be unprofitable in service, so Armstrong Whitworth embarked on a new and improved Argosy 200 which would address the problems of the earlier aircraft. Plans for 10 of the new civil freighter were laid, although there were, as yet, no orders. On 12 August 1961 the company merged with Gloster to become Whitworth Gloster, and on 1 July 1963 was fully absorbed into the parent group under the Hawker Siddeley banner.

The old Shackleton wing used by the Argosy 100 series was built to safe-life principles (30,000-hour life), and was seen as having no further development potential. Accordingly, a new wing structure was

The Argosy C.Mk 1 displayed its rapid deployment capabilities on several occasions, not least when aircraft of No. 114 were dispatched to Zambia to support No. 29 Sqn during the Rhodesian crisis at Christmastime 1965. The Argosy spent a decade in service with the RAF until defence cuts and the virtual elimination of RAF Air Support Command saw their retirement from the late 1960s.

designed, to fail-safe philosophy. The two-spar mass-boom structure of the old wing was changed to a box-spar structure, which allowed the fitment of integral inter-spar fuel tanks in place of bag-type tanks, with a useful consequent reduction in weight. Instead of having two main pin joints to the top of the fuselage, the new wing was structurally integrated into the fuselage structure. The new wing offered better load-carrying, range and fuel consumption figures.

Argosy Mk 200

G-ASKZ was the first and only Argosy 200 to fly, piloted by Eric Franklin from Bitteswell on 11 March 1964. In September, BEA ordered five Argosy 222s (G-ASXL/M/N/O/P), returning its three 102s to the manufacturer. The Series 222 featured the more-powerful Dart 532/I engine and wider door apertures to admit bulkier cargo. They were delivered between 28 January and 16 June 1965 and immediately began work on the airline's freight routes. On 4 July 1965 G-ASXL was flown into an Italian mountainside, miraculously without loss of life, and Argosy 102 G-APRM returned on lease from the manufacturer to cover the shortfall. A more permanent address was found in the shape of a sixth Argosy 222 (G-ATTC), arriving with the fleet in November 1966. This was the last Argosy completed, any other airframe parts being scrapped when there was no chance of any further sales. In December 1967 G-ASXP was written off in a non-fatal landing accident at Stansted. This aircraft was not replaced, as even the second-generation Argosy was proving unprofitable. BEA took the decision to convert its Vanguards to Merchantman freighters, the first entering service in 1969. The final BEA Argosy service was flown on 30 April 1970, and all four survivors were sold to Midwest of Canada.

Civil Argosies continued in service throughout the 1970s and early 1980s with a wide variety of operators. The four Canadian aircraft were principally used on oil field support duties in Alaska. One was leased by Aer Turas for the transport of racehorses, while it and another Canadian machine went to Safe Air of New Zealand to operate on postal and freight contracts. In November 1976 the other two Series 222s were sold to SOACO in Gabon, where they were used to haul construction equipment to rough strips in the bush. The pair then moved to Australia, where IPEC operated an 'air bridge' between Melbourne/Essendon and Launceston in Tasmania.

Of the early aircraft, G-APRM became an engine transport for Rolls-Royce, used to fly Olympus components between Filton and Toulouse in connection with the Concorde programme. G-APRN was used by BAC for a short while to carry support equipment on a sales tour to the Middle East by

The last home operator of the Argosy was Elan Couriers, which operated a pair of aircraft, leased from Air Bridge, on overnight package services between 1985 and 1987.

Above: IPEC Aviation in Australia operated a fleet of three Argosies (a Mk 101 and two Mk 222s) from 1978 into the 1990s. Some of these found their way to Safe Air in New Zealand.

Below: A former Riddle aircraft, this Argosy Mk 101 later flew with Capitol, Zantop and Universal in the USA, before returing to the UK. It is now preserved at East Midlands Airport.

SEPECAT Jaguar T.Mk 2 XX846. Two of BEA's 102s went to Universal to join the ex-Riddle aircraft. This carrier ceased flying in 1972, three aircraft going to Duncan Aviation, one to Nittler Air Transport and four returning to Britain with Sagittair (three) and Shackleton Aviation. The Sagittair machines had a productive career, flying newspapers between Liverpool and Belfast, fruit and flowers from the Channel Islands, and undertaking oil field support work from Aberdeen. Air Bridge Carriers took over these aircraft, although one was sold to BBA in Australia, and subsequently to IPEC (prior to the two Series 222s arriving). Air Anglia operated an Argosy 101 for a time in the mid-1970s.

As recounted earlier, many of the ex-RAF Argosies were sold for scrap or spares, Rolls-Royce buying 11 for the engines. One was overhauled for Philippine Airlines as RP-C-1192, one flew to the Congo in 1977 as 9Q-COA for service with Otrang Range Air Services, and another was sold to Management Jets International in the US. In 1993 there are no Argosies left in service, although a number survive. Two Argosy 101s are preserved at Baginton and East Midlands in the UK, while the RAF Museum has XP411, the sole service T.Mk 2, at Cosford. XN817's damaged airframe remains at West Freugh. A handful of airframes are still extant in the US, although not in flyable condition.

Chapter 5

ATR 42 & 72

•

Avro/Hawker Siddeley HS 748

ATR 42 and 72

The Avions de Transport Régional family of twin-engined airliners is the result of the successful collaboration between Aérospatiale of France and Italy's Aeritalia (now Alenia). Relative new-comers to the regional aircraft market, they have taken it by storm with nearly 400 sales of the stylish airliners now firmly under their belts.

On 29 October 1981 the boards of directors of the French manufacturer Aérospatiale and the Italian company Aeritalia (since renamed Alenia) decided to proceed with a joint project to design and build a twin-turboprop airliner, preliminary agreements on which had been made during the summer of the previous year. On 4 November the partners signed a cooperative agreement for setting up the Groupement d'Intérêt Economique (GIE) to manage the programme on an equal cost-sharing basis. GIE was formally established at Toulouse, France, on 5 February 1982.

Both companies had already independently conducted their own design studies for regional airliners of 30-40 seats capacity, these being the Aérospatiale AS-35 and the Aeritalia AIT 320. Following the establishment of GIE, a French/Italian design group began work on combining the best elements of these projected aircraft into a single machine, to be known as the ATR 42 (ATR for the initial letters of 'Regional Transport Aeroplane' in French and Italian, 42 for the planned passenger capacity of the airliner).

The configuration adopted was for a high wing, T-tail aircraft, with the main landing gear units housed in 'sponsons' on the fuselage side. The powerplant selected was the Pratt & Whitney Canada PW100/2 series turboprop. The design, which was developed using the latest computer-aided design and manufacturing techniques, incorporated advanced technology features in aerodynamics (notably in the Aérospatiale-developed RA XXX 43 wing airfoil section), structures and systems. The ATR 42 was intended to meet US Federal Aviation Administration FAR Part 25 and the forthcoming European Joint Airworthiness Requirements JAR 25 certification standards.

Two models were planned initially, the ATR 42-100 intended for maximum stage lengths of 1126 km (700 nm), with seating for up to 42 passengers and a maximum take-off weight of 14900 kg (32,850 lb); and the ATR 42-200 for stage lengths up to 1258 km (782 miles), accommodation for up to 49 passengers achieved by reducing the capacity of the forward baggage compartment or deleting the galley, and a maximum take-off weight of 15550 kg (34,280 lb).

Two development airframes were built, plus two non-flying airframes for static and fatigue testing. The first prototype (F-WEGA) made its maiden flight from Toulouse on 16 August 1984, and was followed by the second (F-

F-WEGA was the prototype of the new transport family, first flying from Toulouse on 16 August 1984 in smart house colours. The cabin largely contained instrumentation.

Above: The ATR is a 50:50 collaboration of Aeritalia and Aérospatiale, both companies producing sub-assemblies, which are brought together on this, the final line at Toulouse. This assembly line benefitted greatly from experience gained with the Airbus programme.

WEGB) on 31 October, and the first production aircraft (F-WEGC) on 30 April 1985. Certification was granted simultaneously by the French DGAC and Italian RAI authorities on 24 September 1985, while FAA approval followed on 25 October 1985. The first customer delivery took place on 2 December 1985 when the fourth production airframe was handed over to the French carrier Air Littoral, with whom it entered revenue-earning service a week later. Subsequent deliveries were made to the Danish operator Cimber Air in December 1985, and to US launch customer Command Airways in January 1986, Command operating the ATR 42's first services in the United States two months later.

By the time the ATR 42 entered production, the higher capacity -200 version had been adopted as standard, and the greater payload/increased range ATR 42-300 was available optionally.

The production split agreed between the Avions de Transport Régional partners leaves Alenia (Aeritalia) responsible for manufacturing the fuselage, including tail surfaces and landing gear, and all hydraulic, pressurisation and air conditioning systems, while Aérospatiale builds the aircraft's wings and is responsible for final assembly, flight deck and cabin outfitting, engine, electrical, flight control and de-icing systems installation and flight testing.

Cargo adaption

In addition to the commercial passenger variants of the ATR, several other versions have been developed or proposed. The ATR 42F is a commercial freighter version, designed to carry up to 3800 kg (8,377 lb) of cargo over a range of 2010 km (1,250 miles). It features a strengthened cabin floor and a flight-openable cargo loading/airdrop door on the port fuselage side. Only one example has been built, for the Gabonese Presidential Guard, to which it was delivered in 1989. A military freighter version designated ATM 42L was proposed in 1986 as a replacement for the French Armée de l'Air's ageing Nord Noratlas transports. As projected, this version incorporates a large port-side upward-hinged cargo door aft of the flight deck which would permit direct drive-on loading of jeeps, and would enable the aircraft to accommodate three 1.34 × 2.74-m (88 × 108-in) freight pallets, five standard LD3 containers or any aero engine

Above: Italian regional airline ATI (a subsidiary of Alitalia) was an early customer for the ATR 42, receiving its first aircraft in 1986. Currently, the aircraft are on lease to the associate airline Avianova.

Right: Seen with French test registration, this ATR 42-300 was delivered to German regional operator NFD, which had a large fleet of both 42s and 72s. Eleven of the short body transports were in use.

currently installed in French military aircraft. The cargo floor is strengthened to accept heavy loads, with rolling-ball cargo handling systems. The ATM 42L can also carry up to 38 fully-armed troops/paratroops or, in medevac role, up to 27 stretcher cases with four medical attendants. A flight-openable door on each side at the rear of the cabin permits paradropping. As yet no order for this military variant has been placed, and only a mockup of the ATM 42L fuselage has been constructed.

From the outset of the ATR project it was envisaged that the aircraft would be a candidate for 'stretching', to enable it to accommodate more passengers, and initially this aircraft was designated ATR XX and seen as a 54-58 passenger machine. However, when programme go-ahead for this aircraft was announced at the 1985 Paris air show, the configuration had been radically changed, with an increase in fuselage length of 4.25 m (14 ft) providing accommodation for 64, 66, 70 or 74 passengers

Above: Seen during flight tests over the Pyrénées near the ATR final assembly line at Toulouse, this aircraft wears the colours of Air Littoral, the launch customer for the ATR 42. Based at Montpellier and operating services throughout the south of France, the airline was taken over by the French company Euralair in November 1992.

The second prototype ATR 72 demonstrates the sprightly take-off performance of the type at the 1990 Farnborough air show. The stretched member of the family has notched up an impressive tally of sales.

Air Tahiti's fleet is based on one *ATR 42-300*, three *ATR 42-500* and six *ATR 72-200* aircraft, the last two delivered in 1998–9. The airline is based at Faaa, from where it flies services throughout Polynesia.

When Thai Airways adopted the **ATR** for its domestic network, it ordered a pair of each variant. This is an *ATR 72-201*, configured with 64 seats at a seat pitch of 0.81 m (32 in), the lowest-density layout in common usage.

Irish carrier Ryanair was among the multitude to adopt the **ATR** family as its principal turboprop aircraft and the airline established routes from every major city in Ireland to points in the UK. A mix of *ATR 42s* and *72s* was planned, but world-wide recession forced the airline to cut back operations and return its fleet of three *ATR 42s* to the **GPA** leasing company in 1992.

according to seat pitch. Designated ATR 72, the aircraft also features a 2.45 m (8 ft 1 in) increase in wingspan, the outer panels having carbonfibre front and rear spars and skin panels which form an additional 1500 litre (330-Imp gal) fuel tank. The ATR 72's engines are 1611-kW (2,160-shp) Pratt & Whitney Canada PW124/2 turboprops, with single-point pressure refuelling via a port in the starboard main landing gear fairing a standard feature.

The first ATR 72 (F-WWEY) made its maiden flight from Toulouse on 27 October 1988, and was followed by two more prototypes on 20 December and in April 1989. French certification was granted on 25 September 1989, and US FAA approval was obtained on 15 November. Twelve days later the first customer delivery took place when OH-KRA was delivered to Kar-Air of Finland.

In the autumn of 1988 ATR announced development, in conjunction with Avions Marcel Dassault-Breguet Aviation, of a maritime patrol version of either the ATR 42 or ATR 72. Known as the Petrel, this aircraft would be equipped with a mission avionics fit similar to that of the Dassault-Breguet Atlantique 2. The Petrel 42 would be armed with AM39 Exocet air-to-surface anti-shipping missiles, while the

The ATR 72 introduced a significant fuselage stretch to seat about 30 more passengers. This in turn required uprated engines, greater wing span and beefed-up undercarriage to handle the extra weight.

Although the majority of deliveries have been made to operators in Europe and the United States, some ATRs have been sold overseas. Zambia Airways (now incorporated in Aero Zambia) had two for its internal services.

When deliveries are completed, American Eagle will be the largest ATR operator, with 46 ATR 42s and 38 ATR 72s. Three more ATR 72s are due for imminent delivery.

Petrel 72 would be configured for anti-submarine operations and armed with torpedoes. Both variants would have Iguane 360° scanning radar, and undernose forward-looking infrared turret, electronic support measures equipment and sonobuoys.

In April 1991 ATR was refused permission to take over Boeing Canada (formerly de Havilland Canada) by the EC. It was felt that the resultant monopoly between the ATR family and rival Dash 8 would be unacceptable. Undaunted, ATR delivered 62 aircraft that year and by January 1992 the order book stood at 397 firm and 15 options. By the end of 1992 ATR certified the ATR 72-210, an improved hot-and-high version of the existing Series 200 aircraft, powered by PW 127 engines rated at 1849 kW (2,480 shp).

ATR has been successful in selling aircraft in Eastern Europe. First JAT of Yugoslavia then, more recently, LOT of Poland and Czech and Slovakia's CSA have all bought the ATR 72. By June 1998 orders for the ATR 42 and ATR 72 stood at 341 and 221 aircraft respectively, of which 334 and 211 had been delivered. Also by this date the initial products had been complemented by other models. The ATR 42-300 and ATR-200 were joined in 1995 and 1996 respectively by the ATR 42-500 with PW127E engines driving six-bladed propellers, and the ATR 42-400 with PW121A engines, while the ATR 72-200 was supplemented by the ATR 72-210 with uprated PW127 engines and then the ATR 72-210A (later ATR 72-500) with six-bladed propellers, increased weights and a new cabin interior. further improvements to the family result from the introduction of a lighter all-composite tail unit.

Kar-Air of Finland led the way as the ATR72 launch customer, and once flew seven on domestic services (being closely associated with the national airline Finnair). Its aircraft were configured with 66 seats.

Founded in 1967, Pan Am Express operated a network of feeder services into international routes operated from Miami and Philadelphia by its parent Pan American. When the great airline ceased operations on 4 December 1991, the operation was sold to TWA and its 11 ATR 42-300s became part of the Trans World Express network.

Brit Air is another major French regional operator that has adopted the ATR 42 in numbers. Based at Morlaix, the airline undertakes Air France schedules (in full AF colours) in addition to its own services.

Avro/Hawker Siddeley H.S.748

Universally known as the 'Budgie', the dependable 748 was a slow-but-sure seller for the British aerospace industry, appealing primarily to Third World operators who needed a tough, reliable and cheap to operate transport that could operate in primitive conditions. A military version ensured some success with the air forces.

The Avro 748 short/medium range twin turboprop airliner was inspired by the British government's 1957 Defence White Paper which declared that no more manned aircraft would be developed for the Royal Air Force and that future long-term defence of the United Kingdom would be undertaken by guided missiles.

This prompted A. V. Roe & Co. Ltd, which since World War II had been almost exclusively engaged in the manufacture of heavy military aircraft such as Lancaster, Lincoln, Shackleton and Vulcan, to re-enter the civilian market. Design studies began immediately under the direction of J. R. Evans for a short/medium range feeder aircraft with high or low wing, powered by two 1,000-shp turboprop engines, a maximum take-off weight of 18,000 lb and seating 20 passengers. The designation Avro Type 748 was allocated, and by 1958 an initial proposal was being shown to airline operators.

Little interest was shown by potential customers, who indicated that their requirement was for a larger airliner as a replacement for the ubiquitous Douglas C-47/DC-3 and Vickers Viking. They were looking for a rugged, economic aircraft capable of operating from undeveloped airfields and in 'hot and high' conditions. This prompted a complete redesign to cater for a minimum of 36 passengers seated four-abreast, with a gross weight of 33,000 lb and powered by Rolls-Royce Dart engines.

Development of the Avro 748 was officially announced on 9 January 1959, when work was already underway at Avro's Chadderton works near Manchester on the construction of four prototypes, two each for flight trials and static testing. The final configuration chosen was for a low-wing aircraft with a round section fuselage providing accommodation for up to 44 passengers in a pressurised cabin. To maximise ground clearance for the large four-blade propellers and avoid wing spar cut-outs, the two 1,740-shp Rolls-Royce Dart R.Da.7 Mk.514 engines were mounted above the wings, with the main undercarriage legs retracting forwards into fairings ahead of the leading edges.

Farnborough debut

Flown by Avro chief test pilot J. G. Harrison, the first prototype Avro 748, G-APZV, made its maiden flight from Woodford Aerodrome, Cheshire on 24 June 1960 and made its public debut at the Farnborough Air Show that September. The second prototype, G-ARAY, joined the test programme on 10 April 1961 and three months later undertook hot weather trials in Cyprus before flying on to Madrid for high altitude testing.

The first production aircraft, G-ARMV, one of three ordered by launch customer Skyways Coach Air Ltd, flew from Woodford on 31 August 1961, powered by 1,880-shp Dart 514 engines. After an

For many years a staunch supporter of the 748, LIAT (Leeward Islands Air Transport) is based at V.C. Bird International in Antigua, and has a mixed fleet of twin-prop airliners for island hopping services in the Caribbean. Until recently, five 44-passenger 748 Super 2Bs were on strength, serving with Dash Eights, Twin Otters and Islanders.

The Andover C.Mk 1 was a 748 derivative with a raised rear fuselage accommodating a rear ramp for tactical transport duties. After retirement from this role, several C.Mk 1s were converted to E.Mk 3 status (illustrated) for service as radio and navaid platforms with No. 115 Squadron.

Although the 748 sold exceptionally well in the Third World, it never penetrated the lucrative North American market. Air Illinois operated the type from Chicago's lakeside airport.

appearance at the 1961 Farnborough Air Show it made a sales demonstration tour of Jordan and Syria before embarking on a 160-hour route-proving programme, flying between Skyways' base at Lympne, Kent and Beauvais, Lyon and Montpellier in France it then entered revenue service on 1 April 1962 on the carrier's Lympne-Beauvais service carrying coach/air passengers between London and Paris, on which it supplemented long-serving DC-3s. Other customers for the Avro 748 Series 1 included B.K.S. Air Transport, Smith's Industries Aviation Division, and Aerolineas Argentinas, whose first of nine aircraft, *Ciudad de Babia Blanca*, left Woodford on 10 December 1961 on an 11,300-mile, eight-day ferry flight across the North Atlantic to Buenos Aires in the hands of Avro test pilot Colin Allen. Aerolineas' 748 fleet, later increased to 12 aircraft, operated two international routes, to Montevideo in Uruguay and Asuncion in Paraguay, but otherwise flew almost exclusively on domestic Argentinian services from rough provincial airstrips, providing a convincing demonstration of the ruggedness of the new aircraft.

Made in India

In addition to the Woodford production line, Avro had signed an agreement with the Indian government for licence assembly of 748s by Hindustan Aircraft Ltd at Kanpur, who contracted to complete 89 aircraft from British-built components, 72 for the Indian Air Force and 17 for Indian Airlines Corporation. The first of these, a Mark 1 for the IAF, was flown on 1 November 1961.

Meanwhile the second prototype had returned to the factory to have 2,105-shp Dart 531 engines installed as the prototype Avro 748 Series 2, in which form it first flew again on 6 November 1961. With structural strengthening and increased power the aircraft's maximum take-off weight was increased to 43,500 lb and passenger capacity to 52. The Series 2 became the standard production model after 18 Series 1s had been built.

A. V. Roe & Co. Ltd was absorbed into the Hawker Siddeley Group in 1963, and in July the aircraft, the last to bear the Avro title, was redesignated H.S.748. Between March and May of that year

Mounting a MEL MAREC radar under the forward fuselage, the Coastguarder was a maritime surveillance derivative aimed at providing a low-cost patrol platform for a variety of civil/military roles. These included oilfield and fishery patrol, search and rescue and anti-smuggling.

Military use of the 748 has been widespread, the home nation using several for VIP/staff transport duties. This Andover CC.Mk 2 is seen in the markings of No. 32 Sqn, the RAF's staff transport unit. Others previously served with the Queen's Flight for transport of the Royal Family.

Based at Walla Walla in Washington state, Cascade Airways operated a regional network throughout the mountainous north west of the United States. With a large fleet of smaller twins (Beech 99, Swearingen Metro and EMBRAER Bandeirante), Cascade looked to a pair of 748s for high-density operations.

the Series 2 prototype G-ARAY had undertaken an extensive demonstration tour of Europe, Africa, India and the Far East, and early in 1964 repeated the exercise in the Caribbean, South America and Canada, logging 91,263 miles in 413 flying hours on the two tours. Demand for the aircraft mounted, with overseas orders coming from Air Ceylon (1), Austrian Airlines (2), Bahamas Airways (4), LAN-Chile (9), Leeward Islands Air Transport (2), Linea Aeropostal Venezolana (6), Mount Cook Airlines of New Zealand (1), Philippine Airlines (8), Royal Nepal Airlines (2), Thai Airways (3), VARIG of Brazil (10), and at home from Autair and Channel Airways, whose four aircraft were the first to operate at a maximum take-off weight of 44,495 lb. This enabled 62 passengers to be carried on the airline's routes from the south coast of England to Paris and the Channel Islands. In off-season winter months many of the British-operated aircraft were leased abroad, and appeared in the colours of Aerotaxi (Mexico) and Jamaica Air Services as well as those of existing overseas 748 customers.

748 goes military

Following trials in early 1961 for a RAF requirement for a multi-purpose transport with STOL capability, able to operate from rough airstrips, desert or even ploughed fields down to 300 yards in length, an order was placed on behalf of RAF Transport Command for 31 military tactical freighter variants with extended fuselage incorporating rear loading/paradropping doors, and 'kneeling' main undercarriage legs to facilitate on/offloading of vehicles. The prototype G-APZV was returned to Chadderton for conversion to '748MF' (Military Freighter) configuration (also known as the Avro 780), first flying in this guise as G-ARRV on 21 December 1963. RAF aircraft, designated Andover C.Mk.1s, entered service in December 1966 with No. 46 Squadron at RAF Abingdon, and also served in Singapore, Aden and Bahrain, finally retiring in August 1975, although six aircraft modified for radar and navigational aid calibration duties remain on strength while others serve a test purpose.

Four Avro 748 Series 2s with standard fuselages were also ordered for VIP transport duties as Andover CC.Mk.2s, and continue to serve with No. 32 Squadron at RAF Northolt near London for carrying members of the British government and senior RAF officers, and with No. 60 Squadron at RAF Wildenrath, Germany. Two more were supplied to The Queen's Flight at RAF Benson, where they served as transports for the British royal family. Other military customers for the H.S.748 included the Royal Australian Air Force, which took eight aircraft as VIP transports and a further eight for use as navigation trainers, and the Brazilian Air Force, which ordered six for service on 'shuttle' routes between Brasilia and Rio de Janeiro.

In June 1967 Hawker Siddeley announced development of the improved H.S.748 Series 2A with 2,280-shp Dart 532 engines, an additional 260 Imperial gallons of fuel and the increased maximum take-off weight of 44,495 lb as standard. Launch customer for this version was Avianca of Colombia, although the Brazilian carrier VARIG was first to receive the aircraft, the last two of its Series 2s being converted on the production line to the new standard before delivery. Hawker Siddeley flew a Series 2A demonstrator on 5 September 1967, and following the pattern established with previous models, subsequently leased it to a number of potential customers, including SATA, Transair of Canada, Olympic Airways of Greece, Air Cape and Suidwes Lugdiens in South Africa. Zambia Airways, South African Airways and Merpati Nusantara Airlines of Indonesia also leased the aircraft, and subsequently became customers for 748, as did Transgabon, who first leased then bought the demonstrator in 1972. Among other customers for the Series 2A were British Airways, for use on Highlands and Islands routes around Scotland, Air Malawi, Air Gaspé, Chevron Standard Oil and Midwest Airlines of Canada, Rousseau Aviation of France, Mount Cook Airlines of New Zealand, COPA of Panama, SAESA of Mexico, Thai Airways, Botswana Airways, Ghana Airways, the Royal Australian Navy, which took two electronics trainer versions, and Germany's Bundesanstalt für Flugsicherung. The latter took delivery of seven aircraft equipped for airways and navigation aid calibration and flight check duties, for which role two former airline-operated H.S.748 Series 2s were also supplied to the British Board of Trade (later Civil Aviation Authority) Flying Unit based at Stansted Airport. VIP 748s were delivered to the Sultan of Brunei, King of Thailand and the Presidents of Argentina, Brazil, Ecuador, Venezuela and Zambia, while other military operators included the Belgian Air Force (3), Ecuadorian Air Force (5), Colombian Air Force and the Nepal Royal Flight.

Many of the aircraft delivered to military customers featured a large sliding rear freight door and strengthened floor, and in December 1971 a similar option was introduced for civilian aircraft in a convertible passenger/freight version incorporating a cargo door measuring 8 ft 9 in wide adjacent to the passenger door and a strengthened cabin floor capable of supporting a floor loading up to 200 lb/sq ft.

The end and the ATP

Another version of the H.S.748 was introduced in 1977 by British Aerospace, which had taken over the former Hawker Siddeley company that year. Known as the Coastguarder, it was intended for use as a medium-range maritime patrol and search-and-rescue aircraft and could be equipped with a ventrally-mounted surveillance radar and launch chutes for five-man rescue dinghies, flares and flame floats. Additional fuel tanks were installed to extend on-station patrol time to more than nine hours at a range of 200 nautical miles from base, and a variety of internal configurations were offered to accommodate observers and systems operators. A former airline-operated Series 2A, G-BCDZ, was modified to act as a demonstrator and flew widely throughout Europe, Central and South America and the Far East during 1977-78, but no orders for the Coastguarder ensued.

The Coastguarder demonstrator was subsequently rebuilt once again as the prototype for the H.S.748 Series 2B, which featured a four-foot increase in wingspan, modified tail surfaces to reduce drag and 2,290-shp Dart 536-2 engines to enhance hot-and-high performance. The first production Series 2B, G-BGJV, made its first flight on 22 June 1979. The first customer delivery was made in January 1980 to Air Madagascar. Three months later British Aerospace and Rolls-Royce began flight tests of an engine hush-kit for the 748's Dart engines, and these were subsequently offered for retrofit on Series 2Bs and all earlier variants of the aircraft. The Series 2B was marketed in the United States as the Intercity 748.

The final development of the BAe H.S.748 was the Super 748, announced in February 1982 and first flown on 30 July 1984. It incorporated all the features of the Series 2B with the addition of an advanced flight deck, systems improvements, increased baggage capacity, new galley design, more fuel-efficient Rolls-Royce Dart 552 engines which offered a 12 per cent reduction in fuel consumption over the previous powerplants, and engine hush-kits as standard.

Production of the Super 748 terminated in May 1988 with two aircraft sold to Taiwan, bringing total number of BAe H.S.748s built to 382 aircraft, including two prototypes, 164 military aircraft and 89 Indian-produced aircraft, 19 Series 1, 108 Series 2, 100 Series 2A and 34 Series 2B aircraft were built, and 31 Andovers. The 748 was succeeded by the British Aerospace Advanced Turboprop (ATP), originally envisaged as a 'stretched' and re-engined 748. Although it is an entirely new aircraft, the ATP retains the fuselage cross-section and general configuration of its highly successful and long-lived predecessor, which is still in widespread civil and military use throughout the world, 118 civil aircraft remaining in service in 1998.

Affectionately known as the 'Budgie', the 748 played an important part in British Airways' vast organisation, serving with the Highland Division supplying regional transport in Scotland. Flying alongside the ATP, these provided a lifeline to the Scottish islands.

Cameroon Airlines operated this Super 748 on domestic services from its base at Douala. The Super introduced numerous refinements, including a revised cockpit and hush-kitted engines.

Developed by Macavia, this version of the 748 is tailored to the fire-bombing role, with a large ventral pannier to hold retardant. The need for new fire-bombers is growing as the elderly reciprocating engines of the current firefighters require ever more costly maintenance.

For a while augmenting the Hercules fleet, the Andover C.Mk 1 proved a useful if limited tactical transport. Many were sold to New Zealand while others became E.Mk 3 calibration aircraft.

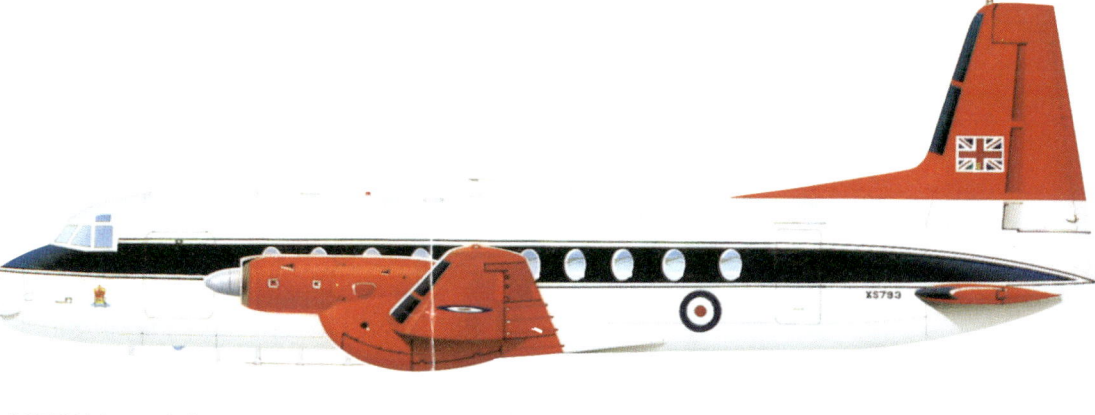

Three Andover CC.Mk 2s were in the service of the Queen's Flight, Royal Air Force. These have been replaced by the BAe 146, the aircraft passing to other staff transport squadrons in the RAF.

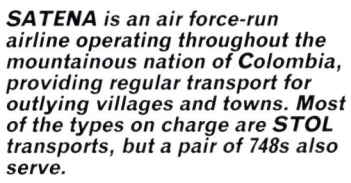

SATENA is an air force-run airline operating throughout the mountainous nation of Colombia, providing regular transport for outlying villages and towns. Most of the types on charge are STOL transports, but a pair of 748s also serve.

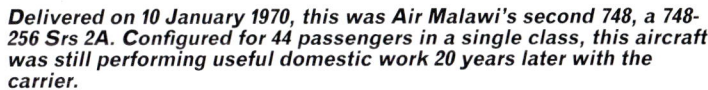

A large batch of 748-234s was delivered to LAN-Chile for that airline's domestic services. The 748 proved ideal for operations in often primitive conditions, being easy to maintain and with the necessary performance for operations from mountainous strips.

Delivered on 10 January 1970, this was Air Malawi's second 748, a 748-256 Srs 2A. Configured for 44 passengers in a single class, this aircraft was still performing useful domestic work 20 years later with the carrier.

British Aerospace (Avro/HS) 748-399 Series 2B (SCD)
Air Niger

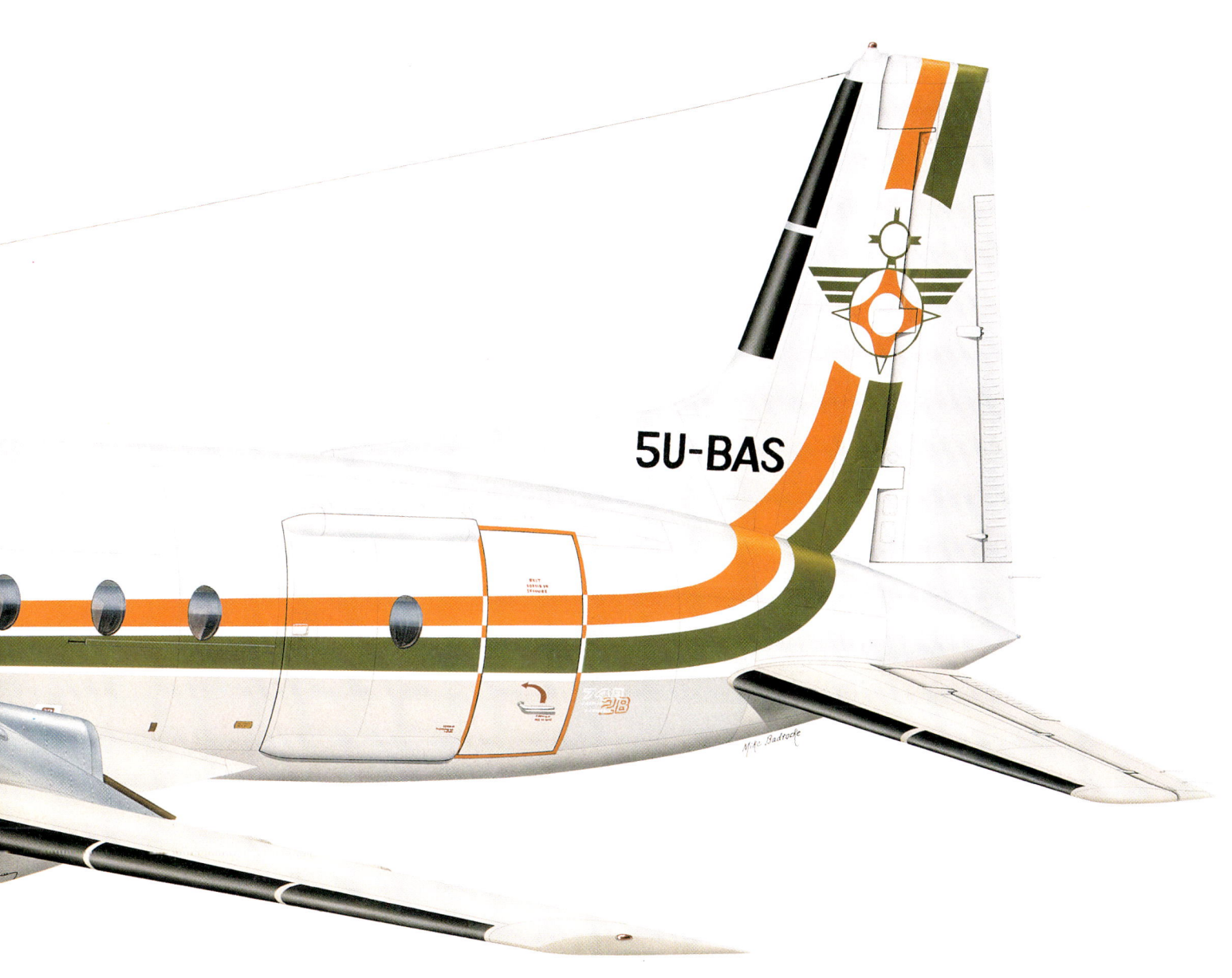

Chapter 6

Ayres Thrush

BAC One Eleven

Beech King Airs

Ayres Thrush

The contribution to modern life made by 'ag aircraft' goes unseen by many, even for a type as long-lived as the Thrush. Designed in the 1950s and remaining in production for decades after, today Leland Snow's Thrush still fulfills diverse tasks.

In 1953, while he was still a student at Texas A & M University, Leland Snow designed the Snow S1 agricultural aircraft, which was among the first aeroplanes to be developed specifically for aerial application work. A large low-wing design with a fixed tailwheel undercarriage and open single cockpit protected by a stout roll-over cage, the S1 was designed for ease of maintenance and cleaning, with a corrosion-proofed tubular fuselage structure with quick-release side panels enabling chemical deposits to be removed easily. Designed also with pilot safety in mind in the hazardous environment of low altitude aerial work operations, its robust structure was configured to absorb energy on impact, greatly reducing decelerative forces and thus the likelihood of serious injury to the pilot in the event of a crash.

The prototype was extensively tested 'in the field' in South America, leading to development of the improved S2, three pre-production prototypes of which were built. These differed from the S1 in having numerous structural modifications to the fuselage and wing.

In 1957 the newly formed Snow Aeronautical Corporation launched production of the Snow S2A, powered by a 179-kW (240-hp) Continental or Gulf Coast Dusting W-670-240 radial engine. A total of 74 S2As was built. Operators' demands for more power and greater lifting capacity resulted in the Snow S2B being introduced with a 335-kW (450-hp) Pratt & Whitney R985-AN1 Wasp Jnr radial engine (17 built), and the externally similar S2C-450 which had a chemical hopper of greater capacity and maximum take-off weight increased by 425 kg (940 lb). A 445-kW (600-hp) Pratt & Whitney R-1340-AN1 was offered as an option on the S2C-600, which had a maximum take-off weight of 2180 kg (4,800 lb), the combined production total of both S2C models being 215 aircraft.

A further increase in maximum take-off weight to 2720 kg (6,000 lb) and standard installation of the 445-kW (600-hp) Pratt & Whitney R-1840 engine were features of the Snow S2D, which was the principal production model when, in 1965, Leland Snow sold his interest in the company to Aero Commander Inc. of Bethany, Oklahoma, manufacturers of the Commander range of twin-engined business aircraft. Snow retained the plant at Olney, where he established Air Tractor Inc. to develop a new range of specialist agricultural aircraft of similar configuration to the S1/S2 series. Production of the Snow S2D was transferred to Aero Commander's factory at Albany, Georgia, where it was known as the Ag Commander S2D and joined the smaller Callair range of agricultural aircraft for which Aero Commander had also acquired manufacturing rights.

Aero Commander's engineers began work on refining Leland Snow's design and, in 1967, when Aero Commander became a division of the North American Rockwell Corporation, the S2R Thrush Commander was introduced. This featured a fully enclosed cockpit for the first

Like Ayres before them, Air Tractor decided on the supremely reliable Pratt & Whitney Canada PT6A engine for their turboprop version of the Snow, the AT-502. The spray bars contain 81 variable nozzles, and the internal glassfibre hopper holds 1900 litres (420 Imp gal) of fertiliser or insecticide.

1967 saw the introduction of the S2R Thrush Commander, a refined version of the Snow which is still the basis for the Ayres designs. Rockwell built the type for 10 years before selling out to Ayres.

Following experience with earlier models, the Snow S2C was the first truly successful Snow design. Although an open cockpit was fitted, with small windshield, a roll-frame was fitted to protect the pilot.

The Snow S2D introduced an enclosed cockpit, and standardised on the 600-hp R-1340-AN1 engine. This was the version in production when Snow was bought out by Aero Commander, becoming the Ag Commander S2D.

time, electrically-operated flaps and a 1515-litre (400-US gal) chemical hopper.

The Thrush Commander continued in production for 10 years until, in November 1977, Rockwell International sold the type certificate for the aircraft and the Albany production facility to Frederick Ayres of Ayres Corporation. Ayres, in conjunction with Serv-Aero Engineering, had previously developed a turboprop conversion kit for the Thrush Commander which replaced the standard radial engine with a 560-kW (750-shp) Pratt & Whitney Canada PT6A, the prototype of which made its first flight on 9 September 1975.

When Ayres acquired design rights it continued production of the piston-engined S2R Thrush (the 'Commander' suffix was dropped) and also launched new production of the S2R-T Turbo Thrush, which had hitherto been available only as an aftermarket conversion. Powered by a choice of 370-kW (500-shp), 505-kW (680-shp) or, most popularly, the 560-kW (750-hp) PT6A-34G driving a three-bladed feathering and reversing propeller, the Turbo Thrush featured a longer nose to compensate for the much lighter weight of the compact turbine engine and a choice of 1515-litre (333-Imp gal) or 1890-litre (415-Imp gal) chemical hoppers. The turbine engine offered many advantages over the radial, including quieter, more 'environmentally friendly' operation (the aircraft was thus known colloquially among ag pilots as the 'Hush Thrush'); improved short field take-off and landing capability; a 454-kg (1,000-lb) increase in payload because of the reduced engine weight; much more ready availability of spares and fuel, both of which could be a problem in remote areas with the radial-engined models; extended engine overhaul intervals, and – most importantly in the agricultural environment where rapid turn-

The Snow designs were for many years the most powerful and 'beefiest' ag planes available in the Western world, able to lift large loads on account of the large power output from the radial engines. This is the 1966 Snow Commander.

Precise flying is required for the agricultural role, as demonstrated here by this Thrush Commander spraying a river. The spray bars are virtually full-span for maximum coverage, and the spray is pumped by a wind-driven generator under the central fuselage.

arounds are essential – the ability to reload the aircraft with the engine running and the propeller stopped, thanks to the PT6A's free-turbine configuration. Two-seat versions of the Thrush and Turbo Thrush were developed in 1979 to enable an engineer/loader or 'flagman' to be carried behind the pilot or, with dual controls installed, to enable trainee ag pilots to gain experience on the aircraft. Two seat ver-

Rivalling the Ayres concern is Air Tractor, a company established by Leland Snow to continue his own refinements to the original Snow design. Shown here at breathtaking altitude is the AT-301, powered by the R-1340 radial of 600-hp.

sions, known as 'DCs' (dual cockpit), are readily identifiable by their extended cockpit fairings which taper down to the base of the fin.

While Ayres was developing its PT6A-powered Turbo Thrush, Marsh Aviation of Mesa, Arizona, was working on a similar turbine conversion involving installation of a Garrett TPE331-1-101 engine, derated from its full 587 kW (788 shp) output to 447 kW (600 shp). Two prototypes were built and the Marsh S2R-T Turbo Thrush received US Federal Aviation Administration type certification in September 1976, since when some 80 Thrush Commanders and Thrushes have been converted to Garrett power by Marsh and are operating in Africa, Europe, South America and the United States. The Marsh conversion is still available, though only small numbers have been converted in recent years.

Choice of engines

Ayres, while continuing to produce the standard P&W Wasp-powered S2R, offered a number of alternative piston engine installations. These included the S2R-R3S with Polish-built 447-kW (600-hp) Pezetel PZL-3S

Specification
Rockwell Thrush Commander
Type: single-seat agricultural aircraft
Powerplant: one 600-hp Pratt & Whitney R-1340-AN-1 or -S3H1 nine-cylinder radial aircooled piston engine.
Performance: max level speed 225 km/h (140 mph); cruising speed 177 km/h (110 mph); stalling speed, flaps up normal weight 89 km/h (55 mph)
Dimensions: wing span 13.54 m (44 ft 5 in); length overall 8.95 m (29 ft 4½ in); height 2.79 m (9 ft 2 in)
Weights: empty, equipped 1791 kg (3,950 lb); maximum take-off weight 3130 kg (6,900 lb)

Typical of the family, this is one of the many Thrush Commanders built by Rockwell's agricultural division, virtually all of which were finished in yellow house colours, this colour being chosen for high conspicuity. Key features of the aircraft were its broad wing which allowed a good load to be lifted but with a low enough wing loading for rapid manoeuvrability. Large control surfaces ensured that this was more than adequate for agricultural work. Note the levelled scale for the hopper contents on the fuselage side.

On the front line in the battle against drugs are the Turbo Thrush NEDS aircraft of the US State Department. This aircraft is seen in Belize, where a major campaign is under way against the cultivation of marijuana. These aircraft often operate under the protective cover of armed Belize Defence Force Britten-Norman Defenders.

radial and the Leo-Thrush with British-manufactured 484-kW (650-hp) Alvis Leonides engine, though neither found particular favour with operators.

Ayres' current production models with piston engines are the S2R-600 with P&W R-1340 Wasp and the S2R-R1820/510 Bull Thrush which is powered by a massive 894-kW (1,200-hp) Wright R-1820 Cyclone, and is claimed to be the world's most powerful agricultural aircraft. Both these models, and the turbine variants described below, share a common airframe structure comprising welded chrome molybdenum steel tube fuselage with quick-release alloy skin panels and stainless steel undersurface skinning, all-metal cantilever wings of two-spar construction with alloy skinning and optional wingtip extensions to increase lifting surface, and light alloy strut-and-wire braced tail surfaces. The cockpit is sealed against chemical ingress and protected by a steel tube roll-over structure with a second, forward-facing seat and dual controls optional. The Thrush has a 1515-litre (333-Imp gal) chemical hopper while the Bull Thrush's hopper has a capacity of 1930 litres (424 Imp gal), and the more powerful aircraft has 719-litre (158-Imp gal) fuel tanks against the 333-litre (73-Imp gal) tanks of the Wasp-powered variant. A variety of spraybooms, atomisers and high-volume spreaders for liquid or solid agricultural chemicals is available, and the aircraft may be adapted for aerial firefighting.

Combat Thrush

In addition to the three standard agricultural versions of the Turbo Thrush currently in production (with 372-, 506- and 558-kW/500-, 680- and 750-shp PT6A-11AG, -15AG and -34AG engines respectively), Ayres has developed two special mission variants of the aircraft. The first of these, developed for a US State Department requirement, was the S2R-T65/400 Turbo Thrush NEDS (Narcotics Eradication Delivery System). Powered by a 1025-kW (1,376-shp) Pratt & Whitney Canada PT6A-65AG turboprop driving a large diameter, five-bladed Hartzell propeller, the Turbo Thrush NEDS features an armoured two-seat cockpit with dual controls, armour-protected engine compartment, self-sealing 63-litre (14-Imp gal) auxiliary fuel tank mounted in a bulletproof structure in addition to the Turbo Thrush's standard wing tanks, and a comprehensive suite of VHF and HF mission avionics. Nine Turbo Thrush NEDS have been delivered to the State Department's International Narcotics Matters Bureau for use on Operation Roundup drug eradication missions against marijuana and poppy crops in such countries as Belize, Burma, Colombia, Mexico and Thailand. The Turbo Thrush NEDS is usually operated by two crew in case one member gets injured by groundfire from armed drug growers, and often operates in conjunction with additional fire support from accompanying aircraft. A chemical called Roundup is carried in the aircraft's 1515-litre (333-Imp gal) hopper and sprayed on drug crops to make the plants overfertilise, grow too rapidly, then wilt and die.

The latest special mission variant of the Turbo Thrush is the V-1-A Vigilante low-cost surveillance and close-support version of the Turbo-Thrush NEDS, which Ayres developed in co-operation with the US State Department and US Army Electro-Optical Survivability Program. The prototype, N3100A, first flew in May 1989. As well as the armoured cockpits and self-sealing auxiliary fuel tank of the NEDS, the Vigilante features four NATO-standard stores hardpoints under each wing and three tandem hardpoints under the fuselage, on which can be carried a variety of stores including 0.50-calibre, 7.62-mm and 20-mm gun pods, 2.75-in air-to-ground rockets, 500-lb bombs, Stinger missiles, land and sea mines, torpedoes, anti-tank missiles and electronic countermeasures equipment. Surveillance systems such as forward-looking infra-red, low light level television, IR linescan and associated datalinks can also be installed, with a systems operator occupying the rear seat where a high resolution monitor is provided. Onboard recorders can take data from two surveillance systems simultaneously and transmit them to ground stations in real time or as secure frozen imagery, with a range of up to 100 nm (185 km; 115 miles). In 1989 Ayres flew the Vigilante on night surveillance demonstrations over Arizona and Texas for the US Border Patrol, successfully detecting groups of illegal immigrants.

One further development of the versatile Turbo Thrush is the Turbo Sea Thrush devised by Terr-Mar Aviation Corporation of Vancouver, Canada. It is an amphibious fire-bomber conversion mounted on Wipline 6000 floats. A retractable probe enables the Sea Thrush's 1703-litre (375-Imp gal) hopper to be filled with water in 15 seconds while the aircraft is planing across a lake or open sea surface; the entire load can then be dumped on a fire site via a gravity-operated drop door set into the base of the hopper.

Not to be confused with the PT6A Ayres Turbo Thrush conversion, Marsh Aviation at Mesa, Arizona, converts piston-powered Thrushes to turbine power by adding the Garrett TPE331. A handful of aircraft are still being converted.

BAC One-Eleven

One of a crop of airliners with rear-mounted engines, the One-Eleven has proved a success for the British aviation industry, its successive variants selling in fair numbers for several years. The programme was later transferred to Romania, where modest numbers of aircraft were completed. New powerplants were one method of development to keep the sturdy airframe viable for service in the 1990s.

The British Aerospace Corporation One-Eleven was designed as a jet successor to the highly successful Vickers Viscount turboprop airliner. Its origin can be traced back to 1956 when Hunting Aircraft Ltd at Luton Airport began a design study for a small 48-seat short-haul jet airliner designated Hunting H-107. The H-107 was to have been a cruciform-tail design powered by two rear-mounted Bristol Orpheus 12B turbojet engines and would have had a maximum range of 1,000 miles.

In September 1958 the H-107 design was reconfigured around the Bristol Siddeley BS.61 or BS.75 turbofan engines which were then in development, with cabin capacity expanded to 48-56 passengers in four-abreast seating. Wind tunnel tests were conducted and static test articles and a full-scale mock-up of the aircraft was built, but no further progress was made before, in 1960, Hunting Aircraft was acquired by the newly established British Aircraft Corporation which had merged the Bristol, English Electric and Vickers companies. Work proceeded on the H-107 under the combined efforts of the former design teams of Hunting and the Weybridge Division of Vickers, and after assessing potential airline interest the decision was taken to further expand the airframe to carry 65 passengers in a five-abreast layout, with range optimised at 600 miles. Two Rolls-Royce Spey turbofans were selected to power the aircraft, which was renamed BAC-111, but later known alternatively as the One-Eleven.

In April 1961 production of an initial batch of 20 BAC-111s was authorised, and on 9 May 1961 British United Airways became launch customer with an order for 10 aircraft of the basic Series 200 model. Vital penetration of the United States market, in which the Viscount had been so successful, occurred in October 1961 when Braniff International ordered six BAC-111s with six options, followed by an order from Mohawk Airlines.

The BAC-111 featured a circular cross-section all-metal fuselage incorporating a pressurised cabin seating 79 passengers in all-coach or economy layout, or 65 in a mixed-class configuration, slightly swept low wings with Fowler flaps and upper surface airbrakes/spoilers, a T-tail configuration with variable-incidence tailplane, and retractable tricycle landing gear with twin wheels on each unit. Passenger entry was via a forward door on the port side and a ventral airstair door below the tail. The prototype, G-ASHG, was powered by two 10,410-lb st thrust Spey Mk 506-14 turbofans, and made its maiden flight from the British Aircraft Corporation factory at Hurn Airport, Bournemouth, on 20 August 1963, at which time orders stood at 60.

The flight test programme for the BAC-111 suffered a major setback on 22 October 1963 when the prototype was lost in a crash which took the lives of seven crew, including test pilot and former world airspeed record-holder Mike Lithgow. The cause was established as a deep stall condition arising from the aircraft's T-tail and rear engined configuration. Modifications were made to provide powered elevator controls, a stick pusher and altered leading-edge

When British Overseas Airways Corporation (BOAC) and British European Airways (BEA) merged to form British Airways, the latter brought a large fleet of One-Elevens with it. Others were picked up later so by the mid-1980s the fleet stood at 34, with most being of the Series 500 model (illustrated).

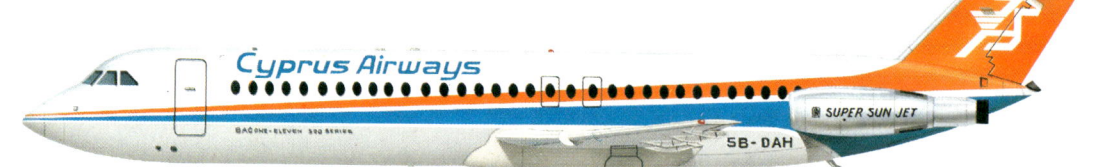

Cyprus Airways came late to the BAC One-Eleven, its Series 537 aircraft being delivered in 1977/78. They operated in Europe and the Mediterranean, but were supplanted by the A310 and A320.

In the Middle East the One-Eleven did not have the sales that might have been expected, but Gulf Air did operate the type for a while before replacing it with the Boeing 737. The airline serves the Gulf states of Oman, Bahrain, Qatar and the United Arab Emirates.

camber to prevent the aircraft entering high angle-of-attack attitudes. The remedy delayed considerably the BAC-111's development programme, but the data gathered as a result of the tragic crash proved invaluable to other manufacturers then developing T-tail jet airliners. British certification of the aircraft was finally granted on 6 April 1965 – a year later than the originally-forecast first delivery dates – and US FAA approval came on 15 April. BUA commenced scheduled services from Gatwick Airport to Genoa on 9 April, and Braniff put the aircraft into service on its Corpus Christi, Texas-Minneapolis/St Paul route on the 25th of the month. BUA inaugurated its BAC-111 'Citijet' service from Gatwick to Glasgow, Edinburgh and Belfast in January 1966.

Prior to certification of the BAC-111 Series 200, the British Aircraft Corporation had announced its intention to develop two further versions of the aircraft: the increased payload/range Series 300 with RR Spey Mk 511 turbofans of 11,400 lb st each mounted in longer engine nacelles, a centre section fuel tank, strengthened wings and undercarriage to cater for an 8,500-lb increase in gross weight to 87,000 lb, and improved heavy-duty brakes; and the generally similar Series 400 which would incorporate modifications to meet US FAA certification requirements for two-crew operation. In July 1965 American Airlines, one of the United States' 'big five' carriers, ordered 15 BAC-111 400s, with options on a further 15, marking a major breakthrough in US sales. The independent carrier British Eagle was first to order the Series 300.

Two Series 400 development aircraft, G-ASYD and G-ASYE, made their first flights from Hurn Airport on 13 July and 16 September 1965 and were joined by American Airlines' first Series 401 on 4 November, during which month G-ASYE conducted a world-wide sales tour which attracted orders from Canada, Central and South America, the Philippines, Spain, the United States and West Germany. FAA type approval was granted on 22 November 1965. American Airlines commenced passenger services on short-haul routes on 6 March 1966. Production of the BAC-111 Series 200, 300 and 400 totalled 56, nine and 69 respectively.

Even before the first BAC-111 flew, BAC had considered the possibility of a 'stretched' version of the aircraft, but it was not until British European Airways expressed an interest in such a model that work began on the Series 500. This aircraft was a 97-119 seat version with a fuselage stretched by 8 ft 4 in forward of the wing and 5 ft 2 in to the rear; a 5 ft increase in wingspan; strengthened landing gear and wings permitting a gross weight of 91,000 lb (later raised to 104,500 lb), and uprated 12,000 lb st Rolls-Royce Spey Mk 512-14 engines. BEA ordered 18 'Super One-Elevens' on 27 January 1967 and a week later the Series 400 development prototype was ferried to Hurn Airport for conversion as the aerodynamic prototype for the Series 500, flying again on 30 June 1967. The first production aircraft flew on 7 February 1968 and United Kingdom certification was

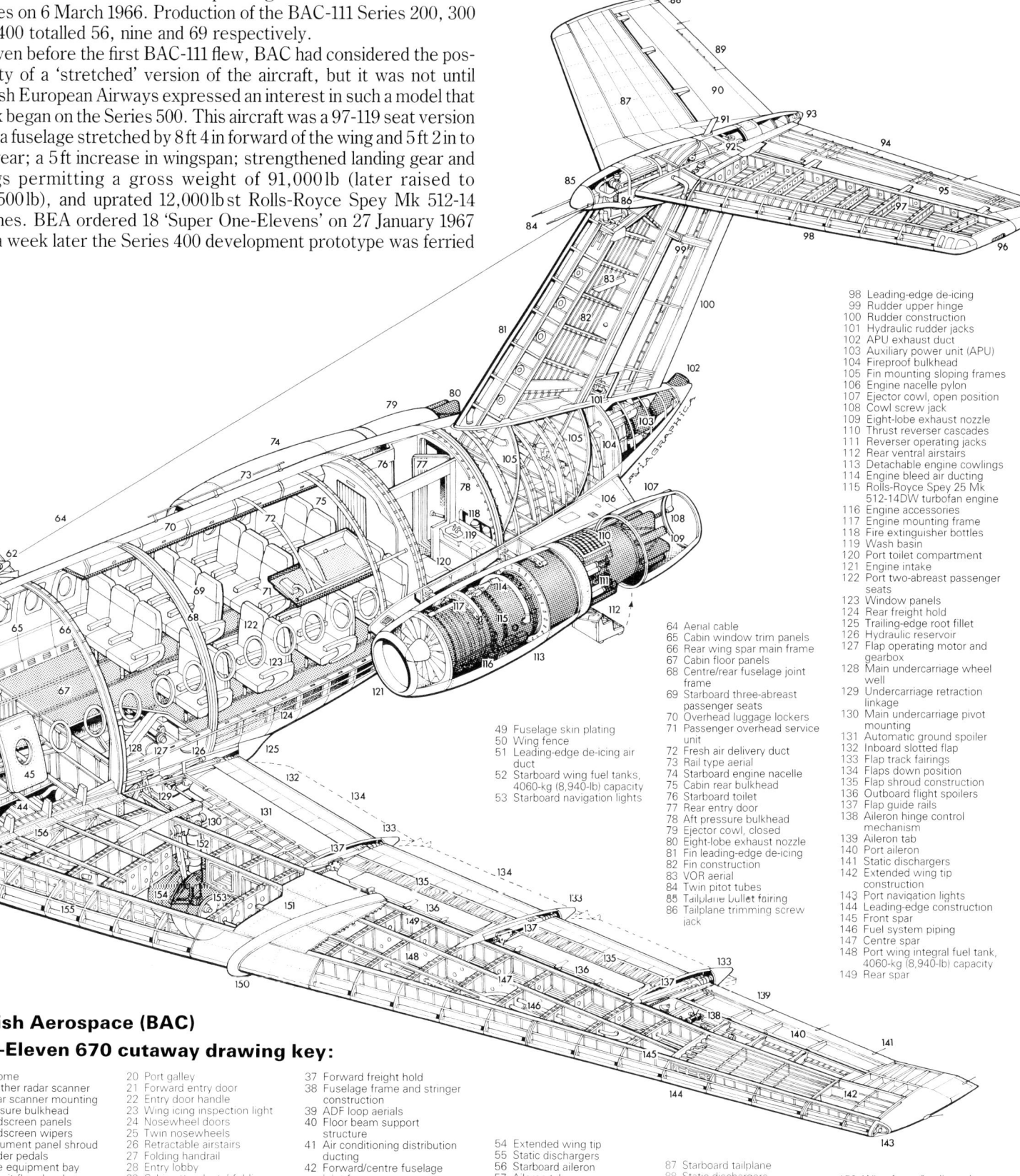

British Aerospace (BAC)
One-Eleven 670 cutaway drawing key:

1. Radome
2. Weather radar scanner
3. Radar scanner mounting
4. Pressure bulkhead
5. Windscreen panels
6. Windscreen wipers
7. Instrument panel shroud
8. Rudder pedals
9. Nose equipment bay
10. Cockpit floor level
11. Control column
12. Pilot's seat
13. Co-pilot's seat
14. Cockpit roof construction
15. Supernumerary crew seat
16. Cockpit bulkhead
17. Radio rack
18. Starboard galley
19. Cockpit door
20. Port galley
21. Forward entry door
22. Entry door handle
23. Wing icing inspection light
24. Nosewheel doors
25. Twin nosewheels
26. Retractable airstairs
27. Folding handrail
28. Entry lobby
29. Cabin attendants' folding seats
30. Starboard service door
31. Cabin bulkhead
32. Wardrobe
33. Communications aerials
34. Forward cabin seating
35. Window panel skin doubler plate
36. Freight hold door
37. Forward freight hold
38. Fuselage frame and stringer construction
39. ADF loop aerials
40. Floor beam support structure
41. Air conditioning distribution ducting
42. Forward/centre fuselage joint frame
43. Front wing spar main frame
44. Ventral air conditioning plant
45. Port emergency exit
46. Wing centre section fuel tank, 3170-kg (6,984-lb) capacity
47. Seat rail support beams
48. Starboard emergency exit window
49. Fuselage skin plating
50. Wing fence
51. Leading-edge de-icing air duct
52. Starboard wing fuel tanks, 4060-kg (8,940-lb) capacity
53. Starboard navigation lights
54. Extended wing tip
55. Static dischargers
56. Starboard aileron
57. Aileron tab
58. Aileron hinge control mechanism
59. Spoilers open position
60. Spoiler jacks
61. Flap screw jacks and gearboxes
62. Flap track fairings
63. Starboard outboard slotted flaps, open position
64. Aerial cable
65. Cabin window trim panels
66. Rear wing spar main frame
67. Cabin floor panels
68. Centre/rear fuselage joint frame
69. Starboard three-abreast passenger seats
70. Overhead luggage lockers
71. Passenger overhead service unit
72. Fresh air delivery duct
73. Rail type aerial
74. Starboard engine nacelle
75. Cabin rear bulkhead
76. Starboard toilet
77. Rear entry door
78. Aft pressure bulkhead
79. Ejector cowl, closed
80. Eight-lobe exhaust nozzle
81. Fin leading-edge de-icing
82. Fin construction
83. VOR aerial
84. Twin pitot tubes
85. Tailplane bullet fairing
86. Tailplane trimming screw jack
87. Starboard tailplane
88. Static dischargers
89. Elevator tab
90. Starboard elevator
91. Communications aerial
92. Elevator control rods
93. Tail navigation light
94. Port aileron tabs
95. Port aileron construction
96. De-icing air outlet louvres
97. Tailplane construction
98. Leading-edge de-icing
99. Rudder upper hinge
100. Rudder construction
101. Hydraulic rudder jacks
102. APU exhaust duct
103. Auxiliary power unit (APU)
104. Fireproof bulkhead
105. Fin mounting sloping frames
106. Engine nacelle pylon
107. Ejector cowl, open position
108. Cowl screw jack
109. Eight-lobe exhaust nozzle
110. Thrust reverser cascades
111. Reverser operating jacks
112. Rear ventral airstairs
113. Detachable engine cowlings
114. Engine bleed air ducting
115. Rolls-Royce Spey 25 Mk 512-14DW turbofan engine
116. Engine accessories
117. Engine mounting frame
118. Fire extinguisher bottles
119. Wash basin
120. Port toilet compartment
121. Engine intake
122. Port two-abreast passenger seats
123. Window panels
124. Rear freight hold
125. Trailing-edge root fillet
126. Hydraulic reservoir
127. Flap operating motor and gearbox
128. Main undercarriage wheel well
129. Undercarriage retraction linkage
130. Main undercarriage pivot mounting
131. Automatic ground spoiler
132. Inboard slotted flap
133. Flap track fairings
134. Flaps down position
135. Flap shroud construction
136. Outboard flight spoilers
137. Flap guide rails
138. Aileron hinge control mechanism
139. Aileron tab
140. Port aileron
141. Static dischargers
142. Extended wing tip construction
143. Port navigation lights
144. Leading-edge construction
145. Front spar
146. Fuel system piping
147. Centre spar
148. Port wing integral fuel tank, 4060-kg (8,940-lb) capacity
149. Rear spar
150. Wing fence/leading-edge fillet
151. Machined wing skin panels
152. Main undercarriage leg strut
153. Automatic wheel brakes
154. Twin mainwheels
155. Leading-edge de-icing air duct
156. Wing attachment joint strap
157. Wing root ventral fairing

113

The Series 475 introduced hot-and-high performance, vital to the internal routes in several South American countries. Faucett of Peru flew this aircraft on such services, where the rough- and short-field performance was called into play many times.

granted on 15 August. Deliveries to BEA began two weeks later, with passenger services commencing on 17 November from Manchester Airport. BEA subsequently introduced the Super One-Eleven on its internal West German services from Berlin.

Improved 500

The BAC-111 Series 500 was subsequently recertificated with water injected 12,550-lb st Spey 512-14DW turbofans, gross weight rising to 99,650 lb, or 104,400 lb with a supplementary fuel tank in the rear of the aft baggage hold, and proved to be numerically the most successful version of which 84 were delivered, serving, in addition to BEA (by then British Airways), with British Caledonian Airways, British Midland Airways, Court Line, Bahamasair, Germanair, Sadia of Brazil, Bavaria and Philippine Airlines.

The final British development of the BAC-111 was the Series 475 'hot-and-high' version which combined the standard fuselage of the Series 400 with the Series 500's extended wings and Spey Mk 512-14DW power plants, and a strengthened undercarriage with low-pressure tyres to permit operation from poor runway surfaces. The Series 400/500 development aircraft G-ASYD was once again converted to serve as the prototype, flying in this guise for the first time on 27 August 1970, followed by the first production aircraft – destined for Faucett of Peru – on 5 April 1971. Nine Series 475 aircraft were built.

Executive transport and freighter versions of the BAC-111 were also offered. More than 40 corporate/VIP aircraft were eventually delivered, including modified ex-airline airframes. Among such operators were Tennessee Gas, the Royal Australian Air Force, Brazilian Air Force and the Sultan of Oman's Air Force, which took delivery of three Series 475 with quick-change passenger/cargo interiors. The One-Eleven Freighter was fitted with a 10 ft by 6 ft 1in upward-opening hydraulically actuated loading door in the port forward fuselage, a quickly-removable freight floor overlay and a cargo handling system.

Quiet One-Eleven

To extend the working life of the BAC-111, BAC developed a 'hush kit' for its Spey engines comprising an intake and bypass duct linings, extended acoustically-lined jetpipes, and a six-chute jet mixing exhaust silencer, with which the hard-working Series 400/500/475 prototype once again served as guinea pig, flying 'hush kitted' for the first time on 14 June 1974.

The 'hush kit' reduced the aircraft's 90 EPNdB noise 'footprint' by 50 per cent, and was claimed to make it as environmentally acceptable as a twin turboprop. The Romanian Airline Tarom became the first operator of hush-kitted BAC-111s when five Series 500s were delivered in this configuration during the summer of 1977.

In May of that year British Aerospace, as BAC had by then

United Kingdom military use is restricted to aircraft used for equipment test purposes and this single One-Eleven Series 479FU used by the Empire Test Pilot School at Boscombe Down.

In the United States the One-Eleven proved popular during its early career, with large orders from Braniff, American Airlines and Mohawk. Since then other operators have used the type, including Florida Express. The airline has since merged with Braniff, which suspended operations.

Far East interests for the One-Eleven lay largely with Philippine Airlines, although its first experience with the type was unhappy. The first aircraft delivered was a Series 402 but was lost three years later. Currently the carrier operates 10 Series 500s, including two ex-Germanair aircraft.

Military use of the One-Eleven has been limited, one of the users being the Sultanate of Oman Air Force. It operates three Series 475s with No. 4 Squadron, these having been fitted with a fuselage-side upward-hinging cargo door.

become, concluded an agreement with Interprinderea de Avionane Bucuresti of Bucharest, Romania, for the licence-manufacture of One-Eleven 475 and 500s, to be known as Rombac 1-11s. Three complete aircraft – a Series 487 freighter and two 525s (as had been traditional with Vickers and BAC, BAe used individual model numbers for each customer within the basic Series range) – were delivered prior to the seven-stage transfer of technology programme which was completed in 1986. The first Romanian-assembled aircraft made its maiden flight on 18 September 1982 and was handed over to the State airline Tarom on 24 December, entering service the following January. Rombac (now Romaero) completed 19 aircraft by 1995 as nine Series 495 (Series 400 with the wings and engines of the Series 560), one Series 497 (equivalent to the Series 475) and nine Series 560 (Series 400 with stretched wings and fuselage).

The most recent development in the BAC-111 story comes from the United States where, in February 1986, the Dee Howard Company of San Antonio, Texas, signed agreements with British Aerospace and Rolls-Royce for a re-engining programme for Series 400, 475 and 500 aircraft under which they will be fitted with 15,100-lb st RR Tay 650 Turbofan engines. A prototype conversion of a corporate BAC-111 Series 400 has been completed by Dee Howard and flew during 1989. Replacement of the aircraft's Spey engines reduces balanced field length by as much as 30 per cent under 'hot-and-high' operating conditions and provides an IFR, with reserves range of up to 3,370 miles in a corporate-configured aircraft. The modification included new engine nacelles and thrust reversers, and enabled the One-Eleven to meet current stringent noise requirements. The programme later fell into abeyance.

BAC One Eleven Series 518
G-AXMH 'Halcyon Sun'
Court Line

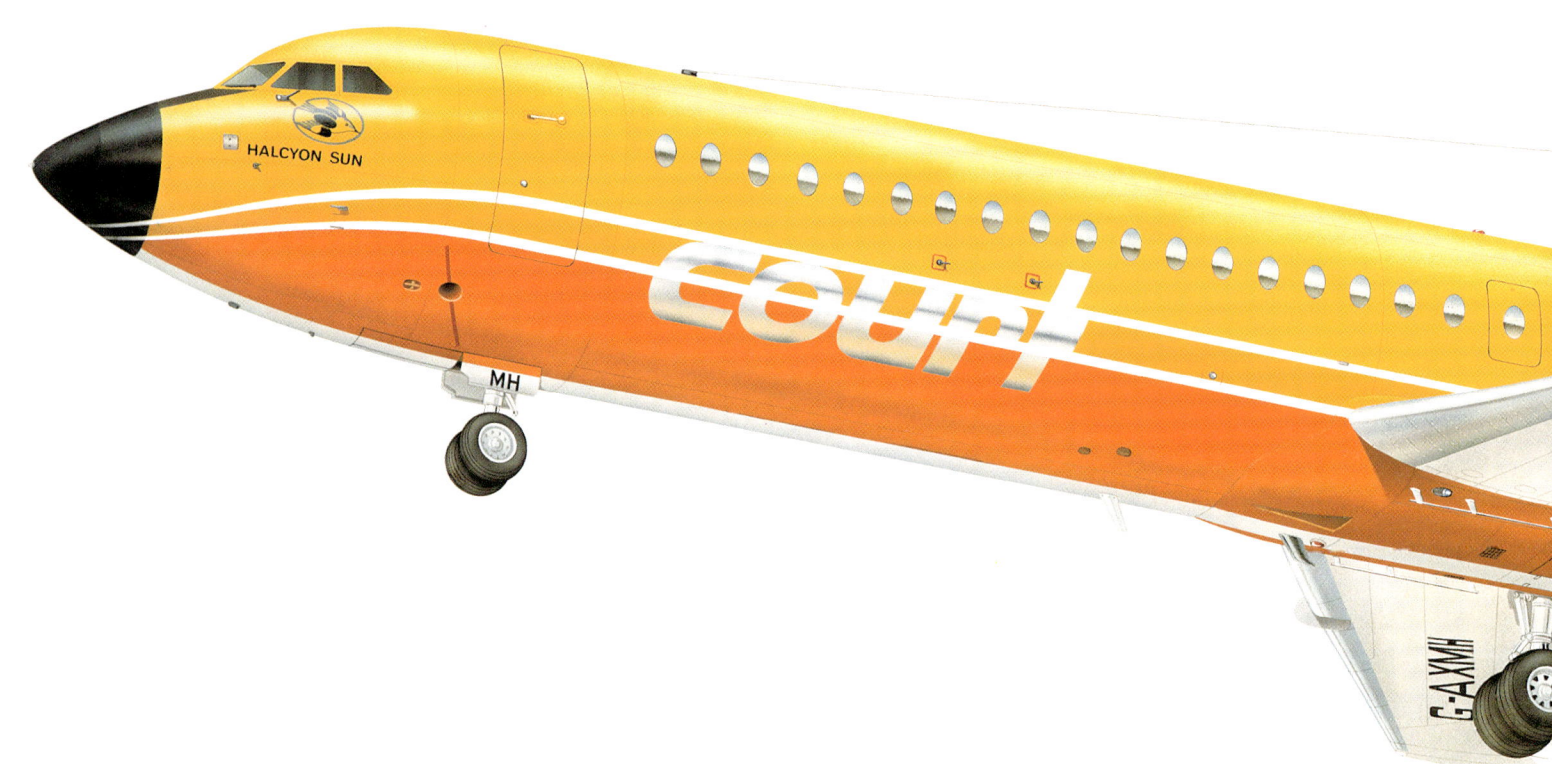

Beech King Airs

Walter Beech's aircraft company already had considerable experience with piston-engined twins when it came to the conclusion in the early 1960s that turboprop powerplants were the way ahead. In the years that followed, the aptly named King Air has become the long-reigning monarch of the lucrative 'biz-prop' market.

Beech (now Raytheon Beech) has been manufacturing twin-engined business and military liaison aircraft since as far back as 1937 when the radial-engined Beech Model 18 first flew. Large numbers were supplied to the US and Allied forces during World War II as C-45 Expeditors, and in peacetime development and manufacture of the Model 18 and Super 18 series continued until 1969.

Beech had meanwhile developed the Model 50 Twin Bonanza, which was also sold to the US Army in substantial quantity as the L-23/U-8 Seminole. Driven initially by a military requirement for a more spacious cabin, the Twin Bonanza evolved via the 1960 L-23F/U-8F Seminole into the 'cabin class' Model 65-80 Queen Air.

When, in 1961, Beech first began planning a 482-km/h (300-mph) twin-turboprop business aircraft, its attention focused on the projected Model 120, a new design which it proposed to power with French-built Turboméca engines. However, a US Army requirement for a turboprop replacement for its U-8 fleet led to trial installation of 373-kW (500-shp) United Aircraft of Canada (now Pratt & Whitney Canada) PT6A-6 free turbines in a modified Queen Air airframe which first flew (as the one-off Beech Model 87) in May 1963 and accumulated 10 months of flight testing before delivery to the Army at Fort Rucker, Alabama as an NU-8F.

In the interim Beech decided to pursue this logical development of the Queen Air for commercial customers, abandoned the Model 120 and announced the Model 90 King Air, the prototype of which (N5690K) made its first flight from Wichita, Kansas on 24 January 1964. Externally the 4218-kg (9,300-lb) maximum take-off weight King Air 90 resembled the contemporary piston-engined Queen Air 80 save for its engine installation and the round rather than square cabin windows demanded by its 23-kPa (3.4-psi) supercharger-driven pressurised six/eight seat cabin. With a maximum speed of 450 km/h (280 mph) the King Air was some 48 km/h (30 mph) faster than its piston-engined sibling, and despite doubts both within and outside Beech about the marketability of a turboprop business aircraft at that time, the type proved an instant success when customer deliveries began in early 1965.

One hundred and twelve King Air 90s had been delivered to civilian customers by the following year, when the Model A90 (first flown on 5 November 1965) appeared, featuring uprated 410-kW (550-shp) PT6A-20 engines driving fully-reversing propellers, an improved 32-kPa (4.6-psi) pressurisation system providing a sea level environment up to 3048 m (10,000 ft) and a 2438-m (8,000-ft) cabin altitude at 6400 m (21,000 ft), and a 136-kg (300-lb) increase in maximum take-off weight. This model was succeeded in 1968 by the similarly-powered King Air B90, which had its wingspan increased from 13.75 m (45 ft 10½ in) to 15.3 m (50 ft 3 in) (as on the then-current Queen Air A80), a recontoured rear fuselage section, balanced control surfaces and a further 23-kg (50-lb) increase in gross weight.

Many of the world's major airlines operate high-powered turbine aircraft as part of their flight crew training programme. Japan Airlines chose the sleek and speedy King Air C90A to initiate its future pilots.

The first Beech King Airs were the product of many years experience in building executive twins, marrying the fuselage of the proven Queen Air with Pratt & Whitney PT6 turboprops. America's Federal Aviation Authority was one of the first customers to recognise the type's merits.

Most common of the King Air 90 family is the C90, which has continued to sell steadily since its introduction in 1971. Originally a B90 fuselage incorporating the systems of the Model 100, the C90 has been continually refined, resulting in the most recent King Air C90-1.

Development of the Series 90 King Air continued with the 1971 Model C90 whose cabin pressurisation system introduced engine bleed air rather than supercharger operation; the 1972 E90 which had 507-kW (680-shp) PT6A-28 engines derated to 410 kW (550 shp) for improved high-altitude performance, affording it a cruising speed of 459 km/h (285 mph) at 4876 m (16,000 ft) and a service ceiling of 8419 m (27,620 ft) (against 7803 m/25,600 ft for the C90); and the 1982 C90-1, which had 410-kW (550-shp) PT6A-21 engines, a strengthened tailplane, and an increase in cabin pressurisation to 35 kPA (5 psi), enabling the pressure equivalent of a 1829-m (6,000-ft) cabin altitude to be maintained to 6096 m (20,000 ft).

This, in turn, was superseded by the Model C90A, which was first flown on 1 September 1983 and became available early the following year. The C90, which as the C90B is the production 'short body' King Air, introduced new low-drag, high-efficiency 'pitot cowl' nacelles whose engine air inlet area is approximately half that of the preceding models, and combined with sealed inlet ducting increases ram air to the aircraft's 410-kW (550-shp) PT6A-21 engines by 30 per cent. As a result, maximum speed has increased by 22 km/h (14 mph) to 448 km/h (277 mph). Other improvements incorporated in the C90A include aerodynamically-faired exhaust stacks, exhaust-heated engine intake lips, a hydraulically rather than electrically-actuated landing gear retraction/extension system, and an automatic rudder boost system to lighten rudder pedal loads during engine-out asymmetric flight.

Below: All Model 90s have been superseded by the King Air C90A and current C90B. The revised cabin can house up to eight, though a layout of four seats is more normal. This aircraft is one of a number flown by Bombardier subsidiary Canadair.

Six years after the first King Air 90 flew, Beech announced the King Air 100. This incorporated a significant stretch of the original fuselage of almost 2 m which allowed two extra windows to be fitted. The initial model was soon followed by the A100 with additional fuel tankage.

Though outwardly it appears much the same as previous models, the King Air E90 was appreciated by its pilots. It was powered by the same engines as the King Air 100, Pratt & Whitney PT6A-28s, which gave the smaller aircraft a noticeable increase in performance.

The most significant external change to the King Air 90 series airframe appeared on the F90, first flown on 16 January 1978. This seven- to 10-seat model combined the fuselage and wings of the E90 with the T-tail of the Super King Air 200 (see below) and had 559-kW (750-shp) PT6A-135 turboprops driving slow-turning four-bladed propellers which greatly reduced cabin noise levels. Production deliveries began in 1979 and continued through 202 airframes until 1983 when the F90-1 appeared, with PT6A-135A engines and drag-reducing 'pitot cowl' nacelles yielding a 28-km/h (18-mph) increase in cruise speed. Thirty-three F90-1s were built before Beech ceased production of the 'short-body' T-tail King Airs in 1985.

Bigger brother

The King Air's airframe was an obvious candidate for 'stretching' to provide more passenger capacity. Unveiled in May 1969 as the head of the Beech 'kingdom' was the King Air 100, which combined the wing, tail unit and 507-kW (680-shp) PT6A-28 engines of the Model 99 Airliner (itself a Queen Air/King Air derivative, but outside the scope of this article) with a Series 90 fuselage lengthened by 1.3 m (4 ft 5 in) to accommodate a maximum of 13 passengers in a 32-kPA (4.7-psi) pressurised cabin. Beech built 89 Model 100s in 1969/70, then introduced the A100, which had a 408-kg (900-lb) increase in maximum take-off

What was to become a bench mark in business aircraft first flew in 1972 when the Beech 200 Super King Air took to the air at the company's Wichita home. Hundreds of this fast, spacious and very graceful design now serve in all corners of the world.

In 1975 the Beech King Air B100 became the first of the family to dispense with the ubiquitous PT6 turboprop. Instead it was powered by a pair of Garrett-AiResearch TPE-331 engines and remained in production until 1984, by which time 137 had been built.

The most major stylistic change in the King Air 90 came with the introduction of the F90, with its distinctive 'T-tail'. Owing an obvious debt to the Super King Air, the F90 also incorporated four-bladed props, not then a feature of its larger sibling.

weight to 5216 kg (11,500 lb), an additional 363 litres (96 US gal) of fuel, and four-bladed propellers of reduced diameter. The A100 was joined from 1975 by the B100, which shared a common airframe but was powered by two 629-kW (840-shp) Garrett AiResearch TPE331-6-251B/252 fixed-shaft turboprops flat-rated to 533 kW (715 shp) each. Beech ceased production of the King Air A100 in 1979 after 157 had been built; the B100 stayed in the range until 1983, with 137 manufactured. Five A100s were delivered to the US Army in 1971 as U-21F transports, joining the service's fleet of U-21/RU-21 utility/electronic surveillance and special missions aircraft, which are essentially unpressurised versions of the Series 90 King Airs.

The definitive 'biz-prop'

In the same year that the King Air 100 appeared, Beech began work on the Model 101, later renamed Super King Air 200. Sharing the same fuselage as the Model 100, this aircraft had increased wingspan through insertion of centre-section 'plugs' housing auxiliary fuel tanks; 633-kW (850-shp) PT6A-41 turboprops; a 42-kPa (6-psi) cabin pressurisation system maintaining sea level to 4212 m (13,820 ft); an eight- to 13-seat cabin; maximum take-off weight of 5670 kg (12,500 lb); and – its most distinctive feature – a 'T-tail' which raised the tailplane and elevators out of the wings' turbulent downwash. Two prototypes were built, making their first flights from Wichita in the hands of Beech test pilot Bud Francis on 27 October and 15 December 1972. FAA certification of the Super King Air 200 under FAR Pt 23 was granted on 14 December 1973, and deliveries began two months later.

The first prototype Super King Air 200 (N38B) was later used as a testbed for turbofan engines. Designated PD 290 (Preliminary Design 290), it had two Pratt & Whitney JT15D-4 turbofans installed in overwing nacelles, and made 103 flights between 1 March 1975 and 30 September 1977 before development was terminated.

The current Series 200 production model is the Super King Air B200, introduced in March 1981. Generally similar to the Model 200, it has 633-kW (850-shp) PT6A-42 turboprops which improve climb and high altitude performance, a 45-kPa (6.5-psi) pressurisation system, and a 272-kg (600-lb) increase in maximum zero-fuel weight. Numerous cabin, flight deck, avionics and systems improvements have also been incorporated since the B200 first entered production.

Military duty

In addition to delivering more than 300 Super King Air 200s to the US forces as C-12/UC-12 transports and RC-12 special missions platforms, Beech has developed a number of specialist versions of the aircraft for commercial customers. Among these were two Model 200Ts delivered to the French Institut Géographique National in 1977 which were equipped with twin Wild RC-10 aerial survey cameras and Doppler navigation systems; wingtip tanks which increase fuel capacity from 2059 to 2460 litres (544 to 650 US gal), giving a mission endurance in excess of 10 hours; and high-flotation landing gear for operation from unimproved runway surfaces. Fixed or removable

The striking red and white scheme of US Navy T-44As disguises the fact that they are basically King Air E90s. More than 60 were ordered by the Navy as multi-engined trainers and they survive today with two squadrons at NAS Corpus Christi in Texas.

The Super King Air has gained a significant number of military orders, its speed and comfort making it a popular choice as a staff transport. Under the designation C-12, Super King Airs serve, in varying colour schemes, with all branches of the US armed forces.

tiptanks, high-float undercarriage, and a 1.35 m (4 ft 4 in) square cargo door at the rear of the cabin on the port fuselage side are available as options on current production Super King Air B200s.

In 1979 the manufacturer announced availability of the Maritime Patrol 200T, which has tiptanks, strengthened landing gear, bubble observation/photographic windows in the rear cabin, flight-openable hatches for dropping survival equipment, a 360° search radar mounted in a ventral radome and advanced mission avionics. Customers for this aircraft have included the Japanese Maritime Safety Agency, which operates 12, Algerian Ministry of Defence, Malaysian Department of Civil Aviation and the navies of Peru and Uruguay.

King Air commuter

Announced in January 1988, the Model 1300 Commuter is a regional airliner version of the Super King Air B200 designed for operation on long routes with light passenger loads. Accommodation is provided for 13 passengers, with provision for a 0.37-m³ (13 cu ft) nose baggage compartment made possible by adoption of panel- rather than remote-mounted avionics, while an optional ventral cargo pod can carry a further 206 kg (455 lb) of baggage or freight. The Model 1300 is further distinguished from the standard Super King Air B200 by its dual outward-canted ventral strakes which enhance stability at low airspeeds and high angles of attack. Mesa Airlines of Farmington, New Mexico was launch customer for the aircraft, taking delivery of the first of 10 Model 1300 Commuters on 30 September 1988.

In October 1981 Beech began test flying a Super King 200 modified with 782-kW (1,050-shp) PT6A-60A engines in nacelles extended forwards and equipped with 'pitot cowl' air intakes and aerodynamically faired exhausts. This aircraft also incorporated extensions to the inboard wing leading-edges and hydraul-

The Super King Air 300 represents yet another example of Beechcraft's efforts to keep their designs competitive, the most obvious outward sign of which is the introduction of four-bladed Hartzell propellers. Comparatively few were produced before the type gave way to the Model 350.

ically-actuated landing gear and had a maximum take-off weight of 6350 kg (14,000 lb). A production prototype, designated Super King Air 300, made its first flight in September 1983 and FAA certification was obtained on 24 January 1984. Deliveries of the Super King Air 300 began in the spring of 1984. Customers included the Federal Aviation Administration, which ordered 19 specially-equipped Super King Air 300s for airborne flight inspection duties. In 1988 Beech introduced a lightweight Model 300LW version of the aircraft for the European market with a maximum take-off weight of 5670 kg (12,500 lb).

Ultimate model

At the 1989 National Business Aircraft Association Convention in Atlanta, Georgia, Beech introduced the Super King Air 350, the first example of which (N120SK) had made its first flight unannounced a year earlier. This aircraft has replaced the Super King Air 300 in production and differs from it in having a 0.6-m (2-ft) long fuselage stretch achieved by 'plugs' forward and aft of wing, whose span is increased by 0.45 m (18 in) with 0.6-m (2-ft) high NASA winglets at the tips. The stretched fuselage provides standard seating for eight passengers in a 'double club' cabin arrangement; three more passengers can be accommodated on optional seats in the rear of the cabin, bringing total capacity to 11 in two-crew operations or 12 with a single pilot. Aerodynamic and control system improvements have also been incorporated. The first customer delivery of a Super King Air 350 was made on 6 March 1990.

By early 1993 Beech had delivered 1,907 King Air 90s (not including T-44s), 1,623 King Air 200s (not including C-12s), 245 King Air 300s and 101 King Air 350s to civilian and military customers, making the King Air by far the world's most successful and widely used twin turboprop business/utility aircraft.

Many Super King Air 200Ts have been sold to date. This maritime patrol version is in service with several nations, most noticeably Japan, where the Maritime Safety Agency operates a fleet of 12.

Above: Since 1991 Beechcraft has had the freight door-equipped *Super King Air 350C* on offer, which shares a common airframe with the basic Model 350. First customer was the Namibia-based Rossing Uranium company.

Below: In the face of growing competition from small 'biz-jets', Beech has produced what may be the ultimate *Super King Air*, the Model 350. Over 1 m longer than the 200, it seats 13 and has prominent 'winglets'.

Chapter 7

Beech Model 18
•
Boeing 247
•
Boeing 307 Stratoliner
•
Boeing 314 Clipper

Beech Model 18

With the Model 17 'Staggerwing', Walter H. Beech had already created a classic. He was to follow this success with one of the best known and most versatile twins of all time. The Beech 18 was a lesson in design; fast and stylish yet efficent and capable. It was the world's first 'executive' aircraft and made a valuable contribution to the Allied war effort in myriad trainer versions.

Beech Aircraft Corporation, founded at Wichita in 1932 by Walter Beech, revolutionised the pre-war private aircraft market with its five-seat Model 17 'Staggerwing' biplane which was faster than many contemporary military pursuit types.

Anxious to capitalise on this success, Beech, his chief designer Ted Wells and project engineer Dean Burleigh began work in November 1935 on an entirely new twin-engined aircraft intended for use by businessmen as a high-speed executive transport offering airline standard comfort, but still able to operate from the small grass airfields that abounded in the United States at the time. Designated Beech Model 18, the aircraft was an all-metal low-wing design with twin fins and rudders, looking not unlike a smaller version of the contemporary Lockheed 10 Electra. It was powered by two 238.8-kW (320-hp) cylinder Wright R-760-E2 radial engines housed in tight cowlings with small 'blisters' to clear the rocker covers and driving two-bladed Hamilton Standard controllable-pitch propellers. The wing airfoil section was NACA 23013, with electrically operated plain flaps; elevators, ailerons and rudders were fabric covered. The tailwheel landing gear was retractable, with electrical actuation. The cabin seated six passengers, with two crew in the cockpit.

The prototype Model 18 (NX15810) made its first flight on 15 January 1937, flown by James Peyton, a Trans World Airlines pilot hired by Beech to do initial flight testing of the company's first twin-engined aircraft, with Beech test pilot H. C. Rankin as co-pilot and Curtiss-Wright's Robert Johnson as flight test engineer. Like the 'Staggerwing', the Model 18 was an innovative design with excellent performance, having a maximum speed of 325 km/h (202 mph) and cruising speed of 309 km/h (192 mph) at 1828 m (6,000 ft) yet could approach to land at just 88.5 km/h (55 mph). With standard fuel capacity of 606 litres (160 US gal) it could fly more than 1206 km (750 miles) at cruise power settings, closer to 1600 km (1,000 miles) with optional auxiliary fuel tanks.

Certification of the Model 18 was obtained on 4 March 1937, after which the prototype undertook a sales tour of the United States that included demonstrations by the well-known racing pilot Louise Thaden. Nonetheless, sales, at $37,500, were sluggish. The Eythl Corporation bought the prototypes, as it had the first Model 17. The first production aircraft (CF-BGY) was a Model S-18A seaplane mounted on Edo 55-7170 floats for bush operations in Canada by Starratt Airways of Hudson, Ontario.

New sales pitch

Slow sales of the Model 18A prompted Beech to offer the lower-powered Model 18B, which was certificated on 29 October 1937 with two 213-kW (285-hp) Jacobs L-5 radials. The Jacobs engines were cheaper to buy and to maintain than the Wrights, enabling the Model 18B to be sold for $33,500, and sales began to

The classic Beech 18 was the first really successful light twin, paving the way for what has become a highly lucrative market. This is a Super E18S, the first variant to feature a raised cabin roof and squared-off wingtip extensions.

Where it all began: the first Beech Model 18 in flight over Wichita in 1937. Just as the earlier Model 17 'Staggerwing' had ushered in a new era of single-engined lightplane transport, so the Model 18 revolutionised the light twin market.

accelerate. The 246-kW (330-hp) Jacobs-engined Model 18D followed in 1938; all three variants were structurally identical, and were offered with a choice of retractable wheel or fixed ski or float landing gear and a variety of interiors from spartan no-frills freighter or bush configuration to luxurious custom-fitted executive configuration whose fabrics and trim were often selected personally by Walter Beech's wife Olive Ann.

Responding to operators' demands for more power, in August 1938 Wells and Burleigh began work on the Model 18S, powered by two 335.7-kW (450-hp) Pratt & Whitney Wasp Jnr engines. The first Model 18S (NC19452) was completed in January 1939 and publicly launched two months later. It had enlarged vertical tail surfaces to cope with the additional power of the Wasps, smooth cowlings, gross weight increased from 2948 kg (6,500 lb) to 3266 kg (7,200 lb), maximum speed up to 386 km/h (240 mph), and was priced at $63,400. In July 1939 Walter Beech flew the first model to California for a sale demonstration tour to celebrate his 25 years in aviation, and in January 1940 he won the McFadden Trophy race flying from St Louis to Miami in four hours 37 minutes.

A winning combination

The combination of P & W's renowned Wasp engines and Beech's rugged airframe proved a fortuitous marriage. Commercial orders came in at an unprecedented rate for the 'Twin Beech' (like the 'Staggerwing', no official name was ever given to the Model 18), but more importantly the aeroplane's excellent performance and versatility caught the attention of the military. In 1940 the US Army Air Corps placed an order for 14 Model B18Ss (the then-current production model) for high-altitude photographic work. Designated F-2 in USAAC service, these aircraft were equipped for day and night vertical and oblique aerial photography, and were followed by a further 44 F-2As and F-2Bs with different variants of the Wasp engine, paving the way for thousands of military Model 18s during World War II. The US Navy followed with an order for five JRB-1 Voyagers, which were equipped with a raised cupola above the cockpit from which an observer controlled unmanned reconnaissance drones. The first Model 18S was modified for flight testing of the cupola structure before being converted back to standard configuration and sold to an oil drilling company in Tulsa, Oklahoma, in June 1940, by which time the improved Model C18C was the standard commercial model, of which the USAAC ordered a further 11 for use as liaison and light transport aircraft under the designation C/UC-45B.

By the time commercial production of the Model 18 ceased following America's entry into World War II, Beech had built a total of 113 aircraft, and were soon mass-producing military variants, all based on the civilian Model B18S

Landing at Laredo AAF, Texas, this is an AT-11 Kansan. In addition to training bombardiers and navigators, the AT-11 was also used for gunnery training, hence the rear fuselage turret. A forward-firing gun was sometimes mounted in the nose.

airframe and powered by 335.7-kW (450-hp) Pratt & Whitney R-9850AN-1 or -AN-3 Wasp Jnr engines. At one time production lines were advancing so rapidly that aircraft were being fitted with wooden wheels to enable them to be moved to storage areas pending delivery of rubber tyres.

Military workhouse

Army Air Force variants included C-45 Expeditor transports, 350 of which were supplied to the British Royal Air Force and 80 to the Royal Navy under lend-lease agreements out of a production run totalling more than 1,400; AT-7 Navigator navigation trainers which were equipped with drift meters, work tables and compasses in the cabin and an astrodome in the roof through which trainee navigators could learn the art of taking 'star shots' for celestial navigation; and AT-11 Kansan bombardier trainers. The AT-11 was the most modified of these variants, having a clear plexiglass 'bombardier' nose with Norden bombsight for trainee bomb aimers, dorsal blisters, and underfuselage racks for up to 10 45-kg (100-lb) practice bombs. A total of 1,560 AT-11s was delivered to the USAAF, operating mostly in the south-western states where good weather enabled almost continuous training of bomb aimers. After the war many AT-11s found favour with commerical operators for survey and aerial photographic work. US Navy variants included JRB-3 and JRB-4 transports; SNB-1 gunnery and bombing trainers which equated to the Army's AT-11 but had an electrically-operated turret with twin 0.30-in calibre machine-guns mounted amidships; and SNB-2 navigation trainers and general purpose aircraft. Total wartime production of military 'Twin Beeches' is thought to have exceeded 5,000 airframes.

Perhaps the most radical of the conversions was the Dumod Liner. This featured a much stretched fuselage to accommodate up to 15 passengers, while a third fin was added to cope with longitudinal stability problems.

While still in the full flush of wartime production, Beech launched the first post-war civilian Model 18 in 1944 when the US War Department granted permission for some manufacturers to supply private aeroplanes to selected customers who needed them for the conduct of government contract business. This was the Model C18S, a seven to 10 passenger light transport and utility aircraft which was similar to the military C-45F/JRB-4 and received its type certificate on 23 September 1944. Among companies which received C18Ss, most of which were C-45Fs diverted from military lines and modified to civilian standard, were Bell Aircraft, General Motors, Goodyear Aircraft, Firestone Rubber, Ford Motor Company, Phillips Petroleum, Standard Oil and Time Magazine.

Among the many weird and wonderful Beech 18 developments and experimental versions was this tri-motor. The outboard engines are standard Continental TS10-520s in place of the radial engines, while the central engine is the Tiara turbocharged unit on flight test.

When war ended Beech replaced the C18S with the D18S, which was very similar externally to previous models but featured longer, more streamlined engine nacelles and a

Low operating costs and ready availability made the Beech 18 and its derivatives extremely popular with small airlines running operations on limited budgets. In addition to passenger transport, the type has been heavily used for light cargo work, particularly around the Caribbean and Latin America. Due to its sturdiness, range and expendability, many Beech 18s were used for smuggling narcotics into the United States, their cheap purchase price representing a tiny fraction of the value of their cargo and allowing them to be abandoned on landing.

Specification
Beech Model E18S Super 18
Wing span: 15.17 m (49 ft 8 in)
Length: 10.75 m (35 ft 3 in)
Height: 2.92 m (9 ft 6 in)
Wing area: 33.53 m^2 (361 sq ft)
Powerplant: two Pratt & Whitney R-985 Wasp Jnr, 335.7-kW (450-hp) each
Passenger capacity: 7-9
Empty weight: 2707 kg (5,970 lb)
Maximum take-off weight: 4128 kg (9,300 lb)
Maximum speed: 376 km/h (234 mph) at 1005 m (3,300 ft)
Crusing speed: 346 km/h (215 mph) at 3048 m (10,000 ft)
Service ceiling: 5943 m (19,500 ft)
Maximum range: 2413 km (1,500 miles)

Volpar of Van Nuys, California, were involved in many Beech 18 conversions, many involving turboprop power. This Swiss aircraft was used for agricultural spraying with Pilatus Aerial Spraying Company.

reinforced wing centre-section that enabled the aircraft to operate up to a maximum take-off weight of 3855 kg (8,550 lb) or 3969 kg (8,750 lb) if its Wasp Jnrs were fitted with Hamilton Standard Hydromatic constant-speed full-feathering propellers, with consequent gains in useful load. Maximum speed was 370 km/h (230 mph) at 1524 m (5,000 ft). When Beech's Wichita factory returned to full production of civilian aeroplanes in February 1946 the company already had orders for hundreds of D18Ss, which sold for $61,500 basically equipped, and built 296 that year – a record for commercial 'Twin Beeches'. A rare variant of the D18 series was the D18C/CT, powered by two 391.6-kW (525-hp) Continental R-9A radial engines and intended for use as a small airliner seating eight to nine passengers. Only 31 were produced.

Yet more versions

Beech also began remanufacturing US Navy SNBs to D18S specification with stronger wing centre-section truss assemblies, disc brakes, Hamilton Standard Hydromatic propellers, new engine nacelles, revised cockpit layouts and Sperry autopilots. These aircraft were redesignated SNB-4, -5 and -5P; a total of 2,263 SNBs was refurbished over a 10-year period between 1947-57 at Beech's Herington, Kansas, plant, which performed similar updates on more than 2,000 USAF C-45s, AT-7s and AT-11s as C-45Gs and C-45Hs. In 1951 Beech also began production of 283 D18Ss for the Royal Canadian Air Force for use as communications aircraft and navigation trainers.

In its heyday the D18S, of which 1,035 were produced, became the flagship of many corporate fleets in the United States, for it was one of the few purpose-built business twins available until the early 1950s, and was also widely used for small feeder airliners, charter companies and cargo carriers throughout the world. Surplus military aircraft further boosted commercial Twin Beech numbers. The USAF retired its last C-45s in November 1963; US Navy SNBs, by then redesignated TC/UC-45Js, remained active at least until the late 1960s. Military variants also served with virtually every nation in Central and South America, with most Western European air arms and everywhere the US mutual aid programme operated.

Stylish 'Super'

In 1954 Beech introduced the new Super E18S which was the first to incorporate major structural and external changes to the classic 'Twin Beech' lines. It featured a cabin roof raised by 15 cm (60 in) to increase headroom, four large cabin windows on each side, an airstair door, optional three-bladed propellers with spinners, a 1.2-m (4-ft) increase in wingspan with squared-off tip extensions that improved single-engine climb rate, and an increased gross weight of 4218 kg (9,300 lb). Production began in August 1954 and continued until 1960, with 451 delivered. France's Armée de l'Air was among few military operators of this variant.

The E18S was superseded by the G18S which had three-bladed propellers as standard, a further 90-kg (200-lb) increase in gross weight, deeper cockpit windows with a two-piece windscreen, and a large 'panoramic' cabin window on each side of the cabin. A total of 154 G18Ss were produced.

1962 marked the 25th year of Model 18 production, and the introduction of the final variant, the Super H18. On this model gross weight increased to 4490 kg (9,900 lb), providing a useful load of more than 1905 kg (4,200 lb) – the highest for any Model 18. Fuel capacity was increased, new lightweight propellers and electric cowl flaps installed and, for the first time as a factory-new option, tricycle landing gear was available. This 'Volpar Mark IV' landing gear had been developed by Volpar Corporation of Van Nuys, California, and had previously been available only as a retrofit for Model 18 owners who preferred the ease of handling of a nosewheel to the standard tailwheel configuration which gave the 'Twin Beech' a reputation for tricky ground handling, particularly during the landing roll in crosswinds. The Super H18 cost $179,500, but despite a high price and somewhat anachronistic appearance compared to more modern piston and turboprop twins, even within Beech's own ranges, it continued to sell in modest numbers throughout the 1960s. The last three of 149 Super H18s built, and the last of more than 8,000 Model 18s of all types produced, left the Beech factory on 26 November 1969 bound for Japan Air Lines, where they served as multi-engine trainers for the carrier's pilots.

Substantial numbers of Beech 18s remain active throughout the world, particularly with small cargo carriers, and the type is also now being recognised as a collectable 'classic' or 'warbird'. The ubiquitous 'Twin Beech' has also inspired many conversions and modifications. Among these are the Wasp Jnr-powered Pacific Airmotive PacAero Tradewind with tricycle landing gear and a sweptback single fin in place of the standard twin fins and rudders; several models of the Hamilton Westwind with Pratt & Whitney Canada PT6A-27/278/34, Garrett AiResearch TPE331 or Lycoming LTP-101 turboprops and stretched fuselages accommodating up to 17 passengers; and the Volpar Turbo 18 and Turboliner, both with tricycle gear, TPE331 turboprops and fuselage stretches, the latter a 17-seater first flown on 12 April 1967.

After World War II, many wartime Beech twins became available for the civil market. This New Zealand-registered aircraft is an ex-military AT-11 Kansan, the variant with a bombardier nose for bombing training. Such aircraft were naturally popular with survey companies.

Boeing Model 247

William Boeing's Model 247 airliner is assured of a place in aviation history, despite the fact that it sold in comparatively small numbers and had a surprisingly short service life with the airlines. By today's standards it may look small and unremarkable, but in the 1930s it heralded a major advance in aircraft construction technique.

In the early 1930s the Boeing Airplane Company built two advanced aircraft that revolutionised design thinking of the era. The Model 200 Monomail, now regarded as the first modern air transport, was a streamlined single-engined mail and cargo carrier which combined all-metal monocoque construction, cantilever low wing and fully retractable undercarriage, putting it years ahead of contemporary designs when it first appeared in 1930. Likewise the B-9 twin-engined bomber designed for the US Army, with its streamlined structure and faired-in engine nacelles, far outstripped the performance of single-engined pursuit aircraft of the day.

Despite the technological advances that each represented in its field, neither the Monomail nor the B-9 was successful in attracting large production orders. The effort had not been wasted, however. While the B-9 was still on the drawing board Boeing president Philip G. Johnson and vice-president Claire Egtvedt launched development of an airliner which would incorporate the revolutionary features of both aircraft. It was to be the first all-metal streamlined passenger transport in the United States.

The specification circulated to airlines in 1932 was indeed revolutionary. The new Boeing Model 247 was to be a twin-engined aircraft (contemporary airliners such as Boeing's own Model 80 biplane and the Fokker and Ford monoplanes were all three-engined) powered by two 410-kW (550-hp) Pratt & Whitney S1D1 Wasp radials driving fixed-pitch three-bladed propellers and housed in ring-cowl nacelles faired into its cantilever low wing. The main undercarriage legs were designed to retract into the lower surface of the wing, leaving just part of each wheel exposed. Retraction was electrically actuated, with a manual emergency winding system. Seating was provided for 10 passengers in adjustable, reclining armchairs spaced 101 cm (40 in) apart, each with a wide window, dome light, individual reading light and the latest heating and ventilating equipment. Soundproofing would further reduce noise in a cabin freed of the clatter of the nose- and underwing-mounted engines of the trimotors. At the back of the cabin was a lavatory and a 1.7 m^3 (60 cu ft) mail/baggage compartment, while the streamlined nose, freed of the need for an engine mount and ancilliary equipment, housed advanced radio gear and up to 180 kg (400 lb) of mail and baggage. Both baggage compartments had external access doors, enabling them to be loaded and unloaded without interference with passengers – an important feature for quick turnarounds at air terminals.

The ex-crop sprayer Model 247 saved from an ignoble fate, seen after its immaculate restoration by the Pacific Aviation Historical Foundation at Boeing's Renton home. It wears the livery it carried during its early service with United Airlines.

A pair of Model 247s was delivered to Deutsche Lufthansa, the only other airline customer for the 247 apart from United. The cabin held 10 passengers in some style, and in United service often carried a 'SkyGirl' to tend to passengers in flight.

The cockpit offered similar advanced features including dual directional gyros, compasses, altimeters, rate-of-climb indicators, turn-and-bank indicators and artificial horizons. Night flying equipment was standard, comprising navigation and landing lights and parachute flares, and leading-edge de-icing boots were available as an option.

First customers

In 1932 Boeing Air Transport, National Air Transport, Pacific Air Transport and Varney Air Lines, which made up the Boeing Air Transport System, placed an order for 70 Model 247s straight from the drawing board. The first of these, NC13301, made its first flight on 8 February 1933 and immediately proved the wisdom of Johnson and Egtvedt's decision to break free from the traditional concept of airliner design. Though offering a lower payload than the tri-motors it was intended to replace, the Boeing 247 cruised 80-110 km/h (50-70 mph) faster, and so aerodynamically clean was the design that at maximum take-off weight it could climb with one engine shut down, something no contemporary twin-engined aeroplane could manage.

Seen at Boeing's Renton plant, this is the prototype Model 247 after conversion to 247E standard with full-cowled engines, rearward-sloping windscreen and other improvements. It had entered service with United Air Lines on 30 March 1933, only seven weeks after the first flight!

Fifty-nine Model 247s were delivered to United Air Lines, as Boeing Air Transport had become, during the first year of production. Two more (D-AGAR and D-AKIN) were sold to the German carrier Deutsche Lufthansa. The 30th production airframe was modified as an executive transport for the Pratt & Whitney Division of United Aircraft Corporation, a former part of the Boeing conglomerate. With tightly cowled 14-cylinder 465-kW (625-hp) Pratt & Whitney Twin Wasp Jnr engines, a maximum speed of 320 km/h (198 mph), 1100 litres (290 US gal) of fuel giving a range of 1045 km (650 miles), and a six-passenger luxury interior, NC 13300 was quite a different aircraft from the standard airline machine, and was given the one-off designation Model 247A. It first flew on 14 September 1933 and was delivered to UAC on 4 November, serving with the company as a corporate transport and testbed until after World War II.

Still wearing its airline colours, the first Model 247 was returned to Boeing for trials which led to the 247D. Along the way different fin shapes were tried, although the final configuration reverted to the original shape. When returned to airline service, this aircraft became the 247E.

Redesign

During 1933 United Air Lines returned its first Model 247 to the Boeing factory to serve as a testbed for a proposed Model 247D. Modifications included trials of several new fin and rudder shapes, although the final choice was identical to the original in outline, but with a fabric-covered rudder in place of the metal-skinned original. Fabric also replaced metal skinning on the elevators. Other improvements incorporated on the 247D included geared Pratt & Whitney S1H1G Wasp engines driving Hamilton Standard controllable-pitch propellers, full NACA cowlings, an additional 245-litre (65-US gal) fuel capacity, a streamlined rearward-sloping windscreen replacing the undercut, forward-raked cockpit windows of the Model 247, and a 338-kg (744-lb) in-

A special 'one-off' version was the Model 247A, developed as a corporate transport for Pratt & Whitney. The interior seated six in sumptuous comfort, while it was the first variant to introduce full cowlings around the engines.

crease in maximum take-off weight. The Model 247D cruised at 304 km/h (189 mph) – faster than the top speed of the first model – and had a maximum speed of 320 km/h (200 mph), while range was increased from 780 to 1200 km (485 to 745 miles) and service ceiling from 6250 to 8290 m (20,500 to 27,200 ft). The modified UAL aircraft was given the designation Model 247E and was returned to the company.

The first production Model 247D was originally destined for Lufthansa, but the sale was not confirmed and it was sold as an executive transport and 'flying office' to the Phillips Petroleum Company, bearing the registration NC 2666 to reflect Phillips' famous '66' brand of aviation gasoline. The remaining 13 of United Air Lines' launch order for 70 were also completed as Model 247Ds, and UAL's entire fleet was progressively brought up to 247D standard.

Fast ships

United's Boeing 247s reduced the flight time for its coast-to-coast passenger and mail services from 27 to 19½ hours, and for two years the Boeing reigned supreme on the airways until the equally swift but much larger Douglas transports eclipsed it, but not before Boeing had been awarded the coveted Guggenheim Medal for "successful pioneering and advancement in aircraft manufacturing and transport." UAL had sold 36 of its 247s to other airlines in the USA and abroad by January 1938.

One of UAL's Boeing 247Ds was to achieve particular fame. It was leased for four months to the flamboyant racing and airshow pilot 'Colonel' Roscoe Turner and his partner, former barnstormer, stunt flyer and expert aerial navigator Clyde Pangborn, to compete in the 1934 MacRobertson Air Race from London to Melbourne, Australia. Registered NR 257Y (a major sponsor was Heinz foods, of '57 Varieties' fame; another was Hollywood movie makers Warner Brothers, in whose honour the aircraft bore the name *Warner Bros Comet*), the 247D was stripped of all non-essential equipment and had eight auxiliary fuel tanks installed in its cabin, bringing total capacity to 4260 litres (1,125 US gal) although race organisers made Turner seal off four 340-litre (90-gal) tanks before the start of the race to keep the Boeing within approved weight limits. Other modifications included new fuel and oil line plumbing, larger oil tanks, a standby vacuum system and special low frequency and short wave radio equipment which Turner and Pangborn used both for air-to-ground communications en route and to file news copy to the North American News Agency.

After test flying in the United States NR 257Y was dismantled and shipped to England for the race, starting off second among the contestants who left RAF Mildenhall in Suffolk on the morning of 20 October 1934. Turner and Pangborn flew via Athens, Baghdad (where

After service with UAL and the Royal Canadian Air Force, this 247D was transferred to the RAF, becoming the first Boeing aircraft in British military service. It performed sterling work at Defford developing instrument landing systems.

they nearly came to grief when Turner inadvertantly landed downwind), Karachi, Allahabad, an unscheduled landing for fuel at Alor Setar, Singapore, Koepang, Darwin, Charleville and a forced-landing at Bourke in New South Wales with engine trouble, arriving at Melbourne in an elapsed time of 92 hours 55 minutes, 38 seconds, and a flight time of 85 hours, 22 mins 50 seconds. The Boeing was actually placed third in the Speed section of the race, but was awarded the second prize of £1,500 because the Dutch crew of the KLM Douglas DC-2 *Uiver* which preceded it opted for first prize in the Handicap section, which they had won; winning two prizes was not per-

Perhaps the most unusual variant was the 247Y, which was fitted with long-range tanks, twin forward-firing machine-guns and a rear gun station. It was delivered to China.

After the war, Model 247s were used by small operators for many years, usually on general cargo transport duties. This aircraft was used as a crop-sprayer until 1978, when it was purchased for restoration.

mitted under race rules. After the race the Boeing was shipped back to the United States and rebuilt to standard airline configuration. It served with United Air Lines until January 1937, when it was sold to the Union Electric Company as an executive transport, later flying as a research aircraft with the US Civil Aeronautics Authority for 14 years until 17 July 1953 when it was presented to the Smithsonian Institute. Restored in the colours it wore during the MacRobertson Race, it is now displayed at the National Air & Space Museum in Washington, DC.

The seventh production Model 247D, originally delivered to UAL as NC 13366, was returned to the factory in January 1937 and converted into the unique Model 247Y. With passenger capacity reduced to six, and range increased by the installation of four of the auxiliary fuel tanks removed from Roscoe Turner's MacRobertson race aeroplane, the aircraft was also equipped with two .50-calibre machine-guns in a fixed mount at its nose and a single gun on a flexible mount at the rear of the cabin. It was delivered to a customer in China, though little is known of its purpose or fate.

After the United States' entry into World War II, 27 Boeing 247Ds were impressed into service with the US Army Air Corps as C-73 transports, crew trainers and instrument trainers. A number were modified while in USAAC service with military-standard Pratt & Whitney R-1340-AN1 Wasp engines, cowlings from North American AT-6 Texan trainers and two-bladed propellers. They were all struck off military charge in 1944 and sold to small airlines, charter operators and private owners, to begin new lives in many parts of North and South America hauling all manner of cargo, but none surely so unusual as the 26 new-born babies ferried three-to-a-crib aboard a 247 flying from Quebec City to Chicoutimi, Canada, in December 1944.

Eight former airline Boeing 247Ds also saw service with the Royal Canadian Air Force. One of these, carrying the serial number DZ203 and nicknamed *Adaptable Annie*, was used by the Telecommunications Research Establishment at RAF Defford for early trials of blind landing equipment, and played a pioneering role in research leading to development of the instrument landing system (ILS) which remains in use today as the primary bad weather approach aid.

The most famous of the Model 247s was this aircraft, an ex-United machine used in the MacRobertson race from London to Melbourne. Flying under the special racing registration NR257Y, it was flown to third place by Roscoe Turner and Clyde Pangborn.

Featuring military-standard Wasp engine with a two-bladed propeller, this is one of 27 Model 247s impressed by the USAAF as C-73s. Note the nose-mounted antenna fairing which must have impeded the forward view somewhat.

Boeing 307 Stratoliner

In 1938 $315,000 could buy you the pinnacle of contemporary airliner design, the Boeing Stratoliner. In fact, the aircraft was almost an adapted bomber, drawing heavily on the B-17 but utilising an entirely redesigned circular fuselage. In the event, the Model 307's airline career was cut short by the advent of war, but it established Boeing as an innovative manufacturer and pointed the way to its great post-war designs.

As production of early Boeing Model 299 (B-17) Flying Fortresses was getting under way in 1935, Boeing's president Claire Egtvedt consulted chief engineer Ed Wells about the possibility of pressurising transport aircraft cabins to permit operation at higher altitudes. Wells had begun to look at this possibility for the B-17, and had concluded that its existing fuselage design could not be readily adapted to pressurisation. He proposed a new fuselage of circular cross-section, mated to the B-17C's wings, tail surfaces and engines.

Boeing's management was keen to develop a

Four Stratoliners were built for launch customer Pan American Airways, gaining the designation PAA-307. One of these, the prototype, crashed and the others were delivered with the revised tail layout, which became standard. They were named **Flying Cloud, Comet,** *and* **Rainbow.**

new transport to rival those planned by Douglas and Lockheed, and gave Wells the go-ahead to draw up a specification for the aircraft. The factory designation S-307 was allocated to the design, and, reflecting its high-altitude operation, the name Stratoliner was copyrighted for it – only the third occasion on which a Boeing design had been given a name.

It was not until 1937, after Pan American Airways president Juan Trippe had placed a launch order for four aircraft, that construction of the prototype Stratoliner began. Wells had opted for a bulbous, almost airship-like fuselage for the aircraft. With a circular cross-section of nearly 3.6-m (12-ft) diameter the Stratoliner was effectively the first 'wide-bodied' airliner. Its supercharger-driven pressurisation system maintained a comfortable 2438-m (8,000-ft) level cabin pressure at 6096 m (20,000 ft), enabling it to operate above the worst of the weather and thus providing a smoother ride for passengers. The aircraft was designed to accommodate 33 passengers in four curtained compartments and a single row of nine reclining 'snoozer' chairs, and crew of six (including, for the first time on an American commercial aircraft, a flight engineer). Its cabin interior was convertible to an overnight sleeper configuration with 16 sleeping berths. Dressing rooms or 'charm rooms' were provided fore and aft, and there was a galley for in-flight hot meals service.

The airframe was of all-metal monocoque construction. The wings, 671-kW (900-hp) Wright GR-1820-G102A two-stage supercharged radial engines, and vertical and horizontal tail surfaces were all identical to those of the Flying Fortress, although span was increased by 1.06 m (3 ft 6 in) because of the larger fuselage cross-section, and broad-chord

Right: This aircraft was diverted from TWA by Howard Hughes, initially for long-distance record-breaking. After being thwarted in these efforts, the aircraft was fitted as his personal transport and named **The Flying Penthouse.**

Howard Hughes sold his **Flying Penthouse** *in 1948, by which time the aircraft had been re-engined with R-2600 Twin Cyclones with characteristic shorter cowls. It was bought by a Texan millionaire, and was eventually damaged beyond flyable repair in a hurricane.*

engine cowlings were fitted. Fuel capacity was 6435 litres (1,700 US gal). Leading-edge de-icing, soundproofing and air-conditioning and autopilot were standard equipment. Boeing set the price of the Stratoliner at $315,000.

Into the air

The prototype Boeing 307 Stratoliner, NX19901, made its first flight on 31 December 1938, flown by test pilot Julius Barr. High-altitude flight testing began the following month. Tragedy struck in March 1939 when Barr, Stratoliner chief engineer Jack Kylstra, project engineer N. D. Showalter, six other Boeing employees and two representatives from the Dutch airline KLM took off from Boeing Field at Seattle for a demonstration flight. While cruising at about 1525 m (5,000 ft) near Mount Rainier in the Cascade Mountains, the Stratoliner was seen to enter a spin, during recovery from which it apparently was over-stressed by one of the Dutch pilots and wings and tail surfaces separated. All aboard perished in the crash.

Right: The prototype Model 307 made its first flight on the last day of 1938, distinguished by the original small fin inherited from the early B-17 variants (as were the wings, tailplane and engines). The aircraft tragically crashed with many key project personnel aboard three months later.

Below: The major second-line operator of the Stratoliner was Aigle Azur of France, which took six aircraft from TWA and Aérovias Ecuatoriana. The fleet was sold to the Allied Control Commission in 1966, which retained two aircraft as late as 1973 on flights between Saigon, Hanoi and Vientiane.

As a result of the accident Welwood E. Beall, who replaced Jack Kylstra as chief engineer on the Model 307, designed new vertical tail surfaces with the broader fin and rudder and large dorsal strake which were later adopted on the B-17E and subsequent Fortresses. The crash of the prototype did not significantly slow development of the aircraft, and certification was obtained on 13 March 1940, after which Pan American took delivery of the remaining three of its order. Named *Clipper Flying Cloud*, *Clipper Comet* and *Clipper Rainbow*, they were based at Miami, Florida, and served on PAA's Latin American routes, bringing a new era of comfort and sophistication to airline travel, just as the earlier Boeing 247 had. It had taken only five years to progress to the far more complex Model 307.

Left: Another French operator of the Stratoliner was Airnautic. Perhaps the strangest user was Air Laos, which had machines leased from the French companies for services from Saigon during the Vietnam War.

Below: NC19902 was finally delivered to Pan American as Clipper Rainbow, bearing the definitive fin. Pan Am and TWA flew their aircraft intensively in the time before the US entered the war, illustrating the capabilities of the type. More would have been sold if the conflict had not intervened.

Boeing's second Stratoliner customer was Transcontinental & Western Air (TWA), which ordered six but only took delivery of five. TWA's aircraft were sufficiently different from Pan American's to warrant a new designation – SA-307B – and a new type certificate, which was issued on 4 May 1940. The principal external changes to the TWA aircraft were narrow-chord cowlings and external flap hinges. TWA's second Stratoliner for the airline, NX19906, undertook a much-publicised pre-service entry tour of TWA's domestic routes wearing the spurious identity NX1940 to herald the arrival of the '1940 Airliner'. The airline's pressurised coast-to-coast service was inaugurated in July 1940 when one of the Stratoliners flew from Los Angeles to New York in a record 11 hours 45 minutes; Los Angeles-bound flights out of New York took about two hours longer due to prevailing westerly winds.

Only one other Stratoliner was built. Actually the first of TWA's intended fleet of six, NX19904 was acquired in typically eccentric fashion by multi-millionaire Howard Hughes. Having tried unsuccessfully to buy a Stratoliner directly from Boeing in 1938 and learning that all production was committed to PAA and TWA, Hughes bought control of TWA and diverted one aircraft to his personal ownership. He intended to use it to break his own three-day, 18-hour around-the-world record set in a Lockheed 14 in July 1938.

Millionaire modification

Uniquely, the Hughes Stratoliner had the smaller B-17C-type fin and rudder of the prototype, and was powered by four 1193-kW (1,600-hp) Wright GR-2600 Twin Cyclone double-row radial engines housed in long-chord cowlings like those of the PAA aircraft. Hughes installed eight auxiliary fuel tanks in its stripped-out cabin, increasing the standard fuel capacity of 6435 litres (1,700 US gal) to 15103 litres (3,990 US gal) – sufficient to circle the globe with just four stops.

The first leg of the trip was to be New York-Berlin non-stop, but the Nazi invasion of

Below: The five TWA Stratoliners were impressed into USAAF service as C-75s, and given the serials 42-88623 to 88627. In olive drab camouflage they were used mainly on the transatlantic route with high-value cargo and important passengers. The fleet was named Apache, Cherokee, Comanche, Navajo and Zuni.

Poland and the outbreak of World War II frustrated Hughes' plans, and he then had the Stratoliner outfitted as a luxury personal transport at a cost of $250,000. The Stratoliner was fitted with accommodation for six sleeper passengers, a full-size bar, kitchen, two lavatories, ladies' powder room, speakers for inflight entertainment, and panelled ceilings with indirect lighting. At the same time a standard Stratoliner fin and rudder were installed. The aircraft remained grounded between September 1939 and May 1947, when Pan American removed the monstrous Double Cyclones and fitted standard single-row Wright radials. Hughes then sold the aircraft to Texas millionaire Glen McCarthy, but it was used little by him and had logged only 500 hours total time when it was next sold, in 1963. Shortly afterwards it was wrecked by a hurricane. Improbably, its fuselage was converted into a luxury houseboat named *The Londonaire* by a new owner in Florida.

High times

Before the outbreak of World War II both PAA and TWA achieved high utilisation rates with their Stratoliners, the latter flying an accident-free total of 7.2 million km (4.5 millions miles). In wartime, TWA's five aircraft were impressed into military service with the United States Army Air Force's Air Transport Command. Designated C-75 by the military and wearing olive drab and grey camouflage, they were operated by their civilian crews for training USAAC pilots in long-range four-engined operations, and also for carrying passengers and cargo on transatlantic routes. Among passengers who travelled aboard the C-75s were Generals George C. Marshall, Dwight D. Eisenhower and H. H. 'Hap' Arnold, and Admiral J. H. Towers.

During their military service the Stratoliners made 3,000 ocean crossings and logged 45,000 flying hours, though not entirely without in-

Between 1951 and 1954 this aircraft was operated by Aérovias Ecuatoriana CA on services mainly between Ecuador and Miami. The aircraft later went on to serve with Quaker City Airways.

cident. On one occasion the crew of a 'friendly' destroyer mistook a C-75 for a Luftwaffe bomber and opened fire on it, damaging its tail with a 20-mm cannon shell.

Their war work done, the Stratoliners were returned to the Boeing factory in 1944 for modification before being handed back to TWA. Wings and engine nacelles were replaced by B-17G components, a B-17G tailplane was installed some 0.9 m (3 ft) further aft than the original, 895.2-kW (1,200-hp) Wright GR-1829-G666 engines replaced the 671.4-kW (900-hp) Cyclones, a B-29 electrical system was installed, the pressurisation system was removed and passenger capacity increased to 38, with a new maximum take-off weight of 20412 kg (45,000 lb). The modified aircraft were redesignated SA-307B-1s and returned to operation on TWA's 'coach' class services until 1951, when four of them were sold to the French carriers Aigle Azur and Airnautic. Three survived into the 1960s and were operated by Air Laos out of Saigon, South Vietnam, during the early years of the Vietnam War. One was inadvertently shot down by an American fighter, and the other two were damaged beyond repair on the ground. An ex-PAA Stratoliner which operated in South Africa later became the private transport of the President of Haiti, and is the only known surviving example of the pioneering pressurised airliner, preserved on behalf of the National Air and Space Museum at the Pima County Museum at Tucson, Arizona.

Above: The surviving Stratoliner is seen here during the period 1950-54, when it served with *Continental Charters*. At one point during its chequered career, it was the presidential transport for Haiti, and at another flew in South Africa.

Right: This was the first production Stratoliner, which originally flew with the same fin as the prototype. Following the crash of the latter, the aircraft trialled dorsal fin arrangements to prevent the inadvertent spinning that had caused the loss of the first aircraft.

Below: In 1944 the five C-75s were returned to Boeing for modification with B-17G components and uprated engines, and were delivered back to *TWA* in April/May 1945 in this unpainted condition. They served for another six years with the airline.

Boeing Clipper

'Clipper' is a name that is legendary. Synonymous with aviation, it became Pan American's radio call sign (like British Airway's 'Speedbird'), instantly recalling the great days of the mighty Boeing Model 314. In a time when flying-boats alone flew the Atlantic, Pan Am's majestic Clippers were the ultimate vehicle to Europe, and were also used on the airline's lengthy Pacific services.

Inspired by the success of Pan American Airways' Martin M-130 and Sikorsky S-42 four-engined flying-boats in pioneering airline routes across the Pacific, in 1935 the Boeing Airplane Company embarked on designing what was to be the world's largest commercial aircraft of the day.

The Boeing Model 314 was the brainchild of Wellwood Beall, a former instructor at the Boeing School of Aeronautics in Oakland, California who had moved to the company's headquarters at Seattle, Washington, where he headed the sales team. Beall sketched out his ideas, proposing to use the wings, engines and horizontal tail surfaces of the experimental Boeing XB-15 bomber which was then under construction (and which led ultimately to the B-17 Flying Fortress), mated to a huge monocoque hull with accommodation for up to 74 day passengers or 34 on transoceanic flights with sleeping berths. Beall's aim was to create a luxurious flying boat, a 'Pullman of the Sky', that would rival the Martin and Sikorsky boats and offer Pan Am a 600 mile increase in range and the ability to fly non-stop across the Atlantic.

His proposal was accepted by Boeing's chief engineer Robert Minshall and design work began on the aircraft. In February 1936 Pan American invited Boeing to submit its design for evaluation, and on 21 June the airline placed a $3 million order for six aircraft, with options on a further six. So large was the aircraft that although construction of the hulls of two Model 314s could be undertaken simultaneously within Boeing's Seattle plant, final assembly had to take place at an outdoor wooden dock equipped with special heavy-duty cranes. The first Model 314, NX18601, was completed in May 1938 and on the last day of the month was floated down the Duwamish river into Puget Sound in preparation for its first flight, with Boeing's celebrated test pilot Eddie Allen at the controls. Unfortunately all 56 spark plugs had fouled on the Model 314's 1120-kW (1,500-hp) Wright R-1820-G102A Double Cyclone radial engines, and the first flight had to be postponed. On subsequent days Allen made high-speed taxi runs across Puget Sound, and on 7 June made a 38-minute first flight. At this time the Model 314 had a single fin and rudder. After landing on Lake Washington Allen reported to Beall that this gave insufficient direc-

This rare colour photograph shows a Boeing Model 314 in flight. The huge boats were used primarily by Pan Am, which put them into service on both Pacific and Atlantic routes.

tional control both on water and in the air and that he had to steer the aircraft with differential power. "It was like herding a reluctant buffalo," he remarked after landing.

Boeing engineers quickly rebuilt the aircraft with a fixed central fin and twin outer fins/rudders, which solved the problem of directional control, and the triple-fin configuration was adopted for production aircraft. Civil Aeronautics Authority certification was granted in January 1939 and the first aircraft, named *Honolulu Clipper,* was delivered to PAA. The remaining five of the initial order had all been delivered by 16 June. Two aircraft were assigned to the Pacific route, flying from San Francisco to Hong Kong. The other four were East Coast based, for transatlantic services. When Pan Am confirmed its order for the second batch of six aircraft, uprated 1190-kW (1,600-hp) Wright GR-2600 Double Cyclones were installed, with additional fuel tankage and accommodation for another three passengers. These aircraft were designated Model 314A; the first six flying-boats were also subsequently converted to this standard.

First launch

On 3 March 1939 America's First Lady, Mrs Eleanor Roosevelt, formally christened *Yankee Clipper,* which subsequently left Baltimore at the end of the month on a route-proving flight to Europe by way of the Azores, returning home via Bermuda. On 20 May it was again *Yankee Clipper,* flown by Captain Arthur La Porte, that inaugurated PAA's first scheduled airmail service to Europe, flying New York-Lisbon-Marseilles. That same day Captain Joseph Chase took *Honolulu Clipper* to Hawaii carrying PAA officials and journalists, but it fell to *Dixie Clipper* and Captain R. O. D. Sullivan to carry the first paying passengers, when the Boeing 314 inaugurated transatlantic passenger flights on 28 June on the New York-Lisbon route, while *Yankee Clipper* launched PAA's North Atlantic service between New York and Southampton on 8 July.

High standards

The Boeing 314 was an immediate success with passengers and crew alike. The aircraft's 5.8-m (19-ft) deep hull was divided into two decks, the upper one containing the flight deck and quarters for the aircraft's 10-strong crew.

Technicians prepare the first Boeing 314 for flight at Puget Sound, from where it made the type's first flight on 7 June 1938. The single fin gave insufficient authority both in the air and on the water, resulting in the rapid fitment of the triple-fin arrangement.

The large wing contained internal companionways through which flight engineers could gain access to the rear of the engine nacelles in flight for minor running repairs. The lower deck, which had large viewing windows throughout its length, featured sumptuous fully-carpeted passenger accommodation which included a lounge with davenport chairs that converted into sleeping berths at night, a dining salon that doubled as a recreational area after meals, a bridal suite, and separate dressing rooms and lavatories for men and women. Service and cuisine were of the highest order, passengers on the inaugural transatlantic flight aboard *Dixie Clipper* enjoying a dinner of breast of chicken and strawberry shortcake prepared in flight by chefs from the Lord Baltimore Hotel. Special kitchens were later set up at La Guardia Field, adjacent to PAA's Marine Terminal, where chefs recruited from top New York hotels oversaw preparation of meals for *Clipper* passengers.

Wartime troubles

Following the outbreak of war in Europe PAA curtailed its transatlantic services, but the Clippers continued to fly the Pacific. On 7

With the Clipper at rest some idea can be gained of the large lower deck for passenger accommodation. Offering sumptuous comfort, the aircraft was an immediate success with the passengers, despite its less than impressive speed performance.

December 1941 *Anzac Clipper* was en route to Hawaii when the Japanese attacked Pearl Harbor, but it landed safely at the island of Hilo. *Pacific Clipper,* under the command of Captain Robert Ford, had already arrived at Auckland, New Zealand. To avoid crossing what had become a war zone, Ford elected to return not

The second Model 314 cruises the Pacific coastline in standard pre-war Pan American colours. Stability on the water was provided by the large sponsons, which gave additional lift in the air, but taxiing was difficult without wingtip floats. Note the water rudder at the rear of the keel.

to his base at San Francisco, but to PAA's East Coast base at New York, flying 55520 km (34,500 miles) via Australia, the Dutch East Indies, India, Arabia, Africa and across the South Atlantic.

After the United States entered World War II two of PAA's *Clippers* were immediately impressed into service with the US Navy, and three requisitioned by the US Army Air Corps. The USAAC applied the designation C-98 to the aircraft, but found few uses for them and these too were handed over to the Navy, all five continuing to be flown by civilian Pan American crews. Three of the newly-built Model 314As were purchased by the British government and handed over to British Overseas Airways Corporation in 1941. Camouflaged, and carrying the registrations G-AGBZ *Bristol*, G-AGCA *Berwick*, and G-AGCB *Bangor*, they were used to maintain essential transatlantic services.

All of the Boeing 314s served honourably during the war. In 1943 *Dixie Clipper* carried President Franklin D. Roosevelt – the first American President to travel by air – to the Casablanca Conference. Clippers flew spares and support equipment to General Claire Chennault's famous 'Flying Tigers' in China, supported General Montgomery's North Africa campaign, and ferried equipment to England prior to the Allied invasion of France in June 1944. British Prime Minister Winston Churchill flew on *Atlantic Clipper* to the United States for an urgent meeting, and travelled several times across the Atlantic on BOAC's aircraft. Returning from a meeting with Roosevelt in Bermuda aboard *Berwick* he took a turn at the controls, and claimed to have greatly enjoyed the experience. It was on this homeward flight that *Berwick* came close to being fired upon by Royal Air Force coastal patrol aircraft whose pilots failed to identify the unfamiliar Boeing flying boat.

Single mishap

Tragically, the Model 314's excellent safety record was marred on 22 February 1942 when *Yankee Clipper*, completing her 240th Atlantic crossing and having flown more than one million miles, crashed while attempting to land on the River Tagus near Lisbon, Portugal, killing 24 of the 39 passengers and crew aboard. They were to be the only fatalities in 3,650 transoceanic flights and more than eight million miles logged by PAA's nine Clippers during seven years of operation, in which each clocked some 18,000 flying hours.

In peacetime, the emergence of new land-plane airliners, such as the Douglas DC-4, Lockheed Constellation and Boeing's Stratocruiser, heralded the end of commercial flying-

Few sights are as majestic as a huge flying boat thundering across the water on its take-off run. Bristol, *shown here, was one of three purchased by* BOAC *to maintain transatlantic services during the war.*

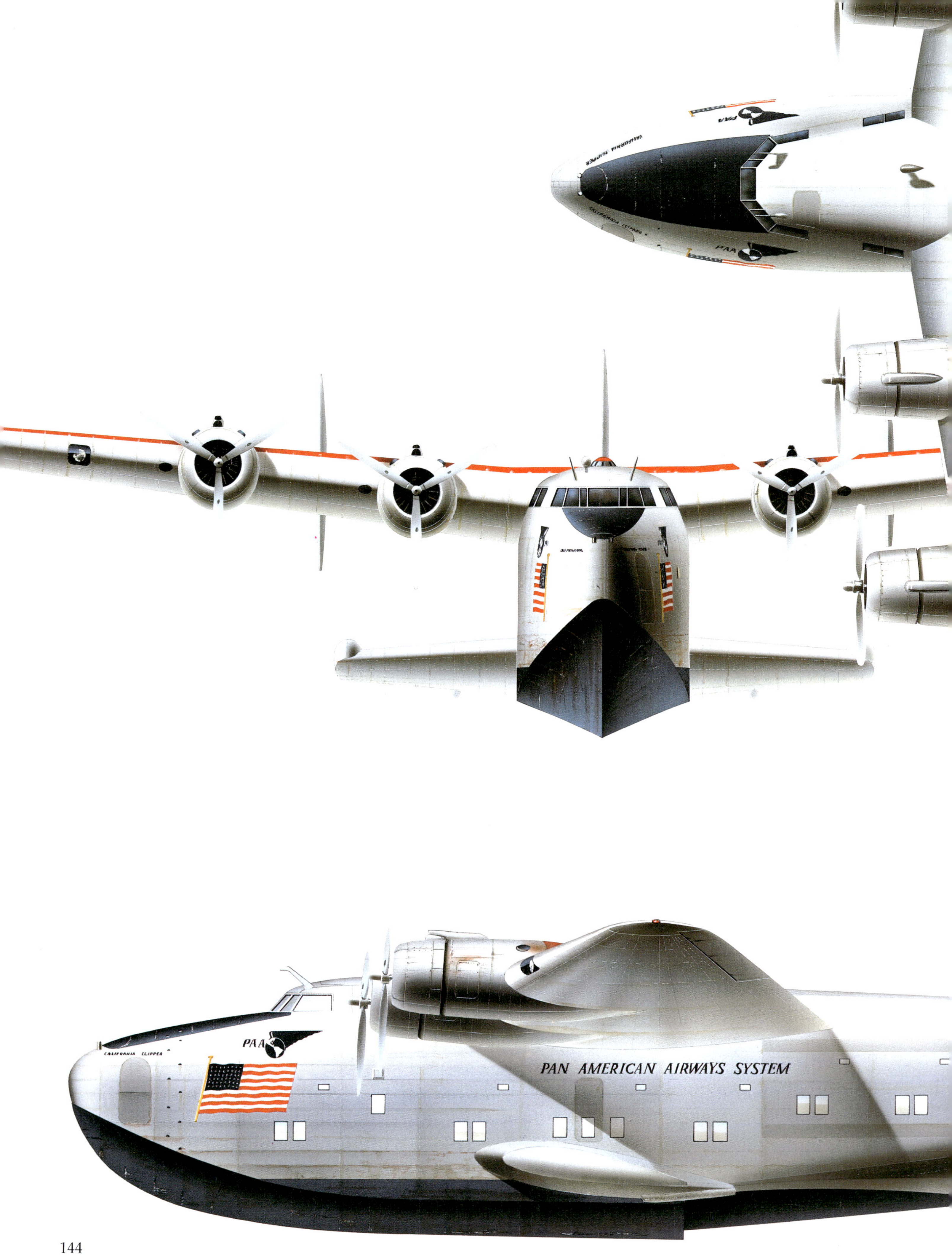
PAA
PAN AMERICAN AIRWAYS SYSTEM

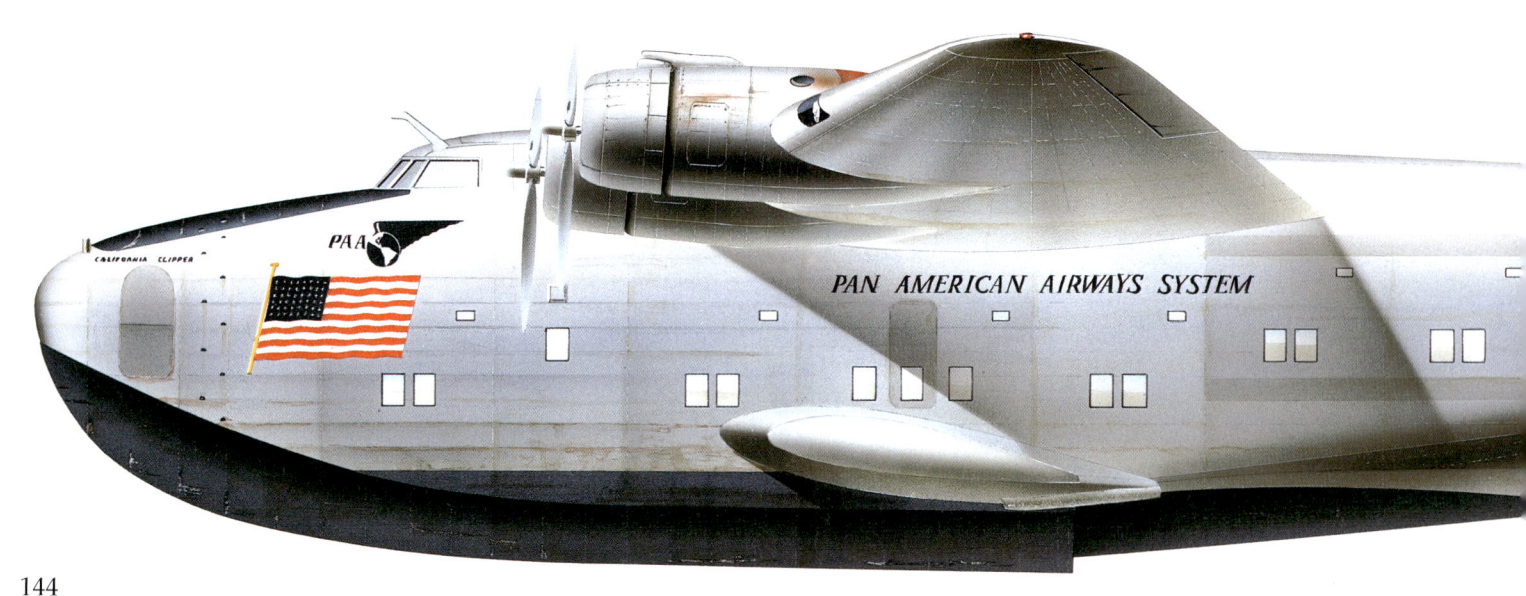

California Clipper was the second aircraft built, delivered to Pan American in early 1939 for service on the Pacific route from San Francisco to Hong Kong. Ten crew flew the mighty craft, comprising two pilots on the 'bridge', navigator, flight engineer, radio operator, ship's master (the officer of the watch who commanded the aircraft but did not fly it) and relief crew, who were provided with bunks behind the 'bridge' and in the nose. A spiral staircase led down from behind the 'bridge' to the main passenger deck, while another staircase led down into the nose and the mooring compartment. The flight engineer had access in flight to the rear of the engine nacelles via walkways in the thick wing section. On long transpacific voyages, the *C*lipper could accommodate 34-40 passengers in sleeper accommodation, but for day journeys up to 74 could be carried.

boat services, and the five surviving PAA Boeing 314s were withdrawn from service in 1946 and put up for sale by the War Assets Administration. Three were dry-docked. *Pacific Clipper* and *Honolulu Clipper* were sold to Universal Airlines, but neither was to survive long. On its inaugural flight with its new owner *Honolulu Clipper* suffered engine problems and was force-landed on open water. The aircraft-carrier USS *Manila Bay* went to its assistance and rescued passengers and crew, but during the rescue operation the ship collided with the flying boat, which subsequently had to be sunk by naval gunfire, since it was deemed to be a hazard to shipping. It is a tribute to the strength of Boeing's structure that it took 1,300 rounds of 20-mm ammunition to send the *Clipper's* hulk to the bottom. *Pacific Clipper* was also damaged in an accident, dismantled and sold for scrap.

BOAC's three Boeing 314s were returned to the United States in 1947, having made 596 Atlantic crossings, flown 6.9 million km (4.3 million miles), and carried 40,042 passengers in wartime service. They were sold to the General Phoenix Corporation of Baltimore, but *Capetown* (formerly *Bangor*) never reached America. Forced down in the Atlantic by fuel exhaustion while on its delivery flight in October 1947, she too was struck and holed by her rescuer, the Coast Guard cutter USS *Bibb*, and like *Honolulu Clipper* had to be sunk. *Bristol* suffered a similar fate. Sold to a clergyman who called himself Master X and claimed he would use the flying boat to fly to the Soviet Union for Cold War peace talks with Stalin, she sank at her moorings in Baltimore Harbor during a storm and was scrapped. By 1951 all remaining Boeing 314s had suffered a similar fate.

As the most reliable and capacious long-range aircraft of their day, the Clippers were widely used by wartime leaders, particularly Churchill, who made several transatlantic journeys in both Pan American and BOAC machines. This aircraft for Pan Am is seen during engine tests.

Specification: Boeing 314A
Wingspan: 46.3 m (152 ft 0 in)
Length: 32.3 m (106 ft 7 in)
Height: 8.4 m (27 ft 7 in)
Wing area: 266 m² (2,867 sq ft)
Passenger capacity: 40-70
Empty weight: 21800 kg (48,000 lb)
Payload: 4000 kg (8,750 lb)
Maximum take-off weight: 38100 kg (84,000 lb)
Cruising speed: 302 km/h (188 mph)
Max speed: 340 km/h (210 mph)
Service ceiling: 4085 m (13,400 ft)
Max range: 5930 km (3,685 miles)

Clippers flew throughout the war on many important transport services. This blue-grey disruptive camouflage was applied to reduce conspicuity over the oceans.

Beached for maintenance at Baltimore, *Bristol* now wears camouflage and wartime British civil insignia on the rear fuselage. Large tractors were necessary to beach the *Clipper*, pulling the boat out of the water on a sturdy trolley.

Chapter 8

Boeing 707

•

Boeing 720

•

Boeing 727

•

Boeing 737

•

Boeing 737 300/400/500 series

•

Boeing 737 600/700/800 series

BOEING 707

The Comet may have been first, but when it came to airliners, few would deny it was the Boeing 707 that was the real inspiration behind the jet revolution. Its bold design incorporated the latest in 1950s technology, with a sharply swept wing and powerful, fuel-efficient engines. After establishing itself as the world's No. 1 airliner, it became a by-word for reliability and economy. Even in the 1990s airframes are still sought to act as civil freighters or sophisticated military electronics platforms.

Though the British de Havilland Comet was the pioneer jetliner, it paid the penalty of being first. It remained an attractive exception to the mainstream of civil air transport, and its main effect was to show the enormous passenger appeal of a vehicle that got there in half the time after a trip that, by comparison with the noisy, vibrating piston-engined machines, was like being wafted by an angel. Could this rather small and supposedly uneconomic machine threaten the almost total grip of US industry on the world transport-aircraft market? The answer was clearly to meet the competition with an American jet, but how could it be financed? The only way seemed to be with federal aid, but after arguing the matter throughout 1949 the US Congress eventually in 1950 threw out a bill to fund a US jet prototype.

Douglas and Lockheed studied the problem intently and published a few brochures, but only Boeing (very much an also-ran in the civil transport aircraft business) had actually built large modern jets. By the autumn of 1950 the Seattle-based company could see that to meet its global range needs, the US Air Force would have to use in-flight-refuelling not only with its Boeing B-47 bombers but also with the giant new Boeing B-52. Boeing was building the tankers as well, in the form of KC-97s; but these were piston-engined, and to hook up with them the bombers had to slow down and lose almost half their altitude. Surely, reasoned Boeing, the answer was a jet tanker?

Boeing began by proposing a swept, jet-propelled KC-97 in March 1951. After prolonged argument the idea was rejected by the USAF on 17 August 1951. Yet Boeing was convinced that both the airlines and the US Air Force would eventually buy jet transports, and that the tanker and the civil liner could be basically the same design. But in the absence of airline orders, government prototype funds or even US Air Force interest, Boeing had to do it the hard way. It had to tighten its belt and produce a prototype with its own money. After the most searching evaluation, the company's board met on 22 April 1952 (exactly one week after the successful first flight of the B-52) and took the big decision. It would cost not less than $15 million.

One essential ingredient was the Pratt & Whitney JT3 engine, the lightweight commercial version of the fuel-efficient J57 used in the B-52. But, whereas the giant bomber used eight of the 4536-kg (10,000-lb) thrust engines, the transport would need just four, hung in single pods below and ahead of the 35°-swept wing. The fuselage would not be that of the C-97 but larger, and with a more streamlined nose tipped by radar. (Even creating the radome would mean solving a completely new set of technical problems.) Gross weight worked out to 86184 kg (190,000 lb), and though the military model would have an interior configured for cargo and fuel, the commercial pas-

Seen here at Mauritius, this Boeing 707-465 typified hundreds of the type which carried on flying with third-world operators long after their service with major airlines was over. Many are still in service today.

This Boeing 707-358C was one of the last of the type to be built. With Boeing hull number 20897 and registration ST-AFA, it was delivered new on 17 June 1974 to Sudan Airways, with whom it has since operated on the trunk route to London with the individual name **Blue Nile.**

senger aircraft could seat 130. Because it could fly at 966 km/h (600 mph), the jetliner promised to do three times the work of either the military KC-97 or a commercial type such as the Douglas DC-7 or Lockheed Super Constellation. But the airlines and US Air Force never showed more than polite interest as the prototype took shape at the Renton (Seattle) plant.

In fact the correct Boeing model number for the prototype was 367-80, and the company-owned prototype was to become popularly known as the Dash-80, but that was because the years of study had been numbered as suffixes to the same Model 367 number of the original piston-engined C-97. Numbers in the 500 series were reserved for gas turbine engines and in the 600 series for Boeing missiles, and aircraft numbers began again at 700. When the Model 367-80 finally became a flyable aircraft it was given the new designation Model 707. Subsequently Boeing deliberately capitalised on this memorable sequence by making subsequent jet transports the Models 717, 727, 737, 747, 757, 767 and 777.

707 emerges

The designation Model 717 was assigned to the proposed USAF tanker, but by 1954 the Korean War had ended, money was tight, and strong forces favoured a cheaper turboprop tanker, or even modified Convair B-36s or B-47s. There was still nothing certain about the programme when the Dash-80, painted in a rich company livery of chocolate-brown and chrome yellow, rolled out in a ceremony on 15 May 1954. The question marks loomed larger when, during taxi tests six days later, the left main gear smashed its way up through the wing and left the vital prototype lying crippled on its left outer engine pod. It was not until 15 July 1954 that Tex Johnston and Dix Loesch were able to fly the aircraft that was to keep the United States the world leader in civil transports.

By this time the USAF had told Boeing it wanted a new-build jet tanker, a great relief after the company had already spent not only $15 million on the prototype but half as much again on design and tooling for production machines. In October 1954 the first tanker order came through, for 29 aircraft, launching a gigantic programme of KC-135 and C-135 versions. This underpinned the commercial Model 707, but in 1954 Boeing did not enjoy the same clout with the airlines as mighty Douglas, and the announcement of the DC-8 on 5 June 1955 meant Boeing was going to have to fight every inch of the way. Moreover, on a run of 50 aircraft Boeing could see no way of pricing the 707 below $5.5 million, far above what the airlines would pay.

Nor was this all. To meet the competition of the DC-8, Boeing decided to do one of the costliest modifications possible: change the body cross section. It remained a figure-8 with smoothly faired sides but the upper lobe was increased in width by 10.16 cm (4 in) to 3.556 m (140 in), beating Douglas by 5.08 cm (2 in) and enabling a triple seat unit to be installed on each side of the aisle, for up to 150 passengers. A new form of fatigue-proof window structure was devised, with two small windows per seat row and chemically machined panels running the whole length of each side of the fuselage to double up on the original skin and frames. Boeing learned a great deal from the British investigation into Comet fatigue problems, as did other US builders. Fuel capacity was increased, with various arrangements of flexible cells inboard and integral tanks outboard, and the first models offered were the Model 707-120 series with a length of 44.04 m (144 ft 6 in) and the special Model 707-138 which was 3.048 m (10 ft) shorter. The standard launch engine was the JT3C-6, rated at 6124-kg (13,500-lb) thrust with water injection, and fitted with a large noise-suppressing nozzle with 20 separate tubes (which soon became covered with soot from wet take-offs, when not only noise but black smoke were emitted on an impressive scale).

Predictably the first airline customer was Pan American, which bought 20 Model 707-121s; but, to Boeing's consternation, it also signed for 25 DC-8s on the same day (13 October 1955) in a $296 million deal. Later in the same month United chose 30 DC-8s, and Douglas announced a longer-ranged DC-8 with the big JT4A engine. Boeing had to respond with a long-range Model 707, and unlike Douglas (whose aircraft were all the same size) decided to make it larger. Braniff, in fact, was odd man out in buying the Model 707-220, which was the original size but had the JT4A, rated at 7167-kg (15,800-lb) thrust, for sprightly take-off at hot-and-high airports on routes to South America. But it was American's order for 30 Model 707-123s on 8 November 1955 that clinched the go-ahead and took Boeing to the 50 mark. Subsequently Boeing at least level-pegged Douglas. In the UK BOAC, having said it had no interest in jets and so getting the Vickers VC 7 cancelled, began talking with Boeing only four months later.

Rolls-Royce power

BOAC, like PanAm, was interested in Boeing's bigger longer-range model, the Model 707-320 Intercontinental. This had a new high-efficiency wing of 3.53 m (11 ft 7 in) greater span, a fuselage 2.57 m (8 ft 5 in) longer seating up to 189 passengers, much greater fuel capacity, and an initial gross weight of 141520 kg (312,000 lb).

N70700 *lifted off from* **Renton** *on 15 July 1954 to open the world-wide era of jet transportation. After many years of valued research by Boeing, this historic aircraft was rebuilt in its original form.*

History was made when the Boeing 707-121 of Pan American World Airways opened the 'Big Jet' era on 26 October 1958 with a scheduled service from New York to Paris.

Boeing 707-436
G-ARC
British Overseas Airways Corporation

Cathay Pacific's colourful livery adorns Boeing airframe no. 18888, delivered in May 1965 as a 707-351C to Northwest Orient and sold in August 1974 to the Hong Kong carrier, with registration VR-HHE.

Boeing 707 variants

Model 367-80: company prototype; first flown 15 July 1954 and subsequently used in numerous research programmes
Model 707-120: four 6124-kg (13,500-lb) JT3C-6 engines; initial production version, with fuselage increased in width and length, gross weight raised to 102060 kg (225,000 lb) and finally 116575 kg (257,000 lb)
Model 707-120B: four 7711-kg (17,000-lb) JT3D-1 engines, aerodynamic improvements for Mach 0.91 cruise
Model 707-138: short-body model for QANTAS, 3.048m (10ft) shorter
Model 707-220: as Model 707-120 but four 7167-kg (15,800-lb) JT4A-3 engines
Model 707-320: first intercontinental version; all-round increase in size; four JT4A engines (various ratings); gross weight 141520 kg (312,000 lb)
Model 707-320B: aerodynamic improvements, four 8165-kg (18,000-lb) JT3D-3 engines; includes **VC-137C**; optional gross weight 151321 kg (333,600 lb)
Model 707-320C: as Model 707-320B but equipped for passengers (up to 202) or cargo
Model 707-420: as Model 707-320 (not Model 707-320B) but with 7945-kg (17,500lb) or 8165-kg (18,000 lb) Rolls-Royce Conway 508 or 508A turbofans
Model 720: derivative of Model 707 series with lightweight structure and shorter fuselage tailored to short/medium-range operations
Model 720B: turbofan-engined version of Model 720
VC-137A: USAF version of Model 707-120 for VIP transport
VC-137B: designation of VC-137A after re-engining with JT3D; gross weight 117025 kg (258,000 lb)
VC-137C: Presidential (often 'Air Force One') aircraft, as Model 707-320B with various special equipment; gross weight 146055 kg (322,000 lb)
707 Tanker/transport: surplus 707 airliners converted to tanker configuration with wing-mounted hose drum units and optional fuselage unit
KE-3A: tanker version for Royal Saudi Air Force with CFM56 engines

Launch engine was the JT4A at the increased rating of 7620 kg (16,800 lb), but Rolls-Royce's Conway bypass turbojet (a turbofan of very low 0.3:1 bypass ratio) fitted perfectly and offered greater power, lower installed weight and much better fuel consumption, and was selected by a small minority of airlines including BOAC and Lufthansa. The considerably greater capability of the Intercontinental quickly made this the standard type of Model 707, while the original size was developed into a new shorter-range family initially marketed with the same Model 717 number as the tanker but in 1959 renumbered as the Model 720.

The first production Model 707, Boeing No. 17586, was flown at Renton on 20 December 1957, but was actually numbered as the second of the initial batch of 20 Model 707-121s for PanAm, N708PA, the second production machine having the specially chosen registration N707PA. Flight development was generally trouble-free,

Robert F. Six, President of Continental Airlines, had a happy experience with the Vickers Viscount and was delighted when Boeing produced the 720B, which likewise suited difficult hot/high altitude on his routes.

Boeing 707-320C cutaway drawing key

1 Nose cone
2 Weather radar scanner
3 Glide-slope aerial
4 Forward pressure bulkhead
5 Pitot head
6 Nose frames
7 Windscreen panels
8 Eyebrow windows
9 Overhead console
10 First Officer's seat
11 Captain's seat
12 Forward frame
13 Twin nosewheels
14 Nosewheel doors
15 Nosewheel box
16 Drag struts
17 Navigator's table
18 Observer's seat
19 Navigator's seat
20 Navigator's overhead panel
21 Flight Engineer's seat
22 Flight Engineer's instrument panels
23 Flight deck entry door
24 Crew coat closet
25 Crew toilet
26 Crew galley/buffet
27 Spare life vest stowage
28 Radio (emergency) transmitter
29 Life raft stowage (2)
30 VHF aerial
31 Smoke and fume-proof curtain
32 Forward entry door (0.61m × 1.83m)
33 Escape slide stowage
34 Forward underfloor freight hold
35 Cabin floor level
36 Six cargo pallets (total 125.3m³)
37 Ball transfer mat (five segments
38 Door actuator rams
39 Main cargo door (raised)
40 Engine intakes
41 Secondary inlet doors
42 Turbocompressor intakes

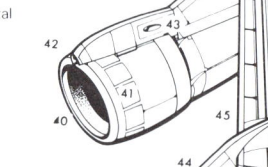

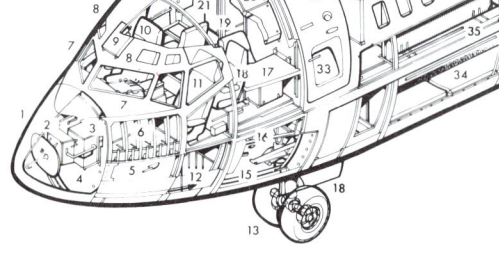

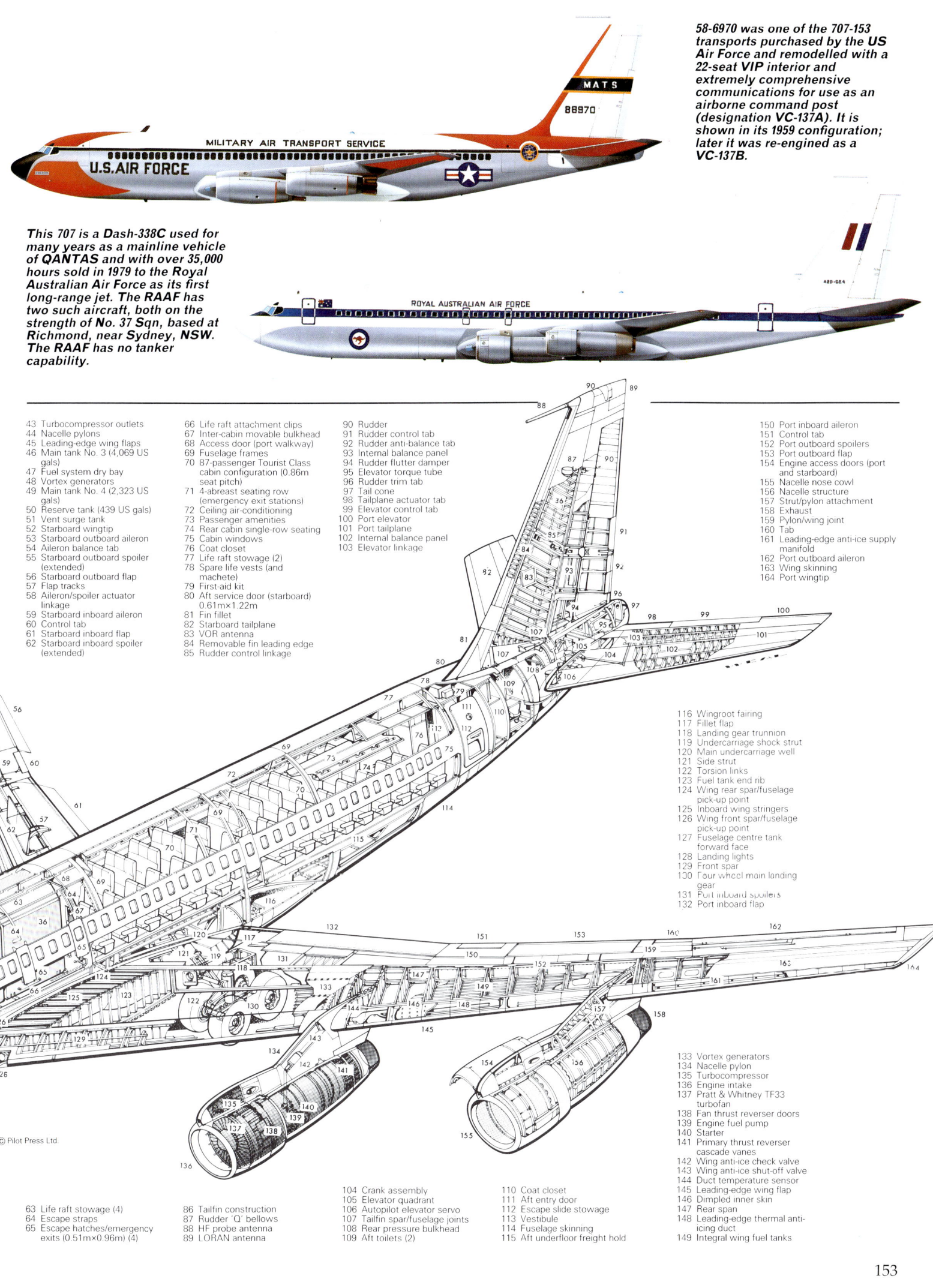

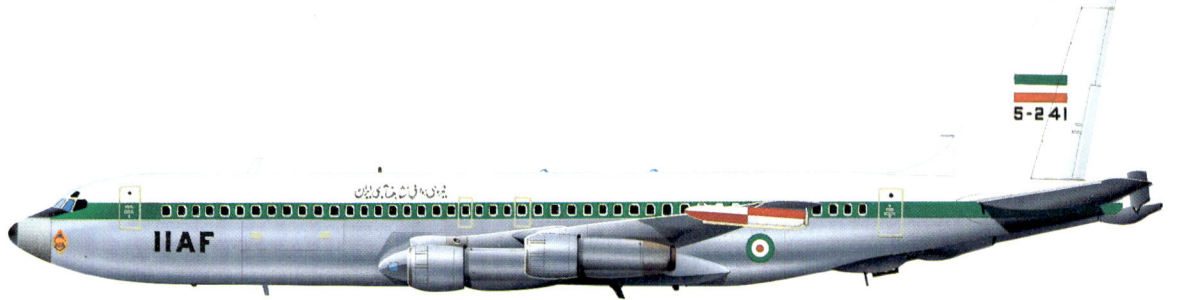

Right at the very end of 707 production, in 1976-7, came a batch of 14 extremely well-equipped 707-3J9C aircraft for the Imperial Iranian air force. They combined special communications for global operation with triple-point air refuelling, with tip-mounted hose/drogue installations and a Boeing high-speed boom driven by an operator in the rear fuselage. All were delivered before late 1977.

Although BOAC had taken a sizeable batch of Model 420s with Conway engines, the fleet was augmented in the late 1960s by Model 320Bs. These went on to serve with British Airways.

and FAA certification was awarded on 23 September 1958. The world airline industry was poised on the brink of a new era. Some thought the prospect disastrous; one important expert said it was 'an industry gone mad'. Bristol, trying to sell Britannia turboprops, had insisted the Model 707 was technically impossible, and would need six engines. Even jet enthusiasts admitted that the Boeing monster was almost impossibly big, heavy, capacious, expensive to buy, needed runways longer than those available, could gobble up passenger-miles at a rate that seemed astronomic, and might indeed bankrupt the airlines (either those that bought it or those that stayed with propellers). The voices which predicted that what had become known as The Big Jets would result in a gigantic expansion of the air-transport business were muted and uncertain. Those who predicted an era of unknown profitability, and freedom from reliance on government subsidies, were hardly heard at all.

PanAm opened scheduled services between New York and Paris on 26 October 1958. New York and London and other European capitals soon followed. On these routes the Model 707-121 was marginal. It had not been designed for the North Atlantic, and the flight crews had to learn fast about correct take-off procedures and how to get the most air-miles per pound of fuel. Had noise certification been in force the operations would have been impossible. As it was, on westbound flights a refuelling stop was invariably needed in Iceland or one of the other wartime-built fields further west. But these were the first faltering steps of a revolution in global travel. On 25 January 1959 American began services with the Model 707-123 between New York and Los Angeles, and here there was no real problem.

Spurred by competition from the DC-8, CV-880 and European types, Boeing had been forced to embark on a programme of building ever-better jetliners. The original risk of $15 million had been left far behind. Now the risks were beginning to approach $100 million. What was still uncertain was whether or not the company would survive. The Model 707 was selling in dozens, but would it ever sell in hundreds? Pratt & Whitney met the competition of the Conway with a startlingly simple modification to the JT3C which replaced the first three stages of the compressor with two stages of enormous blades called a fan; and they coined the name turbofan, more descriptive than 'bypass turbojet' and easier to sell. Called the JT3D, the new engine began life at 8172-kg (18,000-lb) thrust and offered much better fuel economy and dramatically reduced noise, as well as release from water injection. The result was a second generation of Model 707s characterised by a B suffix.

Before these were available, the big Model 707-320 Intercontinental flew as the 16th off the Renton line on 11 January 1959. It was certificated on 15 July the same year and entered service with PanAm a month later, sweeping away the Model 707-121 from transatlantic routes with a replacement aircraft designed for the job, and able to carry a considerably greater payload non-stop, even westbound. The UK's certification was held up while the ARB (today the CAA) studied the handling and stability in adverse circumstances and finally insisted on a greater fin area. At first an underfin was added, with a tail bumper incorporated, but later this was replaced by a much taller vertical tail which was retrofitted on almost all Model 707s and related military models. This cleared the way for the Conway-engined Model 707-420 family, which was approved in February 1960.

Often used to fly Air Force One missions, with the President aboard, USAF no. 72-7200 was built as a 707-353B and delivered on 15 November 1972 as the VC-137C replacing the older 62-6000. Operated by the 89th Military Airlift Wing, it has TF33 (JT3D) engines and is equipped with special communications for direct contact with US diplomatic and military centres in all parts of the world.

Boeing 720

By the late 1950s Boeing knew it had a success on its hands. While few could have predicted just how significant the Models 707 and 717 would become, the manufacturer was confident enough to propose yet another member of the family, the Model 720. Described by others as simply a spoiling tactic to see-off unwelcome competition, the Boeing 720 is still deserving of an auspicious mention.

It is perhaps fitting that the father of the United States', if not the world's, greatest aviation industrial concern first took to the air on Independence Day – July 4th. The year was 1914 and the bearer of the best-known name in aviation manufacture was William Boeing. He was 32 at the time, a successful timberman and boat-builder, and this first joy-ride sparked an interest in aviation that was to produce his first aircraft in 1916. Under his direction the company expanded, mainly constructing types under licence, including the D.H.4. This provided experience for aircraft such as the Model 40, a commercial mailplane, and the PW-9/P-12 fighter family. By the time Boeing resigned from his company in 1934, it had introduced all-metal cantilever monoplanes in the form of the Monomail and Model 247.

In the late 1930s most of Boeing's work was consumed by the epoch-making B-17 bomber, and later with the B-29 Superfortress, an aircraft which made one of the greatest single leaps in aircraft technology of all time. Further development of the B-29 produced the improved B-50 bomber and the Model 367 Stratocruiser and its Model 377/C-97 military counterpart. These piston-engined transports were highly popular, for their double-bubble pressurised fuselages provided ample internal volume.

Another pioneering aircraft of Boeing origin followed, the B-47 Stratojet, which matched the extraordinary thrust of the new jet engines with swept-wing technology, giving it high subsonic performance with exceptional range/load characteristics. The B-47 layout was extrapolated into the monster eight-jet B-52. From the B-17 onwards, the military bomber programme had made Boeing the leader in large aircraft production. For many reasons this industrial might had largely unimpressed the civil market: even the classic Stratocruiser could only achieve a production run of 56. At the same time, the collective national pride was severely dented by Britain, which flew the de Havilland Comet as the world's first jet-powered airliner, well ahead of any US design.

Against this backdrop of a low company standing with the airlines, and a national failure by surrendering the lead in commercial aircraft to the Europeans, Boeing risked all its considerable power and position by launching a company-funded prototype for a jet transport with both civil and military applications. This single aircraft, the Model 367-80 (universally known as 'Dash Eighty') was to become one of the most significant ever built, for it not only

Braniff International was in the vanguard of 707 operators, the only user of the 'hot-and-high' Series 220. It also purchased five 720s for its east coast routes, this particular example being delivered on 22 March 1961. It was sold to AirClub International in September 1973.

United Airlines was the launch customer for the Model 720, using its aircraft on domestic services from July 1960. This aircraft, the third built, was the first to be delivered to any customer, joining United on 30 April.

directly bore two hugely popular families of aircraft, but also established the most successful aircraft dynasty ever known: the Boeing jetliners.

'Dash Eighty' first flew on 15 July 1954, and mirrored the considerable technological expertise of the company. As might be gathered from its '367' designation, the new aircraft was based loosely on the Stratocruiser, although only the double-bubble fuselage section was apparent (and even this had the 'crease' between upper and lower lobes faired over). The wing was pure Boeing magic, long and slender with podded engines, features inherited from the B-47. With one stroke Boeing had set the standard configuration for large airliners that is still adhered to today.

Development of the 367-80 at first followed the trusted military route, resulting in the Model 717 KC-135 Stratotanker and variants, of which 820 were built. However, Boeing knew the real money lay with the airlines, and the 'Dash Eighty' design began its transformation into a jet airliner. After a costly fuselage diameter enlargement, the design was frozen and orders from Pan American and American Airlines allowed Boeing to begin construction of what had become the Model 707, and what was to become the world's standard long-range airliner through the 1960s and early 1970s.

20 December 1957 saw the first 707 take to the air, and on 23 August 1958 Pan Am flew the first revenue-earning service of a US jet-powered transport. On 26 October, the 707 inaugurated full transatlantic services, initially to Paris. These early 707-120s proved marginal on the transatlantic services, virtually always having to make refuelling stops at Gander or Keflavik on westbound journeys and occasionally at Shannon on eastbound sectors.

Conversely, 707-120s delivered to American Airlines for their New York-Los Angeles route, and those leased by Pan Am to National for New York-Miami proved to be a great success where the range performance was not critical. TWA followed suit, and along with American became the leading customer for the 707-120. Pan Am meanwhile introduced the 707-320 Intercontinental, which had a longer fuselage, wider wings, JT4A engines and increased fuel, allowing the 707 to conquer the oceans with ease and establish itself as the major international jet airliner, subsequently developed with JT3D turbofans as the 707-320B and -320C.

Smaller sibling

Early in the development of the 707, Boeing had looked at the possibility of creating a short/medium-range member of the new airliner family, tailored to the sectors not requiring the transcontinental range of the 707-120. Successes on domestic routes with the early 707s further spurred development of this economy version, which at first was designated 707-020 to reflect its parentage. Model 717-020 was also applied before Model 720-020 was assigned, the entirely new number adopted to illustrate the considerable design changes compared with the mainstream 707 family.

Key design motivations behind the Model 720 were weight-saving and improved field performance, the latter to allow the aircraft to use regional airports that might be deemed marginal for the heavy 707. Power for the new airliner was provided by the Pratt & Whitney JT3C-7 turbojet, which each produced 55.62 kN (12,500 lb) thrust. The 57.85-kN (13,000-lb) JT3C-12 was an alternative. With the significantly reduced weight, this gave sparkling take-off performance compared to the ground-hogging 707s. Without a need for long range (if airlines needed it they simply bought the 707!) fuel capacity was cut dramatically, although the actual figure varied from customer to customer between 44857 litres (11,850 US gal) and 56137 litres (14,830 US gal).

Length of the Model 720 was 41.68 m (136 ft 9 in), pitched between the 40.99 m (134 ft 6 in) of the short-body Series 138 707 developed for Qantas and the 44.04 m (144 ft 6 in) of the standard Series 120 and 220. To further confuse, the 707 prototype was the shortest of the family at 38.96 m (127 ft 10 in), the 707-320 the longest at 45.60 m (152 ft 11 in) and the Model 717/C-135 at 41.53 m (136 ft 3 in)! The 720's relatively short fuselage was aimed at sectors with less capacity requirements than the bigger variants, but could be filled with higher-density seating if required, the cabin featuring continuous floor tracks and movable partitions for rapidly reconfigured interiors. The reduced passenger load allowed the deletion of one of the 707's three air conditioning packs, with consequent savings in weight.

Change in shape

Other savings came from the overall philosophy: metal of thinner gauge was used for some areas, and a lighter undercarriage structure was fitted, both possible because of, and contributing to, the weight-reducing process. However, the greatest changes in the Model 720 occurred in the wing design. Although of the same span as the 707-120 (0.35 m/14 in wider than the 367-80), the wing was of different planform. More area was added to the inboard section between the fuselage and the inboard engine nacelle, giving a slight kink to the leading edge. Additional high-lift flaps were added to the leading edge, a development later applied to 707 models. These measures further improved take-off performance.

The first 720 was completed and flown with the 707's original short fin, but production aircraft featured the taller main fin and small ventral fin as first required by the British CAA for the Conway-powered 707-420 and subsequently made standard on all 707/720 models. The tall fin was added to help directional stability at low speeds, while the small ventral fin incorporated a bumper to protect the rear fuselage in the event of over-rotation. A related feature introduced to the whole family by the 720 was the power-assisted rudder, required to overcome potential problems in an engine-out situation at take-off.

The main cabin of the 720 was 29.41 m (96 ft 6 in) long and 3.56 m (11 ft 8 in) wide, offering a floor area of 93.18 m^2 (1003 sq ft). Underneath the cabin floor were two baggage holds, fore and aft of the wing carry-through and main wheel well area. The forward hold could accommodate 19.48 m^3 (688 cu ft) while the rear hold offered 19.54 m^3 (690 cu ft). A crew of two pilots and a flight engineer was standard, with an optional fourth station for a navigator on the port side.

N7201U was the first Model 720, a Series 022 destined for launch customer United Airlines. Its maiden flight occurred from Renton on 23 November 1959, and certification trials began on 18 January, two further aircraft helping to accomplish this by early July, with 442 hours flown. United had received its first aircraft on 30 April, and five others during May and June, so that when certification was granted the training and proving process had been virtually completed.

On 5 July 1960, United flew its first 720 service, between Los Angeles, Denver and Chicago, and on 8 July began Los Angeles-Seattle services. United's aircraft were configured for 105 seats, but second operator American Airlines opted for a fit of 98, of which nearly a half were first-class. American's first schedule was Cleveland-St Louis-Los Angeles on 31 July, after receiving its first aircraft on the 24th. Confusingly it called its aircraft '707-Astrojets'.

Re-engined version

Shortly after the first JT3C-powered 720s entered service, Boeing rolled out the Model 720B. This benefited from the work the company had done in co-operation with American Airlines on the 707-120B, which fitted the Pratt & Whitney JT3D-1 engine of 75.65 kN (17,000 lb) thrust. This was the first engine to be labelled a turbofan, and in addition to the welcome increase in power, was also far more efficient. This stunningly simple yet effective adaptation of the JT3C core engine gave the engine nacelle a different appearance, with a wider-diameter front cowl round the fan blades. At the same time the tailplane was increased in span to 13.21 m (43 ft 4 in) to handle the additional thrust.

Eighty-nine of the 154 Model 720s were to 'B' standard, and not surprisingly American Airlines led the way, even returning the 10 original Model 720s to Boeing for modification to the new standard, a practice adopted by

Based at Minneapolis-St Paul, Northwest was among the group of US domestic carriers to order the type. Its Series 051B aircraft were delivered between May 1961 and July 1964, and were sold in the early 1970s, notably to Maersk, Monarch and Olympic.

N720W was the second of a pair of Model 720s delivered to Pacific Northern, which later served with Western Airlines before passing to Alaska Airlines in May 1973. Note the slender pods for the original JT3C engines.

With a fleet of 27 new Series 047Bs, and three second hand aircraft, Western Airlines was numerically the most important 720 operator, the type forming the bulk of its fleet throughout the 1960s. The last Western 720 service was flown on 6 January 1980 after nearly 20 years in service.

several other carriers later. The first 720B flew on 6 October 1960, and American received its first machine on 3 February 1961. The extra thrust allowed 140 passengers to be carried, and a later increase in thrust to 80.1 kN (18,000 lb) with the JT3D-3 pushed the load to 165 in six-abreast high-density configuration.

For a while, the Model 720 sold well. US users accounted for the vast majority of sales, these being made to United (29), American (25), Western (27), Eastern (15), Northwest (17), Continental (8), Braniff (5) and Pacific Northern (2). Foreign customers were Aer Lingus (3), Lufthansa (8), Avianca (3), Saudia (2), Pakistan International (4), El Al (2) and Ethiopian (3). Completing the list was a single transport (Series 061) for the Federal Aviation Administration. The final aircraft from the line was N3617, a Series 047B for Western Air Lines, delivered on 30 September 1967.

Like Ethiopian, Avianca of Colombia initially purchased three 720-059Bs new from the manufacturer and then added three more second hand machines. Two of the latter were from Lufthansa, the third coming from Western Airlines.

Ethiopian Airlines was among the original purchasers with three machines, its aircraft being designated 720-060B. A similar number of ex-Continental aircraft were subsequently added to the fleet.

Foreign interest in the 720 was not large, but did augment the major sales to the large US operators. Israeli national carrier El Al took a pair of Series 058Bs.

When it appeared, the 720 had surprisingly little competition in its chosen market. Douglas did not produce a corresponding DC-8 variant but Convair (General Dynamics) did try with the 880. Boeing found the competition unwelcome, and reportedly sold its 720s at 'bargain-basement' prices to kill off the unwelcome competition. The 880 did achieve 65 sales, with TWA and Delta as chief sponsors, but each one was sold at a considerable loss. While the 720 was killing the 880, Convair should have heeded the warning, for the follow-on 990 was even more disastrous, selling only 37 examples.

What restricted the Model 720 to its own meagre (by Boeing standards) number of sales was that it was really only an interim type pending introduction of a new design tailored entirely to the short/medium-range market. That design was the Model 727, which began delivery in late 1963 and soon overtook the 707/720 as the world's best-selling airliner, itself since overtaken by the 737.

Central American operator Belize Airways assembled a fleet of five ex-United 720-022s in the 1970s, four of which are seen here. The type served until 1980, when the company's activities were suspended and the aircraft went into storage at Miami. All had been broken up by April 1983.

Typical of the smaller operators which adopted the 720 after its mainstream days were over was Air Malta, which picked up a pair of 720-040Bs leased from PIA in April 1974 (illustrated, still wearing Pakistani registration) and four -047Bs from Western. The final service was flown in November 1989.

Although it was pipped by United Airlines as the launch customer for the 720, American Airlines was the prime mover behind the 'B' modification which added JT3D turbofans. N7527A was America's first 720, originally delivered with JT3Cs but seen here after retrofit with the more powerful powerplant.

By the early/mid-1970s, the US domestic fleet was disappearing fast, the exception being Western, which operated a total of 30 aircraft, and which flew its last 720 service in 1980. Model 720s flooded the second hand market, and went to a multitude of eager purchasers. Even the two US international giants, Pan Am and TWA, flew the type for a short time, while some of the original customers added to their fleets. Some aircraft passed through many hands after their days with mainstream operators were over.

Second careers

Owners of used aircraft included respectable national carriers, notably those of less important nations, small and short-lived airlines operating on even smaller budgets, corporations who flew the aircraft on executive transport duties, leasing companies and, above all, a host of inclusive-tour charter airlines and travel clubs. One well-known ex-American 720-023B was re-registered as N1R and used by the Los Angeles Dodgers baseball team as a squad transport between fixtures. What follows is not an exhaustive list of secondary owners, but does serve to illustrate the diversity of companies which flew the type, and the extent to which the fleet was dispersed around the globe. Approximate numbers follow each user:

AeroAmerica (10), Aero American Leasing (1), Aero Specialities (3), Aerotal Colombia (1), Air Club International (6), Air Lanka (1), Air Malta (6), Air Niugini (1), Air Viking (1), Air Rhodesia/Zimbabwe (3), Alaska Airlines (3), Alia (2), American International (1), Ariana (1), Atlanta Skylarks (2), Belize Airways (5), BWIA (1), Calair Flug (5), Cavanagh Communities (1), Club American (1), Conair (10), Contemporary Entertainment (1), Eagle Air (3), Ecuatoriana (3), Hispaniola Airways (1), Jet Aviation (3), Jet Set Travel Club (1), Kenya Airways (1), Korean Airlines (2), LA Dodgers (1), Maersk (5), Middle East Airlines (18), Monarch (7), Olympic Airways (7), Pan American (9), Pan Aviation (1), Sierra Leone Airlines (1), Somali Airlines (1), Templewood Aviation (1), T.L. Corporation (1), Trans Caribbean (1), Trans European Airways (2), Trans Polar Airlines (3), Trans World Airlines (4) and Voyager 1000 (2).

Desert resting place

Today the Model 720 is a rare beast indeed. Many B-models were purchased by the US Air Force and ferried to Davis-Monthan AFB, Arizona, where they were sacrificed, along with surplus 707s, to provide JT3D (military designation TF33) turbofans and wide-span tailplanes for the USAF's tanker conversion programme. Elderly JT3C (J57)-powered KC-135As emerged from this as KC-135Es for many years' more service with the Air National Guard and Air Force Reserve.

Without doubt the most spectacular end to befall any 720, and indeed any airliner, occurred on 1 December 1984, when the sole Series 061, N113 of the FAA, was deliberately crashed in the interests of airliner safety development. Under remote control, and tracked by NASA's most sophisticated telemetry equipment, the 720 completed one circuit before landing on a specially-constructed gravel strip at Edwards AFB, California. Blades were set alongside the strip to slice in to the aircraft's wings and rupture the fuel tanks, which were filled with an experimental anti-misting fuel, which supposedly would not ignite. Passengers and crew were simulated by 75 dummies, all highly instrumented to record the effects of the impact. As the aircraft approached its landing, it drifted off course, and had to be swung round to compensate. Consequently, it landed short, but hit the blades nonetheless. As the tanks ruptured, the aircraft fireballed and slewed along the runway in a fiery mass of flames, the slithering inferno filmed from just about every angle. Eleven seconds after impact it came to rest and, remarkably, the flames extinguished themselves, illustrating the potential of the new fuel.

Military survivor

Among the handful of survivors into the 1990s, one was the only 720 believed to have entered full military service. Purchased in November 1971, this ex-Northwest 720-051B flew with the VIP Squadron of the Republic of China Air Force at Sungshan Air Base Taipei. Fitted with a luxurious interior, this was the principal transport of the Taiwanese president. Another notable survivor was an ex-American 720-023B, which after service with MEA was bought in 1985 to serve as an engine test bed with Pratt & Whitney, wearing the Canadian registration C-FETB, the last three letters in recognition of its new role.

Often overlooked in histories of commercial aviation, the Model 720 was an important stepping-stone to the establishment of Boeing's position in the marketplace. Not only did it bring jet travel to the shorter sectors for many influential airlines, it also played its part in killing off Convair's attempt to become the third horse in the jetliner race. As with all good aircraft, it also kept on going long after its design role and market had disappeared.

One of the most remarkable events in aviation history occurred on 1 December 1984, when the ex-FAA Boeing 720 was remotely-controlled to a deliberate crash at Edwards AFB during airliner safety trials. Carried onboard were 75 instrumented dummies to simulate passengers, and supposedly anti-misting fuel, which should not have ignited in the impact!

BOEING 727

What makes a winner in the cut-and-thrust of commercial aircraft design? The answer is much hard work, adequate research funding and a keen eye on the demands of a difficult market. Boeing gambled much on its neat Model 727, but the sales proved once again that the company had found the perfect answer to the airlines' needs.

Well before the introduction of its successful Model 707 and Model 720 series the Seattle-based Boeing Corporation was engaged in the preliminary design of a short- to medium-range and medium-capacity air transport to suit the growing needs of the US domestic and inter-city markets. The intention was to produce a high-performance utility jet transport with low approach speeds and short field properties combined with independence in ground support services, to crack the market below those occupied by the high-capacity intercontinental Douglas DC-8, Convair 880 and Boeing 707/720 series, and above the feederliner spheres held by the Sud-Aviation Caravelle and the Douglas DC-9. The market for such an aircraft was certainly there, but as always the go-ahead for production was dependent on the interest, and more vitally, the orders from any interested party and, in addition to the needs of the US domestic market, there was an opening for high-performance medium-range air transports in the export environment. And here, noted the Boeing concern, the new de Havilland D.H.121 Trident with its unique tri-jet configuration promised to be a keen and determined contender. By the late 1950s the growth in the commercial aviation market was on the upsurge; the cry was for extra seating capacity on aircraft plying the high-density routes in the United States and Europe. The old piston-engined Douglas DC-3s, Douglas DC-6s and Lockheed L-749s, and the turboprop Lockheed L-188 Electras (although relatively new), were cheap to operate but too slow, while the new Caravelles had not the required capacity.

The design of the Boeing Model 727, as the new type was named, started in February 1956, and the parameters laid before Boeing's preliminary design group were exacting. Whereas other companies, notably de Havilland (later merged with the Hawker Siddeley group), were vying for high cruise Mach numbers to reduce seat-air mile costs, Boeing wanted this factor on its Model 727, but combined with field operating characteristics that, in such a large aircraft, at first seemed impossible to attain. Put basically, this meant a high power/weight ratio for sprightly acceleration and take off and, of course, a highly adaptable and efficient wing. The wing design of the Model 727 broached new horizons, and its unique system of lift-augmentation flaps and slats allied with spoilers was to become the format in the mighty Model 747. By the time that design work had been finalised, on 18 September 1959, the wing structure had developed into a low-wing format of 3° dihedral, a thickness/chord-ratio of between 8 and 9 per cent, with 2° incidence and special Boeing aerofoil sections: sweepback at quarter-chord was 32° and was less than that of the Model 707. But if the basic wing was of routine design, then the high-lift and lift-dump devices most certainly were not: on the trailing edge of the wing massive triple-slotted flaps, totalling 36.04 m² (388 sq ft) with 40° setting, were combined with four leading-edge slats on the outer two-thirds of the wing, and three Krueger leading-edge flaps on the inner one-third portion. These were joined by seven spoilers on each upper wing surface (0-40°) which doubled as airbrakes and/or roll augmentation spoilers. This beautiful wing was the key behind what was to be the phenomenal success story of the Boeing 727: it gave the aircraft immense utility. Clean, the aircraft was as fast and efficient as the best; dump the flaps and the Kruegers and even at maximum landing weight the pilot could land on any small municipal or rural field in the United States.

Taking shape

Allied with the excellent field performance, the new Model 727 had lively performance and good fuel economy with the adoption of the 6350-kg (14,000-lb) thrust Pratt & Whitney JT8D-1 turbofan in August 1960. The fuselage upper portion was identical to that of the Model 707/720 series: this saved some $3 millions in jig and tool costing, standardised the flight deck layouts and gave the aircraft intercontinental six-abreast cabin accommodation for the passengers. In addition much stress was laid upon independence of operation: the Model 727 needed nothing on the ground if a stop-go transit were required, having a Garrett-AiResearch GTC85 auxiliary power unit for electrics, pneumatic starting and cabin conditioning,

This Boeing 727-256, construction no. 20595, first flew on 23 October 1972 and was delivered to Iberia on 11 May the following year. Named Vascongadas, it then operated with a further 34 of the same type on Iberia's European and North African Services.

Boeing 727-277 of Ansett Airlines, Australia in the livery adopted in 1981. It was first flown as N8278V and was the 1753rd example of 1832 built. It entered service on 20 June 1981. The airline's considerable fleet of Model 727 aircraft had declined to just two by mid-1988.

Air Charter of France operated seven Boeing 727-200 aircraft powered by three Pratt & Whitney JT8D-15 turbofans. The airline also leased additional aircraft from Air France, Air Inter and EAS when required.

Supplementing the Lockheed C-130 in the long-range transport role, the Royal New Zealand Air Force operated three ex-United States Boeing 727-100Cs, one of which was used for spares.

an airstair on Door 1 Left and a ventral staircase to the rear. With a very high maximum landing weight, as a result of wing stressing and landing gear strength, the Model 727 could take on fuel at the originating station, fly several transits, and gain quick turn-arounds and the on-schedule departure that is so vitally important. All these facets were built in during the course of very thorough research and design, entailing 150 studies of which 68 underwent some 1,500 hours of tunnel-testing. Construction go-ahead was given in August 1960, with Boeing acting on the good faith of Eastern Air Lines and United Airlines. Actually, it was not until 5 December 1960 that these operators placed their orders: 20 for United with another 20 on option, and 40 for Eastern. By February 1963 four Model 727s were on the lines at Renton.

At 11.33 (local) on 9 February 1963 the first flight was made from Renton when Lew Wallick, Boeing's senior experimental test pilot, lifted N7001U off after a run of 914 m (3,000 ft); the co-pilot was Dick Loesch and the engineer M. K. Shulenberger. N7001U weighed in at 58968 kg (130,000 lb), carried 20200 litres (5,500 US gal) of fuel and 7258 kg (16,000 lb) of test equipment, and flew for 2 hours 1 minute before Wallick put her down in 610 m (2,000 ft) on Paine Field's limited concrete. To the assembled press, Wallick said that 'She behaved as expected, even better than expected in many respects.' And indeed few problems cropped up during subsequent testing. The second Model 727 (N72700) flew on 12 March, and by the end of the month four Model 727s were undergoing thorough flight trials from Paine, Seattle, Edwards AFB, Denver and Albuquerque. By mid-May N7001U had completed 430 hours on flutter and structural damping tests up to Mach 0.9; 320 hours had been completed by N72700 on systems and braking; 180 hours on Model 727 no. 3, including handling high-g pull-ups, side-slipping and even barrel-rolls; while 313 hours had been completed in furnishing and air-conditioning on Model 727 no. 4. The order book was filling: 25 to American Air Lines, 40 to United, 10 to TWA, 12 to Lufthansa, and four to the Australian TAA and Ansett-ANA.

Into service

FAA certification for the production Model 727-100 was signed on 20 December 1963. Analyses of performance showed that the Model 727-100's parameters were 10 per cent in excess of Boeing's original guarantees: it was faster with better specific fuel consumption offered by the JT8D-1s, slower on the approach, and used reduced field lengths, being quite capable of operations from 1525-m (5,000-ft) runways at maximum weights. In the 1,100 hours of tests since 9 February 1963, dives in excess of Mach 0.95, take-offs at 72576 kg (160,000 lb), take-offs on two engines, and maximum-energy stops in less than 274 m (900 ft) had been made. Significantly, it was at this juncture that the Model 727's arch rival was being readied; on 19 December 1963 the Trident 1 (G-ARPF) completed crew training to be scheduled for its first service with BEA in April.

Already a battle royal was in progress between Boeing with its Model 727 and Hawker Siddeley with the Trident 1C and 1E for the lucrative foreign export markets. In November 1963 Boeing 727 N7003U completed its world tour started on 17 September of that year: out-bound from Montreal it plied the route to Tokyo via the Azores, Rome, Beirut, Karachi (where an uneasy meeting with Trident 1 G-ARPE took place), Calcutta, Bangkok and Manila. Thence the route went via Manila to Australia, followed by visits to Johannesburg and Nairobi, and back via Europe to the States. At home Eastern Air Lines signed up its first Model 727-100 on 22 October 1963 with the intention of starting scheduled services between Miami and New York (La Guardia) in January. The airline flew its first service with the Model 727-100 on 1 February 1964 from Miami to Philadelphia with a Washington (National) transit. Five days later United Air Lines started its schedules: daily Denver/San Francisco shuttles were started, with the line commencing operations on the high-intensity New York-Los Angeles-San Francisco-Seattle routes. In the meantime the salesmen were busy.

Lufthansa's first Model 727-100 (D-ABIB) was rolled out in January 1964, the first of 12 for crew training and the start of the Europa-Jet schedules intended for a start in April (first operation on 16 April 1964); on 10 May the West German operator flew its first Frankfurt-Heathrow service, and by July 1964 six of its Model 727-100s were serving the European and Middle Eastern network.

The announcement of the decision taken by Japan Air Lines and All Nippon to choose the Model 727-100 instead of the Trident came as a bitter blow to Hawker Siddeley. Both airlines looked upon the Trident 1C and 1E with favour, but wanted the stretched Trident 1F.

Boeing Advanced 727-200 cutaway drawing key

1. Radome
2. Radar dish
3. Radar scanner mounting
4. Pressure bulkhead
5. Windscreen panels
6. Instrument panel shroud
7. Back of instrument panel
8. Rudder pedals
9. Radar transmitter and receiver
10. Pitot tube
11. Cockpit floor control ducting
12. Control column
13. Pilot's seat
14. Cockpit eyebrow windows
15. Co-pilot's seat
16. Engineer's control panel
17. Flight engineer's seat
18. Cockpit door
19. Observer's seat
20. Nosewheel bay
21. Nosewheel doors
22. Twin nosewheels
23. Retractable airstairs (optional)
24. Handrail
25. Escape chute pack
26. Front entry door
27. Front toilet
28. Galley
29. Starboard galley service door

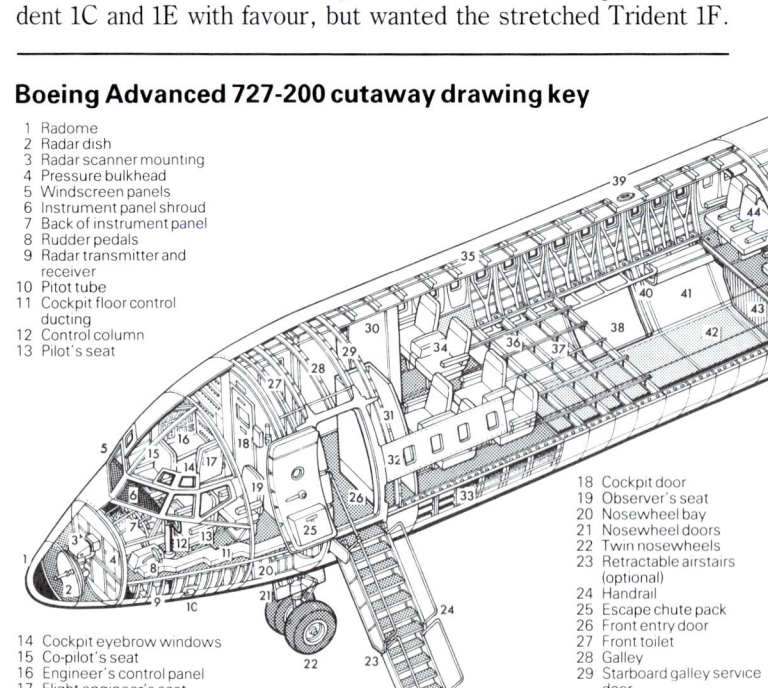

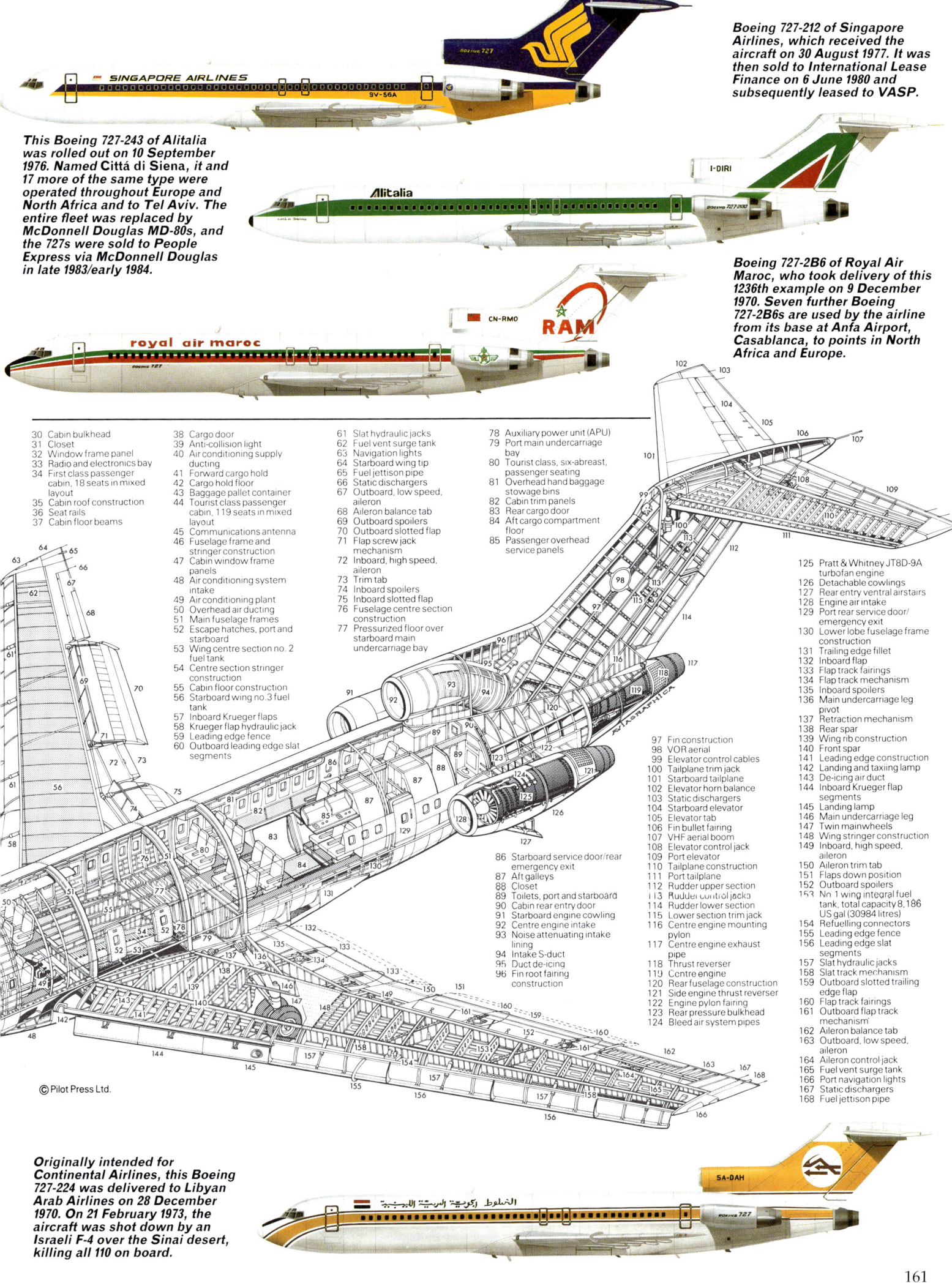

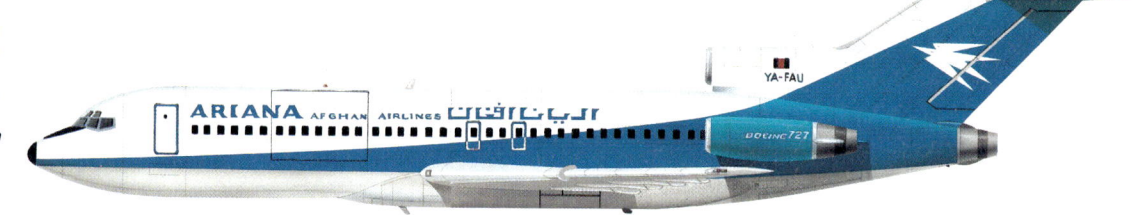

A Boeing 727-200 of Alaska Airlines, which used them largely for services within Alaska and to Seattle. They had 21 such aircraft, plus four of the shorter-fuselage Model 727-100s.

Boeing 727-113C of Ariana Afghan Airlines. The aircraft first flew on 30 December 1969 and was delivered to Ariana on 15 January 1970, enabling the airline to commence jet services to London and Moscow. Currently Ariana (now Bakhtar) operates two Boeing 727s.

But the latter was not due for certification until the spring of 1966, and JAL and All Nippon's intention was to start internal services by April of that year: the Model 727-100 was for them the obvious choice with deliveries promised for October 1965. All Nippon was swung by the decision of its big brother, and no doubt by the poor showing of the Trident from Osaka's 1890-m (6,200-ft) runway in hot weather: in July and August when seasonal temperatures approached 40°C the Trident's poor field operating characteristics would have precluded break-even loads. On 15 May 1964 JAL signed a purchase and loan agreement for six Model 727-100s, the deal being worth $37.5 millions with spares.

A diversity of options

Despite the sales success of the Model 727, the deliveries and options of some 200 were still about 100 short of the break-even figure. On 22 July 1964 Boeing announced its promotion of the Model 727-100C convertible cargo/passenger model: brochures gave operation from a 1525-m (5,000-ft) runway with carriage of 13608-kg (30,000-lb) payloads over a distance of 3058 km (1,900 miles), or the carriage of eight pallets (16670 kg/36,750 lb) over 2414 km (1,500 miles). The St Paul-Minneapolis based Northwest Orient signed for three as the first customer. Identical to the Model 727-100 except for heavier flooring and floor beams and the cargo-door of the Model 707-320C, the Model 727-100C gave operators the option of flying passengers by day and freight by night, thus enhancing utilization; galleys and seats were quickly removable, hatracks could be stowed, and the aircraft could be changed to passenger/cargo or all-cargo configuration within two hours.

By April 1967 the Model 727 was the most widely used commercial jet airliner in service: in that month SABENA took delivery of the 400th aircraft while a total of 586 was on order, in comparison with 564 Model 707s. In June 1967 Pan American placed an order for the milestone 600th, and the aircraft was on option or in service with 32 carriers. On 27 July 1967 Wallick took the first Model 727-200 (N7270C) off Renton's runway and, after a flight of 2 hours 10 minutes, landed at Paine Field for FAA inspection. Certification followed on 30 November 1967 after 457 hours of test flying. Announced on 5 August 1965, the Model 727-200 was the stretched version offering 163 seats, up to a maximum of 189: the fuselage was lengthened by 3.05 m (10 ft) both forward and aft of the main landing gear wheel-well, with localised structural strengthening. Three JT8D-9s powered the Model 727-200, with options for 6804-kg (15,000-lb) thrust JT8D-11s, or 7031-kg (15,500-lb) thrust Pratt & Whitney JT8D-15s. In the meantime Boeing had given the market the Model 727-100QC with palletised passenger seats and galleys, and advanced cargo-loading techniques; ramp weight was increased to 77,112 kg (170,000 lb). Luxurious furnishings and advanced communications were also available on the Model 727-100 Business Jet, the first being ordered by International Telephone and Telegraph on 13 November 1970. The Advanced Boeing 727-200 was announced on 12 May 1971, with a ramp weight of 86638 kg (209,000 lb), and deliveries started in June of the next year. With increased fuel capacity and the JT8D-15s, the aircraft offered a range 1287 km (800 miles) greater than that of earlier models.

By the late 1970s, Boeing was looking at a new design to continue the sales success of the 727, and this emerged as the Model 757. With deliveries of the new aircraft firmly established, the Model 727 line finally closed. The last of 1,832 aircraft, a 727-200F freighter for Federal Express, was rolled out at Renton on 14 August 1984. Another remarkable success story for Boeing had ended, with the 727 becoming the most popular airliner ever. Since then it has lost this position to the 737, but nothing can be taken away from the marvellous achievements of the 727 design and marketing team.

Thirty-five Boeing 727-256 aircraft provided the mainstay of Iberia's short/medium-range operations. Each aircraft could carry up to 189 passengers over a range of 2965 km (2,465 miles).

BOEING 737

A true 'bread-and-butter' aircraft of the airline world, the Boeing 737 had a rough start to its career, but has quietly and efficiently outsold every other airliner in the world, and has always been the short-medium hauler by which all others are measured.

Boeing started serious studies of a new twin-jet 100-seat airliner in 1964. The new Model 727 was already proving very successful, but rivals Douglas and BAC had both launched smaller jets around the same time, and were enjoying good sales. Douglas was already talking about a new version of its DC-9 which would be almost as big as the 727 and would certainly take some sales away from Boeing. Of the four biggest US airlines, Delta had already bought DC-9s, and American had ordered One-Elevens, leaving Eastern and United in the market. The European national carriers were interested as well. The most important of these was Lufthansa, with a strong domestic system as well as its inter-European routes. Lufthansa had also been the first European customer for the Model 727.

Boeing weighed up the pros and cons of going into this market very seriously. Douglas and BAC were already there, but most airlines wanted a slightly bigger aircraft than the current versions and Boeing's rivals would have to do some redesign work as well. Moreover, Boeing's engineers had finished work on the Model 727, and projects for a supersonic transport and a huge military freighter were still awaiting final go-ahead decisions. Also a lot of Model 727 experience could 'read across' to the design of the twin. The picture was promising enough for the start of a design study, in May 1964, on what was named the Model 737.

By mid-summer, the Model 737 design looked quite unlike the One-Eleven and DC-9. Its engines were under the wings, instead of on the rear fuselage, and it had a wider cabin, seating six passengers abreast instead of five. Boeing felt that its design had several advantages. Putting the weight of the engines under the wings reduced the bending load on the structure and made it possible to build a lighter wing. The shorter body, without the weight of the engines on the tail, was also lighter. All in all, the 737 layout saved more than half a ton in empty weight, or the equivalent of six passengers and their bags. It also meant that many components in the fuselage and nose could be similar to the same pieces of the 727.

There were other innovations. The engines were fitted snugly against the wing, instead of being carried on pylons. If the nacelle was long and slim enough, Boeing had discovered, this arrangement would not cause too much drag at high subsonic speed – it had been used on Boeing's first jet, the B-47. Eliminating the pylon saved weight and also meant that the landing gear could be shorter and lighter. In turn, this meant that the Model 737 could have a low, convenient floor level like the DC-9. An unusual feature of the landing gear was that there were no doors over the mainwheels, which were designed so that they sealed the wells.

Boeing crisis

The Model 737 proved attractive to the airlines, but so too did the new DC-9-30, which would be available a year earlier. The second half of 1964 saw a closely fought battle for the crucial launching customers. Boeing wanted to sign up either United or Eastern before committing itself to the programme but, in February 1965, Eastern ordered DC-9-30s. Lufthansa immediately told Boeing that it would order DC-9s unless Boeing made a quick decision, so the US company launched the programme with a 21-aircraft order from the West German airline. This was an unprecedented risk for any US jetliner builder. But they need not have worried. Two months later, United signed for no fewer than 40 Model 737s. Its aircraft were heavier, with a 193-cm (76-in) longer fuselage to seat 12 more passengers. This was the Model 737-200, the Lufthansa model being the Model 737-100.

Resplendent in Aloha Airlines 'Funbird' scheme, this is one of the 17 737-200s that serve the carrier. The aircraft are used on services around the Hawaiian islands, the airline's base being at Honolulu.

The Boeing Surveiller aircraft is a maritime reconnaissance version, equipped with side-looking airborne multi-mission radar (SLAMMR) equipment in two fairings above the rear fuselage. This one serves the Indonesian air force.

Incorporating a gravel-field kit to allow operations from low-standard airfields is this Air Zaïre Boeing 737-298C Advanced. This modification has been adopted by several African nations, consisting of vortex dissipators beneath the engine and deflector plates on the undercarriage to avoid stone damage.

However, initially the Model 737 did not do as well as Boeing had hoped. The US airline pilots' union ruled that it had to be flown by a crew of three, although the DC-9 had been accepted as a two-man aircraft, and this made the Model 737 more costly to operate for the big US carriers. This ruling was later rescinded, but too late for the Model 737; Western was the only other customer among the big US airlines. Douglas was also successful in Europe, beating Boeing to orders from KLM, SAS, Swissair and Alitalia. Sales were not the only problem. The first Model 737 flew in April 1967, just nine months before the deadline for first deliveries, and it quickly became apparent that the design had much more drag than the designers had predicted, and that the thrust reversers were not working. To cut drag, Boeing designed new nacelle/wing fairings (after a long 'try-and-fly' process) and a row of blade-like vortex generators was added to the rear fuselage. The problem with the thrust reversers was that they were too near the wings, and the tailpipes of the engines were extended by 102 cm (40 in) to cure it.

The 'fixes' were implemented before delivery, but the problems hurt the new jet's image. The Model 737-100 never proved very attractive, and apart from Lufthansa's aircraft only a few more were built. The bigger and more economical Model 737-200 became the standard version. Options offered from the start included a convertible passenger-freight version, with a side freight door ahead of the wing. This variant could be fitted with a movable bulkhead for mixed passenger/freight flights, and there was also a quick-change QC version with palletised passenger seats.

Narrow-bodied jets

The Model 737 went into production alongside the Model 727. By mid-1969, Boeing had sold more than 240 of the type, but McDonnell Douglas had sold twice as many DC-9s. What changed the picture, paradoxically, was a crisis at Boeing. In 1969, the company's orders nosedived just as the first of the new Model 747s was to be delivered. Boeing's response was to cut back on Model 747 production, and to throw its weight behind its narrow-body jets in the hope of expanding sales. In the case of the 737, Boeing decided to go for the markets outside Europe and North America, where few airlines had yet selected a twin-jet transport. A new version, the Advanced Model 737, was designed to meet the needs of such operators. It was to be more flexible than the original version, carrying its full payload over greater distances and able to use smaller airfields. It was heavier, was available with more powerful engines, and had more fuel tankage. The flaps were redesigned to reduce the landing speed, and a new anti-skid system was fitted to cut the landing roll. Boeing also offered an optional gravel-runway kit, consisting of a large nose skid and air-jets to keep stones out of the engines. A 'wide-body look' interior, with overhead baggage bins and other refinements, was offered and, slightly later, Boeing standardised on 'hush-kitted' sound-absorbent engine nacelles. The first Advanced Model 737 was delivered to All Nippon Airways in May 1971. By December, Boeing's 737 salesmen had won six out of six head-on DC-9/Model 737 contests in that year. They included a US Air Force order for 19 aircraft, to be delivered as T-43A navigation trainers.

While sales slumped to a low of 13 aircraft in 1972, they recovered well. Even during the lean years of the mid-1970s, Boeing sold an average of 40 Model 737s a year, consistently outselling the DC-9. In the process, the Model 737 took the lead in three growing markets. As intended, the Advanced Model 737 proved attractive to 'third-world' operators, and the Model 737 became a familiar sight in the Middle and Far East, and in Africa. (At the same time, governments in many of the same countries bought Model 737s for their own VIP use.) In Europe, the holiday-charter business had matured to the point where airlines such as Britannia Airways could afford brand-new aircraft; the Advanced Model 737 proved ideal, with 130 seats and plenty of range. The third important set of new customers was the US 'regional' airlines such as Southwest of Dallas and Denver-based Frontier, which were likewise graduating into their first jet equipment.

Sales were steady but unspectacular until 1978, when the Model 737 market exploded. No fewer than 145 sales were logged, for a number of reasons. Boeing took two large 'one-off' orders, from British Airways and Lufthansa (which bought Model 737-200s to replace its old model 737-100s); sales were up throughout the industry; and the US government had abolished the rules which had stopped the small but efficient regional airlines from competing on many routes. The regionals bought Model 737s to hasten their expansion, and three of them (USAir, Southwest and Frontier) are now among Boeing's biggest Model 737 customers, apart from United and Lufthansa.

Best-selling airliner

In 1980, the Model 737 replaced the Model 727 as the world's fastest selling airliner. It is now selling better than ever, well over 100 orders per year between 1978 and 1989. The 1,000th order was booked in 1982. In June 1987 the 737 passed the record of 1,832 set by its stablemate the 727 to become the world's best-selling airliner. The 2,000 mark was passed in March 1988, the 2,500 mark in May 1989, and the 3,000 mark was later passed. This is surely not too bad for an airliner launched with such desperate anxiety on the basis of a single small order by a foreign airline!

The basic Model 737 has been continuously developed. The more powerful JT8D15-17 engines were made available in the mid-1970s, and the maximum take-off weight was gradually raised. The Model 737 could then easily haul a full 130-passenger payload from Scandinavia to the Canaries, year-round. An improved version

The 737 has found favour with many smaller airlines in the Third World. TACA of El Salvador is typical of these smaller lines, operating ten 737s (and two 767s) on services in Central America.

The success of the 737 has been phenomenal, and has achieved renewed vigour with the launch of the -300, -400 and -500 series. TAP started 737-200 operations in 1983, and has since added the Series 300 to its fleet.

of the basic engine, the JT8D-15A, has been available since late 1982 and offers a 5 per cent improvement in fuel consumption. The flight-deck systems have been steadily refined, with the addition of a digital autothrottle, colour radar, a performance management system (continuously computing and displaying the most efficient engine and speed settings for the aircraft) and an optional head-up flight guidance display. Thanks to its basic simplicity, efficient control system and ample power, the Model 737 has won a reputation as one of the world's most pleasant airliners to fly.

A specialised new version of the 737 was launched in early 1981 with the Indonesian government order for three aircraft. Named Surveiller, it is designed to patrol the 320-km (200-mile) exclusive economic zones (EEZ) established by many nations. It is equipped with a Motorola SLAMMR (side-looking airborne multi-mission radar) in the rear fuselage, with its two 4.9-m (16-ft) antennae on each side of the fin. The radar equipment and its two-man control station occupy the rear of the cabin. Cruising at normal heights and speed, the Surveiller can provide a high-resolution radar picture over a very large area; Boeing says that it can cover a given area in less than half as many flying hours as a specialised EEZ patrol aircraft such as the Dassault-Breguet Falcon 20G. Another advantage is that the aircraft is available as a passenger or freight transport. Tests of the Model 737 with SLAMMR began in mid-1983.

While the DC-9-30 is now out of production, the Model 737's rival lives on in the shape of the highly developed MD-80 family, launched in 1978 and put into service in 1981. At that time, Boeing was fully occupied with the Model 767 and Model 757, but in early 1979 the company formed a small design group to study ways of developing a new airliner, smaller than the Model 757, from the Model 727 or Model 737. In little more than a year, Boeing had a design study for a stretched and re-engined Model 737-300, and the programme was launched with an order from USAir in March 1981.

Engine development

Creating the Model 737-300 took some ingenuity, because the basic goals of lower noise and better fuel economy meant high-bypass-ratio engines, and there was no room for these under the wing. General Electric and SNECMA, however, agreed to develop a new version of their CFM56 engine with a smaller fan, and with all the accessories mounted on the sides of the fan case rather than underneath. Boeing's designers slung the new CFM56-3s well out in front of the wing, lengthened the nosewheel leg to provide ground clearance, and stretched the rear fuselage a little more than the nose, to counterbalance the engines. The new engines are efficient,

The US domestic airlines have huge fleets, and USAir is no exception. Boeing 737s are the backbone of the fleet, available in -200 (illustrated) -300 and -400 versions. The airline is based at Pittsburgh, Washington and Charlotte. A new colour scheme was adopted in 1989.

but were heavier than the JT8Ds. The longer fuselage and greater payload also added to the landing weight, but Boeing was under customer pressure to make sure that the Model 737-300 would be able to land anywhere a Model 737-200 could go. The new version featured extended wingtips and a redesigned slat which cut 9.7 km/h (6 mph) off the approach speed, and an extra spoiler panel was also introduced to help braking. In the course of development, careful attention to weight-saving and the introduction of more composites and advanced aluminium alloys (as used on the newer Boeings) shaved some 600 kg (1,450 lb) off the original empty operating weight, though of course the Model 737-300 is a bigger and heavier aircraft than the Model 737-200.

Whereas the Dash-200 seats 120-130 in an all-passenger configuration, the -300 seats 128-149. This, combined with slightly reduced fuel burn, results in fuel consumption per seat-mile being reduced by 25-30 per cent, combined with a dramatic reduction in noise. Range depends on the customer's choice of gross weight; at 56472 kg (124,500 lb) it is similar to a Dash-200 but at 62822 kg (138,500 lb) it is considerably greater. From the recognition viewpoint the 737-300 can be identified by its totally different engines, of much greater diameter, with flattened undersides, hung on pylons entirely ahead of the wing, instead of having nozzles and reversers aft of the trailing edge. Less obvious is the fact that there are new digital avionic systems, including a full range digital autothrottle and a laser IRS (inertial reference system). The cockpit, however, is not of the modern 'glass' type but retains dial instruments.

In June 1986 Boeing was sufficiently encouraged by booming Dash-300 sales to offer a further stretch, the 737-400. This has a fuselage stretched by two plugs, one of 1.83 m (6 ft) ahead of the wings, and one of 1.22 m (4 ft) aft. This stretch, which requires a tail bumper to avoid scraping primary structure, puts up passenger accommodation to 146-170, this bringing it virtually up to 727-200 level! Underfloor hold volume is 38.93 m³ (1,735 cu ft). Engines can be CFM56-3B-2s, rated at 9980 kg (22,000 lb), or CFM56-3Cs rated at 23,500 lb. Outer wings and main gears are strengthened to match an MLW (maximum landing weight) of 56245 kg (124,000 lb), considerably higher than the MTOW of the first 737!

The 737-400 first flew on 19 February 1988. Like the -300 it proved a great success, but in January 1989 a new Dash-400 of British Midland suffered engine difficulties and crashed near East Midlands airport. In June 1989 the British CAA grounded 737-400s on the British register, and would require the -3C engines to be derated to the 9980 kg (22,000 lb) thrust level of previous versions. This seemed to remove the variant's problem, and no other Model 737-400 aircraft have been lost to the same problem.

On 20 May 1987 Boeing announced the 737-500, the last of the 'Classic' series. This traded accommodation for range, having a fuselage closely similar to that of the 737-200, seating 108–130.

Designated T-43A for USAF operation, the Boeing 737 design was selected as a navigation trainer, the cabin equipped with numerous navigator stations. Most fly with the 323rd Flying Training Wing at Mather AFB in California, although a few are used for transport purposes.

Boeing 737-200 cutaway drawing key

1. Hinged nose cone
2. Search radar
3. Glide-slope aerial
4. Forward pressure bulkhead
5. Instrument panel shroud
6. Windscreen sections
7. Sliding side windows
8. Eyebrow windows
9. First officer's seat
10. Overhead panel
11. Centre console
12. Captain's seat
13. Flight kit stowage
14. Circuit breaker panel
15. Nose gear deflector housing
16. Twin nosewheels
17. Nosewheel doors
18. Nose gear gravel deflector
19. Steering cylinders
20. Lock
21. Drag strut
22. Fixed side windows
23. Second observer's seat (optional)
24. First observer's seat (stowable)
25. Wall circuit breaker panel
26. Dome light
27. Flight deck door
28. Forward galley
29. Service door (starboard) 30 in by 65 in (76 by 165 cm)

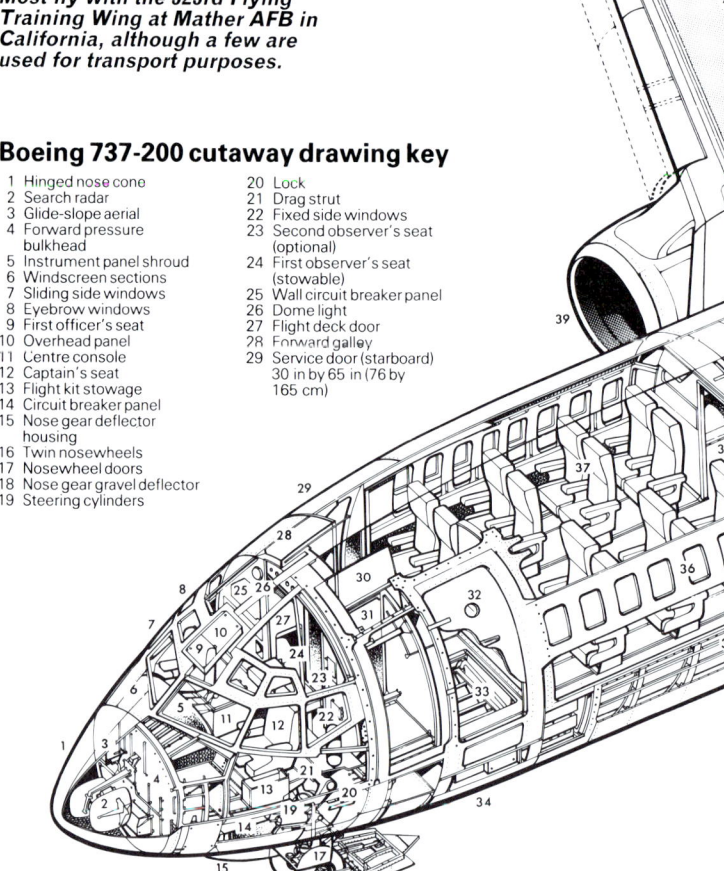

Otherwise it incorporates every new feature of the -300 and -400. Customers have a choice of CFM56-3B-1 ratings from 9390 kg (18,500 lb) (derated) to 9070 kg (20,000 lb), and various weights up to 60554 kg (133,500 lb). With the more powerful engine and maximum fuel, the Dash-500 can carry 108 passengers and baggage no less than 5552 km (3,450 miles). Here again Boeing had a winner.

Today the 737 is more than 30, and being built at a high production rate. Customers have a choice of three body lengths, and as most of the rest of the aircraft is common, they can place orders for TBD (to be determined) aircraft, leaving the exact model until later. So in June 1998 the sales totals for the 'Classic' 737 were: 1,144 (all delivered) for the -100 and -200, and 1,992 (1,895 delivered) for the remainder of the series, namely the -300, -400 and -500.

Boeing 737 series 300, 400 and 500

Boeing's 737 is now the best-selling airliner ever, having surpassed the same company's 727. What allowed the 737 to achieve this was a radical improvement to keep the design viable in today's market, centred around the adoption of high bypass ratio turbofans offering considerably reduced fuel burn compared to the earlier jets. A family of three differing fuselage lengths is now available, catering to the needs of any potential customer.

On 9 April 1967, the prototype of a new airliner flew for the first time, one which was to shatter sales records in every area. This was the Boeing Model 737, and before the end of the year aircraft were being delivered to airlines to begin an illustrious career that, over 20 years on, is still going from strength to strength. That the aircraft is as popular now as during its early years is entirely due to a major update programme to keep the basic airframe viable for service into the next century.

Boeing production began with the 737 Series 100, although only 29 were built for Lufthansa, Avianca and Malaysian before the Series 200 was introduced. This had a lengthened fuselage, seating up to 130 passengers in high-density configuration, matching far better the airlines' requirements than the shorter 737-100. With Pratt & Whitney JT8D turbofans, the 737-200 was a winner, proving efficient, reliable and (for its day) fuel efficient. Without any detriment to safety, it was a no-frills, bread-and-butter aircraft, tailored to serve in the legions of short/medium-haul aircraft that constitute the larger part of an airline's day-to-day operations. Above all it was competitively priced, for its development costs had been kept low by using many components and features of the best-selling 727, not least of which was the fuselage section and cockpit, themselves derived from the long-range 707.

Not surprisingly, the aircraft sold by the hundred, and by the time Series 200 production came to an end in the summer of 1988, 1,144

Nowhere has the new generation of Boeing 737s been more widely appreciated than in or by the inclusive tour charter sector. Offering unrivalled economy over sectors between northern Europe and the Mediterranean and North Africa, they have been chosen by most of the major carriers. Germania has had most of its order for 24 aircraft.

JT8D-powered aircraft had been built. Allowing the production line of this variant to close was the arrival in production of a new variant, tailored to meet the same market demands, but this time of the 1980s and 1990s. This was the Series 300, and it has since spawned a family of variants which cover the spectrum of short/medium-range, low/medium-density requirements.

By the late 1970s, it had become obvious to Boeing officials that fuel economy and noise pollution would be two of the biggest problems during the following years for commercial aviation, and to tackle them both would require the use of high-bypass ratio turbofans. These fans were already in service on the wide-body jets, and showing vastly reduced fuel-burn/power ratios compared with the old technology low-bypass ratio fans like the JT8D. However, at the time the smallest new technology engines were still developing over 178 kN (40,000 lb) thrust – far too much for an airliner in the 737 class.

The answer lay in the CFM International CFM56, the first of a new breed of small fans in the 89-kN (20,000-lb) thrust class. CFM International was in fact a joint company established by the engine giants General Electric of the United States and SNECMA of France. Formed in 1974, the company was established solely to develop and produce a small fan for commercial use, based on a core engine derived from the F101 engine developed for the B-1 bomber, although both companies had been investigating the possibilities of such an engine as far back as the late 1960s.

The first CFM56 ran at GE's Evendale plant on 20 June 1974, and first flew in a McDonnell Douglas YC-15 experimental transport. Most development work, however, was conducted in France using a modified Caravelle, although later a Boeing 707 was completely re-engined to validate inflight operations. Throughout tests, the CFM56 showed lower than expected fuel burn, and impressively low noise output.

The first use of the engine came in 1979, when the 106.8-kN (24,000-lb) thrust CFM56-2 was picked to re-engine DC-8-60s of United, Delta and Flying Tigers, so turning them into DC-8-70s. The KC-135R re-engined version of the USAF tanker also uses this version.

Development of this engine gave Boeing the confidence to begin work on the new 737, this beginning in early 1980. CFM was promising a reduced thrust engine (CFM56-3) with reduced compressor blade diameter, tailored to meet Boeing's needs. Design proceeded apace. The 737-300 retained about 70 per cent commonality with the Series 200, and used basically the same wing with only minor improvements such as an extended wingtip. Certainly the most radical feature was the lengthening of the fuselage by 2.64 m (8 ft 8 in) by the addition of two plugs forward and aft of the wing carry-through structure. This allowed the new model to take advantage of the increase in power to provide additional passenger and baggage capacity. To cater for longitudinal stability problems, a dorsal fin fillet was added.

Anticipating the successful testing of the engine, and with the promise of large orders, Boeing announced the production go-ahead in March 1981, the first metal being cut mid-way through the following year. The CFM56-3 first ran in March 1982 and in the following February flight tests began using a Boeing 707. The first engines were delivered to Boeing which installed them on the prototype 737-300.

This installation caused something of a problem. When the 737 had first been designed, its slim JT8D engines had been suspended directly from the wing structure, eliminating the heavy and draggy pylon and allowing the undercarriage to be considerably shorter (and consequently lighter). With the much fatter CFM56, there was insufficient ground clearance for the engines to be attached without modification. Boeing and CFM circumvented the problem with a simple expedient: by moving all the associated engine equipment from underneath the engine to the sides, the nacelle need only be as deep as the compressor blade. This neat solution gives the characteristic squashed shape to the 737-300's nacelle.

With the adoption by Boeing of the CFM56 for its new breeds of 737, mutual success was assured. The engine was later used to

Aer Lingus has been a staunch supporter of the Boeing 737 since its first Series 200 was delivered in 1969. It is the only airline currently operating all three of the 'Classic' aircraft. Its fleet once included 10 Series 500s, six Series 400s and two Series 200s (illustrated).

Boeing's short-haul family in flight: the prototype Series 500 leads a Series 300 (United) and a Series 400 (KLM) during a flight over the Rocky Mountains. The three variants offer seating from 100 to 150, in mixed-class configuration, and all can be flown on the same pilot rating, offering great versatility to airlines.

Boeing 737-300/400s formed the bulk of the Air Europe fleet, these being used for charter and limited schedule work. Seven of the Series 400s (shown) were used, laid out for charter work, with 172 seats in one class. Note the two escape hatches over the wing, compared with only one in the Series 300.

power the Airbus A320 (CFM56-5), although with this aircraft the engine met its first competition. This came from the IAE V2500, a similar fan developed by a rival international collaboration headed by Pratt & Whitney and Rolls-Royce. A long battle between the two engines has continued up to the present.

On 24 February 1984, the CFM56-3 took the first 737-300 aloft, and even then it was assured of a healthy sales book. A second aircraft joined the flight test programme on 2 March to speed the certification process, this being achieved on 14 November. By the 28th of the month the first delivery was being made to Southwest Airlines, an all-737 operator based at Dallas and Houston in Texas. It flew its first revenue-earning service on 7 December.

The Boeing sales team manoeuvred easily between selling the Series 200 and the new version, and orders came flooding in. By 30 June 1995 these covered 1,054 aircraft, of which 905 had been delivered. Many 737-200 operators, and not a few new ones, came to Boeing for the new fuel-efficient aircraft. Douglas could only offer updated versions of the JT8D-powered MD-80 until the new technology fan-powered MD-90 became available. Many of the US majors bought the 737-300 in huge quantities, while across the globe influential majors, charter carriers and small national airlines alike adopted the type for short/medium-haul work.

Although interiors are configured to suit individual airlines (with maximum seating 149 passengers), the 737-300 has so far been offered with only minor variations. Optional extra fuel tankage is available offering an exceptional range of 4554 km (2,830 miles) at its maximum take-off weight of 62822 kg (138,500 lb), carrying 141 passengers. A VIP/executive version is available, typically seating 30 in luxurious surroundings.

The two-man crew have the latest in avionics and systems to aid them, including a flight management computer system which ties in autopilot, navigation and performance functions. An inertial reference system has laser gyros in place of the gimbal type. INS and Omega navigation is optional, and new aircraft have a full electronic flight instrumentation system (EFIS). Another welcome option is Boeing's windshear detection and guidance system. Digital autothrottle is fitted as standard. With regulatory permission, the 737-300 can undertake extended-range operations over water or undeveloped land more than the statutory one-hour flying time to the nearest airport.

With the Series 300 proving to be a great success, Boeing looked at the aircraft for further development, and opted to produce a family of variants offering different range/load capabilities to meet any potential needs. The first variant to be developed was the 737-400, a lengthened version offering greater capacity but with negligible reduction in range. The first details of this type were announced in June 1986, the first aircraft was rolled out on 26 January 1988, and made its first flight on 19 February. A second aircraft followed on 25 March and a third in June.

As might be expected, the 737-400 is considerably longer than the Series 300, having a 1.83 m (6 ft 0 in) plug forward of the wing and 1.22 m (4 ft 0 in) plug aft. To maintain performance, the uprated CFM56-3B-2 (97.86 kN/22,000 lb thrust) or CFM56-3C (104.5 kN/23,500 lb thrust) engines are fitted. To cope with a maximum take-off weight raised to 68039 kg (150,000 lb), the undercarriage and outer wings have been beefed-up. Another factor of the lengthened fuselage is the risk of hitting the rear fuselage on the runway at

Pakistan International Airlines is one of many worldwide that have adopted the 737 fan family to provide economical and reliable domestic and short range international services. The airline has six on strength, configured for 16 first-class passengers and 96 in economy.

Aéromaritime was a charter subsidiary of UTA, once France's largest independent airline. The basis of the passenger charter fleet was the Boeing 737-300, of which six were on strength. The airline also operated Super Guppy outsize-freight transports on behalf of Airbus.

rotation, so a tail bumper is fitted as standard. Internally the 737-400 offers a typical mixed-class accommodation for 146 passengers, although high-density charter operators regularly carry 170. The FAA certification awarded on 2 September 1988 covers the aircraft for up to 188 passengers, more than most long-range 707s would have carried back in the 1960s!

Piedmont Airlines was the launch customer, and it received its first aircraft on 15 September 1988. Since that time the airline has been absorbed into the huge USAir concern, itself an early customer for the 737-400. Sales have been brisk, particularly among charter operators who have found the increased capacity ideal for their type of work. By June 1995 orders stood at 431, with 383 aircraft delivered. Doubts were cast over the new aircraft and its engines after a CFM56-3C-engined 737-400 of British Midland crashed while attempting an emergency landing at East Midlands Airport in England, but despite a brief grounding, the aircraft were soon plying their trade again.

Further development of the Series 400 has resulted in an increased weight version. With revised avionics, increased fuel capacity and local strengthening of undercarriage and wing structures, this version has a maximum ramp weight of 68265 kg (150,500 lb). Power comes from further uprated engines, in this case the CFM56-3C-1 of 111.2 kN (25,000 lb) thrust.

To complete the 'Classic' family of Model 737s, Boeing needed to produce an aircraft of the same capacity as the original Series 200. Originally known as the 737-100 during the early design studies, this was announced on 20 May 1987 as the 737-500. Combining the features of the 737-300 and -400, the -500 has a reduced overall length of 31.0 m (101 ft 9 in), comparing with that of the 737-200 of 30.53 m (100 ft 2 in). Internal accommodation is for 108-132 passengers.

Power for the new variant comes from the CFM56-3B-1 of 88.97 kN (20,000 lb) thrust, or the same engine derated to 82.29 kN (18,500 lb) thrust, depending on customer preference. Various maximum take-off weights are available, the heaviest being 60554 kg (133,500 lb). With auxiliary fuel tanks and at this weight, a 737-500 with 108 passengers has a still-air range of 5552 km (3,450 miles), making it easily the longest-legged of the family.

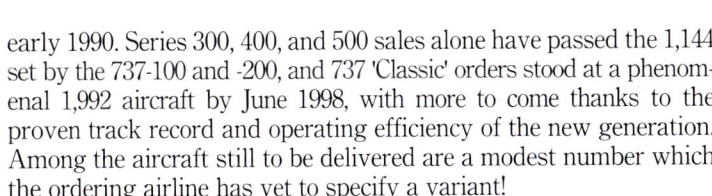

The first 737-500 flew on 30 June 1989 backed by launch orders from Southwest Airlines and Braathens SAFE (Norway), with first deliveries in March 1990. By June 1995, 288 machines had been delivered out of orders for 357 aircraft. With the Series 500 in production, the Series 200 production came to an end, and with it Boeings long and fruitful relationship with the Pratt & Whitney JT8D engine.

CFM International's new fan engine, and Boeing's redesign of the aircraft, have revitalised the Model 737. In June 1987 the Series 300 sales allowed the type to pass the 1,831 mark set by the Boeing 727 as the world's most-popular airliner, the 1,832nd Boeing 737 flying in

The lengthened fuselage, dorsal fin fillet and wide-diameter nacelles of the Series 300 are readily apparent on this Air Belgium 737-300. In the event, the airline took delivery of a Series 400 for its Mediterranean charter work.

early 1990. Series 300, 400, and 500 sales alone have passed the 1,144 set by the 737-100 and -200, and 737 'Classic' orders stood at a phenomenal 1,992 aircraft by June 1998, with more to come thanks to the proven track record and operating efficiency of the new generation. Among the aircraft still to be delivered are a modest number which the ordering airline has yet to specify a variant!

Lufthansa was one of the launch customers for the entire 737 family, buying Series 130s. It subsequently became a major operator of the Series 230 before it adopted the 737-330 for short/medium haul routes. The first was delivered in 1986, and they are to be augmented by Series 530s.

British Midland is a fast-growing scheduled carrier, largely flying domestic routes in the UK. Seven 737-300s (illustrated) are now in service, together with five 737-400 and 12 737-500 aircraft. It was a BMA 737-400 that crashed tragically at East Midlands Airport raising speculation concerning the CFM56 turbofans.

Boeing 737
'Wilhelm C.Schonta'
KLM

BOEING 737 'Next Generation'

With the initial –100 and –200 variants of its 737 twin-turbofan airliner replaced in production from 1983 by the three variants of the 737 'Classic' series, Boeing began to overtake the rival Douglas DC-9 as leader in the world market for short-haul jet transports, and 'Classic' orders eventually reached 1,992 aircraft by 1998. By then, however, Boeing had already embarked on the development of the next sequence of 737 variants, the 'Next Generation' aircraft optimized for longer-range but quieter operation.

There are four basic variants in the 'Next Generation' of the Boeing 737 family, and before any of these was formally launched, Boeing designated the 'Next Generation' as the 737X and from 1991 undertook an exhaustive assessment of its options to ensure that airlines received exactly what they needed and could afford. The company solicited in-depth assistance from more than 30 current and/or possible operators of the new types, and it was only in June 1993 that Boeing's board authorized marketing of the new types. The decision to place the new series in production was taken in November of the same year, the month in which Boeing received an order from Southwest Airlines for 63 examples of the 737-700. The order was of importance in itself because of its size, and also signalled the nature of the market enthusiasm for the types that would materialize over the following years. Thus, by June 1998, only slightly under five years after the 'New Generation' variants had been committed for production, Boeing's order book stood at 940 aircraft, of which 53 had been delivered.

All-round improvements

The key features of the 'Next Generation' aircraft are a new and larger wing, a combination of higher cruising speed and greater range, and an increasingly important combination, in these much more ecologically-minded times, of lower noise and reduced emissions. These features allow the aircraft to operate into smaller, urban, airports at times that would otherwise have been denied to them. The new wing of the 'Next Generation' aircraft is some 25 per cent larger than that of the preceding 'Classic' series as a result of an increase in span of some 4.88 m (16 ft 0 in) and in chord of some 0.46 m (1 ft 6 in), and this additional area also translates into a significant increase in the wing's volume, allowing the carriage of a 30 per cent greater fuel load in the integral tankage inside the torsion box that is the wing's main structural member. This increased tankage gives the aircraft of the 'Next Generation' series a transcontinental range capability even though it retains the ability to operate with the same runways, taxiways, ramps and gates and the smaller, lighter and shorter-ranged aircraft of the 'Classic' series.

Unaltered powerplant

In combination with the new wing, the aircraft of the 'Next Generation' series have an enlarged tail unit for maintenance of adequate stability and control on the longitudinal (pitch) and directional (yaw) planes. The powerplant is based on the same CFM International CFM56 turbofan as used on the later variants of the 'Classic' series, but in this instance the CFM56-7 (originally CFM56-3XS) unit that combines the core of the CFM56-5 with the improved low-pressure compressor of the CFM56-3 and a fan with a diameter of 1.55 m (4 ft 11 in). The CFM56-7 is derated from its nominal thrust for improved engine reliability, lower fuel consumption and reduced emissions, and its noise level on the ground is trimmed by the use of a new diffuser duct. Further decreases in ground-level noise are derived from the introduction of a cooling vent silencer on the APU installation, a new fan and duct for the ECS (Environmental Control System), and a new cooling fan for the electrical and electronics systems.

Flight deck of EFIS type

The 737 has an all-new flight deck for a crew of two, incorporating the same type of Honeywell CDS (Common Display System) as used on the altogether larger 777 airliner, with an EFIS (Electronic Flight Instrumentation System) based on six liquid-crystal displays. It is worth noting, however, that these displays are programmable, and as a result the 737 aircraft of the 'Next Generation' variants can be used to emulate the electro-mechanical displays of the 737 'Classic' aircraft as well as the EFIS of the 757 and 767, and the PFD-ND (Primary Flight Display – Navigation Display) of the 747-400 and 777.

Revised designation system

Whereas the three variants of the 'Classic' series were numbered in order of appearance, the four variants of the 'New Generation' series are numbered in ascending order of size. This means that the 'New Generation' aircraft do not correspond completely with those of the 'Classic' series, and the four variants are thus the 737-600 (originally known as the 737-500X) that supersedes the 737-500, the 737-700 that replaces the 737-300, the 737-800 (originally 737-400X Stretch) that is successor to the 737-400, and the 737-900 that is directly equivalent to no variant of the 'Classic' series but which, as a stretched development of the 737-800, gives Boeing a type to compete directly with the A321 from the rival Airbus consortium that is now Boeing's main competitor in the world airliner market.

Baby of the family

Smallest of the 'Next Generation' types, the 737-600 had by October 1998 secured orders for 62 aircraft, which first flew in January 1998 and of which the first was delivered to SAS in September of the same year against a total order for 55 aircraft. The other current 737-600 customers are Air Algerie and Air Tunis with commitments for three and four aircraft respectively. The 737-600 has an overall length of 31.24 m (102 ft 6 in) for the carriage of up to 108 passengers in a two-class arrangement, and is powered by two CFM56-7 turbofans each rated at 9979 kg (22,000 lb) thrust for a maximum take-off weight of 63504 kg (140,000 lb).

First into production

The 737-700, which was the first variant to enter production, had by October 1998 secured orders for 273 aircraft, of which 66 had been delivered. The first 737-700 flew in February 1997 and deliveries to the lead customer, Southwest Airlines, began in December of the same year. Orders for the 737-700 have currently been placed by one African airline, 10 Asian, Australasian and Middle Eastern operators, nine European airlines, and six American operators. The 737-700 has an overall length of 33.63 m (110 ft 4 in) for the carriage of up to 128 passengers in a two-class arrangement, and is powered

With four different fuselage lengths, offering a useful spread of passenger capacities in different seating arrangements, the 737 'Next Generation' series offers airlines the possibility of selecting the size of small-capacity short/medium-range airliner that exactly matches its route-network profile.

by two CFM56-7 turbofans each rated at 10886 kg (24,000 lb) thrust for a maximum take-off weight of 67586 kg (149,000 lb).

Enter the 737-800

The 737-800 first flew in July 1997 and deliveries to Hapag-Lloyd, which has ordered 21 aircraft, began in April 1998. Total orders for the 737-800 by October 1998 were 420 aircraft from two African airlines, nine Asian, Australian and Middle Eastern operators, 16 European airlines and five American operators. The 737-800 has an overall length of 39.47 m (129 ft 6 in) for the carriage of between 164 passengers in a two-class arrangement and 189 passengers in a one-class arrangement, and is powered by two CFM56-7 turbofans each rated at 11884 kg (26,200 lb) thrust for a maximum take-off weight of 76432 kg (168,500 lb).

The 737-900 was launched as a result of a November 1997 order from Alaska Airlines for 10 aircraft, and the type had by October 1998 notched up orders for a total of 40 aircraft from one Asian, one European and two American airlines. The 737-900 has a fuselage some 2.63 m (8 ft 7 1/2in) longer than that of the 737-800 for an overall length of 42.1 m (138 ft 1 1/2in), increasing the maximum seating capacity by 15 to a maximum of 177 passengers in a two-class

arrangement, although it should be noted that requirements for emergency exits mean that the type's maximum capacity cannot exceed the 737-800's total of 189. The first 737-700 is due to fly in 2000 with deliveries beginning in the following year, and the type is powered by two CFM56-7 turbofans for a maximum take-off weight of 74236 kg (163,660 lb).

Certification and service

Certification of the 737-700 by the Federal Aviation Administration was delayed by some two months by the need to implement structural changes to the tailplane, and certification of the variant by the Joint Airworthiness Authorities followed in February 1998 as this pan-European organization evaluated the revised design of the emergency exits to satisfy its more rigorous evacuation requirements. The exits over the wing had to be redesigned to enable the planned 149/189-passenger exit limits to secure JAA certification in the 737-700 and 737-800 as, without the required modifications, the two variants would have been limited to 140 and 180 passengers respectively. The new exit design incorporates a fast-acting automatic hatch hinged along its upper edge, and this is now installed as standard to all 'Next Generation' variants.

Variation on a theme

Boeing has also developed a derivative of the 737-700 with a side cargo door to meet a US military requirement for a successor to the C-9 Nightingale transport version of the DC-9, and this could pave the way for a similarly equipped civil transport. A BBJ (Boeing Business Jet) dedicated corporate transport derivative of the 737 'New Generation' has also been developed as the combination of the 737-700's fuselage and the 737-800's strengthened wing and powerplant.

Above: The 737 'Next Generation' series is also notable for its wing, increased in area by some 25% by comparison with that of the preceding series, and its enlarged horizontal tail surface.

Left: Launched in September 1994 and originally known as the 737-400X Stretch, the 737-800 can carry up to 189 passengers over transcontinental ranges.

Specification
Boeing 737-700

Type: short/medium-range small/medium-capacity airliner

Powerplant: two CFM International CFM56-7B turbofans each rated at 118,84 kg (26,200lb) thrust

Dimensions: wing span 34.31 m (112 ft 7 in); length 33.63m (110 ft 4in); height 12.50 m (41 ft 0 in); wing area 124.95 m³ (1,345.0 sq ft)

Weights: empty 37,930 kg (83,620 lb); maximum take-off 67,586 kg (149,000 lb)

Performance: maximum cruising speed Mach 0.78-0.8; maximum cruising altitude 12,495 m (41,000 ft); range 5,556 km (3,452 miles)

Overleaf: SAS is the national airline group of Denmark, Norway and Sweden, and is a major operator of Boeing airliners including the 737-600 for service on its extensive network of European routes.

Chapter 9

Boeing 747

●

Boeing 747-400 series

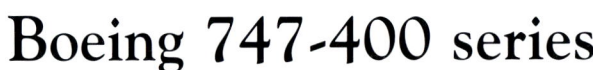

BOEING 747

While Douglas merely stretched the DC-8, Boeing went back to a clean sheet of paper to create a civil airliner much larger than any previously built. Immediately dubbed the 'Jumbo Jet', it revolutionised today's air transport world. Two great advances it introduced, the wide-body fuselage and high bypass ratio turbofans, meant it would move more traffic with less noise and at much lower cost.

The entire history of The Boeing Company (previously The Boeing Airplane Company) has been a succession of the boldest possible decisions to go ahead and build a new transport or bomber far more advanced than any in existence, in advance of definite orders and risking far more than the net worth of the company. Perhaps the biggest gamble of all came in 1966 when the company went ahead with the Boeing 747. Admittedly, that time it did have an order, from Pan American; but the risk on that occasion was probably the greatest any industrial company has ever accepted. In the event, the colossal gamble paid off, and Boeing changed the face of air transport for ever.

This single type of aircraft, for years the most powerful, the heaviest and most capable ever constructed (not forgetting the old airships and giant flying boats of the past), transformed air transport by more than doubling the load of passengers or cargo that could be carried, by more than quadrupling the available payload volume, and not least by introducing a wholly new form of propulsion giving jet speed but with much reduced fuel consumption and less than one-tenth as much noise.

New generation jetliners

Boeing Commercial Airplane Company, the subsidiary responsible for civil transport aircraft, began to study a new-generation jetliner in the early 1960s for long-range high-traffic routes. Rival Douglas was in the throes of eking out DC-8 production by a dramatic process of stretching in the Super Sixty series. Boeing eventually chose not to reply with a super-stretched 707 but to build a totally new and much larger aircraft with a wider fuselage; it was to usher in a new era of so-called wide-body transports. Such an aircraft was possible because, at the same time, General Electric and Pratt & Whitney were developing completely new large turbofan engines for the US Air Force CX-HLS competition for a giant military freighter (won by Lockheed with its C-5A).

With aircraft of this size, new configurations of fuselage became possible. Boeing engineers exhaustively studied double-bubble schemes with two pressurised tubes full of passengers, either side-by-side or superimposed, as well as a single giant tube with two decks inside. In the event the decision was taken to use a single tube but with only one enormous passenger floor, extending right to the nose. Underneath there was room for the capacious bays for electronics, electrics, air-conditioning and hydraulics, as well as cargo holds of unprecedented size. If necessary, galleys and other passenger-service items could be at the lower level. The flight deck was placed above the ceiling of the passenger deck, forming a slight blister above the nose and extended to the rear in an upper deck for typically 32 passengers. The main deck could seat up to 500 passengers in high-density 10-abreast seating (3+4+3), but 350 was judged a more likely number with an ultra-luxurious first-class section (seating passengers in pairs along the sides of the nose) extending right up to the radar filling the tip of the nose.

Boeing went for a high cruising speed with the Model 747, adopting an advanced design of wing with the exceptional sweep-back angle of 37½° at 25 per cent chord (one-quarter of the way back from the leading edge). The leading edge was given a typically ambitious high-lift system, with three sections of Krueger flap hinged down from the under surface of the wing inboard of the inner engines, five sections of novel variable-camber flap between the engines, and five more sections between the outer engines and the tip on each side. The variable-camber flaps resembled traditional slats but comprised flexible skins carried on pivoted links in such a way, that as they were hydraulically extended ahead of and below the leading edge they arched into a curve to give maximum control of airflow at high angles. On the trailing edge were placed enormous triple-slotted flaps, with each section running on steel tracks with prominent fairings projecting behind the trailing edge. Above the wing on each side were added six sections of aluminium honeycomb spoiler, four outboard for control in flight (augmenting the roll power of the conventional trailing-edge low-speed ailerons) and two ground spoilers inboard to destroy lift after landing and thus increase braking power. High-speed ailerons were added at the trailing edge behind the inboard engines where the flaps could not be used.

Unprecedented scale

Though the engine installation looked conventional, being arranged on four widely-separated wing pylons as in the model 707 of 15 years earlier, they were on a scale never before seen except on the military C-5A. In fact Boeing picked the loser in the C-5A engine competition, Pratt & Whitney, whose JT9D was offered as a robust

N7470 was the very first 'Jumbo', seen here on its 9 February 1969 maiden flight during which the landing gear was not retracted. Even at the time there was a healthy customer list, as shown by the 28 airline symbols painted on the fuselage. The aircraft was retained by Boeing for trials.

Stretching the upper deck of the 747 produced the 747-300 (formerly 747 SUD), adding more seats and also improving the aerodynamics to increase cruising speed slightly. Qantas has six of these for operations on its extensive global network.

and reliable turbofan of the new high-bypass-ratio type with a thrust of 18598 kg (41,000 lb). Extremely difficult engineering problems had to be solved in hanging the engines, in arranging a fan-duct reverse and hot-stream spoiler, and in reducing drag. Another tough engineering problem was the landing gear, eventually solved by using four main gear four-wheel bogies, two of them on tall inward-retracting legs pivoted to the wings and the other pair on forward-retracting units pivoted to the fuselage, the four retracted bogies lying together in a large bay amidships under the floor. All flight controls are hydraulically powered, the rudder and elevators being divided into equal-size halves, no tabs being used anywhere. The APU (auxiliary power unit) for ground air conditioning and electric power was placed in the extreme tail of the fuselage.

Not only did Boeing have to build the 747: the company also had to build a new factory to make it, and the new plant at Everett, swiftly created in a 780-acre clearing in a forest, is the largest building (in cubic capacity) in the world. Together with many other programmes Boeing's commitments were awesome, and employment in 1968 peaked at 105,000, compared with 60,000 at the peak in World War II. The risk on the model 747 easily topped one billion dollars, but thankfully orders kept rolling in, and when the first aircraft off the production line (Ship RA001) emerged from the new plant on 30 September 1968 a total of 158 orders had been gained from 26 airlines.

Engine snags

With so new and complex an aircraft it would have been surprising if there had been no snags, but in fact the difficulties, mainly centred on the engines, were prolonged. In crosswinds the engines were difficult to start and ran roughly, and distortion (so-called ovalisation) of the casings caused the blades to rub in a way that had not been apparent in more than two years of ground testing and flight de-

To keep the 747 design viable in the face of new technologies, Boeing have introduced the 747-400, with longer wings (and winglets), increased range performance, updated avionics and a two-man crew. Northwest was the launch customer for the type.

velopment using a B-52. Pratt & Whitney had to devise a Y-shaped frame to hang the engine differently, and eventually produced a new version of JT9D that avoided the problem. But the first flight was delayed until 9 February 1969. This aircraft was retained by Boeing for many development purposes. The first to be delivered was handed to Pan American on 12 December 1969, and that airline – after an unprecedented amount of engineering and training effort and investment in new ground facilities – finally got the Model 747 into service on the New York-London route on 22 January 1970.

Popularly called the 'Jumbo Jet', the Model 747 hit the headlines as well as the pockets of its customers and the world's airport authorities. For a while it appeared almost to be premature, because traffic did not grow as expected and load factor (proportion of seats filled) was often low. With great courage, Boeing continued production at maximum rate, and both the orders and the variants continued to grow. From the outset Boeing had organised a vast manufacturing programme with major sections of airframe made by subcontractors: Northrop made the main fuselage sections, for example, and Fairchild Republic built the flaps, ailerons, slats and spoilers. With such large structures, exceptional precision is needed if there are not to be problems when the parts all come together at Everett. One unusual technique used in assembly is to mount completed aircraft on air-cushion pads so that they can be easily moved in any direction across a smooth concrete floor.

The initial version was designated Model 747-100, and 167 of this model were sold. In the course of production, not only did Pratt & Whitney introduce a succession of improved variants of the JT9D engine, offering power increased in stages from 18598-kg (41,000-lb) to 24040-kg (53,000-lb) thrust, but other engines became available. General Electric offered the lighter and equally powerful CF6-50 series, chosen by several airlines, while Rolls-Royce's RB.211-524 series gradually came from behind and in later years won several important Model 747 order competitions as a result of its unequalled low fuel consumption – a matter of special importance to long-haul aircraft. The Rolls-Royce engine is distinctively shorter and neater than the installed US engines, and is claimed also to offer lower drag, though it is heavy.

On 11 October 1970 Boeing flew the first Model 747-200, with greater fuel capacity and gross weight increased from 334751 kg (738,000 lb) to an initial 351535 kg (775,000 lb), subsequently increased to 37,1945 kg (820,000 lb) and then 37,7840 kg (833,000 lb). Many tests have been made at weights up to more than 38,5560 kg (850,000 lb), almost 45,360 kg (1000,000 lb) heavier than any other aircraft. The basic passenger version is the Model 747-200B, while the Model 747-200F is a dedicated cargo version with no windows, a hinged nose for straight-in loading of pallets and containers (up to 29 standard 3.05-m/10-ft ISO containers, plus 30 lower-lobe containers for up to 112946 kg/249,000 lb of cargo) with a computerised, powered loading system which enables two loaders to load or unload the aircraft in less than 30 minutes. The Model 747-200B Combi is a mixed-traffic aircraft with a movable (or removable) bulkhead separating the passenger sections from any desired length of cargo area. The Model 747-200C Convertible can be converted from all-passenger to all-cargo use, and vice versa.

Further variants

In 1978 Boeing introduced the Model 747-100B as an updated replacement for the Model 747-100 with strengthened structure and more powerful and efficient engines. The Model 747SR (short-range) is a variant of the Model 747-100B with structure further reinforced for frequent take-offs and landings and higher performance because of its lighter weight from reduced tankage, despite a payload of up to 516 passengers.

From the Model 747-100B Boeing took another bold decision in September 1973 in going ahead with the Model 747SP (Special Performance). Though 90 per cent of the parts are unchanged, the other

Four 747-200s were bought by the US Air Force to serve as National Emergency Airborne Command Posts. Designated E-4B, these aircraft have an extensive communications suite and inflight-refuelling capability, enabling them to carry the President and his battle staff aloft in the event of a nuclear attack.

10 per cent represented very substantial engineering effort and investment, and the Model 747SP came out looking strikingly different from its predecessors. Intended for ultra-long routes, the 747SP has a fuselage 14.35 m (47 ft 1 in) shorter, which in turn demanded a vertical tail 1.52 m (5 ft) taller, with double-hinged rudders, and a tailplane increased in span by 3.05 m (10 ft). The flaps were changed to the single-slotted variable-pivot type, without the prominent fairings, and there are many smaller changes. Despite having gross weight lower than other versions, the 747SP has extremely long range, demonstrated by a delivery flight by South African Airways which, with 50 passengers and a heavy load of spare parts, flew nonstop 16560 km (10,290 miles) to Cape Town and arrived with 2½ hours of fuel remaining. Captain Walter H. Millikin of Pan American has flown round the world in a 747SP in 1 day 22 hours 50 minutes (average 809 km/h (503 mph) as well as round the globe in an unusual route which crossed both poles in flying San Francisco-London-Cape Town-Auckland-San Francisco.

Next came the Model 747SUD (Stretched Upper Deck) which is the opposite of the SP in providing increased passenger capacity. Since before the first flight Boeing had been studying double-deck versions, and eventually the market appeared right for a modest move in this direction. An order by Swissair came in the summer of 1980 for the 747SUD with the upper deck extended behind the flight deck a 7.01 m (23 ft). This increases the upper-deck seating from 32 to 69 in standard economy configuration (3+3). A new straight stairway is added at the rear (other 747s have a spiral staircase forward,

The world's most capable commercial cargo-carrier is the 747F variant, equipped with nose door and a large side door for easy loading of bulky cargo. Cargolux specialises in freight operations, flying four of these 747-200F aircraft as well as four 747-400Fs.

and if this is removed the 747SUD can seat a further seven passengers) and a new emergency exit is added, as well as extra windows. The SUD is now available on all models except the Model 747SP and is now known as the 747-300. The latest model is the 747-400, which marks a regeneration of the basic design. Incorporating state-of-the-art avionics, its new wings add a considerable amount to the range performance.

Military customers

Several 747s have been sold to military customers, a fleet which at one time numbered 11 having been supplied to the former Imperial (now Islamic) Iranian air force for air-tanker, electronic-warfare and other strategic roles. Saudi Arabia took delivery of a very special Rolls-engined Boeing 747SP with extremely comprehensive navigation and communications equipment fitted out mainly by E-Systems. The same Texas-based company teamed with another Boeing subsidiary, Boeing Aerospace Company, to provide the US Air Force with the E-4, among the most costly aircraft ever built.

Known as the AABNCP (Advanced Airborne National Command Post), the E-4 is basically a Boeing 747-200B carrying extremely comprehensive communications (command/control) and other equipment in order to serve as the HQ of the US National Military Command System and the operational HQ of SAC (Strategic Air Command) in time of crisis. The E-4s replace various types of EC-135. Originally two E-4A aircraft were delivered, in 1974-5. These had JT9D engines (later replaced by General Electric F103-100s, military versions of the CF6-50) and carried equipment taken mainly from the earlier Boeing EC-135s. In April 1975 the first definitive E-4B was delivered in testbed configuration, the fully equipped E-4B following in 1978. All four aircraft are now E-4Bs, with SHF (super-high frequency) aerials in a large dorsal blister, a 1,200-kVA

The enormous cost and huge capacity of the Boeing 747 give it few military applications, but the Imperial Iranian Air Force assembled a fleet of 16 before the Islamic revolution. Most can be inflight-refuelled, and have been used as tankers and command posts in addition to their regular transport role.

electrical system, advanced environmental controls, nuclear and thermal shielding, and LF/VLF (low or very low frequency) system with trailing wire 'several miles long', flight-refuelling receptacle and unrefuelled endurance of 73 hours. The E-4B interior is divided into many areas, the main deck having an NCA (National Command Authority) work area, conference room, briefing room, work area for a battle staff of more than 60.

Boeing 747 Variants

Boeing 747-100: original production version with JT9D-1 or -3 engines 334751-kg (738,000-lb) maximum weight

Boeing 747-100B: advanced derivative of model 747-100 with strengthened structure and weight up to 341555-kg (753,000-lb); 21297-kg (46,950-lb) JT9D-7 engines, or CF65, or RB.211-524

Boeing 747-200B: long-range model with weights up to 377840-kg (833,000-lb) and increased fuel capacity

Boeing 747-200B Combi: convertible passenger/cargo model with cargo side door aft of wing; all-passenger or up to 12 pallets/containers

Boeing 747-200C Convertible: equipped with basic freight floor and loading system for conversion from all-cargo to all-passenger, or a mix. Weight up to 377845-kg (833,000-lb); same powerplant options as Model 747-200B

Boeing 747-200F: dedicated cargo aircraft with upwards-hinged nose, normally windowless, capable of delivering 90702-kg (200,000-lb) of palletised (or containerised) cargo over a range of 7600km (4,720 miles)

Boeing 747-300: stretched upper deck for 37-44 extra passengers

Boeing 747-400: advanced model with extended wings, winglets, two-man cockpit, stretched upper deck and updated avionics. Range considerably increased over Series 300.

Boeing 747SR: high-capacity short-range aircraft with structural changes, gross weight 273520-kg (603,000-lb) or 333396-kg (735,000-lb)

Boeing 747SP: special-performance long-range model with short body, enlarged tail and weights 273520-kg (603,000-lb) to 318427-kg (702,000-lb); various engines, 289-440 passengers

Boeing 747-200 cutaway drawing key

1. Radome
2. Radar dish
3. Pressure bulkhead
4. Radar scanner mounting
5. First class cabin, typically 32 seats
6. Windscreen
7. Instrument panel shroud
8. Rudder pedals
9. Control column
10. Flight-deck floor construction
11. First-class bar unit
12. Window panel
13. Nose undercarriage bay
14. Nosewheel door
15. Steering mechanism
16. Twin nosewheels
17. Radio and electronics racks
18. Captain's seat
19. Co-pilot's seat
20. Flight engineer's panel
21. Observer's seats
22. Upper deck door, port and starboard
23. Circular staircase between decks
24. Cockpit air conditioning duct
25. First-class galley
26. First-class toilets
27. Plug-type forward cabin door, No. 1
28. First-class seats
29. Cabin dividing bulkhead
30. Anti-collision light
31. Cabin roof construction
32. Upper deck toilet
33. Upper deck seating, up to 32 passengers
34. Window panel
35. Air conditioning supply ducts
36. Forward fuselage construction
37. Baggage pallet containers
38. Forward under-floor freight compartment
39. Communications aerial
40. Upper deck galley
41. Meal trolley elevator
42. Lower deck forward galley
43. No. 2 passenger door, port and starboard
44. Air conditioning system intake
45. Wing-root fairing
46. Air conditioning plant
47. Wing spar bulkhead
48. Fresh water tanks
49. Forward economy-class cabin, typically 141 seats
50. Wing centre section fuel tank, capacity 17,000 US gal (64345 litres)
51. Centre section stringer construction
52. Cabin floor construction
53. Fuselage frame and stringer construction
54. Main fuselage frame
55. Air distribution duct
56. Air condition cross-feed ducts
57. Risers to distribution ducts
58. Machine main frame
59. Satellite navigation aerial
60. Starboard wing inboard fuel tank, capacity 12,300 US gal (36555 litres)
61. Fuel pumps
62. Engine bleed-air supply
63. Krueger flap operating jacks
64. Inboard Krueger flap
65. Starboard inner engine
66. Starboard inner engine pylon
67. Leading edge Krueger flap segments
68. Krueger flap drive mechanism
69. Krueger flap motors
70. Re-fuelling panel
71. Starboard wing outboard fuel tank, capacity 4,420 US gal (16730 litres)
72. Starboard outer engine
73. Starboard outer engine pylon
74. Outboard Krueger flap segments
75. Krueger flap drive mechanism

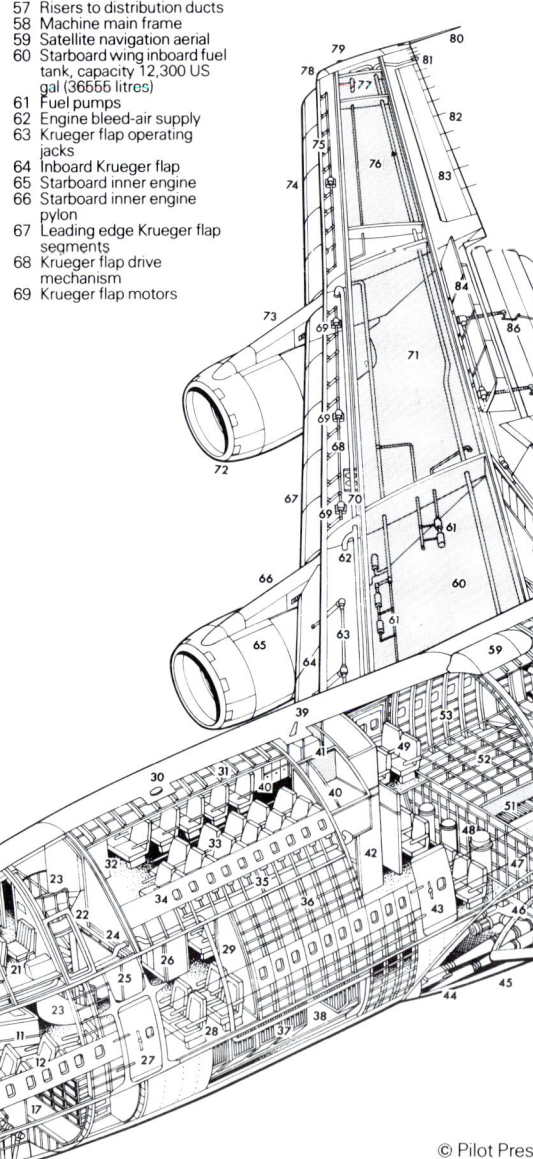

Produced for very long-range routes, the 747SP was introduced in 1975, featuring a shortened fuselage and enlarged vertical tail. Iran Air was one of the major customers for the type, and still flies four.

© Pilot Press Ltd

The 747SP was not one of Boeing's most successful ventures, with only small numbers being built. Only a handful of carriers needed the enormous range offered by the type, one of them being China Airlines of Taiwan. The airline still has four SPs, as does the communist mainland.

Next to KLM, Avianca of Colombia is the world's oldest established airline. It operated this 747-100 on services mainly to Europe, but has now replaced it with a 747-200. This aircraft had a lucky escape at Madrid in 1982 when tyres burst on take-off. Another Avianca 747 was not so lucky a year later when it attempted to land at the same airport, striking ground with the loss of all on board.

76 Extended range fuel tank, capacity 800 US gal (3028 litres) each wing
77 Surge tank
78 Starboard wing tip
79 Navigation light
80 VHF aerial boom
81 Fuel vent
82 Static dischargers
83 Outboard, low-speed, aileron
84 Outboard spoilers
85 Outboard slotted flaps
86 Flap drive mechanism
87 Inboard, high-speed, aileron
88 Trailing edge beam
89 Inboard spoilers
90 Inboard slotted flap
91 Flap drive mechanism
92 Centre fuselage construction
93 Starboard undercarriage bay housing
94 No. 3 passenger door
95 Wing-mounted main undercarriage bay
96 Flap driver motors
97 Undercarriage beam
98 Fuselage-mounted main undercarriage bay
99 Main undercarriage jack
100 Floor panels
101 Spat rails
102 Cabin window trim panels
103 Centre cabin economy-class seating, typically 82 passengers
104 Nine-abreast seating
105 Air distribution ducts
106 No. 4 passenger door, port and starboard
107 Centre cabin galley
108 Overhead baggage racks (with doors)
109 Main air supply duct
110 Rear cabin galley
111 Rear cabin seating, typically 114 passengers
112 Economy-class seating
113 Overseat baggage racks
114 Cabin roof panels
115 Control cable runs
116 Rear fuselage construction
117 Rear cabin seats
118 Rear cabin toilets
119 Wardrobes
120 Rear pressure dome bulkhead
121 Fin root fairing
122 Starboard tailplane
123 Static dischargers
124 Starboard elevator
125 Fin leading edge construction
126 Fin spar construction
127 Fin-tip fairing
128 VOR aerial
129 Static dischargers
130 Upper rudder segment
131 Lower rudder segment
132 Rudder jacks
133 Tailcone fairing
134 APU exhaust
135 Auxiliary power unit (APU)
136 Port elevator inner segment
137 Port elevator outer segment
138 Static dischargers
139 Tailplane construction
140 Elevator jacks
141 Tailplane sealing plate
142 Aft fuselage frames
143 Fin attachment
144 Tailplane centre section
145 Moving tailplane jack
146 APU air duct
147 No. 5 passenger door, port and starboard
148 Rear fuselage window panel
149 Rear under-floor freight hold
150 Freight and baggage pallet container
151 Fuselage frame and stringer construction
152 Trailing edge fillet
153 Fuselage-mounted undercarriage pivot
154 Trailing edge beam
155 Port inboard slotted flap
156 Flap tracks
157 Flap track fairings
158 Inboard spoilers
159 Flap drive shaft
160 Flap down position
161 Fuselage-mounted main undercarriage bogie
162 Wing spar and rib construction
163 Wing root attachment plate
164 Front spar
165 Engine bleed air supply pipe
166 Leading edge ribs
167 Landing lamps
168 Inboard Krueger flap
169 Krueger flap motor and drive
170 Wing mounted main undercarriage leg
171 Four-wheel main undercarriage bogie
172 Main undercarriage side brace
173 Wing-mounted undercarriage jack
174 Wing skins
175 Wing stringer construction
176 Inboard engine mounting beam
177 Pylon attachment strut
178 Port inner pylon construction
179 Heat exchanger
180 Engine intake
181 Rolls-Royce RB.211-524B engine
182 Engine driven gearbox
183 Outer fan ducting
184 Core engine exhaust
185 Integral fuel tankage
186 Inboard, high-speed, aileron
187 Aileron jack
188 Outboard slotted flap
189 Flap track fairing
190 Flap down position
191 Outboard spoilers
192 Flap tracks
193 Flap track mounting beams
194 Wing spar and rib construction
195 Leading edge construction
196 Krueger flap segments
197 Krueger flap mechanism
198 Outboard engine mounting beam
199 Port outer engine pylon
200 Heat exchanger air duct
201 Port outer engine cowlings
202 Thrust reverser cascades
203 Thrust reverser cowling door, open
204 Door operating jacks
205 Outer Krueger flap segments
206 Krueger flap mechanism
207 Outer wing construction
208 Aileron jacks
209 Outboard, low-speed, aileron
210 Static dischargers
211 Fuel vent
212 Wing-tip fairing
213 Navigation light
214 VHF aerial boom

Boeing 747
Z5-SPE
South African Airlines

BOEING 747-400

One of the most difficult tasks in aviation is deciding when to improve a winning design. With the early generations of 747, Boeing had cornered the entire high capacity, long-haul airliner market, but while these still perform excellent service around the world, a thoroughly updated version offers reduced operating costs and far better range, two features appreciated by the high-fliers of the airline world.

Some measure of the advancement of aviation during the 20th century can be gauged from the fact that the main cabin of the Boeing 747 is longer than the distance covered by the Wright Brothers' first powered flight. The 747 is now the standard long-haul carrier for the world's airlines, no other manufacturer competing with Boeing in this large capacity market. However, many of the first 747s in service are 20 years old, so Boeing have introduced a considerably improved model to revitalise the 747 design and maintain its place in the market: the Series 400.

Of course, Boeing had introduced the Series 300 with extended upper deck, newer technology engines and a revised interior, but this had been based on existing 747 technology, and although it looked considerably different externally, did not constitute a massive improvement over the original models. It was a younger brother: what was needed was a new generation.

Using the 747-300 as a starting block, Boeing set about a thorough modernisation of the design, resulting in an aircraft that differs little in appearance from its predecessor, but which is vastly more capable. Proposals for a new model began to take definite shape in April 1985, when the 747-400 designation was adopted, and the airlines were regularly consulted for their own inputs into the final configuration. Major areas attended to were the aerodynamics, powerplants, fuel system, undercarriage and cockpit.

Most notable of the new features are the revised wingtips. A 1.83m (6ft) section has been added to the basic wing, and grafted on to the end are large 1.83m (6ft) winglets, canted outwards at 29°. The effects of these improvements are to reduce drag around the wingtips, thereby reducing the fuel burn, and to improve take-off characteristics and cruising altitude. Increasing overall span to 64.31m (211ft 0in) the winglets are made from graphite epoxy and aluminium. Also improving aerodynamics is a revised wing-fuselage fairing.

New materials

Structural improvements include the use of new materials for the wing skins, stringers and lower spar chords, these saving some 2721kg (6,000lb) of weight per aircraft compared to the Series 300. Also making a reduction in weight is the adoption of BF Goodrich carbon brakes for each of the 16 mainwheels, this resulting in a net saving of 816kg (1,800lb), despite the use of slightly larger wheels.

Power comes from engines in the 258kN (58,000lb) thrust class,

Launch customer for the 747-400 was Northwest Airlines. Surprisingly, only two of the US majors have ordered the type so far, the other being United. Both carriers have extensive Pacific networks.

747-400s near completion at the Boeing-Everett plant. This factory was built specifically to handle the 747, the company's other sites not having the volume to cater for such a large aircraft.

housed in low-drag nacelles. Different nacelle shapes identify the three major engines offered for the type; the Pratt & Whitney PW4000 series, General Electric CF6-80C2 or Rolls-Royce RB211-524G/L. The latter has a full-length cowl of smooth contours. The pylons and nacelles offer lower drag and weight, and are common with those used on the Boeing 767, allowing greater flexibility for operators who fly both types.

Auxiliary power comes from a new Pratt & Whitney Canada APU, replacing the Garrett unit of earlier variants and offering a 40 per cent cut in fuel burn. While many systems remain the same, others have been digitalised to increase reliability and save on maintenance. The fuel system now has automatic management, with digital measurement, and includes an additional 12492 litres (2,748 Imp gal) in the tailplane. Unlike the latest Airbus aircraft, this cannot be used for high speed trim but does offer a considerable improvement in range.

Range increase

Indeed, the range increase is one of the Series 400's major advantages. Combining the new aerodynamics, fuel-efficient engines and extra fuel raises the range with a full load to 13528 km (8,406 miles), allowing several previously unattainable important city pairs to be linked non-stop. On shorter journeys, the payload/range character-

Specification
Boeing 747-400

Type: long range passenger/freight transport

Powerplant: four high bypass ratio turbofans in the 258.0 kN (58,000 lb) thrust class. Available with Pratt & Whitney PW4056, General Electric CF6-80C2 or Rolls-Royce RB211-425G

Performance: maximum level speed at 9150 m (30,000 ft) 980 km/h (609 mph); take-off distance to 11 m (35 ft) 3353 m (11,000 ft); landing run at maximum landing weight 2134 m (7,000 ft); range with 412 passengers and baggage with reserves 13528 km (8,406 miles); ferry range 15569 km (9,673 miles)

Weights: empty 178661 kg (393,880 lb) for RB211 version; maximum take-off 362875 kg (800,000 lb) basic or up to 394625 kg (870,000 lb) optional; max payload 64011 kg (141,120 lb) for RB211 version; maximum fuel 175392 kg (386,674 lb) for RB211 version

Dimensions: span 64.31 m (211 ft 0 in); tailplane span 22.17 m (72 ft 9 in); length 70.66 m (231 ft 10 in); height 19.33 m (63 ft 5 in)

Boeing 747-400 cutaway drawing key:

1. Radome
2. Weather radar scanner
3. Front pressure bulkhead
4. Scanner tracking mechanism
5. Wardrobe
6. First class cabin, 30 or 34 seats at 62-in (1.57-m) pitch
7. Nose undercarriage wheel bay
8. Nosewheel doors
9. Twin nosewheels
10. Hydraulic steering jacks
11. Nose undercarriage pivot fixing
12. Underfloor avionics equipment racks
13. Cabin window panels
14. First class bass unit
15. Flight deck floor level
16. Rudder pedals
17. Control column
18. Instrument panel, five-CRT electronic flight instrumentation system (EFIS)
19. Instrument panel shroud
20. Windscreen panels
21. Overhead systems switch panel
22. First officer's seat
23. Captain's seat (two-crew cockpit)
24. Observer's folding seats (two)
25. Starboard side toilet compartment (two)
26. Cockpit bulkhead
27. Crew rest bunks (two)
28. Upper deck window panel
29. Conditioned air distribution ducting
30. Forward main deck galley unit
31. Plug-type forward cabin door, No. 1 port and starboard
32. Business-class passenger seating, 24 seats typical at 36-in (91-cm) pitch
33. Fuselage lower lobe skin panelling
34. Baggage/cargo pallet containers
35. Forward underfloor cargo hold, capacity 2,768 cu ft (78.4 m³)
36. Forward fuselage frame and stringer construction
37. Upper deck doorway, port and starboard
38. Cabin roof frames
39. Anti-collision light
40. No. 1 UHF communications aerial
41. Upper deck passenger cabin, 52 business-class seats or 69 economy-class seats
42. Lower deck sidewall toilet compartment
43. No. 2 passenger door, port and starboard
44. Air conditioning system heat exchanger intake ducting
45. Ventral flush air intakes
46. Faired wing root leading-edge fillet
47. Ventral air conditioning packs, port and starboard
48. Wing spar bulkhead
49. Economy class seating
50. Staircase to upper deck
51. Fresh water tanks
52. Wing centre section fuel tankage, capacity 16,990 US gal (64315 l)
53. Centre section stringer construction
54. Floor beam construction
55. Front spar/fuselage main frame
56. Upper deck lobby area
57. Curtained bulkhead
58. Galley units
59. Starboard wing inboard main fuel tank, capacity 12,546 US gal (47492 l)
60. Fuel pumps
61. Engine bleed air supply ducting
62. Krüger flap operating mechanism
63. Inboard Krüger flap segments
64. Starboard inner Pratt & Whitney PW4256 engine nacelle
65. Inboard nacelle pylon
66. Leading-edge Krüger flap segments
67. Pressure refuelling connections, port and starboard
68. Krüger flap drive shaft
69. Krüger flap rotary actuators
70. Starboard wing outer main fuel tank, capacity 4,482 US gal (16966 l)
71. Starboard outer engine nacelle
72. Outer nacelle pylon
73. Starboard wing reserve tank provision, capacity 534 US gal (2021 l)
74. Outboard Krüger flap
75. Krüger flap drive mechanism
76. Outer wing panel dry bay
77. Vent surge tank
78. Wing-tip extension
79. Starboard navigation (green) and strobe (white) lights
80. Starboard winglet
81. Fixed portion of trailing edge

Among the most radical changes offered by the 747-400 is the high-tech cockpit, dominated by six large screen displays. The few dials that remain merely duplicate screen information in case of a systems failure with the displays.

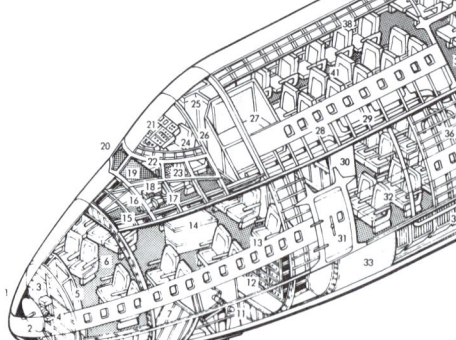

istics have improved to the point of allowing up to four times the payload to be carried on sectors at the edge of the 747-300's range capabilities.

Perhaps the biggest change to the new variant was made to the cockpit. The previous generation of 747s had a cramped and traditional cockpit for three crew (two pilots and flight engineer). The Series 400 introduces an all-glass, two-man cockpit, the reduction in crew members resulting from increased sophistication and reduced pilot workload. Permanent stations for observation or relief crew are provided.

Dominating the cockpit are six large CRT displays, four facing the pilot and two stacked between them. The pilot displays are the primary flight display (PFD) to the left, which presents the aircraft's flying characteristics at the time, and the navigation display (ND) to the right, which has navigation functions displayed. The two central displays are used for the engine indication and crew alert system (EICAS), which presents engine information and any faults that may develop with the aircraft's system.

The PFDs are backed-up by standard electro-mechanical instruments to provide basic flight information such as airspeed, altitude, heading and vertical speed, should all PFD functions be lost, although the ND can be used if one of the PFD screens becomes in-

82 Fuel vent
83 Static dischargers
84 Outboard, low-speed, aileron
85 Outboard four-segment spoilers
86 Outboard triple-slotted Fowler-type flap, down position
87 Flap screw jacks and segment linkages
88 Flap drive shaft
89 Inboard, high-speed, aileron
90 Inboard triple-slotted flap, down position
91 Inboard two-segment spoilers/lift dumpers
92 Flap screw jack
93 Auxiliary trailing-edge wing spar
94 Cabin air distribution ducting
95 Extended upper deck rear bulkhead
96 Upper deck floor beam construction
97 Air system cross-feed ducting
98 Conditioned air risers
99 Machined wing spar attachments main frames
100 Central flap drive motors
101 Wing-mounted outboard main undercarriage wheel bay
102 Undercarriage mounting beam
103 Central keel section
104 Pressure floor above wheel bay
105 Centre fuselage frame and stringer construction
106 Dual navigation aerials
107 Cabin wall trim panelling
108 Seat mounting rails
109 Main cabin floor panelling
110 Fuselage-mounted inboard, main undercarriage wheel bay
111 Hydraulic retraction jack
112 Cabin window panel
113 Overhead conditioned air distribution ducting
114 Economy-class seating, 302 to 410 passengers at 34-in (86-cm) pitch
115 Overhead stowage bins
116 Sidewall toilet compartments, port and starboard
117 Central cabin galley
118 No. 4 passenger door, port and starboard
119 Rear cabin passenger seating
120 Rear cabin galley
121 Rear cabin air supply ducting
122 Fuselage sidewall stowage bins
123 Control cable runs
124 Central overhead stowage bins
125 Cabin roof panels
126 Ten-abreast economy-class seating
127 Rear fuselage frame and stringer construction
128 Rear cabin seating
129 Access ladder to upper deck rest area
130 Overhead cabin-crew rest area, six bunks and four seats typical
131 Rear pressure bulkhead
132 Fin root fillet
133 Starboard trimming tailplane
134 Static dischargers
135 Starboard elevator
136 Fin leading-edge construction
137 Two-spar fin box construction
138 Fin-tip fairing
139 VOR localiser aerial
140 Static dischargers
141 Upper rudder segment
142 Lower rudder segment
143 Rudder hydraulic actuators
144 Tailcone frame construction
145 Pratt & Whitney Canada PW901A auxiliary power unit (APU)
146 Tail navigation and strobe lights (white)
147 APU exhaust
148 Port elevator inboard segment
149 Port elevator outboard segment
150 Static dischargers
151 Port trimming tailplane construction
152 Elevator hydraulic actuators
153 Long range tailplaned fuel tank, capacity 3,300-US gal (12492 l)
154 Tailplane sealing plate
155 Aft fuselage framing
156 Fin root attachment joint
157 Tailplane centre section
158 Tailplane trim screw jack
159 APU high pressure air supply duct
160 Lower deck rear cabin toilet compartments
161 No. 5 passenger door, port and starboard
162 Rear fuselage window panel
163 Underfloor bulk cargo hold, capacity 1,000 cu ft (28.3 m³)
164 Rear fuselage bagage/cargo hold, capacity 2,422 cu ft (68.6 m³)
165 Baggage/cargo pallet
166 Fuselage lower lobe frame and stringer construction
167 Wing root trailing-edge fillet, composite construction
168 Fuselage-mounted main undercarriage pivot fixing
169 Trailing-edge auxiliary spar
170 Undercarriage leg breaker strut
171 Wing-mounted main undercarriage pivot fixing
172 Hydraulic retraction jack
173 Four-wheel inboard main undercarriage bogie
174 Flap drive shaft
175 Flap guide rails
176 Inboard spoilers/lift dumpers
177 Port inboard triple-slotted flap
178 Flap track fairings
179 Flap down position
180 Aileron hydraulic actuator
181 Inboard, high-speed, aileron
182 Outboard triple-slotted flap
183 Outboard flap tracks
184 Outboard spoilers
185 Flap track fairings
186 Flap down position
187 Outboard, low-speed, aileron
188 Aileron hydraulic actuators
189 Static dischargers
190 Fuel vent
191 Fixed trailing edge section
192 Port winglet
193 Winglet composite construction
194 Port navigation (red) and strobe (white) lights
195 Outboard leading-edge
196 Krüger flap segments
197 Krüger flap drive mechanism
198 Outer wing panel rib construction
199 Wing bottom skin access panels
200 Rear spar
201 Outboard engine mounting rib
202 Port outer nacelle pylon
203 Thrust reverser cowling door, open
204 Reverser cascades
205 Outboard engine nacelle
206 Rolls-Royce RB211-524G alternative engine installation
207 Full length nacelle cowling
208 Internal exhaust stream mixer duct
209 Central leading-edge Krüger flap segments
210 Krüger flap drive mechanism
211 Leading-edge rib construction
212 Wing panel spar and rib construction
213 Flap track mounting beams
214 Inner engine pylon mounting rib
215 Wing stringers
216 Wing skin panelling
217 Wing-mounted main undercarriage leg strut
218 Pylon attachment strut
219 Four-wheel outer main undercarriage bogie
220 Nacelle pylon construction
221 Engine bleed air pre-cooler
222 Core engine, hot stream, exhaust duct
223 Fan air, cold stream, exhaust duct
224 Ventral engine accessory equipment package
225 Pratt & Whitney PW4256 turbofan engine
226 Engine intake
227 Detachable engine cowling panels
228 Bleed air de-iced intake lip
229 Inboard Krüger flap segments
230 Krüger flap motor and drive shaft
231 Three-spar wing torsion box construction
232 Inboard wing ribs
233 Bolted wing root attachment joint strap
234 Front spar
235 Engine bleed air ducting
236 Leading-edge nose ribs
237 Twin landing lamps
238 General Electric CF6-80C2 alternative engine installation

During its flight trials, the first 747-400 demonstrated the capabilities of the type by becoming the heaviest aircraft then to have left the ground. Flying from Moses Lake, Washington, N401PW weighed in at 404815 kg (892,450 lb) at lift-off, some 23564 kg (51,950 lb) past the previous record set by a 747-200. Since that time the 747-400 has been easily surpassed by the Antonov An-225.

operative. So sophisticated is the 747-400 that the number of gauges, switches and lights has dropped from 971 on the Series 300 to just 365. This number is also considerably fewer than for the cockpits of the 737 and 757/767 twins.

EICAS functions are equally impressive. The top screen is usually used for engine function display, while the lower screen is used for alerting faults. A combined central maintenance computer stores information which can be down-loaded to technicians for quick and easy maintenance. This function can be used by the crew (but only in the cruise) to access various aircraft systems in flight. A built-in test equipment (BITE) function is also incorporated to allow quick monitoring of the systems by maintenance engineers.

A full range of navigation equipment is fitted, and the aircraft is controlled by a Collins auto-pilot, fully digitalised and triplicated. A Sperry/Honeywell system handles the flight management. Autoland capability is for Cat 3b, allowing operations down to 75m (245 ft) runway-visual range with no decision height constraints.

Pacific Rim airlines have been among the vanguard of Series 400 operators, finding its extra range a great advantage in the region. Thai Airways International has 12 of the type.

This view shows to good effect the extended wingtips and winglets, introduced to provide greater aerodynamic efficiency. British Airways requires 58 such aircraft, of which 37 have been delivered.

Back in the cabin, the 747-400 has had a facelift. New materials are used in the fittings to improve fireworthiness and smoke levels, while the larger overhead bins and new trim provide a completely different appearance to the interior. Customer options for internal configurations have been greatly aided by revisions to the cabin systems. Cabin plumbing allows for a total of no less than 26 possible toilet locations, while the air system is now divided into five sectors rather than three in the previous variants. Hughes has devised a digital advanced cabin entertainment system which dispenses with wire or air tube connections.

Another innovation is the crew rest area situated above the rear cabin. With 747-400s regularly performing long endurance sectors, flight attendants routinely rest for a while during the cruise. Instead of curtaining off potential revenue-earning seats, resting crew on the 747-400 can occupy the 'attic', which can be configured for a varying number of bunks or seats.

Potential options for internal configuration are almost limitless, but the standard interior provides for about 410-420 passengers in three classes. Economy seats are 10-abreast (3-4-3) in the rear cabin, with first- and business-class sharing the forward and upper decks, the latter typically offering 52 business-class seats in 3-2 arrangement. One arrangement exemplified by Boeing has 30 first class seats and 479 economy-class, including 69 on the upper deck in six-abreast seating.

Unsurprisingly, the 747-400 is also offered in -400M Combi versions, trading in some passenger seats for cargo space (typically seven pallets). Combined with the extensive underfloor cargo/baggage hold, this allows the Combi to carry a huge amount of cargo in addition to maintaining a large-scale passenger service. A large freight door is provided in the port rear cabin for admitting the pallets. Dedicated freight interiors are available with the 747-400F variant.

After its 'Big Top' service was established using the 747-300, Singapore Airlines has now introduced the 'Megatop', using 747-400s to link the South East Asian city with the world's major commercial centres.

Programme history

Go-ahead for the Series 400 design was issued in July 1985, and construction of the first aircraft was initiated mid-way through the following year. Northwest Airlines was the launch customer, ordering 10 in October 1985. The carrier was a natural for the new variant, for much of its business came from the Pacific, where the additional range performance offered by the Series 400 could be put to good use. It was also no stranger to Boeing products, being in the company's 'top five' customers and is currently the world's second-largest 747 operator (after Japan Airlines).

The prototype 747-400, N401PW, was rolled out at the Paine Field, Everett plant on 26 January 1988, the same day as the 737-400 was rolled out at the Renton plant. First flight occurred on 29 April 1988, under Pratt & Whitney PW4056 power. Three further aircraft joined the certification programme, one of which was powered by GE engines and one by Rolls-Royce.

Certification was gained for the PW4056-powered version on 9 January 1989, and for the other two powerplants in the following spring. Northwest's first aircraft was received by the airline on 26

A KLM 747-400 sets off on a test flight. The Dutch airline has never been slow to adopt new technology to maintain its prominent position in the marketplace, and it is looking to replace all its older generations of 747 with the -400. Many of the aircraft on order are Combi versions with mixed passenger/freight interiors.

Boeing 747-436
G-BNLE 'City of Newcastle'
British Airways

A cloud of spray marks the departure of this 747-400 destined for China Airlines, the Taiwanese national carrier. The mainland Communist state has also bought the -400 to serve with Air China (CAAC).

January, and on 31 January began route-proving flights between the airline's headquarters at Minneapolis, Minnesota, and Phoenix, Arizona. After crew training was completed, services began on the prestige non-stop New York-Tokyo service in April 1989.

As described below, the 747-400 found a wealth of customers, most of whom are current 747 users wishing to upgrade and expand their long-haul fleets. Labour relations problems caused a slippage of deliveries, but the 747-400 is now an increasingly common sight at the world's major airports. Surprisingly, only Northwest and United have ordered the type from among the US majors. KLM led the

Not unnaturally, British Airways chose the Rolls-Royce RB211-524G turbofan to power its 747-400s. These are the most distinctive of the available engines for they have a smooth contour cowling.

order book for the Combi variant, while Air France was the first to order the dedicated freighter.

Long-range problems

In service the 747-400 has proved every bit as reliable as its predecessor, although some operators have expressed concern over range performances. More concern has been raised over long endurance flights conducted by a two-man crew, and several nations are introducing three-crew operations for long-haul flights to combat fatigue-induced loss of crew performance. Among the milestones for the aircraft was a non-stop delivery flight of an RB211-powered QANTAS aircraft from London to Sydney. Carrying 23 persons, the aircraft covered the 17953 km (11,156 miles) in 20 hours 8 minutes, setting a new distance record for a commercial airliner.

By the end of June 1998, Boeing had amassed a total of 581 orders for the 747-400, of which 433 had been delivered. These raised total

747 orders to 1,305, easily exceeding the 1,009 airframes of the 707 and 720 family (including military aircraft) that up to then held the record for the most produced four-jet transport. Up to the same time, deliveries of all 747 models had reached a total of 1,157 aircraft, including 724 of the 747 'Classic' series, comprising the -100, -200, -300 and SP.

As the world's largest 747 operator, Japan Airlines were an early and large customer for the 747-400. Taking off from Boeing Field, Seattle, this JAL aircraft demonstrates the airline's new colours.

long-haul fleet during the 1990s. Not that Boeing is resting on its laurels, for the design team is already studying the next step – an even larger 747 variant.

Sales breakdown

Details of the June 1998 747-400 sales book are far too extensive for a full listing here, but the main highlights (with a total for deliveries and firm orders in brackets) of the aircraft, with various General Electric, Pratt & Whitney and Rolls-Royce turbofan engines, include Air China (14), Air France (13), Air India (6), Air New Zealand (8), All Nippon (22), Asiana (14), British Airways (58), Cargolux (11), Cathay Pacific (21), China Airlines (12), EVA Air (15), Japan Airlines (50), KLM (20), Korean Air (33), Lufthansa (29), Malaysia Airlines (21), Northwest Airlines (14), QANTAS (24), Saudi Arabian Airlines (5), Singapore Airlines (53), South African Airways (7), United Airlines (51) and Virgin Atlantic Airways (6).

Boeing's aims of revitalising the 747 have been achieved, and the company can expect many further sales as new customers take on ever-larger operations, and old ones return for rejuvenated equipment. Several major airlines are looking to form an all-Series 400

A large and impressive advert for any nation, the 747-400 is now the world's great airliner status symbol. On a practical level, it offers unrivalled economy, capacity and range.

Chapter 10

Boeing 757
•
Boeing 767
•
Boeing 777
•
Boeing C-97

Boeing 757

Towards the end of the 1970s, Boeing was faced with a problem: how to keep its highly successful Model 727 at the top of airliner sales. Subsequent improvement studies to the basic airframe resulted in an aircraft which had little in common with its predecessor, but which has kept the Boeing sales challenge alive. Designated the Model 757, the new aircraft is taking its place in ever-increasing numbers in the ranks of the world's major airlines.

Until the Boeing 727, launched in 1962, no commercial transport – not even the DC-3 – had notched up 1,000 sales. This rear-engined trijet reached the stupendous total of 1,831, the last thousand being gained despite the existence of the A300B offering greater comfort, far better fuel efficiency and much less noise. Even though no airlines in the early 1970s seemed to want to buy the quiet widebody, Boeing could see it would eventually have to improve the 727, and the first idea was the 727-300 with fuel-efficient engines and a stretched body. By 1976 this had become the 7N7, with just two engines hung under a wing more like that of the A300B, with a deeper aft-loaded profile and less sweep. Size of the 7N7 ranged from 125 passengers, with 10-tonne thrust CFM56 or JT10D engines up to 180 seats with the 13600-kg CF6-32.

Increasingly attention focused on the biggest 7N7 versions, and Rolls-Royce came into the picture with the new RB211-535 engine of an initial 14515 kg (32,000 lb) thrust. It was obvious that, instead of trying to sell engines for European A300B and A310 aircraft, Rolls-Royce was eager to get aboard the new Boeing, and it was backed to the hilt by British Airways, up to very recent times the only European flag carrier to prefer new American aircraft exclusively. Indeed in 1976–78 great efforts were made by Boeing, British Airways and Rolls-Royce to get the British aircraft industry to build 757 wings instead of joining Airbus Industrie (these efforts fortunately failed).

In 1979 the decision was taken to drop the T-tail and mount the horizontal tail on the fuselage. A little later the old 727 'cab' – giving commonality with the 707, 727 and 737 – was discarded in favour of the wider nose and flight deck of the 767. By this time Boeing was already in full development, the go-ahead being announced in early 1978 and British Airways and Eastern placing launch orders on 31 August 1978, with the 535C engine flat-rated at 16965 kg (37,400 lb). This power was matched to the increased gross weight of 99792 kg (220,000 lb), providing for up to 233 passengers over ranges initially announced as 3700 km (2,300 miles). Later, despite slightly reducing fuel capacity to 42597 litres, ranges were extended to far beyond 7000 km (4,400 miles).

British Airways signed its contract on 2 March 1979, followed by Eastern three weeks later. On the latter date, 23 March 1979, Boeing announced committal to production. No foreign risk-sharing partner was brought in on the airframe, though contracts were

Delta Air Lines was an early and major customer for the 757, choosing the type to operate alongside 767s and TriStars on its US domestic routes. This aircraft at Atlanta is followed by the similar but larger 767.

Three Boeing 757-2M6s were delivered to Royal Brunei Airlines in 1986. These flew on services linking the nation with other Far East destinations, including Australia. Power comes from the RB211-535E4 – accommodation is in three classes for 148.

placed with subcontractors for 200 ship-sets of parts covering 53 per cent of the aircraft. In the event some of the biggest subcontracts, with Rockwell, Avco Aerostructures and Fairchild, were terminated. Those that remain include Menasco (US) for landing gears, Shorts (UK) for inboard flaps, CASA (Spain) for outboard flaps, Hawker de Havilland (Australia) for wing ribs, Grumman (US) for spoilers, LTV (US) for the rear fuselage, fin and tailplane, Heath Tecna (US) for the graphite/Kevlar fairings over the flap tracks and wing roots, Schweizer (US) for the wingtips and Rohr (US) for the pylon struts.

Sales lapse

Initially Boeing was very disappointed in sales of the 757. Thanks to its narrow body, it had been planned to offer the lowest fuel burn per passenger-kilometre of any jetliner, yet no new customers appeared until in April 1980 three were signed for by Transbrasil and three by Aloha. Both selected the CF6-32 engine, in which General Electric proposed to collaborate with Volvo Flygmotor of Sweden. Still there were no sales until at last, in November 1980, came the breakthrough Boeing had been waiting for: 60 for Delta. The giant US line deferred its choice of engine, and when in December 1980 it announced its decision it was for yet a third engine, the Pratt & Whitney PW2037. This was the ultimate result of the long-running JT10D programme, in which MTU of West Germany and Fiat of Italy had always participated. Starting at the 10-tonne level, it had progressed to 17350 kg (38,250 lb) for the 757, and Pratt & Whitney was determined to sell it as the most fuel-efficient in the world, promising financial compensation to Delta if the PW2037 failed to beat the Rolls-Royce 535 by 'between 7 and 8 per cent'. So attractive did this sound that at the end of 1980 American Airlines announced selection of the PW2037 for its 757s, before it had even announced it would buy the Boeing aircraft!

In January 1981 Rolls-Royce announced the 535E4, rated at 18190 kg (40,100 lb), to power future 757s. Notable for introducing a totally new fan, with very wide snubberless blades made from thin diffusion-bonded titanium skins on a honeycomb core, the E4 promised to give the PW2037 a real run for its money. The E4 also fits a new nacelle with a short and simple integrated rear nozzle. In the same month GE announced it would no longer compete in this market, and never built the CF6-32. Aloha and Transbrasil subsequently picked the PW2037, though this meant delayed delivery.

Boeing centred the 757 programme at its central Renton (Seattle) plant, headquarters of the Boeing Commercial Airplane Company. Here the first 757 was rolled out on 13 January 1982, to make its first flight on 19 February. It was the first time a major US airliner had been launched with a foreign engine, but development of the 535C had gone so much better than prediction that everyone (except P&W) was delighted. Scheduled service began with Eastern on 1 January 1983, and British Airways on 9 February. Initially only the 757-200 was offered, with a fuselage length of 46.96 m (154 ft 10 in), a choice of two engines and a choice between regular or long-range fuselage capacity for a maximum weight of 104325 or 113395 kg respectively. First flown in August 1998, the 757-300 has a fuselage lengthened by 7.1 m (23 ft 1 in) for between 243 and 280 passengers.

Narrow-body fuselage

Though the nose and cab section are similar to those of the appreciably wider 767, the main tube section of fuselage is almost identical to that of the 707, 727 and 737, with a ruling cabin width of 3.53 m. This compares with 3.07 for the MD-80, 3.42 for the BAe 146, 3.696 for the A320 and 5.28 for the A310. Boeing has offered nine interior arrangements for 178 to 239 passengers, seated basically 3+3 with a

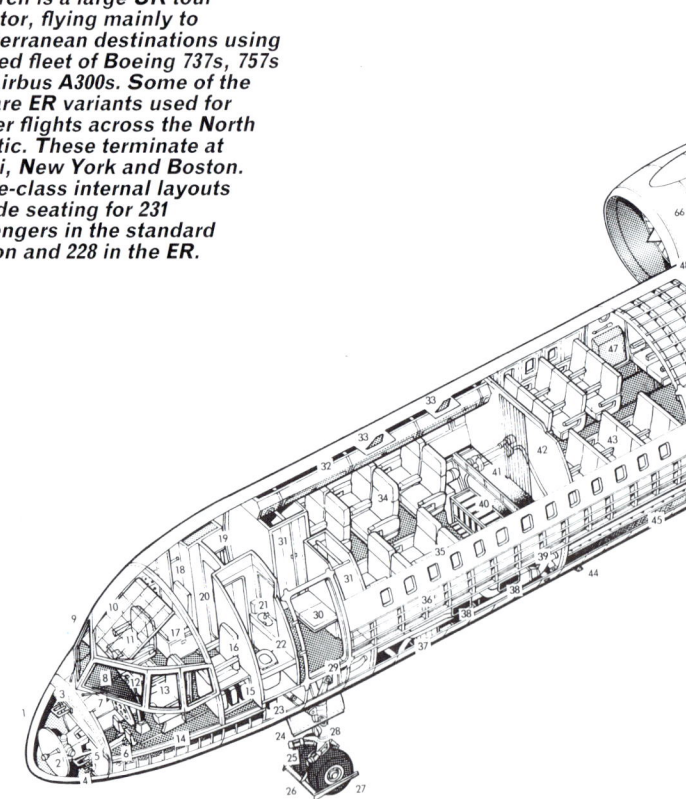

Monarch is a large UK tour operator, flying mainly to Mediterranean destinations using a mixed fleet of Boeing 737s, 757s and Airbus A300s. Some of the 757s are ER variants used for charter flights across the North Atlantic. These terminate at Miami, New York and Boston. Single-class internal layouts provide seating for 231 passengers in the standard version and 228 in the ER.

The 757 has proved very popular with the European tour operators. Air 2000 is another UK company, flying from Gatwick and Glasgow. Thirteen 757s are on strength, five of them ER versions for the US, Caribbean and African charters.

Boeing 757-200 cutaway drawing key

1. Radome
2. Weather radar scanner
3. VOR localiser aerial
4. ILS glideslope aerials
5. Front pressure bulkhead
6. Rudder pedals
7. Windscreen wipers
8. Instrument panel shroud
9. Windscreen panels
10. Cockpit roof systems control panels
11. First officer's seat
12. Centre console
13. Captain's seat
14. Cockpit floor level
15. Crew baggage locker
16. Observer's seat
17. Optional second observer's seat
18. Coat locker
19. Forward galley
20. Cockpit door
21. Wash basin
22. Forward toilet compartment
23. Nose undercarriage wheel bay
24. Nosewheel leg doors
25. Steering jacks
26. Spray deflectors
27. Twin nosewheels
28. Taxiing and runway turn-off lamps
29. Forward entry door
30. Cabin attendant's folding seats
31. Closets, port and starboard
32. Overhead stowage bins
33. DABS aerials
34. First-class cabin four-abreast seating, 16 seats
35. Cabin window panels
36. Fuselage frame and stringer construction
37. Underfloor radio and electronics compartment
38. Negative pressure relief valves
39. Electronics cooling air ducting
40. Radio racks
41. Forward freight door
42. Curtained cabin divider
43. Tourist-class six-abreast seating, 162 seats
44. Ventral VHF aerial
45. Underfloor freight hold
46. Passenger entry door, port and starboard
47. Door mounted escape chutes
48. Upper VHF aerial
49. Overhead air conditioning distribution ducting
50. LD-W cargo container (seven in forward hold)
51. Graphite composite wing root fillet
52. Landing lamp
53. Air system recirculating fan
54. Air distribution manifold
55. Conditioned air risers
56. Wing spar centre-section carry-through
57. Front spar fuselage main frame
58. Ventral air conditioning plant, port and starboard
59. Centre section fuel tank
60. Floor beam construction
61. Centre fuselage construction
62. Starboard wing integral fuel tank; total system capacity 10,880 US gal (41 185 l)
63. Dry bay
64. Bleed air system pre-cooler
65. Thrust reverser cascade doors, open
66. Starboard engine nacelle
67. Nacelle pylon
68. Fuel venting channels
69. Fuel system piping
70. Pressure refuelling connections
71. Leading edge slat segments, open
72. Slat drive shaft
73. Guide rails
74. Overwing fuel filler cap
75. Vent surge tank
76. Starboard navigation light (green) and strobe light (white)
77. Tail navigation strobe light (white)
78. Starboard aileron
79. Aileron hydraulic jacks
80. Spoiler sequencing control mechanism
81. Outboard double-slotted flaps, down
82. Flap guide rails
83. Screw jacks
84. Outboard spoilers, open
85. Spoiler hydraulic jacks
86. Inboard flap outer single-slotted segment
87. Inboard spoilers
88. Starboard main undercarriage mounting beam
89. Cabin wall trim panels
90. Rear spar/fuselage main frame
91. Flap-drive hydraulic motor (electric motor back-up)
92. Port mainwheel bay
93. Pressure floor above wheel bay
94. DF loop aerials
95. Cabin roof lighting panels
96. Port overhead stowage bins, passenger service units beneath
97. Mid-section toilet compartments (two port, one starboard)
98. Emergency exit doors, port and starboard
99. Rear freight door
100. APU battery and controls
101. Rear cabin seating
102. Overhead stowage bins
103. Starboard rear galley unit
104. Fin root fillet
105. Fin construction
106. Fin "logo" spotlight
107. Starboard tailplane
108. Starboard elevator
109. HF aerial couplers
110. Leading edge HF aerial
111. Fin tip aerial fairing
112. Tail VOR aerials
113. Static dischargers
114. Rudder
115. Rudder hydraulic jacks
116. Honeycomb rudder panel construction
117. APU intake plenum
118. Tailcone
119. APU exhaust
120. AiResearch GTCP 331-200 auxiliary power plant (APU)
121. Port elevator
122. Elevator hydraulic jacks
123. Honeycomb panel construction
124. Static dischargers
125. Tailplane construction
126. Fin "logo" light
127. Tailplane sealing plate
128. Fin support frame
129. Tailplane centre-section
130. Tailplane trim control jack
131. Rear pressure bulkhead
132. Aft galley
133. Rear entry door, port and starboard
134. Underfloor freight hold
135. LD-W cargo containers (six in rear hold)
136. Ventral VHF aerial
137. Roller tray cargo handling floor
138. Graphite composite wing root fillet
139. Port inboard double slotted flap
140. Main undercarriage mounting beam
141. Undercarriage left side strut
142. Hydraulic retraction jack
143. Inboard spoilers
144. Flap hinge linkage
145. Inboard flap single slotted outer segment
146. Flaps down position
147. Flap track fairings
148. Outboard double slotted flap
149. Outboard spoilers
150. Aileron hydraulic jacks
151. Port aileron honeycomb construction
152. Tail navigation strobe light (white)
153. Port navigation light (red) and strobe light (white)
154. Vent surge tank
155. Port leading edge slat
156. Slat guide rails
157. Drive shaft
158. Port wing dry bay
159. Ventral access panels
160. Port wing integral fuel tank
161. Wing rib construction
162. Wing stringers
163. Wing-skin plating
164. Four-wheel main undercarriage bogie
165. Main undercarriage leg strut
166. Inboard wing ribs
167. Bleed air ducting
168. Inboard leading edge slat
169. Engine mounting pylon
170. Detachable engine cowlings
171. Port engine intake
172. Intake de-icing air duct
173. Rolls-Royce RB.211-535C turbofan engine (General-Electric CF6-32 optional fit)
174. Engine accessory gearbox
175. Oil cooler
176. Fan air exhaust duct
177. Hot stream exhaust nozzle

203

*The prototype Boeing Model 757 (registration **N757A**) was demonstrated at the 1982 Farnborough air show in its Boeing house colours. Although there were several studies, the 757 was initially offered in only one configuration, the Series 200. This obviously suited airline customers at the time, and the longer Series 300 was launched only in the autumn of 1996.*

Above: Despite a strong home lobby canvassing for the national airline to buy the Airbus, British Airways signed up as one of the launch customers for the 757, its first aircraft entering service in February 1983. Together with its charter subsidiary Caledonian, it operates 51 aircraft, with six more still to be delivered.

central aisle. Customers can have three doors each side plus four overwing emergency exits or four doors each side. In most configurations there is a galley at the front on the right and another at the rear on the left, and toilets at the front on the left and two or three either amidships or at the rear. Some customers offer coat hanging closets.

Technically the 757 is extremely conventional. The wing has a span of 38.05 m, rather less than originally planned, and a quarter-chord sweep of 25° (the same as the 737 and much less than other Boeing jetliners). The leading edge is fitted with full-span powered slats, though these have a gap at the engine pylon, there being a single large slat inboard and four sections outboard. On the trailing edge are inboard and outboard flaps mounted on faired tracks. All are double-slotted except the outer portion of each inboard flap which is single-slotted to avoid interference with the jet wake. Outboard are powered all-speed ailerons of quite long span. Ahead of the flaps are two inboard and four outboard spoilers on each wing, the innermost being a ground lift dumper only. The other 10 spoilers are opened together as speed brakes or differentially for roll, augmenting the ailerons. The horizontal tail is pivoted to serve as the longitudinal trim control, carrying the graphite-composite elevators. Likewise the fixed vertical tail carries a graphite rudder.

Laser technology

In its systems the 757 did break some new ground, notably in the use of laser-light gyros in the navigation IRS (inertial reference system). It is also, as it had to be, one of the new breed of all digital aircraft. There is simply no way Boeing could launch a jetliner in the late 1970s with the analog-type avionics of the 707, 727 and most 737s and 747s. The flight deck, very like that of the 767, is a mix of old and new, the new parts including basic electronic flight instrument displays and an EICAS (engine indication and crew alerting

Below: A member of the Airlines of Europe group that failed in 1991, Air Europe was a growing charter and limited schedule operator, flying mainly tour flights to the Mediterranean and Atlantic islands. At one time the operator had 27 757s in service or on order for a dramatic increase in both scheduled and charter capacity.

The US giant Northwest purchased a total of 73 PW2037-powered 757s to begin the replacement of its domestic 727 fleet. By early 1998, 48 out of 73 aircraft had been delivered. Unusually the airline also operates the rival Airbus A320.

system). The overall flight management is a generation earlier than that of the A320, but it does offer automatic trajectory guidance, terminal navigation, thrust management and an optional feature offering some protection against windshear. An excellent Garrett APU (auxiliary power unit) is mounted in the tailcone, the ECS (environmental control system) packs are under the floor in the centre section, and there are two ram-air turbines to provide emergency electrical and hydraulic power in flight. Each main gear is a four-wheel bogie retracting inwards, with Dunlop wheels, tyres and carbon brakes. The steerable twin-wheel nose gear retracts forwards, all landing-gear doors being of Kevlar.

Super-reliable powerplant

From the start of service the 757 did all that was asked of it, and the 535C did more. It was soon evident that this was the most troublefree engine in history, almost never suffering an inflight shutdown. Over the first four years of service the engine-caused removal rate was only 0.051 per thousand flight hours, described as 'many times better than the previously claimed industry best'. In October 1984 the advanced E4 entered service, and quickly established not only a reputation for reliability which is, if anything, even better, but also a reduction in fuel burn of over 10 per cent. The rival PW2037 followed into service on 1 December 1984, and – though Pratt & Whitney continues to claim the 'lowest fuel consumption' – the unequalled combination of economy, reliability and low cost of ownership resulted in the British engine being selected by every one of the 11 next customers for the 757. Two further advantages for the 535E4 are that the 757 powered by this engine has been officially measured as 'the quietest jet with over 100 seats' and that this is the only version FAA-approved for EROPS (extended-range operations) such as on the North Atlantic. Several airlines, including several British charter operators, use the E4-powered aircraft on services to North America.

Freighter version

An important customer to have picked the Pratt & Whitney engine (in the uprated PW2040 version at 18915 kg) was United Parcel Service, which ordered the 757-200PF (Package Freighter) announced in January 1986. This is a dedicated freighter, with a windowless fuselage and a large cargo door in the port side. It can carry 15 standard 2.24 × 3.18 m containers on the main deck. Other versions on order are the Corporate 77-52, the extended-range version first ordered by Royal Brunei, and the 757-200 Combi which has both the cargo door and passenger windows. As ordered by Royal Nepal Airlines, the Combi typically carries two containers plus 123-167 passengers, depending on the interior arrangement.

By the end of June 1998 Boeing had received orders for 920 aircraft of the 757 family, of which 807 had been delivered. Sixty-three customers had ordered the type. There is no doubt that, on the strict criterion of fuel-burn per seat-kilometre, it has no rival, though its advantage over the A310 is extremely small. On the whole the 757 is an outstandingly good aircraft which has found more competition than it expected and which has, so far, shown no sign of coming near the 1,831 sales of its predecessor.

Eastern Airlines was the other launch customer for the 757, its aircraft powered by RB211-535C engines. The airline was also the first to put the 757 into service, this occurring on 1 January 1983. The fleet once numbered 25, all configured in a two-class layout for 185 passengers.

Boeing 757-236
G-BNSE
Air Europe

One of the most remarkable features of the Boeing 757 was that it was launched with a non-US engine, the Rolls-Royce RB211-535 turbofan. The type is currently offered with Pratt & Whitney engines as well.

Boeing 767

In the late 1970s, Boeing began work on a pair of airliners that would carry their sales thrust through the latter part of the century. One emerged as the narrow-body 757, while the other was the similar, but wide-bodied, 767. This is now a major feature of the world airliner market, offering extraordinary range performance for a twin-engined aircraft.

Following the end of World War II, the commercial airliner market was dominated by the Douglas and Lockheed companies, the former building on the phenomenal success of its DC-3, to produce successively more capable aircraft under the DC- designation. In response, Lockheed produced matching variants of its Constellation design, both lines culminating in the ultimate in piston-engined airliners, the DC-7C and L-1649 Starliner.

Throughout this period, only one aircraft had really challenged this supremacy, the Boeing Stratocruiser, and despite the success of this design, Boeing had largely been concerned with military aircraft, producing hundreds of tankers and bombers for the US Air Force during the massive build-up of the American strategic forces.

Boeing joined, and eventually dominated the airliner club, with the Model 707, an epoch-making airliner that successfully married swept-wing technology with lightweight, powerful turbojets. At this juncture, Lockheed dropped from the competition, leaving Douglas as the only rival to Boeing's meteoric rise in the market. The 707 initiated a memorable sequence of aircraft designations, each of which proved as successful and as revolutionary in its own way. The 717 designation was used by the KC-135 tanker series, built in huge numbers for the US Air Force. The 727 was a revolutionary tri-jet that brought jet power to the smaller fields and short-haul routes, a tradition continued with the twin-jet 737. Then came the 747, ushering in a new era of wide-body transports powered by high bypass ratio turbofans.

With its considerable geographic isolation, Australia's QANTAS needs long-range aircraft to serve its route network. Both regular and stretched versions of the 767 are in use, flying principally on the Pacific routes. Early in 1998 the airline had seven 767-238ERs and 19 767-328ERs, with two more of the latter on order.

Flying nearly three years after the 747 had first been delivered, the European Airbus A300 offered a different kind of competition to Boeing's future dominance. It entered an area where rival products were nothing more than paper projects, and consequently established itself rapidly as a world-leader. Technological excellence was at the heart of the Airbus success, which was continued with the A310. Both Boeing and Douglas were caught out, neither having any aircraft to compete in this twin-engine, medium-haul, high-density market.

Boeing's answer lay in the 7X7 programme. This project was under discussion and study for 10 years, and during a long gestation had several different configurations, including the use of a T-tail, overwing engines and many other features. In July 1978, the company eventually announced the launch of the aircraft as the 767, in the same month as Airbus announced the go-ahead of its A310. Also proceeding roughly in parallel with the 7X7 was the 7N7, a narrow-body airliner intended to replace the 727, but sharing many of the features with its larger brother. Despite its redesignation as the 757, this aircraft actually followed the 767 into service.

Boeing had eventually settled on a conventional configuration for the 767, and although this mirrored that of the much smaller 737, it also looked remarkably like the Airbus it was in direct competition with. Critics, especially in Europe, were quick to point out this fact, but to be fair Boeing had pioneered the configuration nearly 20 years earlier! Also it must be remembered that there is every chance that two superb design teams will arrive at the same answer if it happens to be the right one.

Where the Boeing design differed was that it had a smaller fuselage and bigger wing. The latter was designed to provide higher altitude cruise performance, and also, not surprisingly, higher gross

weights for future enlarged versions. What was surprising was the fuselage, for although the narrow diameter of the 767 (compared to the A300 and A310) produced less drag and consequently greater range, it could not accept standard cargo/baggage containers and the standard eight-abreast seating was rather cramped. In broad terms, the Airbus products were better in load-carrying performance, but inferior in terms of range and altitude.

Despite the shortcomings of the design, the 767 was virtually assured of a large market at home, where the giant domestic carriers were crying out for a suitable type to transfer large numbers of passengers between US cities at economic rates. Nevertheless, this market provided the 767 with fewer sales than might once have been expected, for the Airbus products had made small yet notable inroads into the xenophobic US domestic market. Eastern and Pan Am were important Airbus customers, and these have been joined in recent years by several more, particularly since the smaller A320 has been available.

Notwithstanding this European competition, the 767 at once began to pile up the orders. An order of 30 from United Air Lines had sparked the final go-ahead for the airliner on 14 July 1978. Other giants such as American Airlines, Delta Air Lines, Trans World Airlines and USAir also now operate the type, although the latter's aircraft were picked up when the carrier absorbed Piedmont.

Construction of the prototype began on 6 July 1979. From the outset the 767 was designed with two different engines in mind, according to customer preference. These were the products of Pratt & Whitney (JT9D) and General Electric (CF6). The first aircraft (N767BA) flew on 26 September 1981, with Pratt & Whitney power. The next three aircraft from the line also featured this powerplant. General Electric engines powered the fifth aircraft, and this flew for the first time on 19 February 1982.

FAA certification followed an intensive flight trial period, received for the 213.5-kN (48,000-lb) thrust Pratt & Whitney JT9D-7R4D powered aircraft on 30 July 1982, allowing the first delivery to be made to United Air Lines on 19 August. The General Electric powered aircraft, with CF6-80A engines of the same power, received its certification on 30 September, 22 days after United began

Boeing's 7N7 and 7X7 projects came to fruitition at the start of the 1980s to usher in a new family of airliners to hold Boeing's dominant position in the market. Here the prototypes of the 757-200 (7N7) and 767-200 (7X7) fly together, illustrating the similarities between the types.

revenue-earning services. Delta received the first CF6-powered 767 on 25 October, with services beginning on 15 December.

These 767-200s, with the others that rapidly followed, were soon showing excellent economy over medium-haul routes, Delta's aircraft being configured with 18 passengers in first-class and 186 in economy while United opted for a 24/180 split. Both operators would join the ranks of 757 operators, flying the smaller aircraft alongside the 767 from major hub centres.

As recounted earlier, the large wing of the Boeing 767 made the basic Series 200 a natural step for considerable development. Aimed particularly at overseas customers, the 767-200ER was introduced, this offering a higher gross weight (156490 kg/345,000 lb) and increased fuel capacity for a considerable range increase. Two further increases in weight and fuel capacity were also developed under the 767ER designation, the heaviest weighing in at 175540 kg (387,000 lb). Design range was extended from under 6000 km (3,730 miles) for the basic versions, to 12611 km (7,836 miles) in the heaviest ER variant.

Ever more powerful engines, offering better fuel economy, became available for the type, covering a range of Pratt & Whitney and

Pratt & Whitney and General Electric have provided virtually all the 767 engines to date, but with the delivery in early 1990 of the first of British Airways' 767-336s, Rolls Royce joined the club with the RB211-524H. Twenty-eight aircraft were in service or on order early in 1998 to serve on high-density short/medium-haul routes.

Although United and Delta were the launch customers for the Boeing 767, other US majors followed them in adopting the type. American has a large fleet of 71, with a further 12 on order. These are split between Series 200 (illustrated) and Series 300/300ER aircraft.

The inclusive-tour market has not proved quite as lucrative for the 767 as once might have been hoped, but there are some notable customers. The UK's Britannia Airways has a fleet of 24 serving mainly Mediterranean destinations. It was the first European operator of the type.

General Electric products. The Rolls-Royce RB211-524G turbofan was added to the list, this being the most powerful powerplant yet available, providing 269.6 kN (60,600 lb) thrust for take-off.

While constantly increasing thrust led to increasing take-off weights and so greater fuel capacity and range, the large wing also enabled an increase in overall size. In February 1983, Boeing announced the go-ahead for the 767-300, retaining the basic structure of the Series 200 but with a lengthened fuselage. By placing a 3.07 m (10 ft 1 in) plug forward of the wing, and a 3.35-m (11-ft) plug aft, Boeing achieved a far more capacious aircraft with little change to the systems and construction of the aircraft. Necessary modifications were a strengthened landing gear, and greater metal thickness in key areas of the fuselage and lower wing skin to cope with greater structural loads.

Powerplant options for the lengthened 767 were, and are, the same as for the smaller jet. The first aircraft flew on 30 January 1986, powered by JT9D-7R4D fans. This was certificated on 22 September 1986, at the same time as the CF6-80A2-powered version. Three days later deliveries commenced, Japan Airlines being the first customer, specifying JT9D-powered aircraft for its domestic and Far East routes. Those for home use were completed with a single-class interior for 270 passengers. A major order for the stretched variant was gained in August 1987 when British Airways ordered RB211 powered aircraft.

In January 1985, development began down a familiar road with the announcement of the 767-300ER, this being available in three different gross weight variants, the heaviest topping the scales at an awesome 184615 kg (407,000 lb). American Airlines led the field in ordering this version, the first of its CF6-80C2B6-powered aircraft entering service in 1988.

ER versions in both short and long fuselage configurations have proved popular overseas, especially to small operators seeking new jet equipment but without the capacity requirement for a Boeing 747. Those in geographic isolation have found the incredible range performance of the type perfect for covering the longest sectors, Air Mauritius and QANTAS being good examples. Indeed, during a delivery flight on 17 April 1988, one of the Air Mauritius machines covered the 14044 km (8,727 miles) between Halifax, Nova Scotia, and its new home non-stop, in a flight lasting 16 hours 27 minutes. This sets a world distance record by a twin-engined commercial aircraft. In contrast, and in answer to some of the 767's early critics, the longest-ranged A310 can only manage 9175 km (5,700 miles).

Although other airliners are usually built in small numbers for the military, so far only one 767 has found itself in military service. This is a unique aircraft known as the 767 AOA, the initials standing for Airborne Optical Adjunct. In July 1984, Boeing Aerospace received a contract from the Strategic Defense Command of the US Army to evaluate the use of airborne optical sensors to detect and warn of in-

The prototype 767 received a new lease of life when it was adopted as the platform for the US Army's Airborne Optical Adjunct programme, carrying infra-red sensors in a giant fairing above the fuselage to detect incoming ICBM warheads as part of the Strategic Defense Initiative. Aerodynamic destabilisation caused by the fairing is offset by twin ventral strakes.

Boeing 767-2Q8(ER)
S7-AAS 'Aldabra'
Air Seychelles

coming ICBM warheads, as part of the Strategic Defense Initiative (SDI) programme. The aircraft used for the programme is the very first 767, N767BA, still wearing its civil registration and Boeing house colours.

Sensors for the programme are housed in a giant canoe fairing mounted above the aircraft's forward fuselage, faired into the fuselage immediately above the cockpit. Two large windows open to allow the sensors to peer through. The fairing extends back to about level with the rear of the wing/fuselage join. To offset the loss of longitudinal stability caused by the fairing, two ventral strakes have been added under the rear fuselage.

The principal sensor of the AOA platform is an infra-red camera, delivered by one of Boeing's sub-contractors, Hughes Aircraft Company, in July 1988. The sensor flew during 1989. Earlier, computers from the Space and Strategic Avionics Division of Honeywell had been delivered for data processing. In use, the 767 AOA flies on ranges over the Pacific, operating at standard airliner altitudes. Between 10 and 15 personnel are carried to monitor and operate the equipment. Targets are inert ICBM warheads.

Further development of the 767 occurred in 1997 when Boeing launched the 767-400ER for service from 2000, initially with Delta Air Lines and Continental Airlines. At a maximum take-off weight of 204120 kg, the 767-400ER has a fuselage lengthened by 6.4 m (21 ft) for 303 two-class passengers, an updated flight deck, revised landing gear, and a longer-span wing with raked tip extensions.

Another development was the 767-X, a new, much larger twin that would challenge the market currently catered for by the Airbus A330/A340 and McDonnell Douglas MD-11. This was to have been based on the 767 fuselage cross-section, hence the designation, but this was shelved in favour of a completely new design, the 777. The maximum take-off weight for this new model began at some 247000 kg (544,535 lb), and among the features planned for the type were folding wing tips to facilitate taxiing, approach to terminal gates and hangarage. Power was provided by offerings from the three main manufacturers in the form of the General Electric GE90-76B, Pratt & Whitney PW4077 and Rolls-Royce Trent 877, each rated at some 342 kN (76,885 lb) thrust.

Boeing pushed through the development of the 777 with all possible speed, but in the meantime the 767 led the Boeing challenge in this lucrative market. By the end of June 1998, 767 orders stood at 828 aircraft of all variants, and of these 706 had been delivered.

Good news for smaller airlines requiring a long-range aircraft but without the need for a 747, the 767ER (Extended Range) introduced more fuel and a higher take-off weight. Ethiopian Airlines was the first customer for the ER, receiving the first of three aircraft in May 1984.

Boeing 767-300ER cutaway drawing key

1. Radome
2. Weather radar scanner
3. VOR localiser aerial
4. Front pressure bulkhead
5. ILS glideslope aerials
6. Windscreen wipers
7. Windscreen panels
8. Instrument panel shroud
9. Electronic flight instrumentation system CRT displays
10. Control column
11. Rudder pedals
12. Nose undercarriage wheel bay
13. Cockpit air conditioning duct
14. Captain's seat
15. Direct vision opening side window panel
16. Centre console
17. First officer's seat
18. Overhead systems control panel
19. Crew wardrobe
20. Observer's seat
21. Cockpit bulkhead doorway
22. Second observer's seat
23. Twin pitot heads
24. Angle of attack transmitter
25. Nose undercarriage hydraulic steering jacks
26. Twin nosewheels
27. Nosewheel leg doors
28. Waste system vacuum tank
29. Forward toilet compartment
30. Forward galley unit
31. Starboard side service door
32. Galley compartment curtain
33. Door latch
34. Forward entry door
35. Door-mounted escape chute
36. Underfloor avionics equipment racks
37. Avionics cooling air systems
38. Skin heat exchanger
39. Fuselage frame and stringer construction
40. Cabin window panels
41. Six-abreast first-class seating compartment (26 seats)
42. Overhead baggage bins
43. Curtained cabin divider
44. Cabin wall trim panelling
45. Anti-collision beacon
46. Centre electronics equipment
47. Negative pressure relief valves
48. Forward freight door
49. Total freight hold volume, 3,770 cu ft [106,7 m³]
50. LD-2 baggage container; 16 in forward hold
51. Cooling air ground connection
52. Cabin pressure relief valves
53. Mid cabin door, optional
54. Conditioned air delivery ducting
55. Seven-abreast tourist-class seating, [224 seats] alternative layouts for up to 325 passengers
56. Srs 300 stretched forward fuselage segment, 121 in [3,07 m]
57. Starboard engine nacelle
58. VHF aerial
59. Sidewall toilet compartments, port and starboard
60. Conditioned air distribution manifold
61. Wing spar centre-section carry-through
62. Front spar attachment fuselage main frame
63. Ventral air conditioning pack, port and starboard
64. Centre section fuel tankage; total capacity, 20,101 Imp gal [24,140 US gal/91 400 l]
65. Floor beam construction
66. Overhead rear cabin air distribution ducting
67. Starboard wing integral fuel tank
68. Starboard engine thrust reverser cascades, open
69. Inboard slat segment
70. Nacelle pylon
71. Fixed portion of leading edge
72. Dry-bay

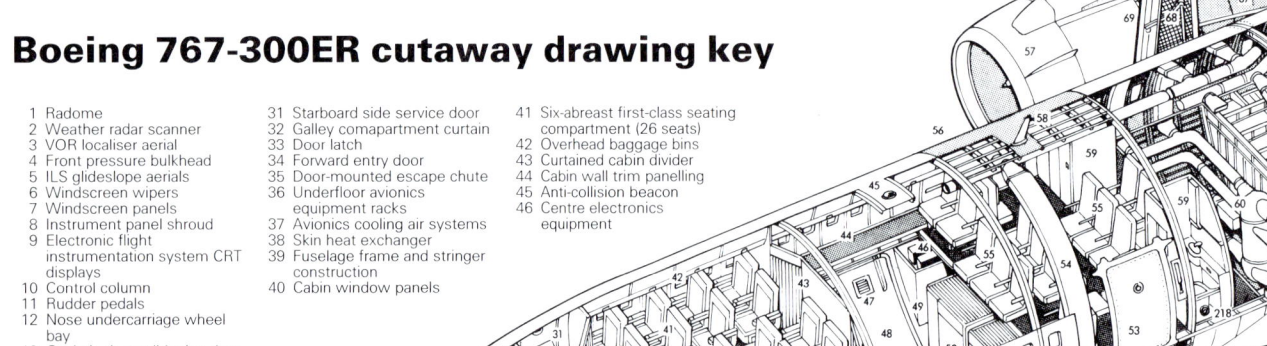

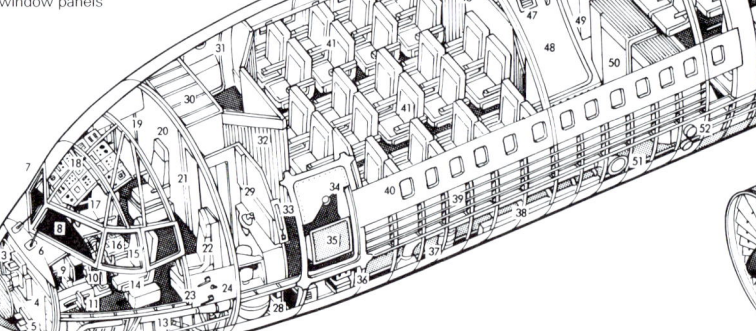

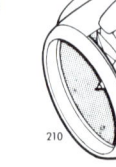

Boeing 767

73 Wing stringers
74 Wing skin panelling
75 Fuel system piping
76 Fuel venting channels
77 Leading-edge slat torque shaft
78 Rotary actuators
79 Outboard leading-edge slat segments, open
80 Starboard navigation light, green
81 Strobe light, white
82 Rear position light, white
83 Static dischargers
84 Outboard, low-speed, aileron
85 Aileron hydraulic actuators
86 Single-slotted outboard flap segment, down position
87 Flap hinge fairings
88 Flap hinge control linkages and rotary actuators
89 Outboard spoiler panels, open
90 Spoiler hydraulic jacks
91 Flap drive torque shaft
92 Inboard, high-speed, aileron
93 Aileron tandem hydraulic actuators
94 Inboard double-slotted flap segment, down position
95 Flap hinge linkages
96 Inboard spoilers/lift dumpers
97 Starboard main undercarriage hydraulic jack
98 Cabin wall insulating blankets
99 Pressure floor above wheel bay
100 Rear spar attachment fuselage main frame
101 Starboard wheel bay hydraulic reservoir
102 Central flap drive motor
103 Port main undercarriage wheel bay
104 Mainwheel door
105 Undercarriage bay rear pressure bulkhead
106 Centre cabin passenger seating
107 Passenger service units
108 Cabin wall stowage bins
109 Cabin roof trim/lighting panels
110 Centre stowage bins
111 ADF aerials
112 Rear fuselage stretch section, 132-in [3.35-m]
113 Rear freight hold door
114 LD-2 baggage containers, 14 in rear hold
115 Rear cabin seat rows
116 Centre toilet compartment
117 Sidewall toilet compartments, port and starboard
118 Starboard service door
119 Rear galley units
120 Fin root fillet
121 Fin spar box construction
122 Fin ribs
123 'Logo' light
124 Starboard tailplane
125 HFG aerial coupler
126 Television aerial
127 Fin tip aerial fairing
128 Tail VOR aerial
129 Static dischargers
130 Rudder
131 Triplex rudder hydraulic actuators
132 Graphite composite rudder construction
133 Rudder mass balance weights
134 APU intake duct
135 Firewall
136 APU intake plenum
137 Airesearch GRCP 332 auxiliary power unit [APU]
138 Tailcone
139 APU exhaust
140 Two-segment elevators
141 Elevator triplex hydraulic actuators
142 Graphite composite elevator construction
143 Static dischargers
144 Tailplane rib construction
145 Fin 'logo' light
146 Tailplane sealing plate
147 Tailplane hinge point
148 Spar box centre-section carry-through
149 Tailplane trim control screw jack
150 Fin attachment main frames
151 Rear fuselage frame and stringer construction
152 APU air delivery duct
153 Rear pressure bulkhead
154 Curtained rear lobby
155 Up-and-over door balance mechanism
156 Cabin attendants' folding seat
157 Rear cabin doorway
158 Underfloor bulk cargo hold, capacity 260 cu ft [7,3 m³]
159 Bulk cargo hold door
160 Rear main freight hold
161 Roller conveyor floor loading panel
162 Fuselage lower lobe frame and stringer construction
163 Emergency external lighting
164 Mid cabin emergency exit door
165 Wing root trailing-edge fillet
166 Air turbine driven emergency hydraulic pump
167 Flap synchronisation motor
168 Main undercarriage mounting beam
169 Hydraulic retraction jack
170 Inboard spoilers/lift dumpers
171 Flap hinge control linkage
172 Inboard double-slotted flap segment
173 Flap hinge fairing
174 Inboard, high-speed aileron
175 Flap down position
176 Outer single-slotted flap
177 Outboard spoilers
178 Flap hinge fairings
179 Graphite composite control surface construction
180 Outboard low-speed aileron
181 Static dischargers
182 Rear position light, white
183 Strobe light, white
184 Port navigation light, red
185 Port leading-edge slat segments
186 Vent surge tank
187 Rear spar
188 Wing rib construction
189 Rear spar
190 Leading-edge slat rib construction
191 Slat guide rails
192 Pressure refuelling connections
193 Slat drive shaft and rotary actuators
194 Port wing integral fuel tank
195 Four-wheel main undercarriage bogie
196 Main undercarriage leg strut
197 Undercarriage leg breaker strut
198 Port wing dry bag
199 Pylon attachment links
200 Port nacelle pylon
201 Rolls-Royce RB211-524G turbofan engine
202 Engine turbine section
203 By-pass air duct
204 Multi-lobe exhaust mixer
205 Combined exhaust nozzle
206 Pratt & Whitney PW4060 alternative powerplant
207 RB211 full-length engine nacelle
208 Thrust reverser cascades
209 Engine accessory equipment gearbox
210 Acoustically lined air intake
211 Nacelle strake
212 Inboard leading-edge slat segment
213 Inboard wing panel integral fuel tank
214 Wing root attachment rib
215 Leading-edge slat drive motor
216 Slat down position
217 Landing and taxying lamps
218 Wing inspection light
219 Kevlar composite leading-edge root fillet
220 General Electric CF6-80C2B6 alternative powrplant

BOEING 777

Neatly filling the capacity niche between the same company's 747 and 767 transports, the Boeing 777 is an all-new wide-body design that builds on Boeing's tried-and-tested concepts for twin-turbofan airliners but adds advanced features such as a choice of the 'big three' turbofans, a structure making extensive use of advanced materials, and the latest avionics including a 'fly-by-wire' control system.

The interest expressed by numerous airlines in an airliner with a capacity between those of its 747-400 and 767-300 airliners led Boeing at the end of 1986 to start on the task of defining the type that would satisfy this perceived need. The result was the 777 (originally 767-X), a transport aeroplane seen within the company as a rival to the European Airbus A330 and A340, and also to another American type, the McDonnell Douglas MD-11. The project definition phase led to the company's decision to offer the type on 8 December 1989, its receipt of the launch order (United Airlines for 34 aircraft) on 15 October 1990, and its commitment to the production of the new type on 29 October 1990.

Two engines and long range

What the Boeing design team had to create as it set to work on the 777 was an airliner with a twin-turbofan powerplant, high capacity and performance that included the ability to move its substantial payload over a long range at high cruising speed and with the lowest possible consumption of fuel. This demanded an airframe of the cleanest possible configuration and lowest structure weight compatible with safety and durability requirements. Boeing therefore adopted the CAD/CAM (Computer-Aided Design/Computer-Aided Manufacture) concept using the Dassault/IBM CATIA system for fully digital definition of the product so that all its millions of components and assemblies, large and small, could be communicated between the various elements of the design team and the manufacturing effort that was started virtually concurrently with the design process. This effort involved no fewer than 238 teams, and the exclusive use of digital design and communications ensured that the maximum accuracy (and therefore the lowest possible cost and delay) was ensured during the essentially simultaneous design, manufacture and test elements of the programme, in both its structural and systems elements. So comprehensive and accurate was the result – thought to be intrinsic to the whole process – that there was no mock-up of the 777.

Risk-sharing development programme

With a programme of the size required for the 777, Boeing sensibly acknowledged that a large measure of subcontracting was essential, as it is in all large-scale aircraft-manufacturing efforts, and also decided to reduce its own overheads and financial risks by bringing on board the programme a number of risk-sharing partners, in this instance in Japan, across the North Pacific from Boeing's home in Washington state. On 21 May 1991 Boeing signed an agreement with Fuji, Kawasaki and Mitsubishi, bringing these three proven companies into the 777 programme as risk-sharing partners responsible for about 20 per cent of the 777's structure, including the central and rear fuselage barrel sections, tail cone, doors, wing-root fairing and landing-gear doors.

Subcontracted manufacture

Boeing also contracted more directly for the construction and delivery of many other parts of the 777's airframe. Thus while Boeing itself is responsible for the flight deck and forward cabin, main structural elements of the wing and tail unit, and engine nacelles, the leading edges of the wing and tail unit, moving parts of the wing, landing gear, floor beams, nose landing-gear doors, wing tips, dorsal fin and nose radome were initially entrusted to a team of subcontractors including Rockwell, Northrop Grumman and Kaman in the USA as well as Alenia in Italy, EMBRAER in Brazil, Shorts in the UK, Singapore Aerospace Manufacturing, Hawker de Havilland and AeroSpace Technologies of Australia in Australia, Korean Air and also a number of smaller subcontractors. Boeing of course kept under its own control the final assembly process and the testing of completed aircraft.

Boeing rolled out the first 777, completed at the plant of its Everett Division (the wide-body element also responsible for the 747 and 767 while the Renton Division controls the manufacture of narrow-body types such as the 737 and 757) on 9 April 1994, and this initial aircraft made its maiden flight on 12 June of the same year. The first machine was followed in the same year by the second aircraft on 15 July, the third on 2 August, the fourth on 28 October and the fifth on 11 November. These aircraft were all powered by two Pratt & Whitney PW4000-series turbofans, while the sixth aeroplane, which was the first for British Airways, was also the first to be completed (on 2 February 1995) with an alternative powerplant, in this instance two General Electric GE.90 turbofans. By the middle of the same month, the Boeing test programme with the six completed aircraft had amassed some 1,950 hours in the air during 1,000 flights. On 19 April 1995 the Federal Aviation Administration and its European counterpart, the Joint Airworthiness Authorities, granted type certification to the 777 in its initial form with Pratt & Whitney engines after the type had accumulated 3,235 hours in the air and 2,340 flight cycles. This high-intensity effort had also successfully involved a 1,000-cycle ETOPS (Extended-range Twin-engined Operations) qualification by the fourth aircraft: this ETOPS qualification allowed the type to undertake overwater flights in which its route took it no further than 180 minutes' flying time from a diversionary aircraft after the failure of a single engine. For this qualification, gained on 30 May 1995 and absolutely essential for airlines wanting to operate the new airliner on long over-water routes such as those across the Atlantic and Pacific Oceans, the fourth aircraft was fitted with high-time engines (specified to mirror

The 777 is the latest civil air transport from the Boeing stable, and also the company's first airliner with a 'fly-by-wire' system. As such, the type was conceived in response to the inroads of Airbus into Boeing's traditional market.

the real world in which the type would be operating) and completed eight single-engined diversions each lasting 180 minutes.

ETOPS qualification programme

Flight crews from United Airlines were involved, together with Boeing personnel, in the last 90 of these ETOPS cycles, which were undertaken between airports on the airline's route network. This allowed the airline to begin the training of its crews during the final stages of the certification and qualification process, and as a result the airline received its first 777 on 17 May 1995 and began revenue-earning services on 7 June of the same year with a service between London in the UK and Washington, DC, in the USA. It is worth noting that the first and second 777 airliners with GE.90 engines, both of them intended for British Airways, undertook a similar 180-minute ETOPS qualification programme for GE.90-powered machines in the course of a 1,750-hour programme of 1,260 cycles, with the last 90 including British Airways personnel. The aircraft were based at London and, to provide realistic experience of British Airways' route network and operating conditions, flew to points such as Abu Dhabi in the Persian Gulf as well as Washington, DC, and Newark, New Jersey, on the east coast of the USA. Certification of the 777 with the GE.90 powerplant was gained in the autumn of 1995 with deliveries to British Airways following almost immediately after this.

Layout of classical simplicity

The third engine type qualified on the 777 is the Rolls-Royce Trent 800, and after the engine had been first flown on a 747-100 test bed machine late in March 1995, qualification of the 777/Trent combination began in May with a Boeing-owned aircraft that was supplemented from August by the first of the Trent-powered aircraft ordered by Cathay Pacific. This qualification programme called for two aircraft to complete 1,220 cycles (including a 1,000-cycle ETOPS element) in the course of 1,700 flight hours on routes such as those linking Hong Kong (Cathay Pacific's base) and Singapore and Sydney. The flight test programme was completed in March 1996 after some 4,900 cycles in the course of 7,000 flight hours.

In overall terms the 777 is a low-wing cantilever monoplane of mixed construction, and among its basic features are an all-new wing characterised by a quarter-chord sweep angle of 31.6° and an advanced airfoil section to permit a fuel-efficient cruising speed of Mach 0.83 but yet possessing sufficient thickness to allow the design of an efficient structure with considerable internal volume for fuel. The wing was designed without the type of drag-reducing winglets then coming into fashion, but with a long span as a means of maximizing take-off and payload/range performance, and a large area for an optimum combination of a high cruising altitude and a low approach speed. Given the 777's considerable span, an option offered from the beginning was provision for the outer 6.48 m (21ft 3in) of each half-span to fold up into the vertical position as a means of reducing overall width on the ground, and thereby facilitating the use of current airport gates.

Wide-body fuselage

The fuselage is cylindrical and of a width greater than that of the 767 to allow the incorporation of a twin-aisle seating arrangement with passengers located between six and ten abreast. Another key element of the design of the accommodation section was the creation of overhead baggage lockers and toilet facilities that would permit rapid change between specific configurations.

Right: All Nippon Airways was the second customer for the 777-200, of which it contracted for 18 examples on 19 December 1990. Since that time the airline has also ordered the higher-capacity 777-300.

Boeing's first 'fly-by-wire' airplane

As noted above, a major innovation in the 777 is the 'fly-by-wire' electrically signalled flight-control system, the first in a Boeing airliner and produced in belated response to Airbus's use of such a system in rival aircraft. All the control surfaces are hydraulically moved in response to electric commands from the 'fly-by-wire' system which controls the slats, flaps, spoilers, inboard flaperons, outboard ailerons, trimming tailplane, elevators and rudder by means of Teigin Seiki America actuators. The complete system offers protection to the flight envelope and also stabilisation and autopilot inputs, with the standard control columns and rudder pedals on the flight deck also moved to offer the two pilots a physical appreciation of the automatic system's functioning.

The flight guidance commands are, in normal mode, created by triply redundant Rockwell Collins digital autopilot/flight directors, with the resulting 'fly-by-wire' control laws and flight envelope protection commands supervised via the three GEC-Marconi Avionics digital primary flight computers, each containing three 32-bit processors. Commands to the powered control units are generated by three Lear Astronics and Teigin Seiki actuator control units, which have a fourth but analogue channel signalled from the sticks and pedals on the flight deck. Thus in the normal operating mode the 777 is flown via the autopilots, primary flight computers and actuator control electronics, with the first degraded (or secondary) mode employed if inertial units and standby attitude sensors all become disabled and the pilots take manual control through the primary flight computers. The second degraded (or direct) mode bypasses the main 'fly-by-wire' system by use of the direct analogue

link between cockpit and actuator control electronics, and the final stand-by system is mechanical control of tailplane incidence for the pitch axis and two wing spoiler panels for lateral control.

Key elements of the 'fly-by-wire' control system are longitudinal control by a system that in effect makes the airliner hold a given airspeed and respond to any departure from that speed by a variation in pitch attitude, the suppression of trim changes resulting from any change in configuration, return to a bank angle of 35° if the pilots exceed that angle and then release the controls, prevention of the aircraft exceeding the limiting airspeed and stalling, countering of any asymmetric thrust, and adjustment of the variable feel system to warn the pilots of any approach to the limits of the flight envelope in manually controlled flight.

Extensive use of composite materials

In structural terms the 777 is fabricated mostly of high-strength aluminium alloys, with 10 per cent of the structure weight provided by composite materials. The latter included a carbon/toughened resin composite for tailplane skins, fin torsion boxes and cabin floor beams, CFRP (Carbon-Fibre-Reinforced Plastics) for the ailerons, flaps, elevators, rudder, engine nacelles and landing gear doors, hybrid composites for the wing-root fairings, and GFRP (Glassfibre-Reinforced Plastics) in the fixed leading edge of the wing, wing aft panels, fore and aft panels of the tailplane and fin, engine pylon fairings and radome.

The 777's landing gear is of the fully retractable tricycle type, with each of the two main units (designed collaboratively by the Franco-American teaming of Menasco and Messier-Bugatti) carrying six-wheel bogies with steering rear axles that are automatically engaged by the steering angle of the twin-wheel nose unit: the tyres on the nose and main units are of Michelin and Goodyear manufacture respectively. The use of six-wheel bogies removed the need for a centreline third main unit under the fuselage and also simplified the system used to work the Bendix Carbenix 4000 mainwheel brakes.

The twin-engined powerplant used in the models of the 777 series (see below) uses variants of three different types of engine. The 777-200 A Market at basic and first alternative take-off weights has two Pratt & Whitney PW4074, or General Electric GE.90-75B or Rolls-Royce Trent 875 units, while the 777-200 A Market at second alternative maximum take-off weight is powered by two PW4077, or GE.90-76B or Trent 877 units, these engines being rated typically at between 33566 kg (74,000 lb) thrust and 34927 kg (77,000 lb) thrust. At all three of its maximum take-off weight options the 777-200 B Market is powered by two PW4084, or GE.90-85B or Trent 884 units typically rated at 38102 kg (84,000 lb) thrust, or PW4090 units rated at more than 40824 kg (90,000 lb) thrust. Growth versions of these engines later made available are the PW4098 rated at 44453 kg (98,000 lb) thrust, GE.90-95B rated at 43137 kg (95,100 lb) thrust and Trent 892 rated at 40790 kg (89,925 lb) thrust.

The fuel for these engines is carried by integral tanks in the wing's torsion box and also in the centre section of the wing, with the reserve tank, the surge tank and the fuel vents and jettison pipes all inboard of wing fold. The main, centre and reserve tanks have a total capacity of 117348 litres (31,000 US gal; 25,813 Imp gal) in the A Market types and of 169208 litres (44,700 US gal; 37,220 Imp gal) in the B Market types.

With a two-man flight crew and a various number of cabin crew, the 777-200 can carry between 305 and 440 passengers in a cabin whose twin-aisle cross section, which is between those of the 747 and 767, was selected to provide the greatest choice of class and seating configurations between six and ten abreast. The galleys and toilet facilities can be installed at a selection of fixed locations in the front and rear cabins, or can otherwise be fitted within large installation footprints in which they can be moved in 2.5-cm (1-in) increments for attachment to pre-positioned mounting, plumbing and electrical fit-

tings. The overhead baggage bins can be removed without disturbing the ceiling panels, overhead ducting and support structure, and the advanced cabin management system incorporates a digital sound system.

Under the cabin floor are cargo compartments with mechanical handling systems and able to carry all LD containers and/or pallets of 2.24- or 2.44-m (88- or 96-in) width. A typical underfloor freight load is 33 LD-3 containers and 600 cu ft (16.99 m^3) of bulk cargo. The underfloor accommodation can also be used for a crew rest module with four bunks, two business-class seats and stowage volume. This module requires only an electrical connection and a hatch in the cabin floor for access.

Advanced avionics on the flight deck

As can be imagined, the 777's systems are complex and comprehensive, and include an Allied Signal air drive unit (using air tapped from the engines, an APU or a ground supply) to power the central hydraulic system, an AiResearch cabin air supply and pressure control system, a Sundstrand variable-speed constant-frequency AC electrical power generating system with two 120 kVA integrated drive generators, an APU-driven generator and an Allied Signal ram air turbine system, an Allied Signal GTCP331-500 auxiliary power unit, a Hamilton Standard air-conditioning system, a Smiths Industries ultrasonic fuel gauge system, and a Smiths Industries electrical load management system.

Aircraft information and flight management systems

The 777's avionics are equally complex and comprehensive, including Honeywell weather radar as standard. Other elements of the avionics are the main navigation system, based on a Honeywell ADIR (Air Data and Inertial Reference) system and containing six ring-laser gyros, backed by a standby system based on the SAARU (Secondary Attitude and Air Data Reference Unit) containing interferometric fibre optic gyros and responsible for producing a secondary flight director attitude display, airspeed and altimeter. Both of these systems are, of course, linked to the aircraft's digital database, as too are the Bendix/King TCAS (Traffic-alert and Collision-Avoidance System), the Honeywell/Canadian Marconi GPS (Global Positioning System) navigation satellite sensor, and the optional Honeywell/Racal multi-channel satellite communications system.

The dual Honeywell AIMS (Aircraft Information Management System) incorporates the equipment required to collect, format and distribute onboard avionics information, including the FMS (Flight Management System), engine thrust control, digital communications management, flight deck display operation and aircraft condition monitoring.

Five-screen EFIS-type instrumentation

The flight deck instrumentation is centred on a five-screen EFIS (Electronic Flight Instrumentation System) using Honeywell colour flat-panel displays of the liquid crystal type and providing two primary flight displays, two navigation displays and one engine indication and crew alerting system display. Three multi-purpose control and colour display units on centre console provide interface with the AIMS, which handles flight management, thrust control, communications control and all systems information.

Poewerplant options

The 777 was initially offered in two versions known as the 777-200 A Market and 777-200 B Market, which became the 777-200 and 777-200IGW respectively before the latter was finally designated as the 777-2300ER, with the latter suffix standing for Extended Range. The 777-200 variant has a maximum take-off weight of 229,522 kg (506,000 lb) and alternative maximum take-off weights of 233,604 kg (515,000 lb) and 24,2676 kg (535,000 lb), and the variant's maximum payload is 54886 kg (121,100 lb) including between 375 and 400 two-class passengers, or between 305 and 328 three-class passengers or between 418 and 440 economy-class passengers depending on the use of nine- or ten-abreast seating in tourist class and seven- or eight-abreast seating in business class. With 375 passengers, the ranges for the 777-200 in its three maximum take-off weight options are respectively 7278 km (4,522 miles), 7778 km (4,833 miles) and 8862 km (5,506 miles).

The 777-200ER has a maximum take-off weight of 263,088 kg (580,000 lb), or 267,624 kg (590,000 lb) or 286,90 kg (632,500 lb), and the variant's maximum payload is 54659 kg (120,500 lb) with the same passenger capacities as the 777-200. With 305 passengers, the ranges for the 777-200ER in its three maximum take-off weight options are respectively 11167 km (6,939 miles), 11,667 km (7,250 miles) or 13,584 km (8,441 miles).

The 777-200 C Market was proposed at one time as an ultra-long-range model capable of flying a typical payload to a range of (12,964 km (8,055 miles).

The 777-300 was initially known as the 777 Stretch, and was conceived as successor to the 747 'Classic' series. Revealed at the Paris Air Show in June 1995, the 777-300 was formally launched by Boeing later in the same month as a development of the 777-200 with structural strengthening of the fuselage, inboard wing sections and landing gear to allow the lengthening of the fuselage by 10.13 m (33 ft 3in) for the carriage of up to 368 passengers in a typical three-class layout or up to 550 passengers in a single-class layout. The 777-300 has a maximum take-off weight of between 158,737 kg (352,750lb) and 300,000 kg (661,375 lb) with a maximum payload of 313,500 kg (142,500lb).

Early commitments

The commitments that persuaded Boeing to launch the 777-300 totalled 31 aircraft, and comprised 10 aircraft with derated PW4098 engines for All Nippon Airways for service from June 1998, seven aircraft (converted from 777-200 orders) with Trent 800 engines for Cathay Pacific Airways with delivery from May 1998, eight aircraft (including four conversions from 777-200 orders) with PW4098 engines for Korean Airlines with delivery from June 1998, and six aircraft with Trent 800 engines for Thai Airways International for service from September 1998.

In overall terms, the 777-300 is designed to carry the same number of passengers as the 747 'Classic' aircraft but at only 67 per cent of the fuel cost and 60 percent of the maintenance cost. The first 777-300 flew in October 1997, and the first deliveries were in fact made to Cathay Pacific in May 1998. Up to the end of June 1998, Boeing held orders for 392 examples of the 777 airliner, and of these 141 had been delivered.

With two variants of the 777 in production and service, Boeing is now focusing its design attention on improved ultra-long-range performance for the next variants of the 777 family under the temporary designations 777-200X and 777-300X. These would have still higher maximum take-off weights, increased fuel capacities, and engines with thrust levels of more than 40,824 kg (90,000 lb). This allows a range of 16,092 km (10,000 miles) in the case of the 777-200X, and more than 12,070 km (7,500 miles) in the case of the 777-300X.

As currently envisaged, the 777-200X and 777-300X have maximum take-off weights of 333,010 kg (734,150 lb) and 323,927 kg (714,125 lb) respectively, but Boeing is considering the possibility of increasing these figures by between 4,500 kg (9,900 lb) and 6,781 kg

Previous pages: This publicity photograph nicely contrasts Boeing's 247, that made its maiden flight in February 1933, with the second example of the 777 in the colours of United Airlines.

Above: The Rolls Royce Trent 800 was the first engine to be certificated at 40,824kg (90,000 lb)thrust. Here a Cathy Pacific 777 is seen during a flight test for commercial operations, which began in January 1996.

(14,950 lb) as a means of enhancing the types' payload/range performance. Achievement of these objectives would demand another one or two percent of power from the engines. Boeing believes that while the variants have the right payload/range capability, they need better take-off performance. As a result the company is investigating the possibility of adding a third engine in the tail of the 777-200X and 777-300X. This unit would also serve as the APU. As such, the additional power of what is being dubbed an APTU (Auxiliary Power and Thrust Unit) would supplement the main engines for take-off and the first stages of the climb, but would otherwise provide the capabilities of a standard APU. Dual-role engines under active consideration in the thrust range between 3,175 kg (7,000 lb) and 6804 kg (15,000 lb) include Rolls-Royce Allison AE3007, BMW Rolls-Royce BR.710 and General Electric CF34-8.

Drag reduction possibilities

Boeing is also examining lift/drag improvements for better take-off, better braking and enhanced thrust, and is reconsidering the use of larger raked tips for the wing, such as those used on the 767-400, or winglets.

Specification
Boeing 737-700

Type: long-range high-capacity airliner

Powerplant: two General Electric GE.90-75B turbofans each rated at 34,655 kg (76,400 lb) thrust, or Pratt & Whitney PW4077 turbofans each rated at 34,927 kg ((77,000 lb) thrust, or Rolls-Royce Trent 875 turbofans each rated at 35,335 kg (77,900 lb) thrust

Dimensions: wing span 60.93 m (199 ft 11in); length 63.73 m (209 ft 1in); height 18.51 m (60 ft 9in); wing area 427.8 m^3 (4,605.0sq ft)

Weights: empty 135,581 kg (298,900 lb); maximum take-off (242,676 kg) (535,000 lb)

Performance: maximum cruising speed Mach 0.87; economical cruising speed Mach 0.83; maximum certificated altitude 13,135 m (43,100 ft); range 8,926 km (5,546 miles) with 375 two-class passengers

BOEING C-97 and Stratocruiser

The giant rounded nose and fin tower above the observer. As the engines cough to life the power shakes the ground, while the huge propellers thrash the air. The Boeing Stratocruiser is a true classic of aviation, serving in civil and military guises and proving its reliability, versatility and longevity in both. Here is the story of the aircraft, from its early development through to the last survivors still flying today.

During World War II the urgent need to concentrate on combat aircraft inevitably slowed the development of transport aircraft. One example is the Avro York, based on the Lancaster, which flew in 1942 but hardly got into production until 1945. In the same way Boeing designed a transport version of the B-29 Superfortress in 1941-2, and received a US Army Air Force order for three XC-97 prototypes in December of the latter year, but was unable to complete one until mid-1944. Farsighted Boeing could see that this might put the firm at a competitive disadvantage after the war, because though Lockheed's Constellation and Douglas's DC-4 and DC-6 were smaller and basically older, they would be immediately available.

In 1942, however, Boeing had to shelve thoughts of the airline market and concentrate on production for the USAAF. By 20 July of that year it completed an engineering summary for the Model 367, and this was soon accepted. Changes from the bomber were mainly confined to the fuselage. While keeping the lower part of the B-29 fuselage (with cargo/baggage holds instead of bomb bays) the Model 367 added a much larger new upper lobe with a diameter of 3.35 m (11 ft), which introduced what became known as the double-bubble cross-section seen on most of today's airliners (though today the sides are filled in to reduce drag). At the time the fuselage of the first XC-97 looked enormously large, and gave a whale-like appearance with a bluff bow that belied the aircraft's extremely high speed.

The first prototype, which received the type name Strato- freighter, made its maiden flight at Seattle on 15 November 1944. Handling was as good as that of a B-29, and speed and mission capability were more than amply demonstrated on 9 January 1945 when the same ship, AAF no. 43-27470, flew from Seattle to Washington DC with 9072-kg (20,000 lb) payload in 6 hours 4 minutes at the remarkable average of 616 km/h (383 mph). By this time the 1641-kW (2,200-hp) Wright R-3350-23 engines had been replaced by R-3350-57As rated at 1734 kW (2,325 hp), but Boeing had plans for a much more powerful version, paralleling the XB-44 (later B-50) bomber. The USAAF ordered six YC-97s broadly similar to the three prototypes, as well as four of the more powerful type: three YC-97As and a YC-97B.

The YC-97As introduced the 28-cylinder Pratt & Whitney R-4360 Wasp Major engine, initially rated at 2237-kW (3,000-hp), which in fact fitted inside a slimmer cowling than the R-3350. The nacelles were of different outline, and new four-bladed propellers were fitted with broad square-tipped paddle blades formed by welding together thin steel sheets. To meet the need for holding a straight course with an outer engine stopped the vertical tail was made much taller, with provision for folding, and other changes included the addition of com-

USAF 53-230 is one of the surviving machines, immaculately cleaned and pictured as a 'museum bird' in 1982. A KC-97L, it is preserved in the livery of its last operator, the 134th Air Refuelling Group, Tennessee Air National Guard. Based at Knoxville, this group today flies the KC-135A. Note the black-top (all aspect) doghouse fairing.

USAF 48-399 was the third of the first batch of 50 production aircraft, designated C-97A and with maker's designation 367-4-19. Features included extra outer-wing fuel tanks, APS-42 weather radar and special provision for rapid air-dropping of stores, usually by parachute, through the open rear clamshell doors.

United Air Lines was one of the civil customers to specify rectangular passenger windows, on their Model 377-10-34. This example, N31225 Mainliner Hawaii, was sold to BOAC in early 1955 and operated as G-ANTX Cleopatra; on 1 March 1958 she flew the first London-Kano-Accra-Lagos service. She was finally used for Guppy parts.

plete thermal de-icing and a switch to the new 75ST alloy for the structure of the wings and certain other parts. A total of 26670 litres (5,867 Imp gal) of fuel was accommodated in a new fuel system with 35 nylon bladder cells, three in the centre section and 16 in each outer wing. There was provision for 18598 kg (41,000 lb) of cargo which could be loaded via a giant rear ramp door and positioned by an electric hoist on rails running the full 23.8 m (78 ft) length of the hold. Trucks or small armoured vehicles could drive on board, and in the troop carrier role no fewer than 134 seats could be installed, or 83 litters (stretchers) and four medical attendant seats. The single YC-97B (45-59596) was valuable as the prototype of what Boeing hoped would be a commercial version. A VIP personnel transport, it had rows of circular windows, upper-deck seating arranged 3 + 2 for 80, an upper-deck galley, and (a feature that was to be a memorable attraction of the civil machine) a spiral stairway down to a lower-deck lounge. The rear ramp door was removed, but in addition to a full passenger load the YC-97B could carry 7711 kg (17,000 lb) of cargo and baggage in lower-deck holds.

The first of the six YC-97s, last of the Wright-engined Boeings, flew on 11 March 1947. These machines transformed the Air Transport Command (later MATS, Military Air Transport Service) route to Hawaii. In 1948 the first of the YC-97As flew at an average daily utilisation of between nine and 12 hours on the Berlin Airlift, greatly relieving fears that the Wasp Major and its twin new-design General Electric turbochargers would prove unreliable. Boeing had few worries over the viability of the C-97, and in 1948 the newly-created USAF ordered 27, soon increased to 50. These first production Stratofreighters had Hamilton Standard propellers, and were visibly distinguished by a chin excrescence housing APS-42 weather radar.

Commercial successes

Boeing was much more concerned about the civil airline model which, though it was virtually the same as the YC-97B, was given the different type number Model 377. The airlines, naturally conservative, appreciated the speed, range and payload of the Model 377, but none wanted to buy it. The minus factors were the new and very complicated engines and propellers, and the sheer size (which had always frightened even the busy trunk-route operators) and the vagueness of timing and price. It was for the same reasons that Pan Am, the first prospect, had lately declined to buy the DC-7, the civil version of the giant C-74 Globemaster. How could Boeing get started? It was a time of giant plants lying empty or ticking over, with long queues at state unemployment benefit windows. In one of his first major decisions in late 1945, the new Boeing president, lawyer Bill Allen, decided to build 50 Model 377s at company risk. The name was obvious: Stratocruiser.

Though the commercial risk was minimised by the close similarity between the Model 377 and the already funded YC-97B, the relief was enormous when Pan Am signed for 20 aircraft in June 1946. Other orders followed: four for SAS, eight for AOA (American Over-

The first Boeing XC-97 prototype photographed before the adoption of the CS-470 fuselage 'buzz number'. Designated Model 367-11, the aircraft was powered by four Wright R-3350-23 1640-kW (2200-hp) piston engines. Note the short tail.

American Overseas Airlines acquired seven Boeing Stratocruisers, the first, Flagship Great Britain, being delivered in 1949. Subsequently serving with Pan American World Airways, the aircraft was written off in Tokyo in 1959.

A Boeing Airplane Company (as it then was) photograph, this is an historic aircraft, the very last of the entire C-97/Stratocruiser tribe to be built (it was the 592nd **KC**-97G, and should have had block number 150, but this was never assigned to the final two aircraft). Further aircraft, 53-3817/3824, were cancelled.

Both of the converted Boeing KC-97Gs are shown after installation of Pratt & Whitney YT34-P-5 turboprop engines. Redesignated YC-97J, the aircraft were evaluated by the MATS 1760th Air Transport Group.

seas Airlines), 10 for Northwest, six for BOAC and seven for United. The Pan Am order, worth $24,500,000, was said at the time to be the biggest purchase of civil airliners in history. The first Stratocruiser was the very next aircraft on the line after the YC-97B, and it made its first flight on 8 July 1947. After much more than the planned amount of development the Model 377 was certificated and opened its Pan Am service on 7 September 1948. Passenger reaction was highly positive. The Stratocruiser was the largest civil transport flying, and certainly the most capacious with seating (not on Pan Am) for 100 on the main deck and a possible 14 additional seats in the lower-deck cabin (which Pan Am and some other operators used as a cocktail lounge, the seats not being sold). What Pam Am did do was adopt the main-deck sleeper facilities for 28 (up to 45 in other versions), and BOAC did the same, the bunks swinging down from the upper sides of the lofty ceiling which followed the curvature of the fuselage.

Pan Am's fleet of 'clipper ships' eventually grew to 27 by adding the prototype plus six of the AOA aircraft. All were retrofitted with more powerful turbochargers, and 10 had extra tankage for Atlantic flights, the eastbound flights invariably being nonstop in something like 11 hours. BOAC's very first delivery, on 15 October 1949, was nonstop from New York in only 10 hours 15 minutes, and the obvious lack of a British aircraft in the same class muted the criticism at yet

Throughout the 1970s Spain's Ejercito del Air (air force) operated three KC-97Ls (ex-USAG Air National Guard). They were assigned to 123 Sqn in 12 Wing to support the F-4C Phantoms at Torrejon. In 1976 the first three KC-130 tankers replaced them in regular use, but they were not scrapped immediately. Designation was **TK-1**.

Boeing Model 377 Stratocruiser cutaway drawing key

1. Windscreen panels
2. Instrument panel shroud
3. Back of instrument panel
4. Rudder pedals
5. Control column handwheel
6. Cockpit eyebrow windows
7. Overhead switch panel
8. Co-pilot's seat
9. Centre control console
10. Pilot's seat
11. Nosewheel steering wheel
12. Cockpit floor level
13. Underfloor control runs
14. Nose undercarriage pivot fixing
15. Nosewheel leg door
16. Twin nosewheels
17. Steering control jacks
18. Torque scissor links
19. Nosewheel rear strut
20. Retraction screw jack
21. Retraction motor
22. Folding observer's seat
23. Radio operator's station
24. Engineer's swivelling seat
25. Flight engineer's instrument panel
26. VOR aerial
27. Cockpit roof glazed hatch
28. Radio and electronics rack
29. Aerial lead-in
30. Crew toilet
31. Cabin bulkhead
32. Flight deck doorway
33. Navigator's seat
34. Chart table
35. Crew access ladder to lower deck
36. Nosewheel well

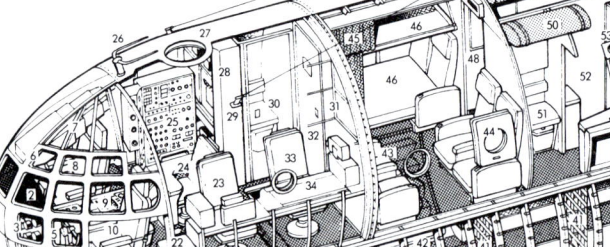

37. Fuselage nose section joint frame
38. D/F loop aerials
39. ADF sense aerial
40. Forward cargo hold
41. Baggage restraint nets
42. Starboard side baggage door
43. First-class seating compartment, eight seats
44. Forward emergency exit window
45. Privacy curtaining
46. Fold-down double bunks, upper and lower
47. HF aerial mast
48. Ladies' cloakroom and toilet, starboard side
49. Main water tank
50. Drinking water tank
51. Toilet
52. Men's cloakroom
53. Wing inspection light
54. Wash basins
55. Drinking water dispenser
56. Magazine rack
57. Upper lobe section joint frame
58. Wing spar centre section carry-through
59. Centre section fuel cells
60. Fuselage centre section construction
61. Emergency exit windows, port and starboard
62. Fuel filler cap
63. Inner wing panel fuel cells; total fuel capacity 7,790 US gal (29450 litres)
64. Starboard inner engine nacelle
65. Intercooler air flap

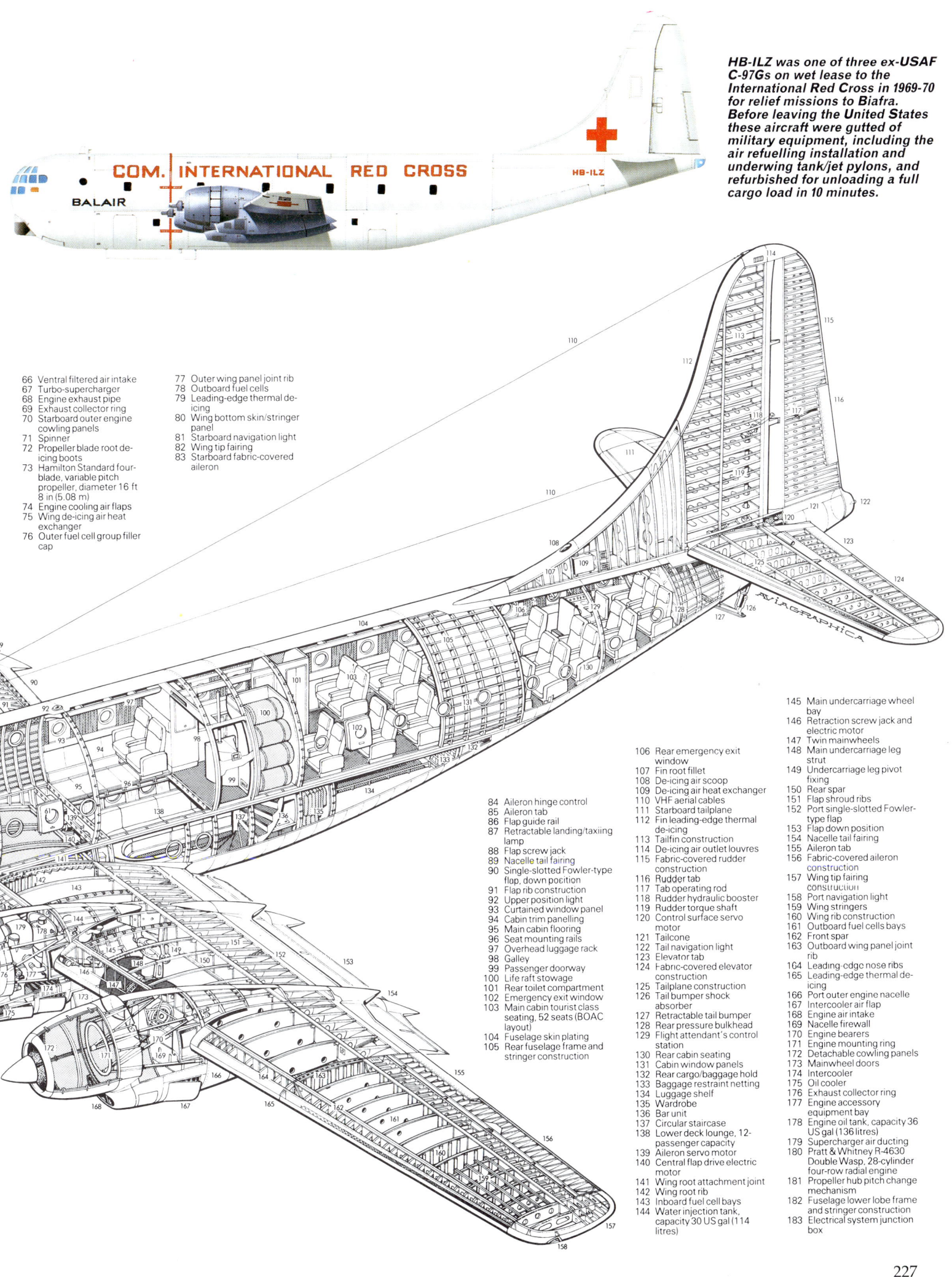

4X-FPV is still in Israel, though no longer airworthy. It was one of five purchased by IAI in February 1962 and rebuilt for service in Israel. Many more Israeli 'Strats' were former C-97s, but 4X-FPV 'Arbel' was ex-BOAC and served as a crew trainer and cargo carrier, with side and rear clamshell doors.

USAF 52-2632 was a KC-976-120-BO, one of the 592 basic G series with full provision for tanking, air refuelling and all forms of troop and cargo transport. Normal interior equipment provided for 96 equipped troops or 69 litter (stretcher) casualties, without the removal of transfer tanks and boom operator's station.

again buying a foreign-built machine. The BOAC aircraft repeated the names of the Short 'Empire' flying-boats, and the British fleet grew to 17, at first by taking over before delivery the four ordered by SAS, and later by adding six after service with United, plus one from Pan Am in 1954, to make good the absence of Comet 1s and 2s. Originally delivered unpainted, the BOAC aircraft were improved in appearance by the newly fashionable white tops introduced in 1951 to reduce cabin temperature on hot airfields. The BOAC Stratocruisers gave excellent service until 1958-9.

Northwest and United specified Douglas-style square passenger windows. Accommodation differences are listed under Variants. From 1959 the two chief operators were Transocean (USA) and RANSA (Venezuela), while in 1962 five bought by the infant Israel Aircraft Industries started an Israeli connection that eventually involved 15 aircraft, including ex-USAF C-97s, used for a host of military and para-military duties after extensive local rebuilds.

Outbreak of war in Korea in June 1950 opened the funding floodgates, and brought orders for C-97s far surpassing Boeing's wildest dreams. These came on top of gigantic orders for the B-47 and B-52 jet bombers, and Boeing was hard-pressed to find enough floorspace and manpower. The entire C-97 and Stratocruiser programme was eventually moved to the Renton plant, while Seattle and Wichita were assigned other tasks. By far the biggest requirement was for inflight-refuelling tankers for SAC, largely in support of Boeing's own B-47, and this led to the tests with 'Flying Boom' refuelling on three KC-97As followed by the production KC-97E, KC-97F and KC-97G. The massive run of 592 of the last brought the total up to 888 military C-97s in all, the last going out of the door on 18 July 1956.

In typical Boeing style, the aircraft was developed out of all recognition. Internal fuel capacity was originally a little over 26498 litres (7,000 US gal), but from the KC-97E tanker the addition of a

The second Boeing Stratocruiser to be completed and flown, NX1039V was used by Boeing for certification testing before passing to Pan American, with whom it served for three years.

With its serial number prefaced by the O (for obsolete), 52-0901 was a USAF KC-97L, originally delivered as a KC-97G-110-BO and rebuilt by Hayes with J47-GE-25A booster jet pods in place of the original long-range underwing tanks, the extra communications 'doghouse' aerial blister above the fuselage and a complete systems update.

27258-litre (7,200-US gal/5,996-Imp gal) group of tanks on the main floor in the fuselage brought the total up to a remarkable 56750 litres (14,990 US gal/12,483 Imp gal), almost all of which could be transferred through the patented Boeing Flying Boom worked by an operator lying in the rear fuselage. All mainstream tankers were designed for rapid conversion to transports, with removal of the fuselage tanks, boom and operator station. The final model, the KC-97G, transferred part of the fuel to giant fixed underwing tanks, leaving enough room in the fuselage for a large cargo payload without any modification.

The vast SAC tanker force, 20 aircraft to every Bomber Wing, gradually became all-jet as the Boeing KC-135A replaced the KC-97 Stratotanker from 1957. Many of the well-liked KC-97s soldiered on as pure transports, but as late as 1964 a considerable number serving with the Air National Guard, some of them in Vietnam, were fitted with the jet booster pods taken from KB-50s to give extra height and speed, thus becoming KC-97Ls. The only other C-97s with anything like the same performance were the two YC-97Js fitted with the 4250-kW (5,700-hp) Pratt & Whitney T34-5 turboprops.

Guppy re-builds

Also included in the Variants list are the Guppy family of rebuilds which, amazing as it would have seemed in 1944, prove that, in adding a giant upper lobe to turn the B-29 into the C-97, Boeing was not really trying. The man behind the grotesquely swollen Guppy series was Jack Conroy, who in 1961 formed Aero Spacelines Inc. to produce special aircraft to transport giant rocket stages for the US space programme, notably the S-II stages of the Saturn V (Apollo) launcher. First, he took an old Stratocruiser and got On-Mark Engineering to splice in an extra 5.08 m (16 ft 8 in) section behind the wing. After flight-testing this, he boldly built on a gigantic new upper fuselage lobe able to carry rocket stages with a diameter of 6.02 m (19 ft 9 in). The result, flown on 19 September 1962, looked impossibly corpulent, but it flew well. Someone said 'It looks like a pregnant guppy', and the name stuck, the official type designation becoming B-377PG.

Nor was even this the limit. NASA wanted to deliver by air the much larger S-IVB stage, the second stage of the same Saturn V vehicle. To fly this monster load Aero Spacelines created the Super Guppy, or B-377SG. This was substantially redesigned with a new tail, longer-span wings and a colossal fuselage 9.4 m (30 ft 10 in) longer and with the upper lobe able to accept loads with a diameter of 7.62 m (25 ft). As a starting point Conroy selected the powerful YC-97J, with T34 engines. The Super Guppy flew on 31 August 1965, and romped through its Certification to join NASA on its rocket airlift. There followed a succession of later Guppy freighters, including the Commercial Super Guppy with Allison 501 turboprops, the Mini Guppy for carrying oil-rig parts and other heavy but less bulky loads, and the Allison-engined Commercial Mini Guppy. The Commercial Super Guppy replaced the original detachable tail by a neat swing-tail hinged open for cargo loading.

In 1970 the first Guppy 101 introduced a powered swing nose for quick loading of giant airframe parts. This was followed by the Guppy 201, the longest and largest of all, with a length of 43.84 m (143 ft 10 in) and Allison 501-D22C turboprops. The first was used in the USA to carry the major fuselage and wing parts of the DC-10 and L-1011 TriStar from the various builders to the prime contractor's assembly plant. In 1971 the French carrier UTA began to operate a Guppy 201 under contract to Aérospatiale to bring to Toulouse all the parts of the prototype Airbus A300B. Gradually, as the Airbus Industrie production rate has increased, so has this aircraft grown into a fleet of four, the last two having been built at le Bourget in 1980-3 by UTA Industries in conjunction with Tractor Aviation (successor to Aero Spacelines Inc.). Who, in 1944, could have dreamed the bulky Boeing would grow in capacity from 141.6 m³ (5,000 cu ft) to an incredible 1415.8 m³ (50,000 cu ft) 40 years later?

The ultimate conversion of the original Stratocruiser design is illustrated by this Super Guppy 201 operated by Aéromaritime, a subsidiary of UTA, from le Bourget near Paris.

Boeing Model 377 Stratocruiser
G-ALSA 'Cathay'
BOAC

Chapter 11

Bristol Britannia

Bristol Freighter

Bristol Britannia

Often referred to as the 'Whispering Giant', the Britannia should have captured almost the whole market for long-haul airliners in the early 1950s. Instead it suffered many problems, almost all of which might have been avoided in a more professional environment, and made only a very brief impact on the world airline scene.

Today's airliners, such as the McDonnell Douglas DC-9 and DC-10 (far bigger and more complex than the Britannia) entered scheduled service well inside a year from their first flight. Bristol's Britannia took almost five years, and this had a crippling effect on the programme because during those five years the long-haul operators entered the jet era. Yet the aircraft itself was excellent, popular with crews and passengers – just like the BAC (Vickers) VC10, which in the same way suffered from abysmal behaviour on the part of the all-powerful airline for which both aircraft were created.

The story starts on 23 December 1942 when, with commendable foresight, the British government set up the first Brabazon Committee to recommend what new types of civil transport should be designed for the era after World War II. Nine types were suggested, of which the most important seemed to be the trunk-route machine in the 45360-kg (100,000-lb) class, to compete with the Lockheed Constellation and Douglas DC-6. Avro schemed a turboprop to meet this 'Brabazon III' requirement, but it then turned into a curiously arranged jet (Avro 693) and was finally cancelled in 1947. Thus the UK was left with such machines as the Avro Tudor, Handley Page Hermes and converted bombers, which were totally uncompetitive.

Despite this, in December 1946 BOAC issued a requirement for an MRE (Medium-Range Empire) aircraft even less ambitious than the Brabazon III specification. Bristol suggested licence-building Constellations with Centaurus engines, but dollar expenditure was not allowed, and eventually Specification 2/47 was issued, calling for a new design. Bristol's Type 175 looked best, but a 32-seater with four Centaurus was thought overpowered, so it was enlarged to 46856 kg (103,300 lb), with 164.9 m² (1,775 sq ft) of wing and up to 48 seats. BOAC declined to risk a production order, but on 5 July 1948 the Ministry of Supply at last got the project moving with an order for three prototypes. The compound-diesel Napier Nomad and turboprop Bristol Proteus were by this time viewed as possible later alternatives to the Centaurus, and late in 1948 BOAC began to show interest in the turboprop.

The Proteus had been designed as a very advanced and economical engine for the Bristol Brabazon II and Saunders-Roe Princess, and because both these giant aircraft had leading-edge inlets the Proteus was designed with reversed-flow layout, the air being ducted to inlets around the rear of the engine, then travelling forwards through the compressors and back through the slim combustion chambers and out through the turbines. It was the most difficult engine in history, and when Dr Stanley Hooker joined Bristol in January 1949 he was horrified. Almost every part of the engine was in trouble, even at half the planned 2237 kW (3,000 hp), and it was much too heavy. In mid-1950 Hooker was appointed chief engineer and he had already decided the Proteus had to be redesigned. Most unfortunately, because of its other applications the new Proteus 3 had to retain the reverse-flow layout; had Hooker known that both the Brabazon and Princess would soon be cancelled he would have put the inlet at the front, and avoided years of difficulty on the Britannia. None of this could be foreseen, and when the Proteus 3 ran in May 1952 it looked much like the old engine, though it was shorter, 454 kg (1,000 lb) lighter and gave 2819 kW (3,780 hp) plus 535 kg (1,180 lb) of thrust.

Type 175

Meanwhile the aircraft designers under Dr A. E. Russell were busy creating an outstanding Type 175, with a highly pressurised fuselage of 3.66 m (12 ft) diameter, a superb wing enlarged to

The introduction of the Bristol Britannia gave RAF Transport Command its first turboprop transport aircraft. The type was widely used on long-range strategic missions all over the world, forming the basis for the rapid deployment of the Army's United Kingdom Strategic Reserve. Twenty Britannia C.Mk 1s were delivered, followed by three C.Mk 2s with a large cargo door.

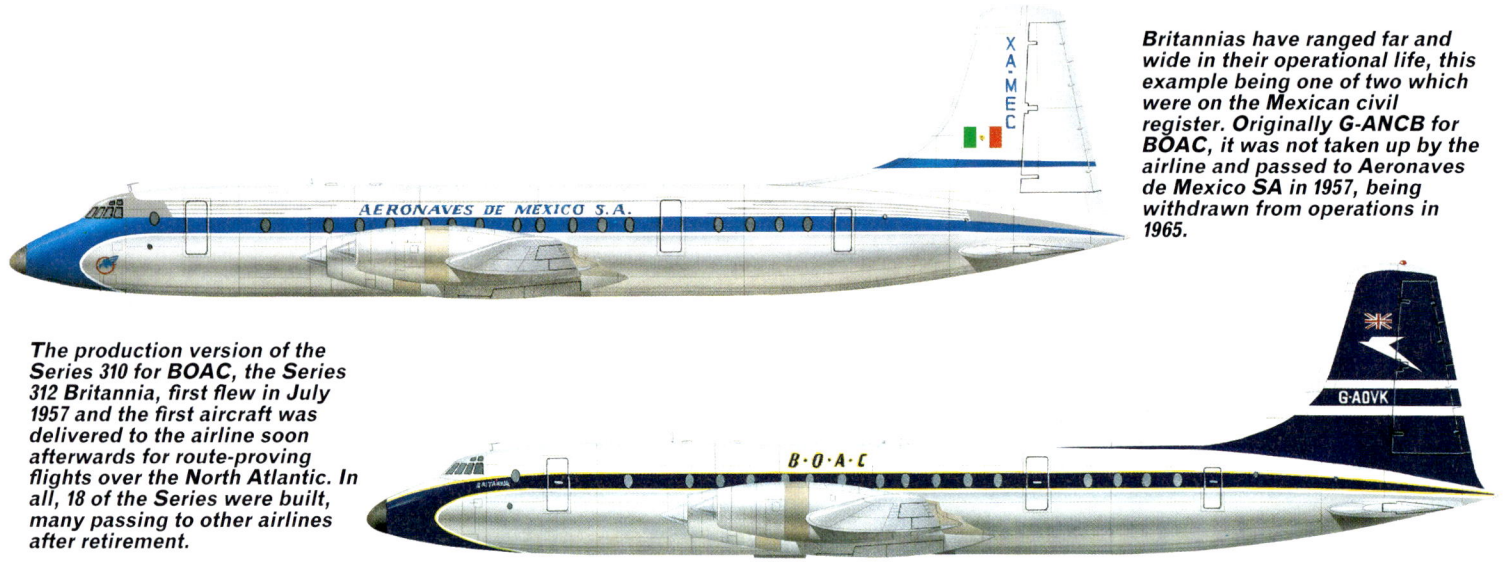

Britannias have ranged far and wide in their operational life, this example being one of two which were on the Mexican civil register. Originally G-ANCB for BOAC, it was not taken up by the airline and passed to Aeronaves de Mexico SA in 1957, being withdrawn from operations in 1965.

The production version of the Series 310 for BOAC, the Series 312 Britannia, first flew in July 1957 and the first aircraft was delivered to the airline soon afterwards for route-proving flights over the North Atlantic. In all, 18 of the Series were built, many passing to other airlines after retirement.

190.9 m² (2,055 sq ft) with large double-slotted flaps, nacelles reduced in diameter (from elimination of the Centaurus, which also removed the third prototype) and Messier bogie main landing gears, which because of the jetpipes had to retract backwards. BOAC argued against integral fuel tankage but did accept Redux adhesive bonding (also used in the de Havilland Comet) and unique flying controls in which the pilot moved nothing but small servo tabs along the trailing edges, which in turn moved the main surfaces. As Ultra produced an electrically signalled engine control system, a common joke in the 1950s was that the Britannia was like other aircraft except that the pilot's controls were not connected to the control surfaces and the throttles were not connected to the engines!

Right from the start

A. J. 'Bill' Pegg was in command when G-ALBO, the first prototype (later designated Britannia 101), made its maiden flight from Filton on 16 August 1952. The only visible modifications needed were to fit extended upturned wingtips and take the jetpipes from the tops of the nacelles to exhaust aft at the trailing edge. An unusual decision was to interconnect the rudder and the inboard tabs on the ailerons. Indeed, the Britannia was almost right from the start, and it looked as if BOAC's fleet of 25, for which the airline finally signed on 28 July 1949, would enter service as planned in 1954. Unlike the Comet, which appeared exotic but risky, the Britannia virtually had the world market at its feet. Obviously tremendously capable, it was also planned to have outstanding economics, and most of the top airlines sent delegations to Bristol in 1953-4.

On 4 February 1954 the visiting team was from KLM, and it flew in the second aircraft (G-ALRX), first flown in December 1953 and for the first time powered by the redesigned engine. Hooker had retained only one part of the original Proteus, the propeller reduction gear. High over Herefordshire the main input gearwheel stripped its teeth; the turbine, suddenly freed from load, instantly over-speeded and exploded, fragments passing through the oil tank and starting a fire. Pegg headed back to Filton, but the fire was so intense he feared the wing spars might go, and so did a belly-landing on the mudflats of the Severn. Prompted by Dr Bob Plumb, Hooker had already produced a set of helical gears, and these cured the problem (an added quick-acting fuel cut-off activated by turbine overspeed was never needed in millions of hours of flying). But G-ALRX had been damaged by salt water and cut to pieces by retrieval cables, and this seriously delayed the programme. A little later, in May 1954, test pilot Walter Gibb had a hair-raising half-roll in G-ALBO following flap drive failure, and this left Bristol without a flyable aircraft until the first production Britannia 102, G-ANBA, flew on 15 September 1954. G-ANBA thus had to do work originally allocated to the prototypes, and Bristol proved incapable of coming anywhere near the agreed delivery dates for either engines or aircraft.

Back in 1952 BOAC had discussed a cargo version, the Series 200, with a stretched fuselage. From this stemmed the passenger Series 300 and mixed-traffic Series 250, and by 1955 design had been completed for the long-range Series 310 with integral tanks in the outer wings increasing fuel capacity from 30322 to 38577 litres (6,670 to 8,486 Imp gal). The Proteus Mk 755 was developed up to 3322 kW (4,455 ehp), easily matching the increased gross weight of 83915 kg (185,000 lb), and the result was an aircraft still of tremendous interest to long-haul operators. In 1954 Canadair took a licence for the Britannia as the basis for a maritime patrol aircraft, as related later,

Three Britannia 313s were sold to El Al, the first being delivered in September 1957. They were swiftly put into operation on the Tel Aviv-New York service, cutting journey times. This aircraft had just completed the 9817 km (6,100 miles) in 14 hours 56 minutes.

Representative of the later variants is this Canadian Pacific (CPAL) Series 314. Six of the variant were delivered to the airline, one setting a record time of 11 hours 44 minutes for the 7564-km (4,700-mile) Vancouver-Tokyo route on 20 September 1958.

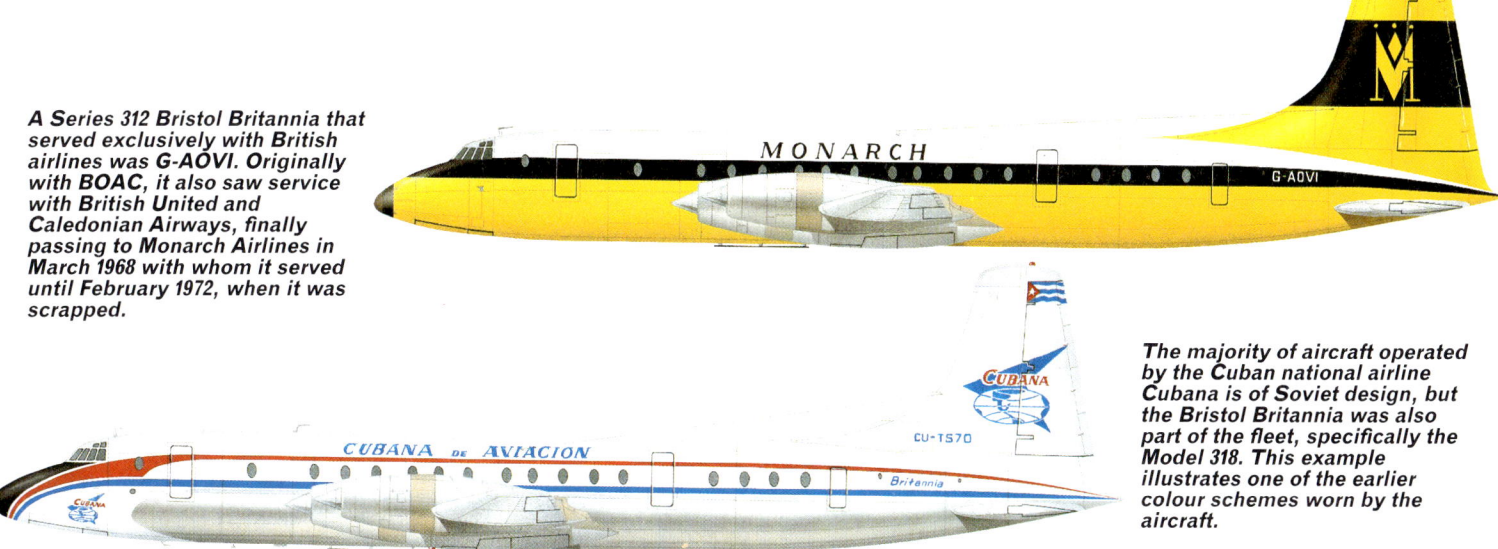

A Series 312 Bristol Britannia that served exclusively with British airlines was G-AOVI. Originally with BOAC, it also saw service with British United and Caledonian Airways, finally passing to Monarch Airlines in March 1968 with whom it served until February 1972, when it was scrapped.

The majority of aircraft operated by the Cuban national airline Cubana is of Soviet design, but the Bristol Britannia was also part of the fleet, specifically the Model 318. This example illustrates one of the earlier colour schemes worn by the aircraft.

while to try to get some production going a second-source line was set up at Shorts at Belfast. The Belfast line was started with seven Series 302 aircraft for BOAC; the airline wanted the extra fuel capacity of the Series 310 and this was provided in the last five of this batch, which thus became Series 305s. In the end BOAC did not have to take any Series 302s or 305s, but finally took 15 Series 102s, the only short-body Britannias, and 11 long-haul Series 312s. With painful slowness deliveries of the Series 102 got under way in 1955, at the increased take-off weight of 70307 kg (155,000 lb) compared with 58967 kg (130,000 lb) for G-ALBO and 63503 kg (140,000 lb) for G-ALRX, and 55792 kg (123,000 lb) for landing, with eight crew and 90 passengers. The fourth Series 102 was for some reason diverted into having 1st-class seating and a lounge bar and stayed on the ground when it might have been doing development flying, 2,000 hours of which had been demanded, with BOAC crews assisting. At last two aircraft with Certificates of Airworthiness were handed over at Heathrow on 30 December 1955, to continue 'route proving'. In March 1956 one aircraft encountered flameouts with all four engines whilst in cumulo-nimbus cloud over Uganda. The problem was the build-up of enormous amounts of ice and slush in the bend of the reverse-flow inlet; when this broke away it put the flame out. Dr Hooker, at the newly formed company Bristol Aero-Engines, quickly had glow-plugs fitted which caused instant relight, and pointed out that the trouble was confined to the tropics and that it could be avoided by (on the rare occasions when this was necessary) selecting a slightly different altitude or route. BOAC totally refused to accept any of this and spent two years apparently doing all it could to magnify the problem, to the extent of searching for severe ice in cold climates. All this came within an ace of bankrupting Bristol, and eliminated almost all the interest shown previously by world airlines.

The one exception was El Al of Israel, which simply ordered three long-range Series 313s, and from 19 December 1957 put them into widely publicised service between New York and Tel Aviv. On the first trip hundreds of observers waited to see if 4X-AGA would land at Rome to refuel; it did not, but set a world airline record at 9817 km (6,100 miles) non-stop, flown at an average of 645 km/h (401 mph). This rather shamed BOAC into getting a move on, but by this time the world's major carriers were thinking only of the Boeing 707 and Douglas DC-8.

Series 301 crash

The sole Series 301 (G-ANCA) was painted in Capital livery but was not sold, and tragically crashed at Bristol on 6 November 1957 through, it was thought, autopilot failure. The two Series 302s, without long-range tanks, went to Mexico. Other marks are listed in the variants table. By far the most important long-term operator was the Royal Air Force which, with the civilian ministry (which had various names) used 26 with extreme success over many years. Civil Britannias flew under the flags of Argentina, Belgium, Burundi, Canada, Cuba, Czechoslovakia, Ghana, Ireland, Israel, Kenya, Liberia, Mexico, Pakistan, Spain, Switzerland, Uganda, the USA and Zaïre, as well as with a host of British independent lines.

Hooker's team had developed a superb new turboprop, the Orion engine, which not only gave better fuel efficiency for less weight, and avoided the reverse-flow layout, but also maintained constant sea-level power of 3840 kW (5,150 hp) to a height of 6095 m (20,000 ft), giving tremendous high-speed cruise performance. Bristol planned the Britannia Series 400 around this engine, one of which began flight test in G-ALBO in August 1956. For the longer-term, collaboration with General Dynamics led to the Type 187 'thin-wing Britannia' with a double-bubble two-deck fuselage, which promised unrivalled seat-

A successful design with effective load-carrying capabilities, the Canadair-built CL-44D has served with many airlines. Typical was Seaboard World, a transatlantic freight-carrying operator which used the type in the 1960s.

An airline that has operated both the Britannia and the CL-44 is Aer Turas of Eire. This aircraft, one of the stretched CL-44Js, was used mainly for bloodstock transportation.

Canadair initially produced the CL-28 Argus maritime patrol aircraft based on the Britannia fuselage. Following this, it built a military transport, the CL-44, which was used by the RCAF on long-range duties under the designation CC-106 Yukon. This example served with No. 437 Squadron.

mile costs with 200 passengers at 805 km/h (500 mph) over very long ranges. In its wisdom the Treasury said it would support only the rival Rolls-Royce Tyne engine, and the Orion was cancelled in 1958.

Canadian manufacture

Canadair, another General Dynamics company, used only the basic airframe in the CL-28 ocean patrol aircraft, 33 of which were built in 1957-60 as the Argus. Canadair's licence extended to a transport version for the RCAF (now CF), and this was most competently engineered at Montreal using the Tyne engine, with the fuselage stretched to 41.63m (136 ft 7 in) and weight increased to 92986kg (205,000lb). Large cargo doors were added on the left side and there were many other changes, the resulting aircraft first flying on 15 November 1959. Designated CL-44-6, it became the CC-106 Yukon in service, and the 12 built served with No. 437 Squadron and (VIP nos 15929 and 15932) with No. 412 Squadron until replaced by Boeing 707s in 1970-1. They then began a new life with commercial operators.

Market research showed that as big jets had taken over the passenger market, there was a place for a pure freighter. Canadair therefore developed the first commercial swing-tail freighter, the CL-44D. The opportunity was taken to introduce many further updates, the most visible of which was the new flight-deck windshield with 40 per cent more window area but only seven panes in-

stead of 14. The first CL-44D-4 flew on 16 November 1960 and 23 were delivered to the main US cargo airlines. Carrying 30 tonnes over transatlantic range with a direct cost of just over 4 cents per tonne-mile, they were the most efficient cargo aircraft of their day, and a few are still operational.

A further four (CL-44D-4-8) were sold in passenger configuration to Loftleidir of Iceland, which used them with 189 seats on the New York-Luxembourg route. The fourth CL-44D-4-8 was modified before delivery into the CL-44J with fuselage stretched to 46.34 m (152 ft 0.5 in), and with seating at the same pitch for 214 passengers. This proved so successful that the other three were subsequently rebuilt to CL-44J standard.

One CL-44D-4-2, N447T of Flying Tiger, was bought in 1968 by Jack Conroy, whose company rolled it out again in 1969 as the CL-440, with a giant fuselage 4.24 m (13 ft 11 in) wide. As *Skymonster* it served with Transmeridian on outsize-freight duties before being sold to a British operator, HeavyLift, for a few more years.

Bristol Britannia 312 cutaway drawing key

1 Radome
2 Weather radar scanner
3 Radar tracking mechanism
4 Radar receiver and transmitter
5 ILS aerial
6 Nosewheel bay bulkhead
7 Nosecone construction
8 Forward pressure dome
9 Control column linkages
10 Rudder pedals
11 Nose undercarriage wheel bay
12 Hydraulic retraction jack
13 Nosewheel leg strut
14 Twin nosewheels
15 Steering linkage
16 Nosewheel leg doors
17 Pitot tube
18 Cockpit floor level
19 Nosewheel steering control
20 Control column handwheel
21 Instrument panel
22 Instrument panel shroud
23 Windscreen panels
24 Overhead switch panel
25 Cockpit eyebrow windows
26 Co-pilot's seat
27 Pilot's seat
28 Navigator's station
29 Aerial mast
30 Cockpit roof escape hatch
31 Periscope sextant mounting
32 Radio racks
33 Radio operator's station
34 Cabin air system vent valve
35 Forward underfloor cargo hold
36 Air system ducting
37 Escape chute stowage
38 Cabin attendant's folding seat
39 Crew entry door
40 Forward galley
41 Wardrobe compartment
42 Entry door, open position
43 Cabin bulkhead
44 Forward passenger cabin
45 Six-abreast tourist-class seating; maximum seating capacity 133 passengers in all tourist-class layout
46 Forward toilets, port and starboard
47 Underfloor freight hold
48 Floor beam construction
49 Cabin wall trim panels
50 DF loop aerials
51 Radio aerial mast
52 Starboard inner engine nacelle
53 Air cooled transformer rectifier
54 Inner wing fuel cells; total fuel system capacity 8,486 Imp gal (38576 litres)
55 Starboard outer engine nacelle
56 Detachable engine cowling panels
57 Annular engine air intake

The ability to carry large cargo items or palletised freight by loading through the hinged rear fuselage proved an extremely attractive operating feature. The CL-44D was the world's first cargo aircraft with this innovation.

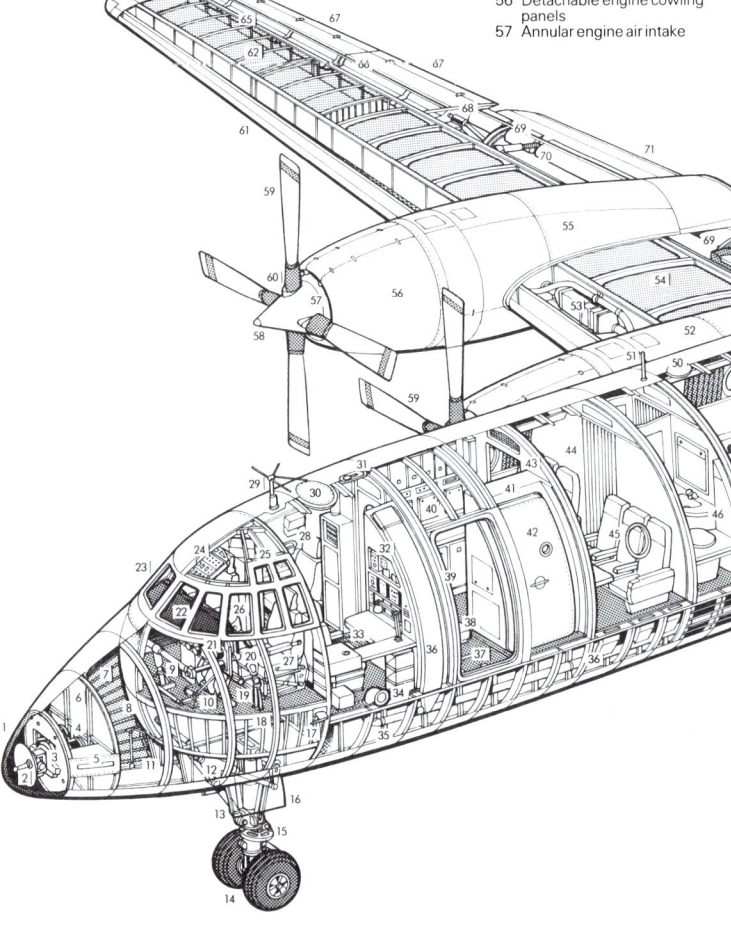

Two CC-106 Yukons were used by No. 412 Squadron, RCAF, for VIP duties. Along with the 10 aircraft of No. 437 Squadron, these differed from later civil aircraft by having normal cargo doors and no swing-tail. They were replaced by Boeing 707s (designated CC-137) in 1971.

N447T started life as a CL-44D-4 serving with Flying Tigers until sold to Conroy Aircraft, who fitted an outsize fuselage for transporting large items. Called the CL-44O and named Skymonster, it has flown for the British companies Transmeridian and HeavyLift.

58 Propeller spinner
59 de Havilland Hydromatic variable pitch, reversible four-bladed propellers
60 Propeller blade root de-icing boots
61 Hot-air leading edge de-icing
62 Outer wing integral fuel tank
63 Starboard navigation light
64 Wing tip fairing
65 Aileron balance
66 Starboard two-segment aileron
67 Aileron servo tabs
68 Tab control linkage
69 Flap guide rails
70 Flap screw jack
71 Starboard double-slotted flap, down position
72 Life raft stowage
73 Jet exhaust nozzle
74 Fuselage centre section construction
75 Wing spar attachment fuselage main frames
76 Wing centre section carry-through
77 Central flap drive motor
78 Emergency exit hatch, port and starboard
79 Cabin floor panelling
80 Cabin air vents
81 Removable cabin divider
82 Overhead luggage racks
83 Passenger service units
84 Tourist class seating
85 Emergency exit hatch, port and starboard
86 Four-abreast first-class seating; 93-passenger capacity in mixed-class layout
87 Curtained window panel
88 Rear bar/galley units
89 Fuselage frame and stringer construction
90 Aft cabin seating
91 Starboard emergency exit door
92 Fin root fillet
93 Starboard tailplane
94 Starboard elevator
95 Fin electric mat leading edge de-icing
96 Fin construction
97 Fin tip fairing
98 Anti-collision light
99 Rudder construction
100 Rudder servo tabs
101 Rudder tab control gearbox
102 Tailcone
103 Tail navigation light
104 Elevator servo tabs
105 Port elevator construction
106 Elevator horn balance
107 Tailplane construction
108 Electric mat leading edge de-icing
109 Fin/tailplane attachment bulkhead
110 Elevator hinge control
111 Tailplane centre section carry-through
112 Rear pressure bulkhead
113 Aft toilet compartments (three)
114 Cloakroom
115 Dressing room
116 Cabin rear bulkhead
117 Port emergency exit doorway
118 Escape chute stowage
119 Rear cabin air ducting
120 Rear underfloor cargo hold
121 Passenger entry door, open
122 Escape chute stowage
123 Wing root fillet construction
124 Life raft stowage
125 Inboard flap segment
126 Heat shrouded jet pipe
127 Exhaust nozzle
128 Main undercarriage wheel well
129 Nacelle tail fairing
130 Life raft stowage
131 Port outboard double-slotted flap, down position
132 Trailing edge flap shroud ribs
133 Flap rib construction
134 Aileron servo tab control mechanism
135 Port aileron construction
136 Aileron servo tabs
137 Wing tip fairing
138 Port navigation light
139 Wing rib construction
140 Leading edge corrugated inner skin
141 Leading edge de-icing hot air duct
142 Front spar
143 Port wing fuel cell bays
144 Engine nacelle construction
145 Port outer engine nacelle
146 Engine bay firewall
147 Oil cooler
148 Engine bearers
149 Annular air intake
150 Four-wheel main undercarriage bogie
151 Main undercarriage leg strut
152 Hydraulic retraction jack
153 Undercarriage leg pivot fixing
154 Wing de-icing system air ducting
155 Cabin air heat exchanger
156 Bristol Proteus 755 turboprop
157 Engine oil tank, capacity 9 Imp gal (41 litres)
158 Engine flame tubes
159 Engine mounting ring
160 Propeller reduction gearbox
161 Port inner propeller
162 Propeller hub pitch change mechanism

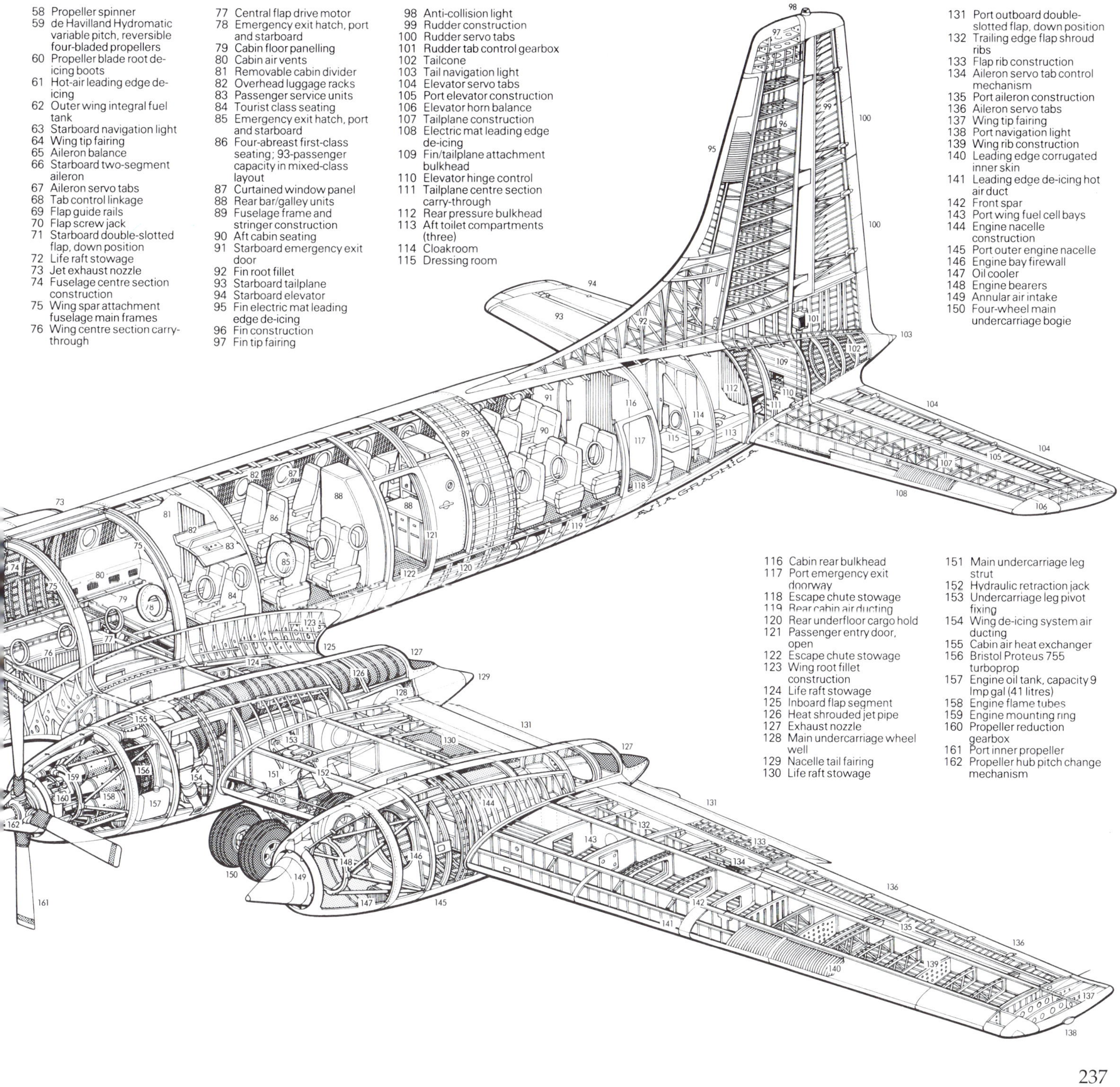

Bristol Freighter

Slow, noisy yet possessed of a certain dignity, the B.170 Freighter had its roots in wartime needs but soon became a servant of the rising demand for civilian air transport. Best known perhaps for its cross-Channel car-ferry services, Freighters served in modest numbers around the world, with some surviving until quite recently.

As World War II progressed in the Far East plans were laid by the Bristol Aeroplane Company for a military transport aircraft capable of operating from short jungle airstrips. Victory for the Allies in the Pacific precluded the need for the aeroplane, but in peacetime it quickly became apparent that such a machine would have a role to play in the revived civil air transport industry.

Bristol's chief designer, A. E. Russell, and his team quickly adapted their preliminary plans for the military aeroplane, producing instead the Bristol Type 170. It was designed with structural simplicity, rugged construction, low-maintenance requirements, and low acquisition and operational costs in mind. A high wing configuration was chosen, with the three-crew cockpit set above the rectangular-section fuselage, leaving its 15-m (49-ft) long, 2.4-m (8-ft) wide, 66.8-m³ (2,360-cu ft) cargo hold free of obstructions. Two 1250-kW (1,675-hp) Bristol Hercules 632 radial engines were chosen to power the aircraft, mounted in 'power eggs' attached to underslung nacelles and driving four-bladed propellers. In the airframe Russell eschewed use of expensive alloys in favour of steel, and employed the minimum of machined parts and double-curvature sections, aiming for an airframe which could be serviced and repaired 'in the field' without specialist tools. For the same reasons of low cost and simplicity he opted for a fixed tailwheel undercarriage.

Two versions of the Bristol 170 were planned. The Series I freighter had a strengthened floor and hydraulically-operated sideways-opening clamshell nose doors to permit direct loading of vehicles or heavy freight, of which it could carry up to four and a half tons. The Series II Wayfarer was a 34-passenger version without nose doors but with additional windows in the nose section.

G-AGPV, the first of four prototypes, was a Freighter, but lacked the clamshell doors and had small round porthole cabin windows, replaced by square window apertures in all succeeding aircraft. It first flew, with Bristol's test pilot Cyril Uwins at the controls, on 2 December 1945 and was thus the first new post-war British transport aeroplane to take to the air. Early test flights revealed a number of handling shortcomings which were corrected by increasing the span of the tailplane by 1.2 m (4 ft) and relocating it from its original position atop the rear fuselage to a mid-point, after which Uwins reported that the Freighter happily flew

Freighter Mk 31Ms continued to ply their trade around the world into the 1970s, with Australia a veritable hotbed of activity. Air Express Ltd gathered a fleet of aircraft from sources such as the Pakistan air force and Trans Australia Airlines. Visible here is the crew entry stairs in the lower fuselage.

'hands off'. The second prototype (G-AGVB) was built to Wayfarer configuration, making its first flight on 30 April 1946. The Bristol 170 was awarded an unrestricted British certificate of airworthiness on 7 May 1946 and two days later the Wayfarer prototype, bearing the colours of Channel Islands Airways, made a proving flight to Jersey, commanded by another celebrated Bristol test pilot, A. J. 'Bill' Pegg. Within six months the aeroplane had made 358 flights to the Channel Islands and carried 10,000 passengers.

In the meanwhile, Bristol had completed a production-standard prototype Freighter I with cargo floor and nose doors, G-AGVC making its maiden flight on 23 June 1946. Six weeks later it left the Bristol company airfield at Filton for a demonstration and sales tour which was to keep it out of the United Kingdom for two years, journeying first across the North Atlantic for demonstrations in Brazil, then to Canada for leased service with Canadian Pacific Airways, and finally meat-hauling with Linea Aéropostale Venezolana in Venezuela. The first prototype Freighter was handed over to the Royal Aircraft Establishment at Boscombe Down for service trials, while the fourth aircraft built, a Wayfarer IIA, also acquired military marks with the Telecommunications Research Establishment at Defford, where both were eventually to serve until demobilised in November 1958.

The first production Wayfarer IIA was due to follow the Freighter I across the Atlantic for sales tours, but on 4 July 1946, en route from Bathurst, Gambia to Natal and Buenos Aires, compass problems caused its crew to miss their planned landfall and they were forced to ditch in the Atlantic 193 km (120 miles) from land after the Bristol 170's fuel was exhausted.

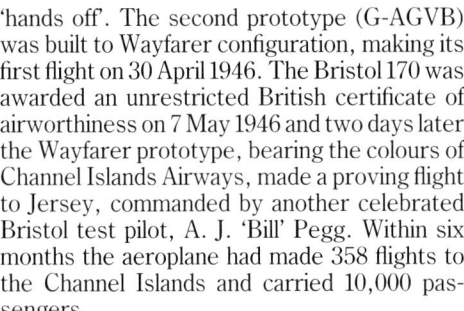

The principal identification feature of the Freighter 31 was the addition of a dorsal fin. Among the customers was the Royal Canadian Air Force, which added glazing in the lower nose of its three aircraft.

Proving flights

Other early production aircraft also undertook sales tours of Europe and the Middle East, and were leased for proving trials with such British carriers as Airwork, Skytravel, British Aviation Services and Air Contractors, during which they proved invaluable in carrying such diverse loads as servicemen and their families on trooping flights and leave trips, fruit and vegetables, and racehorses and cattle.

The Argentine air force became the first customer to buy the aircraft, ordering 30 Freighter IAs (later reduced to 15) in 1946. The first of these flew from Filton on 24 October and all 15 had been delivered by the summer of 1947. The survivors of this fleet were withdrawn from use during 1959. Customer interest was principally in the Freighter version, with little demand for the passenger-carrying Wayfarer. Only 16 of those were built, all but four subsequently being converted to Freighter configuration.

SAFE of New Zealand established itself as a major Freighter user, and undertook some modifications to its aircraft. This Freighter 31 has had the majority of its windows blocked out, while a specialist passenger capsule was developed by the airline.

The first major modification introduced on the Bristol 170 was the fitting of rounded wingtips, which increased wingspan from 29.8 m (98 ft) to 32.9 m (108 ft) on the 'New Freighter' or Freighter XI G-AIFF, which made its debut at the 1947 Society of British Aircraft Companies show at Radlett. Only two Freighter XIs were built before installation of 1260-kW (1,690-hp) Bristol Hercules 672 engines driving 4.3-m (14-ft) diameter propellers, and an increase in maximum take-off weight to 12190 kg (40,000 lb), created the Freighter 21, G-AIFF again serving as prototype and demonstrator. Several versions were offered: the standard Freighter 21 had an all-cargo configuration with a soundproofed cockpit; the Mk 21A was a

Without doubt the best-known Freighters were the car ferries of Silver City Airways. These plied the Channel carrying two cars, their occupants and additional passengers. The later 32 could add another car in the lengthened nose section.

The prototype Bristol 170 was completed to Freighter standard, with clamshell front doors. The cockpit was mounted high above the freight deck to allow direct entry to the capacious cabin.

G-AIFF served as the prototype for two successive marks of Freighter. The Freighter XI introduced these extended wings with rounded tips, while in a later guise (Freighter 21) more powerful Hercules engines were fitted.

*Military customers of the Freighter appreciated the ease with which vehicles could be loaded through the clamshell doors. This is a **No. 1 Squadron, RNZAF**, aircraft seen in Thailand in 1962 during a logistic support mission in aid of the growing war effort in Vietnam.*

mixed-configuration aircraft with accommodation for 16-20 passengers and freight; and the Mk 21E had a movable bulkhead and airliner-standard full cabin soundproofing, enabling it to be converted quickly to all-freight, mixed, or 32-36 passenger Wayfarer 22A standard. The Freighter 21E made its debut at the 1948 SBAC show at RAE Farnborough.

Overseas commercial customers for the Freighter 21 came from Australia, France, the Middle East, Saudi Arabia, Spain and West Africa, but the biggest single order to date for the aircraft was placed by the Pakistan air force, which required 30 specially-equipped Mk 21Ps with Wayfarer-style windows in their clamshell nose doors.

It was in England that the Freighter found a role with which it was to become synonymous. In the summer of 1948 Freighter I G-AGVC was modified for Silver City Airways to carry two family cars in the forward cargo hold and 15 passengers in a separate cabin at the rear. On 13 July it made the first 25-minute flight between Silver City's base at Lympne, Kent and Le Touquet, inaugurating the carrier's soon-to-be-famous cross-Channel car ferry service which proved immensely popular as an alternative to slow and uncomfortable sea crossings. The airline soon employed a large fleet of

The revival of Instone Air Line in the 1980s saw two ex-New Zealand Freighter 31s flying again in the United Kingdom on ad hoc freight duties. Later the airline had a single Douglas DC-6, operated by Air Atlantique.

Probably the last Freighter in service was this Freighter 31, returned to New Zealand after service with Instone. It was on the strength of Hercules Airways, which ceased trading in 1989.

Freighters on the route. Five of these, and two more operated by Airwork, were pressed into service on the Berlin Airlift in 1948-49, logging 287 sorties between them.

Unfortunate losses

The success of the Freighter was marred by two fatal crashes. G-AIFF, the prototype Mk 21, was lost in the English Channel on 6 May 1949, while Silver City's G-AHJJ crashed at Llandow, Glamorgan in the following March. Investigators concluded that both had suffered structural failures in their fins due to over-stressing from rudder hard-overs during single-engine climbs. As a result, Bristol developed a modification for all existing aircraft, and also developed the new Model 31 which featured a large dorsal fin strake, 1477-kW (1,980-hp) Hercules 734 engines and a 19960-kg (44,000-lb) maximum take-off weight, and was designated Freighter 31. The first two Mk 31s were extensively evaluated by military test pilots at Boscombe Down.

The Bristol 170 proved popular with many customers. Most aircraft were to Freighter standard, as the Wayfarer passenger version was largely overshadowed by more efficient airliner types. This aircraft was used by BEA for mail and freight work.

Production of the Mk 21 ceased after 92 aircraft had been built. It was superseded by the new Mk 31, and a military variant, the Freighter 31M, which achieved substantial sales success, with the Pakistan air force once again being the major customer with an order for 38 aircraft. Other military operators of the Freighter 31M included the Burmese air force with a single example, Iraqi air force with two, Royal Canadian Air Force with three operated in support of RCAF units in Europe, and the Royal New Zealand Air Force which took delivery of seven. Although it was superseded by the Mk 32 (see below), production of the Freighter 31 did not formally cease until early 1958, when the final two of 93 production aircraft were delivered, one to the British carrier Dan Air, the other to Straits Air Freight Express (S.A.F.E.) of New Zealand. S.A.F.E. became a major operator of the Freighter, buying 10 surplus Pakistan air force Mk 31Ms as well as aircraft from Iberia Airways of Spain. The carrier also leased several RNZAF machines for its all-cargo services across Cook Strait between New Zealand's North and South Islands. S.A.F.E. developed its own 'Cargon' self-loading system for the Freighter, and also devised a unique 20-seat 'capsule' compatible with the system for passenger services on the Wellington-Chatham Island route.

In 1953 the Freighter production line was relocated from Filton, which was being prepared for manufacturing the new Bristol Britannia turboprop airliner, to the company's plant at Weston-Super-Mare. That same year the final version of the Bristol 170 emerged as the Freighter 32 or 'Super Freighter'. It featured an elongated nose extended by 1.6 m (5 ft 4 in) to permit three cars to be carried, a taller fin and rudder and deeper dorsal strake to compensate for the additional side area of the new nose, and accommodation for 25 passengers aft of the car deck. Twenty Freighter 32s were built. Fourteen were supplied to Silver City Airways, which on 13 July 1954 opened a new purpose-built cross-Channel air terminal at Lydd-Ferryfield Airport in Kent.

In addition to its cross-Channel service to Le Touquet, which formed part of the London-Paris coach-air link and was served by specially modified Mk 32 G-AMWA fitted out with 60 passenger seats and no car deck as a 'Super Wayfarer', Silver City had developed Freighter services to Cherbourg and the Channel Islands and to Northern Ireland. By the tenth anniversary of Silver City's pioneering 1948 first cross-Channel service, its Freighters and Super Freighters, affectionately dubbed 'Fly-

The second 170 was a Wayfarer, identified by the lack of nose doors and additional cabin windows. The machine was used by Channel Island Airways in the late 1940s for passenger services to the mainland.

ing Nissen Huts', had made 125,000 Channel crossings and carried 215,000 cars, 70,000 motorcycles and 759,000 passengers.

Six more Freighter 32s, including the 214th and last Bristol 170 to be built, were delivered to Air Charter Ltd of Southend Airport for use on its burgeoning Southend-Calais Channel Air Bridge and Southend-Ostend Air Bridge car ferry services. So good was business in 1958, when all Freighter production had ceased, that Air Charter Ltd had two of its Mk 31s modified to Mk 32 standard to provide increased capacity. One of the company's Super Freighters wore the colours of the Belgian airline SABENA, with which it operated the Ostend service jointly.

Long-lived Mk 32s

In October 1962 Silver City and Air Charter Ltd merged to form British United Air Ferries, and later British Ferries, operating a combined fleet of 24 Freighters of all marks. The new carrier began withdrawing its ageing Freighter 31s the following year as they became due for major inspection, since it was an airworthiness requirement following a fatal crash in Ireland caused by inflight wing failure that wing booms had to be replaced after every 25,000 landings. While this had been done several times on the hard-used aircraft, it was no longer economic. The Mk 32 Super Freighters continued to give good service, some operating in the colours of BAF's French associate Compagnie Air Transport, until the fleet was run down from 1967. Most had been scrapped by 1970, although the last two delivered continued to serve with Midland Air Cargo of Coventry Airport until 1975.

Small numbers of Freighter 31s survived into the 1980s, chiefly in Britain and New Zealand. The aptly named Hercules Airways remained faithful to the B.170 until 1989. In Britain, Instone Air Line retained a pair of hard-working Mk 31s which was ultimately sold in the late 1980s to Trans Provincial Airlines of British Columbia, Canada. Founded in 1982, Trans Provincial built up a fleet of three aircraft between 1987 and 1989, with a large stock of spare parts to support operations. Ideally suited for cargo hauling in the harsh northern climes, the Bristol Freighter seems sure to have found its final home there, doing what it still does best.

Silver City took the lion's share of the Freighter 32s built, these not only adding a lengthened nose for an extra car but an extended fin to offset the consequent reduction in directional stability. This particular aircraft was an all-passenger 'Super Wayfarer'.

Several Freighters were sold to France, including this 21. At least one was used in the war in Indo-China for military freight work under civilian contract. Long-nose 32s later flew the Channel car ferry run with CAT.

The Bristol 170 was designed to reflect a no-nonsense approach to the requirements. Any needless complexities were left out to allow ease of maintenance under primitive conditions, hence the fixed undercarriage. The expansive wing won no prizes for performance, but could lift a good load from a small strip, while the undercarriage was sturdy enough to withstand operations from most surfaces. The high-wing configuration allowed a flat floor to the main cargo cabin, and also kept the nose as low to the ground as possible for ease of access.

Specification
Bristol Freighter Mk 21
Wingspan: 32.9 m (108 ft 0 in)
Length: 20.8 m (68 ft 4 in)
Height: 6.6 m (21 ft 8 in)
Wing area: 138 m^2 (1,487 sq ft)
Powerplant: two 1261-kW (1,690-hp) Bristol Hercules 672 radial engines
Passenger capacity: 15-34
Empty weight: 12013 kg (26,484 lb)
Maximum take-off weight: 18144 kg (40,000 lb)
Cruising speed: 266 km/h (165 mph)
Service ceiling: 6400 m (21,000 ft)
Maximum range: 790 km/h (490 miles)

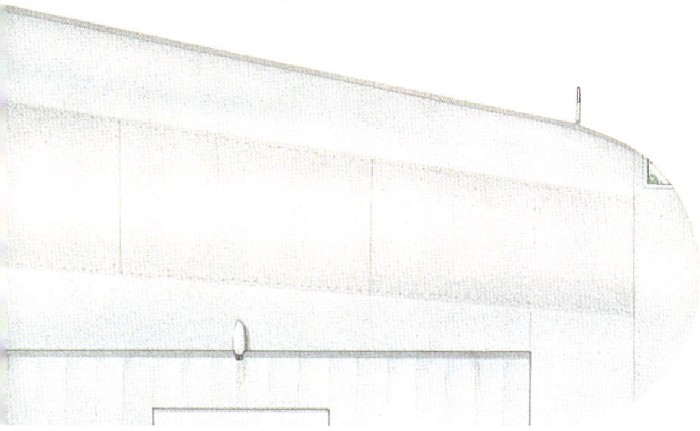

Specification
Bristol Super Freighter Mk 32
Wingspan: 32.9 m (108 ft 0 in)
Length: 22.4 m (73 ft 4 in)
Height: 6.6 m (21 ft 8 in)
Wing area: 138 m^2 (1,487 sq ft)
Powerplant: two 1261-kW (1,980-hp) Bristol Hercules 734 radial engines
Passenger capacity: 23
Empty weight: 13365 kg (29,465 lb)
Maximum take-off weight: 18144 kg (44,000 lb)
Cruising speed: 266 km/h (165 mph)
Service ceiling: 7470 m (24,500 ft)
Maximum range: 790 km/h (490 miles)

Chapter 12

British Aerospace BAe 125

British Aerospace BAe 146

BAe 125

For more than 30 years the 125 business jet has been at the forefront of its market, continuing development matching advances made by its competitors. Initially a de Havilland design, the 125 then switched to Hawker Siddeley and then British Aerospace, and in its current incarnation is the Raytheon Hawker. The latest variants offer ever greater range, efficiency and comfort.

In February 1961 the de Havilland Aircraft Company formally announced that it was developing a 'mini' jetliner successor to the D.H.104 Dove twin-engined feedliner and business aircraft. Designated D.H. 125 Jet Dragon after the pre-war D.H.84 Dragon twin-engined biplane airliner, the aircraft was to seat six/eight passengers in a fully pressurised and air-conditioned executive cabin and be powered by two 13.7-kN (3,000-lb st) Bristol Siddeley Viper 20 turbojets mounted in pods on the aft fuselage, which would give it a cruising speed of up to 800 km/h (500 mph) and a range of 2413 km (1,500 miles).

The chosen configuration was a cylindrical section fuselage offering a cabin 0.3 m (one foot) wider than that of the Dove and with the same headroom as the four-engined Heron; slightly swept one-piece low wing with large double-slotted flaps; and swept fin and tailplane of cruciform layout. An initial production batch of 30 'Jet Dragons' was laid down at de Havilland's factory at Hawarden, near Chester, but two prototypes, of slightly smaller overall dimensions than the production standard airframe, were hand-built in the company's experimental workshops at Hatfield, Herfordshire. The first of these, G-ARYA, was rolled out in July 1962, at which time the name Jet Dragon was dropped. It made its 56-minute maiden flight on 13 August 1962 and was followed by the second aircraft on 12 December. Both aircraft were flown extensively for the certification programme, in which they were later joined by the first production aircraft, which was powered by two Viper 520 engines and featured a 0.91-m (3-ft) increase in wingspan and a fuselage stretch of 1.20 m (3 ft 11 in).

Including the prototypes, eight examples of the D.H. 125 Series 1 were built. The first customer delivery was made to West German company Krupp GmbH of Essen on 26 February 1964 – an early por-

Like other British aircraft, the 125 found favour in the Middle East, where it is a popular business/VIP transport. This Series 700 flies with a Saudi Arabian finance company, the interior fitted with a luxury executive suite.

Many of the early 125 variants are still flying on executive duties. This Saudi machine is a Series 1B, powered by the Rolls-Royce Viper 522 engine. The short fuselage and rounded nose profile are typical of the early aircraft, most of which were built by de Havilland.

British Aerospace 125 Series 800 cutaway drawing key

1. Radome
2. Lightning conducting strips
3. Weather radar scanner
4. Scanner tracking mechanism
5. ILS glideslope aerial
6. Nosewheel bay
7. Nosewheel doors, closed after cycling of undercarriage
8. Ice detector probe
9. Auxilary hydraulic reservoir
10. Flush VOR aerial
11. Nose compartment construction
12. Windscreen rain repelling air blower
13. Front pressure bulkhead
14. Rudder pedals
15. Sloping cockpit framing
16. Static ports
17. Trim control handwheel
18. Instrument panel, with five-tube EFIS
19. Instrument panel shroud
20. Stand-by compass
21. Curved windscreen panels, electrically de-iced
22. Overhead switch panel
23. Co-pilot's seat
24. Circuit breaker panels
25. Pilot's seat
26. Cockpit side window panels
27. Control column
28. Nose undercarriage pivot fixing
29. Nosewheel leg strut
30. Twin nosewheels
31. Torque scissor links
32. Nosewheel steering jack
33. Nosewheel leg door
34. Cockpit floor level
35. Pitot head
36. Chart case
37. Door latch
38. Entry door/airstairs
39. Folding handrail
40. Entry lobby
41. Galley units
42. Underfloor avionics equipment racks
43. Baggage compartment
44. Forward fuselage frame and stringer construction
45. Baggage compartment folding door
46. Main cabin bulkhead doorway
47. Folding table
48. Aft facing passenger seat
49. Seat mounting rail
50. Wing-root leading-edge fairing
51. Inspection light
52. Wing spar centre-section beneath fuselage structure
53. Footwell
54. Cabin window panels
55. Swivelling passenger seats, eight-seat standard layout
56. Cabin wall trim panelling
57. Window blind
58. Starboard emergency exit hatch
59. Starboard landing/taxiing lamps
60. Leading-edge stall strip
61. Starboard wing integral fuel tank; total fuel capacity 1,248 Imp gal (5674 l)
62. Fuel system piping
63. Starboard wing fence
64. Leading-edge de-icing strip
65. Wing stringers/tank venting ducts
66. Fuel filler cap
67. Wing-tip surge tank
68. Starboard navigation and strobe lights

tent of the export sales success which the aircraft still enjoys today. Like its piston-engined forerunner, the D.H. 125 was expected to be popular in the United States, where business jets were beginning to replace elderly converted World War II bombers and transports as the flagships of corporate fleets.

With the substitution of 13.9-kN (3,100-lb st) Viper 521 and 522 engines permitting a 2200-kg (1,000-lb) increase in maximum take-off weight, and a reduction in the number of cabin windows from six to five, the aircraft was redesignated Series 1A (for FAA-certificated aircraft destined for the North American market) and Series 1B (for customers in the rest of the world. These suffixes were then retained for all subsequent models. First deliveries to the United States were made in September 1964, and thereafter D.H. 125s for US customers were ferried across the Atlantic in 'green' condition, in primer paint with no interior furnishings and only a 'strap-on' avionics pack, for outfitting and completion to customer requirements by American specialists. A total of 77 D.H. 125 1A and 1Bs were built. One British-operated charter aircraft achieved notoriety on 30 June 1967 when it was hijacked over the Mediterranean while carrying the former Prime Minister of the Congo, Moise Tshombe.

In 1963 the Royal Air Force ordered 20 HS 125s (de Havilland having been absorbed into the Hawker Siddeley Group, although the 'D.H.,' prefix continued to be used for some time in US sales campaigns). These aircraft were Series 2s, similar to the 1B but with lower powered Viper 520 engines, and designated Dominie T. Mk 1 in service use as navigation trainers. The first was flown on 30 December 1964 and most remain operational in 1991. Later models have also been supplied to the RAF as the BAe 125 CC.1, CC.2 and CC.3 for duties with No. 32 Squadron, the VIP transport unit based at RAF Northolt near London.

By mid-1966 HS 125 production had been increased to one per week and the improved Series 3A and 3B introduced, featuring 15.7-kN (3,600-lb st) Vipers 522s, improved air conditioning and auxiliary power units, a further increase in gross weight to 21,700 kg (47740 lb) and numerous detail changes. Thirty-nine HS 125 Series 3s, 3As and 3Bs were manufactured, including two specially equipped aircraft for use by the Australian airline QANTAS as crew trainers, with flight decks identical in layout to those of the carrier's Boeing 707 airliners. The addition of a flush-fitting 509-litre (112-gallon) ventral fuel tank faired into the underside of the fuselage and a small triangular ventral fin created the longer range Series 3A-RA and 3B-RA, 36 of which were built, including 10 for the Brazilian air

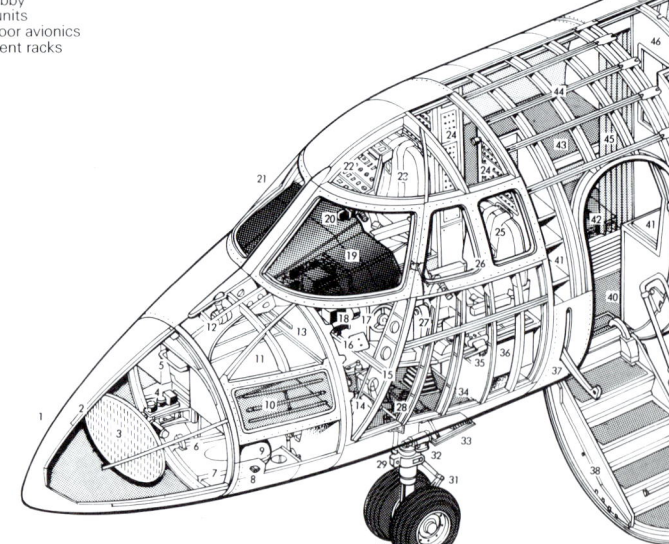

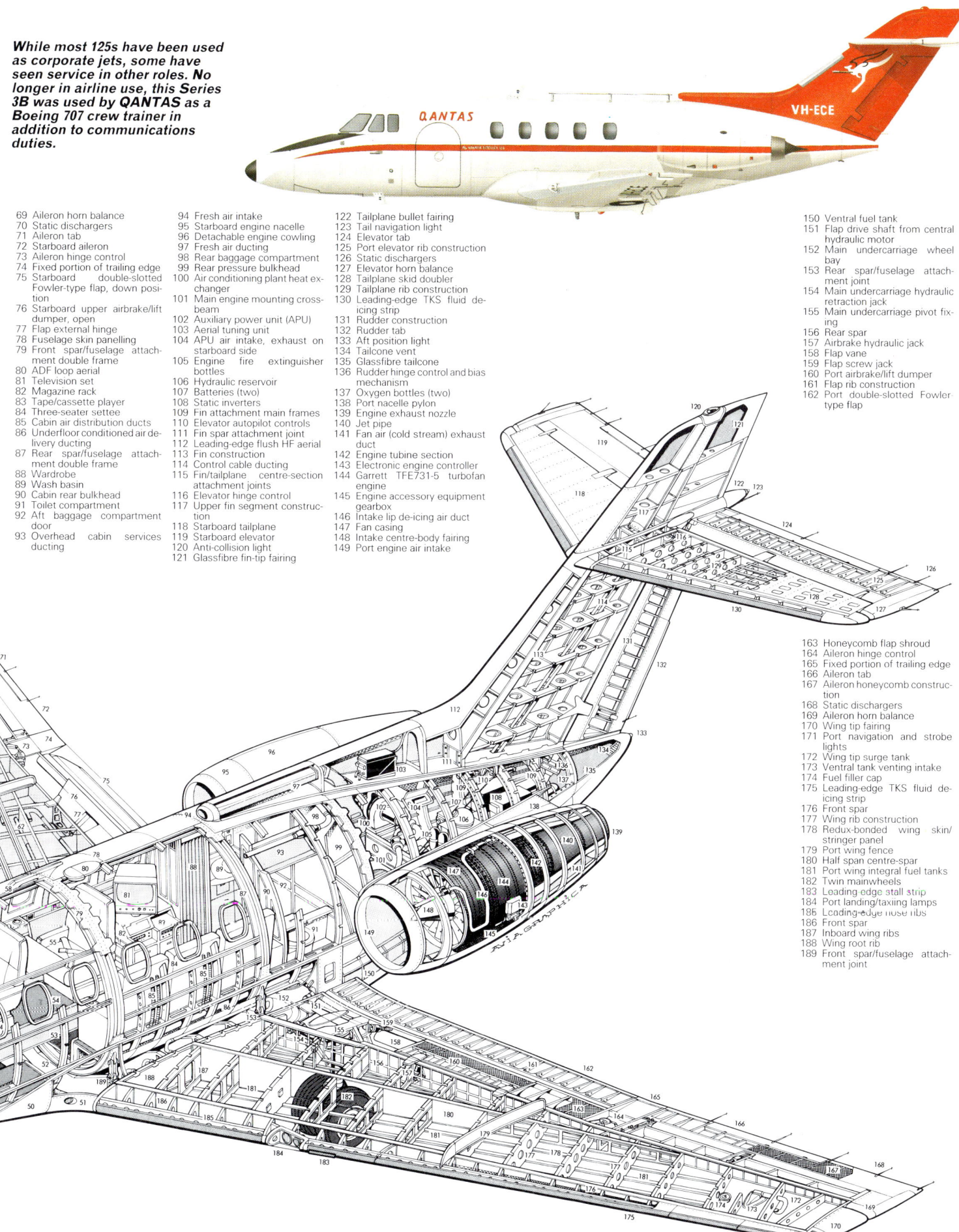

While most 125s have been used as corporate jets, some have seen service in other roles. No longer in airline use, this Series 3B was used by QANTAS as a Boeing 707 crew trainer in addition to communications duties.

69 Aileron horn balance
70 Static dischargers
71 Aileron tab
72 Starboard aileron
73 Aileron hinge control
74 Fixed portion of trailing edge
75 Starboard double-slotted Fowler-type flap, down position
76 Starboard upper airbrake/lift dumper, open
77 Flap external hinge
78 Fuselage skin panelling
79 Front spar/fuselage attachment double frame
80 ADF loop aerial
81 Television set
82 Magazine rack
83 Tape/cassette player
84 Three-seater settee
85 Cabin air distribution ducts
86 Underfloor conditioned air delivery ducting
87 Rear spar/fuselage attachment double frame
88 Wardrobe
89 Wash basin
90 Cabin rear bulkhead
91 Toilet compartment
92 Aft baggage compartment door
93 Overhead cabin services ducting
94 Fresh air intake
95 Starboard engine nacelle
96 Detachable engine cowling
97 Fresh air ducting
98 Rear baggage compartment
99 Rear pressure bulkhead
100 Air conditioning plant heat exchanger
101 Main engine mounting cross-beam
102 Auxiliary power unit (APU)
103 Aerial tuning unit
104 APU air intake, exhaust on starboard side
105 Engine fire extinguisher bottles
106 Hydraulic reservoir
107 Batteries (two)
108 Static inverters
109 Fin attachment main frames
110 Elevator autopilot controls
111 Fin spar attachment joint
112 Leading-edge flush HF aerial
113 Fin construction
114 Control cable ducting
115 Fin/tailplane centre-section attachment joints
116 Elevator hinge control
117 Upper fin segment construction
118 Starboard tailplane
119 Starboard elevator
120 Anti-collision light
121 Glassfibre fin-tip fairing
122 Tailplane bullet fairing
123 Tail navigation light
124 Elevator tab
125 Port elevator rib construction
126 Static dischargers
127 Elevator horn balance
128 Tailplane skid doubler
129 Tailplane rib construction
130 Leading-edge TKS fluid de-icing strip
131 Rudder construction
132 Rudder tab
133 Aft position light
134 Tailcone vent
135 Glassfibre tailcone
136 Rudder hinge control and bias mechanism
137 Oxygen bottles (two)
138 Port nacelle pylon
139 Engine exhaust nozzle
140 Jet pipe
141 Fan air (cold stream) exhaust duct
142 Engine tubine section
143 Electronic engine controller
144 Garrett TFE731-5 turbofan engine
145 Engine accessory equipment gearbox
146 Intake lip de-icing air duct
147 Fan casing
148 Intake centre-body fairing
149 Port engine air intake
150 Ventral fuel tank
151 Flap drive shaft from central hydraulic motor
152 Main undercarriage wheel bay
153 Rear spar/fuselage attachment joint
154 Main undercarriage hydraulic retraction jack
155 Main undercarriage pivot fixing
156 Rear spar
157 Airbrake hydraulic jack
158 Flap vane
159 Flap screw jack
160 Port airbrake/lift dumper
161 Flap rib construction
162 Port double-slotted Fowler type flap
163 Honeycomb flap shroud
164 Aileron hinge control
165 Fixed portion of trailing edge
166 Aileron tab
167 Aileron honeycomb construction
168 Static dischargers
169 Aileron horn balance
170 Wing tip fairing
171 Port navigation and strobe lights
172 Wing tip surge tank
173 Ventral tank venting intake
174 Fuel filler cap
175 Leading-edge TKS fluid de-icing strip
176 Front spar
177 Wing rib construction
178 Redux-bonded wing skin/stringer panel
179 Port wing fence
180 Half span centre-spar
181 Port wing integral fuel tanks
182 Twin mainwheels
183 Leading-edge stall strip
184 Port landing/taxiing lamps
185 Leading-edge nose ribs
186 Front spar
187 Inboard wing ribs
188 Wing root rib
189 Front spar/fuselage attachment joint

249

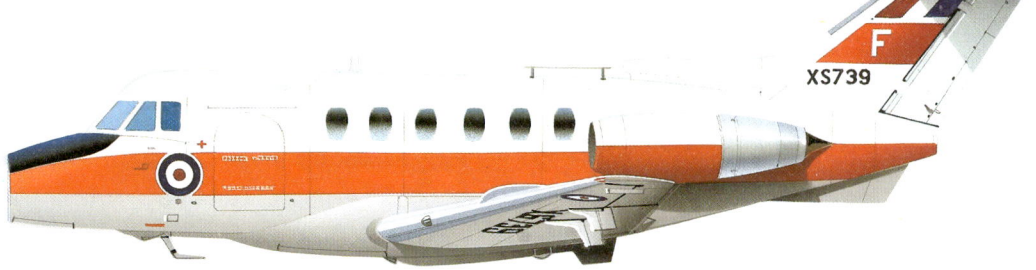

The 125 Series 2 covered an order for 20 aircraft for the Royal Air Force. Known in service as the Dominie T.Mk 1, these now veteran aircraft are used on navigation training sorties, flying with 6 Flying Training School from Finningley and normally carry two students and an instructor.

Hawker Siddeley developed the Series 600, which introduced a fuselage stretch to increase maximum passenger load to 14 and to revive the type on the biz-jet market. This aircraft was operated by the Nigerian Government on VIP tasks.

force for VIP and navigational aid checking duties.

In September 1968 Hawker Siddeley announced the HS 125 Series 4, which was marketed as the Series 400. It featured uprated 15.7-kN (3,600-lb st) Viper 522 turbojets, a redesigned cabin seating seven passengers as standard, flight deck improvements, an integral airstair entrance door and a further increase in maximum take-off weight to 23,300 kg (51260 lb). The popularity of the HS 125-400 in the United States prompted Hawker Siddeley to enter into an agreement with Beech Aircraft Corporation in December 1969 whereby the jet was marketed in the USA as the Beechcraft-Hawker BH 125, although Beech did not play any part in manufacture and none of the planned joint Beech-HS familiy of business jets ever materialised.

In all, 69 HS 125-400As and 47 -400Bs were built, the production rate at Hawarden peaking at seven per month.

Competition from the Lear Jet and Dassault Falcon 20 slowed sales at the turn of the decade, but drawing on input from Beech, Hawker Siddeley undertook a major revision of the HS 125 in 1971 and introduced the Series 600, first flown in prototype form on 21 January. This version was the first to incorporate major external changes, with a fuselage lengthened by 0.95 m (3 ft 1 in) to provide executive class accommodation for up to eight passengers, or up to 14 in a high-density layout. Other changes included more powerful 16.7 kN (3,750-lb st) Viper 601-22 engines with optional 'hush kits', a taller fin with extended dorsal fin containing a ventral fuel tank, aerodynamic improvements to the wing/fuselage junction, a sleeker nose profile, restyled flight deck with new avionics and annunciator panels, and a gross weight of 25,000 kg (55060 lb). Production of the HS 125-600 began in 1972 and continued until 1976, by which time 33 600As and 39 600Bs had been delivered.

New generation

It was succeeded by the Series 700, first announced on 12 May 1976. The prototype, converted from a Series 600, made its first flight on 28 June 1976, followed by the first production aircraft on 8 November. The Series 700 was the first turbofan-engined HS 125, powered by two 16.46-kN (3,700-lb st) Garrett AiResearch TFE731-1-3-1H turbofans which offered much improved fuel economy and increased range over the pure jet variants and could be fitted optionally with Aeronica thrust reversers. With maximum payload of 2,355 kg (5181 lb) the series 700 had a range of 3929 km (2,442 miles), against 3121 km (1,940 miles) for the Series 400. Other improvements included substitution of countersunk rivetting in place of much of the mushroom head rivetting on preceding models, redesign and enlargement of the ventral fin and fairings, the addition of windscreen wiper fairings to the flight deck windows, provision for single-point pressure refuelling, and entirely new interior furnishings. These including the use of figured walnut veneer and leather trim, a luxury toilet compartment, high quality cabin sound system, and a new range of colour co-ordinated cabin decors. UK certification of the HS 125 Series 700 was obtained on 7 April 1977. Initially it had been intended to keep both the pure jet Series 600 and turbofan Series 700 in parallel production, but the superior performance of the Garrett-engined aircraft quickly established it as the sole production model, of which 215 were built.

Always up-to-date

The production version from the mid-1980s of what is now the British Aerospace 125, is the BAe 125-800, first flown as a prototype on 26 May 1983 and certified on 30 May 1984. The Series 800 is powered by two 19.13-kN (4,300-lb st) Garrett TFE731-5R-1H turbofans, which afford substantial improvements in take-off, climb and 'hot-and-high' performance and maximum speed, and provide adequate with-reserves range for non-stop coast-to-coast operation in the United States or Transatlantic flights. Many detail changes were made in developing the Series 800 airframe, including a curved windscreen; redesigned outboard wing panels of increased span for reduced drag, greater fuel capacity and enhanced aerodynamic efficiency; a larger ventral fuel tank; extended fin leading edge; Dee Howard thrust reversers as an option; and sequentially actuated nosewheel doors. The flight deck incorporates an 'all-glass' instrument panel featuring a Collins EFIS-85 five-tube electronic flight

An important step in 125 development was made by the Series 700, which dispensed with the Viper turbojets and introduced the quieter and more fuel-efficient Garrett TFE731 turbofans. These engines also became available as retrofits to earlier variants.

The RAF's communications squadron is No. 32 based just outside London at Northolt. The mainstay of its equipment is the 125, available in three versions. Four are CC.Mk 1s (Series 400), two are CC.Mk 2s (Series 600) and six are CC.Mk 3s (Series 700). This is one of the latter.

instrument system with centrally-mounted multifunction display, while the cabin area has been redesigned to offer greater headroom and width.

In addition to the two most recent production models, earlier 125s have been retrofitted with turbofan power plants by British Aerospace and Garrett General Aviation Services of Los Angeles. The Garrett aircraft, retrofitted with new Garrett TFE731-3 turbofans enabling the aircraft to comply with FAA FAR Pt 36 Stage III noise regulations, is known as Garrett 731 Hawker. The modification package also includes a dual Collins EFIS-85B system, Collins ATS-80 autopilot, Collins WXR-270 colour weather radar, Universal UNS-1JR flight management system and a new seven-seat passenger interior. The first aircraft modified under the now-discontinued Garrett programme was a former D.H. 125-400A delivered to its new owner in the Philippines in 1985.

Prestigious sales

The British Aerospace 125 has been an outstandingly successful aircraft, particularly in the highly competitive American market, which has accounted for 60 per cent of total sales. In the 24 years since the first deliveries were made, the value of export sales has totalled some £1.5 billion. In the summer of 1988 the aircraft established a sales record for any British-manufactured civil jet aircraft when the 700th BAe 125 was sold. BAe 125s are in service in 40 countries worldwide, and have been adapted for many roles beyond VIP/executive transport, including air ambulance, military liaison, aircrew training, maritime surveillance and navaid calibration. The aircraft is operated by many international corporations, heads of state and Royal families, and by foreign military services including those of Argentina, Botswana, Brazil, Eire, Ghana, Malawi, Malaysia and South Africa. Recent military customers include the Royal Saudi Air Force which ordered four 125-800s for VIP transport duties as part of the 'Al Yamamah' programme, the United States Air Force who ordered six to fulfil the Combat Flight Inspection and Navigation (C-FIN) role, replacing Sabreliner and JetStar aircraft. The 125s, designated C-29A by the USAF, are outfitted by the Sierra Research Division of LTV Electronics as flight inspection aircraft for checking and calibrating airways and airfield navigation aids and installations. First delivery was made on 24 April 1990. In August 1989 the Japan Air Self Defense Force ordered three 125-800s to carry out flight inspection duties from December 1992. These will also be fitted out by Sierra Research.

In 1988 the company revealed it was developing the BAe 1000, an extensively redesigned and re-engined 125-800 featuring increased headroom, a bigger cabin and extended range, primarily to compete with the Falcon 50 and later the Falcon 2000 in the long-range business jet market. Officially launched at the NBAA Convention in October 1989, the 1000 incorporates a 0.84 m (2 ft 9 in) fuselage stretch, an additional cabin window, extra fuel and other detail improvements. By the time the BAe 1000 made its first flight on 16 June 1990, the company was claiming orders for 10 aircraft. Joint CAA/FAA certification is scheduled for autumn 1991, followed by customer deliveries by the end of the year. Engines are two Pratt & Whitney Canada PW305 turbofans delivering 23.13 kN (5,200 lb) thrust, and equipped with thrust reversers.

In August 1993 a US company, the Raytheon Company, bought BAe Corporate Jets, and in the following reorganisation the 125-800 and 125-100 became the Hawker 800 and Hawker 1000, names under which both variants are still marketed.

The UK earth-moving equipment manufacturer Bamford operate this Series 800, displaying the increased wing span and curved windscreen that characterise this variant. The 800 won a six-plane order to replace the USAF's calibration fleet of Sabreliners and JetStars and are designated C-29A.

British Aerospace 146

With the world increasingly conscious of noise pollution, aircraft such as the BAe 146 have an ever-important part to play in commercial aviation. With four fuel-efficient and very quiet jets, the 146 can slip unnoticed into any suburban airfield. This powerplant also gives it exceptional performance, a fact not lost on Third World airlines with small and primitive airfields.

Gradually the BAe 146 is winning friends all over the world. It is doing so partly by being the quietest jet, and almost the quietest aircraft of any type, in the sky. It is pleasing travellers with an exceptionally comfortable interior, a perfect view from every seat, and by bringing jet speed to routes where no jet could go before. And it is pleasing airline accountants by being extremely reliable and burning very little fuel.

The odd things are that, when it was launched, some supposed experts said it was ridiculous to use four engines. Fokker tried to get the project stopped on the grounds that it competed with their F28 (which cannot fly from many airports today served by 146s). And, at the very start, it almost never went ahead at all!

This is all the more remarkable when it is realised that the 146 was preceded by perhaps the longest period of study and refinement of any aircraft in history. The original objective was to replace the DC-3 with a modern short-hauler able to operate from any field that a DC-3 could use. The work began in 1958, and nobody could have foreseen that it was to take 20 years to get the final design launched!

The first project, the DH.123, looked like a slightly smaller F27, with two Gnome turboprops. Next came a series of DH.126s with twin aft fans, looking like enlarged 125 bizjets [business jets]. Next came ideas for jet 748s, culminating in the DH.131 which used a shortened 748 fuselage (and T-tail). By the mid-1960s there were few DC-3s to be replaced, and the market needed bigger aircraft. A series of HS.136 designs followed, looking like small 737s, followed by a series of HS.144s with the RR Trent engines moved to the rear (and looking very like the F28). In 1970 Rolls-Royce crashed, eliminating the Trent, and the Hatfield team studied the whole range of available engines including the M45, RB.410 and 415, GE TF34 and Spey before deciding that the best answer, by a clear margin, was four Lycoming (today Allied Signal) ALF502 engines. A further detailed market survey showed that a field length of 914 m would satisfy the entire market.

Choice of the ALF502 automatically meant four engines. This was no problem. The capital cost, fuel burn and installed weight of the four engines was in every case better than for any of the bigger engines for a twin. An incidental advantage is that the quickly disconnected modules which make up the ALF502 are individually small and light enough to be man-handled and, if necessary, stowed under the floor in the 146's cargo compartments. Its extremely low noise level, absence of smoke, and mature core with (in 1989) over 5 million hours behind it were all further giant 'plus' factors.

All the previous projects had had low wings, but when Hatfield added together the requirements – four turbofan engines, with inlets high off the ground breathing undisturbed air, a large range of per-

Under the designation 146-QT, BAe has developed the aircraft as a quiet, efficient freighter, capable of carrying six standard 2.74×2.24 m pallets or up to nine standard LD3 containers. The freight door is in the rear, allowing seats to be retained forward if required.

Air Wisconsin was the first US customer for the BAe 146, although it now operates its aircraft under the United Express banner.

PSA was another US carrier which found the 146 ideal for its shuttle-type operations between noise-conscious airports. Its aircraft later flew with USAir.

missible centre of gravity position, and powerful high-lift flaps for STOL performance – it was inevitable that the wing should be on top of the fuselage. This imposed no penalties, gave the passengers a good view, left an uninterrupted upper surface to give high lift, unaffected by the wide fuselage below, and made it possible to put the fuselage close to the ground so that (as a customer option) passengers could board via a short airstairs at each end of the cabin, making the aircraft independent of airport jetways or stairways. The high wing led to fuselage-mounted main landing gears, and Dowty Rotol designed a gear that is extremely neat, offers a wider track than that of a C-130 (giving excellent stability on rough strips or in crosswinds) and weighs less than that of typical low-wing airliners of similar take-off weights.

The wing itself was made quite thick (over 15 per cent at the root) and given hardly any sweepback (a mere 15°, much of which is due to natural taper). This fits in well with the planned short field length and Mach 0.7 maximum operating Mach number, and it enables the required high lift coefficients to be achieved with a plain leading edge without slats or drooping flaps. At the same time the tabbed Fowler flaps along the trailing edge are very powerful. To get the horizontal tail out of the wing wake it was put on top of the large vertical tail (large because of the 146's ability to fly slowly and with an outer engine shut down). This T-tail arrangement also made it easy to fit powerful airbrakes forming the tail end of the fuselage. These airbrakes, together with lift dumpers above the wing and powerful wheel brakes, made it possible not to need engine thrust reversers.

Obviously the rudder had to be fully powered, but all other flight controls were made manual, though with extremely refined qualities. There are conventional ailerons, backed up by small outboard roll-control spoilers, and the tailplane is fixed. The tailplane and all control surfaces are made by Saab of Sweden. A Smiths SEP.10 flight control system is fitted, with an autopilot cleared for ILS-coupled approaches down to Cat II minima (runway visibility down to 400 m and decision height down to 30 m). The stall protection system, which operates with a stick shaker, gives a 'soft' response (which varies according to the sub-type of aircraft) and is fired under

British Aerospace 146-100 cutaway drawing key

1. Radome
2. Weather radar scanner
3. Radar mounting
4. ILS aerial
5. Oxygen bottle, capacity 1812 l (400 Imp gal)
6. Sloping front pressure bulkhead
7. VOR flush aerial
8. Nose undercarriage wheel bay
9. Nosewheel leg strut
10. Twin nosewheels
11. Pitot tube
12. Rudder pedals
13. Instrument panel
14. Windscreen wipers
15. Instrument panel shroud
16. Windscreen panels
17. Overhead switch panel
18. First Officer's seat
19. Centre control pedestal
20. Control column handwheel
21. Side console panel (area navigation system)
22. Cockpit floor level
23. Captain's seat
24. Direct vision window/flight deck emergency exit
25. Folding observer's seat
26. Flight deck bulkhead
27. Air conditioning ducting
28. Starboard galley unit
29. Forward service door
30. Main cabin divider
31. Port side forward toilet compartment
32. Forward entry door
33. Door latching handle
34. Escape chute stowage
35. Underfloor radio and electronics equipment bay
36. Machined doorway cut-out main frames
37. Nose section/forward fuselage skin joint strap
38. Door frame support structure
39. Entry vestibule
40. Cabin attendant's folding seat
41. Six-abreast passenger seating
42. VHF aerial
43. D/F loop aerial
44. Cabin wall trim panels
45. Air conditioning ducting
46. Forward cargo hold door
47. Forward underfloor cargo hold
48. Seat rail support structure
49. Fuselage keel construction
50. Pressurisation air control valve
51. Fuselage/front spar attachment main frame
52. Floor beam construction

So far military usage has been restricted to a pair of aircraft for the United Kingdom Queen's Flight, dedicated to transporting royalty under the designation CC.Mk 2. Two earlier CC.Mk 1 examples had been evaluated for this role and subsequently sold on.

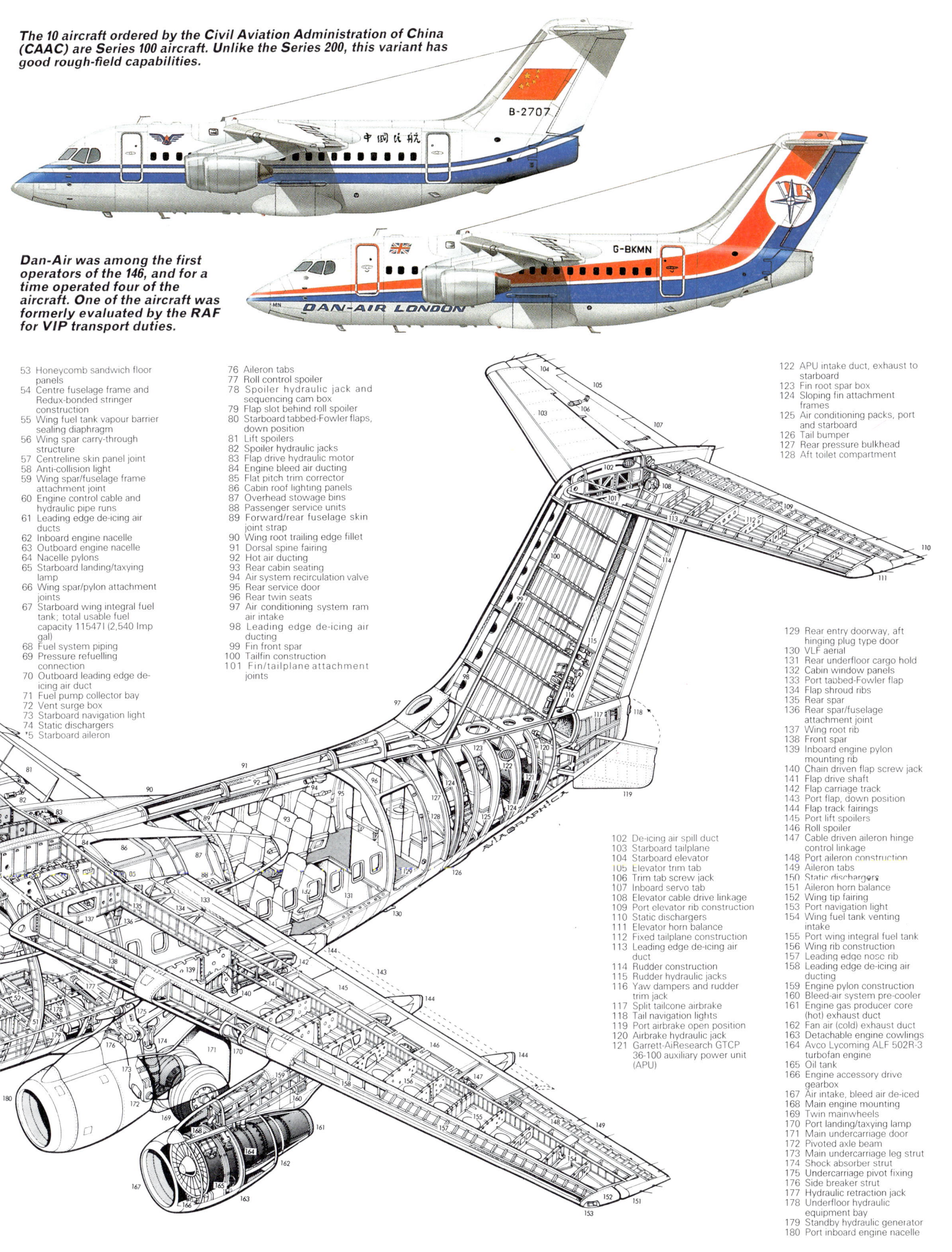

computer control which takes into account the rate at which the boundary of safe flight is approached. It is doubtful if any previous aircraft has had such safe handling qualities, and agility even at speeds below 200 km/h is exceptional for so large an aircraft.

Aircraft systems

Each engine is installed in a simple pod, hung well ahead of and below the wing. There are three integral tanks in the main wing box, which is supplied by Textron Aerostructures of Nashville, Tennessee. Normal capacity is 11,728 litres, increased to 12,901 by using auxiliary tanks in the wing root fairings. Engine bleed air is used to de-ice the leading edges of the wings and tail, and also to energise the cabin pressurisation system, which has a maximum differential of 0.45 bar. The hydraulic system at 207 bars powers almost all ancillary services and also drives an emergency electric generator. In the tail end of the fuselage is the Garrett gas-turbine auxiliary power unit, which among other things can provide air-conditioning on the ground.

From the outset the Hatfield designers were determined to make the 146 as comfortable as possible. The fuselage diameter was fixed at 3.56 m, sufficient for an internal cabin width of 3.38 m in early versions and 3.42 m for the stretched 146-300. This compared with, for example, 3.07 m for a Fokker F28 or Fokker 100 and 3.1 m for a DC-9 or MD-80. Cabin height was also made generous, and in service the 146 has proved to have a totally flexible interior able to adapt to any desired seating or cargo configuration. It was a design objective to provide five-abreast seating with a general level of comfort not less than that in a 747 – in fact, the 146 mock-up was initially furnished with actual 747 seats. The initial production aircraft, the 146-100, is normally equipped for 71 seats in 13 rows of 3+2 and three pairs of seats at the rear. Some operators, however, have used the width to fit triple seat units on each side of the aisle, giving 82 passengers at 84-cm pitch and 93 at 74-cm pitch. In the standard layout there is a forward galley and two toilets.

Hawker Siddeley collected an excellent team of designers, systems engineers and marketing staff to launch the HS.146, and on 29 August 1973 the British government announced that the project was going ahead with the government providing half the estimated launch cost of £92 million as a risk-sharing investment. Unfortunately various factors conspired to wreck this plan. At just about this time the Middle East oil price began rising in a way nobody had dreamed of, triggering off a worldwide crisis and economic recession which hit the manufacturing industry and the airlines especially hard. (One might have thought it would hasten the replacement of fuel-hungry airliners, but the operators did not see it that way.) Inflation soared, making nonsense of all the carefully refined 146 launch and operating costs. Hawker Siddeley was to be swallowed up in the nationalised British Aerospace, and the shareholders would do far better if the 146 were to be shelved and the money not spent. In October 1974, therefore, the programme was terminated by the Hawker Siddeley board. This brought a vigorous reaction from workers at Hatfield, who immediately expressed their total belief in the aircraft and tried to find ways to keep work going. Eventually the board agreed to let work tick over with a trickle of (mostly government) funding.

New tasks for a new company

British Aerospace was formed in 1977. One of the first tasks undertaken by the civil marketing team was to review the 146 and re-assess the potential market. During the delay of almost four years quite a lot had been done to refine the design, carry out tunnel testing and build systems test rigs, and manufacture production tooling. Virtually all the market analysis yielded positive answers, and the BAe board recommended a re-launch. Government approval was obtained on 10 July 1978. This time there were no hold-ups, and a big manufacturing organisation was quickly set up. Hatfield builds the forward fuselage and flight deck and handles assembly and flight test. Textron in the USA builds the wing box. Saab in Sweden makes movable control surfaces and the tailplane. Brough makes the fin and flaps, Bristol the centre fuselage, Manchester the rear fuselage, Hamble the flap track fairings and Prestwick the pylon struts. In 1987 output had to be stepped up from 28 to 40 per year, as a result of which a second assembly line was set up at Woodford – the first Woodford-built 146 flying on 16 May 1988.

The first 146, appropriately registered G-SSSH, was flown on 3 September 1981. Type certification followed in early 1983, and Dan-Air began scheduled services on 27 May 1983. It was typical of the 146 that one of the first Dan-Air destinations should be Innsbruck, never previously served by scheduled jetliners. Production of the 146-100 continued for customers with severe field-length or noise problems; one customer being HM The Queen's Flight. Later aircraft have thicker wing skins giving the option of greater weights.

146 is at its best shuttling sizeable numbers of passengers between regional airports and major route centres, and on such flights the type has been welcomed by US carriers. The super-quiet Avco Lycoming ALF 502R turbofans meet the most stringent of noise restrictions.

Pacific Southwest Airlines was a major user of the BAe146, flying Series 200s along the California/Nevada corridor. The airline merged with USAir in 1989 under the latter's name, which was changed in February 1997 to US Airways.

It was increasingly obvious that some operators would gladly trade the 146-100's amazing field capabilities for increased payload. Accordingly the fuselage was stretched by five frames (a length of 2.39 m) to produce the otherwise almost identical 146-200. Compared with the -100 the -200 has maximum weight increased from 38102 kg to 42184 kg, payload increased from 8845 kg to 10478 kg, and seating capacity increased to a maximum of 111 (3+3 at a pitch of 74 cm).

200 into service

The first -200 was flown on 1 August 1982, and the first scheduled service was flown on 27 June 1983, only a month after the start of -100 operations. Almost immediately the -200 became the baseline standard aircraft, and the launch pad for a cargo version (the 146 QT Quiet Trader), a VIP executive version (the Statesman), and four military versions, the STA (Small Tactical Airlifter), the MSL (Military Side Loader), the MT (Military Tanker) and the BAe 146M (a major revision with a lowered floor, rear ramp doors and tandem main landing gears).

At the time of writing, the Statesman and QT are in service, but nothing has been said of launch customers for the military variants. The giant second arms sale to Saudi Arabia announced in July 1988 was said to include 'unspecified numbers of military versions of the 146', but nothing firm is known of this contract. Certainly, it would take a major buy of this nature to launch a considerably revised military version such as the extremely attractive 146M. In contrast, the STA and MSL are only minor adaptations of the QT, though they could have a flight-refuelling probe above the flight deck.

As for the QT, this did so well when evaluated by the TNT global package and cargo group that this Australian conglomerate reached an agreement with BAe to take five years' production of QTs, a total of 72 aircraft committed in stages. TNT's growing fleet has so far been largely confined to Europe, where they operate a gruelling night-long schedule involving extremely rapid turnarounds and making use of the fact that the 146 is not noisy enough to be subject to any curfews.

While the Dash-200 does need a paved runway, unlike the -100, it still has a field length much shorter than other jetliners (1500 m). The market suggested stretching the 146 fuselage a second time, still keeping field length shorter than (for example) almost every local-service airport in the USA. The result was the 146-300, first flown on 1 May 1987. This was planned to have extended wings with winglets, more powerful engines and other new features, but in the event the best view of the market was 'Leave the 146 as it is but stretch it further'. The -300 has a 2.46-m plug in the forward fuselage and a 2.34-m plug in the rear fuselage, and a thicker centre-fuselage skin which allows maximum weight to rise to 43090 kg.

After a brief dalliance with ATR in the Aero International (Regional) consortium, resulting in the designation Avro International Aerospace RJ (standing for Regional Jet), the 146 has become the Avro RJ or RJ Avroliner. Updated with LF507 turbofans and digital avionics, the family comprises the RJ70 for 70 to 94 passengers, the RJ85 for 85 to 112 passengers and the RJ100 for 100 to 128 passengers, these models being equivalent to the 146-100, -200 and -300 respectively.

The large Ansett concern in Australia has provided BAe with its largest prospective order, with up to 72 146-QTs for use by TNT Couriers or for sale and lease by Ansett Worldwide Aviation Services. Ansett W.A. (illustrated) also has five in service.

British Aerospace 146-300A
Air Wisconsin/United Express
Appleton Wisconsin

Chapter 13

Business Jets

Canadair CL-44

Canadair CL-215

Canadair CL-600 Challenger

Business Jets

Modern jet aircraft are an essential tool of big business. The growth in the company use of jets has been an important facet of the general aviation scene since the 1960s, and there are now many thousands of corporate jets operating around the world. Here we review the major types in use.

In the early 1950s jet transport aircraft were a novelty; the de Havilland Comet had stolen the headlines as the first pure jet airliner, and the Boeing 707 was arriving in airline service with Pan American titles painted boldly along its white roof. It was, however, not too early for other manufacturers to see the great advantage offered by the jet engine in terms of higher speed for military communications aircraft. At Lockheed and at North American studies were put in hand which resulted in the JetStar and the Sabreliner high-speed staff transports for the US Air Force. These two types were built to a high-level military specification and their application for civil purposes was expected to be limited because of the high capital cost involved. However, American business was not slow in appreciating that speed and convenience were very good reasons for investing in one of these new aircraft. Both types soon found their way into the general aviation fleet which had thus far been dominated by Convairs 440s, Douglas Dakotas and converted World War II piston-engined bombers.

Now the business jet fleet is quite a bit different from that of those early years, when only an elite group of big corporations could afford such luxury. The range of choice stretches from the light jets with a practical five-passenger capacity to the large number of airliners which have been converted as long-range executive aircraft and offer a very high standard of luxury for flights from Jeddah to London or New York to Mexico City. Over 30 Boeing 707s have been converted for this type of role, and numerous BAC One-Elevens, Boeing 727s and Boeing 737s are also employed in transcontinental business transportation. Perhaps the most prestigious aircraft in use anywhere are the two Boeing 747SPs which are used by the Saudi Arabian royal flight. The original order came from the late King Khalid, who issued an impressive specification which included a special communication system which would link the aircraft to the American hospital where his heart pacemaker had been fitted. Sadly, the aircraft are no longer required to provide this support, but they do serve as very high standard equipment for the regular travel necessary to this principal oil producer.

The battlegrounds for business jet salesmen have changed over the years, but the 1980s saw the heavy intercontinental business jets in the forefront, with the light and medium jets facing a temporarily saturated market. For many years the Grumman Gulfstream II (known universally as the 'G-2') was the Rolls-Royce of corporate aviation. This position started to be eroded in 1976, when the first Dassault-Breguet Falcon 50 was flown and then in 1978 when the first Canadair Challenger appeared. The Falcon 50 was the latest in a line of business jets which had started with the Mystère 20 (Fan Jet Falcon) 8/10-seater in 1963. Avions Marcel Dassault had great success in the United States with the Fan Jet Falcon and sold examples to prestige names such as IBM, General Electric and Pepsi Cola. In Europe the Falcon 20 was bought by Fiat, Philips and Zanussi. With such success it seemed natural to design a larger aircraft, and after building a prototype of the much larger one-off Falcon 30 they decided to use a slightly enlarged Falcon 20 fuselage combined with three engines and greatly increased fuel capacity to offer as the Falcon 50. Backed by the reputation for reliability established by the smaller aircraft, the Falcon 50 has reached a production total in excess of 190 units, of which over two-thirds have been sold in the US.

Dassault created a replacement for the Falcon 20, the Falcon 2000 wide-body, first announced in 1989 as the Falcon X. The type flew in March 1993, and is sold as a transcontinental type with up to 19 passengers and two General Electric/Allied Signal CFE738 turbofans.

*For many companies a converted airliner offers a cost-effective way of owning a comfortable long-range business aircraft. Some 42 BAC One-Elevens were at one time flying in this role, including eight on the Saudi register alone. This **BAC One-Eleven 401** owned by **Caw Corporation** is registered in the **Cayman Islands**.*

Canada's stake in the executive jet market is the Canadair Challenger, here represented by the Model 601, the current production design with advanced technology winglets and General Electric turbofan powerplants. Maximum accommodation is for 19, though various executive layouts are available.

A direct competitor of the Gulfstream III and Canadair Challenger at the top end of the corporate market, the Dassault Falcon 900 was first flown on 21 September 1984. This intercontinental business jet is powered by three Garrett TFE731-5A turbofans and first customer deliveries were made in December 1986.

Quiet Challenger

The CL-600 Challenger from Canadair, now part of Bombardier, was originally the design of William P. Lear (creator of the Learjet). It has changed its form greatly since the design studies, and is now a mature widebody jet powered (in original form) by a pair of Avco Lycoming ALF 502 turbofans. It is very much the modern concept of the 'good neighbour' aircraft with very quiet engines meeting the regulatory pressures to reduce jet noise at airports all around the world. Many operators of the Challenger fly from airports such as White Plains and Teterboro in New Jersey state, where strong environmentalist lobbies exist. The inherent fanjet design is a spur to existing operators of older business jets to modify their aircraft with new engines or 'hush kits' in order to meet the approaching noise threshold rules. The Challenger was available in two versions: the basic CL-600 and the long-range CL-601 which can be identified externally by the Whitcomb winglets at the outer tips of the wings.

The CL-601 is now powered by General Electric CF34 turbofans. A coup for Canadair was an order from the West German government for seven aircraft to replace its fleet of VIP JetStars and Hansa Jets. Today, only variants of the CL-601 and longer-range CL-604 are produced alongside the Regional Jet airliner version of the Challenger.

For many company chairmen, however, there will never be any alternative to the Gulfstream. Designed by Grumman Aerospace as a jet version of the successful Gulfstream I turboprop, the long-range 19-passenger aircraft dominates the corporate big jet market with more than 600 in service worldwide. Following 258 G-1159 Spey-powered Gulfstream IIs down the production line, around 200 of the improved G-1159C Gulfstream III variant were sold. These featured a stretched fuselage with a new nose housing a redesigned cockpit, improved wings with leading edge extensions fitted with the now fashionable NASA winglets, and improved Spey turbofans. Grumman sold the design to Allen E. Paulson's American Jet Industries in 1978, and this company in turn became the Gulfstream American Corporation (later Gulfstream Aerospace). In August 1985, Paulson sold Gulfstream to Chrysler but in 1990, after only five years, he bought the company back. In the meantime 47 Gulfstream IIs had been modified into G-IIBs by fitting the G-III wing, and development of the current version, the G-1159C Gulfstream IV, got underway. Announced in 1983, the first G-IV flew in September 1985 and more than 280 have been delivered to date. With a fuselage

Manufactured by Gulfstream Aerospace at Savannah, Georgia, the Gulfstream III was regarded as the ultimate corporate jet and was bought by companies such as Shell, Coca-Cola and Honeywell. This example is one of nine Gulfstream jets operated by Saudia and used for the transport of members of the Saudi Royal Family.

Dassault-Breguet has been building the Falcon 20 since 1965. Falcons are in service in 33 countries with military examples used by Canada, France, Iran, Belgium and Australia. The Falcon 200 has a nine-seat interior and is powered by a pair of Garrett ATF3-6A-4C turbofan engines.

The top-of-the-line model from Gates Learjet Corporation is the Model 55 Longhorn, which has a maximum passenger capacity of 11 together with two crew members and is readily identifiable by the prominent wingtip winglets. Over 80 per cent of the Model 55s built are flying in the United States.

The most popular of the light business jets, the seven/nine-seat Cessna Citation I is powered by two Pratt & Whitney Canada JT15D-1B turbofans and a version is certificated for single-pilot operation. The stretched Citation II was adopted by the US Navy as the T-47A.

stretched by a further 54 in and fitted with a sixth window and a glass (EFIS) cockpit, a modified wing, and powered by Rolls-Royce Tay 610 turbofans, the long-distance Gulfstream IV can carry its passengers in exceptional comfort over 4,300 miles cruising at Mach 0.80. Its endurance of over nine hours has led to a series of special mission variants primarily for military use, some with cargo doors.

Falcon 900

When launched, an unknown quantity among the heavy jet contenders was the Falcon 900, which was unveiled by Dassault at its rollout ceremony in May 1984. Larger than the Falcon 50, the new aircraft has a wide fuselage with many cabin windows. In its high-density arrangement the Falcon 900 can be fitted with a forward cabin with 12 seats and an aft stateroom equipped with a four-seat couch and two armchairs. Most aircraft are delivered with 12 or 15 seats, and exceptionally low noise levels are delivered by the three Garrett TFE731 turbofans fitted to the aircraft. Without doubt the Falcon 900 is a strong contender in this sector in the 1990s.

The medium-sized pure jet and turbofan types were somewhat more sensitive to the worldwide recession of 1979 to 1983. The older models, namely the Dassault Falcon 20, British Aerospace BAe 125, IAI-1124 Westwind and the Sabreliner series (built successively by North America, North American Rockwell and Rockwell Sabreliner Division) faced declining sales volumes in the prime United States market and were given facelifts by their manufacturers in order to bring them up to modern standards. Gates Learjet and Cessna came in as new contenders in this section with their Learjet Model 55 and Citation III respectively, but this occurred as sales volume fell and fewer aircraft were sold than the initial predictions forecast. Unfortunately, it is vital for any manufacturer to succeed with the American customer. Cessna delivered over 180 Citation IIIs and probably needs many more sales to break even, even though the initial customers included such companies as Conoco and Sperry Corporation.

Numerically, the British Aerospace BAe 125 (and its de Havilland/Hawker Siddeley predecessors) is a substantial part of the medium jet fleet. Over 750 are currently in service and more than 40 per cent of these are the turbofan-powered Models 700 and 800. The current Model 800 represents a most worthwhile improvement over previous models with a brand new wing design, revised nose profile, TFE731-5 turbofans and the attraction of a 4828-km (3,000-mile) VFR range made possible by a revised rear fuselage shape and increased fuel capacity.

In order to capitalise on their market success, British Aerospace launched the development of the BAe 1000 in 1988 and two prototypes are now flying. Intended to occupy the middle sector of the 'bizjet' market alongside

The Model 1124 Westwind, manufactured by Israel Aircraft Industries, was derived from the Aero Commander Jet Commander design, which was purchased by the Israeli company from North American Rockwell in September 1967. The Westwind II model carried winglets on the tip fuel tanks.

The original HS.125, designed by de Havilland, was first flown in August 1962 and, after many examples were sold to the USA and Europe, it was upgraded with Garrett TFE731 turbofans to become the HS.125-700. The BAe 125-800 version shown here has a longer wing, new nose, deeper rear fuselage and new interior.

the Falcon 900, the longer range 15-passenger stretched and re-engined variant of the 125-800 has a bigger cabin, and first deliveries are due next year.

It is in the light jet field that the greatest competition exists. One of the very early models was the Learjet 23, which first flew in October 1963 and has attained a reputation as a very fast business transport. More that 1,700 examples of the Learjet are in service currently; this total includes a substantial number of early production aircraft, although some of these are being forced into retirement as mandatory wing requirements make further service uneconomic. The emphasis on low noise impact at local airports has, again, had its effect on the Learjet series, which moved from the turbojet-powered Models 24 and 25 to the TFE731 fanjet-powered Model 35 and its long-range sister, the Model 36.

Cessna competition

In numerical terms, the Citation family from Cessna comes a close second to the Learjet fleet. With passenger accommodation ranging from the five-seat CitationJet to the 12-seat Citation V, these aircraft are somewhat slower than the Learjets, but have rather larger cabins. They also offer special versions which allow single-pilot operation, which can be a significant advantage to small users who wish to operate satisfactorily with a single crew member. Turbofans make the Citations very quiet in the approach and take-off phases of operation, and a number of military deliveries have been made. Second-generation light bizjets are the Citation VII and Citation X, still in development and expected to fly in 1992. Dassault has also competed in this sector since 1970 with its seven-seat Falcon 10. A late arrival was the twin-jet Diamond introduced by Mitsubishi, who hoped to find a market among former operators of the successful MU-2 turboprop. However, the entire Diamond project was sold to Beech in 1985 and the type is now marketed as the Model 400A Beechjet. This has found favour with the USAF, which ordered more than 200 T-1A Jayhawk versions in February 1990 for training transport and tanker crews. Most recent entrant in the light twin-jet category is the six-place Swearingen/Jaffe SJ30 Fanjet (known until September 1989 as the Gulfstream SA-30 Gulfjet), which flew for the first time in 1990. Powered by two Williams/Rolls-Royce JF-44 turbofans, first customer deliveries are scheduled for 1992.

A fair number of business jet models have fallen out of production over the years, but the fleet still contains examples of the Aero Commander Jet Commander (now represented in current production by the Israel Aircraft Industries' Westwind and Astra), the Sabreliner, the JetStar and the HFB320 Hansa. All give very good service and are often taken for refurbishment with new main spars and new turbofan engines. Some 78 per cent of all current business jets are registered in the United States, with a further nine per cent appearing with European companies. With shorter distances in Europe the business jet often appears less viable than in the United States, but many well-known names such as Shell and Rank Xerox are enthusiastic operators, and companies such as Ford and Philips run virtually scheduled services linking their factories and offices in many cities across Europe.

For many years the Japanese manufacturer Mitsubishi built the MU-2 turboprop business aircraft. The turbofan-powered MU-300 Diamond I was introduced in 1982 as a direct competitor for the Cessna Citation. Main airframe assemblies were built in Japan and completed at Mitsubishi's factory in San Angelo, Texas.

Canadair CL-44

Inspired by – indeed based on – the Bristol Britannia, Canadair's CL-44 retained the former's stylish good-looks, coupled with less problematical engines. Intended from the outset as a freighter, the CL-44 excelled in its chosen role far out-living the last Britannias and continuing to work to this day.

The Canadair CL-44's birth may be traced back to 1952, when the Canadian air force was looking for a long-range maritime reconnaissance and anti-submarine patrol aircraft to replace its ageing Avro Lancasters, and for a long-range troop/freight transport aircraft. Airframe commonality and local production were key features in the requirement, which led in 1954 to a licence-agreement between Canadair and the Bristol Aeroplane Company in England for Canadian development and manufacture of modified versions of the Bristol Type 175 Britannia four-engined turboprop airliner.

The maritime reconnaissance aircraft was the first to be developed, under the designation CL-28 Argus. It employed the wings, tail surfaces and landing gear of the Britannia, revised to permit substitution of North American- rather than British-standard specification materials and to provide weapons hardpoints, mated to a completely new unpressurised fuselage incorporating a weapons bay. Four 2760-kW (3,700-hp) Wright R-3370-TC981 EA-1 turbo-compound piston engines were substituted for the Britannia's Bristol Siddeley Proteus turboprops. The prototype Argus made its first flight from Canadair's Cartierville plant on 28 March 1957 and was followed by 12 more Argus Is and 20 Argus 2s, all of which served only with the RCAF.

Work had been proceeding on development of the long-range transport aircraft, which received the designation CL-44. In the spring of 1957 Canadair was awarded a contract for eight CL-44s for the RCAF, and was additionally contracted for one prototype of a rear-loading all-cargo version of the aircraft.

The CL-44 also employed the wings, horizontal and vertical tail surfaces and undercarriage of the Britannia, with local strengthening to cater for a higher maximum take-off weight. The Britannia 253's fuselage was lengthened by 3.76 m (12 ft 4 in) and the cabin pressurisation system was retained. Canadair proposed three engine options for the CL-44: Bristol BE25 Orion, Pratt & Whitney T34 or Rolls-Royce Tyne. The RCAF selected the Orion, but when this engine development programme was cancelled by Bristol Siddeley in 1958, the 4103-kW (5,500-shp) R-R Tyne II was substituted for the first production variant, designated CL-44-6 and known in Canadian military service as the CC-106 Yukon. This version had two side-loading freight doors forward and aft of the wing on the port fuselage side and was capable of carrying 27525 kg (60,680 lb) of freight. The prototype (RCAF l5501) made its first flight from Cartierville, Montreal in the hands of test pilots W. S. Longhurst and 'Scotty' McClean on 15 November 1959. Twelve CL-44-6/CC-106s were delivered to the RCAF during 1960/61. Nine were configured for all-cargo operations and served with No. 437 Squadron, while three were outfitted as personnel/VIP transports and were operated by No. 412 Squadron.

The RCAF disposed of its Yukons from 1973. All found ready buyers among commercial cargo carriers, mostly in Latin America where operators included Aerocondor of Colombia, Aerolineas Nacionales de Ecuador (ANDES), Aeronaves del Peru, Aerotransportes Entre

Preceding the CL-44 civil airliner was the CL-28 Argus, which reverted to piston power for the long endurance maritime patrol role. In Royal Canadian Air Force service it was known as the CP-107.

*This beautiful portrait captures a **CC-106** Yukon (service designation) over a typical **C**anadian scene. The Yukons served for over 10 years before replacement by **B**oeing 707s, and subsequent sale into the civil market.*

*Flying Tiger Line was the biggest **CL-44** customer, taking a dozen. Its interests lay mainly in the Pacific and Far East regions, and it appreciated the capacity, speed and range offered by the aircraft on its cargo routes.*

Rios and Transporte Aéreo Rioplatense of Argentina, and TACA of Salvador. Elsewhere, other operators of former RCAF Yukons included Batchair, Beaver Enterprises and Canhellas in their native Société Générale d'Alimentation in the Congo.

While developing the Yukon, Canadair's design team took note of commercial operators' observations that the side-loading doors did not take fullest advantage of the aircraft's capacious cargo hold, and hit upon the idea of providing a hinged rear fuselage and tail section which could be swung aside to permit direct loading/unloading of bulky freight, including palletised loads which could be rolled straight into the fuselage

CF-MKP-X was **C**anadair's first civil aircraft, used as a company demonstrator. The ninth aircraft from the production line, it was fitted from the outset with the swinging tail, as evidenced by the large hinge fairings on the starboard rear fuselage.

from trucks. This so-called 'swing tail' was not new - it had been employed as long before as 1917 to facilitate shipboard stowage of the British Parnall N.2A Panther fleet reconnaissance spotter - but it had never been successfully adapted to a transport aircraft.

Thus, for the CL44D4 (also known for marketing purposes as the Canadair Forty-Four), which was externally similar to the RCAF's CL-44-6 apart from deletion of the rear side-loading cargo door and installation of uprated 4275-kW (5,730-shp) Rolls-Royce

*Right: With the company designation **CL-44-6**, this was the first prototype of the series, tailored to a military specification from the **RCAF**. Note the fore and aft cargo doors, which also contained smaller passenger entry doors.*

R.Ty.12 Tyne 515/10 engines, Canadair devised a hydraulically-actuated system whereby the entire tail unit, from the fin's dorsal fairing aft, was hinged on the starboard side and opened and closed by two actuators housed in the dorsal fin. The system took 90 seconds to open fully the tail section, and was designed to operate in crosswinds up to 48 km/h (30 mph) and to be capable of holding the tail open in 96-

*Below: The **CL-44** became a regular sight in Britain as several **CL-44D**s were used by cargo operators Transglobe, Trans Meridian and Tradewinds. These were bought when the Flying Tiger Line/Seaboard fleets were replaced by Douglas DC-8s.*

Slick Airways was the third of the US cargo carriers to opt for the CL-44-D4, buying four. This was their first aircraft, having first flown as CF-NYL-X. It was delivered to the airline on 17 January 1962.

Millon Air was among the last users of the CL-44. In 1992 the last survivors were operating with Tradewinds International Airline, which had six CL-44D4s flying from its Greensboro, North Carolina, base.

TF-LLH was one of Loftleidir's four CL-44Js, originally built as the first CL-44D4 with a swing tail. The Canadair 400 designation would have been applied to new-build machines to this standard, if any had been ordered.

A small number of small operators kept the CL-44 in service on cargo operations through the 1980s. Air Express International was based at Atlanta, Georgia, flying two ex-Trans Meridian CL-44-D4-8s (themselves ex-FTL).

km/h (60-mph) gusts. When closed, the tail section was secured by eight automatic locks and cabin pressure was maintained by an inflatable seal. Control runs to the rudder, elevators and tail trim tabs were automatically disconnected and reconnected during the tail-swinging, which was controlled from an on-board operator's station in the tail adjacent to the joint line. The swinging tail section was also part of the hold, and could carry up to 1360 kg (3,000 lb) of cargo.

The first CL-44D4 (CF-MKP-X), crewed by the same pilots who had first flown the CL-44-6 prototype, made its maiden flight on 16 November 1960, and received US Federal Aviation Administration approval seven months later. Canadair received initial orders for 17 CL-44D4s, later increased to 23, from three American carriers: Flying Tiger Line (12 CL-44D4-2s), Seaboard & Western (seven CL-44D-ls) and Slick Airways (four CL-44D-6s). The first delivery was made to Flying Tiger on 31 May 1961. At that time these airlines were the world's leading all-cargo carriers and quickly exploited the 100-mph speed advantage and much lower operating costs which the Canadair design offered over contemporary piston-engined cargo types such as the Douglas DC-6 and Lockheed Constellation. Flying Tiger Line estimated that its 12 CL-44s had the cargo capacity of more than 30 Constellations and operated at

Loftleidir's requirement for additional passenger capacity resulted in the CL-44J conversion of the CL-44D4, with a lengthened fuselage to accommodate up to 214 passengers. The first conversion was registered CF-SEE-X for trials before delivery to the Icelandic airline.

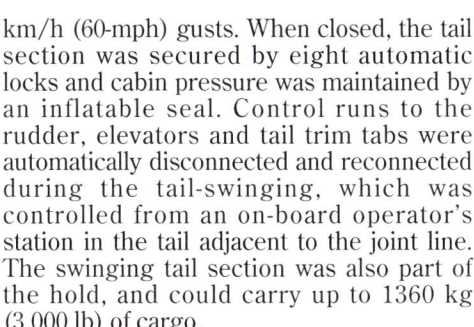

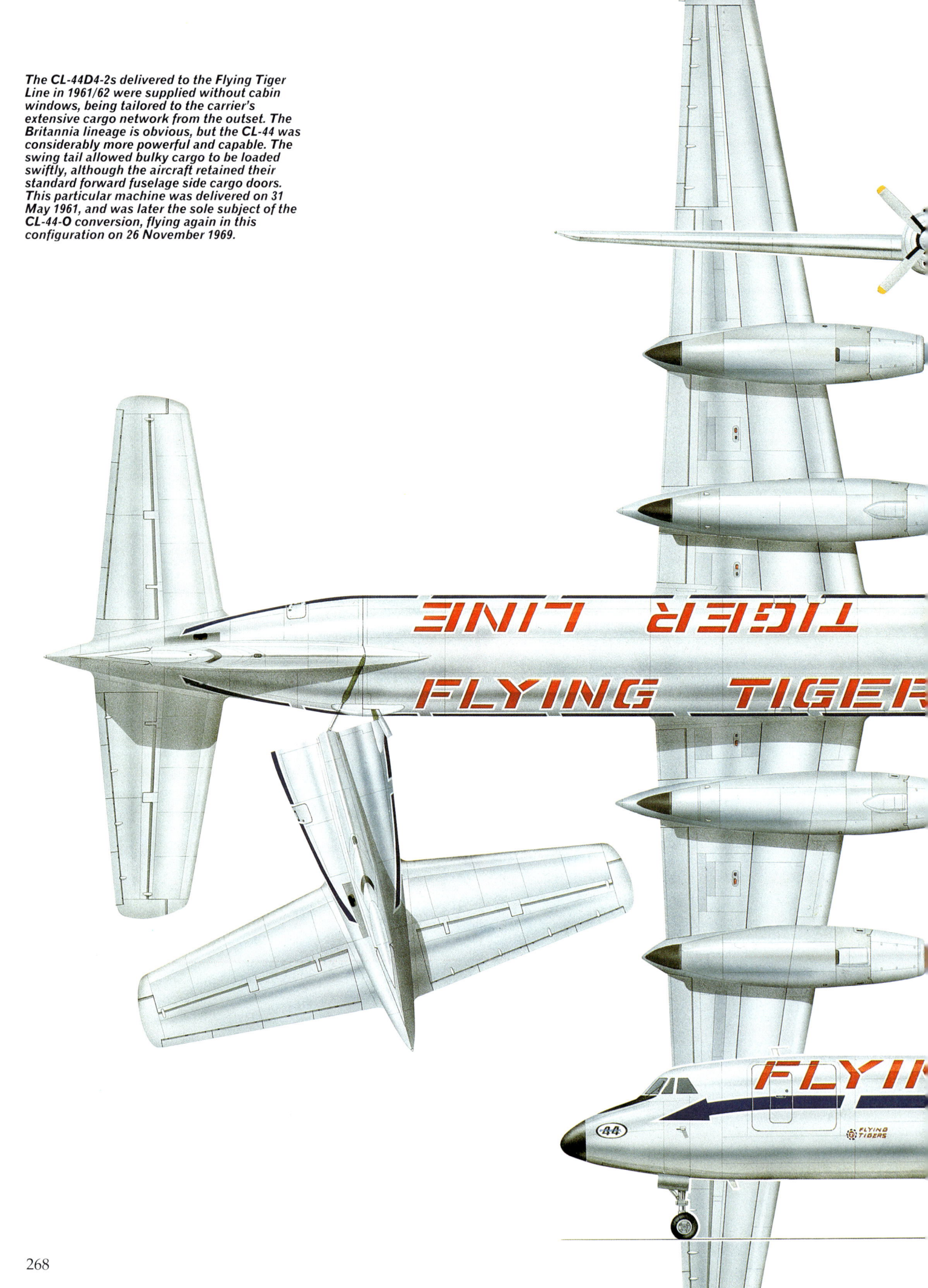

The CL-44D4-2s delivered to the Flying Tiger Line in 1961/62 were supplied without cabin windows, being tailored to the carrier's extensive cargo network from the outset. The Britannia lineage is obvious, but the CL-44 was considerably more powerful and capable. The swing tail allowed bulky cargo to be loaded swiftly, although the aircraft retained their standard forward fuselage side cargo doors. This particular machine was delivered on 31 May 1961, and was later the sole subject of the CL-44-O conversion, flying again in this configuration on 26 November 1969.

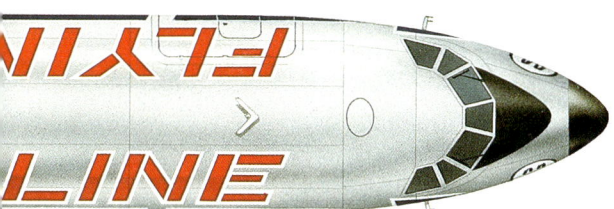

Specification
Canadair CL-44D4
Wingspan: 43.38 m (142 ft 3½ in)
Length: 41.73 (136 ft 10¾ in)
Height: 11.76 m (38 ft 7 in)
Wing area: 192.76 m² (2,075 sq ft)
Powerplant: four 4275-kW (5,730-shp) Rolls-Royce Tyne Mk. 515/10 turboprops
Empty weight: 40349 kg (88,952 lb)
Maximum take-off weight: 95256 kg (210,000 lb)
Passenger capacity: 178
Maximum speed: 621 km/h (386 mph)
Cruising speed: 464 km/h (288 mph)
Service ceiling: 9144 m (30,000 ft)
Maximum range: 8991 km (5,587 miles)
Range with maximum payload: 5246 km (3,260 miles)

By far the most important customers for the CL-44 were the two US cargo carriers Seaboard World and Flying Tiger Line, who were later to merge under the FTL title. Seaboard bought seven of the CL-44D4-1 variant for their scheduled freight network, which largely served Europe from the United States.

The Conroy CL-44-0 conversion retained the swing tail but featured an outsize cabin courtesy of the Guppy-specialist Jack Conroy. The aircraft last flew with UK-based HeavyLift, named Skymonster.

less than one-half the cost per ton/mile.

Canadair also developed a tail-loading military transport version of the aircraft, known as the CL-44G, for the Canadian Department of Defence, but none was ordered. Proposed variants of the CL-44D4 were presented to British Overseas Airways Corporation, Pan American World Airways and the United States Air Force, but also failed to achieve sales.

The only other customer for new-build aircraft was Loftleidir, the national airline of Iceland, which placed an order for four aircraft to be completed as 160 to 178-passenger CL-44D4-8 airliners for its low-fare transatlantic flights. These retained structural provisions for the swing tail, but without the hydraulic equipment.

The final development of the design was the CL-44J, created to meet Loftleidir's requirement for more passenger capacity. Canadair took the fourth and last of the CL-44D4-8s on order for Loftleidir (and the last of 39 CL-44 airframes built) and converted it to meet the carrier's specification by lengthening its fuselage by 4.6 m (15 ft 2 in). This was achieved by means of two 'plugs', one of 3.07 m (10 ft 1 in) forward of the wing and another 1.55 m (5 ft 1 in) long aft of it, enabling cabin seating to be increased to a maximum of 214. Although it had the same maximum take-off weight as the shorter fuselage variant, the CL-44J's increased maximum landing weight of 79380 kg (175,000 lb) required strengthening of the wings and landing gear. Originally flown in standard CL-44D4-8 form on 17 March 1965, the modified aircraft (CF-SEE-X) flew again as the prototype CL-44J on 8 November of that year. Loftleidir's three CL-44D4-8s were later also converted to CL-44Js, but plans to put the stretched version of the aircraft into series production as the Canadair 400 did not materialise.

One other variant of the CL-44 has appeared, a hybrid modification of a former Flying Tiger Line CL-44D4. Conversion to high-volume cargo carrier was carried out by Conroy Aircraft Company of Santa Barbara, California, whose founder Jack Conroy had pioneered the Guppy, Pregnant Guppy, Mini Guppy and Super Guppy conversions of the Boeing 377 Stratocruiser and KC-97 Stratofreighter for carrying outsize cargo. The basic CL-44D airframe was retained, but immediately aft of the flight deck the entire upper fuselage above floor level was removed and replaced with an enlarged circular section 'bubble top'. Though much less voluminous than the Guppies, conversion provided a cargo hold with a constant internal height of 3.45 m (11 ft 4 in) and a diameter of 4.24 m (13 ft 11 in), and retained the aircraft's swing-tail loading facility. Redesignated CL-44-O, the Conroy conversion was first flown from Santa Barbara on 26 1969 by the famous Lockheed company test pilot Herman 'Fish' Salmon, and was put into service transporting Rolls-Royce RB.211 engines and pods from England to California in support of Lockheed L-1011 TriStar production. This unique aircraft still survives and after several changes of owner is now operated by the London/Stansted Airport-based outsize air freight specialists Heavylift Air Cargo.

Following disposal by Flying Tiger, Seaboard and Slick, CL-44s served with many new owners around the world, notably in the United Kingdom with Tradewinds Airways, Transglobe Airways and Transmeridian Air Cargo, which between them operated 15 of the 27 aircraft built. Eight aircraft were still in service in 1998.

Originally delivered to Seaboard World on 15 June 1962, N228SW was later leased to UK national flag-carrier BOAC for its cargo services. Having returned to Seaboard, it was written off in December 1966 at Da Nang while involved in government contract work.

Canadair CL-215

Fighting fires from the air is mainly the domain of converted bombers and transports, but a number of operators have chosen the only purpose-built aircraft available for the job: the CL-215. Operating from any available stretch of water, the Canadian flying-boat can scoop vast amounts of water into its belly while skimming the surface.

Everyone knows that big flying-boats are a thing of the past. So too are big aircraft powered by piston engines. How strange then, that a factory in the suburbs of Montreal, Canadians were building big piston-engined flying-boats up to 1989, when the fifth production batch (increased from 29 to 45 aircraft) was completed for an overall total of 125 before the production line switched to an updated version with turboprop power.

The reason for all this is that the Canadair CL-215 is unique. It is the only aircraft designed from the start as a water bomber for putting out forest fires, and other types of fire. It is also an extremely versatile aircraft, able to fulfil a variety of other tasks. As it is an amphibian it can operate from short airstrips as well as stretches of water, and it was designed to be simple to maintain.

Canadair, at that time a subsidiary of the US giant General Dynamics, began considering such an aircraft in the early 1960s. By 1965 the firm decision had been taken to go ahead, encouraged partly by the promise of an order for no fewer than 20 from the Province of Quebec (in which the Canadair factory is situated) and partly by the belief that such an aircraft was needed. The fact that nobody else wanted to build such an aircraft could have been construed as a deterrent, but in fact it was judged to be a good sign, and Canadair has been proved right.

Basically the CL-215 was designed as a simple and tough aircraft that could operate day or night in all climates, from airfields or reasonably calm water. It was envisaged above all as a water bomber to fight forest fires, but with the added capability of carrying passengers or cargo, or of flying coastal patrol missions, or of rescuing people at sea (a job otherwise done only by helicopters with much shorter range). Canadair asked the long-retired aircraft designer Ed Heinemann (creator of dozens of famed Douglas military, naval and commercial aircraft) to serve as consultant on the basic design, and this paid off.

Of course the airframe was made all-metal, and well able to withstand harsh use. As there was no requirement for high speed the wing was more like that of 30 years earlier, with an extremely deep section and rectangular plan shape, carrying big hydraulically driven slotted flaps but manual ailerons. The hull, or fuselage, did not have to be especially large, because in the basic firefighting role the load of water occupies quite a modest volume. In the nose is the flight deck with two pilots side-by-side. In patrol and similar missions there is provision for a navigator, flight engineer and two observers. On the left side there are large square doors at front and rear, ladders being needed when the aircraft is parked on land. At the rear is a very large tail, again with fully manual control surfaces but made so big to give good stability and control power at the low airspeeds at which the CL-215 often operates.

Radial reliability

Piston engines were chosen mainly because the CL-215 was designed to operate almost always at very low level, where turboprops would be relatively inefficient. The Pratt & Whitney R-2800 Double Wasp was the obvious choice. Together with the three-blade Hamilton Standard propellers, this engine has been in worldwide civil and military service for over 40 years, though it is fading from the scene rapidly and its 115-grade petrol is becoming a scarce commodity. Each engine has 18 cylinders, with a capacity of 45.9 litres. Take-off power is 2,100 hp. Canadair planned to offer a CL-215-109 version powered by R-2800-CB17 engines boosted to 2,500 hp with water injection, but this was eventually dropped. In the original design two fuel tanks, each made up of six flexible cells, accommodated a total of 4,105 litres, but the production aircraft have the increased capacity of 5,910 litres. This is enough for a ferrying range of at least 2500 km, or an endurance of over 10 hours.

In the basic firefighting mission by far the biggest advantage of the CL-215, shared by hardly any rivals, is that by swooping down and making a run at very low level over any stretch of calm water it is possible to refill the water tanks in flight. Of course for the greatest effect plain water is not used. Instead the water is mixed with special

Not all CL-215s are used for firefighting, the pair operated by the Royal Thai Navy being employed on coastal patrol and rescue duties. Some of the Spanish aircraft fulfil a similar function, while those of Venezuela are used for transport duties.

Much of the former Yugoslavia is forested and hot in summer, making it prone to fires. To fight these, the former Yugoslav air force acquired the CL-215. The aircraft is currently the only dedicated firefighting design available, although conversions of surplus airliners/military types are popular.

chemicals to form a foam which can blanket a large area of fire and starve it of oxygen. This is especially valuable in the case of oil fires, which are often very difficult to extinguish.

There are two main water tanks, arranged left and right in the centre of the fuselage, each with a capacity of 2,673 litres. They can be pumped full on the ground through a hose, and then dumped over the fire through a large drop door, resembling a bomb door, in the planing bottom under each tank. It is then possible to refill the tanks by extending left and right retractable probes under the aircraft and lowering these into the water of a lake, river or calm sea as the aircraft flies very low over the surface at 130 km/h. With practice a pilot can judge exactly what throttle opening to use to keep airspeed correct despite the sudden quite severe drag of the probes entering the water. Using this method it takes only 10 seconds to refill the tanks, compared with two minutes using maximum-pressure hoses on the ground.

Tank versatility

Thus there are several ways of using the water tanks. They can be filled on the ground with fresh or salt water, or with a formulated chemical retardant. They can be filled with water and then have a concentrated foaming agent mixed on board. In repeated passes over a fire the CL-215 can swoop over either fresh or salt water, and if necessary mix chemicals with each fresh load to give a foam-type extinguishing mixture (though of course eventually the foaming agent tanks, which are much smaller than the main ones, will have to be refilled back at base).

One of the few parts of the CL-215 that was modified between original design and the emergence of the production aircraft was the landing gear. Of course, for water operation the aircraft alights on the planing bottom of the hull, or fuselage, which has the usual step, or discontinuity, about half-way along to help the aircraft unstick from the water on take-off. To keep the aircraft upright on the water there is a fixed stabilising float under each wingtip. One normally rests on the water, but on take-off the pilot uses aileron to bring the wings level, keeping both floats out of the water. On land the CL-215

Canadair CL-215 (Spanish SAR) cutaway drawing key:

1 Nose radome
2 Scanner
3 Snubbing post
4 AVQ-21 radar transceiver
5 Walkway
6 Mooring cleat
7 Water towing fitting
8 Air-oil shock strut
9 Land towing lug
10 Twin nosewheels
11 Nosewheel doors
12 Nosewheel door hinges
13 Nosewheel retraction strut
14 Engine controls
15 Rudder pedal assembly
16 Mooring compartment access (tunnel under control pedestal)
17 Mooring compartment hatch
18 Mooring hatch window
19 Windshield wiper
20 Glide slope antenna
21 Overhead instrument panel
22 Co-pilot's seat
23 Instrument panel shroud
24 Control columns
25 Pilot's seat
26 Pilot's repeater radar scope
27 Control cables
28 Underfloor electrical bay
29 Engine controls
30 Elevator controls
31 Bulkhead
32 Flight-engineer's seat (folding)
33 Cockpit roof window
34 HF transceiver pack
35 Emergency exit hatch
36 Loop and sense antennae (ADF)
37 Cabin heater installation
38 Navigator rack
39 Navigator's seat
40 Crew entry door (emergency exit)
41 Webbing safety curtain
42 Chine spray suppressor
43 Underfloor water tank space
44 Three-tier litter (or four-man bench seat)
45 Fuselage frames
46 Navigator's chart table (with folding extension)
47 Rudder controls
48 UHF/IFF antenna

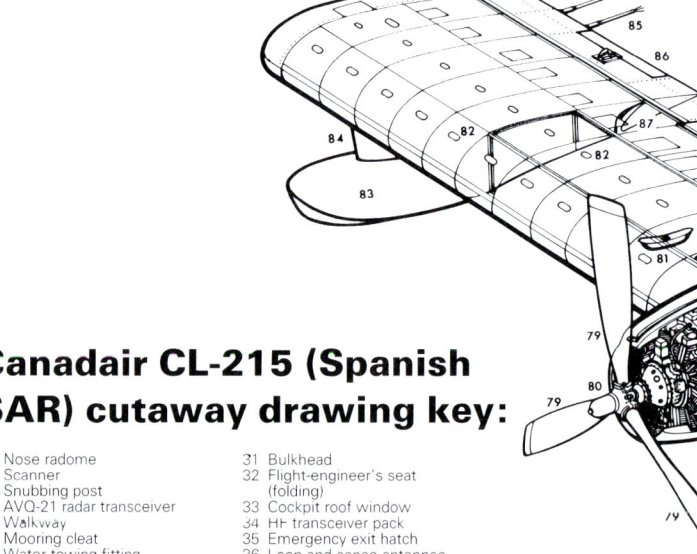

Another major CL-215 operator is Greece which, like Spain, entrusts the fire-fighting operation to the air force. No. 355 Mira flies 11 CL-215s for this task. Although sales are limited in contrast to other types, considering the specialised role of the aircraft they have been excellent.

49 Maintenance walkway
50 Life-raft/droppable stores (optional location)
51 Roof fairing members
52 Engine controls
53 Firex discharge indicators
54 Cabin window
55 Three-point electrical outlet (blankets)
56 Glass-fibre cabin water tank sections (removable)
57 Ground water fill point (port and starboard)
58 Water tank door (open)
59 Mainwheel door
60 Landing gear 'A'-frame bearing housing
61 Landing gear retraction/extension actuator
62 Port landing gear well
63 Battery and jar
64 Water tank overflow
65 Fuel system booster pump
66 Wing/fuselage attachment frames
67 Wing centre-section
68 Hydraulic reservoir (3.5 Imp gal/15.9 l)
69 UHF/VHF homer receiver and antenna
70 Cabin heater fuel regulator
71 Fuel system filter
72 Fuel system water drains
73 Ten bladder-type fuel cells in wing, total capacity 1,266 Imp gal (5,755 l)
74 Fuel lines
75 Engine firewall
76 Tacho generator
77 Carburettor
78 Pratt & Whitney R-2800-83-AM engine

extends tricycle landing gear, with hydraulic retraction and hydraulic steering for the twin nosewheels. The nose gear retracts backwards into a compartment under the cockpit, which is closed by twin doors strong enough to bear the heavy water loads during sea take-offs. The single main wheels extend out on linkages on each side of the fuselage. They were to have had a tyre pressure of 6.68 kg/cm^2 and be housed in recesses in the side of the fuselage. In the produc tion aircraft it was found more useful to reduce the tyre pressure to only 5.4 kg/cm^2, to enable the aircraft to operate from unpaved strips. As the landing gear is retracted, the main legs fold into the fuselage, where they are covered by fairing doors, but the wheels are left out in the slipstream. This does not matter much at CL-215 speeds, though if the aircraft is hurrying to a fire at its maximum cruising speed of 291 km/h there is an appreciable extra drag.

79 Hamilton Standard three-blade propeller
80 Propeller hub
81 Marker beacon antenna
82 Inspection panels
83 Starboard wingtip float
84 Float pylon
85 Static dischargers
86 Starboard aileron tab
87 Aileron profile
88 Starboard aileron
89 Starboard flap (outboard)
90 Oil tank (30 Imp gal/136.5 l)
91 APU
92 Nacelle construction
93 Starboard flap (inboard)
94 Centre-section stringers
95 Galley unit location (starboard side)
96 Aileron controls
97 Hydraulic system flap control and jettison valves
98 Inspection panels
99 Extended electrical distribution centre (MDC): 2 transverters
100 Landing gear downlock actuator
101 Landing gear folding strut
102 Landing gear 'A'-frames
103 Air-oil shock strut
104 Port mainwheel
105 Oil cooler
106 Oil cooler outlet flap
107 Engine drains
108 Fuel cells (see item 73)
109 Wing/nacelle joint
110 Observer's seat (port and starboard)
111 Circular observation port
112 Droppable stores
113 Extended overhead handrail/static line
114 25-man life-raft
115 Three-tier litter unit (securing provision only)
116 VHF(FM) antenna
117 DME antenna
118 Emergency exit (starboard)
119 VHF (comm) antennae
120 HF wire antenna
121 Fuselage dorsal skinning
122 Mounting provision for rescue ladder (aft starboard wall)
123 Fin spar/fuselage attachment
124 Downed aircraft locator antennae (port and starboard)
125 Fin front spar
126 Tailplane attachment points
127 Elevator controls
128 Leading-edge de-icing boots
129 Starboard tailplane
130 Elevator balance
131 Static dischargers
132 Starboard elevator
133 Elevator tab
134 Elevator tab control linkage
135 Rudder geared trim tab actuator
136 VOR/ILS antenna
137 Anti-collision beacon
138 Mooring light
139 Rudder balance
140 Static dischargers
141 Rudder upper hinge
142 Rudder geared trim tab
143 Rudder spring tab
144 Elevator torque tube
145 Elevator spring tab
146 Spring tab control linkage
147 Elevator trim tab
148 Trim tab control linkage
149 Elevator outboard hinge
150 Elevator balance
151 Stafoam tip stiffeners
152 Tailplane construction
153 Radio altimeter antennae
154 Aft position light
155 Radio altimeter receiver
156 Elevator downspring
157 Rudder spring tab actuator
158 Rudder controls
159 Water towing fitting
160 Aft skinning
161 Angled fuselage frames/fin members
162 Downed aircraft locator
163 Aft bulkhead
164 Bilge pump stowage
165 Potable water containers (2)
166 Toilet
167 Wash basin
168 Aft entry/rescue door (removable sill)
169 Flap actuating mechanism
170 Port flap structure
171 Flap hinge fairings
172 Fuel vent system discharge and ram air scoop
173 Leading-edge construction
174 Wing box assembly
175 Bulkhead-type rib
176 Landing light (port and starboard)
177 Pitot heating tube
178 Pitot head
179 Open strut-braced rib
180 Aileron control
181 Aileron geared tab actuator
182 Aileron hinge
183 Aileron trim tab actuator
184 Aileron trim tab
185 Aileron geared tab
186 Static dischargers
187 Port aileron
188 Aileron profile
189 Port navigation light
190 Flux valve access panel
191 Flux valve
192 Stafoam tip stiffeners
193 Docking/beaching tie-down fitting
194 Float pylon structure
195 Port wingtip float

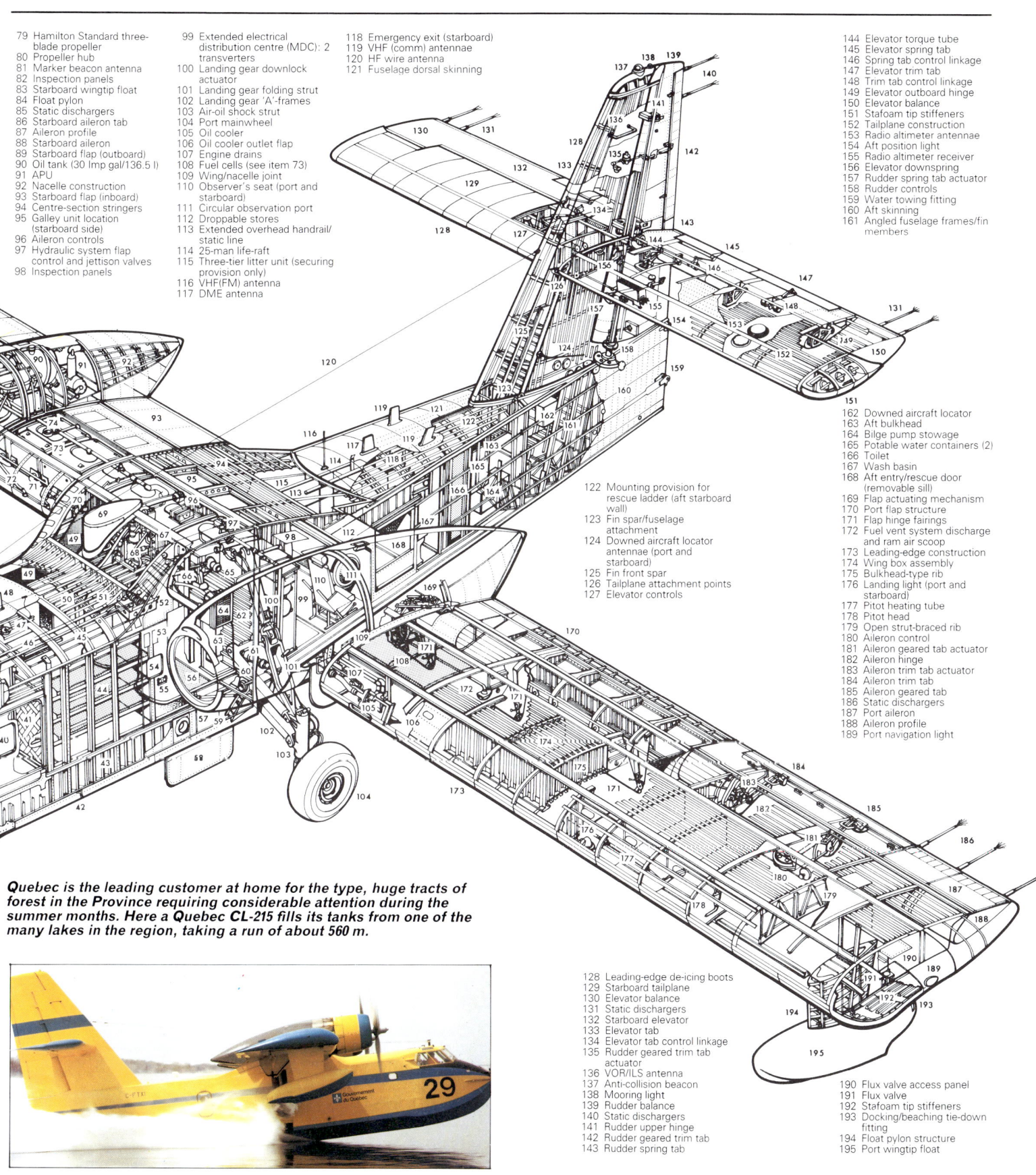

Quebec is the leading customer at home for the type, huge tracts of forest in the Province requiring considerable attention during the summer months. Here a Quebec CL-215 fills its tanks from one of the many lakes in the region, taking a run of about 560 m.

Specification
Canadair CL-215
Type: firefighting and utility transport amphibian
Powerplant: two Pratt & Whitney R-2800-CA3 18-cylinder radial engines each developing 1566 kW (2,100 hp)
Performance: cruising speed 291 km/h (181 mph); maximum rate of climb at sea level 305 m (1,000 ft) per minute; take-off run to 15 m (50 ft) 811 m (2,660 ft) from land, 800 m (2,625 ft) from water; stalling speed 123 km/h (76 mph); landing run from 15 m (50 ft) 732 m (2,400 ft) on land, 835 , (2,740 ft) on water; range 2095 km (1,300 miles)
Weights: empty 12160 kg (26,810 lb); maximum take-off (land) 19731 kg (43,500 lb), (water) 17100 kg (37,700 lb); maximum payload (water bomber) 5443 kg (12,000 lb)
Dimensions: span 28.60 m (93 ft 10 in); length 19.82 m (65 ft 0½ in); height (on land) 8.92 m (29 ft 3 in); wing area 100.33 m² (1,080 sq ft)

The Canadair **CL-215** is designed around the swift extinguishing of forest fires at a practical cost. At the heart of the machine are the water tanks, which can accommodate 5346 litres of water. These are filled as the aircraft skims across a lake or river, two scoops projecting into the water from the hull and using ram pressure to fill the tanks in about 10 seconds. Once the filling is complete, the scoops are retracted and the aircraft climbs away. Over the drop site, large doors are opened to dump the full load in a second. The new turboprop-powered **CL-215T** features two sets of doors to allow the water to be dumped in salvo (all at one time) or in train (one set preceding the other), this giving a variable spread pattern.

The first CL-215 made its maiden flight from Cartierville airport on 23 October 1967, and made its first take-off from water on 2 May 1968. Flight testing unearthed few problems, and with five aircraft flying the initial customers in Quebec and in France (Sécurité Civile) began operations at the start of the forest-fire 'season' in spring 1969. At the start it was found that to make a water pick-up completely filling the tanks a clear run of 1.7 km was needed, measured from a height of 15 m on the inward run to the same height on the climb-out. Gradually it was found possible to take water on board even faster, cutting the run time from 12 to 10 seconds, and the clear distance needed today is only 1.2 km. In most areas there are numerous stretches of water available offering a 1.2 km run, and there is no problem in making repeated pick-ups.

In 1969 Canadair proudly announced that it ought to be possible to make 75 water drops in a single day. In fact the impressive figure of 100 has often been exceeded. The record is believed to be held by a Yugoslav CL-215 which in a single gruelling day made 225 drops on fires, putting down over 1.2 million litres! Another remarkable fact is that the tanks have been replenished repeatedly from the Mediterranean in waves 2 m high.

Customers were found in Canada, France, Greece, Italy, Spain, Yugoslavia, Thailand and Venezuela. A proportion of the Spanish and Thai aircraft are equipped for search and rescue, and the Venezuelan machines are utility passenger transports with 26 seats. In the cargo role a payload of 2,268 kg is allowed for, though total payload is 3,864 kg (compared with 5,443 kg in the water-bomber role). Yugoslavia purchased four sets of a liquid spray system which enables the CL-215 to apply oil dispersants and pesticides, without interfering with the basic fire-extinguishing role.

Turboprop power

Even at the launch of the programme Canadair was wondering whether the choice of piston engines was right, and the increasing difficulty of finding high-octane petrol and R-2800 spares has steadily increased the pressure for a turboprop version. The trouble is, new turboprops cost many times more than reconditioned R-2800s. Even the original aircraft is too expensive for many impoverished would-be customers. For example in April 1988 Canadair demonstrated the CL-215 to several states in the USA with forest fire problems. After getting over the shock of discovering that the aircraft was almost universally unknown in the USA, the manufacturer found that everywhere the CL-215 went it made a deep impression and was urgently wanted, but nobody had $6 million. Several forestry departments suggested that perhaps several states could club together to share the cost. This is remarkable, bearing in mind that the

A large operator of the CL-215 is the French Sécurité Civile, which contends with many forest fires during the summer months. Most occur in the south of the country. Here the full 5346 litres is released on a fire in the area.

Canadian state of Quebec alone has 19, as does Spain, and France and Greece have 15 each. (This demo tour also displayed the CL-215's toughness, when the aircraft carried on after slicing through a large poplar tree.)

It is surely amazing that the CL-215 failed to achieve a single sale in the USA, and this market thus played no part in the planning for the CL-215T (from 1991 the CL-415) turboprop. After years of study this went ahead in August 1986. Canadair determined to make the 415 a fully versatile aircraft, while keeping the basic design unchanged. The company guesses that water bombing would account for only 44 per cent of sales of the turboprop, with civil transport accounting for 23 per cent. The other 33 per cent is expected to be made up of sea patrol, search and rescue and maritime combat missions such as anti-ship or anti-submarine.

The chosen engine is the Pratt & Whitney Canada PW123AF, rated at 2,380 hp and driving a four-blade Hamilton Standard propeller with reversible pitch. Propeller diameter is 3.97 m, compared with 4.34 m, which assists the aim of reducing noise and vibration inside the new version. Fuel capacity is unaltered, though of course aviation turbine fuels such as JP-1, JP-4 and JP-5 are available virtually everywhere. Among the very few other changes is replacement of the 36 ampere-hour lead/acid battery by two much lighter nickel/cadmium batteries with a combined capacity of 80 ampere-hours.

The biggest advantage of the 415 is that the engines are lighter, enabling payload to be increased to 6,123 kg of water or 4,790 kg of cargo. First flight was in April 1989; deliveries started a year later. There is a kit to convert the 215 to partial 415 standard as the 215T.

The prototype CL-215T demonstrates its water-dropping capability. At the same time as the PW123 turboprops are added, the aircraft has upturned wingtips to improve manoeuvrability at low speeds and increase aerodynamic efficiency. New-build aircraft are now being delivered, while conversions of existing radial-powered aircraft are available.

Canadair Challenger

The design for the Challenger business jet fell into Canadair's hands almost by accident. Now flown by some of the world's biggest companies, the CL-600 family has grown to become an accomplished long-range aircraft.

In 1974 William P. Lear, creator of the Lear Jet series of business jets, began design work on a new aircraft project, having previously been involved with an expensive and ultimately unsuccessful attempt to develop a steam-powered automobile.

The aircraft was known as the LearStar 600, thus taking up the name previously applied in the 1950s to Lear's executive conversion of Lockheed Lodestar twin-engine transport – the '600' referred to a projected cruising speed of 600 mph (966 km/h). It was inspired by research conducted by Dr Robert Whitcomb of the National Aeronautics and Space Administration into a radical new 'supercritical' wing aerofoil design that, by delaying the formation of shock waves at high Mach numbers, could reduce drag and enhance range and specific fuel consumption while permitting lower landing speeds. Initially Lear planned to design a supercritical wing for the Lear Jet, but con-

The second of three Canadair Challengers is seen during a test flight. Notable features are the 'X' suffix to the registration to indicate the aircraft's test status, and the air data boom projecting from the nose for accurate measurements during flight trials.

cluded that the efficient wing, combined with a new-generation turbofan engine, would be best exploited by an entirely new aircraft that would fulfill business jet market needs for two decades or more.

Working with aerodynamicist Larry Heuberger, Lear created a preliminary design for a twin-turbofan jet powered by the new, high bypass ratio Avco Lycoming ALF 502D engine, and offering three possible configurations: 14-passenger business jet with a range in excess of 6437 km (4,000 miles) cruising at 966 km/h (600 mph) at 13716 m (45,000 ft); 30-passenger commuter airliner with 805-km (500-mile) unrefuelled range; and an all-cargo transport.

Having sold Lear Jet to the Gates Rubber Company, Lear had only his small Reno, Nevada-based Lear Avia facility, which had neither space nor capital to launch full-scale production of the aircraft, so he offered the design to major companies such as Beech, Cessna and Gates Learjet, and sought financing from banks and investors to expand his own company to manufacture and market the LearStar, but was unsuccessful in interesting any of them.

However, in April 1976, the Canadian government-backed Canadair Limited of Montreal, which was looking for a project to be a centrepiece for the revitalised company, took an option on the LearStar 600 which, if exercised, would pay Lear more than $7 million for design and marketing rights to the aircraft once a prototype had been built and certificated, with a royalty on every aircraft delivered thereafter. Lear was hired to act as consultant on continued development of the design, while Canadair technical and marketing staff began their own studies. Canadair was due to give a final decision within a year, subject to the confirmation of Lear's performance estimates, a satisfactory order book, and Canadian government approval and financial backing.

It quickly became apparent that relationships between Lear and Canadair were not going to be easy. Canadair's marketing studies suggested that a 51-cm (20-in) increase in fuselage diameter was needed to provide a full 1.8-m (6-ft) 'stand up' cabin. "Corporate executives don't want to sit in the prenatal position in an executive mailing tube," Canadair's marketing consultant James Taylor remarked. More fuel was needed to ensure the

Above: Doug Adkins gets air under the wheels of the Challenger for the first time on 25 May 1978, at Canadair's Cartierville factory. The futuristic shape and promised performance of the aircraft had already accounted for a healthy order book at the time of its first flight.

Below: The principal structural member of the Challenger is the one-piece primary wing structure, built up from two spars and skins with integrally-milled stringers which form a strong but light torsion box.

Above: As the CL-600, the Challenger entered the marketplace hoping to capture much of the long-range corporate market. Eighty-three were built but the performance figures were disappointing, so hastening the development of the CL-601.

projected maximum range, and since this was entirely stored within the wing, an increase in span and root thickness followed. These revisions in turn increased gross weight from the projected 10886 kg (24,000 lb) to 14742 kg (32,500 lb), demanding the more powerful Lycoming ALF 502L engine to maintain take-off and climb performance. Wind tunnel tests also showed Lear's choice of a low-mounted tail to have disadvantages, so Canadair opted for a T-tail instead.

All of these proposed changes enraged Lear, who felt that his exceptionally clean aerodynamic lines had been destroyed. Conflict quickly mounted. Lear took to calling Canadair's proposal 'Fat Albert'. Finally a compromise was agreed: Canadair and Lear Avia would each undertake a redesign around the 274-cm (108-in) diameter cabin which the Canadians were adamant was essential, and Canadair's president Fred Kearns would make the final decision on which aircraft was to go on to prototype stage. Lear lost. His new Allegro was rejected by Canadair, which forged ahead with its own design. Go-ahead was authorised on 29 October 1976, and in March 1977 Canadair announced that the new aircraft was to be designated CL-600 (for 'Canadair Lear') and called Challenger. At that time the company held orders for more than 50 aircraft, which were then priced at Can$4.275 million.

Production jigs were used for prototype construction to speed development, and on 25 May 1978 the first of three pre-production Challengers (C-GCCR-X) was rolled out of Canadair's Cartierville, Montreal plant. Systems installation work took a further five months. The aircraft made its first flight on 8 November 1978 – only two years after the go-ahead – with Canadair's chief test pilot Doug Adkins at the controls. Initial tests were performed with Avco Lycoming ALF 502H

The current major production model of the Challenger is the CL-601-3A, powered by the General Electric CF34-3A turbofans. Compared to the -1A engines originally fitted to the CL-601, these engines offer better 'hot-and-high' performance.

Just as the Canadian Forces found the Challenger ideal for the electronic warfare mission, so the Canadian civil aviation authorities have found it perfect for the flight testing of landing and navigation aids. On board is ample capacity for the sensitive calibration equipment.

engines pending availability of production standard 33.4-kN (7,500-lb) static thrust ALF 502Ls. On 27 December, with 27 flights totalling 50 hours logged, the Challenger was flown to the Flight Systems Test Center at Mojave, California, so that testing could continue uninterrupted through the harsh Canadian winter.

By early March 1979 a number of changes had been incorporated in the aircraft as a result of initial testing and ground vibration studies, including reduced elevator span and a modified fuel system. ALF 502L engines have also been retrofitted. On 17 March the second pre-production Challenger flew at Montreal and joined the first at Mojave, followed by the third on 14 July. The first production aircraft, actually the sixth airframe, also joined the test and development programme after making its first flight on 21 September 1979. In addition to the flying prototypes, two non-flying airframes were built for static testing and fatigue testing.

The flight test programme proceeded swiftly, but was marred by the loss of the first pre-production aircraft in the Mojave desert on 3 April 1980 after a stall-spin recovery parachute failed to jettison after it had been deployed to recover from deep stall tests; two pilots and a flight engineer successfully bailed out of the aircraft, but the captain was killed.

Canadian Department of Transportation certification was obtained on 11 August 1980 after some 1,300 hours of flight testing. US Federal Aviation Administration approval was granted on 7 November. Pending completion of testing, both authorities initially limited the Challenger to a gross take-off weight of 14969 kg (33,000 lb) and a maximum speed of 573 km/h (356 mph), with flight into known icing conditions and use of the ALF 502Ls' thrust reversers prohibited.

As anticipated, the Challenger found its major market in the United States, although orders were also forthcoming from the Middle East and Europe, where the Saudi Arabian-funded TAG Aeronautics Corporation handled sales. Twelve Challenger 600s were delivered to the Canadian Department of National Defence, six of these being equipped for electronic support/countermeasures training under the designation CE-144A with 414 Squadron (now 434) at CFB North Bay, and six as government/VIP transport CC-144s with No. 412 Squadron at CFB Uplands.

During 1980 Canadair announced two new versions of the Challenger. One was to be a 'stretched' CL-610 Challenger E variant, having fuselage plugs fore and aft of the wing totalling 2.7 m (8 ft 9 in) which would increase maximum passenger capacity from 19 to 24. Extended wingspan, new high-lift leading-edge

A change of powerplant was the most significant new feature introduced by the CL-601, but the most obvious was the addition of drag-reducing winglets. These were subsequently retrofitted to many CL-600s, leaving the different shape of the engines as the principal recognition feature between the two variants.

devices, 40.7-kN (9,140-lb st) General Electric CF34-1A turbofans in place of the Avco Lycoming powerplants, and an increase in fuel capacity to 13044 litres (3,446 US gal) would give this model a maximum with-reserves range of 7686 km (4,150 nm; 4,776 miles) at a cruising speed of 476 kt (879 km/h; 546 mph) while carrying a 907-kg (2,000-lb) payload. The second version, designated CL-601, was to retain the basic Challenger airframe but with the CF34 engines.

In the event, the stretched Challenger E failed to materialise, but poor initial performance by the ALF 502-powered Challenger, particularly with regard to meeting projected payload/range, speeded up development of the General Electric-powered CL-601 and led to the design of wingtip 'winglets' to enhance range performance through drag reduction. The ninth production Challenger 600 was experimentally fitted with a pair of 1.2-m (4-ft) high winglets and made its first flight in this configuration on 13 November 1981, by which time wind tunnel and computer simulation tests had led Canadair to take the decision to make winglets standard on the Challenger 601. Subsequently, winglets were offered as a retrofit on the Challenger 600 (known as the Challenger 600S), and to date 67 of the total 83 Model 600s manufactured have been so modified.

Canadair used a converted CL-600 airframe as the basis for the Challenger 601 prototype (C-GCGT-X) which first flew from Montreal on 10 April 1982, and was followed by the first production model on 17 September. The 601 was adopted as the standard production model; Model 600 production is officially only suspended pending receipt of sufficient orders, but no Lycoming-powered Challenger has been built since the mid-1980s. Canadian and US certification were obtained on 25 February and 11 March 1983 respectively for the initial

Above: Another military customer was the German Luftwaffe, which bought seven CL-601s for VIP work with the Flugbereitschaftstaffel (FBS) at Köln-Bonn. These are regular visitors at European airports, ferrying German government officials.

Challenger 601-1A version. This had 40.7-kN (9,140 lb st) CF-34-1A turbofans, 9278 litres (2,451 US gal) fuel capacity in four tanks located in the centre-section, wings and under the cabin floor, and a 62 kPa (9.0 psi) cabin pressure differential against the 64 kPa (9.3 psi) differential in the Model 600, reducing its maximum certificated operating altitude from 13716 m (45,000 ft) to 12497 m (41,000 ft).

Deliveries of this version began immediately after certification. On 23 August 1983 a Challenger 601 set a new non-stop straight-line world distance record for its class, flying from Calgary, Alberta to London's Heathrow Airport – a distance of 3,790 nm (7019 km; 4359 miles) – in nine hours, four minutes. Among Challenger 601 deliveries, again mostly to corporate customers in the United States, Middle East and Europe, were four for the Canadian government as VIP transports (still referred to as CC-144s) and flight inspection aircraft, seven to the special air missions wing (FBS) of the German Luftwaffe for VIP duties, passenger/cargo and air ambulance duties, three for

Above: Seen above Utah's Lake Powell, this CL-601 displays its new CF34 engines which cured the disappointing range/payload performance of the Challenger's early career. This has seen the popularity of the type increase, most notably in the United States.

Right: The prototype Challenger (construction number 1001) poses for the camera at its Cartierville home. It was lost in April 1980 during the flight test programme at Mojave, California, when its spin-recovery parachute failed to open.

Above: As with most executive jet designs, the Challenger has been adopted in small numbers by some military users. Leading the way was the home country, which operates the type in CC-144 (transport) and CE-144 (ECM training) variants. Three aircraft are to be modified to CP-144 standard for maritime patrol duties.

The People's Republic of China adopted the Challenger as a government transport in 1986, when three CL-601-1As (illustrated) were delivered. Two years later the fleet was swelled by the addition of two further aircraft, this time the uprated CL-601-3A version.

the People's Republic of China and two for the Royal Malaysian air force (now withdrawn). Because of its wide-body configuration (which was much against original designer Bill Lear's conception) the Challenger has proved popular as an air ambulance/medevac aircraft, Model 600s and 601s having been modified to enable them to provide inflight intensive care for up to eight patients while flying non-stop over a range of 5552 km (3,450 miles). A prominent user of the Challenger in this role is the Zurich-based Swiss Air-Ambulance Service, which was among the first Challenger operators and has both models in its fleet.

The Challenger 601-1A was superseded from 1987 by the 601-3A, which remains in production as the current 'standard' Challenger. First flown on 28 September 1986 and certified in Canada and the USA in April 1987, it features a number of improvements over the -1A variant. Its upgraded General Electric CF34-3A engines, rated at 40.7 kN (9,140 lb st) with automatic power reserve, 38.5 kN (8,650 lb st) without APR, are flat-rated at 21° Celsius for enhanced climb and 'hot-and-high' performance, and the flight deck is now 'all glass', incorporating a fully integrated flight guidance and flight management system with Honeywell SPZ-8000 four-tube electronic flight instrument system (EFIS). Other improvements include a power-assisted passenger door and twin landing lights in the nose.

From 1989 Canadair has offered an extended range option on the Challenger 601-3A/ER, available on new-build aircraft and as a retrofit to earlier -1As and -3As, which increases range from 3,430 nm (6352 km; 3,942 miles) to 3,650 nm (6760 km; 4,200 miles) based on standard US National Business Aircraft Association criteria for maximum fuel, five passenger operation with IFR reserves. The -3A/ER Challenger features a conformal tailcone fuel tank holding 697 litres (184 US gal) in place of the existing tail fairing, increasing overall length by 46 cm (18 in) and adding 113 kg (250 lb) to the aircraft's empty weight. Minor modifications are necessary to the landing gear to accommodate increased maximum ramp/take-off weights of 20299/20230 kg (44,750/44,500 lb). The ER modification was certificated on 16 March 1989 and is a common option on new aircraft, while more than 50 modification kits have also been sold.

Another variant of the aircraft, announced in June 1989, was the Challenger 601-S, intended as a 'no frills' variant for operators not seeking intercontinental range. The Challenger 601-S had no auxiliary fuel tank, a simplified avionics fit, and a standard 12-passenger interior. None were ordered and the line has been discontinued. By late 1998 Challenger deliveries to corporate owners had passed 275 aircraft, and the latest model in production was the Challenger 604 with CF34-3B engines, greater fuel capacity and a more advanced avionics package and systems for a range of 4,603 miles (7408 km).

Although it is essentially a new aircraft, the Canadair Regional Jet, or CRJ, is based on Challenger airframe components, notably the outer wings and fore and aft fuselage sections – between which plugs totalling 6 m (20 ft) in length have been added to provide a 50-passenger cabin – and tail unit. As such the CRJ is an extension of the abandoned CL-610 Challenger E concept. The CRJ prototype made its first flight on 10 May 1991 and by June 1998 some 240 such aircraft (out of orders for 376) had been delivered. There are three CRJ variants, and the CRJ-700 was due to make its first flight in 1999 as a stretched version with 72-seat capacity. The Global Express is a large corporate transport developed from the Challenger and CRJ concepts.

Chapter 14

Cessna Citation

Cessna Family

Concorde

Cessna Citation

While the business jet is now both a common sight and a de rigueur corporate possession, this was not always the case. The first such aircraft, the Lockheed Jetstar for example, were too constrained by their military roots to become popular, and it fell to Cessna to take the lead in forging this new market.

The Cessna Aircraft Company was the first of America's 'big three' general aviation aircraft manufacturers to produce a business jet, leap-frogging directly from an all piston-engined line into the jet age on 7 November 1968, when it announced development of the Fanjet 500.

Although the Fanjet 500 was to be Cessna's first business jet, the company had long experience of jet manufacture, having produced some 1,774 T-37 jet trainers/attack aircraft for the United States Air Force. In the mid-1950s Cessna had proposed a five-seat executive transport version of the T-37, known as the Model 407, but the prototype never flew and the project was not pursued.

The Fanjet 500 was an eight-seat pressurised twin-jet powered by two 8.9-kN (2,000-lb) Pratt & Whitney Canada JT15D-1 turbofans. Cessna intended it to be quieter, easier to fly and to maintain, more fuel-efficient and cheaper to buy and to operate than any other business jet then on the market, and able to operate from any airfield used by medium-sized twin piston-engined or turboprop business aircraft.

The prototype Fanjet 500 (N500CC) made its first flight from Wichita, Kansas on 15 September 1969, at which time Cessna announced that the aircraft's name would be changed to Model 500 Citation. During the course of the Citation's flight test programme a number of changes were made to the airframe, including lengthening of the forward fuselage, increasing the area of the fin and rudder, relocation of the engine nacelles further aft, and resiting of and addition of dihedral to the tailplane, while maximum take-off weight was increased from 4310 to 4700 kg (9,500 lb to 10,350 lb). A production-standard Citation (N502CC) was flown for the first time on 1 July 1971, and Federal Aviation Administration certification to FAR Part 25 was granted on 9 September, with customer deliveries beginning immediately.

Despite the scepticism of Cessna's competitors in both turboprop and jet markets who dubbed the aircraft, with its distinctly unjet-like unswept wings and tailplane, 'Nearjet', the Citation proved a

Now the world's best selling biz-jet, it was only a matter of time before the economical Citation II was made an even more attractive proposition through its certification for single pilot operations, as the Model 551.

1968 saw the foundation of a dynasty by the Cessna Aircraft Company, with the unveiling at its Wichita factory of the new eight-seat twin-engined Fanjet 500. Restyled and renamed, it became the Cessna 500 Citation.

great success. Its very low noise levels and excellent short-field performance enabled it to operate freely, day and night, into airports from which other business jets were excluded by either environmental or performance limitations.

During the early 1970s the Citation was certified in Australia, Austria, Belgium, Canada, Denmark, Germany, Japan, Italy, the Netherlands, Spain, Sweden, Switzerland, the United Kingdom, Yugoslavia and numerous African states. Maximum take-off weight was increased progressively to 4920, 5215 and eventually 5375 kg (10,850 lb, 11,500 lb and 11,850 lb), and from February 1976, Rohr thrust reversers became available as an option, further enhancing its landing field performance.

Production of the original Citation ceased in 1976 after 350 had been built. It was succeeded by the Model 500 Citation I, which incorporated a 1.03-m (3-ft 4-in) increase in wing span and uprated 9.79-kN (2,200-lb) JT15D-1A turbofans. The Citation I was certificated on 15 December 1976,

deliveries commencing six days later. A single-pilot version known as the Model 501 Citation I/SP was certificated to FAA Part 23 on 7 January 1977 and, apart from some equipment changes, was identical to the two-crew model. Production of the Citation I was terminated in the summer of 1985 after a total of 691 Model 500/501s had been delivered.

Stretched Citation

On 14 September 1976 Cessna announced development of the first 'stretched' Citation derivative, the Model 550 Citation II, which featured a 1.12-m (3-ft 9-in) increase in fuselage length to accommodate up to 10 passengers and two crew; increased-span high aspect ratio wings; increased fuel and baggage capacity; 11.1-kN (2,500-lb) Pratt & Whitney Canada JT15D-4 engines; and a maximum take-off weight of 6033 kg (13,300 lb). The first Citation II (N550CC) made its maiden flight on 31 January 1977 and was certificated by the FAA in the following March. A single-pilot Model 551 Citation II/SP variant was also available, with maximum take-off weight limited to 5670 kg (12,500 lb).

Production of the Citation II continued through 503 aircraft until, in the summer of

1984, Cessna introduced a number of improvements to the airframe, resulting in a change of designation to Model S550 Citation S/II. These modifications included a new supercritical airfoil section for the wing to reduce high-speed drag, while retaining the Citation's excellent low-speed handling and short-field performance; a modified wing/fuselage juncture fairing; a new inboard wing leading-edge cuff extension which increased lift and provided more room for fuel; graphite composite Fowler flaps and ailerons; recontoured drag-reducing engine pylons; aileron and speed brake gap seals; a 'weeping' wing TKS glycol anti-icing system for wing leading edges; and JT15D-4B turbofans that provided greater thrust at high altitudes. Maximum take-off weight went up to 6486 kg (14,300 lb). The S/II's cabin interior was also redesigned to provide more headroom, width and baggage volume, while a CRT-based Sperry electronic flight instrument system (EFIS) became available as a flight deck option. The Citation II prototype, N550CC, was modified as the initial S/II, flying in this configuration for the first time on 14 February 1984. Certification was obtained during the following July.

Navy orders

Among customers for the Citation S/II was the US Navy, which ordered 15 aircraft under the company Model 552 and military designation T-47A for use in its Undergraduate Naval Flight Officer Training Systems Upgrade (UNFO/TSU) programme. These aircraft differ from the commercial S/II in having uprated 1315-kg (2,900 lb) JT15D-5 engines and shorter wing span to increase climb rate and maximum speed, and are also equipped with a nose-mounted Emerson APQ-159 radar. The first T-47A was flown from Wichita on 15 February 1984, and was FAA-certificated (the aircraft

Eight years after its 1969 launch, and with 350 of the basic Model 500 Citation delivered, Cessna introduced the 501 Citation I/SP. Designed for single-pilot operations, it sold in almost as great numbers.

The inevitable development of the small eight-seat jet came in 1977, with the first flight of the noticeably longer Model 550 Citation II. With room for up to 12 passengers, it was intended as a transcontinental biz-jet.

The Citation I and II were outwardly very conventional in appearance: straight-winged and with a narrow cabin. The only notable features were the hexagonal windows in the fuselage and the unusual cockpit frames on some Citation IIs.

operate under civilian registration and are flown by civilian pilots) on 21 November 1984. Deliveries were completed during 1985. The Cessna T-47s, which replaced some North American T-39 Sabreliners in the US Navy inventory, were used to train radar intercept officers and other 'back-seaters' in air-to-air and air-to-surface radar interception techniques.

Five specially-equipped Citation S/IIs have also been delivered to the Chinese Academy of Sciences Airborne Remote Sensing Unit for aerial survey work. A Citation S/II delivered in January 1988 was the 1,500th Citation manufactured. The Citation S/II remains in current production, as does the aircraft that it succeeded. Production of the Model 550 Citation II was resumed in September 1985 as a result of customer demand. Total deliveries of all Citation II, II/SP and S/II models exceed 800.

Citation III and VI

In 1976 Cessna began studying plans for a Citation III, initially seen as a further development of the straight-wing Citations, but with increased power and passenger capacity to enable it to compete with medium-size business jets. Several configurations were mooted, including a three-engined Model 700, but when the new aircraft finally emerged as the Model 650 it was an entirely new, advanced design which had nothing in common with its stablemates save for the Citation name.

Key features of the Citation III were its swept-back wing with NASA-developed supercritical airfoil section, eight to 11-seat passenger cabin with stand-up headroom,

Citation IIs soon became common the world over and, like the Model 500, the basic Citation 550 was improved as the S550, through the fitment of uprated Pratt & Whitney JT15D-4B engines and an aerodynamically cleaner wing.

swept-back T-tail configuration and 16.24-kN (3,650-lb) Garrett TFE731-3B-100S turbofan engines. The use of the fuel-efficient Garrett powerplants and state-of-the-art wing section enabled Cessna to promise true intercontinental range capability for the new aircraft, the prototype of which (N650CC) made its first flight at Wichita on 30 May 1979, followed by the second prototype on 2 May 1980. A two-year flight test, development and certification programme led to FAA approval being granted under the FAR Part 25 Transport category on 30 April 1982. The first three production Citation IIIs undertook extensive demonstration tours throughout 1982. Although deliveries did not begin until the following spring (golfer Arnold Palmer received the first customer aircraft), Cessna quickly built up production of the Citation III, delivering the 100th aircraft to Martin-Marietta Corporation of Georgia in March 1986. By mid-1993 a total of 203 Citation IIIs had been delivered.

A developed version of the Citation III, with longer range, improved short-field performance, increased cabin room and revised fuel system was announced in the autumn of 1989 as the Model 670 Citation IV. Lack of customer appeal caused Cessna to abandon the development of this aircraft in favour of two further derivatives, the Model 650 Citation VI and Model 660 Citation VII, both of which were announced during 1990. The Citation VI, which is being produced on a common production line with the Citation III from the 200th airframe onwards, is a simplified, lower-cost version of the aircraft sold with standard factory fit avionics and cabin interior, but is otherwise identical. The first Citation VI was rolled out on 2 January 1991; deliveries began during March and now total 23.

Citation VII

The Citation VII shares the Citation III's airframe but is powered by 17.8-kN (4,000-lb) Garrett TFE731-4R-2S turbofans which improve the aircraft's hot-and-high performance, and has a maximum take off weight

Cessna abandoned the projected Citation IV in favour of the 1987 Model 560 Citation V. Their first new design in some time, it owed a lot to the earlier Citation S/II but incorporated years of avionics and engineering improvements, in addition to a longer fuselage. The 100th aircraft was delivered to Cartier's in April 1991.

of 10183 kg (22,450 lb). An engineering prototype made its first flight in February 1991. FAA certification was gained in January 1992 after a 400-hour, 225-flight test programme. By May 1995 some 53 of these aircraft had been built.

Citation V

At the NBAA Convention in New Orleans, Louisiana, in October 1987, Cessna announced the next development in its line of straight-wing Citations, the Model 560 Citation V, which had first flown in prototype form (N560CC) two months previously. This aircraft is a derivative of the S/II with a fuselage nearly two ft longer to increase passenger capacity to a maximum of 12, or to enable a fully enclosed toilet/vanity area to be installed in the cabin in an eight-passenger configuration. Two baggage compartments with a combined capacity of 385 kg (850 lb) are located outside of the main cabin area. The Citation V also features a tailplane of increased span, a new anti/de-icing system using engine bleed air to heat the inboard leading edges of the wing, and low-profile 'silver' inflatable de-icer boots on outer wing and tailplane leading edges. Powerplant is two 12.9-kN (2,900-lb) Pratt & Whitney Canada JT15D-5A turbofans, and maximum take-off weight is 6804 kg (15,900 lb).

Following a 1,029-hour certification test flight programme flown by two prototypes, the Citation V was approved by the FAA on 9 December 1988. The first customer delivery was made in April 1989, and by July 1995 some 300 Citation V and avionically upgraded Citation V Ultra aircraft had been delivered.

In 1989, at the National Business Aircraft Association Convention in Atlanta, Georgia, Cessna unveiled a successor to the original Citation: the Model 525 CitationJet, a six-seat 'entry-level' business jet which would fly higher, faster and further than the original Model 500, taking off from shorter runways and offering more cabin space and comfort while burning less fuel. Target specifications and performance figures

The 650 Citation VII is a heavier and more powerful development of the old Citation III. Its uprated Garret engines provide a reassuring amount of extra thrust for operations from hot-and-high airfields.

A sleek and speedy design, the Citation III deserved to do better in the marketplace. The supercritical wing was designed with help from NASA, and in a dive the aircraft has reached Mach 0.90 during test flights.

Left: While retaining obvious family features like the long nose, low undercarriage and twin-engine layout, 1979's swept-wing Model 650 Citation III was a major advance by Cessna, taking the family into the realm of the long-range jet.

Above: 1990 saw the launch of two aircraft based on the Citation III. The Model 650 Citation VI shares the same model number as its predecessor, but its cockpit is less expensively equipped and the cabin interior more spartan.

include a maximum take-off weight of 4536 kg (10,000 lb), a cruising speed of 379 kt (700 km/h; 435 mph) (up 30 kt on the earlier aircraft), and a 10 per cent increase in range, despite carrying 317 kg (700 lb) less fuel.

Externally, the CitationJet's most obvious new feature is its T-tail, but it is in fact an entirely new aircraft sharing little in common with the Citation I save for its fuselage cross-section, windscreen, cabin door and nosewheel assembly. Its natural laminar flow (NLF) wing section (developed by NASA and Cessna) and highly fuel-efficient 8.45-kN (1,900-lb) Williams/Rolls-Royce FJ44-1A turbofans are the keys to meeting Cessna's performance goals. The engineering prototype CitationJet (N525CJ) made its first flight on 29 April 1990, and was followed by a pre-production prototype (N525CC) on 20 November 1991. These two aircraft undertook a 1,000-hour flight test programme leading to FAA FAR Part 23, Amendment 36 certification in December 1992. Cessna holds orders for more than 100 CitationJets, believing there is a market for more than 1,000 over a 10-year period.

Citation X

The latest, and perhaps ultimate, development in the Citation line is the Model 750 Citation X (ten) 12-passenger high-speed long-range business jet, announced at the 1990 National Business Aviation Association Convention, although engineering work on the aircraft had commenced five years previously. The Citation X has been designed to become the world's fastest production business jet. Though sharing some of the external features of the Citation III, it will have an entirely new high aspect ratio wing with sharp sweepback (37°) and new-generation 31.1-kN (7,000-lb) Allison GMA 3007A turbofans flat rated to 26.7 kN (6,000 lb) each with full-authority digital engine control (FADEC) and target-type thrust reversers. Maximum speed is expected to be Mach 0.90, with a full-fuel range of 3,300 nm (6111 km/9835 miles) with IFR reserves, permitting US transcontinental flights east-west or west-east regardless of prevailing wing conditions, and intercontinental capability. Other key features of the Citation X will include cruising speeds of Mach 0.86/0.88, a maximum operating altitude of 15544 m (51,000 ft), take-off runway length of 884 m (2,900 ft), landing field length of 884 m (2,900 ft), and a maximum take-off weight of 14060 kg (31,000 lb).

Wind Tunnel tests of Citation X models began in the summer of 1991, and the first aircraft flew in December 1993. FAA certification followed in 1995 and delivery to customers started in 1996.

With the Model 525 CitationJet, Cessna got right back to basics with a straight-forward, cheap and easy-to-operate small business jet. Making its first flight from Wichita's Mid-Continent Airport on 29 April 1991, the CitationJet is also being offered for the USN/USAF JPATS competition.

CESSNA
Single-engined family

Visit a private or club airfield virtually anywhere in the world and the odds are that you will find at least one example of the aircraft that made Cessna the light aircraft market leader in the postwar years. Here we examine the single-engined aircraft upon which they built their success.

World War II brought the opportunity to Cessna for enormous expansion in its production capacity. From being a small-volume builder of high-wing touring aircraft, the company had to meet huge wartime demands for the T-50 (UC-78 Bobcat), the Waco CG-4A transport glider and major components for the Douglas A-26 and Boeing B-29 bombers. By 1944, however, Cessna could foresee peace over the horizon, and with it the prospect of many wartime aviators wanting to purchase light aircraft. Their immediate solution was the Model 190 based on the prewar Model 165 Airmaster, and the prototype first flew on 7 December 1944. In production form, the Cessna 190 was an all-metal aircraft with a 179-kW (240-hp) Continental R-670-23 radial engine and, together with the 224-kW (300-hp) Jacobs-engined Cessna 195, this model was first delivered to customers in July 1947.

The main opportunity, however, was for a reasonably-priced two-seater for sale to the homecoming GIs. Cessna therefore produced the Model 140. This was a strut-braced high wing monoplane with a fixed tailwheel landing gear which embodied slim main gear legs of spring steel. Powered by a 63-kW (85-hp) Continental C85-12 flat-four engine, the prototype (NX41682) first flew on 28 June 1945 and was an immediate success. Together with its economy partner, the Model 120 (which lacked flaps, electric starter and extra side windows) the aircraft sold some 3,950 examples in 1946.

The sales boom was short-lived and, from the heady days of 1946 when a total of over 12,000 light aircraft was delivered by United States general aviation companies, the market plummeted to an output of only 2,550 units in 1949. Nevertheless, Cessna enlarged its range of models with the design of a new four-seat version of the Model 140. The prototype was first flown in September 1947 and, in its initial version, the new Model 170 featured a wider, longer fuselage than the Model 140 and was powered by a 108-kW (145-hp) Con-

The workhorse of the outback, Cessna's U.206 Stationair started life as the Super Skywagon and could be fitted with a fixed tricycle undercarriage, skis, amphibious floats or, as on this Canadian aircraft, standard water floats. A larger tail unit was fitted when the Stationair operated from water.

Cessna's original Model 150 trainer, introduced in 1959, had a square tail and a 74.5-kW (100-hp) Continental O-200A engine. Over the years the tail has been swept back, an all-round vision cockpit provided and tubular steel undercarriage fitted, and the ultimate Cessna 152 was powered by a 82-kW (110-hp) Lycoming engine. Production was centred at both Wichita and at Reims.

tinental C145-2, but still employed constant-chord fabric-covered wings. With the Model 170A, however, the aircraft was fitted with a completely new all-metal wing which had inboard sections of constant chord and tapered outer panels. This was to become the standard wing design used for virtually all subsequent single-engine Cessnas. The Model 120 had been discontinued in 1949, but the company retained the Model 140A for a further couple of years, and also gave it wings similar to those used on the Model 170A.

Observation post for the Army

During the 1950s Cessna was faced with new military demands brought about by the onset of the Korean War. A two-seat liaison aircraft was required and a prototype of the Cessna 305A (NX41694) was built and flown in December 1949. Essentially a Cessna 170A with a cut-down rear fuselage, all-round vision cockpit enclosure and a 157-kW (210-hp) Continental O-470 engine, the Model 305 was ordered for the US Army as the L-19 Bird Dog with production starting in 1950. The US Navy received 60 similar aircraft designated OE-1, and Cessna also built a specialised instrument trainer as the L-19A-1T. This was developed into the Model 305B (TL-19D) with a constant-speed propeller and increased gross weight; the similar Model 305C (L-19E) was produced for a number of overseas nations under the MDAP programme and, in particular, was delivered to the French army's ALAT.

In April 1959 Bird Dog production was finished after a total of 3,394 had been built, but the aircraft was far from obsolete and it really came into its own with the advent of the Vietnam War. The US Air Force took over a large number of US Army O-1s for use by forward air controllers, in which form aircraft subsequently passed into the hands of the air forces of Vietnam, Laos, Thailand and Kampuchea.

In 1952, the high-lift flaps of the Bird Dog had been introduced to the civil Model 170B, providing considerable improvements in performance, but Cessna had its sights set on a brand new high-performance four-seater. This materialised as the Model 180 prototype (N41697), which was rolled out in January 1952. Outwardly it was similar to the Model 170 but it had a large squared-off vertical tail unit and, with its 168-kW (225-hp) Continental O-470-R, it achieved a remarkable 266 km/h (165 mph) maximum speed. The Cessna 180 proved to be an outstanding utility aircraft, and was frequently fitted with floats or skis for operation in very isolated terrain. It was so successful that Cessna brought out a six-seat variant, the Model 185 Skywagon, which was powered by a 194-kW (260-hp) Continental IO-470-F engine and had an enlarged tail and extra rear cabin side windows. First deliveries were made in the 1961 model year. The Model 185 and its military derivative, the U-17, have changed little over the years and were still in production when a construction halt was called to piston-engine models in 1986/87.

Recognising that the Cessna 180 had potential as an up-market private and business aircraft, the design team at Wichita fitted it with a tricycle landing gear to create the Model 182. This was introduced in 1956 and two years later Cessna offered the alternative Skylane, which was a Model 182A with an all-over paint scheme, improved upholstery and landing gear spats. Eventually, virtually all of the Cessna 182s delivered were the Skylane variant and the utility Cessna 182 was only available by special order. In a parallel development, the Cessna 170B was also changed, in 1956, to become the Cessna 172. In this case the modification was not only the conversion to a tricycle landing gear but also the fitting of a squared-off tail unit of contours generally similar to that of the Model 182.

In its marketing of light aircraft, Cessna was increasingly following the practices of the American automobile industry. To create volume sales the models were changed every year (and given a new letter suffix to the designation) following the philosophy that well-off private owners would want to trade in their old model frequently for the latest variant with the most up-to-date modifications. The paint scheme was changed each year and, when large fins were being featured on the 1958 range of motor cars the Cessna colour scheme incorporated similar 'go faster' fins running across the tail unit. Later, between 1964 and 1970, the paint design for all models in the product line received a similar theme each year, and it was quite possible to tell the year of construction of any Cessna by looking at the motif painted into its tail design.

Annual changes were not merely cosmetic, however. In 1958 an alternative version of the Model 172 with a larger 130-kW (175-hp) Continental GO-300A geared six-cylinder engine was announced, and this became known as the Cessna 175. 1960 also saw the Model

The precursor of all Cessna's modern single-engined aircraft, the C-34 first flew in June 1935, powered by a 108-kW (145-hp) Warner Scarab radial engine. Forty-two were built during 1935-36 and the C-34 gained fame for winning the 1936 Detroit Trophy Race. The postwar Cessna 190 and 195 were, essentially, all-metal versions of the C-34.

Originally delivered to Belgium in 1947, OO-ACE was an example of the Cessna 120 two-seater built in quantity immediately after the war. A total of 7,724 of this type and its better-equipped cousin, the Model 140, emerged from Wichita before production ceased in 1951, and they are much sought after today.

The Greek air force is only one of the users of the T-41B military trainer version of the Cessna 172. The T-41A Mescalero was provided by the US Air Force to civilian contractors for the air cadet training evaluation programme, and the type also serves with the air forces of Peru, Ecuador, Colombia, Turkey and Honduras.

Demand for a higher-powered version of the ubiquitous Cessna 180 resulted in the 194-kW (260-hp) Model 185 Skywagon, initial examples of which were delivered in early 1961. It had seats for five passengers and could be made even more productive by the fitting of the optional underfuselage cargo pannier illustrated.

172A, Model 175A and Skylane/Model 182C being fitted with swept vertical tails. Having adopted the de luxe option for the Skylane and Model 182, Cessna did the same with the Model 172 and Model 175 by bringing in the Model 175A Skylark in 1960 and the Model 172B Skyhawk a year later. By 1962, the designers had moved on a few more steps and proceeded to cut down the rear fuselages of the Model 172D and Model 182F to give all-round vision cabins to these models. In addition to this alteration, the Model 175 was given a completely new designation, Model P172D Skyhawk Powermatic.

Introduction of the Cardinal

By 1977 Cessna was convinced that the Model 172 needed a major redesign if it were to remain competitive. The result was the Model 172J, and the prototype (N3766C) was completed and flown in early 1966. It had a cantilever wing, broader-chord tail unit and a low-profile streamlined cabin enclosure. It was powered by a 112-kW (150-hp) Lycoming O-320 engine, and Cessna aimed to finish production of the Model 172H and introduce the new aircraft in the 1968 model year. At the last moment it was decided to build both models in parallel and the Model 172J became the Model 177 (Cardinal in de luxe form). The Model 177 was maintained in production until 1978, and in 1971 was joined by the Model 177RG Cardinal RG which had fully retractable main landing gear generally similar to that of the large Cessna Centurion. It was a measure of the public loyalty to the Model 172 Skyhawk that it outlived the Cardinal in production and deliveries exceeded 35,500.

Basic trainer

Between 1952 and 1959, Cessna had concentrated on the four-seat single-engine market, and the demise of the Model 140 had left the company without a two-seat trainer in the product line to offer as a replacement for the ageing Piper Cubs, Ercoupes and Cessna 120s in the training fleet. Thus in mid-1957 the company flew the first Cessna 142 (N34258). This was a side-by-side high-wing cabin two-seater with a fixed tricycle landing gear. The designation was soon changed to Model 150, and a total of 683 aircraft powered by the 75-kW (100-hp) Continental O-200A engine was produced in 1959, the first year of production. By the end of 1963 the company had sold over 2,200 Cessna 150s, and the 1964 Model 150D introduced the first major change, namely a cut-down rear fuselage and all-round vision cabin. A swept tail came in with the 1966 Model 150F and, in 1970, the aerobatic Model A150K Aerobat went into production beside the standard trainer version at Cessna's Winfield, Kansas, factory. The traditional Cessna spring steel landing gear was replaced by tubular steel on the 1971 Model 150L and Model A150L and then, in 1978, the trusty Continental engine was replaced by the 82-kW (110-hp) Lycoming O-235-L2C and the aircraft became the Cessna 152.

By the mid-1950s attention turned to a new top-of-the-line model

Originally built during the Korean War era as a two-seat artillery spotting aircraft, the Model 306 (L-19) Bird Dog was later pressed into service as the main Forward Air Control platform and used in the Vietnam conflict. Bird Dogs were also used by the US Navy and many overseas armies, with the French ALAT being the largest user.

ZP-PKG is a basic Cessna 182 delivered in 1980 to a customer in Paraguay. Most of the Model 182s ordered were the de luxe Skylane version which featured an all-over paint scheme, wheel fairings and improved upholstery and trim. Over 22,000 Skylanes were delivered, including 20 for the Venezuelan army.

The Model 177 Cardinal was Cessna's answer to the need to modernise the popular Model 172, and it featured a streamlined airframe and cantilever wing structure. The Cessna 177RG Cardinal RG pictured here had a retractable tricycle undercarriage similar to that of the Model 210 and was in production between 1971 and 1978.

Cessna's Model 207 Stationair 8 had the option of a turbocharged engine. Its quick-change cabin could be fitted with seven passenger seats or used for freight transport. The Stationair 8 is also popular for sport parachuting, which is facilitated by the removal of its large double doors.

which would offer greater speed, comfort and avionics capability than the Skylane. While retractable landing gear had almost exclusively been fitted to low-wing aircraft hitherto, the Cessna engineers saw no reason why they should not fit this type of gear to the larger high-wing aircraft in the product line. Accordingly, a standard four-seat Cessna 182 (N1296) was used as a development aircraft and flew for the first time in 1956 with the retractable wheels and a 194-kW (260-hp) fuel-injected Continental IO-470S engine. A new engine cowling with a large chin fairing was required, and the higher engine power demanded a large dorsal fin fairing be added to the vertical tail. In fact the tail went through numerous changes and production Model 210s had an elegantly swept fin and rudder. The landing gear mechanism was complex, with the main gear legs twisting and folding back into fully enclosed wells under the fuselage centre section. The production Model 210, by now fitted with down-turned wingtips, was introduced as a 1960 model and, by September of that year, no less than 542 aircraft had been delivered. The 1961 Model 210A was given extra cabin side windows and the Model 210B, announced in October 1961, followed the Model 172 and Model 182 in having a cut-down rear fuselage and rear cabin window. A 213-kW (285-hp) fuel-injected Continental IO-520-L was installed in the 1964 Model 210D, which also gained a full cantilever wing in place of the strut-braced wing of previous models, together with the option of a Continental TSIO-520-R turbocharged engine permitting the aircraft to be flown at altitudes of up to 9205 m (30,200 ft). The Centurion and Turbo Centurion remained in production until 1987 and, apart from detailed improvements, were very similar to the versions sold in the late 1960s. Deliveries exceed 8,500.

Model 210 influence

The Cessna 210 influenced Cessna's product policy in a number of ways. Firstly, the landing gear retraction system allowed the company to offer the Cardinal RG and later, in 1977, the Skylane was also given the retractable gear option, the Cessna 172 following, in 1979, as the Cutlass RG. Cessna was also anxious to use the Model 210 airframe to create a larger range of upper-class singles. First to appear was the Model 210-5 (known as the Model 205) which was, essentially, a 1963 Model 210C with fixed landing gear and a six-seat utility interior. In 1964, when the power of the Model 210 was increased, the Model 205 was also given a 213-kW (285-hp) engine and thereby became the Model 206 Super Skywagon. Large double doors were fitted on the starboard side, but for passenger use Cessna also offered the Model P206 Super Skylane, from 1965 onwards, and this dispensed with the cargo doors but featured a luxury standard of internal trim. From 1971, both the utility and de luxe models were marketed as one type (with different trim options) under the designation Model U206E Stationair and, as with the Model 210, turbocharging was available with effect from the 1966 model year. The company did, however, retain the strut-braced wing rather than giving the Stationair the new Centurion cantilever unit.

Stretching the Model 206 to seven seats was the next action of the ever-creative Cessna design department, and this resulted in their flying the Cessna 207 (N1970F), which was first offered as a 1969 model. This had additional fuselage sections let in ahead of and behind the existing Model 206 cabin section to provide an extra row

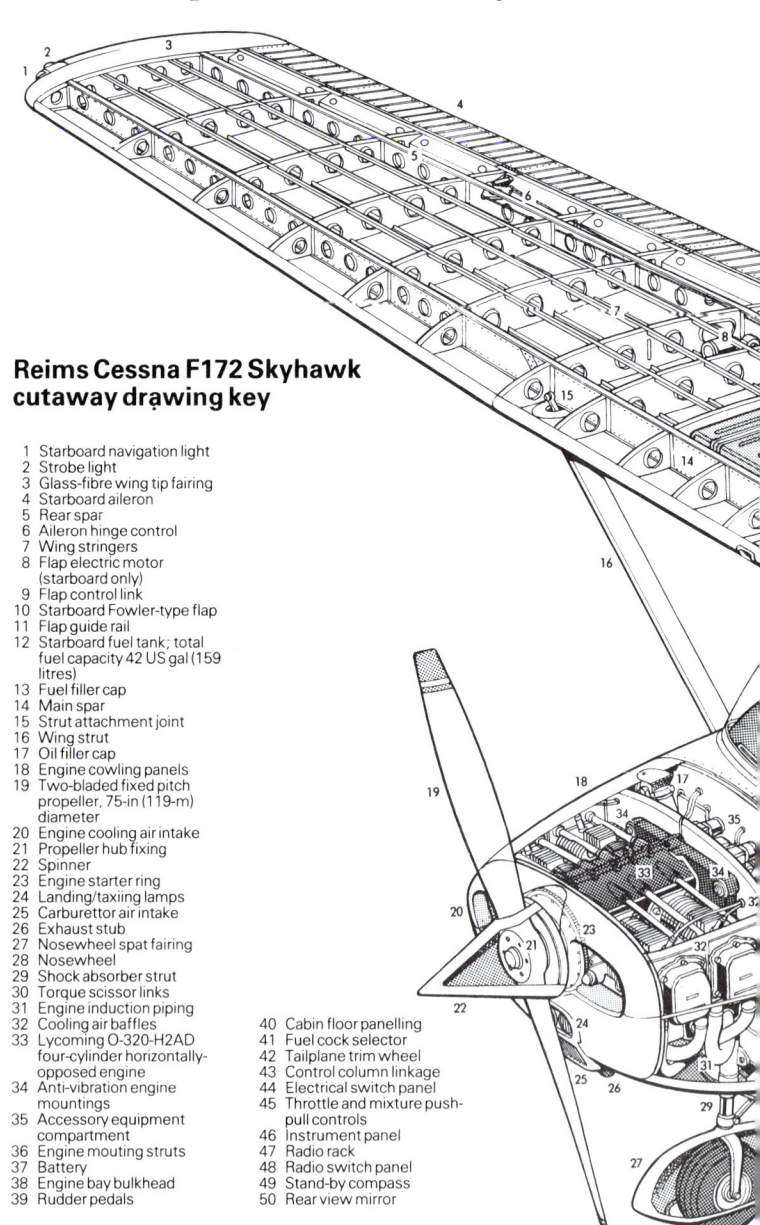

Reims Cessna F172 Skyhawk cutaway drawing key

1. Starboard navigation light
2. Strobe light
3. Glass-fibre wing tip fairing
4. Starboard aileron
5. Rear spar
6. Aileron hinge control
7. Wing stringers
8. Flap electric motor (starboard only)
9. Flap control link
10. Starboard Fowler-type flap
11. Flap guide rail
12. Starboard fuel tank; total fuel capacity 42 US gal (159 litres)
13. Fuel filler cap
14. Main spar
15. Strut attachment joint
16. Wing strut
17. Oil filler cap
18. Engine cowling panels
19. Two-bladed fixed pitch propeller, 75-in (119-m) diameter
20. Engine cooling air intake
21. Propeller hub fixing
22. Spinner
23. Engine starter ring
24. Landing/taxiing lamps
25. Carburettor air intake
26. Exhaust stub
27. Nosewheel spat fairing
28. Nosewheel
29. Shock absorber strut
30. Torque scissor links
31. Engine induction piping
32. Cooling air baffles
33. Lycoming O-320-H2AD four-cylinder horizontally-opposed engine
34. Anti-vibration engine mountings
35. Accessory equipment compartment
36. Engine mouting struts
37. Battery
38. Engine bay bulkhead
39. Rudder pedals
40. Cabin floor panelling
41. Fuel cock selector
42. Tailplane trim wheel
43. Control column linkage
44. Electrical switch panel
45. Throttle and mixture push-pull controls
46. Instrument panel
47. Radio rack
48. Radio switch panel
49. Stand-by compass
50. Rear view mirror

The Cessna 210 was the first production high-wing light aircraft to be fitted with a retractable undercarriage, the main gear legs having a complex twisting action to allow them to fold into the lower fuselage. Later Model 210 Centurions, such as N3654C, had all-round vision cockpits and full cantilever wings.

of seats and a baggage compartment between the windshield and the engine firewall. The engine was a beefy 224-kW (300-hp) IO-520-F fuel-injected Continental or its turbocharged equivalent, the TSIO-520-M. Initially, the Model 207 was named the Skywagon 207, but it became the Stationair 7 (or Turbo Stationair 7) in 1978 and then, in 1980 when the seating capacity was raised to eight, it was re-titled the Stationair 8.

Perhaps the ultimate Cessna single was unveiled in November 1977 when the company brought out the Pressurized Centurion. Announced as the only production pressurised single in the world (the Mooney M22 having been discontinued in 1970) the Model P210N was based broadly on the standard Cessna T210N, but had a cabin pressure differential giving 3695-m (12,120-ft) cabin pressure at 7010 m (23,000 ft). It was equipped with a Continental TSIO-520-P engine with a 231-kW (310-hp) take-off power rating and could be identified by its four small cabin side windows on each side (as opposed to the two windows of the standard Centurion). Maximum speed of the Model P210 had reached 235 mph (378 km/h), and a fully equipped model could be delivered for $107,330. It was a measure of the impact of world inflation that the same model was priced at $182,750 by the time the 1982 model was announced, an increase of some 70 per cent.

Overseas production

Many of Cessna's single-engine models have been built overseas. In February 1960 Cessna bought a major interest in Avions Max Holste, later renamed Reims Aviation, and established a French production line to build Cessna aircraft for sale in Europe and the Middle

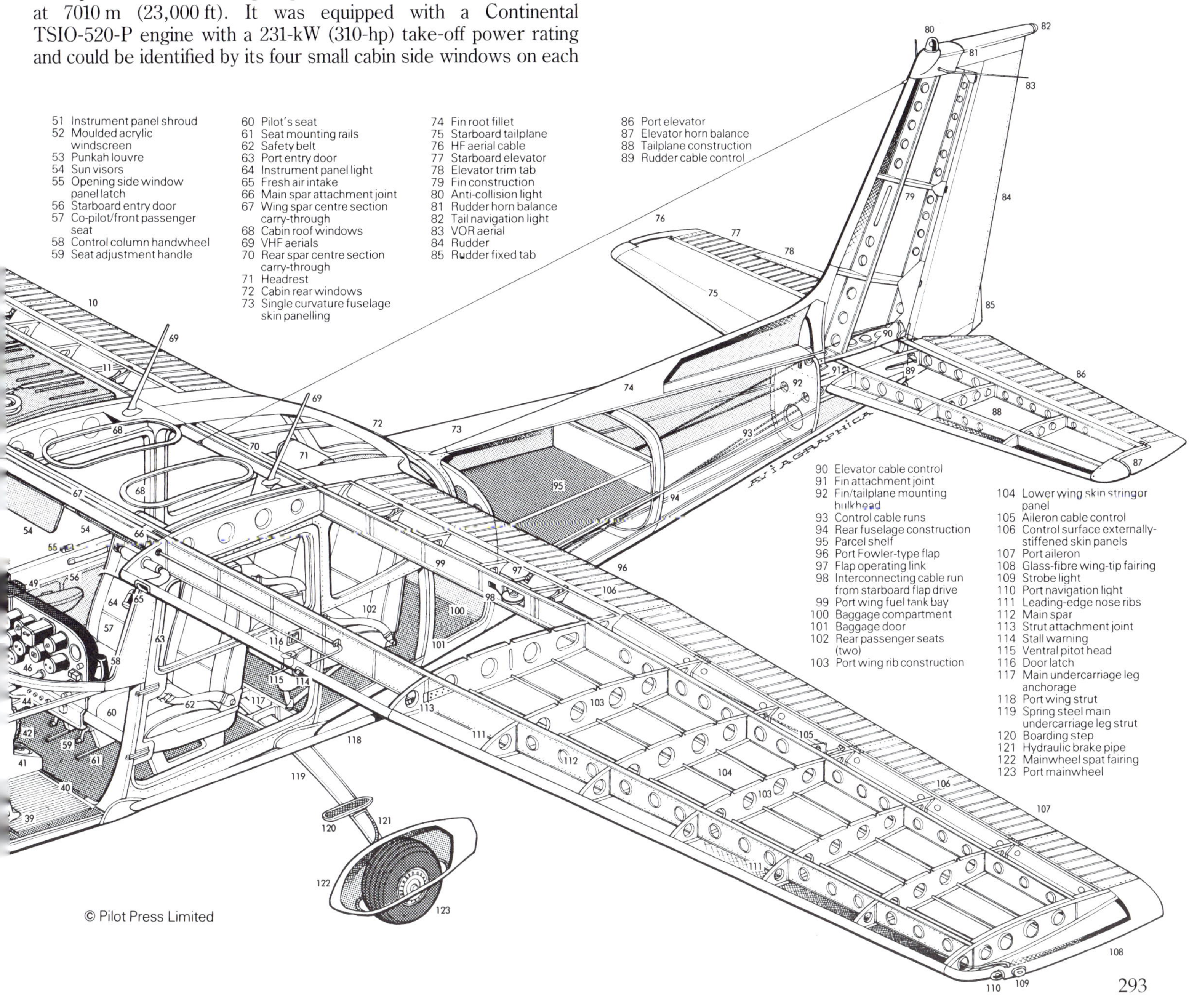

51 Instrument panel shroud
52 Moulded acrylic windscreen
53 Punkah louvre
54 Sun visors
55 Opening side window panel latch
56 Starboard entry door
57 Co-pilot/front passenger seat
58 Control column handwheel
59 Seat adjustment handle
60 Pilot's seat
61 Seat mounting rails
62 Safety belt
63 Port entry door
64 Instrument panel light
65 Fresh air intake
66 Main spar attachment joint
67 Wing spar centre section carry-through
68 Cabin roof windows
69 VHF aerials
70 Rear spar centre section carry-through
71 Headrest
72 Cabin rear windows
73 Single curvature fuselage skin panelling
74 Fin root fillet
75 Starboard tailplane
76 HF aerial cable
77 Starboard elevator
78 Elevator trim tab
79 Fin construction
80 Anti-collision light
81 Rudder horn balance
82 Tail navigation light
83 VOR aerial
84 Rudder
85 Rudder fixed tab
86 Port elevator
87 Elevator horn balance
88 Tailplane construction
89 Rudder cable control
90 Elevator cable control
91 Fin attachment joint
92 Fin/tailplane mounting bulkhead
93 Control cable runs
94 Rear fuselage construction
95 Parcel shelf
96 Port Fowler-type flap
97 Flap operating link
98 Interconnecting cable run from starboard flap drive
99 Port wing fuel tank bay
100 Baggage compartment
101 Baggage door
102 Rear passenger seats (two)
103 Port wing rib construction
104 Lower wing skin stringer panel
105 Aileron cable control
106 Control surface externally-stiffened skin panels
107 Port aileron
108 Glass-fibre wing-tip fairing
109 Strobe light
110 Port navigation light
111 Leading-edge nose ribs
112 Main spar
113 Strut attachment joint
114 Stall warning
115 Ventral pitot head
116 Door latch
117 Main undercarriage leg anchorage
118 Port wing strut
119 Spring steel main undercarriage leg strut
120 Boarding step
121 Hydraulic brake pipe
122 Mainwheel spat fairing
123 Port mainwheel

© Pilot Press Limited

Specification
Reims Cessna FA.152 Aerobat
Type: aerobatic basic trainer
Powerplant: one 80-kW (180-hp) Avco Lycoming O-235-N2C piston engine driving a fixed-pitch metal propeller
Performance: maximum speed at sea level 200 km/h (125 mph); cruising speed at 75 per cent power 194 km/h (121 mph); typical range at 75 per cent power and 2591 m (8,500 ft) altitude with standard tanks 574 km (310 nautical miles) or 3 hours; service ceiling 4480 m (14,700 ft); take-off distance over 15.2-m (50-ft) obstacle 408 m (1,340 ft); stall speed, flaps down and power off 80 km/h (50 mph)
Weights: maximum ramp weight 760 kg (1,675 lb); maximum take-off 757 kg (1,670 lb); standard empty in basic configuration 513 kg (1,131 lb); standard fuel capacity with long-range tanks 148 litres (39 gal)
Dimensions: span 10.11 m (33 ft 2 in); length 7.34 m (24 ft 1 in); height 2.60 m (8 ft 6 in)
Crew and accommodation: one instructor and one pupil in side-by-side seating

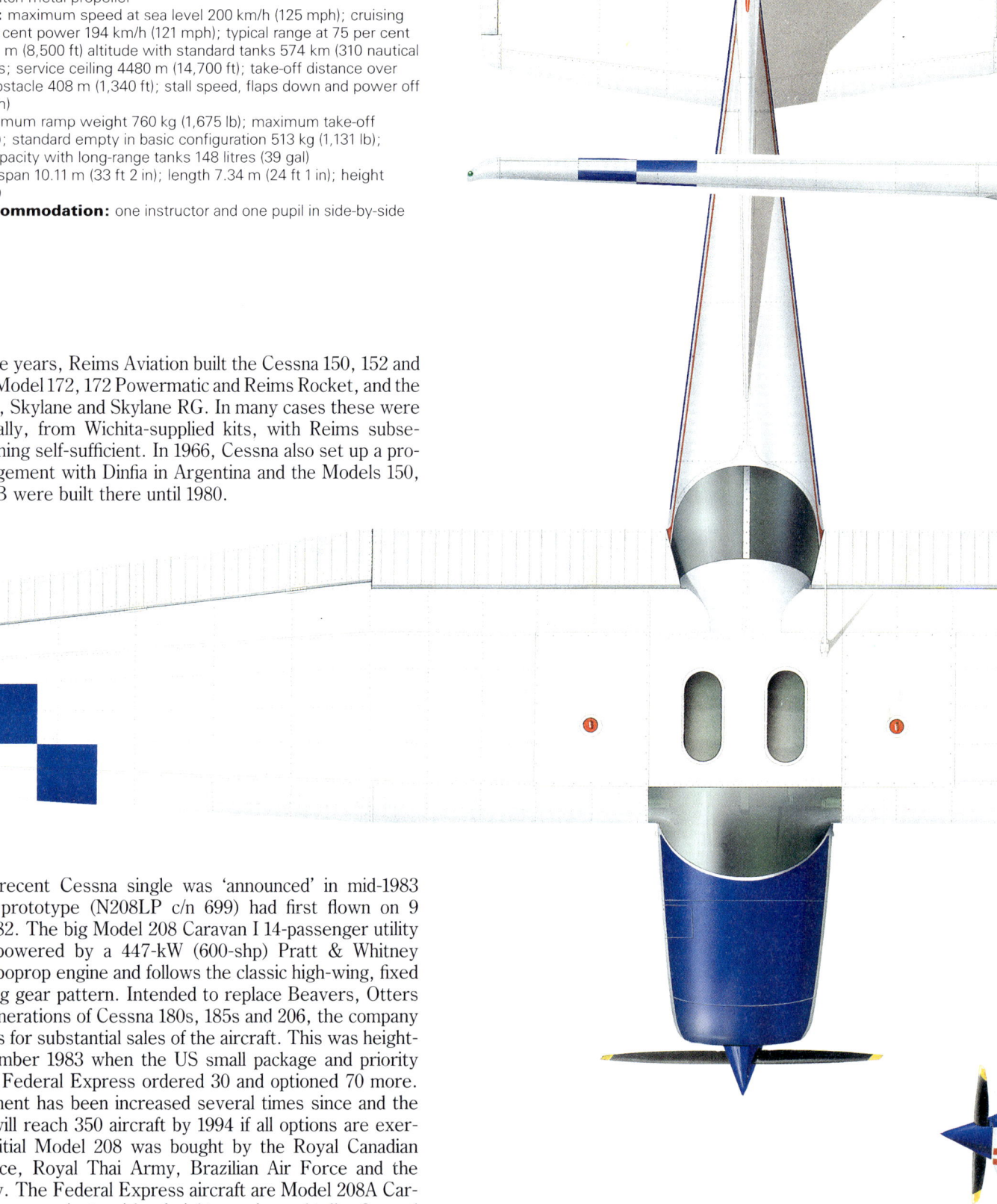

East. Over the years, Reims Aviation built the Cessna 150, 152 and Aerobat, the Model 172, 172 Powermatic and Reims Rocket, and the Model 177RG, Skylane and Skylane RG. In many cases these were built up, initially, from Wichita-supplied kits, with Reims subsequently becoming self-sufficient. In 1966, Cessna also set up a production arrangement with Dinfia in Argentina and the Models 150, 182 and A188B were built there until 1980.

The most recent Cessna single was 'announced' in mid-1983 although the prototype (N208LP c/n 699) had first flown on 9 December 1982. The big Model 208 Caravan I 14-passenger utility transport is powered by a 447-kW (600-shp) Pratt & Whitney PT6A-114 turboprop engine and follows the classic high-wing, fixed tricycle landing gear pattern. Intended to replace Beavers, Otters and earlier generations of Cessna 180s, 185s and 206, the company has high hopes for substantial sales of the aircraft. This was heightened in December 1983 when the US small package and priority freight airline Federal Express ordered 30 and optioned 70 more. This commitment has been increased several times since and the FedEx fleet will reach 350 aircraft by 1994 if all options are exercised. The initial Model 208 was bought by the Royal Canadian Mounted Police, Royal Thai Army, Brazilian Air Force and the Liberian Army. The Federal Express aircraft are Model 208A Cargomaster all-cargo variants with a freight pannier and taller fin, and with windows and starboard rear door deleted, together with the Model 208B Super Cargomaster with a bigger pannier and fuselage stretched by 1.22 m (4 ft) aft of the wing. Cessna also markets the Model 208B as the Grand Caravan, fitted with seats for 14 passengers, windows and a starboard rear door. Sales of the Caravan I had passed 800 by the end of 1998.

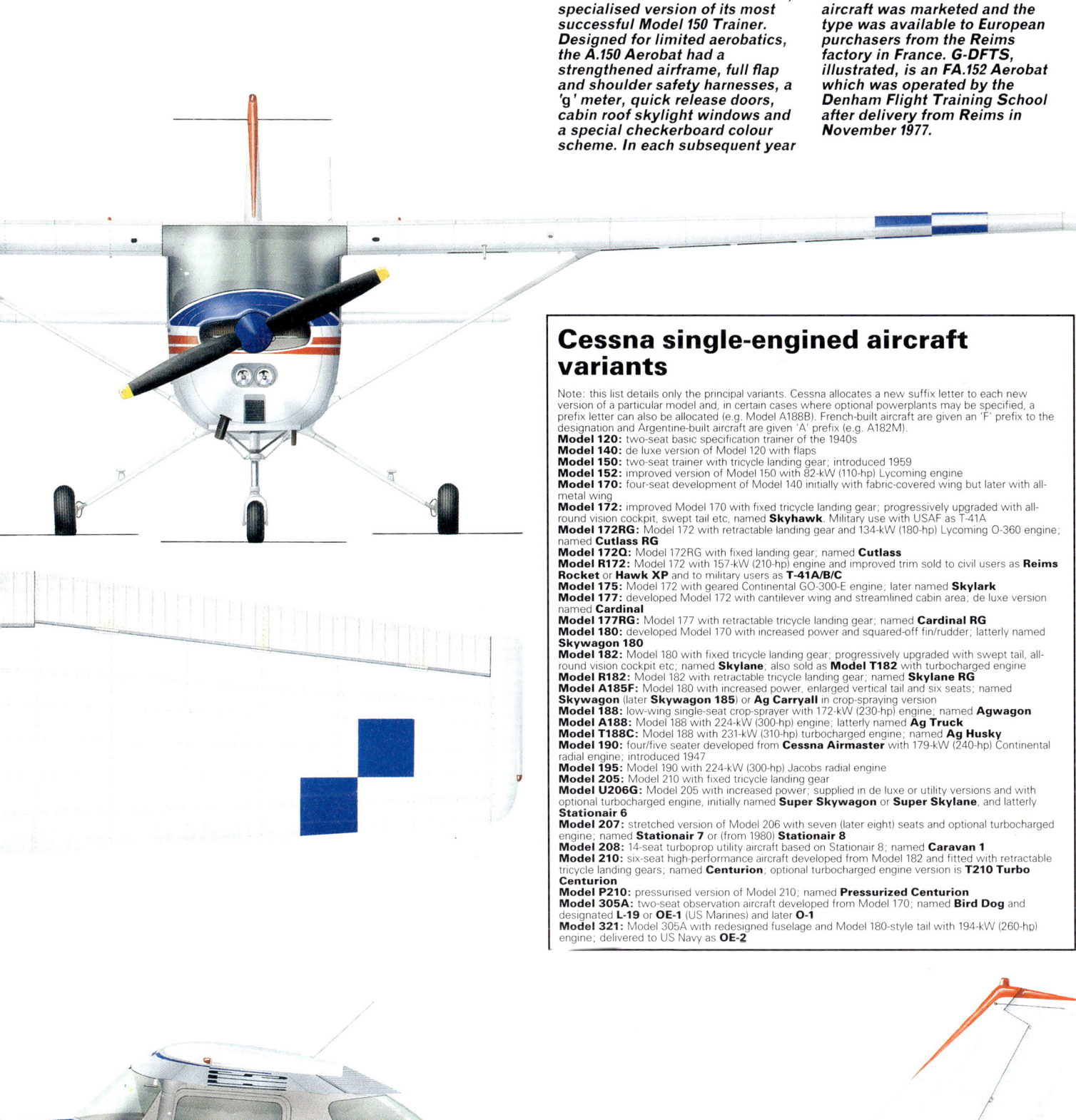

In 1970 Cessna introduced a new, specialised version of its most successful Model 150 Trainer. Designed for limited aerobatics, the A.150 Aerobat had a strengthened airframe, full flap and shoulder safety harnesses, a 'g' meter, quick release doors, cabin roof skylight windows and a special checkerboard colour scheme. In each subsequent year an Aerobat version of the basic aircraft was marketed and the type was available to European purchasers from the Reims factory in France. G-DFTS, illustrated, is an FA.152 Aerobat which was operated by the Denham Flight Training School after delivery from Reims in November 1977.

Cessna single-engined aircraft variants

Note: this list details only the principal variants. Cessna allocates a new suffix letter to each new version of a particular model and, in certain cases where optional powerplants may be specified, a prefix letter can also be allocated (e.g. Model A188B). French-built aircraft are given an 'F' prefix to the designation and Argentine-built aircraft are given 'A' prefix (e.g. A182M).

Model 120: two-seat basic specification trainer of the 1940s
Model 140: de luxe version of Model 120 with flaps
Model 150: two-seat trainer with tricycle landing gear; introduced 1959
Model 152: improved version of Model 150 with 82-kW (110-hp) Lycoming engine
Model 170: four-seat development of Model 140 initially with fabric-covered wing but later with all-metal wing
Model 172: improved Model 170 with fixed tricycle landing gear; progressively upgraded with all-round vision cockpit, swept tail etc, named **Skyhawk**. Military use with USAF as T-41A
Model 172RG: Model 172 with retractable landing gear and 134-kW (180-hp) Lycoming O-360 engine; named **Cutlass RG**
Model 172Q: Model 172RG with fixed landing gear; named **Cutlass**
Model R172: Model 172 with 157-kW (210-hp) engine and improved trim sold to civil users as **Reims Rocket** or **Hawk XP** and to military users as **T-41A/B/C**
Model 175: Model 172 with geared Continental GO-300-E engine; later named **Skylark**
Model 177: developed Model 172 with cantilever wing and streamlined cabin area; de luxe version named **Cardinal**
Model 177RG: Model 177 with retractable tricycle landing gear; named **Cardinal RG**
Model 180: developed Model 170 with increased power and squared-off fin/rudder; latterly named **Skywagon 180**
Model 182: Model 180 with fixed tricycle landing gear; progressively upgraded with swept tail, all-round vision cockpit etc; named **Skylane**; also sold as **Model T182** with turbocharged engine
Model R182: Model 182 with retractable tricycle landing gear; named **Skylane RG**
Model A185F: Model 180 with increased power, enlarged vertical tail and six seats; named **Skywagon** (later **Skywagon 185**) or **Ag Carryall** in crop-spraying version
Model 188: low-wing single-seat crop-sprayer with 172-kW (230-hp) engine; named **Agwagon**
Model A188: Model 188 with 224-kW (300-hp) engine; latterly named **Ag Truck**
Model T188C: Model 188 with 231-kW (310-hp) turbocharged engine; named **Ag Husky**
Model 190: four/five seater developed from **Cessna Airmaster** with 179-kW (240-hp) Continental radial engine; introduced 1947
Model 195: Model 190 with 224-kW (300-hp) Jacobs radial engine
Model 205: Model 210 with fixed tricycle landing gear
Model U206G: Model 205 with increased power; supplied in de luxe or utility versions and with optional turbocharged engine, initially named **Super Skywagon** or **Super Skylane**, and latterly **Stationair 6**
Model 207: stretched version of Model 206 with seven (later eight) seats and optional turbocharged engine; named **Stationair 7** or (from 1980) **Stationair 8**
Model 208: 14-seat turboprop utility aircraft based on Stationair 8; named **Caravan 1**
Model 210: six-seat high-performance aircraft developed from Model 182 and fitted with retractable tricycle landing gears; named **Centurion**; optional turbocharged engine version is **T210 Turbo Centurion**
Model P210: pressurised version of Model 210; named **Pressurized Centurion**
Model 305A: two-seat observation aircraft developed from Model 170; named **Bird Dog** and designated **L-19** or **OE-1** (US Marines) and later **O-1**
Model 321: Model 305A with redesigned fuselage and Model 180-style tail with 194-kW (260-hp) engine; delivered to US Navy as **OE-2**

CONCORDE

One of the greatest challenges in the history of aviation, Concorde is also among the very greatest technical triumphs. Though its aim was to bring people together, it created a powerful opposition lobby which, added to spiralling fuel costs, ensured that the SST revolution was not to happen.

In transport, speed is usually a saleable commodity. There has never been any question that, provided other factors were reasonable, it was logical to build faster and faster airliners. The leader in this process since World War II has been the UK. In 1946 it began with the design of the de Havilland D.H.106 Comet, which came within an ace of becoming the first-generation jetliner for the world market and should also have been the basis for the world-beating next generation. In the middle of the 1950s, when the Comet programme had faltered, talks were held between the Controller of Aircraft, M.B. (later Sir Morien) Morgan, and leaders of the British aircraft industry on the possibility of building an SST (supersonic transport). The logical thing was to form a committee, and in November 1956 the STAC (Supersonic Transport Aircraft Committee) met under Morgan's chairmanship.

The STAC embraced not only government technologists, top men from engine and airframe companies and numerous officials, but also nominees from civil airlines. Nobody then could possibly foresee that opposition to such a project would assume the proportions of a national crusade, or that for political reasons the USA and many other countries would delay for years admitting such an aircraft to its airports or even permit it to fly overhead, and certainly the price of fuel was not expected to be multiplied tenfold. All that could be seen clearly was that the technical problems were of an order of magnitude greater than those of other aircraft. Just one immutable problem was that, for any given amount of wing lift, the drag is more than doubled as the vehicle accelerates from Mach 0.9 to beyond Mach 1.

Three options

It was obvious that a supersonic airliner would need a novel configuration, with a wing of very low aspect ratio and extremely long body of minimal cross section. The STAC studied three main answers. One was an area-ruled aircraft with a curious M-wing, initially swept forward and then back to the tips, cruising at Mach 1.2 (1285 km/h/800 mph) for 2414 km (1,500 miles). Second was a slender delta, very much like the eventual Concorde, but cruising at Mach 1.8 (1930 km/h/1,200 mph) for 5635 km (3,500 miles). Third was a steel/titanium monster cruising at Mach 3 (3220 km/h/2,000 mph) for 5635 km (3,500 miles). The third idea was deemed technically too difficult and costly, though for a further 14 years it consumed over $2,000 million in the USA. In March 1959 the STAC recommended Mach 1.2 short-range or Mach 1.8 transatlantic, and

Singapore Airlines was an enthusiastic potential partner on the British Airways Far Eastern route, and for a time some Concordes were painted in Singapore colours on one side of the fuselage. However, overflying rights became a burden on the costs and benefits of the operation, forcing Singapore to withdraw.

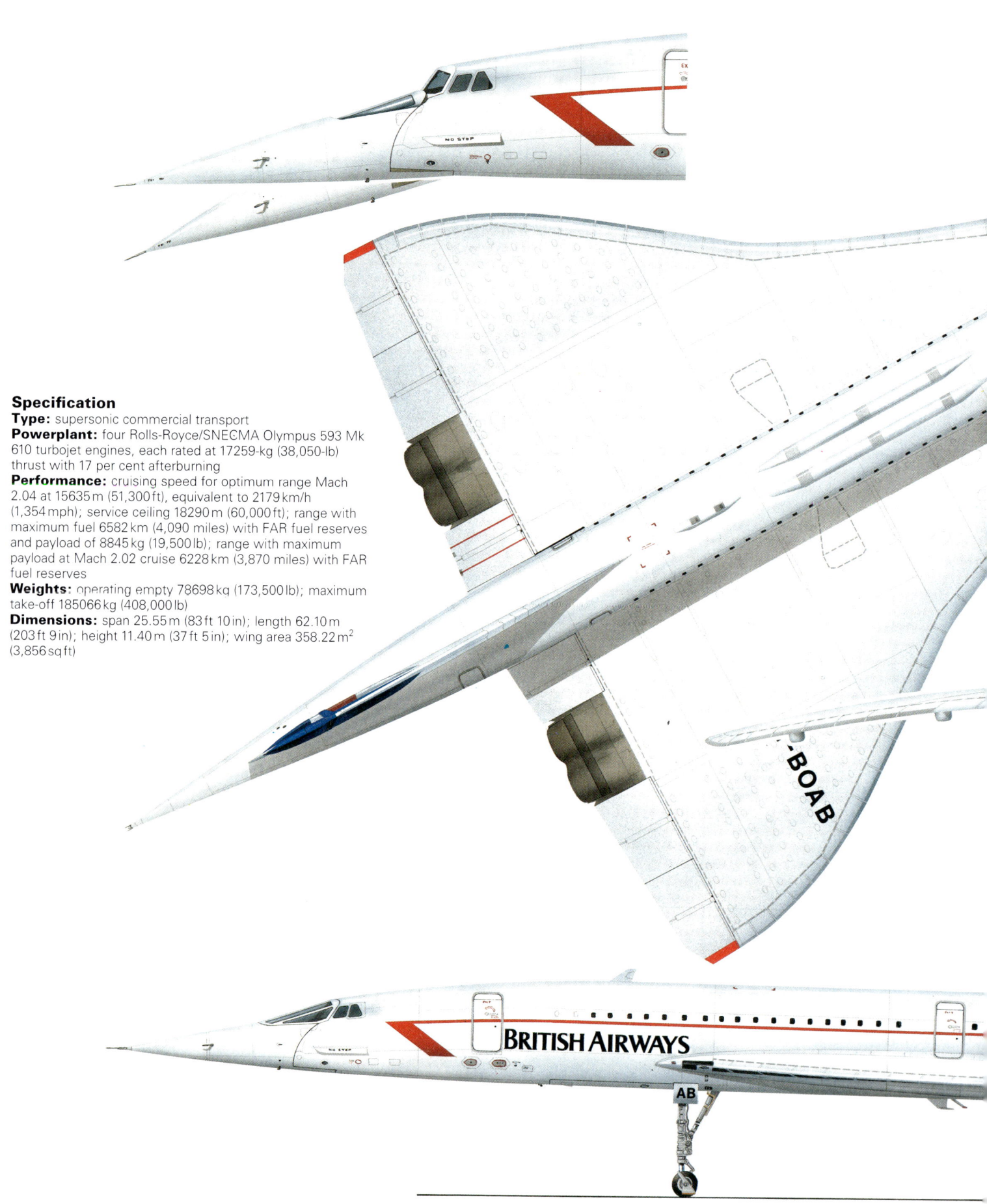

Specification
Type: supersonic commercial transport
Powerplant: four Rolls-Royce/SNECMA Olympus 593 Mk 610 turbojet engines, each rated at 17259-kg (38,050-lb) thrust with 17 per cent afterburning
Performance: cruising speed for optimum range Mach 2.04 at 15635 m (51,300 ft), equivalent to 2179 km/h (1,354 mph); service ceiling 18290 m (60,000 ft); range with maximum fuel 6582 km (4,090 miles) with FAR fuel reserves and payload of 8845 kg (19,500 lb); range with maximum payload at Mach 2.02 cruise 6228 km (3,870 miles) with FAR fuel reserves
Weights: operating empty 78698 kg (173,500 lb); maximum take-off 185066 kg (408,000 lb)
Dimensions: span 25.55 m (83 ft 10 in); length 62.10 m (203 ft 9 in); height 11.40 m (37 ft 5 in); wing area 358.22 m^2 (3,856 sq ft)

The elegantly simple lines of Concorde, optimised for an economical cruising speed of just over Mach 2, tend to disguise the extreme complexity of both aerodynamics and systems of this pioneering SST (supersonic transport). The ogival wing has cambered leading edges, and it creates powerful vortices upon which the aircraft rides at cruising speed. The four underslung jets are fed with carefully-controlled air, involving a complex series of intake ramps. With its long nose and high angle of attack the crew have no forward vision at low speeds, so the nose droops to two positions for take-off and landing. Forward of the windscreen is a retractable 'visor' which is lowered to improve visibility.

the Ministry of Aviation awarded contracts for detailed studies of the latter aircraft. By late 1960 it was clear not only that the slender delta, with an ogival (curved) leading edge, was the best shape but also that its aerodynamic efficiency improved up to about Mach 2.2. This not only reduced journey time but improved propulsion efficiency, because the latter depends on the ram pressure generated in the engine inlet and this rises rapidly with increasing Mach number.

By 1960 Bristol Aircraft, then in the process of merging into the British Aircraft Corporation (BAC), had schemed the Type 198, a 130-passenger transatlantic SST looking much like today's Concorde but powered by six Bristol Siddeley Olympus engines. In the summer of 1961 the Ministry decided a 172368-kg (380,000-lb) six-engine aircraft was too ambitious, and asked for a 100-seater weighing 113400 kg (250,000 lb) with four engines. By this time Sud-Aviation in France, likewise in the throes of becoming part of a giant group (SNIAS or Aérospatiale), had studied Mach 2 SSTs and decided on a very similar configuration but aimed at short ranges as a 70/80-seat successor to the company's Caravelle. At the Paris Air Show in June 1961 Sud displayed a model of its idea, named Super Caravelle. Some months previously BAC had, on government insistence, put out feelers to possible foreign partners, and Sud-Aviation was the only positive response. The first formal meetings were held during the Paris Air Show, and a month later at Weybridge.

The similarity between the British and French proposals was amazing; almost the only difference apart from size, weight and range was that Sud thought it could get away without using a hinged 'droop snoot' nose, while BAC considered such a feature essential for adequate forward view during the nose-high landing. After many further talks, an intergovernment agreement of 29 November 1962 formally launched the project, with government funding on a 50/50 basis and the main effort shared between BAC and Aérospatiale on the airframe and Bristol Siddeley (from 1966 Rolls-Royce) and SNECMA on propulsion, using as a basis an enlarged Olympus called the Olympus 593. The British partner was charged with developing a transatlantic version, while the French adhered to a short-haul model with a ventral stairway instead of fuel in the rear fuselage. A Committee of Directors was appointed to run the airframe, and another to manage the engine, and there were separate British and French government contracts to each national industrial group.

Widespread talents

Suffice it to say that such an arrangement will never be repeated, but it was made to work by the goodwill and towering stature of the main engineers involved. Sir George Edwards, who, when head of Vickers-Armstrongs (Aircraft) in 1958 had predicted the entire future course of events with extreme accuracy, was the architect of the whole project. Top designers at Bristol were Sir Archibald Russell and Dr Bill Strang, while leaders at Aérospatiale were Pierre Satre and Lucien Servanty. Collaboration extended down through the main airframe and engine teams to the hundreds of major suppliers of systems and equipment, much of which had to be specially designed to meet new requirements.

Gradually the French accepted the idea of a transatlantic SST,

Aérospatiale/BAe Concorde cutaway drawing key

1 Variable geometry drooping nose
2 Weather radar
3 Spring pot
4 Visor jack
5 'A'-frame
6 Visor uplock
7 Visor guide rails and carriage
8 Droop nose jacks
9 Droop nose guide rails
10 Droop nose hinge
11 Rudder pedals
12 Captain's seat
13 Instrument panel shroud
14 Forward pressure bulkhead
15 Retracting visor
16 Multi-layer windscreen
17 Windscreen fluid rain clearance and wipers
18 Second pilot's seat
19 Roof panel
20 Flight-deck air duct
21 3rd crew member's seat
22 Control relay racks
23 1st supernumery's seat
24 2nd supernumery's folding seat (optional)
25 Radio and electronics racks (Channel 2)
26 Radio and electronics racks (Channel 1)
27 Plug-type forward passenger door
28 Slide/life-raft pack stowage
29 Cabin staff tip-up seat
30 Forward galley units (port and starboard)
31 Toilets (2)
32 Coats (crew and passengers)
33 Twelve 26-man life-rafts
34 VHF1 antenna
35 Overhead baggage racks (with doors)
36 Cabin furnishing (heat and sound insulated)
37 4-abreast one-class passenger accommodation
38 Seat rails
39 Metal-faced floor panels
40 Nose-wheel well
41 Nosewheel main doors
42 Nosewheel leg
43 Shock absorber
44 Twin nosewheels
45 Torque links
46 Steering mechanism
47 Telescopic strut
48 Lateral bracing struts
49 Nosewheel actuating jacks
50 Underfloor air-conditioning ducts
51 Nosewheel door actuator
52 Nosewheel secondary (aft) doors
53 Fuselage frame (single flange)
54 Machined window panel
55 Underfloor forward baggage compartment (6.72 m³/237 cu ft)
56 Fuel lines
57 Lattice ribs
58 No. 9 (port forward) trim tank
59 Single-web spar
60 No. 10 (port forward) trim tank
61 Middle passenger doors (port and starboard)
62 Cabin staff tip-up seat
63 Toilets
64 Emergency radio stowage
65 Provision for VHF3
66 Overhead baggage racks (with doors)
67 Cabin aft section
68 Fuselage frame
69 Tank vent galley
70 No. 1 forward collector tank
71 Lattice ribs
72 Engine-feed pumps
73 Accumulator
74 No. 5 fuel tank
75 Trim transfer gallery
76 Leading-edge machined ribs
77 Removable leading-edge sections
78 Expansion joints between sections
79 Contents unit
80 Inlet control valve
81 Transfer pumps
82 Flight-deck air duct
83 No. 8 fuselage tank
84 Vapour seal above tank
85 Pressure-floor curved membranes
86 Pre-stretched integrally machined wing skin panels
87 No. 8 wing tank
88 No. 4 forward collector tank
89 No. 10 starboard forward trim tank
90 No. 9 starboard forward trim tank
91 Quick-lock removable inspection panels
92 Spraymat leading-edge de-icing panels
93 Leading-edge anti-icing strip
94 Spar-box machined girder side piece
95 No 7 fuel tank
96 No. 7a fuel tank
97 Static dischargers
98 Elevon
99 Inter-elevon flexible joint nozzles/reverser buckets
100 Combined secondary nozzles/reverser buckets
101 Nozzle-mounting spigots
102 Cabin air delivery/distribution
103 Inspection panels
104 Cold air unit
105 Fuel-cooled heat exchanger
106 Fuel/hydraulic oil heat exchanger
107 Fire-suppression bottles
108 Main spar frame
109 Accumulator
110 No. 3 aft college tank
111 Control linkage
112 'Z'-section spot-welded stringers
113 Riser to distribution duct
114 Anti-surge bulkheads
115 No 6 (underfloor) fuel tank
116 Machined pressurised keel box
117 Fuselage frame
118 Double-flange frame/floor join
119 Machine pressure-floor support beams
120 Port undercarriage well
121 Mainwheel door
122 Fuselage/wing attachments
123 Main spar frame
124 Mainwheel retraction link
125 Mainwheel actuating jack
126 Cross beam
127 Forked link
128 Drag strut
129 Mainwheel leg
130 Shock absorber
131 Pitch dampers
132 Four-wheel main undercarriage
133 Bogie beam
134 Torque links
135 Intake boundary layer splitter
136 Honeycomb intake nose section
137 Spraymat intake lip de-icing
138 Ramp motor and gearbox
139 Forward ramp
140 Aft ramp
141 Inlet flap
142 Spill door actuator
143 Intake duct
144 Tank vent gallery
145 Engine front support links
146 Engine-mounting transverse equalizers
147 Oil tank
148 Primary heat exchanger
149 Secondary heat exchanger
150 Heat exchanger exhaust air
151 Rolls-Royce/SNECMA Olympus 593 Mk 610 turbojet
152 Outer wing fixing (340 high-tensile steel bolts)
153 Engine main mounting
154 Power control unit mounting
155 No. 5a fuel tank
156 Tank vent
157 Transfer pump
158 Port outer elevon control unit fairing
159 Static dischargers
160 Honeycomb elevon structure
161 Flexible joint
162 Port middle elevon control hinge/fairing
163 Power control unit twin output
164 Control rod linkage
165 Nacelle aft support link
166 Reverse-bucket actuating screw jack
167 Retracter silencer lobes ('spades')
168 Primary (inner) variable nozzle
169 Pneumatic nozzle actuators
170 Nozzle-mounting spigots
171 Port inner eleveon control hinge/fairing
172 Control rod linkage
173 Location of ram-air turbine (RAT) in production aircraft
174 Accumulator
175 Vent and pressurisation system
176 Forged wing/fuselage main frames
177 Ground-supply air-conditioning connection
178 Control mixing unit
179 Control rod (elevon) linkage
180 Aft galley unit
181 Rear emergency doors (port and starboard)
182 Wingroot fillet
183 Air-conditioning manual discharge valve
184 Automatic discharge/relief valve
185 First-aid oxygen cylinders
186 Rear baggage compartment (door to starboard)
187 Rear pressure bulkhead
188 Fin support frames
189 No. 11 aft trim tank
190 Machined centre posts
191 Shock absorber
192 Retractable tail bumper
193 Tail bumper door
194 Tank overflow and pressure relief lines
195 Tile cone bulkhead
196 Fuel jettison
197 Monergol-powered emergency power unit (pre-production aircraft only)
198 Tail cone
199 Rear navigation light
200 Rudder lower section
201 Servo control unit fairing (manual stand-by)
202 Fixed rudder stub
203 Multi-bolt fin-spar attachment

especially after a large meeting of possible airline customers had shown in 1963 that the feeling of the market was that the proposal was not bold enough. A powerful body of opinion held that the Anglo-French SST would be outmoded by the promised later American SST to carry some 250 passengers at Mach 3. It needed steady nerves to stick to the belief that 100 passengers at about Mach 2 was right. Named Concorde in 1963, the Anglo-French machine grew from 118 843 kg to 129 730 kg (262,000 lb to 286,000 lb), increasing the seating for full range from 90 to 100, but the engine team then redesigned the Olympus, as the Mk 593B, to give much greater power, and this allowed the Concorde to follow the pressures of the market and grow to 14 874 kg (326,000 lb) with 118 seats. In early 1965 design

Above: *Air France's routes to South America looked promising, lying as they did across the Atlantic and Sahara desert. Crippling fuel costs ruined the route despite the appeal and reliability of the service. Like British Airways, Air France now serves only North America.*

204 Fin construction
205 Fin spar
206 Air-conditioning ducting
207 HF antennae
208 Finroot fairing

209 Leading-edge structure
210 Servo unit threshold bellcrank
211 Servo control unit fairing
212 VOR antenna
213 Rudder upper section

was frozen, and construction of prototypes 001 and 002 began. In parallel a colossal research programme was funded, including a complete airframe thermal rig at Farnborough and the Handley Page H.P.115 and BAC.221 research aircraft.

Though part of the agreement was that there should be an assembly line in both countries, producing odd-numbered aircraft at Toulouse St Martin and even-numbered at Filton (Bristol), there was no duplication in actual manufacture. Thus BAC was assigned the nose, tail and engine installations, and Aérospatiale the wings, centre fuselage and landing gears. France actually had about 60 per cent of the airframe because the UK had most of the engine, while systems were shared more or less evenly and included a number of items from the USA.

Complex wing production

Aerodynamically the design is that of a tailless ogival delta. The wing has continuous subtle curvatures, with strong conical camber giving pronounced leading-edge droop outboard, but the basic thickness is extremely low, only 3 per cent inboard and 2.15 per cent outboard of the engines. Wing sections were made at Bouguenais, Toulouse, St Nazaire and Marignane, and the section outboard of the engines was made by Dassault at Bourges. Flight control is by six elevons, two of them inboard of the engines, each driven by a Dowty Boulton Paul tandem jack in the 281.2-kg/cm^2 (4,000-lb/sq in) hydraulic system. One of the drawbacks is that at take-off and landing these surfaces cannot be used to increase wing camber and thus lift and drag, but the problem of trimming out the change in centre of pressure (the point through which the resultant lift force acts) was solved very neatly without causing any drag. Most of the 119695 litres (26,330 Imp gal) of fuel is housed in integral tanks in the thin wing and under the passenger floor, but by using extra tanks at the extreme front of the wing and in the tail of the fuselage, it was found possible to shift the centre of gravity of the aircraft to match the shift in the centre of pressure. During transonic acceleration the contents of the forward tanks are pumped into the rear trim tank and main tankage. At the end of supersonic cruise the contents of the rear trim tank are pumped forward into the front trim and main tanks.

Ruling structural material is an aluminium alloy developed in the UK as RR.58 and produced in France under the designation AU2GN. The engines, however, are almost entirely of ferrous alloys, titanium alloys, Waspalloy or high-nickel alloys, and they are fed by extremely large ducts leading from fully variable sharp-edged inlets with electric anti-icing, and with front and rear variable upper wall ramps and controllable doors in the underside through which air can be admitted or expelled. In the course of development the engine was further increased in power, given a new jetpipe combining afterburner, variable nozzle and reverser, and also vaporising combustors which eliminated visible smoke. The four-wheel main gears fold inward and have Dunlop carbon/carbon brakes, the first service application of such brakes in the world and indicative of the unprecedented severity of the rejected take-off of the fully loaded aircraft. There is no braking parachute or airbrake, and leading edges are fixed. Systems, however, are advanced and complex, though the high hydraulic pressure had been used on the Bristol Britannia. The system marking the biggest jump in complexity was un-

British Airways' third Concorde sits on the tarmac at Melbourne in Australia. The airline and manufacturers spent much time and trouble proving the route to Australia, but met with such opposition from overflown countries and residents local to destination airports. The aircraft was banned from landing at Sydney's airport.

doubtedly the engine inlets, followed by environmental control, with the high cabin pressure-differential of $0.75\,kg/cm^2$ ($10.7\,lb/sq\,in$) and the fuel used as a heat-sink.

By 1966 major pieces of structure were on thermal/fatigue test, the engine was running with its variable exhaust system and the main flight simulator was in use. Prototype 001 was rolled out at Toulouse on 11 December 1967, but it was the following August before it taxied and the first flight was delayed until 2 March 1969, the pilot being André Turcat. No. 002 flew in command of Brian Trubshaw from Filton a month later. Airline pilots first flew the prototypes in November 1969, and from the first there were no major problems connected with the aircraft. The problems stemmed from protesters, who considered any SST a menace to the environment (apparently on grounds mainly of noise), and from soaring increase in development cost, half of which was the result of the childishly low initial estimate, which made no allowance for the progressive increases in size and capability of the aircraft, and half the result of inflation, for which no allowance at all had been made.

Contracts signed

By 1971 the prototypes were making long overseas trips, and in December that year the first pre-production aircraft, with visibly different vizor, longer forward fuselage and extended tail, made its first flight from Filton to the UK test base at Fairford. On 28 July 1972 British Airways signed with BAC for five aircraft and Air France signed with Aérospatiale for four. There had previously been options by Pan Am and purchase agreements by China, but these were never taken up. Prototype 001 was retired to a museum at Le Bourget in October 1973, two months before the first flight by the first production aircraft. By 1975 there was hardly a major city that had not been visited by at least one Concorde, and very intensive route proving was showing a remarkable reliability even at this early date. Scheduled services began on 21 January 1976, British Airways flying London-Bahrain with aircraft 206 and Air France flying Paris-Dakar-Rio with 205. Services to Washington were begun by both airlines on 24 May 1976. Prototype 002 was retired to Yeovilton on 4 March 1976 and 01 to Duxford on 20 August 1977. Category III autoland was cleared in passenger operation on 1 September 1978.

Though passenger operations could not have been more successful, both in terms of passenger appeal and reliability (to the extent that in 1981 Concorde was the most punctual type in British Airways service according to the monthly report to the airline board, averaging 94 per cent), political troubles and rising fuel costs crippled global plans for using the aircraft properly and had by late 1982 reduced utilisation to an extremely low figure. A joint British Airways/Braniff service from Washington on to Dallas was suspended in June 1980, and the British Airways service to Singapore followed five months later. In April 1982 Air France discontinued services to Caracas and Rio, and by summer 1982 the only scheduled services were (British Airways) twice daily to New York and three times weekly to Washington, and (Air France) 11 times a week to New York, of which two continue to Washington and two to Mexico City. A small amount of additional flying is made up by charters to business companies and enthusiast groups.

The main traffic routes are transatlantic, and by the mid-1980s they had brought British Airways some 1,300,000 passengers and Air France about 795,000, totalling just over two million. The aircraft has sustained reliability in excess of 93 per cent on sectors up to 6440 km (4,000 miles) in length, total service flight time being over 136,000 hours on more than 40,000 scheduled departures. Small changes, for example to the inlet lips and rudder trailing edge, have significantly improved operating economics, but profitability remained marginal. Both airlines then announced that they had entered an era of profitability with the type, and British Airways also looked at ways to increase its Concorde operations, including running a cargo service in partnership with Federal Express, open-

This remarkable formation was put up for Concorde's tenth anniversary in British Airways revenue-earning service. BA and Air France each operate seven, and the aircraft operate at a profit. However, this operating profit is more than offset by the huge research and development cost of the type, which could never be recouped.

ing passenger service to Miami and a nonstop route to Lagos, Nigeria.

At the government level much has been done to reduce costs by trimming out the vast research and support effort, which would be needed only if there were to be a successor. British Minister for Industry Norman Lamont held formal discussions with French Transport Minister Charles Fiterman in May 1982 at which the entire project was reviewed, the decision having been taken earlier that to cancel the project would cost far more than it would save. Ways were studied to share the remaining burden more evenly, but though both British Aerospace and Aérospatiale have studied second-generation aircraft, with proven lift/drag ratio of 10 compared with seven for Concorde, and with much better engines – and the French partner has even exhibited an impressive model – there is little chance of any go-ahead. But this reflects only the current mood. In the longer term vehicles which effectively shrink the planet are surely bound to come.

With the afterburners of its four Olympus jets glowing, a British Airways Concorde begins another journey, leaving behind the yellow-tinged smoke trail unique to the type. In addition to its regular services to New York and Washington, Concorde is regularly used for charters and one-off 'specials' that feature other activities, such as New York theatre visits or Nile boat rides.

Chapter 15

Conroy/Aérospatiale Guppy

Convair Twins CV 240/640 Series

Convair 880/890

Conroy/ Aero Spacelines GUPPY

In the league of aviation oddities, the Guppy family stands near the top of the first division. Seemingly defying the laws of aerodynamics, the swollen-bodied aircraft have proved eminently successful over the years, transporting rocket sections and airliner components between construction and assembly sites.

The late John M. 'Jack' Conroy was a former United States Army Air Force bomber and transport pilot. In the 1950s Conroy was flying for a commercial airfreight company and also serving with the Air National Guard flying Boeing C-97 Stratofreighters, the military transport variant of the Stratocruiser airliner.

At that time large numbers of Stratocruisers were being retired from airline service with British Overseas Airways, Pan American World Airways and Northwest Airlines following the introduction of turboprop and early jet airliners, and many of these were purchased by aircraft broker Lee Mansdorf and stored at Van Nuys, California, where Conroy lived. The Stratocruisers were available for sale at low cost, and the obvious potential of the aircraft's immense double-deck fuselage for the carriage of freight was not lost on Conroy. Simultaneously the National Aeronautics and Space Administration, NASA, was encountering problems with transporting massive stage sections of Saturn rockets for the Apollo space programme from manufacturers in California to test sites in Louisiana and Mississippi, and to the launch complex at Cape Canaveral, Florida. So big were the Saturn rocket stages that they could not travel by road or rail, and had to be shipped by barge along a tortuous, costly and time-consuming sea route via the Panama Canal.

Conroy and Mansdorf believed that a Stratocruiser could be modified to incorporate a large diameter freight hold capable of carrying outsize cargo such as NASA's rockets. Conroy resigned his airline job, mortgaged his home, borrowed US$15,000 and began work on his

Until quite recently Airbus Inter Transport had four Guppy 201s, this being F-GDSG, the first of the two UTA-modified aircraft. Airbus was so impressed with the type that it bought the modification rights.

projected conversion, which he presented in outline to NASA officials at Marshall Space Flight Center, Huntsville, Alabama, in 1960.

First, the Pregnant Guppy

NASA was sceptical of the idea, but Dr Werner von Braun offered the use of the Administration's wind tunnel facilities for detailed design studies which led to the first B377PG Pregnant Guppy (so named in jest by a NASA official on first seeing the drawings for the bloated aircraft) converted from a former PAA Stratocruiser. The aircraft retained the airliner's wings, engines, tail surfaces and cockpit section, with a fuselage increased in length by 5.08 m (16 ft 8 in). Above the lower deck level a bulbous upper section was added which provided an internal diameter of 6.02 m (19 ft 9 in) and a total volume of 826.5 m^3 (29,187 cu ft). The entire rear fuselage and tail section detached aft of the wing trailing edge for loading. With no hangar at Van Nuys Airport large enough to accommodate it, the prototype Pregnant Guppy was assembled in the open, and was first flown by Conroy, co-pilot Clay Lacy and flight engineer Bob D'Agostini on 19 September 1962. Despite the addition of some 2272 kg (5,000 lb) of structural weight, they found that the Pregnant Guppy handled little differently to a C-97 and was only 8 km/h (5 mph) slower. Feasibility studies commissioned from Boeing and Douglas by NASA had predicted that the additional drag of the Guppy's immense 'bubble top' would reduce its cruise speed by at least 64.4 km/h (40 mph) and make handling difficult, particularly in the event of engine failure. To prove them wrong Conroy shut down two engines when Dr von Braun was aboard the prototype; the rocket scientist failed to notice until the feathered propellers were pointed out to him.

The Aero Spacelines B-377MG Mini Guppy retained the piston engines and tail unit of the original Stratocruiser, and incorporated a less bulged fuselage than other members of the family. This single example served with Aero Union of Chico, California, alongside a fleet of Douglas DC-4s.

Once certificated by the Federal Aviation Agency, the Pregnant Guppy was quickly pressed into service by NASA in the summer of 1963. The US Department of Defense awarded Conroy's newly-created Aero Spacelines Corporation an exclusive operating contract for transporting Saturn rocket sections, on the strength of which Conroy bought 25 surplus Stratocruisers and C-97s to cannibalise.

Serving the space programme

It quickly became apparent that the Pregnant Guppy was not large enough to meet all of NASA's needs. A second aircraft was modified, using parts from four C-97 airframes. This time the fuselage length was increased by 9.4 m (30 ft 10 in), wingspan by 4.57 m (15 ft) by means of a centre section 'plug', and the internal height of the cargo hold increased to 7.77 m (25 ft 6 in) and width to just 0.015 m (6 in) less. The area of the fin and rudder was increased and the left-hand side of the forward fuselage hinged so that the entire nose could be swung sideways for loading. Four 5220-kW (7,000-shp) Pratt & Whitney T34-P-7WA turboprop engines replaced the C-97's standard Pratt & Whitney R-4360-B6 piston engines. The B377SG Super Guppy made its first flight on 31 August 1965. With a 33.17 m (108 ft 10 in) long, 1,410 m^3 (49,790 cu ft) capacity cargo hold it was then the world's largest aircraft in terms of volume, and the only aeroplane capable of carrying the massive third stage of the Saturn V launch vehicle.

Between them the Pregnant Guppy and Super Guppy carried 85 per cent of the Saturn and Apollo programmes' hardware, flying 11 of the 13 Apollo launchers, and command, service and lunar modules. In 1978 NASA purchased the Super Guppy for logistical support duties on the Space Shuttle programme. A planned six-engined Guppy was never built, nor was the 'Colossal Guppy' which would have been based on a Boeing B-52 Stratofortress bomber with a 12.1 m (40 ft) diameter cargo hold and a total payload of 90720 kg (200,000 lb).

The giant bluff nose profile of the Guppy dwarfs the occupants of the flight deck. The extensive glazing and undernose radar fairing show signs of the aircraft's Boeing 377 Stratocruiser ancestry.

The first Aero Spacelines product was the B-377PG Pregnant Guppy. A 5.08 m (16 ft 8 in) plug aft of the wing lengthened the fuselage, while the large upper fuselage bubble allowed the carriage of items over 6.02 m (19 ft 9 in) in diameter. It first flew on 19 September 1962 before flying for NASA.

NASA also considered a Conroy proposal for a craft called Virtus which would have served as a test launch vehicle and transporter for the Space Shuttle. Conroy proposed using two B-52 fuselages mated to a straight wing and high tail. The Shuttle Orbiter would have been carried beneath Virtus for air launching or ferrying from the landing site at Edwards Air Force Base in California to the launch site at the Kennedy Space Center in Florida, and a 10.6 m (35 ft) diameter cargo hold would also have enabled Virtus to carry outsize items of Shuttle equipment such as its external fuel tank. In the event, NASA chose to use a converted Boeing 747 as the Space Shuttle 'mother ship'.

Following completion of the Super Guppy, Jack Conroy moved Aero Spacelines from Van Nuys to Santa Barbara where he built another Pregnant Guppy with auxiliary turbojet engines in underwing pods, and created the B377MG Mini Guppy, which had a shorter fuselage than the Super Guppy, an internal cargo hold diameter of 4.70 m (15 ft 5 in), a swing tail for loading, and reverted to piston-engine power. It first flew on 24 May 1967.

The definitive Guppy

The Mini Guppy was succeeded by the Guppy 201, first flown on 24 August 1970, followed by a second example of the type exactly a year later. The Guppy 201 was dimensionally similar to the Super Guppy, with a 43.84 m (143 ft 10 in) long fuselage, but was powered by four 3663-kW (4,912-shp) Allison 501-D22C turboprops, and was effectively the 'production' version of the earlier model, though still created from surplus Boeing C-97 airframes. So was the smaller Guppy 101, also powered by Allison 501 engines, the sole example of which was first flown on 13 March 1970. With the original Mini Guppy it was sold to Allen Paulson, then president of American Jet Industries, for use in ferrying salvaged aircraft. The Guppy 101 was later employed by Mr Paulson's Gulfstream American (now Gulfstream Aerospace) operations for carrying aircraft components between the company's facilities in Georgia, Oklahoma and California.

While work was proceeding on the Guppy 201 and 101, Jack Conroy devised one further 'oversize' cargo aircraft conversion, based on a former Flying Tiger Line Canadair CL-44-D4 cargo aircraft. The basic CL-44 airframe was retained, but immediately aft of the flight deck the entire upper fuselage above floor level was removed and replaced with an enlarged circular section pressurised 'bubble top'. Though much less voluminous than the Guppy, the CL-44 conversion provided a cargo hold with a constant diameter of 4.24 m (13 ft 11 in) and retained the Canadair aircraft's swing tail loading facility. Redesignated CL-44-0, the Conroy conversion was first flown from Santa Barbara on 26 November 1969 by the famous Lockheed company test pilot Herman 'Fish' Salmon, and was used to transport Rolls-Royce RB.211 engines and pods from England to California for Lockheed L-1011 TriStar production. This unique aircraft still survives and is now operated by the London–Stansted Airport-based outsize air freight specialists Heavylift Air Cargo.

The two Guppy 201s built by Aero Spacelines were delivered to Airbus Industrie at Toulouse in 1971 and 1972 for use in support of the European Airbus programme. So successful were they in this role that Airbus Industrie acquired production rights and technical material from Tracor Aviation – to which Jack Conroy had sold the Guppy project on his retirement from Aero Spacelines – and commissioned UTA Industries at Le Bourget Airport, Paris, to manufacture two more Guppy 201s, which were delivered in 1982 and 1983.

The aeroplane's aeroplane

The four Guppies of the Airbus Skylink were used to bring together components of the Airbus family of airliners. They were operated until 1989 for Airbus Industrie by UTA subsidiary Compagnie Aéromaritime d'Affrêtement SA, providing an around-the-clock schedule shuttling completed wing, fuselage, tail and other Airbus components from Airbus partners

Seen shortly after it arrived in France, F-BPPA was the second Guppy 201, previously registered N212AS. The first, N211AS, became F-BTGV after delivery to Aeromaritime. The Aero Spacelines title is still visible on the bulged fuselage.

Until 1989 when Airbus Industrie took over the direct management of the aircraft, the four Guppy 201s were operated by Aeromaritime, a subsidiary of UTA and based at Puteaux, although operated from Toulouse. They were used to transport large Airbus airframe components from consortium members' factories to the assembly line.

The giant fuselage cross-section of the Guppy 201 enabled it to carry bulky as well as heavy freight. The Airbus aircraft entered service in 1972, after which they became vital to the multi-national programme. The hold could accommodate items 7.77 (25 ft 6 in) in height.

The entire nose section of the Guppy 201 swung through 110° to admit the cargo direct to the hold. Here the second Airbus A320 forward fuselage is delivered from MBB's Hamburg plant to Toulouse. Note the flat floor area at the bottom of the hold.

Aérospatiale in Nantes and Saint Nazare, Belairbus in Belgium, British Aerospace in Bristol and Chester, CASA in Madrid, Fokker in Amsterdam and Messerschmitt-Bölkow-Blohm in Bremen and Hamburg to the final assembly plant at Toulouse. They were then operated directly by Airbus. With their swing-out

Airbus Industrie's four aircraft were the Guppy 210 variant, the largest of the family. These feature 3663-ekW (4912-shp) Allison 501-D22C turboprops and increased wingspan. The enlarged fuselage has a cargo capacity of some 24006 kg (52,925lb).

nose section, which opens through 110 degrees via a self-contained powered wheel, 7.77 m (25 ft 6 in) diameter cargo hold and certificated gross weight of 77110 kg (170,000 lb), the Airbus Skylink Guppies can easily accommodate the heaviest regular load – a pair of British-made A310 wing boxes weighing 50909 kg (50 tons). Each manufacturer has custom-made transporters/loaders to ensure precise loading of assemblies which are often a tight fit even in the cavernous Guppy.

Who could have imagined that an aircraft based on an early post-war piston-engined airliner would be playing such a vital role in the creation of high-technology fly-by-wire jet transports nearly half a century later?

Specification
Aero Spacelines Guppy 201
Type: outsize cargo-carrying aircraft
Powerplant: four Allison 501-D22C turboprops, each rated at 3663 kW (4,912 shp)
Performance: cruising speed at 7620 m (25,000 ft) 407 km/h (253 mph); maximum cruising speed 463 km/h (288 mph); initial rate of climb 457 m (1,500 ft) per minute; service ceiling 7620 m (25,000 ft); range with maximum payload with IFR reserves 813 km (505 miles)
Weights: empty 45359 kg (100,000 lb); maximum take-off 77110 kg (170,000 lb); maximum payload 24494 kg (54,000 lb)
Dimensions: span 47.62 m (156 ft 3 in); length 43.84 m (143 ft 6 in); height 14.78 m (48 ft 6 in); wing area 182.52 m² (1,965 sq ft)

Convair Twins CV 240/640 Series

One of the most successful, long-lived and attractive 'families' of American aircraft has been the Convair twins. Initial designs were drawn up at the end of World War II and, while the first CV 240 was a good aircraft, it was soon succeeded by new versions in a production run that lasted many years.

In early 1945 American Airlines issued a specification for a new twin-engined airliner required as a post-war replacement for its fleet of Douglas DC-3s. Consolidated Vultee Aircraft Inc (Convair) began work on a prototype, known as the Convair Model 110, which featured an unpressurised 30-passenger cabin, low wings and two 1342-kW (2,100-hp) Pratt & Whitney R-2800-SC13G Double Wasp radial engines mounted in nacelles incorporating twin overwing 'augmentor' exhaust stacks in which cooling air and exhaust gases were mixed and discharged through a venturi to provide 'aspirated cooling' and a small measure of additional thrust. This remained a common feature of all subsequent piston-engined Convair twins, and gave them a distinctive rasping exhaust note. The Model 110 incorporated integral airstairs in the underside of the rear fuselage and, unusually for airliners of the day, an entry door forward of the wing on the port side.

The sole prototype Model 110 NX90653 made its first flight from Convair's plant at San Diego on 8 July 1946, flown by company test pilots Russ Rogers and Art Bussy. But even before it had flown market studies had shown that airlines wanted a larger aircraft, and design was already under way on a 40-seat pressurised development designated Model 240 (for 'two engines, 40 passengers') and named Convair Liner, for which American Airlines placed a launch order for 100 aircraft, at an initial price of $316,000 each. Other early customers for the Model 240 included Pan American, United Airlines, Western Air Lines, Ethiopian Airlines, Garuda Indonesian Airlines, KLM Royal Dutch Airlines and Trans-Australia Airlines.

The prototype Model 110 was assigned to development flying for the Model 240, which featured a fuselage 1.12 m (3 ft 8 in) longer and 10 cm (4 in) less in diameter than that of the Model 110, accommodating up to 10 rows of seats in four-abreast configuration with a central aisle in a pressurised cabin – the first such to be incorporated in a twin-engined commercial transport aircraft. The Model 240's windows were square, rather than the round 'porthole' type of the Model 110. The airframe was of all-metal construction apart from some glassfibre used in non-structural fairings, and 1714-kW (2,300-hp) Pratt & Whitney R-2800-CA15 or water-injected 1788-kW (2,400-hp) -CA18 engines were chosen to power the new model, driving three-bladed Hamilton Standard or Curtiss Electric 'paddle blade' propellers. Engine servicing was facilitated by large 'orange peel' cowlings that hinged up or down to give access to the entire powerplant installation. Fuel was contained in two 1890-litre (500-US gal) tanks outboard of the engine nacelles. Four door configurations were offered, featuring combinations of front/rear, port/starboard passenger/cargo doors with optional ventral airstair.

This aircraft is a Rolls-Royce Dart-powered Convair 600. This designation was applied to modified 240s, the Dart being fully compatible with the short fuselage design.

This beautiful study depicts a Convair 240 on a test flight before delivery to the Central Air Transport Corporation of China. The aircraft were delivered via Europe to Hong Kong, and were used for regional services in the south of China and to Taiwan.

The most notable improvement introduced by the Model 340 was an increase in fuselage length to permit more seats to be fitted, and corresponding increases in wing span and power to cope with increased high altitude performance demands.

American Airlines launched the Convair 240 with an order for 100 aircraft, and was a major operator of the type as a replacement for DC-3s on regional routes. The aircraft were all named in the airline's 'Flagship' series.

There was no Model 240 prototype as such. The first aircraft, N90849, was built on production jigs and tooling and made its first flight on 16 March 1947 in the hands of Sam Shannon and Russ Rogers. Following a 16-week test programme the aircraft was approved for operation at a maximum take-off of 18900 kg (40,500 lb). First customer delivery was made on 23 February 1948 to American Airlines, with whom the Model 240 entered revenue service on 1 June. Production was brisk. The 100th Convair 240 was delivered to KLM in January 1949. In 1950 approval was granted for an increase in gross weight to 19000 kg (41,790 lb) and fuel capacity was increased to 5680 litres (1,500 US gal) by the addition of tanks in the outer wing panels. The first production aircraft to this standard were delivered to Garuda and Ethiopian Airlines, the latter's aircraft also being fitted with Aerojet 14AS-1000 Jet Assisted Take-Off (JATO) units under the fuselage for use from 'hot-and-high' airports on the carrier's African routes.

One hundred and seventy-six commercial Model 240s were built, five of which were retained for testing or sold to private operators, while the remainder entered airline service. Apart from those carriers already noted, other buyers of new Convair 240s included Continental and Northeast Airlines, Civil Air Transport Corporation of China, FAMA of Argentina, Orient Airways of Pakistan, SABENA and Swissair. Further aircraft were built for the United States Air Force: 48 unpressurised T-29A, 105 T-29B and 119 T-29C 'flying classrooms' for navigator training, 92 T-29D bombardier trainers, and 26 C-131A Samaritan casualty evacuation aircraft and VC-131A VIP transports, bringing total production of the Model 240 to 566 aircraft.

Towards the end of the Model 240 production run Convair considered terminating the entire project, since sales were slow. However, the success of the rival Martin 4-0-4 prompted a redesign of the aircraft, initially known as the Model 240A but announced on 28 November 1950 as the Model 340. Launch customer was United Airlines, which ordered 30. The Model 340 featured a 1.37 m (4 ft 6 in) increase in fuselage length to allow an extra row of four seats to be installed, wingspan was increased to 32.1 m (105 ft 4 in) to improve high altitude performance, flaps redesigned in two sections, more powerful 1860-kW (2,500-hp) P & W R-2800-CB16/17 engines installed in

Five Convair 580s are still on the books of the Federal Aviation Administration for flight inspection duties, serving in a mixed fleet which centres on Sabreliners, King Airs and newly-delivered BAe 125-800s (briefly operated by the USAF as C-29As).

The NC-131H was a heavily-modified aircraft with Allison engines. Operated by the Cornell Aeronautical laboratory, it simulated the flying characteristics of various large aircraft, flown from the additional cockpit in the nose.

nacelles that were 18 cm (7 in) longer than those of the Model 240 and driving larger Hamilton Standard Hydromatic propellers that were interchangeable with those used by United's four-engined Douglas DC-6s, and an additional 2840 litres (750 US gal) of fuel capacity provided. Other modifications included longer undercarriage legs, an improved pressurisation system, and standardisation of all passenger doors and the integral airstairs on the port side of the aircraft to ensure compatibility with airport docking facilities.

Widespread use

The prototype Convair 340, based on a T-29A airframe, made its first flight on 5 October 1951, again flown by Sam Shannon. Certification was granted on 27 March 1952, with first delivery to United four days later and first revenue service flown on 16 November. Among other customers for the Model 340 were Braniff, Continental, Delta, Hawaiian and National in the United States, Aeronaves de Mexico, Alitalia, All Nippon of Japan, Ansett of Australia, AVENSA of Venezuela, Cruzeiro do Sul of Brazil, Finnair, Garuda Indonesian, JAT of Yugoslavia, KLM, LACSA of Costa Rica, Lufthansa, Philippine Air Lines and Saudi Arabian Airlines. A total of 209 civilian Model 240s were built, including 12 sold to corporate operators as executive transports. Like the Model 240, the 340 also attracted military orders: 36 C-131B 48-seat transports and 33 C-131D and VC-131D 44-seat transports for the USAF and 37 R4Y-1 passenger/cargo/medevac transports for the US Navy.

In 1954 Convair announced development of the Model 340B, intended as a competitor for the turboprop Vickers Viscount. A testbed aircraft, flown on 6 October 1955 by Sam Shannon, incorporated the modifications planned for the new model, which was soon renamed Model 440 Metropolitan. It featured nose-mounted weather radar with 240-km (150-mile) range (optional but specified by nearly all customers), the radome increasing overall length 71 cm (28 in) although cabin length remained as on the 340, but with the option of 52 seats with an extra cabin window on each side. Other improvements included better cabin soundproofing, more streamlined engine nacelles and a new thrust augmentation system in which the twin exhaust pipes of earlier models were replaced by a single rectangular outlet at the rear of the wing trailing edge. Maximum take-off weight was increased to 22270 kg (49,100 lb), and later approved up to 22540 kg (49,700 lb).

The first production Convair 440 flew on 15 December 1955. Certification was quickly obtained on 30 January 1956, when the first customer delivery took place to the Swiflite Corporation (Cities Services Oil) for use as a corporate transport. Continental Airlines inaugurated airline services with the Model 440 on 8 March 1956. Other US airline customers for the Metropolitan were Braniff, Delta, Eastern, Hawaiian, Mohawk and National, but almost half the 153 Model 440s built for airlines went to Europe, where they were operated by Alitalia, Condor-Flugdienst, Finnair, Iberia, JAT, Kar-Air of Finland, KLM, Lufthansa, Sabena, Scandinavian Airline System and Swissair. Military customers included the Australian Air Force and German Luftwaffe, with two each, the Italian Air Force (one) and the USAF and US Navy (C-131D/E and R4Y-2 respectively). These, and aircraft sold for private use, brought Model 440 production to 199 aircraft.

Convair also offered upgrade kits to bring earlier Model 340s up to 440 standard, and some 100 aircraft were so converted.

The Convair Liner series had been designed from the outset to be suitable for turboprop power, and in 1950 the Allison Division of General Motors Corporation re-engined the first production Model 240 with two 2050-kW (2,750-shp) Allison 501-A4 (T38A) turboprops. Known as the Model 240-21 Turbo-Liner, this aircraft was the first American turboprop transport to fly, but development of the engine for commercial use was not proceeded with and no production aircraft were built. The one-off Turbo-Liner was restored to standard piston-engined configuration.

The first commercial turboprop conversion of the Convair Liner was completed by the British firm D. Napier and Son Ltd of Luton in 1955, when it installed two 2280-kW (3,060-shp) Napier Eland N. E1.1s in a Convair 340. This aircraft, registered G-ANVP, made its first flight on turboprop power on 9 February 1955, and proved capable of cruising at 500 km/h (315 mph) at 4870 m (16,000 ft). Certification of the 'Napier Eland Convair Series 2' was obtained on 21 October 1957 and PacAero Engineering Corporation of Santa Monica, California, was appointed by Napier to continue development of a 2542-kW (3,412-shp) N. E1.6 Eland 504 engine conversion using the testbed aircraft. This modification was certificated on 22 August 1958. In December of that year a second conversion was completed on a Convair 440 airframe, and on 1 July 1959 this aircraft entered service with Allegheny Airlines on a three-month lease, the first turboprop aircraft to fly with a local service carrier in the United States. Allegheny ordered five Eland conversions of aircraft in its fleet, these receiving the unofficial name Convair Model 540 Cosmopolitan.

'Canforce' Convairs

When production of the piston-engined Model 440 ceased Convair transferred all jigs and tooling to Canadair Ltd in Montreal, Canada, in anticipation of new production aircraft being built there with Eland turboprops. In the event this did not happen immediately, but Canadair did convert three unsold 440s to 540 standard as CL-66Cs, the first flying on 2 February 1959 and remaining with the manufacturer as a demonstrator. The other two were leased to Quebecair. Canadair subsequently received an order for 10 new-build CL-66Bs from the Royal Canadian Air Force. These were delivered to No. 412 Squadron between July 1960 and March 1961 under the RCAF designation CC-109 Cosmopolitan. Following the take-over of Napier by Rolls-Royce in 1962 work on Eland conversions of Convairs ceased, and all but the RCAF aircraft reverted to piston-engined configuration.

Meanwhile, in 1957, Allison had performed a second Convair Liner turboprop conversion, installing 2790-kW (3,750-shp) Allison 512-D13 engines in a military YC-131C airframe which had been loaned by the USAF. Successful completion of a 1,000 flight hour test programme known as Operation Hourglass resulted in the 1958 launch of a conversion programme for the Allison 'Prop Jet Super Convair', based on Model 340 and 440 airframes (the Model 240 was not structurally suitable).

PacAero was once again chosen as prime contractor for performing the modifications. Apart from the engine installation, these included increasing tailplane span by 1 m (3 ft 4 in), fin/rudder height 30 cm (1 ft), raising fuel capacity to 11000 litres (2,908 US gal), revising the cockpit layout, providing accommodation for up to 56 passengers, and installing additional cabin soundproofing. The first PacAero-modified 'Super Convair' flew from Burbank near Los Angeles on 19 January 1960 and received FAA certification on 21 April. Initial deliveries were all to corporate operators, the first going to Allison's parent company General Motors in May. Frontier Airlines was the first commercial carrier to take delivery of the modified aircraft, which it dubbed 'Convair 580', though this name was never officially adopted by the manufacturer. Frontier began operating its Allison-powered Convairs in June 1964 and was followed by Allegheny, AVENSA, Lake Central and North Central Airlines. The USAF also ordered four conversions of VC-131Ds operated by the VIP transport

When powered by Darts, the 340/440 was redesignated the Model 640. Caribair had seven 340s modified, and flew the first revenue-earning service of the 640. These aircraft were unique in retaining JATO capability.

The Model 440 with extended nose radome for weather radar was especially popular in Europe where many of the major airlines purchased the type for inter-city routes. SABENA, an earlier purchaser of the 240, bought 12 for services from July 1956.

New life was breathed into the Convair twin family with the adoption of turboprop power. After the Eland-powered 540 came the Allison-powered 580, which was the most successful of the turbine conversions. Prinair operated the type in the Caribbean from Puerto Rico into the late 1980s.

Thirty-six Model 340s were bought for the US Air Force under the C-131B designation. They were used as transports for many years, although six were diverted for temporary tests with missile tracking equipment.

unit 89th MAW at Andrews AFB, Maryland, and a fifth C-131 was later converted as the NC-131H Total In-Flight Simulator (TIFS), which remained in use until 1986, performing valuable service in developing advanced flight control systems for new transport aircraft and for the Space Shuttle. Eight Eland-powered RCAF CC-106s were also retrofitted with 3020-kW (4,050-shp) Allison 501-D36 engines in 1967. In all 175 Convair Liners were converted to Model 580 configuration before the programme ceased in July 1969, and many remain in service with cargo carriers, third level airlines and charter and corporate operators.

In 1964 a third turboprop engine modification was launched by Convair, based on the 2250-kW (3,025-shp) Rolls-Royce RDa10/1 Dart 542-4 engine developed for the Japanese NAMC YS-11 airliner. Unlike the Allison powerplant, this engine was deemed suitable for installation in any of the piston-engined Convair Liners, converted aircraft initially receiving the designations 240D (for Dart), 340D and 440D. These were later changed to Model 600 (for converted 240s) and Model 640 (340s

and 440s).

The first Model 600 made its maiden flight from San Diego on 20 May 1965, and was followed by the first Model 640 on 20 August. Certification of both types was received at the end of 1965. The Dart conversion was relatively straightforward, involving few modifications aft of the engine firewalls, and was therefore offered as a factory modification or as a retrofit kit for local installation by the operator's own engineers. Some structural strengthening was necessary on the Model 240/600, and realignment of the nacelle centrelines was required to provide adequate clearance for the Darts' 4-m (13-ft) diameter Dowty-Rotol four-bladed propellers, but even much of the original engine cowling structure was able to be fitted around the Rolls-Royce turboprops without modification. Optional equipment included an AirResearch auxiliary power unit in the starboard engine nacelle, dual air compressors for enhanced cabin pressurisation at the higher altitudes flown by turboprops, a forward cargo door, and increased fuel capacities of 7570 litres (2,000 US gal) for the Model 600 and 11150 litres (2,945 US gal) for the Model 640.

Distinctive Darts

Launch customers for the Model 600 were Central Airlines, whose first of 10 Convair-converted aircraft was delivered in September 1965 and entered service on 30 November, and Trans-Texas Airways, which converted 25 of its fleet using kits supplied by Convair. The first Model 640 order came from Hawaiian Airlines, which introduced the first of its fleet of eight on 23 December 1965, a day after the first 640 service had been operated by Caribair, which had its seven-strong fleet of Model 340s converted but uniquely retained the JATO installation deemed necessary on the piston-powered aircraft. Other customers for Dart conversions included Air Algerie of Algeria, Martin Air Transport of the Netherlands and SATA of Switzerland. A total of 39 Convair 600s and 31 Convairs 640s was produced. Five ex-Hawaiian Airlines aircraft were subsequently modified by American Jet Industries to 640F (Freighter) configuration with heavy duty cabin floors and 274 × 185 cm (108 × 73 in) cargo doors on the aft port fuselage side, for operation by Zantop International. Similar cargo versions of other turboprop Convairs have also been developed, the type proving popular with small load haulers in its twilight days.

The latest chapter in the protracted and highly successful story of the Convair Liner began in 1984 when Allison, creators of the Model 580 'Super Convair', launched an updated version powered by two 2990-kW (4,000-shp) Allison 501-D22G Series II engines. These, it was claimed, would offer an increase in cruising speed to 600 km/h (374 mph), enhanced engine-out performance, a 40 per cent increase in hot-and-high payload capacity and a 40 per cent reduction in operating costs through the greater fuel efficiency of the new-generation engines. Known as the Super 580, and marketed by the Super 580 Aircraft Company of Carlsbad, California, this modification also includes a redesigned flight deck with optional 'all-glass' cockpit featuring electronic flight instrumentation system (EFIS) displays, upgraded cabin interior and extensive systems improvements. The first Super 580 conversion was performed by Hamilton Aviation at Tuscon, Arizona, and flown for the first time in March 1984, but no details have been released of further conversions. A stretched version with accommodation for up to 78 passengers in modified Convair 340, 440, 580 or 640 airframes has also been proposed as the super 580ST, for which tooling design has been completed, but no conversions have yet been made.

The T-29A was a version of the Convair 240 produced for the US Air Force to act as a navigation trainer. After their training duties were over many were reconfigured for transport or VIP carriage. This aircraft is seen in CT-29A configuration, with three-tone tactical camouflage.

For general transport and air ambulance duties, the US Navy bought the Model 340, designated the R4Y-1. These had the Model 440-style weather radar nose, and were redesignated C-131F in 1962.

Convair 880/990: The Broken Dream

Elegant and fast, the Convair jetliners will be remembered primarily for the considerable amount of money they lost their manufacturer. However, the aircraft themselves were well-made and efficient, and their speed made them the hottest of the jetliners until the advent of Concorde. Nevertheless, they were too small to make economic sense compared to the rival Boeing and Douglas designs, which sold so well they swept the Convairs from the market.

During the 1950s Convair ranked as one of America's aerospace giants. With a successful range of twin-engined airliners, two successive front-line interceptors (the F-102 Delta Dagger and the F-106 Delta Dart), the gigantic intercontinental B-36 Peacemaker bomber and even more impressive Mach 2 B-58 Hustler, Convair seemed to have a finger in every aeronautical pie.

The success of Britain's de Havilland Comet soon showed that people were willing to pay for the speed and comfort offered by high-flying jets, and that the jet airliner would be the aircraft of the future. A series of Comet accidents shocked Britain's aerospace industry into inactivity, giving America a chance to catch up.

The first US jetliner was Boeing's Model 707, and this was quickly followed by the slightly larger, longer range Douglas DC-8. The air transport industry was booming and every operator seemed to be trading-in its piston-engined aircraft in favour of the new jets. Production of the 707 and DC-8 could hardly keep pace with demand, and it was inevitable that Convair would not ignore the opportunity of producing a prestigious and profitable product, and would build America's third jetliner.

In fact Convair's ill-fated jetliners were built in small numbers, and the project lost Convair some $450 million, in 1962 dollars! This was the largest loss made on a single manufacturing project, amounting to 10 times what the Bristol Brabazon cost the British government, and four times what Lockheed lost on the ill-fated Electra. It wasn't even as if the giant Convair Corporation could somehow afford this loss, as there was no taxpayers' money involved and the amount represented one quarter of the value of the company! And yet there

The first, third and fourth Convair 880s line up at the San Diego factory. The prototype is in the foreground, distinguished by the long instrumentation boom projecting from the nose.

Civil Air Transport of Taiwan received this, their only Convair 880, in June 1961. Its most remarkable feature was a colourful dragon motif on the nose. It was sold on to Cathay Pacific in 1968.

was little wrong with the Convair jetliners themselves, apart from their rather high seat/mile costs, and they won a well-deserved reputation for being the preferred choice of passengers and pilots alike. So what went wrong?

The eccentric Hughes

The man who suggested that Convair build a jet airliner was Howard Hughes, the controlling shareholder of TWA and owner of the Hughes Tool Company, the organisation from which TWA leased all its aircraft. Convair should perhaps have known better than to deal with the increasingly eccentric and reclusive Hughes, with whom it had conducted fruitless negotiations in 1950 concerning the sale of some Convair 240s and 340s.

These negotiations had usually been conducted in the early hours of the morning, with Convair representatives being picked up by car and whisked to 'secret' locations, such as the municipal rubbish dump at Palm Springs, for torchlit discussions with Hughes. These bizarre business meetings were frequently interrupted when Hughes decided that everyone present would be taken to watch a private screening of one of his movies. Unsurprisingly, these meetings resulted in no sales.

On 22 March 1955 Convair and TWA held a first meeting to dis-

Cathay Pacific became one of the major Convair 880 operators, eventually picking up nine from a variety of sources. All were Model 22Ms, the version with more powerful engines for short-field performance.

cuss the proposed Model 18 jetliner. Hughes himself had once been an aircraft designer, and he insisted on having some input into the design of the new aircraft, with disastrous consequences. The most damaging Hughes requirement was the insistance on five-abreast seating, which significantly reduced the capacity of the aircraft and condemned it to suffering appreciably higher seat/mile costs.

There was strong internal opposition from the military side of Convair, who pointed out that only a blunder of huge proportions by Douglas or Boeing would leave any market for the proposed jetliner. This advice was ignored, and dismissed as being the product of jealousy over the resources that might be devoted to a new civil project. There was a power vacuum at the top of Convair's parent, General Dynamics, caused by the fatal cancer of the charismatic and hard-headed Jay Hopkins and his replacement by the ineffectual Frank Pace. This allowed the Convair Civil Division's optimism to run wild, and in April 1956 the launch of the new jetliner, now known as the 'Skylark 600', was announced.

Within three months, the new Convair jetliner had been re-named again, becoming the Model 22 (which stuck) Golden Arrow (which didn't). In a June 1956 press release announcing orders for 30 of the new aircraft by TWA, and 10 by Delta, Convair proudly stated that: 'The new airplane will bring top-speed jet travel luxury to scores of cities where existing airports cannot accommodate any other jet transport . . . which will be able to get in and out of 5,000-ft runways, which are the common maximum in many airports and which are too short for the operation of any other commercial jet airliners.' It should have been obvious that virtually all these runways would be lengthened to allow the operation of the best-selling Boeing 707, and in any case, the Convair was never able to meet this requirement.

Gilding the lily

The June 1956 release went further, explaining the new name: 'The name Golden Arrow is derived not only from the airplane's commercially unmatched speed, but from its planned unique and sensational employment of metal color. The Golden Arrows are planned to be the first airplanes to include exterior metal which is shimmering gold in color rather than the conventional silver associated with airplanes over the years. This, it is believed, would not only greatly enhance the beauty of the airplane, but would afford it an unmistakeable stamp of recognition.'

Four months later, Convair had realised the production difficulties that anodising the aircraft skin would cause, and the plan for gold-coloured aircraft was quietly dropped, along with the name. It was too late to stop production of various anodised gold interior fittings, which found their way into every Convair jetliner built. The final name selected for the new Convair was the Convair 880, supposedly reflecting the designed cruising speed of 880ft per second (600 mph). By this time the proposed Pratt & Whitney J57 engines had been dropped in favour of a civil version of the military J79, which powered the Convair B-58 Hustler, designated CJ-805. As a matter of record, this was General Electric's first civilian turbojet engine, but remarkably it was one of the few parts of the troubled programme that went without a hitch.

At the very beginning of its life, the future of the Convair 880 looked bright. They'd sacked the accountant who'd pointed out that the value of bought-in equipment alone was more than the planned $4.5 million price tag, dismissing him as a mentally unstable crank! Two years later he was back on the payroll, but the first Convair 880s were each sold at a huge loss, with the prices not even covering the cost of the engines, radios and sub-contracted airframe parts. With such a low purchase price the Convair 880 looked very attractive, and soon KLM, United and American had expressed a strong interest in the aircraft, despite its foolish five-abreast seating.

This front view of a Swissair 990A Coronado shows to advantage the large 'carrot' fairings on the wing's trailing edge. These carried additional fuel and helped to 'area-rule' the aircraft, improving speed.

The flight test programme went well enough, with only one serious incident. This occurred when the first prototype, which had made its maiden flight on 27 January 1959, lost most of its rudder and a large portion of vertical fin during high-speed flutter testing. Despite this incident, the four prototypes completed the flight test programme within the timescale originally envisaged by Convair, and the FAA issued an unrestricted certificate of airworthiness on 1 May 1959.

TWA problems

The first cloud appeared on Convair's horizon during the flight test programme, when it became apparent that launch customer TWA actually couldn't finance its aircraft. Hughes refused to allow Convair to flight-test his first two aircraft and impounded them, locking them in a guarded hangar. Convair decided to stop work on the remaining 18 TWA aircraft in production, which were left in open storage, where they suffered water damage. Worse was to come when, with the help of the courts, TWA finally shook off control by the Hughes Tool Company and found the finance for all but 10 of its order for 30 Convair 880s.

Work on the 18 remaining aircraft re-commenced, but in the intervening months the paperwork relating to them had disappeared, and there was no record of what modifications had been incorporated in which aircraft. For example it was known that some of the aircraft had been re-wired, and that others need re-wiring. All 18 aircraft had to have their wiring replaced, at enormous cost. TWA finally received its first aircraft in January 1961, a year and a half late. Subsequent purchases took the TWA '880' fleet to 27 aircraft, including one Convair 880M, not far short of the original order.

Fortunately for Convair Delta had no such problems, and had quickly put the new machine into service. Delivery of the first aircraft from San Diego to Miami on 10 February 1960 established a new transcontinental speed record, the first of several records to fall to the type. Delta eventually took delivery of 17 880s. Unfortunately, Delta and TWA were to be the only mainstream US airlines to operate the 880. Capital's order for seven aircraft lapsed when financial problems led to a merger with United, whose own interest in the Convair evaporated when the Boeing 720 arrived on the scene. This was a short-range derivative of the Boeing 707, with a lightweight

This Convair 880 was purchased by the US Navy and designated UC-880. It performed a variety of trials and tests, including air-to-air refuelling. Convair 990s were later used by NASA for trials and Shuttle test work.

Swissair received a total of eight 990A Coronados, the first being delivered on 12 January 1962 and the last being retired in 1975. The aircraft was very popular with crews and passengers alike.

airframe and a new wing, and it was offered to United at a deliberately uneconomic price by Boeing, as a means of killing off the unwelcome competition in the jetliner market.

The first flight of the Boeing 720 in November 1959, and its subsequent slick marketing by Boeing, did not completely kill off the Convair 880, which picked up several important orders. After TWA took 20 of the original 30 aircraft ordered, the Hughes Tool Company actually picked up four aircraft, which were leased to a variety of carriers, including Northeast Airlines and the Nicaraguan carrier LANICA.

Overseas version

With production for domestic carriers totalling only 47 aircraft, Convair introduced a modified version for overseas customers, known as the 880M. Eighteen of these aircraft were built, including nine aircraft for Japan Air Lines, three for VIASA, two for Cathay Pacific, and one each for Taiwan's Civil Air Transport, Alaska Airlines and the FAA. Cathay Pacific later increased its fleet by purchasing VIASA's three aircraft, two aircraft from JAL, and one each from Alaska and Civil Air Transport.

Most of the original 880 operators retired their aircraft during the early 1970s, passing them on to a variety of smaller operators. Some

Part of the large TWA Convair 880 fleet seen in mothballs after the type was withdrawn from service. TWA eventually had 27 of the aircraft, and many were retired to sit in the desert at Mojave.

were converted to freighter configuration, and others flew with a variety of charter operators and travel clubs. Only a handful survive today, including one operated by the US Navy, and at least one in Central America. Large numbers rot in the desert sun at Mojave, ostensibly in storage, but unlikely to fly again. One of the most interesting 880s has actually been preserved in a museum. This is the 38th aircraft built, once used by Elvis Presley to travel around the country, and named *Lisa Marie*. The aircraft sits in the Graceland museum, still in immaculate condition.

Convair is estimated to have lost some $425 million on the 65 880s built. The aircraft was priced far too low, and its high operating costs led to few orders. The Boeing 707 and Douglas DC-8 were slightly slower, and appreciably noisier, but they were larger and carried more people further at a lower cost. Convair would have been well advised to learn the lesson and go back to making military aircraft, but it had one further expensive mistake to make, and this was the Convair 990.

Early interest in the 880 by American Airlines threatened to evaporate when the Boeing 720 arrived on the scene. Desperate to retain this important order, Convair offered to develop a new higher-capacity, longer-range version for American, with new turbofan engines (the first in a civil jetliner), a stretched fuselage, and greater wing area. Foolishly, Convair guaranteed its speed, noise, range and payload figures, and accepted American's condition that 25 tired old piston-engined DC-7Cs should be accepted in part payment, at twice their market value. Finally, Convair offered the new version at a price of only $4.7 million each, which it should have realised was less than the cost of building an additional 880.

In the 1950s it wasn't easy to predict performance from wind tunnel tests and theoretical calculations, but instead of taking the safe option of building a prototype, Convair committed itself to going straight into production. This meant, of course, that if the aircraft did not meet the performance specified all production aircraft would have to be recalled for modifications, or be modified on the line if they hadn't been delivered. Predictably enough, the new Convair did not meet its specified performance.

990 differences

The main features of the Model 30, later designated 990 to elevate it above the original 880, were a new engine, a stretched fuselage, and a new thinner section wing. The new powerplant was a turbofan version of the CJ-805, with an aft-mounted fan, which promised 40 per cent more power, 10-15 per cent less fuel burn, and higher reliability. In order to improve performance the Model 30 was area-

ruled to minimise transonic drag. This was achieved by adding conical anti-shock fairings above the trailing-edge of the wing. These were dubbed 'speed fairings' by Convair, but were more graphically known as 'Küchemann carrots' after the aerodynamicist who invented them.

Despite the performance promised by the new Convair, the order book remained disappointingly empty. American remained the only major customer, with 25 firm orders, with REAL of Brazil ordering three, SAS two, and Swissair seven. The first 990 rolled out at San Diego on 23 November 1960, and made its maiden flight on 24 January 1961. Things soon started to go wrong. When the aircraft entered the high-speed part of its flight test programme serious drag problems were encountered, limiting top speed to 580 mph (40 mph less than the guaranteed cruising speed) and threatening to reduce range to a point where non-stop coast-to-coast operation would be impossible. The outboard engines vibrated badly, and the elevators lacked effectiveness. Modifications to the pylons that supported the engine nacelles soon cured the vibration problems, but the excessive drag was more serious.

Modification agreements

American Airlines was desperate to increase its passenger capacity, and to switch its order from the 990 to another manufacturer would have meant huge delays. Accordingly, the airline agreed to buy 20 of its 990s, 15 of them unmodified, and with a guaranteed cruise of only 584 mph. This scuppered American's plans to operate the 990 in all-first-class configuration, with passengers paying a premium for speed, comfort and the low cabin noise. In turn, Convair undertook to produce a modification package that would restore the original 620-mph cruising speed within one year of certification. Similar agreements were reached with Swissair and with VARIG (which had absorbed REAL), but SAS cancelled its order.

A modification programme was worked out, with new slats and Krueger flaps, new engine nozzle fairings, and new wing root fillets. The $30 million programme was paid for by Convair, and involved 15,000 man-hours per aircraft. The 990 entered service with

After their service with American Airlines was over, some of the 990 'Astrojets' passed on to Middle East Airlines. Two were destroyed during a 1968 Israeli commando raid on Beirut airport.

American in March 1962, as the Astrojet, and the first modified aircraft arrived in 1963. The last five aircraft of the American order were built, and sold to Garuda, and Aerolineas Peruana.

Charter operations

The 990s had a brief life with their original operators, and like the 880s were quickly passed on to charter operators and travel clubs. Spantax, a Spanish charter airline, eventually received 14 990s, with a maximum fleet of 12 aircraft in service at once. The last of these made its last flight in 1987, apparently bringing the career of the 990 to an end. This was not to be. In October 1987, a former 'Denver Ports of Call' aircraft was taken out of storage at Marana and painted up in 'Ciskei International Airways' markings. It was delivered to South Africa, where it re-entered service in February 1988. Finally, another 990 was resurrected by NASA to serve as a testbed for the Space Shuttle undercarriage, with the new undercarriage being installed on the centreline for high impact landing tests.

This 990 was one of the aircraft originally ordered for American Airlines but cancelled due to the type's shortfall in performance. It flew with the Peruvian carrier Aerolineas Peruanas.

Convair 320-21
General Dynamics 990A
Model 30-6 Coranado

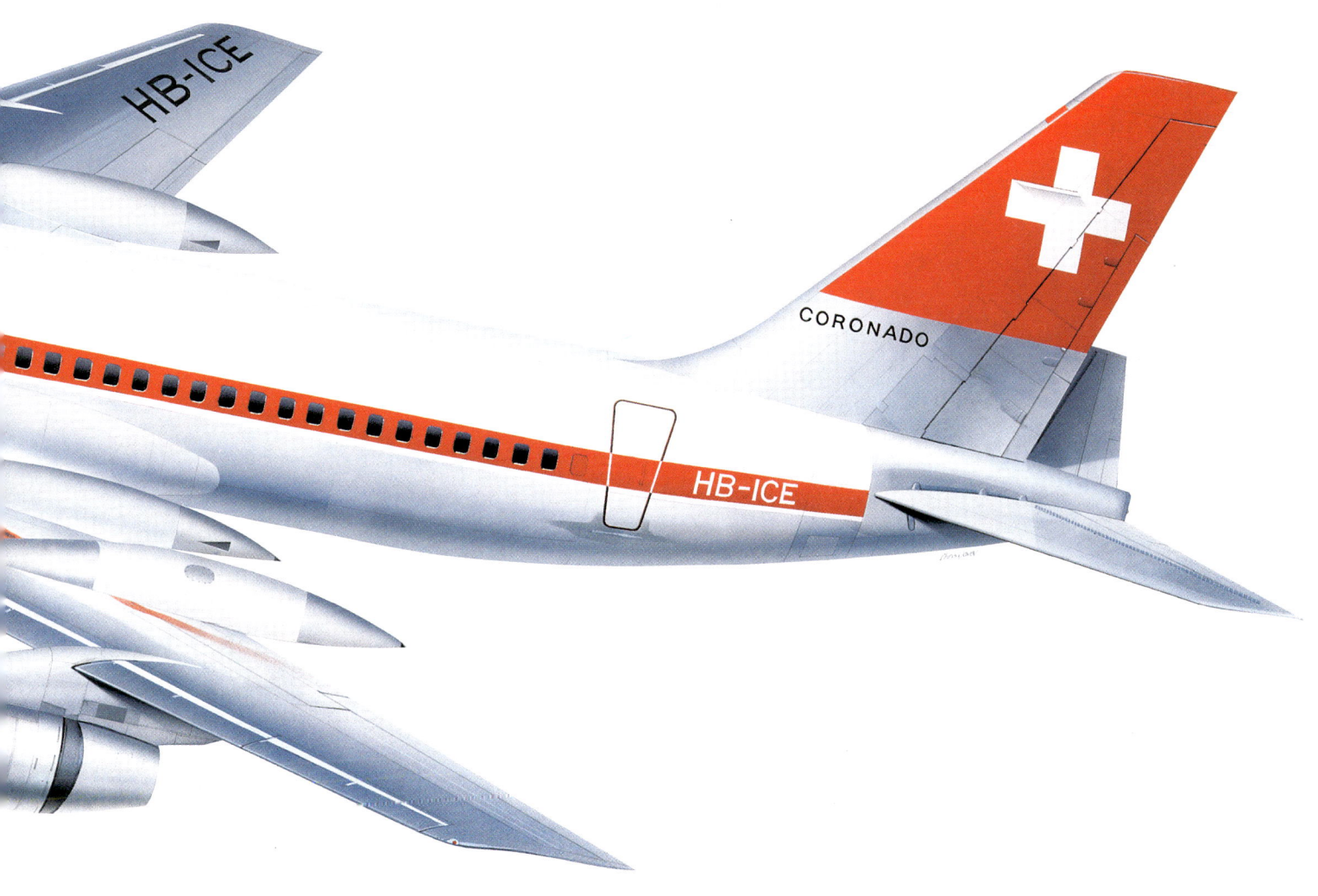

Chapter 16

Curtiss Condor II

Dash 7

Dash 8 Canadian Commuter

Curtiss Condor II

The first Curtiss Condor was a converted bomber, which received a lukewarm reception from US airlines. The company's second attempt resulted in the far more successful 15-seat Condor II. Finding fame as a day passenger transport, a night sleeper and even a polar explorer, the Condor II was later developed as a bomber for China.

Following its order for two production Curtiss B-2 Condor twin-engined biplane bombers in 1928, the US Army gave approval to the Curtiss Aeroplane & Motor Company Inc. to develop the design for commercial purposes. The result was the Curtiss Condor 18, also known as the Model 53 Condor CO Transport, the first of which flew in June 1929. The Condor 18 was essentially a B-2 with a lengthened fuselage modified to accommodate 18 passengers (hence the designation) in three rows of seats, plus three crew in an enclosed cockpit. The aircraft was powered by two 466-kW (625-hp) Curtiss GV-1570 engines, housed in nacelles similar to those of the B-2 bomber but with the rear machine-gun positions faired over. Its fabric-covered metal-framed wings and archaic biplane tail surfaces were identical to those of the military aircraft.

US Civil Aeronautics Board certification of the Condor 18 was granted in August 1929 and a batch of three was completed at Curtiss' Garden City, New York, factory. Three further aircraft were also completed there, but differed substantially from the first batch in having fuselages 0.6 m (2 ft) shorter, dihedral on upper and lower wings instead of just the lower wings, taller fins and rudders, increased span tailplanes and revised engine nacelles of metal monocoque construction.

The Condor 18 was not well received by US airlines, most of which were already operating Fokker and Ford tri-motor airliners. Only Transcontinental Air Transport (TAT) briefly showed interest when it tested two Condors on its routes, but the airline did not take up an option to buy. On 30 May 1930, the 20th anniversary of his historic $10,000 prize-winning flight from Albany, the capital of New York state, to New York City in the *Hudson Flyer* biplane, company founder Glenn Curtiss flew the fifth production Condor 18 over the same route. It was to be his last flight; he died on 23 July that year.

In the face of poor sales prospects all six Condors were stored by Curtiss at Garden City and eventually rebuilt and modified at the company's St Louis, Missouri, factory for sale (at less than cost price) to Eastern Air Transport, which flew them from 1931-34. Thereafter four were bought by barnstormer and transatlantic flier Clarence D.

The only European operator of the Condor was Swissair, which purchased a single AT-32 to operate services between Switzerland and Germany. Among the new features introduced by this service was the carriage of a hostess. Sadly the Condor crashed after just four months of operations.

The BT-32 was the company-developed bomber version of the Condor II, featuring an internal bomb bay, racks under the wings and two upper turrets for defence. This machine was the prototype of the variant, eventually sold to China for use by Chiang Kai Shek as a personal transport.

Chamberlain, who stripped them of their airline interiors to increase seating capacity and used them for itinerant pleasure flying tours throughout the United States in the mid-1930s.

Curtiss' St Louis plant was closed during the depression which began in October 1929, and when time came to reopen the factory a successor to the Condor was chosen to launch the revival of production. In April 1932 a design team headed by George A. Page Jnr was dispatched to Missouri to begin work on a new trunk-line transport aircraft. Curtiss knew that both Boeing and Douglas had new, advanced all-metal monoplane airliners (the Model 247 and DC-1/2 respectively) on their drawing boards. There was neither time nor capital available to produce a rival for these, but the Curtiss management believed a market existed for a simpler aircraft of conventional appearance and construction, yet incorporating such refinements as a retractable undercarriage. Page's brief was to complete the task of designing and building a prototype as quickly as possible in order to beat Boeing and Douglas onto the market.

Initial design work and preliminary wind tunnel testing had been performed at Garden City during 1931, so Page's small team was able to have the prototype Curtiss T-32 (NC 12353) flying on 30 January 1933, just 10 months from the date when they started work at the St Louis plant. The T-32 was a traditional design, the only obviously

The Condor 18 did not interest any of the airlines sufficiently for an order, although this and another aircraft wore Transcontinental Air Transport colours while the airline evaluated the type.

'modern' feature being its electrically-actuated retractable landing gear, which Curtiss claimed as a 'first' among commercial aircraft. It was a short-lived record, for Boeing's first production Model 247 flew for the first time just eight days after the T-32 and was in every respect an aeroplane of its times, just as the biplane Condor was an anachronism even before it flew.

Sturdy construction

The T-32's airframe was principally of chrome-molybdenum steel tubing. Its fuselage comprised a welded truss frame, with aluminium alloy skinning on the nose section. The wings were built up on two welded steel Warren-girder main spars with tubular light alloy ribs and sheet metal leading edges. The tail surfaces were also of welded tube construction, and the entire airframe was fabric-covered. The retractable tailwheel undercarriage was hinged at the rear wing spar position, retracting and extending by means of an electrically-actuated screw-jack and so arranged that, when retracted, the mainwheels protruded below the engine nacelles to provide a measure of protection for the lower fuselage in the event of a wheels-up landing.

Geared Wright Cyclone GR-1820-F11 engines driving fixed-pitch three-bladed Hamilton Standard metal propellers were chosen for the T-32, with fuel contained in four tanks in the upper wing giving a total capacity of 1135 litres (300 US gal). The engines were mounted on rubber bushes to minimise vibration, and all powerplant accessories were installed in a bay behind the engines and ahead of the firewalls, creating self-contained 'power eggs' which facilitated maintenance. The cabin was configured for 12 passengers in six pairs of facing seats with a central aisle, each pair of seats converting to a couch for night sleeper services, with a second sleeping berth hinging down from the cabin ceiling.

Much attention was devoted to passenger comfort, with effective soundproofing, individual hot and cold air controls for each seat, and a toilet/washroom with hot and cold running water. The main luggage compartment was situated below the cabin floor, further enhancing the already spacious cabin provided by the T-32's bulky fuselage. Curtiss believed that while the biplane would inevitably be slower than the sleek monoplanes being developed by their rivals, speed was not such an important asset for sleeper transports, for which role the T-32 was ideally suited. The cockpit was also advanced for its day, incorporating a Sperry artificial horizon, a directional gyro, sensitive altimeter, and airspeed, turn-and-slip and vertical speed indicators mounted in a central 'blind flying panel', as was later to become standard practice.

The T-32's flight test programme was short, and its Type Certificate, ATC 501,

Right: With 'FLY $1' emblazoned under the wing, this Condor 18 was one of four purchased from Eastern in 1934 and used by Clarence D. Chamberlain's flying circus. Countless American citizens must have taken their first-ever flight in the trusty Condors, thanks to Chamberlain's barnstorming.

Below right: Eastern and American led the way for the T-32 Condor II, each buying nine aircraft. Inevitably slower than the Boeing and Douglas products being developed at the same time, the Condor offered far greater comfort and was often used for sleeper operations.

Bottom right: Both Navy R4C-1s were handed over to squadron VJ-7M of the USMC at San Diego, California, to continue the staff transport role. In 19393 they were shipped to Antarctica to support the US bases there, and both aircraft were abandoned in the region during 1941.

was issued on 18 March 1933. To capitalise on the good reputation earned by the earlier Condor 18, Curtiss-Wright's marketing department chose the name Condor II for the new aircraft, though it had little if anything in common with the earlier transport. The aircraft was priced at $55,000. Initial orders for the Condor II came from Eastern Air Transport and American Airways, which each ordered five, later increased to nine apiece. EAT took delivery of the prototype on 23 March 1933, while AA received the first production machine (NC 12354) just over a month later. Deliveries continued at two-week intervals throughout 1933.

'Flying Brooklyn Bridges'

Eastern Air Transport launched its Condor II services on the New York-Miami route, while American Airways used its blue-and-orange-painted aircraft for transcontinental coast-to-coast services. It took five and a half hours for an AA Condor to fly from Newark, New Jersey, to Chicago, and 90 minutes from Newark to Washington DC. The Condors, affectionately dubbed 'Flying Brooklyn Bridges' because of their cat's cradle of bracing wires, were popular

with passengers, more comfortable, much quieter and free of the exhaust fumes which plagued travellers on the Fokker and Ford tri-motors.

A total of 21 Model T-32 Condor IIs was built. Apart from the EAT and AA aircraft, two were built as VIP transports for the US Army Air Corps, which gave them the designation YC-30 and operated them from Bolling Field, Washington, one being used briefly as the personal aircraft of President Franklin D. Roosevelt. The last production Model T-32 (NC 12384), originally destined for AA, was specially modified for Admiral Richard E. Byrd's second Antarctic expedition. Named *William Horlick* in honour of the expedition's principal sponsor, it had a fixed wheeled undercarriage, with provision for the attachment of floats or skis, and extra fuel tanks in its fuselage adding a further 3030 litres (800 US gal) to the Condor's standard capacity. After test flights on wheels the aircraft was delivered in September 1933 equipped with floats and shipped aboard the support vessel SS *Jacob Ruppert*, from which it was launched to make preliminary survey flights as the expedition reached Antarctic approaches. Thereafter it operated on skis from pack ice to Byrd's base camp at Little America on the Ross ice barrier. The *William Horlick* made a number of protracted survey flights over the region during late 1934, playing a vital role in gathering topographical information, but came to grief on 3 February 1935 when it was dropped while being loaded aboard the support ship at the end of the expedition. Damaged beyond repair, the historic aircraft was scrapped on return to the United States.

Improved AT-32 variant

Curtiss-Wright had developed an improved model of the Condor, designated AT-32, which featured 530/537-kW (710/720-hp) supercharged Wright Cyclone SGR-1820-F2/F3 engines driving the then-new Hamilton-Standard controllable (variable-pitch, two-position) propellers and was first flown on 17 March 1934. The enhanced performance of this engine/propeller combination, which was housed in full NACA cowlings and streamlined long-chord nacelles, enabled fuel capacity to be

Specification
Curtiss AT-32B Condor II
Wingspan: 25.0 m (82 ft 0 in)
Length: 14.8 m (48 ft 7 in)
Height: 4.9 m (16 ft 4 in)
Wing area: 112 m² (1,208 sq ft)
Powerplant: two 537-kW (720-hp) Wright SGR-1820-F2 Cyclone radial engines
Passenger capacity: 12
Empty weight: 5550 kg (12,235 lb)
Maximum take-off weight: 7940 kg (17,500 lb)
Maximum speed: 305 km/h (190 mph)
Cruising speed: 270 km/h (167 mph)
Service ceiling: 7010 m (23,000 ft)
Maximum range: 1152 km (716 miles)

American Airways was the major customer for the 'second-generation' Condor, taking 19 of the aircraft in total. The second batch of these consisted of 10 AT-32s, which had uprated engines and variable-pitch propellers. American was impressed with the new variant, and uprated the first batch of nine to the same configuration.

increased to 1420 litres (375 US gal) and maximum take-off weight to go up to 7940 kg (17,500 lb). Take-off, climb and cruise performance were all improved over the T-32. The AT-32 was offered in five variants: the AT-32A and AT-32B were 12-passenger convertible day/night-sleeper models, with a -F3 and -F2 Cyclones respectively; the AT-32C and AT-32D had day-only 15-passenger cabins, with -F2 and -F3 Cyclones respectively; and the AT-32E, which was a de luxe day-only 12-passenger aircraft, accommodation being in single seats rather than the pairs of the other Condors, and with 530-kW (710-hp) -F3 engines.

Swiss operator
American Airways ordered 10 AT-32s, the first of which flew on 17 March 1934, and later had its nine T-32s modified to the new standard, these aircraft then being redesignated T-32Cs. Eastern Air Transport did not place an order for the AT-32 Condor, but had its fleet of earlier aircraft equipped with the new controllable propeller. Apart from American Airways, only one other commercial customer was found for the AT-32. The Swiss national carrier Swissair ordered a single AT-32C (CH-170), which was delivered on 28 March 1934 and became the first Condor to operate in Europe, as well as the last civil Condor manufactured. It entered service on Swissair's Zurich-Berlin route in April, but its operation was short-lived. On 27 July 1934 CH-170 crashed in Germany, killing all on board, following failure of the starboard wing flying wire attachments.

Two further AT-32 Condors were built, both 'E' models delivered in June 1934 to the US Navy as R4C-1 VIP transports. They differed from commercial aircraft in having electrically-actuated Curtiss-Wright variable-pitch propellers in place of the hydraulically-operated two-position Hamilton-Standard units. Initially based at NAS Anacostia, Virginia, and subsequently transferred to the US Marine Corps with VJ-7M/VMJ-2 squadrons at San Diego, California, they – like their YC-30 counterparts in the US Army Air Corps – were little used. In November 1939 both Condors, repainted high-visibility red overall, were loaded aboard the SS *North Star* and shipped, like the *William Horlick* before

them, south to Antarctica, where they were to serve with Admiral Byrd's newly-established US Antarctic Service, operating from West and East Base 2415 km (1,500 miles) apart at Little America and Marguerite Bay. The East Base Condor made a number of flights in late 1940, including one on 30 December which ventured further south along the Weddell sea coast than explorers had previously been. The aircraft was abandoned on Wetkins Island in March 1941 after rescuing 24 scientists trapped by pack ice at Mikkelson Island. The other US Antarctic Service Condor had also given sterling service before it too had been abandoned in Antarctica two months previously after a forced landing following engine failure.

Even before deliveries of the improved AT-32-series Condors began, Curtiss-Wright accepted that the slow, dated biplane's future as a first-line airline transport was severely limited, and they began looking at the possibility of adapting the Condor's excellent load-carrying capabilities to military roles, thus reversing the process which had seen the original Condor 18 developed from the B-2 bomber.

'China Bomber'

A company-funded prototype/demonstrator, unofficially known as 'The China Bomber', was built. Its under-cabin cargo compartment was modified to accommodate bomb racks with 500-kg (1,100 lb) capacity. External bomb racks were installed below the lower wings, and small gun cupolas, each armed with twin 0.30-in Browning machine-guns, were fitted immediately aft of the cockpit and at the rear of the cabin, where additional side glazing was also provided at the gunner's position, with provision for a single 0.30-in gun in the cabin floor. Additional fuel tankage in the fuselage gave this BT-32 Bomber/Transport) Condor a total capacity of 2840 litres (750 US gal). The first BT-32 made its maiden flight on 9 February 1934 and subsequently was taken by sea to China where it was demonstrated by the famed racing and record-breaking flier Frank Hawks. It was damaged in a wheels-up landing incident, but after repair was sold to the Chinese government for use as the personal aircraft of Generalissimo Chiang Kai-Shek.

Colombian floatplanes

Three production BT-32s followed, all mounted on Edo floats for use as trainers, mailplanes and military transports by the Colombian air force, which used them in operations along interior rivers and was still flying two of the Condors on anti-submarine patrols along Colombia's Atlantic coast during World War II. Four more BT-32 landplanes were built to the order of the Bolivian government, but their export was blocked by the US State Department. After passing through the hands of several brokers and dealers they were acquired by Tampa-New Orleans-Tampico Airlines, ostensibly for service on a New York-Buenos Aires route, but were quickly leased to the Peruvian carrier Condor Peruana de Aviacion. Impounded on delivery to Lima on 28 March 1935, the bomber/transports subsequently had all their military equipment stripped and entered charter service supporting mining operations in Peru. One crashed in 1936. The remainder, named *Iquitos*, *Cuzco* and *Madre de Dios*, were later used on scheduled passenger flights within the country. One was scrapped in 1939, the survivors being taken over by the Peruvian air force in 1941. One of these remained flying until 1956 on aerial survey and photo-mapping duties, and was the longest-surviving airworthy example of the breed.

Although considered obsolete even at the time of its first flight, the Condor II proved a robust and durable aircraft, and its capacious cabin was put to good use long after its days with the mainstream airlines were over.

The final Condor development was a freighter version designated CT-32, which had a 1.8 m (6 ft) square cargo loading door on the starboard side of the fuselage, a built-in cargo hoist, and strengthened cabin floor with cargo hold-down points. Such was the voluminous interior of the Condor that moderate-sized vehicles could be loaded. Three CT-32 freighters were built, ending Condor II production after 45 aircraft had been manufactured. The CT-32s were all delivered to the Argentine navy, and were based at Puerto Belgrano and Punta del Indio, the last being scrapped in 1949.

Foreign operators

When the US trunk carriers disposed of their Condors in 1936/37, the aircraft all found new homes abroad. Six former American Airways aircraft, purchased by the Spanish Embassy in Mexico for shipment to the Republican side in the Spanish Civil War as potential bombers, were impounded by the Mexican authorities and were eventually modified to cargo configuration and sold to the China National Aviation Corporation for its services linking Hong Kong with mainland China. Five of these were destroyed on the ground at Hong Kong during a Japanese attack in December 1941; the sixth was captured and operated in Japanese colours. Four ex-Eastern Air Transport T-32s, also converted as freighters, were sold to the British company International Air Freight at Croydon Airport, but after IAT ceased trading in 1938 were taken over by the Royal Air Force for use as navigation trainers. Never flown by the RAF, they were scrapped shortly after the outbreak of World War II. Other Condors proved popular with small airlines and charter companies in Central and South America, the original prototype T-32 serving with owners in Colombia and Nicaragua before ending its days in the late 1940s as an executive transport in Mexico. Only one Condor is believed to exist today, under restoration with a private owner in the United States.

A rare air-to-air photograph of a Condor depicts one of the US Navy's R4C-1 VIP transports. Noteworthy are the retractable undercarriage and the luxury interior for the carriage of high-ranking officers. The aircraft were based at NAS Anacostia, in the Washington, DC, region.

Dash 7

Since the mid-1940s, when the DHC-2 Beaver was introduced, de Havilland Canada has maintained a reputation for building strong and reliable light utility transports. The Beaver was joined in 1951 by the Otter, and then by the Caribou, Buffalo and Twin Otter. Then came the Dash 7, introducing more sophistication to become one of the world's leading STOL airliners.

De Havilland (now Bombardier), Canada's modern commercial production line, really became established in the middle 1960s when the network of 'third-level' airlines started to develop in the USA. Their routes required a reliable 15-20-seat light passenger aircraft which could be certificated under the FAR Part 23 regulations: these kept the maximum take-off weight of the aircraft to under 5670 kg (12,500 lb). The de Havilland solution was the DHC-6 Twin Otter. This was a high-wing STOL machine with a fixed tricycle landing gear and two Pratt & Whitney Canada PT6A-20 turboprop engines. The system's simplicity of the Twin Otter made it attractive to smaller airlines, and it was put into service with many American operators, including Golden West Airlines, Pilgrim Airlines and Suburban Airlines.

The market opened up by the Twin Otter and its competitors (the Beech 99 and Swearingen Metro) soon started to mature and this meant that an 18-seat aircraft was inadequate to fulfil the demand on many routes. The next option for third-level operators was to go up to the 40/52-seat Fokker F.27 which was a 1955 design, and de Havilland could see that there was a strong opportunity for a new 48-seat aircraft with good take-off and landing performance which would appeal to existing Twin Otter users. Thus was born the new DHC-7 which received the obvious but effective name Dash-7.

Design work on the DHC-7 started in 1972, and the first prototype (C-GNBX-X) made its first flight at Downsview, Ontario, on 27 March 1975 in the hands of Robert H. Fowler, chief test pilot for the company. De Havilland pressed forward quickly with test flying of the new aircraft, and brought a second prototype (C-GNCA-X) into the programme to speed up the certification process. The basic model in standard passenger configuration was given the designation Dash-7 Series 100. As it turned out, few alterations were required to the basic design of these prototypes and the Canadian Department of Transport awarded a Canadian type approval certificate on 19 April 1977. This was followed immediately by the United States FAA type certificate under the requirements of FAR Part 25 and by this time de Havilland had already rolled out the first production aircraft (number 3) and was ready to press ahead with Dash-7 deliveries.

The Dash-7 has a high wing, circular-section fuselage and a high T-tail. A strong tricycle landing gear is fitted, and the main legs retract forward into the inboard engine nacelles, while the twin-tyred nosewheel unit retracts backward and is fully enclosed. The aircraft was designed specifically for short-haul routes of around one-hour duration and de Havilland gave special consideration to environmental factors because of the need for many operators to fly from smaller airports which are often close to city centres. As a result, the Dash-7 has demonstrated a particularly low noise level in airport

The first Dash-7 to enter revenue service was N27RM (c/n4), which was initially a de Havilland development aircraft and then went to Rocky Mountain Airways of Denver, Colorado. The Dash-7 represented a considerable step up from the carrier's Twin Otters, and the six aircraft flew from Denver to Vail, Aspen and Steamboat Springs.

Emirates Air Service was another loyal operator of de Havilland products and used four Twin Otter 300s in addition to its single-passenger version of the Dash-7. It was based at Abu Dhabi and flew to Dubai in addition to being used on charter services in the general Gulf region.

arrival and departure patterns and, in testing of the aircraft for approval under the American FAR Part 36 noise regulations, the Dash-7 was flown into Chicago's Meigs Field Airport at noise levels which were virtually inaudible against a background of normal Chicago road traffic.

As with the Twin Otter, de Havilland aimed to give the Dash-7 excellent short-field performance and it has an FAR Part 25 take-off field length of 689 m (2,260 ft) and a landing distance of 594 m (1,950 ft) at maximum take-off gross weight, sea level ISA. This remarkable STOL performance has been achieved by using design technology developed over a number of years by de Havilland on previous designs. In particular, the high-aspect ratio wing (10:1 factor) is fitted with double-slotted large-chord flaps which extend across 80 per cent of the wingspan. The four turboprop engines are positioned well apart to augment the airflow across the wings and flaps with the maximum propeller thrust, and this is especially advantageous in low-speed performance areas. Considerable safety advantages accrue to the Dash-7 in critical 'go-around' situations arising from an aborted landing. It is normal procedure when this occurs for the 45° landing flap configuration to be cleaned up only to the 25° setting, which gives a very high level of lift with minimum drag and the aircraft can climb away at a steep angle and clear the area at altitude.

Short-field landings are particularly effective in the Dash-7 with the combined use of spoilers, propeller pitch control and standard braking. Normal practice is to move the propellers into ground fine pitch control, or betas, as soon as the main wheels touch the ground and this neutralises up to 90 per cent of the wing lift immediately. At the same time the inboard spoilers are deployed. When the nose-wheels touch down, the pilot deploys the outboard spoilers (which are also used for aileron augmentation) and completes the landing run with the anti-skid mainwheel braking system.

Obviously, the main objective for de Havilland was to make the Dash-7 attractive to its 'third-level' airline customers. Therefore, the cabin is pressurised to provide maximum passenger comfort and to allow the aircraft to operate at more economical higher altitudes. The passenger cabin has a 'big aircraft' feel with 81-cm (32-in) pitch four-abreast seating divided by a central aisle and with large overhead lockers for carry-on baggage. The stand-up headroom inside is 1.94 m (6 ft 4.5 in), which is more than adequate to allow passengers and flight attendants to move freely around the cabin. The standard passenger configuration allows for a lavatory and small galley in the rear of the cabin. From an operational standpoint, de Havilland has stressed the ease of operation of the Dash-7 on multi-stop routes. When it is necessary to make intermediate calls a total passenger change can be accomplished in no more than seven minutes from the time the aircraft stops at the gate to the moment it taxies out to the runway again. This is most important to the small operator, who aims to get maximum utilisation from his fleet.

Cargo-carrying capability

An optional cargo/passenger version of the Dash-7 is available and this is designated Dash-7 Series 101. It is fitted with a large upward-opening (71 by 91-in/1.8 by 2.31-m) cargo door on the forward port side between the cockpit and the wing leading edge. This has resulted in the repositioning of some of the forward passenger cabin windows and it is normal for this variant to be fitted with a heavy-duty cargo floor with a roller mat surface. When it is used in the all-cargo role, the Dash-7 has a maximum usable cabin volume of 9.34 m^3 (330 cu ft) and it can be loaded with either palletised freight or up to five LD-3 containers providing a maximum cargo payload of 4990 kg (11,000 lb). The more normal application adopted by several carriers is a mixed passenger and freight arrangement with 18 passenger seats in the rear cabin and three containers in the forward section. However, the passenger capacity can be raised to 34 (with one container) or 26 (with two containers). The cabin floor and cargo sill are designed for direct truckbed loading to keep ground handling equipment to a minimum.

While the bulk of Dash-7 production to date has been for the commuter airlines, de Havilland has a number of specialised applications for the aircraft. One version is the Dash-7R Ranger reconnaissance and surveillance aircraft, which is aimed at the relatively low-cost requirement of many air arms which have a coastal patrol mission to fill but cannot afford the increasingly sophisticated and expensive types on offer. The Ranger can be fitted with Litton LASR-2 airborne search radar-mounted in the forward fuselage with a radome positioned in the belly to give full 360° search. The normal layout would also provide for vertical cameras to be mounted to the rear of the

Seen on a test flight over the countryside near Toronto is the first Dash-7 prototype, C-GNBX-X. The forward cabin window arrangement shows that the upward-opening freight door is not fitted to this aircraft. The convertible passenger/freight version of the Dash-7 has only two windows forward of the propeller line.

One of the larger operators of the Dash-7 is Petroleum Air Services, based at Cairo. A total of five of the type appears in its inventory and this aircraft, SU-CBA, was the first to be delivered and was the 93rd Dash-7 to be built. Petroleum Air Services flies on support contracts for the North African oil companies.

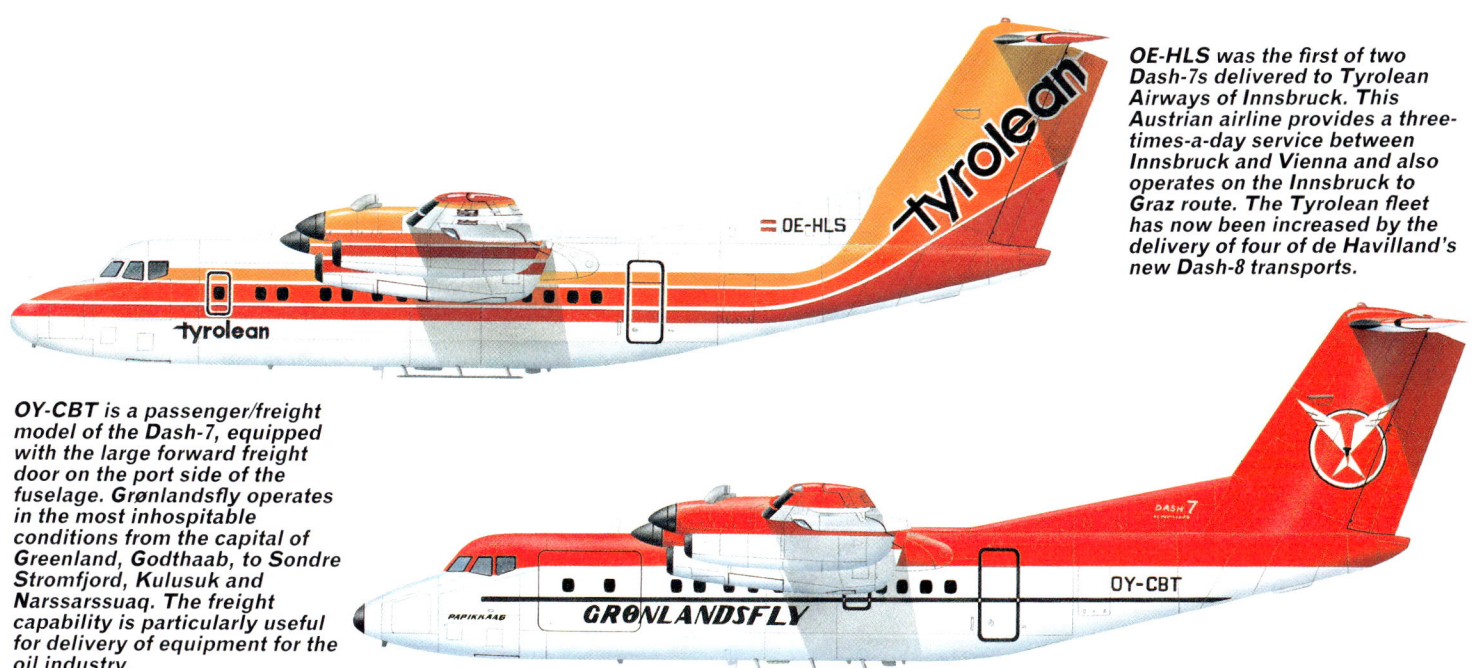

OE-HLS was the first of two Dash-7s delivered to Tyrolean Airways of Innsbruck. This Austrian airline provides a three-times-a-day service between Innsbruck and Vienna and also operates on the Innsbruck to Graz route. The Tyrolean fleet has now been increased by the delivery of four of de Havilland's new Dash-8 transports.

OY-CBT is a passenger/freight model of the Dash-7, equipped with the large forward freight door on the port side of the fuselage. Grønlandsfly operates in the most inhospitable conditions from the capital of Greenland, Godthaab, to Sondre Stromfjord, Kulusuk and Narssarssuaq. The freight capability is particularly useful for delivery of equipment for the oil industry.

radome and for the usual flare chutes and liferaft ejection equipment. Another use for the Dash-7 is as an executive transport but de Havilland would not complete these aircraft itself. Rather, the 'green' airframe would be delivered to Innotech Aviation at Dorval for a custom interior to be installed. No such conversions were delivered, but a typical layout would provide for a forward de luxe cabin with six to 10 executive chairs and a rear standard cabin with 12 airline seats.

The first operator was Rocky Mountain Airways of Denver, Colorado, which placed its initial aircraft in service at the end of 1977 on routes to the local mountain ski resorts. A pair of Dash-7s also reached Greenlandair to supplement and expand the operations carried on for some while by Twin Otters. They were required to fly in the most extreme winter conditions and to land and take-off from very demanding airfields which are almost permanently covered in snow and ice. The Dash-7s link Godthaab, the main base of Greenlandair, with Sondre Stromfjord and Narssarssuaq, and also fly the route from Sondre Stromfjord to Kulusuk. In many cases these airfields are surrounded by mountains and the ability of the Dash-7 to make a very steep approach is critical. In fact, the backbone of Greenlandair's fleet before the arrival of the Dash-7 was the Sikorsky S-61 helicopter which provided the only practical method of entry to many of the airline's destinations. The company also uses the Dash-7s to provide support for the US Air Force strategic radar base at Thule and full use is made of the cargo-carrying facilities of the aircraft.

Short-runway effectiveness

Several of the commuter airlines who have bought the Dash-7 have made use of the 'Separate Access Landing System' concept which has been promoted by de Havilland. This involves the use of limited sections of runways or stub sections of runway which are available only because of the unique ability of the aircraft to manoeuvre onto such sections outside the general flow of the air-

The Canadian Armed Forces used this Dash-7, 132001, for passenger and freight transport between various locations in Europe in support of the Canadian NATO detachment in West Germany. Based at Lahr, the Dash-7 operated with No. 412 Sqn and carried the designation CC-132. Two aircraft were used by this mixed squadron.

Delivered in 1983, LN-WFL was the 84th Dash-7 and serves with Widerøes Flyveselskap on routes in northern Norway. The main routes served are from Bodø to Trondheim and to Tromsø. Widerøe also has a fleet of 12 Twin Otters, which operate on the shorter routes and replaced another de Havilland product, the DHC-3 Otter.

port's traffic. This effectively increases the capacity of any airport, and also gives the Dash-7 operator preferential treatment at peak periods. This system has been operated successfully by Ransome Airlines at New York and Washington National and by Golden West Airways who have used it at San Francisco. In Europe, Maersk Air has been able to use the same techniques at Copenhagen. In Canada, the Dash-7 has been able to develop services into downtown airports at Toronto and Montreal. City Express Airlines has four Dash-7s which it uses to link these two cities with the capital, Ottawa, on a very-high-frequency schedule, and the quick turnround of the aircraft is vital to the achievement of this service.

STOLport operations

In the UK the principal operator to adopt the Dash-7 has been Brymon Airways. Again, Brymon was a large Twin Otter user and purchased five Dash-7s to meet increased-capacity demand. The aircraft has been a key factor in the joint venture between Brymon and the John Mowlem construction group to develop a STOLport in London's dockland. Approval for the 792-m (2,600-ft) runway to serve the City of London was given in May 1985, and the short-field performance and low noise level of the Dash-7 were the basis on which the government was willing to sanction construction of the airstrip. Brymon operates routes from the STOLport to Brussels, Amsterdam, Paris, Rotterdam and Frankfurt, with substantial time savings for central London businessmen. A second DHC-7 operator, London City, also flies from here.

The Dash-7 has also gained a number of military orders. The first of these was for two aircraft for the Canadian Armed Forces, who needed them to transport high-ranking passengers and freight around Europe. These aircraft received the CAF designation CC-132 and were delivered to No. 412 Squadron at Lahr in West Germany. One of the machines is fitted with a rear 22-seat cabin, a central galley area and a forward VIP section for 10 senior passengers. The other aircraft is built to Series 101 standard with the cargo door and was used on the weekly scheduled service operated by the Canadians between their base at Lahr and Gatwick (where it collected passengers from the transatlantic CAF Boeing 707 service). On the return flight, the Dash-7 dropped off passengers at Chievres and Maastricht before coming back to its home base. These have now been replaced by Dash-8s.

With the Dash-7 reaching the end of production, at 114 aircraft in 1986, de Havilland had been seriously considering the possibility of building a stretched version known as the Series 300. In market surveys carried out during 1982 there was a strong indication of interest in such a development but it was decided that the Dash-8 (a smaller twin-engine aircraft) should receive priority. The design study showed the Series 300 to have a 5.64-m (18.5-ft) fuselage stretch providing the capacity to carry up to 78 passengers. It would have been powered by the upgraded PT6A version of the present powerplant and would have had an improved cruising speed and rate of climb while still being able to operate from a 914-m (3,000-ft) runway. The de Havilland designers believed that it would have been possible to reduce the present cost-per-seat of the Series 100 when the Series 300 was introduced and to improve significantly the overall direct operating costs over the normal 241-km (150-mile) stage lengths. With Boeing taking over DHC, all work on the Dash-7 stopped, leaving the Dash-8 as the company's main product.

The first Dash-7 Series 150 is now in flight-test for the Canadian government, which will operate the aircraft on ice reconnaissance duties. Equipment includes SLAR in a port fuselage fairing, and an observation cupola on the upper forward fuselage.

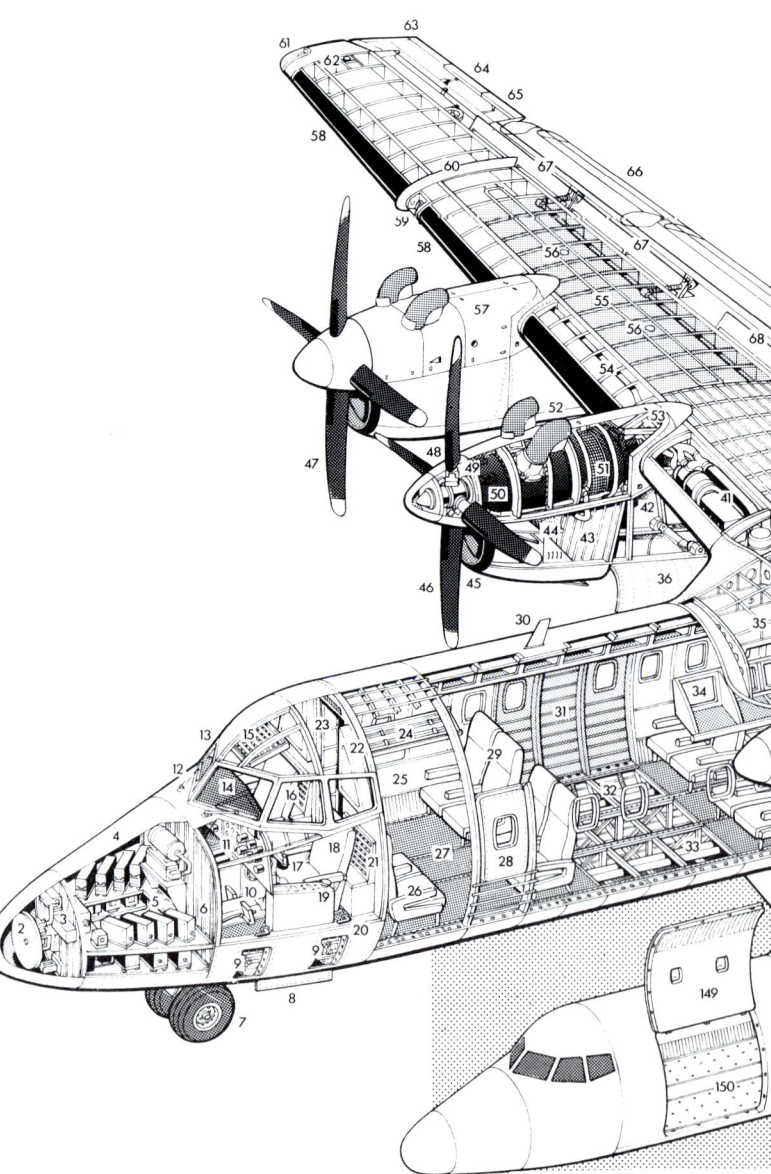

Wearing the distinctive colour scheme of the major Danish domestic operator Maersk Air, this DHC-7-102 is one of two currently operated by the airline. Maersk's DHC-7s are configured in a single-class, 44-seat layout.

de Havilland Canada DHC-7 Dash-7 cutaway drawing key

1. Radome
2. Weather radar scanner
3. Radar transmitter and receiver units
4. Nose electronics compartment
5. Radio and electronics racks
6. Front pressure bulkhead
7. Twin nosewheels
8. Nosewheel doors
9. Control runs beneath cockpit floor
10. Rudder pedals
11. Instrument panel
12. Windscreen wipers
13. Windscreen panels
14. Instrument panel shroud
15. Overhead switch panel
16. Co-pilot's seat
17. Control column handwheel
18. Pilot's seat
19. Nosewheel steering control
20. Pitot tubes
21. Circuit breaker panel
22. Cockpit bulkhead
23. Electrical distribution panel
24. Cabin roof control runs
25. Cabin trim panels
26. Rearward facing seat row
27. Seat attachment rails
28. Emergency exit window panel, port and starboard
29. Four-abreast passenger seating, 50 seats
30. VHF aerial
31. Fuselage frame and stringer construction
32. Floor beam construction
33. Underfloor air conditioning ducting
34. Overhead stowage bins
35. Wing root fairing construction
36. Main undercarriage wheel doors
37. Air system water separators
38. Wing spar box centre section
39. Skin panel joint strap
40. Wing stringers
41. Starboard air conditioning plant
42. Main undercarriage wheel bay
43. Engine compartment firewall
44. Intake debris separator
45. Engine air intake
46. Hamilton Standard four-bladed, reversible pitch propellers
47. Glass-fibre propeller blades
48. Blade root de-icing boots
49. Propeller hub pitch change mechanism
50. Pratt & Whitney Canada PT6A-50 turboprop
51. Engine intake screen
52. Exhaust stubs
53. Engine support link
54. Bleed air piping
55. Starboard wing fuel tanks; total aircraft fuel capacity 1,480 US gal (5602 litres)
56. Overwing fuel filler caps
57. Engine cowling panels
58. Leading edge de-icing boots
59. Landing lamp
60. Wing fence
61. Starboard navigation light
62. Compass flux valve
63. Starboard aileron
64. Geared tab
65. Aileron trim tab
66. Starboard double slotted flaps, down position
67. Roll control spoilers
68. Ground spoilers
69. Flap screw jacks
70. Wing root trailing edge fillet
71. Fuel transfer pipe fairing
72. Starboard service door
73. Rear seat row
74. Buffer unit
75. Starboard baggage door (open)
76. Fin root fillet
77. Refuelling/defuelling pipe
78. Emergency locator transmitting aerial
79. Fin leading edge
80. Fin construction
81. VOR aerial
82. Elevator control rods
83. Tailplane/fin attachment spar box
84. Upper position light
85. Anti-collision light
86. Tailplane leading edge de-icing boots
87. Starboard tailplane
88. Static discharge wicks
89. Elevator trim tabs
90. Elevator spring tab
91. One-piece elevator
92. Elevator horn balance
93. Tailplane construction
94. Rudder hydraulic jacks
95. Trailing rudder
96. Fore-rudder
97. Tail navigation light
98. Rear fuselage vent
99. Tailcone access door
100. Retractable tail bumper
101. Cockpit voice recorder
102. Sloping fin attachment frames
103. Ventral pressure refuelling connection
104. Rear pressure bulkhead
105. Baggage compartment
106. Baggage restraint net
107. Toilet compartment
108. Wash basin
109. Passenger door upper segment
110. Trailing edge wing root fillet
111. Inboard flap track
112. Wing spar/fuselage main frame attachment joint
113. Flap shroud ribs
114. Port wing integral fuel tank bays
115. Lower passenger door segment/airstairs
116. Handrail
117. Nacelle tail fairing
118. Port double slotted flaps
119. Roll control spoilers
120. Port aileron construction
121. Aileron geared tab
122. Static discharge wicks
123. Aileron horn balance
124. Compass flux valve
125. Port navigation light
126. Wing rib construction
127. Leading edge nose ribs
128. Wing fence
129. Wing tank outboard end rib
130. Landing lamp
131. Leading edge de-icing boots
132. Outboard nacelle hydraulics bay
133. Engine nacelle construction
134. Twin mainwheels
135. Engine air intake
136. Front engine mounting
137. Undercarriage breaker strut
138. Main undercarriage leg strut
139. Hydraulic retraction jack
140. Main undercarriage pivot mounting frame
141. Wing tank inboard end rib
142. Bleed air piping
143. Port air conditioning plant
144. Port inner nacelle construction
145. Propeller spinner
146. Oil cooler
147. HF aerial rail
148. Quick-change passenger/cargo version
149. Cargo door
150. 'Ballmat' heavy duty cargo handling floor

© Pilot Press Limited

DASH 8
Canadian Commuter

Renowned for no-nonsense biplanes, the de Havilland Canada (later Boeing and now Bombardier) company had placed its faith in a new market: commuter liners. Although the DHC-8 shows the high-wing STOL traditions of its predecessors it has introduced a new level of style and sophistication.

Today, when it seems almost every industrially developed country is building or developing at least one type of twin-turboprop commuter transport, de Havilland Aircraft of Canada's long background of successful bush and STOL transports gives them an almost unique edge on its rivals. Once owned by Boeing but eventually sold to Bombardier after the rival Alenia/Aérospatiale ATR had been forced out, DHC has produced designs which over the years have tended to get ever larger and more powerful. Until recently the biggest was the four-engined Dash 7, a quiet 50-seater able to operate from runways 600 m (1,970 ft) long, but in terms of fuselage length Dash 7 dimensions have now been overtaken by the Dash 8-300 and, first flown on 31 January 1998, the even longer Dash 8-400 designed to carry up to 70 passengers.

During the 1970s DHC kept studying the prospects for a further product, and from the start it appeared likely that this would be sized to fill the giant gap between the DHC-6 Twin Otter, typically seating 20, and the big Dash 7. By 1979 these studies had crystallised into the Dash-X project, disclosed at that year's Paris air show. By this time some of the multitude of rivals had already begun to show themselves. The Toronto company never wavered in refining the design, however, and on 2 April 1980 it accepted an order from NorOntair for two, designated as DHC-8s, or Dash 8s, showing the design was going ahead. The first of four flight prototypes made its maiden flight on 20 June 1983. Orders came in a healthy stream, reaching 137 by mid-1987 and topping 150 by 1988.

In planning the Dash 8 careful note was taken of changing requirements. There was never any doubt that the aircraft should have a high wing, two turboprops, a pressurised cabin and good STOL qualities. At the same time it was clear that almost all sectors would be flown from reasonable airports, and the field length was pitched at the 1000-m (3,280-ft) level, appreciably longer than for all previous DHC aircraft. This fitted in well with the modest size of wing needed for cruising faster than the Dash 7, in order to get close to jet schedules on short sectors within the USA. At an early date 36 seats was decided upon as the optimum size, and this has continued to be a popular size, though – like almost all its rivals – the Dash 8 has since been developed into a stretched version with a still larger model planned for the future.

In all major respects the Dash 8 is wholly conventional. The aerodynamics are appropriate to speeds below 500 km/h (310 mph), with a traditional wing profile of 18 per cent thickness/chord ratio over the centre-section, the outer panels tapering to slender tips for high effi-

The Dash 8 has emerged as one of the two market leaders in the cut-throat international marketplace of the 36-40 passenger commuter aircraft. Many have found a home in their native country flying feeder services both for Canadian Airlines and Air Canada.

Air Nova, now a subsidiary of Air Canada, flys scheduled commuter services for its parent company from its base at Halifax, Nova Scotia. Its fleet of 14 Dash 8-120s operates alongside British Aerospace 146 jets.

Despite its sophistication, the Dash 8 has sold well in the remoter parts of the globe. Talair was one of the largest carriers in Papua New Guinea, and one had a pair of Dash 8-100 aircraft in service.

ciency. Wing loading of the first Dash 8 version is 275.3 kg/m², compared with 249 for the Dash 7 and 145 for the Twin Otter. The two-spar structure is light-alloy, except for the leading edge and wingtips which are glass-fibre honeycomb composite. The large fairing between the wing and fuselage is a Kevlar honeycomb sandwich.

Flight controls

The leading edge is fixed. Immediately outboard of the engines it contains twin landing lamps on each side. Outboard of these the whole leading edge is fitted with a pneumatic pulsating boot de-icer. During flight test it was found desirable to add a stall strip along the inner part of each outboard leading edge to ensure clean and repeatable stall qualities. On the trailing edge are long-span slotted flaps mounted on tracks, Fowler fashion, to run out and give high lift at the take-off setting of 15°. Outboard are conventional manual ailerons with inset trim tabs. In cruising flight these alone provide roll control. Above the wing are large hydraulically powered spoilers covering most of the span. At speeds between 259 and 380 km/h (160 and 235 mph) the outer spoilers augment the ailerons for roll control. Below 259 km/h (160 mph) all four flight spoilers provide roll control. After landing, ground spoiler/lift dumpers inboard and outboard of the nacelles flick open to kill lift, together with the four flight spoilers further outboard.

Like the Dash 7 and Buffalo, the Dash 8 has a STOL-type tail of T configuration. The enormous vertical surface again follows the two previous designs in having tandem powered rudders to give positive yaw control down to 145 km/h (90 mph). The front rudder is hydraulically powered and drives the rear rudder through pivoted links. The modest horizontal surface comprises a fixed tailplane and manual elevators with mass and horn balances and spring and trim tabs. The tailplane leading edge has a rubber boot de-icer.

Perhaps the biggest single argument in designing the Dash 8 concerned the main landing gear. At first short gears were to be pivoted to the fuselage, retracting into blisters. There is abundant experience with such gears on 30 years of C-130 operation all over the world, and it has been successfully used on such rival aircraft as the ATR42, CN-235 and (in a bigger aircraft) BAe 146. Nevertheless, despite the penalties, the Dash 8 as finally built has landing gears with a track of 7.87 m (26 ft) retracting backwards into the engine nacelles. The extra track, especially in short crosswind landings, was judged well worth the extra weight, complexity and cost, the need for long nacelles dividing the flaps into shorter inner and outer sections and the fact that the gears free-fall against the slipstream. The nose gear has 'steer by wire' electric signalling. Each gear is a product of Dowty of Canada, with twin wheels and brakes by Goodrich, with Hydro-Aire anti-skid control.

The withdrawal of Canadian forces from Europe has meant that the pair of Dash 8s, known by the military as CC-142s and once based at Lahr in Germany, has returned home, leaving Convair CC-109s to soldier on.

*Lufthansa-partner **DLT** operates a large and modern fleet of 'regional' aircraft. It was one of a relatively small number of European carriers to operate Dash 8s, with which the company had only a brief association.*

De Havilland Canada's greatest market for the type was in North America, where it proved to be a reliable workhorse for feeder carriers. Metro Express collapsed only months before Eastern Airlines itself.

After carefully looking at the GE CT7, the choice of engine predictably fell on the home product, the Pratt & Whitney Canada PW120A, rated at 1432 kW (1,800 shp). Each is fed by an S-duct from below and exhausts through a jetpipe passing under the wing to curve sharply up to a nozzle above the rear of the nacelle. The 3.96-m (13-ft) propellers are HamStan products, each with four glass-fibre blades with reversing and ground Beta (direct pilot) control and raw AC de-icing. Most of each engine cowl is supplied by Westland Aerospace, using aluminium, titanium (firewall) and Kevlar/Nomex honeycomb sandwich for external panels.

Hidden features

The fuselage is of almost circular section, with a maximum diameter of 2.69 m (8 ft 10 in). Again the structure is conventional, with adhesively bonded stringers and cutout reinforcement. The tailcone is a Kevlar/glass composite. Almost the only striking feature is the nose, which is a long pointed cone angled downwards so that its upper line almost follows the angle of the pilots' windscreens. The latter are flat, while the large side windows are curved and bear structural loads in flight. Primus 800 colour radar is standard. There is a passenger/crew door forward on the left, with integral airstairs, and a baggage/cargo door 1.27 m (4 ft) wide on the port side behind the wing. The 9.2-m (30-ft) passenger cabin normally has nine rows of 2+2 seats, each row being opposite a window. The rear compartment area is 8.5 m^3 (300 cu ft) and can accept a spare engine or 907 kg (2,000 lb) of cargo (depending on the amount of baggage). There is normally a toilet and small buffet. A movable rear bulkhead enables mixed passenger/cargo operations to be flown, and the Dash 8 can be operated in the all-cargo mode, in which case the payload is 4268 kg (9,409 lb).

Normal fuel capacity is 3160 litres (695 Imp gal), in integral tanks outboard of the engines. Auxiliary long-range tanks can be provided, and the Corporate model (marketed in North America by Innotech of Montreal) has centre-section integral tanks raising capacity to 5742 litres (1,265 Imp gal), sufficient for a range of 3706 km (2,300 miles) in typical business configuration. The Corporate also has an auxiliary power unit, mainly for ground air-conditioning. The ordinary (so-called Commuter) Dash-8 can fly four 185-km (115-mile) sectors with full payload on internal fuel, or a single sector of 1650 km (1,025 miles). Typical cruising speed is 493 km/h (305 mph).

Adaptable airplane

All Dash 8s in service at the start of 1988 were of basic Series 100 type. The only non-standard aircraft were six for the Canadian Department of National Defence. Two of these are CC-142 transports which replaced Dash 7s in Germany; they have rough-field landing gear, strong cargo floors, long-range tanks and special avionics. The other four are CT 142 navigation trainers distinguished by their extended noses. Even more remarkable aircraft are the two E-9As completed by Sierra Research of Buffalo for the

Two Dash 8M-100s were bought by the US Air Force in 1988, to serve as missile-test telemetry aircraft, under the designation E-9A. They are both fitted with an AN/APS-128D surveillance radar under the fuselage.

Announced in 1985, the Dash 8 Series 300 brings the design up to a 50-seat aircraft by means of two fuselage 'plugs' totalling 3.43 m (11 ft 3 in). Orders currently stand at 110, with over 30 delivered.

USAF. These are flying data links operating up to 370 km (230 miles) off the Florida coast, performing radar surveillance and sending back voice and telemetry data during test, training and RPV drone operations. Equipment includes a large electronically steerable phased-array radar in a fairing on the right side of the fuselage, an APS-128D sea surveillance radar in a ventral radome and extensive internal avionics.

Early in the 1980s DHC planned a Dash 8 Series 200, with 1614-kW (2,200-shp) PW122 engines, and displayed a model of a proposed 200M in ASW configuration, outfitted with radar, FLIR, MAD, ESM, INS, sonobuoy processing, comprehensive data management and display systems, and wing and fuselage pylons for weapons and other stores, but this has since given way to the more developed Triton maritime patrol version of the Series 300. Current production centres on the Series 100 and 300, the latter announced in 1985 and first flown (by stretching the first prototype Dash 8) on 15 May 1987. This has 1775-kW (2,380-shp) PW123 engines and fuselage plugs ahead of and behind the wings, which extend the fuselage by 3.43 m (11 ft 3 in) to provide standard seating for 50 passengers. Wingtip extensions increase the span from 25.91 m (85 ft) to 27.43 m (90 ft). Other changes include dual ECS air conditioning packs, a Turbomach T-40 APU (as fitted in earlier Corporate versions), a rear service door on the starboard side and more space for toilets and coat-hanging. Maximum weight is increased from 14968 kg (32,402 lb) to 18642 kg (41,000 lb). Cruising speed is increased to a maximum of 526 km/h (326 mph), but there is little change in other aspects of performance. Deliveries of the Series 300 began in February 1989 immediately following certification.

Bigger and better

In June 1987 DHC, by now a subsidiary of Boeing, disclosed their interest in a further stretched version – the Series 400. This would have additional plugs added to the front and rear fuselage, extending the length by a further 4.67 m (15 ft 4 in) to 29.87 m (98 ft 4 in).

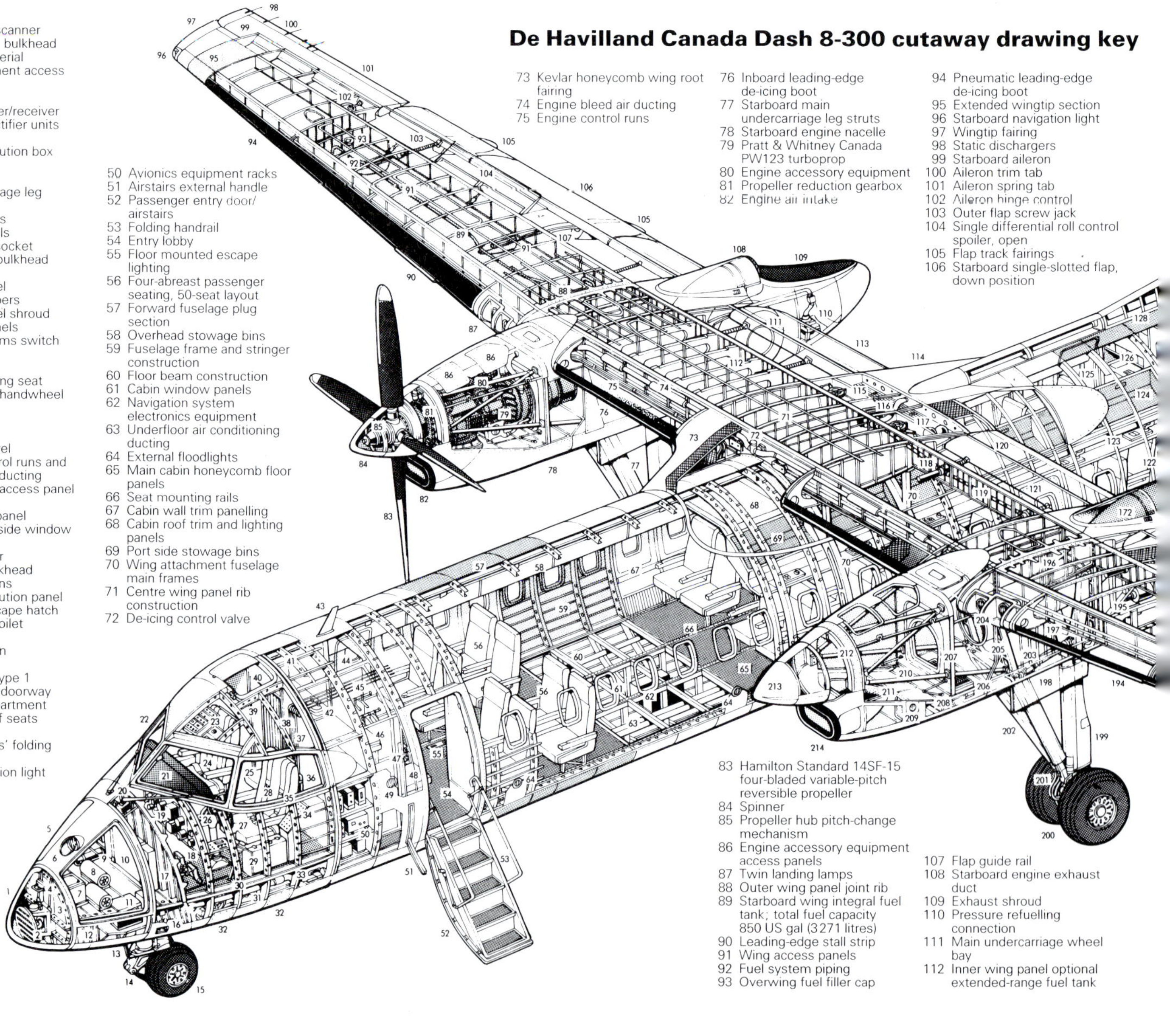

De Havilland Canada Dash 8-300 cutaway drawing key

1 Radome
2 Weather radar scanner
3 Radar mounting bulkhead
4 ILS glideslope aerial
5 Nose compartment access doors
6 Taxiing lamp
7 Radar transmitter/receiver
8 Transformer rectifier units
9 Oxygen bottles
10 Electrical distribution box
11 Battery
12 Nosewheel bay
13 Nose undercarriage leg doors
14 Nosewheel forks
15 Twin nosewheels
16 Ground power socket
17 Front pressure bulkhead
18 Rudder pedals
19 Instrument panel
20 Windscreen wipers
21 Instrument panel shroud
22 Windscreen panels
23 Overhead systems switch panel
24 Co-pilot's seat
25 Observer's folding seat
26 Control column handwheel
27 Pilot's seat
28 Safety harness
29 Document case
30 Cockpit floor level
31 Underfloor control runs and air conditioning ducting
32 Control system access panel
33 Pitot head
34 Circuit breaker panel
35 Curved cockpit side window panels
36 Fire extinguisher
37 Cockpit rear bulkhead
38 Control cable runs
39 Electrical distribution panel
40 Cockpit roof escape hatch
41 Starboard side toilet compartment
42 Wash hand basin
43 VHF aerial
44 Starboard side type 1 emergency exit doorway
45 Wardrobe compartment
46 Aft-facing pair of seats
47 Interphone
48 Cabin attendants' folding seat
49 External inspection light
50 Avionics equipment racks
51 Airstairs external handle
52 Passenger entry door/airstairs
53 Folding handrail
54 Entry lobby
55 Floor mounted escape lighting
56 Four-abreast passenger seating, 50-seat layout
57 Forward fuselage plug section
58 Overhead stowage bins
59 Fuselage frame and stringer construction
60 Floor beam construction
61 Cabin window panels
62 Navigation system electronics equipment
63 Underfloor air conditioning ducting
64 External floodlights
65 Main cabin honeycomb floor panels
66 Seat mounting rails
67 Cabin wall trim panelling
68 Cabin roof trim and lighting panels
69 Port side stowage bins
70 Wing attachment fuselage main frames
71 Centre wing panel rib construction
72 De-icing control valve
73 Kevlar honeycomb wing root fairing
74 Engine bleed air ducting
75 Engine control runs
76 Inboard leading-edge de-icing boot
77 Starboard main undercarriage leg struts
78 Starboard engine nacelle
79 Pratt & Whitney Canada PW123 turboprop
80 Engine accessory equipment
81 Propeller reduction gearbox
82 Engine air intake
83 Hamilton Standard 14SF-15 four-bladed variable-pitch reversible propeller
84 Spinner
85 Propeller hub pitch-change mechanism
86 Engine accessory equipment access panels
87 Twin landing lamps
88 Outer wing panel joint rib
89 Starboard wing integral fuel tank; total fuel capacity 850 US gal (3271 litres)
90 Leading-edge stall strip
91 Wing access panels
92 Fuel system piping
93 Overwing fuel filler cap
94 Pneumatic leading-edge de-icing boot
95 Extended wingtip section
96 Starboard navigation light
97 Wingtip fairing
98 Static dischargers
99 Starboard aileron
100 Aileron trim tab
101 Aileron spring tab
102 Aileron hinge control
103 Outer flap screw jack
104 Single differential roll control spoiler, open
105 Flap track fairings
106 Starboard single-slotted flap, down position
107 Flap guide rail
108 Starboard engine exhaust duct
109 Exhaust shroud
110 Pressure refuelling connection
111 Main undercarriage wheel bay
112 Inner wing panel optional extended-range fuel tank

Seating capacity would go up into the 66-70 seat region, cruising speed to 350 kt and maximum take-off weight to around 21320 kg (47,000 lb). This shows the extent to which the major markets need speed and capacity, with the emphasis on fuel burn per seat kilometre. STOL capability has become less and less important, and the field length for the Series 400 would be around the 1300-m (4,265-ft) mark. Although development work was at one time suspended, it was restarted early in 1990 and by the middle of the year approximately 100 people were employed on definition studies. By then an element of the Bombardier empire, DHC launched the Dash 8-400 in June 1995 with Sextant Avionique and Mitsubishi as partners.

At the start of the programme a Canadian writer compared the Dash 8 with four major rivals, and concluded that the Toronto product would be "the benchmark against which the others are measured." Although the competition has perhaps been tougher than DHC expected, some 567 Dash 8s had been sold by August 1998, and of these 494 had been delivered. Meanwhile, the Franco-German ATR-42/72, the Canadian turboprop's main rival, is only slightly behind in the market place with just over 560 sold, though starting more than a year later. For the future, the Dash 8 will have to stand up to a whole phalanx of new similarly-sized jet commuters including the Canadair Regional Jet, EMBRAER EMB-145 Amazon, the Sino-German MPC.75 and down-sized BAe 146s (the RJs) and Fokkers (the 70). Hopefully, the Dash 8's high cruise speed, bringing only a small disadvantage in block times, together with superior fuel burn characteristics, may save the day for the sleek DHC turboprop.

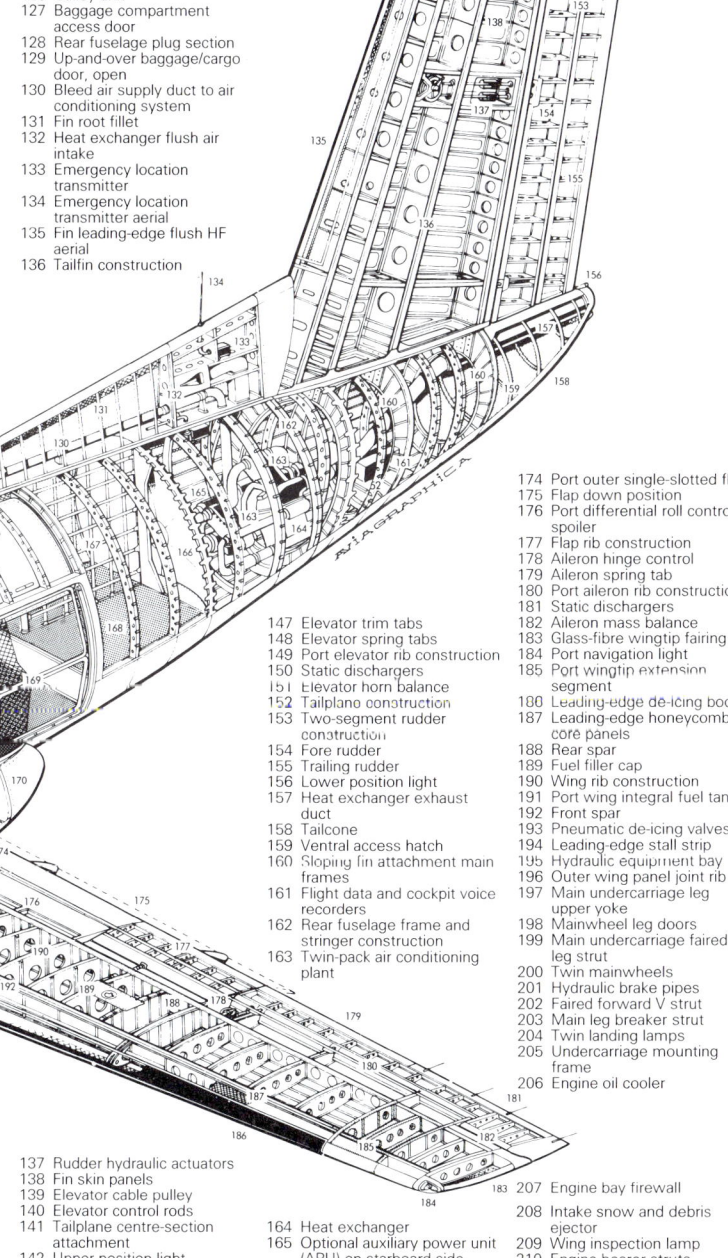

113 Inboard flap segment
114 Trailing-edge wingroot fillet
115 Control cable linkages
116 Flap hydraulic motor
117 Fire extinguisher bottles
118 Flap screwjack housing
119 Flap drive shaft
120 Port inboard single-slotted flap segment
121 Emergency exit window panel, port and starboard
122 Rear cabin window panels
123 Overhead passenger service units
124 Stowage bins
125 Starboard side service door/emergency exit
126 Galley unit
127 Baggage compartment access door
128 Rear fuselage plug section
129 Up-and-over baggage/cargo door, open
130 Bleed air supply duct to air conditioning system
131 Fin root fillet
132 Heat exchanger flush air intake
133 Emergency location transmitter
134 Emergency location transmitter aerial
135 Fin leading-edge flush HF aerial
136 Tailfin construction
137 Rudder hydraulic actuators
138 Fin skin panels
139 Elevator cable pulley
140 Elevator control rods
141 Tailplane centre-section attachment
142 Upper position light
143 Anti-collision light
144 Tailplane leading-edge pneumatic de-icing boot
145 Starboard tailplane
146 Starboard elevator
147 Elevator trim tabs
148 Elevator spring tabs
149 Port elevator rib construction
150 Static dischargers
151 Elevator horn balance
152 Tailplane construction
153 Two-segment rudder construction
154 Fore rudder
155 Trailing rudder
156 Lower position light
157 Heat exchanger exhaust duct
158 Tailcone
159 Ventral access hatch
160 Sloping fin attachment main frames
161 Flight data and cockpit voice recorders
162 Rear fuselage frame and stringer construction
163 Twin-pack air conditioning plant
164 Heat exchanger
165 Optional auxiliary power unit (APU) on starboard side
166 Rear pressure bulkhead
167 Baggage restraint net
168 Baggage/cargo bay floor
169 Baggage door guards
170 Port nacelle tail fairing
171 Exhaust duct shroud
172 Port engine exhaust pipe
173 Honeycomb trailing-edge shroud panels
174 Port outer single-slotted flap
175 Flap down position
176 Port differential roll control spoiler
177 Flap rib construction
178 Aileron hinge control
179 Aileron spring tab
180 Port aileron rib construction
181 Static dischargers
182 Aileron mass balance
183 Glass-fibre wingtip fairing
184 Port navigation light
185 Port wingtip extension segment
186 Leading-edge de-icing boot
187 Leading-edge honeycomb core panels
188 Rear spar
189 Fuel filler cap
190 Wing rib construction
191 Port wing integral fuel tank
192 Front spar
193 Pneumatic de-icing valves
194 Leading-edge stall strip
195 Hydraulic equipment bay
196 Outer wing panel joint rib
197 Main undercarriage leg upper yoke
198 Mainwheel leg doors
199 Main undercarriage faired leg strut
200 Twin mainwheels
201 Hydraulic brake pipes
202 Faired forward V strut
203 Main leg breaker strut
204 Twin landing lamps
205 Undercarriage mounting frame
206 Engine oil cooler
207 Engine bay firewall
208 Intake snow and debris ejector
209 Wing inspection lamp
210 Engine bearer struts
211 Intake duct
212 Forward engine mounting ring frame
213 Port propeller spinner
214 Intake lip de-icing

Speedy turn-round times are essential if aircraft are to earn their keep, not least when they are being relied on to feed larger airlines at 'hub' airports, a role to which the Dash 8 is ideally suited.

The recent foiling of a take-over bid by ATR has secured the future of Dash 8 at its Downsview home, for the time being. However, its then-owner, Boeing, later sold the plant to Bombardier.

Chapter 17

Dassault Falcons

De Havilland Canada Beaver

De Havilland Canada DHC-3 Otter

De Havilland Canada DHC-6 Otter

De Havilland Comet

Dassault Falcons

In the hard-fought arena of executive jet sales, the Dassault Falcon series is unquestionably the best-selling family of biz-jets, with a pedigree that stretches back over several decades. Dassault's stylish Falcons look set to carry their success well into the next century.

In the late 1950s Avions Marcel Dassault, in conjunction with Sud Aviation, began design studies for a light twin jet intended principally as an executive transport. Design and engineering work were carried out under the leadership of Paul de Plante and Paul Chassagne, who proposed an eight/10-seat aircraft which incorporated some of the aerodynamic features of the Mystère IV jet fighter.

Marcel Dassault gave his personal approval to go ahead with prototype construction of the aircraft, designated Mystère 20, in December 1961, and two months later the first metal was cut at Dassault's factory at Bordeaux-Merignac, where the wings were built, and at Sud Aviation's St Nazaire plant, which was responsible for manufacturing the fuselage, tail assembly and engine nacelles.

The prototype Mystère 20, F-WLKB, was rolled out on 1 April 1963 and made its first flight on 4 May, temporarily powered by 14.6-kN (3,300-lb st) Pratt & Whitney JT12A-6 turbojets pending availability of the 18-kN (4,125-lb st) General Electric CF700 engines which had been selected for production aircraft. Among the witnesses to the first flight were senior executives of Pan American World Airways, including Colonel Charles Lindbergh. Pan Am was interested in marketing and operating the aircraft in the United States, and subsequently established a Business Jet Sales Division jointly with Dassault to handle marketing the aeroplane throughout the Western hemisphere, placing an initial order in July 1963 for 40 aircraft, with options on a further 120.

The prototype made its public debut at the 1963 Paris air show and flew again after re-engining with production-standard CF700-2B turbofans on 10 July 1964. By the end of that year it had completed more than 250 flights, and was joined in the test and certification programme on 1 January 1965 by the first production Mystère 20 F-WMSH, which incorporated a number of airframe changes, notably a 100-cm (39½-in) increase in wingspan. An increase of 46-cm (18-in) in fuselage length and twin wheels on the main and nose landing gears units in place of the prototype's single wheels.

Joint French and US certification of the Mystère 20 was granted on 9 June 1965 in a ceremony at that year's Paris air show, where the first production aircraft appeared in 'Pan Am Business Jets' colours. Later that month the French test-pilot Jacqueline Auriol set 1,000 km (621 miles) and 2,000 km (1248 miles) closed-circuit world speed records for its class with the second production Mystère 20 F-BMSS, averaging 859 km/h (534 mph) and 819 km/h (509 mph) respectively, over triangular courses between Istres and Cazaux, France.

Deliveries were quickly under way, those destined for the United States being ferried as 'green' (bare shell) airframes to completion centres where avionics, customised interiors and exterior paint were added before handover to customers. Aircraft for other markets were mostly outfitted at Dassault's Bordeaux

The Dassault executive jet family at the end of the 1980s. In the foreground is a Falcon 100, followed by a Falcon 200, Falcon 50 and finally a Falcon 900. Dassault has never given type numbers to these aircraft, and any designation such as My.20 or Da.100 is entirely unofficial.

The Mystère XX flew on 1 January 1965 and deliveries soon began to the type's American distributor, Pan American Business Jets Inc. Initially it was marketed as the Fan Jet Falcon, before being rechristened Falcon 20.

The Falcon 30 was a one-off stretched Falcon 20, intended as a 30-seat airliner, powered by a pair of Lycoming ALF 502D turbofans. Beginning life as the Falcon 20T, it was felt to be too far ahead of its time.

The shapely Falcon 10 followed the success of the larger Falcon 20. This Mini-Falcon, as it was first known, entered production in 1973 and in the 17 years that followed 193 were sold around the world.

plant or at its wholly-owned subsidiary Europe Falcon Service at Le Bourget Airport, Paris. The name Mystère 20 was used primarily in France; Fan Jet Falcon was adopted in the United States, later evolving into Falcon 20, the initial production version being known as the Falcon 20C because of its CF700-2C engines.

This model was superseded in 1968 by the Falcon 20D, which had uprated 19-kN (4,250-lb st) CF700-2D engines, a 303-litre (80-US gal) increase in fuel capacity, a 402-kg (887-lb) increase in maximum take-off weight to 12400 kg (27,337 lb) and larger wheel brakes. Further improvements were incorporated in the Falcon 20E, with 19.2-kN (4,315-lb st) engines, 13000 kg (28,660 lb) maximum take-off weight, higher zero-fuel weight and high capacity starter/generator. Next came the Falcon 20F. This model, first flown in the summer of 1969, introduced full-span leading-edge slats for enhanced take-off and landing performance,

From 1970 the standard production Falcon was the Falcon 20F. Improvements over the 20E version were its full leading-edge slats and increased fuel tankage. One hundred and thirty-four of this version were sold and many customers moved up to the Falcon 50 when it arrived. Note the non-regulation style of registration on this German example.

and also had a further increase in fuel capacity to 5224 litres (1,380 US gal) and single-point pressure refuelling as standard. Deliveries of the Falcon 20F began in July 1970; it was the first aircraft to receive type approval under US Federal Aviation Administration FAR Part 36 noise regulations.

Ocean guardians

In the spring of 1976 Dassault announced development of a retrofit programme to enable 24.6-kN (5,538-lb st) Garrett ATF3-6-2C turbofans to be fitted to Falcon 20s, with new-production Garrett-powered aircraft to follow later under the designation Falcon 20G. Later that year Falcon Jet Corporation successfully tendered this proposed aircraft for the US Coast Guard's Medium Range Surveillance (MRS) requirement, for which a contract was placed for 41 Falcon 20Gs under the US military designation HU-25A Guardian. The retrofit programme was then abandoned to enable Dassault to concentrate on the USCG aircraft, a development prototype for which (F-WATF) made its maiden flight on 28 November 1977. USCG HU-25As were delivered in 1982-83. A number have since been modified as HU-25B sea pollution monitors and HU-25C narcotics interdiction aircraft. The latter, eight of which have been converted, are equipped with an F-16-type Westinghouse AN/APG-66 radar, turret-mounted Texas Instruments WF-360 forward-looking infra-red (FLIR), and secure communications radios.

A commercial version of the Falcon 20G was known as the Falcon 20H, and was introduced at the 1981 Paris air show. It featured 23.1-kN (5,200-lb st) Garrett ATF3-6A-4C engines, increased fuel tankage in the rear fuselage, redesigned wing root fairings, automatic leading-edge slat extension and numerous systems changes. Certification of the Falcon 20H, which was marketed as the Falcon 200, was achieved on 21 June 1981. Initially the Falcon 20F and Falcon 20H/200 were produced concurrently, but in late 1983 the last 20F came off the production line, and in 1988 the last of 35 Falcon 200s was delivered, bringing total production of Falcon 20/200s of all variants to 508.

The Falcon 20 proved adaptable to many roles beyond its primary task of transporting businessmen, and has been used for airline pilot training, airways and navaid calibration, high altitude aerial photography and mapping, electronic countermeasures training, medevac, military transport and airborne sensing. A fleet of 21 Falcon 20s is operated by the British company FR Aviation to provide target-towing and anti-shipping threat simulation.

Cargo carrying has also been an important role for the aircraft, and the first conversion of a Falcon 20 to this role led directly to the foundation of what has since become one of the

Five Falcon 10MER (Marine Entrainement Radar) aircraft were delivered to France's Aéronavale, though one was lost in an accident in January 1980. They operate with 57 Escadrille from Landivisiau.

The last 33 Dassault Falcon 10s off the line were known as Falcon 100s. The most obvious difference between the two is the 100's additional fourth window on the starboard side. Its gross weight was also increased.

world's leading overnight package and airfreight operations – Federal Express Corporation. FedEx's founder, Fred Smith, acquired an Arkansas business jet maintenance and sales company called Little Rock Airmotive in 1969 and was contracted by Pan Am Business Jets to develop an all-cargo prototype of the business jet, known as the Falcon 20 Cargo Jet, the first of which flew on 28 March 1972. The conversion involved replacement of the aircraft's standard cabin door with an upward-opening hydraulically-operated cargo door 188 cm (74 in) wide and 152.4 cm (60 in) high, and installation of a 7-m (23-ft) long aluminium honeycomb cargo floor with tie-down points, providing a 14-m^3 (500-cu ft) flat-floor cargo hold. Smith subsequently formed Federal Express and acquired 33 Falcon Cargo Jets as the nucleus of a fleet which nightly served 76 major airports throughout the United States, bringing packages to a computerised sorting centre at Memphis, Tennessee. The Cargo Jet conversion can be applied to any Falcon 20 model, and following Federal Express's success was offered by AMD as a factory option.

Range of options

Two engine replacement programmes have been developed for the Falcon 20 series. Garrett General Aviation Services, in conjunction with Dassault, announced in May 1987 the availability of TFE731-5AR turbofans to replace the aircraft's standard GE CF700s. The modification also involves installation of new engine pylons with improved aerodynamics, and Falcon 900 engine nacelles. Converted aircraft are known as 731 Falcons, and are given a -5 suffix to their factory designations (thus Falcon 20C-5) to denote the engine change. The modification results in increases in range of up to 50 per cent, depending on model, and enables the retrofitted aircraft to meet FAA FAR Part 36 Stage 3 noise standards. Dee Howard TR-5020 thrust reversers are an option with the TFE731 retrofit. The first 731 Falcon conversion was performed by Dassault and flew on 7 October 1988; more than 40 Falcon 20s have since been modified for customers by Garrett and Falcon Jet Service in the USA, Europe Falcon Service in France and TransAirCo in Switzerland.

Volpar Aircraft Corporation of Van Nuys, California offers a modification to install 23-kN (5,225-lb st) Pratt & Whitney Canada PW305 turbofans with Rohr Industries thrust reversers in Falcon 20s. The modified Volpar Falcon PW300-F20, first flown in prototype form on 12 February 1991, cruises 80 km/h (50 mph) faster than the standard CF700-powered Falcon 20, has a maximum range of 4184 km (2,600 miles) with IFR reserves, and also meets FAR Part 36 Stage 3 noise criteria. Certification and first customer deliveries are expected during 1992.

In June 1969 Avions Marcel Dassault announced development of a smaller business jet known as the Mystère 10 MiniFalcon, which was to be a scaled-down version of the Falcon 20 having about 75 per cent of its power and weighing about 30 per cent less, with accommodation for four/eight passengers and two crew. Subsequently renamed Falcon 10, the prototype (F-WFAL) made its first flight on 1 December 1970, powered by two 13-kN (2,954-lb st) General Electric CJ610 turbojet engines. Following modifications to increase the angle of wing sweepback and alter its angles of dihedral and incidence, the aircraft resumed test flying on 7 May 1971 and set an FAI 1,000-km (621-mph) closed circuit speed record of 926 km/h (575 mph) on 1 June. The second and third prototypes, powered by production-standard 14.3-kN (3,230-lb st) Garrett AiResearch TFE731-2 turbofans, made their first flights on 15 October 1971 and 14 October 1972 respectively.

Falcon 10 in service

The first production aircraft was flown on 30 April 1973; certification was granted by the French authorities on 11 September 1973, followed by FAA certification nine days later, enabling customer deliveries to begin in November. Final assembly of the Falcon 10 took place at AMD's Istres plant, using airframe components supplied by Potez (fuselage), CASA of Spain (wings), IAM of Italy (nose assembly and tail unit) and Latécoère (doors, fin and other small components). In 1981 Dassault introduced a 224.9-kg (496-lb) increase in maximum take-off weight, a fourth cabin window on the starboard side, a larger rear cabin baggage compartment and optional five-tube Collins EFIS-85 avionics on the aircraft, which was then redesignated Falcon 100. Production of this smallest member of the Falcon family came to an end in September 1990 when the 226th aircraft was delivered to Mexican charter operator Aerpersonal.

Although most of the Falcon 10/100s built were sold to commercial customers, primarily in North America, six were delivered to the French navy as Mystère-Falcon 10MERs. These aircraft, five of which remain in service, are used as instrument, navigation and radar trainers, for communications duties, calibration of shipboard radars and for medevac missions.

In May 1985 an experimental Falcon V10F was flown with a set of wings manufactured from resin-impregnated carbonfibre under a French government-sponsored research and development programme in which Dassault

Dassault never abandoned the Mystère tag on its biz-jets, and even when the Falcon 20 evolved into the Falcon 200 it was often referred to by the manufacturer by its French title.

The US Coast Guard bought 41 Falcon 20Gs as HU-25 Guardians. The aircraft have evolved into three versions, the most radical of which is the long-nosed APG-66-equipped HU-25C Interceptor.

Dassault's move into the long-range jet market came with the Falcon 50 of 1976. It challenged Gulfstream's domination of the scene, and 190 had been sold by the time the Falcon 900 was introduced.

and Aérospatiale co-operated. After completing 40 test flights the aircraft received French certification on 19 December 1985 and subsequently entered service in the charter fleet of Europe Falcon Service at Le Bourget, Paris.

Avions Marcel Dassault first announced development of larger versions of the Falcon 20 in 1971, when it revealed the Falcon 20T feederliner project, intended as a straightforward 'stretched' variant of the business jet to carry 29 passengers in high-density commuter airliner seating. This concept was soon abandoned, and an entirely new larger diameter fuselage with a 30/40-seat cabin created for the Falcon 30/40. A prototype, powered by two Lycoming ALF 502D turbofans, made its maiden flight on 11 May 1973, but after two years of flight testing further development was considered uneconomic and the project was abandoned.

Since production ended, the Falcon 20 has become a prime candidate for re-engining, with several conversions on offer. Aircraft fitted with new Garrett TFE731 turbofans are known as Falcon 20-5s.

Meanwhile, Dassault had been working on a way to extend greatly the Falcon 20's long distance capability, aimed at almost doubling the early model's 3704-km (2,301-mile) range. This was achieved by the development, using computer-aided design (CAD) techniques, of an entirely new wing featuring a highly efficient supercritical airfoil section and sophisticated high-lift devices; a new fuselage employing area ruling; and substitution of three fuel-efficient 16.45-kN (3,700-lb st) Garrett AiResearch TFE731 turbofans for the more powerful CF700s of the twin-engined Falcons. The result was the Falcon 50, an eight/10 passenger aircraft with a maximum range in excess of 6482 km (4,028 miles).

The prototype Falcon 50 (F-WAMD) flew for the first time on 7 November 1976 and was joined in the test programme by a second prototype (F-WINR) on 18 February 1978, and a pre-production aircraft four months later. French certification was achieved on 27 February 1979, with FAA approval following a week later. On 31 March 1979 a Falcon 50 established a straight-line world distance record for its class, flying 6099 km (3,790 miles) non-stop across the United States.

The Falcon 50 remains in production. Total deliveries exceeded 265 aircraft by late 1998, including two operated by the Armée de L'Air's Groupe de Liaisons Aériennes Ministérielles for Presidential transport duties, three for the Italian air force, three for the Portuguese air force, and VIP aircraft for the heads of state of Djibouti, Iraq, Jordan, Libya, Morocco, Portugal, Rwanda, South Africa, Spain, Sudan and the former Yugoslavia.

The 1983 Paris air show was selected by AMD for the announcement of the next major step in the Falcon programme – a three-turbofan intercontinental range business jet to rival the Gulfstream IV then in the initial stages of development in the United States. The Falcon 900 is similar in general configuration to the Falcon 50, but of bigger overall size, particularly its deep-section fuselage which houses a 19-seat cabin. Extensive use of CAD/CAM techniques was made in designing the aircraft, and carbonfibre and Kevlar composites figure largely in its structure.

The first Falcon 900 (F-GIDE), named *Spirit of Lafayette* flew on 21 September 1984, powered by three 20-kN (4,500-lb st) Garrett TFE731-5AR-1C turbofans. A second development aircraft (F-GFJC), which joined the flight test programme on 30 August 1985, demonstrated the aircraft's intercontinental capability in the following month when it flew non-stop from Le Bourget to the Falcon Jet Corporation's headquarters at Little Rock, Arkansas, a distance of 7973 km (4,954 miles), subsequently making a non-stop return trip from Teterboro, New Jersey to the French flight test centre at Istres. French and US certification of the Falcon 900 were received a week apart at the end of March 1986. Customer deliveries began in the following December, and had reached 120 aircraft by the end of 1992. Military and governmental customers include Algeria, Australia, France, Malaysia, Nigeria, Qatar, Spain, Syria and the UAE. Two specially-configured long-range maritime surveillance versions of the Falcon 900 were delivered to the Japan Maritime Safety Agency in 1989. These aircraft are equipped with bulged observation windows, drop hatches for launching sonobuoys, flares and markers and a ventral search radar installation. Civilian Falcon 900s have been sold in 17 countries.

In 1991 AMD announced a new version of the aircraft, designated Falcon 900B, which is powered by uprated 21.1-kN (4,750-lb st) TFE731-5B turbofans whose five and a half per cent per engine power increase enables the 900B to direct-climb fully-loaded to an initial cruising altitude of 11887 m (39,000 ft), operate with reduced take-off run at maximum payload in hot-and-high runway conditions, and provides a 185-km (115-mile) increase in maximum payload range with IFR reserves. The TFE731-5B engine upgrade is available as a retrofit to earlier Falcon 900s.

New generation

At the 1989 Paris air show AMD first announced the latest addition to its business jet range, then provisionally designated Falcon X, but subsequently renamed Falcon 2000 in October 1990 when the development programme was officially launched with an order for five aircraft from the Swiss executive jet charter company Aeroleasing. The Falcon 2000 airframe is based substantially on that of the Falcon 900, retaining its cabin cross-section with a 1.98-m (6-ft 6-in) reduction in fuselage length, but is a twin-engined aircraft powered by two 25.5-kN (5,727-lb st) General Electric/Garrett CFE738 turbofans. The first Falcon 2000 has a maximum passenger capacity of 12 in executive configuration, and was rolled out for the first time on 4 March 1993, with French and US certification gained in November 1994 and February 1995 respectively. In terms of its performance, the Falcon 2000 has a range of 5556 km (3,452 miles) with IFR reserves, cruising at Mach 0.83-0.85 with eight passengers. The Italian manufacturer Alenia is a 25% risk-sharing partner on the development and production of the Falcon 2000, with responsibility for the rear fuselage sections and engine nacelles. By late in 1994 Dassault had secured commitments for a total of 60 aircraft within the context of a total market then estimated at more than 300 such aircraft in 10 years.

The French military has been a loyal customer for Dassault's Falcons and was an early buyer of the Falcon 900. ET 1/60 GLAM (Groupe de Liaison Aériennes Ministérielles) currently has two aircraft on charge.

de Havilland Canada Beaver

In the Beaver, DHC succeeded in building an aircraft that was equally at home in the desert and in the snows of Canada. It can operate from land or water and, despite its age, it will certainly be flying into the next century.

Among the de Havilland Canada company's plans for post-war activity was a proposal for a rugged light transport aircraft to service the anticipated boom in mineral, oil and timber exploration in the Canadian north, and to replace such types as the Noorduyn Norseman and Fairchild 82 which were then being used by bush operators to link isolated communities otherwise dependent on boats, railroads or even dog teams.

Many experienced Canadian bush fliers were surveyed for their requirements, and preliminary design work was undertaken before war's end by DHC's chief designer Wsiewolod Jakimiuk and aerodynamicist W. Z. Stepniewski. The appropriate name 'Beaver' was chosen at an early stage, after the industrious animal indigenous to Canada, but development was postponed while the company concentrated on design and prototype construction of the DHC.1 Chipmunk trainer.

Initial plans were for an aircraft that bore a resemblance to the pre-war D.H.80 Puss Moth light aircraft. However, when design work resumed in the spring of 1946, R. D. Hiscocks replaced Stepniewski as aerodynamicist on the project, and he redesigned the wing, changing its aerofoil from an RAF section to one of his own creation, and replaced the original two-spar/V-strut arrangement with a single spar/single strut configuration, and to enhance take-off performance devised a combined flap/aileron system whereby the aircraft's ailerons would droop 15° when the flaps were lowered fully.

During the preliminary design phase DHC liaised closely with a potential customer, Ontario Provincial Air Service, and refined the aircraft's configuration on advice from that company, introducing features such as fuselage rather than wing fuel tanks to avoid awkward over-wing refuelling, drop hatches in the cabin floor, hatches in the rear cabin bulkhead to enable long items of cargo such as drilling rods to be carried, and wide doors that would enable 45-US gal oil drums to be rolled directly into the cabin.

It had been proposed to power the Beaver with a 216-kW (295-hp) or 246-kW (330-hp) DH Gipsy Queen engine, in the belief that a combination of de Havilland airframe and engine would prove a strong selling point, but doubts about the power output and suitability of these new engines for bush operations led to

Operations of the Beaver are constrained by few types of terrain thanks to its amazing STOL performance and ability to match its undercarriage to the chosen landing site. Many flew on floats, including this Royal Canadian Mounted Police aircraft.

a change to the proven, reliable, rugged and ubiquitous Pratt & Whitney R-985 Wasp Junior radial engine driving a Hamilton Standard two-bladed constant-speed propeller.

Detail engineering design began in September 1946, and construction of a prototype was started at DHC's Downsview factory on 15 January 1947. It was a robust and simple 'no-nonsense' aeroplane of all-metal stressed-skin construction, the airframe stressed to +3.5g. The slab-sided fuselage was a semi-monocoque structure with strengthened cabin floor for heavy load-carrying and room for up to eight quickly-removable seats in two-three-three configuration. Two wide entrance doors were provided for the main cabin, with a separate door for the pilot, and three fuel tanks for total capacity of 360 litres (95 US gal) were mounted in the belly. The parallel chord wings had heavy dural spar beams and alclad skinning, with external stiffeners on flaps and ailerons; fin and rudder bore the hallmark of the classic 'DH' shape. The cantilever landing gear was strong and simple, employing rubber-doughnut and oleo shock struts, with a heavy-duty castoring tailwheel. Float, ski, wheel/ski and amphibious landing gear were optional – an essential requirement for the off-airport environment which was to be the Beaver's natural habitat. All-up weight was 2040 kg (4,500 lb), later increased to 2213 kg (5,100 lb) on production aircraft, which provided a payload of 612 kg (1,350 lb) with sufficient fuel to fly 756 km (470 miles) with reserves. The Beaver's short take-off qualities were remarkable: with a full load, in still air it could lift off in under 183 m (600 ft).

For 20 years de Havilland Canada built the Beaver I, the basic production version with Pratt & Whitney Wasp Junior engine. Production of all Beaver variants reached a creditable 1,631.

'Beaver One'

The prototype DHC.2 Beaver, CF-FHB-X, made its first flight on 16 August 1947 in the hands of DHC operations manager and chief test pilot Russ Bannock. Early flight trials convinced de Havilland Canada that it had a winner in the Beaver, though just how successful it was to become could not have been apparent at that time. Production was launched almost immediately, with virtually no changes to prototype configuration save for some refining of the windscreen shape to improve airflow. A Canadian Department of Transport type certificate was issued on 12 March 1948 and deliveries began to commercial customers in Canada and the United States, where certification was granted simultaneously with Canadian approval. Among early operators were the Royal Canadian Mounted Police and the US Border Patrol, along with numerous small airlines, charter and cargo carriers throughout Canada and in Alaska, where Beavers have always been favoured and continue to serve in large numbers today.

In 1949 DHC achieved the sales breakthrough which was to make the Beaver Canada's most successful export at that time and establish the company among the world's leading manufacturers of STOL utility and transport aircraft. In June of that year Russ Bannock, as head of military sales, demonstrated a production Beaver to the United States Air Force at Elmendorf AFB, Alaska, where the USAF's Search and Rescue Command had a requirement for a small number of utility aircraft. A planned purchase of 22 Beavers for this role was frustrated by the Buy American Act which in peacetime prohibited the purchase of military equipment from outside of the United States without competitive bidding, so beginning in December 1950 fly-off trials were conducted at Wright Field, Ohio, between the Beaver and 13 other types, of which its principle rival was the Helio Courier. The Beaver won the competition, and repeated this feat at Fort Bragg, North Carolina, in May 1951 when the US Army held a similar competition.

Combat Beaver

Legislation was enacted to enable a foreign military purchase to be made, following which an initial order was placed for six Beavers for trials with the US services. Designated YL-20, these were standard civilian aircraft straight from the Downsview production line. They were followed by a further 982 US Army/USAF Beavers, which differed from their

Although the performance increase offered by the Turbo Beaver was slight, the improvement in reliability, maintenance requirements and economy were significant. The low weight of the engine required the use of a longer nose to maintain the centre of gravity.

Australia was a major military operator for the Beaver, using its aircraft principally for support of Antarctic exploration with mixed ski/wheel undercarriage.

This Beaver was procured by the UK Ministry of Supply prior to a 41-aircraft order for the British Army, which flew the aircraft until May 1960 on operational missions as the Beaver AL.Mk 1. It saw much service in Northern Ireland.

civilian counterparts in having instrument and avionics changes, different cabin seating configurations and four skylight windows in the cabin roof panel. These aircraft were designated L-20A (later U-6A) in US military service and were widely deployed throughout the world, seeing combat service in the Korean and Vietnam Wars and operating on wheels, skis, floats and, in Vietnam, amphibious floats. They fulfulled a wide variety of roles from liaison, troop transport and cable laying to medevac, psychological warfare and forward air control. President Eisenhower used an L-20A for his personal transport during a 1952 visit to the Korean theatre. The last Beaver delivery to the US Army took place in 1960. At least one was transferred to the US Navy and was operated by the USN Test Pilot's School at NAS Patuxent River, Maryland.

The L-20 was redesignated as the U-6 in 1962, and continued in military service in many roles. This pair is seen in South Vietnamese markings during the South East Asia war, the loud speaker panel on the fuselage side indicating their psychological warfare role.

The Beaver's success in America soon led to orders from other military operators, and saw them wearing the insignia of the air arms of Argentina, Australia, Cambodia, Chile, Colombia, Cuba, the Dominican Republic, Finland, Ghana, Kenya, the Netherlands, New Zealand, Oman, Peru, the Philippines, South Korea, South Vietnam and Zambia. But, next to the United States, the single biggest military order came from Britain, which ordered 41 Beaver AL.1s for service with the Army Air Corps. These aircraft served worldwide with the AAC, the last finally retiring from surveillance duties in Northern Ireland in May 1990 when they were replaced by Pilatus Britten-Norman Turbine Islanders. The Argentinian, Australian and New Zealand air forces used wheel/ski-equipped Beavers to support expeditions to Antarctica, and RAAF crews were so impressed by the aircraft's performance on South Pole exploration that they named a lake, glacier and island after it.

In 1953 DHC developed a new variant known as the Beaver II, which first flew on 10 March, piloted by George Neal and bearing the registration CF-GQE-X. Powered by a British-

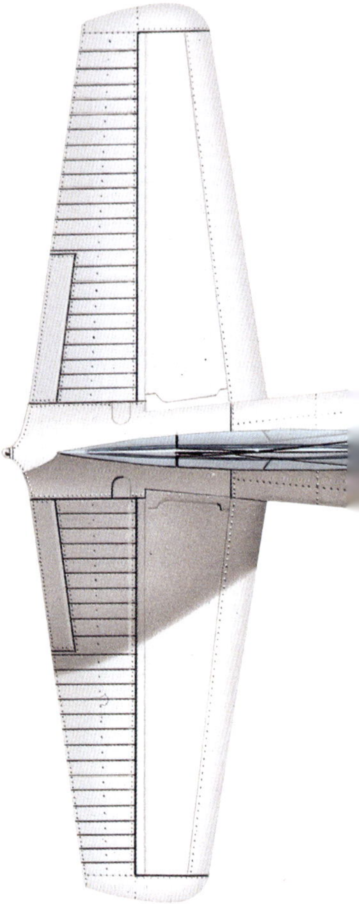

Specification
Wingspan: 14.62 m (47 ft 9 in)
Length: 9.24 m (30 ft 3 in)
Height: 2.74 m (8ft 9 in)
Wing area: 23.2 m² (249 sq ft)
Powerplant: 335.7-kW (450-hp) Pratt & Whitney R-985 Wasp Jr
Passenger capacity: 7
Empty weight: 1,294 kg (2,852 lb)
Maximum take-off weight: 2313 kg (5,100 lb)
Maximim speed: 257 km/h (160 mph)
Service ceiling: 5486 m (18,000 ft)
Maximum range: 1252 km (780 miles)

Beaver variants

Beaver I: major production version with 335-kW (450-hp) P&W R-985 Wasp Jr engine. Designated L-20A/U-6A in US military service
Beaver II: one prototype only, with 410-kW (550-hp) Alvis Leonides 502/4 engine and larger fin and rudder
Turbo Beaver III: turboprop version with 410-kW (550-shp) Pratt & Whitney Canada PT6A-6 engine, lengthened fuselage with 10-seat cabin and larger fin/rudder

A classic amongst bushplanes, the de Havilland Canada Beaver was instrumental in opening vast tracts of remote land for exploitation. Designed primarily for the hostile environment of the Canadian north, the Beaver was from the outset designed with ski or float undercarriage as options. With large areas of remote land under their jurisdiction, the Royal Canadian Mounted Police were natural customers for the type, using them for rapid transport across large distances of hostile terrain. The float undercarriage allowed a landing to be made on any of the myriad lakes that cover the Canadian interior.

The largest military customer was the US Army, which procured nearly 1,000 examples, although over 200 were used by the US Air Force. These flew on a bewildering array of liaison and light transport missions.

Beavers found favour with many overseas operators, where rough terrain or ill-prepared airstrips required the use of a tough STOL aircraft. Aerotaxi of Colombia served a mountainous region.

Ski operations were second nature for the Beaver, especially during the snowy winters in the Canadian bush. This aircraft is actually the Beaver prototype, seen during ski testing.

made 410-kW (550-hp) Alvis Leonides 502/4 radial engine which DHC hoped would make the Beaver more attractive to 'soft currency' nations, it differed externally in having a longer cowling, three-bladed propeller and a taller fin and rudder, which marred the classic DH profile. After initial flight testing the Beaver II was despatched to the parent de Havilland company in England for use as a demonstrator, and bearing the British marks G-ANAR made its debut at the 1953 SBAC Show at Farnborough. After periods of temporary trials duties with the Royal Air Force and British Army it was transferred to engine manufacturers Alvis Ltd before returning to Canada in August 1971. No production Beaver IIs were manufactured.

By 1961 DHC was considering an obvious development of the Beaver – conversion to turbine power. Good though the faithful P&W Wasp was, turbine engines offered greater reliability and economy and longer time between overhauls, and turbine fuel was often more readily available than avgas in the remote areas of the world where Beavers usually found themselves operating. De Havilland Canada studied the possibilities of re-engining with 395-kW (530-shp) Turboméca Astazou II or 410-kW (550-shp) Pratt & Whitney PT6 turboprops, finally opting for the latter engine in late 1962 when it was learned that Beaver operator Pacific Western Airlines of Vancouver was planning to market its own PT6 conversion.

Building a better Beaver

Design work on the Turbo Beaver II began in June 1963. Because the turbine engine was much lighter than the piston powerplant, it had to be mounted further forward to maintain the aircraft's centre of gravity. This enabled an extra cabin section to be inserted, moving the cockpit ahead of the wing and enabling two more passengers and additional fuel to be carried. The prototype Turbo Beaver, temporarily fitted with the fin and rudder of the Beaver II, made its first flight on 31 December 1963, flown by Bob Fowler and Jock Aitken. Production aircraft, powered by the PT6A-6 engine, had swept fins and rudders of new design. Never sprightly, the Beaver gained little in performance from the turbine engine, the Turbo Beaver's maximum speed of 273 km/h (170 mph) adding only 16 km/h (10 mph) to that of the Wasp-engined aeroplane, though rate of climb increased by 50 m (165 ft) per minute and

The prototype Turbo Beaver displays the revised lines of the type. In addition to the lengthened nose, the new aircraft introduced an angular and swept-back fin of larger area to cope with the extra power of the engine.

A versatile configuration for the Beaver had wheeled floats for amphibian operations. Each float had a nosewheel which swung round and under the float, and a semi-retractable mainwheel. Water rudders also swung down for water operations.

New Zealand was another nation to use the Beaver for Antarctic support missions, this example having ski-only undercarriage. Named City of Auckland, this aircraft is seen over the frozen wastes of Antarctica, where the type performed uncomplainingly in harsh and hazardous conditions.

service ceiling by 610 m (2,000 ft), while payload went up to 816 kg (1,800 lb). Deliveries began in January 1965. Sixty Turbo Beaver IIIs were built; retrofit conversions of piston-engined aircraft were planned, but never carried out.

De Havilland Canada delivered the 1,631st and last production Beaver in 1967, having sold 1,077 to military customers and 552 to commercial operators in 62 countries. Never glamorous, Beavers have always been worked hard, serving in such diverse roles as aerial ambulances for remote settlements, paradrop aircraft, spotters for fishing fleets, locust sprayers, border patrollers, indeed any activity where ruggedness, dependability and the ability to carry virtually anything you could get into its cabin were the operator's requirements. Its utility was further extended in the early 1960s when Ottawa-based operators Bradley Air Services developed huge low-pressure 'tundra' tyres which enabled the Beaver to be operated from unprepared rock-strewn surfaces on barren Arctic islands.

Now largely redundant in the armed services, surplus military Beavers are still much sought after by bush fliers and are being remanufactured by Wipaire Inc. of Minnesota, well-known maker of Wiplane seaplane floats. Wipaire completely strips Beaver airframes, extends the rear cabin area and adds two additional windows on each side, adds a baggage door and tinted skylights, replaces all seats, installs soundproofing, electrically-actuated flaps and customised IFR instrument panels with autopilot and electric trim, replaces the standard two-bladed propeller with a three-bladed Hartzell unit with spinner, reassembles the corrosion-proofed airframe using all new fasteners, cables, pulleys and rods, installs a zero-timed R-985 engine and finishes the aircraft in polyurethane paint. These 'good-as-new' Wipaire Super Beavers are available with conventional wheel landing gear or with Wipline 6000-series floats or amphibious floats, and are set to extend the life of Canada's most famous bush aircraft well into the 21st century.

de Havilland Canada DHC-3 Otter

The DHC-2 Beaver convinced all involved that such versatile aircraft were a necessity, not least in the rugged terrain of its Canadian homeland. The heir to the Beaver's throne was the de Havilland Canada Otter, which became king of the 'bush' aircraft. Bigger and faster than the DHC-2, it retained the ability to operate from ice and snow, lakes and river-banks; in fact, anywhere that people or cargo are needed the Otter can bring them there.

The successful introduction of the DHC-2 Beaver by Ontario Provincial Air Service prompted the operator to inform de Havilland Canada that if an aircraft of comparable performance but with much greater payload could be developed, OPAS would place a launch order for 20.

DHC's Richard D. Hiscocks and Frederick H. Buller were assigned to conduct initial design studies, and produced a configuration generally similar to that of the Beaver, but some 3 m (10 ft) longer both in fuselage dimensions and wingspan, with 50 per cent increases in empty and maximum take-off weights. The single-strut braced high wing featured full-span slotted flaps, the outer portions of which formed the aircraft's ailerons. Accommodation was provided for two crew and nine passengers on folding seats, or up to six stretcher cases in air ambulance role, with provision for loading bulky freight via double doors on the port fuselage side. A strengthened cabin floor was incorporated for cargo carrying, with provision for a drop hatch for cargo or paratroop dropping or for aerial photography. Wheel, float, ski, wheel/ski or amphibious landing gear could be installed and would in interchangeable. Desirable performance parameters included the ability to operate from a 1000-m (3,048-ft) long rough airstrip, and a payload/range capability of 1430 kg (3,153 lb) over 321 km (200 miles) or 1055 kg (2,325 lb) over 1420 km (882 miles) unrefuelled. The powerplant selected was a 447-kW (600-hp) Pratt & Whitney R-1340 S3H1-G Wasp geared radial engine with a pair of exhaust augmentor tubes on each side of the cowling to enhance

Like the Beaver, the Otter was a natural for the Royal Canadian Mounted Police, which operated the type on skis and floats to maintain authority over far-flung communities in the bush. This floatplane has a ventral fin added to counteract the loss in directional stability caused by the floats: not all Otter floatplanes had this.

engine cooling, particularly during high-power, low-airspeed operation. Pratt & Whitney Aircraft of Canada had manufactured 1,149 Wasps under licence, and when the Otter entered production the company developed a modification kit to enable direct-drive R-1340 engines to be converted to the geared type during overhaul.

On 29 November 1950 Factory Instruction 390 was issued authorising construction of one prototype DHC-3 'King Beaver'. This name was changed to Otter before the aircraft, registered CF-DYK-X, made its first flight from Downsview near Toronto on 12 December 1951. Flown by DHC chief test pilot George Neal, the Otter lifted off after a ground roll of less than 182 m (600 ft), which boded well for its ability to meet short field requirements. A second prototype, CF-GCV-X, joined the test programme in 1952.

Proving flights

During the course of its test flight programme a number of changes were made to the Otter's airframe, the most significant being an increase in fin and rudder area and replacement of the original straight-edged fin with a curved dorsal strake, and the addition of a fence on the upper surface of each wing immediately inboard of the ailerons. These were incorporated in the third and subsequent airframes. Canadian certification of the Otter was obtained on 5 December 1952 for operation at an initial maximum take-off weight of 3265 kg (7,200 lb), later increased to 3629 kg (8,000 lb). Customer deliveries began almost immediately, with the fourth production Otter

The US Army was the major military customer for the Otter, designating its aircraft U-1A. The US Navy also operated 16, designated UC-1 (illustrated) and used primarily by VX-6 on Antarctic support duties; these became U-1Bs in 1962. The USAF evaluated one Army machine but did not adopt the type.

CF-DYK was the prototype Otter, the 'X' suffix denoting its experimental status. The rugged lines of the type were inherited from the smaller Beaver, the new machine offering similar performance but with a much greater load.

going to Hudson Bay Air Transport on 11 December.

Among other early deliveries were aircraft for Imperial Oil Ltd, Wardair Canada Ltd, Wideroe's Flyveselskap A/S of Norway, and launch customer Ontario Provincial Air Service, but the most significant order (and a surprising one since it had not been a customer for the Beaver) came from the Royal Canadian Air Force, which ordered a total of 66 Otters in three batches under the RCAF designation CSR-123. The first RCAF Otter was handed over to the Central Experimental and Proving Establishment at Rockliffe, Ottawa, on 28 March 1953. RCAF Otters subsequently saw much service in remote parts of northern Canada, operating off land and water in liaison, resupply, search-and-rescue and general 'hack' roles with seven RCAF auxiliary squadrons, six reserve squadrons and five communications and rescue flights, and also wore United Nations colours in trouble spots such as the Belgian Congo and Middle East, not finally retiring from Canadian military service until late 1982.

The biggest customer

In the summer of 1953 DHC was invited to participate in the US Army's Operation Skydrop II exercise at Fort Bragg, North Carolina, an exercise intended to evaluate the cargo-carrying capabilities of helicopters and fixed-wing aircraft. So impressive was the Otter's performance in operating with minimal maintenance from short airstrips (on one occasion it bettered a helicopter of similar power by clearing a 50-foot obstacle carrying twice the helicopter's payload) that the US Army placed an initial order for six evaluation aircraft, designated YU-1. These entered service with the 521st Aviation Company, Corps of Engineers on survey duties in Alaska, where they proved the ruggedness and low-maintenance dependability of the type. Orders for a total of 190 Otters followed, accounting for about one third of total production and making the US Army the largest operator of the aircraft, which subsequently served in all theatres of US Army operations, including Europe and Vietnam. An Army U-1A Otter was used as a tanker aircraft to refuel the Piasecki H-21 Shawnee which made the first non-stop transcontinental flight across the United States by helicopter. US Army Otters remained in large-scale service until the mid-1970s. The last pair, operated by the 'Golden Knights' freefall parachute display team, did not retire until 1981, when they were replaced by Beech U-21 twin turboprops. A small number of former US Army U-1As were supplied to foreign air arms in Latin America and South East Asia under American military aid programmes following retirement from US service.

Other military users

The US Navy was also a customer for the Otter, operating a total of 14 UC-1/U-1B aircraft, including four loaned by the RCAF/United Nations which were shipped to McMurdo Sound, Antarctica, in December 1955 to take part in Operation Deep Freeze I. Operated on wheel/skis by VX-6 squadron, all four had been written off in accidents by the summer of 1956, but four new UC-1s were taken to Antarctica for Operation Deep Freeze II, remaining on station for the following year's expedition during which two of the USN Otters made the first wheeled landing on the continent at Marble Point Airstrip.

US Navy Otters served in Antartica until 1966, suffering only one fatal accident. They were joined on that continent by a single example operated by the Royal Air Force on behalf of the British Commonwealth Trans-Antarctic Expedition. On 6 January 1958 the RAF Otter became the first single-engined aircraft to cross the continent non-stop, flying from Shackleton to Scott Base in 11 hours.

Following the lead set by the RCAF and US Navy, a number of other military operators ordered Otters new from the Downsview pro-

*The rugged structure, **STOL** capability and good load-carrying ability made the **O**tter a favourite for relief operations in the **T**hird World and for military liaison duties. This **C**anadian machine flew with the **UN** contingent in the **S**inai.*

*In civilian service the **O**tter proved extremely adaptable to the harsh conditions under which it was flown. The **C**anadian bush was an excellent testing ground for the type, where it even fought fires.*

Among the more colourful Otters was this aircraft operated by Wardair, which began as a local carrier but which rose to be a major transatlantic operator. During the winter months, Canadian Otters donned skis to continue services to remote settlements, while in summer floats were more appropriate for landing on the many lakes that dot the Canadian interior. The fuselage side door allowed the admission of sizeable cargo, and the Otters were regularly used to transport vital machinery.

Specification
de Havilland Canada DHC-3 Otter
Wingspan: 17.68 m (58 ft)
Length: 12.75 m (41 ft 10 in)
Height: 3.96 m (13 ft)
Wing area: 34.8 m² (375 sq ft)
Powerplant: one 447-kW (600-hp) Pratt & Whitney S3H1-G Wasp radial piston engine
Passenger capacity: 9-11
Empty weight: 2398 kg (5,287 lb)
Maximum take-off weight: 3629 kg (8,000 lb)
Maximum speed: 257 km/h (160 mph)
Cruising speed: 222 km/h (138 mph)
Service ceiling: 4008 m (16,400 ft)
Maximum range: 1520 km (945 miles)

Specification
Vardax Vazar Dash 3
Wingspan: 17.68 m (58 ft)
Length: 14 m (46 ft)
Height: 3.96 m (13 ft)
Wing area: 34.8 m² (375 sq ft)
Powerplant: one 559.5-kW (750-hp) Pratt & Whitney Canada PT6A-135 turboprop engine
Passenger capacity: 9-11
Empty weight: 1905 kg (4,201 lb)
Maximum take-off weight: 3629 kg (8,000 lb)
Maximum speed: 267 km/h (166 mph)
Service ceiling: 4876 m (16,000 ft)
Maximum range: 1203 km (748 miles)

The Otter proved equally at home on skis as it did on floats or wheeled undercarriage. This was the sole aircraft on British military charge, an ex-US Navy UC-1 procured for the 1956 Trans-Antarctic expedition. It was later re-registered as NZ8601 of the Royal New Zealand Air Force.

A major stepping stone between the Otter and the Twin Otter was this experimental aircraft, which began life as a STOL research platform. The trolley-style undercarriage simulated a tricycle undercarriage which was fitted to the definitive DHC-6 Twin Otter.

duction line. These included the Royal Australian Air Force, which purchased two in 1961 for use by No. 1 Air Trials unit in support of operations at the Woomera Weapons Research Establishment, ferrying range observers and test crews to remote sites and tracking down and retrieving target drones and missiles. They remained in service until 1979, when they were disposed of to a civilian operator in their homeland, Canada.

Among other military customers for the Otter were the Indian Air Force, a major operator with 27 delivered new in 1957-58, a further five supplied from surplus RCAF stocks in 1963 and five others acquired in 1966; the Royal Norwegian Air Force, which operated a total of 10 Otters between 1954-68, using them on wheels and amphibious floats for communications duties throughout the country; the Ghana Air Force, which took delivery of 12 aircraft; the Union of Burma Air Force with nine; Indonesia with four; Argentina with two, which were used to support Argentinian Antarctic operations; and the Chilean Air Force, whose five Otters were delivered by air, flying in a group from Toronto to Santiago in 1957.

An airline asset

Several foreign airlines became early customers for the Otter. Wideroe's Flyveselskap A/S used Otter seaplanes for domestic services in northern Norway. Philippine Airlines' six Otters served a network of short-stage routes among the country's 730 inhabited islands, frequently operating from roughly-prepared strips no more than 365 m (1,200 ft) long. They reduced to minutes journey times which otherwise might take days on foot or a week by infrequent coastal freighters, and provided vital feeder links with PAL's trunk routes, as well as serving in search-and-rescue and medical evacuation roles in remote areas unsuitable for larger and heavier aircraft.

In Australia, Qantas Empire Airways took delivery in 1958 of four Otters, three landplanes and one amphibian, which were based in Papua New Guinea, operating in 14-seat 'high density' configuration out of Lae and Port Moresby. The Qantas Otter operations were taken over by Trans Australia Airlines in 1960, in 10-seat configuration, finally ceasing in October 1966.

Otter production ended in 1967, by which time 466 had been built, of which 359 went to military operators, and 107 to civilian customers, the majority of whom were Canadian. The Otter did the job it was designed for extremely well, frequently in harsh weather conditions and the inhospitable terrain of bush and outback, where its trouble-free day-in, day-out service earned it a reputation as "an aeroplane designed for the pilot who comes into town once a year for annual overhaul of his aircraft and a shave for himself!"

Among the multitude of roles for which the Otter proved adaptable was that of aerial firefighter. Launch customer Ontario Provincial Air Service developed a fire-bomber conversion for its Otter seaplanes which involved installation of an 363-litre (80-Imp gal) water/fire retardant tank on a rotating mount above each float. The tanks were filled automatically during 'touch-and-go' landings on a lake, and the contents dispersed by rotating the tank over a fire site. The twin tank installation was later replaced by a single 954-litre (210-Imp

A problem facing Otter operators was the increasing difficulty of support for the original R-1340 engines. Airtech in Canada provided an answer in the form of a PZL ASz-62IR radial, rated at 746 kW (1,000 hp). This uncowled aircraft wears Peruvian colours.

The re-engined DHC-3/1000 Otter offers considerably improved take-off and load-carrying performance compared to the original R-1340 engined aircraft. Airtech's original Otter re-engining scheme used the PZL-3S engine offering the same power (447 kW/600 hp) as the original powerplant.

Civilian and military customers alike with the need for a tough go-anywhere transport bought the Otter. The Indian Air Force took 32 in all, using them to supply primitive air strips and in the mountainous regions of the country.

gal) streamlined tank mounted centrally, and later still by integral float tanks.

More bizarre were the STOL research modifications made by de Havilland Canada and the Canadian Defence Research Board to an RCAF Otter. Large flaps were added inboard of the ailerons, a tall parallel chord fin and rudder and longer, dihedralled tailplanes installed, and the aircraft was mounted on a skeletal seaplane-type chassis with four wheels, to simulate tricycle landing gear. After initial flight trials in this configuration the large flaps were removed and a General Electric J85-GE-7 turbojet engine installed in the rear fuselage. With adjustable exhaust nozzles this powerplant was intended to give the pilot a means of controlling rate and angle of climb and descent – a crude form of vectored thrust – greatly enhancing STOL performance. In a final development the aircraft's standard P & W Wasp radial engine was removed, the nose faired over and two Pratt & Whitney Canada PT6 turboprops installed on the Otter's wings. This research programme provided useful data for subsequent development of the DHC-6 Twin Otter.

De Havilland Canada investigated, but chose not to proceed with development of a single turboprop conversion of the Otter, as they had with its smaller stablemate the Beaver, but independent companies have engineered turbine engine conversions of the aircraft. The first of these, a Garrett TPE331 conversion, was devised by Western Rotorcraft Ltd of Seattle, Washington, in 1972, but was not a success. Ray Cox, an experienced Arctic Otter pilot and mechanic, formed Cox Air Resources of Namao near Edmonton, Canada, to develop a

Pratt & Whitney Canada PT6A conversion. The modification, which involved replacing the Wasp radial with a 494-kW (662-shp) PT6A-27 driving a four-bladed Hartzell reversible-pitch propeller, proved problematical and required substantial re-engineering to retain the aircraft's centre of gravity position with the much lighter turbine engine. The Otter's 190-litre (42-Imp gal) rear fuel tank had to be moved forward, and a further 181-litre (40-Imp gal) tank installed forward of the engine firewall. First flown on 26 September 1978, the prototype DHC-3-T Turbo Otter, CF-MES-X, had the same maximum take-off weight as its piston-engined counterpart, but gained a 317-kg (700-lb) increase in useful load thanks to the lighter weight of the PT6A engine.

A second PT6A re-engining project was initiated by Vardax Corporation of Bellingham, Washington during 1986. Serv-Aero Engineering Inc. of Salines, California completed prototype modification of one Otter, re-engined with a 560-kW (750-shp) PT6A-135 turboprop, which improved performance and offered a 331-kg (730-lb) increase in useful load. Known as the Vardax Vazar Dash 3, this turbine Otter conversion has received US Federal Aviation Administration certification, and about a dozen have so far been modified by Vardax.

The great advantages that the turboprop engine brings to the inherent qualities of the Otter are its light weight and greater fuel economy over a piston engine.

Since 1958 Tyee Airways have been flying from their Vancouver base in British Columbia. Their youngest Otter is now well over 30 years old, and the oldest aircraft of the four in service dates from 1953.

Typical Twin Otter country: a DHC-6 performs a steep approach to a rough mountain strip in South America. Operating effectively and safely into this sort of airfield has been the type's forte since its first flight on 20 May 1965.

de Havilland Canada DHC-6 Twin Otter

With its Beaver and Otter aircraft, de Havilland Canada had established itself as the world's leading supplier of bushplanes. Combining the proven ruggedness of DHC's design with the reliability and power of two turboprops produced the Twin Otter. Excellent STOL performance, good load-carrying ability and the capability to land on virtually any surface, including snow and water, made it the world's standard go-anywhere light transport.

The de Havilland Canada company, with a long history of manufacture of successful utility and 'bush' aircraft behind it, began detail design work in November 1963 on a twin-engined development of the DHC-3 Otter. An experimental Otter powered by two Pratt & Whitney Canada PT6 turboprops had earlier shown great promise in a Defense Research Department test programme, and this engine was selected for the new design, which was designated DHC-6.

De Havilland Canada's design philosophy was to create an aircraft which could operate unsupported in the most inhospitable environment, from land, water, ice or snow, offering good short take-off and landing characteristics, low operating costs, simple maintenance requirements and employing as many components and as much of the production tooling of its single-engined forebear as possible.

The Otter's wing section was retained, with increased span, as was the fuselage cross section and many other components, providing a surprising degree of commonality with the earlier aircraft, despite an external appearance suggesting that the Twin Otter was an entirely new aeroplane. The chosen configuration was for a conventional strut-braced high wing monoplane of all-metal construction with double-slotted trailing edge flaps and drooping ailerons to enhance STOL performance, cruciform tail surfaces with slightly swept fin, and fixed tricycle undercarriage with float and ski options. The unpressurised cabin was designed to accommodate 20 passengers in individual seats with a centre aisle. Two 579-shp PT6A engines were selected to power the initial pre-production batch of three aircraft.

Work began on the prototype in July 1964, and this aircraft first flew from the DHC plant at Downsview, Ontario on 20 May 1965, at which time a preliminary production batch of 15 was laid down, with plans for a further 45 aircraft to be built at the rate of two a month. The pre-production DHC-6 Twin Otter Series 1 was certificated by Canadian Authorities in April 1966 and received US Federal Aviation Agency type approval three months later. In July 1966 the first production standard Series 100 went into service with launch customer Trans-Australia Airlines and with the Ontario Department

Not surprisingly, the Twin Otter's roomy fuselage, sturdy construction and splendid short-field performance have led it to be one of the major types in use by the relief organisations. This aircraft operates in the drought-stricken area of Sudan for UNICEF.

Under the designation UV-18B, USAF operates a pair of Twin Otters. These fly with the Air Force Academy at Colorado Springs, Colorado, providing cadets with parachute training. The other US military Twin Otters are two UV-18As for the Army.

of Lakes and Forests. The Series 100 was powered by PT6A-20 turboprops of similar power output to the PT6A-6.

Sales of the Twin Otter came from unexpected sources. De Havilland Canada had designed the aircraft primarily with the needs of bush operators in mind, but the aircraft quickly attracted great interest from airlines operating feeder and commuter services, and by military forces. One hundred and fifteen Twin Otter Series 100s were built before the Series 200 was introduced in April 1968, featuring an extended nose housing a larger baggage compartment (not offered on seaplane variants) and an expanded rear baggage area.

Certification was granted on 25 April 1968 of the Twin Otter Series 300. This further improved model, which lasted in production until late 1988, when the Twin Otter line closed, has a large two-part cargo/passenger door on the port side of the fuselage, a passenger door to starboard, separate crew doors, fuel capacity increased to 378 US gallons as standard with

From its base at Goose Bay, Labrador Airways (later known as Lab Air) operated a mix of bush types on trips round the rugged territory. Four Twin Otters were operated, including this Series 100, identified by the shorter nose. The enlarged nosewheel allows the aircraft to operate from rough, icy strips.

Like many other light transports, the Twin Otter forms the basis of a coastal patrol aircraft, complete with surveillance radar in the nose and a searchlight under the wing. This aircraft operates for the Senegalese government.

optional wingtip tanks holding a further 92 gallons, an increase in maximum take-off weight of 454 kg over earlier variants, and is powered by 620-shp Pratt & Whitney Canada PT6A-27 engines. After a further 115 Series 200s had been built, Series 300 deliveries began in the spring of 1969, the first of this model and the 234th Twin Otter off the line going to the Arabian-American Oil Company (Aramco). The Twin Otter 300S was a special version developed for the Canadian Ministry of Transport for operation on Air Canada subsidiary Air Transit Canada's STOLport service between Ottawa and Montreal. Six were delivered between July and November 1973. The Series 300S differed from standard production aircraft in having upper wing spoilers, high-capacity brakes with anti-skid systems, improved engine fire protection and electrical and hydraulic systems, 11-seat cabins with airline-style interiors, and a sophisticated avionics fit including area navigation and ILS systems designed for steep glideslopes necessary for operation into 610 m long by 30 m wide paved strips. Air Transit's Twin Otters flew half-hourly schedules between the STOLports, pioneering city centre to city centre air travel.

Sea search

During 1977 a maritime patrol version of the Twin Otter 300 was developed for Greenlandair Charter to operate on ice patrol and surveillance duties. It incorporated a Litton LASR-2 search radar in a chin radome, Omega navigation system, four observers' stations with bubble windows, flight openable doors for supply dropping, two additional fuselage fuel tanks, total capacity 1214 kg, and 'finlets' on the tailplane similar to those installed on floatplane variants to offset the additional nose area of the radome. The aircraft was delivered in January 1978. Six months later de Havilland Canada delivered its 600th Twin Otter and increased production rate to five aircraft per month.

The Twin Otter's rugged airframe has proved readily adaptable to a number of special roles. Eight were delivered to the Canadian Armed Forces as CC-138s for search and rescue duties, frequently operating on wheel/ski undercarriages in remote locations. Ten were sold to the Royal Canadian Mounted

Designed in Canada, it is no surprise that the Twin Otter is hardened to all climates. During the southern summer months, this aircraft operates on Antarctic survey work for the United Kingdom. It is registered in the Falkland Islands.

Inset right: Away from the jungles, deserts, ice-caps or mountains, the Twin Otter is still a superb performer, and in fact large numbers serve as feederliners in Europe and America. Its STOL performance is useful in the traffic patterns of large airports, allowing them to slip in between big jets.

Police, and 10 of 12 delivered to the Peruvian Air Force operate on floats with Grupo Aéreo No. 42 based at Iquitos, serving isolated native settlements along the River Amazon. One series 300 flew in support of the British Transglobe Expedition. Another serves the British Antarctic Survey Team.

Baggage pannier

Optional equipment developed for the Twin Otter aircraft includes a ventral pannier capable of carrying up to 272 kg of baggage, a cabin-mounted fabric membrane chemical retardant tank with a capacity of 1,818 litres for aerial fire-fighting missions, and oversize low pressure 'tundra' tyres for soft field operation.

Among the most extensively modified examples were two aircraft delivered in April 1979 and the autumn of 1981 respectively to China and Kenya for geophysical survey work. These aircraft were equipped with a Scintrex Tridem airborne electromagnetic system housed in two wingtip pods. Three widely separated frequencies are transmitted from one pod and received by the other, comparison of returns enabling metal and mineral deposits to be located. The Twin Otters also carried a long noseboom housing a proton magnetometer, a VLF electromagnetic system in the nose baggage compartment, a radiometric spectrometer sensor in the rear cabin, and a strip camera and sophisticated Doppler navigation system for accurate positioning during survey runs.

Among many military operators of the Twin Otter are the air forces of Argentina, Canada, Chile, Ecuador, Ethiopia, France, Jamaica, Norway, Panama, Paraguay, Peru, and the United States, where the US Army National Guard also operates two Twin Otters under

Loganair operates a sizeable network throughout Scotland and the northern islands. Several feederliner types are used, ranging in size from the Fokker F27 to the Britten-Norman Islander. Four Twin Otters once operated on the more difficult sectors.

Left: The DHC-6-300MR is the maritime reconnaissance version, seen here in prototype form. Four pylons under-wing can accommodate rocket pods and guns for light attack. Other options include infra-red linescan and cameras for reconnaissance, and Omega/VLF long-range navigation equipment.

the designation UV-18A.

On 16 July 1982 de Havilland Canada announced development of a dedicated multirole military version of the aircraft, designated Twin Otter 300M. This was offered in three versions. A basic military transport model had a standard interior with side-facing seats accommodating 15 combat-equipped troops and a jumpmaster, and was quickly convertible to 20-passenger, passenger/freight, or sixstretcher medevac configuration. A counter insurgency variant was similar, but had provision for self-sealing fuel tanks, armour protection, a light machine-gun mounted at a flight-openable cabin door on the port side of the fuselage, and four underwing hardpoints for a variety of stores including light bombs and rocket launchers. A maritime reconnaissance version, first flown in 1982, was designated Series 300MR, and featured a chin-mounted 360° search radar, infra-red linescan equipment, a searchlight mounted beneath its starboard wing, tactical/search operator's console, underwing stores hardpoints, handheld data-annotating camera, observers' bubble windows, auxiliary fuel tanks in the cabin, flare chutes, and crew rest stations. Twin Otter 300M variants have maximum take-off weights increased to 6350 kg.

The Twin Otter remains in service throughout the world with commuter airlines – many of which might never have been established but for de Havilland Canada's foresight in creating the perfect vehicle for their needs – and with armed forces and cargo/bush operators for whom the rugged twin turboprop's undemanding reliability has won it a reputation which will survive long after the 844th and last Twin Otter left the Downsview factory during 1988.

de Havilland Comet

In May 1952 a de Havilland Comet left London for Johannesburg to open the jet age for the world's airlines. So far ahead of all competition was the Comet, that no rival even had any firm idea what to build. However, this lead was controversially lost during a series of setbacks, but the Comet recovered to become a successful and safe airliner.

At the height of World War II, in late 1941, the UK took the farsighted step of appointing a special group of experts to study what commercial transport aircraft should be planned for use after final victory. In May 1943 the Second Brabazon Committee had its first meeting, and towards the end of that year it issued a series of recommendations, one of which, called Type IV, was for a jet-propelled transport. It is paradoxical that, whereas in the USA several large jet bombers were then being planned, without the slightest thought of a jet airliner, in the wartime UK the reverse was true. A major factor was that in the pre-war era the United States had come to dominate the world airliner market, and it was rightly felt that the only way to compete after the war would be to take bold technological leaps.

The various Brabazon recommendations quickly became associated with particular companies, and Type IV went to de Havilland. This enterprise had both an Aircraft Company, with a design team led by R. E. Bishop and R. M. Clarkson, and an Engine Company whose top engineers were F. B. Halford and J. S. Moult. In early 1944 both de Havilland teams studied the Type IV, and after many possibilities had been examined they were given authority to go ahead with the D.H.106, in February 1945. Even then the design remained fluid, ranging from a twin-boom Vampire-like machine with three Goblin engines, via a canard to a tailless swept-wing machine scaled up from the D.H.108. British Overseas Airways Corporation (BOAC) found it hard to make up its mind what was wanted, and the D.H.106 hovered between a 14-passenger short-hauler and a non-stop North Atlantic mailplane carrying a few sacks of letters plus two VIP passengers.

Fortunately de Havilland managed to persuade BOAC and the government that a more conventional transport would be useful; the company also recognised that transatlantic range was unattainable. The design at last went ahead in September 1946, when the Ministry of Supply ordered two prototypes. BOAC had said, in 1944, it would need 25, but it had second thoughts and the first production order, in January 1947, was for eight. British South American Airways then ordered six, but this airline was merged into BOAC which cancelled the BSAA order but increased its own purchase to nine. The name Comet was announced in December 1947.

Because of the dramatically advanced nature of the new Comet, parallel research programmes were launched to provide a basis of proven hardware. The new Ghost engine was tested in the outer nacelles of Avro Lancastrian VM703 and in a special high-altitude de Havilland Vampire TG278. The Lockheed Servodyne-powered flight controls were exhaustively tested at Farnborough on Lancaster PP755, while control components were flown on a D.H.108 and a Hornet. Even the streamlined nose was tested on an Airspeed Horsa glider, mainly to verify pilot vision in rain. Construction of the prototypes moved rapidly, and the first, unpainted and bearing Class B registration G-5-1, was rolled out at Hatfield on 25 July 1949, and flown by John Cunningham two days later.

A revolution in design

Though the Comet was less radical than it might have been, it nevertheless broke new ground in almost every part of its aerodynamics, structure, propulsion and systems. Basically, it had a low

RAF Comet C. Mk 2 aircraft served No. 216 Sqn flawlessly from 1956 until April 1967, fully modified with reskinned fuselages with oval windows and many other changes for military operations. XK697, the 32nd Comet and intended to be G-AMXJ, was named Cygnus from October 1959. The legend later read Air Support Command.

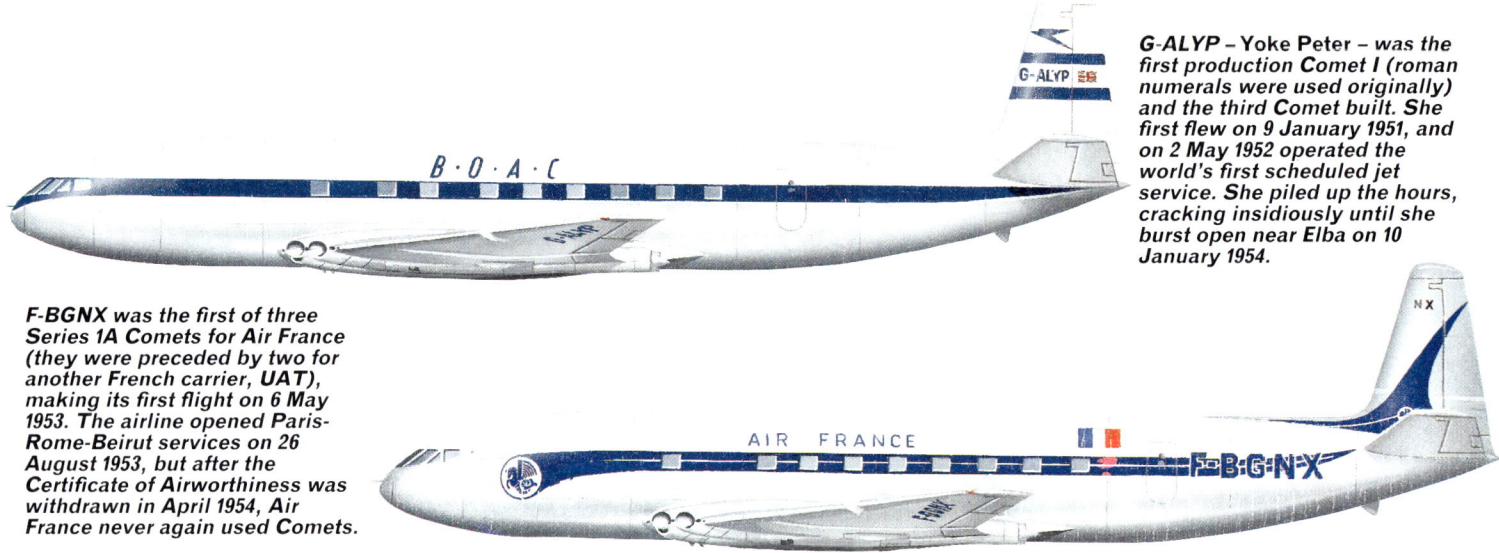

G-ALYP – Yoke Peter – was the first production Comet I (roman numerals were used originally) and the third Comet built. She first flew on 9 January 1951, and on 2 May 1952 operated the world's first scheduled jet service. She piled up the hours, cracking insidiously until she burst open near Elba on 10 January 1954.

F-BGNX was the first of three Series 1A Comets for Air France (they were preceded by two for another French carrier, UAT), making its first flight on 6 May 1953. The airline opened Paris-Rome-Beirut services on 26 August 1953, but after the Certificate of Airworthiness was withdrawn in April 1954, Air France never again used Comets.

wing of substantial area whose 20° sweep resulted mainly from taper; the tail was unswept. The three widely separated spars passed through the fuselage under the floor, the inboard part of the wing on each side being bulged to house the 2291-kg (5,050-lb) thrust Ghost 50 turbojets, with plain oval inlets in the leading edge and long jetpipes projecting just behind the trailing edge. Large plain flaps were fitted, with split flaps under the jetpipes. The leading edge was fixed, with a small fence well outboard, ahead of the inner end of the powered aileron. Narrow perforated airbrakes could be raised ahead of the outer flap sections. Thickness ratio was 11 per cent, and the first Comet had a lower stalling speed and was easier to fly than most contemporary airliners with piston engines!

With a diameter of 3.05 m (10 ft), the circular tube fuselage was pressurised to 0.58 kg/cm^2 (8.25 lb/sq in), to give a 2440-m (8,000-ft) interior when flying at 12190 m (40,000 ft). This was double the pressure of any previous airliner, and a further factor was that outside temperatures of minus 70°C had a profound effect on materials throughout the aircraft. Not least of the innovations was the use of Redux metal-to-metal bonding throughout the structure. In effect a glueing process, Redux had never before been used on such a scale or on such highly stressed parts. The de Havilland engineers conducted extensive tests to prove the strength of many critical parts, and also pressure-tested large fuselage sections. One of the latter exploded, the shattered pieces making it hard to pinpoint the source of the original crack, a fact that was later to be very significant. Its main result was to cause the company to switch to underwater pressure testing, where failures did not cause such explosive release of stored energy.

Secrecy worthy of a military project was maintained throughout. This applied especially to the quite simple engine installations, which were backed up by a neat stored-liquid rocket motor, a 2268-kg (5,000-lb) thrust de Havilland Sprite, between the jetpipes on each side to maintain take-off thrust in hot or high conditions. In sharp contrast with previous British aircraft, the Comet was planned to house all the fuel that could be packed in, virtually the whole wing apart from the engine and landing-gear bays forming a giant sealed integral tank with a capacity of 27503 litres (6,050 Imp gal), some three times that of a Lancaster. Another new feature was pressure fuelling, the large hose being attached in the underside of the wing to feed at the rate of one ton per minute. The cabin pressurisation boldly took bleed air from the engines, while other bleed air was used to de-ice the wings and tail.

Overall technical risk was very great, and it is greatly to the credit of de Havilland that the company went ahead on the basis of an order for eight aircraft at a fixed price of £250,000 each. It rightly judged there would be many more orders once this revolutionary airliner had entered service. Passenger accommodation was on the basis of pairs of seats on each side of the aisle, and at the initial gross weight of 47627 kg (105,000 lb) it was possible to carry 36 passengers, eight in facing pairs in a 'smoking room' forward and 28 in the main cabin at what today seems the princely seat pitch of 1.14 m (45 in), the rearmost seats being in line with the jet nozzles. The limited underfloor space resulted in the main baggage bay being above the floor, aft of the capacious flight deck laid out for two pilots, an engineer and a navigator.

Inevitably, every overseas trip by the prototype broke records, and it was established that full payload could be carried over a range of 2816 km (1,750 miles) cruising at 788 km/h (490 mph). The top of the fuselage was painted white, BOAC livery applied and civil registration G-ALVG bestowed, the certificate of airworthiness being issued as early as 21 April 1950. In December of that year the ungainly single-wheel main gears were replaced by neat four-wheel

One of the first air-to-air photographs ever taken of a Comet, G-ALVG, is seen here in late 1949, after the addition of BOAC's Speedbird badge and a Union Jack on the fin. Later, G-ALVG had the full BOAC white-top livery then current, though it never served with the airline. It was scrapped after fatigue testing.

First of two Series 1A aircraft for CPA, CF-CUM had 31823 litres (7,000 gal) of fuel and water-methanol injection for full-load take-offs. After the destruction of her sister in a take-off crash she was passed to BOAC, where she served as G-ANAV. Photographed on 15 August 1952, she was the 13th Comet to be built.

G-APYC was the 37th of the 'New Comets' and was originally flown from Chester as SX-DAK of Olympic Airways, the only customer for the short-span 4B version other than BEA. She is depicted in Channel livery (1970-2), and finally found her way (like most surviving Comets) to Dan-Air in April 1972.

SU-ALC was the 39th of the 4/4B/4C series, being delivered as the first of nine 4Cs for Misrair. She is shown after repainting in the livery of UAA, Misrair's successor, with whom she was lost, after 10 years of use, in an accident near Tripoli on the second day of 1971. (Biggest export buyer was Aerolineas.)

bogies (which could not retract in G-ALVG), and these were standard on the production machines. The latter also did not have the take-off rockets, though provision for them was retained.

The first production aircraft flew on 9 January 1951, and an unrestricted passenger-carrying certificate of airworthiness was awarded on 22 January 1952, ready for the start of regular service on 9 May. The initial route was from London to Johannesburg via Rome, Beirut, Khartoum, Entebbe and Livingstone, the end-to-end time being just under 24 hours. Thus, as Concorde was to do 25 years later, the Comet halved the effective size of the world; but in addition it utterly transformed air travel, the traditional thumping and lurching progress through bad weather being replaced by a totally effortless and virtually silent progress far above any turbulence, vibration or any kind of unpleasant environment. The world airline industry was naturally cautious, but could not overlook the fact that passengers who had flown in BOAC's Comet disliked any other kind of air travel.

Thus other customers slowly began to emerge, and de Havilland upgraded the basic aircraft to Comet 1A standard with weight raised to 52163 kg (115,000 lb), with extra fuel, 44 seats and water/methanol injection to maintain hot/high engine thrust. The first sale was to Canadian Pacific Airlines, followed by Air France, UAT and the Royal Canadian Air Force. For the next generation de Havilland switched to the Rolls-Royce Avon engine in the Comet 2, and early Avons flew the Comet 2X prototype on 16 February 1952. BOAC asked for 11 Comet 2s, using almost the same airframe as the Series 1 but with a 0.91-m (3-ft) stretch giving an extra passenger window on each side. After them came British Commonwealth Pacific, Japan Air Lines, LAV and Panair do Brasil, as well as the Comet 1A customers Air France, UAT and CPA. Suddenly de Havilland ran out of available capacity, because every inch of floor space was already bursting with production on many other types, and Shorts at Belfast was brought in with a duplicate Comet 2 assembly line.

Developments and setbacks

At the 1952 Farnborough Air Show de Havilland announced the Comet 3. The power of later Avon engines enabled gross weight to jump to 65772 kg (145,000 lb), transforming the 44-seat 2816-km (1,750-mile) Comet 1 into a really capable and efficient airliner that could carry 76 passengers 4345 km (2,700 miles), and at slightly higher speed. With a fuselage stretch of 5.639 m (18 ft 6 in), the Comet 3 also looked a truly modern jetliner, and the Hatfield company was deluged with enquiries. Sales were quickly made to BOAC, Air India and Pan Am, and plans were drawn up for a third Comet assembly line at the company's Chester factory. The Comet 3 prototype flew on 19 July 1954, by which time six BOAC Comet 2s had been completed, and shown their capability by flying non stop to Khartoum, 4930 km (3,064 miles) in 6½ hours. But other things had happened by this time that crippled the programme.

The first accident came on 26 October 1952 when G-ALYZ failed to become airborne at Rome and was damaged beyond repair, though without injury to the 42 passengers and crew. On 3 March 1953 exactly the same thing happened to CF-CUN of CPA making a maximum-weight take-off in hot air at Karachi, and on this occasion all on board were killed. The cause was hauling back on the yoke too early, causing the aircraft to proceed down the runway with the wing stalled, the high drag preventing acceleration and the inlet angle reducing thrust. The answer was to droop the leading edge and refrain from hauling the aircraft off until a proper rotation speed had been reached.

More serious was the inflight destruction of G-ALYV whilst climbing out of Calcutta on the first anniversary of BOAC services on

Originally planned for 58 passengers, the Comet 4 was late in life regularly seating 106. G-APDR was the 18th of 19 Series 4 aircraft ordered by BOAC. She is seen here on pre-delivery testing in July 1958. Later she was sold to Mexicana and ended her days in 1972 at the Standard Fire School.

The 45th Comet 4 series was this 4C for Middle East Airlines of the Lebanon. OD-ADR, along with sisters ADQ (no. 46) and ADS (no. 48), was destroyed in the Israeli attack on Beirut Airport on 28 December 1968. Altogether Comet 4, 4B and 4C aircraft flew almost exactly 2,000,000 hours, ending in December 1979.

2 May 1953. Tragically there was plenty of monsoon turbulence, and no enquiry was made to see whether or not the structural break-up might have been due to a weakness in the aircraft. But when the famed G-ALYP, which had flown the first scheduled service, disappeared on 10 January 1954 outward bound from Rome, with no radio transmission from the crew, the Comet 1 was grounded and all available aircraft were subjected to a detailed examination. Nobody knew what flaw to look for, but a month later a Royal Navy salvage team found the wreckage near Elba and began the laborious process of recovering every fragment. There were many theories about what had happened, and after various small modifications the Comet 1s resumed services, but on 8 April G-ALYY disappeared whilst climbing out of Rome in the opposite direction. This time the Comets stayed grounded, causing great difficulties to BOAC, Air France, UAT and the RCAF.

There followed the biggest technical investigation up to that time, culminating in the certain knowledge that G-ALYP's fuselage had ripped open from a fatigue crack started at the corner of one of the rectangular cut-outs for the ADF aerials in the top skin. Another fuselage, that of G-ALYU, was tested in a giant water tank and after 1,830 simulated flights also ripped open, from the corner of a passenger window. Why de Havilland should have used straight-sided cut-outs in the most highly pressurised fuselage then attempted remains a mystery. For many months the whole Comet project hung in the balance. All export orders were terminated, and all Comet 1As were returned, except for those of the RCAF which without fuss were suitably modified with stronger skins and oval cut-outs and served with No. 412 Squadron until late 1964. The other Comet 1s and Comet 1As were used up in test programmes or simply scrapped. The Comet 2s were rebuilt with oval cut-outs in thicker skins and delivered from June 1956 to RAF Transport Command, whose No. 216 Sqn achieved a perfect record in intensive operation until April 1967. The assembly lines at Belfast and Chester were abandoned.

Two Comet 2s were passed to Rolls-Royce for use in developing the more powerful Avons being used in France's Sud Caravelle, and for BOAC route-proving, while three others with the original fuselage and with pressurisation removed served as electronic-warfare 'ferret' platforms with RAF No. 51 Squadron at Wyton. But the big question was whether de Havilland could ever succeed in selling a completely redesigned derivative of the Comet 3, or whether the image was so damaged that the Comet had become unsaleable to the airlines and to the public. The answer came in February 1955 when BOAC announced it would buy 19 Comet 4s, similar to the Series 3 (which from the start had oval windows) but with even greater fuel capacity and range, with a weight up to 70762 kg (156,000 lb), which later grew to 73483 kg (162,000 lb). This at last resulted in a capable mainliner, but it was clear the new model would emerge into a world dominated by the Boeing 707 and Douglas DC-8. Thus, for narrow competitive reasons, the Comet 4's service life began on the North Atlantic on 4 October 1958, beating Pan Am's Model 707s into service by 22 days, though the type had never been intended for this route. BOAC's own Model 707-420s replaced the Comet on the Atlantic after little over a year. Other Comet 4s were sold to Aerolineas and East African.

The BOAC fleet of 19 Comet 4s demonstrated that at last de Havilland had created a safe airliner, and were sold until 1965, and even then there were plenty of customers for the second-hand airliners. The first export order, in 1956, was for four Series 4 and 10 of a new clipped-wing long-body short-haul model, the Series 4A, for Capital Airlines of Washington DC. The Series 4A was specially strengthened for high speed at low levels, and it was a major blow when Capital was taken over by mighty United, which cancelled the order. But by this time BEA had become interested, and (to the disgust of the French, who said the Caravelle did the same job on two of the identical engines instead of four) in April 1958 the British airline ordered six Comet 4Bs, later increased to 14. Olympic ordered four. This version had the clipped wings and even better payload, and without effort could carry over 100 passengers for 5390 km (3,350 miles), which a Caravelle certainly could not do. They put up a fine record, and de Havilland ended the programme by merging the long body with the large pinion-tanked wing to produce the Comet 4C.

The Comet 4C was the most successful model of all, being bought initially by Mexicana, Misrair (Egyptair), Aerolineas Argentinas, MEA, Sudan and Kuwait. Altogether 30 of this series were made, 23 of them at Chester, bringing total production to 113, the final examples being made speculatively and finally going to King Ibn Saud of Saudi Arabia, five for No. 216 Squadron RAF, one for the A&AEE at Boscombe Down and the last two being passed to Hawker Siddeley Manchester for use as starting points for the first Nimrods, a derived type.

Some ex-BOAC Comet 4s were snapped up by the British Ministry of Aviation for research and trials purposes, notable examples being G-APDF which became the much-modified XV814, and

Not included in the two million hours flown by Comet 4 variants, are the RAF and Ministry versions. These are three of the latter. XV814, ex-BOAC G-APDF, has a dorsal fin to balance the equipment gondola; XN453 was a Comet 2E G-AMXD used on navaid research; XP915 was the Mk 3B G-ANLO, used on blind-landing research.

de Havilland Comet 4 cutaway drawing key

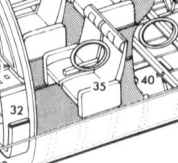

1 Radome
2 Radar scanner
3 Front pressure bulkhead
4 Windscreen framing
5 Windscreen wipers
6 Instrument panel coaming
7 DME aerial
8 Rudder pedals
9 Cockpit roof construction
10 Co-pilot's seat
11 Control column
12 Pilot's seat
13 Engineer's control panel
14 Emergency escape hatch
15 Radio rack
16 Engineer's work table
17 Engineer's swivelling seat
18 Navigators' seats
19 Navigator's worktable
20 Nosewheel bay construction
21 Nosewheel leg strut
22 Twin nosewheels
23 Nosewheel door
24 Crew entry door
25 Crew's wardrobe
26 Forward galley
27 Galley supplies stowage boxes
28 Radio and electrical equipment bay
29 Forward starboard toilet compartment
30 Forward port toilet compartment
31 Wash basin
32 Air conditioning duct
33 Toilet servicing panel
34 Cabin window panel
35 First class cabin seats
36 Twin ADF loop aerials
37 Air conditioning grilles
38 Floor beams
39 Forward freight and luggage hold
40 Freight hold door
41 Control cable runs
42 Fuselage keel construction
43 Overhead hat rack
44 Cabin dividing bulkhead
45 Air distribution duct
46 Emergency escape window
47 Air conditioning plant
48 Hydraulics bay
49 Starboard wing integral fuel tanks
50 Flow spoilers
51 External fuel tank
52 Tank bumper
53 Fixed slot
54 Outer wing fuel tanks
55 Navigation light
56 Wing tip fuel vent

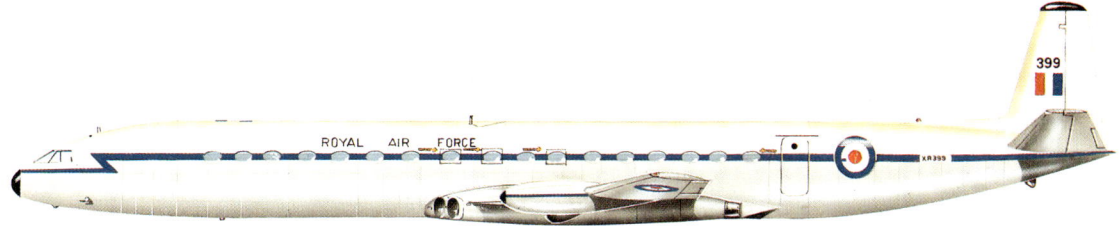

XR399 was built as a Comet C.Mk 4 for the RAF, almost at the end of the Chester line and first flown in March 1962. Usually fitted with 94 aft-facing seats, XR399 was the last of the five C.Mk 4s (basically 4Cs) which operated without any problems with No. 216 Sqn from RAF Lyneham. In 1975, with commendably few hours, she was sold to Dan-Air, being re-registered as G-BDIX.

Variants

D.H.106 Comet: two prototypes (G-ALVG and G-ALZK) with Ghost 50 engines and large single mainwheels; provision for Sprite rockets
Comet I: nine production aircraft for BOAC with bogie main gears, Ghost 50-1 engines and no rockets (G-ALYP/S and G-ALYU/Z)
Comet 1A: 10 aircraft with more fuel (31395 litres/6,906 Imp gal) and water/methanol injection for CPA, AF, UTA and RCAF
Comet 1XB: two RCAF (later CAF) Series 1As (nos 5301/5302) after structural rework in 1957
Comet 2X: one aircraft (G-ALYT), test-bed with Avon 501 engines
Comet 2: planned production civil transport with slightly longer fuselage (29.26 m/96 ft instead of 28.35 m/93 ft) and Avon 503 engines; 36 intended for BOAC and seven other operators; none delivered
Comet 2E: two ex-Comet 2s (G-AMXD and G-AMXK) used by BOAC for route-proving with modified airframes and Avon 524s in the outer positions; G-AMXD later rebuilt as radar and navaid laboratory for use by RAE as XN453
Comet 2(RAF): three aircraft (G-AMXA, G-AMXC and G-AMXE) transferred to RAF as XK655, 659 and 663; later designated **Comet R.Mk 2** for use as Elint platforms
Comet C.Mk 2: 10 aircraft begun as BOAC Series 2 but rebuilt for RAF; XK669 *Taurus*, XK670 *Corvus*, XK671 *Aquila*, XK695 *Perseus*, XK696 *Orion*, XK697 *Cygnus*, XK698 *Pegasus*, XK699 *Sagittarius*, XK715 *Columba* and XK716 *Cepheus*; XK669 and XK670 were initially **Comet T.Mk 2** trainers
Comet 3: one prototype (G-ANLO) with stretched fuselage and Avon 522 engines; later Avon 523, then (1958) rebuilt as Series 3B
Comet 3B: rebuild of G-ANLO as prototype of Series 4B with clipped wing, Avon 525 and other changes, finally rebuilt as blind-landing and avionics laboratory for RAE/BLEU with serial XP915
Comet 4: production fleet for BOAC, total 19 (G-APDA/DT) with revised airframe, Avon 524 engines and increased fuel
Comet 4A: planned short-haul version for Capital, not delivered
Comet 4B: short-haul version for BEA (14, G-APMA/MG, G-ARCO/CP, G-ARGM, G-ARJK/JN) and Olympic (SX-DAK/DAL/DAN/DAO) with Avon 525B engines and clipped wing
Comet 4C: final production model with long fuselage and large wing, total 30; last two completed as XV147 Nimrod with Avons and XV148 Nimrod with Speys, XV148 finally being avionics test-bed at Pershore/Bedford

G-APDS which grew a Nimrod AEW radome. Most, however, went to other airlines, all over the world. Malaysia Singapore purchased seven, but by far the most important operator after 1966 was the UK's Dan-Air. Having no doubts of the reliability and practicability of the now ageing Comet, the British independent purchased no fewer than 47 of various (Series 4, 4B and 4C) types, though never more than one-third were in use at any one time. Dan-Air really showed what Comets could do, even after the 1973 fuel crises trebled the price of kerosene, and it put 106 seats in its Series 4s and 119 into the Series 4Bs and 4Cs. Its scheduled and inclusive-tour Comet services finally ceased on 3 November 1980.

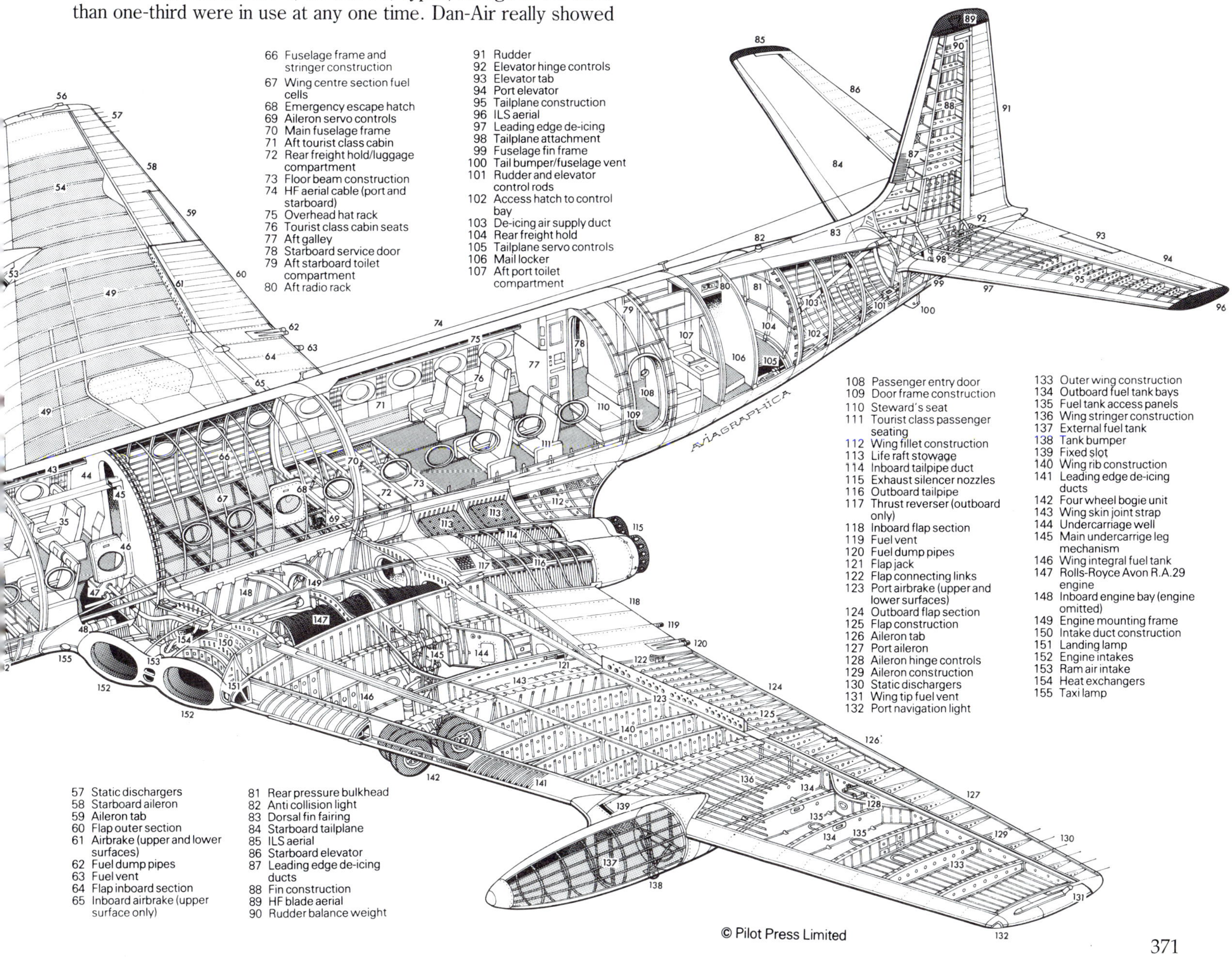

- 57 Static dischargers
- 58 Starboard aileron
- 59 Aileron tab
- 60 Flap outer section
- 61 Airbrake (upper and lower surfaces)
- 62 Fuel dump pipes
- 63 Fuel vent
- 64 Flap inboard section
- 65 Inboard airbrake (upper surface only)
- 66 Fuselage frame and stringer construction
- 67 Wing centre section fuel cells
- 68 Emergency escape hatch
- 69 Aileron servo controls
- 70 Main fuselage frame
- 71 Aft tourist class cabin
- 72 Rear freight hold/luggage compartment
- 73 Floor beam construction
- 74 HF aerial cable (port and starboard)
- 75 Overhead hat rack
- 76 Tourist class cabin seats
- 77 Aft galley
- 78 Starboard service door
- 79 Aft starboard toilet compartment
- 80 Aft radio rack
- 81 Rear pressure bulkhead
- 82 Anti collision light
- 83 Dorsal fin fairing
- 84 Starboard tailplane
- 85 ILS aerial
- 86 Starboard elevator
- 87 Leading edge de-icing ducts
- 88 Fin construction
- 89 HF blade aerial
- 90 Rudder balance weight
- 91 Rudder
- 92 Elevator hinge controls
- 93 Elevator tab
- 94 Port elevator
- 95 Tailplane construction
- 96 ILS aerial
- 97 Leading edge de-icing
- 98 Tailplane attachment
- 99 Fuselage fin frame
- 100 Tail bumper/fuselage vent
- 101 Rudder and elevator control rods
- 102 Access hatch to control bay
- 103 De-icing air supply duct
- 104 Rear freight hold
- 105 Tailplane servo controls
- 106 Mail locker
- 107 Aft port toilet compartment
- 108 Passenger entry door
- 109 Door frame construction
- 110 Steward's seat
- 111 Tourist class passenger seating
- 112 Wing fillet construction
- 113 Life raft stowage
- 114 Inboard tailpipe duct
- 115 Exhaust silencer nozzles
- 116 Outboard tailpipe
- 117 Thrust reverser (outboard only)
- 118 Inboard flap section
- 119 Fuel vent
- 120 Fuel dump pipes
- 121 Flap jack
- 122 Flap connecting links
- 123 Port airbrake (upper and lower surfaces)
- 124 Outboard flap section
- 125 Flap construction
- 126 Aileron tab
- 127 Port aileron
- 128 Aileron hinge controls
- 129 Aileron construction
- 130 Static dischargers
- 131 Wing tip fuel vent
- 132 Port navigation light
- 133 Outer wing construction
- 134 Outboard fuel tank bays
- 135 Fuel tank access panels
- 136 Wing stringer construction
- 137 External fuel tank
- 138 Tank bumper
- 139 Fixed slot
- 140 Wing rib construction
- 141 Leading edge de-icing ducts
- 142 Four wheel bogie unit
- 143 Wing skin joint strap
- 144 Undercarriage well
- 145 Main undercarriage leg mechanism
- 146 Wing integral fuel tank
- 147 Rolls-Royce Avon R.A.29 engine
- 148 Inboard engine bay (engine omitted)
- 149 Engine mounting frame
- 150 Intake duct construction
- 151 Landing lamp
- 152 Engine intakes
- 153 Ram air intake
- 154 Heat exchangers
- 155 Taxi lamp

© Pilot Press Limited

One of the best-looking Comet variants, the short-span 4B was derived for British European Airways from the still-born 4A for the US operator Capital Airlines. Longer than all other versions, it was planned for short-haul European operations, but natural growth in certificated weight endowed it with such range that BEA Airtours and the final operator of the type, Dan-Air, found it an excellent vehicle for long charters and inclusive-tour flights. The aircraft shown was the fifth of BEA's initial batch of six, to which eight more were later added, the only other initial customer being Olympic with four. Interesting features include the bogie main gears with dimpled tyres retracting into bays bulged on the underside, the fatigue-resistant windows and ADF aerials (two black ovals above the forward fuselage), outward-swept jetpipes with Greatrex-type noise-reducing nozzles, and reversers on the outer engines only.

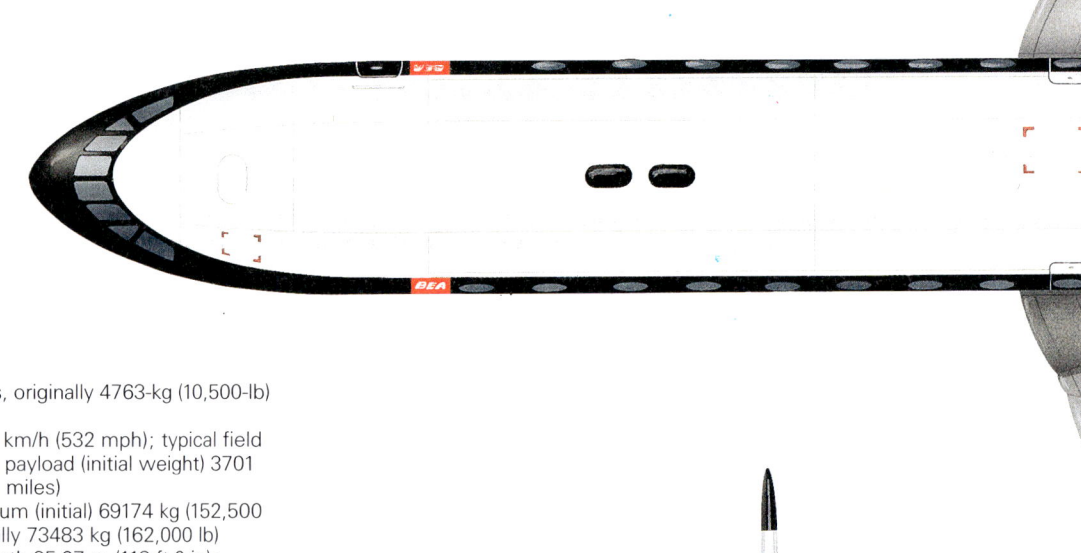

Specification
D.H.106 Comet 4B
Type: medium-range passenger transport
Powerplant: four Rolls-Royce Avon turbojets, originally 4763-kg (10,500-lb) Mk 542; later Mk 525B
Performance: maximum cruising speed 856 km/h (532 mph); typical field length 2134 m (7,000 ft); range with maximum payload (initial weight) 3701 km (2,300 miles), (final weight) 5391 km (3,350 miles)
Weights: empty 33483 kg (73,816 lb); maximum (initial) 69174 kg (152,500 lb) subsequently 70762 kg (156,000 lb) and finally 73483 kg (162,000 lb)
Dimensions: span 32.87 m (107 ft 10 in); length 35.97 m (118 ft 0 in); height 8.69 m (28 ft 6 in); wing area 191.28 m² (2,059 sq ft)
Accommodation: flight crew of three or four and normal seating for (initially) 101 passengers, (Dan-Air) 119 passengers

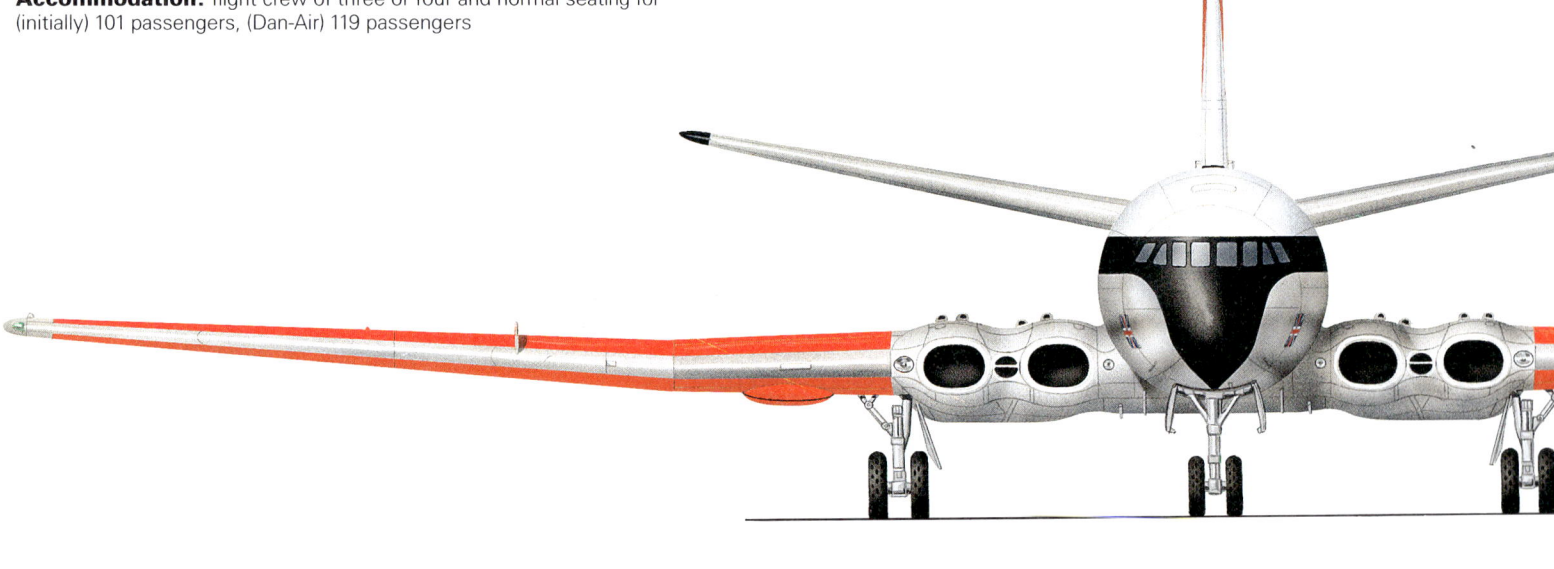

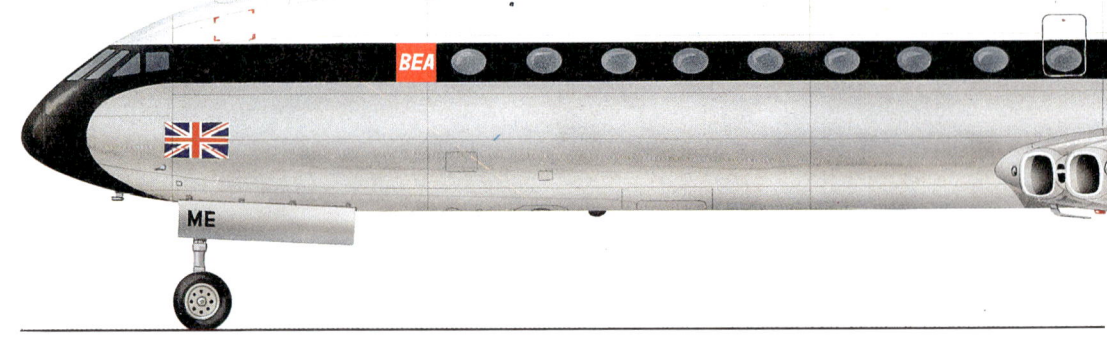

Chapter 18

De Havilland Dove & Heron

De Havilland Dragon

De Havilland's Moths

de Havilland Dove and Heron

The aptly named Dove was the first aircraft to be built by de Havilland after World War II. It would become a popular and long-lived type, finding favour around the world. Later both it and its larger sibling, the D.H.114 Heron, became the subjects of several modification programmes which saw them change almost out of all recognition to the original design.

In 1943 the Brabazon Committee issued specifications for a number of civil aircraft types that it anticipated would be needed by British airlines in peacetime. Among these was a proposal for a 'Type 5B', intended as a replacement for the de Havilland D.H. 89 Dragon Rapide twin-engine biplane feeder-liner.

Responding to this proposal, the de Havilland Aircraft Company formed a design team headed by R. E. Bishop, who had created the Mosquito fighter bomber and was later to design the Comet jet airliner. Bishop's D.H. 104 design was for an all-metal low-wing twin with retractable nosewheel undercarriage, powered by two 246-kW (330-hp) D.H. Gipsy Queen 70 supercharged engines driving three-blade D.H. Hydromatic featherable and reversible propellers and seating two crew and eight passengers in two-abreast configuration with a central aisle, or up to eleven passengers with less space for baggage and no lavatory. Built to Air Ministry Specification 26/43, the D.H. 104 was a much more sophisticated machine than its biplane forerunner, and appropriately for an aircraft that flew just six weeks after the fall of Japan, the name Dove was chosen.

The prototype Dove, G-AGPJ, made its first flight from Hatfield Aerodrome in the hands of Geoffrey Pike on 25 September 1945, which by happy chance was the 25th anniversary of the founding of the de Havilland company. After a month of test and development flying the Dove returned to the workshop for installation of an enlarged dorsal fin to improve engine-out handling, later acquiring a graceful curving dorsal fillet that retained the distinctive fin and rudder shape that had been a D.H. hallmark since the Moth biplanes of the 1920s. The prototype was passed to the British Ministry of Supply in 1946 and served in military markings as a development aircraft until 1954, when it was sold to the Portuguese government.

A second prototype joined the development programme in early 1946 but was destroyed in a crash near Hurn Airport, Bournemouth on 14 August 1946 while on loan to the British Overseas Airways Corporation's Development Flight. The fourth production Dove G-AHRJ made its public debut at the Society of British Aircraft Company's display at Radlett Aerodrome in September 1946.

By that time large scale production of the Dove Mark 1 had been launched at Hatfield, but domestic sales were slow. Although the Dove, with a cruising speed of 265 km/h (165 mph), was much faster than the Rapide and safer in the event of engine failure, its high purchase price and operating costs were beyond the means of most British charter operators in the austere years immediately following the Second World War.

Early deliveries went mostly to overseas operators, including Airlines of Western Australia, Airways (India) Ltd, Eagle Airlines of Persia, the Iraq Petroleum Transport Company, the Belgian carrier Sabena, SATA of the Azores, South African Airways, Sudan Airways, Union of Burma Airways, and West African Airways Corporation. Others went to Brazil, Chile and Ethiopia, while Argentina became a major operator of the type with an initial batch of 20 delivered, followed by a further 50. Unique among exports was the 15th production aircraft, which went to de Havilland Canada in

Doves and Herons were used on nascent commuter airline operations, heralding the later rapid growth of such services. Belgiums's SABENA flew Heron 1s.

The third Dove seen on flight test. Early Doves were notable for having an all-glazed canopy, later replaced by a more domed solid cockpit roof. The dorsal fin was fashioned into an elegant curve, recalling earlier de Havilland designs.

Many Doves were given a new lease of life by the Riley conversion, which substituted Lycoming flat-eight engines for the inverted-six Gipsy Queens. Many of these conversions also had a swept fin added. This pair flew with Bristow Helicopters, retaining the original fin.

1947 and was the only Dove floatplane. British operators who took delivery of Dove 1s included Skyways Ltd, whose *Sky Maid,* the fifth production Dove, became the first British aircraft to land at Kathmandu in the Himalayas, where it was inspected by the King of Nepal; Hunting, Olley, and Morton Air Services of Croydon Airport; Cardiff-based Cambrian Air Services, and British Overseas Airways Corporation, which used three for crew training.

Executive transport

In 1948 the six-seat Dove 2 executive transport was introduced, and found a ready market with business operators at home and abroad, particularly in the United States which was to prove an especially lucrative market, with more than 100 aircraft exported there. The Dove's success in the USA over many years was capitalised on by de Havilland and its successors Hawker Siddeley and British Aerospace when developing and selling the D.H. 125 executive jet, conceived as a successor to the Dove for the corporate market and now, as the BAe 125 series, among the most successful business jets in an intensely competitive market.

Also developed in 1948 was the Dove 4, a military liaison and communications aircraft meeting British specification C.13/46. The 48th production Dove was modified as the prototype, and featured reduced cabin seating for seven to make room for a ten-man survival dinghy. An initial batch of 30 Dove 4s were delivered to the Royal Air Force in 1948 as Devon C. Mk 1s, serving with No. 31 (Metropolitan Communications) Squadron at RAF Hendon near London, and overseas for use by British Air Attachés. Nine more Devons were later delivered to the RAF, finally retiring from service in 1984, although one remains in use as the support aircraft for the Battle of Britain Memorial Flight. In 1955-56 13 ex-civilian Doves were supplied to the Fleet Air Arm for service with 781 Naval Air Squadron based at RNAS Lee-on-Solent in Hampshire, and designated Sea Devon C. Mk 20s. Other military Doves/Devons were delivered to the air forces of the Belgian Congo, Ceylon, Egypt, Ethiopia, India, Iraq, Ireland, Jordan, Lebanon, New Zealand, Pakistan, South Africa and Sweden.

In 1951 demand for the Dove was running at such a high level that the production line was transferred from Hatfield to a larger site at Hawarden, Chester. Few major changes were made to the Dove during early production. From 1952 onwards 270-kW (340-hp) D.H. Gipsy Queen 70-4 engines became standard, and designations changed to Dove 1B and 2B (or 1A and 2A for aircraft destined for America, the 'A' suffix tradition for U.S. models continuing throughout Dove production and on through the current BAe 125 range). The following year 300-kW (380-hp) Gipsy Queen 70 Mk 2 engines became available, model numbers changing to Dove 5 and 6 respectively, and maximum take-off weight was increased to 3991 kg (8,800 lb) providing a 20 per cent increase in payload on a typical 804-km (500-mile) flight. Uprated Gipsy Queen engines were also retrofitted to earlier models to bring them up to Marks 5/6 standard.

The only significant external change to the aircraft in fifteen years of production was the installation of asymmetrically-shaped elevators to eliminate buffeting, the port elevator having a square-cut tip while the starboard one retained the original rounded profile.

The ultimate Dove

The final production models of the Dove were the 8-11 passenger Mark 7 feeder liner and the six passenger Dove 8, which was marketed as the Dove Custom 800 in the United States. Powered by 298-kW (400-hp) Gipsy Queen 70 Mk 3 engines and having a maximum take-off weight increase to 4060 kg (8,950 lb), these models were distinguishable from earlier Doves by the enlarged oil cooler intakes on top of their engine cowlings, thrust augmentor exhausts, raised cockpit roof line (actually a Heron cockpit canopy) which provided five inches more headroom for the crew, and square-cut tips to both elevators. The prototype Dove 8 executive model first flew in February 1960 and made its public debut at that year's Farnborough Air Show. First customer delivery took place on 24 January 1961 when excavator manufacturers J. C. Bamford re-

This Heron 1 (RMA Sir James Simpson) served with BEA on its Scottish routes, linking the mainland with the outer Hebrides. Two such aircraft performed sterling work on this island service.

Military use of the Heron was not as widespread as that of the Dove, Jordan being one of the nations to purchase the type for staff transport and communications duties. A handful flew in British colours, the last serving as an Admiral's Barge with the Royal Navy.

New Zealand National Airways was the first customer for the Heron, and this aircraft is seen on a demonstration tour. The Heron retained many Dove features for ease of design and manufacture, although the Heron 1 had fixed undercarriage.

ceived G-ARJB *Explorer*; the aircraft is now preserved at JCB's factory at Rocester, Staffordshire, its place having been taken by a succession of BAe 125s. Other British corporate operators of the Dove 8 included the National Coal Board, British Insulated Callenders Cables and the Dowty Group, whose aircraft was flown by the famous test pilot and world speed record breaker Neville Duke.

End of the line, for now

Dove production finally ended on 20 September 1967 when the 542nd aircraft was delivered, although the 541st aircraft was technically the last built, being assembled at Baginton Airport, Coventry in early 1968 from Chester-manufactured components.

Apart from various Gipsy Queen engine retrofits bringing earlier models up to Marks 5/6 or 7/8 standard, the Dove inspired two major modification programmes in the USA. The first of these was developed in 1963 by Jack Riley of Riley Aeronautics Corporation, Fort Lauderdale, Florida and was known as the Riley Turbo Executive 400. It featured 300-kW (400-hp) flat-eight Lycoming IO-720-A1A supercharged flat-eight engines in new 'slimline' nacelles, a large sharply-swept fin, and restyled flight deck and cabin interior. Seventeen Doves were converted to this standard, including four undertaken by McAlpine Aviation at Luton Airport, although only one of the British-converted Riley Doves had the swept fin modification.

Super Dove

The most ambitious modification of the Dove was undertaken by Carstedt Air Inc. of Long Beach, California, which stretched the fuselage by 2.22 m (7 ft 3 in) by means of 'plugs' inserted fore and aft of the wings, enabling 18 passengers to be accommodated. The cockpit position was also moved forward, and the cabin roof line lowered. Two 447-kW (605-shp) Garrett-AiResearch TPE331 turboprop engines were installed, driving three-blade feathering and reversing Hartzell propellers, and an additional 850 litres (187 Imp gal) of fuel provided in tanks outboard of the engine nacelles. The result of this extensive re-engineering was known as the Carstedt Jet Liner 600, the first of which made its maiden flight on 18 December 1966. Only a small number of Jet Liners were built, principally for US third-level carrier Apache Airlines of Phoenix, Arizona which operated four. A similar conversion started in England by Dove operator Channel Airways of Southend, Essex was abandoned before completion.

The immediate success of the Dove in 1945 prompted de Havilland to consider a larger four-engined version to replace the pre-war D.H. 86 as the Dove had the Dragon Rapide, but lack of market prospects caused the idea to be shelved until 1949, when detailed design work began on the D.H. 114. Simplicity was the key to designer W. A. Tamblin's thinking. The D.H. 114 incorporated many Dove components such as outer wing panels and forward and rear

CF-DJH was a unique Dove, for it was the only one to be fitted with floats. It was the 15th production aircraft, and was bought by de Havilland Canada.

In addition to their regular RAF communications role, the Devon was also used for trials work, and transport in association with Royal Aircraft (now Aerospace) Establishment work. Two of these aircraft (VP959 and XM223) were used until recently on sonobuoy-dropping work at West Freugh.

fuselage sections, linked by a 'stretched' 2.62-m (8-ft 6-in) cabin section seating 14-17 passengers. A fixed tricycle undercarriage was chosen for ease of maintenance, and 187-kW (250-hp) Gipsy Queen 30 Mk 2 engines driving two-blade variable pitch propellers selected to power the aircraft, for which the name Heron was chosen.

Big brother

The prototype Heron, G-ALZL, was first flown from Hatfield by Geoffrey Pike on 10 May 1950, and completed 180 hours of test flying during which the only major external change made was to add dihedral to the originally flat tailplane. After completing hot-and-high tropical trials in the Sudan and Kenya, G-ALZL was transferred to de Havilland's sales department in May 1951 and was demonstrated in the colours of many airlines, including British European Airways and Japan Air Lines.

The first production Heron Mark 1 was delivered to New Zealand National Airways in April 1952, and was followed by six further aircraft before, like the Dove, Heron production was concentrated at Chester. The seventh production aircraft, G-AMTS, was also the prototype Heron Mark 2, which incorporated a retractable undercarriage resulting in a 32-km/h (20-mph) increase in cruising speed and significant fuel savings. It first flew on 14 December 1952.

Production of the two marks of Heron continued concurrently, early fixed-undercarriage Mark 1s going to All Nippon Airways for services between Osaka and Tokyo; Braathens S.A.F.E. Air Transport in Norway; Butler Air Transport in Australia; Cambrian Airways; Garuda Indonesian Airways; Jersey Airways, which became the major British user of the type on its services from the south of England to the Channel Islands; PLUNA in Uruguay; and TAS in Brazil. British European Airways took delivery of two Heron 1s in February 1955, replacing Dragon Rapides on 'highlands and islands' services between Glasgow and the beach airstrips at Barra and Benbecula.

The retractable undercarriage Heron 2 proved popular as a VIP and executive transport, examples going to the Belgian Congo, Saudi Arabia and South Africa for this purpose, while four were supplied to the RAF's Queen's Flight between 1955-61 for use by the British Royal Family. Other military operators were to include the air forces of Ceylon, Ghana, Iraq, Jordan and West Germany.

Few structural changes were made to the Heron airframe, save for the adoption of a wider-chord rudder in 1955, the introduction of optional fully-feathering propellers on the Mark 2C, and an increase in maximum take-off weight of 6123 kg (13,500 lb) on the mark 2D. Total production of the Heron was 148 aircraft, of which 52 were fixed-undercarriage Mark 1s.

Like the Dove, the Heron became the basis for several conversions in later life, beginning in 1965 when Toa Domestic Airlines of Japan had five of its Mark 1s re-engined by Shin Meiwa with 195-kW (260-hp) Continental IO-470D 'flat sixes'. These converted aircraft were renamed Tawrons. Connellan Airways of Alice Springs developed a similar Lycoming IO-540-G1A5 conversion for its fleet of aircraft used on services throughout Australia's Northern Territories, and Riley Aeronautics produced a turbosupercharged Lycoming IGO-540-B1A5 conversion known as the Riley Turbo Skyliner.

Mr Saunders' idea

The most extensive conversion of the Heron was that developed by David Saunders of Saunders Aircraft Corporation of Montreal, Canada. Largely engineered by Aviation Traders (Engineering) of Southend, England, the aircraft was known as the Saunders ST-27. It featured a fuselage stretch of 3.16 m (10 ft 4 in) which enabled 23 passengers to be carried, strengthened wings, provision for nose-mounted weather radar, and replacement of the Heron's four Gipsy Queen piston engines with two 560-kW (750-shp) Pratt & Whitney Canada PT6A-34 turboprops. The ST-27 prototype, then powered by 530-kW (715-shp) PT6A-27 engines, made its first flight on 28 May 1969, followed by a production prototype in April 1970. Twelve ST-27s were converted. A thirteenth became the prototype ST-27B, which was later redesignated ST-28 and first flew in July 1974. This incorporated structural and systems redesigns to enable it to meet more stringent certification requirements, and had four-bladed propellers, a wider chord rudder, larger cabin windows, integral airstairs and greater fuel capacity. Because the supply of surplus Heron airframes was dwindling, Saunders planned to build production ST-28s from scratch using new jigs and tooling supplied by Hawker Siddeley Aviation. The company started work on an initial batch of 15, but financial problems caused the project to be abandoned in 1976.

The most radical of the Dove conversions was the Carstedt Jet Liner 600, which featured a considerably lengthened fuselage and turboprop power in the form of two Garrett AiResearch TPE331s. A few of these conversions saw service with regional airlines.

Conversions of Herons roughly mirrored those of the Dove, the main upgrade being to fit more modern and reliable engines. Shin Meiwa created the Tawron for TDA by using Continental engines on the Heron 1.

Dove 4 was the company designation for the aircraft acquired by the Royal Air Force for communications and embassy support work. Named Devon in RAF service, the aircraft had a long and fruitful career, the last examples flying with the Battle of Britain Memorial Flight and RAE West Freugh.

de Havilland Dragon

Producing more than twice as many transport aircraft as any other British manufacturer, the de Havilland Aircraft Company carved its niche in aviation history. The Dragon family of biplane transports formed a high proportion of the company's output; with their reliable pedigree and overall performance attracting both civil and military customers, the three major designs logged over 30 years of operations, with some aircraft still flying today.

During the late 1920s and early 1930s de Havilland built up a remarkable reputation for its range of light biplanes based on the D.H.60 Moth. The company had built the eight-passenger D.H.61 Giant Moth and the seven-passenger D.H.66 Hercules for service with Imperial Airways and for use in other parts of the Empire, while the needs of small operators were met by the D.H.83 Fox Moth.

One of the main Fox Moth operators was Hillman's Airways. It had been founded by Edward Hillman, who moved into aviation as a natural extension of his motor coach business. This colourful character was responsible for pressing de Havilland to design a larger 10-seat twin-engined aircraft to meet the expanding demand for his airline's services. Hillman's Airways was accordingly the first operator of de Havilland's new light airliner, the D.H.84 Dragon. The Dragon I was actually rather smaller than Hillman's initial specification and provided accommodation for a pilot and six passengers in a slab-sided fuselage built of spruce and plywood. The biplane wings were built with outer panels which folded backwards just outboard of the engines and the shape of the wings themselves was somewhat complex. Two 97-kW (130-hp) de Havilland Gipsy Major 1 engines were mounted on the lower wings with the main landing gear legs directly mounted to the engine firewall bulkheads and, in some cases, the wheels were enclosed with streamlined spats to give some drag improvement. A swept-back fin and rudder assembly was used, perpetuating the de Havilland 'trade-mark' which had become familiar on the sport biplanes, and the pilot was housed in an extensively glazed cockpit in the extreme nose of the aircraft.

The prototype Dragon (E-9, c/n 6000) made its first flight at Stag Lane on 24 November 1932. The period of flight testing was incredibly short compared with present-day experience and the prototype was soon resprayed with the registration G-ACAN and delivered by Mr Hillman during December 1932. Four production Dragons followed in quick succession and were an immediate success. Hillman's Airways offered very low fares (by the device of keeping overheads down and paying the pilots and the bus drivers the same wage rates) and this soon resulted in two further Dragons being ordered and the existing machines being converted to eight-passenger configuration by eliminating the rear baggage compartment.

Many other small airlines started to order the Dragon for passenger services and carrying mail, and for ambulance operations in Scotland and other isolated areas. Another bus company, the Scot-

Owned and operated by the Prince of Wales in the 1930s, the immaculate D.H.84 Dragon 1 wore a red, blue and silver colour scheme. Powered by two 97-kW (130-hp) Gipsy Major engines, this biplane had seating for between six and 10 passengers.

*The prototype D.H.84 Dragon made its first flight with the test registration E-9, but was re-registered **G-ACAN** within a few weeks. It joined the first customer, Hillman Airways, at its base at Maylands Aerodrome during December 1932 and was later joined by a further five Dragons.*

tish Motor Traction Co. Ltd, had established an integrated system of coach and air services and used the Dragon for this purpose and the type entered service with Highland Airways and Aberdeen Airways, which were merged into Scottish Airways in 1938. The Dragon II was introduced after 62 of the initial version had been built, and this later model was identifiable by the individually framed cabin windows which replaced the continuous glazing of the Dragon I. Quite a number of Dragons were operated on floats (particularly in Canada), which necessitated an increase in side area and these aircraft were fitted with an extended vertical fin which stayed on even when they reverted to the normal wheeled landing gear.

An early purchaser of the Dragon was the Iraqi air force, which had probably prompted the D.H.84 development in the first place. It ordered eight of the D.H.84M version, and these were delivered to Baghdad during 1933 to be used for patrolling the rather turbulent tribally-controlled areas of the country. A gunner's open position was provided on top of the fuselage just aft of the wings and the pilot was able to take aggressive action using two machine-guns fitted one on each side of the nose of the Dragon. The D.H.84M was also fitted with a seaplane-style extended fin. Other air forces which ordered the type were the Danish army air force (two aircraft), the Portuguese air force (three aircraft) and the Turkish and Irish military forces which purchased standard civil Dragons as general-duties transports.

Long-distance record attempt

The most famous of all Dragons was G-ACCV (c/n 6014) named *Seafarer* and owned by Jim and Amy Mollison. As Amy Johnson, Mrs Mollison had become famous through her epic flight to Australia in 1930 and subsequent flights from England to South Africa. Together, the Mollisons planned to use *Seafarer* in an attempt on the world long-distance record, which would start with a flight to New York. The Dragon made its first attempt to leave Croydon on 8 June 1933 but was so heavily loaded that it failed to evade a ditch and the aircraft nosed over. The aircraft was repaired and flown to Wales where it successfully took off from Pendine Sands and, after 39 hours in the air, arrived over Bridgeport, Connecticut. Attempting to land downwind *Seafarer* overshot and was damaged beyond repair, the Mollisons ending up in hospital. The remains of G-ACCV were later cannibalised to be used in the Dragon II *Seafarer II* which was used by Ayling and Reid to fly the Atlantic from Canada to England in August of the following year.

In 1934, de Havilland had rushed through development of a new four-engined airliner designated D.H.86. Aimed at the requirement for a fast 10-seater which could operate the Singapore to Australia section of the proposed Croydon to Brisbane passenger service, the D.H.86 (Express Air Liner, as it was known) used much of the design layout of the Dragon but was scaled up, had a completely new wing cellule and was powered by four of the new 149-kW (200-hp) de Havilland Gipsy Six inline engines. There was also a degree of pressure from existing Dragon operators for a more comfortable and faster replacement for the D.H.84, and this led to the company scaling down the D.H.86 and fitting it with two Gipsy Six engines to produce the D.H.89 Dragon Six. Outwardly, the Dragon Six was not a great deal different from the Dragon, but it used the completely new tapered wings without the folding mechanism of the previous model and the maximum speed of 206 km/h (128 mph) which had been achieved by the Dragon I rose to 253 km/h (157 mph) because of improved streamlining and the higher engine power. The D.H.89 prototype (E-4, c/n 6250) was first flown from Hatfield on 17 April 1934. It had a fully faired landing gear, which was a major contributor to drag reduction, and the nose cone was tapered upwards to give a new profile to the cockpit area.

Rapide operators

Once again, Mr Hillman was a keen purchaser of the new model, which was now named the Dragon Rapide, and the first three production examples were delivered to Hillman's Airways between July and September 1934. A further four Dragon Rapides were delivered to this operator, and these were finally transferred to British Airways when Edward Hillman sold his business in December 1935. Another Rapide operator was Railway Air Services Ltd, which had been formed by the principal railway companies in 1934 and had begun flying on routes from Croydon to the Isle of Wight, Plymouth, Birmingham and Liverpool. The expansion of mail services to Glasgow and Belfast prompted the company to buy the D.H.86 and also a total of eight D.H.89s. Other British operators of the Dragon Rapide included Olley Air Service Ltd, Northern & Scottish Airways Ltd, Jersey Airways Ltd and British Continental Airways Ltd. All of these small air carriers used the Dragon Rapide intensively, and in all weather conditions, right up to the outbreak of World War II in 1939.

The D.H.89 built up a good export reputation, 41 of the first 100 aircraft being sold to overseas customers. Several of the Middle East petroleum companies bought Rapides for communications with the desert oil extraction sites, and Canadian companies, including Quebec Airways and Canadian Airways Ltd had fleets of them, often fitted with the extended fin and twin floats. Once the Rapide production line had built up to a fair volume, de Havilland introduced the D.H.89A version which featured split trailing-edge flaps fitted just outboard of the engine nacelles on the lower wings, and optional metal propellers. Many of the earlier D.H.89s were retrospectively modified with the flap system, and production of the D.H.89A started at G-AERN (c/n 6345) which was sold to West Coast Air Services and delivered in March 1937. At about this time, also, the Rapide was given an effective cabin heating system, the rear cabin

This Canadian-registered D.H.84 Dragon 2 displays a floatplane configuration, a feature applied to at least one Dragon Rapide in later years. British production of the Dragon 1 & 2 totalled 115 aircraft, with additional aircraft being built in Australia pushing the overall figure to 202.

*The five-seat **D.H.90 Dragonfly** was a small, luxury development of the Rapide but was constructed with a pre-formed plywood fuselage. One of the 67 examples is **G-AEDU** which was originally in Mozambique as **CR-AAB**. It was restored as **G-AEDU** in 1979, and in June 1983 it was sold to an owner in Louisville, Kentucky, as **N190DH***

***G-AGTM** is a **D.H.89A Dragon Rapide** (c/n 6746) which was originally delivered during the war as a Dominie for RAF use. When the war was over it was civilianised and, in 1953, sold to Lebanon as **OD-ABP**. It subsequently flew in Jordan and, in 1964, returned to England and is shown in its Rothman parachuting livery.*

windows were enlarged and optional provision was made for a toilet in the rear of the cabin.

Several Rapides had been sold abroad for military purposes during 1936 and 1937. The Imperial Iranian war ministry bought three, the National Government of China had five (delivered in December 1937) and the Spanish government ordered three for police work in its North African territories. These were, in fact, never used in North Africa and they joined the Republican forces when the Spanish Civil War broke out. Three ex-civil D.H.84s and approximately 10 D.H.89s also reached one side or the other. These were all aircraft which were sold by dealers who had purchased them in England and were able to break the blockade which both the British and French governments established in the summer of 1936. Both the Nationalists and the Republicans were desperate to obtain aircraft, and a number of Airspeed Envoys, Monospars, Fokker transports and Rapides were despatched, although a fair number never reached their destinations. The Rapides were fitted in Spain with open gun turrets and there is no evidence that any of these machines survived the war.

Coastal patrol version

In 1935, the British Air Ministry carried out evaluations to see if the Rapide would be suitable for the coastal patrol role set out in Specification G.18/35. One example (K4772, c/n 6271) was built as a D.H.89M with the extended fin and a dorsal upper turret which housed a Lewis Mk III machine-gun, and it was tested at Martlesham Heath and Gosport but the patrol contract was awarded to the Avro 652 (which became the Anson). Despite this setback, two Rapides were ordered as communications aircraft and several were ordered as navigation trainers by Airwork Services Ltd, which had gained substantial contracts from the Air Ministry during the build-up to the outbreak of war. As 1938 gave way to the fateful 1939, the Rapides in use at the School of Navigation at Shoreham received a coat of camouflage paint, and the Rapide went into battle.

Up to the start of the war on 3 September 1939, de Havilland's civil Rapide output had numbered about 180 aircraft and this gave way to military production at Hatfield as the company started to receive direct Air Ministry orders for D.H.89s equipped as W/T and navigation trainers. These were designated D.H.89B Dominie Mk I and were externally identical to the civil model with the exception of the large loop antenna on the roof just behind the cockpit and, of course, the camouflage paint scheme. Dominies entered service with the air navigation schools, particularly No.6 Air Observers' Navigation School (AONS) at RAF Staverton and No. 7 AONS at RAF Perth (Scone). They were also delivered as the D.H.89B Dominie Mk II for general communications duties and employed as the mainstays of station flights both with the RAF and the Royal Navy. The Air Transport Auxiliary (ATA) took over responsibility for the ferrying of mili-

*Originally known as the Express Air Liner, the **D.H.86** introduced tapered wings and a faired undercarriage housing in addition to the Gipsy Six air-cooled powerplant. This **D.H.86B** also displays the Zulu-shield fins added to the tailplane tips on this model. Sixty-two aircraft were built.*

*Rejected by the RAF in favour of the Avro Anson for general reconnaissance, the **D.H.89B Dragon Rapide** was ordered for communications and training duties as the Dominie Mk I & II. This example served with No. 2 Radio School and has the direction-finding loop on the upper fuselage to facilitate the training of wireless operators.*

tary aircraft from airfield to airfield in May 1940 and used Dominies for positioning crews all over the United Kingdom and Europe. It also set up a Medical Section which used Dominies with up to two stretchers fitted into the modified D.H.89 cabin to ferry medical cases from isolated locations to its central hospital at White Waltham.

The arrival of the war had brought complete suspension of all civil air services. Once the military priorities had been sorted out some of the aircraft used by the airlines were put back in service with civil markings to run priority routes under the control of the Associated Airways Joint Committee (AAJC). Typical of the Rapides so employed were G-ACPP and G-ACPR of Great Western & Southern Air Lines, which were put to work flying three return flights each day between Land's End (St Just) and the Scilly Isles. Most of the Dragons and Rapides on the civil register, however, were impressed into military service with standard service markings and serials. Many of them suffered accidents during their war service, with the result that few were still airworthy when peace came in 1945.

The overall total of Dragon Rapides built by de Havilland was 728 and the production line was closed in 1945 to make way for new types. The last 346 aircraft were built by the Brush Coachworks at Loughborough and, in common with many other aircraft, the D.H.89 was subject to some cancellation of Air Ministry contracts when peace arrived. In fact, the last batch of approximately 100 Rapides was built for post-war civil sale. These were known as the Rapide Mk 2, which was fitted with six passenger seats, and the Rapide Mk 3 which had eight passenger seats in addition to the pilot. The Rapide Mk 4 was a conversion of the standard Dominie with the Gipsy Queen 3 engines replaced by Gipsy Queen 2s with constant-speed propellers, and the single example of the Rapide Mk 5 had manually-operated variable-pitch propellers and a special version of the Gipsy Queen 3 engines. The final post-war variant was the Rapide Mk 6 which, again, was a modification of earlier aircraft with fixed-pitch Fairey X5 propellers which provided performance equivalent to that of the Rapide Mk 4 but with less mechanical sophistication.

Post-war use

The majority of post-war Rapides were Dominies, which had survived to be declared surplus and sold at bargain prices to various overhaul companies such as W. A. Rollason, Field Aircraft Services, and Lancashire Aircraft Corporation. These Dominies initially supplemented the fleets of the AAJC operators to expand the wartime route network which stretched from Shetland (Sumburgh) to Guernsey and the Scilly Isles. Eight principal airlines came into this group, including Olley Airways, Railway Air Services, Great Western & Southern Air Lines, and Scottish Airways with a fleet of 25 Rapides, four D.H.86 Expresses and one Dragon. There was also Allied Airways (Gandar Dower) Ltd, which had five Rapides and a Dragon and had always resisted the pressure to fit into the AAJC grouping. The independence of these companies was to be short-lived, however, because the policy of the Attlee government was to nationalise air transport. British European Airways came into being on 1 January 1946 and most of the independent companies' Rapides were taken over. The numbers diminished over the years and the type was finally phased out when the Scilly Isles route became operated by Sikorsky S-61N helicopters and the Highlands and Islands service was passed on to de Havilland Herons.

Rapides found their way into the hands of small air taxi companies such as Air Kruise. They were popular joy-riding machines in the hands of Skegness Air Taxi and of Airwork Services, who used to offer Saturday pleasure flights from Turnhouse and Renfrew for £1 per passenger. They also joined a fair number of companies as business aircraft and they flew with Saunders-Roe, The Windmill Theatre Transport Company, Midland Metal Spinning Co. Ltd and Fox's Glacier Mints, to name but a few. As a general workhorse, the Rapide found employment with Fairey Aviation who used several for survey work, and with many parachute schools and clubs. As time went by, the number of Rapides diminished but around a dozen are

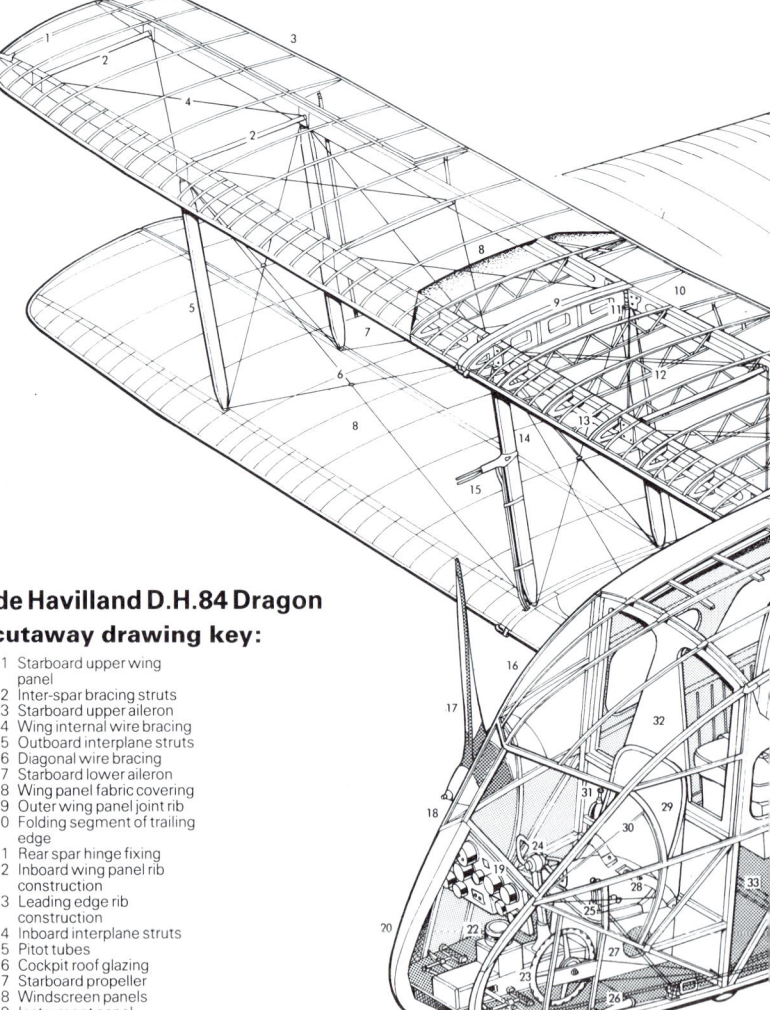

de Havilland D.H.84 Dragon cutaway drawing key:

1 Starboard upper wing panel
2 Inter-spar bracing struts
3 Starboard upper aileron
4 Wing internal wire bracing
5 Outboard interplane struts
6 Diagonal wire bracing
7 Starboard lower aileron
8 Wing panel fabric covering
9 Outer wing panel joint rib
10 Folding segment of trailing edge
11 Rear spar hinge fixing
12 Inboard wing panel rib construction
13 Leading edge rib construction
14 Inboard interplane struts
15 Pitot tubes
16 Cockpit roof glazing
17 Starboard propeller
18 Windscreen panels
19 Instrument panel
20 Moulded plywood nose section
21 Rudder pedals
22 Compass
23 Tailplane trim control handwheel
24 Control column
25 Engine throttle and mixture control levers
26 Venturi tube
27 Seat support box
28 Safety harness
29 Pilot's seat
30 Direct vision opening side window panel
31 Fire extinguisher
32 Cockpit bulkhead
33 Main cabin flooring
34 Fuselage side panel framework
35 Diagonal frame member
36 Cabin window panels
37 Passenger seats (six)
38 Cabin roof framing
39 Ventilating air duct
40 Cabin wall trim panels
41 Leading-edge ventilating air scoop
42 Wing spar centre-section carry-through
43 Wing root attachment rib
44 Diagonal engine nacelle bracing struts
45 Passenger entry door
46 Cabin roof escape hatch
47 Ventilating air duct
48 Starboard side baggage compartment, capacity 50 cu ft (1.42 m³)
49 Starboard wing folded (ground handling) position
50 Skin panel external stiffening strip
51 Fuselage upper longeron
52 Horizontal spacers
53 Fin/tailplane root fillet
54 Starboard tailplane
55 Fin construction
56 Sternpost
57 Rudder horn balance
58 Fabric covered rudder construction
59 Tailplane upper bracing wires
60 Rudder operating lever
61 Elevator hinge control

G-AGSH, *a Dragon Rapide Mk 6, is seen here in the markings of the Royal Air Force Sport Parachute Association. The predominantly red colour scheme reflected its use by the 'Red Devils' parachute team as the jump-ship at many air displays.*

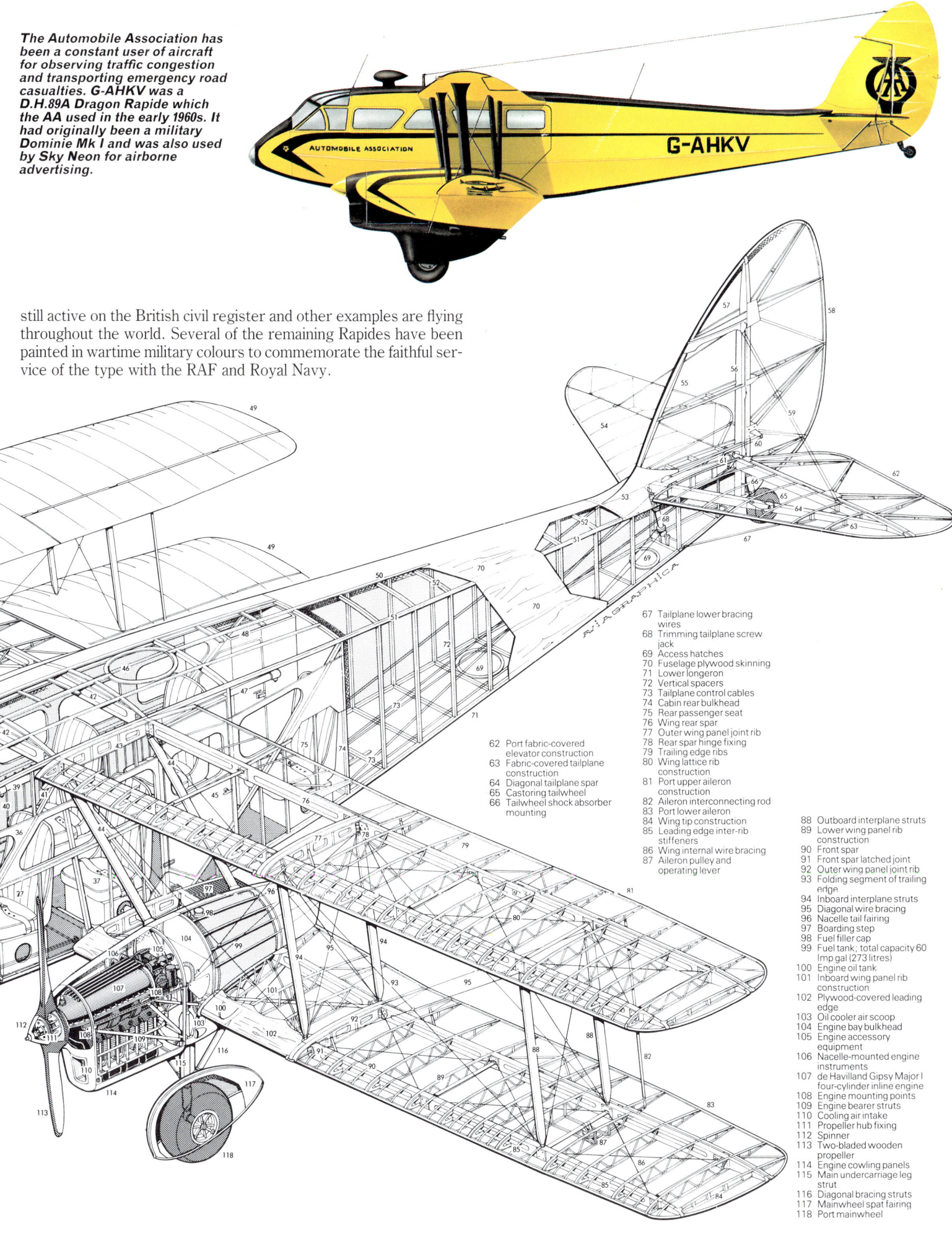

The Automobile Association has been a constant user of aircraft for observing traffic congestion and transporting emergency road casualties. G-AHKV was a D.H.89A Dragon Rapide which the AA used in the early 1960s. It had originally been a military Dominie Mk I and was also used by Sky Neon for airborne advertising.

still active on the British civil register and other examples are flying throughout the world. Several of the remaining Rapides have been painted in wartime military colours to commemorate the faithful service of the type with the RAF and Royal Navy.

62 Port fabric-covered elevator construction
63 Fabric-covered tailplane construction
64 Diagonal tailplane spar
65 Castoring tailwheel
66 Tailwheel shock absorber mounting
67 Tailplane lower bracing wires
68 Trimming tailplane screw jack
69 Access hatches
70 Fuselage plywood skinning
71 Lower longeron
72 Vertical spacers
73 Tailplane control cables
74 Cabin rear bulkhead
75 Rear passenger seat
76 Wing rear spar
77 Outer wing panel joint rib
78 Rear spar hinge fixing
79 Trailing edge ribs
80 Wing lattice rib construction
81 Port upper aileron construction
82 Aileron interconnecting rod
83 Port lower aileron
84 Wing tip construction
85 Leading edge inter-rib stiffeners
86 Wing internal wire bracing
87 Aileron pulley and operating lever
88 Outboard interplane struts
89 Lower wing panel rib construction
90 Front spar
91 Front spar latched joint
92 Outer wing panel joint rib
93 Folding segment of trailing edge
94 Inboard interplane struts
95 Diagonal wire bracing
96 Nacelle tail fairing
97 Boarding step
98 Fuel filler cap
99 Fuel tank; total capacity 60 Imp gal (273 litres)
100 Engine oil tank
101 Inboard wing panel rib construction
102 Plywood-covered leading edge
103 Oil cooler air scoop
104 Engine bay bulkhead
105 Engine accessory equipment
106 Nacelle-mounted engine instruments
107 de Havilland Gipsy Major I four-cylinder inline engine
108 Engine mounting points
109 Engine bearer struts
110 Cooling air intake
111 Propeller hub fixing
112 Spinner
113 Two-bladed wooden propeller
114 Engine cowling panels
115 Main undercarriage leg strut
116 Diagonal bracing struts
117 Mainwheel spat fairing
118 Port mainwheel

383

de Havilland's Moths

A product of the 1920s, the classic Moths represented a successful venture to enable the public to enjoy the sport of flying, and sowed the seeds of private aircraft ownership that were to blossom throughout the inter-war years. The basic design led to a hugely successful series of aircraft.

The de Havilland Moths were more than a family of light aircraft; they represented the manifestation of Captain Geoffrey de Havilland's feeling that there was a dormant market among the British middle class to enjoy a participation sport in private flying. The same motivation had prompted the *Daily Mail* to sponsor competitive trials at Lympne in 1923 for light aeroplanes (by latter-day standards they would be termed 'ultra-lights'), the objective being to encourage production of an aeroplane so simple and inexpensive as to lie within the budget of small flying clubs or within the means of 'the man in the street'. Although a number of ingenious little aircraft were entered, de Havilland realised that, being limited to engines of less than 1100-cc (67.13-cu in) capacity, the aircraft must be underpowered and probably unreliable in the hands of pilots with limited flying experience.

Accordingly the works at Stag Lane set about producing a scaled-down version of the D.H.51 three-seater, the new aircraft being a two-seat single-bay biplane, tough enough to withstand the rigours of instructional work and not so cramped as to be uncomfortable during cross-country flying. Powered by a four-cylinder 45-kW (60-hp) engine specially designed by Major F. B. Halford and named the A.D.C. Cirrus I, the new D.H.60 was a model of robust simplicity, its fuselage being in effect a plywood box built round spruce longerons and stiffened by vertical and horizontal members screwed to the plywood.

Named Moth in recognition of Geoffrey de Havilland's renown as a lepidopterist, the first D.H.60 (G-EBKT) was flown at Stag Lane by 'DH' himself on 22 February 1925. On that day was set a pattern for private flying that was to be emulated the world over until World War II. Efforts to canvass support from government sources were successful when Sir Sefton Brancker, Director of Civil Aviation, announced the founding of five Air Ministry-subsidised Moth-equipped flying clubs: the Lancashire Aero Club, the London Aeroplane Club, the Newcastle Aero Club, the Midland Aero Club and the Yorkshire Aeroplane Club. By the end of 1925 20 Cirrus Moths had been completed, of which 16 had been delivered to the clubs and two to private owners. 1926 brought forth 35 more Moths, of which 14 were exported, nine of them to Australia; Major Halford produced a 63-kW (85-hp) Cirrus II engine which was fitted in a Moth to be piloted by Geoffrey de Havilland in the King's Cup Race, but a broken oil pipe let through Hubert Broad into first place in a Cirrus I Moth. Another famous British sporting pilot, Neville Stack, flew the second Moth prototype (G-EBKU), powered by the first production Cirrus II, from Croydon to India where it attracted much interest as it embarked on six months' joyriding in the sub-continent.

The so-called 1927-model Moths, of which 150 were produced, were all powered by the Cirrus II, and one of these (G-EBPP) was shipped to Australia where Major Hereward de Havilland set up an agency in Melbourne to assemble imported Moths. Numerous outstanding lightplane flights were made by enthusiastic private owners, typical being Lady Bailey's altitude record for light aircraft of 5268 m (17,283 ft) on 5 July of that year, and Lieutenant R. R. Bentley's two extraordinary return flights between the UK and Cape Town totalling 83125 km (51,652 miles) in G-EBSO. Six aircraft, powered by 56-kW (75-hp) Armstrong Siddeley Genet radials, were bought by the Air Ministry for use at the Central Flying School.

1928 introduced the 67-kW (90-hp) Cirrus III engine and a split-axle landing gear which, on account of the new cross-braces between the landing gear, resulted in this model being termed the D.H.60X Moth. By the end of the year 403 Moths of all types had been produced and production was running at 16 per week; moreover, licences to build Moths had been negotiated with the Finnish Government Aircraft Factory and Veljekset Karhumäki O/Y (also of Finland) and the General Aircraft Co. of Sydney, Australia.

Meanwhile stocks of war-surplus Renault engine components, employed in the Cirrus engines, had been exhausted. In 1926 'DH' asked Major Halford to design a new replacement engine; the first example of this, the legendary Gipsy, was completed in July 1927 and within 16 weeks a production shop at Stag Lane was built. After trials

G-ATBL was a D.H.60G Gipsy Moth which was exported to Switzerland. It returned to England to be restored, one of quite a handful still flying in their country of origin. It is seen here at Old Warden, a perfect setting for such a historic aircraft.

The longest-surviving D.H.60 Moth, G-EBLV, was the sixth production example and was one of two Moths supplied to the Lancashire Aero Club on 29 August 1925, the first of the government-sponsored flying clubs initiated by Sir Sefton Brancker. This aircraft was painstakingly restored by de Havilland apprentices in 1951.

in an early Moth (G-EBOH), the new engine gained its certificate and appeared in the production D.H.60G Gipsy Moth. One of these, flown by W. L. Hope, won the 1928 King's Cup Race at 169 km/h (105 mph). Records tumbled, and many memorable flights followed: 'DH' himself in his own Gipsy Moth (G-AAAA) established a new lightplane height record at 6090 m (19,980 ft) on 25 July 1928, and Hubert Broad remained aloft for 24 hours (using extra fuel tanks in the fuselage); DH's son Geoffrey flew a Gipsy Moth for 600 hours in nine months during 1929, covering 82075 km (51,000 miles) with only routine inspections; when tested afterwards the cost of replacement parts was only just over £7! (The cost of a Gipsy Moth, ex-works, was a mere £475.)

Overseas success

Gipsy Moth-equipped flying clubs sprang up throughout the UK, as well as in Australia, Canada, Colombia, Finland, India, Kenya, New Zealand, Singapore, South Africa, Spain, Sweden and Switzerland. The type was also flown commercially in the United States, Argentina, Austria, the Bahamas, Belgium, Brazil, China, France, Germany, Mexico, Mozambique, the Netherlands East Indies, New Guinea, Norway, Nigeria, Peru, Portugal, Poland, Southern Rhodesia, the Straits Settlements and Yugoslavia. Two of countless outstanding flights should be mentioned: Amy Johnson flew *Jason*, a secondhand Gipsy Moth (G-AAAH), from Croydon to Darwin, Australia, between 5 and 24 May 1930 to complete the first solo England-Australia flight by a woman; and Francis Chichester, the first man to make the same flight, had completed the journey five months earlier in *Madam Elijah*, a Gipsy Moth (G-AAKK). The Gipsy Moth continued in production until 1934, by which time a total of 595 had been produced in England, in addition to 40 built by Morane-Saulnier in France, 18 by the Moth Aircraft Corporation of Massachusetts, USA, and 32 by the Larkin Aircraft Supply Co. Ltd of Melbourne, Australia.

A derivative of the wooden Gipsy Moth, the D.H.60M (sometimes known as the Metal Moth), was introduced in 1928 to meet overseas demands for a strengthened version, particularly in remote areas where damage to the structure could not easily be repaired. Interest in this version came initially from Canada, the first aircraft built being shipped there to be evaluated with wheel, ski and float landing gears, after which an order for 50 aircraft was placed by the RCAF, and others for flying clubs throughout North America. Metal Moths were also shipped to the agents Arnhold & Co. of Shanghai for sale to China. Others, fitted with light bomb racks, served on No. 1 Squadron, Royal Iraqi Air Force, helping to quell the Kurdish rebellion of 1932. In RAF service some D.H.60Ms flew with the CFS and with No. 5 FTS at Sealand, as well as numerous station flights. Two of the civil D.H.60Ms (G-AALG and G-ABDG) were owned by the then-Prince of Wales and the Duke of Gloucester. The all-comers England-Australia record was lowered by C. W. A. Scott in G-ACOA to 8 days, 20 hours and 47 minutes, Scott arriving in Darwin on 28 April 1932. Jean Batten established a new women's record in just under 15 days in May 1934.

The introduction of the inverted Gipsy III in 1931 led to the D.H.60G-III Moth, whose prototype (G-ABUI) first flew in March 1932. This was followed by orders from all over the world. The third production aircraft was fitted with a special high-compression Gipsy IIIA which developed 99 kW (133 hp); flown by Hubert Broad, it finished fifth in the 1932 King's Cup Race at 211.5 km/h (131.34 mph). One long-distance flight by a D.H.60G-III is worthy of mention, that by the Portuguese pilot Carlos Bleck, who flew from Lisbon to Goa (Portuguese India), averaging 169 km/h (105 mph) over the 9655 km (6,000 miles) in February 1934. From the 58th airframe onwards the designation was changed to Moth Major to mark a change to the standard 97-kW (130-hp) Gipsy Major, although externally the aircraft was indistinguishable from the earlier type (the cooling fins of the Major's cylinders tapered inwards towards the crankcase, while those of the Gipsy III did not). Among British sporting owners of Moth Majors was the Duchess of Bedford; at the age of 72 the old lady took off for a solo flight in her G-ACUR over the North Sea on 23 March 1937, but was never seen again.

Military trainer

Last of the classic D.H.60s, and a link between them and the famous Tiger Moth *ab initio* trainer, was the D.H.60T Moth Trainer of which 63, including two prototypes, were produced, prompted by an order for 10 aircraft for the Swedish air force in 1931. Light training armament (such as camera gun, light reconnaissance camera and practice bombs, etc) were specified, and subsequent batches were supplied to the Egyptian air force, the Brazilian army and navy air arms, and Iraq.

By numerical designation the next Moth was the D.H.61 Giant Moth, but this had not only first flown in 1927 but was not strictly a Moth in anything but name. It was a large, two-bay, six/eight-passenger cabin biplane, indeed a small airliner. Powered by a direct-drive 336-kW (450-hp) Bristol Jupiter VI radial, the prototype (G-EBTL) was first flown by Hubert Broad in December 1927, before being delivered to Australia as G-AUTL in February of the following year. Five other Giant Moths went to Australia during the next 15 months, as well as two to Canada.

Also in 1927 were produced in great secrecy two small monoplanes, designated D.H.71 and intended for high-speed lightplane research. Given the name Tiger Moth (in no way to be confused with the later D.H.82), G-EBQU and G-EBRV were each powered initially by 63-kW (85-hp) A.D.C. Cirrus II engines, but the former was soon re-engined with Major Halford's new 101-kW (135-hp) Gipsy engine. Both were entered for the 1927 King's Cup Race, but

The prototype D.H.60 Moth G-EBKT, which was first flown by Captain Geoffrey de Havilland on Sunday 22 February 1925 at Stag Lane. It differed from subsequent production aircraft in having an unbalanced rudder and exhaust pipe on the starboard side. Moving the pipe to the port side allowed access to an aft baggage locker.

The light aeroplane that made the greatest impact on sporting aviation throughout the world in the 1920s was the D.H.60 Gipsy Moth, and it was flown in almost every country that supported flying facilities. Shown here with Australian registration, the Gipsy Moth was used by flying clubs at Adelaide, Brisbane, Longreach, Melbourne, Perth and Sydney.

The classic Tiger Moth was licence-built in Canada, New Zealand, Australia, Portugal, Norway and Sweden. The example depicted here was one of 20 D.H.82As produced by AB Svenska Järnvagsverkstaderna, Lindingö, as the Sk 11A for the Flygvapen.

one was scratched and the other withdrew from the race having averaged 267 km/h (166 mph). Shortly afterwards, with wings reduced in span from 6.86 m (22 ft 6 in) to 5.79 m (19 ft 0 in), the first aircraft was flown by Broad to establish a world Lightplane Class III speed record over a 100-km (62.1-mile) closed circuit at 300 km/h (186.47 mph) on 24 August 1927. This machine ended its days in Australia in September 1930 when its engine cut on take-off, sending it crashing into a street in Mascot. The other D.H.71 was eventually stored among the rafters of a hangar at Hatfield, where it was destroyed on 3 October 1940 when a lone Junkers Ju 88 bombed the de Havilland factory.

Much closer to the true Moth tradition was the D.H.75 Hawk Moth, a four-seat cabin high-wing monoplane for which a new engine had been developed by Major Halford by inclining two Cirrus Is to mate on a common crankshaft as a V-8. The new engine, the Ghost, provided power of only 148 kW (198 hp), and after the prototype Hawk Moth had flown on 7 December 1928 the first two D.H.75A production aircraft were powered by 179-kW (240-hp) Armstrong Siddeley Lynx geared radials. Only five other aircraft are believed to have been built, of which one, designated D.H.75B, was fitted with a 224-kW (300-hp) Wright Whirlwind R-975 radial.

The next Moth, the D.H.80 Puss Moth, was also a high-wing cabin monoplane, this time a 'club three-seater' with the pilot seated forward of a two-place bench seat. This attractive little aeroplane remained in production for three years from mid-1930 until a total of 350 had been built, of which roughly half were exported. A spate of early accidents (in one of which the famous sporting pilot Bert Hinkler was killed on 7 January 1933) was diagnosed as the result of wing failure following control flutter, cured by mass-balanced ailerons and strengthened wing struts. Jim and Amy Mollison (the latter was Amy Johnson before her marriage) were the outstanding Puss Moth pilots; the former in G-ABXY *The Heart's Content* made the first solo east-west crossing of the North Atlantic on 18/19 August 1932; he also became the first to fly from England to South America, the first to make a solo east-west crossing of the South Atlantic and the first to make crossings of both the North and South Atlantic.

Birth of a legend

The single D.H.81 Swallow Moth was a two-seat low-wing monoplane which, first flying on 24 August 1931 in the depths of the Depression, attracted no commercial interest. The next in line, however, the D.H.82, was to eclipse even the magnificent D.H.60. The D.H.60T Moth Trainer had been in use by the RAF for some years when Air Ministry Specification 15/31 was issued, calling for an improved version. Eight pre-production aircraft, using D.H.60 fuselages, were produced, but with the top wing centre-section moved forward and the outer sections swept back 48.26 cm (19 in) at the tips, thereby avoiding large changes in the centre of gravity position; this change in effect created the basis of the famous D.H.82 Tiger Moth. The first 35 production aircraft to Specification T.23/31 for the RAF were designated Tiger Moth Mk Is and entered service with the CFS and FTS, followed by two float-equipped aircraft to Specification T.6/33, but large-scale production started at Stag Lane early in 1934 with the issue of T.26/33, only to be switched to de Havilland's new factory at Hatfield later in that year. Known in the RAF as the Tiger Moth Mk II, the new version departed from the Moth's traditional use of fabric and stringers behind the cockpits in favour of plywood decking, as well as provision for a blind-flying hood over the rear cockpit. Further contracts as well as Specification T.7/35 were received in 1935, so that pre-war production amounted to 1,150 at Hatfield, 227 by the de Havilland factory at Toronto, one by de Havil-

Easily distinguishable from the Gipsy Moth by the prominent fuselage stringers in the rear fuselage, the D.H.60M (sometimes called the Metal Moth) employed welded steel tube construction in answer to calls for a more robust structure, particularly by foreign air forces.

Exuding nostalgia for RAF pilots of a former generation, this photo shows a de Havilland-built Tiger Moth being flown solo in 1941. So sensitive was the little trainer that it was possible to induce yaw simply by putting one's hand into the slipstream as if to signal one's intention to turn!

land at Wellington, New Zealand, and three by the DH Technical School. The vast majority had entered RAF service, but many had also gone to foreign owners and air forces as well as British flying clubs. On the outbreak of war most of the civil aircraft in the UK were impressed, the civilian schools providing the aircraft for the service's Elementary Flying Training Schools. In 1941 production of the de Havilland Mosquito at Hatfield forced the company to move Tiger Moth production to Morris Motors at Cowley, Oxford, where a total of 3,216 examples was produced before manufacture ended on 15 August 1945. Many of these Tiger Moths were shipped overseas where they provided the ab initio flying equipment of the schools that constituted the great Empire and Commonwealth Air Training Plan. Indeed, there were precious few wartime RAF pilots whose first experience of flight was not in the cockpit of a 'Tiger'.

Post-war Tiger

After the war, surplus Tiger Moths flooded into the civil market, while others continued to serve in the RAF for a number of years until replaced by the Percival Prentice and de Havilland Chipmunk in the *ab initio* training phase. Among the famous exponents of Tiger Moth flying was the Tiger Club, whose aircraft (*The Bishop, The Archbishop, The Deacon* and *The Canon* single-seaters) became well known for their aerobatics at air shows. Between 1957 and 1959 Jackaroo Aircraft Ltd converted a number of Tiger Moths to four-seat cabin aircraft, this variant being known as the Thruxton Jackaroo.

A derivation of the D.H.82, but one that had little but its designation in common with it, was the D.H.82B Queen Bee radio-controlled gunnery target of which most, equipped with Short float landing gear, served with the Royal Navy or with RAF anti-aircraft co-operation units at home and overseas.

The D.H.83 Fox Moth which followed the original Tiger Moth in 1932 was a four-passenger single-bay biplane produced to meet the growing demands of small feeder line operators whose businesses were springing up all over the world in the early 1930s; a total of 153 was built by de Havilland in the UK, Canada and Australia, many of the Fox Moths actually operating scheduled services, particularly in India. The type was flown on wheel, float and ski landing gear.

Two other famous members of the Moth family were the D.H.85 Leopard Moth and the D.H.87 Hornet Moth, the former a high-wing three-seat cabin aircraft introduced in 1933, and the latter a cabin biplane that first flew in the following year. Both were ideal 'taxi' aircraft, being operated by such well known companies as Olley Air Services and Air Taxis Ltd. A Leopard Moth, flown by Jimmie Broadbent, set a new record for the Australia journey, landing at Lympne on 3 May 1937 only 6 days 8 hours and 25 minutes after leaving Darwin.

Finally, the D.H.94 Moth Minor of 1937 may be seen as having created the link between the classic pre-war Moth and the post-war Chipmunk, being a tandem two-seat low-wing monoplane. More than 100 Moth Minors had been produced by the beginning of the war when production was drastically slowed, and eventually transferred to Australia. At a price of only £575 ex-works, the Moth Minor was obviously the ideal private owner's aircraft which, but for the war, would have perpetuated an unbroken tradition that had originated as Geoffrey de Havilland's dream a decade and a half earlier. Total production of all Moths was 10,794.

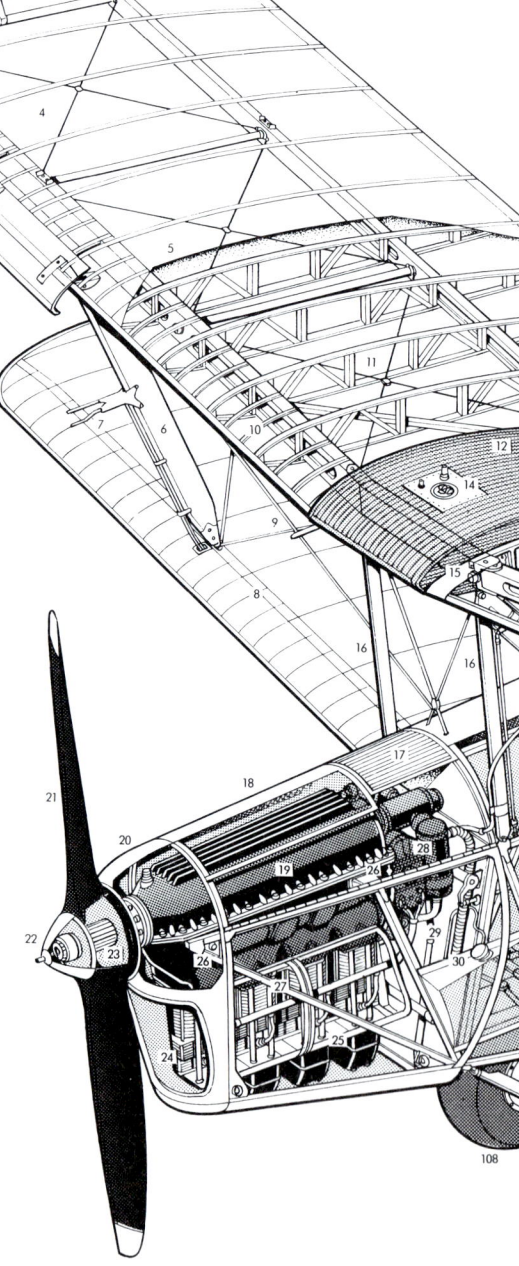

1. Starboard navigation light
2. Automatic leading edge slat, open
3. Slat hinges
4. Starboard upper and wing panel
5. Wing fabric covering
6. Starboard interplane struts
7. Pitot static tubes
8. Starboard lower wing panel
9. Diagonal wire bracing
10. Leading-edge stiffening ribs
11. Lattice rib construction
12. Fuel tank, capacity 22.8 Imp gal (104 litres)
13. Fuel contents gauge
14. Filler cap
15. Main spar attachment joint
16. Centre section 'N' struts
17. Cowling step
18. Detachable engine cowlings
19. de Havilland Gipsy Major four-cylinder inline engine
20. Fixed nose cowling
21. de Havilland two-bladed fixed-pitch wooden propeller
22. Spinner
23. Splined propeller shaft
24. Cooling air intake
25. Internal air duct
26. Engine mounting points
27. Engine bearer struts
28. Accessory equipment
29. Sloping engine bay bulkhead
30. Oil filler cap
31. Forward rudder pedals
32. Oil tank, capacity 2.5 Imp gal (11.4 litres)
33. Control column
34. Instrument venturi
35. Front pilot's (instructor's) seat
36. Tailplane trim control lever
37. Rear rudder pedals
38. Seat safety harness
39. Engine throttle lever
40. Instructor's instrument panel
41. Slat locking lever
42. Rear view mirror
43. Fuel tank sump and gravity feed pipe
44. Centre section root rib

The Thruxton Jackaroo was a post-war four-seat cabin conversion of the Tiger Moth undertaken by Jackaroo Aircraft Ltd, starting in 1957, G-AOIR (shown here), being the fourth produced that year. Most were powered by the 108-kW (145-hp) Gipsy Major IC in place of the more usual 97-kW (130-hp) Gipsy Major I of the Tiger.

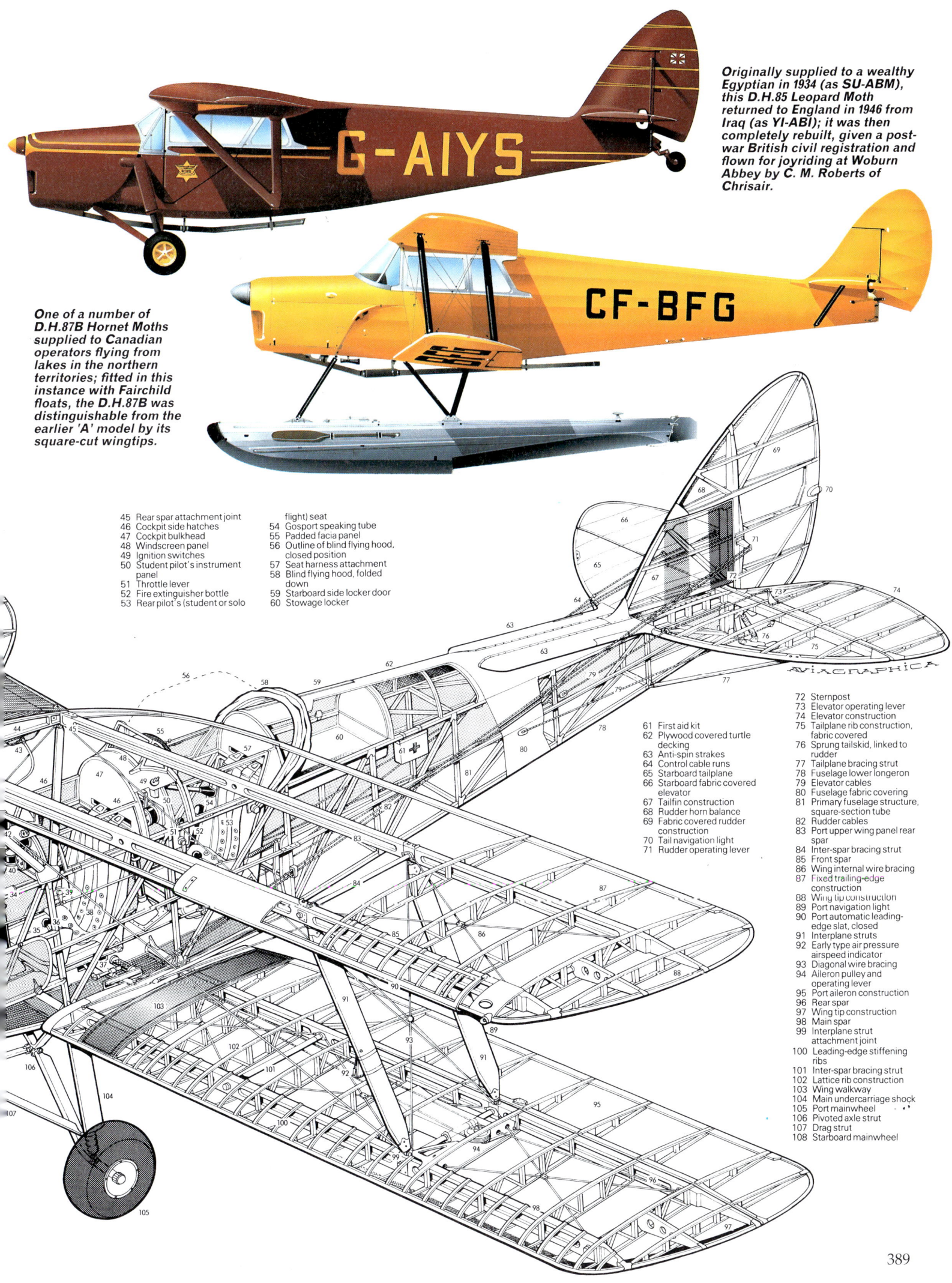

Originally supplied to a wealthy Egyptian in 1934 (as SU-ABM), this D.H.85 Leopard Moth returned to England in 1946 from Iraq (as YI-ABI); it was then completely rebuilt, given a post-war British civil registration and flown for joyriding at Woburn Abbey by C. M. Roberts of Chrisair.

One of a number of D.H.87B Hornet Moths supplied to Canadian operators flying from lakes in the northern territories; fitted in this instance with Fairchild floats, the D.H.87B was distinguishable from the earlier 'A' model by its square-cut wingtips.

45 Rear spar attachment joint
46 Cockpit side hatches
47 Cockpit bulkhead
48 Windscreen panel
49 Ignition switches
50 Student pilot's instrument panel
51 Throttle lever
52 Fire extinguisher bottle
53 Rear pilot's (student or solo flight) seat
54 Gosport speaking tube
55 Padded facia panel
56 Outline of blind flying hood, closed position
57 Seat harness attachment
58 Blind flying hood, folded down
59 Starboard side locker door
60 Stowage locker
61 First aid kit
62 Plywood covered turtle decking
63 Anti-spin strakes
64 Control cable runs
65 Starboard tailplane
66 Starboard fabric covered elevator
67 Tailfin construction
68 Rudder horn balance
69 Fabric covered rudder construction
70 Tail navigation light
71 Rudder operating lever
72 Sternpost
73 Elevator operating lever
74 Elevator construction
75 Tailplane rib construction, fabric covered
76 Sprung tailskid, linked to rudder
77 Tailplane bracing strut
78 Fuselage lower longeron
79 Elevator cables
80 Fuselage fabric covering
81 Primary fuselage structure, square-section tube
82 Rudder cables
83 Port upper wing panel rear spar
84 Inter-spar bracing strut
85 Front spar
86 Wing internal wire bracing
87 Fixed trailing-edge construction
88 Wing tip construction
89 Port navigation light
90 Port automatic leading-edge slat, closed
91 Interplane struts
92 Early type air pressure airspeed indicator
93 Diagonal wire bracing
94 Aileron pulley and operating lever
95 Port aileron construction
96 Rear spar
97 Wing tip construction
98 Main spar
99 Interplane strut attachment joint
100 Leading-edge stiffening ribs
101 Inter-spar bracing strut
102 Lattice rib construction
103 Wing walkway
104 Main undercarriage shock
105 Port mainwheel
106 Pivoted axle strut
107 Drag strut
108 Starboard mainwheel

Chapter 19

Dornier Do X

Dornier Flying Boat

Douglas DC2

Dornier Do X

*Only three were built, but the Do X remains one of the truly great aircraft of the world. Its construction was entirely conventional and its achievements were few and plagued with problems, yet its sheer size made it a breathtaking example of the aircraft constructor's art. Its manufacturer refused to call it a flying-boat, preferring instead the much grander and more appropriate title of **Flugschiff** — a flying-ship!*

As the world moves towards the 21st century, aviation is characterised by ever-increasing complexity in matters of design and technology as the boundaries of flight are pushed back. Yet it is only some 80 years since the world began to take notice of frail little machines which carried the hopes and dreams of the pioneering fathers of powered flight. One such person was Dr Claudius Dornier, a brilliant German designer whose forte was flying-boats. He produced civil and, with the onset of World War I, military aircraft, each a little more daring and capitalising on the successes of their predecessors.

But though successful, these machines were relatively small and limited in passenger accommodation. For Dr Dornier the dream was a multi-engined passenger-carrying flying-boat which could operate on commercial transatlantic routes. And soon the dream began to turn into reality. In the early 1920s a special design bureau was established at Manzell to make a full feasibility study, and by late 1926 a full-scale wooden mock-up had been ordered.

This grand design was christened the Dornier Do X – a reference to the unknown factor – and was constructed in the old Zeppelin shed at Friedrichshafen. Certainly this giant was unlike anything seen before, yet it relied heavily on tried and trusted technology. Successful testing of the mock-up led to an airworthy example being commissioned with construction commencing at the Altenrhein factory on the shores of Lake Constance in December 1927. Some 18 months later the largest aircraft of its time in the world was rolled out and readied for flight testing. Dr Dornier was to acknowledge the sheer size of the design by referring to it not as a flying-boat but as a flying-ship.

The design was based on a two-step hull of concave cross-section and a transverse step to facilitate take-off from water. A fabricated

Riding on Lake Constance, the first Do X shows the enormous plank-like wing which carried it aloft. This also provided a good measure of ground effect lift over the water, which augmented the power of the 12 engines ranged across the wing centre section.

During the Do X's Europaflug *tour of European countries, it visited Calshot. During its stay it played host to the Prince of Wales, seen here being welcomed by Dr Claudius Dornier.*

keel structure formed the 'skeleton' around which the flat-plate fuselage was built. Internally the fuselage was divided into three decks (the first aircraft to feature such a layout): a lower, operating deck which housed the fuel system and freight/baggage; the middle deck given over to luxurious accommodation for some 66-70 passengers; and the upper, crew deck which incorporated, from fore to aft, the flight crew station, commander's bridge, engine room and wireless/auxiliary machine room.

Though impressive in all respects, the fuselage was totally eclipsed by the massive main wing structure, atop which sat six rows of paired Bristol Jupiter 525-hp air-cooled engines. Untapered in both plan and depth except at the very tips, the wing relied for lift on its sheer area and camber, there being no flaps, slots or other lift devices fitted. The entire slab-like construction was supported by a trio of parallel struts from each of the large side sponsons.

And so it was that early in the morning of 12th July 1929 the Do X entered the waters of Lake Constance under the command of Dir Schulte-Frohlinde, with chief test pilot Richard Wagner at the controls. The initial taxiing trials went well, so well that it was decided to lift the 35-ton flying-boat off the lake surface in the first of three short test flights that morning. As the flight testing continued a top speed of 214 km/h (133 mph) was achieved, while take-off at an all-up weight of 51.5 tons was timed at approximately 130 seconds.

But such successes masked a fundamental problem that was becoming all too clear to Dr Dornier and his team. The massed ranks of engines were simply not producing enough power, with an average output of 231.16 kW (310 hp) compared with the calculated minimum cruising output requirement of 298.28 kW (400 hp). There was no choice but to seek alternative engines for the licence-built Jupiters, and soon work was underway to fit liquid-cooled Curtiss GV-1570 Conquerors, each of which boasted an output of 458.6 kW (615 hp) at 2,500 rpm. They were fitted to the existing mountings to save time and money, but the original vertical nacelles below the

One of the Italian Do Xs shows off the huge thickness of the wing, and bracing to the fuselage sponsons. The Italian aircraft were powered by Fiat A-22R engines in place of the Conquerors of the first machine.

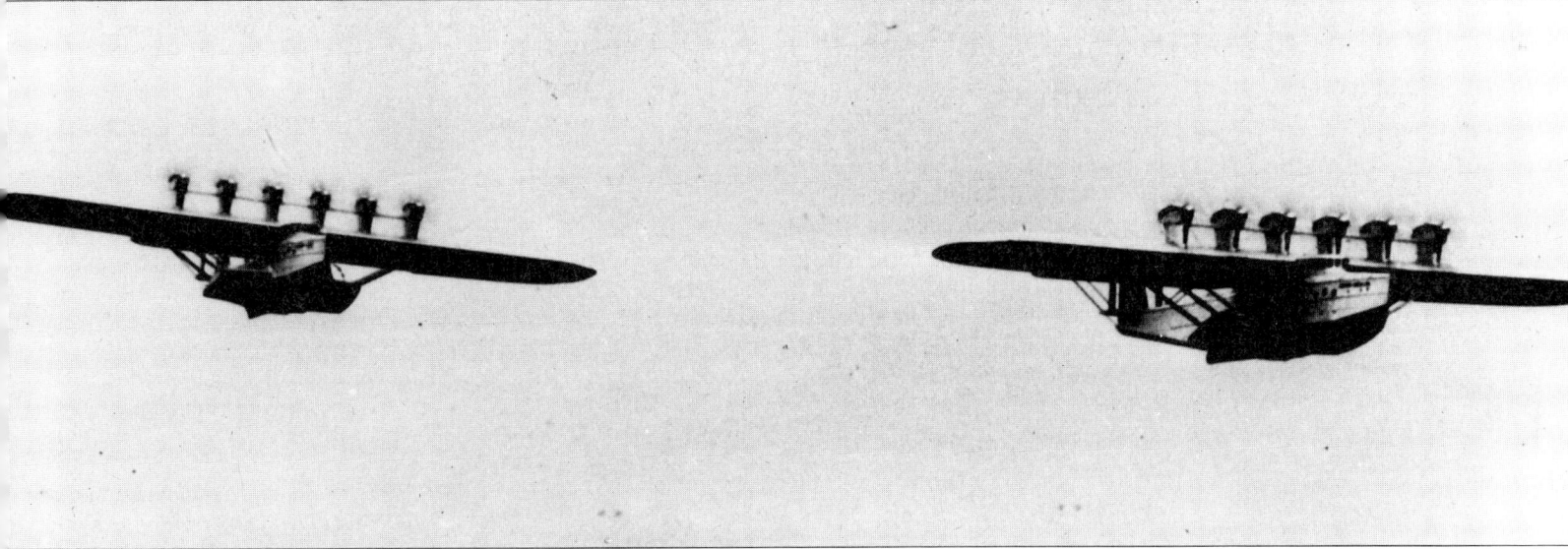

*Photographed in 1932, this rare shot shows both Italian machines together. Intended for passenger transport with **SANA**, the pair were found unsuitable and handed over to the military.*

engine bodies gave way to open-strut assemblies. In this re-engined form the Do X took to the air for the first time on 4th August 1930, but the improvements in performance were limited, primarily because the new engines weighed some 1088 kg (2,393 lb) more than the Jupiters.

Trials in Germany with the re-engined aircraft (sometimes referred to as the Do X1A) were completed in October 1930, but in order to prove its performance capabilities further the decision was taken to mount a transatlantic flight which would include experience in a variety of climatic and operational conditions. Unfortunately the German media built up what was essentially further flight testing into a record-breaking attempt, and soon the magnificent machine was in the limelight for all the wrong reasons.

The outward leg began on the morning of 5 November 1930, the thirteen-man crew guiding the Do X to the west of the Rhine. In the next few days a number of photo-sessions was flown and courtesy visits made to Rotterdam in the Netherlands and Calshot in Great Britain. The route then turned south for Portugal via points in France and Spain, but at Lisbon a major setback occurred when the entire port wing fabric covering was destroyed in a fire. Though the damage was confined to one wing, it took some six weeks for new material to arrive and a further two weeks for repairs to be completed. For the German media this delay and the generally leisurely pace of the flight hardly equated with what it saw as a record-breaking attempt, and the image of Dornier's magnificent design was badly tarnished.

Worse was to follow when the time came to depart Portugal. The two-month delay for wing repairs meant that it was early February before the flight could resume, and by then the wintery weather over the North Atlantic was too severe. The decision was taken to follow a longer, slower route over the South Atlantic, but on 2 February 1931 the take-off from Gando Bay was abruptly halted by heavy seas. Such was the force that the starboard wing and three fuselage main frames were severely overstressed and the aircraft was once again hauled onto land for repairs that would take three months to complete.

Bearing the Lufthansa title on the nose (although the airline showed little interest in the type), the Do X is seen after its return from North America. In 1934 the aircraft went to a Berlin museum.

The sumptuous interior was intended for 65-70 passengers carried in great comfort. In an early flight, however, the Do X had lifted off from Lake Constance with 10 crew, 150 passengers and nine stowaways!

When the Do X did at last take to the air again another change of plan had been made. The route now headed for Portuguese Guinea to enable hot-weather trials to be conducted, and a relatively uneventful flight seemed to signal a change in fortunes. But reality soon dashed all such hopes. Certainly the weather was hot, but it was also extremely humid and when the time came to depart for the actual transatlantic crossing, the heavy, fuel-laden aircraft just could not get into the air. One attempt after another failed and it soon became apparent that the aircraft was all but marooned. In desperation the luxurious interior fittings were discarded, along with all non-essential equipment and even two crew members who had to travel overland, and eventually a successful take-off was made for a short hop to the Cape Verde Islands.

Dornier Do X cutaway drawing key

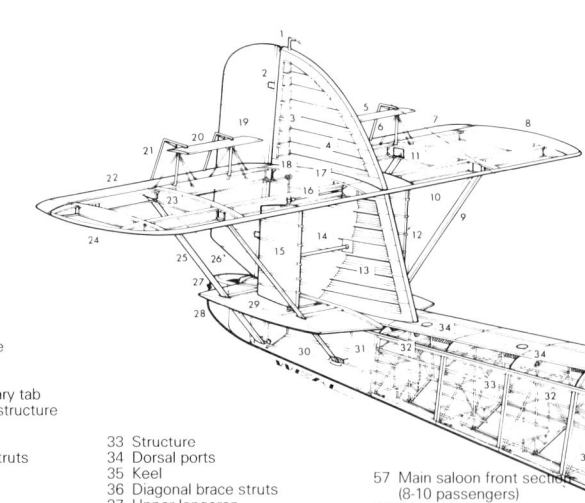

The flight deck of the Do X had positions for two pilots, navigator, ship's captain and engineers. The main navigation equipment was carried here, while engineers had inter-spar access to the engines.

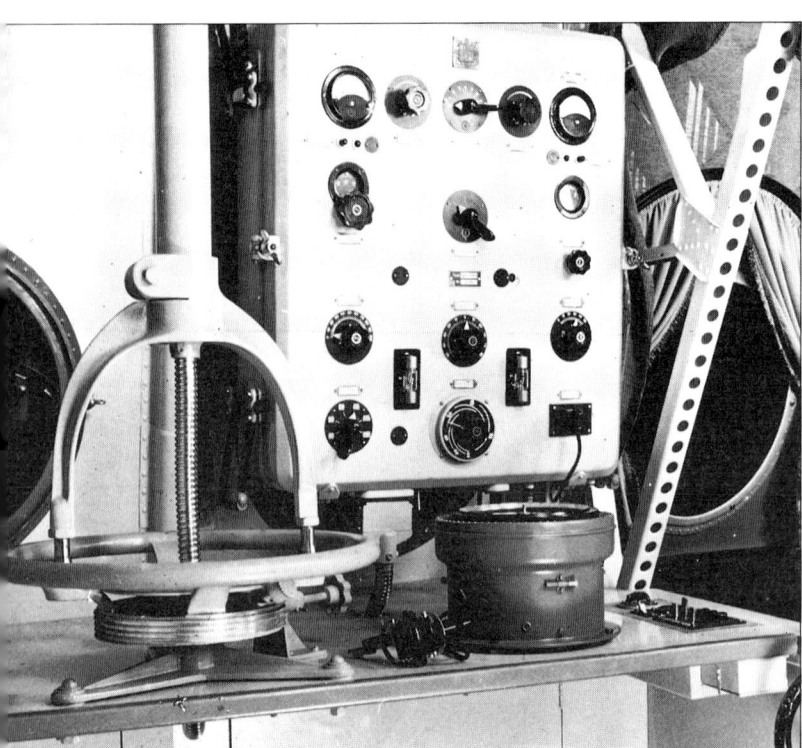

1. Aerial attachment
2. Rudder upper hinge
3. Rudder post
4. Tailfin structure
5. Port elevator auxiliary tab
6. Tab support/hinge structure
7. Port elevator
8. Port tailplane
9. Tailplane support struts
10. Strut brace wires
11. Strut/tailplane spar attachment
12. Port auxiliary rudder surfaces
13. Corrugated tailfin skin
14. Auxiliary rudder connection
15. Starboard auxiliary rudder surface
16. Rudder actuating link
17. Tailplane spar attachment
18. Elevator control linkage
19. Rudder
20. Starboard elevator auxiliary tab
21. Tab support/hinge structure
22. Starboard elevator
23. Strut/tailplane spar attachment
24. Starboard tailplane
25. Tailplane support struts
26. Strut brace wires
27. Rear navigation light
28. Stern structure
29. Stub tailplane
30. Tailplane strut/fuselage attachment
31. Metal skinning
32. Fuselage aft frames
33. Structure
34. Dorsal ports
35. Keel
36. Diagonal brace struts
37. Upper longeron
38. External strakes
39. Accommodation aft bulkhead
40. Aft fuselage access
41. Aft baggage/cargo hold
42. Portholes
43. Dorsal loading hatch
44. Galley (port)
45. Toilets and washrooms (starboard)
46. Passenger accommodation aft bulkhead
47. Aftermost passenger cabin (8 seats)
48. Starboard wing aerial mast
49. Bulkhead door
50. Passenger cabin (8 seats)
51. Main saloon aft bulkhead
52. Main saloon rear section (8-10 passengers)
53. Bulkhead
54. Main saloon centre section (8-10 passengers)
55. Sponson/fuselage support struts
56. Bulkhead
57. Main saloon front section (8-10 passengers)
58. Cloakrooms
59. Main entry doors (port and starboard)
60. Vestibule
61. Ladder to upper crew deck
62. Floor hatch
63. Doorway to flight engineer's station
64. Main engine control banks
65. Wing crawlway access
66. Star outer engine support struts
67. Engine access ladder
68. Soundproofed wireless room
69. Underslung engine radiator
70. Entry hatch (through rear main spar)
71. Auxiliaries (generator and plant) compartment
72. Wing/fuselage decking
73. Four-blade pusher propellers

*I-ABBN was the Do X3 built for Italy and named **Alessandro Guidoni**. Note the large auxiliary tab section mounted above the ailerons and elevators which reduced stick forces.*

Seven months to the day since its departure from Lake Constance, the Do X completed the 2,324-km (1,254-mile) crossing from the Cape Verde Islands to Natal on the Brazilian coast. The Americas had been reached, but the goal was now to reach the United States. In Brazil this flying shell had its interior refitted before departing for New York. A broken crankshaft and thirteen-day wait for the engine to be changed seemed a minor problem compared with past events! Finally, on 27 August 1931 a huge crowd in New York caught its first glimpse of the impressive German flying-boat.

A complete overhaul followed, but the nine-month stay in the USA was really too long to sustain public interest. Ironically the return to Europe took just six days from Newfoundland, but this successful leg received limited publicity. The Do X would be remembered for the monumental problems of the outward leg, a fact reflected in an almost total lack of airline interest. Though the Deutsche Lufthansa title was worn with pride on the bows, no orders were ever placed by the German flag-carrier. But that was not quite the end of the story.

Two further examples were built and flown to Italy for a proposed airline link between Genoa, Naples, Rome and Tripoli. Powered by Fiat A-22R liquid-cooled engines, the aircraft were designated Do

74 Curtiss Conqueror tandem paired engines
75 Exhaust manifolds
76 Aileron tab support/hinge structure
77 Port aileron auxiliary tab
78 Four-blade tractor propellers
79 Port wingtip
80 Port navigation light
81 Front main spar structure
82 Nose ribs
83 Engine support strut attachments
84 Engine access/inspection upper surface walkway
85 Port wing leading-edge fuel tank (66 Imp gal/300 l capacity)
86 Wing leading-edge/flight deck fairing
87 Commander's station
88 Flight deck side windows
89 Navigator's chart table
90 D/F loop control
91 Cockpit entry door
92 Port sponson/wing strut
93 Flight deck roof strakes
94 Antenna
95 First pilot's seat
96 Water rudder control handwheel
97 Second pilot's seat
98 Control wheels
99 Cockpit hinged front window panels
100 Central console
101 Flight deck forward bulkhead
102 Rudder pedal assembly
103 Flight deck floor support frames
104 Forward passenger cabin (8 seats; optional sleeper berths)
105 Smoking room
106 Bulkhead door
107 Bar
108 Forward upper frames
109 Nose hatch
110 D/F loop
111 Forwardmost compartment (see equipment)
112 Forward collision bulkhead
113 Access hatch
114 Bow frames
115 Bow structure
116 Mooring lug
117 Strengthened bow post
118 Chine
119 Towing eye
120 Anchor
121 Hull lower frames
122 Solid support frame
123 Keel
124 Lower hull portholes
125 Main passenger deck floor level
126 Mooring bollard
127 Auxiliary fuel tanks (2)
128 Pumping room
129 Starboard sponson leading edge
130 Sponson front spar
131 Passenger entry non-slip walkway
132 Main fuel tanks (2)
133 Sponson/wing support struts
134 Sponson main spar member
135 Squared-off sponson end rib
136 Starboard wing leading-edge
137 Hull step
138 Sponson attachment
139 Planing hull structure
140 "Vee" bottom frames
141 Water rudder
142 Starboard aileron auxiliary tab
143 Aileron tab support/hinge structure
144 Aileron hinge fairing
145 Starboard aileron
146 Aileron tube
147 Aileron control linkage
148 Rear main spar
149 Wing structure
150 Turnbuckle
151 Centre main spar
152 Front main spar
153 End rib structure
154 Starboard identification light
155 Interspar rib structure
156 Trailing-edge rib structure
157 Corrugated wing skinning
158 Starboard wingtip
159 Spar end section structure
160 Starboard navigation light

In its original form, the Do X was powered by the Siemens-built Bristol Jupiter radial. These proved to be inadequate for the task, producing less-than-stated power and having overheat problems.

X2 and Do X3 respectively, but performance fell well short of expectations and both were destroyed. Interestingly the remains of one have been located at the bottom of an Italian lake, and maybe one day they will be brought to the surface.

For the famous D-1929 the end was tragic indeed. After her epic flight she was passed to the Deutsche Versuchsanstalt für Luftfahrt for testing. In 1934 she was dismantled and taken to the Aviation Museum in Berlin where she proudly took her place amongst the other exhibits. But in the closing weeks of World War II an Allied bombing raid destroyed the Museum and its exhibits. The Do X was completely destroyed, and so ended one of the most interesting chapters in German aviation history.

After a long stop in Lisbon for repairs following a fire, another in the Canaries and frustrating excursions to Africa and South America, the Do X eventually arrived in New York on 27 August 1931.

Dornier Flying-Boats

In April 1983 Dornier flew the Do 24TT, a turboprop derivative of the pre-war Do 24. This had been the culmination of a long and impressive line of all-metal flying-boats, putting Dornier at the top of the class in this discipline. Here we look at the designs which led to the Do 24.

Professor Dr. Claude Dornier was probably the world's greatest designer of flying-boats in the pre-1930 era, and also one of the greatest exponents of the modern metal aeroplane. His very first design, the RS I of 1915, was not only constructed almost entirely from steel and aluminium alloys but was also the biggest aeroplane then built, with a span of 43.5 m (142 ft 8.6 in). It was the first of a series of RS (*Riesenflugzeug See*, or giant seaplane) aircraft that had to finish with the 1918 Armistice.

As well as the RS series, Dornier's company, Zeppelin-Werke Lindau GmbH, had almost completed the prototype of a smaller but still very capable flying-boat, the Gs I. Powered by two 201-kW (270-hp) Maybach Mb.IV water-cooled engines arranged in tandem at the centre of a rectangular monoplane wing, the Gs I had a slim and very efficient hull, stabilising sponsons of the kind favoured by Dornier, and a twin-finned biplane tail. The four-seat enclosed cockpit in the bows gave an appearance like the head of a serpent, but as a tough transport flying-boat it had no equal when it flew on 31 July 1919. It was later flown from Lindau to the Netherlands, but the fact remained that the machine contravened the rules imposed by the Allies and the Allied Control Commission had it sunk off Kiel in 1920.

Dornier went ahead with small aircraft, such as the Delfin, Libelle and Komet, but he never gave up his work on big and powerful flying-boats. His wartime design for the Gs II, rather larger than the Gs I with 96 m² (1,033.4 sq ft) of wing instead of 80 m² (861.1 sq ft), and a laden weight raised from 4315 kg (9,513 lb) to 5700 kg (12,566 lb) appeared to be just what the infant airlines and air forces needed. Under the noses of the hated Control Commission, Dornier set up an Italian subsidiary, Società di Costruzioni Meccaniche di Pisa, at Marina di Pisa. He carried on designing at Lindau, the works there being renamed Dornier Metallbauen GmbH in 1922, and licensed the big Gs II machines to the Italian subsidiary, even buying two 224-kW (300-hp) Hispano-Suiza engines from Paris for the prototype of the improved GS II. Dornier redesignated it as the Type J, and named it Wal (whale). The prototype flew on 6 November 1922.

Dornier was correct in his assessment of the market. The Wal was ideally suited to both commercial and military operations for the next 15 years, having very satisfactory performance and payload, a robust all-metal structure and excellent reliability. Airline Wals set new standards in interior appointments, and with glass portholes resembled staterooms in luxury yachts. The many military versions were fitted with four small stub wings along the upper part of the hull from which bombs and other stores could be hung, and there were

This Wal was one of the later examples of the 8000-kg (17,637-lb) gross weight civil family, in this case with BMW VI engines. These later variants had the same 96 m² (1,033 sq ft) wing area, but rounded tips and span increased from 22.5 m (74 ft) to 23.2 m (76 ft). D-2069 Monsun (monsoon) served with Deutsche Lufthansa. Note the giant D/F loop aerial.

This regular Military Wal was one of a substantial number operated by the Spanish navy. Many earlier Spanish Wals had been built by CMASA, but the example shown was a Dornier-built Do J IId, with BMW VI engines. It is shown during the Civil War whilst serving with 1-G 70 Grupo at Puerto de Pollensa, in Majorca.

The Luftwaffe Do 18D reached the peak of its brief career in 1939-40, when the type equipped four Staffeln (squadrons) of the Küstenfliegergruppen (coastal groups). This example served at Kamp, on the Pomeranian coast, with 2/KüFlGr 906. By 1941 the Do 18Ds were being converted as Do 18H crew trainers, the more powerful Do 18Gs then becoming Do 18Ns.

The 10000-kg (22,046-lb) gross weight Wal was the last and largest version, dating from 1933. Span was increased to 27.2 m (89 ft 2.9 in) and wing area to 112 m² (1,206 sq ft). This example was one of those that made 328 scheduled crossings of the South Atlantic, operating from the ships Westfalen and Schwabenland. Note the length of the hydraulic catapult.

various schemes for defensive guns.

Of course, with the benefit of hindsight, the Wal can be criticised. Tandem engines are inefficient, can lead to cooling problems and can cause severe stresses in the rear propeller. The Wal's very broad wingtips were aerodynamically poor, and the high induced drag in-

The original Do X 12-engined monster is seen here after its refit with Curtiss Conqueror water-cooled engines. It is shown after its long trip to the Americas in brief service with Lufthansa. It then went to the DVL research organisation before becoming the biggest exhibit in the Berlin air museum, where it was destroyed by bombs.

evitably reduced the attainable range. The large 'park bench' balancing surfaces attached to the ailerons were a scheme outmoded by 1930, and the open cockpit of most Wals would have been disliked by the softer pilots of later times. But in overall terms no other marine aircraft came near it. Altogether at least 300 and possibly more than 320 Wals were built, including between 157 and 177 in Italy (156 by SCMP and its successor CMASA of 1929, a Fiat subsidiary), at least 56 by Dornier at Friedrichshafen from 1932, 40 by CASA in Spain, about 40 by Aviolanda in the Netherlands and three by Kawasaki in Japan. It is doubtful if any aircraft has left the original factory in so many different variations. There were four major and nine minor variations in span, six quite different designs of nose and cockpit, eight different designs of tail, at least 17 different basic types of engine, and a host of gross weights from 4000 kg (8,818 lb) to more than 10000 kg (22,046 lb).

Though some early customers continued to specify the Hispano, many of the pre-1925 Wals were powered by the 268-kW (360-hp) Rolls-Royce Eagle IX, and this rather outdated V-12 was fitted to a Wal that set 20 world records for speed/height/range with payloads up to 2000 kg (4,409 lb) on 4, 9, 10 and 11 February 1925. Other Wals, with three different types of engine, made crossings of the North and South Atlantic, flew from Europe to near the North Pole and to various places in the Far East, and also flew round the world. Airline Wals often had an enclosed cockpit (always from 1931) and began as six- to nine-seaters, with space for loose cargo. In 1931 Dornier began production of the '8-tonne Wal' first flown the pre-

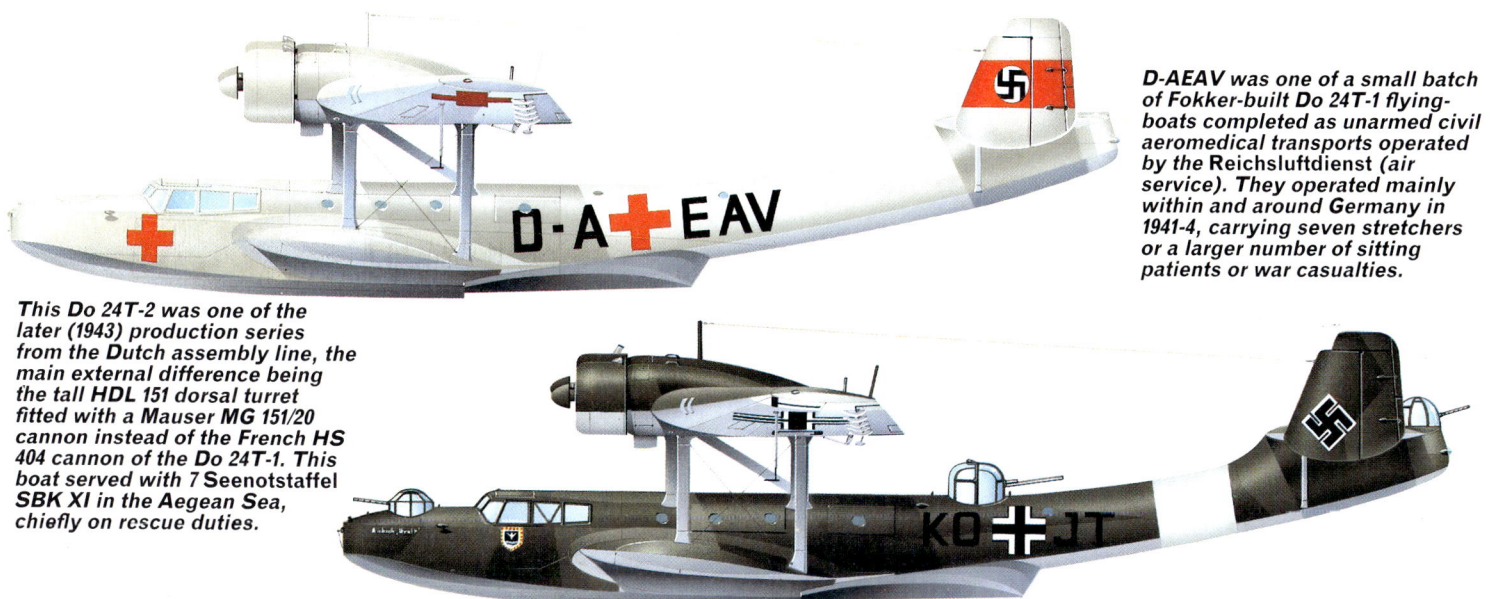

This Do 24T-2 was one of the later (1943) production series from the Dutch assembly line, the main external difference being the tall HDL 151 dorsal turret fitted with a Mauser MG 151/20 cannon instead of the French HS 404 cannon of the Do 24T-1. This boat served with 7 Seenotstaffel SBK XI in the Aegean Sea, chiefly on rescue duties.

D-AEAV was one of a small batch of Fokker-built Do 24T-1 flying-boats completed as unarmed civil aeromedical transports operated by the Reichsluftdienst (air service). They operated mainly within and around Germany in 1941-4, carrying seven stretchers or a larger number of sitting patients or war casualties.

vious year with 515-kW (690-hp) BMW VI engines and a 96-m² (1,033.4-sq ft) wing. This became the standard model, though again with many variations, and it led to the 8.5-tonner in 1933, and the corresponding Militär-Wal 33 which was produced for the Luftwaffe as the Do 15. In the same year appeared the biggest version, the 10-tonne Wal, with 112 m² (1,205.6 sq ft) of wing, which was operated by Deutsche Lufthansa from the depot ships *Westfalen* and *Schwabenland* and made 328 crossings of the South Atlantic in regular mail service.

Different enough to have a different name, Superwal, the Type R enlarged version first flew on 30 September 1926. This was constructed at another Dornier factory, at Manzell on the Swiss shore of the Bodensee (Lake Constance). The prototype of this much bigger boat had two Rolls-Royce Condors of 485 kW (650 hp) each, and another had two 597-kW (800-hp) Packard A-2500s, but most other Superwals had two push/pull pairs, the most common engines being the Siemens-built Bristol Jupiter, Pratt & Whitney Hornet, and, for the CASA-built military Superwals, the 373-kW (500-hp) Hispano-Suiza.

Just as the RS I had been the biggest aeroplane of its day, so did Dornier's liking for size manifest itself in an extreme form in 1926 when work began at Manzell on the Model X. This monster flying-boat was the ultimate scaled-up Wal, with no fewer than six push/pull pairs of engines, still the largest number ever used to propel a single aircraft. The Do X first flew in the hands of Richard Wagner on 12 (often reported as 25) July 1929, powered by Siemens Jupiters. Planned as a long-range passenger machine with 66 to 72 seats, it took off on 31 October 1929 with 10 crew, 150 passengers and nine stowaways, a total not exceeded for 20 years. On 4 August 1930 the Do X emerged after a rebuild with water-cooled Curtiss Conqueror engines with plain struts instead of auxiliary wings between the nacelles, the US engines giving extra power (477 instead of 391 kW/640 instead of 525 hp) and avoiding customs problems on a visit to the USA in 1930-1. The Italian air force used two of these monsters with 433-kW (580-hp) Fiat A 22R engines.

On 26 September 1939 a Blackburn Skua flown by Lieutenant B. S. McEwan scrambled from HMS *Ark Royal* and shot down a shadowing Do 18D, the first aircraft shot down by British forces in World War II. The Do 18 was the natural successor to the Wal, designed in early 1934 to meet both a requirement for an updated Militär-Wal and a more modern South Atlantic boat for Lufthansa. The prototype Do 18a flew on 15 March 1935, powered by tandem 403-kW (540-hp) Junkers Jumo 5 diesel engines giving great range and endurance. Similar to the 8-tonne Wal in size and weight, the Do 18 was naturally much improved aerodynamically, with a tapered wing with rounded tips and 'double wing' flaps and ailerons, a central

D-ABYM Aeolus was the Do 18 V3 (third prototype). Powered by tandem 447-kW (600-hp) Jumo 205C diesel engines, all three were later designated as Do 18E transports and used from 1936 on Lufthansa's routes on the North, and later South, Atlantic. Just before joining the airline, on 10-11 July 1936, Aeolus made a 30-hour 21-minute test flight over the Baltic.

wing pylon incorporating the cooling radiators, and a well-streamlined hull of graceful form. Small numbers of several versions were produced for civil applications, and in summer 1936 deliveries began from the Weser company of the military Do 18D series, most of which had 447-kW (600-hp) Jumo 205C diesels and were used for armed reconnaissance with coastal units. Delay in development of the Blohm und Voss BV 138 led to continued production of the Do 18 into World War II, the total of all versions being in the region of 160 including 75 Do 18Ds.

The last aircraft that can be said to have descended from the Wal was the Do 24, a considerably larger boat that at last abandoned tandem engines, though it adhered to the use of sponsons for stability on the water. It had its origins in the requirement of the MLD, the Dutch naval air force, for a successor to its mainstay, the trusty Wal. In 1934 the MLD refined its ideas and opened talks with Dornier, soon deciding that the Do 18 was not quite big enough. The MLD wanted not only long range and more than two engines but also the ability to operate safely at great distances from land and, if necessary, alight on quite rough seas. During 1935 Dornier went ahead with the design of the three-engine Do 24 purely for the MLD, charging the Dutch government a fee for the design process as well as selling a licence to Aviolanda, which had finished building Wals in 1931. Four prototypes were put in hand, the Do 24 V3 being the first to fly and being powered, like the Do 24 V4, with the engine specified by the customer for commonality with its Martin 139WH-1 bombers, the Wright Cyclone F52.

Successful trials

A wholly modern stressed-skin boat, the Do 24 hung its shapely hull under the rectangular centre section on two pairs of inverted-V struts, with a fifth diagonal strut to react to the pull of the engines (by contrast with the big streamlined pylon of the Do 18). The tail had twin fins on the ends of the wide-span tailplane, and there was provision for gun turrets in the bow, amidships and at the extreme tail. The MLD also specified wing bomb racks, for up to 12 stores of 100 kg (220 lb) each, and a heavy load of mission equipment. Flight trials, including operations from very heavy seas, showed that the Do 24 possessed exceptional strength and good performance. In autumn 1937 the MLD signed for 60, to be licence-built in the Netherlands, but in the event Dornier built 12 (including the Do 24 V3 and V4 prototypes), with the designation Do 24K-1, using the 652-kW (875-hp) F52 engine, while Aviolanda, De Schelde and other Dutch companies undertook to build the other 48, designated Do 24K-2, powered by the 746-kW (1,000-hp) Cyclone G102.

All the Do 24K-1s had been delivered by the outbreak of war, but when Germany invaded the Netherlands on 10 May 1940 only 25 of the Do 24K-2s had been despatched. Three were captured intact, and when these were tested at Travemünde the Luftwaffe discovered they were superb boats, and Dutch production was quickly restarted. Under the control of the Weser company the Do 24K-2s were modified without bomb gear but with MG15 nose and tail guns and captured Hispano 20-mm cannon in the amidships turret, as well as large hatches and comprehensive internal furnishings for the air/sea rescue mission. Designated Do 24N-1, the production boats entered Luftwaffe service from August 1941, 11 being delivered by November when the stock of engines was exhausted. Production then continued with the Do 24T-1 with the 746-kW (1,000-hp) BMW Bramo Fafnir 323, a total of 170 in all being supplied. Such was the demand for this fine boat that a second source was organised at the former Potez-CAMS factory at Sartrouville, France, where 48 additional Do 24s were built.

From early 1943 most were of the Do 24T-2 sub-type with extra radio and sometimes radar, as well as an MG 151 replacing the Hispano cannon amidships. Very similar boats designated Do 24T-3 were prepared for Spain, 12 being flown to Majorca in 1944.

Dornier Do 24T cutaway drawing key

1. Bow navigation light
2. Towing/mooring ring
3. Fore hull structure
4. Retractable mast
5. Ammunition magazine racks
6. Spent cartridge chute
7. Bow compartment (see equipment stowage)
8. Turret mechanism
9. Bow 7.92-mm MG 15 machine-gun
10. Bow turret
11. Removable turret dome
12. Nose decking
13. Nose mooring lug
14. Mooring rope stowage
15. Bulkhead
16. Crawlway
17. Rudder pedal assembly
18. Bulkhead door frame
19. Instrument panel shroud
20. Co-pilot's control column
21. Compass
22. Windscreen panels
23. Cockpit roof glazing
24. Sliding entry panels
25. Hatch runners
26. Flight deck windows
27. Navigator's station
28. Co-pilot seat
29. Pilot's seat
30. Seat adjustment lever
31. Side-mounted control column
32. Floors support frame
33. Radio-operator's position
34. 'Vee' bottom hull structure
35. Fuselage/sponson fairing
36. Access panels
37. Sponson nose ribs
38. Sponson main fuel cells (four, approx 77 Imp gal/350 litre capacity each)
39. Sponson abbreviated fuel cell (36.6 Imp gal/180 litre capacity)
40. Port sponson mooring lug
41. Rib reinforcement
42. Sponson main spar/fuselage frame
43. Sponson/forward wing strut lower atachment
44. Fuel collector tank (46 Imp gal/210 litres)
45. Access panels
46. Fuselage hull step
47. Sponson rear spar/fuselage frame
48. Sponson/aft wing strut lower attachment
49. Mooring lug
50. Non-slip sponson walkway
51. Forward midships bay (two-stretcher accommodation; 28 oxygen bottles stowed along centre-line walkway)
52. Forward bay hinged entry hatch
53. Sponson/forward wing strut
54. Fuselage main frame bulkhead
55. Medical attendant: station
56. Fuselage/wing strut attachment fairing
57. Port engine nacelle intake
58. Fuselage porthole
59. Starboard crew entry door
60. Bulkhead
61. Radio equipment installation
62. Roof external strakes
63. Three-blade VDM metal propellers
64. Spinners
65. Starboard sponson
66. Roof (starboard) entry hatch
67. Centre engine nacelle intake

Continuing the story of the Do 24 into the 1980s, this ex-Do 24T-2 has acquired a new wing and triple turboprop powerplants whilst retaining the original hull and tail units. Currently undergoing flight testing and research into composite materials and corrosion problems, a commercial future for the Do 24TT remains to be realised.

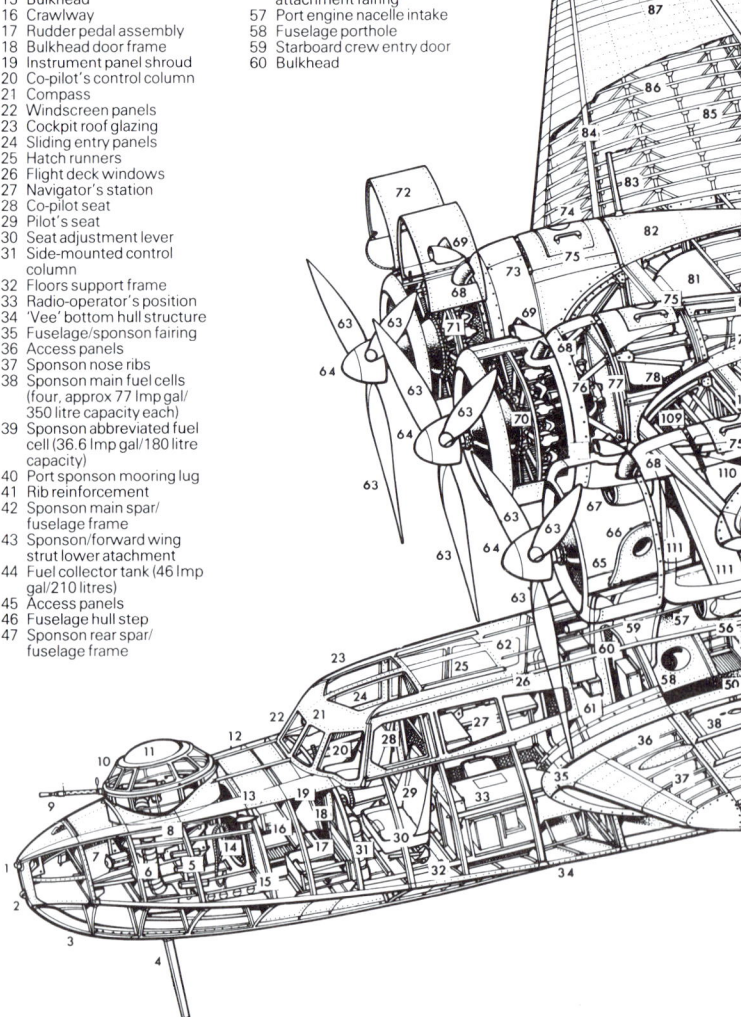

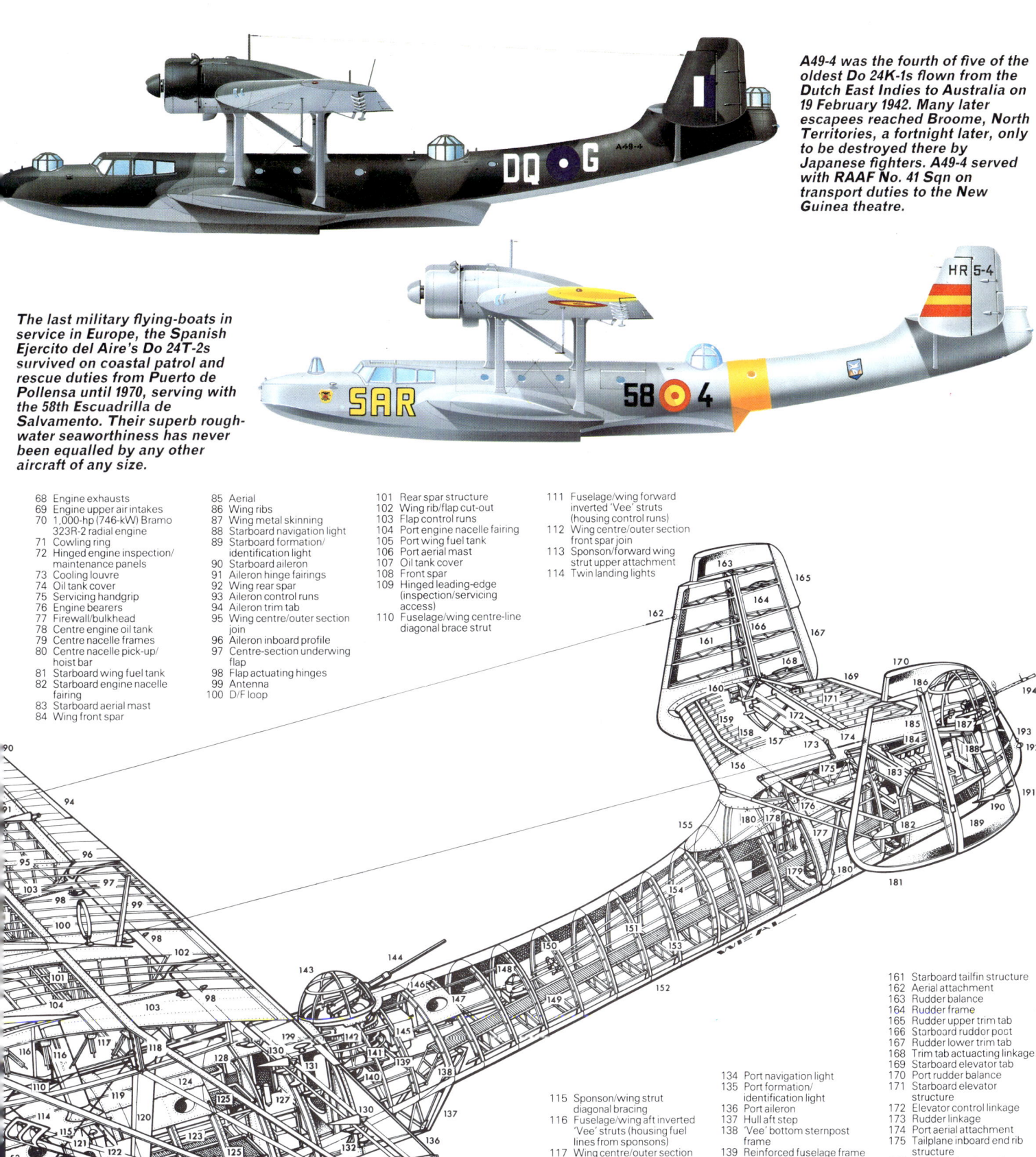

A49-4 was the fourth of five of the oldest Do 24K-1s flown from the Dutch East Indies to Australia on 19 February 1942. Many later escapees reached Broome, North Territories, a fortnight later, only to be destroyed there by Japanese fighters. A49-4 served with RAAF No. 41 Sqn on transport duties to the New Guinea theatre.

The last military flying-boats in service in Europe, the Spanish Ejercito del Aire's Do 24T-2s survived on coastal patrol and rescue duties from Puerto de Pollensa until 1970, serving with the 58th Escuadrilla de Salvamento. Their superb rough-water seaworthiness has never been equalled by any other aircraft of any size.

68 Engine exhausts
69 Engine upper air intakes
70 1,000-hp (746-kW) Bramo 323R-2 radial engine
71 Cowling ring
72 Hinged engine inspection/maintenance panels
73 Cooling louvre
74 Oil tank cover
75 Servicing handgrip
76 Engine bearers
77 Firewall/bulkhead
78 Centre engine oil tank
79 Centre nacelle frames
80 Centre nacelle pick-up/hoist bar
81 Starboard wing fuel tank
82 Starboard engine nacelle fairing
83 Starboard aerial mast
84 Wing front spar
85 Aerial
86 Wing ribs
87 Wing metal skinning
88 Starboard navigation light
89 Starboard formation/identification light
90 Starboard aileron
91 Aileron hinge fairings
92 Wing rear spar
93 Aileron control runs
94 Aileron trim tab
95 Wing centre/outer section join
96 Aileron inboard profile
97 Centre-section underwing flap
98 Flap actuating hinges
99 Antenna
100 D/F loop
101 Rear spar structure
102 Wing rib/flap cut-out
103 Flap control runs
104 Port engine nacelle fairing
105 Port wing fuel tank
106 Port aerial mast
107 Oil tank cover
108 Front spar
109 Hinged leading-edge (inspection/servicing access)
110 Fuselage/wing centre-line diagonal brace strut
111 Fuselage/wing forward inverted 'Vee' struts (housing control runs)
112 Wing centre/outer section front spar join
113 Sponson/forward wing strut upper attachment
114 Twin landing lights
115 Sponson/wing strut diagonal bracing
116 Fuselage/wing aft inverted 'Vee' struts (housing fuel lines from sponsons)
117 Wing centre/outer section join rib
118 Sponson/aft wing strut upper attachment
119 Midships dorsal decking
120 Sponson/aft wing strut
121 Fuselage air ventilation plant
122 Fuselage main frame bulkhead
123 Centre-line walkway
124 Aft bay hinged entry hatch
125 Aft midships bay (four × stretcher accommodation)
126 Hull 'Vee' bottom structure
127 Galley/hot plate (port)
128 Blanket/survival clothing cupboard (starboard)
129 Fuselage midships mooring lug
130 Aileron hinge fairings
131 Bulkhead door
132 Aileron underwing mass balances
133 Wing rear spar
134 Port navigation light
135 Port formation/identification light
136 Port aileron
137 Hull aft step
138 'Vee' bottom sternpost frame
139 Reinforced fuselage frame
140 Turret support
141 Compressed air bottles
142 Turret ring
143 Dorsal turret
144 20-mm MG 151 cannon
145 Toilet
146 Peat bag (toilet sanitary refill)
147 Porthole
148 Master compass
149 Aft fuselage centre-line catwalk
150 Tail surface control rod linkage
151 Fuselage aft frames
152 Fuselage skinning
153 Ventral stringers
154 Control rods
155 Fuselage/tailplane fairing
156 Tailplane front spar
157 Elevator control rod
158 Elevator hinges
159 Tailplane ribs
160 Tailplane rear spar/tailfin attachment
161 Starboard tailfin structure
162 Aerial attachment
163 Rudder balance
164 Rudder frame
165 Rudder upper trim tab
166 Starboard ruddor post
167 Rudder lower trim tab
168 Trim tab actuacting linkage
169 Starboard elevator tab
170 Port rudder balance
171 Starboard elevator structure
172 Elevator control linkage
173 Rudder linkage
174 Port aerial attachment
175 Tailplane inboard end rib structure
176 Tailplane front spar/fuselage frame attachment
177 Fuselage aft main frame
178 Control linkage
179 Stern mooring rope stowage
180 Tailplane brace struts
181 Port tailfin
182 Tailplane/fin attachment
183 Port elevator hinge
184 Rear turret ammunition stowage
185 Port rudder post
186 Rear turret
187 Turret ring
188 Tail gunner's armour plating
189 Port rudder
190 Rudder lower trim tab hinge fairing
191 Port rudder lower trim tab
192 Tail navigation light
193 Tail mooring ring
194 7.92-mm MG 15 stern machine-gun

401

Douglas DC-2
First of the Tough Old Birds

Overshadowed by its illustrious successor, the Douglas DC-1/2 was nevertheless a true pioneer of commercial aviation, and scored a considerable success in its own right. Of course it also provided the inspiration for the classic DC-3, and established the 'Douglas Commercial' airliner dynasty which continues today.

Following the fatal crash on 31 March 1931 of a Transcontinental & Western Air Fokker F-10A tri-motor airliner, which killed the famous Notre Dame University football coach Knute Rockne, the US Bureau of Air Commerce enforced stringent inspection requirements on commercially-operated wooden-structure aircraft in the United States. These inspections proved costly and time-consuming, and prompted TWA Vice-President of Operations Jack Frye to draw up a specification for a new all-metal airliner.

TWA's requirement was for a three-engined monoplane to be powered by engines of 500-550 hp, with a cabin seating 12 passengers, a cruising speed of 150 mph, maximum speed of at least 185 mph, landing speed of not more than 65 mph, a minimum service ceiling of 21,000 ft and a range of at least 1,080 miles. The aircraft would also have to be capable of taking off on the power of two engines from any airport served by the airline's transcontinental routes, which included the 4,954-ft high field at Albuquerque, New Mexico, where temperatures frequently exceeded 90°F, creating a taxing high density/altitude combination.

On 2 August 1932 TWA circulated its requirements to the Consolidated, Curtiss, Douglas, General Aviation and Martin companies, and 10 days later representatives of the Douglas Aircraft Company in Santa Monica presented TWA with their proposals for the Douglas Commercial One. Despite the fact that the proposed DC-1 was a twin-engined design, over several weeks of intense negotiation Douglas was able to convince TWA that the aircraft could meet or exceed all the airline's requirements ("I am 90 per cent sure the DC-1 can meet the specified performance – it is the other 10 per cent that's keeping me awake at night," chief designer Arthur Raymond reported to company founder Donald Douglas), and on 20 September 1932 TWA contracted to buy the first DC-1 for $125,000, with options on a further 60 aircraft at a cost per aircraft of $58,000 without engines.

The DC-1, powered by two 690-hp Wright SGR-1820-F nine-cylinder radial engines, made its maiden flight from Clover Field, Santa Monica, on 1 July 1933, piloted by Douglas's chief test pilot and Vice-President, Sales, Carl A. Cover, and went on to meet all TWA's requirements, including a single-engine flight from take-off to landing over the airline's 'hot-and-high' route between Winslow,

Where it all began: the one and only Douglas DC-1. First flying on 1 July 1933, the aircraft was used for tests and demonstrations. It later surfaced in Spain, flying reconnaissance and staff-transport missions for the Republicans.

Arizona, and Albuquerque, New Mexico. By then TWA had placed a firm initial order for 20 improved DC-2s, later increased to 31 aircraft, but took delivery of the sole DC-1, which the airline used as a flying laboratory and for occasional revenue flights prior to delivery of the new aircraft.

The DC-2 differed from the DC-1 in having a fuselage 2 ft longer, enabling the cabin to accommodate 14 passengers. TWA specified the more powerful 710-hp Wright Cyclone SGR-1820-F3 engine for its aircraft. The DC-2 was a low-wing monoplane of all-metal construction employing a semi-monocoque fuselage, cantilever wing using Douglas-Northrop cellular multi-web construction with a parallel chord centre section built integrally with the fuselage, cantilever tail surfaces also of multi-cellular construction, and forward-retracting tailwheel undercarriage, the mainwheels protruding from the engine nacelles when retracted to minimise damage to the aircraft's undersurfaces in the event of a wheels-up landing. The passenger cabin provided adjustable seats on rubber mounts to minimise vibration, with ventilation and steam heating. A forward compartment between the passenger cabin and flight deck, in line with the engines, could hold up to 1,000 lb of cargo and mail, with an additional cargo and baggage compartment aft of the cabin. Cabin entry was by a single door on the port side, with a buffet to the rear of the door and a toilet at the rear of the cabin.

The Sky Queen takes to the air

The first DC-2, TWA's *City of Chicago*, made its maiden flight on 11 May 1934 and was delivered to the airline three days later. Contemporary TWA publicity described its new 'Sky Queen' thus: "The silver ship is a low-wing cantilever monoplane. The entire external appearance of the transport is remarkable for its complete freedom from struts and control system parts. In harmony with clean design, the wheels retract into the engine nacelles... The passenger salon is 26 ft 4 in long, 5 ft 6 in wide and 6 ft 3 in high. The great height of the passenger salon permits even the tallest person to walk fully erect in the cabin for its entire length. The compartment is fitted to accommodate 14 passengers in two rows of specially-designed lounge chairs 40 in wide and separated by an aisle 16 in wide. Chairs are deeply upholstered and fully adjustable for reclining or reversing to face the passenger behind. Each seat has a private window and because of the height of the seat above the wing there is excellent vision from all chairs... Flying in the new luxury liner transport is like putting wings on a luxurious living room and soaring in complete security and comfort. The Douglas transport is the crowning achievement in commercial transportation, creating a new ideal in luxurious travel combined with high speed and high performance characteristics coupled with great security."

The DC-2 entered TWA service on the Colombus, Ohio-Pittsburgh, Pennsylvania-Newark, New Jersey, route on 18 May 1934, and a week later began serving Chicago and Newark, breaking the speed record between the two cities four times in one eight-day period. The *Sky Chief* transcontinental service flew from New York to Los Angeles by way of Chicago, Kansas City and Albuquerque. The journey took 16 hours and 20 minutes eastbound and 18 hours westbound, considerably faster than the Boeing 247s of rival United Air Lines – a fact which did not escape the attention of other US carriers.

Orders for a prestige aircraft

Soon, orders had been received from Eastern Air Lines for 14 aircraft to serve East Coast routes to Miami, Florida; from American and Pan American Grace for a total of 14; from Pan American Airlines for 16; and from General Air Lines (later Western Airlines) for four. To promote Eastern's Miami service Captain Eddie Rickenbacker, the carrier's general manager, made a one-day VIP return flight between New York and the Florida city, taking guests for dinner in Miami and returning them the same day. Three other DC-2s were ordered as corporate transports by major American companies.

Abroad, the Dutch company Fokker negotiated licence manufacturing and distribution agreements for the DC-2, as did Nakajima Hikoki KK in Japan and Airspeed in Great Britain. Nakajima took delivery of one complete DC-2 and a kit of parts, and built five more in 1936-37 for operation by Dai Nippon Koku KK on routes between Japan and Formosa. The Soviet trading agency Amtorg also took de-

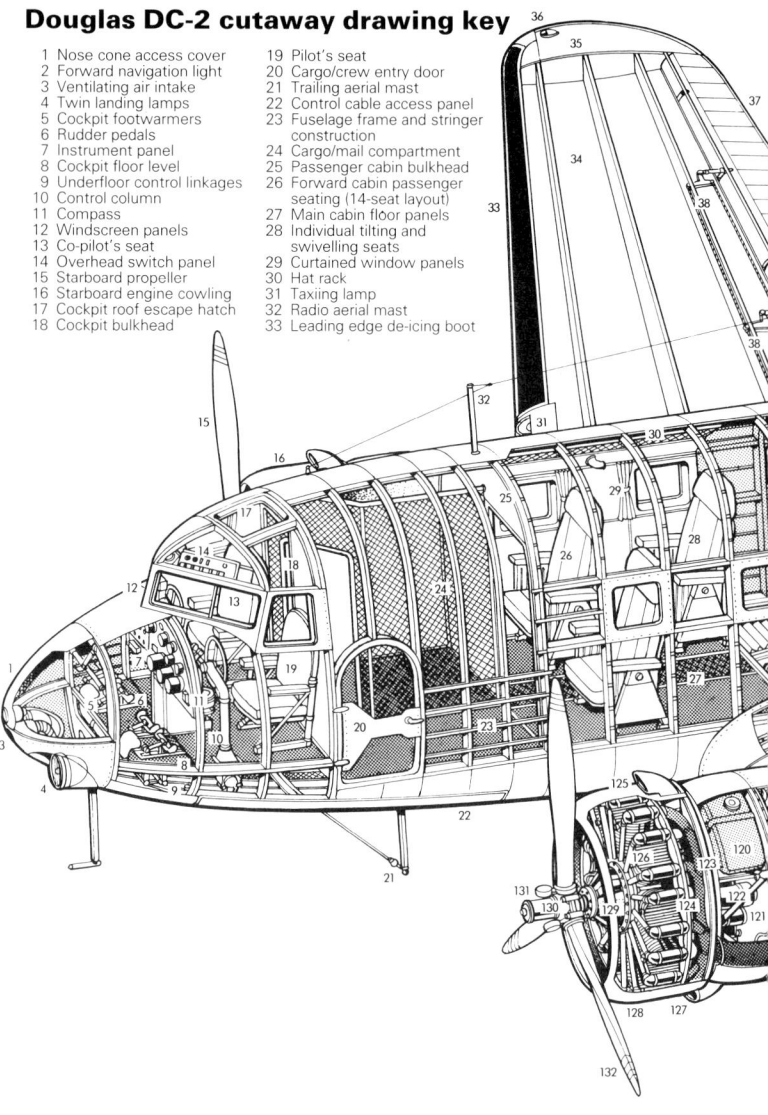

Douglas DC-2 cutaway drawing key

1. Nose cone access cover
2. Forward navigation light
3. Ventilating air intake
4. Twin landing lamps
5. Cockpit footwarmers
6. Rudder pedals
7. Instrument panel
8. Cockpit floor level
9. Underfloor control linkages
10. Control column
11. Compass
12. Windscreen panels
13. Co-pilot's seat
14. Overhead switch panel
15. Starboard propeller
16. Starboard engine cowling
17. Cockpit roof escape hatch
18. Cockpit bulkhead
19. Pilot's seat
20. Cargo/crew entry door
21. Trailing aerial mast
22. Control cable access panel
23. Fuselage frame and stringer construction
24. Cargo/mail compartment
25. Passenger cabin bulkhead
26. Forward cabin passenger seating (14-seat layout)
27. Main cabin floor panels
28. Individual tilting and swivelling seats
29. Curtained window panels
30. Hat rack
31. Taxiing lamp
32. Radio aerial mast
33. Leading edge de-icing boot

Four DC-2s were in Republican service at the start of the Spanish Civil War, but this aircraft was captured by the Nationalists and used mostly as General Franco's personal transport. On one memorable occasion it fought with defensive machine-guns against a similarly-armed Republican DC-2!

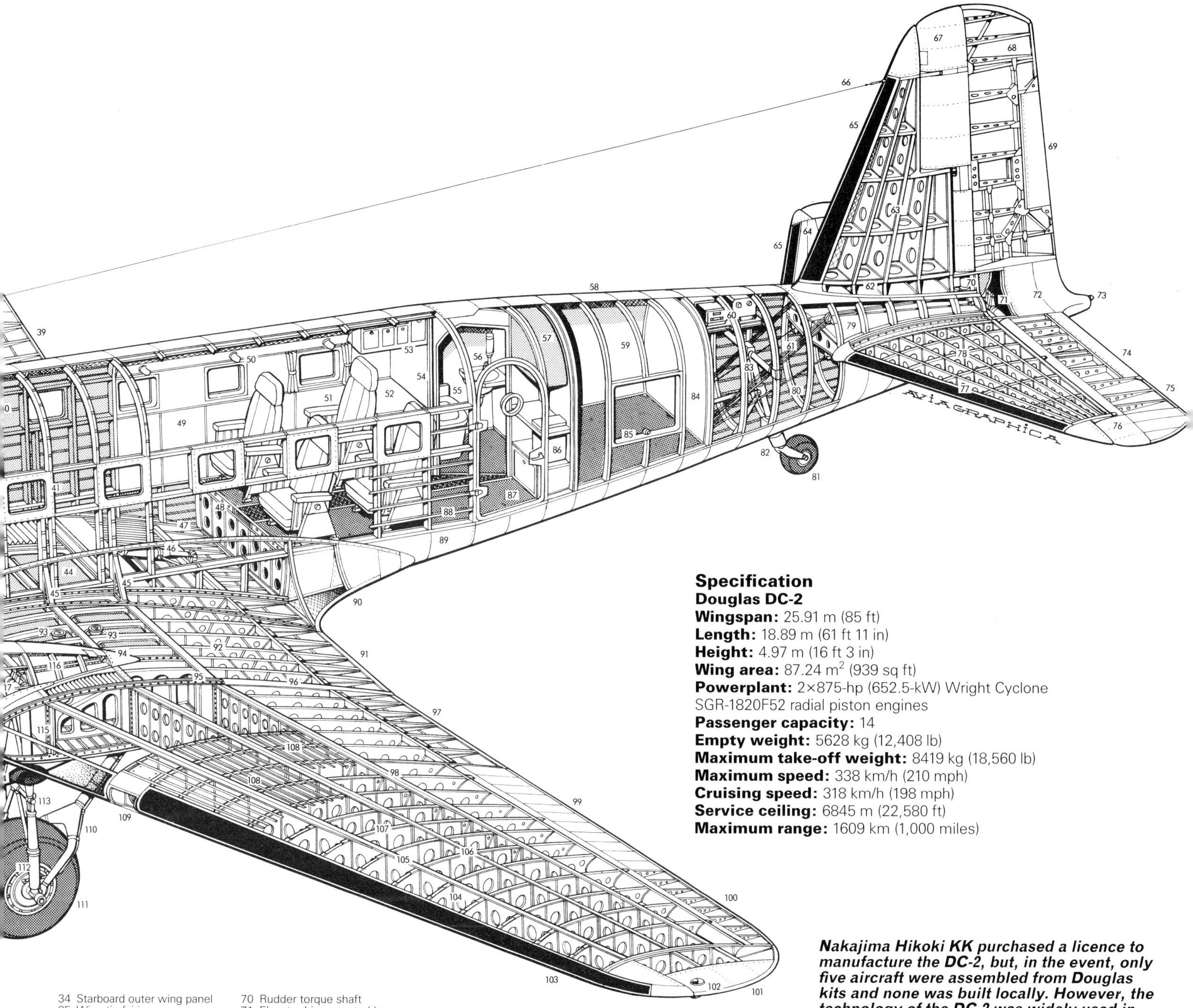

Specification
Douglas DC-2
Wingspan: 25.91 m (85 ft)
Length: 18.89 m (61 ft 11 in)
Height: 4.97 m (16 ft 3 in)
Wing area: 87.24 m² (939 sq ft)
Powerplant: 2×875-hp (652.5-kW) Wright Cyclone SGR-1820F52 radial piston engines
Passenger capacity: 14
Empty weight: 5628 kg (12,408 lb)
Maximum take-off weight: 8419 kg (18,560 lb)
Maximum speed: 338 km/h (210 mph)
Cruising speed: 318 km/h (198 mph)
Service ceiling: 6845 m (22,580 ft)
Maximum range: 1609 km (1,000 miles)

Nakajima Hikoki KK purchased a licence to manufacture the DC-2, but, in the event, only five aircraft were assembled from Douglas kits and none was built locally. However, the technology of the DC-2 was widely used in subsequent military designs.

34 Starboard outer wing panel
35 Wing tip fairing
36 Starboard navigation lights
37 Starboard aileron
38 Aileron hinge controls
39 Split trailing edge flap
40 Centre fuselage construction
41 Cabin window panels
42 Centre wing section corrugated inner skin
43 Port main fuel tank
44 Port auxiliary fuel tank (total fuel capacity 510 US gal/1930 litres)
45 Wing spar attachment joints
46 Flap actuator
47 Centre section flap
48 Floor beam construction
49 Cabin wall trim panels
50 Fresh air louvres
51 Rear cabin seating
52 Steward's table
53 Overhead stowage locker
54 Cabin rear bulkhead
55 Toilet compartment
56 Wash basin
57 Drinking water tank
58 Fuselage skin panelling
59 Baggage compartment
60 HF radio receiver and transmitter
61 Tailplane control cables
62 Fin root attachment
63 Tailfin construction
64 Starboard tailplane
65 Leading edge de-icing boots
66 HF aerial cable
67 Rudder horn balance
68 Fabric-covered rudder construction
69 Rudder tab
70 Rudder torque shaft
71 Elevator hinge control levers
72 Tailcone fairing
73 Tail navigation light
74 Elevator tab
75 Port fabric-covered elevator construction
76 Elevator horn balance
77 Tailplane leading edge de-icing boot
78 Tailplane construction
79 Centre section carry-through
80 Aft fuselage frame-and-stringer construction
81 Tailwheel
82 Shock absorber leg strut
83 Tailwheel strut
84 Baggage compartment bulkhead
85 Baggage door
86 Buffet unit
87 Passenger entry doorway
88 Rear cabin sidewall construction
89 Ventral skin panelling
90 Trailing edge wing root fillet
91 Inboard split trailing edge flap
92 Flap shroud construction
93 Fuel filler caps
94 Nacelle tail fairing
95 Outer wing panel bolted joint
96 Wing panel joint capping strip
97 Outer split trailing edge flap
98 Rear spar
99 Port fabric covered aileron
100 Aileron rib construction
101 Wing tip fairing
102 Port navigation lights
103 Leading edge pneumatic de-icing boot
104 Leading edge nose ribs
105 Front spar
106 Centre spar
107 Wing rib construction
108 Wing stringers
109 Port taxiing lamp
110 Main undercarriage rear strut
111 Port mainwheel
112 Shock absorber leg struts
113 Undercarriage knee joint
114 Exhaust pipe
115 Outer wing panel spar joint
116 Undercarriage bungee cables
117 Main undercarriage leg pivot fixing
118 Hydraulic retraction jack
119 Mainwheel well
120 Oil tank (19 US gal/72 litres)
121 Accessory equipment compartment
122 Engine bearer struts
123 Fireproof bulkhead
124 Exhaust collector
125 Carburettor air intake
126 Wright Cyclone SGR-1820 nine-cylinder radial engine
127 Oil cooler air intake
128 Detachable engine cowlings
129 Propeller reduction gearbox
130 Propeller hub pitch change mechanism
131 Feathering counterweights
132 Three-bladed variable-pitch propeller

The DC-2 was evolved from the DC-1 with a stretched fuselage, and was built to the order of TWA. However, other US airlines were quick to purchase the aircraft, American Airlines being among them. This is one of its 16 DC-2-120s.

livery of a single DC-2, but no licence manufacturing agreement was made with the USSR. Neither Airspeed, which allocated the designation AS.23 to the aircraft, nor Fokker actually built DC-2s, but Fokker enjoyed considerable success in assembling and selling Douglas-built aircraft, eventually delivering 39, including 21 to the Dutch national carrier KLM, five to Swissair, five to Ceskoslovenska Letecka Spolecnost (CLS) of Czechoslovakia, and four to the Spanish airline Lineas Aéreas Postales Españolas (LAPE). KLM, and its overseas subsidiary KNILM, operated DC-2s on European routes and on a long-distance service between Holland and the Dutch East Indies, flying between Amsterdam and Batavia (Jakarta) in stages, the total journey lasting six days. This service, operated twice weekly, was inaugurated on 12 June 1935, with the DC-2s in five-passenger (later eight-passenger) configuration.

Ground-breaking flight

It was one of KLM's DC-2s, PH-AJU *Uiver*, which made aviation history in October 1934 when it took part in the MacRobertson London-Australia Air Race. Crewed by Captain K.D. Partmentier and J.J. Moll, a navigator and mechanic, and carrying three fare-paying passengers and 30,000 airmail letters, *Uiver* won the transport section of the race and came second to the purpose-built de Havilland D.H.88 Comet racer overall, completing the 11,123 miles between RAF Mildenhall and Melbourne in 90 hours 13 minutes and 36 seconds, and firmly establishing the Douglas company as the leading manufacturer of commercial aircraft.

In little more than a year 108 DC-2s were in service in 21 countries, and had logged more than 20 million miles. A contemporary operator survey conducted by Douglas revealed: "United States operators and Pan American Airways in South America report 15,000,000 miles flown in the first eight months at an efficiency of 98.8 per cent . . . TWA, American Airlines, Eastern Air Lines and Pan American Airways report their fleets of Douglas ships are negotiating 52,289 miles a day with 21,499 miles flown at night and 20,790 miles of daylight flying. Of the 26,259,665 miles flown in the US during the first six months this year, 7,286,437 miles were flown by Douglas transports, or 27.7 per cent. However, the 42 Douglas airplanes available for service constituted only 7.6 per cent of the total airplanes in service in the US – a remarkable tribute to the Douglas operating ability. Operating efficiency increased on the airlines using Douglas luxury liners from the first month the airplanes were placed in service. . . In some cases there was a reported 66-per cent gain in average 'air time' per day."

Three principal commercial variants of the DC-2 were developed. The 'standard' DC-2 was powered by Wright Cyclone SR-1820 engines in 710-875 hp variants, each sub-type being identified by a numerical suffix according to customer. When powered by Pratt &

In Europe it was KLM which was best known as a DC-2 operator, although it was by no means the only one, as evidenced by this Swiss Air Lines example. During World War II, Lufthansa operated a sizeable fleet of ex-KLM, -LOT and -Czech aircraft.

Several DC-2s were flown by the China National Airways Corporation before and during World War II. Some of the aircraft had previously served with Pan American, while two were purchased direct from Douglas.

Whitney Hornet engines the aircraft was known as the DC-2A. Two DC-2s operated by Polskie Linie Lotnicze (LOT) in Poland were powered by two 750-hp Bristol Pegasus VI engines, the modification being developed jointly by Douglas and the Bristol company at LOT's request, since the Skoda company was licensed to manufacture Bristol engines, which were in common use in Poland. These aircraft, which were supplied by Fokker, were unofficially designated DC-2Bs and were among 10 DC-2s captured at the outbreak of World War II and operated by Deutsche Lufthansa.

The DC-2 in uniform

The American military forces also ordered DC-2s, five 710-hp Wright-engined variants being delivered to the US Navy as R2D-1 staff transports, while 18 were operated by the US Army Air Corps as C-33 military cargo aircraft, these having enlarged fins and rudders and a cargo-loading door on the port side. In addition, 24 former civilian DC-2s were impressed into service by the USAAF as C-32As, two YC-34 personnel transports were delivered, and a 'DC-2½' was developed from a prototype C-38 which combined elements of the DC-2 and later DC-3. The single C-38 featured two 975-hp Wright R-1820-45 engines and a DC-3 tail section, and led to an order for 35 C-39s which combined the C-33 fuselage with DC-3 wing centre section, tail and landing gear. Two similar aircraft, the C-41 powered by 1,200-hp Pratt & Whitney R-1820-21 engines, and the C-42, powered by 1,000-hp Wright R-1820-53s, were delivered to General Henry H. Arnold, Air Corps Chief of Staff at Bolling Field, Washington, DC, and to the commanding general of the USAAF General Headquarters, respectively.

The Douglas DC-2 not only inspired the ubiquitous and vastly successful DC-3, but more importantly it also established the Douglas Aircraft Company as a major manufacturer of commercial aircraft, a position it retains today within the McDonnell Douglas Corporation.

The DC-2's contribution was recognised at the highest level on 1 July 1936, when President Roosevelt presented Douglas with the prestigious Robert J. Collier trophy – aviation's highest accolade – noting in the citation that: "This airplane, by reason of its high speed, economy and quiet passenger comfort, has been generally adopted by transport lines throughout the United States. Its merits have been further recognised by its adoption abroad, and its influence on foreign design is already apparent."

Numbers of the *DC-2* were purchased by the Royal Air Force for transport work, mostly with **No. 31 Squadron**. Designated *DC-2K*, these aircraft flew in the Middle and Far East. This aircraft wears an Indian civil registration in addition to the RAF serial *DG478*.

The first export *DC-2s* went to the Dutch national airline *KLM*, which promptly put them into service on most of its routes, including one to Batavia. One (PH-AJU/*U*iver) was used in the 1934 MacRobertson Trophy race between Mildenhall, England, and Melbourne, Australia, arriving in second place.

Hanssin Jukka *of the Finnish air force flew one mission as a makeshift bomber before reverting to transport duties. It served throughout World War II and long after, until its final retirement in 1955. After service as a restaurant, it is now in the Finnish air force museum.*

Chapter 20

Douglas DC-4

Douglas DC-6/7

Douglas DC8

Embraer Bradeirante

Fokker F.VII

Douglas DC-4

Flying for the first time in 1942, the Skymaster was to ply the world's airways in both battledress and civil liveries for over 40 years, with some still soldiering on today. As well as its longevity, it achieved fame for ushering in what we now accept as the era of modern air transport.

The DC-4E had features hitherto only dreamed of; tricycle landing gear, four 858-kW (1,150-hp) Pratt & Whitney R-2180 Twin Hornet radials (among the most powerful engines in the world), and a distinctive triple tail low enough for existing hangars. With a wing span of 42.15 m (138 ft 3 in), a length of 29.74 m (97 ft 7 in), and able to accommodate 52 passengers, it was a benign behemoth. Created in partnership between two legendary figures in aviation, W. A. Patterson of United Air Lines and aircraft-builder extraordinaire Donald Douglas, the DC-4E was to usher in a new age. But it was a false start. Uneconomical for US airline routes, the DC-4E was dismantled, sold to Japan in September 1939, and relegated to obscurity. A smaller, better version, the DC-4, then appeared, and it was this which gained immortality, no fewer than 1,245 airframes being manufactured by Douglas over a decade. But it would not, at first, enhance civil transportation, for the Skymasters which followed the one-off DC-4E at first wore olive drab Army garb, and won fame in the crucible of war as a military C-54.

Like many of the world's greatest aircraft, the Skymaster emerged from second thoughts. In May 1939, Douglas' design team under Arthur Raymond did a double-take, reduced its aspirations, and came up with the simpler, more pragmatic DC-4A. United and American Airlines placed orders. Unpressurised (unlike the DC-4E which had introduced full cabin pressurisation and air conditioning) the new transport had its wing span reduced to 35.81 m (117 ft 6 in), its length to 28.63 m (93 ft 11 in) and passenger capacity to 42. Pearl-Harbor, the USAAF's severely limited airlift capacity, and the need to supply fighting men in a global struggle prompted the US War Department to snap up the first 34 machines, now configured with a single tail, and label them C-54. The first (42-20137) took to the air at Santa Monica on 26 March 1942, flown by John F. Martin. The dimensions and outward configuration of the Skymaster were now fixed, but its appearance in airline livery would, perforce, await VJ Day.

The structurally improved C-54A, powered by Pratt & Whitney R-2000-7s yielding 895 kW (1,200 hp), flew on 14 January 1943, and the later C-54B in March 1944. Olive-drab coloured, impressively reliable, respected by maintenance crews, US Army Air Force Skymasters began to be seen in the war theatres, delivering the goods. The USAAF's Air Transport Command used some contract airline pilots and some women pilots for its gruelling long-range hauling jobs. A typical mission, from Grenier Field, New Hampshire to Prestwick would bring up to 7258 kg (16,000 lb) of war matériel across 3675 km (2,285 miles), sometimes in the face of the most arduous weather and navigational obstacles, the perils chronicled by Ernest K. Gann in *Fate is the Hunter*. C-54s appeared in Monrovia, Karachi and Honolulu. A solid and eye-pleasing aircraft, the Skymaster thus played a decisive unsung role in wresting victory from the Axis.

Aircraft 42-10751, converted from a C-54A, was given the designation C-54C, the nickname *Sacred Cow*, and a special electric lift for the wheelchair carrying President Franklin D. Roosevelt as the very first presidential aircraft, before the term 'Air Force One' was coined. First delivered in Janaury 1944, it carried Roosevelt to the important convocation of Allied leaders at Yalta in February 1945. Winston Churchill had to wait longer for his Skymaster, receiving C-54B 43-17126; quickly given RAF serial EX999 in June 1944. The cigar-smoking prime minister enjoyed numerous luxury features in his Skymaster, including electrically-heated lavatory seats, but an

Ferrying war supplies across the stormy North Atlantic was the C-54's main contribution to the Allied war effort. This post-war C-54G-5-DO (45-553) of the 83rd Air Transport Squadron, Military Air Transport Service (MATS) is seen at Grenier AFB, outside Manchester, New Hampshire – one hub of transatlantic operations. The MATS paint scheme included a tail stripe identifying aircraft as belonging to its Atlantic Division, and Arctic red trim on fin and wingtips.

After serving with their initial purchasers as military C-54s or civil DC-4s, Douglas Skymasters moved on to give sterling service with secondary purchasers. OY-OFI appears in the late 1940s paint scheme of Scandinavian Airlines System.

The first DC-4 (later DC-4E), the company's original four-engine design, was a bigger and more complex machine than the DC-4 which became the military C-54 when war broke out and served as an airline workhorse afterwards. The prototype DC-4E (NX18100), seen in pre-war Californian skies, first flew in 1938.

Though initially conceived as a commercial airliner with strong support from US airline companies, the DC-4 Skymaster was pre-empted by the military until VJ Day. Then some machines, like Delta's N37474, went directly from production into civil livery.

elaborate plan for a heavily-armoured protection capsule, to shield him in the event of fighter attack, was deigned impractical and abandoned. During the war, the UK operated a further 10 Skymasters (KL977/986) taken from the US inventory. Most continued the 'round-the-clock' mission of ferrying war supplies, but three served as VIP executive transports at RAF Northolt near London.

Solid performer

The robust Skymaster was such a solid performer that outwardly no important changes were made from one variant to the next. The C-54D with more powerful R-2000-11 engines entered service in August 1944, and 380 were built at Douglas' wartime plant in Chicago. The identically-powered C-54E had increased fuel capacity; 125 came off the Santa Monica line. The final production Skymaster was the C-54G, 162 of which were built in Santa Monica.

A total of 331 Skymasters from the USAAF production run was turned over to the US Navy and US Marine Corps as R5D-1 to R5D-5 variants, and 14 served with the US Coast Guard. Some airframes transferred often from one service to another, an example being C-54D 42-72484 built for the USAAF, turned over to the RAF as KL978, returned to the USA as US Marine Corps R5D-3 92003, and ultimately restored to its USAF identity.

Among experimental variants or proposals in the Skymaster series were the one-off XC-54F (42-72321) with paratroop jump doors, and the long-range XC-54K (45-59602) with 1063-kW (1,425-hp) Wright R-1820-HD single-row radials. The projected paratroop-carrying C-54H, the C-54J staff transport, and its naval R5D-6 equivalent were never built. The sole XC-114 (45-874) was powered by 1208-kW (1,620-hp) Allison V-1710-131 12-cylinder Vee engines and had its fuselage lengthened by 2.13 m (7 ft). The only XC-116 (45-875), also Allison-powered, was used for thermal de-icing experiments. For many years, a US Navy R5D-2 operated at NATC

President Roosevelt's personal aircraft, the VC-54C-1-DO known as the Sacred Cow (42-107451), was originally ordered from Douglas-Santa Monica as a C-54A. In January 1947 it wore flags of the numerous countries it had visited. The aircraft carried Roosevelt abroad only twice before his death in 1945, but took his successor Harry S. Truman to 55 nations. Note the tail support strut.

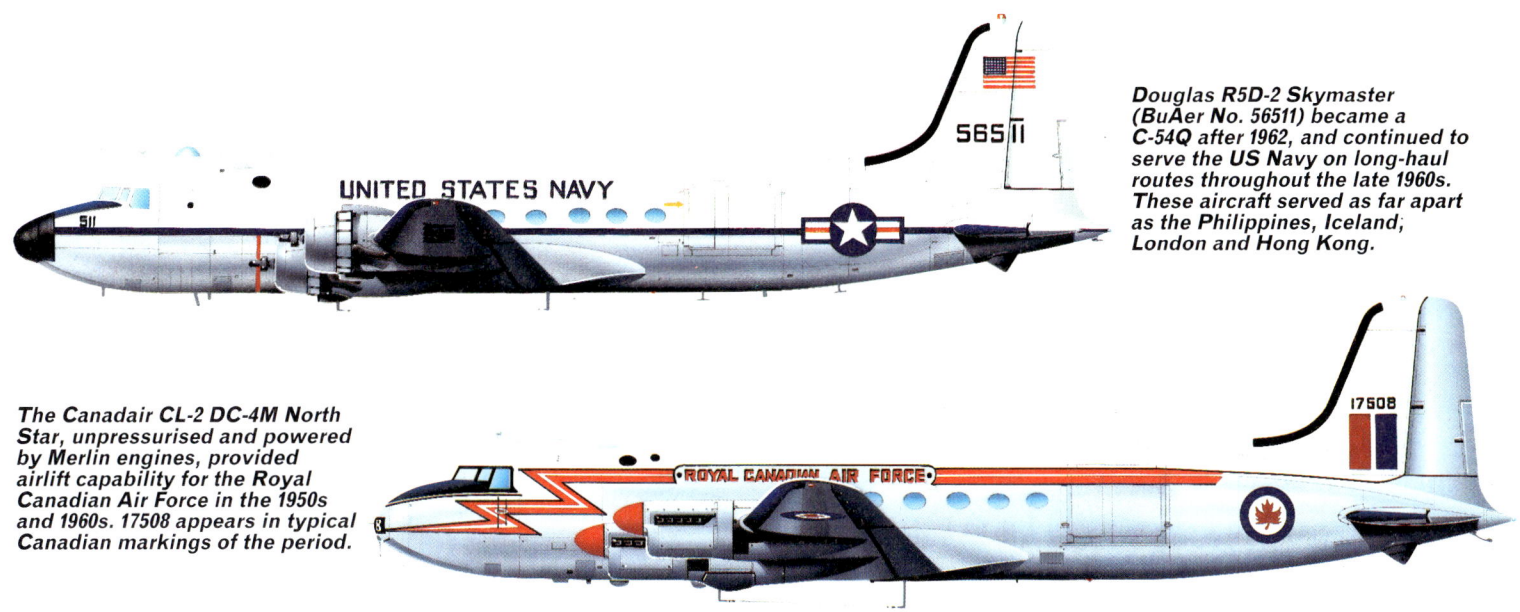

Douglas R5D-2 Skymaster (BuAer No. 56511) became a C-54Q after 1962, and continued to serve the US Navy on long-haul routes throughout the late 1960s. These aircraft served as far apart as the Philippines, Iceland, London and Hong Kong.

The Canadair CL-2 DC-4M North Star, unpressurised and powered by Merlin engines, provided airlift capability for the Royal Canadian Air Force in the 1950s and 1960s. 17508 appears in typical Canadian markings of the period.

Patuxent River, Maryland had four enormous under-wing pods carrying radio and radar gear and a unique, 4.27-m (14-ft) retractable antenna on the dorsal spine of the aircraft.

Apart from the phased-out veteran C-54s which finally shed their olive drab and began to appear in the liveries of the world's airlines after VJ Day, beginning in 1945 Douglas finally turned out 79 machines built from the outset as DC-4 civil airliners. By the end of the war, however, the more advanced DC-6 was on the horizon and, to make things worse, Douglas was competing with itself in the sense that trunk carriers could purchase used C-54s more cheaply than new DC-4s. Initial customers for the civil DC-4 included Air France, Sabena, Western and National, but over decades to come the airliner would serve with many hundreds of carriers. Major airlines normally operated new DC-4s in a 44-passenger configuration, but by the late 1950s and early 1960s the ingenuity of interior designers had crammed no less than 85 seats into the old bird. The C-54/DC-4 was also the basic design which led to the very successful DC-6 and DC-7 'propliners'.

A unique record was set by a USAF machine on 22 September 1947 when a C-54D (42-72461) took off from Curphey, Newfoundland and winged its way across the North Atlantic to RAF Brize Norton on a fully automatic flight: at no point did Captain John D. Wells and his crew place their hands on the controls.

Many of the ex-USAF C-54s and ex-Navy R5Ds went through several airlines and ended their days serving on the cargo routes to and from South America. A fair number are still in service, whilst others still fly as water bombers.

Among the dozens of C-54s which went to the US Navy and Marine Corps under the designation R5D, none was more distinctive than this R5D2-2 (BuAer No. 50851), used for radar tests at the Naval Air Test Center (NATC), Patuxent River, Maryland, just after the war.

For the military Skymaster, the Soviet Union's 1948 decision to seal off Berlin was a new call to arms. Beginning on 26 June 1948, Allied cargo aircraft flew 'round the clock' to supply Berlin, and USAF C-54s from bases around the world were assembled to support the Berlin Airlift. So many aircraft were flying in the constricted Berlin corridor that pilots were never more than seconds away from mid-air collision. The weather was frequently dismal. Pilots of more

With Honolulu's Diamondhead in the background and Pacific waters beneath, this Douglas R5D-4 Skymaster (BuAer No. 90409) of the US Marine Corps has probably just taken off from Hickam Field, Hawaii, and is heading for the American mainland. The R5D-4 became the C-54P after 1962.

A long-serving veteran, C-54G-15-DO (45-633) of the Iowa Air National Guard soldiered on in the early 1960s. The C-54 enjoyed great longevity with Guard and Reserve units even after regular military transport units had moved into the jet age in the mid-1960s.

than 300 C-54s which flew the airlift took frightful risks daily. Berlin was reopened on 12 May 1949, the airlift having been a total success, but the Skymaster soldiered on: a USAF C-54G (45-518) at Seoul's K-14 Kimpo Airport was the first casualty of the Korean War when North Korean Yakovlev Yak-9s strafed it and set it on fire on 25 June 1950. Thirty airframes were converted to serve as MC-54M hospital planes in 1951.

Continuing to serve the US forces well into the 1960s (including a role as support aircraft for the US Navy's 'Blue Angels' flight demonstration team) the C-54 also served in many other air forces, including those of Argentina, Belgium, Brazil, Colombia, Cuba, Denmark, Ethiopia, France, Honduras, Mexico, Peru, Portugal, Saudi Arabia, South Africa, Spain, Thailand, Turkey and Zimbabwe.

Canadian production

Some 71 airframes were added to the total of DC-4 variants in the post-war period by Canadair Limited of Montreal. Trans-Canada Air Lines had proposed a fully pressurised version of the DC-4 and this became the Canadair DC-4M, first flown on 15 July 1946 with four 1286-kW (1,725-hp) Rolls-Royce Merlin 626s. A total of 24 unpressurised examples of the Canadian machines served with the RCAF as the North Star, including service in support of Korean operations in 1950-3, while a 25th fully pressurised Canadair machine, the C-5, was powered by Pratt & Whitney R-2800 engines in its role as a VIP staff transport for the RCAF from 1950 until its retirement in 1967. And 46 civil Canadair-built DC-4M-2s served with British Overseas Airways (as the Argonaut), Trans-Canada, and later other carriers, with Merlin 724s. They raised the total number of Skymaster derivatives to 1,315. A final variant was evolved by Aviation Traders in England, who converted existing airframes by adding 2.64 m (8 ft 8 in) to the DC-4 forward fuselage and adding an hydraulic swing nose to permit direct in-loading of vehicles. Known as the Carvair, an apt contraction of car-via-air, and able to carry five vehicles, this conversion made its first flight at Southend on 21 June 1961. Nineteen Carvairs were operated on cross-channel ferry flights in the early 1060s and have since been used as general transports.

In 1966, Charlotte Aircraft Corp. of Miami, Florida, developed a C-54/DC-4 conversion powered by four 1119-kW (1,500-hp) Wright R-2600 engines. By then the jet age was full-blown and the Skymaster's long life was approaching its useful end.

Douglas DC-4 cutaway drawing key

1. Nose cone
2. Radio homing aerial
3. Fire extinguisher bottles
4. Nosewheel doors
5. Nose undercarriage bay construction
6. Cockpit front bulkhead
7. Pitot tubes
8. Windscreen panels
9. Instrument panel shroud
10. Instrument panel
11. Rudder pedals
12. Nosewheel hydraulic retraction jack
13. Steering jacks
14. Nosewheel
15. Torque scissors
16. Battery bay
17. Cockpit floor level
18. Control column handwheel
19. Captain's seat
20. Opening side window panel
21. First Officer's seat
22. Starboard side crew entry door
23. Navigator's folding chart table
24. Cockpit bulkhead

The C-54 Skymaster laboured long and hard in Arctic climes. This C-54D-5-DC (42-72606), built at Douglas' Chicago plant, still served with the 21st Composite Wing, Alaskan Air Command, at Elmendorf AFB near Anchorage with Arctic trim as late as 1967.

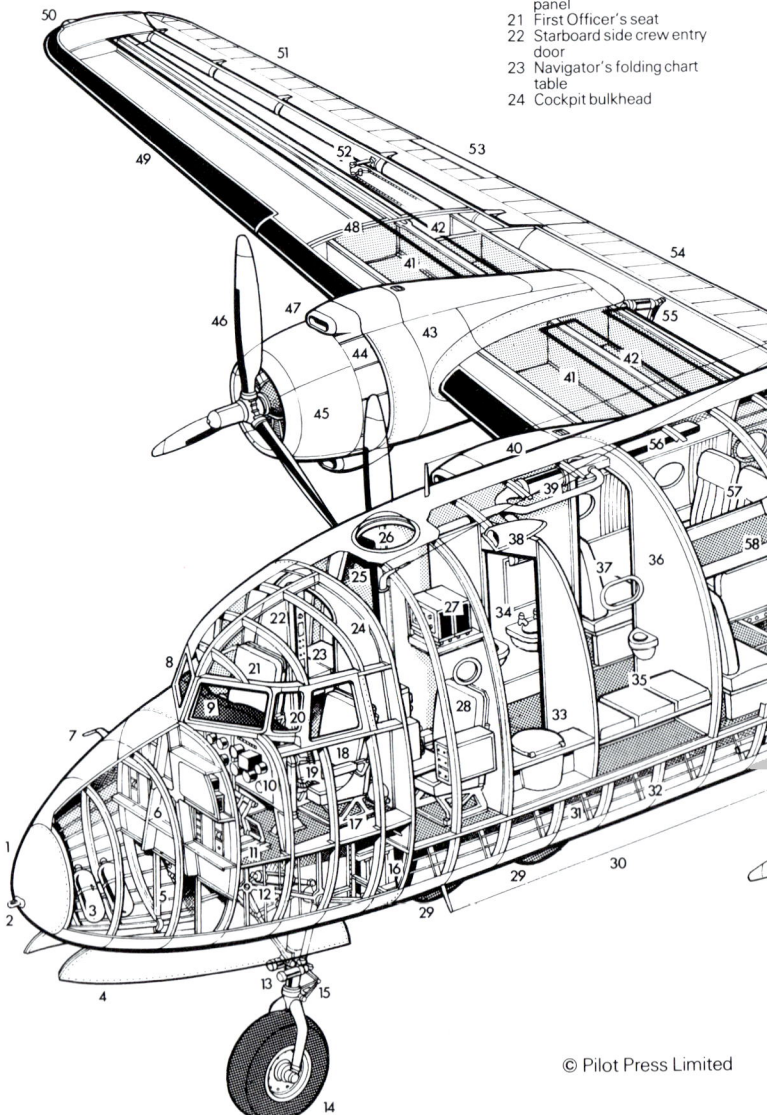

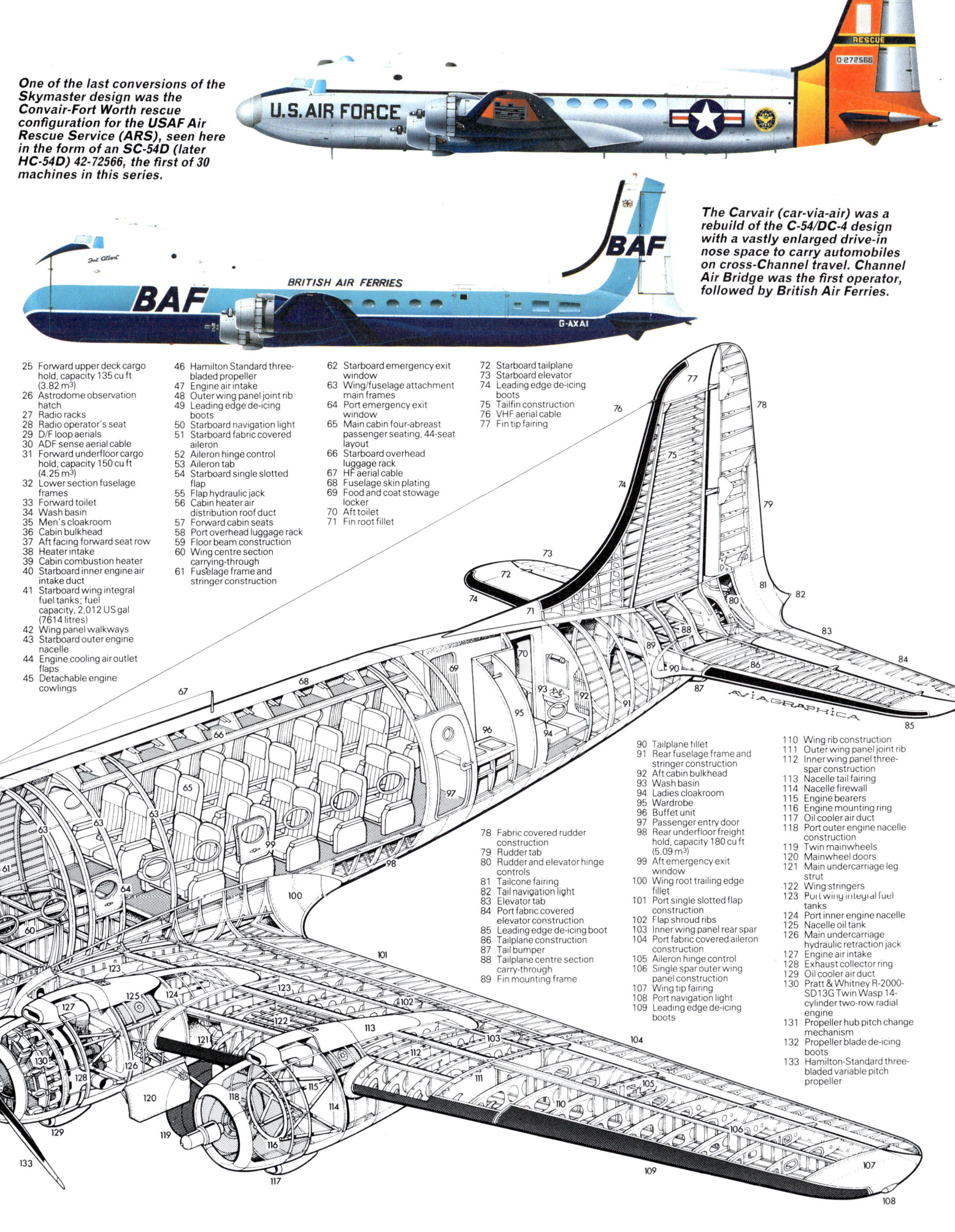

Douglas DC-6/7

The classic C-54/DC-4 provided Douglas with the starting block for a family of ever-larger airliners, blessed with steadily increasing speed and range. These were the DC-6 and DC-7 families, which culminated (with the DC-7C) in the ultimate achievement in propliner technology.

At the end of World War II the Douglas aircraft company found itself seriously behind its two main competitors, Boeing and Lockheed, in the United States transcontinental transport market. Wartime production of the four-engined Douglas DC-4 had topped the 1,000 mark, and with almost 88,000 wartime ocean crossings of the Atlantic and Pacific to its credit for the loss of only three aircraft, the type had more than proved its reliability.

However, the new Boeing 307 Stratoliner and the Lockheed 749 Constellation were not only larger in freight capacity but were also pressurised to carry passengers in comfort at over 3048 m (10,000 ft). So the USAAF, which had flown the DC-4 as the C-54, financed Douglas in the building of a pressurised, larger version. This was designated XC-112A and the prototype first flew on 15 February 1946. Its potential for airline operation was immediately recognised and Douglas continued its development in this direction under the company designation DC-6.

The prototype retained the DC-4's 35.81-m (117-ft 6-in) wing span, but the new, pressurised fuselage was lengthened by 2.06 m (6 ft 9 in) to give increased passenger capacity. Standard seating was for 48 to 52, but a high-density cabin layout could accommodate 86, and square windows – to become an identifying feature of the DC-6 series – replaced the rounded shapes of the DC-4. The initial power plant comprised four 1566-kW (2,100-hp) Pratt & Whitney R-2800-CA-15 Double Wasp radial piston engines.

American Airlines ordered 50 of the civilian version, and the first of these made its maiden flight on 29 June 1946; the following April the DC-6 began to enter service, initially on the airline's New York-Chicago route. The aircraft also satisfied military requirements; the newly independent US Air Force and the US Navy ordered 166, mainly for the Military Air Transport Service (MATS). The USAF version was dubbed the C-118 Liftmaster and could transport 74 passengers, 12247 kg (27,000 lb) of cargo, or 60 stretcher cases, while the US Navy versions were known as R6D-1; four, designated R6D-1Z, were VIP versions. Altogether 176 civilian DC-6s were delivered between 1946 and 1951 to American, United and Braniff airlines.

Perhaps the most famous DC-6 was 29th off the production line. It was fitted with a VIP interior, and operating from Andrews Air Force Base, Maryland, became President Harry S. Truman's personal aircraft. The term Air Force One for the President's executive aircraft was not in use at this time; instead, the DC-6 was designated VC-118, and given the name *Independence*. It had accommodation for 25 passengers, or could be converted to accommodate 12 sleeping, and had a stateroom and offices.

Having met with the approval of both the military and civil establishments, the Douglas company produced an improved version of the DC-6 which first flew on 29 September 1949. With the fuselage further lengthened by 1.52 m (95 ft), the DC-6A, as it was designated, first flew as a windowless, all-cargo version, with two freight doors, one forward, one aft, of the wing on the port side, and strengthened flooring.

Air Atlantique is based at Coventry in England, from where it operates two DC-6As on ad hoc cargo charter work. The DC-6 is expensive to maintain but the purchase price is low.

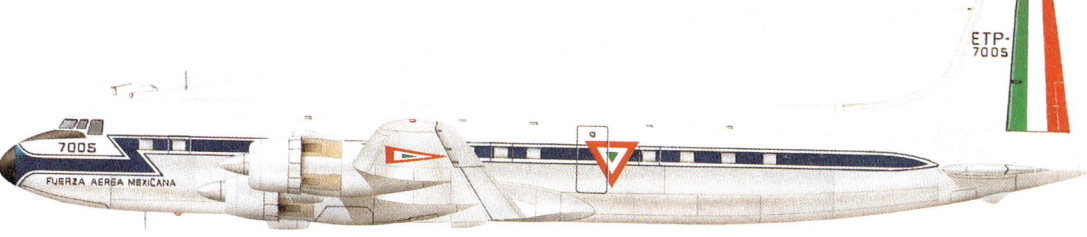
The DC-6 is now all but out of service with the military, a handful of aircraft surviving, mostly in South America. This machine wears the colours of the Mexican air force, with whom it flew general transport duties.

Power was improved with the use of four alcohol-water injection, 1790-kW (2,400-hp) Pratt & Whitney Double Wasp CB 16s. The DC-6A's initial trials were marred by two inflight fires. These, however, were traced to fuel venting into the cabin heater intakes, and as soon as the teething troubles were soothed orders began to come in. The US Navy ordered a total of 65 DC-6As between 1949 and 1951, and the USAF followed suit shortly afterwards. The military versions of the DC-6A, dubbed the C-118A, were capable of carrying 81 fully armed troops with their support equipment. The type proved so resilient that as late as 1962, 40 of the Navy versions were redesignated C-118B and VC-118B – the latter for use by top military brass – and transferred for further use to the USAF.

Slick Airways, an all-cargo civilian operator, ordered the first civilian DC-6As in 1951, and were closely followed by Pan American, which eventually operated a fleet of five. The last, Pan American *Ocean Express*, operated like its fellows on the trans-Atlantic route and came into service with the airline as late as 1958.

Convertible 'Six'

Many smaller operators, civil as well as military, voiced an interest in a 'convertible' version of the DC-6A, and to cater for their needs the DC-6C was brought into being. Although powered by the same type of engines as the DC-6A, and structurally and dimensionally the same, the DC-6C could carry up to 76 passengers, or could be converted to carry a mixed passenger-cargo load, or be stripped down as a pure cargo aircraft to carry nearly 13 tons of freight.

Its principal feature was a moveable bulkhead which could be positioned at four different stations so that the interior arrangement could be adjusted to suit varying passenger-cargo requirements and ratios. The cabin was equipped with lightweight seats which folded up against the wall when not in use, and tourist or coach-type buffet facilities could be removed when not required. Like the DC-6A, the craft's two loading doors swung upwards to allow the operation of twin, self-powered loading elevators, each of which could lift 1820 kg (4,000 lb) from truck-bed height to cabin-floor level, and folded for storage within the aircraft. Seventy-four DC-6C types were built, and like others of the DC-6 series, several were still operating in various parts of the world into the 1990s.

Ultimate 'Six'

The series reached its zenith with the DC-6B, one of the last, finest and most economical piston-engined airliners ever built. Test flown on 2 February 1951, several major airlines used the DC-6B on regularly scheduled passenger routes until 1967 – over a dozen years after the first jet airliners came into regular operation. It remained in production until 1958, during which time 288 were built.

Such had been the success of the series as a whole that orders for

Air Force serial 46-0505 was the sole VC-118, a DC-6 procured as a presidential transport, fitted with 25 seats and 12 bunks and named Independence. *It later joined the general transport fleet.*

Douglas DC-6/DC-7 variants

XC-112A: lengthened and pressurised version of DC-4/C-54. This featured square windows in place of round, and was powered by four Pratt & Whitney R-2800-34 Double Wasp engines, each developing 2,100 hp. Only one (45-873) was built. Length 30.66 m, wing span 35.81 m

YC-112A: projected service test model. Not built

DC-6: first civil version similar to XC-112A with 2,100-hp R-2800-CA15 engines. Seating between 52 and 86 passengers depending on configuration, the DC-6 was delivered between 1946 and 1951. American, United and Braniff were the main recipients. 175 built

DC-6A: strengthened and lengthened civil version with 2,400-hp water-injection R-2800-CB16 Double Wasps. Most were built as cargo-haulers with two large upward-hinging cargo doors fore and aft, a strengthened floor and no windows. Several were later converted for passenger use. Length 32.18 m

DC-6B: first flown on 2 February 1951, the DC-6B was a passenger-only version of the DC-6A with windows and a lighter structure, and lacking the cargo door. The major production version, 288 were delivered between 1951 and 1958. Several were later fitted with a large underbelly tank to fight forest fires

DC-6BF: designation applied to DC-6Bs subsequently converted to freighter configuration, often without cabin windows. Several still serve in Latin America on freight duties. Two DC-6Bs were modified to freight configuration by SABENA with swing-tails for the easy loading of bulky cargo

DC-6C: version of DC-6A with strengthening but completed as quick-change passenger/freight aircraft with windows. A moveable bulkhead could be positioned in any of four stations for mixed passenger-cargo operations

R6D-1: military version of DC-6A for the US Navy. Powered by 2,500-hp R-2800-52W, a total of 65 was delivered. Forty were loaned to the USAF as C-118A. Serial blocks were BuNos 128423/128433 and 131567/131620

R6D-1Z: four R6D-1s (128433, 131572, 131575 and 131597) converted for staff transport duties

C-118A: military version of DC-6A for US Air Force. Powered by four R-2800-52W engines, the C-118A could accommodate 74 troops, 60 stretchers or 12247 kg of cargo. For troop-carrying flights rearward-facing seats were fitted, although these were later changed to forward-facing. One-hundred-and-one were built (51-3818/3835 and 53-3223/3305) and 40 US Navy R6D-1s (50-1843/1844, 51-17626/17661 and 51-17667/17668) were loaned to the USAF. Two further ex-civil DC-6Bs (65-12816 and 66-14467) were procured as C-118As for sale to the Chilean air force. Four aircraft (53-3228, 53-3255, 53-3256 and 53-3290) transferred to US Navy, the latter trio receiving BuNos 152687/152689

MC-118A: little-used designation for C-118A configured for aeromedical evacuation duties

VC-118: single DC-6 (46-505) procured to serve as presidential transport for Harry S. Truman, named *Independence*. Internally configured with 25 seats, 12 bunks and an executive stateroom. Currently on display at Pima County museum in Arizona

VC-118A: at least 15 C-118As (51-3826, 51-3829, 53-3229, 53-3234, 53-3239, 53-3240, 53-3260, 53-3262, 53-3269, 53-3274, 53-3287, 53-3294, 53-3300, 53-3303 and 53-3305) converted for staff transport duties with varying interior fits

C-118B: US Navy R6D-1 redesignated in 1962, when all US military designations were rationalised into one system

VC-118B: US Navy R6D-1Z after 1962

DC-118B: single VC-118B (131572) converted for use in directing pilotless drones

DC-7: direct development of DC-6B to incorporate the 3,250-hp Wright R-3350 turbo-compound engine. The fuselage was lengthened to accommodate one extra row of seats, and with the increase in weight the undercarriage was considerably strengthened. 105 were built. Length 33.24 m

DC-7B: minor improvements to DC-7 including extended engine nacelles to accommodate saddle tanks with extra fuel. 112 were built

DC-7C: final new-build variant with uprated 3,400-hp R-3350 engines, increased wing span with taperless centre-section and a further row of seats in an extended fuselage, raising accommodation to 105. To cater for the longer wings, the vertical tail was enlarged and given a square profile. These improvements gave the DC-7C true intercontinental range, for which it earned the popular name 'Seven Seas'. De-icing equipment and advanced weather radar were installed, along with other new avionics systems. 120 were built. Length 34.22 m, wing span 38.86 m

DC-7D: proposed DC-7C development powered by Rolls-Royce Tyne turboprops of 5,730 shp each. The programme was halted in favour of the jet-powered DC-8

DC-7F: factory freighter conversion of any of the DC-7 passenger series, known as 'Speedfreighter'. Modifications included the fitting of cargo floor (with optional rollers), strengthened fuselage and fore and aft loading doors. Some retained the ability to be quickly reconfigured as 99-seat passenger carriers

When the passenger-carrying days were over, many DC-6s were snapped up by cargo operators, many of whom had their aircraft converted to windowless cargo configuration. Zantop flew this aircraft from its Detroit base, where it now has a large fleet of Electras and Convair 540s.

the DC-6B flowed in as early as 1950, before the prototype was off the factory floor. United was the first airline to take delivery on 11 April 1951, closely followed by American on 29 April. Within the year, five more of the United State's major operators – Pan American Grace Airlines (Panagra), Braniff, National, Delta and Northwestern – had the aircraft among their complement. Between February 1952 and June 1954, the main Pan American company was to re-equip its entire fleet with the type, buying 45 for trans- and intercontinental work.

Revised interior

Although the overall length of the DC-6B fuselage – 32.20 m (105 ft 7 in) – was the same as that of the all-cargo DC-6A, the former's cabin layout offered approximately seven per cent greater payload capacity and 14 per cent greater passenger capacity than the standard DC-6, with only four per cent greater operating costs. The standard domestic dayplane version carried 64 passengers, eight more than the standard DC-6, while the trans-oceanic model with larger galleys, coatrooms and toilet compartments carried 54 passengers. A high-density 92-passenger version with an air-coach type interior for short-haul journeys was also available.

The DC-6B was available with either the 'CB17' or 'CB16' powerplants. The former consisted of four Pratt & Whitney Double-Wasp R-2800-CB17 18-cylinder radial air-cooled engines, each developing 1417 kW (1,900 hp) at maximum continuous cruise, and with 1865 kW (2,500 hp) available for take-off with alcohol-water injection. Each engine had either a Hamilton Standard or a Curtiss Electric full-feathering and reversible aircrew. Standard fuel capacity was 15111 litres (3,992 US gal), with optional capacities of 20550 litres (5,406 US gal) and 20918 litres (5,512 US gal). The maximum take-off weight was 48125 kg (106,000 lb). The aircraft had a maximum speed of 576 km/h (360 mph) at 5520 m (18,100 ft), with a cruising speed of 494 km/h (307 mph) at 7390 m (24,400 ft) and landed at 149 km/h (93 mph) using the reversible airscrews, in a distance of 686 m (2,250 ft) from a height of 15.25 m (50 ft). The empty weight of the standard DC-6B was 24458 kg (54,148 lb) and maximum landing weight was 40043 kg (88,200 lb).

Tourist-class services

The DC-6Bs supplied to Pan American World Airways were specially modified for use on the line's tourist-class services, and the 45 aircraft involved were dubbed 'Super Sixes'. Certificated for a gross take-off weight of 48,580 kg (107,000 lb) – 454 kg (1,000 lb) more than the standard DC-6B – they carried a greater payload, had an increased fuel capacity and an improved performance, cruising at a steady 500 km/h (315 mph) over long distances. The advantages made possible the first all-tourist service on the London-New York

Representing the pinnacle of airliner design with reciprocating engines, the aptly-named DC-7C was a true globe-spanner. However, its success was short-lived, for the long-range jets were soon in service.

101 C-118As were procured for the USAF, and flown mostly on trooping, staff transport or medical evacuation work. For the last role, the aircraft were designated MC-118A and painted with a large red cross.

route, dubbed the 'Rainbow' service by Pan Am, and inaugurated on 1 May 1952 by the 'Super Six' *Clipper Liberty Bell*.

Liberty Bell was fitted with 82 seats, although a more typical tourist arrangement was 88; for almost a decade Pan Am's 'Super Six' clipper service remained a leader of the Atlantic passenger trade.

Local modifications

Although overtaken successively by turboprop and pure jet designs, the Douglas DC-6 series continued in use worldwide, often adopted to local needs. SABENA, for instance, modified two DC-6Bs, giving them swing-tails to simplify the direct loading of bulk cargo. And in the United States, a number of DC-6Bs were provided with a 61356-litre (3,000-US gal) capacity underfuselage tank to carry fire-retardant chemicals. These saw considerable service in the USA and Canada during periods of high fire risk in national timberlands.

Successful as the DC-6 series had been, competition pressed hard on its heels. In 1952 Lockheed announced its L.1049C Super Constellation driven by Wright Turbo-Compound engines, and in October 1953 TWA introduced the first non-stop transcontinental service using the 'Super Connie'.

One month later, American Airlines responded with the DC-7, matching the non-stop capability. The giant Wright R-3550-18EA-1 turbo-compound radial piston engine, delivering up to 2760 kW (3,700-hp) was the key to the DC-7, conceived jointly by Donald Douglas and C. R. Smith of American Airlines. Each of these engines had three exhaust-driven turbines disposed equally around its circumference, given a power output some 22 per cent greater than ordinary piston petrol engines.

More fuselage, bigger engines

Smith produced up-front funding for 25 new airliners each powered by four of the big Wrights, and again Douglas's design team set to work, though initially the DC-7 was yet another variation of the DC-6B. However, the fuselage was again lengthened by 1.02 m

The DC-6 had a long career with the US Navy, a total of 65 R6D-1s being delivered (redesignated C-118B in 1962). This aircraft is an R6D-1Z (VC-118B) configured for staff transport duties.

Douglas DC7
'Clipper Bald Eagle'
Pan American Airways

(3 ft 4 in) to allow for the inclusion of an additional row of seats. Installation of the turbo-compound engines made possible an increase in gross weight of 6895 kg (15,200 lb) and to accommodate this a strengthened undercarriage was fitted. A total of 105 DC-7s was built.

The 112 DC-7B aircraft which followed had only minor changes. These included an extension of the engine nacelles further aft to permit the installation of saddle tanks in the rear of the nacelles, the whole unit being constructed of the new metal titanium. But even with this new fuel capacity, those operators who used the aircraft for transatlantic crossings – including Pan American, which inaugurated non-stop London-New York services with the DC-7B on 13 June 1955 – soon discovered that the fuel capacity was marginal. In fact, flying against the prevailing westerly headwinds with a full load, New York-bound DC-7Bs frequently had to divert for a refuelling stop. At best this was unsatisfactory and at worst potentially disastrous.

More range required

Furthermore, Pan American's aim of making a non-stop Pacific crossing was quite out of the question with the type, and fuelling arrangements had to be made at Hawaii and Wake Island on the flight from California to Tokyo. The designers at Douglas's Santa Monica plant went to work again, and the result was the DC-7C.

From the DC-4 onwards, all Douglas four-engined aircraft had flown on the one, 35.81-m (117-ft 6-in) wing, but greater fuel capacity meant a greater wingspan. Two parallel-chord wing sections were inserted inboard between the fuselage and each inboard engine nacelle. This had the extra advantage of distancing the high noise and vibration levels of the compound-turbo engine from the cabin environment. During the development of the DC-7C, Curtiss-Wright was able to offer a further increase in engine power, which meant that the fuselage could be lengthened yet again by the insertion of a 1.02-m (3-ft 4-in) plug. Accommodation could thus be provided for up to 105 passengers.

The final version of the DC-7C was powered by four 2535-kW (3,400-hp) Wright R-3350-18EA-1 turbo-compound radial piston engines. Maximum speed was 653 km/h (406 mph) at 6615 m (21,700 ft), with a normal cruising speed of 571 km/h (355 mph). Service ceiling was 6615 m (21,700 ft) and the range, with maximum payload, was 7411 km (4,605 miles). Empty weight was 33005 kg (72,763 lb) with maximum take-off weight of 64864 kg (143,000 lb). Span was 38.86 m (127 ft 6 in); length, 34.21 m (112 ft 3 in); height 9.70 m (31 ft 10 in) and wing-area 152.08 m^2 (1,637 sq ft).

The ultimate propliner

Production of DC-7s totalled 120. The '-7Cs' suffix was naturally and popularly corrupted to 'Seven Seas', which was an apt nickname for an aircraft which could now take both the Atlantic and Pacific in its stride. On 11 September 1957, Pan American used the 'Seven Seas' to inaugurate its 'Great Circle' polar route from New York to Seattle.

An improved DC-7D was planned, to be powered by the 4273-eKW (5,730-eshp) Rolls-Royce Tyne turboprop engine, but the emergence of the Boeing 707 and the Douglas company's own purpose-built DC-8 jetliners meant that this remained an unfulfilled project. It was the end of Douglas's line of airscrew-powered long-range transports, and sadly, although some DC-7 series aircraft were taken on by second-rank operators, many were sold for scrap. But it remains a salient fact that, although only two per cent of the population of the USA had ever flown in an aeroplane as late as 1960, some 96 per cent of that group had flown in a Douglas propeller-driven aircraft.

A handful of DC-6s were converted to the firebombing role with a large ventral pannier containing the retardant tanks. France's Sécurité Civile organisation has two fire-fighting DC-6Bs based at Marseille.

Douglas DC-8

Overshadowed by its rival, the Boeing 707, Douglas' first jetliner nevertheless achieved its own brand of success. Whereas Boeing looked to the entirely new and much bigger 747 to keep their place in the market, Douglas opted for applying a massive stretch to the DC-8 with the Super Sixty series, ensuring the aircraft was kept in production into the early 1970s. Although none have been built since, they are still popular.

The designation DC-8 was first applied to a jet transport design in 1952, when Douglas produced a mock-up of a four-jet airliner designed for US transcontinental routes. Its appeal to the airlines, at a time when the US economy was still on a war footing, was limited, and the company dropped the study in the following year. In early 1955, however, interest in jets began to increase. The de Havilland company and the British authorities announced that they had identified the cause of the Comet crashes of 1953-4, and that the new long-range Comet 4 would be in service by 1958 on the North Atlantic. In March, the US Air Force ordered the first batch of military tanker-transport aircraft based on the Boeing 367-80 prototype, ensuring that US jet transport technology would be thoroughly tested before the airlines started operations. Pan American, which would be most directly affected by British competition on the Atlantic, was the first US airline in the jet market. Boeing quickly offered a design based on the 367-80, for delivery in 1957.

Douglas, with no prototype in the air, could not match the delivery date. Studying the market, however, Douglas saw that the 86183-kg (190,000-lb) Boeing was smaller than would be ideal. More powerful engines were under development for military aircraft, and would be available for civil use by the end of the decade. With such engines, it would be possible for a jet to fly the North Atlantic without landing to refuel at Gander, Newfoundland, as the Comet 4 and early Model 707 would have to do. The bigger aircraft could also have a slightly wider cabin, to seat six passengers abreast. This was the thinking behind the new DC-8 design, presented to Pan Am in June 1955.

The DC-8 was clearly superior to Boeing's proposal, and the Seattle company had to respond with a scaled-up, heavier Model 707, which would not be available until 1958. In October 1955, after months of intense negotiations, Pan Am gave a judgement of Solomon, ordering 20 Model 707s and 25 DC-8s. The Model 707s were 'one-stop' transatlantic aircraft, needed to meet British competition in late 1958; the DC-8s were intercontinental models with the new Pratt & Whitney JT4A engine (the civil version of the J75) and would be in service in the following year. Just 12 days later, United Airlines placed the first jet order from a US domestic airline, for 22 lighter, lower-powered DC-8 variants. United cited the fractionally wider cabin of the DC-8 as one of the main reasons for its decision, and Boeing responded by increasing the fuselage diameter of all Model 707s.

As the development of its DC-8 advanced, Douglas adhered rigorously to a philosophy of building a single basic airframe with minimal changes to suit different roles. While Boeing, eventually, produced what were virtually three different aircraft (the original Model 707-120, the Intercontinental Model 707-320 and the medium-haul Model 720), all the Douglas aircraft built before 1967 were identical in exterior shape, had identical flight controls, and had the same electrical, hydraulic and pressurisation systems. Differences were confined to the engines, the fuel systems and the structure. The longer-range versions were heavier, and thicker gauges and stronger materials were used in heavily loaded parts of the airframe such as the upper and lower wing skins, the rear fuselage, the tailplane and the landing gear.

The DC-8 was an advanced aircraft for its day, one of the first large, high-subsonic-speed aircraft to be designed to commercial standards of safety and reliability. The basis of its design was the wing. Like that of the Model 367-80, it was swept, carried the four engines in underwing pods, and was markedly tapered in chord and thickness, providing plenty of room for uncomplicated wide-track landing gear. But Douglas lacked Boeing's unique experience with this layout, and adopted a more conservative wing design. Thus the DC-8's wing was less sharply swept than that of the Model 707, and the leading edge carried no full-span high-lift devices. Instead, two pairs of partial-span slots were built into the wing, inboard of each engine. Because the basic leading-edge shape had to perform at all speeds, rather than being modified by a mechanical device for landing and take-off, its cruise efficiency was inevitably compromised. The aileron system was rather different, too. Instead of a separate inboard aileron for high speeds, the Douglas wing featured a two-section outboard aileron. The inboard section was powered directly by the aircraft hydraulic systems, and was linked to the outer, low-speed section by a torsion-bar linkage. At high speeds, the air loads

The first in a successful airliner family, the prototype DC-8 in full Douglas company livery reveals the less sharply-swept wing in comparison to the Boeing 707. Due to underpowered engines, only 28 of these DC-8-10s were sold.

Through its own operations and those of its charter and inclusive tour subsidiary Scanair, Scandinavian Airlines System was a regular operator of the DC-8. Illustrated is OY-KTN 'Viking', a Series 32 aircraft delivered in July 1960.

overcame the torsion linkage and the outer aileron ceased to operate, limiting the twisting loads on the wing. A notable feature of the wing was that it was entirely free of aerodynamic fences, notches and vortex generators; in this respect it was unique among jetliner wings until the appearance of the Airbus A310 in 1982.

Competition with the 707

The DC-8 was slightly less expensive than the Model 707, and slightly slower, although the speed difference was barely noticeable even on long trips. (Into the 1970s, in fact, the DC-8 was to remain cheaper in the second-hand market than the Model 707.) Both types established similar records in service: after an early flurry of accidents, mostly linked to the vastly greater complexity and performance of the jets, they proved to be safe and reliable aircraft. The industry's ultimate verdict on the two types was that the Boeing was somewhat more efficient, but the structure of the Douglas required much less attention in later life.

Douglas was handicapped, though, by its late start in the jet business. The first DC-8 to fly was the prototype of the transcontinental variant ordered by United; the DC-8-10, which made its first flight in May 1958. It entered service in September 1959. Powered by 6123-kg (13,500-lb) thrust Pratt & Whitney JT3C-6 engines (the same as those used on the smaller Model 707-120), the DC-8-10 was somewhat underpowered; only 28 were sold, and most were later fitted with more powerful engines. By that time, Boeing had not only put the Model 707-120 into service, but had also developed and delivered the Intercontinental Model 707-320, which could carry more passengers than the DC-8, and the Model 707 generally outsold the DC-8 from that time onwards.

The second version of the Douglas type was the DC-8-20, a developed domestic aircraft with the more powerful 7167-kg (15,800-lb) JT4A-3 engine, certificated in early 1960. The intercontinental DC-8 earned its ticket soon afterwards, in two versions, entering service in April 1960. The original DC-8-30 used a higher-rated, water-injected version of the JT4A, while the DC-8-40 was powered by the Rolls-Royce Conway low-bypass-ratio turbofan, which offered better performance without the complication of a water-injection system. Air Canada, Alitalia and Canadian Pacific Airways ordered DC-8-40s.

By the time these versions entered service, their successor was under development. Pratt & Whitney had started development of its own higher-bypass-ratio turbofan engine in early 1958 (after Boeing threatened to use the General Electric CJ805 in the Model 720) and this had matured into the JT3D. More effective in basic design than the Conway, with a larger fan and higher bypass ratio, the JT3D was more efficient than any other commercial engine of its day. It was the powerplant of the DC-8-50, which made its first flight in December 1960 and was certificated six months later. The increased power and higher efficiency of the JT3D gave the DC-8-50 the ability to fly non-stop from the US west coast to Europe, and the type quickly re-

As a result of surpassing the speed of sound in August 1961, this company-owned DC-8 was adorned with a suitable proclamation on the vertical tail surfaces. The aircraft later saw service with KLM and PAL.

placed all earlier models in production. Aerodynamic refinements, including a low-drag leading edge, and improved JT3D-3B engines produced the improved DC-8-55 model, introduced early in production.

A DC-8-40, fitted with the new leading edge, was used for a remarkable record flight in August 1961. It was flown to 15877 m (52,090 ft), carrying ballast equivalent to its full payload, and was dived at a shallow angle. At 12300 m (40,350 ft), still diving, the DC-8 recorded a Mach number of 1.012, becoming the first transport-type aircraft to exceed the speed of sound.

In January 1963 Douglas delivered the first of a series of freighter and convertible versions of the DC-8. The basic DC-8F Jet Trader, also designated DC-8-54, was a convertible cargo/passenger version of the Series 50 aircraft, with a cargo door, a system of rollers, guides and tie-downs built into the floor, and an increased landing weight. It was superseded by the DC-8-55F, with the same improvements as the passenger-carrying DC-8-55. In the late 1970s, McDonnell Douglas converted many older DC-8s into freighters at its Tulsa facility.

Air traffic congestion

The Model 707 and DC-8 stimulated an explosion in the growth of air traffic; so much so, in fact, that by the mid-1960s the problem of congestion on the airways and at airports was beginning to become

McDonnell Douglas/Cammacorp DC-8 Super 71 cutaway drawing key

1. Radome
2. Weather radar scanner
3. Air conditioning system ram air intake
4. Pitot tubes
5. Air conditioning units, port and starboard
6. Front pressure bulkhead
7. Rudder pedals
8. Instrument panel
9. Windscreen rain dispersal air ducts
10. Windscreen panels
11. Overhead systems switch panels
12. Co-pilots's seat
13. Cockpit eyebrow window
14. Pilot's seat
15. Cockpit floor level
16. Air system heat exchanger exhausts
17. Nosewheel door, closed after cycling of undercarriage
18. Landing/taxiing lamps
19. Nosewheel steering jacks
20. Twin nosewheels, forward retracting
21. Nosewheel leg door

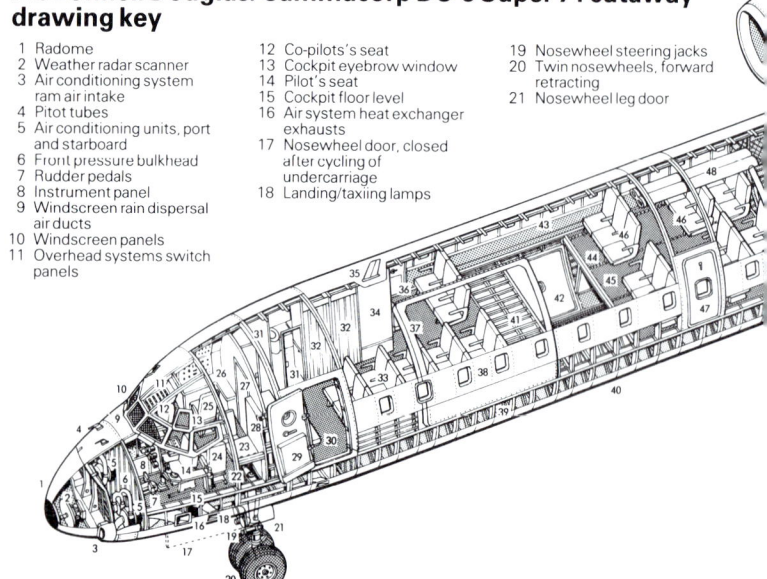

424

With the Series 50 models, Douglas made the DC-8F Jet Trader available in the AF (all cargo) or CF (convertible passenger/cargo) configurations. Shown here is a DC-8-55CF in the colours of Seaboard World Airlines, a major cargo operator which also undertook charter flights for MAC.

22 Nose undercarriage leg pivot point
23 Navigator's station
24 Supernumerary crew seat
25 Flight engineer's station
26 Engineer's instrument panels
27 Cockpit bulkhead
28 Avionics equipment racks
29 Forward entry door, open
30 Cabin attendant's folding seat
31 Toilet compartments (two)
32 Wardrobes (two)
33 Forward cabin seating
34 Galley unit
35 VHF aerial
36 Starboard side service door/emergency exit
37 Outline of freight door (Super 71CF variant) 85 in by 140 in (216 cm by 356 cm)
38 Cabin window panels
39 Fuselage lower lobe frame construction
40 Underfloor freight compartment, 1,290 cu ft (36.53 m³)
41 Floor beam construction
42 Starboard side freight door, 54 in by 63 in (137 cm by 160 cm)
43 Cabin wall trim panelling
44 Seat mounting rails
45 Main cabin floor panelling
46 Six-abreast passenger seating, 251-seat single class layout
47 Forward cabin emergency exit doors, port and starboard
48 Cabin air distribution ducting
49 Bulk cargo door, 36 in by 44 in (91 cm by 112 cm)
50 Cabin wall soundproof lining
51 Optional auxiliary power unit (APU) installation, port and starboard
52 Fuselage skin panelling
53 Fuselage frame and stringer construction
54 Wing inspection light
55 APU exhaust
56 Runway turn-off lamp
57 Forward cargo bay pressure bulkhead
58 Recirculated air ducting
59 Wing attachment fuselage main frames
60 Wing root joint strap
61 Overwing emergency exit hatches, port and starboard
62 Wing centre section construction
63 Pressure floor above wing carry-through
64 Fuselage centre section construction
65 Fuel system piping
66 Variable leading edge slot operating mechanism
67 Starboard wing integral fuel tanks; total fuel system capacity 23,392 US gal (88548 litres)
68 Pressure refuelling connections
69 Starboard engine nacelles
70 Nacelle pylons
71 Outboard variable leading edge slot mechanism
72 Tank dividing rib
73 Outboard wing integral fuel tank
74 Fuel vent piping
75 Wing tip vent tank
76 Starboard navigation lights
77 Wing tip fairing
78 Static dischargers
79 Fixed portion of trailing edge
80 Outboard aileron
81 Aileron tabs
82 Aileron hydraulic actuator
83 Flap operating linkages
84 Starboard spoilers, open
85 Double-slotted Fowler-type flaps, down position
86 Flap exhaust gate
87 Inboard double-slotted flap segment
88 Pressure floor above wheel bay
89 Wheel bay dividing keel member
90 Main undercarriage wheel bay
91 Rear cargo hold pressure bulkhead
92 Overhead stowage bins
93 Cabin ceiling lighting panels
94 Cabin air distribution ducting
95 Aft freight hold door, 54 in by 56 in (137 cm by 142 cm)
96 Air delivery ducts to cabin sidewalls
97 Overhead passenger service units
98 Anti-collision light
99 Aft bulk cargo door, 36 in by 44 in (91 cm by 112 cm)
100 Rear cabin air ducting
101 Rear cabin passenger seating
102 Cabin aft bulkhead
103 Starboard side galley unit
104 Toilet compartment water tank
105 Fin root fillet
106 Leading edge de-icing air duct
107 Tailfin construction
108 Starboard trimming tailplane
109 Starboard elevator
110 VOR aerial
111 HF aerial
112 Fin tip VHF aerial
113 Static dischargers
114 Glass-fibre aerial isolating strip
115 Rudder rib construction
116 Rudder tab
117 Tailcone
118 Elevator hinge control
119 Elevator tabs
120 Port elevator rib construction
121 Port trimming tailplane rib construction
122 Corrugated leading-edge skin doubler
123 Leading edge de-icing air duct
124 Tailplane sealing plates
125 Rudder hydraulic actuator
126 Trimming tailplane screw jack
127 Screw jack motor
128 Tail bumper
129 Sloping fin attachment bulkheads
130 Rear pressure bulkhead
131 Aft toilet compartments (three)
132 Wardrobe
133 Port galley unit
134 Cabin attendant's folding seat
135 Rear entry door, open
136 Rear cabin window panels
137 Lower fuselage skin panelling
138 Fuselage lower lobe frame construction
139 Rear cabin emergency exit doors, port and starboard
140 Rear underfloor cargo hold, 1,270 cu ft (35.97 m³)
141 Wing trailing edge root fillet
142 Inboard double-slotted flap segment
143 Inboard spoilers
144 Main undercarriage hydraulic retraction jack
145 Retraction spring strut
146 Main undercarriage pivot mounting
147 Flap shroud ribs
148 Outboard double-slotted flap segment
149 Flap exhaust gate
150 Flap rib construction
151 Aileron tabs
152 Inboard aileron rib construction
153 Port outboard aileron
154 Fixed portion of trailing edge
155 Static dischargers
156 Wing tip fairing
157 Port navigation lights
158 Hot air leading edge de-icing
159 Wing rib construction
160 Rear spar
161 Wing bottom skin/stringer panel
162 Fuel tank ventral access panels
163 Port wing integral fuel tank
164 Leading edge nose ribs
165 Outboard pylon mounting ribs
166 Pylon attachment joint
167 Nacelle pylon two-spar torsion box construction
168 Engine bleed air pre-cooler
169 Vented exhaust tailcone
170 Core engine (hot stream) exhaust
171 Fan air (cold stream) exhaust
172 CFM International CFM56-2-1C turbofan
173 Main engine mounting
174 Fan casing
175 Engine oil tank
176 Ventral accessory equipment gearbox
177 Intake duct acoustic lining
178 Intake lip hot air de-icing
179 Nacelle strake
180 Engine bleed air ducting
181 Port outboard variable leading edge slot
182 Leading edge de-icing air duct
183 Three-spar torsion box wing construction
184 Wing stringers
185 Wing skin panelling
186 Engine fire suppression bottles
187 Reverse thrust translating cowl, open
188 Reverser cascades
189 Detachable engine cowling panels
190 Inboard nacelle pylon
191 Inboard variable leading edge slot
192 Nacelle pylon attachment joint
193 Pylon mounting double ribs
194 Four-wheel main undercarriage bogie, inward retracting
195 Main undercarriage leg strut
196 Inboard wing integral fuel tanks
197 Wing centre spar
198 Inboard wing ribs
199 Overwing fuel filler cap
200 Front spar
201 Fuel system piping

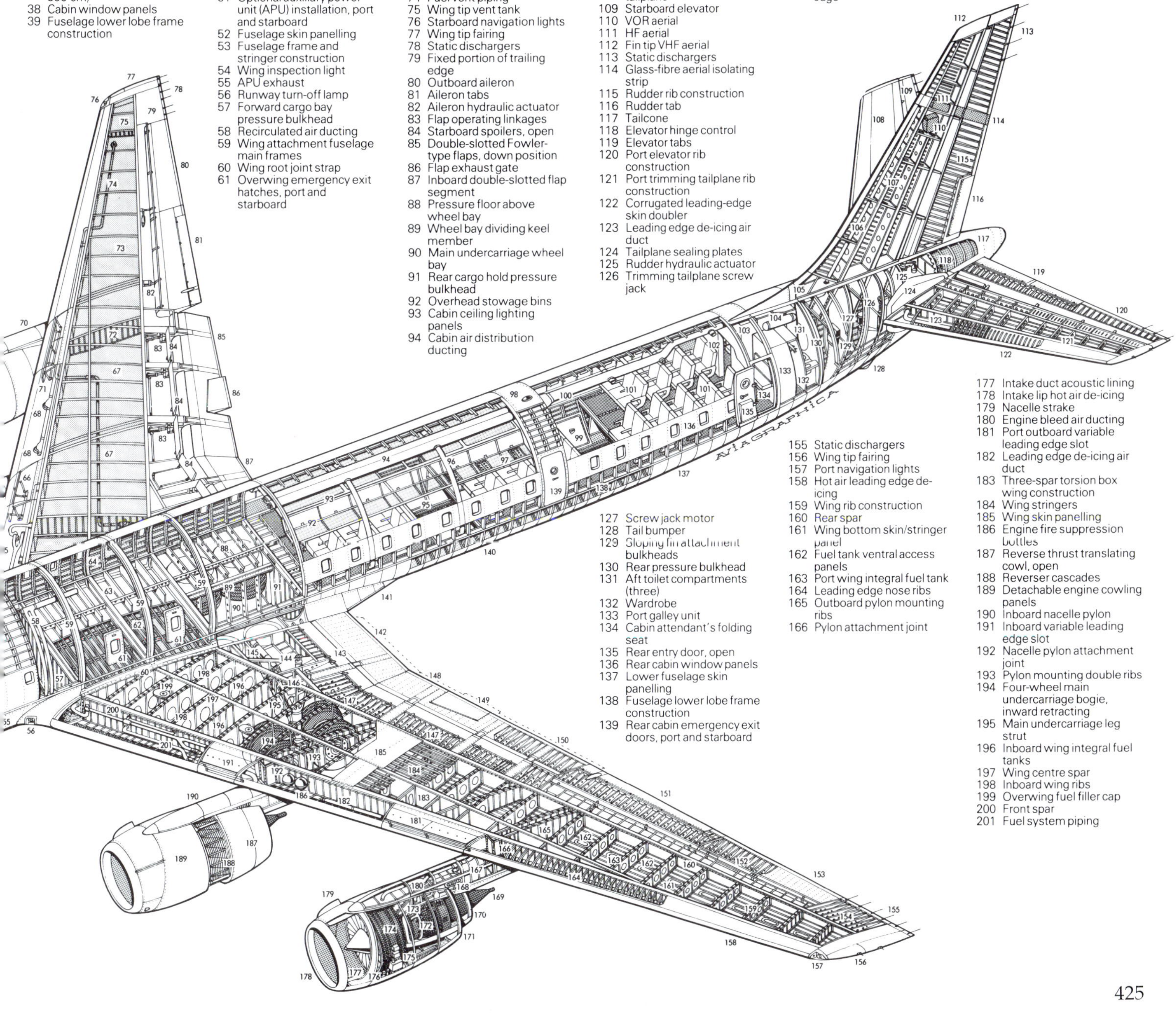

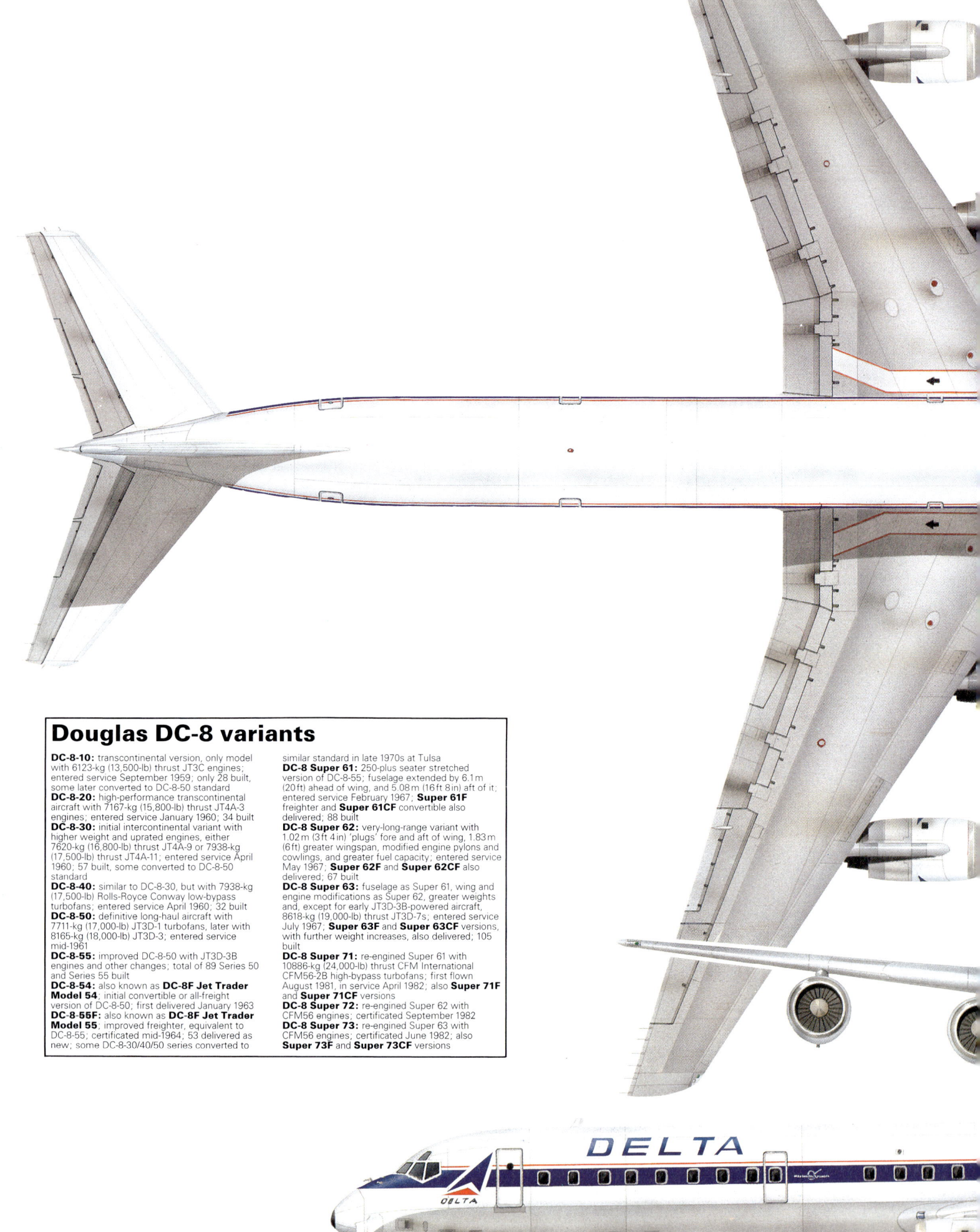

Douglas DC-8 variants

DC-8-10: transcontinental version, only model with 6123-kg (13,500-lb) thrust JT3C engines; entered service September 1959; only 28 built, some later converted to DC-8-50 standard
DC-8-20: high-performance transcontinental aircraft with 7167-kg (15,800-lb) thrust JT4A-3 engines; entered service January 1960; 34 built
DC-8-30: initial intercontinental variant with higher weight and uprated engines, either 7620-kg (16,800-lb) thrust JT4A-9 or 7938-kg (17,500-lb) thrust JT4A-11; entered service April 1960; 57 built, some converted to DC-8-50 standard
DC-8-40: similar to DC-8-30, but with 7938-kg (17,500-lb) Rolls-Royce Conway low-bypass turbofans; entered service April 1960; 32 built
DC-8-50: definitive long-haul aircraft with 7711-kg (17,000-lb) JT3D-1 turbofans, later with 8165-kg (18,000-lb) JT3D-3; entered service mid-1961
DC-8-55: improved DC-8-50 with JT3D-3B engines and other changes; total of 89 Series 50 and Series 55 built
DC-8-54: also known as **DC-8F Jet Trader Model 54**; initial convertible or all-freight version of DC-8-50; first delivered January 1963
DC-8-55F: also known as **DC-8F Jet Trader Model 55**; improved freighter, equivalent to DC-8-55; certificated mid-1964; 53 delivered as new; some DC-8-30/40/50 series converted to similar standard in late 1970s at Tulsa
DC-8 Super 61: 250-plus seater stretched version of DC-8-55; fuselage extended by 6.1 m (20 ft) ahead of wing, and 5.08 m (16 ft 8 in) aft of it; entered service February 1967; **Super 61F** freighter and **Super 61CF** convertible also delivered; 88 built
DC-8 Super 62: very-long-range variant with 1.02 m (3 ft 4 in) 'plugs' fore and aft of wing, 1.83 m (6 ft) greater wingspan, modified engine pylons and cowlings, and greater fuel capacity; entered service May 1967; **Super 62F** and **Super 62CF** also delivered; 67 built
DC-8 Super 63: fuselage as Super 61, wing and engine modifications as Super 62, greater weights and, except for early JT3D-3B-powered aircraft, 8618-kg (19,000-lb) thrust JT3D-7s; entered service July 1967; **Super 63F** and **Super 63CF** versions, with further weight increases, also delivered; 105 built
DC-8 Super 71: re-engined Super 61 with 10886-kg (24,000-lb) thrust CFM International CFM56-2B high-bypass turbofans; first flown August 1981, in service April 1982; also **Super 71F** and **Super 71CF** versions
DC-8 Super 72: re-engined Super 62 with CFM56 engines; certificated September 1982
DC-8 Super 73: re-engined Super 63 with CFM56 engines; certificated June 1982; also **Super 73F** and **Super 73CF** versions

Specification
McDonnell Douglas DC-8 Super 71
Type: long-range passenger transport
Powerplant: four CFM International CFM56-2-1C turbofan engines each developing 10887-kg (24,000-lb) thrust
Performance: maximum level speed 965 km/h (600 mph); cruising speed 854 km/h (531 mph) at 10670 m (35,000 ft); initial cruising altitude 10550 m (34,600 ft); take-off field length 2698 m (8,850 ft); range with maximum passenger payload 7485 km (4,650 miles)
Weights: empty 73800 kg (162,700 lb); maximum payload 30240 kg (66,665 lb); fuel weight 71093 kg (156,733 lb); maximum take-off weight 147415 kg (325,000 lb); maximum landing weight 108860 kg (240,000 lb)
Dimensions: span 43.36 m (142 ft 3 in); length 57.22 m (187 ft 5 in); height 12.93 m (42 ft 5 in); wing area 271.92 m² (2,927 sq ft)

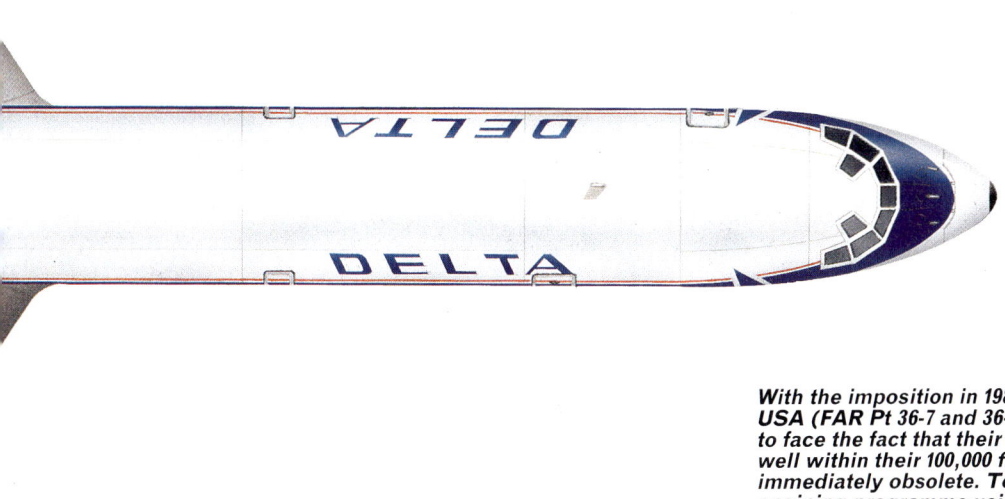

With the imposition in 1985 of new FAA airport noise regulations in the USA (FAR Pt 36-7 and 36-8), the operators of JT3D-powered DC-8s had to face the fact that their aircraft, many only just over 10 years old and well within their 100,000 flight-hours' life, would have been immediately obsolete. To this end, Cammacorp Inc. offered a re-engining programme using the CFM International CFM56-2-1C fanjet for DC-8 Series 60s. The performance figures with the new powerplants have proved very impressive: 60 to 70 per cent true noise reduction; 25 per cent reduction in fuel burn and pollutant levels below environmental requirements; an increase in range up to 10139 km (6,300 miles); and 17 per cent more power. These and other statistics led several operators to have their aircraft modified, including Delta Air Lines, whose first DC-8 Series 71 is illustrated. Delta has re-engined 13 aircraft, with those from the second onwards having the work carried out at the Delta Technical Operations Centre as part of a major overhaul, repaint and wing re-skinning programme. The end result is an aircraft that can continue to operate for many years to come, can fly into more airports, and which offers effective performance and economic operating costs to large and small airlines alike.

Douglas DC-8
Alitalia

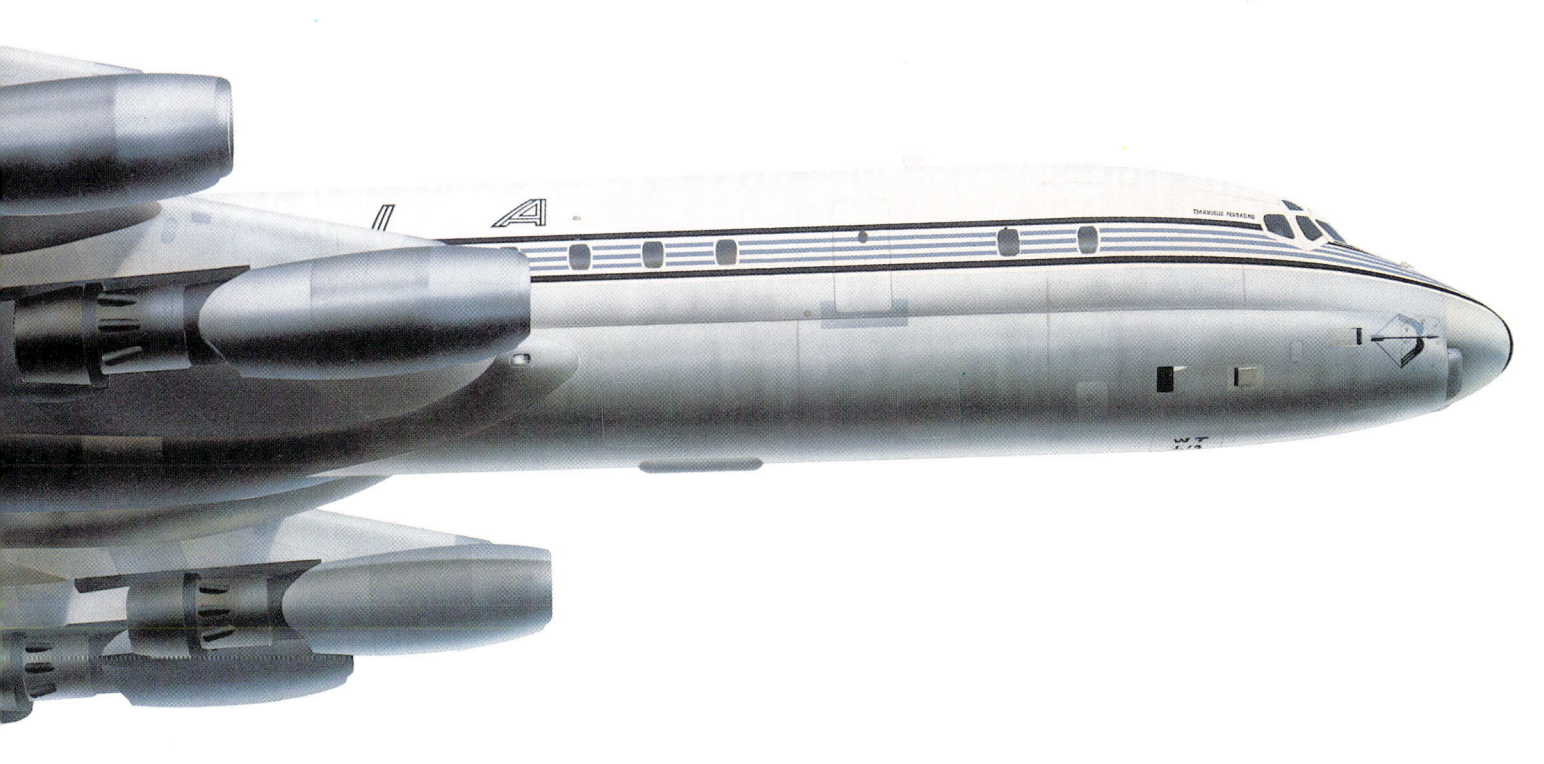

One of the few military operators of the DC-8 is the French air force (others being the Thai air force and US Navy), which operates DC-8-72s and DC-8-55Fs (illustrated) on transport duties. One DC-8 has been modified to SARIGUE standard with a comprehensive electronic surveillance suite.

Surely one of the most striking paint schemes ever to appear on an aircraft, this was the work of the artist Alexander Calder. The aircraft, a Braniff Airways DC-8-62, was hand-painted and made its public debut at the Paris Air Salon of 1973, with Calder reputedly finishing the designs on the engine nacelles hours before its unveiling.

serious. While the first jets had seemed both large and expensive at the time they were introduced, the airlines were now ready to look at larger aircraft. The DC-8, Douglas believed, could be 'stretched' into a large-capacity vehicle for the US transcontinental markets. The company was also conscious of the fact that the existing DC-8 was still less efficient than the competing Model 707-320B, despite the improvements incorporated in the DC-8-55. Thus in April 1965 Douglas announced three new DC-8 variants. The Super 61 was a high-capacity transcontinental aircraft with the same wing and engines as the DC-8-50. This operated at the same weights (apart from the maximum landing weight, which was the same as that of a DC-8F). Its fuselage was stretched by no less than 11.48m (37ft 8in), increasing passenger capacity by 45 per cent. With only a modest increase in operating costs, the type offered vastly better economics than any smaller aircraft, and was ordered by some of the major US domestic carriers.

Second of the new versions was the Super 62. Only slightly longer than the standard aircraft, it was stretched just enough to match the maximum seating capacity of the Model 707-320. But the engine installation was completely redesigned. On the original DC-8 and Model 707, the engine pylons had wrapped around the leading edge and over the top surface of the wing, but it had soon been realised that this incurred a disproportionately large penalty in cruise drag. The Super 62 featured cut-back pylons which did not disturb the line of the leading edge. Also, the engines were housed in full-length cowlings, replacing the distinctive half-cowls of the early JT3D installation. The modification improved the mixing between the fan and core exhausts, improved engine efficiency in the cruise and reduced drag. Finally, the Super 62 was heavier than the DC-8-50, and carried more fuel. The result was the longest range of any airliners in existence. The Super 62 solved a lot of problems for many airlines whose route systems were biased towards greater-than-transatlantic sector lengths.

The third of the new series, logically enough, combined the DC-8-61 fuselage with the Super 62 wing and uprated engines. The Super 63 was heavier, once again, and offered full intercontinental range and excellent economy. All three subtypes were available as convertible aircraft, or as windowless pure freighters, and the Super 63CF featured a further weight increase.

Super 60 development continued throughout the financial crisis that forced Douglas into takeover by McDonnell Aircraft. The first Super 61 flew in March 1966, and all three passenger versions were in service by mid-1967, the Super 63CF following in June 1968. The new aircraft proved more successful than their predecessors. Between 1967 and 1972, Douglas delivered almost as many DC-8s as it had done in the previous eight years, all of them Super 60s.

The problem was that the new wide-body jets were on their way, and sales of the Super 60s began to slump by 1970. Moreover, McDonnell Douglas was developing its own wide-body, the DC-10, and was faced with a difficult decision. Should it do as Lockheed was doing, and build a new production plant for the wide-body, or should it drop the slow-selling DC-8 and use its existing facilities for the new aircraft? The balance was further tipped against the DC-8 by the fact that the company could gain tax benefits from writing off remaining development costs if production stopped. Finally, the company concluded that the DC-8 could be kept in production if the sales teams could gather enough orders to justify a new batch of aircraft. Unfortunately, not enough airlines were ready to buy, and the last DC-8 was delivered in May 1972.

Just over a year later, the 1973 Arab-Israeli War and the consequent surge in oil prices drastically slowed the growth of the airline business, and plans which had depended on the exclusive use of wide-bodies began to seem over-ambitious. The Super 60s were by

With 30 aircraft being converted, United Airlines was the biggest single customer for the re-engining programme. Illustrated is the first of their Super 71s, N8093U, which re-entered service in 1982.

The DC-8, trailing the Boeing 707 in development, was launched by an order from Pan Am for 25 aircraft. Ironically these were to operate alongside the 707 on the airline's routes to Europe during the 1960s. Illustrated is the last DC-8, a Series 33, to be delivered to Pan Am.

One of the most consistent of European DC-8 operators was Alitalia, with over 20 years' experience with the type. This is the first aircraft to enter the Alitalia fleet, a DC-8-42 which was delivered on 28 April 1960. This machine was to see 17 years' service with the airline.

far the best of the narrow-body aircraft, particularly in the freight market, and those airlines which had them tended to keep them. Throughout the 1970s, the Super 60s commanded higher prices on the second-hand market than McDonnell Douglas had been paid for them when they were new. Top prices were paid for the Super 63CF; when they changed hands at all, they did so for more than $12 million. By comparison, the market was flooded with low-hour wide-body jets, sometimes knocked down for a barely higher price than the older, smaller Super 60s.

The end of production usually means the end of major development for any aircraft. The Super 60 was to be an exception, partly because of regulations on aircraft noise drafted in the early 1970s. These made it clear that future commercial aircraft would have to be as quiet in relation to their weight as the wide-body jets, and it followed that new quiet engines, smaller than the first generation of big-fan powerplants, would be needed. The first engine in this category to go ahead was the CFM56, developed by General Electric of the USA and SNECMA of France. Development started in 1971, but seven years later the engine had still to find a customer. The Super 60s were still at a premium on the market, and most of the aircraft were relatively 'young', but they would all be grounded by 1985 unless they were made quieter. The CFM56 was the right size, and would not only reduce noise but also improve performance and reduce fuel consumption. (The Pratt & Whitney JT8D-200, put forward as an alternative, would have been cheaper but offered less dramatic performance gains.) Late in 1978, GE and SNECMA (now joined in CFM International) formed a new company, named Cammacorp, to organise and market a re-engining programme for the Super 60 series.

Photographed in 1968 shortly after delivery is a Philippine Air Lines DC-8-53, one of eight DC-8s used by this operator. Acquisition of the type allowed PAL to resume operations to the west coast of the USA.

The re-engined aircraft are redesignated as Super 70s. Detail design and flight testing were carried out by McDonnell Douglas, under subcontract to Cammacorp, and the new nacelles were designed and built by Grumman. Some aircraft were converted by McDonnell Douglas at Tulsa, but some airlines have done the work themselves, using kits supplied by Cammacorp. All aircraft have Super 62-type pylons and redesigned pressurisation systems. Options offered by Cammacorp included a Garrett auxiliary power unit, extra fuel in under-floor tanks and (from 1985) a modern flight-deck with cathode-ray-tube (CRT) displays. The first Super 71 flew in August 1981; a Super 73 flew later in that year, and the first Super 72 conversion followed soon afterwards. All three variants entered service between early 1982 and early 1983. By the end of 1986 when work came to an end 110 conversions were completed.

The new engines transformed the performance of the aircraft. Simply put, the Super 71 matches the range performance of the intercontinental DC-8-63; the Super 73 can fly as far as the very-long-range DC-8-62, and the Super 72 has a greater range than any other commercial aircraft in the world, including the much bigger, highly specialised Boeing 747SP. Most of the long-body aircraft have been re-engined; the smaller Super 62s have been slower to move, because the conversion works out to be more expensive per passenger seat, and the economics of the modified aircraft are less attractive. A number of Super 72s, however, have been converted for large corporations and for government use: with the optional extra fuel and a light payload, the type has a near-global range of around 16093 km (10,000 miles). The Super 72 can link most of the world's airports non-stop, eliminating the need to ensure security at intermediate refuelling points. Short of the Saudi Arabian government's 747s, the Super 72 is probably the ultimate in personal transportation – a remarkable comeback for an aircraft that started its life in second place.

EMBRAER Bandeirante

Bandeirantes were sold to third-level operators around the world, most appreciating the type for its excellent reliability and efficiency on low-density services. Loganair flew the type on regional services in Scotland.

There can be little question that the EMBRAER Bandeirante has been one of the most successful of the modern light turboprop commuterliners. By the time it finished production in 1989, over 500 had been sold. In addition to its phenomenal civil success, many were delivered to the military for a variety of roles.

EMBRAER is the shortened title of Empresa Brasileira de Aeronautica S.A. As a company it was started on 19 August 1969, and has built not only the Bandeirante but also the Ipanema agricultural aircraft, the Tucano turboprop military trainer and, as a licensee of Piper Aircraft, virtually all the single- and twin-piston-engined Piper models. It was formed around the Bandeirante project, however, with a 51 per cent interest invested by the Brazilian government, and now employs over 9,000 workers at its main production plant at São José dos Campos in the state of São Paulo.

The Bandeirante concept was initiated in June 1965 by Max Holste, the famous French designer who conceived the MH.52, the Broussard and the Super Broussard (which later became the Nord 262). His design team came under the control of the Centro Técnico Aerospacial (CTA) of the Brazilian air ministry, and the original requirement was for a light transport to replace the ageing fleet of Beech Super 18s then in widespread service with the Brazilian air force (FAB). The study received the project designation PAR-6504 and the aircraft emerged as a low-wing monoplane with a retractable tricycle landing gear and two 410-kW (550-shp) Pratt & Whitney Canada PT6A-20 turboprops. The nose profile of the prototype closely resembled that of Holste's original Broussard Major and the Super Broussard prototype.

The first IPD-6504 was built by the production department of the CTA, the Instituto de Pesquisas e Desenvolvimento (IPD) at São José dos Campos and was first flown by Major Jose Ferreira on 22 October 1968. This was followed by a further two prototypes and a static test airframe which was reviewed by the CTA while airworthiness testing was in hand. The IPD-6504 had a number of shortcomings from a technical standpoint. The project definition progressively matured as more thought was given to the true needs of the FAB and to the possible market slot which the aircraft would fit in the US commuter industry. The ungainly engine nacelles of the first aircraft were redesigned to allow the main landing gear units to be fully retracted without part of the wheels being exposed and to permit higher-powered 507.1-kW (680-shp) PT6A-27 engines to be fitted. The rear fuselage was lengthened and there was a major redesign of the cockpit and nose area to give a much more streamlined appearance. The tail was enlarged to compensate for the longer fuselage and higher engine power, and the aircraft was given increased fuel capacity. The redesign also involved replacement of the porthole-style cabin windows with square windows and a revised fuselage structure to allow airstair entry doors on the port side both ahead and behind the wing. Compared with the EMB-100 (IPD-6504) the definitive aircraft had 12 to 16 seats as against the maximum nine passenger seats of the prototype.

Following the first flight of the EMB-110 prototype on 19 August 1972, the FAB carried out its official evaluation and confirmed its satisfaction with the aircraft. The military designation was C-95 and the first three of these joined the FAB in January 1973, a further four being delivered before the end of that year. The initial batches of C-95s were standard transport versions of the aircraft which went to the five general-duties *esquadrao de transporte aéreo* (ETA) units based at various military stations around the country. These were the short-fuselage version of the Bandeirante with a standard 12-seat interior, but the air force

Finnaviation is a subsidiary of the state airline Finnair, and flies regional services between Finland's cities for both passengers and express freight. Bandeirantes were used, but later the Saab 340 was the only type on strength.

Alas Chiricanas is based at David in the Chiriqui Province of Panama. A pair of EMB-110P1 Bandeirantes are on strength for regional services, this being the quick-change 19-passenger or freight version, carrying 1633 kg (3,800 lb) in the latter configuration.

subsequently received the stretched C-95A which was based on the commercial EMB-110P1 and fulfilled a dual role as a cargo transport and paratroop-dropping type. While the C-95 had a standard rear entrance door, the C-95A was fitted with the forward airstair door for crew and troop entry and a large rear fuselage cargo door with an upward-opening mechanism. This cargo door also had an inset sliding paratroop door which permitted free dropping of troops using static lines secured to the cabin roof.

Maritime patrol

Other versions of the Bandeirante soon appeared in the FAB inventory. Four EC-95s were taken on strength for airways aids calibration and navaid checking, and the R-95, of which six were received, was an aerial photogrammetric version with a ventral fairing added to form a camera bay. An electrically-operated sliding door could be opened to give a clear view to the Zeiss RMK A8.5/23 and RMK A15/23 vertical cameras, and the interior of the aircraft was fitted to accommodate three equipment operators and their consoles for precision subject tracking. For search and rescue work the stretched SC-95B was brought into service with the 2° Esquadrao of the 10° Grupo de Aviacao, and this unit also received the specialised EMB-111A maritime surveillance version. Designated P-95, the EMB-111 first came into service in 1978 and was also delivered to the Chilean naval air service as the EMB-111A(N). While it is based on the C-95, the P-95 is considerably modified with more powerful PT6A-34 engines and wingtip tanks which raise the fuel capacity to

The colourful markings on this Bandeirante identify it as a SAR-configured SC-95B. Observers on each side of the fuselage are stationed behind bubble windows, while the main cabin stores rescue equipment and can transport stretcher cases.

2545 litres (560 Imp gal) to give the type an endurance of over eight hours. Three underwing hardpoints are provided for bombs, depth charges or triple rocket packs, and a 50-million candlepower searchlight is fitted to the starboard wing leading edge. The P-95 has a large nose radome which accommodates the antenna for a Cutler-Hammer APS-128 SPAR-1 search radar which is interfaced with a Litton LN-33 inertial navigation system, giving the aircraft a potent search capability. An inward-opening rear door is provided to enable the crew to drop survival equipment or para-rescue teams.

The CTA had granted the Bandeirante a civil certificate of airworthiness on 20 December 1972, and the fourth production aircraft (an EMB-110C) was delivered to Transbrasil Airlines in March 1973. Transbrasil received a further five Bandeirantes and was later merged into the new regional airline network as a part of Nordeste Aéreas Regionais S.A. It now has a fleet of four EMB-110Cs, two EMB-110Ps, six EMB-110P1s and one EMB-110P2 to cover a network in the east of the country from Belo Horizonte and Rio in the south to Salvador and Recife in the north. Other regional airlines were established at the same time under a law passed in November 1975, which divided the internal routes of Brazil into geographical areas. The Bandeirante played a major part in the fleets of all the new airlines. Rio Sul Servicos Aéreos Regionais S.A., for instance, had five EMB-110Ps together with seven Piper Navajo and Navajo Chieftain light piston-engined aircraft to give a mixture of capacity on its sector south of São Paulo. It has since disposed of the piston-engine aircraft and has increased its Bandeirante fleet to six.

Executive transport

Another application for the Bandeirante was the executive and air taxi market. For this purpose EMBRAER offered the EMB-110E and EMB-110J, which had seating for seven and eight passengers respectively. The standard airline Bandeirante is fitted with three-abreast seating comprising a single seat on the port side and a twin bench seat on the starboard side with a narrow aisle in between. These seats are normally finished in hard-wearing leather. The executive version replaces this interior with two groups of seats in a 'club' face-to-face arrangement, and these are fully reclining seats upholstered to a high standard providing a great deal more leg-room than those in the high-density arrangement. An early customer was the air taxi operator Sudeco, while a good number of business organisations purchased

Specification: EMB-111 Bandeirante Patrulha

Wings
Span, over tanks	15.95 m	(54 ft 2 in)
Area	29.10 m²	(313.23 sq ft)

Fuselage and tail unit
Accommodation	two pilots and three to five operators/observers	
Length overall	14.91 m	(48 ft 11 in)
Height overall	4.91 m	(16 ft 1.3 in)
Tailplane span	7.54 m	(24 ft 9 in)

Landing gear
Hydraulically retractable tricycle landing gear with a single wheel on each unit
Wheelbase	4.26 m	(13 ft 11.7 in)
Wheel track	4.94 m	(16 ft 2.5 in)

Weights
Empty, equipped	3760 kg	(8,289 lb)
Maximum take-off	7000 kg	(15,432 lb)
Internal fuel load	1730 kg	(3,814 lb)

Powerplant
Two Pratt & Whitney Canada PT6A-34 turboprops
Rating, each	559 kW	(750 shp)

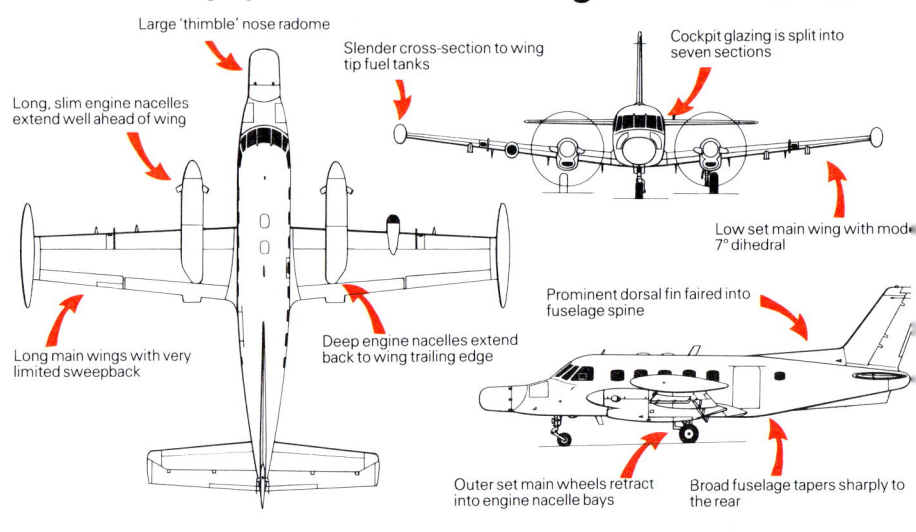

EMB-111A(A) Bandeirulha recognition features

- Large 'thimble' nose radome
- Slender cross-section to wing tip fuel tanks
- Cockpit glazing is split into seven sections
- Long, slim engine nacelles extend well ahead of wing
- Low set main wing with mod 7° dihedral
- Long main wings with very limited sweepback
- Deep engine nacelles extend back to wing trailing edge
- Prominent dorsal fin faired into fuselage spine
- Outer set main wheels retract into engine nacelle bays
- Broad fuselage tapers sharply to the rear

Bandeirante variants

EMB-100: known to the FAB as the **YC-95**, this was the CTA project IPD/PAR 6504 with 410-kW (550-shp) Pratt & Whitney Canada PT6A-20 turboprops, rounded windows and eight passengers; three pre-series aircraft (last two with increased windscreen rake and additional window on each side) and one static test airframe; first flight 26 October 1968

EMB-110 Bandeirante: known to the FAB as the **C-95**, this initial production model has 507-kW (680-shp) PT6A-27s in redesigned nacelles completely enclosing retracted landing gear; first flight 9 August 1972; 12-seat military transport with fuselage length 13.74 m (45 ft 1 in) and maximum weight 5300 kg (11,684 lb)

EMB-110A: known to the FAB as the **EC-95**, this is the navaids calibration variant of EMB-110; up to six passengers/operators

EMB-110B: known to the FAB as the **R-95**, this is the aerial survey variant of C-95 with Zeiss cameras and additional avionics; up to five passengers/operators; two additionally sold (one to Uruguay and one civil) designated **EMB-110B1** with alternative, quick-change 14-passenger capacity

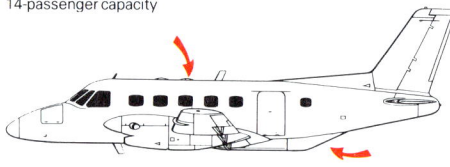

EMB-110C: civil transport with 15 (occasionally 12-16) seats; military exports to Uruguayan air force and Chilean navy, the latter as **EMB-110C(N)** with anti-icing provision

EMB-110E: executive transport variant of EMB-110C; 6-8 seats in VIP interiors; also **EMB-110E(J)** with special equipment

EMB-110K1: known to the FAB as the **C-95A**, this has PT6A-34 engines of 559 kW (750 shp), ventral fin, fuselage length 14.59 m (47 ft 10.5 in), rear cargo door and added passenger/crew door forward; military transport with 1650-kg (3,638-lb) capacity, delivered from May 1977

EMB-110P: civil commuter version of EMB-110K1 with PT6A-27 or PT6A-34 engines and 18 seats; first flown in January 1976; subvariants are the **EMB-110P1** quick-change passenger/freight model; **EMB-110P2** with up to 21 seats, but lacking freight door and having a maximum weight of 5670 kg (12,500 lb)

EMB-110P1K: known to the FAB as the **C-95B**, this is the quick-change model of the EMB-110K1/C-95A with similar payload

EMB-110P1SAR: known to the FAB as the **SC-95B**, this is the SAR version of EMB-110P1K with four bubble windows and accommodation for six stretchers; maximum weight 6000 kg (13,228 lb)

EMB-110P1A: civil variant with **EMB-110P2A, EMB-110P1A/41** and **EMB-110P2A/41** subvariants as P1 etc, but with 10° tailplane dihedral, improved soundproofing and other changes; delivered from December 1983

EMB-110S1: geophysical survey variant of EMB-110C with increased internal wing tankage, magnetometer tail boom, two equipment operators and PT6A-34 engines; one civil model sold

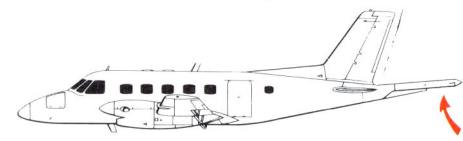

EMB-111: known to the FAB as the **P-95 Bandeirante Patrulha**, this is the maritime surveillance model with short fuselage, PT6A-34 engines, tip tanks and additional internal fuel, large nose radome and four underwing weapon pylons; maximum weight 7000 kg (15,432 lb); standard model **EMB-111A(A)**, or **EMB-111(M)** for Brazil; the Chilean **EMB-111A(N)** has changed equipment and full anti-icing provision

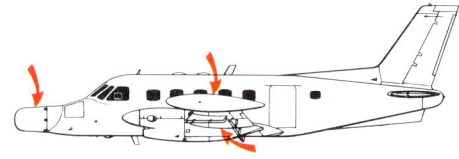

EMB-120 Araguaya: pressurized EMB-110 with mid-tail and PT6A-45s of 875 kW (1,174 shp); construction of prototype (to fly July 1979) abandoned and designation reassigned to EMB-120 Brasilia 30-seat commuterliner

EMB-121 Xingu: Based on EMB-110 wings and engine nacelles with short, T-tailed fuselage seating up to nine passengers; first flight 10 October 1976; 104 built

EMB-123 Tapajos: sized between EMB-121 and Araguaya; PT6A-45 engines; projected first flight in March 1979 but abandoned and designation reassigned in 1986 to 19-seat rear-engined twin-jet, having 60 per cent commonality with Brasilia

EMBRAER EMB-110 Bandeirante cutaway drawing key

1. Nose cone
2. Radar array (Bendix RDR-1200 or RCA AVQ-47)
3. Nosewheel well
4. Nosewheel doors (close after activation)
5. Pitot probe
6. 250W taxi light
7. Nosewheel fork
8. Goodyear 6.50x8 nosewheel tyre
9. Nosewheel oleo (by ERAM)
10. Nosewheel oleo flap
11. External power socket
12. Avionics bay
13. Avionics bay access doors (upward hinged)
14. Bulkhead
15. Plexiglas windscreen side panels
16. 13-mm stressed acrylic windscreen centre panels
17. Instrument panel shroud
18. Variable speed wipers
19. Second pilot's seat
20. Clear vision panel
21. Pilot's adjustable seat
22. Control column
23. Rudder pedals
24. ADF antenna
25. Forward/centre fuselage join
26. Port cloaks/stores
27. Aerial mast
28. Starboard equipment rack
29. Starboard nacelle
30. Spinner
31. Hartzell HC-B3TN-3C/T10178H-8R constant speed propeller
32. Leading edge wing fence
33. Starboard 450 W shielded glare landing light
34. Riveted aluminium sheet wing skin
35. Starboard navigation light
36. Aileron static dischargers
37. Starboard statically-balanced aileron
38. Fuel filler cap
39. Two wing integral fuel tanks each side (total capacity 1690 litres/372 Imp gal)
40. Double-slotted flap
41. Aerial
42. Starboard cabin-air trunking
43. Seven cabin windows (starboard)
44. Emergency exit window (starboard only)
45. Five-a-side cabin seating
46. Riveted aluminium sheet fuselage skin
47. Five cabin windows (port)
48. Floor support structure (stressed for cargo)
49. Front spar/fuselage steel join
50. Rear spar/fuselage steel join

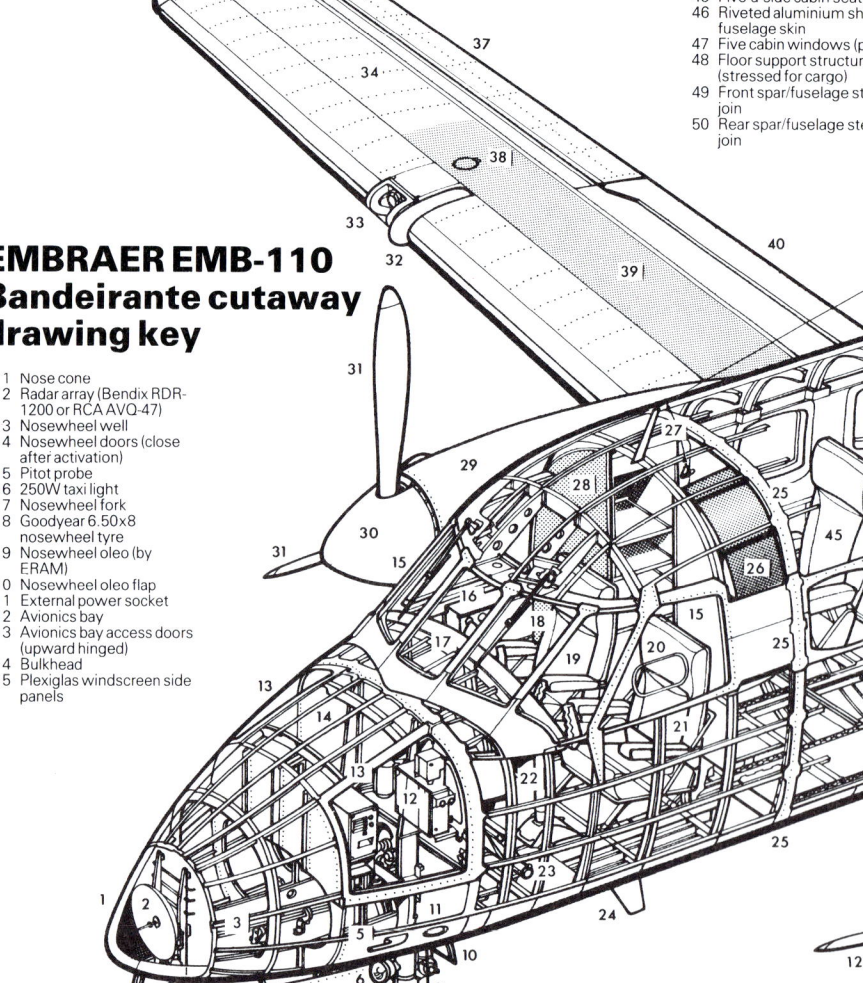

© Pilot Press Ltd

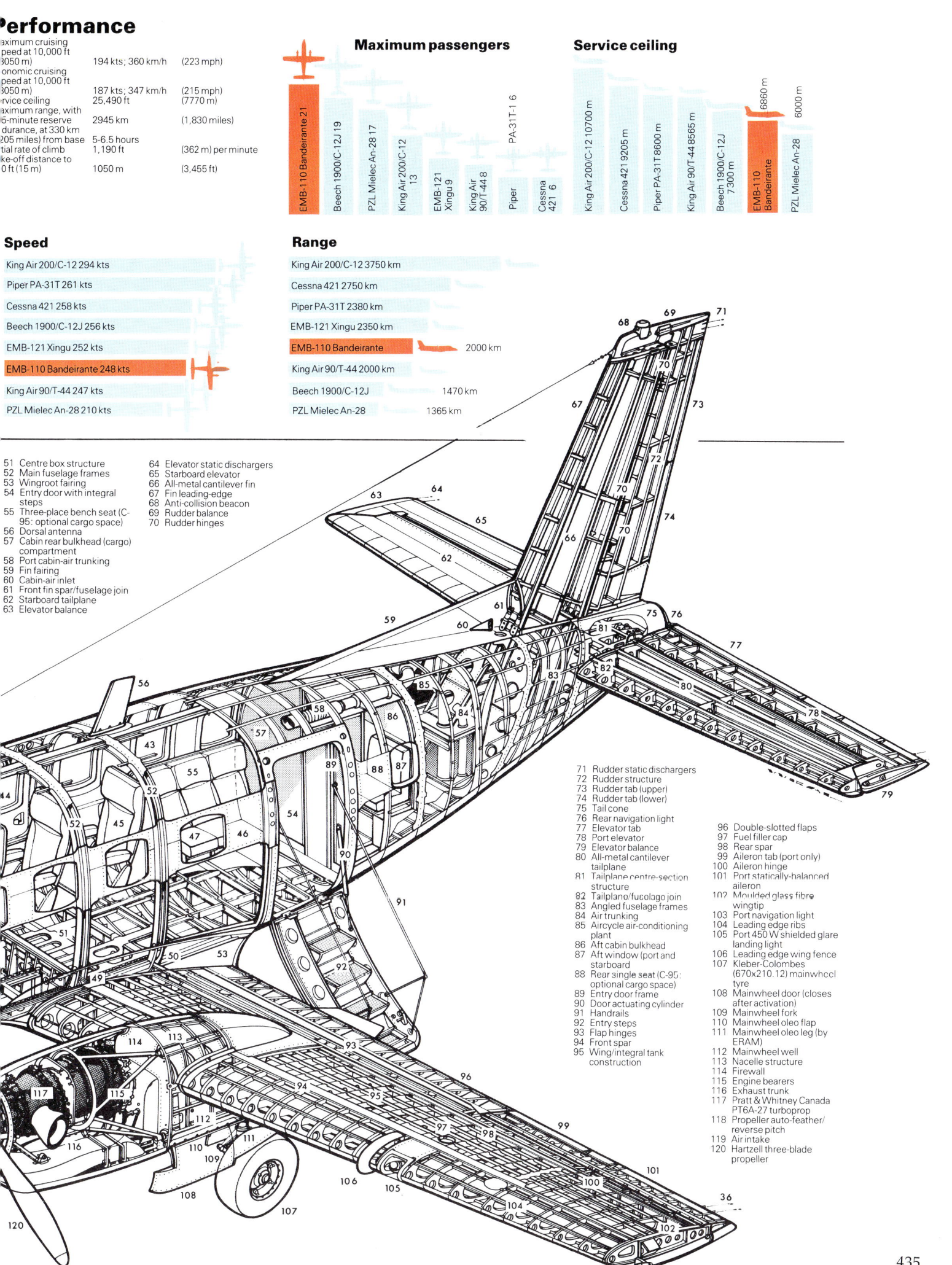

EMBRAER EMB-111A/P-95 Bandeirulha
FORCA AEREA BRASILEIRA
Coastal Command

the aircraft in preference to a Beech King Air or Swearingen Merlin. These corporate users included Furnas Centrais Eletricas S.A., which now operates three aircraft, and a great number of nationalised industries and local government operators such as the Governo do Estado de Pernambuco and the Governo do Estado de Minas Gerais. Several airline aircraft were also fitted with a reduced number of executive style seats in order to increase the comfort of passengers on long stage lengths. One aircraft was placed on contract for Mobil Oil in the north of Scotland and was modified in this way because of the somewhat cramped conditions provided by the twin bench seats, and the rather substantial frames of the oilmen the aircraft had to carry!

American sales

Clearly, the prime market for the commuter versions of the Bandeirante was the United States. Certification by the Federal Aviation Administration was slow in being granted – perhaps as a certain measure of protection for the domestic producers of light turboprops. In 1978, however, a distribution agreement was signed between EMBRAER and Aero Industries of Los Angeles (a company which owned a number of local service air carriers) and the FAA then proceeded to grant the essential type certificate. The Bandeirante continued to face considerable opposition from the American air carriers and manufacturers, reflected in charges of predatory pricing through subsidised interest rates supported by the Brazilian government, but this did little to deter many small airlines from buying the type. By May 1979 operators included Wyoming Airlines with three aircraft, Mountain West Airlines with four aircraft and Transmountain Airlines of Denver, Colorado, with three.

Today there are over 100 Bandeirantes operating with US local service airlines. Many of the carriers operate large fleets which use the capacity of the Bandeirante to fill the gap between the 8/10-seat light aircraft and the larger 30/40-seater turboprops. Provincetown-Boston Airlines was a good example. It was highly unusual in that it had two separate locations for its services; in summer it flew routes in the Boston area covering a triangle which included La Guardia, New York, in the south, Burlington, Vermont, in the north and communities such as Martha's Vineyard, Hyannis and Nantucket in the vicinity of the city of Boston. For the winter season the airline took a large part of its fleet to Florida and operated from Miami and Tampa to destinations such as Sarasota, Punta Gorda, Key West and Daytona Beach. With varying distances and load factors Provincetown-Boston also used a large fleet of Cessna 402 Utililiners for up to nine passengers over one-hour stage lengths. The 15 Bandeirantes came into use for the route lengths of 402 to 483 km (250 to 300 miles) which used the 18 passenger seats and the cargo section and where it was quite prac-

The EMB-111 has been produced in only modest numbers, with Brazil the major user. The APS-128 is backed-up by an optional 50-million candlepower searchlight on the inner starboard wing, and the ability to hit back in anger with up to eight underwing rockets.

ticable for the aircraft to fly at around 2135 m (7,000 ft) and not require pressurisation.

The EMB-110P1 and EMB-110P2 started to gain a foothold in Europe in 1977 when a number of aircraft were sold to French regional carriers. Air Littoral uses six of the EMB-110P2 model covering a route network which radiates from its home base at Montpellier-Frejorgues to some 29 destinations in southern France and in Italy and Spain. The Bandeirantes now share these routes with the Nord 262, Metro II and Brasilias but they were used on operations from Montpellier to Nantes, Nice, Rennes, Valencia and Venice as a regular routine. Another airline (familiar to users of London Gatwick) is Brit Air, which has three Bandeirantes and flies to Rennes and Quimper every weekday.

UK operators

A number of Bandeirantes served with British operators. Many of these were imported to meet the needs of the North Sea oil market, and entered service with newly created air carriers such as Air Ecosse and Air Shetland. The established second-level airlines also found the Bandeirante useful, examples appearing with Loganair, Dan-Air and Air UK. At one time Air UK had six aircraft which it used both on internal services and on short-range European routes: Bandeirantes, for example, replaced the airline's rather elderly Handley Page Heralds on the route from Gatwick to Rotterdam, and also operated between Southend and Rotterdam. The largest fleet

The maritime reconnaissance EMB-111 (P-95) is easily identified by the thimble-like nose which houses an APS-128 search radar, and the large wingtip fuel tanks for increased patrol range. For nations with territorial waters to protect, it is a cheap but effective asset.

user in the United Kingdom was Fairflight, based at Biggin Hill, which ran the Air Ecosse services and chartered Bandeirantes to Mobil and the British National Oil Corporation. They also used one of the quick-change EMB-110P1 aircraft for the Datapost service for the Post Office and this Bandeirante, in an all-red colour scheme, shared the national Datapost service with a Cessna Titan and a Shorts 360 providing an overnight express delivery around the country.

For many years the Allegheny Commuter system worked successfully in the United States. Under this arrangement low-density routes held by Allegheny Airlines, now merged into US Air, were passed as concessions to smaller airlines which then had access to Allegheny's booking system and organisation. This grew to become a general method of entry for these small commuter carriers to the mother airline's organisation even though the routes served were not basically those of Allegheny. In this interesting commuter system the Bandeirante found favour with several operators, and a parallel system operated by Air Florida brought in the four EMB-110P1s of Finair Express, which were painted up in the Air Florida colour scheme.

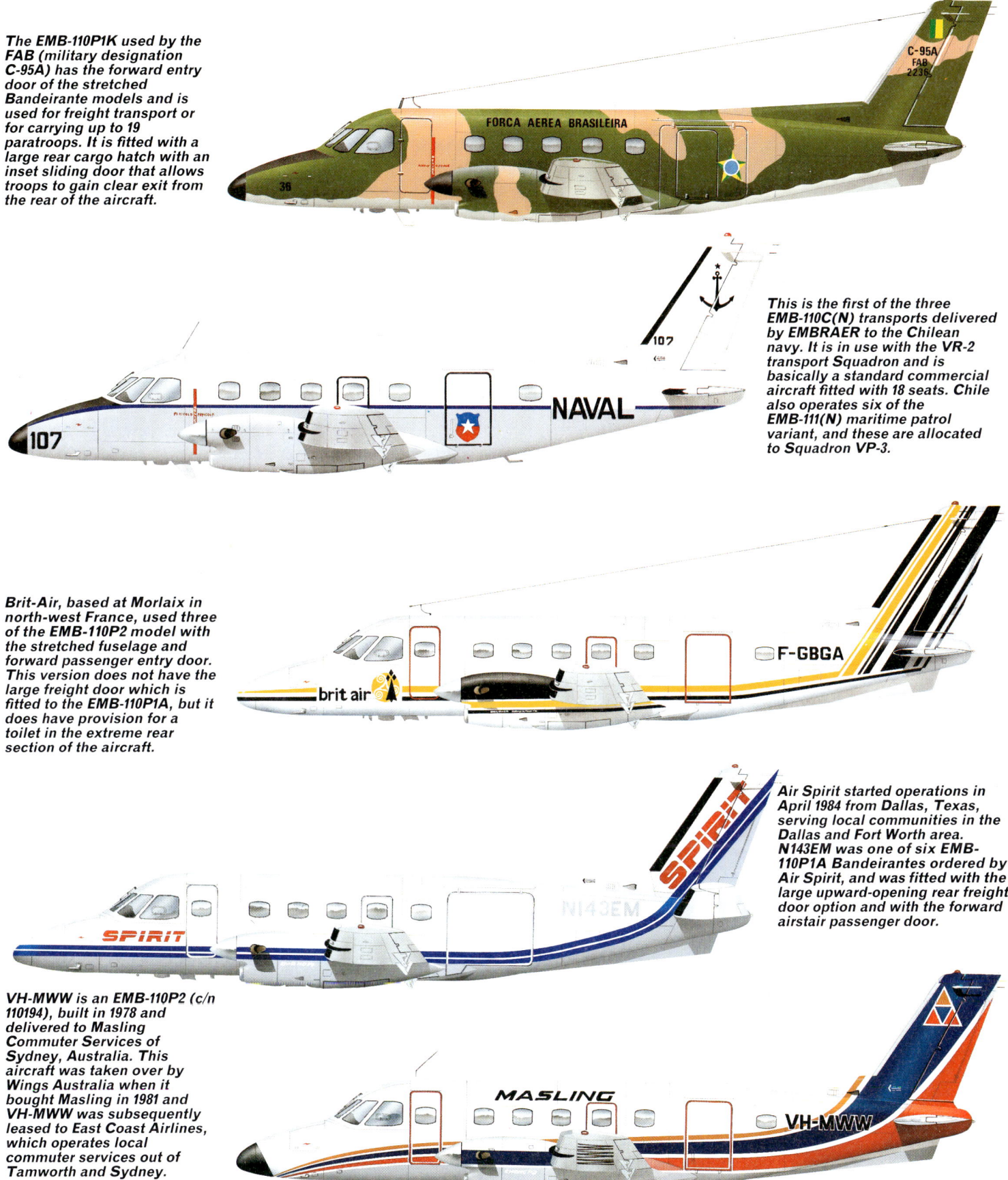

The EMB-110P1K used by the FAB (military designation C-95A) has the forward entry door of the stretched Bandeirante models and is used for freight transport or for carrying up to 19 paratroops. It is fitted with a large rear cargo hatch with an inset sliding door that allows troops to gain clear exit from the rear of the aircraft.

This is the first of the three EMB-110C(N) transports delivered by EMBRAER to the Chilean navy. It is in use with the VR-2 transport Squadron and is basically a standard commercial aircraft fitted with 18 seats. Chile also operates six of the EMB-111(N) maritime patrol variant, and these are allocated to Squadron VP-3.

Brit-Air, based at Morlaix in north-west France, used three of the EMB-110P2 model with the stretched fuselage and forward passenger entry door. This version does not have the large freight door which is fitted to the EMB-110P1A, but it does have provision for a toilet in the extreme rear section of the aircraft.

Air Spirit started operations in April 1984 from Dallas, Texas, serving local communities in the Dallas and Fort Worth area. N143EM was one of six EMB-110P1A Bandeirantes ordered by Air Spirit, and was fitted with the large upward-opening rear freight door option and with the forward airstair passenger door.

VH-MWW is an EMB-110P2 (c/n 110194), built in 1978 and delivered to Masling Commuter Services of Sydney, Australia. This aircraft was taken over by Wings Australia when it bought Masling in 1981 and VH-MWW was subsequently leased to East Coast Airlines, which operates local commuter services out of Tamworth and Sydney.

The Allegheny system was copied by British Caledonian, which saw an opportunity to bring the Bandeirante fleet of Liverpool-based Genair into the BCAL booking net and create commuter services which would be complementary to British Caledonian trunk routes. British Caledonian Commuter Services was able to provide connections with Gatwick from Humberside, Leeds/Bradford, Liverpool, Teesside and Norwich using Bandeirantes painted in a version of the British Caledonian colour scheme. Regrettably, Genair went out of business in July 1984, but the Commuter Services' concept continued with Connectair flying Bandeirantes to Antwerp and other airlines using Shorts 330s within the same system, until BCAL was merged with British Airways in December 1987.

The Bandeirante continued in production until 1989 when just over 500 had been sold. The EMB-110 design was also used as a basis for the short-fuselage EMB-121 Xingu sold to business users and to the French military forces and the FAB. The line of turboprop aircraft was a credit to the engineering and marketing ability of a company which is now a major world manufacturer.

EMB-110 Bandeirante in service

Angola (Força Aérea Populaire de Angola/Difesa Anti-Avioes)
Although equipped with a single Fokker F.27MPA Maritime, based at Luanda, the FAPA/DAA expanded its coastal patrol fleet in 1986 with an order for two EMB-111 Bandeirante Patrulhas.

One of 56 C-95 transport aircraft delivered to the Forca Aérea Brasiliera from 1973 onwards, displaying an alternative to the white and light grey scheme often worn.

Brazil (Força Aérea Brasileira)
Eight models of military Bandeirante have been supplied to the FAB, comprising two EMB-100 prototypes (serialled YC-95 2130/2131); 56 EMB-110 transports (C-95 2132/2176 and 2179/2189) delivered 9 February 1973 to November 1976; four EMB-110A navaids calibration aircraft (EC-95 2177/2178 and 2190/2191) delivered between April and November 1976; six EMB-110B survey aircraft (R-95 2240/2245) in July and August 1977; 20 EMB-110K1 cargo transports (C-95A 2280/2299) delivered September 1977 to September 1978; 30 EMB-110P1K convertible transports (C-95B 2300/2329) delivered in 1980-4; five EMB-110P1KSAR rescue variants (SC-95B 6542/6546) in 1981; and 12 EMB-111A(A) maritime patrollers (P-95 7050-7061) delivered September 1977 to August 1979. A requirement exists for a further six EMB-111s. In May 1982 EMB-111s 7058 and 7060 were loaned to the Argentine navy (Comando de Aviación Naval Argentina), flying briefly with 2ª Escuadra at Puerto Belgrano/Comandante Espora (coded '2-P-201' and '2-P-202') then with 6ª Escuadra at Trelew (as '6 P 201' and '6 P 202') before being returned in the following month, or shortly thereafter. Brazilian units operating the Bandeirante are:

Six R-95 photo survey aircraft (serialled 2240-2245) are in the inventory of the FAB, all having been delivered during mid-1977. All six equip 1°/6 GAv at Recife.

1° Esquadrão/ 6° Grupo de Aviação
Base: Recife
Role: Photo-survey
Equipped: August 1977
Aircraft: R-95 Bandeirante

2° Esquadrão/ 7° Grupo de Aviação
Base: Florianapólis
Role: Maritime surveillance
Equipped: February 1982
Aircraft: P-95 Bandeirante Patrulha

2° Esquadrão/ 2° Grupo de Transporte
Base: Campo dos Afonsos (ex-Galeão)
Role: Light transport
Equipped: 1973
Aircraft: C-95A/B Bandeirante

Escola Preparatória de Cadetes do Ar (Air Cadets' Preparatory School)
Base: Barbacena
Role: Training
Equipped: about 1976
Aircraft: C-95 Bandeirante

Base: Natal
Role: Twin-conversion and transport
Equipped: about 1974
Aircraft: C-95 Bandeirante

Esquadrãos de Transporte Aéreo
Base: 1° ETA – Belem; 2° ETA – Recife; 3° ETA – Galeao; 4° ETA – Cumbica; 5° ETA – Canoas; 6° ETA – Brasilia; 7° – Manaus
Role: Regional communications
Equipped: various dates
Aircraft: C-95 Bandeirante

1° Esquadrão/ 7° Grupo de Aviação
Base: Recife
Role: Maritime surveillance
Equipped: April 1978
Aircraft: P-95 Bandeirante Patrulha

2° Esquadrão/ 10° Grupo de Aviação
Base: Campo Grande
Role: SAR
Equipped: late 1981
Aircraft: SC-95B Bandeirante

1° Esquadrão/ 15° Grupo de Transporte
Base: Campo Grande
Role: Light transport
Equipped: 1981
Aircraft: C-95B Bandeirante

1° Esquadrão/ 5° Grupo de Aviação (Centro de Apliação Tática e Recomplementação de Equipagens, or Tactical and Crew Training Centre)

Grupo Especial de Inspeção e Vigilância (Special Inspection and Surveillance Group)
Base: Rio de Janeiro
Role: Navaid calibration
Equipped: 1976
Aircraft: EC-95 Bandeirante

The large marking on the fin of this C-95 identifies it as an aircraft operated by 6° Esquadrao de Transporte Aéreo based at Brasilia, one of seven such squadrons.

Chilean Navy (Servicio de Aviación de la Armada de Chile)
Three EMB-110C(N) personnel transports, serialled 107-109 and equivalent to the civilian EMB-110C, were supplied to VC-1 (Naval Transport Squadron No. 1) at Quilpué/El Belloto, Valparaiso, in July 1977. Between December 1977 and September 1976, VP-3 took delivery at the same base of six EMB-110A(N)s, 261-266, for surveillance of Chile's extensive coastline.

In addition to the trio of personnel transports supplied to the Chilean navy, six EMB-111A(N) maritime patrol aircraft were supplied from 1977-9 for use by VP-3. These replaced four embargoed ex-US Navy Lockheed SP-2E Neptunes.

Gabon (Forces Aériennes Gabonaises)
The first Bandeirante delivered to Gabon, in mid-1980, was an EMB-110P1 (call-sign TR-KGB) transport assigned to the bodyguard of President Omar Bongo, La Garde Presidentielle. The regular air force then received two more EMB-110P1s (TR-KNA/KNB) from July 1980, plus a maritime surveillance EMB-111A (TR-KNC) in July 1981.

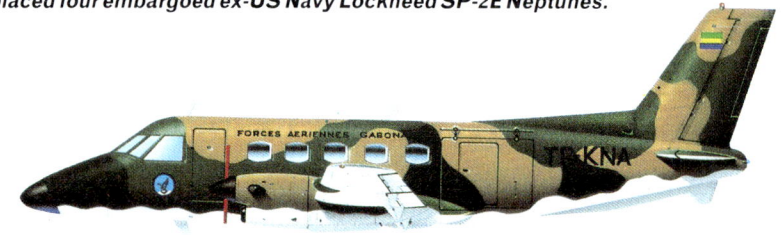

Publicity photographs show the first of Gabon's EMB-110P1s incorrectly registered as TR-KMA, but this has since been corrected to TR-KNA.

Guyana (Air Command, Guyana Defence Force)
A single EMB-110P1, 8R-GFO, was delivered to the GDF in 1983 for general transport duties. It was reportedly out of service in 1986 – whether temporarily or permanently is not known.

Uruguay (Fuerza Aérea Uruguaya)
In September 1975, the air force formed Grupo de Aviación 6 within Brigada Aérea 1 at Carrasco to operate five EMB-110C civil-standard transports, all of which had been delivered by the following November, serialled T-580/584. A later arrival, in August 1978, was a single photo-survey/transport EMB-110B1, T-585. For use by the paramilitary airline TAMU (Transporte Aéreo Militar Uruguayo), the aircraft have the parallel civil registrations CX-BJJ, -BJK, -BJB, -BJC, -BJE and -BKF respectively.

Six Bandeirantes are in service with the Fuerza Aérea Uruguaya, this being the first example delivered during 1975. Both military and civilian transport tasks are carried out, the latter under the auspices of TAMU.

Fokker F.VII

After World War I the best-selling airliners did not come from the Allies but from defeated Germany's Junkers and 'Flying Dutchman' Anthony Fokker, whose German factory had produced the enemy's top fighters. Fokker's high-wing monoplane transports also supported a major industry in the USA.

As Germany collapsed in chaos in November 1918, Anthony Fokker found himself with a factory stuffed with the latest fighters, scores of millions of marks in the bank and the possibility of losing everything. Red revolutionary guards were in control, shooting any industrialist who failed to hand over his money. The Allies were about to commandeer all German aircraft and prohibit further manufacture, except in the smallest and lowest-powered class of sport aircraft. One night Fokker managed to slip in disguise past Red guards (later shot for not recognising him) and reach Berlin, managing by various means to transfer large sums to his native Netherlands. He also hid vast amounts of air material and then in 1920 smuggled six trains, each of 60 box or flat cars, under the noses of Allied officials, from Schwerin to his new factory in Amsterdam!

However, back in early 1919, Fokker and chief designer Reinhold Platz had decided they had better start designing passenger aircraft. The first result was the F.I. There was no argument about the structure. The fuselage was to be made of steel tubes, accurately cut and, where necessary, curved, and welded together in a precision jig, with fabric covering. This was how they had made their fighters. Platz had pioneered unbraced wooden wings whose plywood covering was an early form of stressed skin. By making the wing deep it could be light yet strong enough to need no external bracing. The wing of the F.I was almost a direct scale-up of that of the D.VIII fighter. This was to be mounted in the parasol position above the fuselage, the latter having an impressive row of open cockpits for the pilot and six passengers, the latter in three side-by-side pairs. But before this prototype (the V 44) had got very far, Fokker and Platz took a decision that probably made the difference between failure and success. They decided to redesign the fuselage to accommodate the passengers in comfort, internally.

The result was the V 45 prototype, or F.II, first flown by Fokker's trusty chief pilot and school head, Bernard de Waal, in October 1919. The wing, originally intended for the V 44, was simply bolted directly to the top of the enlarged fuselage. The cabin was furnished for four passengers, and a fifth could sit in the open front cockpit under the leading edge alongside the pilot. Powered by a 138-kW (185-hp) BMW IIIa 6-cylinder inline water-cooled engine, the F.II was slow but, to the passenger, vastly better than the crudely converted military machines produced by British and French companies. One odd feature was the absence of a fin, the weathercock stability being provided by the long flat-sided rear fuselage to which was hinged the diminutive rudder. There was nothing lacking, however, from the tailplane, elevators and ailerons, all controls being fully balanced and the ailerons projecting distinctively beyond the wingtips.

Escape to Amsterdam

De Waal secretly flew the V 45 prototype to the Netherlands on 20 March 1920. Subsequently, the type entered service with several airlines, some being built by Fokker and others under licence by Grulich, the latter having 186.4- or 238.6-kW (250- or 320-hp) BMW engines and usually some wing bracing struts and an enclosed cockpit. They firmly established Fokker as a builder of practical passenger airliners. Despite their wooden wings and fabric-covered fuselages, they proved to be robust and long-lived, and they rather surprisingly remained in scheduled service with several operators (including even Deutsche Lufthansa) well into the 1930s, one still flying in 1940.

Naturally Fokker developed the F.II, moving via the bigger F.III to the F.IV of 1921. The former, powered by a BMW, Eagle, Puma or Jupiter, was a best seller and by far the most important airliner in northern Europe, including the Soviet Union, in 1922-25. The F.IV was much bigger but was still only single-engined, with a 313.2-kW

Of typical Fokker design, with a plywood-covered two-spar wooden wing and fabric-covered welded steel tube fuselage, the four production F.VIIs had accommodation for eight passengers and two crew members.

Sir Charles Kingsford Smith's famous F.VII-3m – the **Southern Cross** – arrived at Brisbane's Eagle Farm Airport on 9 June 1928 following an 11890-km (7,390-mile) flight across the Pacific from Oakland, California. The American registration (1985) was changed to VH-USU for its onward flight to England in 1929.

The Italian airline Avioline Italiane received a batch of three F.VIIa-3ms, including I-BBED (c/n 5059). This was sold by them in 1933 to Societá Aérea Mediterranea and then, in company with the other ALI aircraft, it was taken on charge by the Ala Littoria company and is believed to have been scrapped in 1939.

(420-hp) Liberty. Two were sold to the US Army, which called them T-2s. One gained immediate fame on 2-3 May 1922 by flying non-stop across the USA, from New York to San Diego. This triggered major sales to the US Army and US Navy, which in turn led to an outcry among US manufacturers. Fokker overcame this by setting up a US subsidiary, Atlantic Aircraft Corporation, at Teterborough, New Jersey. Subsequently, Atlantic Fokkers tended to depart from the parent product in design details and engines, until from 1925 Atlantic's manager, R. B. C. Noorduyn (later a famed builder of his own Norseman), developed Fokker transports quite independently.

Birth of 'a real airliner'

In early 1923 Fokker and Ing. Walther Rethel (later famed at Arado) began designing the F.VII, a straightforward machine with a 268.5-kW (360-hp) Rolls-Royce Eagle and six passenger seats. It had a wing of 71.72-m^2 (772-sq ft) area, compared with 39.11 m^2 (421 sq ft) for the F.III and 89.00 m^2 (958 sq ft) for the F.IV. It worked well enough, but the way to improve it was obvious. With a newer and more powerful engine and smaller wing it would be possible to carry more yet fly faster, and thus reduce the operating costs. Only five F.VIIs were built, but the improved F.VIIA was another smash hit. First flown in early 1925, it was given a wing reduced in area to only 58.53 m^2 (630 sq ft), with the ailerons at last accommodated within the outline of the wing. The usual engine was a 335.6-kW (450-hp) Bristol or Gnome-Rhône Jupiter, but one early example had the 298.3-kW (400-hp) Packard A-1500 (improved Liberty). Carrying eight passengers and two crew, the F.VIIA was what a modern observer might call 'a real airliner'. Minor improvements were the fitting of a larger rudder with a fixed fin, and a neater landing gear with vertical shock struts (sprung by multiple elastic cords inside a fairing) pinned under the front wing spar.

In the summer of 1924 the Dutch airline KLM, by then almost entirely Fokker-equipped, issued a specification for an airliner to carry 10 passengers and be able to fly with one engine stopped. It suggested the use of three Siddeley Puma engines. However, Fokker was intensely interested in the tri-motor Junkers G.23 when this flew in November 1924. Most significant of all, in his autobiography he wrote, "When Byrd sought to buy the first tri-motored Fokker he was unknown to me. That plane had been constructed on the basis of my cabled instructions to put three Wright motors on an F.VII." It is therefore odd that Fokker later claimed that he built the F.VIIA-3m tri-motor purely in order to compete in the Ford Reliability Trial in the USA, which began in September 1925. Be that as it may, Fokker had his eye initially on the US market, and in any case was always intensely interested in publicity. So the first F.VIIA-3m was given American engines, 149.1-kW (200-hp) Wright J-4 Whirlwinds, and fitted for the American trial. Fokker wanted the outboard engines on the leading edges, but for reasons of balance and thrust-line axis designer Platz put them in separate nacelles on the vertical main legs, braced to fuselage and wings. This machine was completed with rather crude exhaust stacks wrapped round right in front of the air-cooled cylinders and discharging under the nose and above the wings. In typical Fokker fashion the aircraft was painted with the name in gigantic capitals along the wings and fuselage. It first flew on 4 September 1925, required little alteration, and was dismantled and shipped to the USA.

The machine's appearance was sensational, and it performed almost faultlessly throughout the trial. It proved it could (just) fly with one outer engine stopped, and Henry and Edsel Ford took a keen personal interest. The Fokker won the trial easily, and Edsel Ford bought the aircraft for Commander Richard E. Byrd's scientific expedition over the North Pole. Named *Josephine Ford*, and now with refined rear exhaust collector rings and fitted with skis, it was flown by Byrd and Floyd Bennett over the Pole on 9 May 1926. The aircraft is now in the Ford Museum in Dearborn. Ford later produced

The first prototype of the single-engined Fokker F.VII is seen here making its maiden flight on 11 April 1924. The vertical tail shown in this photograph was later changed for a redesigned version with greater area and a rounded rudder profile. This aircraft was subsequently given the registration H-NACC.

The first three-engined Fokker (c/n 4900) was flown in the 1925 Ford Reliability Tour and was subsequently sold to Edsel Ford. Carrying the name **Josephine Ford**, it made a historic flight from Spitzbergen to the North Pole on 9 May 1926 under the command of Lieutenant Commander Richard E. Byrd and piloted by Floyd Bennett.

F.VIIB-3m PH-AFS was used initially by KLM, who named it **Specht** *(woodpecker)*. It was sold in August 1936 to Crilly Airways, but did not see any further airline service because it was then passed on to the Spanish Nationalist forces, who took it on charge at Burgos and used it in the fight against the Republican army.

SABENA was a prominent *F.VIIB-3m* user, and flew aircraft built under licence by the Belgian manufacturing company SABCA. This particular example, **OO-AGH**, was delivered on 20 September 1932 and flew on European routes until May 1940, when it was taken over by the invading German forces at Haren Airport near Brussels.

his own tri-motor, with Junkers-type all-metal structure with corrugated skin.

From the start, demand for the F.VIIA-3m was almost more than Fokker's Dutch and US plants could handle. Later he did a deal with industrialists in Wheeling, West Virginia, who thought aircraft might replace part of their declining steel industry; soon Wheeling was building a 10-seat tri-motor every five days. Licences were purchased by Belgium, Italy, Poland and the UK, the British licensee being A. V. Roe, which produced several derived designs which included a nice-looking twin, the Avro 642. This was a much more graceful design than Fokker's own derived twin, the F.VIII. The latter was a hefty 15-passenger machine, of which two were made by Manfred Weiss in Hungary.

Legendary flights

Until 1934 Fokker's mainstream airliner developments were tri-motors. In 1926 two Dutch-built F.VIIs were ordered for Sir Hubert Wilkins' Arctic expedition. One was a Liberty-F.VIIA, but the other was a tri-motor of a new and enlarged type which gradually became the standard model, the F.VIIB-3m. This had a bigger wing of 67.54 or 67.72 m² (727 or 729 sq ft), but not as big as that on the original F.VII, which was designed at Fokker's Dutch factory for the American Fokker F.9 (see variants list). Thus the second machine was a unique in-between, for it retained the 1491-kW (200-hp) Whirlwind and fuselage of the F.VIIA. After expedition flying in Alaska this special long-range Fokker was overhauled by Boeing and sold to an Australian, Squadron Leader Charles Kingsford Smith. Re-engined with 176.7-kW (237-hp) J-5 Whirlwinds and boldly painted with the name *Southern Cross*, the Fokker took off from Oakland, California, on 31 May 1928. This was the start of a career that was to make the Fokker the most famous aircraft between the world wars, surpassing even Lindbergh's Ryan. Some of its flights figure in a separate listing of famous Fokker flights. Today it is on display at Eagle Farm, Brisbane.

The 'big-wing' tri-motor had great potential, especially with greater power, and the F.VIIB-3m was formally marketed from 1928, usually with the J-6 series Whirlwind of 223.7-246.1 kW (300-330 hp). The first production example was another 'special' built in late 1927 as the *America* for the Byrd transatlantic flight. Later renamed *Friendship* (a prophetic name for Fokker) it participated in many further epic flights. Several of the enlarged tri-motors were F.VIIA rebuilds, one of the most famed starting life as H-NADP, powered by a Jupiter. This carried a wealthy American on the first KLM passenger flight to the Dutch East Indies (see list of flights). In 1928 it was rebuilt as an F.VIIB-3m with Gnome-Rhône (Bristol) Titans, and was again hired by Black for a trip to Cape Town. It was later fitted with Whirlwinds and almost commuted between Croydon, Cape Town and the Far East, amazing for 1929-30.

Licence-building

The F.VIIB-3m was the world's premier airliner of the 1927-33 period. Fokker and Atlantic built 147, and basically similar machines (with many makes of engine) were built under licence in Belgium, Czechoslovakia, France, Italy, Poland and the UK. Nine of the licensed sub-types were military, four being bombers. Probably the most important tri-motor Fokker bombers, however, were the Avia F.IXs. The Czech Avia company had built 21 F.VIIB-3m aircraft, one a bomber, and it followed with 12 F.IX bombers based on the Fokker F.IX, a much more powerful machine with 372.9-kW (500-hp) Walter-built Jupiters and able to take off at a weight of 9000 kg (19,842 lb). An ex-KLM F.IX later found its way to the Spanish Republicans where it too became a bomber.

Fokker's American team ended in a blaze of glory with 10 giant F.32s, the designation indicating the number of seats. Powered by four push/pull 428.8-kW (575-hp) Pratt & Whitney Hornets, they

The F.VIIB-3m c/n 5028 was delivered for use on the 1928 Antarctic expedition planned by Lieutenant Commander Richard E. Byrd. In the event, a Ford Trimotor was used instead and the Fokker, named **Friendship**, *was fitted with twin floats and, with the great Amelia Earhart as passenger, made a record transatlantic flight.*

The US Army received eight **C-2A** *aircraft, which differed from the normal F.VIIB-3m in having a larger, 10-seat fuselage, a modified cockpit and three 164-kW (220-hp) Wright J-5 engines. This example, named* **Question Mark**, *was used to establish an inflight-refuelled endurance record of 150 hours during January 1929.*

The second F.VIIA-3m (c/n 4917) was delivered by Fokker to the Royal Aircraft Establishment in British military colours in April 1927. It was evaluated as a potential transport type and carried the serial number J7986. This aircraft was later modified to test the experimental Steiger Monospar ST.2 wing design.

were generally judged too big for the market. But back in Amsterdam the traditional Fokker tri-motors just kept on being improved. In 1930 the important F.XII made its appearance, with an 82.96-m² (893-sq ft) wing and 316.9-kW (425-hp) Pratt & Whitney C-series Wasp engines. Normally seating 16 passengers, the F.IX was the KLM main liner and also served with Sweden's ABA and Denmark's DDL, two actually being built in Denmark. This model in turn led to the further-enlarged F.XVIII of 1932. These proved to be the ultimate production Fokker tri-motors. KLM's incredibly experienced crew did marvels with what were becoming outdated machines. F.XVIII PH-AIP *Pelikaan* took 1933 Christmas mail to Batavia in 73 hours 34 minutes flight time, and PH-AIS *Snip* took 1934 Christmas mail to Curaçao in under 56 hours.

Overtaken by technology

By 1932 Fokker was beginning to notice the advantages of all-metal stressed-skin construction, but doggedly stuck to traditional methods. The result in 1933 was the F.XX, which strove to keep up with the times by having 477-kW (640-hp) Cyclone engines and retractable landing gear. For the first time obvious attention was paid to drag reduction, and the new 12-seater could reach 305 km/h (190 mph). But it now had to compete with the Boeing 247 and Douglas DC-1, and with great reluctance KLM decided in 1933 to talk to Douglas. Fokker still kept trying, and his last two transport designs were the big F.XXXVI flown in June 1934 and slightly smaller F.XXII of a year later. This time the designations followed the American F.32 by indicating the maximum number of passengers. Both were traditional high-wing, steel-tube, fabric and wood machines, but with four engines on the wing, 559-kW (750-hp) Cyclones for the '36' and 372-kW (500-hp) Wasps for the '22'. Four of the latter were built, but only two survived until World War II, when they joined the RAF in Scotland where they operated alongside the sole F.XXXVI.

Fokker F.VIIs were pressed into service with both sides in the Spanish Civil War. This example was one of three F.VIIB-3ms originally delivered to Lineas Aéreas Postales Españolas (LAPE) in November 1933 and used by the Republicans. Makeshift external bomb racks were fitted on spars secured to the window frames.

Famous Fokker flights

2-3 May 1923: Lieutenants Macready and Kelly fly non-stop from New York to San Diego in a Fokker T-2 of the US Air Service, the flight time being 26 hours 50 minutes
24 November 1924: Van der Hoop starts the first airline flight from Amsterdam to Batavia in the Netherlands East Indies in a Fokker F.VII of KLM
27 July 1925: M. Grase sets an endurance record of 3 hours 30 minutes 30 seconds with a payload of 1500 kg (3,307 lb) in a Fokker F.VIIA
September 1925: Anthony Fokker wins the Ford Reliability Trial in the Fokker F.VII-3m
9 May 1926: Floyd Bennett flies the Byrd expedition's Fokker F.VII-3m from Spitzbergen over the North Pole
28-29 June 1927: Lieutenants Hegenberger and Maitland of the US Army fly the Fokker C-2 *Bird of Paradise* from Oakland, California, to Wheeler Field, Hawaii, in 25 hours 15 minutes
29 June-1 July 1927: The F.VIIB-3m prototype is flown by Bernt Balchen and Byrd from Roosevelt Field, USA, to Ver-sur-Mer, France, in 43 hours 21 minutes in fog
15-23 July 1927: An F.VIIA of KLM flown by Geysendorffer and Scholte makes the first revenue flight from Amsterdam to Batavia
31 May 1928: Charles Kingsford Smith and his crew in an F.VIIB-3m make the first aerial crossing of the Pacific Ocean from Oakland, California, to Sydney in Australia (via Honolulu, Suva and Brisbane) in 88 hours for the 12555-km (7,800-mile) trip
10 June 1928: Flown by Barnard and Alliott, an F.VIIA sets off for a round trip from the UK to India; the outward leg was delayed by the need for an engine change, but the return leg was accomplished in 4½ days
17-18 June 1928: Amelia Earhart piloted by Wilmer Stultz becomes the first woman to cross the Atlantic Ocean by air, flying from Trepassey Bay to Burry Port in the F.VIIB-3m *Friendship*
10-11 September 1928: Kingsford Smith makes the first aerial crossing of the Tasman Sea from Australia to New Zealand in an F.VIIB-3m; he makes the return flight in October of the same year
January 1929: The Fokker C-2A *Question Mark* is flown by a US Army crew (Spaatz, Eaker, Quesada and Halvorsen) for an inflight-refuelled record flight of 150 hours 40 minutes 15 seconds
10 July 1929: Kingsford Smith and his crew arrive at Croydon from Australia in their F.VIIB-3m after a flight of 12 days 18 hours
24-26 June 1930: Kingsford Smith and his crew fly from Portmarnock to Harbour Grace and New York in their F.VIIB-3m

Fokker F.VIIB-3m cutaway drawing key

1 Starboard wing tip tie-down shackle
2 Starboard navigation light
3 Aileron cables
4 Aileron control horn
5 Starboard aileron
6 Plywood wing skinning
7 Fixed trailing edge construction
8 Rear spar
9 Wing ribs
10 Front spar
11 Leading edge nose ribs
12 Starboard engine nacelle mounting struts
13 Engine instruments
14 Control cable duct to engine nacelle
15 Cooling air louvres
16 Starboard main undercarriage leg
17 Starboard engine
18 Three-blade propeller
19 Exhaust collector ring
20 Wright J-6 Whirlwind nine-cylinder radial engine
21 Engine accessories
22 Engine mounting struts
23 Fireproof bulkhead
24 Oil cooler
25 Centre engine oil tank
26 Oil tank filler cap
27 Cockpit floor level
28 Nose baggage compartment
29 Landing/taxiing lamp
30 Wind driven generator
31 Undercarriage strut mounting
32 Mail locker
33 Rudder pedal bar
34 Elevator control linkages
35 Instrument panel
36 Windscreen panels
37 Co-pilot's seat
38 Control column handwheel
39 Pilot's seat
40 Radio

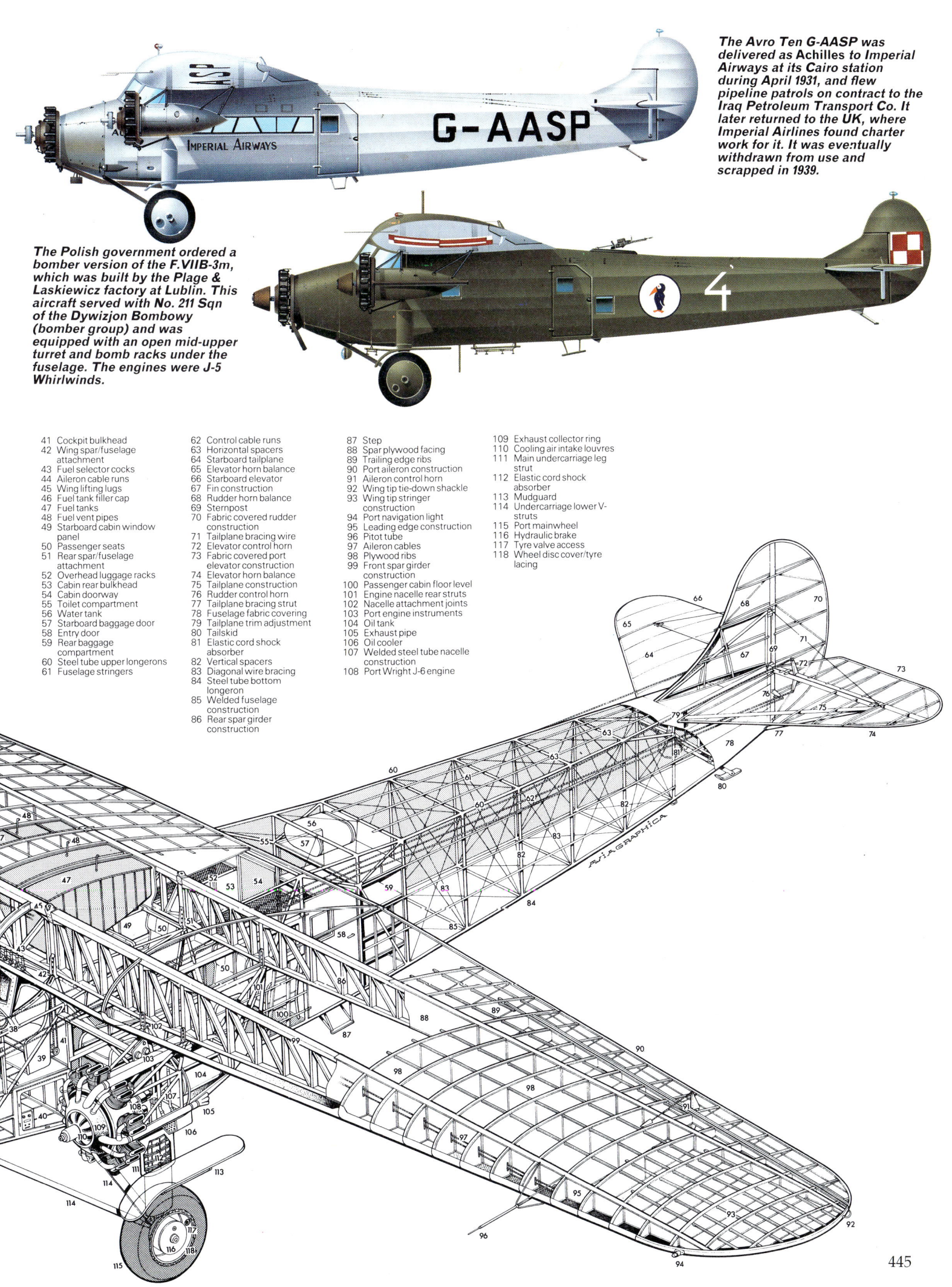

The Avro Ten G-AASP was delivered as Achilles to Imperial Airways at its Cairo station during April 1931, and flew pipeline patrols on contract to the Iraq Petroleum Transport Co. It later returned to the UK, where Imperial Airlines found charter work for it. It was eventually withdrawn from use and scrapped in 1939.

The Polish government ordered a bomber version of the F.VIIB-3m, which was built by the Plage & Laskiewicz factory at Lublin. This aircraft served with No. 211 Sqn of the Dywizjon Bombowy (bomber group) and was equipped with an open mid-upper turret and bomb racks under the fuselage. The engines were J-5 Whirlwinds.

41 Cockpit bulkhead
42 Wing spar/fuselage attachment
43 Fuel selector cocks
44 Aileron cable runs
45 Wing lifting lugs
46 Fuel tank filler cap
47 Fuel tanks
48 Fuel vent pipes
49 Starboard cabin window panel
50 Passenger seats
51 Rear spar/fuselage attachment
52 Overhead luggage racks
53 Cabin rear bulkhead
54 Cabin doorway
55 Toilet compartment
56 Water tank
57 Starboard baggage door
58 Entry door
59 Rear baggage compartment
60 Steel tube upper longerons
61 Fuselage stringers
62 Control cable runs
63 Horizontal spacers
64 Starboard tailplane
65 Elevator horn balance
66 Starboard elevator
67 Fin construction
68 Rudder horn balance
69 Sternpost
70 Fabric covered rudder construction
71 Tailplane bracing wire
72 Elevator control horn
73 Fabric covered port elevator construction
74 Elevator horn balance
75 Tailplane construction
76 Rudder control horn
77 Tailplane bracing strut
78 Fuselage fabric covering
79 Tailplane trim adjustment
80 Tailskid
81 Elastic cord shock absorber
82 Vertical spacers
83 Diagonal wire bracing
84 Steel tube bottom longeron
85 Welded fuselage construction
86 Rear spar girder construction
87 Step
88 Spar plywood facing
89 Trailing edge ribs
90 Port aileron construction
91 Aileron control horn
92 Wing tip tie-down shackle
93 Wing tip stringer construction
94 Port navigation light
95 Leading edge construction
96 Pitot tube
97 Aileron cables
98 Plywood ribs
99 Front spar girder construction
100 Passenger cabin floor level
101 Engine nacelle rear struts
102 Nacelle attachment joints
103 Port engine instruments
104 Oil tank
105 Exhaust pipe
106 Oil cooler
107 Welded steel tube nacelle construction
108 Port Wright J-6 engine
109 Exhaust collector ring
110 Cooling air intake louvres
111 Main undercarriage leg strut
112 Elastic cord shock absorber
113 Mudguard
114 Undercarriage lower V-struts
115 Port mainwheel
116 Hydraulic brake
117 Tyre valve access
118 Wheel disc cover/tyre lacing

445

Chapter 21

Fokker F.27 Friendship

Fokker F.28 Fellowship

Ford Tri-Motor

Gulfstream

Fokker F.27 Friendship

In 1955 a small twin-turboprop aircraft took off from Amsterdam amidst forlorn marketing prospects; no chance of large home sales, and a nagging worry that the company might have got it wrong. Not only did the aircraft exceed all expectations, but an improved version was then placed in production to take the basic design into the 21st century.

After World War II the famous Fokker company had to begin again from scratch. It sold only a handful of S.11, S.13 and S.14 trainers, but was kept going by licence-building Hawker Sea Furies and Gloster Meteors and by overhauling Douglas DC-3s. The last were available for next to nothing all over the world, yet Fokker's board was convinced the 'DC-3 replacement' was the market at which to aim. But how could anyone replace such a cheap and capable workhorse, except at 10 times the price?

Before the war Fokker had designed the F.24 with two R-1820 engines on a high wing, carrying 36 passengers at almost 483 km/h (300 mph). It had been an outstanding concept, the high wing giving a level floor close to the ground and better wing efficiency, as well as perfect passenger view. Indeed, almost all Fokker airliners had possessed high wings, though the argument that a forced landing could crush the fuselage was valid. Fokker had little money in 1948 but managed to send designers to talk with Boeing and Canadair who were looking at the same market. This clinched the basic concept, but in sharp contrast with the North American companies, Fokker could see the advantages of the turboprop, which promised more power, less weight, safer fuel, greater smoothness and, conceivably, better reliability in the fullness of time.

In August 1950 the company produced its brochure on the P.275. A high-wing 32-seater, with eight seat rows, this had a wing of high aspect ratio and quite small area, sized not for short-field performance but for the best possible efficiency in cruising flight. Remarkably, for fuel was then cheap, the design team led by H.C. van Meerten believed airlines would have good runways but would be interested in burning less fuel. For a time the best engine seemed to be the axial-flow Armstrong Siddeley Mamba, but when Vickers chose the Rolls-Royce Dart for the Viscount this was clearly the engine to pick. Not only did Rolls have an unmatched reputation but the fact that the Viscount had four engines and was already on order for BEA meant the tough and simple Dart would gain airline experience rapidly. It was adopted along with the Rotol (later Dowty Rotol) four-bladed 3.51-m (11-ft 6-in) propeller.

By 1952 the P.275 had gained two more seat rows and detail design then went ahead as the F.27. Van Meerten stuck his neck out slightly in copying de Havilland's use of Redux bonding to join major items of metal primary structure, as well as in the extensive use of glassfibre for unstressed parts. A crucial decision was pressurisation; Saab had rejected it on the otherwise excellent Scandia and had failed to find customers. From the start, the new aircraft was planned to have a long-life structure with doublers and crack-stoppers, and the Viscount-type elliptical window was adopted unhesitatingly. Many British firms made major contributions to all the accessory systems, and though Dowty provided the very neat landing gears it was Dunlop who, as well as contributing wheels, tyres and brakes (important for an aircraft that may fly 12 sectors or more each day) persuaded Fokker to use pneumatic accessory power for the brakes, landing-gear retraction and nose-wheel steering. Roots-type blowers, driven by the engines, were chosen for the cabin air, the heat exchangers being in the rear fuselage. The double-slotted

Two classic commercial designs from the Fokker aircraft company: behind the F.VII is the second F.27 prototype, shortly after acquiring the name 'Friendship'. This aircraft was representative of the Series 100 production aircraft, with 32 seats in a 23.2-m (76-ft) fuselage.

As the first recipient of the licence-built Fairchild F.27, West Coast Airlines offered an important breakthrough for Fokker into the lucrative US market. For today's operators of the type, modification kits are available to increase performance. This particular F.27 was the first Friendship to enter service.

447

The fifth Fokker Friendship off the Dutch production line was delivered to Aer Lingus on 19 November 1958. A Mk 100 airliner powered by Rolls-Royce Dart 511 turboprops, this particular aircraft has seen over 25 years' service with various operators, including NZ National Airways Corporation and Air New Zealand.

flaps were driven electrically, and de-icing was by inflatable pneumatic boots.

In 1953, Netherlands' government approval, with some funding, was obtained for two prototypes plus single static and fatigue test airframes. In 1954 Fokker bought Avio-Diepen, and Frits Diepen headed the new Fokker sales team. From potential operators they got much agreement but no orders, and it says much for the Dutch government that it never wavered in its support, in the way certain others would have done. At just this time, Handley Page's salesman, Group Captain 'Bush' Bandidt, was asking operators the world over what was needed, as a result of which that company built a very similar aircraft but powered by four piston engines (649-kW/870-hp Alvis Leonides Majors). The first H.P.R.3 Herald appeared at the 1955 SBAC show, in Queensland Airlines' livery, and to the Americans this seemed a surer bet than turboprops (though nobody liked the Leonides Major). Fokker began discussion on a possible licence agreement with Fairchild in Maryland, and the US company insisted on a switch to piston engines, such as two Cyclone 14s. Fokker not only managed to resist this, but even persuaded Fairchild it was right, and the deal was signed on 26 April 1956.

A big part in clinching the licence was the aircraft itself, the first prototype, PH-NIV, having started wholly successful flight trials on 24 November 1955. It had 1044-kW (1,400-hp) Dart 507 engines and a 22.25-m (73-ft) fuselage, but the second F.27, by this time named Friendship, had a 0.91-m (3-ft) stretch and 1145-kW (1,535-hp) Dart 511s. Field performance was so good the flaps were simplified to the single-slotted type. Viscount experience quickly swept away all talk of piston engines, and airlines suddenly had confidence in trusting up to 40 passengers to just two turboprops. A major role was played by TAA in Australia, who persuaded Fokker to cure a worrying wing cracking problem by fitting thicker skin, get four more seats in, and modernise the interior. TAA placed an order for six F.27s on 9 March 1956, but the first two customers, a few weeks earlier, were Aer Lingus and Braathens SAFE. Fairchild, which used the designation F-27, quickly sold to West Coast, Bonanza and Piedmont.

Sales came quite well for the next two years, but in 1958 fell to a mere eight by Fokker and 16 by Fairchild; but the costly tooling had by then been completed, the production line was in being and the original F.27 had been certificated to what were then new rules (CAR Pt 4B and SR-422) for turbine transports. The first production F.27 flew on 23 March 1958, the first F-27 on 14 April and very few last-minute changes were needed. West Coast flew the first commercial service with an F-27 on 28 September 1958. Aer Lingus received its first two F.27s, named *St Fintan* and *St Fergal*, on 19 November 1958 and flew the first service in the following month. By 1959 orders had again picked up and the production line at Amsterdam has never looked back. Over the years, Fokker brought in international partners, and later almost the entire fuselage (except

Pilgrim Airlines operated a pair of F.27 Mk 100s on routes in the north-eastern USA and into Canada, the Friendships being the largest aircraft in its fleet. It is a measure of the F.27's performance and reliability that early models continue to serve faithfully.

The Royal Netherlands air force has been a consistent operator of the Friendship, the majority of the 14 in service being of Troopship configuration. This machine, operated by No. 334 Sqn at Soesterberg, is an F.27 Mk 300 and is used as a navigational trainer. The air force has recently taken over the two Dutch navy F.27 Maritimes and operates them from Hato in the Dutch Antilles on maritime patrol missions.

Six F.27 Mk 400M Friendships are operated by the Senegambia Air Force from Dakar, this example being fitted with two 950-litre (209-Imp gal) underwing external fuel tanks. This version can accommodate 46 paratroops, 6025-kg (13,283-lb) of freight or 24 stretchers with nine attendants.

Immaculately finished in the tri-colour trim of French carrier Air Inter is this F.27 Mk 500, one of several to have seen service on the airline's extensive internal network linking many of the French cities with the island of Corsica.

the nose) was made by Dassault-Breguet at Biarritz, the outer wings in Belgium by SABCA, and the flaps, ailerons and dorsal fin in West Germany by MBB. The US line, however, came to an end in July 1973 at the 173rd aircraft, despite the introduction of the FH-227 (FH for the 1964-merged company Fairchild Hiller) with a stretched fuselage seating up to 52 passengers. Other variants are listed separately.

Increased range

Several customers requested increased range, and in 1958 the standard model became the Friendship 200 with 1368-kW (1,835-hp) Dart 528 engines and increased fuel capacity. Subsequently this basic machine has been re-engined with the Dart 532-7 and then the 536-7R, both rated at 1700 ekW (2,280 ehp). From late 1982 the engines have been provided with hush-kits which significantly reduce noise and in particular alleviate the piercing scream of the first-stage compressor which is heard mainly from the front. The Dart has the inestimable advantage of over 110 million hours flown by some

An F.27 Mk 600 in the colours of Libyan Arab Airlines; the operator currently has 14 F.27s on strength. Short-haul flights are the type's domain, with many routes geared to the oil industry sites in the desert.

From the licence production of the F.27 by Fairchild arose the Fairchild-Hiller FH-227, a 'stretched' model which increased passenger capacity to 56 and provided a second cargo section in the aft fuselage. This is an FH-227B, which featured heavy-duty landing gear, redesigned windshield and larger propellers.

7,200 engines, but it is the oldest gas turbine in use and this is reflected in high weight and poor fuel economy. Rolls-Royce could have introduced a replacement engine in the 1970s but left it too late, and today a complete new generation of aircraft is flying or projected with engines by General Electric and Pratt & Whitney. Rolls-Royce has no competitor engine, but it has picked around the edges by introducing the Dart 551 mentioned above, which has improved compressors and turbine, and will save at least 10 per cent in fuel. Retrokits are available for bringing older Darts up to this standard.

Fokker, now under dynamic chairman Frans Swarttouw and his deputy Dan Krook, watched the engine situation with disquiet. By the early 1980s it looked as if the F.27 would simply peter out, and the company concentrated on the jet F.28 and attempts to market the F.29 and (with McDonnell Douglas) MDF-100. But the startling and sustained upsurge in local-service turboprops forced the company to think of an F.27 successor, the P.335. The simplest thing was just to replace the engines, but the insistent word from the market was that a new transport ought to do more than this, and in 1983 project P.335 gradually became a new design. The pneumatic landing gear had never been popular, and Fokker decided to replace it with a conventional 3,000-lb/sq in (207-bar) hydraulic system, with a pump on each engine, which also works the flaps. The electrics are completely updated, with a constant-speed IDG (integrated-drive generator) on each engine serving an AC system. Newer de-icer boots are fitted, and the leading edges are among a wide range of additional parts made in advanced composites instead of metal. The landing gear is refined and strengthened, with double-acting main legs and a twin-wheel nose gear steerable 73° left or right. The environmental (cabin air) system is new, and customers can have an APU (auxiliary power unit) in the tailcone for ground conditioning

Fokker F.27 Friendship Mk 500 cutaway drawing key

1. Radome
2. Weather radar
3. Glideslope aerial
4. Nosewheel doors
5. Taxiing lamp
6. Nosewheel forks
7. Steerable nosewheel
8. Nose undercarriage pivot fixing
9. Nosewheel well
10. Radar transmitter/receiver
11. Nosewheel retraction jack
12. Brake reservoirs
13. Front pressure bulkhead
14. Windscreen wipers
15. Windscreen panels
16. Instrument panel shroud
17. Overhead switch panel
18. Co-pilot's seat
19. Electrical equipment switch panel
20. Instrument panel
21. Rudder pedals
22. Cockpit floor level
23. Side console panel
24. Control column handwheel
25. Pilot's seat
26. Fire extinguisher
27. Cockpit bulkhead
28. Radio and electronics equipment rack
29. Control runs
30. HF aerial cable
31. Aerial lead-in
32. Control cable duct
33. Cargo door operating jack
34. Pneumatic system air bottles
35. Crew entry door/emergency exit
36. Port cargo stowage area
37. Starboard cargo stowage area
38. Cargo/baggage restraint net
39. Main cabin bulkhead
40. Door locking handle
41. Cargo door, 92 in (2.34 m) by 72 in (1.83 m) (open position)
42. Forward fuselage frames
43. UHF aerial
44. Upper fuselage unpressurised (outer) skin/control duct
45. Pressurised (inner) skin panelling
46. Curtained cabin window panelling
47. Cabin wall trim panels
48. Main cabin flooring
49. Door surround structure
50. Twin keel beams
51. Floor beam construction
52. Cabin window panels
53. Forward cabin four-abreast seating
54. Wing/fuselage joint drag strut
55. Wing spar/fuselage attachment joint
56. Centre section spar
57. Glassfibre wing root fairing
58. Optional wing bag tanks, increasing fuel capacity by 504 Imp gal (2289 litres)
59. Inboard leading edge de-icing boot
60. Fuel system collector tank
61. Engine fire extinguisher bottles
62. Starboard main undercarriage front strut
63. Engine cowling panels
64. Propeller spinner
65. Dowty-Rotol four-bladed, variable pitch propeller
66. Oil cooler intake
67. Oil cooler exhaust
68. Centre wing panel inner corrugated skin
69. Outer wing panel bolted joint
70. Wing stringers
71. Starboard wing integral fuel tank; normal fuel capacity 1,130 Imp gal (5136 litres)
72. Landing lamp
73. Fuel filler cap
74. Leading edge de-icing boot
75. Pitot tube
76. Starboard navigation light
77. Static dischargers
78. Starboard aileron
79. Aileron tab
80. Aileron hinge control
81. Two-segment outboard slotted flap
82. External flap hinge
83. Nacelle tail fairing
84. Inboard slotted flap segment
85. Centre section rib construction
86. Flap interconnecting shaft and gearbox
87. Inboard flap track
88. Glassfibre trailing edge fillet
89. Overhead hand baggage bins
90. Passenger service units
91. Rear cabin seating
92. Dorsal spine, air system and control cable ducting
93. Starboard side emergency exit doorway
94. Toilet compartment
95. Fin root fillet construction
96. Control cable access panel
97. Fin tailplane attachment joints

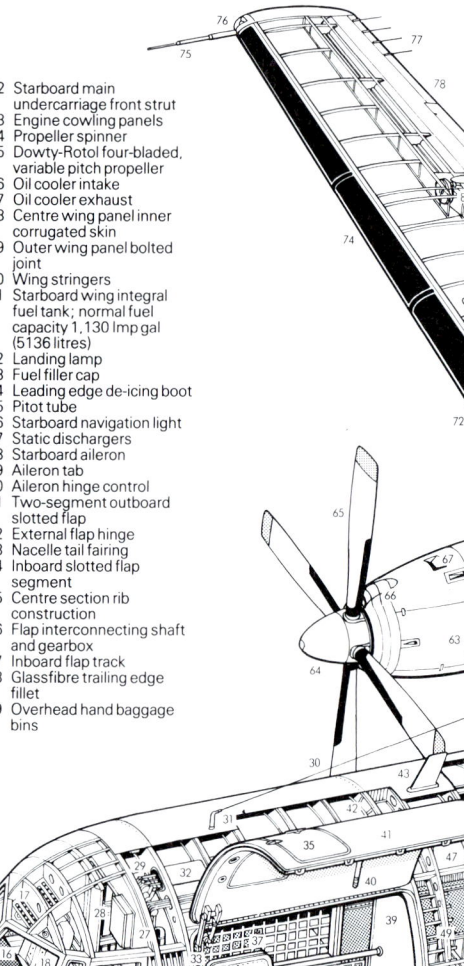

Displaying one of the many roles in which the F.27 has been used is this Dutch air force F.27 Mk 300 Troopship during a paratroop drop mission. The model can be configured for cargo or combined cargo/passenger operations, and is fitted with a reinforced floor.

The F.27's duties include those of an executive and corporate aircraft. Shown here is an F.27 Mk 525CRF, as operated by ARAMCO (Arab American Oil Company), which used these customised aircraft for business in the Middle East.

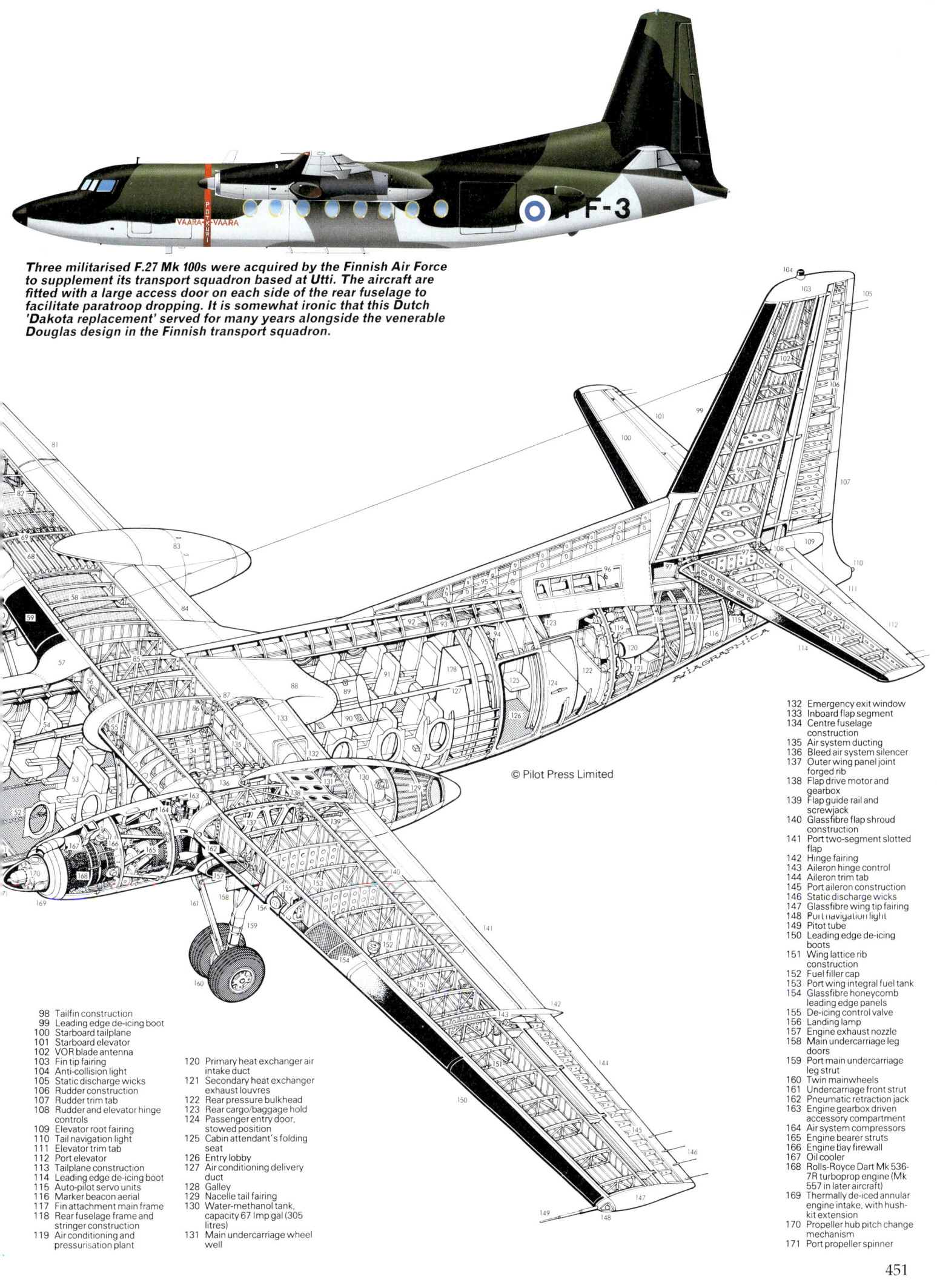

Three militarised F.27 Mk 100s were acquired by the Finnish Air Force to supplement its transport squadron based at Utti. The aircraft are fitted with a large access door on each side of the rear fuselage to facilitate paratroop dropping. It is somewhat ironic that this Dutch 'Dakota replacement' served for many years alongside the venerable Douglas design in the Finnish transport squadron.

98 Tailfin construction
99 Leading edge de-icing boot
100 Starboard tailplane
101 Starboard elevator
102 VOR blade antenna
103 Fin tip fairing
104 Anti-collision light
105 Static discharge wicks
106 Rudder construction
107 Rudder trim tab
108 Rudder and elevator hinge controls
109 Elevator root fairing
110 Tail navigation light
111 Elevator trim tab
112 Port elevator
113 Tailplane construction
114 Leading edge de-icing boot
115 Auto-pilot servo units
116 Marker beacon aerial
117 Fin attachment main frame
118 Rear fuselage frame and stringer construction
119 Air conditioning and pressurisation plant
120 Primary heat exchanger air intake duct
121 Secondary heat exchanger exhaust louvres
122 Rear pressure bulkhead
123 Rear cargo/baggage hold
124 Passenger entry door, stowed position
125 Cabin attendant's folding seat
126 Entry lobby
127 Air conditioning delivery duct
128 Galley
129 Nacelle tail fairing
130 Water-methanol tank, capacity 67 Imp gal (305 litres)
131 Main undercarriage wheel well
132 Emergency exit window
133 Inboard flap segment
134 Centre fuselage construction
135 Air system ducting
136 Bleed air system silencer
137 Outer wing panel joint forged rib
138 Flap drive motor and gearbox
139 Flap guide rail and screwjack
140 Glassfibre flap shroud construction
141 Port two-segment slotted flap
142 Hinge fairing
143 Aileron hinge control
144 Aileron trim tab
145 Port aileron construction
146 Static discharge wicks
147 Glassfibre wing tip fairing
148 Port navigation light
149 Pitot tube
150 Leading edge de-icing boots
151 Wing lattice rib construction
152 Fuel filler cap
153 Port wing integral fuel tank
154 Glassfibre honeycomb leading edge panels
155 De-icing control valve
156 Landing lamp
157 Engine exhaust nozzle
158 Main undercarriage leg doors
159 Port main undercarriage leg strut
160 Twin mainwheels
161 Undercarriage front strut
162 Pneumatic retraction jack
163 Engine gearbox driven accessory compartment
164 Air system compressors
165 Engine bearer struts
166 Engine bay firewall
167 Oil cooler
168 Rolls-Royce Dart Mk 536-7R turboprop engine (Mk 557 in later aircraft)
169 Thermally de-iced annular engine intake, with hush-kit extension
170 Propeller hub pitch change mechanism
171 Port propeller spinner

451

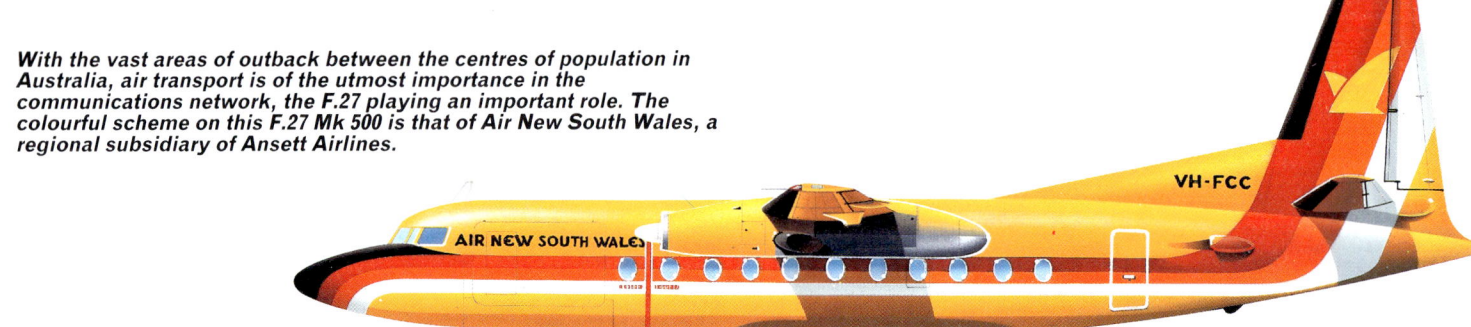

With the vast areas of outback between the centres of population in Australia, air transport is of the utmost importance in the communications network, the F.27 playing an important role. The colourful scheme on this F.27 Mk 500 is that of Air New South Wales, a regional subsidiary of Ansett Airlines.

and electric power. Obviously the engine installations look completely different, but the two features that would strike the observer at once are the futuristic Dowty Rotol six-bladed composite propellers, with digital control, and the Boeing-type row of small passenger windows.

The future takes shape

The crucial decision was engine selection, but after the most careful analysis the choice fell on the Pratt & Whitney Canada PW124, the 1790-kW (2,400-hp) member of the already widely sold PW100 series. Compared with a Dart this offers a cruise specific fuel consumption of around 0.434 instead of 0.578; what this means in practice is that fuel burn is slightly reduced and power is increased, and, as the cruising speed is about 532 km/h (331 mph) instead of 480 km/h (298 mph), the journey is over quicker, thus saving still more fuel. The PW124 and its propellers are much quieter, and handling and economy are improved by digital electronic control. The cockpit is immediately familiar to an F.27 pilot, but has so-called 'state of the art' with improved controls and electronic displays.

Fokker decided to call the new airliner the Fokker 50, its partner being the 100-seat Fokker 100 derivative of the F.28. The first Fokker 50 flew on 28 December 1985, followed by the second on 30 April 1986. The first production machine flew on 13 February 1987. Fokker remained confident that the Fokker 50 would outsell the Fokker 100, such was the switch from jets to propellers, and as a result expected to get a good share of a market foreseen as up to 1,250 aircraft to add to 787 examples of the Friendship series. There were plenty of rivals, one of them the larger British Aerospace ATP with the same engines and seating for up to 72, and to meet this competition Fokker designed all the systems of the Fokker 50 to handle a fuselage stretch of four seat rows, pushing normal capacity from 50 to 66 (though this version appears likely to scrape the tail on take-off).

By 11 June 1989, Fokker had sold 105 Fokker 50s with another 19 on option. The first had been delivered to DLT on 7 August 1987, and production had been set at about 30 aircraft per year. Two maritime variants have been announced: the unarmed Maritime Mk 2 and the armed Maritime Enforcer Mk 2.

In March 1996, however, Fokker went into liquidation and, despite extended talks about rescue packages and the sale of the company, ceased trading in 1997 after the delivery of 212 aircraft of the Fokker 50 series, including four longer-fuselage Fokker 60s.

Photographed in its typical operational environment is a Spanish air force F.27 Maritime, a version specifically tailored to meet the needs of coastal agencies. With an endurance of 10-12 hours, the aircraft is ideal for coastal patrol, fishery protection and SAR duties.

Fokker F28 Fellowship

Having scored a great success with the turboprop F27, it was logical for Fokker to move into the jet market. Here the F28 proved another winner for the Dutch company, offering low-cost, short-range transport from short runways. As such it has proved particularly attractive to Third World customers. In more recent times, the Fellowship spawned a larger brother, the Fokker 100, which continued the sales success of its predecessor.

By the late 1950s Fokker of Amsterdam had achieved a major success with the F27 Friendship, but took note of the prevalent view among airlines, and aviation experts generally, that turboprops would soon be obsolete. The short-haul jet was all the rage. In Britain BEA was looking for a rear-engined jet to fly at 600 mph, and production of such fast and advanced turboprops as the Electra and Vanguard looked like coming to a sudden termination. Clearly, the only thing to do was to build a short-haul jet.

In fact the idea that turboprops were outmoded was totally mistaken. The F27 went on to sell a total of 786, not finishing until 1987, by which time new turboprops were being sold like hot cakes. But to the Dutch company a jet seemed essential, and after careful consideration the F28 Fellowship was announced in April 1962. The closest competition came from the British BAC One-Eleven,

The Stockholm-Arlanda-based Linjeflyg is a well-known Fellowship operator, beginning in 1972 with a trio of Series 1000s. Since then another 17 Series 4000s were added to the fleet, two of the later aircraft being leased rather than bought from Fokker.

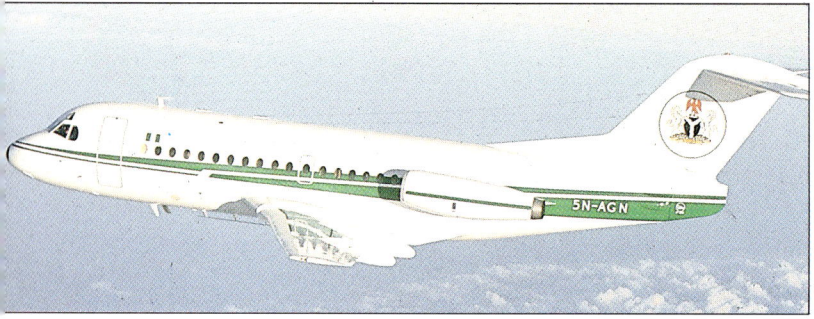

With its smart colour scheme, this Series 1000 aircraft served the Nigerian government on ministerial and presidential transport duties. It was augmented in 1976 by a Gulfstream II and replaced by a Boeing 727.

launched in 1961. Fokker picked the same aircraft layout, with a slightly swept low wing, twin-wheel tricycle landing gear, aft-mounted engines and a T-tail. There were, however, a few subtle differences.

The British jet was aimed at major operators who wanted high speeds and were not too bothered about having to fly from small airstrips. Fokker decided to tailor the F28 to a slightly different market, trading speed and capacity for shorter field length.

Whereas the One-Eleven started with about 89 seats and went on to carry 109, Fokker sized the F28 to carry from 55 to 65 in quite a short cabin. The wing sweep was fixed at only 16°; in fact the trailing edge was made almost unswept. On the other hand powerful double-slotted flaps were fitted, running out on tracks to increase wing area, to give exceptional lift at low airspeeds. These made possible operations from runways as short as 5,000 ft or less, and landings were slowed by very powerful wheel brakes, with lift destroyed by a row of five powerful lift dumpers above each wing, helped by opening airbrakes which when closed formed the entire rear end of the fuselage. Such airbrakes had previously been used on the Buccaneer attack aircraft, and today they are also seen on the BAe 146 jetliner.

Simplified powerplant

Fokker picked almost the same engine as the One-Eleven's, but in a new and slightly simplified version. Rolls-Royce originally called it the Spey Junior, but later marketed as the RB.183. Rated at 9,850 lb each, these were mounted in pods on the sides of the fuselage only just behind the wing, with no water injection and no thrust reversers. Among other things the choice of these engines made the F28 significantly less noisy than the One-Eleven, and this was later to prove a boon in avoiding the need for costly 'hushkits'.

One of the biggest problems facing Fokker was the small size of the home market. This tends to promote a major commercial programme being launched on a collaborative basis, though with Fokker retaining total design authority. For a start the Netherlands Government had to find most of the launch funding, half of it as a repayable loan. In 1964 Fokker signed agreements with companies in West Germany and Britain and was at last ready to go. Fokker, which for a time was linked with a West German company as Fokker-VFW, took on responsibility for assembly, test and marketing, and for manufacture of the nose, centre fuselage and wing-root fairings. The West German partner, VFW-Fokker, built the forward fuselage and complete tail end. MBB in West Germany became responsible for the rear fuselage, including the engine pods and pylons. In Northern Ireland Shorts signed contracts for the complete left and right wings, as well as the landing-gear doors.

In the USA Fairchild-Hiller Corporation was in production with the FH-227 version of the Friendship, and showed immediate interest in the F28. For a time in the mid-1960s the US partner wanted to go ahead with an advanced version designated as the FH-228, with all-American systems and powered by the Rolls-Royce Trent. This extremely advanced three-shaft turbofan promised to be much quieter and more fuel-efficient than the Spey Junior, but eventually Fairchild decided not to go ahead with the FH-228 and Rolls-Royce discontinued the very promising Trent. Instead Fairchild decided merely to market the Fokker-built F28 in the Americas, and, following the original launch order for one aircraft placed by LTU of West Germany in November 1965, Fairchild ordered 10 aircraft for the US market in June 1968. The other early customers, on which the programme rested, were Braathens SAFE of Norway, Itavia of Italy, MacRobertson-Miller (Ansett) of Australia and Martinair of Holland. Together with the LTU and Fairchild orders these totalled just 25 aircraft by mid-1969. This contrasts with US makers who are used to receiving single orders for 50, 80 or even over 100 from their somewhat bigger home market!

Initial disappointment

Fokker built two prototype F28s, which flew on 9 May and 3 August 1967. The home government granted a Certificate of Airworthiness on 24 February 1969 and the first delivery, to LTU, took place on the same day. At that time, with 22 aircraft sold, Fokker was rather despondent. It did not look as if the programme would ever get into three figures, though they knew the aircraft itself was excellent. Among other things the brochure figure for landing speed was 109 mph, only about two-thirds as high as for other short-haul jets. Fokker decided the answer lay in trying harder. Its experienced sales staff visited literally hundreds of possible customers and gradually sold F28s in ones and twos, not only to airlines but also to heads

Sales of the Fokker F28 have mostly been to European, Asian, African and South American nations, although several US operators adopted the type. Empire Airlines no longer operates, but in the early 1980s it flew eight F28 Series 4000s. Note the split tailcone airbrake.

Türk Hava Yollari is the national airline of Turkey, and was an early customer for the F28. Operating Series 1000 aircraft, the first was delivered in December 1972. They have since been sold, the airline now relying on DC-9s, Boeing 727s and A310s.

A major European operator was Air France, with a fleet of 20 being on strength. These belonged to TAT, which took over the aircraft of Air Alsace. Services are flown around France and to neighbouring countries by a mix of Series 1000, 2000 and 4000 aircraft.

of state and to air forces. Meanwhile, back at the factory, the F28 was developed as a family of related aircraft designated as Mk 1000 to Mk 6000 and 6600.

The Mk 1000 was the original version, sized for up to 65 passengers. Originally certificated at 54,000 lb maximum weight, it was later upgraded to 65,000 lb. In 1970 Fokker introduced the Mk 2000, with a fuselage stretched by 2.21 m (87 in) to seat up to 79 passengers. Another variant with the short fuselage was the 1000C (Combi, or convertible), with a big side-loading cargo door and equipped for mixed loads of cargo and passengers.

Chronologically, but not numerically, the next versions were the Mks 5000 and 6000 (first flown on 27 September 1973). These were, respectively, short- and long-fuselage versions of an advanced aircraft with an extra high-lift wing with the span increased from 23.58 m (77 ft 4 in) to 25.07 m (82 ft 3 in), and with slats added along the leading edge. Other changes included the RB.183 Mk 555-1H engine with new noise-reduction acoustic pod liners and silenced nozzles. These versions were cleared to a further increase in weight, but after a while Fokker decided that few customers wanted the slatted wing. The result was the two final versions, the short Mk 3000 and long Mk 4000, which incorporated all the other improved features. Both were cleared to a take-off weight of 33110 kg (73,000 lb) and the Mk 4000 could seat up to 85 passengers.

These continued in production at the rate of about one per month, steadily building up a useful production run almost entirely by finding additional customers for ones and twos. Apart from Garuda of Indonesia and Linjeflyg of Sweden, hardly any customer bought more than 5 or 6, yet when the production line finally closed down in 1986 Fokker had sold 241 Fellowships, to 57 operators in 37 countries. This was a fine achievement, and incidentally outsold the British One-Eleven.

F28 follow-on

Throughout the late 1970s Fokker gave urgent attention to the problem of a successor. There were several possibilities. Rolls-Royce kept offering re-fanned Speys, giving more thrust with much less noise and better fuel economy, but nothing actually went ahead. Fokker continued to produce brochures for Super F28s, with a stretched fuselage and a new wing of so-called supercritical profile, most having a further increase in span. By early 1979 one Super F28 was almost launched, with seating for up to 130 passengers. Agonising doubts about such a long, narrow-body aircraft, allied with the fact that rear engines and a T-tail were going out of fashion, led to further studies. Many customers wanted a basically bigger aircraft, and in 1980 these crystallised as the F29.

The F29 had a fuselage with the same cross-section as the Boeing 727 and 737 (and 757), seating up to 156 passengers in a triple seat on each side of the aisle instead of 3+2. With much greater fuel capacity, the F29 was to weigh 59770 kg (131,770 lb), and the CFM56 or RJ.500 engines were to be hung under the wings, though a T-tail was retained in order to keep the proven effectiveness of the rear-fuselage airbrakes. The F29 was expected to be launched in 1981,

A major customer of Fokker products was Garuda of Indonesia, using the type between Jakarta and many other South East Asian population centres. The fleet once numbered 34 F28 Fellowships, with 12 of the follow-on Fokker 100 on order.

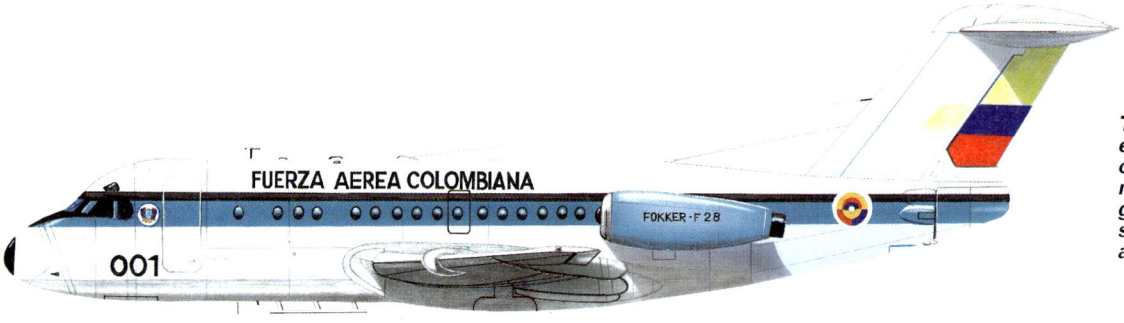

The Fokker F28 Fellowship offers economic and safe operations over good ranges, making it a natural choice for many governments. Colombia is one such nation, and operates this aircraft on VIP duties.

This is the second Fokker 100, proudly displaying the early customers on the forward fuselage. By March 1990 224 sales had been announced, of which 41 aircraft had been delivered.

but by this time Fokker was talking about a joint programme with McDonnell Douglas for an aircraft called the MDF-100.

In the event none of these things happened. Fokker wisely kept both feet on the ground and decided the best solution was a derivative aircraft, which the company and the Netherlands government could afford to launch and which posed the lowest possible risk. At the same time, in 1983, Fokker was coming to the same conclusion regarding a successor to the long-lasting F27. Here again a minimum-change derivative was adopted, designated Fokker 50, the new number being the basic seating capacity. For the new jetliner the designation chosen was Fokker 100, again indicating the basic seating capacity.

Though there were attractive advantages to be gained from a new deep-profile wing, Fokker eventually decided to stay with the original structural wing box, but to add new secondary structure on the front and back. The longer-chord leading edge of the Fokker 100 almost eliminated the kink of the F28 wing and brought important gains in lift, buffet limit and cruising drag. At the rear, the flaps and ailerons were redesigned in detail, and the material changed to carbon-fibre composite. Shorts continues as supplier of the wings.

The fuselage remains pure F28, but again stretched, this time by 5.74 m (18 ft 10 in) to seat 107 passengers in standard layout or up to 119 in high-density. Again MBB is responsible for large sections of the fuselage and tail. The engines, however, are what finally resulted from the refanned Spey studies: the Rolls-Royce Tay.

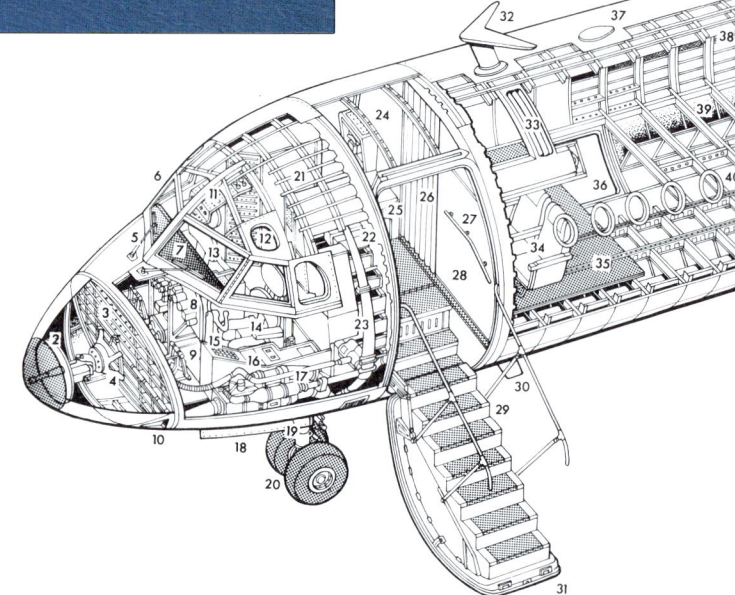

The first production aircraft was delivered to the Ivory Coast government, but airline deliveries began to Swissair on 29 February 1988 with orders for a total of 10 such aircraft.

Though it retains the original high-pressure spool of the RB.183, the Tay gives dramatically more thrust for less noise and improved specific fuel consumption. It completely eliminates the former severe worries about meeting ever more harsh noise legislation being imposed, particularly FAR (Federal Aviation Regulation) 36 Pt III. The Fokker 100 went ahead with the Tay 620-15, rated at 13,850 lb thrust and installed in pods which, together with the effective target-type reversers, are supplied by Grumman (the installation does not differ greatly from that for the Gulfstream IV). With these engines the first Fokker 100 flew on 30 November 1986.

Unfortunately for Fokker both the new aircraft, the 50 and the 100, suffered various development problems which made them late on delivery and cost the company substantial sums. Eventually, deliveries to Swissair began in spring 1988, and after that the Fokker 100 did not look back until Fokker's liquidation in 1997.

Unlike the F28, the Fokker 100 has from the start appealed to major carriers who can buy in substantial numbers. The launch customer was Swissair who signed for an initial eight. Then came KLM with 10, followed by US Air with an initial 20, and the specially-formed leasing company GPA-Fokker 100 Ltd signed for 40. Thus

Fokker F28 Fellowship Mk 4000 cutaway drawing key

1. Radome
2. Weather radar scanner
3. Front pressure bulkhead
4. Radar equipment mounting
5. Windscreen wipers
6. Windscreen frame
7. Instrument panel shroud
8. Back of instrument panel
9. Rudder pedals
10. Ram air intake
11. Cockpit roof control panel
12. Overhead window
13. Co-pilot's seat
14. Pilot's seat
15. Control column
16. Pilot's side console
17. Air conditioning plant
18. Nosewheel doors
19. Nose undercarriage leg
20. Twin nosewheels
21. Cockpit roof construction
22. Radio and electronics' rack
23. Radio rack cooling duct
24. Galley
25. Stewardess' seat
26. Curtained doorway to passenger cabin
27. Handrail
28. Entrance vestibule
29. Entry stairway
30. VHF aerial
31. Main passenger door
32. Upper VHF aerial
33. Air conditioning duct
34. Forward cabin passenger seating
35. Seat rails
36. Freight and baggage hold door
37. ADF loop aerials
38. Fuselage frame and stringer construction
39. Soundproofing panels
40. Underfloor freight and baggage hold
41. Window panels
42. Cabin floor construction
43. Hot air duct
44. Wing centre section front spar
45. HF aerial fixing
46. Leading-edge fence
47. Starboard wing integral fuel tank
48. Fuel filler
49. Starboard navigation light
50. Static discharge wicks
51. Starboard aileron
52. Aileron tab
53. Flap mechanism fairings
54. Starboard outer flaps
55. Outboard spoilers (open)
56. Starboard inboard flaps
57. Inboard spoilers (open)
58. Centre section main fuselage frames
59. Air distribution duct
60. Wing centre section construction
61. Emergency escape windows
62. Mainwheel well pressurised cover
63. Port mainwheel well
64. Cabin window trim panels
65. Rear cabin seating
66. Passenger overhead service panels
67. Overhead luggage bins
68. Cabin rear bulkhead
69. Starboard engine cowling
70. Toilet
71. Wash basin
72. Air intake to APU
73. Fin root fairing
74. Fuselage sloping frames
75. De-icing air duct
76. HF aerials
77. Fin leading-edge de-icing
78. Fin construction
79. Tailplane hydraulic jacks
80. Tailplane de-icing air duct
81. Anti-collision light
82. Starboard tailplane
83. Starboard elevator
84. Tailplane pivot fairing
85. Elevator hinge controls
86. Tailcone fairing
87. Tail navigation light
88. Port elevator
89. Tailplane construction
90. Leading-edge de-icing
91. Tailplane pivot
92. Rudder
93. Rudder hydraulic jack
94. Port airbrake (open)
95. Airbrake jack housing
96. Rear fuselage construction
97. Hydraulic accumulators
98. Exhaust silencer nozzle
99. Engine pylon fairing
100. Bleed air ducting
101. Rolls-Royce Mk 555-15 Spey Junior engine
102. Engine mountings
103. Engine cowlings
104. Auxiliary power unit (APU)
105. Engine mounting beam
106. Air intake
107. Five-abreast passenger seating
108. Underfloor air duct
109. Trailing-edge wing root fairing
110. Inboard flap track fairing
111. Port inboard spoilers
112. Port flaps
113. Flap mechanism fairings
114. Port outboard spoilers
115. Flap construction
116. Aileron tab
117. Port aileron
118. Static discharge wicks
119. Port wingtip
120. Port navigation light
121. Outer wing rib construction
122. Leading-edge de-icing air ducts
123. Leading-edge construction
124. Lattice ribs
125. Wing integral fuel tanks
126. Twin mainwheels
127. Main undercarriage leg
128. Leading-edge fence
129. Undercarriage retraction jack fixing
130. Wing panel bolted joint
131. Corrugated inner wing skin
132. Wing spar attachment frame
133. Leading-edge de-icing air duct

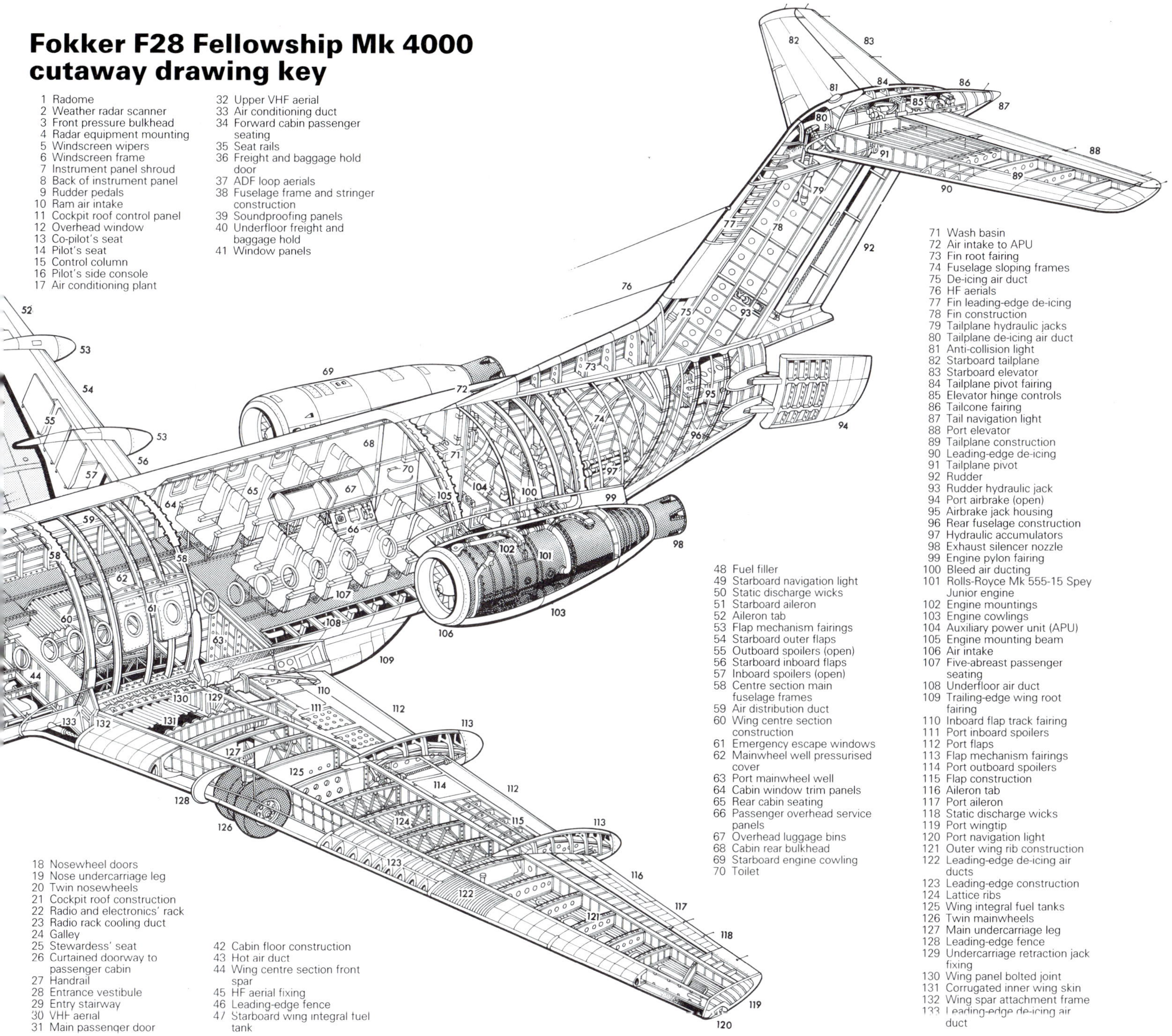

quite quickly, 87 had been sold, plus a further 91 on option. Moreover, US Air wanted a heavier gross weight, and to meet this need Rolls-Royce began in 1986 to develop the Tay 650 rated at 15,100 lb thrust. The heavier version with the uprated engines was certificated on 1 July 1989, and this became the standard model. Fokker offered the type with various interior layouts and a new passenger door, optional wide cargo door for Combi versions, upper-deck avionics and a brightly polished skin.

The development of the Fokker 100, available from March 1995, when the first machine was delivered, was the Fokker 70 with its fuselage shortened by 4.6 m for the carriage of 79 passengers. Other plans were overtaken by Fokker's 1997 failure, and the production line for the Fokker 70 and 100 closed down after the delivery of 48 and 280 aircraft respectively.

Aside from one aircraft for the Argentine air force, AeroPeru was the first South American customer for the Fellowship. Three Series 1000s were delivered in 1972 and 1973, one of which is still in service. The airline flies domestic and short-range international services from Lima.

Few aircraft evoke as much nostalgic interest as the Ford Tri-motor. Consequently they change hands for over a million dollars apiece! Scenic's Model 5-AT basks in the Nevada sunshine at its base at Las Vegas, from where it flies on charters and promotional work. This aircraft has worked all its life, and has not been restored during its 60-year career.

Ford Tri-motor

Popularly called the 'Tin Goose', the Ford Tri-motor combined the high-wing three-engine configuration of the Fokker with the tough, all-metal construction of Junkers. This alone was enough for success in the tough airline world of the 1920s, but who could predict that a handful of these machines would still be active in the 1990s?

In spring 1925 car baron Henry Ford announced a reliability trial for transport aircraft, to start at company headquarters in Dearborn, Michigan, and cover much of the United States. He had decided that he ought to move into the field of civil aviation, and sponsored the contest to see who could build the best airliner. Dutchman Anthony Fokker quickly redesigned his established F.VII to have three Wright engines and brought the first F.VII/3m over to the USA in July 1925, finding little opposition and not only winning the Ford tour but attracting fantastic attention coast-to-coast. Henry's son, Edsel Ford, bought the Fokker for the Byrd North Polar expedition, and also convinced his father the high-wing tri-motor was the best answer; but Fokker was not interested in a licence deal.

The Fokker was made of steel tube, wood and fabric, and after taking a lot of advice Henry Ford became convinced the future lay with what were at the time called 'all-metal ships'. Instead of just copying the Fokker, which is what Fokker himself insisted, incorrectly, was what happened, Ford decided to build an all-metal tri-motor high-wing transport of his own. In August 1925 he bought the Stout Metal Airplane Company, whose founder, William Bushnell 'Bill' Stout, had pioneered the internally braced metal monoplane in the USA. From 1922 Stout had built reliable and aerodynamically clean transports, with Liberty, Wright and other engines, with an outer skin entirely of Alclad (described later).

Ford purchased several of these single-engined transports, and like many other companies jumped on the mail bandwagon made possible by the Kelly Act of 1925, which enabled private companies to bid for air mail routes. On 15 February 1926 Ford's Stout Air Services began flying the Detroit-Cleveland and Detroit-Chicago routes. But Ford could see that the future called for a bigger multi-engined aircraft, and he told Stout to build him a tri-motor.

Stout had the ball at his feet, but he rushed through his Model 3-AT as if timing mattered more than the result. It will probably never be known precisely what happened, but what seems beyond dispute is that the Model 3-AT was a crude and unimpressive lash-up. It may have flown only once before it was destroyed in a fire on the ground, which also destroyed the factory at Ford Airport, on 17 January 1926. Ford and Stout had a violent argument which ended in Stout being fired.

In February 1926 Ford decided to proceed with an improved aircraft, the Model 4-AT, whose design was based on that of the Model 3-AT but differed in almost all details and especially in the engine installations, cockpit, landing gear and fuselage. The small engineering team was headed by Harold Hicks and Tom Towle, but it is difficult today to determine whether they or Stout should receive the main credit for this classic and long-lived aircraft.

The first of the successful Ford tri-motors was flown for the first time on 11 June 1926. Most surviving pictures, and most surviving

Tri-motors nos 69, 74 and 75 were all of the Model 5-AT-CS type with twin floats for water-based operation. It is shown as built, with two engines fitted with ring cowls and registered as a US civil aircraft (NC414H). Later it is believed to have served, with at least one other Model 5-AT-CS, with the Chilean air force.

Possibly Model 5-AT-11, this Tri-motor in the 5-AT series was one of four supplied in 1930 to Cia Mexicana de Aviacion, an affiliate of Pan American. All had uncowled Wasp engines, but after World War II various powerplants appeared on Ford Tri-motors throughout Latin America, including cowled Wasp Juniors from BT-13s.

Fords, are of the larger Model 5-AT variety, but the Model 4-AT was as important as its successor.

Though some of the high-wing transports of the day, including the Fokkers, had their wing mounted above the fuselage, the Ford wing was recessed so that the three spars significantly reduced headroom in the cabin. Each spar was a Warren-braced truss assembled mainly by riveting, and the ribs were likewise built up from rolled strip and sections. The wing was a generation earlier in form than the superb multi-spar stressed-skin wings of John K. Northrop, one of which was the basis of the Douglas DC series in the early 1930s. The Northrop (Douglas) wings proved to have hardly any fatigue problem, but Fords' suffered a certain amount of cracking after flying over 5,000 hours, a figure seldom even approached in commercial service between the world wars.

Rugged structure

Like that of Douglas, the Ford wing was made in three parts, the rectangular centre section being integral with the fuselage. The latter was a capacious box, though unlike that of the Fokkers it had a rounded top which may have reduced drag. The three engines, which in the Model 4-AT were 149-kW (200-hp) Wright J-4 Whirlwinds driving two-blade metal propellers, were uncowled and in most Fords not even provided with cooling airflow baffles or fairings.

In the Model 3-AT the nose engine had been mounted low to balance the thrust lines of the two wing-mounted engines, but in the Model 4-AT the centre engine was mounted exactly on the nose and the other pair brought down so that all thrust lines were at the same level, the wing engines now being carried on the front of separate strut-mounted nacelles well below the wing. The wing engines had short stub exhausts or, from 1927, an exhaust manifold leading to a short pipe on the outboard side of the nacelle. The nose engine, however, sent its gas through a pipe running (in most versions) half way back to the tail, surrounded by a muff open at the front so that, in cold weather, heated fresh air could be admitted to the cabin.

Unlike those of many later metal stressed-skin aircraft, the movable control surfaces, as well as the fixed tail, were metal-skinned. The extreme nose and the top and bottom of the fuselage were often skinned with smooth metal sheet, but almost all other surfaces were of Junkers-style corrugated sheet. Such sheet could be made in thin

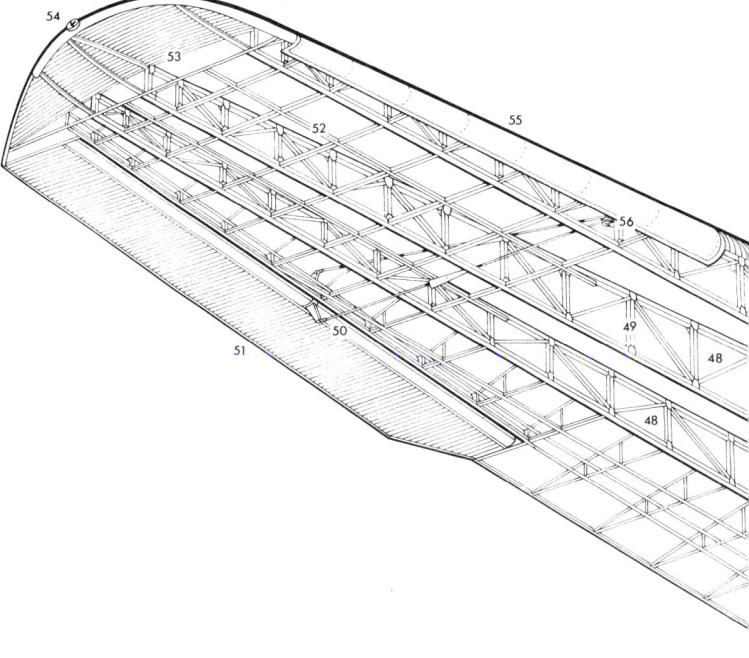

Specification
Ford Model 5-AT-B
Type: passenger transport
Powerplant: three 313-kW (420-hp) Pratt & Whitney Wasp C-series nine-cylinder radial piston engines
Performance: maximum speed 259 km/h (161 mph); cruising speed 198 km/h (123 mph); normal range 708 km (440 miles)
Weights: empty (typical, as built) 3447 kg (7,600 lb); maximum 5738 kg (12,650 lb)
Dimensions: span 23.72 m (77 ft 10 in); length 15.3 m (50 ft 3 in); height (tail down) 3.66 m (12 ft 0 in); wing area 77.57 m² (835 sq ft)
Accommodation: flight crew of two side-by-side plus normal seating for up to 15 passengers, or (later) 13 plus steward, or 1520 kg (3,350 lb) cargo

US Army Air Corps X (NX) 9652 was the one-off XB-906-1 bomber, built at the firm's expense and using what was mainly a Model 5-AT-D airframe (but with a redesigned vertical tail). There were guns above and below at the rear and a bombardier's station under the cockpit. This aircraft disintegrated fatally in a dive in 1931.

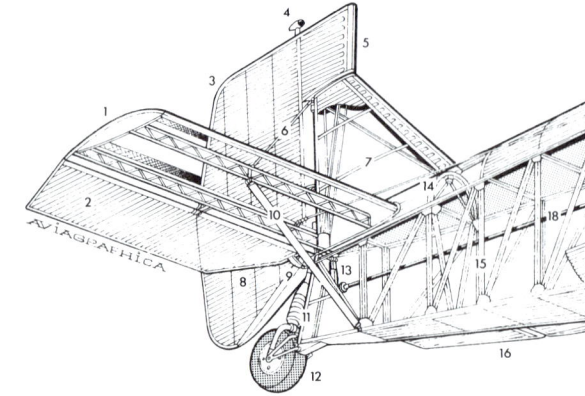

N414H is the last Tri-Motor in commercial service. Owned by Scenic Airways Inc., of Las Vegas, Nevada, it is still in existence and has been used for local pleasure flying (for example, as far as the Grand Canyon). It is used today for promotional and movie work, as well as charter tours.

gauges yet resist local impacts or buckling loads, and it was retained in Junkers aircraft up to the last Ju 52/3m in mid-1944. In fact, it was only after World War II that careful measurements revealed that the drag penalty of corrugated skin was considerable, the air hardly ever flowing in the direction of the corrugations! The one good thing about the skin of the Fords was that, from the first Model 4-AT, the material used was Alclad, duralumin coated with non-corrosive pure aluminium. For good measure the assembled structure was then given a sprayed coating of varnish. This explains the outstanding longevity of these aircraft.

Internal fittings

All the Fords had an oval entrance door on the right side, used for passengers, cargo and also by the two pilots. In the Model 4-AT the latter had the usual open cockpit, a feature of which was a sharply angled Vee windshield with glass panes that were not vertical but sloped outwards from bottom to top, to give a better view for landing. Panels in the windshield could be slid open in really bad weather.

The Model 4-AT had four passenger seats on each side of a central aisle, and was available with a lavatory at the rear. Light personal baggage could be placed in a typical suspended-mesh rack along each side of the cabin above the windows, level with the bottoms of the wing spars. In a few Model 4-ATs further baggage and mail was stowed in drop-down lockers between the spars at the inner end of each outer wing panel, though this was really a feature of later models. Other features by no means common in 1926 were wheel brakes, which enabled a tailwheel to be used instead of a skid, and an electrical system with the battery charged by a generator on the nose engine. This system served navigation lights and, in most aircraft, leading-edge landing lamps and two-way radio. On the other hand, the Fords adhered to the old practice of running the tail control cables along the outside of the fuselage from large rocker arms on the sides of the nose.

One of the first customers was Stout himself, who had formed Stout Air Services to fly schedules from Detroit to Chicago and other cities in the Great Lakes region. But the first and largest customer was a Los Angeles car dealer, Jack Maddux. He first had to be satisfied of the Ford's ability to operate over the mountains of California

Ford Trimotor 5-AT-D cutaway drawing key

1. Starboard tailplane
2. Starboard elevator
3. Rudder
4. Tail navigation light
5. Rudder horn balance
6. Tailplane bracing wire
7. Tailfin
8. Corrugated tailplane skins
9. Elevator hinge strut
10. Tailplane bracing strut
11. Tailwheel shock absorber
12. Tailwheel
13. Tailplane incidence screw jack
14. Fin attachment
15. Rear fuselage construction
16. Port elevator
17. Port tailplane
18. Incidence control shaft
19. Fuselage top decking
20. Corrugated fuselage skins
21. Tail control cable pulleys
22. Flare dispenser
23. Wash basin
24. Step
25. Cabin door
26. Toilet compartment
27. Fire extinguisher
28. Rear cabin seating
29. Cabin roof luggage racks
30. Cabin windows
31. Fuselage strut bracing construction
32. Bottom longeron
33. Starboard mainwheel
34. Shock absorber leg strut
35. Starboard Pratt & Whitney Wasp radial engine
36. Two bladed propeller
37. NACA cowling ring
38. Engine cooling air shutters
39. Engine mounting framework
40. Oil tank
41. Exhaust pipe
42. Engine cowling fairing
43. Engine pylon struts
44. Centre wing panel
45. Wing corrugated skins
46. Spar attachment joints
47. Drop-down mail and baggage lockers
48. Outer wing panel spars
49. Wing spar strut bracing
50. Aileron hinge control
51. Starboard aileron
52. Wing rib bracing
53. Wing tip construction
54. Starboard navigation light
55. Reinforced leading edge
56. Aileron cable pulley
57. Landing and taxi lamp
58. Corrugated leading edge skin
59. Outer wing panel attachment rib
60. Fuel tanks
61. Cabin roof fairing
62. Fuselage main frame
63. Passenger seats
64. Starboard undercarriage swing axle
65. Cabin floor
66. Cabin heater duct fairing
67. Centre engine exhaust pipe
68. Air vents
69. Battery
70. External control cables
71. Co-pilot's seat
72. Cockpit side windows
73. Instrument panel
74. Control column handwheel
75. Cockpit roof windows
76. Sliding windscreen panel
77. Windscreen frame
78. Centre engine fairing air louvres
79. Oil tank
80. Centre engine mounting framework
81. Rudder pedals
82. Exhaust collector ring
83. Centre Pratt & Whitney Wasp engine
84. NACA cowling ring
85. Two-bladed propeller
86. Port Pratt & Whitney Wasp engine
87. NACA cowling ring
88. Engine pylon fairing
89. Port landing and taxi lamp
90. Reinforced leading edge
91. Instrument pitot head
92. Port wing tip
93. Port navigation light
94. Outer wing panel construction
95. Aileron hinge control
96. Port aileron
97. Port mainwheel
98. Mudguard fairing
99. Shock absorber leg fairing
100. Swing axle struts

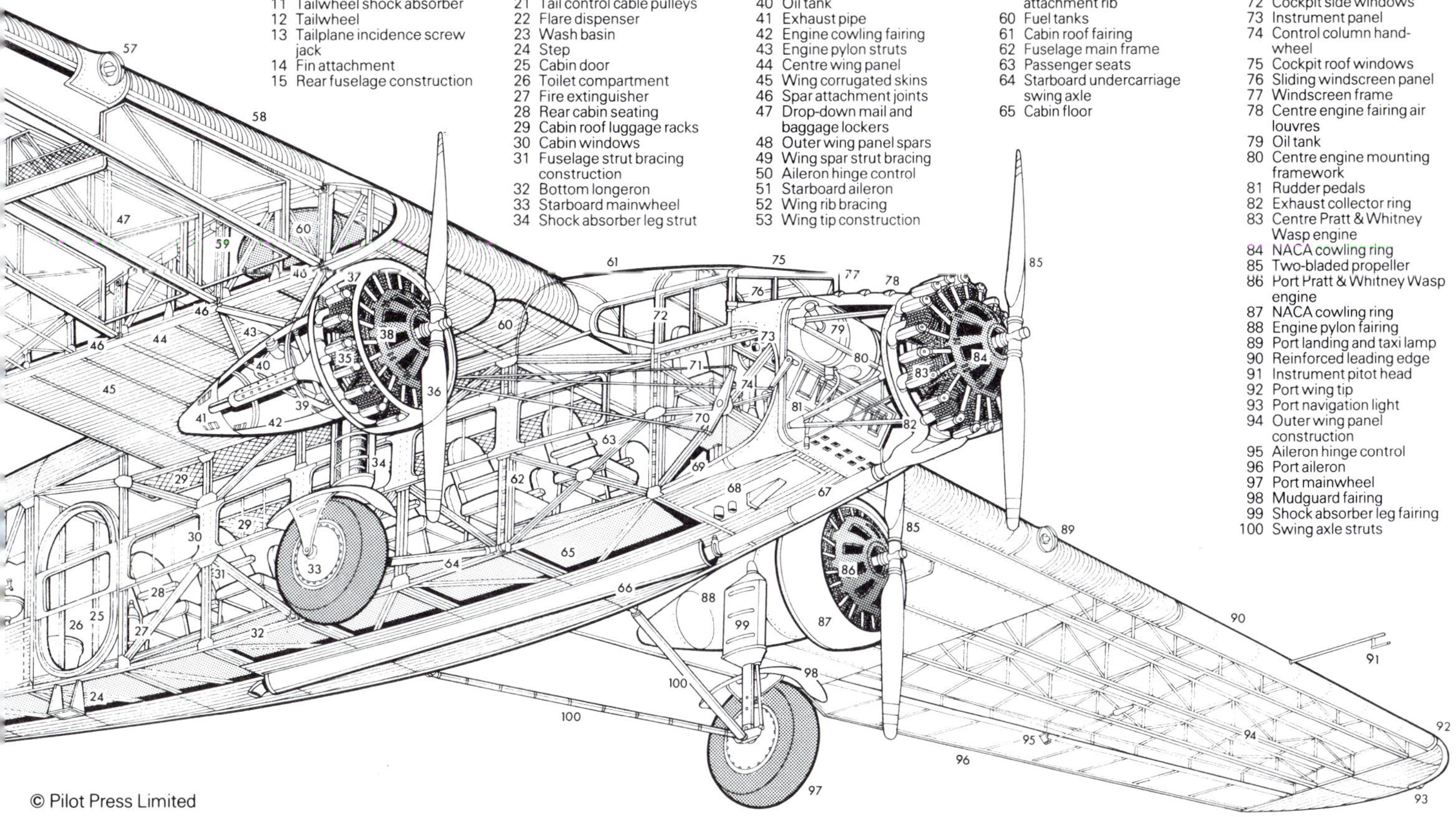

© Pilot Press Limited

The fourth Tri-motor built was the lone XJR-1 ordered by the US Navy in March 1927. A Model 4-AT, it was tested at Anacostia in 1928 and put into service as a cargo and personnel transport before being damaged beyond repair in April 1930. It was the precursor of eight additional Tri-motors in navy/marines service.

and neighbouring States, and of its reliability; then he eventually bought 16 of different versions for Maddux Air Lines.

Subsequent Fords had an enclosed cockpit with sliding side windows, and various other refinements, and almost all were more powerful. As listed in the separate variants section, there were almost as many sub-types as there were aircraft built, partly because there were numerous rebuilds and conversions. The initial Model 4-AT-A eight-seater entered service with Ford's own air service on 2 August 1926. Ford Airport at Dearborn had by this time become the first in the world with paved runways and full electric lighting, and later Ford not only made other improvements to the buildings but also set up one of the world's first training schools for commercial flight crews.

Though the Model 4-AT had been the biggest all-metal aircraft then built in the USA, and possibly in the world, it was rightly judged too small and the Model 4-AT-B of 1927 had a span increased from 20.97 m (68 ft 10 in) to 22.53 m (73 ft 11 in), matching the greater power of 164-kW (220-hp) J-6 Whirlwind engines. Though the fuselage was almost the same as before, the seating limit went up to 12. There followed various sub-types of Model 4-AT, differing chiefly in engine arrangement. Almost all the Fords were originally built with spatted main wheels, and their modern image did much to promote sales.

When the Ford was a new and very modern aircraft, most had Wright engines in the 149/164-kW (200/220-hp) bracket, but production of this Model 4-AT family came to an end (except for the single Model 4-AT-F) in 1929. By this time the much more powerful, and further enlarged, Model 5-AT family was selling even better than its predecessors, and output of Model 5-ATs reached a remarkable four per week in 1929, before the Wall Street crash.

More engines, more sales

Thanks to the use of the Pratt & Whitney Wasp engine, the Model 5-AT was a much more capable aircraft. First flown in mid-1928, the Model 5-AT had a further enlarged wing, the fuselage deepened to

The Model 4-AT-3, the third Tri-motor to be built, typified the early production machines with enclosed cockpit, revised cabin windows and many other changes, but still with wire wheels with mudguards. It operated on the Ford Freight Line from late 1926 with 149-kW (200-hp) Whirlwind J4 engines. Gross weight was 4173 kg (9,200 lb).

After the false start with the Model 3-AT the Stout Metal Airplane division built the first true ancestor of the Tri-motors in the Model 4-AT-1, flown on 11 June 1926. The assistant chief engineer was a young man of Scottish ancestry named James S. McDonnell, a recent graduate from MIT, whose later aircraft went faster.

give more headroom under the wing, Townend-ring cowled wing engines (often a cowled nose engine also), and a structure strengthened for operation at greater weights. No fewer than 117 were built in 1929-31, many of them having float or ski landing gear. Many were built with the wing baggage/mail lockers, and others were thus equipped as a modification. The cabin could be equipped to seat up to 17, with eight seats on the right and nine on the left, even with a rear lavatory.

Airline success

Military models are listed under variants. The original buyers of the 198 airline Fords were: American, British Columbia, CLASSA (Spain), Colonial Western, CMA (Mexico), Curtiss Flying Service, Eastern, Ford, Jefferson, Maddux, Mamer, Mohawk, NAT (National), Northwest, NYRBA (New York, Rio & Buenos Aires), Pacific, Pan American, Pan American Grace, Pennsylvania, Pitcairn, Queen City (Ohio), Rapid, Robertson, SCADTA (Colombia), SAFE (Southwest Air Fast Express), Spokane, Stout, TAT (Transcontinental Air Transport) and Universal Flyers. After 1930 many Fords came on the second-hand market, finding ready buyers. Standard Oil was the first executive customer, buying its first aircraft new, while Royal Typewriter wrote on the fuselage 'capacity 210 portable typewriters'. Several were imported into England, where the aerodrome at Ford, Sussex, was taken over because of its name and turned into

Most famous of all Tri-motors, the Model 4-AT-15 was donated by Henry and Edsel Ford for the Antarctic expedition of Commander Richard E. Byrd. Pilot Bernt Balchen found it needed more range, so the span was increased to 22.55 m (74 ft), tankage was augmented and a powerful Cyclone was put on the nose. It overflew the South Pole.

This unique Ford was Model 6-AT-1, basically a Model 5-AT-C fitted with low-powered Wright Whirlwind J6 engines, sold as a sea/ski-plane to the Royal Canadian Air Force in 1929. It is pictured with only the last two letters of its military registration, G-CYWZ. Its main duty was forest patrol and dusting, followed by radio research.

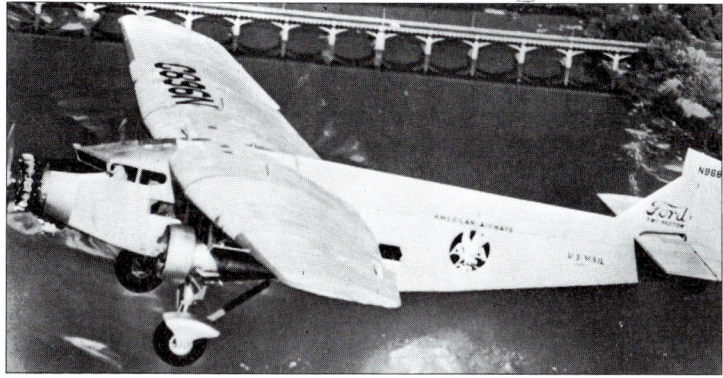

One of the almost priceless survivors, Ford Model 5-AT-39 served with American Airways in 1933-5. Then followed many years of trucking all over the Americas, from Chile to Alaska. In 1962 the still highly airworthy ship was bought by American, completely refurbished and used for publicity and promotional purposes.

A8840 was the US Navy RR-4, basically a Model 5-AT-C with the enlarged wing and three 335-kW (450-hp) Wasp engines with Townend ring cowls. The Tri-motor in the US Naval Aviation Museum at NAS Pensacola is painted and presented as this aircraft, but its true identity is not certain. Today it has no spats.

a maintenance base for what was vainly hoped would be many European tri-motors. In 1934 W. S. Shackleton of Piccadilly sent four to New Guinea to carry mining freight alongside various Junkers, and one aircraft used by the Guinness family even found its way into No. 271 Squadron RAF in 1940, in camouflage and with serial number X5000.

Large numbers served with the US forces, RAAF, RCAF, Colombian air force and Republican Spain, but the most long-lived were the many Fords bought by Latin American operators, notably TACA (then of Honduras) which once had 30. The outstanding single example was Model 5-AT-11 (the eleventh Model 5-AT), which passed through the hands of Pan American and TACA and in August 1945 was bought for $4,500 by TATSA of Mexico to operate between Mazatlan and a tiny strip at the bottom of a canyon serving a mine at Tayotita. It logged a further 5,376 hours on this single-handed task, carrying over 65,000 passengers and 7,390 short tons of freight, without incident. In 1966, with almost 23,000 hours on its books, it was bought by Island Airlines of Port Clinton, Ohio, before being passed on to Scenic Airlines at Las Vegas, which used it for sightseeing over the Grand Canyon. A landing accident badly damaged it, but it is now undergoing restoration in San Diego to bring it back to flying condition.

US Army Air Corps aircraft 31-401 was the first of three Tri-motor Model 5-AT-Ds with the Army designation C-4A. It was the 91st Tri-motor, delivered in 1931. Like most Army Fords it had Townend rings round its R-1340 Wasps and spats over the mainwheels. The Army bought 13 Fords, seven being less-powerful C-3As.

Ford 5-AT Tri-motor
Pan American Airways

The Gulfstream Story

For over 40 years the 'high end' of the corporate transport market, both prop- and jet-driven, has been dominated by a single company. While its name has changed over the years, its aircraft themselves have won universal acclaim, known simply as 'Gulfstreams'.

There are some things in life, be they cars or vacuum cleaners, that have become far better known by their manufacturer's name than any make or model designation. Such names have fallen into everyday usage. If any aircraft can be said to have achieved similar recognition, the most likely candidates can be found in the prestige world of executive jets. Bill Gates' Learjet, through its stylish lines and startling performance, is one such example. At the other end of the spectrum there is another exemplar, one whose name has become synonymous with quality and prestige – Gulfstream.

The story of the masters of the big biz-jet begins at Baldwin, Long Island, with the founding of the Grumman Aircraft Engineering Corporation in 1930. Grumman became known for their naval fighter designs, and during World War II its F4F Wildcat and F6F Hellcat established themselves as two of the finest carrierborne aircraft of their day. The company had little experience of civil aircraft, but in 1946 it produced the G.73 Mallard, a development of the pre-war G.21 Goose amphibian. The disappearance of a military market in the late 1940s was coupled with an increasing demand for commercial aircraft, not least in the emerging market for corporate transports. What aircraft existed were almost entirely converted bombers or rapidly ageing small airliners from the late 1930s.

Grumman's plans were interrupted somewhat by the Korean War, but by the mid-1950s (spurred by the increasingly popular 'missile' theory that predicted an end to all manned military aviation) the company had decided to be the first to build a dedicated executive aircraft. Initial thoughts of building another flying-boat were soon discounted, and the new design was targeted squarely at replacing the DC-3, two of which were used by the corporation itself.

Taking its TF-1 Trader COD/utility transport (itself a development of the S2F Tracker anti-submarine aircraft) as a starting point, Grumman envisioned a similar-looking 12-seater with the same high wing and mid-set tailplane. While the Trader and Tracker were powered by Wright R-1820 piston engines, the new aircraft would be fitted with a pair of turboprops instead. The obvious choice was Rolls-Royce's newly developed RDa.7 Dart engines. In 1956 Grumman began accepting $10,000 deposits to secure a delivery position. Hand-in-hand with an extensive marketing survey (as Grumman would be risking a lot in offering an aircraft with a projected cost of $500,000) came an extensive redesign. A Dart-powered Trader looked increasingly unrealistic, as the original wing had been designed for high-endurance, low-altitude flight, far from Grumman's envisaged environment. Customer response underlined the need for a 560-km/h (350-mph) aircraft with a stand-up cabin, capable of covering up to 3500 km (2,200 miles). By 1957, when the go-ahead was given to project engineer B. J. Harriman, the eventual low-wing configuration was evolving and 250 engineers were assigned to the newly christened Design 159. The G.159 would become far better known as the Gulfstream, yet the origins of the name are somewhat hazy. It was chosen by Leroy Grumman himself, with an eye on the aircraft's transatlantic range, and proved to be a durable one.

On 14 August 1958 the first aircraft (N701G) emerged from its Bethpage home for its maiden flight at the hands of Carl Alber and Fred Rowley. This proved to be one of history's less routine first flights, as an electrical failure shut down the fuel system, forcing a rapid emergency landing before the engines stopped turning. The second aircraft followed on 11 November, delayed by a strike at suppliers, the Pittsburgh Plate Glass Co. The flight test programme was undertaken largely at Stewart Field, Florida, and led to FAA certification to CAR 4B, Part 40 and SR422A 'Special Requirement, Turbine-Powered Airplane' standards, permitting full airline-style operations up to 12192 m (40,000 ft).

One of Gulfstream's latest offerings is the GIV-SP (Special Performance), which boosts the payload of the baseline Gulfstream IV with no increase in empty weight. All GIVs built after September 1992 are to this standard.

The first Gulfstream (N701G), not formally a prototype, undertook most of the early test flying and was later fitted with a cargo door and used as a company aircraft by Grumman.

The unmistakable Academe joined the USN in 1967. Current examples are fitted with an undernose 'TRAM ball'. VA-42 and -128 operated the eight surviving TC-4Cs.

Below: Not strictly within the remit of this feature, in 1979 Grumman launched the Gulfstream I-C, a stretched, commuter version of the original GI. Extending existing aircraft by 3.25 m (10 ft 7 in) to seat up to 34 passengers, Grumman hoped it would be a cheap and quick way to corner the feeder-airliner market. Only eight conversions were undertaken and the project was abandoned in 1984.

preceding years the type had been certified for 24-passenger operations, and featured an optional 1.57 m x 2.08 m (62 in x 82 in) aft cargo door plus optional wing slipper fuel tanks outboard of the engines, and seen a general rise in fuel capacity and operating weights. Grumman had always hoped for military sales of the aircraft and planned a swing-tail freighter, casevac, anti-submarine (the PF-1) and nav-aid calibration versions. During the early 1960s the US Navy repeatedly sought funding for a transport and training version of the Gulfstream that eventually led to the purchase of nine TC-4C Academes from 1967 onwards. These aircraft were fitted with a large, bulbous nose radome that housed the attack radar for the A-6 Intruder, another Grumman product. The cabins were equipped with four A-6 'backseat' consoles and a replica cockpit to train student pilots and B/Ns. Their systems and avionics have now been upgraded to A-6E/TRAM standard. A single VC-4A was acquired by the United States Coast Guard as a VIP transport.

The jet dimension

The Gulfstream was then competing with the first generation of biz-jets: the Lockheed JetStar, North American Sabre, Dassault Falcon and Hawker-Siddeley 125. All of these aircraft had their advantages, but none could match the G.159 for comfort or range. A jet-powered Gulfstream was a logical progression, and Grumman began studies into a swept (25°) wing aircraft, using essentially the same cabin as the G.159, but powered by two Pratt & Whitney JTF10 or Rolls-Royce Spey Junior turbofans. An essential consideration was to retain the original aircraft's 1220-m (4,000-ft) runway performance capability. In tandem with this new aircraft arose studies of a HU-16 Albatross development, powered by General Electric T64 engines, with a hydro-ski hull. Attention was more focused on the new Gulfstream, and the spur was provided again by Rolls-Royce in the shape of the 49-kN (11,000-lb) RB-163-25 Spey engine. Cruising at Mach 0.75 at 13100 m (43,000 ft), the T-tailed G.1159 was to have a maximum range of 3,190 nm (5907 km/3670 miles) with 30-minute reserves. In late 1964 Grumman announced that it would proceed with the programme on the receipt of 50

The first customer delivery (aircraft number four) was made to the Sinclair Oil Company, entering service in October 1959. With firm orders standing at over 30 (early critics of the programme had forecast sales of 15 at best), the production rate was set at three per month; however, the break-even figure was a more daunting 125 to 150 aircraft.

In 1965 the 150th Gulfstream was delivered to Anheuser-Busch, at St Louis. In the

$420,000 deposits. In the event, it went ahead, in May 1965, with substantially fewer than this, although five months before the first flight the desired half century was reached.

Carl Alber, along with Bob Smith, again took charge of another maiden flight as the first Gulfstream II (N801GA) left the ground for a 45-minute hop on 2 October 1966. The first four production aircraft were involved in the flight test programme that was rewarded with an FAA Type Certificate on 19 October 1967. Potentially the Gulfstream II, or GII as it would become increasingly known, could carry 30 passengers, as it was certified to CAR 4b transport standards, but more importantly it could undertake Category II landings. These are bad-weather/poor-visibility arrivals, with a decision height below 60 m (200 ft) and a Runway Visual Range not less than 550 m (1800 ft). As Gulfstream I production came to an end at 200 aircraft in 1969, over 50 GIIs had been delivered. The G.1159 soon established itself as the fastest biz-jet available. In March 1975 one aircraft set a FAI speed record for its class, covering the distance from Cairo to London in four hours, 49 minute and 16.5 seconds at an average speed of 729.59 km/h (453.35 mph).

Space research

A more unusual task came with the selection of the G.1159, in December 1973, as a flying simulator for NASA's Space Shuttle. Two GIIs were modified as Shuttle Training Aircraft (STA) by adding fly-by-wire controls in a replica Shuttle cockpit (left-hand seat only), thrust reversers, speed brakes and direct-lift flaps, but the most obvious change was two large fins under the fuselage centre for lateral control. Both aircraft were used to make dead stick landings at White Sands and Edwards AFBs, from 10668 m (35,000 ft). NASA also obtained a third GII, in 1986, for use as an engine testbed, fitted with an eight-bladed Hamilton Standard propfan in a nacelle above the port wing.

To comply with FAR Part 25 and 36 noise abatement regulations, a hush kit for the GII was developed and fitted as standard from aircraft number 166 onwards, and a more radical change came in the tip-tank-equipped Gulfstream II(TT). The modification was certified in September 1976 and led to a slight improvement in performance and handling due to the increased wing area. Available as an option from aircraft number 199 onwards, the GII(TT) offered a 400-nm (740-km/460-mile) increase in range at Mach 0.72 and a 230-nm (425-km/265-mile) at Mach 0.80. In fact, Grumman considered

Left: The Gulfstream II marked a radical development of Leroy Grumman's original aircraft and finally established the company, outside the United States, as a supplier to the corporate elite.

Right: For nearly 10 years the Gulfstream III inherited the GIIs crown as the ultimate biz-jet. The new aircraft boasts a distinctive, restyled nose with only a single side cockpit window instead of the G.1159's two.

Above: Adding the Gulfstream III wing to the Gulfstream II to produce the GIIB brought about a dramatic improvement in the latter's performance and a not-insubstantial increase in their re-sale value also. This picture of the first aircraft illustrates where the fuselage has been strengthened to carry the new wing.

Right: Another GII development, this time a production-line improvement, was the tip-tanked GII(TT). Several versions of tank were tested at first, and the final maximum gross weight of the aircraft was 29940 kg (66,000 lb).

the tip-tanks to be standard equipment on all aircraft from No. 206 onwards, and customers had to specify their exclusion (at a saving of $30,000 on the then list price of $5.1 million).

In 1969 Grumman had reorganised its company structure, leading to the formation of a dedicated subsidiary in charge of Gulfstream production. In 1973 the American Aviation Corporation, builder of light aircraft such as the AA-1 Lynx and AA-5 Tiger, was bought out, resulting in the formation of the Grumman American Aviation Corporation. A change was also coming in its plans for the Gulfstream (which now found itself related by marriage to the Agcat crop sprayer).

In a move to further improve its product line, Grumman American began to study and market a stretched GII, the Gulfstream III. The plan called for a radically reprofiled 26.6 m (87.3 ft) aircraft with a wing span of 25.7 m (84 ft). The wing would be of central importance to the new design, as it was planned to develop a new supercritical airfoil with end-plate fins, or 'winglets', as designed by Dr Richard Whitcomb at NASA. A supercritical wing is one that functions most efficiently at specific subsonic speeds, the disadvantage of this being that as the aircraft approaches its critical Mach number, airflow over its surface becomes both sub- and supersonic, leading to buffeting and performance fall-off. Despite retaining the familiar Spey engines, the new GIII would have an estimated base price of some $6.4 million, not including avionics or interior outfitting. Though the manufacturer confidently predicted sales of more than 300 over a 15-20-year production life, the mid- to late 1970s saw a dramatic down-turn in the market and just before the 1977 Paris air show Grumman American announced that R&D costs would price it out of the market, and the GIII was cancelled.

A worthy successor

A less adventurous aircraft was proposed instead; there was no question of ignoring a successor to the GII and, furthermore, 40 customers had handed over their deposits. The new wing design was abandoned, but 1.52-m (5-ft) high winglets were retained on what was essentially a refined GII wing. With a cabin 61 cm (2 ft) longer than the Gulfstream II, the G.1159A Gulfstream III would have a more pointed nose with redesigned cockpit windows. In July 1978 the Grumman Corporation announced that it was negotiating to sell its controlling stake in Gulfstream American to American Jet Industries, headed by Allen E. Paulson. Grumman would continue to aid GIII development, and thus receive a royalty on the first 200 built, but control of the aircraft passed on to a new company named for their single product, the Gulfstream American Corporation.

In addition to simple transport duties, USAF C-20s would fly (classified) wartime Presidential Successor Support System missions to collect and safeguard (surviving) government leaders.

A major redesign saw the Gulfstream IV completed with 30 per cent fewer parts in its advanced wing than the GIII, reduced weight and significantly increased fuel capacity.

A major new auto-pilot – the Sperry SPZ-800 EFIS – was a developed for the GIII, and Rolls-Royce further improved the fuel burn of the Spey Mk.511-8 turbofans (offering the same thrust as those on the GII). The 14-18-seat aircraft was a large one, 90 per cent the size and weight of the original BAC One-Eleven. Gulfstream increased the proposed range with eight passengers, three crew and baggage from 3,600 nm (6667 km/4142 miles) to 3,760 nm (6963 km/4327 miles), with an accompanying increase in ramp weight. Long-range cruise speed would be Mach 0.775 as opposed to the GII's Mach 0.75. The Gulfstream II also had a 1.3 g manoeuvring capability at high altitude that provided a Mach range of 0.62 to 0.82 without encountering a buffet boundary, a generous margin between design limitations and available performance. In September 1979 Allen Paulson attended the roll-out of the first Gulfstream III (N300GA), referring to it as "a new definition of the ultimate", with 52 orders already secured. A new production line at Savannah, Georgia, had been established specifically for the

*With deliveries approaching 220 by the close of 1993, the **G**ulfstream IV is easily the company's best selling design, though its development costs were equally high. **G**ulfstream hopes it will still be selling at the end of the century.*

GIII and it was from here that the type first flew on 2 December 1979.

The 258th and last Gulfstream II was delivered in April 1980, and later that month Gulfstream hammered home its supremacy in the long-range biz-jet field with a record-breaking non-stop flight from Savannah to Hanover, Germany, for that year's air and trade show, covering the 7347-km (4565-mile) distance at an average speed of 831 km/h (516 mph). That same aircraft (N303GA, the third production example) set an altitude record of 15850 m (52,000 ft) on a journey from Geneva to Washington, DC, the following May, after a European tour that saw the establishment of seven other records. As a result of the GIII development programme, Gulfstream offered a wing retrofit to existing GII customers. Any owner of the later production 29620-kg (65,300-lb) gross weight version could have their existing fuselage mated with the wingletted GIII wing, improving the efficiency, range and value of their still almost-new aircraft. Forty-three aircraft were thus transformed into Gulfstream IIBs. These aircraft are often referred to as G.1159As, with the Gulfstream III becoming sequentially the G.1159B. Today, Gulfstream is offering Stage 3 hushkits for GIIs and GIIIs, and (to offset the penalties imposed by their increased weight) in association with Seattle-based Aviation Partners has certified for the GII an aerodynamic improvement kit featuring new blended winglets. By late 1993 six kits had been delivered with 16 on order. Installation is being undertaken by Gulfstream and Marshalls of Cambridge as the Gulfstream IIW.

Military machinations

For the first time Gulfstream developed dedicated military versions. The SMA-3 (Support Missions Aircraft) was a GIII configurable for maritime patrol, search and rescue, medevac, cargo or passenger transport duties. The GIII's speed, range and capacity were all instrumental in the aircraft's suitability for these roles. So far the only taker has been Denmark, which operates two aircraft on patrol/SAR duties modified with a large cargo door to starboard and the facility to air-drop stores and marker flares. The SMA-3 lead to the SRA-1 (Surveillance Reconnaissance Aircraft), a dedicated Elint-configured aircraft with an optional pod-mounted sideways-looking airborne radar, synthetic-aperture radar, long-range cameras and ESM gear. There was also the option of underwing weapons hardpoints if the customer so desired, but none were forthcoming and the SRA-1 never entered production. The most significant military users of the GIII are the US armed forces, which operate it under the designation C-20. Between them the USAF (C-20A,

B, C), USN (C-20D) and US Army (C-20E) operate a total of 17 as VIP transports

In early 1983 the Gulfstream went public, offering seven million shares to raise working capital for a new project. That new project would be the Gulfstream IV (G.1159C), which (although the GIII was selling well and had a bright future) was already under consideration at Savannah. A driving consideration behind its design was the imminent introduction of new international environmental regulations that would eventually catch up with the reliable, but loud (and smoky) Spey engine. The answer again lay with Rolls-Royce, and Gulfstream contracted with them for the delivery of 200 new Tay Mk.611-8 turbofans. Even before a mock-up had appeared Gulfstream had obtained 45 firm orders, and it seemed the company could do no wrong. Its financial position was far from secure however, and in 1985 the Chrysler Corporation (in competition with Ford) announced it had acquired Gulfstream Aerospace in its entirety.

The GIV represented another leap forward in range capability. As the GIII had bettered the GII by some 37 per cent, so the GIV brought about a 22 per cent improvement over its predecessor, carrying eight passengers over 4,600 nm (8519 km/5294 miles). The GIV would also be longer (by 1.37 m/4.5 ft), faster and fitted with the most modern of 'glass cockpits', using the Sperry SPZ-8000. The first aircraft (N404GA) was rolled out on 11 September 1985, followed by a first flight eight days later and FAA certification on 22 April 1987. That year the second prototype captured the record for a westbound (against the prevailing winds) circumnavigation of the globe from the Paris air show at Le Bourget. In 1988 the eastbound record (held by a Boeing 747SP) fell too. Production of the GIII halted at 198 (plus two GII-derived prototypes) in September 1988. Gulfstream has again offered a dedicated military version of the GIV, the SRA-4. It offers a similar range of mission fits to the SRA-3, but capitalises on the new airframe/engine combination. Once again the US military has become a customer for a transport version, this time designated C-20F

The next chapter in the Gulfstream story is the ultra-long-range GV. First flown late in 1995, the type is selling very well despite its high purchase price.

Above: The last word in sophistication, the GIV boasts a six-screen (four for the pilot) Honeywell EFIS cockpit incorporating all the flight and navigation displays with a colour weather radar and the potential for MLS and GPS.

Above: The SRA-4 is intended for electronic warfare support, and more active surveillance and reconnaissance duties. Sweden has so far been the only announced customer.

(USAF), C-20G (USN) and C-20H (USAF). The Swedish air force has ordered three aircraft designated Tp102s, two of which will replace Caravelles on Elint-gathering duties.

SSBJ and the Gulfstream V

Plans for a supersonic business jet, once thought to be a vital future project, are on hold. Gulfstream teamed up with Russian fighter-manufacturer Sukhoi, in 1988, to jointly develop the latter's S-21 design, until 1992. The difficulties of working with a Russian firm proved too much, and Gulfstream withdrew from the project.

In 1990 Allen Paulsen, in association with Forstmann Little and Company, bought Gulfstream back from Chrysler, installing William C. Lowe as Chairman and CEO. At the 1992 Farnborough air show Mr Lowe unveiled the Gulfstream IV-SP (Special Performance), which offered an increase in payload/range capability over the standard GIV of 53 per cent. The price remained unchanged. By late 1993, 228 GIVs of both versions had been built, but the GIV-SP is only a stepping stone to Gulfstream's latest offering, the Gulfstream V.

At the 1992 Farnborough show (rather than the traditional venue of the NBAA show) Mr Lowe returned to announce that Gulfstream would begin delivery of its biggest aircraft yet in 1996. Stretched by a further 2.13 m (6 ft 10 in) over the GIV, the Gulfstream V boasts a new wing developed by Vought, along with a larger horizontal tail. The $29.5 million GV represents the world's first ultra-long-range biz-jet, with an unrefuelled range of 6,300 nm (11667 km/ 7,250 miles), placing a non-stop New York-Tokyo or London-Singapore sector within reach. Cruising at 15545 m (51,000 ft) and Mach 0.90, the GV will fly higher and further than any other business jet. Its engines will be the BR 710 turbofan, being jointly developed by Rolls-Royce and BMW. Gulfstream is paying for the $260 million development from GIV revenue and admits that it will be no easy proposition. For the first time, the company has formed several risk-sharing partnerships with Vought, Fokker, Shin Meiwa, Rolls-Royce/BMW and Honeywell. The initial 26 production aircraft have been sold at a unit price of $29.5 million, with the next batch costing $1 million more – a total of 40 has so far been spoken for. While there are competitors such as Canadair's Global Express and the Dassault 9000, Gulfstream has completed tunnel testing and are again first in what is likely to be a small market. First metal was cut in October 1993 and the first flight was recorded late in 1995. Twenty-five GIVs were built in 1993, and by 1995 annual production had risen to 15 GIVs and 30 GVs. With the GV on line, Gulfstream had become, for the first time, a 'two-aircraft' company, with this situation due to last until well into the next century as demand for both types is buoyant.

Chapter 22

Hamilton Metalplane

Handley Page Herald

Handley Page HP-42

Hawker Siddeley Trident

Heinkel He 70

Hamilton Metalplane

The inter-war years in America were ones of phenomonal growth in the still-young world of aviation. The vast expanses of the United States were finally opened up by civil aircraft and one of the most significant, if often ignored, pioneers was Thomas Hamilton's Metalplane. At a time when aircraft were still largely composed of wood, the new Metalplanes were of highly advanced and sturdy construction.

It is beyond contention that one of the great revolutions in air transport between the two World Wars was the introduction of the all-metal aircraft. Ironically, the aircraft which did most to usher in this revolution is today virtually forgotten, unmentioned in most reference books, and unknown to most aviation historians and enthusiasts, despite the fact that the names of the company which made it, and its designer, remain famous to this day.

Thomas Hamilton's Milwaukee, Wisconsin-based Hamilton Metalplane Company eventually became the mighty Hamilton Standard, the world's leading propeller maker, while Hamilton's young assistant James S. McDonnell went on to design the Phantom, Banshee and Voodoo jet fighters.

Thomas Hamilton was one of America's great aviation pioneers. On 28 May 1916, the 16-year-old Tom went solo in a biplane he had built himself, and was the man responsible for introducing William Boeing to aviation. During World War I, Hamilton organised propeller production at a number of factories in the USA and Canada, and was contracted to build floats for US Navy seaplanes. Hamilton was unimpressed by the durability of wooden floats, and their propensity to leak, so he designed his own metal floats. These were remarkably advanced, incorporating a drag reducing step and spray suppressors.

From the design and production of metal floats it was a short step to the design and production of a 'Metal Airplane' (invariably shortened to Metalplane), and Hamilton began work on the H-18 *Maid in Milwaukee* in 1925. The aircraft first flew in 1926, and carried six passengers in the fuselage, with the pilot sitting up front in an open cockpit. As a floatplane the aircraft took second place overall in the Third National Air Tour of 1927, winning the efficiency contest. Later that year it won the transport division of the Spokane Air Races.

Unfortunately, the Hamilton H-18 was too far ahead of its time, and its circular section fuselage was too expensive to build. Accordingly, Hamilton, ably assisted by Professor John Ackerman and James S. McDonnell, set about re-designing the aircraft with a cheaper-to-manufacture slab-sided metal fuselage, similar to that adopted by the contemporary Stout and Ford Tri-Motors. This consisted of an aluminium alloy framework, covered by a corrugated Alclad Duralumin skin. The re-designed aircraft, flown in early 1928, was de-

A true aeronautical pioneer in every sense, the Hamilton was credited with a string of notable achievements in the development of US domestic aviation. Among these was the establishment of the world's first rail-air joint service which saw Metalplanes transferring mail and passengers at the rail head for onward transportation.

Hamilton's first 'Metalplane' was the H-18 Maid In Milwaukee, *seen here with conventional wheeled undercarriage. The circular fuselage was at the time considered too expensive for production, despite the aircraft's efficiency.*

The H-18 was given float undercarriage for the 1927 Third National Air Tour, gaining a creditable second place. Hamilton had earlier developed metal floats for aircraft, which leaked less than wooden ones.

signated the H-45, and was generally known as the 'Silver Streak' (or as the 'Silver Dan' or 'Sea Dan' when fitted with floats).

A quality aircraft

The new aircraft enjoyed a high specification, with a ground-adjustable Hamilton propeller, Bendix brakes, navigation lights, cabin heaters and even what was coyly described as a 'passenger comfort station'. While the original H-18 had been powered by a 164-kW (220-hp) Wright J.4 Whirlwind (and later a J.5) the H-45 was given a 301-335-kW (410-450-hp) Pratt & Whitney Wasp nine-cylinder radial engine, which dramatically increased performance. Despite all this (undreamed of luxury to air travellers of the time) the price of the H-45 was set at only $23,000 (later $24,500), making the new aircraft a bargain. Orders soon began to come in, and the 1928 brochure was optimistic:

"Just as the coming of the airplane marked a new era in the methods of travel so the Hamilton Metalplane marks a new era in aircraft development. Its enthusiastic acceptance has been a splendid tribute to its advanced design – to the greater strength, safety and service of Hamilton all-metal construction. From Alaska to the Tropics, the brilliance of Hamilton performance has become a by-word. Great fleets of Hamilton Metalplanes fly their way between the great cities of the country, bringing into familiar contact the people of widely separated communities, and strengthening their faith in flying by a new standard of safety, speed, reliability and economy.

"The Hamilton Metalplane is constructed entirely of metal, non-corrosive Alclad Duralumin, because metal has proved its superiority in every respect over other materials – because it is weather proof, fire-resisting and non-rusting, because it makes possible a measure of safety and security that can be achieved in no other way – and because durability and continued service are the heart of airplane value. In every detail of its construction, the Hamilton Metalplane is destined to remain America's outstanding single-motored transport. It is luxurious in its appointments, it has unusual speed and power to conquer the unexpected, it is quick and easy to handle, it is low in operating costs and depreciation and high in efficiency. It is, purely and simply, the product of an inevitable combination of sound, proven engineering and acute vision into the future."

Place in history

The Metalplane was a genuinely revolutionary aircraft, its low operating costs pleasing operators, and the 'magic' of all metal construction re-assuring passengers, whose confidence in the aeroplane had been shaken by the structural failures of the Ford Tri-Motor's wooden wing. The Hamilton did point the way to the future, and did change the face of air transport. To talk of great fleets of Metal-

planes was something of an exaggeration, however, since production totalled only 29 aircraft, including the later H-47.

Though the Wasp-engined H-45 was an impressive performer by the standards of the day, the up-engined H-47 marked an even greater step forward. Powered by the 373-391-kW (500-525-hp) Pratt & Whitney Hornet, the H-47 had improved payload and performance, giving a useful competitive edge. At least one aircraft was powered by a 391-kW (525-hp) Wright Cyclone, but such an installation was

This view displays the broad wing which gave the Metalplane its economic performance. The use of metal in the structure allowed the wing to be clear of much external bracing, consequently reducing drag. However, the corrugations did cause extra drag as the airflow across the wing rarely flows exactly chord-wise.

rare. The H-47 was known as the 'Silver Eagle' in landplane configuration, and as the 'Silver Swan' when fitted with floats. The latter were quickly interchangeable with a conventional undercarriage, while skis or special fat 'doughnut' low-pressure tyres could also be quickly fitted. This gave the aircraft useful flexibility and allowed it to operate from virtually any type of surface. Both the H-45 and the H-47 were later modified with greater wing area.

The first and most important Metalplane customer was Northwest Airways, whose Hamilton flights were known as 'Silver Streak' services, to differentiate them from the Stinson Detroiter ('Blackbird') and Ford Tri-Motor ('Grey Eagle') services. The H-45 entered service some months before the Ford 5-AT Tri-Motor, but was not the first all-metal airliner in service in the USA, this honour falling to the Stout 2-AT Tri-Motor. Contemporary Northwest advertising usually concentrated on the Ford Tri-Motor, which, with its three engines and accommodation for 14 passengers, was generally perceived to be more advanced and sophisticated than the single-engined, eight-passenger Hamilton. In fact, the rather longer range of the Hamilton made it in many ways a more useful machine to Northwest, and the

The restored H-47 in flight. A second Hamilton survives, but in poor condition. It is hoped that Northwest may buy the aircraft and spend the money necessary to restore it.

fleet played a vital (if largely unsung) part in the airline's growth. Two of the first H-45s were delivered to Northwest in September 1928, and initial trials, carrying a full load of fuel and sandbags, were highly successful. One evaluation pilot reported that the "ship can be landed without difficulty in any field which the Stinson can be landed, and has a better take-off". Another was more direct: "The more experience I have with the airplane, the better I like it."

Northwest's operations

One of the two H-45s was later lost, in February 1938, in a hangar fire, but Northwest had already taken seven of the more powerful H-47s, so the fleet remained a large one. Northwest's Hamiltons clocked up many firsts and achievements during their long period of service. In 1927 Northwest's Transcontinental service had to shut down in winter because of the severe weather encountered in Minnesota. With Hamiltons, however, Northwest was able to run the service all-year round, using Federal skis in the worst weather. On 1 September they established the first co-ordinated air-rail service in the USA, providing connections with the railways connecting Chicago with the West Coast and the East. In 1938, a Metalplane was used to open the service to Billings, Montana, carrying Chief Big Man and several of his Crow Indian braves. "If the Indians are not afraid of the white man's bird, nobody else should be," trumpeted Northwest's marketing men. Two years later the Hamilton established Northwest's first international service, opening a route to Winnipeg, Manitoba, Canada.

Less significantly, Northwest's Hamilton pilots were among the first commercial pilots to be issued with a uniform (with six instead of four buttons to differentiate them from ground staff, before a 'Wings' insignia was devised). This was usually hidden beneath a voluminous leather coat, since the draughty cockpit was less warm than the cabin, with its efficient little heaters.

Even the arrival of Lockheed Electras didn't mean the end of the Hamilton in Northwest service, which continued in use on various passenger routes and for carrying the all-important mail. Northwest finally sold its last Hamilton in December 1941, after 13 years of service, and by which time DC-3s were on charge, which made the old H-47 look decidedly ancient and rather obsolete.

Two views of an H-47, in use as a company demonstrator. In fact, Boeing's Hamilton Division ran a small airline itself, equipped with Metalplanes. For its day the Metalplane was innovative, introducing all-metal structure for greater durability, safety and strength.

Other Metalplanes served with Twin Cities-Chicago Airline, Universal Airlines (and with American Airways which absorbed Universal in January 1930), and with Coastal Air Freight, Condor Air Lines and Isthmian Airways. The latter company cheekily advertised its short trans-Panama hop as America's fastest transcontinental service! Boeing's Hamilton Division even used the aircraft as the equipment of its own small airline.

Out in the wilds

The Hamilton was popular as a bushplane, serving with distinction with various commercial and governmental operators in Canada, Alaska, and Central America. The aircraft was particularly popular in the Arctic, and the famous explorer Carl van Eissen was killed in the crash of a Hamilton Metalplane. The last examples of the type survived in use until well into the 1950s. One Metalplane actually gained a military designation (VC-89) and briefly wore US military markings and the serial 42-79546. Impressed during World War II, this former Panamanian Transportes Aereos Gelebert aircraft was taken on charge by the USAAC and was briefly used as a VIP transport in the Canal Zone, serving from 11 August 1942 until 24 August 1943.

Immaculately restored by Jack Lysdale of Lysdale Flying Service, with help from Northwest employees, this is the only airworthy Hamilton left, painted in the colours of its most important user (although this example never actually flew with Northwest). A museum place is being sought for this historic machine.

A line-up of five Northwest Hamiltons. The pair nearest the camera are H-45s; the other three H-47s. In Northwest service the landplane Hamiltons were called 'Silver Eagles', and operated the 'Silver Streak' service.

The Hamilton Metalplane's tremendous success had quickly attracted the interest of William Boeing. He bought the company to form the Hamilton Metalplane division of Boeing, swallowing the company whole, design staff, factory and all. Hamilton and McDonnell were soon put to work designing a more advanced Metalplane, which eventually became the classic Boeing Model 247, the world's first truly modern airliner, and Metal-

Northwest were by far the most important operators of the Metalplane, flying them alongside Ford Tri-Motors and Stinson Detroiters on their routes in the north of the United States. Shown here are two H-47s, together with a Model 5-AT Tri-Motor and a Waco Taperwing.

plane production ceased after only 29 had been built. Their studies also led to the Model 307 Stratocruiser.

Remarkably, two Hamilton Metalplanes survive today. The first was originally discovered in 1947 at Boeing Field, Seattle, where it had been inactive for some years, having flown some 5,183 hours and 30 minutes in a long and arduous career with the Ontario Provincial Air Service of Canada (who acquired the aircraft new) and Joe Crosson's North West Air Service. It was bought by Harry McKee, a Northwest Airlines pilot. Rebuilding by Northwest volunteers began in January 1955 but was abandoned in March 1955, after demands on hangar space and workers' time increased. The aircraft was eventually purchased by Jack Lysdale, operator of Fleming Field at South St Paul. He re-commenced restoration in 1972, completing it in 1975, having replaced two thirds of the corrugated skin, the leading edges, the windows, and the whole of the interior. No drawing or blueprints were available, so Lysdale and his team had to make their own. The completed aircraft is a remarkable testimonial to their skill and professionalism.

Aloft again

The aircraft first flew on 16 August 1975, in immaculate Northwest colours, and later that month Lysdale flew it to the annual meeting of the Antique Airplane Association at Blakesburg, Iowa, where it won the grand champion prize, best in monoplane and transport classes, outstanding workmanship and several other prizes. The aircraft flew some 44 hours and 30 minutes before December 1978, when Lysdale grounded the aircraft. This flying included the production of an orientation film for new NWA employees. He hopes to get the aircraft into

Montana to the West Coast at Seattle: Metalplanes were instrumental in establishing these routes, and had to operate in often harsh and primitive conditions. Here an H-45 receives engine maintenance.

the Aero-Space museum at Boeing Field, Seattle.

The second extant Hamilton was donated to Lysdale's restoration project by Wien Alaska after it was discovered by Lysdale's son Steve, a pilot with Alaskan Airlines. Used as a spares ship, the carcass was donated to the Antique Airplane Association's Air Power Museum at Ottumwa, Iowa, in December 1979, for restoration as a static exhibit. No restoration work was carried out and the aircraft was passed first to the Wings of Yesterday Museum at Santa Fe in 1984, and then to Austen Sky Services at Stead Field, Reno.

Austen Sky Services have a small collection of historic aircraft, including a Howard 500, a Lockheed 10, a Waco, and a Cessna 140, and is run as a labour of love by Fred Austen and Jay Carter (pilots with American Airlines) and Fred Patterson (a DC-10 pilot with World Airways). They reluctantly had to decide that to restore the aircraft was beyond their means (it will cost over $100,000 even to restore it to static display condition) and put it up for sale. Pessimistically described as a 'basket case' and only '80 per cent complete' it is missing its horizontal stabiliser, doors, tyres, wheels and engine, but could be restored. Parts have been located in Alaska and in Panama, so perhaps Northwest will step in to save this historic machine, which represents an important part of the company's past. This airframe actually served with Northwest until 1939, when it was purchased by Wien Alaska, who used it until it suffered a minor accident.

The sound structure and performance of the Metalplane led to its adoption by bush operators for many years after its main airline service ended. This may have been the last in commercial service, operated by Wien Alaska at Fairbanks. In front of its hangar are Cessna Airmasters.

Handley Page Herald

The H.P.R.-7 Herald was one of history's many attempts to find a replacement for the irreplaceable Douglas DC-3. It began life with four piston engines and so missed the boat when Fokker introduced its twin-turboprop F27, stealing the Herald's market almost entirely. Despite this, the Herald achieved some notable sales and a large number were still in service during the early 1990s, when the type had something of a renaissance.

Having inherited from the defunct Miles Aircraft Ltd the commercially unsuccessful H.P.R.1 Marathon light airliner, in the spring of 1952 Handley Page (Reading) Ltd began work on another four-engined design intended to meet the needs of short-haul carriers and serve as a replacement for the ubiquitous Douglas DC-3.

The design team was based at Handley Page's factory at Woodley, near Reading, which had also been taken over from Miles, and was headed by chief designer E. W. Gray. Taking as the basis for his design the projected Miles M.73, Gray projected a larger aeroplane designated H.P.R.3 which would utilise the high-mounted outer wings, nose and triple-fin tail unit of the Marathon mated to a new fuselage of circular cross-section to permit cabin pressurisation, which was seen as an essential requirement for future civil transports. Four Alvis Leonides Major 701/1 14-cylinder twin row radial piston engines of 447 kW (600 hp) were selected to power the aircraft. A Marathon was modified to serve as a flying testbed for the engine installation.

The announcement by Dutch manufacturer Fokker in September 1952 of its DC-3 replacement, the twin-turboprop F27 Friendship, prompted some renewed thinking at Handley Page, and consideration was given to future development of the H.P.R.3 as a turboprop. Market research among airlines in Asia, Central and South America and Australasia, where the company perceived its best sales prospects to be, suggested that carriers in the less developed parts of the world were suspicious of the relatively new and unproven turboprop, and sought the simplicity of operation and servicing of piston engines.

In the meanwhile, a single fin and rudder configuration was adopted on the personal instructions of Sir Frederick Handley Page, and on 18 January 1954 approval was given for the construction of two piston-engined prototype H.P.R.3s, to be manufactured at Woodley. The name Herald was adopted later that year.

The first prototype (G-AODE) was completed in mid-July 1955 and taken by road to the Handley Page factory at Radlett, Hertfordshire for final fitting out, making a 30-minute first flight from there in the hands of company chief testpilot Squadron Leader H. G. Hazelden on 25 August. By then Handley Page had received orders for a total of 29 Heralds from Australian National Airways, Lloyd Aereo Colombano, and Queensland Airways, in whose colours the prototype was painted. British charter operator Air Kruise (Kent) Ltd placed an order for six in September 1955, when the Herald made its public debut at the Farnborough air show.

A month later Handley Page gave the go-ahead for production of 100 aircraft in four batches of 25 under the designation H.P.R.4. These aircraft were to incorporate structural improvements to permit installation of two turboprops when a suitable powerplant became available, and were to be configured for 36-44 passengers. The second prototype H.P.R.3 (G-AODF) made its first flight from Woodley on 14 August 1956, and again made its debut at that year's SBAC show at Farnborough, wearing a company colour scheme.

Although the Leonides Major engines had proved reliable in more than a year's test fly-

Many Heralds were still in service in the mid-1990s, with Channel Express, British Air Ferries, Janes Aviation, Aerosource and Aérovias. This machine still carries passengers, configured for 50 seats. It was built as a Series 206 for Canadian airline Eastern Provincial.

The prototype H.P.R.3 was flown in the colours of Queensland Airlines, one of the small group of carriers to order the type. Three of the four original customers were lost due to buy-outs, including the Queensland order.

The second Herald was built as an H.P.R.3 with Leonides power, and was also modified to Dart power, flying in this configuration for the first time on 17 December 1958. It subsequently became the prototype for the Series 200 Herald.

ing, no further customer interest was shown in the piston-engined Herald, and having lost the ANA, Queensland and Air Kruise orders when the airlines were taken over, and with the Colombian contract looking increasingly uncertain because of currency exchange difficulties, in May 1957 Sir Frederick Handley Page decided to abandon the four-engined Herald in favour of the twin-engined H.P.R.7, which would be powered by 1568-kW (2,105-shp) Rolls-Royce Dart 527 turboprops. His decision had been influenced not only by the poor market showing of the Leonides-powered Herald against the rival Dart-powered Fokker Friendship, but also by the reliability, economy and ease of operation which the Dart had exhibited in three years of service in the highly successful Vickers Viscount.

The prototype Herald was grounded in June 1957 for work to commence on conversion to turboprop power. No simple task, this involved design of new engine nacelles which were mounted 76 cm (30 in) further outboard than the innermost nacelles of the H.P.R.3 to provide clearance for the Darts' larger diameter four-bladed propellers; the forward fuselage of the aircraft was lengthened by 51 cm (20 in) to improve the centre of gravity location and also to move the crew entry and baggage doors well forward of the propeller arcs; a large dorsal fairing was added to the fin and aerodynamic changes made to the tailplane's leading edge; the cabin windows, which had been circular 'portholes' on the Leonides-powered aircraft, were changed to a vertical oval shape; and a new fuel system was developed which comprised a 2650-litre (700-gal) flexible bag type tank in the wing centre section and a 606-litre (160-gal) integral tank in each outer wing panel.

Dart into the air

Thus modified, G-AODE flew again from Woodley on 11 March 1958, test pilot Hazelden reporting a marked improvement in handling and performance. Flight time built rapidly over the ensuing weeks, enabling a provisional certificate of airworthiness to be issued on 22 April, with the prospect of full approval by year's end and start of production in February 1959.

The new Dart Herald was due to make its first public appearance at the 1958 Farnborough air show, where Handley Page was modestly confident that its order book would begin to fill, especially since the British aircraft was to be sold at a lower price than the Dutch Friendship. On 30 August, with eight passengers on board, Hazelden took off from Woodley to position G-AODE to Farnborough in readiness for the opening of the show. The aircraft had completed 200 hours of test flying and had only minor checks left to be completed before gaining its C of A. En route to the Hampshire aerodrome, while cruising at 1829 m (6,000 ft), the starboard Dart engine caught fire, quickly burning through the engine mounts and falling from the aircraft, complete with nacelle. Despite partial loss of aileron control, Squadron Leader Hazelden made a masterly forced landing in a field near Godalming, Surrey; all occupants escaped unhurt, but the fire, which had come close to destroying one wing and tailplane, had damaged the Dart Herald prototype beyond economic repair.

The second H.P.R.4 was already being converted, and flew again on 17 December 1958, spending much of the following summer on extensive sales tours of India, the Middle and Far East, South America and Australasia during which it was away from England for a total of 265 weeks and visited 44 countries. The first sale of the aircraft was made in June 1959 when the Ministry of Civil Aviation placed an order

Military interest in the Herald was restricted to Jordan and Malaysia. Shown here is the first of two Herald 207s for the Royal Jordanian air force, prior to its delivery on 22 January 1963. The aircraft were handed over to the Jordanian airline Alia in December of that year.

for three Dart Heralds which were to be leased to British European Airways for use on its Scottish Highland and Islands services.

Handley Page laid down an initial batch of 10 Dart Heralds, the first of which, G-APWA, was a company demonstrator to replace the ill-starred G-AODE. It made its maiden flight on 30 October 1959 and immediately underwent certification acceptance tests leading to the issue of the Dart Herald's C of A on 25 November, more than a year later than first anticipated. In BEA colours, it made a second demonstration tour of South America in the following spring, during which it helped with flood relief operations in Brazil and demonstrated its ability to take off fully laden from Rio de Janeiro and climb straight ahead over the 436-m (1,430-ft) Sugar Loaf Mountain situated two miles from the end of the runway – a feat which no contemporary aircraft in its class could accomplish.

In September 1960 Jersey Airlines ordered six Dart Heralds after the manufacturer had offered to develop a higher-capacity version with a 107-cm (42-in) fuselage extension and a 907-kg (2,000-lb) increase in maximum take-off weight to enable up to 56 passengers to be carried. This version was designated Series 200 and, following its launch, no more 'short body' Dart Heralds, subsequently designated Series 100, were ordered. The demonstrator G-APWA, and BEA's first Series 100, G-APWB, were leased to Jersey Airlines in the summer of 1961 pending delivery of its first Series 200, G-APWE, on 15 January 1962. The second prototype, meanwhile, had once again been modified as the prototype/demonstrator Series 200 and appeared at the 1961 SBAC show equipped with a nose radome for weather radar and wearing the colours of Canadian operator Maritime Central Airways, which had ordered four, two of which were destined for its subsidiary Nordair.

Royal duties

During 1962 G-APWA, which had been equipped with a 21-seat executive interior and 1249-litre (330-gal) pylon-mounted underwing fuel tanks, was used for HRH Prince Phillip's 30577-km (19,000-mile) tour of South America, during which the Duke personally flew the aircraft for 99 of the 127 hours logged on the trip.

Although production of the Dart Herald was building up at Handley Page's Cricklewood and Radlett factories to which it had been moved from Woodley, sales were still slow in coming. Aerolinee Itavia of Italy ordered five Series 203s, the Royal Jordanian air force took delivery of two Series 207s for passenger/cargo/paratroop use (subsequently transferring them to Alia Royal Jordanian Airlines), Arkia of Israel acquired five Series 209s, Globe Air of Switzerland took delivery of four Series 210/211s, Bavaria Fluggesellschaft of West Germany ordered two Series 213s, SADIA SA Transportes Aéreos of Brazil received six Series 214s and Air Manila of the Philippines ordered a single Series 215. These, with the Canadian aircraft delivered to Nordair and Eastern Provincial Airlines, which had taken over Maritime Central, accounted for all 36 of the Dart Herald Series 200s built.

In April 1963 the Royal Malaysian air force placed an order for four military Series 401s, increased two months later to eight aircraft. These had a strengthened cabin floor, underwing hardpoints for external stores and a flight-openable door for paratroop or supply dropping. The first RMAF aircraft, FM1020, made its maiden flight from Radlett on 28 September 1963, and was delivered in November. All eight had been delivered to No. 4 Squadron, RMAF at Kuala Lumpur by November 1964.

During the course of Dart Herald development Handley Page had projected a number of variants of the aircraft which were not proceeded with. Among these were the H.P.R.8 Car Ferry, proposed in 1959 to meet a Silver City Airways requirement for a high-capacity aircraft to replace Bristol Freighters on its busy Lydd-Le Touquet 'fly-drive' services. This aircraft would have had a 9-m (30-ft) centre-section extension to its wings, a flattened oval-section fuselage and front clamshell doors to enable cars to be driven directly on to a vehicle deck ahead of the passenger cabin. An all-passenger version with a panoramic observation lounge replacing the clamshell doors was also mooted, but neither attracted an order, Silver City opting instead for Aviation Traders Ltd's Carvair conversion of the Douglas DC-4.

The H.P.124, designed in 1960 as a replacement for the Royal Air Force's ageing Vickers Valettas in the tactical transport role, was also Dart Herald-based, incorporating an upswept rear fuselage with loading ramp. Although the prototype Dart Herald 200 (which had not been

The Herald prototype in its original form, with four Leonides radial engines and small tail. The fitment of radials was chosen as a result of a misconception concerning small airlines' mistrust of turboshaft propulsion, but the Fokker F27 proved exactly the opposite.

G-APWA was the first production Dart Herald, which was leased to various airlines by the manufacturer, including Jersey Airlines, Autair (illustrated) and SADIA. It was also used by HRH Prince Phillip for a South American trip.

The Herald did not sell well, most of its market being taken by the F27 Friendship, restricting total production to just 50 aircraft. Arkia of Israel took five aircraft, the first of which (illustrated) was delivered on 17 April 1964.

In terms of sales the Herald was unsuccessful, but it proved durable. Many returned to service with UK domestic operators in their later lives, Air UK being an operator. This aircraft had originally flown with Arkia.

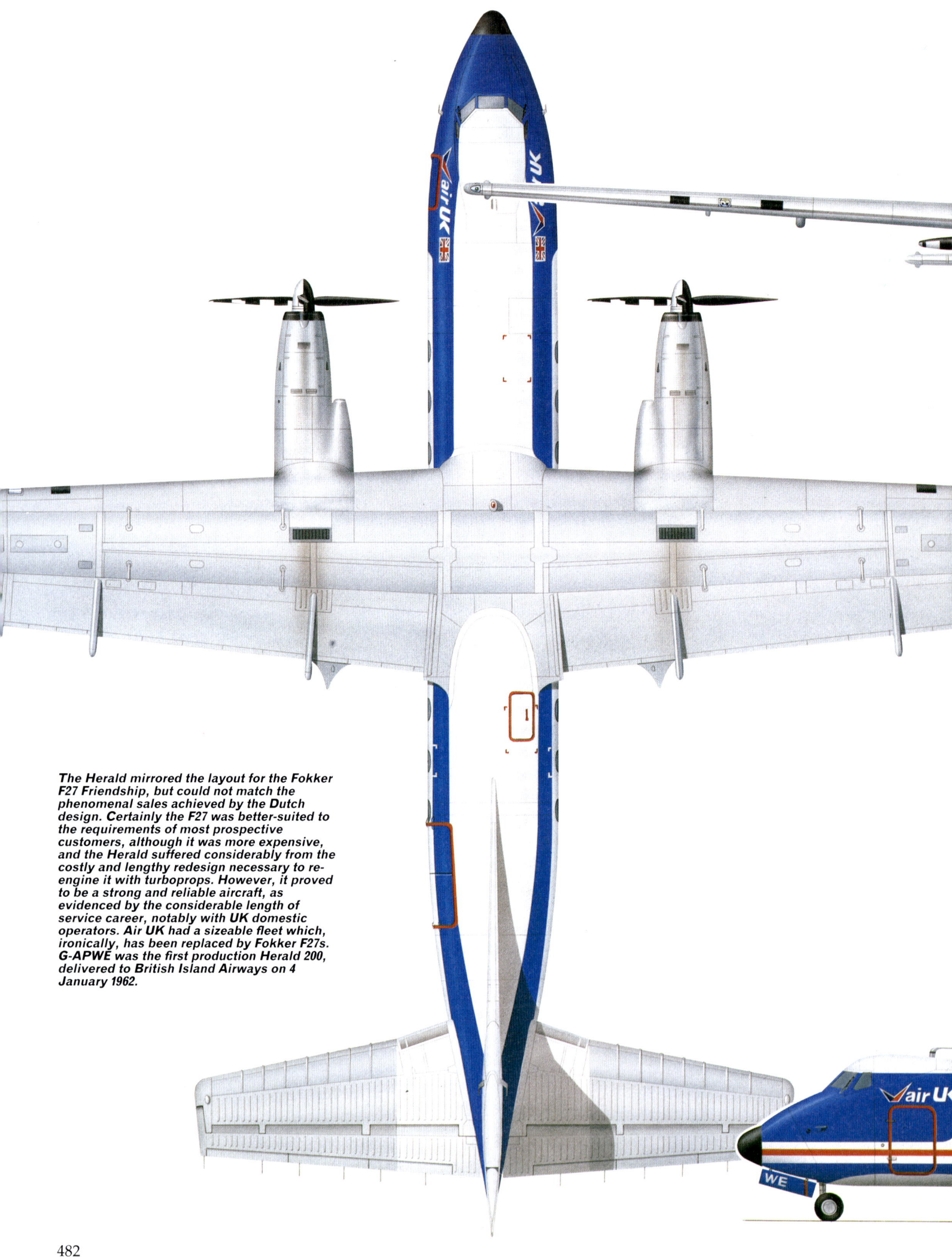

The Herald mirrored the layout for the Fokker F27 Friendship, but could not match the phenomenal sales achieved by the Dutch design. Certainly the F27 was better-suited to the requirements of most prospective customers, although it was more expensive, and the Herald suffered considerably from the costly and lengthy redesign necessary to re-engine it with turboprops. However, it proved to be a strong and reliable aircraft, as evidenced by the considerable length of service career, notably with UK domestic operators. Air UK had a sizeable fleet which, ironically, has been replaced by Fokker F27s. *G-APWE* was the first production Herald 200, delivered to British Island Airways on 4 January 1962.

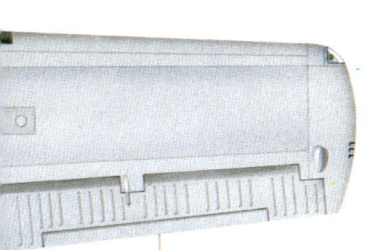

**Specification
Handley-Page H.P.R.7 Dart Herald
Series 200**
Wingspan: 28.9 m (94 ft 9½ in)
Length: 23 m (75 ft 6 in)
Height: 7.34 m (24 ft 1 in)
Wing area: 82.3 m² (886 sq ft)
Powerplant: two 1570-kW (2,105-shp) Rolls-Royce Dart Mk 527 turboprops
Passenger capacity: 50-56
Empty weight: 11703 kg (25,800 lb)
Maximum take-off weight: 19505 kg (43,000 lb)
Maximum speed: 239 kt (170 km/h; 274 mph)
Cruising speed: 230 kt (164 km/h; 264 mph)
Service ceiling: 8504 m (27,900 ft)
Maximum range: 1,530 nm (2834 km; 1761 miles)
Range with maximum payload: 608 nm (1126 km; 700 miles)

The second Herald is seen wearing British European Airways colours, the UK carrier being the first customer for the type, buying three Series 101 aircraft for its Scottish services. They were sold to Autair in 1966.

Re-registered as G-ARTC, this is the second aircraft after modification to Series 200 standard, in which it first flew on 8 April 1961. Maritime Central Airways had ordered a pair of Herald 200s, but these were not delivered to the airline.

Heralds served with French operators Aigle Azur and Europe Air Service, this aircraft wearing the colours of the former. Its Guatemalan registration denotes its sales to Aérovias. It was originally delivered to Brazilian carrier SADIA.

modified to this configuration) fared well in comparative trials against the Hawker Siddeley 748MF in July 1961, it became clear that an order for the aircraft was dependent on Handley Page merging with either the British Aircraft Corporation or Hawker Siddeley, and this fiercely independent Sir Frederick resolutely refused to do. The RAF order went to Hawker Siddeley, whose aircraft entered service three years later as the Andover C.Mk 1.

Similarly stillborn were the Herald-based HP.125 vertical lift military transport, and the HP.127 Jet Herald and HP.129 Mini Herald. These latter two aircraft were civilian jet transports powered by two 17.13-kN (3,850-lb) Rolls-Royce RB.183/1 Spey Junior turbojets mounted in underwing nacelles. The HP.127 would have carried up to 70 passengers in a fuselage stretched by 3.35 m (11 ft) compared to that of the Dart Herald, while the Mini Herald would have carried 30 passengers in a smaller airframe. Models were exhibited at the 1962 Farnborough air show, but neither proceeded beyond the design study stage.

Export variations

Less radical but also abortive developments of the Dart Herald were the Series 300, which was a Series 200 incorporating minor modifications to meet US Federal Aviation Agency certification criteria; Series 500, with 2417-kW (3,245-shp) Rolls-Royce Dart R.Da.12 engines, upper wing spoilers and nosewheel brakes; and the Series 700. This aircraft, 10 of which were ordered in April 1964 by the Brazilian airline VASP, was to have been powered by 1731-kW (2,320-shp) Dart R.Da.9s and would have carried an additional 1060 litres (280 gal) of fuel in integral wing tanks, with a maximum take-off weight of 20412 kg (45,000 lb) and cabin accommodation for up to 60 passengers. A 68-passenger version with a 1.5-m (5-ft) fuselage stretch was designated Series 600, and a proposed military version of this was dubbed the Series 800. None was built. VASP, unable to obtain financing from the Brazilian Treasury for Dart Herald 700s, ordered Japanese NAMC YS-11s instead, and the six Series 700 airframes which had been started on Handley Page's Radlett production line were scrapped.

Production of the Dart Herald ended in 1969 when Handley Page went into liquidation. Fifty aircraft, including the converted H.P.R.3 prototypes, had been built, plus one static test airframe.

The Dart Herald never achieved the sales success for which Handley Page had hoped, but in service with such carriers as Air UK, British Island Airways (the major operator of the type in later years) and British Air Ferries it gave sterling service, particularly on holiday routes to the Channel Islands. Bournemouth International Airport-based Channel Express is now the principal operator of the surviving Dart Heralds.

The prototype Herald was the first of the marque to feature turboprop power, the installation of Darts necessitating a major redesign of wing structure and an additional plug in the fuselage. The changes made the Herald a much more sprightly performer.

Handley Page H.P.42

In an age when plastic aeroplanes (with plastic food) have become the norm for air travellers, it is easy to forget that flying was once a spectacle and an adventure. The H.P.42 was perhaps the greatest symbol of those days, an imposing design that seemed to be composed entirely of spars, rigging and engines. It was with such aircraft that Imperial Airways carried the flag to the outposts of the British Empire around the world.

In 1928 Imperial Airways, the between-wars British civil flag carrier, invited tenders from aircraft manufacturers for a new fleet of airliners to serve on its Empire airmail route to India. Three- and four-engined aircraft were specified, each in versions suitable for service in Europe ('Western') and the East ('Eastern'). They were to feature "the greatest safety, highest possible paying load capacity and lowest cost of operation", to have a stalling speed of not more than 83 km/h (52 mph), to be able to be flown 'hands-off' with any one engine shut down and to have ranges of at least 485 and 800 km (300 and 500 miles) respectively for Western and Eastern versions.

Handley Page Ltd tendered for the contract and received an order for four each of the four-engined Eastern and Western aircraft, priced at £22,300 and £23,000 each without engines. The design team at Handley Page's Cricklewood factory in North London was headed by George Volkert, assisted in the early stages by Harold Boultbee. They opted for an all-metal airframe with sesquiplane wing configuration, fabric covered, employing streamlined-strut Warren girder bracing. Engines, either Armstrong Siddeley Jaguar or Bristol Jupiter air-cooled radials, were to be mounted uncowled in a 'two-up, two-down' cluster, their thrust lines as close together as propeller diameter permitted to minimise asymmetric trim loads.

The fuselage comprised a Duralumin monocoque for the main cabin with corrugated metal skinning, with a fabric-covered welded steel tube rear section, and a triple-finned biplane tail unit. To increase ground angle and bring the cabin door as near to the ground as possible for convenient ease of entry and exit a slightly 'kinked' fuselage profile was adopted, later earning the H.P.42 the nickname 'flying banana'.

In July 1929 a full-scale wooden fuselage mock-up of the aircraft was exhibited at the Aero Exhibition at Olympia in London, and attracted much attention for the luxury of its cabin furnishings, which included Pullman-style passenger seating with individual controls for heating and ventilation (though, as crews were later to discover, the H.P.42's cockpit was notoriously cold and draughty), inlaid wood panelling, a lavatory, galley and, thanks to the centre-section of the lower wings being sharply anhedralled to permit wing attachment above the passenger cabin, an unobstructed downwards view of the passing landscape through the large cabin windows.

Initially the two versions of the aircraft were given different designations: H.P.42 for the Eastern version, which was to be powered by unsupercharged Bristol Jupiter XIF engines, and H.P.45 for the Western variant with A.S. Jaguars. When the design was finalised Imperial Airways opted for Jupiters in both variants – XIFs for the Eastern model and supercharged XFBMs for the Western version. Although the manufacturer retained the separate designations, Imperial Airways and most other sources referred to the aircraft as the H.P.42E and H.P.42W.

The first H.P.42 completed was an Eastern model, G-AAGX named *Hannibal*, reflecting Imperial's new policy of adopting aircraft names from the pages of history and mythology. *Hannibal* began taxiing tests at Handley Page's airfield at Radlett, Hertfordshire, on 31 October 1930, and made its first short 'hop' on 10 November, completing its first flight away from the airfield three days later in the hands of test pilots Squadron Leader A. England and Major J. L. Cordes. A demonstration for press and public was arranged on 17 November, during which Cordes took off in less than three times *Hannibal's* own length, then flew a sprightly exhibition under a low cloud base which included flying with three of the four engines throttled back to little more than idle thrust.

G-AAGX was the first of eight H.P.42s to be built, and made its first flight from Radlett on the last day of October 1930. The huge wing area bestowed superb low-speed qualities on the type, and its short field performance was breathtaking by modern standards.

Hengist *began life as an H.P.42W as seen here, but was later converted to Eastern standard for service on the India route. It was the first H.P.42 to be lost when it was burned in a hangar fire at Karachi.*

The prototype H.P.42 was soon delivered to Imperial Airways, with whom it received the name **Hannibal**. *It went into service initially on the Paris route, but was later switched to the Eastern routes for which it was designed.*

Hannibal was sent to RAF Martlesham Heath on 17 April 1931 for certificate of airworthiness trials which lasted five days. On 5 June 1931 the H.P.42's certificate of airworthiness was granted and *Hannibal*, on only its 35th flight, was delivered to Croydon Airport and officially handed over to Imperial Airways. Next day is flew to Hanworth Air Park for inspection by members of both Houses of Parliament before commencing proving flights on Imperial Airways' Silver Wing service to Le Bourget Airport, Paris. From 11 July *Hannibal* began carrying fare-paying passengers on Paris proving flights. A month later, on 8 August, a loose engine fairing detached, hitting the port lower propeller and destroying it; the propellers of both upper engines were severely damaged. Captain F. Dismore carried out a skilful forced landing in a small field near Tonbridge, Kent, but while he was trying to avoid telephone wires the aircraft's tail struck a tree stump and was broken off. No-one was hurt in the accident, and the H.P.42 was quickly recovered by road for repair at Croydon. A second Eastern machine, G-AAUE *Hadrian*, had meanwhile been delivered.

What's in a name?

The remaining H.P.42s were Eastern models *Horsa* (originally to have been named *Hecate*, which was thought to have unfortunate connotations) and *Hanno* and Western models *Heracles*, *Horatius*, *Helena* and *Hengist* (which was to have been *Hesperides* until Imperial Airways' staff complained that it would be too onerous to have to spell it out on their paperwork!). Delivery continued throughout 1931. The last, *Helena*, arrived at Croydon on 31 December, nine months later than scheduled, the delay costing Imperial Airways financial penalties almost totalling its value.

A classic image of colonial air transport, an H.P.42E refuels in the Raluchistan desert with ensign flying. Despite many minor incidents, the H.P.42s never failed to get their Imperial passengers to their destinations.

The H.P.42W seated 38 passengers in considerably less comfort than those on the H.P.42Es. Note the primitive leading edge slats which helped give the type its low-speed flying qualities.

The H.P.42Es carried 24 passengers, 12 in each forward and rear cabin. The centre-section of the fuselage adjacent to the engines housed the lavatory, and mail and baggage holds carried mail and baggage. H.P.42Ws seated 20 and 18 passengers respectively in the fore and aft areas, with reduced baggage space. A flight crew of three comprising captain, first officer and radio officer was carried, with one or two cabin attendants.

The aeroplanes were soon a familiar sight at Croydon Airport, where, thanks to their excellent short field performance with a take-off run of just 200 m (650 ft) and a landing speed of 104 km/h (65 mph), they often operated directly from the Customs apron in front of the terminal rather than from the grass airfield. But the H.P.42's forte – slow speed – was also its Achilles' heel: cruising at a sedate 152 km/h (95 mph), it could be severely affected by strong headwinds. One captain reported taking more than an hour to cross the narrowest stretch of the English Channel during a gale. Nonetheless, on 20 May 1932 Imperial Airways' redoubtable Captain O. P. Jones took part in a publicity stunt which involved racing *Heracles* against the *Flying Scotsman* between London and Edinburgh. The H.P.42 just barely beat the train. The following month *Heracles* was involved in a less happy incident when, joyriding at the Royal Aeronautical Society's garden party at Hanworth in the hands of Captain 'Cockney' Rogers, it broke through the grass surface of the aerodrome and subsided into a drainage ditch, damaging its lower port wing.

Business express

For a premium return fare of £6.15.0 the Silver Wing service proved extremely popular with businessmen, and by 1932 Imperial Airways had won 58 per cent of the Croydon cross-Channel traffic. Passengers checked in at the Airways Terminal at London's Victoria Station in the morning and were deposited 're-freshed and relaxed' on rue Lafayette in the heart of Paris a mere three hours and 34 minutes later, after a flight which lasted two hours and 15 minutes.

'Speed without hurry' was the boast of Imperial Airways, who proudly, if inaccurately, pointed out that a complete service of meals – breakfast, lunch and tea – was provided on the flight between London and Paris. C. G. Grey, editor of *The Aeroplane* magazine, replied to

Heracles was the first of the H.P.42W (H.P.45 company designation) for the Western routes. The principal route was Croydon-Paris, although other European cities were also served.

Helena was the last of the H.P.42Ws, and survived until August 1940. Serving with No. 271 Squadron, RAF, it was used for domestic communications flights until a landing accident at Donibristle. Part of the cabin became a Fleet Air Arm squadron crew room!

this boast that he "had no idea the distance was so great, or alternatively that Imperial Airways' big four-engined airliners took so long to get there".

Smoking on board was prohibited, as a passenger found to his cost when he lit an after-lunch cigar aboard *Heracles* one day. Imperial Airways prosecuted him and he was fined £10

487

for the offence. That apart, Imperial Airways took the view that its passengers were always right: one senior H.P.42 captain was dismissed for rudeness to a particularly exasperating passenger. Captains were supposed to be models of diplomacy, but many of the early Imperial Airways' pilots were 'rough diamond' former Royal Flying Corps NCO pilots whose turn of phrase did not always fit the airline's ideal. One apocryphal story has it that passengers departing from Croydon aboard a flight to Paris were shocked to overhear an H.P.42 captain make several requests for his radio officer to lower the ensign that fluttered from the aircraft's cockpit whenever they were on the ground. Frustrated at the lack of response from 'sparks', the captain bellowed a decidedly non-standard command: "Mister, pull that bloody rag in!"

Imperial journey

The Eastern H.P.42Es flew out of their base at Cairo to Karachi and Delhi, night-stopping en route. Two stewards served four- (or even seven-) course meals with fresh vegetables and fine wines, from the H.P.42s' galleys. No other airline could match the standard of comfort and service offered by Imperial. The complete through route from London to Delhi took six and a half days, so a close rapport grew up between crew and passengers on such a protracted journey. It was not unusual for the captain to entertain his charges to dinner, and on occasions when landing at one of Imperial's unattended desert fuel stores became necessary, he was not above asking passengers to help fill the H.P.42's fuel tanks for the onward flight.

On 22 August 1935 *Horsa's* passengers, en

Pictured in front of the famous terminal building at Croydon Airport, Hanno was an 'Eastern' model, which plied its trade to India from Cairo via desert refuelling strips. With only 24 passengers, the service was sumptuous on this route.

route from Basra to Bahrain in darkness, were subjected to an unexpected and unwelcome sojourn in the desert when the Bahrain station failed to receive a signal that the H.P.42 had departed and did not light its landing flares. *Horsa's* crew overflew and ran out of fuel nearly 100 miles south of their destination. A forced landing was made with no damage or injury. The passengers and crew were rescued next day, hot and thirsty but otherwise unharmed, by a search party from RAF Shaibah.

On European routes the hard-working H.P.42Ws served Paris, Brussels, Cologne, Basle and Zurich, and were employed between European journeys for joyriding over London and day trips to events such as the Grand National horse race. Utilisation was remarkably high. In 1935 *Heracles* flew on all but four days of the year, and on 23 July 1937 became the world's first commercial aircraft to complete one million flying miles. In eight years this H.P.42 flew a total of 2,121,000 km (1,318,000 miles) and carried more than 160,000 passengers, in an age when air travel was still largely restricted to the middle and upper classes.

In all, Imperial Airways' fleet of H.P.42s each logged more than 12,000 flying hours and jointly flew nearly three million miles (seldom at more than 160 km/h; 100 mph!) without harming a passenger. The aeroplane's service was not entirely without incident, however. Following *Hannibal's* damaging forced landing at Tonbridge it was more severely damaged in a gale at Galilee the following year. Repairs took six months on that occasion. *Hengist*, which inaugurated Imperial's Australian air mail service from Croydon on 8 December 1934, was converted to Eastern standard and eventually burned out in a hangar fire at Karachi in May 1937.

The remaining seven H.P.42s continued to give good service until the outbreak of World War II, when three Croydon-based machines, *Hanno*, *Heracles* and *Horatius*, were immediately turned over to the National Air Communications service at Whitchurch and Exeter for ferrying vital supplies to the British Expeditionary Force in France. On 7 November 1939 *Horatius*, unable to land at Exeter in stormy weather, was forced to put down on a golf course at Tiverton, Devon, and was destroyed in the attempt.

Wartime duties

The Eastern aeroplanes were then still based at Cairo, but after the retreat at Dunkirk were recalled to England. *Hannibal* disappeared in the Gulf of Oman between Jask and Sharjah on 1 March 1940 while en route, with the loss of four crew and four passengers. *Hadrian*, *Horsa* and *Helena* were safely ferried back to join the National Air Communications pool at Whitchurch, and all five remaining H.P.42s served on communications flights to Scotland and Northern Ireland until 19 March when *Hanno* and *Heracles* were wrecked by a gale at Whitchurch. Impressed into Royal Air Force service and camouflaged, the three survivors were allocated to No. 271 Squadron and continued to serve, albeit briefly. In August 1940 *Helena* was damaged in a heavy landing at RNAS Donibristle and was eventually dismantled there, the forward cabin being turned into a temporary office for Fleet Air Arm pilots. That same month *Horsa* was destroyed by fire after a forced landing in Cumberland while flying between Manchester and Stornoway. Like those destroyed earlier at Whitchurch, the last survivor, *Hadrian*, also succumbed to a gale while parked at Doncaster Airport in December 1940 and was blown from its picketings and wrecked.

In its later years the Handley Page H.P.42 may have seemed an anachronism, a fabric-covered, drag-ridden biplane in an age of sleek all-metal monoplane airliners. But when it entered service it was the first four-engined airliner in regular passenger service, and in nine years of plying busy European routes and lengthy, leisurely journeys throughout the British Empire, these slow and steady 'aerial galleons' won passengers' affections and earned jealously-guarded reputations for safety, service, dependability and comfort which have seldom been matched by any other aircraft in the history of commercial air transportation.

G-AAUE Hadrian was the second H.P.42 for Imperial Airways, again an Eastern machine. These had two cabins, one forward of the wing structure and one aft, 12 passengers were carried in each cabin, and there were large windows for an excellent view of the outside world.

Hawker Siddeley Trident

Envisaged in a day when short-range jetliners were an entirely unknown quantity, the Trident's clear ability should have guaranteed it success. The manufacturers then ignored the market outside Britain and allowed their American competitors to steal the march on them. Despite its long service career afterwards, this was a loss from which the Trident would never recover.

In July 1956 British European Airways published an outline specification of a requirement for a short/medium-range jet airliner to operate from 1829-m (6,000-ft) runways carrying 100 economy-class or 70-80 mixed-class passengers over stage lengths of 482-1931 km (300-1,200 miles).

Three British aircraft manufacturers responded to the BEA requirement, submitting their designs in the summer of 1957. These were the Avro 740, Bristol 200 and the de Havilland 119. The latter aircraft was projected as a four-engined machine powered by Rolls-Royce RA29 Avon turbojets. This design was later revised as the D.H.120, which was intended to meet both BEA's short/medium haul needs and the British Overseas Airways Corporation's requirement for a long-range intercontinental jet airliner. Combining such vastly different requirements in one airframe proved impossible (the BOAC specification was met by the Vickers VC-10), and de Havilland again revised its BEA proposal, this time to a three-engined configuration as the D.H.121. This was to be powered by three 61-kN (13,790-lb) thrust Rolls-Royce RB.141/3 Medway turbofans, two mounted in external pods at the rear of the fuselage, and the third in a buried installation at the base of the fin – a configuration stipulated by the airline. The D.H.121 was to have a 111-seat cabin and a maximum range of 2896 km (1,800 miles).

It was a government requirement that whichever manufacturer won the BEA order should merge with another company to fulfill the contract. Accordingly, de Havilland and the Fairey and Hunting companies set up The Aircraft Manufacturing Co. Ltd (Airco), reviving a name last used during World War I, and in February 1958 it was announced by the British Government that the Airco D.H.121 had been selected on its technical merit and that detailed negotiations could now begin between the airline and the manufacturing consortium. As these proceeded BEA changed its specification, and by 1959 the new jet airliner had been redesigned and scaled down by a team headed by C. T. Wilkins to carry 97-103 passengers over a maximum range of just 1300 km (810 miles). Because of the unavailability of the Medway engine, the selected powerplants had also been changed to 43.8-kN (9,850-lb) thrust Rolls-Royce RB.163 (later named Spey) turbofans.

The redesigned aircraft met BEA's requirements exactly, having been tailor-made for the airline, but it was no longer attractive to other carriers, particularly overseas customers, whose route profiles, load factors and range requirements were quite different from those of the British airline. Unwisely, de Havilland ceased all marketing operations with export customers until BEA's specification had been met. This was to be a key factor in the aircraft's ultimate lack of sales and market failure in competition with the rival Boeing 727, from which the British design could have capitalised on its significant time advantage and probably shared part of the 727's outstanding success had de Havilland not devoted all of its efforts to meeting the needs of just one carrier.

BEA was well satisfied, however, and on 2 August 1959 the airline signed an order for 24 aircraft with options on a further 12, 'straight

Pakistan International Airlines operated a quartet of improved performance Trident 1Es. One was lost in a crash and the survivors were sold to the Chinese air force.

Iraqi Airways took delivery of three of the 14 Trident 1Es built, in 1966. They could carry up to 115 passengers and were withdrawn from use in 1977.

Most dedicated of all Trident users was China's CAAC, which shared its aircraft with the air force. This is one of their 18 Trident 2Es.

from the drawing board'. No prototype was to be built. By this time the Airco consortium had been disbanded following de Havilland's merger with the Hawker Siddeley Group, and the D.H.121 became an exclusively de Havilland project once more, design work proceeding at the company's factory at Hatfield, Hertfordshire.

The forefront of design

As the final design for the D.H.121 emerged, it became clear that it was in every respect a modern, aerodynamically efficient design. Particularly noteworthy was the clean wing design, a by-product of rear-mounted engines. The wing leading edges had 35° of sweepback – the greatest sweepback angle adopted on any commercial aircraft at that time, drooped leading edges, Krueger flaps, double-slotted trailing-edge flaps, split ailerons, of which only the inner portions were activated at high cruising speeds, and spoilers which doubled as lift-dumpers or airbrakes. All flying controls were powered, and operated by three completely independent hydraulic circuits, one pressure system being driven by each engine. This triplex flight control system, and tri-jet configuration inspired the name chosen for the aircraft in September 1960: Trident.

To further enhance the cleanliness of the aircraft's wing design, its main undercarriage legs, each of which carried two pairs of wheels, were pivoted about the main wing spar so that they swivelled through 90° and extended 15 cm (6 in) before retracting inboard into the fuselage rather than into the lower surface of the wings; the twin nosewheels were offset to port and also retracted sideways. All fuel tanks were contained in the wing, and thanks to a maximum landing weight only 2268 kg (5,000 lb) less than maximum take-off weight it was possible for the Trident to make a stop within a short time from its full-fuel departure and still have sufficient fuel to fly several more short route segments before refuelling. As demanded by BEA, provision was made for future fully automatic 'blind' landings through a Smith's Flight Control System with triple redundancy. Cabin accommodation was finalised at 75 passengers in four/six abreast mixed-class or 95 in six-abreast, all coach class seats.

Structural and water tank pressure testing of a static test airframe began late in 1959 while work proceeded on construction of the first of BEA's Tridents, G-ARPA, which was rolled out at Hatfield on 4 August 1961. Flight trials of the definitive 46.0-kN (10,400-lb) thrust RB.163-1 Spey turbofan began in October using an Avro Vulcan V-bomber as testbed, paving the way for the Trident's first flight on 9 January 1962, piloted by de Havilland's famed chief test pilot John Cunningham and co-pilot Peter Bugge. No less than three Tridents made the type's first public appearance at the 1962 Farnborough air show, and by early 1963 four were engaged in the intensive test programme which culminated in the Air Registration Board awarding the aircraft its full certificate of airworthiness on 18 February 1964.

Making money

The Trident flew its first revenue service for BEA on 11 March, and began operating full scheduled services on 1 April. Although the carrier had ordered 24 Tridents, only 23 were to enter service. On 3 June 1966 the penultimate production aircraft, G-ARPY, crashed at Felthorpe, Norfolk during its first flight, a victim of the 'deep stall' phenomenon which beset early rear-engined T-tail jet airliners and business jets. Test pilots Peter Barlow, George

The Trident pioneered modern 'blind' landing aids and, in 1971, a Trident III became the first commercial airliner to land in Category (Cat) IIIA zero-visibility weather conditions. The Trident 1 (illustrated) was certified for less demanding Cat II landings.

Three Tridents, two 1Cs (illustrated) and a single 1E, were delivered to Kuwait Airlines. One was lost in a crash on approach to Kuwait International in 1966.

With the break-up of CAAC, the honour of maintaining Trident operations fell to China United Airlines. All nine remaining aircraft carried air force serials.

Errington (a long time veteran of Airspeed Aircraft and de Havillands) and two other crew-members died in the crash.

The second production Trident, G-ARPB, was retained by the manufacturer for Autoflare and Autoland trials, and made its first fully-automatic landing at the Royal Aircraft Establishment airfield at Bedford on 5 March 1964. The first fully-automatic landing on a revenue-earning service was performed by G-ARPR arriving at London-Heathrow from Paris on 10 June 1965, while G-ARPB made the world's first fully-automatic landing by a civil aircraft in true zero visibility on 4 November 1966, when it landed at a fog-bound Heathrow with John Cunningham in command, but not actually handling the controls.

BEA's Trident 1s were all eventually upgraded to Trident 1C standard with 43.8-kN (9,850-lb) thrust Spey Mk.505-5F engines, an additional 3785-litre (1,000-US gal) fuel tank in the wing centre-section and provision for up to 109 passengers.

BEA took delivery of its first Trident 2 in February 1968. All other customers referred to it as the Trident 2E. Though no larger than preceeding versions, it offered an increased fuel capacity and maximum take-off weight.

Because its specification had been so finely tuned to the needs of BEA, the Trident 1 attracted little interest from other airlines. To broaden its appeal Hawker Siddeley developed an export version designated Trident 1E which featured uprated 50.7-kN (11,400-lb) thrust Rolls-Royce RB.163-25 Spey 511-5, extended wings of 1.58 m (5 ft 2 in) greater span with full-span leading-edge slats in place of the Trident 1's dropped leading edges, increased fuel capacity and an increase in maximum take-off weight to 58060 kg (128,000 lb). Cabin accommodation was increased to 115 passengers.

Kuwait Airways was launch customer for the Trident 1E with an order for three, the first of which flew from Hatfield on 2 November 1964. The second of the airline's Trident 1Es was destroyed in a non-fatal landing accident at

The Trident 3B was a high-capacity short-haul development of the Trident 1. The fuselage was stretched by 2.0 m (6 ft 5 in) to seat up to 180 passengers. Twenty-six were ordered initially by BEA, with the first entering service on April Fool's Day, 1971.

With the arrival of British Airways on the scene in 1973 the Trident fleet just adopted a new colour scheme and kept on flying. This is one of the former BKS-Northeast Trident 1Es which was withdrawn from use in 1980. The Trident 3s lasted almost another 10 years.

The final Trident variant, again developed specifically to meet a BEA requirement, was the Trident 3B. This was optimised for medium-range inter-city routes, and was essentially a Trident 2E with a 5.0 m (16 ft 5 in) fuselage stretch to enable up to 180 passengers to be carried. Wing span was unchaged, but wing area, incidence and the span of the flaps were increased, fuel capacity reduced by 12 per cent, and maximum take-off weight increased by 3311 kg (7,300 lb). To provide extra thrust for take-off from short airfields or at high maximum operating weights, a 23.2-kN (5,250-lb) thrust Rolls-Royce RB.162-86 auxiliary turbojet engine was installed at the base of the fin below the rudder. The first Trident 3B, G-AWYZ, made its maiden flight on 11 December 1969, initially without the booster jet installed, and made its first four-engine take-off on 22 March 1970. BEA ordered 26 Trident 3Bs, which it called Trident Threes, the type entering service on 1 April 1971. The Trident 3B was equipped with full Autoland systems, approval for 'zero-zero' Category IIIA automatic landings being granted in December 1971, enabling the aircraft to operate with a runway visual range of 90 m (295 ft) and a decision height of just 3.5 m (12 ft). By that time more than one million BEA passengers had (most unknowingly) been delivered to their destinations automatically by the Smiths Autoland equipment installed in the carrier's earlier Tridents. Final Trident production comprised two 'Super 3Bs' for CAAC of China, these aircraft having 152-seat cabins, extra fuel tanks and a further increase in maximum take-off weight to 71668 kg (158,000 lb). The first Super 3B was flown on 9 July 1975, and delivery of the two CAAC aircraft brought Trident production to an end after a run of just 117 aircraft – 24 Trident 1s, 15 1Es, 50 2Es, 26 3Bs and two Super 3Bs.

Trident swansong

Between 1977-79 BEA's Trident fleet was plagued with fatigue problems after cracks were discovered in the wing structures of Trident Twos and Threes. All were remanufactured to have their wingspans reduced by 0.9 m (3 ft) and Kuchemann wingtips removed, thus reducing bending movements and enhancing fatigue life.

Two BEA Trident Ones were lost in accidents. G-ARPU was destroyed on the ground at Heathrow Airport on 3 July 1968 when an Airspeed Ambassador went out of control on landing and crashed into it; G-ARPI, which was damaged in the same incident but repaired, was lost shortly after take-off from Heathrow on 18 June 1972 after entering a 'deep stall'. G-ARPS had to be withdrawn from use after a fire was deliberately started in its cabin while it was parked at Heathrow in the summer of 1969. The remainder of the Trident One fleet served BEA and its successor British Airways until 1975 when the fleet was withdrawn and mostly flown to Prestwick Airport for storage and eventual scrapping.

British Airways ceased all Trident operations on 31 December 1985, leaving only CAAC of China and Air Charter Service of Zaire (which acquired four of BA's trident Threes) operating the type into the later 1980s.

Kuwait Airport in June 1966, just a month after delivery. Iraqi Airways also took delivery of three Trident 1Es, and was the first operator to put the type into service after handover on 1 October 1965. Pakistan International Airlines accepted the first of four Trident 1Es in March 1966. Five further aircraft were built, remaining unsold at the Hatfield factory until October 1967 when the Southend Airport-based carrier Channel Airways Ltd contracted to buy all five, which were to be modified to high density configuration with accommodation for 139 passengers, four additional emergency exits and a maximum take-off weight of 61500 kg (135,580 lb). This version was known as the Trident 1E-140. Only two were actually delivered to Channel Airways, in May and June 1968. Two more went to BKS Air Transport, later renamed Northeast Airlines, for its scheduled services between London Heathrow and Newcastle, and the fifth, the last of 15 Trident 1Es built, was delivered to Air Ceylon on 16 July 1969. Channel Airways' two Trident 1Es were later transferred to Northeast in 1972, when the two remaining Kuwait Airways aircraft were sold to Cyprus Airways. Pakistan International Airlines' Trident 1Es were sold to CAAC, the national carrier of the People's Republic of China, in 1970.

When Hawker Siddeley developed the Trident 1E, which was intended primarily for export, it also proposed a stretched, 125-passenger Trident 1F to British European Airways. Although it did not opt for the longer fuselage, on 26 August 1965 BEA placed an order for 15 aircraft, redesignated Trident 2E. This revised version featured a further small increase in wingspan created by the addition of low-drag Kuchemann wingtips, structural strengthening using titanium instead of steel in some key areas, additional fuel tank in the fin, 53.0-kN (11,930-lb) thrust Spey 512-5W engines, and a cabin configured for 97 mixed class passengers for BEA, although it could accommodate up to 149 in high-density layout. The first Trident 2E, which was known as the Trident Two in BEA service, made its maiden flight on 27 July 1967. Scheduled services began on 18 April, the additional range conferred by the extra fuel capacity enabling BEA to fly the type London-Middle East non-stop for the first time.

Cyprus Airways took delivery of two Trident 2Es in September 1969, but the major customer for this variant was CAAC of China, which ultimately ordered 33, the first of which left Britain for delivery to Peking on 13 November 1972.

Heinkel He 70

During the 1930s American aircraft companies were making great advances in passenger transport design. Europe's nationalistic manufacturers, particularly those in Hitler's Germany, could not stand idly by. Heinkel's answer was the sleek He 70, built initially for Lufthansa's European routes. So impressive was its performance that it soon dropped its civilian guise and joined the emerging Luftwaffe as a bomber.

In February 1932 Deutsche Lufthansa commissioned Ernst Heinkel AG and Junkers to design new airliners with which the airline could compete against the fast Lockheed Orion low-wing monoplane then entering service with Swissair on the Zurich-Munich-Vienna route.

The initial specification called for a maximum speed of 285 km/h (177 mph) and for this requirement Heinkel began work on the He 65 fixed undercarriage monoplane, but within a few months it became clear that this would not be adequate to compete with Swissair's Lockheeds, and an entirely new design, the He 70, was begun by Heinkel's designers Siegfried and Walter Günther.

The chosen configuration was a cantilever low-wing monoplane incorporating every known aerodynamic refinement. The powerplant selected was a liquid-cooled 469-kW (630-hp) BMW VI 6,OZ 12-cylinder Vee engine which used ethylene-glycol rather than water for cooling. The higher evaporation rate of the ethylene-glycol enabled a small retractable radiator to be used, further reducing aerodynamic drag. The fully-cowled engine installation blended smoothly into the streamlined contours of the fuselage, which comprised a semi-monocoque duralumin structure built up of hollow frames and longitudinal stringers and covered with metal skinning on which all rivets were countersunk, resulting in a perfectly smooth finish.

The fully-enclosed cockpit featured pilot's and radio operator's seats slightly offset to the port side, and was smoothly faired into the upper fuselage decking and had a sliding canopy for access. A fireproof bulkhead separated the cockpit from the passenger cabin, which contained two pairs of facing seats for the four passengers, each having arm and headrests. A separate window was provided for each passenger and the forward pair could be opened on the ground for ventilation. The cabin was heated and was boarded via a flush-fitting door on the starboard side of the fuselage. A baggage compartment was provided in the aft area of the cabin, where there was also an access 'manhole' permitting internal inspection of the rear fuselage structure.

The He 70's wing and tail surfaces had the elegant elliptical planform which was to become a trademark of the Günther brothers' Heinkel designs. The wing was a one-piece structure built up on two spruce-and-plywood box spars with closely spaced ribs and smooth plywood skinning, and featured an inverted gull-wing section in the root area to eliminate tail buffeting from disturbed airflow. Split trailing edge flaps were fitted. The landing gear was of the tailwheel type, each main unit comprising a tripod arrangement with an oleo leg, hydraulically retracted outwards into the lower surface of the wing, with a fixed oleo-sprung tailskid. The result of the Günthers' careful attention to streamlining was an aeroplane of striking beauty and cleanliness of line.

The third prototype He 70 had the company designation He 70C, but in Lufthansa service was known as the He 70B. Originally registered D-3114, it was later D-UBAF and christened **S***perber. When the He 70Gs were taken on charge in 1935,* **S***perber was released from its airline duty.*

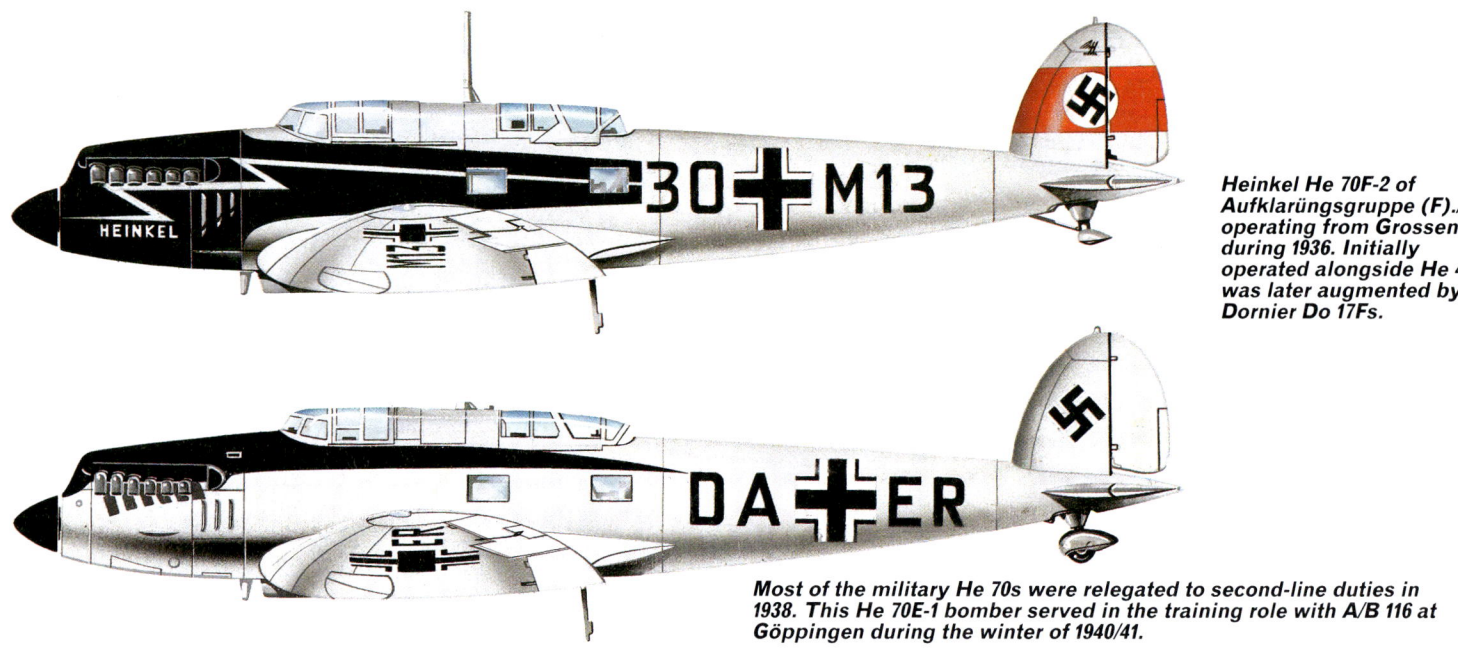

Heinkel He 70F-2 of Aufklärüngsgruppe (F)./123 operating from Grossenhain during 1936. Initially operated alongside He 45s, was later augmented by Dornier Do 17Fs.

Most of the military He 70s were relegated to second-line duties in 1938. This He 70E-1 bomber served in the training role with A/B 116 at Göppingen during the winter of 1940/41.

Preliminary design work was completed by June 1932, and construction of a prototype and an initial batch of production aircraft began immediately. The prototype Heinkel He 70V-1 made its maiden flight from the Heinkel works at Warnemünde to Travemünde airfield on 1 December 1932, flown by Flügkapitan Werner Junck. The efficiency of the design was quickly demonstrated when, early in 1933, the He 70 V1 set eight records for speeds over distances of 100-2000 km (62-1,243 miles) with payloads of 500 and 1000 kg (1,102-2,204-lb) achieving a speed of 357 km/h (222 mph) over a 100-km (62-mile) distance, and an absolute maximum speed of 377 km/h (234 mph) – faster than most contemporary fighter aircraft.

When the first production-standard He 70A appeared it bore the name *Blitz* (Lightning), which was subsequently unofficially adopted as the Deutsche Lufthansa fleet name. The He 70 made its international public debut at the Paris Salon in November 1933, and was widely acclaimed, doing much to re-establish Germany's prestige in aviation.

The Heinkel He 70A entered service with Deutsche Lufthansa on 15 June 1934, and was joined by the generally similar He 70B, C, and D variants, the latter He 70D differing from earlier aircraft in having a more powerful 559-kW (750-hp) BMW VI 7, 3 engine. The definitive variant of the commercial aircraft was the He 70G, which had a longer fuselage, its cockpit relocated on the fuselage centreline, and increased maximum take-off weight. One of these, a Heinkel He 70G-1, was delivered in 1935 to the Rolls-Royce company at Hucknall, England for trials installation of a Rolls-Royce Kestrel V engine, and served for nine years as a flying testbed, successively re-engined with a 555-kW (745-hp) Kestrel XVI and a 659-kW (885-hp) Peregrine engine, with which it attained a maximum speed of 483 km/h (300 mph).

Military developments

The excellent performance of the He 70 prompted its use by the then-clandestine Luftwaffe, which operated a number of aircraft for high-speed liaison duties during 1934. Meanwhile specialised military variants were under development, these being the He 70E-1 high-speed light bomber, which was similar to the He 70D but had a 7.9-mm MG 17 machine-gun mounted in the rear cockpit and could carry a 300-kg (661-lb) bomb load, and the He 70F-1 long-range reconnaissance aircraft.

Eighteen He 70F-2s were despatched to Spain in the autumn of 1936 for use by the Condor Legion reconnaissance wing, A/88, and

This unmarked aircraft is the first prototype, the He 70a. It was retained for trials by Heinkel, making its first flight from Warnemünde on 1 December 1932. Later trials were undertaken from Travemünde.

The He 70B second prototype received the registration D-3 for competitive trials with the Ju 60 and a W34, and went on to set a number of speed-with-load records. It then passed to Luftansa in whose service it was redesignated He 70A and named Blitz.

A close-up of the He 70 nose area shows the carefully streamlined cowling and the neat radiator installation. Also noticeable is the slight anhedral of the wing root area, giving rise to an inverted gull-wing configuration.

some remained operational in Spain until the early 1950s, but the type was progressively phased out of Lufthansa and Luftwaffe service during 1938, being replaced in military roles by the Dornier Do 17F-1.

Prior to that, a new variant was developed during 1937 for export to Hungary. This aircraft, designated Heinkel He 170, differed from the He 70 principally in having a 678-kW (910-hp) Gnome/Rhône 14K Mistral Major 14-cylinder aircooled radial engine enclosed by a circular cowling which destroyed the smooth lines of the original design. Twenty He 170s were delivered to the 1st Independent Long-Range Reconnaissance Group of the Hungarian air force, entering service in Carpatho-Ruthenia in March 1939. The Hungarian He 170s first saw action against Russia on 26 June 1941, but were quickly withdrawn from front-line service because their defensive armament of two light machine-guns left them vulnerable to fighter attack. The He 170 had a maximum speed of 270 km/h (168 mph).

The ultimate development of the He 70 series was the He 270, a single prototype of which was first flown during 1938. Powered by an 875-kW (1,175-hp) Daimler Benz DB 601A engine driving a three-bladed airscrew, the He 270 had a maximum speed of 460 km/h (286 mph) at 4000 m (13,120 ft) and was intended for use as a light bomber/reconnaissance aircraft. It was armed with one forward-firing and two rearward-firing MG 15 machine-guns and could carry a bomb load of 272 kg (600 lb), but was not proceeded with. Total production of the He 70/170/270 series was 324 aircraft, including 28 commercial airliners.

*The potential of the He 70 was so great to the nascent Luftwaffe that three aircraft for Lufthansa were completed as military machines. To compensate, three He 70D-0s were supplied to Lufthansa, these being the first such production machines. This one was **D-UBIN** Falke.*

*Perhaps the most remarkable He 70 was **G-ADZF**, ordered by Rolls-Royce in 1935 as an engine test-bed and used extensively by that company to develop the very engines that would be used to such effect against the Luftwaffe four years later. It sadly ended its days in 1945 on the scrap-heap.*

Heinkel He 70F-2 cutaway key

1. Tail navigation light
2. Tail cone
3. Elevator tab control cable
4. Elevator tab
5. Starboard elevator
6. Elevator hinges
7. Starboard tailplane
8. Tailskid
9. Tailskid support strut
10. Tailskid leg cuff
11. Leg attachment
12. Elevator control linkage
13. Fuselage and frame
14. Elevator torque shaft
15. Rudder linkage
16. Rudder lower hinge
17. Rudder lower structure
18. Port elevator tab
19. Rudder tab
20. Rudder tab control cable
21. Rudder structure
22. Rudder centre hinge
23. Aerial attachment
24. Tailfin
25. Port elevator
26. Tailfin forward spar
27. Port tailplane
28. Fuselage frame/tailfin attachment
29. Tailplane attachment
30. Tail surface control cables
31. Fuselage structure
32. Fuselage frames
33. Longerons
34. Control cable pulleys
35. Fuselage skinning
36. Wingroot fairing
37. Bomb-aimer's position
38. Gliding ventral hatch
39. Provision for vertical camera
40. Radio equipment
41. Bombsight installation
42. Gunner's step
43. Gunner's folding seat
44. Window surround
45. Magazine stowage racks
46. First-aid kit
47. MG 15 machine-gun stowage
48. Flexible 7.9 mm MG 15 machine-gun
49. Ring sight
50. Canopy sliding section
51. Canopy fixed sections
52. Aerial
53. D/F loop assembly
54. Aerial mast
55. Port flap
56. Flap hinge
57. Flap actuating rod and linkage
58. Main boxed ribs
59. Intermediate ribs
60. Underwing inspection/access ports
61. Aileron control rod linkage

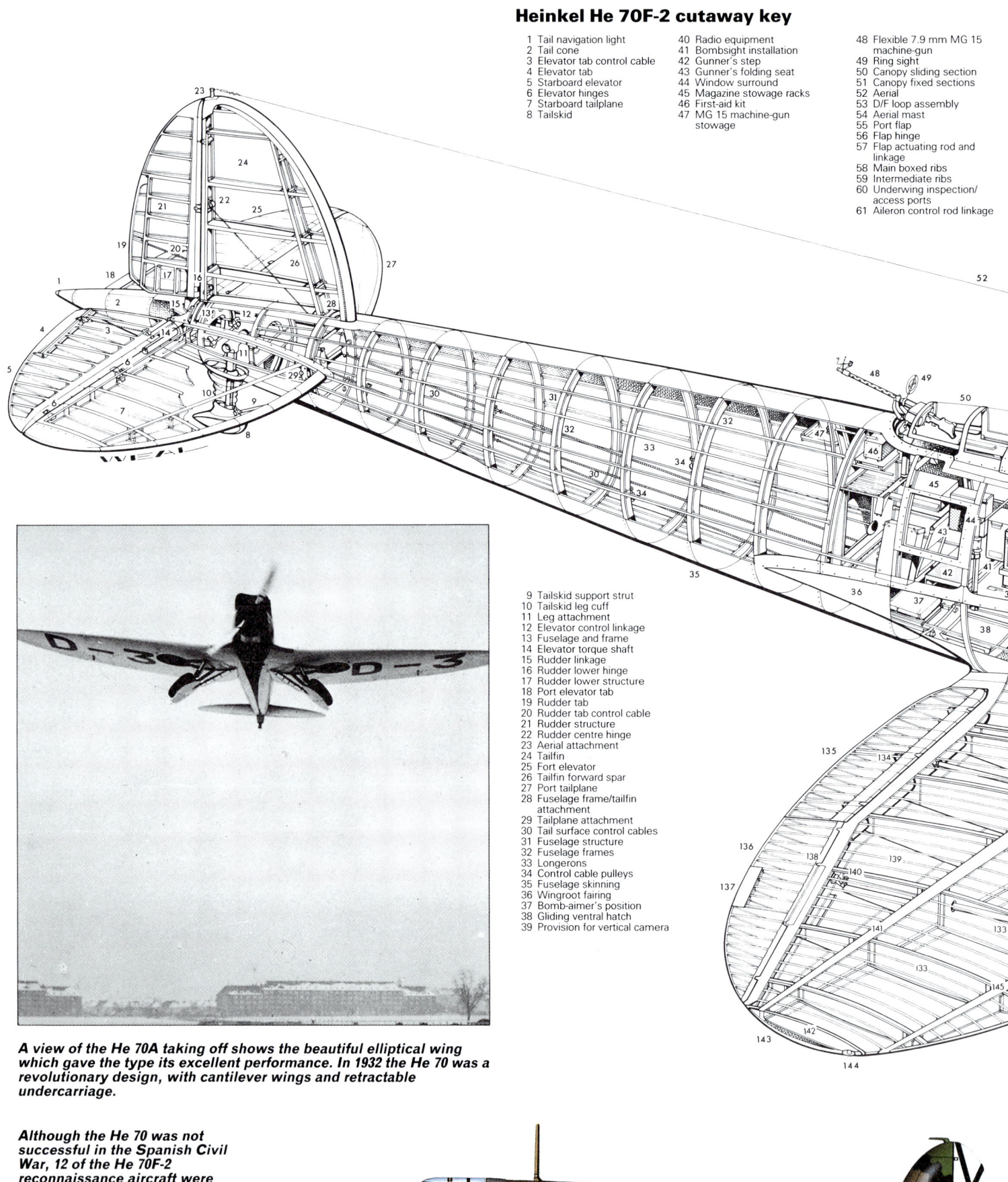

A view of the He 70A taking off shows the beautiful elliptical wing which gave the type its excellent performance. In 1932 the He 70 was a revolutionary design, with cantilever wings and retractable undercarriage.

Although the He 70 was not successful in the Spanish Civil War, 12 of the He 70F-2 reconnaissance aircraft were transferred to the Spanish Nationalists for service with Grupo 7-G-14 in the Bilbao region.

Eleven of the 12 Spanish He 70F-2s survived the Civil War, and remarkably flew on into the 1950s on a variety of communications and liaison tasks. This example served with Escuadrón 101.

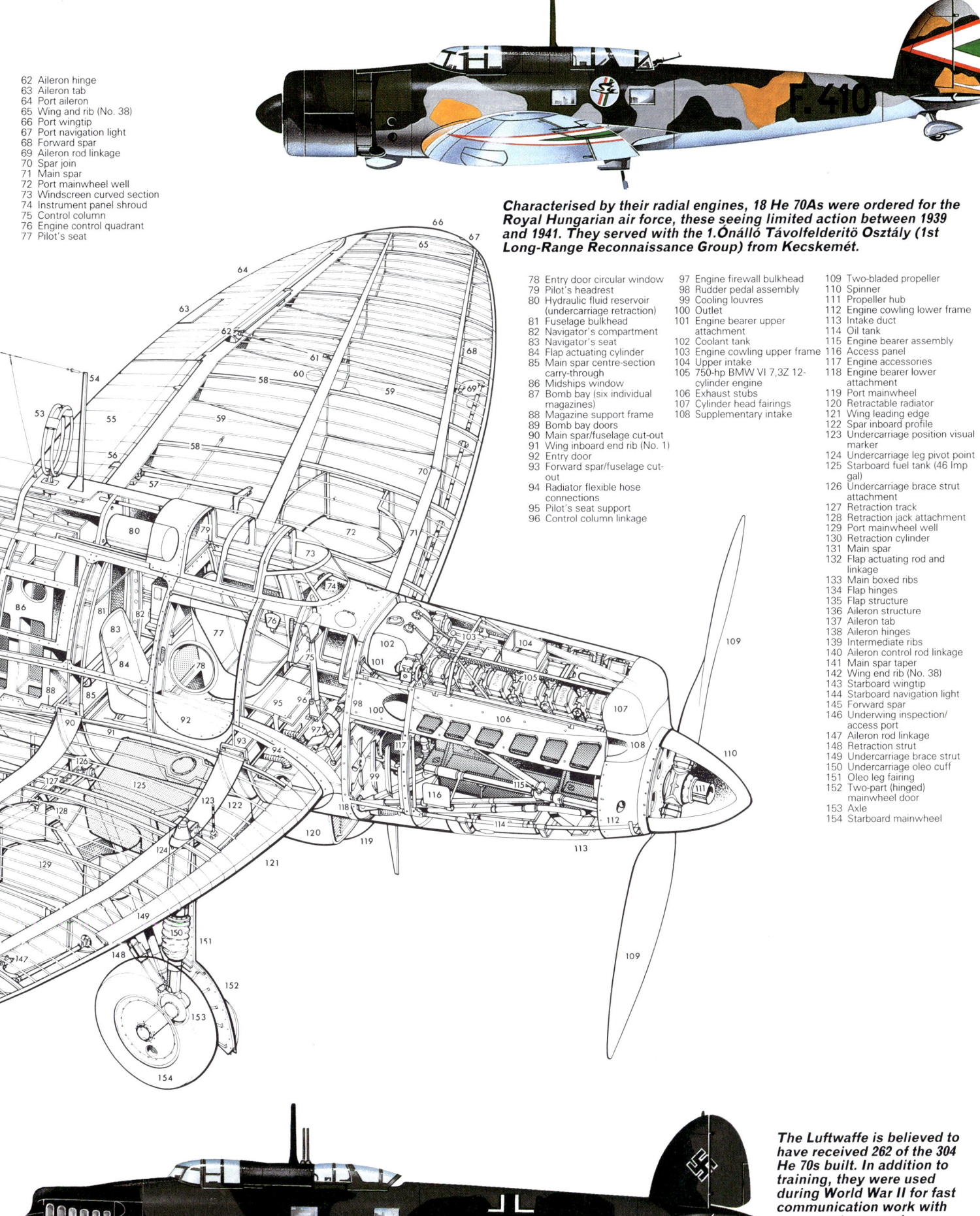

The heart of the He 70's remarkable speed performance lay in its shape, and this represents one of the high points in the art of aeronautical design. Everything about the aircraft suggested speed, and it delivered the goods although it was not very manoeuvrable. For Lufthansa in the mid-1930s it was used mainly on internal routes, linking major cities with high-priority passengers. It also took the mail, even as far as Seville on the South American route. Luftwaffe interest was naturally high, and they received three early aircraft destined for Lufthansa. This is one of them, originally built as an He 70B but completed as an He 70C military communications aircraft.

Specification
Heinkel He 70G-1
Wingspan: 14.78 m (48 ft 6 in)
Length: 11.48 m (37 ft 7 in)
Height: 3.1 m (10 ft 2 in)
Wing area: 36.5 m² (393 sq ft)
Powerplant: one 469-kW (630-hp) BMW VI V-12 piston engine
Passenger capacity: five
Empty weight: 2300 kg (5071 lb)
Maximum take-off weight: 3310 kg (7297 lb)
Maximum speed: 355 km/h (221 mph)
Cruising speed: 310 km/h (193 mph)
Service ceiling: 6000 m (19,685 ft)
Maximum range: 800 km (497 miles)

Chapter 23

Islanders & Trislanders

●

Ilyushin 11-12 & 11-14

●

Ilyushin IL-18

Islanders and Trislanders

One of Britain's greatest success stories of the post-war years did not emanate from a main manufacturer, but from a small company established by two aeronautical engineers, John Britten and Desmond Norman. Their BN-2 Islander design, and its many developments, has sold widely around the world, where its toughness and dependability have proved ideal for a wealth of small operators, both civil and military.

John Britten and Desmond Norman, both graduates of the de Havilland Technical School, formed a partnership in 1955 to develop aerial crop spraying equipment and operate agricultural aircraft in many parts of the world. One of their interests was in a small charter company, Cameroon Air Transport, which operated Piper Apache and Aztec aircraft on scheduled services between Tiko and Douala. The Pipers were not entirely suited to the work, and having discovered that no suitable aircraft was available to small air taxi and charter companies, Britten and Norman decided to design their own aeroplane to fill the market gap. The requirement was for a simple and rugged twin-engined transport seating more than six passengers, with good short-field performance, modest maintenance requirements and low cost – in effect, a modern replacement for the pre-war de Havilland Rapide biplane.

Design work on the Britten-Norman BN-2 began in 1963. Initially a fabric-covered welded steel tube airframe was considered, but this was soon abandoned in favour of an all-metal structure. The final design was for an angular, boxy, high-wing aircraft with fixed tricycle undercarriage, seating 10 people including the pilot in five rows of bench-type seats without a centre aisle, so that three separate entrance doors were needed. Powerplant was two 157-kW (210-hp) Rolls-Royce Continental IO-360-B horizontally-opposed piston engines.

Manufacture of the prototype BN-2 began at Bembridge Aerodrome, Isle of Wight, in September 1964. The prototype, G-ATCT, made its 70-minute first flight on 13 June 1965 flown by Desmond Norman, and within a few days was flown across the English Channel to make its public debut at the Paris air show. Early flight tests revealed some shortcomings in the BN-2's performance, particularly in engine-out climb rate, and as a result wingspan was increased from 13.7 to 14.9 m (45 to 49 ft) and more powerful 194-kW (260-hp) Lycoming O-540-E engines installed, in which form the aircraft flew again on 17 December 1965, but was later destroyed in an accident while on demonstration in Holland.

With a huge network of island destinations, often with short and ill-prepared runways, Olympic Aviation's domestic network was a perfect match for the Islander's talents. After many years' good service, these were replaced by Dornier Do 228s.

The production prototype, G-ATWU, made its maiden flight on 20 August 1966, by which time the name Islander had been chosen for the aircraft. The first production aircraft was flown on 24 April 1967 and British certification was granted on 18 August 1967. Deliveries began immediately to British launch customers Glos-Air and Loganair. Certification in the United States was obtained on 19 December 1967, by which time deliveries had already begun to Britten-Norman's US agents Jonas Aircraft of New York, which had ordered 30 aircraft initially, and was to take half of Britten-Norman's 1968 production to satisfy an eager American market among air taxi, charter and commuter airline operators.

Romanian production line

During 1968 agreement was reached, as part of an offset deal connected with the sale of BAC One-Eleven jet airliners to the Romanian airline TAROM, for Intreprinderea de Reparat Material Aeronautic (IRMA) of Bucharest to build an initial batch of 215 basic Islander airframes which would be flown to the Isle of Wight for installation of interior furnishings, avionics and painting prior to customer delivery. The first Romanian-built Islander, built from a kit supplied by Britten-Norman, made its maiden flight on 4 September 1969, and arrived in England later that month.

The BN-2A, introduced in June 1969 as the basic production model, was similar to the BN-2 but had its maximum take-off weight increased from 2585 to 2720 kg (5,700 to 6,000 lb). It was later marketed as the Islander II and also incorporated an updated interior, aerodynamic improvements to the engine nacelles and undercarriage legs and a larger rear baggage compartment with a side-loading door.

Performance 'pokes'

To improve the 'hot and high' performance of the Islander, Britten-Norman test flew an aircraft powered by 229-kW (300-hp) fuel-injected Lycoming IO-540-K1-B5 engines on 30 April 1970, this aircraft also having conformal wingtip fuel tanks which extended the Islander's span to 16.1 m (53 ft). Deliveries of 224-kW (300-hp) Islanders began in November 1970; the tip tanks were introduced as options on both 194- and 224-kW (260- and 300-hp) variants. In conjunction with Britten-Norman, the US distributor Jonas Aircraft developed a 'bolt on' Riley-Rajay turbocharger modification for the 194-kW (260-hp) variants which increased power output and offered improved cruising speeds and single- and twin-engined service ceilings. A lengthened nose providing a $0.62 m^3$ (22 cu ft) forward baggage compartment was also available as an option.

Meanwhile Britten-Norman was looking at ways to increase the passenger capacity of the Islander, and modified the second prototype with a 0.83-m (33-in) fuselage 'plug' which made room for an

The rugged reliability and easy maintenance have led the Britten-Norman BN-2 design to be adopted by many third world air forces, Belize being a typical example. Shown here carrying torpedoes, the Belizean Defenders are more usually used in operations against drug growers.

The Islander is superbly suited to operations in any areas where terrain or weather prove difficult. Munz Northern Airlines flew the type for many years in Alaska.

Around 30 military air arms have bought various Defender versions, flying these on utility transport duties or in the light attack role. Oman's No. 5 Squadron has eight, based on the BN-2A civil version.

extra row of seats. This aircraft, called the Islander Super, flew from Bembridge on 14 July 1968 and was to have been powered by two eight-cylinder Lycoming O-720 engines, but because of certification problems was never put into production. Instead, as a result of market surveys which indicated that customers required a 50-per cent increase in passenger capacity, a major redesign was undertaken. A 1.98-m (6-ft 6-in) extension was added to G-ATWU's fuselage forward of the wing to increase passenger capacity to 17, and a third engine installed in a nacelle on the top of a new broad chord fin, faired into the fin/tailplane junction – following the contemporary 'tri-jet' trend among large airliner manufacturers. Designated BN-2A Mark III, this aerodynamic prototype first flew on 11 September 1970, departing almost immediately to make its debut at that year's Farnborough Air Show. Among improvements incorporated during initial flight testing was the addition of a fin extension above the tailplane.

The simple plank-like wing and powerful engines provide good STOL capability, while the rugged undercarriage can cope with very rough landing strips. Bush airlines and parachute clubs have found the Islander ideal.

With an eye to stringent noise regulations, the UK firm Dowty Rotol fitted this Islander with experimental ducted fans powered by Lycoming IO-540 224-kW (300-hp) engines. The aircraft first flew on 10 June 1977 and demonstrated very quiet operations.

A new military departure for the Defender is radar-carrier for stand-off surveillance. The flat radome on this aircraft hides the ASTOR (Airborne STand-Off Radar) unit, which would be used to give long-endurance battlefield surveillance.

Another proposed radar use for the Turbine Defender is the AEW role, this resulting in an even more pronounced nose bulge. Thorn EMI's Skymaster is the radar fitted, which can be used either for airborne or maritime target detection.

Britten-Norman (Pilatus) Trislander cutaway drawing key

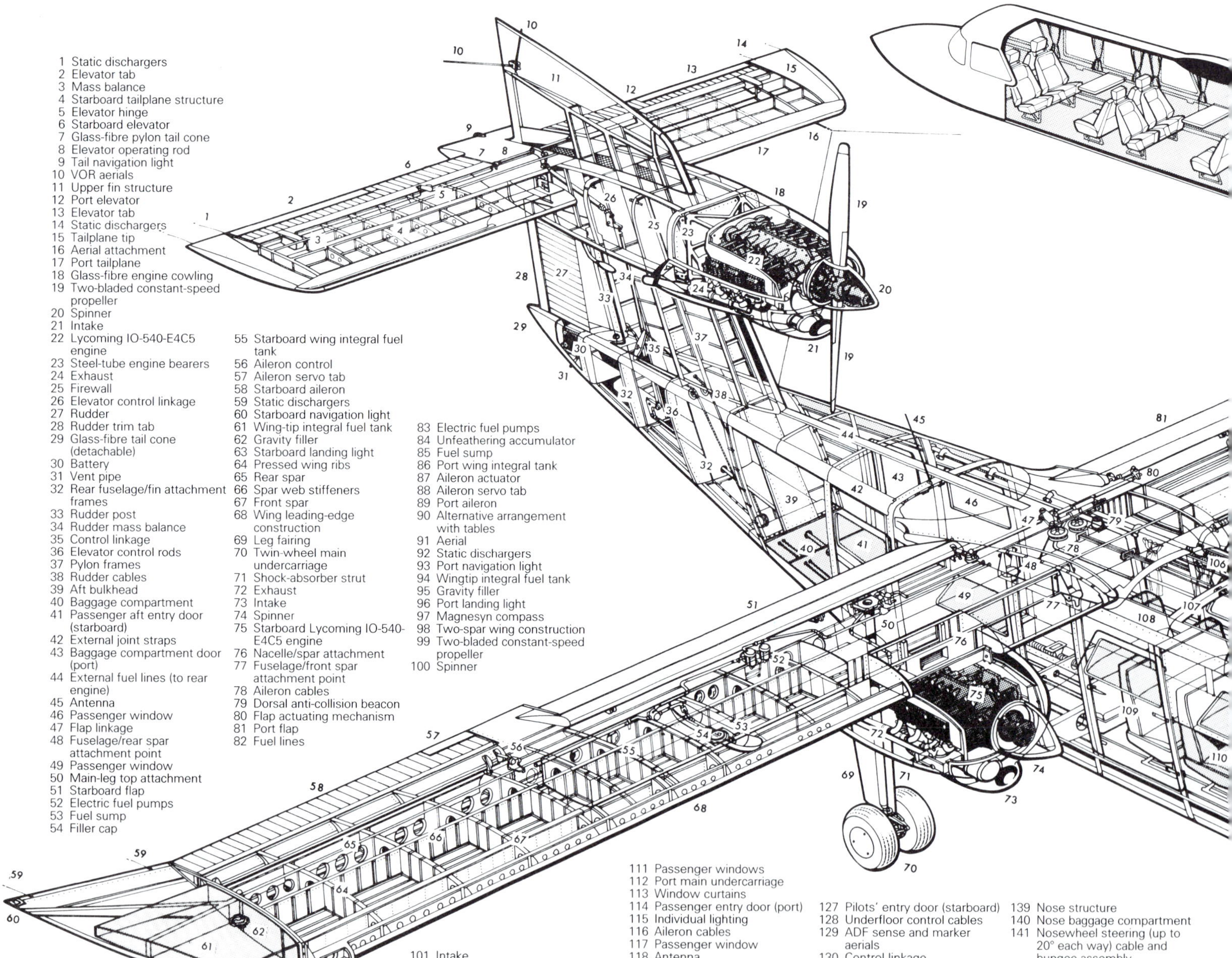

1. Static dischargers
2. Elevator tab
3. Mass balance
4. Starboard tailplane structure
5. Elevator hinge
6. Starboard elevator
7. Glass-fibre pylon tail cone
8. Elevator operating rod
9. Tail navigation light
10. VOR aerials
11. Upper fin structure
12. Port elevator
13. Elevator tab
14. Static dischargers
15. Tailplane tip
16. Aerial attachment
17. Port tailplane
18. Glass-fibre engine cowling
19. Two-bladed constant-speed propeller
20. Spinner
21. Intake
22. Lycoming IO-540-E4C5 engine
23. Steel-tube engine bearers
24. Exhaust
25. Firewall
26. Elevator control linkage
27. Rudder
28. Rudder trim tab
29. Glass-fibre tail cone (detachable)
30. Battery
31. Vent pipe
32. Rear fuselage/fin attachment frames
33. Rudder post
34. Rudder mass balance
35. Control linkage
36. Elevator control rods
37. Pylon frames
38. Rudder cables
39. Aft bulkhead
40. Baggage compartment
41. Passenger aft entry door (starboard)
42. External joint straps
43. Baggage compartment door (port)
44. External fuel lines (to rear engine)
45. Antenna
46. Passenger window
47. Flap linkage
48. Fuselage/rear spar attachment point
49. Passenger window
50. Main-leg top attachment
51. Starboard flap
52. Electric fuel pumps
53. Fuel sump
54. Filler cap
55. Starboard wing integral fuel tank
56. Aileron control
57. Aileron servo tab
58. Starboard aileron
59. Static dischargers
60. Starboard navigation light
61. Wing-tip integral fuel tank
62. Gravity filler
63. Starboard landing light
64. Pressed wing ribs
65. Rear spar
66. Spar web stiffeners
67. Front spar
68. Wing leading-edge construction
69. Leg fairing
70. Twin-wheel main undercarriage
71. Shock-absorber strut
72. Exhaust
73. Intake
74. Spinner
75. Starboard Lycoming IO-540-E4C5 engine
76. Nacelle/spar attachment
77. Fuselage/front spar attachment point
78. Aileron cables
79. Dorsal anti-collision beacon
80. Flap actuating mechanism
81. Port flap
82. Fuel lines
83. Electric fuel pumps
84. Unfeathering accumulator
85. Fuel sump
86. Port wing integral tank
87. Aileron actuator
88. Aileron servo tab
89. Port aileron
90. Alternative arrangement with tables
91. Aerial
92. Static dischargers
93. Port navigation light
94. Wingtip integral fuel tank
95. Gravity filler
96. Port landing light
97. Magnesyn compass
98. Two-spar wing construction
99. Two-bladed constant-speed propeller
100. Spinner
101. Intake
102. Engine cowling
103. Oil filler access
104. Cowling hinge line
105. Fresh air inlet and trunking
106. Fresh air cabin ducting
107. Fuselage frames
108. External capping strip
109. Passenger entry door
110. Bench seat (eight, side-to-side)
111. Passenger windows
112. Port main undercarriage
113. Window curtains
114. Passenger entry door (port)
115. Individual lighting
116. Aileron cables
117. Passenger window
118. Antenna
119. Rear-view mirror (port only)
120. Ceiling panel (starter switches)
121. Aileron cable turnbuckles
122. Fuel gauges
123. Ceiling panel (power supply)
124. Pilot's seat
125. Control yoke
126. Co-pilot's (or passenger's) seat
127. Pilots' entry door (starboard)
128. Underfloor control cables
129. ADF sense and marker aerials
130. Control linkage
131. Engine control pedestal
132. Instrument panel shroud
133. Windshield
134. Bulkhead
135. Rudder pedals
136. Heater and blower installation
137. Radio tray
138. Nose baggage compartment door
139. Nose structure
140. Nose baggage compartment
141. Nosewheel steering (up to 20° each way) cable and bungee assembly
142. Forward frame
143. Glass-fibre detachable nose cone (weather scanner optional)
144. Nosewheel leg attachment
145. Upper torque link bolt
146. Nosewheel leg shock absorber
147. Steerable/self-centring nosewheel

504

Mauritania is typical of third world military operators, fitting its Defenders with underwing rocket packs. Nine were delivered in total, but three are thought to have been lost.

A simple way of extending the Islander design to cater for greater loads was to stretch the fuselage and add a third engine on top of the fin. Thus was born the Trislander, offering 17 seats in maximum density as opposed to nine in the Islander.

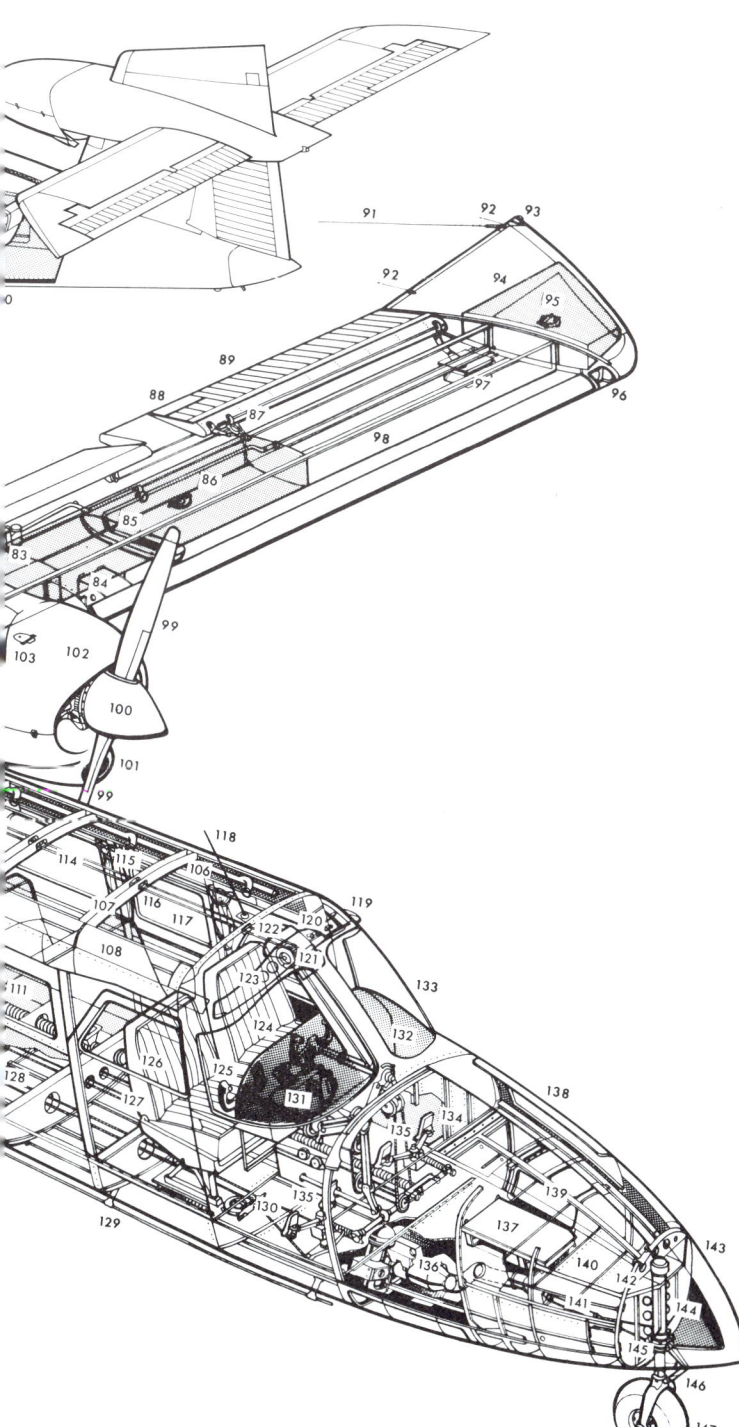

The name Trislander was adopted in January 1971, and UK certification was granted in May, followed by first customer delivery to Aurigny Airlines of Alderney, Channel Islands, on 29 June – only 292 days after the aircraft's maiden flight.

Further developments of the Trislander included the BN-2A Mk III-2 with an extended nose containing a baggage compartment, the BN-2 Mk III-3 with a propeller autofeather system, and the BN-2 Mk III-4, which had a standby rocket system mounted in the tailcone to provide additional thrust in the event of an engine failure on take-off. In June 1982 production rights to the Trislander were sold to the International Aviation Corporation of Florida, USA, which was to complete an initial batch of 12 aircraft, renamed Tri-Commutair, from a set of UK-supplied parts. However, there is no evidence that any of these has yet been completed.

Cheap firepower

At the 1971 Paris air show Britten-Norman introduced a military version of the Islander named Defender, which incorporated wing hardpoints for four NATO-standard weapons/stores pylons which could carry 7.62-mm machine-gun pods, 113- or 226-kg (250- or 500-lb) general purpose bombs, rocket launchers, wire-guided missiles or 227-litre (60-US gal) droptanks. Provision was also made for the installation of sophisticated electronics and medevac equipment. A Maritime Defender variant incorporated a nose-mounted search radar and was intended for coastal and fishery patrol, oil rig protection and search-and-rescue duties. Among operators of the Defender are the armed forces of Abu Dhabi, Belgium, Belize, Botswana, Ciskei, Ghana, Guyana, Haiti, Hong Kong, Jamaica, Malagasy, Malawi, Mauritania, Mexico, Oman, Panama, Rwanda, Seychelles, Surinam, Zaire, Zimbabwe and, of course, the UK.

Despite the great success of the Islander, of which production was then running at 12 per month, by late 1971 Britten-Norman was in financial difficulties and was forced into receivership. In August 1972 the Belgian Fairey Group took over the company, and a new production line was established at Fairey S.A.'s Gosselies plant, to which all UK Islander and Trislander production was progressively transferred. In September 1974 the Islander broke the British sales record for multi-engined commercial aircraft, passing the previous total of 548 held by the de Havilland Dove. At this time a further licence production agreement was signed with the Philippine Aerospace Development Corporation (PADC) of Manila, so that Islanders were then being manufactured or assembled in Belgium, Romania and the Philippines.

Development work continued to be undertaken at Bembridge, although by February 1976 the two company founders had left Fairey-Britten-Norman to start new ventures. On 6 April 1977 the prototype Turbo Islander made its first flight. This aircraft was powered by two 498-kW (600-shp) Lycoming LTP101 turboprops,

Aurigny Air Services of the Channel Islands ... operator. This is one of the original aircraft ... nose for extra baggage capacity. Note the ... the rear fuselage to prevent the aircraft tipping back ... loading.

The Trislander was not as popular with military users as the Defender, Botswana being one of the few purchasers. Already a user of the Defender, the southern African state received two Trislanders in 1984.

which provided sparkling performance, but doubts about engine reliability and the extensive modifications required to the wing structure caused it to be shelved early in the test programme.

Buy-out from Switzerland

Financial difficulties within the Fairey Group brought a further crisis in late 1977 which cast doubts about Fairey-Britten-Norman's future, but in July 1978 agreement was reached with the Swiss company Pilatus for a further take-over, and the company was renamed Pilatus Britten-Norman Ltd.

Under the new management, production was re-started at Bembridge and a comprehensive product improvement plan was launched. The BN-2B version of the Islander was introduced, featuring the 194- or 224-kW (260- or 300-hp) engine options of the BN-2A, but with an increased maximum take-off weight of 2993 kg (6,600 lb), a redesigned flight deck, new passenger cabin interior with improved ventilation, and smaller diameter propellers to reduce cabin noise. Pilatus Britten-Norman celebrated delivery of the 1,000th BN-2 in April 1982 when a BN-2B Islander was handed over to the Cyprus government. The BN-2B remains the standard production Islander, with Defender variants also in production at Bembridge.

Special missions

Pilatus also approved development work on a new turboprop Islander, for which the 298-kW (400-shp) Allison 250-B17C powerplant was selected. The prototype BN-2T Turbine Islander made its first flight from Bembridge on 2 August 1980 and was certificated by the Civil Aviation Authority at the end of May 1981, with US Federal Aviation Administration approval being granted on 15 July 1982. The Turbine Islander has a maximum cruising speed of 315 km/h (196 mph) at 3050 m (10,000 ft) and a VFR range, with standard reserves, of 1349 km (728 nm/838 miles). The first production aircraft was delivered in December 1981; by the end of 1990 more than 30 were in operation, including Turbine Defenders, which are offered in a variety of coastal patrol, search-and-rescue, medevac, paradropping, ASW, AEW and Elint variants. Recent deliveries of BN-2Ts have included two for the Dutch National Police Force, which are equipped with special police communications equipment, weather

A Trislander destined for the Seychelles Islands (re-registered S7-) flies along the southern English coast on a test flight. The use of several doors along the fuselage allows the seats to be accessed without the need for an aisle, thereby maximising use of the internal capacity.

radar, and flight-openable rear cabin doors with mounts for cameras and infra-red sensors; and seven aircraft for the British Army have replaced de Havilland Canada Beavers and are believed to be equipped with electronic sensors for use in a signals intelligence gathering (Sigint) role in Northern Ireland. Others are with the RAF and Royal Navy.

The Islander story has been one of remarkable sales success despite serious financial setbacks. More than 1,250 Islanders and their derivatives have flown with 385 or more operators in some 125 countries – a fine testimony to the foresight of John Britten and Desmond Norman.

Ilyushin Il-12 and Il-14

After the war the Soviet Union was faced with the same problem as the West: how to replace its irreplaceable DC-3s (or Lisunov Li-2s)? The answer was first the Il-12, followed later by the Il-14. While neither of these types was a great leap forward from what they had replaced, they were built in large numbers in several nations and became a versatile family of transport aircraft.

During World War II the Soviet state aircraft industry produced nearly 3,000 licence-built Douglas DC-3 transports as Lisunov Li-2s for military and civil use. These, together with 'genuine' DC-3s supplied by the United States, formed the mainstay of Aeroflot's aircraft fleet during the immediate post-war years.

Plans for a replacement for the Li-2 were laid in 1943, when the Kremlin issued a requirement for a modern transport aircraft. Despite high priority being given to production of the 'Shturmovik' ground attack aircraft at the time, the Ilyushin Design Bureau under Sergei Vladimarovich Ilyushin formed a sizeable design team to undertake the project, which was given the designation Il-12. The Ilyushin design competed with two others, the Bartini T-117 and Yermolayev Yer-2-ON, both of which failed to meet requirements, and the Il-12 was selected for production.

Based to a large extent on the well-tried structure of the Li-2, Il-12 was an all-metal low-wing design with tricycle undercarriage, and bore a resemblance to the later Swedish Saab Scandia and American Convair CV-240. The fuselage was circular in section, of multi-stringer semi-monocoque construction with mostly flush-riveted skin. The wings were built up on three spars, with a Clark YH airfoil on the centre-section, and K-4 section on the outboard tapered wing panels, which had a narrower chord at their root end than the corresponding centre-section panel. Hydraulically-actuated split flaps were fitted; all control surfaces were of metal construction with fabric covering.

The prototype Il-12 was first flown on 9 January 1946 by test pilot V. K. Kokkinaki and his brother Konstantin. Early testing revealed a sensitive centre of gravity position so close to the location of the main landing gear legs that a tail prop was necessary to support the unloaded aircraft on the ground, and a lack of adequate directional control in the air which was solved by adding a dorsal fin.

Production was launched at Khodinka in 1946, with 1380-kW (1,850-hp) Shvetsov ASh-82FNV radial engines replacing the Charomskii diesels originally planned. The engines drove four-bladed constant-speed AV-9E-91 propellers, which were copies of US-built Hamilton Standard units. The Il-12, which was given the NATO codes name 'Coach', entered service with Aeroflot on 22 August 1947, and within nine months was being used on most of the airline's domestic and international routes. Aeroflot's Il-12s were equipped to carry a crew of four (pilot, co-pilot, navigator and radio operator) and 27 passengers, in nine rows of seats, although performance limitations restricted capacity to at most 21 passengers, often fewer on long-distance routes.

Europe's last active Il-14s were the Polish air force's few East German-built examples. Some were fitted for photo-mapping missions, while others retained a classic 1950s Eastern Bloc VIP interior complete with up-to-the-minute formica table tops.

A rare photograph shows the Ilyushin Il-12 'Coach' prototype (SSSR-L1380) in flight, in its original form prior to fitment of a dorsal fin. The latter was necessary to cure a directional stability problem which surfaced during early flight tests.

At rest, this Aeroflot Il-12 displays the prop under the tail needed to prevent the aircraft tipping up on the ground. A feature inherited from the Li-2/DC-3 for Il-12T cargo versions was the large rear door incorporating a smaller passenger door.

From virtually every angle, the Il-12 showed the obvious influence of the Douglas DC-3 design (built in the Soviet Union as the Lisunov Li-2). The 'Coach' largely replaced the Li-2/DC-3 on Aeroflot internal routes.

Of the 1,500 or so Il-12s built, several were exported for use by other Communist airlines. This 'OK Liner' served with CSA, the Czech national airline.

Both East Germany and Czechoslovakia built the Il-14 under licence, this being a Czech-built Avia-14. Avia took over all production from 1958, ending production in 1962.

Many air forces used the Il-14 for utility transport and other roles. This Polish aircraft is in VIP transport configuration. Both the Polish and Soviet air forces refitted 'Crates' for electronic intelligence gathering missions.

Early service experience led to a number of improvements, including a strengthened nosewheel, and anti-icing equipment for propellers, windscreen, tailplane and engines. These improvements added some 260 kg (570 lb) to the aircraft's maximum take-off weight and reduced engine-out performance to such an extent that passenger capacity had to be limited to 18. An all-cargo Il-12T version was also developed, with capacity for up to 3628 kg (8,000 lb) of freight. In addition to Aeroflot, civilian Ilyushin Il-12s were also operated by CAAC of China, the Czech carrier CSA, LOT of Poland, TABSO of Bulgaria and TAROM of Romania.

Versatile versions

A military assault transport variant, designated Il-12D, had large double cargo doors on the port rear fuselage side and could carry up to 26 fully-equipped troops or 20 paratroops; or in medevac role, 16 stretcher cases and six sitting patients. Other variants included one equipped with skis and additional anti-icing measures for use in Soviet Arctic and Antarctic exploration expeditions. Total production of the Il-12 series is believed to have exceeded 1,500 aircraft.

While production deliveries of the Il-12 continued, design work was under way on a developed version to overcome the payload and performance deficiencies from which the aircraft suffered.

The many necessary changes resulted in an almost complete redesign, with only the configuration of the fuselage and undercarriage remaining similar, although these too were modified on the Il-14, which made its first flight on 20 September 1950. An entirely new, higher aspect ratio, square-tipped wing had been designed. Unlike that of the Il-12 it had a straight taper from root to tip, using an SR-5 airfoil section throughout, this being a slightly deeper section than those used on the Il-12. Slotted flaps were fitted. The fin and rudder were also square-tipped rather than rounded, and increased in area to enhance directional control and reduce forces in asymmetric flight. Since performance limitations had prevented all nine rows of seats ever bing used in the Il-12, the number of cabin windows was reduced to seven on each side in the Il-14. Uprated 1417-kg (1,900-hp) ASh-82T ('Transportnii') engines were installed in aerodynamically refined nacelles which had twin overwing exhaust stacks with exhaust augmentation in an installation closely copied from that of the CV-240. New AV-50 propellers with much-reduced feathering times were fitted to improve engine-out handling.

The Il-14 underwent a protracted test programme to try to resolve performance deficiences that left it still well behind contemporary Western transports of similar power, but was finally approved for production in late 1953. The production standard Il-14P (NATO code name 'Crate') entered service with Aeroflot on 30 November 1954 and was initially configured to carry only 18 passengers, although theoretically it should have been able to accommodate up to 40 in 10 rows of four. A cargo version was designated Il-14G.

In 1956 the improved Il-14M ('Modifikatsiya') was introduced. This aircraft had a fuselage lengthened by 1 m (3 ft 3 in), enabling 24 passengers to be accommodated (later increased to 32) and also incorporated revised

Seen at London's Heathrow Airport, this Il-14 brought former Prime Minister Georgi Malenkov to inspect British power stations in 1956. Wearing Soviet air force markings, the aircraft had a luxury interior for VIP transport.

This Czech Avia-14 was used for engine tests, a Walther turboprop being carried under the nose. Illustrating the success of this arrangement, the Il-14 has both its main engines shut down with props fully feathered.

wingroot geometry, reduced structural weight, increased maximum take-off weight and increased fuel tankage. A freighter variant with 2.2 m (7 ft 2 in) wide double cargo doors was designated Il-14T. Other versions included the Il-14D ('Desantnyi') assault transport and the Il-14F aerial survey and mapping variant. Production at Khodinka continued until 1 January 1958, with the total built for civilian and military use estimated at 2,200.

'Crates' abroad

In addition to Soviet production the Il-14 was also produced by East German state manufacturer VEB Flugzeugwerke at Dresden, and Avia in Czechoslovakia, who received drawings and tooling assistance from Ilyushin in 1955, and took over production jugs and tooling from the Soviet manufacturer in 1958. VEB built 80 Il-14Ps for service with the DDR carrier Interflug as 18-26 passenger airliners or as all-cargo aircraft. Avia built some 50 Avia-14s (equivalent to the Il-14P) before introducing the improved 32-passenger Avia 32 (equivalent to the Il-14M). In 1958 Avia introduced the Avia-14-32A, which was structurally strengthened to permit operation at a higher gross weight, and later developed the unique Avia-14 Salon which differed from all previous models in having a pressurised cabin with circular windows, and a revised fuel system incorporating wingtip fuel tanks. The Salon could accommodate 36-42 passengers. Production of Avia-14 variants in Czechoslovakia ceased in 1962, and totalled around 200.

The Ilyushin Il-14 was the first Soviet aircraft to be widely exported, and saw service with the civilian carriers and military forces of most Eastern Bloc and many Middle Eastern and Third World nations. A number were configured as VIP transports and donated to heads of state, among them President Nasser of Egypt, Prime Minister Nehru of India, President Tito of Yugoslavia and the Shah of Iran.

Ilyushin Il-18

Unsophisticated and crude by Western standards, the Il-18 'Coot' became a mainstay of the Soviet civil aviation scene, and did much to expand Aeroflot's services. Still in limited civil service today, the Il-18 has also provided an excellent basis for special mission variants with both civilian and military applications. Here we describe the development of this important turboprop aircraft and its derivatives.

Sometimes the supremely methodical Russians cause confusion by giving the same designation to two completely different aircraft. On 30 July 1947 an impressive passenger airliner in the class of the Boeing 377 Stratocruiser made its first flight. It was judged too big and complex to fit into Aeroflot's primitive fields, and never went into production. But just six years later the Soviet civil operator, biggest airline in the world, issued a requirement for an aircraft in the same class, but carrying slightly more passengers (75) and powered by turboprops. The task was assigned to the design bureau of S. V. Ilyushin, which repeated the same designation as that used for the piston-engined aircraft: Il-18.

The prototype, named *Moskva* and with registration L-5811, made its first flight on 4 July 1957. In almost every respect it looked an efficient and modern aircraft, one of the few adverse features being the obvious joint between the hemispherical front of the pressurised fuselage and the projecting unpressurised nose, into which the twin-wheel steerable nose gear retracted. Apart from that, and the smoky trails from the four 2980-kW (4,000-ehp) Kuznetsov NK-4 turboprops (and the NK-4 was no worse in this regard than Western turboprops of the same era), the Il-18 looked to be something new in the world of aviation: a Soviet civil transport every bit as good and modern as those produced in the USA and Britain.

When the Il-18 was designed the only aircraft in the same class that already existed was the Bristol Britannia. The Ilyushin designers followed the same layout, with four engines mounted high on a tapered wing of high aspect ratio with double-slotted flaps and manually driven ailerons, a pressurised fuselage of circular section of 3.5-m (138-in) diameter, a single-finned tail again with manual controls, and jetpipes taken back above the wing to nozzles over the trailing

Long after their days on scheduled routes, a few Il-18s are still on charge with Balkan Bulgarian Airlines, and are used for general passenger work, including charters. Some freight is also carried.

This Il-18 served in a VIP/staff transport function with the Democratic and Popular Republic of Algeria. It has since been withdrawn, its role being handled by a mix of Gulfstream IIIs, King Air 200s and a Fokker F27.

The Il-20 'Coot-A' is based on the Il-18 airframe but packed with electronic surveillance equipment. It is the chief Soviet Sigint (signals intelligence) gatherer, and is regularly intercepted by Western fighters.

edge, flexible fuel cells between the wing spars except in the outer wings where integral tankage was used, four-bladed reversing propellers and four-wheel bogie main landing gears retracting into the inner nacelles.

A quality design

Apart from the fact that the main gears retracted forward instead of backward, the only significant difference between the British and Soviet aircraft was that the latter was slightly smaller and lighter, being designed for operation over shorter sectors. With maximum payload the range was only 2500 km (1,550 miles), while with maximum fuel it was possible to fly 5000 km (3,100 miles). These distances were fully adequate for Aeroflot's network – indeed, they compared very favourably with the Li-2 (DC-3) and Il-12 which the Il-18 was designed to supplement and, on major routes, to replace – and they were also long enough for the new transport to be of great interest to various export customers, which was again a new achievement for the Soviet Union.

Early in the design process the Comet disasters and subsequent investigation made a deep impression on the Ilyushin team. The entire structure was carefully inspected and in numerous places redesigned in detail, to give fail-safe or safe-life qualities. This work involved adding more than 150 double skins or duplicate load paths, which added 270 kg (595 lb) to the airframe weight. During the testing of the prototype it was decided to add a pressure bulkhead behind the flight deck, to be locked in flight. Apart from this, few modifications were needed.

Best in the world

In line with common Soviet practice, several flights were submitted as world records to the Fédération Aéronautique Internationale, and duly homologated. In each case the pilot was Ilyushin chief test pilot Vladimir Kokkinaki. These flights were: a speed of 719.4 km/h (447.07 mph) over a 2000-km (1,242-mile) circuit with various payloads up to 15000 kg (33,070 lb) on 19 August 1959; a speed of 693 km/h (430.5 mph) over a 5000-km (3,107-mile) circuit with various payloads up to 10000 kg (22,046 lb) on 2 February 1960; a climb to 12471 m (40,915 ft) with a payload of 15 tonnes on 14 November 1958; and a climb to 13154 m (43,156 ft) with a payload of 10000 kg (22,046 lb) on 15 November 1958. The Il-18 won a Lenin Prize for Ilyushin, Kokkinaki and nine others.

Cubana operated the Il-18 from 1965, and had Il-18Ds and Il-18Vs on strength. This aircraft is an Il-18D, first delivered in 1967, and is configured for 100 passengers.

Production at the GAZ-30 plant at Moscow Khodinka was authorised in November 1958. This was a commendably fast development, contrasting very sharply with the protracted development of Soviet transports of the 1970-90 period. The initial production aircraft, Nos 1-20, had NK-4 engines, 75 passenger seats, a flight crew of five, two rather small underfloor baggage/cargo holds and an unpressurised rear hold accessible from the outside only. MTO (maximum take-off) weight was 57200 kg (126,102 lb). Take-off run was about 1193 m (3,750 ft), and cruising speed typically 625 km/h (388 mph).

In 1953 the NK-4 turboprop was transferred to the design collective of A. G. Ivchyenko at Zaphorozhye. Here it was redesignated as the AI-20, refined in detail and prepared for production. By the end of 1958 Zaporozhye was supplying for Il-18 production the complete powerplant comprising the 2939-kW (3,945-ehp) AI-20K engine, all accessories and cowlings, and the 4.39-m (14-ft 9-in) four-bladed AV-68I propeller. This enabled production to switch from the 21st aircraft to the Il-18B, with MTO weight of 59200 kg (130,511 lb). Normal seating was increased to 84, and another change was that, whilst retaining traditional gravity filling from remote airports, pressure-fuelling sockets were added in the inner nacelles.

A few Il-18B aircraft had been delivered when Aeroflot services began on 20 April 1959. Two years later, in spring 1961, production was transferred to the Il-18V, with a re-arranged cabin able to seat up to 100 passengers, though it was not possible to move the rows of call buttons, lights and fresh-air outlets which were arranged in three compartments at a fixed seat pitch. In 1964 the Il-18I introduced the

The principal sensor of the Il-20 'Coot-A' is the huge SLAR fairing under the forward fuselage. Additional fuselage-side fairings contain infra-red and electronic sensors, while the airframe is liberally festooned with other electronic antennae.

Right: As might be expected, the nations of the Warsaw Pact were early customers for the Il-18, all the major airlines receiving the type. Malev had several Il-18Vs (illustrated) and a long-range Il-18D.

AI-20M engine, rated at 3166 kW (4,250 ehp). This enabled MTO weight to be again increased, this time to 6400 kg (141,093 lb). The extra weight enabled 6300 litres (1,386 Imp gal) of additional fuel to be carried in four fuel cells between the wing spars under the cabin floor. This enabled maximum payload to be carried for 3700-km (2,300-miles), or a distance of 6500 km (4,040 miles) to be flown with maximum fuel. Take-off run, however, rose to around 1371 m (4,500 ft). The designation Il-18I was thought confusing, so before this version entered service in 1965 it was changed to Il-18D. This proved to be the ultimate Il-18 variant, and it was produced alongside the Il-18Ye (written in Russian Il-18E), which was identical except that the extra fuel cells were not fitted. Most D and Ye aircraft had the rear pressure bulkhead moved to the rear so that the pressurised cabin swallowed up the previous unpressurised rear hold. This enabled high-density seating to be provided for 122 passengers.

Out of production but into uniform

About 800 Il-18 aircraft were built, production being completed at GAZ-30 in about 1970. About 25 were equipped for military and government use, usually as VIP transports, one being used on 1 May 1960 to fly Francis Gary Powers, the shot-down U-2 pilot, to Moscow. About 100 were exported, early customers including Czechos-

lovakia, East Germany, Hungary, China and Ghana. Later customers came from Guinea, Mali, Bulgaria, Romania, Mauritania, Poland, Cuba, Egypt and Yemen.

To meet the need of the AV-MF, the Soviet naval air force, for a large shore-based ASW and maritime patrol aircraft, the Il-18 was converted into the Il-38, in precisely the way that the very similar Electra was converted into the P-3 Orion. The first conversion took place in about 1965, and the production Il-38 probably entered service in about 1968. Though the airframe remained sensibly unaltered

North Korea's CAAK airline (now called Chosonminghang Korean Airways) still operates this Il-18D, the aircraft having been delivered in 1968. The airline is regularly called upon to undertake government duties.

An Aeroflot Il-18 demonstrates some agility during a test flight, and also the clean design which typified the type. The powerful engines provided a considerable amount of jet thrust, in addition to that from the propellers. In 1959 the type set numerous turboprop class records.

there was an almost unbelievable shift in CG (centre of gravity) so that, while the fuselage ahead of the wing was made shorter, the aft fuselage was made 26 ft longer, with a further increase in overall length from 35.9 m to 39.57 m (117 ft 9 in to 129 ft 10 in) resulting from the MAD (magnetic-anomaly detection) device projecting behind the tail. Apart from the MAD the most prominent new feature was the search radar in the forward fuselage, with the rotating antenna in a bulged radome under the forward fuselage. Unlike the P-3 little provision was made for weapons, there being no evidence of external pylons to back up the two shallow weapon bays ahead of and behind the wing centre section. Fuel capacity, engines, cabin pressure and virtually all systems remain unchanged, though the passenger windows and doors are deleted and replaced by a few small portholes, the door being on the right side at the rear.

Total production of the Il-38 was not greater than 100. In 1990 about 59 were still in service with the AV-MF, and these are very active and often encountered over the Baltic and North Atlantic. Some have been detached to operate from bases in Egypt, Libya and, more recently, the PDR of the Yemen and Syria. Three used aircraft were purchased by India and equip Indian Navy Squadron No. 315 based at Dabolim, Goa.

Cargo Coots

The Il-38s were all built as new aircraft, but by the mid-1970s increasing numbers of ex-Aeroflot Il-18s became available for conversion to other duties. Following general Soviet practice of never throwing anything away if it could be put to use, every one of these aircraft has been converted for other purposes. Probably the largest number have been rebuilt at the GAZ-402 plant at Moscow Bykovo, as Aeroflot freight aircraft. Several detail features of this conversion were copied from the British conversion of Vanguards into Merchantman freighters. The main changes have been to fit a strengthened floor, mainly of titanium and thick panels of aluminium alloy, with ball mats, rollers and multiple tiedowns. At the rear is the new power-driven side door for loading pallets and other bulky loads, with a width of 3.5 m (138 in). Cargo conversions began in 1974 and were still going on in 1989, probably 200 aircraft having been thus rebuilt. Each conversion starts the Il-18 off again with a 'zero time' airframe. Designation of the freighter is Il-18T.

Variations on a frame

There are many other versions, rebuilt from the original Il-18 series aircraft. One is the Il-18DORR, described in the Soviet media in 1986 as a long-range ocean-fisheries reconnaissance aircraft. Another, of unknown designation, is exemplified by SSSR-75442, which in 1989 passed through Shannon bound for Cuba. One of four operated by the Soviet Meteorological Institute, it was covered in bumps, blisters, dipole antennae, canisters, 'bomb doors' and other devices, the largest being what appeared to be a radar blister behind the nose landing gear.

Il-18s were supplied to many Soviet clients. Among them was Communist China, which still has a small number on charge. Their remaining service is diminishing as Western types are procured.

Much larger numbers of ex-Aeroflot Il-18s have been rebuilt as testbeds for various systems and equipment, mainly of an avionic nature, and as Il-20 or Il-22 aircraft for the armed forces, chiefly the VVS (air force). The Il-20, first seen in the West in 1978, is the most important Elint (electronic intelligence) aircraft of the VVS. Unlike the Il-38, the basic airframe is unchanged, the rebuild being confined to gutting the fuselage and fitting totally different crew stations and equipment. Many of the passenger doors and windows, and service doors, are left unchanged. Internally, though still pressurised, the interior is occupied by a flight crew of four and emission crew which numbers from eight to 11. Outside there are usually from 14 to 18 sensors, of which by far the largest is a SLAR (side-looking airborne radar), the antennae for which face to left and right along the sides of a container slung under the forward fuselage and measuring 10.25 m (33 ft 8 in) long and 1.14 m (45 in) high. On each side of the forward fuselage is a fairing 4.39 m (14 ft 5 in) long and 0.86 m (34 in) deep, containing infra-red linescan and cameras. Two very large plate antennae are mounted on the centreline above the fuselage aft of the flight deck, while eight further antennae or Elint receivers are mounted along the underside of the rear fuselage. Several further antennae are flush with the skin of the aircraft.

Senior Soviet spyplane

The primary mission of the Il-20 is electronic 'ferret' sensing, whilst patrolling near NATO or other fleets or, less often, off the shores or frontiers of potentially hostile countries. The SLAR, cameras, IRLS and some other sensors are used in the general reconnaissance role, concerned with items other than foreign electronic emissions. There is some variation in the equipment fitted to Il-20 aircraft, and some have been seen which are probably civilian Il-18 testbeds for Elint or reconnaissance systems. One such aircraft, seen in 1981, is SSSR-75431, with fewer sensors or antennae than standard Il-20s but equipped with a very large streamlined blister under the aft fuselage superficially resembling the J-band radar seen on some Tupolev ocean patrol aircraft. There is evidence that this particular, civil-registered aircraft was one of the many Il-18s now in use as testbeds. It continues to carry Aeroflot titles, though that does not necessarily mean it is still serving with the airline.

The Il-20 repeats the designation of an earlier Ilyushin, the civil transport versions of the Il-28 bomber. So too does the Il-22 repeat an earlier designation, in this case that of the Soviet Union's pioneer four-jet bomber of July 1947. Today's Il-22 is an Il-18 rebuilt as an airborne command post for the higher command of the Soviet armed forces. Details were not available in the West as this was written, but this is just one of five known types of rebuild which by late 1990 had been identified in the West yet remained in classified literature. NATO has assigned the reporting names 'Coot-A' to 'Coot-E' to these aircraft, all the original Aeroflot Il-18 versions being lumped together under the general name of 'Coot'. The reporting name for the Il-38 is 'May'.

Two turboprop types that dominated Soviet aviation in the 1970s sit outside Dnepropetrovsk airport terminal in 1974. Both An-12 (background) and Il-18 were powered by the Ivchenko AI-20 engine.

Chapter 24

Ilyushin IL-62

Ilyushin IL-76

Ilyushin IL86 & IL96

Ilyushin Il-62

While the Il-62 has been often dismissed as a 'carbon-copy VC10', it has had a longer career and a far greater impact in its own field than Vickers' matching four-jet ever did. The cornerstone of Aeroflot's international routes from the mid-1960s onwards, it still carries the flag of its homeland around the world.

The Il-62 was the first jet-propelled long-range transport to be designed in the Soviet Union. Western observers naturally at first thought it a 'copy of the VC10', but in fact the resemblance is skin-deep.

The air services of the former Soviet Union – the world's largest country – tended to be of a basically short-haul nature, serving local networks. This was almost entirely because the airfields were poorly equipped, by Western standards. Except for Moscow, where international services were operated from good runways, most cities were limited by their airports to small propeller aircraft. Even as recently as the 1960s, the main overseas trunk routes were operated by the rather ungainly Tu-114 turboprop, related to the Tu-95 'Bear' bomber. By 1957, the year in which the Tu-114 first flew, it was recognised that Aeroflot would soon have need of a large long-haul jetliner. The contract for what became the Il-62 was placed with the design bureau of S. V. Ilyushin the following year, 1958. The GVF (civil air fleet) specification required the carriage of 150 passengers and baggage at 900 km/h (559 mph) over a range of 8000 km (4971 miles), plus advanced navigation and blind-landing systems to permit sustained operations in bad weather. Ilyushin and his design staff carefully considered six configurations, before deciding on four engines grouped in horizontal pairs on the rear fuselage. A special powerplant was needed for the Il-62, and the task was assigned to the bureau of N. D. Kuznetsov. The result was the NK-8 two-shaft turbofan, with a bypass ratio close to unity, very similar to the later versions of Rolls-Royce Conway.

This engine was not ready in time for the first prototype Il-62, SSSR-06156 (the Cyrillic letters appear to read CCCP in Roman script), nor for at least one of the following four aircraft, which comprised a second prototype and three pre-production aircraft. No. 06156 was fitted with Lyulka AL-7 turbojets, which at 74 kN (16,535 lb) take-off thrust were inadequate for test flying at maximum weight, and whose greater fuel consumption made range figures pointless.

In many respects, such as the use of very large forged frames and machined skin panels, even the prototype was well up with the 1960 state of the art. The large wing, swept 35° at the quarter-chord line, was designed as a centre-section integral with the fuselage, with span sufficient to carry the main landing gears, and long outer

Poland's national carrier LOT received its first six Il-62s slowly, between 1972 and 1976. These were eventually replaced by Il-62Ms during the 1980s. Named, in the main, after famous countrymen, LOT's 'Classics' were retired in 1991/92, and stored at Warsaw-Okecie.

UAA came into being in 1961, after the union of Syria and Egypt to form the United Arab Republic in 1958, and the merger of the two states' airlines. To replace existing Egyptair Comets, Il-62s were leased as a temporary measure pending the arrival of more favoured Boeing 707s.

panels. The basic wing torsion box had four spars to the aileron and two from there to the tip. It was sealed to form three integral tanks in the centre-section and two in each outer panel, total capacity being 100000 litres (21,998 Imp gal). A large track-mounted slotted flap was mounted on each outer panel and on each side of the centre-section. Ahead of the outer flap on each wing was inserted a two-section spoiler, hydraulically driven in unison for use as an airbrake, for rapid let-down, and as a lift dumper for use after touchdown. Outboard were large two-section ailerons, divided into three sections, the outermost being plain, the middle having a spring-loaded servo tab and spring tab, and the inner a spring servo tab. The ailerons were manual, the flaps electric and the spoilers hydraulic.

The fuselage was a near circular-section tube, the height being fractionally greater than the width of 3.75 m (12 ft 3.5 in). Pressure differential was set at 0.63 kg/cm2

Czechoslovakia's CSA had a significant Il-62 fleet of its own, but this Aeroflot example was leased to cover excess demand during the summer of 1969.

(8.96 lb/sq in), an exceptional value. The main environmental system machinery was mounted in the inner wing ahead of the front spar, with ram inlets in the wingroots. In the tailcone was mounted a TA-6 gas-turbine APU (auxiliary power unit). The small elliptical passenger windows closely resembled those of the VC10 in structure. A new feature for Soviet transports was that the passenger floor was made of a metal/foam sandwich. Plug-type doors were provided ahead of and behind the front cabin on each side, ahead of the wing. The plan to fit an aft ventral door with stairs was not accepted.

Massive tail

The bold T-tail comprised two sections of manual rudder with a yaw damper, spring servo tab and trim tab, and manual elevators with four trim tabs (two automatic and two manual) hinged to a pivoted tailplane powered by an irreversible hydraulic drive for trimming. The three units of the landing gear were powered hydraulically. The steerable twin-wheel nose gear retracted forwards and the four-wheel main bogies folded inwards to be housed in the wingroots and under the passenger floor. The hydraulic disc brakes were controlled by electric inertia anti-skid units. In the rear fuselage of the prototype was an emergency braking parachute. An unusual feature was that when the aircraft was parked a vertical strut with two small wheels was extended down to support the rear fuselage during loading operations.

Apart from the AL-7 engines, the early prototypes differed from production aircraft in having DC (direct current) electrical systems. Externally it had conspicuous axial fairings along the lower sides of the nose, along the top of the forward fuselage and from the mid-fuselage to the fin. First flown in January 1963, this aircraft soon showed that various modifications were needed. The flight-test programme was protracted, no true production Il-62 appearing for nearly four years. The most obvious changes introduced during this time were the switch to the Kuznetsov engines, redesign of the outer wings with extended-chord drooped leading edges, addition of six small fences (later removed) and elimination of the four axial fairings along the fuselage.

Wearing the 1960s-style Aeroflot scheme this Il-62 has the extended spine sometimes seen on early aircraft, but is not to be confused with the communications antenna of VIP aircraft.

overall pressure ratio and a bypass ratio of 2.42 (compared with 1.02) it promised considerably better fuel economy. For example, the specific fuel consumption in cruising flight was about 0.695 instead of 0.78. Thus, despite being slightly heavier, the overall weight of engines plus fuel for any long sector was many tons lighter. With the help of the Central Aero and Hydrodynamic Institute, the Ilyushin bureau designed a completely new nacelle with lower drag. The thrust reversers, still over the outer engines only, were of the clamshell type instead of the internal cascade type, giving increased performance and enabling approach speed to be reduced. Range was further increased by adding a fuel tank in the fin, of 5300-litre (1400-gal) capacity.

Improved Il-62M

The Ilyushin bureau called the re-engined aircraft the Il-62M. The opportunity was taken to introduce many further upgrades, by far the most important being an almost completely new suite of avionics, far more comprehensive than in any previous Soviet transport. In particular, the use of dual SAU-1T autopilots, each with its own monitoring, enabled flight trajectory to be automatic from just after take-off until 30 m (100 ft) from the end of the landing runway. The pilots' yokes were redesigned to improve view of the instruments, and roll control at all speeds was enhanced by redesigning the spoiler system so that it could work asymmetrically.

The production line for the Il-62 was established at Kazan, capital of the then Tartar SSR, now Khazakstan, which is today by far the world's largest aircraft production complex (the Il-62 always being a very minor part of the work). From the start, the production aircraft had NK-8-4 engines rated at 99 kN (22,273 lb) thrust, with reversers on the outer engines. Hot bleed air was piped to de-ice the airframe leading edge and engine inlets. Except for the first few aircraft, the electrical power system was changed to raw AC (alternating current), generated by a 40-kVA alternator on each engine. Internally the fuselage was furnished for a flight crew of five and up to 186 passengers, all in 3+3 seat rows. In all normal configurations there are two cabins, with the galley between them. In the highest-density arrangement the forward cabin seats 72 and the rear cabin 114.

Aeroflot was able to introduce the Il-62 on experimental internal services from 10 March 1967. Serious international services, replacing the Tu-114, began on 15 September of the same year, initially from Moscow to Montreal. Subsequently, the Il-62 replaced the Tu-114 on all other routes, including those to New York, Cuba and Tokyo. Modest export sales were soon made, particularly to the airlines of the Eastern Bloc countries which had little choice but to accept the sole long-range airliner available to them. Operators of the basic Il-62 have included Aeroflot, Cubana, Interflug, CSA, Choson Minhang, LOT, Egyptair, CAAC and TAROM.

By the time scheduled service had started a new engine had been chosen. Although the original NK-8-4 had been uprated to 103 kN (23,150 lb) take-off thrust, and had established quite a good record, the Soloviev D-30KU represented newer technology. Thanks to a higher

TAROM's usage was confined to three Il-62s and a pair of Il-62Ms delivered between 1973/1975 and 1977/1978 respectively. Several of the Romanian aircraft were leased out to other 'friendly' airlines such as LOT and Cubana.

Old and new sit on the tarmac at Prague's Ruzyne airport. When CSA adopted a new colour scheme for the 1990s, it repainted many of its Il-62s. With the coming of the A310 the airline's international routes have been increasingly taken away from Ilyushin's workhorse.

Despite the arrival of the Il-86 and the even more recent appearance of the Il-96 (now officially in service with Aeroflot Russian Airlines), the Il-62M remained the mainstay of Aeroflot's long-range services. On its transatlantic routes the Il-62M reigned supreme, far out-ranging the short-legged Il-86. Even so, the scheduled run from Moscow to Lima, Peru, was a marathon flight incorporating stops at Luxembourg, Shannon, Gander and Havana, before reaching its destination. Most of the Eastern European COMECON nations replaced their early Il-62s with Il-62Ms, and aircraft were also delivered to TAAG Angola, Linhas Aereas de Mocambique and CAAC. Since the break-up of the Soviet Union, these aircraft have now been largely (although not completely) replaced by far more efficient Western types, such as the Airbus A310 and Boeing 767. Equally as a result of the liberalisation of the air transport market in Eastern Europe, there is a healthy trade in secondhand aircraft, and Il-62s are being snapped up by new airlines for their price and familiarity. When LOT withdrew its entire fleet in 1991/92 they were largely bought up by Air Ukraine and are now regular sights on the North Atlantic airways again. Uzbekistan Airways took over six of the former Interflug aircraft, three of which were also pressed into Luftwaffe service with the FBS.

Increased seating

The passenger cabin was rearranged to seat up to 174, while provisions for cargo and baggage were completely rethought. Loading and unloading time was dramatically reduced by arranging for the underfloor holds to house baggage and cargo in containers, with mechanised loading and unloading. Another improvement was to enhance the provisions for emergency egress, and the on-board emergency and rescue equipment carried.

The first Il-62M, SSSR-86673, at first styled Il-62M-200, made its maiden flight in 1970, and appeared at the Paris air show in 1971 and 1973. Redesignated as simply the Il-62M, the improved aircraft entered Aeroflot service in 1974, and progressively took over all the airline's very long routes. Thanks to the extra fuel and new engines, take-off weight could be raised from 162 tonnes (357,150 lb) to 165 tonnes (363,760 lb), so that range with maximum payload was increased from 6700 km (4,160 miles)

Below: The Il-62M and MK remains the principal long-haul type for Aeroflot Russian International Airlines. While some old-style examples remain, most have now adopted the Russian flag and new 'RA-' registrations.

Right: Air India adopted closer ties with Aeroflot for a period during the late 1980s, leading to the leasing of several aircraft including this Il-62M in 1988. This aircraft is now back in regular service in Russia.

North Korea's Chosonminhang is a shadowy Il-62 operator whose aircraft have a very definite secondary military role, as evidenced by their ambiguous 'civil' registrations.

to 7800 km (4,846 miles). From the flight crew's viewpoint, the M was a markedly superior and more modern aircraft.

Nevertheless, there is nothing so good that it cannot be improved, and after 1975 the Soviet Union, and Aeroflot in particular, was keenly feeling Western competition with what was then superior technology. Ilyushin had handed over the reins of his growing design bureau to Genrikh Novozhilov. Despite a huge workload on many other types, including the new Il-76 and -86, Novozhilov decided on one further major upgrade of the Il-62, which was clearly going to be Aeroflot's flagship on long international routes for a further 10 years at least. The Il-62MK was therefore produced and, although externally virtually identical to the Il-62M, it is a new aircraft and cannot be produced by modifying existing machines. This is largely because the MK has a restressed structure to permit operation at take-off weights up to 167 tonnes (368,170 lb). Redesign extended to the complete spars and machined skins of the main wing torsion box and to several minor parts of the structure. The main landing gears were modified with improved high-strength materials, low-pressure tyres on wheels moved further apart to give a greater footprint area and higher-capacity brakes with a new anti-skid system. Braking is enhanced still further by arranging for all spoiler sections to be increased in size and to deploy fully as soon as the oleos are compressed on landing. The interior was subjected to a major revision. Despite rearranging the passenger seats to permit 195 to be carried, the aisle was widened to give greater freedom of movement and permit passenger service carts (trolleys) to be used. Overhead enclosed baggage bins were installed, together with a new furnishing scheme giving indirect lighting and a 'wide-body' look. Maximum payload was increased from the previous 23 tonnes (50,700 lb) to 25 tonnes (55,115 lb).

Winding down

Production of the Il-62MK continued at Kazan until 1990, being completed with the 244th aircraft (delivered to Cubana). In addition to those operators named earlier, aircraft of this family were sold to Choson Minhang of North Korea and to Cubana. From 1985 all surviving Il-62M and MK aircraft were being put through an update cycle that installs a triplex INS (inertial navigation system) and completely rebuilds the engine pods to reduce noise and emissions. These aircraft do not meet future noise legislation but they will continue in service until at least the early 2000s as in 1999 there were still 124 of the aircraft in service. The Il-62 deserves its place in aviation history as the first Soviet (now Russian) jet to offer reasonably comfortable travel over long international sectors.

A change of colour scheme and new plans for the future did not spell the immediate end for TAROM's Il-62s. However, the arrival of the A310-300 in the fleet did finally mark the end of their careers.

Ilyushin Il-76

Bearing a passing resemblance to the West's StarLifter, the Il-76 'Candid' is considerably more versatile, and although it does not have the load-carrying ability, it has much better performance. In addition to a wide variety of trooping, para-dropping and cargo-hauling missions, it has been adapted to several special roles.

The Il-76 was the first major project of the great Ilyushin OKB (experimental construction bureau) in which S. V. Ilyushin himself played no direct part. The design was directed by G. V. Novozhilov, the work beginning in 1965 to meet an important requirement of both the civil air operator Aeroflot and the military air transport service V-TA for a replacement for the An-12BP, the most numerous heavy cargo transport.

Among the numerical requirements was the ability to carry a 40-tonne cargo for 5,000 km in less than six hours. It also had to be able to operate from short unpaved airstrips, maintain reliability in the extreme climatic conditions of the hot southern Steppes and deserts and northern Siberia, and be easy to service.

To a considerable degree the design was influenced by the contemporary Lockheed C-141, though the Soviet aircraft has a much less constricted fuselage cross-section and, because of its more powerful engines, much better field performance. Compared with passenger jets the wing was given less sweep (25°), because short field performance was much more important than high cruising speed. The wing has multiple spars and the left and right wings are attached to the fuselage-width centre section with 4° anhedral. The main wing box in each wing is divided into three sealed integral tanks, the total capacity being 81,830 litres.

Each wing carries conventional outboard ailerons, which are powered but with manual reversion, inboard and outboard triple-slotted flaps, five sections of powered slat and eight sections of powered spoiler used for roll-augmentation, fast letdown and lift dumping after landing. The T-type tail has a powered trimming tailplane and hydraulically boosted and aerodynamically balanced tabbed elevators and rudder, with manual reversion.

The engines are four Soloviev D-30KP turbofans, each rated at 12 tonnes thrust and fitted with a clamshell reverser. The engine installations are particularly neat, and have less weight and drag than those of the much less powerful NK-8 engines of the early Il-62. The landing gear has high flotation for operation at maximum weight from soft strips, using 20 wheels. The main gear comprises tandem units on each side, each having a single axle with four wheels with 1300 × 480-mm tyres inflated to a pressure selectable in flight between 2.5 and 5 kg/cm^2. Each unit folds straight in and up, the axle turning so that the four wheels lie in a row at 90° to the axis of the fuselage. The nose gear again has a single axle with four wheels, with tyres 1100 × 330-mm, hydraulically steerable and retracting forward. All 20 wheels have anti-skid hydraulic brakes. All engines drive generators and hydraulic pumps, and one of each is driven by the auxiliary power unit in the left main landing gear blister.

From the start the Il-76 was comprehensively equipped. All leading edges have hot-air de-icing, and all forward-facing windows have electro-thermal de-icing. Hot bleed air protects the engine inlets. In the nose is a weather radar, with a navigation and mapping radar in a large blister under the nose. Despite the presence of these two

Full-span leading edge slats and flaps over 75 per cent of the trailing edge bestow excellent short-field performance on the Il-76. Take-off run is reported to be as little as 850 m, with landing run considerably less. The 16-wheel main undercarriage ensures good rough-field capability.

At least two Il-76MD 'Candid-Bs' have been delivered to Cubana (CU-T1258 and CU-T1271). These are unusual for this model as they have no tail turret. They are used for joint airline/air force operations.

radars it was still found possible to provide a cabin for a navigator in the nose in the traditional Soviet style with all-round glazing to give a perfect view throughout the forward hemisphere. This is useful not only for old-fashioned contact (map reading) navigation but also for guiding the pilot during missions involving heavy dropping or putting down sticks of paratroops. Other flight crew comprise two pilots, radio operator and supernumerary on the capacious flight deck, and two freight handlers.

Aeroflot once had over 120 Il-76s, often on military-type operations. Both Il-76T and Il-76M are in service, this being one of the latter, complete with tail turret. Some Aeroflot aircraft are used to support Arctic exploration, including one stationed in Mozambique.

Pressurised cabin

The main cargo cabin measures 20 m long (24.5 m including the rear ramp), with ruling width of 3.4 m and height of 3.46 m. Usable volume is 235.3 cubic metres. Unlike most earlier Soviet aircraft with a rear ramp door, the entire interior is pressurised, normal pressure being 4.5 kg/cm^2. On each side of the forward fuselage is a forward-hinged main door, while two windows on each side serve as emergency exits. At the rear is the main ramp door, which hinges downwards, aft of which is a central upward-hinged door and left/right clamshell doors, all four doors being powered hydraulically, the main ramp door being able to be raised with a 30-tonne load resting on it. The entire cargo floor is made up of tough titanium panels, with anchorages and optional or folding roller/ball panels to facilitate the movement of pallets and containers. Cargo is easily loaded and unloaded by overhead hoists of 2.5 tonnes and 3 tonnes capacity, which can be positioned anywhere in the interior under control of a computer which governs the load distribution for the best centre of gravity position.

Loads can include a very wide range of wheeled or tracked vehicles, 12 m standard ISO freight containers, or various standard pallets of 2.44 m width and length either 1.46 m or 2.99 m. The hold can be equipped for troop or paratroop transport (normally up to 140 troops or 125 paras), or special modules can be inserted (for example, three modules each with 30 seats). Other modules can be added for stretcher patients and attendants. Paratroops can leave via the open rear ramp door, though it should be stressed that by far the most important role of the Il-76 is cargo carrying. The normal

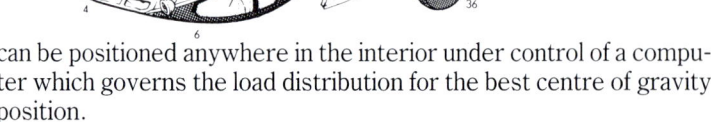

A major operator of the Il-76 is Iraq, which operates this Il-76MD on cargo flights (seen here at London's Heathrow Airport). Although wearing nominal Iraqi Airways' colours, the empty gun turret in the tail belies the use of the aircraft by the air force.

Ilyushin Il-76MD/TD 'Candid-B' cutaway drawing key

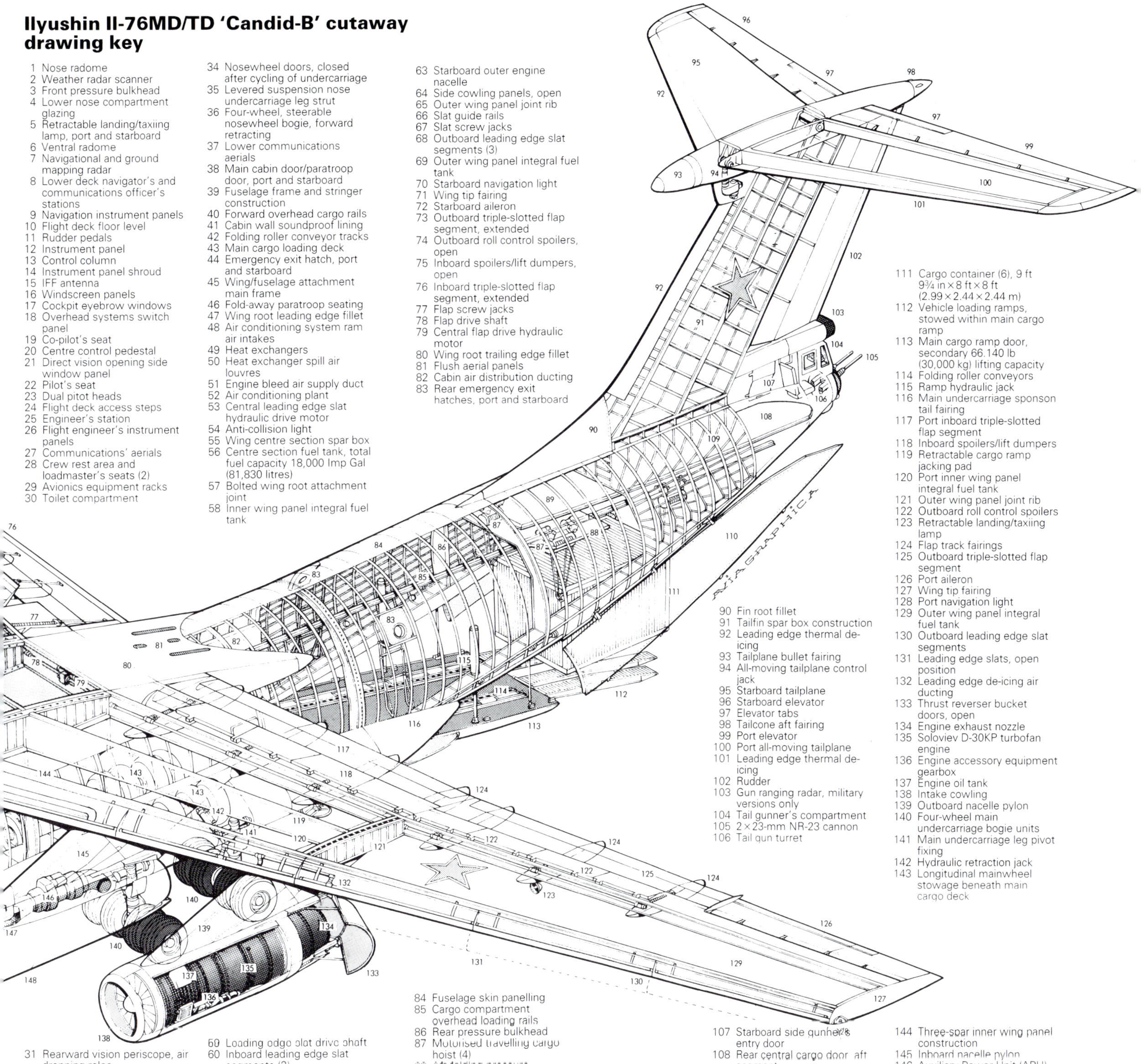

1. Nose radome
2. Weather radar scanner
3. Front pressure bulkhead
4. Lower nose compartment glazing
5. Retractable landing/taxiing lamp, port and starboard
6. Ventral radome
7. Navigational and ground mapping radar
8. Lower deck navigator's and communications officer's stations
9. Navigation instrument panels
10. Flight deck floor level
11. Rudder pedals
12. Instrument panel
13. Control column
14. Instrument panel shroud
15. IFF antenna
16. Windscreen panels
17. Cockpit eyebrow windows
18. Overhead systems switch panel
19. Co-pilot's seat
20. Centre control pedestal
21. Direct vision opening side window panel
22. Pilot's seat
23. Dual pitot heads
24. Flight deck access steps
25. Engineer's station
26. Flight engineer's instrument panels
27. Communications' aerials
28. Crew rest area and loadmaster's seats (2)
29. Avionics equipment racks
30. Toilet compartment
31. Rearward vision periscope, air dropping roloc
32. Crew entry door
33. Underfloor equipment stowage racks
34. Nosewheel doors, closed after cycling of undercarriage
35. Levered suspension nose undercarriage leg strut
36. Four-wheel, steerable nosewheel bogie, forward retracting
37. Lower communications aerials
38. Main cabin door/paratroop door, port and starboard
39. Fuselage frame and stringer construction
40. Forward overhead cargo rails
41. Cabin wall soundproof lining
42. Folding roller conveyor tracks
43. Main cargo loading deck
44. Emergency exit hatch, port and starboard
45. Wing/fuselage attachment main frame
46. Fold-away paratroop seating
47. Wing root leading edge fillet
48. Air conditioning system ram air intakes
49. Heat exchangers
50. Heat exchanger spill air louvres
51. Engine bleed air supply duct
52. Air conditioning plant
53. Central leading edge slat hydraulic drive motor
54. Anti-collision light
55. Wing centre section spar box
56. Centre section fuel tank, total fuel capacity 18,000 Imp Gal (81,830 litres)
57. Bolted wing root attachment joint
58. Inner wing panel integral fuel tank
59. Leading edge slat drive shaft
60. Inboard leading edge slat segments (2)
61. Nacelle pylons
62. Starboard inner engine nacelle
63. Starboard outer engine nacelle
64. Side cowling panels, open
65. Outer wing panel joint rib
66. Slat guide rails
67. Slat screw jacks
68. Outboard leading edge slat segments (3)
69. Outer wing panel integral fuel tank
70. Starboard navigation light
71. Wing tip fairing
72. Starboard aileron
73. Outboard triple-slotted flap segment, extended
74. Outboard roll control spoilers, open
75. Inboard spoilers/lift dumpers, open
76. Inboard triple-slotted flap segment, extended
77. Flap screw jacks
78. Flap drive shaft
79. Central flap drive hydraulic motor
80. Wing root trailing edge fillet
81. Flush aerial panels
82. Cabin air distribution ducting
83. Rear emergency exit hatches, port and starboard
84. Fuselage skin panelling
85. Cargo compartment overhead loading rails
86. Rear pressure bulkhead
87. Motorised travelling cargo hoist (4)
88. Aft folding pressure bulkhead door, open
89. Central rear cargo door, open position
90. Fin root fillet
91. Tailfin spar box construction
92. Leading edge thermal de-icing
93. Tailplane bullet fairing
94. All-moving tailplane control jack
95. Starboard tailplane
96. Starboard elevator
97. Elevator tabs
98. Tailcone aft fairing
99. Port elevator
100. Port all-moving tailplane
101. Leading edge thermal de-icing
102. Rudder
103. Gun ranging radar, military versions only
104. Tail gunner's compartment
105. 2 × 23-mm NR-23 cannon
106. Tail gun turret
107. Starboard side gunners entry door
108. Rear central cargo door, aft segment
109. Rear fuselage framing
110. Port cargo loading door, open
111. Cargo container (6), 9 ft 9¾ in × 8 ft × 8 ft (2.99 × 2.44 × 2.44 m)
112. Vehicle loading ramps, stowed within main cargo ramp
113. Main cargo ramp door, secondary 66.140 lb (30,000 kg) lifting capacity
114. Folding roller conveyors
115. Ramp hydraulic jack
116. Main undercarriage sponson tail fairing
117. Port inboard triple-slotted flap segment
118. Inboard spoilers/lift dumpers
119. Retractable cargo ramp jacking pad
120. Port inner wing panel integral fuel tank
121. Outer wing panel joint rib
122. Outboard roll control spoilers
123. Retractable landing/taxiing lamp
124. Flap track fairings
125. Outboard triple-slotted flap segment
126. Port aileron
127. Wing tip fairing
128. Port navigation light
129. Outer wing panel integral fuel tank
130. Outboard leading edge slat segments
131. Leading edge slats, open position
132. Leading edge de-icing air ducting
133. Thrust reverser bucket doors, open
134. Engine exhaust nozzle
135. Soloviev D-30KP turbofan engine
136. Engine accessory equipment gearbox
137. Engine oil tank
138. Intake cowling
139. Outboard nacelle pylon
140. Four-wheel main undercarriage bogie units
141. Main undercarriage leg pivot fixing
142. Hydraulic retraction jack
143. Longitudinal mainwheel stowage beneath main cargo deck
144. Three-spar inner wing panel construction
145. Inboard nacelle pylon
146. Auxiliary Power Unit (APU)
147. APU air intake
148. Port inboard engine nacelle

maximum payload is 40 tonnes, though for record purposes loads of up to 70 tonnes have been carried.

The prototype, SSSR-86712, made its maiden flight on 25 March 1971, and two months later made a most favourable impression at the Paris Air Salon. Test flying under Eduard Kuznetsov, carried out mainly from Khodinka, showed the need for very few modifications and it was clear that the Ilyushin OKB under its new leader had produced a winner. There followed one civil pre-production Il-76 and four pre-production military Il-76M transports, the latter having comprehensive V-TA communications, special navaids and electronic-warfare systems with six passive receivers and active jammers giving all-round coverage, as well as an Ilyushin rear turret with twin NR-23 cannon.

Four Il-76s were delivered to Syrianair for freight duties. Two are Il-76Ts and two are Il-76Ms (illustrated). These are occasional visitors to European cities to collect cargo.

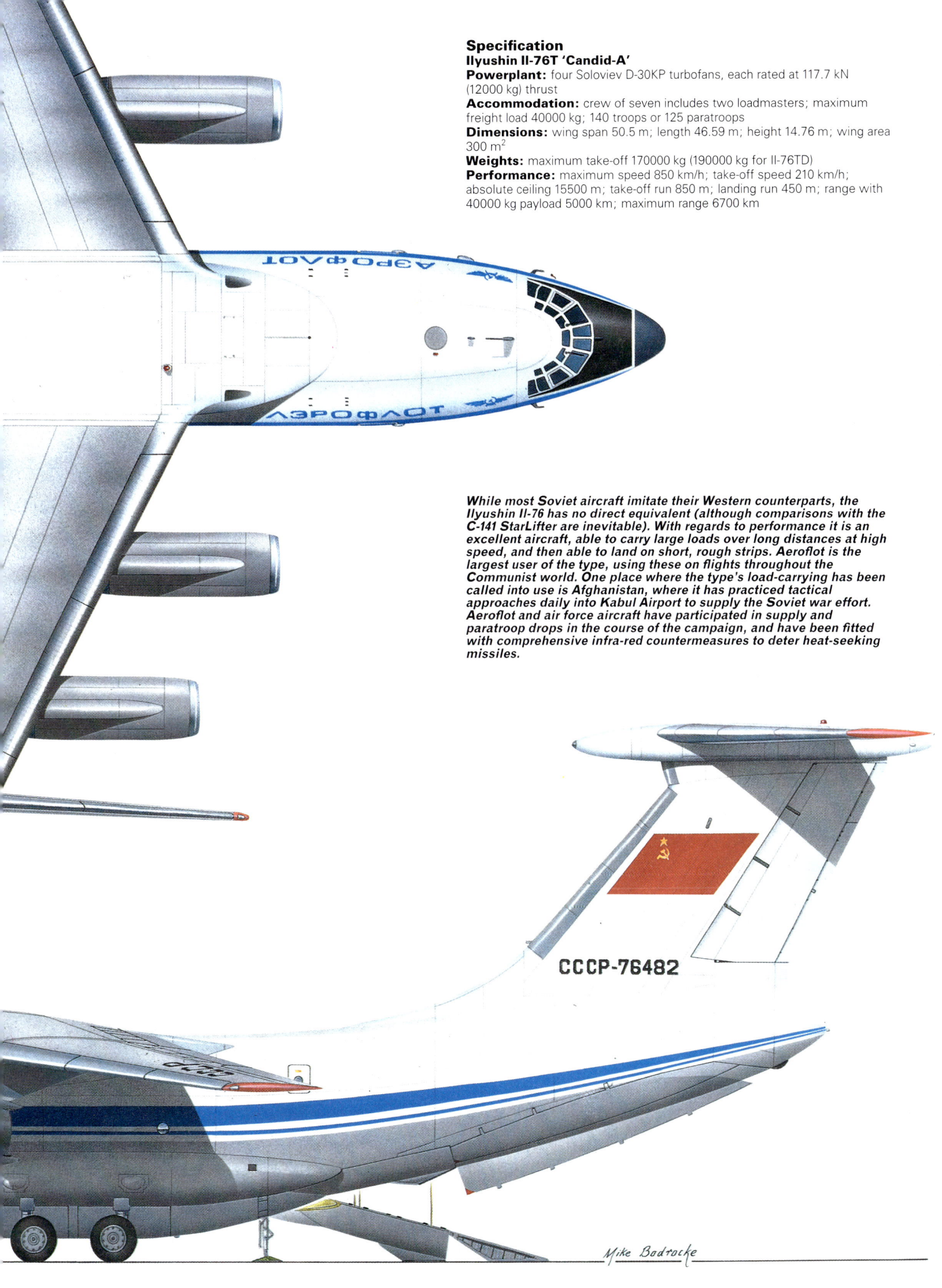

Specification
Ilyushin Il-76T 'Candid-A'
Powerplant: four Soloviev D-30KP turbofans, each rated at 117.7 kN (12000 kg) thrust
Accommodation: crew of seven includes two loadmasters; maximum freight load 40000 kg; 140 troops or 125 paratroops
Dimensions: wing span 50.5 m; length 46.59 m; height 14.76 m; wing area 300 m^2
Weights: maximum take-off 170000 kg (190000 kg for Il-76TD)
Performance: maximum speed 850 km/h; take-off speed 210 km/h; absolute ceiling 15500 m; take-off run 850 m; landing run 450 m; range with 40000 kg payload 5000 km; maximum range 6700 km

While most Soviet aircraft imitate their Western counterparts, the Ilyushin Il-76 has no direct equivalent (although comparisons with the C-141 StarLifter are inevitable). With regards to performance it is an excellent aircraft, able to carry large loads over long distances at high speed, and then able to land on short, rough strips. Aeroflot is the largest user of the type, using these on flights throughout the Communist world. One place where the type's load-carrying has been called into use is Afghanistan, where it has practiced tactical approaches daily into Kabul Airport to supply the Soviet war effort. Aeroflot and air force aircraft have participated in supply and paratroop drops in the course of the campaign, and have been fitted with comprehensive infra-red countermeasures to deter heat-seeking missiles.

Initially requiring 20, India has since added another four Il-76s to its order. These are to replace Antonov An-12s in service with two squadrons, the first of which is No. 44 'Mountain Geese' Squadron.

Libyan Arab Airlines has 20 'Candids' for transport duties, although like those of Iraq and Syria, these are more usually flown on military-related flights. Most of the fleet is Il-76Ts, but at least four Il-76Ms are on strength.

This is doubly interesting because airborne early-warning and control control is the mission of a new Il-76 version whose designation was finally revealed as A-50 only after it had been dubbed 'Mainstay' by NATO. First reported in 1977, this version differs in many respects from the basic transport, apart from the obvious fact that its fuselage is completely given over to the very powerful surveillance radar, communications and display consoles, with the antennae mounted in a streamlined 'rotodome' which rotates on top of a tall pylon above the rear fuselage.

'Mainstay' close-up

The first 'Mainstay' to be encountered in service with the PVO (air defence organisation) was seen over the Kola Peninsula by a P-3 Orion of the Royal Norwegian AF in December 1987. The rotodome appears to be severely blanketed by the huge wing close beneath it. No fewer than 13 bumps, blisters and antennae could be seen, though few have yet been publicly identified. The forward fuselage is slightly longer than that of the transport versions, and while there are several new or modified avionics blisters there is no glazed navigator's station in the nose. Above the nose, immediately ahead of the cockpit windows, is an inflight-refuelling probe, which as noted previously is seldom seen on the transport versions. The tail turret is removed, and replaced by major electronic warfare installations designed to protect the vulnerable aircraft as it orbits round its racetrack pattern in operational service. Another distinctive feature is the addition of a ram-air inlet at the base of the fin, probably to cool liquid cooling radiators which dump excess heat from the main radar transmitter and other avionics.

According to the US Department of Defense, 'Mainstay' can detect missile-sized targets even when these are flying at very low altitude over land or water. Certainly it is a very much more capable aircraft than the Tu-126 'Moss' which it is replacing in service, which was alleged to have very limited look-down capability, especially over land. The 'Mainstays' operating over the Kola Peninsula and the surrounding Arctic region are said to be used mainly for direction of the growing force of Su-27 long-range all-weather interceptors in this region.

Tanker derivative

Yet another specialised version of the Il-76 is an air refuelling tanker, known as the Il-78 with the NATO reporting name 'Midas'.

Again this aircraft has been under development since before 1980, although few details are yet available. What is beyond question is that it uses the probe/drogue method, with three hose-drum units, one on the port side of the rear fuselage and two in streamlined pods under the outer wings. The transfer fuel capacity will probably be about 130,000 litres, very considerably more than the limit for the Myasischchyev M-4 'Bison' which the new Ilyushin aircraft has replaced. The M-4 also lacked three-point capability, having only a single hose-drum unit in the fuselage.

According to the US department of Defense the 'Midas' tanker version reached initial operational capability in late 1987. Probably about 50 of this version will be built, and at least 20 of the 'Mainstay' surveillance and control version. Indeed, modest exports of either or both versions remain a possibility, especially to the Indian Air Force. These pushed total production of all Il-76 versions past the 800 mark by early 1994, with production still continuing today.

Altogether the IL-76 has already proved an extremely successful

This view of an A-50 'Mainstay' highlights the rear-fuselage mounting for the rotodome, and the large air intake at the base of the fin for cooling the onboard equipment. The rotodome contains search radar and IFF equipment.

From certain angles the Il-76 resembles the Lockheed StarLifter, but it is considerably better-performing, and possesses true rough-field performance. It has proved popular in service, with considerable manoeuvrability for an aircraft of its size.

Featuring a slightly lengthened fuselage, the A-50 'Mainstay' is the airborne early warning version, complete with rear fuselage rotodome and associated equipment forward of the wing root.

Iraqi Airways/Iraqi air force operate 30 'Candids', all with the tail turret. Early deliveries were Il-76Ms, but more recent machines have been Il-76MDs with improved Soloviev D-30KP-1 engines and extra fuel.

'Mainstay' is now fully operational, this example being intercepted by the Royal Norwegian Air Force. Time on station can be vastly increased by inflight refuelling, a role in which the Il-76 is also being developed, under the NATO reporting name 'Midas'.

programme. Compared with other recent Soviet aircraft it was developed with commendable speed and absence of difficulty, and its performance in extremely harsh environments has been exemplary. The Indian Air Force spent many months in detailed evaluation before committing a large sum to purchase of this big Soviet aircraft, but was so pleased with the first few delivered that it soon increased its order from 20 to 24, and expects in due course to purchase an additional quantity. The number delivered to Aeroflot has so far been modest, but this is probably because of the great pressure to deliver military variants. At the last Paris Air Salon Mr. Genrikh Novozhilov commented that development of the Il-76 family was continuing, and that further improved versions or special-purpose variants might appear in the course of time.

Record breaker

Series production at GAZ-243 at Tashkent began in 1975. Aeroflot found that tonne-kilometre direct costs were at least 25 per cent lower than for the An-12, and it was announced that this made the Il-76 competitive with river transport even during the summer when all rivers are open. In July 1975 one of the first production aircraft set 25 world records. Some have since been broken, but the Il-76 still holds a record for having lifted a payload of 70 tonnes to a height of 11,875 m. Another record still held is a speed of 816 km/h round a 5000-km circuit with a payload of 40 tonnes.

Early in production the basic series version changed to the Il-76T, with fuel in the centre section and gross weight increased from 157 to 170 tonnes. In 1982 the standard versions were again upgraded to the 76TD and military 76MD. The engines in current aircraft are D-30KP-1s, which are flat rated at 12 tonnes at ambient temperatures up to ISA+23°C. Maximum take-off weight is increased yet again to 190 tonnes, and a further increase of 10 tonnes in fuel capacity gives about an additional 1200 km of range with any payload up to the new maximum of 48 tonnes. The Il-76MD can be fitted with an inflight-refuelling probe, though this is not often seen.

Transport production

By spring 1990 about 650 Il-76s of all transport versions had been delivered, with production continuing. Of these about 405 have been supplied to V-TA regiments, and more than 120 to Aeroflot. Curiously, manufacture has been so locked-in on the 76M and MD that many of these have gone to Aeroflot and to civilian export customers, the guns naturally being removed. Purchasers include Iraqi Airways (25-plus, at least one being shot down), Jamahiriyan Air Transport of Libya (19), Syrianair (two Ms and two Ts), Bakhtar Afghan (at least one) and Cubana (at least two).

Customers for the M and MD include the air forces of Czechoslovakia, India, Iraq and Poland. The Indian Air Force calls the 76MD the Gajaraj. Nos 25 and 44 Squadrons operate 24 Gajarajs, and it has been reported that some may be rebuilt as airborne early-warning platforms.

India's transports are Il-76MD 'Candid-B', needing the extra rating of the improved engines for hot-and-high performance. In service with the Indian Air Force, the Il-76 is known as the Gajaraj, and is often employed on trans-Himalayan flights from the Soviet Union, carrying heavy machinery.

Ilyushin Il-86 & 96

Ilyushin intended the Il-86 to be the sucessor to the Il-62 on Aeroflot's long-haul routes. It didn't quite make the grade, falling short in range and turning out to be not quite the 'airbus' that was promised. Despite these difficulties, it is still in regular service and its problems have been addressed with the new Il-96.

As the world's largest country the Soviet Union had a natural tendency towards bigness, but in the field of air transport this has repeatedly led to errors. Such aircraft as the ANT-14, ANT-20 and 20bis, Tu-70 and the first (piston-engined) Il-18 were unsuited to either the traffic offered or the available airfields. Only one was built of each type. The first large Soviet airliner to be built in numbers was the Tu-114, a huge turboprop derived from the Tu-95 'Bear' bomber, but this was again not a fully successful vehicle for main trunk routes. As soon as possible, it was replaced by the Il-62 jet.

During the 1960s it became increasingly evident that Aeroflot, the national airline, would soon need larger equipment in what was being loosely called the "airbus" class. Such a belief was suddenly sharpened by the announcement in 1966 of the Boeing 747, which was not only the first so-called 'wide-body' airliner but also the first to be powered by giant HBPR (high bypass ratio) turbofan engines. No such engine existed in the Soviet Union and, surprisingly, nothing was done at that time to create one. On the other hand, the design bureau of S. V. Ilyushin was picked over two rival teams and assigned the task of creating a new "airbus" passenger transport with a wide body and large capacity. This aircraft was to fly short/medium sectors, and was to be in service in time for the Moscow Olympic Games in 1980.

Ilyushin himself had retired, and his successor, Genrikh V. Novozhilov, announced the Il-86 at the 1971 Paris airshow. Soon afterwards preliminary details were published, showing what was essentially an Il-62 with a fuselage of increased diameter and an improved wing. The T-tail and four 117.7-kN (26,455-lb) Soloviev D-30KP engines were retained. Normal seating was to be provided for 380, eight-abreast. A special feature of the Il-86 was to be a double-deck layout making it completely independent of stairways or passenger jetties (loading bridges). Passengers would simply walk to the aircraft and climb one of three integral staircases to the lower deck. Here they would stow their luggage and hang up coats and hats (important in Soviet winter). Then they would climb fixed internal stairways to the main deck on which the seats were mounted.

A tremendous amount of study and argument followed. By the summer of 1972 the Il-86 had been totally redesigned, with a conventional tail and the engines spaced out in single pods under the wing. The wing and fuselage were completely redesigned, though fuselage diameter remained 6.08 m (19 ft 11.4 in). Rather less floor area remained available for seating, so even with a nine-abreast layout

In Aeroflot service it was claimed that the Il-86 'Camber' was the airline's largest passenger-carrying aircraft, with a seating capacity for 350 passengers in nine-abreast seating. However the standard layout for international routes was fixed for 28 passengers in first-class and 206 eight-abreast in two economy classes – hardly a great improvement over the Il-62's 168 seats.

Among the earliest routes assigned to the Il-86 was the long trip to Mineralnye Vody, where the second aircraft to be accepted by Aeroflot is seen in March 1981.

(3+3+3) the maximum capacity was published as 350. A further change was the decision to use three main landing gears, each with a four-wheel bogie, two retracting inwards and the third, on the centreline, retracting forwards.

Slightly behind schedule the first of two prototypes, SSSR-86000, made its first flight on 22 December 1976. It was flown by the bureau's chief test pilot A. Kuznetsov, from Khodinka, formerly Moscow Central Aerodrome, the small airfield on which the Ilyushin bureau is situated. By this time many further changes had been incorporated, notably a switch to the Kuznetsov NK-86 engine rated at 127.5 kN (28,660 lb). These engines were developed from the NK-8-4 which were fitted to the original form of Il-62. In the new aircraft each was fitted with a noise-attenuating nozzle incorporating thrust-reversers. The extra power enabled fuel capacity and gross weight to be increased. Maximum weight was increased to 206 tonnes, a figure later increased again to 208 tonnes, or 458,560 lb. A new

The prototype Il-86 (SSSR-86000) was taken aloft for the first time on 22 December 1976 with A. Kuznetsov at the controls. The flight occurred five years after the announcement of the type's development.

feature in the Il-86 was that all leading edges were de-iced by a system of giant electrical pulses, claimed to require "500 times less energy than a conventional system".

The airframe was almost entirely conventional light-alloy structure, though the floors were mainly carbon-fibre panels and metal honeycombs. The wing, swept at 35° at the quarter-chord line, was fitted with full-span slats, double-slotted flaps, conventional ailerons and surfaces ahead of the flaps serving as spoilers, airbrakes and lift dumpers. A shallow fence was added across the wing behind each engine pylon. The tail comprised a powered tailplane used for trimming, powered elevators in four sections and two portions of powered rudder. The tailcone, to house an APU (auxiliary power unit) in production aircraft, was used in the first prototype to accommodate a parachute system for test-flight purposes.

Photographed at Tashkent on an early proving flight, this Il-86 demonstrates the two main doors to the baggage holds. In total, the Il-86 can lift a payload of 42000 kg (92,600 lb) for a maximum take-off weight of 208000 kg (458,560 lb) on a good runway.

Portions of the fuselage of the prototypes were made at Kharkov, the tail unit at Kiev, the main landing gears at Kuibyshyev and the wings at Voronezh. Though the first aircraft was assembled at Khodinka, all subsequent Il-86 aircraft were assembled at Voronezh, and large sections of the production airframe were subcontracted to Poland, PZL-Mielec receiving the contract for the slats, flaps, engine pylons, fin and tailplanes. Aircraft SSSR-86002, the first production Il-86, made its first flight on 24 October 1977. The first to be accepted by Aeroflot was SSSR-86004, on 24 September 1979. The Il-86 did not assist in carrying traffic for the Moscow Olympics, scheduled services beginning on 26 December 1980. The first international service, to East Berlin, was operated from 3 July 1981.

Despite the very long time taken over design and development, the Il-86 proved to be seriously deficient in range. Aviaexport continued to quote "Design performance", including cruising speed of 900 to 950 km/h (559-590 mph) and a range with 40 tonnes payload (less than the maximum figures of 42 tonnes) of 3600 km (2,235 miles). Yet the former East German airline Interflug, which was the partner on the first international services, published the maximum range as 2500 km (1,553 miles). Despite this the Il-86 is a useful vehicle, not least because of its ability to move 350 troops and their equipment and offload them at any airfield with a good enough runway, without needing any ground equipment. By the start of 1991 Mielec had delivered no fewer than 263 shipsets to Voronezh, though the Soviet authorities merely said that it was "hoped to produce 100 Il-86s by 1995".

There is no doubt that the short range of the Il-86 has proved a major disappointment, and one wonders at the Mielec plant's score of 263

The sixth production aircraft is unloaded alongside an Antonov An-12 'Cub'. Roughly the same size as the Western Hercules, it gives some impression of the enormity of the Il-86.

shipsets delivered and production continuing. One way of increasing range was to fit newer and more efficient engines, and in 1990 one Il-86 was fitted with the Soloviev PS-90A, the most efficient Soviet HBPR engine with a thrust of 156.9 kN (35,275 lb). But the Ilyushin design bureau, perhaps feeling their pride had been hurt, started almost from scratch in 1980 with the Il-96. Ilyushin himself had died in 1977, and Novozhilov was appointed General Constructor. He had hoped the Il-96 could merely be a modified form of Il-86, with a new wing carrying the specially developed PS-90A engines, but in fact the 96 turned out to be a new design. Almost the only components left unaltered, or only slightly modified, were major sections of the fuselage (though this was made much shorter) and the four units of the landing gear.

A new breed

The Il-96 wing has no part in common with that of the 86. Span is greatly increased, the profile is completely new (of rooftop or supercritical form), and the arrangement of movable surfaces closely follows that of Western civil jetliners, such as the 747. On the trailing edge are (from root to tip) double-slotted flaps, a small all-speed aileron in line with the inboard engine, two single-slotted flaps, and low-speed outboard ailerons. Ahead of the flaps are airbrakes inboard and combined airbrakes and roll-spoilers outboard. On the leading edge are full-span powered slats. There are no fences, but on the wingtip are what Airbus call wingtip fences and most people call winglets. Major parts of the wing secondary and movable structures are of composite materials.

With the break-up of the Soviet Union and the fragmentation of the monolithic Aeroflot, many new airlines have sprung up. One such carrier is TransAero, in which the Sukhoi aircraft company has an interest. This Il-86 is named Moscow and still retains its SSSR registration prefix.

The PS-90A engine marked the first time the Soviet Union had aimed at performance and efficiency fully comparable with any Western airline engine. Its bypass ratio is 4.8 and maximum pressure ratio no less than 35.5, resulting in cruise specific fuel consumption of around 0.58 lb/h/lb/st. compared with over 0.62 for typical Western engines. Contributing to this engine's efficient performance is a FADEC (full-authority digital electronic control). This

The Il-86 has become a stalwart on Aeroflot's high density international routes – this example is seen landing at London-Heathrow. The type has also played a large part in expanding internal operations between the main population centres.

Most Il-86s still fly with Aeroflot. The international arm of the airline has been renamed Aeroflot Russian International Airways and many aircraft now wear Russian flags and registrations.

The second Il-96 SSSR-96001 is heavily involved in the certification programme of the type. The first Il-96 (SSSR-96000) is undergoing conversion to Il-96M standard.

is partnered in the aircraft by completely new avionics, including triplex FBW (fly by wire) flight controls. The tail unit is only superficially similar to that of the Il-86, the new surfaces having higher aspect ratio, different profiles and composite leading and trailing edges.

Apart from the new wing, engines and systems, the chief differences in the Il-96 are the fuel capacity and the way the fuselage is used. The integral tankage in the very large wing accommodates 152620 litres (33,572 Imp gal) of fuel, double the capacity available in the Il-86. Combined with substantially reduced drag and a major reduction in engine specific consumption this results in a dramatic increase in range, to 7500 km (4,660 miles) with the maximum payload of 40 tonnes and to 11000 km (6,835 miles) with 15 tonnes. Normal cruising speed is reduced, to 850-900 km/h (528-559 mph). Field length and general flight performance were little altered, and in fact despite the massive increase in fuel capacity the power loading (weight divided by engine thrust) was appreciably reduced.

The fuselage had the same diameter as before, and was generally made in the same way, but internally it was rather different. Gone were all the internal and external staircases. Instead the fuselage resembles that of Western airliners. The lower deck is entirely given over to the wing, to systems, to the galley and trolley lifts to the upper deck, and to cargo, part general (loose) and part palletised or in LD3 containers. The upper deck is almost entirely for the passengers, who enter at this level via three doors on the left side. Thus, the Il-96 is totally dependent on the stairways and loading bridges at the airports. This has meant a considerable amount of upgrading at airports within the Soviet Union. As the fuselage is shorter (51.15 m/167 ft 10 in, instead of 56.1 m/184 ft 0½ in) seating capacity is reduced. The first production version is designated the Il-96-300, signifying 300 passengers, but in fact most 96s put into service will be 235-seat mixed-class.

The first of three flying and two static test prototypes, SSSR-96000, was flown from Khodinka on 29 September 1988. By the start of 1991 the Polish Mielec factory had delivered 19 shipsets of the same components made for the Il-86. It had been hoped to certificate the Il-96 in 1990, but despite a reduction in the programme, certification came only in December 1992. By June 1998 only 10 aircraft had been delivered to the first operators including Aeroflot, which was once said to have ordered between 70 and 100 examples of the Il-96. Current firm orders stand at only 41 aircraft including 31 yet to be delivered.

In 1993 the Ilyushin bureau plans to fly the stretched Il-96-350, now redesignated Il-96M. As its original designation suggests, this is to be a 350-seater, the fuselage having a 3 m (9 ft 10 in) plug ahead of the wing and a 2.5 m (8 ft 2 in) plug aft. Most other Il-96 features will be unchanged, and the range with full payload is estimated to be 7000 km (4,350 miles).

Future plans

In addition to the Il-96M, the Ilyushin bureau has also been studying a twin-engine version, originally designated Il-96MD and more recently re-styled the Il-90. This will have an almost identical airframe to the Dash-300, but initial versions at least will be powered by two Lotarev D-18T engines, uprated to deliver between 274.7-313.6 kN (61,750-70,500 lb) of thrust, or if export/import licences are granted, the Pratt & Whitney PW2337 or Rolls-Royce Trent turbofans in the same power bracket. A further stretched version is the projected twin-deck Il-96-550, powered by four ultra high by-pass engines in the 176.0-211.2 kN (39,700-48,500 lb) thrust class, and able to accommodate 190 and 360 passengers in the upper and lower cabins respectively.

SSSR-96005 operates from Moscow's Sheremetyevo airport on route proving trials. It gained Rossia titles and flag to carry President Yeltsin to the USA on his state vist in 1992, but was actually never used.

535

Chapter 25

The Jet Commander, Westwind, Astra Story

Jetstream

The Jet Commander, Westwind, Astra Story

What has emerged as a major family of business and corporate transports began life as a jet-powered partner of the successful Aero Commander series. In subsequent times the design became Israeli property, and has been considerably developed into today's Astra, one of the most impressive performers on the biz-jet stage.

In 1961 the Aero Commander division of the Rockwell Standard Corporation announced development of a new executive jet at its Bethany, Oklahoma, factory under the design leadership of the late Ted R. Smith, who had been responsible for the Aero Commander series of twin-engined business aircraft which had first entered production 10 years earlier. Known as the Model 1121 Jet Commander and bearing a family resemblance to its piston-engined forebears, the prototype made its first flight on 27 January 1963, and was followed by a second development aircraft (actually the third airframe manufactured – the second was retained for static testing) on 14 April 1964. Powered by a pair of 1292-kg (2,850-lb) thrust General Electric CJ610-1 turbojets mounted in pods on the aft fuselage, the Jet Commander provided pressurised accommodation for two crew and seven/eight passengers and featured an unswept mid-wing of NACA 64A212 airfoil section with swept, cruciform tail surfaces.

The Jet Commander's structure was all-metal with a semi-monocoque flush-rivetted fuselage with pressurised fail-safe cabin and baggage compartment; flush-rivetted wing with a two-spar fail-safe structure, electrically operated double-slotted Fowler-type trailing edge flaps, and hydraulically operated spoilers and lift dumpers on the upper wing surface. Ailerons, elevators and rudder were manually operated, with a variable-incidence tailplane

When Rockwell and North American merged, US laws decreed that one of the new company's executive jets must go. Israel Aircraft Industries picked up the Jet Commander and began developing the type. The first major variant was the 1123 Commodore Jet, with a further increase in fuselage length.

Redesigning the aerofoil section for greater high Mach number efficiency and adding Whitcomb winglets to the tip tanks to reduce induced drag produced the Westwind 2. This retained the cabin and powerplant of the Westwind 1.

The 1124N Sea Scan is a maritime surveillance model based on the civil Westwind and equipped with a Litton seach radar in the nose and Omega navigation system. Other equipment is optional. Only the Israeli navy uses the type.

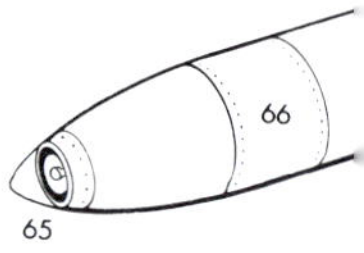

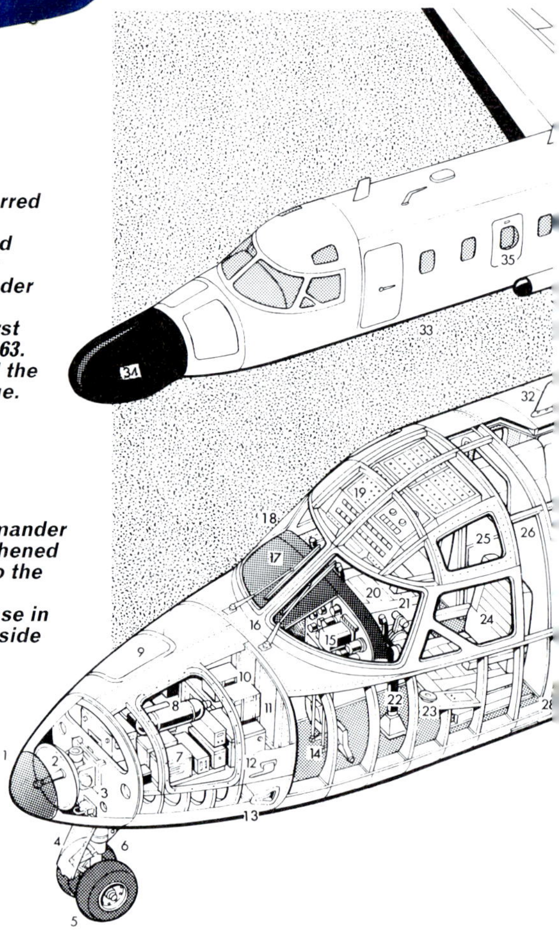

The success of the Commander twin-turboprop family spurred the company on to develop a jet-powered executive. The result was the Jet Commander 1121, the prototype of which (illustrated) first flew on 27 January 1963. Only this aircraft had the original short fuselage.

Production Jet Commander 1121s featured a lengthened fuselage compared to the first prototype, and a corresponding increase in windows on the port side from four to five. The 1121 sold well if not spectacularly, low cost being one of its main attributes.

for pitch trim and electrically operated trim tabs in the ailerons and rudder for roll and yaw trim. Pneumatically-operated de-icing boots were installed on the leading edges of all flying surfaces.

Jet Commander's early debut

The first production aircraft made its first flight on 5 October 1964, and following certification by the US Federal Aviation Adminstration, customer deliveries began in January 1965. The Jet Commander made its European public debut at the 1965 Paris Air Show. Early operators of the Jet Commander included American golfers Arnold Palmer and Jack Nicklaus and German publishing magnate Axel Springer. The Jet Commander, which sold for $750,000 fully equipped, was produced at the rate of eight per month. Among exports were aircraft for customers in Canada, France, Mexico, Sweden, Switzerland and a number of South American countries.

In 1967 Rockwell Standard acquired North American Aviation, manufacturer of the Sabreliner series of military/corporate light jet transports, and under US anti-trust laws was required to divest itself of one of the two business jet programmes. The decision was taken to retain the Sabreliner, and tooling, production and marketing rights for the Jet Commander were sold to Israel Aircraft Industries, which also marketed the last 49 of 150 Jet Commanders manufactured by Aero Commander in Oklahoma.

Initially IAI began manufacture of the Jet Commander 1121A virtually unchanged apart from an increase in fuel capacity and gross weight, while the 1121B featured uprated engines. Two prototype Model 1122s were built which led to the development of a new Model 1123, known as the Commodore Jet. This featured a fuselage lengthened to accommodate 10 passengers, 1410-kg (3,100-lb) thrust CJ610-9 turbojets, and wingtip fuel tanks in addition to the standard 'wet wing' tanks. The 1123 Commodore Jet was first flown on 28 September 1970 and was later known as the Westwind Eleven-23. Two were delivered to the Heyl Ha'Avir (Israeli Air Force) and were used as VIP transports during the Yom Kippur War in October 1973. Thirty-six Westwind Eleven-23s were built before the Model 1123 was superseded by the Model 1124 Westwind, first flown on 21 July 1975. The Model 1124 represented a major change to the aircraft, incorporating fuel-efficient 1678-kg (3,700-lb) thrust Garrett AiResearch TFE731-3-1-G

IAI Westwind 2 cutaway drawing key

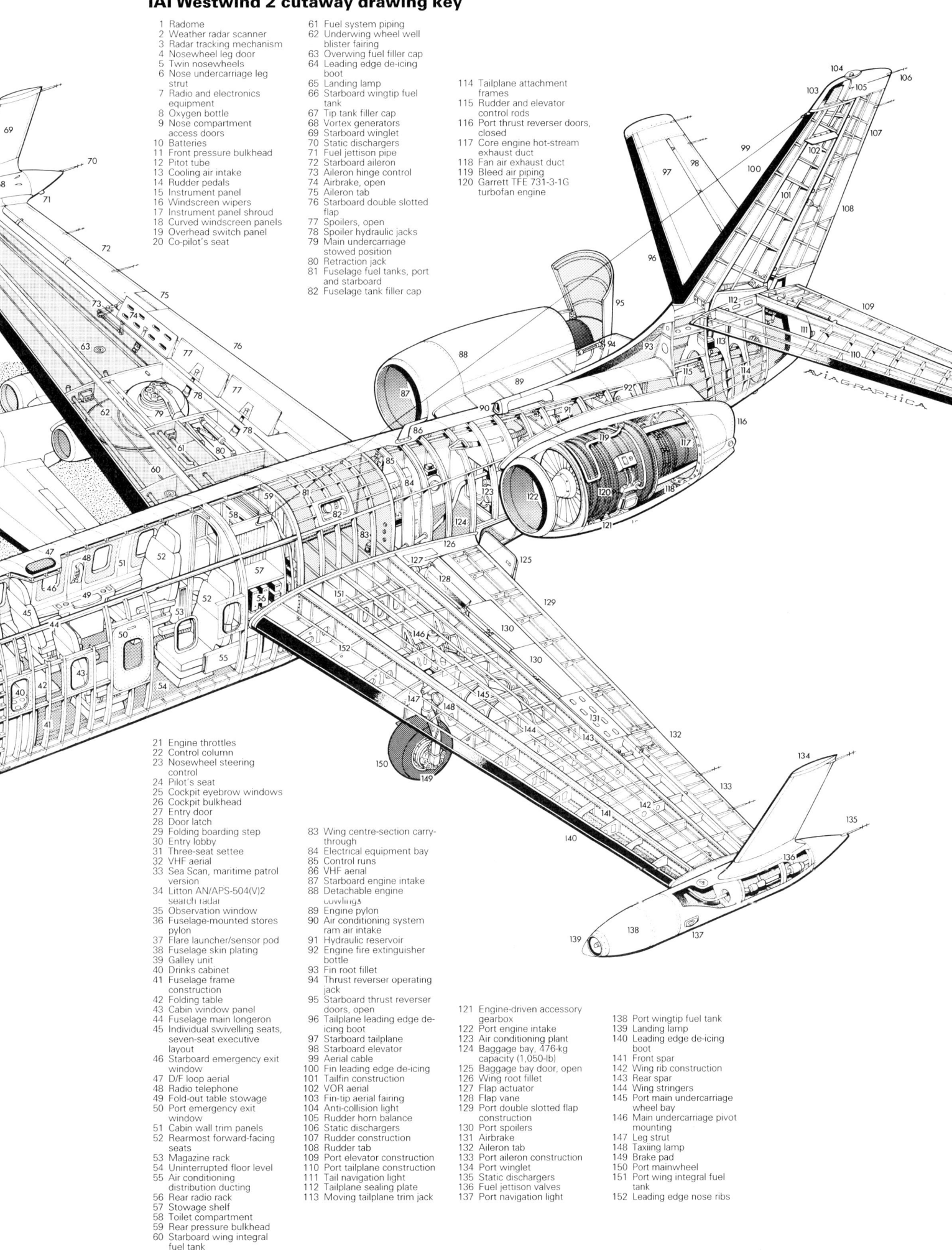

1. Radome
2. Weather radar scanner
3. Radar tracking mechanism
4. Nosewheel leg door
5. Twin nosewheels
6. Nose undercarriage leg strut
7. Radio and electronics equipment
8. Oxygen bottle
9. Nose compartment access doors
10. Batteries
11. Front pressure bulkhead
12. Pitot tube
13. Cooling air intake
14. Rudder pedals
15. Instrument panel
16. Windscreen wipers
17. Instrument panel shroud
18. Curved windscreen panels
19. Overhead switch panel
20. Co-pilot's seat
21. Engine throttles
22. Control column
23. Nosewheel steering control
24. Pilot's seat
25. Cockpit eyebrow windows
26. Cockpit bulkhead
27. Entry door
28. Door latch
29. Folding boarding step
30. Entry lobby
31. Three-seat settee
32. VHF aerial
33. Sea Scan, maritime patrol version
34. Litton AN/APS-504(V)2 search radar
35. Observation window
36. Fuselage-mounted stores pylon
37. Flare launcher/sensor pod
38. Fuselage skin plating
39. Galley unit
40. Drinks cabinet
41. Fuselage frame construction
42. Folding table
43. Cabin window panel
44. Fuselage main longeron
45. Individual swivelling seats, seven-seat executive layout
46. Starboard emergency exit window
47. D/F loop aerial
48. Radio telephone
49. Fold-out table stowage
50. Port emergency exit window
51. Cabin wall trim panels
52. Rearmost forward-facing seats
53. Magazine rack
54. Uninterrupted floor level
55. Air conditioning distribution ducting
56. Rear radio rack
57. Stowage shelf
58. Toilet compartment
59. Rear pressure bulkhead
60. Starboard wing integral fuel tank
61. Fuel system piping
62. Underwing wheel well blister fairing
63. Overwing fuel filler cap
64. Leading edge de-icing boot
65. Landing lamp
66. Starboard wingtip fuel tank
67. Tip tank filler cap
68. Vortex generators
69. Starboard winglet
70. Static dischargers
71. Fuel jettison pipe
72. Starboard aileron
73. Aileron hinge control
74. Airbrake, open
75. Aileron tab
76. Starboard double slotted flap
77. Spoilers, open
78. Spoiler hydraulic jacks
79. Main undercarriage stowed position
80. Retraction jack
81. Fuselage fuel tanks, port and starboard
82. Fuselage tank filler cap
83. Wing centre-section carry-through
84. Electrical equipment bay
85. Control runs
86. VHF aerial
87. Starboard engine intake
88. Detachable engine cowlings
89. Engine pylon
90. Air conditioning system ram air intake
91. Hydraulic reservoir
92. Engine fire extinguisher bottle
93. Fin root fillet
94. Thrust reverser operating jack
95. Starboard thrust reverser doors, open
96. Tailplane leading edge de-icing boot
97. Starboard tailplane
98. Starboard elevator
99. Aerial cable
100. Fin leading edge de-icing
101. Tailfin construction
102. VOR aerial
103. Fin-tip aerial fairing
104. Anti-collision light
105. Rudder horn balance
106. Static dischargers
107. Rudder construction
108. Rudder
109. Port elevator construction
110. Port tailplane construction
111. Tail navigation light
112. Tailplane sealing plate
113. Moving tailplane trim jack
114. Tailplane attachment frames
115. Rudder and elevator control rods
116. Port thrust reverser doors, closed
117. Core engine hot-stream exhaust duct
118. Fan air exhaust duct
119. Bleed air piping
120. Garrett TFE 731-3-1G turbofan engine
121. Engine-driven accessory gearbox
122. Port engine intake
123. Air conditioning plant
124. Baggage bay, 476-kg capacity (1,050-lb)
125. Baggage bay door, open
126. Wing root fillet
127. Flap actuator
128. Flap vane
129. Port double slotted flap construction
130. Port spoilers
131. Airbrake
132. Aileron tab
133. Port aileron construction
134. Port winglet
135. Static dischargers
136. Fuel jettison valves
137. Port navigation light
138. Port wingtip fuel tank
139. Landing lamp
140. Leading edge de-icing boot
141. Front spar
142. Wing rib construction
143. Rear spar
144. Wing stringers
145. Port main undercarriage wheel bay
146. Main undercarriage pivot mounting
147. Leg strut
148. Taxiing lamp
149. Brake pad
150. Port mainwheel
151. Port wing integral fuel tank
152. Leading edge nose ribs

IAI followed the 1121B Commodore with the first of its improved designs, the similar-looking 1123. Thirty-six of these were built before production switched to the 1124 (illustrated). This version mated the long fuselage of the 1123 with TFE731 engines.

The IAI 1124 became the Westwind I with the addition of long-range fuel tanks, and sales increased accordingly. Again IAI found their main market in the US. One-hundred-and-sixteen were built before production shifted to the winglet-equipped Westwind II.

turbofan engines, which offered an increase in range over the CJ610-powered model of some 33 per cent. Fifty-three IAI 1124s were built.

The IAI 1124 Westwind I, introduced from serial number 240 onwards in 1978, was an improved version differing principally in having an optional 317.5-kg (700-lb) increase in fuel capacity in a removable tank mounted in the forward luggage compartment, a five-per cent increase in cabin volume brought about by relocating avionics and lowering the floor in the aft toilet compartment, RCA Primus 400 colour weather as standard, and improved fuel and environmental systems. The United States was the most important marketing area for the Westwind, most of which were ferried across the Atlantic in 'green' condition for completion, cabin outfitting, avionics installation and external painting. Atlantic Aviation of New Jersey acted as distributor for the jet. Four civilian-registered IAI 1124 Westwinds were operated for a time by Rhein-Flugzeugbau from Mönchengladbach for target-towing duties on behalf of the West German armed forces.

Military market

Parallel with manufacture of the commercial Westwind, IAI developed a dedicated maritime reconnaissance version known as the 1124N Sea Scan. The Sea Scan prototype, converted from one of the Israeli Air Force Model 1123s, made its first flight in August 1978. This, and two more 1124Ns manufactured between 1979-81, were delivered to the Israeli navy and are used to guard the coast of Israel against sea landings by terrorists. The Sea Scan is equipped with underwing hardpoints for external stores/weapons pylons, bubble windows in the cabin to provide a vertically downwards view for observers, a nose-mounted Litton APS-504(V)2 360° search radar, VLF/Omega navigation equipment, a VHF/UHF communications suite, operators' consoles in the cabin, and provision for a variety of mission sensors including identify friend or foe, magnetic anomaly detectors and low light level television. Weapons capability includes torpedoes and Gabriel III anti-shipping missiles. The Sea Scan has a low-altitude search range of 2110 km (1,311 nm) at 914 m (3,000 ft), covering an area of 1665 m^2 (103,496 sq ft), or at high altitude it can range 3910 km (2,430 miles) with an eight-hour endurance while cruising at 816 km/h (506 mph) at up to 13720 m (45,000 ft) and loitering on-station at speeds as low as 259 km/h (160 mph). IAI have undertaken studies for a second-generation Sea Scan which would have signals intelligence (Sigint), anti-submarine warfare (ASW) and anti-shipping air-to-surface capability, but no development programme for such an aircraft has yet been announced, nor have earlier Sea Scans been sold outside of Israel.

The all-new Westwind 2

On 24 April 1979 the prototype of the 1124A Westwind 2 made its first flight from Lod Airport. This version was a development of the Westwind 1 aimed at improving 'hot and high' runway performance, range and specific fuel consumption. A major change was the introduction of an IAI-developed 'Sigma' wing of supercritical airfoil section, with swept-back NASA/Whitcomb winglets above the tiptanks. Other improvements included a flat cabin floor replacing the 'trenched' floor of the earlier model, affording increased headroom; overhead passenger service units, and an airline-standard flushing toilet. In standard form the Westwind 2 was delivered with accommodation for two crew and seven passengers, six on individual tracking and swivelling seats and one on a single-place divan. A fully-enclosed toilet and flight-accessible pressurised rear baggage compartment were provided, while two further baggage areas were located in the rear fuselage, accessible on the ground via a door on the port side. Optional seating arrangements were available for up to 10 passengers. Standard avionics included full dual IFR instrumentation comprising Collins VHF navcom, DME, ADF and flight director/autopilot/flight management systems, and WXR-300 colour weather radar. A wide range of optional avionics, including VLF/Omega, was available to customer specification.

The Westwind 2 was certificated by the Israeli Civil Aviation Authority on 11 December 1979 and by the US Federal Aviation Administration on 17 April 1980. The first customer delivery, of the prototype aircraft, was made to an operator in Colombia on 16 May 1980. More than 50 Westwind 2s were delivered before production ceased in 1987 in favour of the IAI 1125 Astra, which is virtually a new aircraft, sharing only the Westwind's engine nacelles and tail unit.

To give the type a distinct character, the 1123 was renamed Westwind and further developed into the 1124 version with turbofan power. Minor improvements introduced the Westwind I, for which this is the US demonstrator, resplendent in a Pierre Cardin design.

The most recent development of the basic Jet Commander is the 1125 Astra, which in fact only retains the tail and engine nacelles of the Westwind. A deeper fuselage offers greater comfort, and the new swept-wing gives superb range and speed performance. First flight was on 19 March 1984.

Jetstream

The Jetstream has led a charmed life of many incarnations. The design has passed through the hands of several manufacturers, failed, been resurrected and until very recently has formed one of British Aerospace's most viable products.

In 1962 Britain had a thriving civil aircraft industry. Under the auspices of the British Aircraft Corporation, the new One-Eleven jet was taking shape and taking on the Americans, while the VC10 continued to roll off the Weybridge line. The venerable Britannia was still in production, as was the Viscount, and the new Vanguard was entering service. Over at Hawker Siddeley attention had moved on from the Comet to the Trident, which was again ahead of its US competitors. The Dove was still in demand and great hopes were invested in an important new small jet design, the D.H.125.

The fact that many of these aircraft can be associated with other aircraft companies, such as Bristol's Britannia for example, was indicative of the major shake-up that had occurred in British aviation by that time. The great names of the pre- and post-war years had been merged into larger, less well-defined units in a government-inspired attempt to rationalise the industry and improve competitiveness abroad. This proved an unpopular move with many, but even more unpopular (in official eyes) were the few manufacturers who chose to stay independent of the new giants.

Chief among these was the Radlett-based firm of Handley Page. Founded in 1909 by Frederick Handley Page (who in 1942 became Sir Frederick), the company was a pioneering firm, specialising in the design and manufacture of heavy bombers during and between both world wars. By the mid-1940s, with only the Lincoln and Canberra bombers receiving official backing, the company was forced into the transport market, first with the Hastings and later with the Hermes airliner. It acquired Miles aircraft in 1948 (developing the unsuccessful Marathon), and then moved on to the Rolls-Royce Dart-powered Herald. When the pressure came to merge with other companies under the Sandy's plan of 1957, Handley Page refused, and it began to lose government orders to its rivals.

In 1962 Sir Frederick died, and chairmanship of the company passed on to George Russell. While the Herald was still in limited production, Handley Page had to content itself primarily with the overhaul of Victors for the RAF. The company found itself

The Jetstream 31 became a favourite among US feeder networks, such as Continental Express. During the late 1980s, Air New Orleans operated Jetstreams, under the Continental banner, alongside BAe 146s.

Peregrine Air Services was formed in Aberdeen in 1969, and grew along with the city's oil business. In December 1982 it was operating Piper Aztecs and Cessna Titans when they took delivery of this, its first Jetstream 31, the fourth production aircraft.

involved more and more with non-aircraft products such as radiators, brewery equipment and portable hangars. After three years, in a move that was to see the company emerge again as a successful aircraft manufacturer, an announcement was made in August 1965 that work had commenced on the H.P.137 Jetstream.

Turboprop inspiration

Handley Page had observed with interest the trend, particularly in the United States, towards re-engining existing piston-driven aircraft with turboprop engines. Companies such as the Riley Aeronautics Corporation had successfully converted significant numbers of Doves and Herons by fitting them with Garrett AiResearch TPE331s or Pratt & Whitney PT6s. Yet, thus far, no-one had attempted to supply that market with a purpose-built 12/20-seat passenger aircraft intended to cater for short to medium routes of less then 1600 km (1,000 miles). The impetus behind the Jetstream was the recent promulgation by the Federal Aviation Authority in the United States of its FAR (Federal Airworthiness Requirement) Part 23 ruling. This permitted the operation of any commercial aircraft weighing up to 5670 kg (12,500 lb) to be flown by a single pilot, thus cutting airlines' operating costs by a significant margin. Handley Page began work on the Jetstream, lead by designer Mr C. F. Joy, and optimised it at this weight, but with the eventual intention of increasing it to two-crew operation at 6350 kg (14,000 lb).

It was Jack Riley's Fort Lauderdale firm that led the way for the Jetstream. The aircraft's name had been chosen in deference to the American distributors who wanted something striking and easily remembered. After specifications had been circulated among prospective customers, the Riley Aeronautics Corporation announced a firm order for 20 aircraft. It was, however, an aircraft that thus far existed only on paper. In 1966 the go-ahead was given to proceed with four prototypes, to be paid for entirely out of the manufacturers' coffers. In a surprising move, the Ministry of Aviation then offered the firm £1,500,000 towards the launch cost, with the caveat that they would have some input in the design and would levy a duty on every production aircraft sold. This would repay the government's loan after 125 Jetstreams had been sold. The offer was accepted, and a new production line was established next to those for the Heralds and Victors at the company's Colney Street factory.

In August of that year BAC completed tunnel-tests on a one-sixth model of the aircraft. In its construction the Jetstream broke no new ground. A twin-engined, low-wing monoplane, it spanned 15.8 m (52 ft) and measured 14.3 m (47 ft) in length. The wing was unswept, used a NACA-developed 63A aerofoil and had 7° dihedral. Each wing had an integral fuel tank of 738 litres (195 Imp gal) as part of the box spar. Like the wing, the tailplane was tapered and unswept. The fin was swept by 45°, which was said by many to be purely for effect, as it contributed greatly to the type's rakish appearance. The cylindrical cabin was designed from the outset to be 'stand-up' with a ceiling height of 1.8 m (6 ft). Passengers in the 18-seat, three-abreast cabin were provided with seven elliptical windows on each side, the fourth one to starboard serving as the emergency exit. Situated at the rear of the cabin, the door was downward opening and incorporated folding stairs. Extensive use of chemical milling on the skin was used to save weight wherever possible. Turboméca's diminutive 514-kW (690-ehp) Astazou XIV turboprop was chosen as the initial powerplant, a choice which would later dog the design, and drove Hamilton-Standard reversible-pitch propellers.

First flight fright

A furnished mock-up was displayed at the 1966 SBAC show at Farnborough, and it was planned to have the aircraft's public debut at the following year's Paris show. In the event, the first aircraft (G-ATXH) was not rolled out at Radlett until the end of June 1967, and its Paris date was missed. On 18 August 1967, at the hands of pilots John Allam and Harry Rayner, the 100-minute maiden flight was made. Events were marred only by a partial seizure of the starboard wheel-brake on touch-down, which was dealt with by the hasty application of full reverse pitch on the port engine. By the end of November the aircraft was flying with larger and definitive 625-kW (840-ehp) Astazou XIV engines, and from the outset these were fitted to the second aircraft (G-ATXJ), which first flew on 28 December. Winter flight tests were moved to France and the two Jetstreams were joined by a third (G-ATXI) on 8 March 1968. In May of that year the fourth and final development aircraft (G-AXTK) was ferried to the US for use by the distributors, International Jetstream Corporation (IJC). Formerly the Riley Jetstream Corp., IJC already held 56 options for the aircraft, while the UK firm CSE were distributing around Europe, Asia and Africa with an undertaking to sell 20 aircraft a year for five years.

Handley Page decided to undertake only military sales of the Jetstream directly, and it was a military customer that placed the first big order (or so it seemed). The USAF had a requirement

No. 750 Sqn, Royal Navy, operates the Astazou-powered Jetstream T.Mk 2 as a navigation trainer from its Yeovilton home. Formerly RAF T.Mk 1s, they replaced the Sea Prince T.Mk 1s in navy service in 1977.

for a 'CX' light turboprop transport and at first the Beech 99 was selected for the 80-aircraft contract. Then, in December 1967, the decision was overturned in favour of a batch of 11 TPE331-powered Jetstreams. This new version, the Jetstream 3M, would be a FAR 25 certified 6577-kg (14,500-lb) aircraft, known as the C-10A in USAF service. With a number of improvements, such as a strengthened floor and a wider door for para-dropping, the C-10 required a serious development effort. Furthermore, the initial contract payment of £2,400,000 was only just enough to cover the manufacturing costs. However, Handley Page anticipated further USAF orders, and felt that a Garrett-powered Jetstream would have good civilian prospects too. Two aircraft werere-engined and Handley Page invested heavily in the project, but no further USAF money was forthcoming, and the British firm's bill soon exceeded £5,500,000. A new factory had been built in Ayrshire for Scottish Aviation to manufacture the wings, and components were also imported from Canada. But by the time the Jetstream had entered production 'proper', it was encountering weight problems. Then the ARB delayed certification, and bills for components were rolling in, without aircraft rolling out.

Success and failure in the United States

The Jetstream began to sell in small numbers to non-airline clients, and then, suddenly, North American sales improved as both Sun Airlines' and Cal-State AirLines' positive early experience saw a further 30 aircraft ordered by the US and Canadian distributors. The aircraft was still falling short of its original range/payload promises, however, and the only way out of this was to install more powerful Astazou XVI engines. All this took time and money, two luxuries that, increasingly, the company did not have. Then the USAF cancelled its C-10s and the whole Jetstream project seemed doomed. Fortunately, International Jetstream Corporation stepped in to save the day, and with financiers Mercantile Trust of St Louis, established a new company, Handley Page Limited, to continue building Jetstreams. Work on the Astazou XVI-powered aircraft, soon to become known as series 200s, progressed at Filton, and further Jetstreams were ferried to US customers between September 1969 and February 1970. Then towards the end of February the cash ran out again, and technically Handley Page ceased to exist. Further disaster

First flown in February 1969, this was the fourth production Handley-Page Jetstream Mk 1. Delivered to Bavaria Fluggesellschaft in October of that year, it crashed after colliding with powerlines in March 1970.

then struck with the crash of Bavaria Flug's single aircraft, killing all on board including the airline's managing director. Restrictions were placed on the 26 aircraft in the US and Canada, limiting their flying hours to only 1,000. Many were already approaching that point and were set to be grounded. Most were stored, and Handley Page finally disappeared from the aviation world, leaving behind 40 complete and 33 incomplete airframes.

The Jetstream, though, did not disappear. Former Handley Page pilot Bill Bright founded Jetstream Aircraft Ltd in 1971 and acquired all the existing spares and airframes. Production was re-started at Sywell with the intention of supplying Jetstream Is from existing parts and new-build Jetstream 200s. Bill Bright flew the Jetstream in the well publicised Daily Express air race of 1971, then took the aircraft to Australia for climate testing. Three new Jetstreams appeared at the 1971 Paris air show, and a Sywell-built aircraft became the first to be sold in Europe by Jetstream Aircraft Ltd, to Air Wasteels of Metz. Another large military order loomed, this time for a replacement for the RAF's Varsity multi-engined pilot-trainers. The Jetstream emerged the winner, and with the help of Scottish Aviation (who handled all the construction) 26 aircraft were delivered to the RAF as Jetstream T.Mk 1s. Essentially 742-kW (996-shp) Astazou XVI.D-powered Jetstream 200s, they were equipped with Sperry-Rand STARS avionics and Marconi AD.370BA DF gear. The first aircraft, built at Prestwick, was delivered to No. 5 FTS on 16 May 1974, and training courses on the type commenced in July

The Jetstream 31 was a staple of Dutch airline Netherlines, which operated six aircraft until their replacement by the rather larger Saab 340. Netherlines was absorbed by KLM-subsidiary NLM, forming KLM Cityhopper.

of that year. After a crash in November all aircraft were grounded until the cause was traced to turbulent airflow caused by the nacelle panel joints and the fine-pitch setting on the props. No sooner had these problems been rectified than No. 5 FTS was forced to disband, due to a drastic contraction of the RAF's transport fleet, and its aircraft were placed in storage at St Athan. Fourteen of these were later converted for Royal Navy use as T.Mk 2s for observer training. Ultimately, 12 RAF aircraft re-entered use with No.6 FTS in 1977. Production of civil Jetstreams then moved to Leavesden and, while some aircraft were sold to corporate users, Scottish Aviation (who had acquired all rights from Jetstream Aircraft) refused to undertake any further production without orders for 15 aircraft. By the beginning of 1975 some 80 aircraft had been completed, including those for the RAF and USAF (the C-10s were largely built and then scrapped on the production line).

British Aerospace and the Jetstream 31

In 1975 the British aviation industry was nationalised and in the shake-up that followed Scottish Aviation was combined with BAC and Hawker Siddeley to form British Aerospace. On 5 December 1975 BAe announced a new 18/19-seat airliner to be powered by 700-kW (940-ehp) Garrett TPE331-10 engines, which would be built at Prestwick and named the Jetstream 31. At last the excellent airframe was freed from its troubled marriage with the mismatched Astazou. The prototype (G-JSSD) first flew on 28 March 1980, and production aircraft would be available in airline, executive shuttle or corporate configurations. Military versions were also foreseen. British Aerospace decided to go ahead with the project, having received 13 orders and nine options from five customers and a further 40 letters of interest. The Jetstream Commuter would carry up to 19 passengers and baggage and was optimised to operate three 160-km (100-mile) sectors without having to refuel. The first production aircraft (G-TALL) took to the air on 18 March 1982 to partner the prototype in the flight test programme. It was then refitted with a 12-seat corporate interior for use as a demonstrator. The intended production rate was a minimum of 25 aircraft per year.

By June 1982 the Jetstream 31 had obtained its UK certification. At that year's Farnborough show BAe announced the Jetstream 31EZ, which would be fitted with an underfuselage surveillance radar, a quick-change interior, military-standard navigation and communications equipment, searchlight, increased fuel capacity and bulged observation windows. The 31EZ was to be a dedicated maritime patrol version, in the same vein as the BAe 748 Coastguarder, which in the event failed to gain any orders either. Two years later the 10-seat Corporate Jetstream was unveiled at Farnborough. Outfitted by the Arkansas Modification Center, it featured six fully-reclining chairs with a four-seat divan, and a Sperry-equipped EFIS cockpit. That year saw the first sale of a BAe Jetstream to a private client, and the company was keen to offer the type as the perfect complement to its BAe 125 jet. In the same year the Royal Navy ordered four Garrett-powered Jetstream T.Mk 3s to serve with No. 750 Sqn at Culdrose. The Navy already operated Astazou-powered Jetstream T.Mk 2s, but the new aircraft also differed through the addition of a belly-mounted Racal ASR 360 multi-mode search radar. These aircraft brought the Jetstream 31 order book up to 42 by mid-1984. A further military order came from the Royal Saudi air force, which ordered two aircraft equipped with Tornado IDS avionics for navigator training.

In fact, sales continued ever upwards. By 1987 179 Jetstream 31s had been delivered, not only to its primary market of the United States, but also among Europe's burgeoning regional airlines in Norway, Sweden, Italy, West Germany, the Netherlands and, of course, the UK. At the Paris air show of that year BAe announced their first improvement on the basic civil Jetstream 31, the Jetstream Super 31. Sometimes referred to as the Jetstream 32, this aircraft was fitted with 670-kW (1,020-shp) TPE331-12UAR engines, and the first was delivered to Wings West in the USA. It partnered the Jetstream 31 from aircraft number 205 on the line, and finally replaced it from number 230 onwards. In 1990 the Jetstream passed the 300 sales mark with a delivery to Pan Am Express and by 1992 the type had amassed nearly 3,000,000 flying hours. The Jetstream 31 has found particular favour in the US among regional and feeder airlines and has been operated by Presidential Airways, Westair, Metro Express, Chaparral Airlines, Express Airlines, Jetstream International Airlines, North Pacific Airlines, Wings West, Nashville Express and Air Midwest Eagle. The most important customer has been the American Eagle network, which operates 75 Jetstreams 31s of both versions. By October 1993 384 Jetstream 31 and Jetstream Super 31s had been delivered.

Jetstream for the 1990s

With its American market very much in mind, BAe announced a radical new addition to its Jetstream family in 1989. Though still bearing the family name and sleek lines, the 29-seat Jetstream 41 is a significant step up from its forebears. A mock-up appeared at the 1989 Paris air show and several manufacturers are involved in producing the components for the finished aircraft. Pilatus manufacture the tail and ailerons in Switzerland, while Gulfstream in Oklahoma have been contracted to supply 200 wing sets. In the UK, Slingsby provide composite components, while Field Aircraft outfit the interiors and handle the electrical

Until its demise, one of the earliest and largest Jetstream users in the United States was Eastern Airlines, which named its newly established feeder carrier Metro Express, after the aircraft.

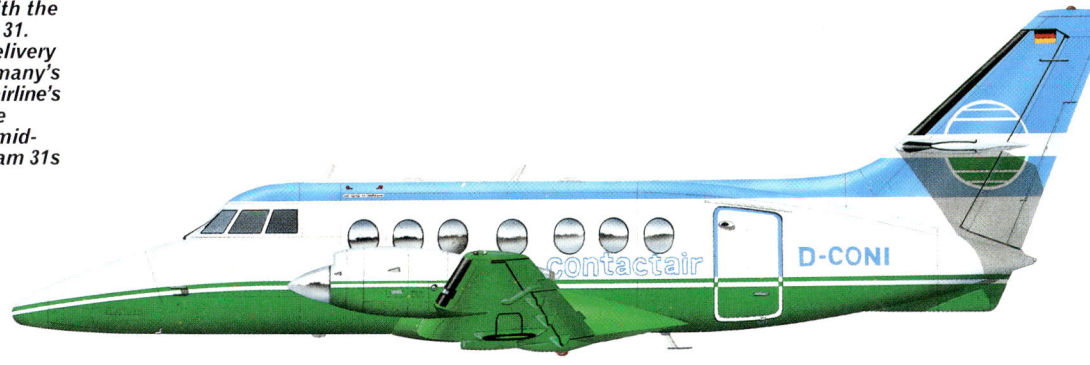

Stuttgart-based Contactair joined the Jetstream club in 1982 with the delivery of its first Jetstream 31. The airline would later take delivery of three more aircraft as Germany's sole Jetstream operator. The airline's fleet expanded to accommodate seven DHC-8 Dash 8s, but in mid-1993 two of its 16-seat Jetstream 31s were still in use.

looming and avionics installation. In outward appearance the Jetstream 41 differs from the Jetstream 31 chiefly through the addition of two fuselage plugs which stretch the airframe by a total of 4.87 m (192 in). Under the centre section is a 0.99 m³ (35 cu ft) external baggage hold. In addition, the rudder and flaps have increased chord, the engines have new nacelles and five-bladed composite props, the main passenger door is in the forward fuselage, and a second type III emergency exit has been added to the starboard side. The flight deck is equipped with a Honeywell EFIS system, with four CRT displays. Powered by a pair of 1117-kW (1,500-shp) TPE331-14 engines and some 2721 kg (6,000 lb) heavier than its sibling, it still retains a high degree of commonality with the Jetstream 31. Maximum cruising speed is 295 kt (547 km/h; 340 mph) and maximum range with 29 passengers and IFR reserves is 681 nm (1263 km/785 miles). The Jetstream 41 is 19.25 m (63 ft 2 in) in length and spans 18.29 m (60 ft). Intended to compete with the Dornier Do 328, Saab 340B, Dash 8-100 and EMB-120 (all of which are larger), BAe promises the best break-even load-factor in its class, with acquisition costs being considerably lower.

Several existing Jetstream operators had signed for the Jetstream 41 by the time the first aircraft was rolled out in June 1991. The first aircraft (G-GCLJ) took to the air for 2 hours and 45 minutes on 25 September of that year, with 115 commitments received by BAe. It was followed into the flight test programme by G-PJRT in February 1992 and G-OXLI in the following March. The unusual registrations were all significant. The first two were a tribute to John Larroucau (Jetstream Senior Vice President) and Jimmy Thompson (Head of Prestwick Operations). The third represented 'Go 41'. The first delivery took pace on 25 November 1992 to Loganair and Manx Airlines, part of the Airlines of Britain group, who ordered five and three aircraft respectively. Other early customers were Sun-Air (two) and United Express-affiliate Atlantic Coast Airways (17). Current deliveries, as of October 1993, stand at 22.

Still a growing family

Production of the Jetstream family is now undertaken at Prestwick by Jetstream Aircraft Limited, a BAe company, and are part of the trend towards spinning off British Aerospace's aircraft manufacturing 'components'. In an attempt to boost non-existent sales, the (Avro) ATP has been re christened the Jetstream 61 in an attempt to bathe it in the smaller aircraft's reflected glory (and order book). Even more recently Jetstream announced plans for 52-seat Jetstream 51 and 78-seat Jetstream 71 T-tailed variants, but by May 1997 sales were slipping and BAe announced that production would cease in 1997 with the completion of the 100th airframe, ending a programme that had also seen the delivery of 446 earlier-generation Jetstream aircraft.

Possessed of larger engines and a longer fuselage the Jetstream 41 bridges the gap between the Jetstream 31 and Jetstream 61 (ATP). The first three prototypes were involved in climatic testing, hence the desert and snowman logos on the tails and engine cowlings. This aircraft is intended to open a new chapter in the Jetstream story, leading to the Jetstream 51 and 71.

Chapter 26

Junkers F13

Junkers G38

Learjet

Junkers F 13

The development of commercial aviation has been marked by several landmark aircraft designs, the Junkers F 13 being one such innovative type. Produced in large numbers for home and overseas consumption, the all-metal F 13 not only laid the foundations for subsequent airliners, but also introduced design features that would typify Junkers aircraft for many years.

In the early 1900s Professor Hugo Junkers was principally associated with the manufacture of engines, but in 1910 he was granted a patent which was to have far-reaching consequences for the development of modern aeroplanes. The patent referred to 'making the ratio between the supporting capacity of an aeroplane and the resistance to forward motion as large as possible, (by arranging) the non-supporting parts of the machine in casings having their upper and lower surfaces shaped so as to act as supporting planes while reducing head resistance to a minimum'.

Thus was born the cantilever monoplane concept. Professor Junkers built his first cantilever-winged monoplane, the J 1, in 1915. It featured not only a wing devoid of external bracing, but smooth tin-plate skinning over its entire metal airframe, thus earning it the nickname 'Tin Donkey'. Later developments led to the adoption of the load-bearing corrugated skinning which was to become a hallmark of Junkers Flugzeug und Motorenwerke AG designs in the between-wars era.

The Junkers F 13 was the result of Junkers' cantilever wing experience explored during World War I, and a far-sighted realisation that civil aircraft were to become big business in the years following the war.

According to legend, on the day that the Armistice and Peace Treaty of Versailles were signed forbidding aircraft construction in Germany until 1922, Professor Junkers gathered his technical staff together at Dessau and ordered work to begin on a new civil transport aeroplane based on the J 10 two-seat low wing attack monoplane, which had first flown on 4 May 1918.

The design philosophy adopted by Junkers was for economics and reliability rather than outright high performance. At first the design team, led by Dipl Ing Otto Reuter, planned a derivative of the J 10, designated J 12. A hastily-converted J 10 with a rudimentary closed cabin for one passenger or a small amount of freight inaugurated Junkers' airline service between Dessau and Weimar in March 1919, thus becoming the first all-metal aeroplane in commercial service.

The new design would have been a two-passenger aircraft powered by a 160-hp Mercedes engine, but the concept was abandoned before a prototype was ever built, probably because of the impracticality of operating airline services with an aeroplane of such limited capacity.

Instead Reuter opted for an entirely new design, the J 13, which had an enclosed four-passenger cabin set behind two open but

Among the military users was the Swedish air force, which designated the aircraft Tp 1 in service. This example is an air ambulance of the early 1930s. Sweden's F 13s served from 1928 until 1943.

The coat of arms on the tail of the 'Dz' registration show this aircraft to be from the Free State of Danzig. Kasuar *served as a mailplane with* Danzigar Luftpost.

centrally divided cockpits for its pilots. The J 13 incorporated Junkers' strong corrugated duralumin skinning over the entire structure and was powered by a 160/170-hp Mercedes D.IIIa engine. Construction of the first aircraft started on 10 February 1919 and was completed on 20 June. This prototype, named *Annelise* after the wife of Prince Leopold of Anhalt-Dessau, made its first flight in the hands of Junkers test pilot Monz on 25 June 1919. Subsequently on that same day it carried a full complement of six people, three hours' fuel supply and 88 lb of ballast simulating luggage to an altitude of 6,560 ft. On 13 September 1919 *Annelise*, with eight people aboard, set a record by climbing to an altitude of 22,145 ft, though the achievement was never officially sanctioned because Germany was not a member of the Fédération Aéronautique Internationale.

The production version of the J 13 was later designated F 13 and was to become one of the most significant contemporary designs, a true ancestor of the all-metal transport aircraft which were to emerge during the 1920s and 1930s. The Junkers F 13's airframe was immensely strong and durable. The cantilever wing of characteristic thick airfoil section was formed around nine tubular duralumin spars braced by duralumin tubes to form a girder, each panel bolted to the similarly constructed centre section which formed an integral part of the fuselage. A strong metal box formed the cabin area, which was comfortably furnished with individual seats for the front passengers and a two-seat bench at the rear equipped – uniquely for its day – with seat belts, and with three large windows on each side.

Early production F 13s were powered by the Mercedes engine or the 185-hp BMW III and IIIa, but the F 13's airframe was progressively strengthened to take increasingly more powerful engines such as the 230/265-hp Junkers L2, 250/300-hp BMW IV and Va, 280/385-hp Junkers L5 liquid-cooled six cylinder powerplants, and Armstrong Siddeley Puma, Jaguar, Jaguar Major, Bristol, Gnome-Rhône or Siemens Jupiter IV and Pratt & Whitney Wasp and Hornet air-cooled radial engines. The early F 13s had a wingspan of 47 ft 5¾ in, but this was increased to 48 ft 7½ in, and following introduction of a long-span 58 ft 2¾ in wing for float-equipped F 13s, this was adopted as standard. In all, some 60-70 distinct sub-variants of the F 13 are known to have been developed, incorporating airframe improvements which included strengthened wings and fuselages, enlarged fin and rudder, balanced control surfaces and a fully-developed cock-

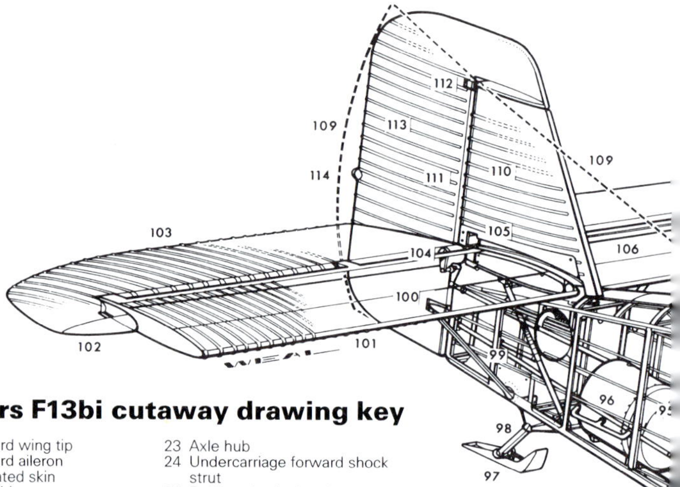

Junkers F13bi cutaway drawing key

1 Starboard wing tip
2 Starboard aileron
3 Corrugated skin
4 Aileron hinges
5 Outer wing structure
6 Interspar brace struts
7 Corrugated leading-edge
8 Aileron control linkage
9 Aileron actuating hinge
10 Trailing-edge corrugation
11 Rib end section
12 Wing lower surface Duralumin spars (four)
13 Wing upper surface Duralumin spars (four)
14 Leading-edge spar
15 Wing inner/outer section joints
16 Joint capping strake
17 Entry step (port and starboard)
18 Wing walkway steps
19 Walkway support frame
20 Starboard wing triangular fuel tank, capacity 35.6 Imp gal (162 l)
21 Undercarriage rear shock-strut
22 Starboard mainwheel (soft-ground tyre)
23 Axle hub
24 Undercarriage forward shock strut
25 Rubber shock absorbers (2.9 in/74 mm travel)
26 Undercarriage brace strut
27 Vee strut
28 Engine firewall/bulkhead
29 Engine bearer assembly
30 Bearer support strut
31 Nose frames
32 Auxiliary (tropical) radiator
33 Port mainwheel
34 Service hand/footholds
35 Nacelle face panels
36 Propeller shaft
37 Hub
38 Junkers two-blade metal propeller
39 Radiator intake
40 Service handhold
41 Radiator
42 BMW IV in-line engine
43 Exhaust manifold
44 Cowling panel
45 Rearward hinged nacelle upper panel
46 Alternative exhaust outlet assembly
47 Port wing multi (9)-spar structure
48 Corrugated underwing surfaces
49 Port wingtip
50 Port aileron hinges
51 Aileron control rod
52 Port aileron
53 Aileron actuating hinge
54 Pilot's windscreen
55 Pilot's cockpit coaming cut-out
56 Control wheel column
57 Instrument panel
58 Cockpit floor
59 Control linkage
60 Rudder pedal assembly
61 Fuselage forward frames

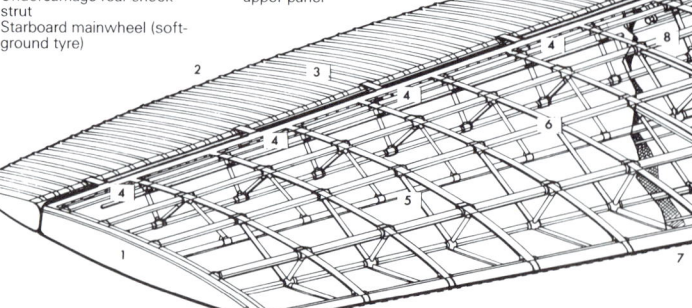

Some ex-Lufthansa machines were passed to the Luftwaffe to act as pilot trainers. This was one such aircraft, flying with the Flieger Übungsstelle at Böblingen in 1934.

pit. Wheel, ski and float undercarriage were offered.

Some 60 F 13s were put into service by Junkers-Luftverkehr, the airline company established in 1921 by the Junkers company. That year the newly-formed Deutsche Lufthansa took over most of Junkers-Luftverkehr's fleet and operated them on 45 domestic passenger and cargo services, some remaining in service until 1938.

Junkers F 13s were employed by nearly all major European airlines, including Olag in Austria; Danziger Luftpost in Germany; Aero

This Junkers F 13W was owned by the famous Norwegian aviatrix Miss Gidsken Jakobsen, who flew several Junkers aircraft, all of them floatplanes. With floats attached the F 13 needed an increased rudder area to maintain longitudinal stability.

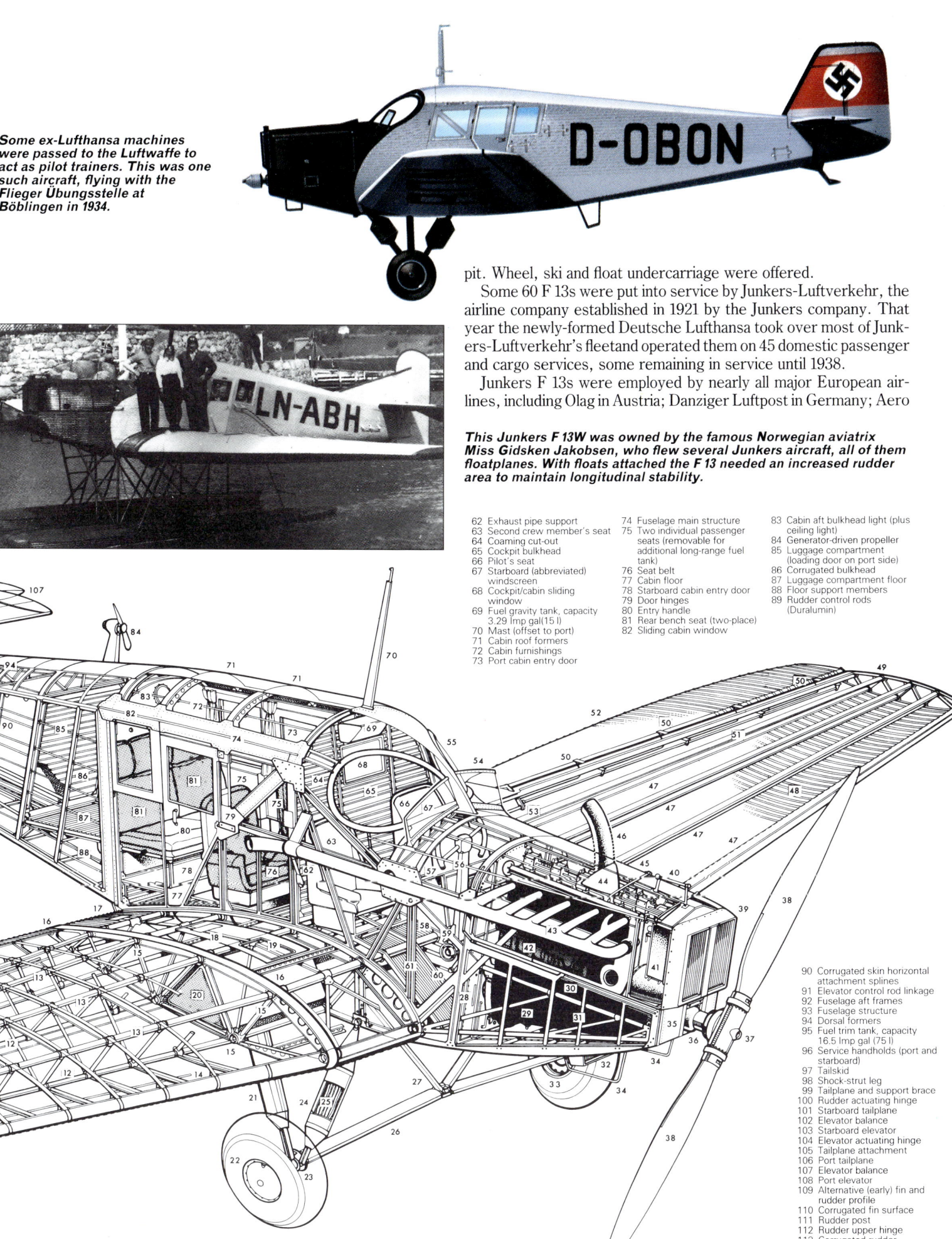

62 Exhaust pipe support
63 Second crew member's seat
64 Coaming cut-out
65 Cockpit bulkhead
66 Pilot's seat
67 Starboard (abbreviated) windscreen
68 Cockpit/cabin sliding window
69 Fuel gravity tank, capacity 3.29 Imp gal (15 l)
70 Mast (offset to port)
71 Cabin roof formers
72 Cabin furnishings
73 Port cabin entry door
74 Fuselage main structure
75 Two individual passenger seats (removable for additional long-range fuel tank)
76 Seat belt
77 Cabin floor
78 Starboard cabin entry door
79 Door hinges
80 Entry handle
81 Rear bench seat (two-place)
82 Sliding cabin window
83 Cabin aft bulkhead light (plus ceiling light)
84 Generator-driven propeller
85 Luggage compartment (loading door on port side)
86 Corrugated bulkhead
87 Luggage compartment floor
88 Floor support members
89 Rudder control rods (Duralumin)
90 Corrugated skin horizontal attachment splines
91 Elevator control rod linkage
92 Fuselage aft frames
93 Fuselage structure
94 Dorsal formers
95 Fuel trim tank, capacity 16.5 Imp gal (75 l)
96 Service handholds (port and starboard)
97 Tailskid
98 Shock-strut leg
99 Tailplane and support brace
100 Rudder actuating hinge
101 Starboard tailplane
102 Elevator balance
103 Starboard elevator
104 Elevator actuating hinge
105 Tailplane attachment
106 Port tailplane
107 Elevator balance
108 Port elevator
109 Alternative (early) fin and rudder profile
110 Corrugated fin surface
111 Rudder post
112 Rudder upper hinge
113 Corrugated rudder
114 Tail navigation light

The Junkers F 13 was produced in many versions, mainly with differing engines. The largest user was Deutsche Lufthansa, which was formed in 1926 as the German national airline. Most of its aircraft had previously served with Junkers-Luftverkehr, an airline started in 1921 by the F 13's manufacturer. F 13s served DLH until 1938, when they were transferred to Luftwaffe training schools and pleasure flying companies.

Specification
Junkers F 13da
Wingspan: 17.75 m (58 ft 2¾ in)
Length: 9.60 m (34 ft 6 in)
Wing area: 44.0 m² (473.61 sq ft)
Powerplant: 1 × 280-hp Junkers L2
Passenger capacity: 4
Empty weight: 1150 kg (2,535 lb)
Maximum take-off weight: 1730 kg (3,814 lb)
Maximum speed: 171 km/h (106 mph)
Cruising speed: 140 km/h (87 mph)
Service ceiling: 4000 m (31,120 ft)
Range: 949 km (590 miles)

The Junkers F 13 sold well in Europe, where most of the major airlines bought the type. One of the major users was Ad Astra of Switzerland, the forerunner of today's Swissair.

This Junkers F 13ke served as a trainer with the Würzburg DVS, a pilot training school. Named Königsmilan, it is shown as it appeared in 1932-3.

O/Y in Finland, which operated F 13 floatplanes; Aero-Express in Hungary; Aerolloyd, Aerolot and LOT in Poland; ABA in Sweden, also with floatplanes which inaugurated the first passenger flight by a Swedish operator from Stockholm to Helsinki in 1924; and Ad Astra Aero of Switzerland, and carriers in Belgium, Czechoslovakia, Estonia, Portugal, Romania and Spain.

Junkers sold F 13s far afield from Europe. They served in Afghanistan, Australia, Bolivia, Brazil, where the national airline VARIG continued flying two as late as 1948, Canada, China, Colombia, New Guinea and South Africa. Several were operated in Persia with machine-gun emplacements built into the tops of their cabins.

F 13s were particularly successful in South America, pioneering difficult routes in isolated areas, their rugged all-metal airframes standing up well to the harsh conditions and often rudimentary maintenance facilities. Four F 13s were used as air-taxis by a British company based at Croydon Airport. Eight, redesigned Junkers-Larsen JL 6s in anticipation of US production by the John M. Larsen Corporation, were delivered to the United States where they served the US Post Office on mail routes between New York-Chicago-Omaha and New York-San Francisco, though none was ever manufactured in America. Five F 13s were built in the USSR and two in 1931 by Transadriatica in Venice, one of these being powered by a 410-hp Fiat A.20 engine.

A total of 322 Junkers F 13s is thought to have been built before production ceased in 1932, although the exact figure is uncertain. The aircraft not only played a vital part in the growth of commercial air transport and in the development of all-metal transport aircraft, but led directly to a series of Junkers designs which were essentially little changed from the original 1919 concept. The 1924 Junkers G 23 was a scaled-up F 13 powered by three Junkers engines, and was the first civilian all-metal multi-engined monoplane. The derivative G 24 emulated the F 13's commercial success, as did the single-engined Junkers W 33 and W 34 which were little more than refined F 13s powered, respectively, by the 310-hp Junkers L5 or 420-hp Gnome-Rhône Jupiter VI engines. One specially-modified W 33, named *Bremen*, was equipped with long-range fuel tanks and flown by Hermann Koehl, Gunther von Heunfeld and James Fitzmaurice from Baldonnel, near Dublin, to Greenly Island, Labrador on 12-13 April 1928,

The Junkers F 13 was a true milestone in the history of aviation, being the father of the modern metal airliner. Design concepts such as the cantilevered wing and corrugated metal skin led through a classic series of Junkers aircraft to the Ju 52, the greatest of them all.

making the first east-west transatlantic air crossing. Junkers W 34s remained in large-scale service with the Luftwaffe in communications and training roles throughout World War II. The ultimate development of the design concept initiated with the F 13 was seen in the Junkers Ju 52. Originally flown on 13 October 1930 as a single-engined design, it was to emerge as the legendary tri-motor Ju 52/3m *Tante Ju* transport aircraft which rivalled the Douglas DC-3 as the most significant civil passenger aeroplane of the 1930s, and in military guise formed the backbone of the Luftwaffe's air transport fleet.

The F 13 was as at home on floats as it was on wheels. In remote parts of the world this was a particular boon to opening up the airways, particularly through South America, where the type was very active.

JUNKERS G 38

Only two G 38 aircraft were built, but they were so impressive that even today they are regarded with awe. Featuring an enormous wing of typical Junkers design, the G 38 had a double-deck fuselage for accommodating passengers, while cabins were also incorporated in the leading edge of the wings! Airline service with Lufthansa was sporadic, yet one of the aircraft soldiered on to don military camouflage and to serve in World War II.

In 1909, a year in which flying machines were gossamer-like creations scarcely heavier than their single 'aviators', Professor Hugo Junkers of Dessau prepared drawings for a transport aircraft in a somewhat different class – indeed, it might be considered bold even today. Passengers, crew, engines, fuel, cargo and almost everything else were housed inside a single giant monoplane wing. This wing was a cantilever, devoid of external bracing, and it was to be made of aluminium alloy so that the stressed skin could bear the main loads. At the rear was a vestigial fuselage in order to carry a tail for stability and flight control. This amazingly advanced aircraft was patented in 1910.

Junkers never had a chance to build it, though by 1915 he was building very advanced all-metal cantilever monoplanes. He never lost sight of his idea for a huge "flying wing" transport, however. Chief designer Karl Plauth was killed in a crash in 1927, and his successor, Dipl Ing Ernst Zindel eagerly asked Junkers if work could begin on such a machine. By 1928 the design was on paper, and though for practical reasons it did have something of a fuselage projecting both to the front and rear, it was not far removed from the original conception of 1910. It was still, in 1928, an incredible project to undertake, though by that time the technology of both airframe structures and powerplants had advanced to the point at which it was capable of realisation.

Designated as the G 38, two examples were built. The finance was found partly by the Junkers company, partly by the German Air Ministry (RLM) and partly by the airline Deutsche Lufthansa (DLH).

Junkers could hardly have expected any more. Indeed, it is remarkable that the airline agreed to pay a share of the costs at all, because, like the contemporary Do X, the G 38 was far bigger than normal airliners of the day, and did not really look like a profitable vehicle.

Even today the wing of the G 38 seems impressive. The span was 44 m, and the area 300 m². At the root the chord was 10.8 m, and the maximum thickness no less than 2.02 m. Thus, it was simple to walk upright within the wing. The structure was similar to smaller Junkers aircraft, with multiple tubular spars, diagonal bracing throughout and corrugated skin of a gauge thinner than would have been necessary without the stiffening corrugations. Catwalks were arranged along the wing to a point just beyond the outboard engines. The latter were installed inside the wing, where mechanics could reach all parts of the engine exterior in flight.

As originally built the G 38, registered D-2000, was powered by four Junkers water-cooled engines. The inners were L 55 vee-12 engines rated at 650 hp and driving through a reduction gearbox and extension shaft to a four-bladed wooden propeller of 4.5 m diameter. The outboard engines were L 8 inlines rated at 400 hp and driving via a reduction gearbox and extension shaft to a two-bladed propeller of the same diameter. Petrol was carried in 14 aluminium tanks in each

The second example was designated G 38b and featured a deep, double-deck fuselage for up to 34 passengers. A new wing had full-span slotted flaps/ailerons. The powerplants remained the same, with the two smaller engines outboard.

wing, aft of the catwalk, giving the remarkable flight endurance of 20 hours. Between the engines in each wing was a machine room, with various instruments and controls including all those associated with the fuel and coolant systems. The cooling water was pumped round a closed circuit which incorporated vertical-matrix (car type) radiators which could be cranked up and down along vertical guides. At full power and at low speeds, as at take-off and during the initial climb, the radiators were normally fully extended beneath the wing. As speed increased, they could be progressively retracted, until at economic cruising powers they were mainly raised inside the machine room, giving minimal cooling drag.

Design beyond its time

Almost all the rest of the airframe was aluminium alloy, apart from fabric-covered control surfaces. The biplane tail unit had a single central fin and three rudders, and a span of 9.38 m. The outboard trailing edge of the wing carried ailerons of very long span, extending almost as far in as the outer engines. Landing gear comprised tandem main wheels with vertical shock-absorber legs braced to the fuselage, the whole main gear being encased in a gigantic spat, and a single tailwheel. The four main wheels had compressed-air brakes, the system being supplied from a small diesel auxiliary power unit in the left wing. This compressed-air system was also used for starting the main engines, and, in emergency, for discharging fire-extinguisher fluid.

The cockpit seated two pilots side-by-side with giant circular control wheels typical of the period. The cockpit was enclosed by an aluminium framework holding multiple Plexiglas panes, all immediately above the upper surface of the leading edge. The leading edge of each giant root rib was immediately beside each pilot. Immediately further outboard, in the forward part of each wing, was a small cabin for two passengers. A further 26 could be seated in 13 rows of two single seats along the length of the fuselage, all at one level. Under the floor and in the rear part of the wing roots were compartments for up to three tonnes of cargo and baggage. As originally built, the forward cabin immediately aft of the cockpit was illuminated only by four small roof skylights and could be equipped for four sleeping passengers. The intended crew comprised two pilots, two engineers (one in each machine room), a wireless operator and a steward. Occasionally an additional member was carried to serve as captain.

Most of the history of the G 38 consisted of design changes and modifications. The first aircraft made its maiden flight at Dessau on 6 November 1929. Either before or immediately after this date the inboard engines were replaced by L 88s, only slightly larger than the L 55s but uprated to 800 hp. At about the same time the giant wheel spats were removed, appreciably reducing weight and having little effect on speed, which was modest. Cruising speed had been 160 km/h, but with the L 88 inboard engines it rose to 180 km/h.

In 1930 the Dessau factory built a second G 38, originally registered D-2500. Though obviously another G 38, it was actually a complete redesign. The wing was changed in several respects, the most obvious alteration being the double-kinked trailing edge with full-span "double wing" surfaces which served as flaps and as ailerons. These made a surprising difference to field performance. Stalling speed was brought down to only 78 km/h. Take-off at maximum weight over a 20-m obstacle took only 575 m, compared with 765 for the original machine, in spite of the fact that gross weight was increased from 21 to 24 tonnes. Landing distance over a 20-m obstruction was reduced from 615 m to only 455. Handling at low speeds was significantly crisper.

The fuselage of the second aircraft was deeper than that of the original, Zindel and his team having discovered from experience that Junkers' all-wing conception was not attractive to passengers. Each wing-root cabin was rearranged to seat three passengers, not two. The new fuselage seated two passengers in the nose, with a marvellous view (better than that of the pilots). The amidships cabin aft of

*The radiator grilles, wing root observation cabin and nose lights are clearly visible in this view of the **G 38**. Although only two aircraft were built, the **G 38** did enter service with **Deutsche Lufthansa**, although its use was restricted.*

*The giant Junkers **G 38** first prototype wheels over the German countryside during an early test flight, graphically illustrating the giant wings of the type. **N**ote the radiator grilles under the wings.*

the cockpit was now lofty enough to have windows at the sides, immediately above the wing roots. It seated 11, as did a second cabin immediately to the rear and at a lower level. The double entrance doors opened into the rear of this cabin. Aft of this was a four-seat compartment for smokers. At the extreme rear was a toilet, then very unusual in airliners, and a second toilet was located in the root of the right wing. On the opposite side, just ahead of the forward cabin, was a pantry from which the steward could dispense food and drink which had been previously heated or cooled as appropriate.

The other major new feature of the second aircraft was that the tail unit had three fins and three rudders. The engines comprised L 88s inboard and L 8s outboard, as in the re-engined first aircraft.

Debut with Deutsche Lufthansa

The original G 38 was delivered to DLH in June 1930; it was the largest landplane in service in the world. On 1 July 1931 it began a rather sporadic career in regular revenue service which eventually saw its appearance at Frankfurt am Main, Munich, Königsberg, Rome, London and Stockholm. It was joined by the second aircraft on 1 September 1931, and, unlike its predecessor, D-2500 did maintain regular services on particular routes which, however, changed

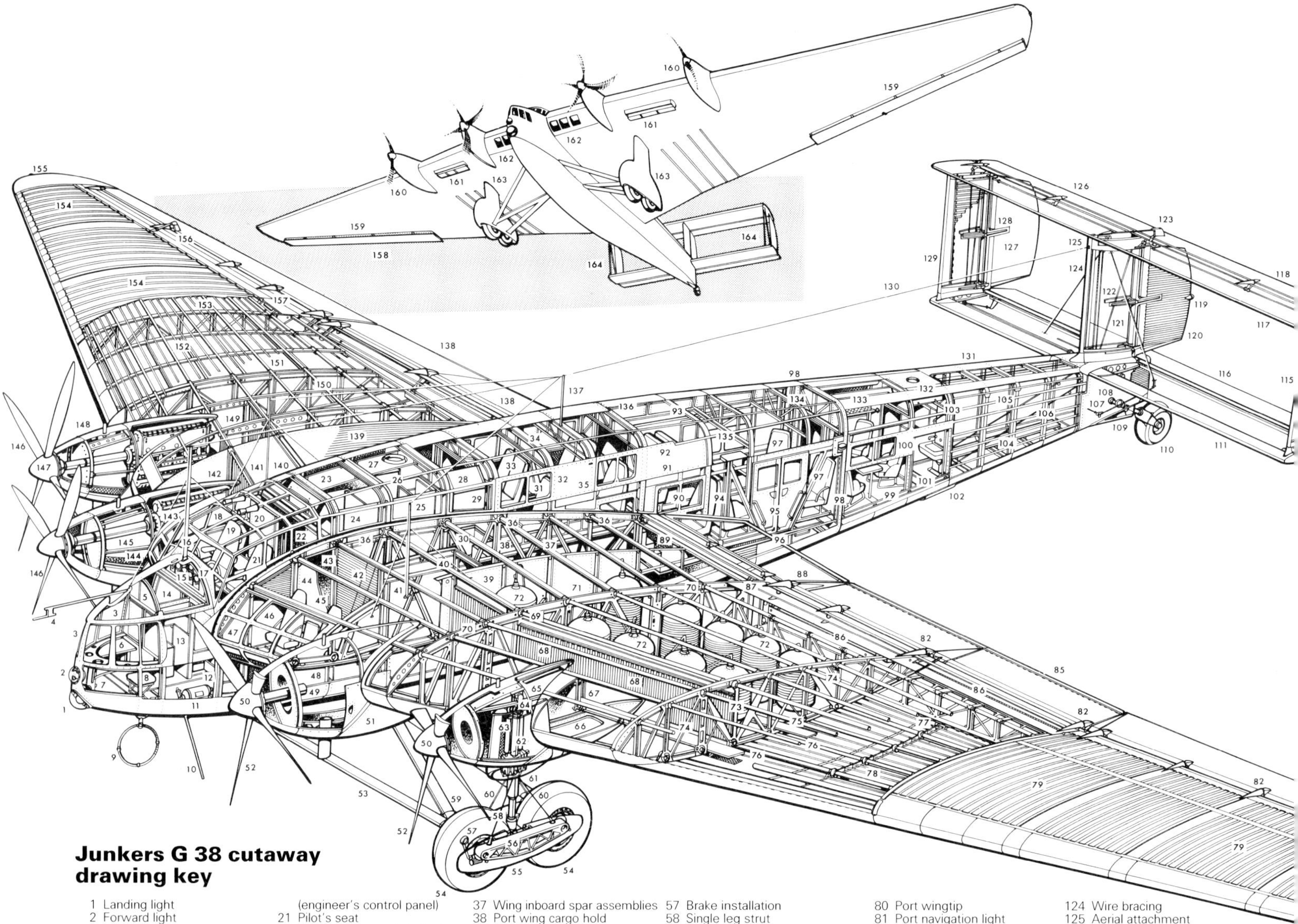

Junkers G 38 cutaway drawing key

1. Landing light
2. Forward light
3. Nose glazing
4. Pitot head
5. Bulkhead
6. Navigator's table
7. Nose observation station
8. Navigator's seat
9. D/F loop
10. Trailing aerial lead
11. Aerial motor
12. Port gangway (nose/flight deck)
13. Wireless operator's station
14. Flight engineer's station
15. Instrument panel
16. Forward aerial mast
17. Pilot's control wheels
18. Flight deck windows
19. Co-pilot's seat
20. Flight deck rear wall (engineer's control panel)
21. Pilot's seat
22. Engineer's central station/gangway
23. Starboard cupboard space
24. Buffet area
25. Kitchen
26. Washroom/toilet
27. Dorsal porthole
28. Port cupboard space
29. Fuselage frame
30. Stairwell to upper passenger cabin
31. Single seats (port)
32. Upper cabin windows
33. Double seats (starboard)
34. Upper passenger (non-smoker) cabin
35. Frame assembly skinning
36. Wing/fuselage multi-joint attachments
37. Wing inboard spar assemblies
38. Port wing cargo hold
39. Partition wall
40. Door to cargo hold
41. Port wing aerial mast
42. Port wing passageway
43. Fuselage/wing cross corridor
44. Port wing observation cabin
45. Three seats
46. Wing nose glazing
47. Nose rib profile
48. Wing lower spar/fuselage join
49. Propeller shaft
50. Spinners
51. Port inner nacelle
52. Four-bladed propellers
53. Port undercarriage forward strut
54. Port twin mainwheels
55. Axle fairing
56. Elektron axle assembly
57. Brake installation
58. Single leg strut
59. Diagonal brace strut
60. Stays
61. Undercarriage rear strut
62. Elektron shock-absorber loom
63. Twin upper legs
64. Undercarriage/wing rib attachment
65. Port outer nacelle
66. Engine removal/access bay
67. Support frame
68. Flight mechanic's inner walkway
69. Wing inner/centre section join
70. Lattice rib assembly
71. Spar joints
72. Individual fuel tanks
73. Rib structure
74. Wing centre/outer section join
75. Upper spar assemblies
76. Internal service walkways
77. Lower spar assemblies
78. Inner stringers
79. Corrugated wing panels
80. Port wingtip
81. Port navigation light
82. Flap hinge fairings
83. Flap actuator
84. Flap outer section (serves as aileron)
85. Flap centre section
86. Control runs
87. Linkage
88. Flap inner section
89. Upper/rear passenger cabin stairwell
90. Rear (lower) passenger cabin windows
91. Fuselage longeron
92. Double seats (starboard)
93. Frame assemblies
94. Single seats (port)
95. Entry door
96. Door support frame
97. Individual armchairs
98. Fuselage construction break
99. Four-seat smoker cabin
100. Aft cabin windows
101. Aft washroom/toilet
102. Ventral skinning
103. Aft bulkhead
104. Service walkway
105. Control runs
106. Fuselage aft structure
107. Tailplane bracing wire attachment
108. Tailwheel attachment assembly
109. Tailwheel leg fairing
110. Tailwheel
111. Port lower tailplane
112. Elevator horn
113. Port outer rudder
114. Rudder hinge
115. Port outer fin
116. Lower elevator
117. Port upper tailplane
118. Upper elevator
119. Tail light
120. Centre rudder
121. Centre fin
122. Rudder hinge
123. Upper elevator actuator
124. Wire bracing
125. Aerial attachment
126. Elevator hinge fairing
127. Starboard outer rudder
128. Rudder hinge
129. Starboard outer fin
130. Aerial
131. Dorsal decking
132. Aft washroom/toilet dorsal porthole
133. Smoking cabin luggage rack
134. Partition bulkhead
135. Starboard window frames
136. Lower/upper passenger cabin partition
137. Midships aerial mast
138. Starboard flap sections
139. Corrugated wing skinning
140. Starboard wing passageway
141. Starboard wing aerial mast
142. Starboard inner nacelle
143. Starboard inner engine (750 hp Junkers Jumo 204 Diesel)
144. Starboard wing nose observation cabin windows
145. Propeller shaft
146. Four-bladed propellers
147. Spinner
148. Starboard outer engine nacelle
149. Flight mechanic's station
150. Fuel bays
151. Wing upper spar assemblies
152. Stringers
153. Wing rib structure
154. Corrugated skin panels
155. Starboard navigation light
156. Flap actuator
157. Flap hinge fairings
158. Original G 38 configuration; note items 159-164
159. Single section ailerons
160. Two-bladed outer propellers
161. Underwing radiators
162. Wing nose cabin lower glazing
163. Undercarriage spat fairings
164. Inverted V outer tailfins struts

The second G 38 was later impressed into the Luftwaffe for transport duties, by which time it had acquired four L 88a engines, each driving a four-bladed propeller. It is seen here at Celle in 1941, still wearing its name Generalfeldmarschall von Hindenburg.

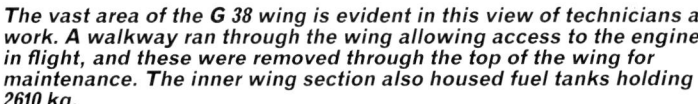

Passengers in the G 38 flew in some comfort. This is the view looking aft in the lower cabin. Beyond the small flight of stairs is the aft smoking cabin. This front lower cabin accommodated 11 passengers, as did the upper front cabin.

A major operation was undertaken to remove an engine of the G 38, these having to be lifted out of the wing by crane and pulley. This is one of the outboard L 8 engines, which produced 440 hp at their maximum setting. The inboard L 88s produced 700 hp.

The vast area of the G 38 wing is evident in this view of technicians at work. A walkway ran through the wing allowing access to the engines in flight, and these were removed through the top of the wing for maintenance. The inner wing section also housed fuel tanks holding 2610 kg.

The G 38ce in flight, complete with L 88a engines driving four-bladed propellers. In addition to the main cabins, passengers were also housed in the leading-edge observation cabins, and on occasion two more were seated in the nose, previously occupied by the navigator.

The surviving G 38 served with KGrzbV 172 on transport duties, initially in Norway and later in the Balkans. It was unfortunately destroyed by the RAF in an air raid on Tatoi, Greece, on 17 May 1941.

from time to time. There was some evidence that the second G 38 did pick up a little traffic from rivals on international routes because of its size and consequent feeling of safety. From the publicity viewpoint, the media became polarised around the cabins in the wings, though as already noted by the 1930s this had reluctantly been dropped by Junkers and was never resurrected. Digressing slightly, until after World War II artists' impressions of fictitious giant airliners of the future invariably featured passenger cabins along the leading edge, even though the only actual experience of such accommodation had not been wholly positive.

In 1932 both G 38s were returned to Dessau and fitted with four L 88a engines, all driving four-bladed propellers made up from a pair of two-bladed propellers arranged at 90° one behind the other. The first aircraft was then redesignated as the G 38a and the second as the G 38ce. When they returned to the airline they also carried names, the G 38a becoming *Deutschland* and the G 38ce *Hindenburg*, the latter being changed to *Generalfeldmarschall von Hindenburg* in April 1933. A little later the registration numbers were changed to letters, in conformity with new national policy, the 38a becoming D-AZUR and the 38ce D-APIS. The rudders, previously doped silver, also began to display national insignia. The 38a began by showing the national black/white/red horizontal stripes facing outwards and the new swastika flag facing inwards on the outboard rudders. The 38ce soon displayed the swastika flag in all four locations.

In 1934 the 38a was completely rebuilt almost to the same standard as its sister, at the same time receiving propellers with all four blades in the same plane, and many minor changes including the addition of a very long braced pitot/static head projecting ahead of the starboard wing.

Radical re-engining

One of the biggest changes made to the 38a in 1934, and carried out on its sister in the same year, was to remove all the L 88a engines and replace them with Junkers Jumo 204 diesels. The designations of the two aircraft were not changed. Junkers had been developing diesel engines since just after World War I, and had settled on the two-stroke type using opposed pistons oscillating towards and away from each other inside extremely long cylinders, with crankshafts at the top and bottom of each engine and a long row of gears taking the drive to the single propeller shaft. The objectives were, primarily, excellent fuel economy and, secondly, the ability to run on heavy fuel oil, which was considered to improve safety. The G 38 had never been intended to have particularly long range, but fitting the diesel engines almost doubled the range, from 1900 km to 3500 km. Each rated at 750 hp, the Jumo 204 engines also increased the cruising speed from about 180 to 208 km/h. They were reasonably quiet in operation, but neither Junkers nor the airline was ever able to eliminate the four trails of black smoke that thereafter followed the monsters. Later-model Jumo 205 and 207 diesels were a little better in this respect.

In 1936 the G 38a, D-AZUR, crashed during a take-off at Dessau. Though it was repairable, DLH and Junkers did not bother. The cost would have been considerable, and by this time the G 38 had become an obsolescent aircraft by comparison with the outstanding Ju 52/3m and the speedy He 70, Ju 86 and Ju 160.

A short time in uniform

The G 38ce, however, soldiered on, although its utilisation was not particularly intensive. In September 1939 the airline may even have been rather relieved when it was impressed into the Luftwaffe. The aircraft itself was hardly altered, apart from being equipped to Luftwaffe standards with radio and operational equipment, and painting in service colours of 70/71 (dark green and black green) camouflage above and colour 65 (very light blue) on the underside and rudders. It was assigned to KGrzbV 172 and received unit code letters GF-GG. All this made it quite a handsome-looking aircraft, and it saw plenty of use. In April 1940 it took a complete military brass band to Oslo Fornebu during the invasion of Norway. In July it appeared at Paris Le Bourget and several other airports in the newly-occupied countries, and in the spring of 1941 it made quite long flights to support the campaign in Greece. On 17 May 1941 it was destroyed in an RAF bombing attack on Athens Tatoi aerodrome. This time there was no question of repair.

What has received little publicity is that Junkers schemed a bomber version of the G 38, as the K 51, though this was never built. However, the design attracted the interest of the Imperial Japanese Army, and Mitsubishi were instructed to obtain a licence. The Japanese company modified the design into the Ki-20, or Army Type 92 Super Heavy Bomber, powered by four 800-hp Jumo 88a engines and carrying a bombload of up to 5 tonnes. At 25488 kg it was heavier than the airliner and cruised at little over 160 km/h. The crew numbered 10, and five gunners manned eight 7.7-mm guns and one 20-mm cannon. Six of these giants were built by Mitsubishi in 1931-35, and used to form a special heavy bomber unit. So far as is known this did not take part in raids on China, and the Ki-20 was relegated to transport duties before Pearl Harbor in December 1941.

The second G 38 was also re-engined with Jumo 204 engines (in 1935), gaining the registration D-APIS in 1934. This view shows to good effect the 'double-wing' flaps and ailerons that characterised many of Junkers' designs.

LEARJET
Best-selling Biz-jet

Learjet claims that its aircraft have appeared in more movies and TV programmes than any other aircraft in history, and the Learjet name is one of the few in aviation recognisable enough to find its way into a rock lyric. Behind this popular image is the hard fact that the Learjet has been the world's best-selling business jet since the concept was invented, despite a turbulent corporate history.

A distinction borne by the Learjet is that it is the only aircraft in large-scale production outside the Soviet Union to carry its designer's name. The late William P. Lear Sr was no stranger to achievement before the Learjet flew. He had made his mark in electronics, and won the Collier Trophy in 1950 for designing the first autopilot suitable for jet use. In 1953, Lear became directly involved in the nascent market for business aircraft by converting Lockheed 18 airliners into Learstar transports.

Piston-engined transports such as the Learstar and the British de Havilland Dove introduced many US companies to executive aircraft. With jet airliners in service, the corporate operators did not need to be sold very hard on the advantages of higher speeds and much higher cruising altitudes. Small, high-subsonic jets held no technical mysteries. Between 1957 and 1962, these factors resulted in the appearance of a swarm of new jets from many of the world's major aircraft companies.

In the late 1950s, Bill Lear conceived a different kind of business jet, based on minimum size, maximum power-to-weight ratio, high performance and low cost. It would weigh under 5670 kg (12,500 lb), putting it into the less restrictive light-aircraft category (Part 3 of the Federal Aviation Regulations) for certification purposes, and greatly simplifying development. The aircraft would be designed and built in Europe, where costs were lower, and the bare airframes would be shipped to the USA for completion and delivery. Lear chose Switzerland as a European base, forming the Swiss-American Aircraft Company (SAAC) in April 1960.

At that time, the Swiss aircraft company FFA (Flug-und Fahrzeugwerke AG) was attempting to save its P-16 attack fighter from cancellation. Lear was impressed by the P-16, and particularly by its unswept, thin-section, high-aspect-ratio wing. It had been designed to combine high subsonic performance with low landing speeds, and used an innovative and efficient system of construction with few ribs, multiple spars and thick skins. Lear enlisted the help of FFA chief designer Dr. Hans Studer in outlining SAAC's new business jet around a modified P-16 wing. Because the new jet was the third twin-engine aircraft which Dr. Studer had designed, it was named the SAAC-23.

The American designer Gordon Israel was instrumental in producing the final shape of the SAAC-23, its characteristic wedge-shaped nose recalling Israel's Grumman Tigercat and Panther. Compared with its contemporaries, the SAAC-23 was striking for its small size and slenderness, accentuated by remarkable detail cleanliness. The big two-piece windscreen, with a single slender pillar, remains one of the most dramatic single styling features in aviation. The SAAC-23 was small inside as well, with just 1.37 m (4 ft 6 in) of headroom. The wing lost some of its relationship to that of the P-16 during design, being slightly thicker and longer in span, with a fixed leading edge rather than nose flaps. The dense structure left little room for fuel, so the SAAC-23 had tip tanks. The empennage started as a T-tail with a very short, fully powered rudder and elevator, and was changed first to a larger design with manual controls and a cruciform layout to save weight. The final T-tail, with a swept tailplane, was adopted to increase control power and allow a higher cruising Mach number.

General Electric provided a non-afterburning, civil-certificated version of its J85 engine, designated CJ610, for the new aircraft. The J85 was already in production for the USAF's Northrop T-38A Talon

The Gates Learjet in its various forms has found its way into the inventories of several air arms around the world. Typical are two Model 25Bs operated by the Peruvian air force for survey work on behalf of the Directorate General of Photographic Air Survey and Mapping from Las Palmas.

The high-speed and fighter-like handling of early Learjets suited them to a number of roles. A great deal of target-towing for the Swedish air force is carried out by the highly modified Model 24Bs of Swedair. A hit-miss indicating system is carried under the forward fuselage, and a dart target is carried under the tail.

supersonic trainer, and its most remarkable feature was its excellent 7:1 thrust-to-weight ratio. As development progressed, experience with the T-38 permitted a 20 per cent increase in power to match the higher speed potential of the redesigned tail.

The only contemporary business jet to match the SAAC-23's design cruising speed was the much more sophisticated, heavier and more expensive Dassault Mystère 20. Lear's aircraft would also have the highest cruising altitude of any business jet, and a higher initial climb rate than a North American F-100 fighter.

Systems were kept simple, to reduce costs and improve reliability. All controls except the flaps and spoilers were manually operated; the fuel system used 'jet pumps' with no moving parts, and the hydraulics were backed up with lighter electric and compressed air power rather than being fully duplicated. Many of the electronic systems (including autopilot, primary flight instruments and gyro reference systems) were of Lear's own low-cost design, and were to be built by an associated Lear company.

Plans called for the first five SAAC-23s to be completed in Switzerland, and shipped to the USA for final assembly and finishing. FFA was to build the wings and tip-tanks, and the rest of the airframe was to be manufactured by Heinkel in Germany. Engines, electronics, landing gear and other system components were to be imported from the USA. International aircraft development and manufacture, as opposed to licence production, is commonplace now, but it was not so in 1961; the SAAC-23 programme began to bog down in unforeseen administrative complications.

Lear was not the man to be patient with such problems. Late in 1962, he transferred the entire programme, tooling, parts and all, to Wichita, Kansas, and opened shop as the Lear Jet Corporation. Completion of the first Lear Jet Model 23, as the aircraft was then designated, was inevitably delayed by the exodus from Europe, but the entire programme was back on track by the middle of 1963. The Model 23 rolled out in September and flew on 7 October 1963.

The second aircraft flew in March 1964, and the test programme was carried through to FAA certification with record speed, despite the loss of the first aircraft in a non-fatal take-off accident in June 1964. A few minor changes were needed: the leading-edge profile was revised to improve the stalling behaviour, and other modifications were made to improve wing aerodynamics. Like many other T-tail designs, the Learjet was fitted with a 'stick-shaker' and 'stick-pusher' system to help the pilot avoid a locked-in deep stall. With the third aircraft joining the test programme in May 1964, and a final, accelerated effort, flying totalled 300 hours when the type certificate was issued under Part 3 regulations at the end of July.

The 1961 target price for the SAAC-23 was $350,000, but with the move to the USA and the investment of $4 million in the new factory, this figure gradually increased. Also, the standard specification was steadily improved, with an accompanying increase in price. The first production aircraft were offered at $575,000, fully equipped and furnished, a clear $150,000 less than its cheapest competitor.

Lear had been correct in concluding that speed and altitude were what business jets were all about. The Learjet not only offered more of both for the money than any competition, but looked the part as well. That, in a nutshell, was the reason why 94 Model 23s had been ordered by early 1965, putting the fledgling company in the lead of the business jet market. By the middle of the year, Lear was building 10 Model 23s a month.

The price of success

The Model 23 was only 18 months into its production life before it was replaced by the externally identical Model 24. The new model had been designed to meet the more stringent FAR Part 25 rules for public-transport aircraft. Changes were individually minor but significant in sum; an example was a strengthened, birdproof windscreen, not needed under FAR Part 3 but required by Part 25. Production of the Model 24 started in March 1966.

The new certification basis cleared the Learjet for weight growth beyond 5670 kg (12,500 lb), opening the way to development of the stretched Model 25; this version flew in August 1966, and was certificated in October 1967. It was heavier than the Model 24, had a 50 per cent bigger cabin and a slightly shorter range.

Despite the Learjet's success and, to some extent, because of it, all was not well with the programme. Production had been accelerated so quickly that aircraft were being delivered 'off the top of the learning curve': workers had not had the time to become familiar with their jobs, and more manhours went into each aircraft as a result. The market was highly competitive, and Lear's bigger rivals could better afford to cut prices than the small new company. Lear had also chosen the traditional way of selling private aircraft: a dealer network. But dealers' 'territories' became meaningless when they could cross the continent in a single day. Competition between Learjet dealers put even more pressure on the Learjet's price, while its competitors were distributed through centralised factory sales organisations.

Meanwhile, Lear added new divisions and new activities to the company. But Learjet production was simply not profitable enough to carry the load of the subsidiaries, and the balance sheets began to look steadily worse. In April 1967, Bill Lear sold a controlling interest in his company to the Gates Rubber Company of Denver, and disputes over the company's continuing failure to turn its results around culminated in his resignation in April 1969. At the end of the year, the company's name was changed to Gates Learjet Corporation.

A Gates Learjet marketing division was set up in Denver, but the 800 km (500 miles) between Denver and Wichita proved too long for effective management, or for the development of a working re-

The first Learjet Model 23 makes its maiden flight from Wichita on 7 October 1963. Relatively few changes were needed during development of the original Model 23, which, unlike its contemporaries, was certificated under light-aircraft regulations. The single large, oval window on each side distinguished the early Learjets, but was dropped later due to changing airworthiness requirements.

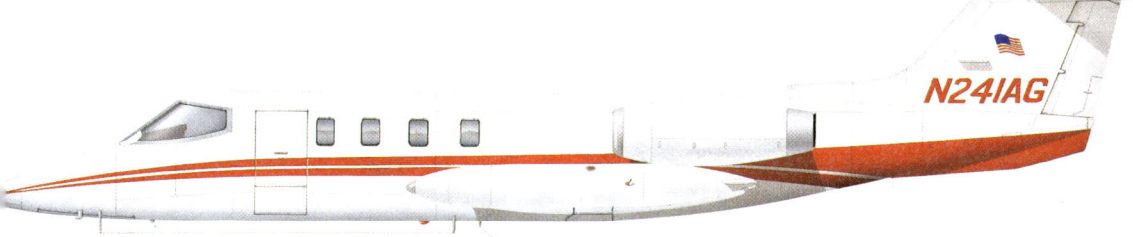

The Model 25C marked Learjet's move into the market for transcontinental-range jets; it was based on the original stretched Model 25, but the cabin length was reduced to make room for extra fuel in the centre section.

Freedom's Way, a standard Model 36, was used by golfer Arnold Palmer for a round-the-world goodwill flight in 1976, celebrating the 200th anniversary of the foundation of the United States.

lationship between the sales force and the manufacturing facility. Losses continued, and in 1971 Gates Rubber tried to sell the entire operation once more. Eventually, in October 1971, Harry Combs was appointed president of Gates Learjet. Combs immediately began a complete overhaul of the company.

The foundation for the company's economic recovery after 1972 was the aircraft itself, which was still selling as strongly as ever, particularly in export markets. New and improved models continued to appear. In early 1969 the Model 24 was replaced by the Model 24B, with higher take-off and zero-fuel weights, airframe anti-icing equipment and other changes. Only a year-and-a-half later, though, Gates Learjet announced a complete range of new models. The Model 24B was superseded by the heavier Model 24D, which was distinguished by the lack of a tailplane bullet, and three small cabin windows on each side, replacing the characteristic single large window of earlier Model 23s and Model 24s. A parallel type was the Model 24C, identical except for a 5670 kg (12,500 lb) take-off weight limit which allowed it to be operated under less stringent FAA operating rules. Intended to compete with new small jets like Cessna's Citation, it enjoyed little success. New versions of the stretched aircraft, incorporating the same features as the Model 24D, were the Model 25B and the extended-range Model 25C. The 25C had an extra fuel tank in the fuselage, and the internal length of the cabin was reduced to make room for it, cutting capacity to six passengers.

While most of these changes were designed to make the Learjet more attractive in the marketplace, a great deal of development was directed at a much more serious problem: Learjets were, from the start of operations, involved in a disturbingly large number of accidents. Like most aircraft, the Learjet is not in itself dangerous. But the early models were demanding aircraft, ill-suited to an inexperienced, under-trained or over-confident pilot. As the cheapest, fastest and most fashionable of business jets, Learjets and marginally adequate pilots seemed to be drawn together much too often.

Aerodynamic and systems modifications helped with the problem: the British authorities, significantly, issued their first certificate to a Learjet model in 1974. But one of the biggest steps forward did not affect the aircraft at all. In January 1972, FlightSafety International, one of the world's best regarded flight training organisations, established a training centre in Gates Learjet's Wichita facility. Since that time, FlightSafety has conducted all factory training for Learjet pilots and maintenance people, and FlightSafety-trained Learjet operators have demonstrated an excellent safety record.

Improved powerplant

With improving financial results, Gates Learjet could afford to move into the next stage of development: the incorporation of a more efficient, quieter engine. Garrett AiResearch had started work on such an engine, carefully sized to replace turbojets such as the CJ610, Viper and JT12, in the late 1960s. It was a shrewd move, because concern for the environment in general, and about the problem of aircraft noise in particular, was rapidly increasing. Garrett's new TFE 731-2 turbofan engine made its first flight on a Learjet 25 test-bed in May 1971.

The first production Learjets to use the TFE 731 were unveiled in August 1973. Both were based on the stretched Learjet 25, with increased wingspan and a further small fuselage stretch to balance the extra weight of the engines. The Model 35 was a transcontinental eight-seater, while the Model 36 had an extra fuselage fuel tank and shorter cabin, like the Model 25C, endowing it with enough range for transatlantic operations. Both were certificated in July 1974. They were more efficient than the Model 25B/C, and had greater range, but the pure-jet types were faster, had better altitude performance and were cheaper to buy and maintain.

Work on improving the basic Learjet design continued, both inside and outside the company. In October 1975, Gates Learjet announced that the entire range was to incorporate a modification to the wing leading edge camber, combined with other features including a better stall warning system, for improved handling and low-speed performance. New designations identified aircraft fitted with the Century III package, and certification of the new variants was com-

The Learjet's high-altitude capability has made it particularly attractive to military users such as the Fuerza Aérea Argentina. The service's six Model 35As can be used for liaison or survey/reconnaissance duties, with a camera pack faired into the lower fuselage.

Apparent in this view of a Swiss-registered Longhorn 55 are the new-technology 'winglets' and the wide-body fuselage intended to offer a 'stand-up' cabin with a maximum capacity of 10 passengers. Four versions of the Longhorn 55 are now on offer.

pleted during 1976. Meanwhile, two independent companies (Dee Howard and the Raisbeck Group) were working on a similar package for retrofit to older Learjets. Since mid-1979, all Learjets have been fitted with additional wing modifications, the so-called Softflite package, which further improve low-speed handling and performance. Softflite can be retrofitted to any Learjet with the Century III wing. Another Learjet first was recorded in April 1977, when the CJ610-powered Model 24 and Model 25 were certificated for a cruise altitude of 15545 m (51,000 ft), the first commercial aircraft to be cleared to cruise at such an altitude.

In mid-1977 Gates Learjet announced the development of a new family of Learjet developments. The latest aircraft would represent a clean break with the original design philosophy, featuring a much enlarged cabin with stand-up headroom. Power would be provided by uprated versions of the TFE 731, and lift by a highly modified version of the basic Learjet wing. This was the first on any production aircraft to incorporate the wing-tip vertical surfaces known as 'winglets'. Devised by Dr. Richard Whitcomb of NASA, winglets are designed to work in the vortex flow which exists at the tip of the wing, turning the flow to provide a small forward thrust, or drag reduction. The wingspan was also considerably increased, and the wingtip fuel tanks were eliminated. The new wing, named 'Longhorn' by the company, was tested on a Learjet 25 in August 1977, and Gates Learjet decided to offer a version of the test aircraft commercially while development of the definitive Learjet 50 series continued. This was marketed as the Longhorn 28 and 29 (with different fuel capacity and internal cabin length) and both versions were certificated at the end of 1978. Only a few Longhorn 28s and Longhorn 29s were sold, and the version is now out of production. A successor is the Learjet 31, combining the 28/29 wing with the 35/36 fuselage and TFE731 turbofans. It has large ventral strakes known as 'Delta Fins'. The prototype first flew on 11 May 1987, shortly before the company was renamed Learjet Corporation after Gates sold out.

Longhorn 50

The first Longhorn 50 was flown in April 1979, and the type was certificated in March 1981. As originally planned, three versions were to be offered: the large-capacity Model 54, with a smaller fuselage fuel tank; and the Model 55 and Model 56, basically similar but with a higher gross weight and fuel capacity for the latter model. Demand for the Model 54 proved limited, however, so the Model 55 became the basic model, with extended-range (Model 55ER) and long-range (Model 55LR) sub-types.

With so many Learjets in service, the type continued to prove attractive for the modification industry. In the late 1970s, the Dee Howard company developed a performance improvement system for the Learjet 24. The most important of many changes in the XR package was a swept, extended-chord leading-edge glove which both reduced drag and increased internal fuel capacity. In 1980, Gates Learjet and Dee Howard agreed to offer the same modifications on new Learjets, and the resulting Learjet 25G was certificated in early 1982.

The Learjet has been used for a number of roles apart from corporate transport. Its high operating altitude makes it a useful photo-survey platform, and a number of South American governments have acquired suitably equipped Model 24s. The Learjet is also in service with the Finnish air force, primarily as a target tug. Special variants on offer include the EC-35A electronic warfare platform, PC-35A maritime patroller, RC-35A and RC-36A photo survey variants, UC-35A utility machine and U-36A ECM, target tug and fleet support craft for the JMSDF. Other Learjets have been modified as air ambulances (the large entry door of the 20/30 series is a great advantage) to carry patients to their home countries or for special treatment when they are too sick or injured to travel by normal means. Howard Hughes died in one.

Military buyer

Both 1980 and 1981 set all-time records for Learjet deliveries, but like other companies Gates Learjet has been affected by the 1982-3 recession. However, sales received a tremendous boost in September 1983, when the US Air Force announced that it would lease 80 Model 35As to replace elderly Rockwell (North American) CT-39s. The decision doubled the company's planned production for 1984-5, and placed it in a uniquely favourable position to handle the foreseen recovery in the market. The US Air Force aircraft are designated C-21A, and were delivered between April 1984 and the end of 1985 eventually being bought outright in September 1986.

The Learjet line is currently headed by the Longhorn 55. Still in production at Wichita are the Learjet 35/36 and the Learjet 31. In 1990, the turbulent corporate history took yet another turn when Learjet was bought by Bombardier Aircraft, who also own the de Havilland Canada and Shorts concerns. Under Bombardier, the company has produced the 6/9-passenger Learjet 60, from 1993 the successor to the Learjet 55 with Pratt & Whitney Canada PW305 turbofans and a stretched fuselage. The company's latest offering is the 8/10-passenger Learjet 45 with Allied Signal TFE731-20 turbofans.

With original production starting in 1966, the Model 25D was produced for over 20 years, although there was a temporary cessation from August 1982 to January 1983. Illustrated is a Mexican-registered aircraft.

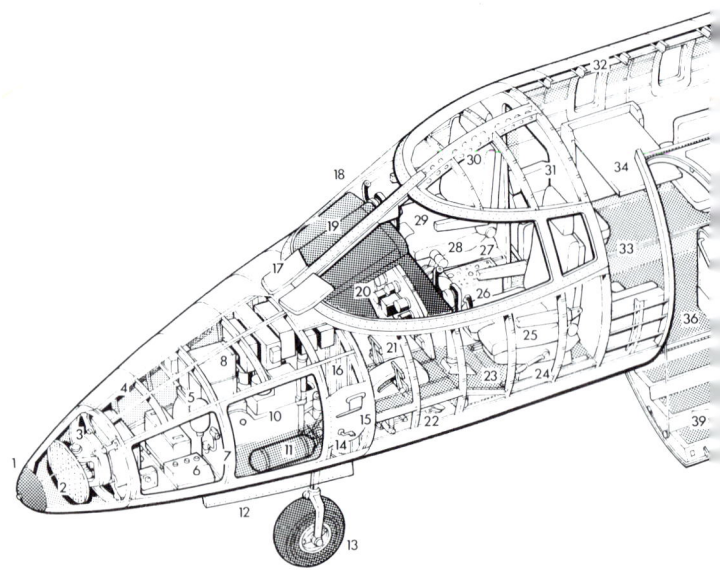

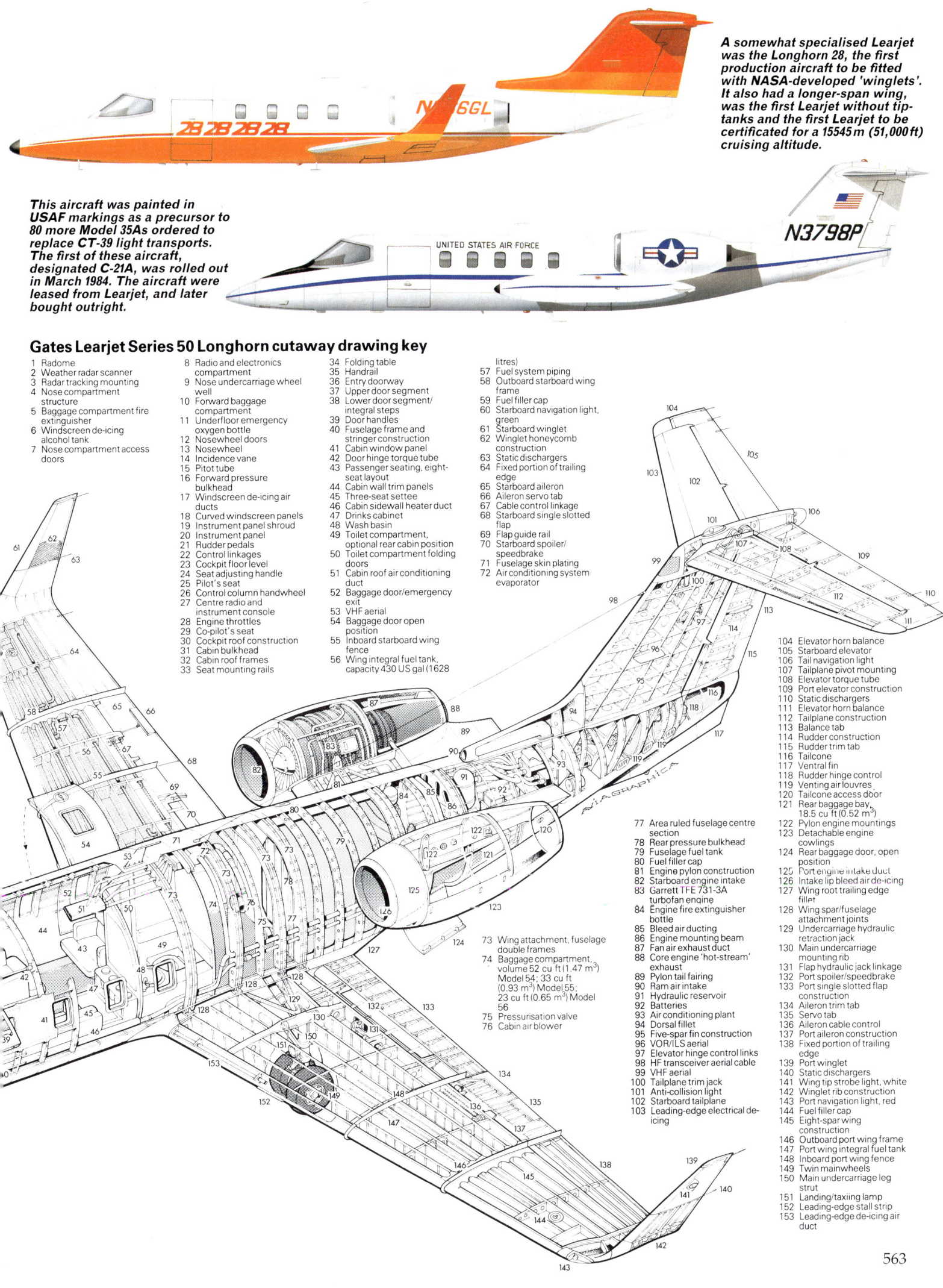

Chapter 27

Let L-410/610

Lockheed Constellation

Lockheed L-188 Electra

Lockheed 'Singles': The Plywood Bullets

Lockheed Tristar

Lockheed Twins

LET L-410/610

Czechoslovakia has a well-established aviation industry that was building aircraft when many European countries saw them at best, as frivolous novelties. Things were not helped after the war when Soviet-inspired stagnation set in, but LET's L-410 was a success story: a capable small transport which sold well. It has been followed by the LET 610, a larger and more modern aircraft, which is currently seeking a place in the 'new world order' of aviation.

Czechoslovakia had for many years produced the standard jet trainers for all former Warsaw Pact nations except Poland, as well as a range of attractive general-aviation machines. In addition, a factory at Kunovice (Uherské Hradiste) has established a solid reputation with a light twin-turboprop airliner which has been built in large numbers – 1,000 off the line and plenty more to come. This in turn has supported the development of a larger transport which seems assured of a similar long production run, and could well penetrate world markets.

The original aircraft is the L-410. It was simply designed as a versatile aircraft for local passenger and freight transport but – according to the original draft objective of 1966 – "suitable also for executive, aerial survey, radio/navigation training, ambulance and other duties." Another requirement was the ability to operate from rough unpaved airstrips.

There was never much doubt that the new aircraft, designated L-410 and named the Turbolet, would be a high-wing twin-turboprop. Equally, there was never any doubt that it would be generally simple. Much thought was given to the pros and cons of fixed landing gear, before the final choice was taken to make it hydraulically retractable. All three units have a single wheel on a long arm pivoted at the foot of a fixed leg, sprung by an oleo-pneumatic shock abosorber to give levered suspension with long wheel travel. The nose gear, with hydraulically assisted steering, retracts forwards, while the main units, with manually controlled disc brakes, fold straight inwards to lie inside stub wings in compartments closed by large hinged doors. Operation is hydraulic.

Another point for discussion concerned pressurisation and air-conditioning. In the event no such systems were fitted, yet instead of taking advantage of this by making the fuselage a simple rectilinear box, the decision was taken to make it a streamlined tube with curved sides. Thus, apart from the cost and weight advantage of not having the systems, the only gain was that it simplified the fuselage structure and made it easier to fit a row of large passenger windows and doors at front and rear on the left side, the rear door being a double unit 1.5 m (59 in) wide for loading cargo. Interior width was set at 1.92 m (75.6 in) across the widest part, well matched to 1+2 seating (single seats on the left, double on the right). Normal seating was arranged for up to 19 passengers, but various executive and special-role arrangements were provided for, including an ambulance version for six stretchers and five sitting patients.

Of course, the structure was made of simple aluminium alloy. The wing was made in one piece, 18 per cent thick at the root and 12 at the tip, with a fixed leading edge. On the trailing edge were arranged powerful double-slotted flaps, driven hydraulically, one of the few con-

A single example of the L-410AF was exported to Hungary in 1974. The extended glazed nose housed a downward-looking camera for survey work, and a seat for the navigator/operator. The modification meant that the nosewheel was non-retractable.

Accounting for a huge slice of the L-410 production run are the Aeroflot machines, the L-410MU and L-410UVP. The UVP included many detailed changes, the most notable being an increase in wing span.

cessions to slight complexity. Outboard were added plain manual ailerons, with electric trim. The slightly swept fin carried the horizontal tail above the level of the fuselage, all controls being manual, the elevator tabs being manual and the rudder tab electric. All leading edges were considered to require protection against icing, and at the outset two methods were chosen: TKS fluid or Kléber-Colombes pneumatically-inflated boots.

Homegrown engine

Mainly to power the L-410 the Czech Motorlet factory designed a new turboprop, very much in the class of the Pratt & Whitney Canada PT6A engine, and with the same reverse-flow layout. This new engine, the M 601,

The LET 410 has proved an attractive and reliable light transport, and a good basis for further development. Despite its rounded-section fuselage, the LET 410 is surprisingly unpressurised.

was not ready in time for the prototype L-410, which was accordingly fitted with 715-ehp PT6A-27 engines. The fuel system, however, was standard, with eight bag tanks in the wings with total capacity of 1300 litres (286 Imp gal), filled through four traditional gravity refuelling

This is the first prototype LET 410, which initially flew from Kunovice on 16 April 1969. As its indigenous engines were not available in time, it was powered by Pratt & Whitney Canada PT6As.

points above the wing. Electric de-icing was provided for the engine inlets and for the propellers. Both Hamilton Standard and Hartzell propellers were tested, and offered to customers, with a diameter of 2.49 m (98 in), with three feathering and reversing blades.

First off the line

The prototype, registered OK-YKE, made a successful first flight at Kunovice on 16 April 1969. Designated the XL-410, this carried out the basic handling and performance testing, but was followed by three further PT6A-engined prototypes. The second of these, designated L-410AB, was fitted and later flown with Hartzell four-bladed propellers which resulted in reduced cabin noise and vibration levels. The initial production version, the L-410A, retained the PT6A-27 engine and three-bladed propeller. This version entered service with Slov-Air at Bratislava in late 1971. A total of 27 was built, not including the four prototypes. Other PT6A-powered aircraft were planned, but only

one was built, a single L-410AF photo-survey aircraft. This had a considerably enlarged and glazed nose, housing a vertical camera with a seat for the operator; the latter required that the nose gear be locked down at all times. This aircraft was exported to Hungary in 1974.

Flight testing of the L-410M, powered by the Czech M 601 engine, began in 1973. This engine originally gave 410 kW (550 shp), less than the planned figure, and most of the production M-series aircraft were powered by the enlarged M 601B engine, rated at 515 kW (691 shp) driving an Avia V 508B three-bladed reversing propeller. On hot days a limited supply of injection water was available with which take-off power could be increased to 790 ehp. Aircraft fitted with the M 601B were designated L-410MA, and this version incorporating detail changes required by Aeroflot was designated L-410MU. Between 1976 and 1979 a total of 110 M, MA and MU aircraft were delivered, the majority being of the MU type.

By far the greatest market for the Turbolet was the Soviet Union, where – together with the very similar An-28 – the L-410 was selected as the standard general-purpose light transport for local service. Accordingly, in the mid-1970s the LET design team, initially under Ladislav Smrcek and later Vlastimil Mertl, carried out a complete detail redesign to adapt the aircraft exactly to the requirements of Aeroflot. The result was the L-410UVP, the first of three prototypes which made its initial flight on 1 November 1977. In 1980 the UVP became the first non-Soviet aircraft to be certificated to the Soviet NLGS-2 airworthiness regulations. One of the requirements was the ability to operate for long periods in ambient temperatures ranging from 45°C (113°F) down to −50°C, with the systems still operable at −60°C.

Aeroflot modifications

The engines remained M 601B turboprops, but driving slightly modified VJ8 508B propellers with beta (direct pilot) pitch control. The wing was increased in span from 17.48 m to 19.48 m (57 ft 4 in to 63 ft 11 in), area rising from 32 to 35 m² (355 to 375 sq ft). The fuselage was likewise lengthened by 47 cm (18.5 in) to give an overall aircraft length of 14.47 m (47 ft 5.5 in). The tail surfaces were all enlarged, and the tailplane given moderate dihedral. The wing was fitted with spoilers ahead of the flaps, while immediately in front of each aileron was added a pop-up hinged strip which, during single-engined flight, automatically rises on the side of the running engine to decrease lift and keep the wings level. The elevators and rudder were redesigned with fabric covering, and all leading edges of both the tail and wings were fitted with the French pneumatic de-icing system as standard. Auto propeller feathering was added, together with a Moravan Otrokovice anti-skid system for the wheel brakes. A new option was metal ski landing gear, with a non-stick plastics undersurface. Not least, the cockpit instrumentation and systems were improved and augmented.

All these changes increased the basic empty weight from 3100 kg (6,834 lb) to 3725 kg (8,212 lb), so that even though maximum weight was increased from 5700 kg (11,905 lb) to 5800 kg (12,786 lb) the maximum payload tended to be reduced. Performance was not greatly altered, though cruising speed tended to be slightly lower, at 300 km/h (186 mph). Most production UVPs were equipped with 15 passenger seats, though a few have been freighters, or ambulances (as previously described), or equipped for 14 parachutists and a dispatcher or for 12 firefighters and an observer (who can serve as second pilot).

The L-410UVP was adopted by Aeroflot principally as a light passenger transport to serve short regional routes in outlying areas. In this role it carries 15 passengers and light freight.

Numerous Turbolets have been used by Czech government organisations and factories for light transport, VIP and utility work. This L-410AB is retained by the manufacturer for transport work.

The latest and hottest variant is the L-410UVP-E, with powerful M 601E engines driving five-bladed propellers. STOL performance is exceptional, and range is increased by the addition of wing tip-tanks. This is LET's demonstrator.

Another aircraft flown by LET at Kunovice is this L-410MA, which features the uprated M 601B turboprop. This version is similar to the first for Aeroflot, the L-410MU.

Production of the UVP, in four sub-types, took place in 1980-85, and totalled 495.

Five-bladed propeller

Further refinement led to the UVP-E, which first flew on 30 December 1984 and succeeded the original UVP in production the following year. One of the principal differences in the -E version is that it is powered by Motorlet Walter M 601E engines, each rated at 750 shp and with 809 shp available for short periods with water injection. These uprated engines drive Avia V 510 five-bladed propellers, with reversible pitch, auto-feathering and beta control. The powerplants incorporate new oil coolers and oil/fuel heat exchangers. Fire-extinguishing systems are redesigned, the public-address system is improved and portable oxygen is provided. The wings are restressed to carry a fixed tank on each tip with a capacity of 200 litres (44 gal), increasing range by some 40 per cent. Maximum flap deflection is increased, to hold approach speed roughly unchanged despite increased weights, while spoiler angle is fixed at 25° in flight and 72° on landing, to dump lift. In the rear of the fuselage the toilet and baggage compartment are moved further aft, making room for four additional seats without increasing fuselage length. Thus, standard configurations in the UVP-E include 19 passengers, 18 parachutists or 16 firefighters. Overall capability is greatly enhanced, the additional capacity being matched by the greater power and increased take-off weight of 6400 kg (14,110 lb).

Production of the UVP-E has continued from 1987 to date, the total by 1 January 1990 being 287, and rising to about 340 a year later. Thus, by 1991 total production of all versions of the L-410 had reached 990, with the 1,000th flying early in the year and ongoing requirements for several hundred more, almost all of them from Aeroflot. Small numbers have been produced of special variants, including at least one additional machine for Hungary equipped with magnetometers and other sensors, on the rear fuselage and in wingtip pods, and with a large observation blister in the side of the cockpit, for photogrammetry mapping.

The new arrival

The solid success of the various UVP versions in Aeroflot service has now led to LET being funded to produce a larger transport as a possible replacement for the An-24 and Yak-40, in the 40-seat class. The result is the L-610, the prototype of which, OK-130, made its first flight on 28 December 1988. Though it naturally followed the layout established with the L-410, it was not only bigger and more powerful but also differed in several major respects. The most obvious change is the adoption of a T-type tail. Another obvious change is that, though both span and length are increased, the ratio of fuselage length to wing span has greatly increased. Less obvious is that the fuselage is a pressurised tube of circu-

Slov-Air is the main internal aviation organisation of Czechoslovakia, and was the first recipient of the LET 410. The initial aircraft involved were L-410As and ABs powered by PT6As.

In service with the Polish coast guard in small numbers is the LET 410UVP-K2. Carrying the service's anchor emblem on the nose, these aircraft can be fitted with a camera for survey work. Like all LET 410s it is capable of operating in extreme weather.

lar section, with external width increased from 2.08 m to 2.7 m (106.3 in), matched to 2+2 seating (a double seat on each side of a central aisle). Thus, normal seating is increased to 40, in 10 rows of four.

Maximum pressure differential is 30 kPa (4.35 lb/sq in), a modest level but good enough for cruising heights in the region of 7200 m (23,620 ft), which is the brochure optimum, and to maintain sea level pressure to about 3000 m (10,000 ft). On-board systems are completely redesigned and upgraded. Hydraulic pressure is increased from 14700 to 21000 kPa (2,133 to 3,045 lb/sq in), electric power is massively increased and generated as raw AC by two 25 kVA alternators, and there is an APU (auxiliary power unit) in the tailcone. The main engines are Motorlet Walter M 602 turboprops, each rated at 1360 kW (1,822 shp) and driving an Avia V 518 five-bladed propeller. This would have been the standard engine for aircraft for Eastern bloc countries, but in 1991 LET flew an L-610 powered by 1300-kW (1,750-shp) General Electric CT7-9B engines, and this would probably be the preferred engine for the rest of the world market.

New wing

The flexible bag tanks of the L-410 are replaced by the simplest possible integral tanks, one area in each wing (left and right) being sealed to have a capacity of 3500 litres (770 Imp gal). All leading edges have pneumatic boot de-icing. All flight control surfaces have gone back to being metal-skinned, with manual control. The narrow pop-up strips ahead of the

The LET 610 was intended, like its smaller brother, to serve in large numbers with Aeroflot. Given that no such order is now likely, the L-610 has become the L-610G with Western avionics and engines (the latter the General Electric CT7-9D rated at 1305 kW/ 1,750 shp). LET is now controlled by the Ayres Corporation of Georgia, USA, which is looking for alternative markets.

The LET 410UVP-E is the air ambulance version and can be configured to carry six stretchers and five sitting patients.

ailerons for asymmetric flight are deleted, and the flaps are of Fowler type. Bought by Ayres in 1998, Let secured certification of the L-610G late in 1998.

Lockheed Constellation

Though it initially suffered from protracted technical problems, the 'Connie' became the best-loved piston-engine airliner of all and, in its makers own words, was 'Queen of the Skies'. Finally, while commercial Connies were slowly rotting away, military examples of a dozen species were working unnoticed around the clock.

In its day the Lockheed Constellation was the biggest, most powerful and most expensive of all airliners. But it avoided joining the list of unsuccessful giants, because at first its capacity was not so great as to frighten the airlines. The 'Connie', as it was affectionately known, was made possible by the development of engines of great power, and this power was used for speed, and to lift fuel for long range whilst cruising in pressurised comfort at high altitude. Once the basic type was established, Lockheed met the demand for greater capacity by introducing one of the first and greatest of all 'stretching' programmes to yield the 'Super Connie', seating up to 100 or more.

But in 1938 this could not be foreseen. On 23 June of that year the McCarran Act had transformed US commercial aviation, and the manufacturing industry had rationalised into the same three names that dominate it today, Boeing, Douglas and Lockheed. The first two had already built the big four-engined DC-4 and pressurised Model 307 Stratoliner respectively, but Lockheed was overloaded with small twins and the military Hudson and P-38, and its promising Excalibur remained a succession of mock-ups. This was tough, because Lockheed had a strong leaning towards powerful, fast aircraft, and had pioneered pressure cabins with the XC-35 flown in May 1937. Nothing much could be done until suddenly on 9 June 1939 the company was visited by the famed Howard Hughes, who had secretly bought most of the stock of TWA, and Jack Frye, whom he had appointed president.

Hughes had lately given Lockheed a giant boost by flying a Model 14 airliner around the world in record time. TWA was in severe trouble with money and route competition, and Hughes urged the development of a new super-luxury transport that could fly nonstop coast-to-coast across the continental USA. The specifications sounded out of this world (empty weight 24132 kg/52,300 lb, four 1641-kW/2,200-hp engines, cruising speed over 483 km/h/300 mph and the ability to fly from New York to London nonstop), and all Lockheed's rivals gave Hughes a thumbs-down. But Lockheed's Bob Gross called in his top designers, Hall L. Hibbard and Clarence L. 'Kelly' Johnson, and said, "Come up with something". Long meetings took place at a secret Hughes place on Romaine Street, Hollywood, and before the end of the year a formal meeting took place at which Hughes asked the price and was told $425,000. He rocked back and forth like an Indian at a pow-wow and said, "Hell, TWA can't pay, the damn airline's broke. Go ahead and build 40. I'll have to pay for them myself."

C-69 for the US Army

Wartime pressures delayed the Model 49 Constellation, but eventually Lockheed hired pilot Eddie Allen to take the first one aloft on 9 January 1943. No commercial production was allowed after Pearl Harbor, and TWA's idea of a commercial lead with a super

Known to Lockheed as the Model 749-79-38, the C-121A was a properly designed military transport, the C-69s having mainly been converted airliners. USAF no. 48-616 is seen here in MATS (Military Air Transport Service) markings, flying as a PC-121A passenger transport. Engines were 1864 kW (2,500-hp) R-3350-75 Duplex Cyclones.

Left: This aircraft was a Model 049 originally ordered pre-war by TWA but completed as a C-69-5. It wears typical post-war grey camouflage, having discarded the wartime olive drab. After USAF service it finally went to TWA.

Right: Several Constellations found their way to the small cargo operators of Latin America, one being Argo of Dominica. This was an L-749A.

dream ship evaporated as Pan Am came in alongside; and in the event all production went to the US Army. But even in olive drab the C-69, as it was now called, was quite something. Its wing was a scaled-up version of the wing of the P-38, with giant area-increasing Fowler flaps. The fuselage was curved like the body of a fish and ended in a triple tail. The circular-section cabin seated 64 passengers, though Hughes was bemused that the US Army could take out all the luxury and still contrive to make the C-69 heavier than the Model 49. All flight-control surfaces were hydraulically boosted. Not least, the height of the Constellation off the ground was unprecedented. In two respects, however, the civil registered NX67900, or Lockheed ship no. 1961, was conventional: it had normal cockpit windows instead of the once-planned perfectly streamlined nose, and the engines were also fairly conventional instead of being in perfectly streamlined nacelles with reverse-flow cooling from inlets in the leading edges.

During the war the USAAF received 22 Constellations, comprising nine ex-TWA and 13 of a contract for 180 signed in 1942. At VJ-Day the military contract was cancelled. Lockheed shut down the Burbank plant for five days to plan its future. It almost decided to start again with an even newer Constellation, but finally elected to buy back surplus government tooling, parts, materials and unfinished C-69s. This resulted in the commercial Model 049 having an 18-month lead over the DC-6 and Stratocruiser, and even more over the Republic Rainbow, and within nine days 103 Constellations valued at $75.5 million had been ordered by eight airlines. TWA at last got the first of 27 Model 049s in November 1945. CAA certification followed on 11 December. Commercial services followed in early February 1946, TWA flying the New York-Paris and Pan Am the New York-Bermuda routes.

The USAAF sold its surplus C-69s to airlines in 1946, and the in-service record built up so fast that by July 1946 over 200 million passenger miles had been flown without anyone suffering injury. This was despite numerous engine fires and both engine and propeller failures, but on 11 July 1946 a TWA training flight crashed at Reading, Pennsylvania, because the pilots could not see from the smoke-filled cockpit. A six-week grounding followed, in which 95 modifications were made to the powerplant and systems. There was light at the end of the tunnel (and the DC-6 and other rivals were grounded too): on 19 October 1946 the first definitive post-war Model 649 took to the air, so luxurious and with such good air-conditioning and sound-proofing it was called the 'Gold Plate Connie'. Eastern worked on the specification and was first to use it, one of its features being a Speed-pak external cargo pod under the belly.

Post-war civil success

Ten airlines then bought nearly 100 Model 749s in which 2014-kW (2,700-hp) engines enabled gross weight to rise to 46267 kg (102,000 lb), with a long-range tank in each outer wing to give an extension of range of 1609 km (1,000 miles) without any reduction in payload. Payload/range was further enhanced by the Model 749A at 48535 kg (107,000 lb), with Curtiss paddle-blade propellers. The USAF again adopted the Constellation, buying 10 Model 749s as C-121s of various sub-types, one of which was General MacArthur's *Bataan* and the other General Eisenhower's *Columbine*. Later another of this batch (48-610) was used as *Columbine II* when Eisenhower became President. The US Navy, which in 1945 had used two R7O-1 transports off the US Army line, purchased two Model 749s as its first dedicated radar picket (AEW) aircraft, initially designated PO-1W and later WV-1. Initially flown in June 1949, these were the first aircraft in the world bought from the start as high-flying radar stations, if one discounts modified single-engined Grumman TBM-3 Avengers. Their giant radars were served by aerials (antennae) above and below, but despite the grotesque appearance the addition of extra height to the fins resulted in a very tractable aircraft, and the success of these first two examples led to massive orders of no fewer than 27 distinct subsequent versions for various electronic purposes.

The unexpected USAF and USN sales helped to carry the commercial line through a bad patch, and spurred Lockheed into the striking 'stretch' which turned the Constellation into the Model 1049 Super Constellation. By 1950 the whole programme, under Carl M.

Wearing US Army olive drab, the prototype Constellation takes to the air at Burbank on 9 January 1943.

The final variant of the Super Constellation was the L-1049H, which featured a mixed passenger/cargo configuration with rapid role conversion as a prime feature.

Right: KLM operated several Connies, and was one of the first to operate the Turbo Compound L-1049C. This is one of the later L-1049Gs, complete with tip tanks.

Left: The most successful Connie in terms of sales was the L-1049G, which offered a range of 6470 km (4,020 miles). Lufthansa flew its first intercontinental service with an L-1049G on 8 June 1955 between Hamburg and New York.

Haddon, was poised for its second generation, backed by Lockheed Aircraft Service which had begun at Idlewild in 1949 spurred by the Berlin Airlift, in which C-121s flew almost 6 million passenger miles from Westover AFB to Rhein-Main. But the big news was the L-1049, in which Hibberd injected rational sense with a straight passenger tube of constant section, it having been realised that an airliner in which the body section varied continuously was a mistake.

The first L-1049, the old no. 1961 rebuilt, flew on 13 October 1950. There were many minor changes, but the most obvious one, apart from the 5.59 m (18 ft 4 in) extra length, was the switch from portholes to rounded square windows, removing what had become another outdated feature despite the better fatigue resistance of rings to squares. TWA was behind the Super Constellation, though it was pipped to the post by Eastern which got the new transport, able to seat 99, on the Miami run on 15 December 1951. The largely new engine installations worked well, as did the improved de-icing systems and the larger cockpit windows, integrally stiffened wing skins machined in the newly opened 'Hall of Giants' at Burbank, and better environmental systems.

The L-1049B was a cargo model with integrally stiffened floor and two large loading doors. Lockheed was requested by the US Navy to switch to the complex new Wright Turbo-Compound engine derived from the existing R-3350 used in all previous Constellations, and the 2425 kW (3,250 hp) available from each engine not only promised more speed but also a jump in gross weight to 60329 kg (133,000 lb), representing a further great advance in payload/range. Though the Turbo-Compound engine predictably took a long time to mature, it also totally removed the slight sluggishness which had crept into the original Model 1049, which cruised at barely 483 km/h (300 mph). In addition to 57 US Navy R7V-1s, of which 32 were transferred to the USAF as C-121Gs, Lockheed sold 33 C-121Cs to the USAF. The C-121C was cleared to 61236 kg (135,000 lb), and the others introduced the uprated 2611-kW (3,500-hp) R-3350-34 engine and could take off at 65772 kg (145,000 lb). Subsequently the USAF and US Navy variants of the Model 1049 far outnumbered the commercial versions, as the variants list shows.

Comfort and capability

Seaboard & Western bought a commercial version of the Model 1049B, but the passenger version was the Model 1049C, first put into service by KLM on the New York-Amsterdam route in August 1953. Increasingly the airliner models were fitted with weather radar, which added 0.91 m (3 ft) to the length, as in the military variants. This helped improve passenger comfort, especially on US coast-to-coast trips, and such was the capability of the Model 1049C that from 19 October 1953 TWA at last opened a nonstop service between Los Angeles and New York, rivalling that opened by American with the similarly powered DC-7. This was a time when the British de Havilland Comet was blazing the trail of the airline jet. The Constellation's Mach 0.58 was less impressive, but history was to show that the US industry, greatly aided by the unfortunate collapse of the original Comet programme, was able to fend off this new competition with its fundamentally old-generation piston-engine machines, with the Model 1049 in the forefront. Unit price, typically $1 million for a Model 749, rose to $1.25m with the first Model 1049s, and more than doubled with later Model 1049 versions.

The Model 1049D was an improved cargo aircraft, and the Model 1049E a corresponding passenger version, but most Model 1049E

Iberia was among the last European operators to retire the Super-G, although it continued to use cargo DC-4s long after. This important model had entered service with Northwest in January 1955.

aircraft were actually completed as Model 1049Gs, which were among the most important of all commercial versions. Improved climb ratings from the R-3350-DA3 Turbo-Compound enabled gross weight to reach 62370 kg (137,500 lb), which among other things enabled 600-US gal (2271-litre) tip tanks to be added. The Model 1049G entered service with Northwest in January 1955. A convertible passenger/cargo model was the Model 1049H, of which 54 were built. The very last Constellation was an L-1049H delivered in November 1958.

Piston-engine pinnacle

Before then, one final model had been designed, flown and built in quantity; but the quantity was small, and this was the only version to prove unprofitable to Lockheed, because its development cost over $60 million. The Model 1649A Starliner was launched in 1954 to meet the severe competition of the DC-7C Seven Seas on ultra-long routes, especially the nonstop North Atlantic services. No Model 1049 was really a nonstop Atlantic aircraft (certainly not westbound) and TWA for the last time decided to push Lockheed into a further development of what had by this time become an old aeroplane. In going for long range, Lockheed capitalised on the one thing the Super Constellation could offer in competition with the new Comet 3: lower fuel consumption, though of course it was 115/145 grade gasoline. Fitting Turbo-Compounds had inevitably resulted in severe noise and vibration in an airframe originally planned for about half this power, and it was ultimately decided to out-do Douglas by designing a totally new wing, of great span and high aspect ratio, which would give unrivalled cruising efficiency, house more fuel and enable the engines to be moved well away from the fuselage to reduce noise and propeller vibration.

The Model 1649A first flew on 10 October 1956. It was a beautiful aircraft, and can be regarded as the final pinnacle reached by the piston-engined airliner. The wing box, which had acute dihedral from the roots, was wholly skinned in gigantic machined planks. The body was slightly stretched, and the redesigned nacelles housed the most powerful civil piston engines ever used, driving Hamilton Standard three-blade propellers with synchrophasers to keep their speeds exactly synchronised. TWA, which called the Model 1649 the Jetstream, began services on the North Atlantic on 1 June 1957, but by then the jet era was only weeks away and at last the old Constellation had begun to lose its once powerful appeal. Only 44 of this vastly improved final version were sold, and all were withdrawn in the 1960s while basically older models kept at work, several occasionally trucking odd cargoes around the Caribbean and Dominica.

Far more important in the post-1960 era was the profusion of technically intriguing military versions, packed with unusual devices mainly of an electronic nature, which thundered faithfully all over the

Fitting turboprops to the Constellation was a natural choice, but only a handful received Pratt & Whitney T34s. Four were delivered to the Navy as R7V-2s (company designation L-1249) for high speed transport, and two were transferred to the USAF for tests under the YC-121F designation.

Wearing tactical three-tone camouflage, USAF EC-121Rs flew from Thai bases during the Vietnam war.

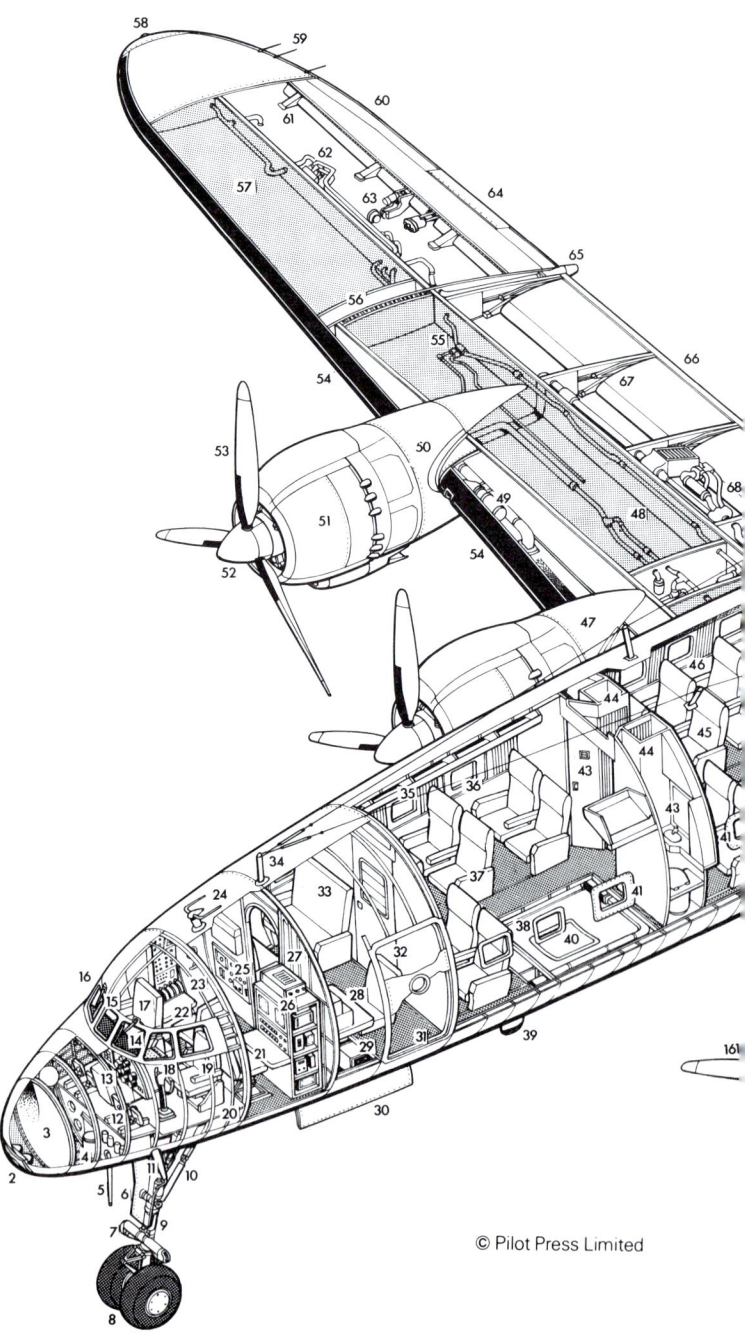

After ten years' service with Air India, this machine passed to the air force for another decade of duty. In addition to transport tasks, the Indian Connies flew maritime patrol missions.

Lockheed L-1049C Super Constellation cutaway drawing key

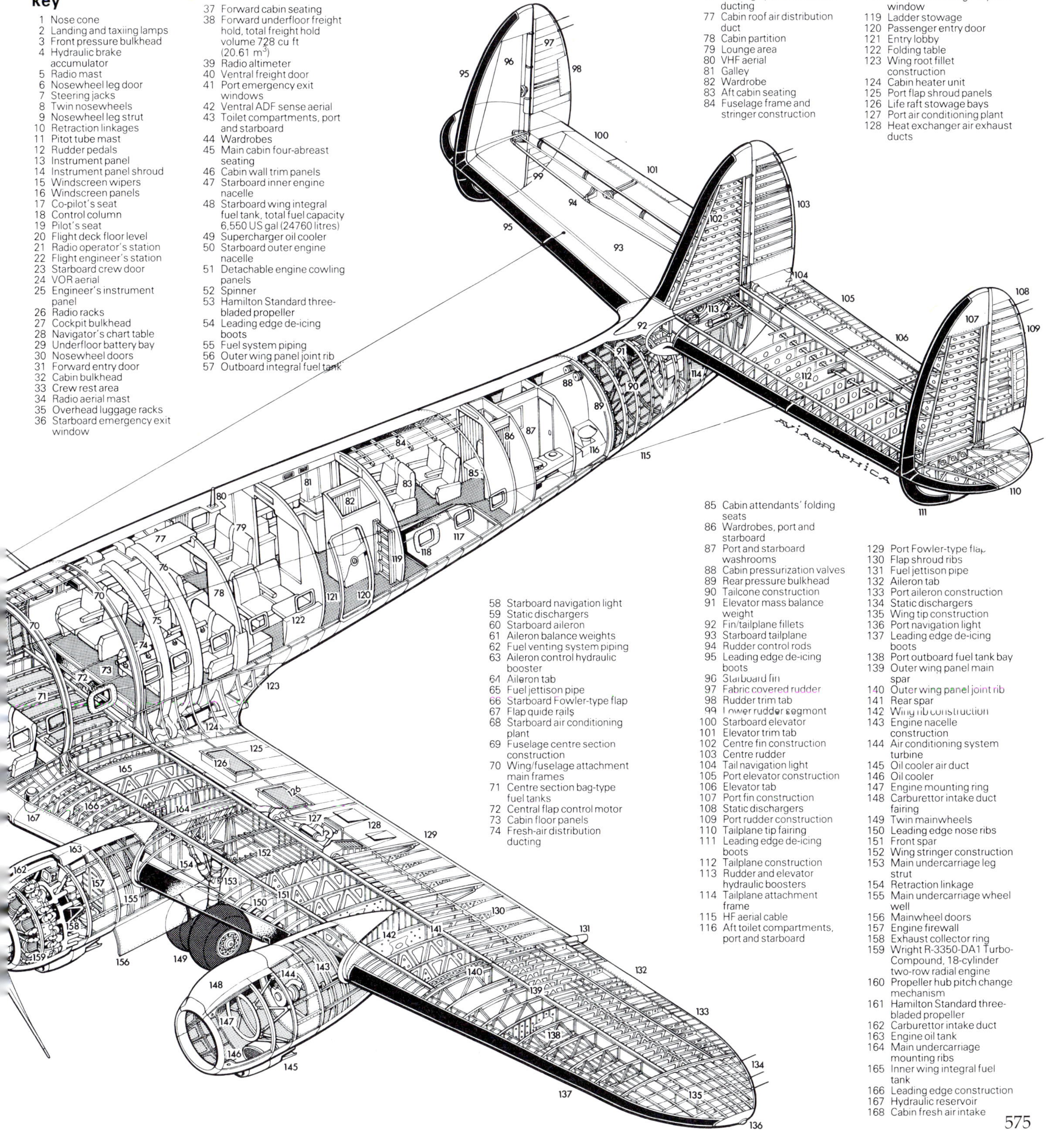

1. Nose cone
2. Landing and taxiing lamps
3. Front pressure bulkhead
4. Hydraulic brake accumulator
5. Radio mast
6. Nosewheel leg door
7. Steering jacks
8. Twin nosewheels
9. Nosewheel leg strut
10. Retraction linkages
11. Pitot tube mast
12. Rudder pedals
13. Instrument panel
14. Instrument panel shroud
15. Windscreen wipers
16. Windscreen panels
17. Co-pilot's seat
18. Control column
19. Pilot's seat
20. Flight deck floor level
21. Radio operator's station
22. Flight engineer's station
23. Starboard crew door
24. VOR aerial
25. Engineer's instrument panel
26. Radio racks
27. Cockpit bulkhead
28. Navigator's chart table
29. Underfloor battery bay
30. Nosewheel doors
31. Forward entry door
32. Cabin bulkhead
33. Crew rest area
34. Radio aerial mast
35. Overhead luggage racks
36. Starboard emergency exit window
37. Forward cabin seating
38. Forward underfloor freight hold, total freight hold volume 728 cu ft (20.61 m³)
39. Radio altimeter
40. Ventral freight door
41. Port emergency exit windows
42. Ventral ADF sense aerial
43. Toilet compartments, port and starboard
44. Wardrobes
45. Main cabin four-abreast seating
46. Cabin wall trim panels
47. Starboard inner engine nacelle
48. Starboard wing integral fuel tank, total fuel capacity 6,550 US gal (24760 litres)
49. Supercharger oil cooler
50. Starboard outer engine nacelle
51. Detachable engine cowling panels
52. Spinner
53. Hamilton Standard three-bladed propeller
54. Leading edge de-icing boots
55. Fuel system piping
56. Outer wing panel joint rib
57. Outboard integral fuel tank
58. Starboard navigation light
59. Static dischargers
60. Starboard aileron
61. Aileron balance weights
62. Fuel venting system piping
63. Aileron control hydraulic booster
64. Aileron tab
65. Fuel jettison pipe
66. Starboard Fowler-type flap
67. Flap guide rails
68. Starboard air conditioning plant
69. Fuselage centre section construction
70. Wing/fuselage attachment main frames
71. Centre section bag-type fuel tanks
72. Central flap control motor
73. Cabin floor panels
74. Fresh-air distribution ducting
75. Air conditioning system overhead ducting
76. Heating system overhead ducting
77. Cabin roof air distribution duct
78. Cabin partition
79. Lounge area
80. VHF aerial
81. Galley
82. Wardrobe
83. Aft cabin seating
84. Fuselage frame and stringer construction
85. Cabin attendants' folding seats
86. Wardrobes, port and starboard
87. Port and starboard washrooms
88. Cabin pressurization valves
89. Rear pressure bulkhead
90. Tailcone construction
91. Elevator mass balance weight
92. Fin/tailplane fillets
93. Starboard tailplane
94. Rudder control rods
95. Leading edge de-icing boots
96. Starboard fin
97. Fabric covered rudder
98. Rudder trim tab
99. Lower rudder segment
100. Starboard elevator
101. Elevator trim tab
102. Centre fin construction
103. Centre rudder
104. Tail navigation light
105. Port elevator construction
106. Elevator tab
107. Port fin construction
108. Static dischargers
109. Port rudder construction
110. Tailplane tip fairing
111. Leading edge de-icing boots
112. Tailplane construction
113. Rudder and elevator hydraulic boosters
114. Tailplane attachment frame
115. HF aerial cable
116. Aft toilet compartments, port and starboard
117. Rear underfloor freight hold
118. Rear cabin emergency exit window
119. Ladder stowage
120. Passenger entry door
121. Entry lobby
122. Folding table
123. Wing root fillet construction
124. Cabin heater unit
125. Port flap shroud panels
126. Life raft stowage bays
127. Port air conditioning plant
128. Heat exchanger air exhaust ducts
129. Port Fowler-type flap
130. Flap shroud ribs
131. Fuel jettison pipe
132. Aileron tab
133. Port aileron construction
134. Static dischargers
135. Wing tip construction
136. Port navigation light
137. Leading edge de-icing boots
138. Port outboard fuel tank bay
139. Outer wing panel main spar
140. Outer wing panel joint rib
141. Rear spar
142. Wing rib construction
143. Engine nacelle construction
144. Air conditioning system turbine
145. Oil cooler air duct
146. Oil cooler
147. Engine mounting ring
148. Carburettor intake duct fairing
149. Twin mainwheels
150. Leading edge nose ribs
151. Front spar
152. Wing stringer construction
153. Main undercarriage leg strut
154. Retraction linkage
155. Main undercarriage wheel well
156. Mainwheel doors
157. Engine firewall
158. Exhaust collector ring
159. Wright R-3350-DA1 Turbo-Compound, 18-cylinder two-row radial engine
160. Propeller hub pitch change mechanism
161. Hamilton Standard three-bladed propeller
162. Carburettor intake duct
163. Engine oil tank
164. Main undercarriage mounting ribs
165. Inner wing integral fuel tank
166. Leading edge construction
167. Hydraulic reservoir
168. Cabin fresh air intake

Carrying search radar and height-finding equipment in giant fairings, the WV-2 was the US Navy's shore-based airborne early warning version, later redesignated EC-121K. Radar picket Connies also served in large numbers with the USAF.

The shape of things to come: the WV-2E was a one-off AEW variant which carried an APS-82 surveillance radar in a rotating rotodome similar to that fitted to the Boeing E-3, Grumman E-2 and Ilyushin Il-78. In 1962 the WV-2E was redesignated EC-121L.

American crew members – the only loss in anger of a US military Constellation – highlights the special role of this aircraft type in reconnaissance, intelligence-gathering, and airborne early warning (AEW).

This role was first undertaken when the ubiquitous 'old 1961' was converted – again! – by Lockheed/Burbank for trials as an AEW platform. Modifications to accommodate radar and other gear gave the Connie a tall, narrow dorsal hump and a large underfuselage bulge. This configuration was first seen operationally on the US Navy's PO-1W and PO-2W (later designated WV-1 and WV-2) weather-reconnaissance and AEW machines based respectively on the 749 and 1049 airliners.

Later, other duties were added to the AEW role, including electronic eavesdropping and countermeasures. The variety of humps, bumps and bulges protruding from various 'spook' Connies in the C-121 series became, itself, an encyclopedia-length subject. A one-off variant, the WV-2E, later called EC-121L, carried its antennae in a saucer-like disc evocative of today's Boeing E-3A Sentry.

In these Constellations, large crews could journey for up to 30 hours far from the Fleet or from shore, relief crews replacing each other while searching for air or sea threats. Initially, with the purpose of extending the range of ground and ship radars, USAF Constellations were used off American coasts by the Air Defense Command. In Vietnam, under the 'Big Eye' and 'College Eye' programmes, USAF Constellations stood off the enemy coast and reported MiG activity to endangered US combat pilots. In South Vietnam, the USAF's EC-121R, camouflaged and devoid of bulges, relayed radio signals to ground station. More than a dozen designations were eventually applied to 'spook' Connies (see Variants list), and today one of the most beautiful of these preserved machines is the US Navy's EC-121K no. 141292, from Squadron VAQ-33, on display at the Air and Space Museum in Florence, South Carolina.

globe until 1980 despite concerted efforts made from 1975 to eliminate everything calling for 115/145 gasoline. In the 1960s Super Constellations of the USAF and US Navy flew more than 600,000 hours, though by this time the basic type ceased to appear in books on US military aircraft. Total production of all Constellation models was 856 (one XC-69, 233 Constellations, 578 Super Constellations and 44 Starliners).

On 15 April 1969 a US Navy 'Connie' piloted by Lieutenant Commander James Overstreet was attacked by two North Korean MiG-17 fighters over the Sea of Japan and shot down. The loss of EC-121M no. 135749 from electronics squadron VQ-1, and of all 31

Lockheed Constellation variants

Model 049-39-10: pre-war TWA/Pan Am aircraft with four 1641-kW (2,200-hp) R-3350-35 after being taken over by USAAF prior to completion; first aircraft (Lockheed no. 1961) completed as **XC-69** in olive drab, civil registration NX67900, later NX25600, later AAF 43-10309; remaining aircraft AAF **C-69-1** (43-10310/10317) and **C-69-5** (42-94549/94561); total 22
C-69C: conversion of 294550 as VIP transport, later **ZC-69C**
XC-69E: conversion of first prototype with 1492-kW (2,000-hp) R-2800 Double Wasp engines
Model 1049 Constellation: initial new-build commercial model with 1641-kW (2,200-hp) R-3350-BA1 engines, gross weight 39010 kg (86,000 lb), later 40824 kg (90,000 lb) and later 43546 kg (96,000 lb); nos 2023/88, total 66
R7O-1: initial US Navy counterpart of Model 049-46, used by VPB-101, BuAer nos not assigned
Model 649: commercial transport with 1865-kW (2,500-hp) R-3350-BD1 engines, gross weight 44453 kg (98,000 lb); total 14, later brought up to L-749 standard
Model 749: long-range tanks in outer wings; gross weight 47628 kg (105,000 lb)
Model 749A had stronger spars and main legs; gross weight 48535 kg (107,000 lb); total 111
C-121A (Model 749A-79-38): 1865-kW (2,500-hp) R-3350-75 engines as USAF passenger transports (48-609/617); at one time designated **PC-121A**, for passenger transport; total 9
VC-121A: VIP conversions of C-121A aircraft, 48-610 as *Columbine II*, 48-613 as *Bataan* and 48-614 as *Columbine I*
VC-121B: special long-range VIP aircraft (48-608); total 1
Model 1049 Super Constellation: stretched version with 2014-kW (2,700-hp) R-3350-CA1 engines, gross weight 54432 kg (120,000 lb); prototype was conversion of no. 1961, registered N6201C; square windows retained but tip tanks deferred until the Model 1049G; total 24 (nos 4001/4024)
R7V-1: US Navy **Model 1049B**, first with 2425-kW (3,250-hp) R-3350-91 Turbo-Compound engines; weather radar lengthened nose 0.91 m (3 ft); total 51 (BuAer nos 128434/128444, 131621/131629, 131632/131659, 140311/140313); see C-121G, C-121J and VC-121E
PO-1W: see WV-1
WV-1: first AEW radar picket version, based on Model 749, height-finder radar above fuselage, plan surveillance radar below; crew 22; BuAer nos 124437/124438, total 2 and ordered as PO-1Ws
WV-2: production AEW model, later named **Warning Star**; based on Model 1049, 2536-kW (3,400-hp) R-3350-34 or -42 Turbo-Compounds, improved radars and data links, crew 26; BuAer nos 126512/126513, 128323/128326, 131387/131392, 135746/135761, 137887/137890, 141289/141333, 143184/143230, 145924/145941, total 142
WV-2E: conversion of BuAer no. 126512 with APS-82 radar using rotating dish 'rotodome'
WV-2Q: conversions (16-plus) as high-power ECM warning, D/F and jamming platforms
WV-3: weather reconnaissance variant, also called Warning Star, same basic airframe as WV-2 but totally different interior and equipment, no tip tanks, crew 8; BuAer nos 137891/137898, plus conversion of 141323, total 9; became **WC-121N**, but two to USAF as **EC-121R**
R7V-2: US Navy **Model 1249** high-speed transports to evaluate turboprop propulsion, restressed airframe for high speeds (703 km/h/437 mph); four P&W YT34-P-12A driving broad paddle-blade propellers; BuAer nos 131630/131631, 131660/131661, total 4
C-121C: standard USAF long-haul passenger transport for MATS derived from Model 1049, weather radar, 2611-kW (3,500-hp) R-3350-34 Turbo-Compounds; 54-151/183, total 33
EC-121C: see RC-121C
JC-121C: two C-121C (54-160, 54-178) and one RC-121C (51-3841) converted as systems (mainly electronics) test aircraft
RC-121C: first USAF AEW radar picket version, based on C-121C airframe with dorsal and ventral radars similar to WV-2; 51-3836/3845, total 10; in 1962 became EC-121C
TC-121C: nine RC-121Cs converted as AEW radar trainers; subsequently became EC-121C
VC-121C: conversions as VIP executive transports total 4 (54-167/168, 54-181/182)
EC-121D: see RC-121D
RC-121D Warning Star: improved long-range AEW&C (airborne early warning and control) version, tip tanks and other changes, equipped second AEW&C Wing (551st), basis for many other versions; 52-3411/3425, 53-533/556, 53-3398/3403, 54-2304/2308, 55-118/139, total 72; most became **EC-121D**
VC-121E: ex-USN R7V-1 (131650) transferred to USAF as 53-7885 as Presidential *Columbine III*
YC-121F: last two R7V-2s transferred to USAF as 53-8157/8158, later re-engined with T34-P-6
C-121G: transferred USN R7V-1s to USAF for MATS use, renumbered 54-4048/4079; total 32
TC-121G: conversions as crew trainers; 54-4050/4052 and 4058, total 4
VC-121G: VIP conversion of 54-4051
EC-121H: rebuild of 551st AEW&C Wing EC-121Ds with SAGE data-links, large airborne computer, new navaids and other equipment; total 42
C-121J: transfer of C-121G (54-4079) back to USN as 140313; later became NC-121K
EC-121J: updates of EC-121Ds with extra equipment (classified); 52-3416 and 55-137, total 2
EC-121K: redesignation of WV-2
YEC-121K: conversion of WV-2 128324 for classified 'Ferret' Elint
JC-121K: conversion of EC-121K 143196 for classified US Army test programme(s)
NC-121K: conversions (21-plus) for special tests, including C-121J and EC-121K 145925 for Project 'Magnet' mapping Earth's field
EC-121L: redesignation of WV-2E
EC-121M: redesignation of WV-2Q
WC-121N: redesignation of WV-3
EC-121P: conversions (13-plus) of EC-121K with ASW sensors and special overwater navaids; 143184, 143189, 143199, 143200 to USAF with numbers as before omitting initial 1
JEC-121P: three of above transfers (189/199/200) later on special USAF test
EC-121Q: conversions (unknown number) of EC-121Ds with augmented and updated AWACS systems
EC-121R: conversions (total 30) for Project 'Igloo White' in Vietnam, serving as airborne data-relay stations (67-21471/21500)
EC-121S: near-total rebuild of five C-121Cs (54-155, 159, 164, 170, 173)
EC-121T: classified conversions (25-plus) of EC-121D, H and J as Elint electronic intelligence platforms
Model 1049C: first commercial model with Turbo-Compounds (2425-kW/3,250-hp R-3350-DA1), strengthened wing, extra fuel; total 49
Model 1049D: cargo version of Model 1049C; total 4
Model 1049E: improved passenger model; total 18
Model 1049G: major upgrade of commercial aircraft with DA3 engines, tip tanks and other improvements; total 104
Model 1049H: corresponding long-range restressed cargo aircraft; total 53
Model 1649 Starliner: ultra-long-range model with new long span wing, slightly stretched body and many other improvements; total 44

Lockheed L-188 Electra

One of the most elegant airliner designs ever, the Electra was Lockheed's great hope in the pre-jet age days. The very best in turboprop design, it promised an unprecedented level of performance both for operators and passengers. Sadly, the type was born under an unlucky star and its career was cut short by accidents. A sizeable proportion of those built survive to this day, still adding a little glamour to the greying world of air transport.

When Capital Airlines placed an order for 60 British-built Vickers Viscount turboprop airliners in 1954, other US carriers were alerted to the coming of the 'jet age' in airline transportation.

Prior to ordering Viscounts, Capital had approached the Lockheed company to ask if it would be interested in developing a turboprop airliner, but at the time none of America's 'big three' domestic carriers – American Airlines, Eastern Air Lines and United Airlines – showed interest in such an aircraft. That situation changed in late 1954 when American issued a specification for a four-engined design with accommodation for 75 passengers and a maximum range of 3218 km (2,000 miles) which would still be economical to operate over shorter routes of 800 to 1300 km (500 to 800 miles).

Initially Lockheed proposed a four-engined, Rolls-Royce Dart or Napier Eland powered version of the projected Model CL-303, a high-wing, twin-engined design which had already been proposed to American for its short-haul routes. When Eastern Air Lines also expressed interest in a four-engined aircraft, but with 85-90 seats, a 4020-km (2,500-mile) range capability and a 560-km/h (350-mph) cruising speed, Lockheed revised its submission, proposing the Model 188, a low-wing aircraft with circular cabin cross-section, powered by four 2796-kW (3,750-shp) Allison 501D turboprops, which were civilian versions of the T56-A1 powerplant then test flying aboard the prototype YC-130 Hercules military transport and a modified Super Constellation testbed.

American and Eastern were impressed with Lockheed's proposal, and placed 'off the drawing board' orders for the new aircraft, American contracting for 35 L-188s on 8 June 1955, while Eastern signed for 40 on 27 September.

The name 'Electra' was chosen for the L-188. The prototype, N1881, made its first flight from the Lockheed Air Terminal at Palmdale, California, eight weeks ahead of schedule and 26 months after the first metal was cut, on 6 December 1957, flown by Lockheed's celebrated test pilot Herman 'Fish' Salmon. By the time it had flown, Lockheed had received orders for 129 Electras, with options on a further 48 aircraft. Besides American and Eastern, other carriers which ordered Electras were Braniff, National Airlines, Northwest Airlines, Pacific Southwest Airlines (PSA) and Western Airlines in the United States, Trans Australian Airlines and Ansett-ANA in Australia, Tasman Empire Air Lines (TEAL) in New Zealand, Garuda Indonesian, and KLM Royal Dutch Airlines.

The L-188A design had been optimised for 66-80 passengers, with six further seats in a lounge, but in high density configuration seating could be increased to 98. A three-crew flight deck and two cabin attendants were standard. Turboprop powerplants offered were the 2796-kW (3,750-shp) Allison 501D-13/13A or 3020-kW (4,050-shp) 501D-15, mounted ahead of the wing in long nacelles with separately-mounted reduction gearboxes and overwing exhausts, the large nacelles giving the Electra the appearance of being 'all engine and little wing'. Four-bladed 'paddle' type Aeroproducts or Hamilton Standard reversing propellers were installed. Fuel tankage was 20630 litres (5,450 US gal), and maximum take-off weight 51267 kg (113,000 lb). To meet the long-range requirements of over-water operators such as Northwest, PSA, Garuda, KLM and TEAL, Lockheed developed the L-188C version, which had provision for an additional 3937 litres (1,040 US gal) of fuel and a maximum take-off weight of 52618 kg (116,000 lb), with accommodation for 74-99 passengers and a range, with a payload of 5080 kg (11,200 lb) of 5633 km (3,500 miles). The designation L-188B was not used officially, but applied to L-188C-configured aircraft for

The first Lockheed L-188A Electra took to the air on 6 December 1957. With the advent of passenger jets, Lockheed attached the title 'Prop-Jet' to its new baby to dampen passenger resistance to 'old-fashioned' propellers. Sadly, this historic aircraft was not preserved. It was sold and later broken up in 1975.

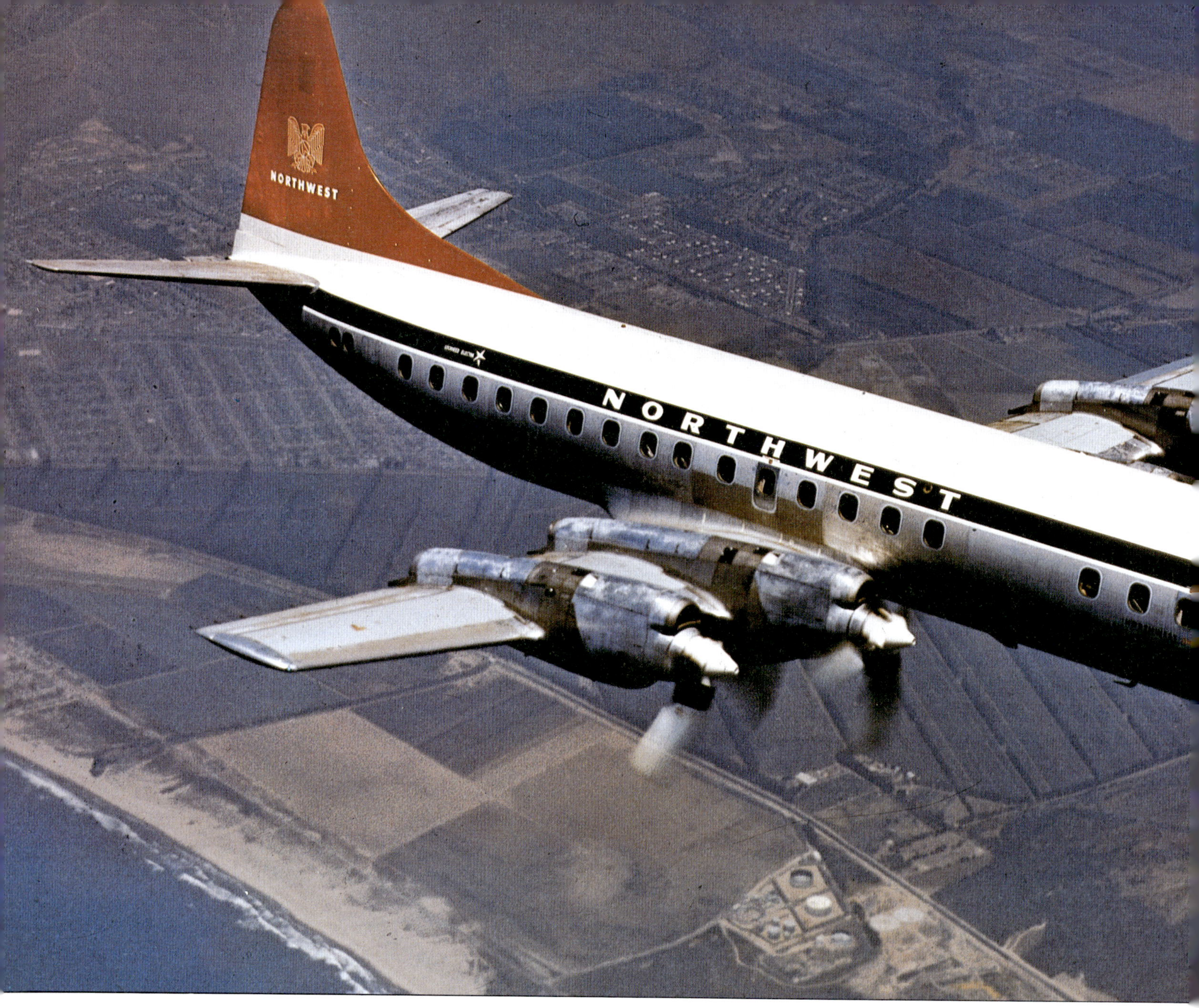
NORTHWEST
NORTHWEST

N5501
FLY-EASTERN'S PROP-JET · ELECTRA

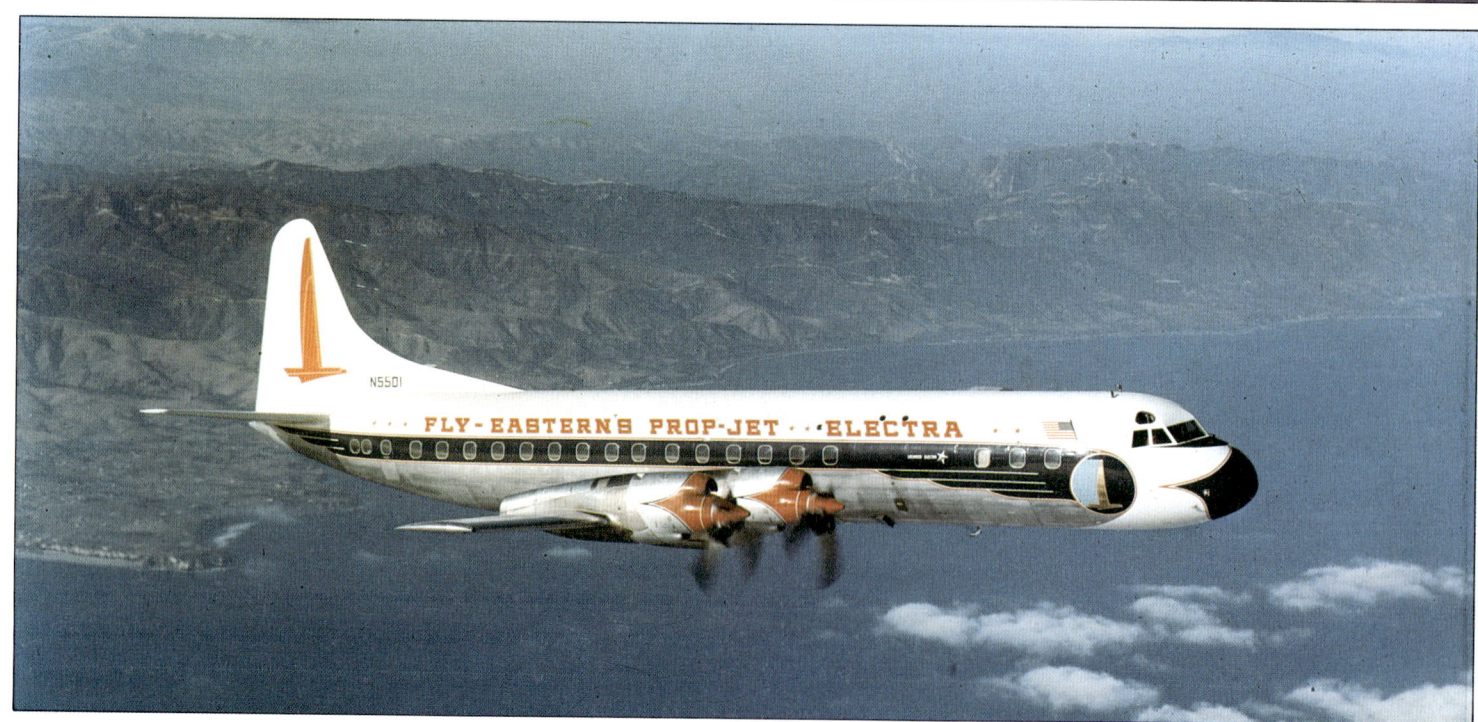

American Airlines ordered the L-188 as a response to the Viscounts operated by rival Capital Airlines. A pilots' strike meant that Eastern snatched the honour of the world's first Electra service from American, however.

'Tiao Khoun Boulom' was the name of one of two L-188A Electras operated by Royal Air Lao in the early 1970s. They served with the airline for little over a year before being disposed of in 1973.

Eighteen L-188Cs served with Northwest Airlines on its domestic routes. The first to be delivered, in 1959, was the third Electra to crash after structural failure, effectively setting the seal on the type's career.

foreign operators, and had track-mounted passenger seats, additional lavatory facilities and a navigator's station.

Three further company-owned development aircraft joined the Electra flight test programme during early 1958, leading to Federal Aviation Agency certification on 22 August 1958 after a swift and trouble free certification programme that boded well for the Electra's service entry and market success, since it was more than a year ahead of its British rival the Vickers Vanguard.

The first customer aircraft (actually the seventh airframe built) was delivered to Eastern Air Lines. Six more were delivered to Eastern and four to American Airlines by the

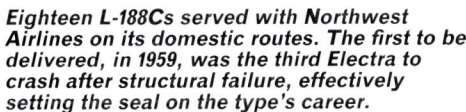

Eastern Airlines heralded the introduction of its first 'Prop-Jet Electras' with the most eye-catching of colour schemes. Eastern was also one of the last American carriers to relinquish the type on passenger services.

end of 1958, but labour disputes between the airlines and their pilots delayed the Electra's service entry until 12 January 1959, when Eastern introduced the new turboprop on its New York-Miami route, followed on 23 January by American's inaugural Electra service between New York and Chicago. Ansett-ANA became the first foreign carrier to operate the type, in March 1959, while KLM introduced the Electra to Europe in December.

Tragedy strikes

The Electra proved an immediate success with its passengers, but its service entry was marred just weeks after inaugural commercial flights by a fatal crash on 3 February 1959. The cause was attributed to pilot error. Deliveries continued throughout the summer of 1959, with the first of the long-range L-188C models, the 57th production Electra, going to Northwest on 19 July 1959. Early service experience with the aircraft led to modification of the engine nacelles to increase the angle of incidence of the engine mounting for reduced cabin noise and vibration levels and to solve a propeller resonance problem. While this programme was being carried out a second Electra, operated by Braniff, was lost in a fatal crash in Texas on 29 September 1959. The Electra had broken up in flight, but the cause of the accident could not immediately be established. On 17 March 1960 a similar inflight break-up occurred to a Northwest L-188C over Indiana.

The loss of three almost-new aircraft in just over a year seriously undermined public and airline confidence in the Electra, and brought calls from the US Congress for the aircraft to be grounded. The Federal Aviation Agency did not comply with this request, but a week after the Indiana crash it imposed a reduction in maximum cruising speed on Electras to 275 kt (510 km/h; 320 mph), subsequently reduced to 225 kt (417 km/h; 260 mph), while the cause of the inflight break-ups was being sought.

After extensive investigations it was discovered that because of a design fault in the Electra's engine mountings, slight damage to a mounting (in a heavy landing, for example) could lead to a 'whirl mode' oscillation developing between the nacelle structure and propeller, causing flexing of the wing and eventual

Zantop has been associated with the Electra since its foundation in 1972 as a contract carrier. It grew out of the demise of Universal Airlines and is still an all-cargo airline. It once had 20 Electras, all L-188AFs and L-188CFs, on its books.

QANTAS' four L-188Cs were some of the last Electras off the production line. The first was named 'Pacific Electra' and all the others had the 'Pacific' prefix also. All four aircraft went on to enjoy long careers in their days after Australian service.

catastrophic failure of the wing at its root. This was positively established as the cause of the Braniff and Northwest accidents.

Lockheed immediately launched a $25 million Lockheed Electra Achievement Program (LEAP) to modify all in-service L-188As and L-188Cs and incorporate these structural changes in aircraft then on the production line. The modifications included adding bracing and stiffeners to the engine nacelles, reinforcing the forward wing spar, strengthening 18 wing ribs in the nacelle areas, and replacing the skinning on some wing panels with heavier gauge metal. The Federal Aviation Agency approved the modifications and re-certificated the aircraft on 5 January 1961, when all speed restrictions were removed from modified aircraft. The first LEAP-modified 'Electra II' returned to service on 24 February 1961.

Although the LEAP modifications completely solved the Electra's structural problems, and enhanced Lockheed's reputation for integrity within the airline industry, public confidence in the aircraft had waned to the point that no new orders were received for Electras after the FAA imposed its speed restrictions. Production terminated on 15 January 1961 after 170 Electras (55 of them long-range L-188C models) had been built, the last one going to Garuda Indonesian. Four were retained by Lockheed for trials and development, and one was acquired by engine manufacturer Allison Division of the General Motors Corporation for use as an engine testbed.

The aircraft's fall from favour with the travelling public, and the emergence of jets such as the Sud Aviation Caravelle, BAC-111 and Boeing 727 on short/medium-haul routes within the USA, ensured that the Electra had a short career with most of its original operators. American Airlines began disposing of parts of its fleet of L-188As as early as 1962, although some of Eastern's Electras continued to serve as backups for jets on the carrier's Boston-New York-Washington D.C. shuttle services until 1977. Pacific Southwest Airlines actually increased its fleet of Electras from three to six, and later still, in 1975, bought back Electras to operate its services to the resort area at Lake Tahoe in the Sierra Nevada, where jet operations were prohibited.

While most major carriers soon abandoned the Electra, it found a ready market among smaller airlines in the United States, South America and the Pacific, and as with its British rival the Vanguard, the Electra proved an ideal vehicle for cargo hauling. Lockheed Aircraft Service Company, a subsidiary of the manufacturer, developed a cargo conversion in 1968 which involved strengthening the cabin floor and installing a 134 × 80 in cargo loading door

A pre-delivery photograph of one of KLM's 'Flying Dutchman' Electras. A total of 12 long-range L-188Cs was delivered to the airline between 1959 and 1960.

on the port side of the fuselage aft of the flight deck and, optionally, a 99 × 80 in door aft of the wing. LAS converted 41 Electras to L-188AF/CF standard, and some 20 similar modifications were performed by other contractors.

Long-term service

Among the most successful operations in the Electra's second career were those of the Brazilian national airline VARIG, which acquired 12 ex-American Airlines Electras in 1962. VARIG used the Electra on its *Ponte Aérea* (air bridge) shuttle services linking Rio de Janeiro's downtown Santos Dumont Air-

Western Airlines was one of the earliest Electra operators, putting its first of 12 aircraft into service on 1 August 1959 on the Los Angeles to Seattle run. The company's Indian's head motif was worn prominently.

Modern-day Electras still abound. Hunting Cargo Airlines may be better known to many as Air Bridge, but the recent name change has not diminished their commitment to the Electra as the ideal freighter.

port, where jet operations were banned, with Sao Paulo-Congonhas. The Electras operated a half-hourly no-reservations service, averaging 37 round trips each weekday and 22 at weekends, with load factors around 80 per cent and annual passenger loads in excess of 1.5 million. Only in 1991 did these aircraft finally give way to jets.

Also in South America, Ecuador's government-sponsored carrier Transportes Aéreos Militares Ecuatorianas (TAME), operated by the Fuerza Aérea Ecuatoriana, acquired five L-188As and one L-188C in 1974/75. These, former American and Western Airlines and Cathay Pacific aircraft, played a key role in developing air services to the Galapagos Islands, and only reluctantly left service in September 1989 when the last of the TAME fleet suffered a landing gear malfunction while en route from Ecuador's capital Quito to Guayaquil, and was damaged beyond economical repair in a forced landing at Taura Air Base.

Other operators of passenger and cargo carrying Electras during the 1970s and 1980s included Air California, Evergreen International, Hawaiian Air, Overseas National Airways, Reeve Aleutian Airways, Southeast Airlines, Transamerica Airlines and Zantop International Airlines in the United States; Aerocondor, Copa, LACSA, Lineas Aereas Paraguayas, SAHSA, TACA and Transportes Aéreos Nacionales SA of Honduras in Latin America; Fred Olsen Flyveselskap in Norway, and Mandala Airlines of Indonesia. A number of Electras were also operated as corporate transports or by travel clubs. Two were acquired by Nordair of Canada for use on ice survey flights on behalf of the Canadian government.

In addition to its commercial transport roles, the Electra, while not a financial success for Lockheed, provided the airframe from which the highly successful P3V-1/P-3 Orion maritime patrol and anti-submarine warfare aircraft was developed, the third production L-188A serving as aerodynamic prototype for the Orion, in which guise it was first flown on 19 August 1958. After development trials with Lockheed and the US Navy it was transferred to the National Aeronautics and Space Administration (NASA) in 1967 and employed until the late 1970s on aerodynamic and astronomical data gathering flights, bearing the identity NASA 927.

Ongoing operations

A surprising number of Electras are still extant, with more than 75 around the world. Of these 56 were still airworthy in 1988, of which 7, 19 and 30 were in Africa, Europe and the Americas respectively. A major US operator was Zantop, with 20 flying from Detroit-Willow Run, Michigan, and Macon, Georgia. Zantop is the world's largest Electra operator but, like most, all its aircraft are freighters. Only Anchorage-based Reeve Aleutian Airways still offers American passengers the chance to fly on an Electra on its services to the Aleutian Island chain, in the Barents Sea, and elsewhere. South America is no longer the bastion of Electra users it once was, but some linger on with the Argentinian navy, Honduran air force and Paraguayan air force, in addition to civil users. Sao-Tomé based TransAfrik is a typical African Electra owner. In Indonesia Mandala Airlines was the only other carrier in the world to offer this great aircraft in the original passenger-carrying form.

There are currently more Electras flying in Europe than anywhere else in the world, as the phenomenal growth in the overnight parcel delivery business has created a gap that the Electra is uniquely qualified to fill. The long-time Norwegian operator Fred Olsen now has four aircraft, and a similar number was operated by Falcon Cargo of Sweden. The Electra received British certification relatively late in its career, and Hunting Air Cargo (ex-Air Bridge Cargo) and Channel Express each had four aircraft, rising to the current levels of six and eight aircraft respectively. Amerer Air and Atlantic Airlines have three and four aircraft respectively.

Lockheed's 'singles': The Plywood Bullets

The Vega was indeed a milestone aircraft. Not only was it a technological marvel, it was also used on many famous flights, including the first solo circumnavigation of the globe. While spawning an important family of derivatives, it laid the foundation for two of the world's most important aircraft manufacturers: Lockheed and Northrop.

Following the collapse of their Loughead Aircraft Manufacturing Company in 1921, the brothers Allan and Malcolm Loughead, and their friend and chief engineer/designer John K. (Jack) Northrop, pursued other careers but met occasionally to discuss future aircraft designs, from which emerged a concept for a high-speed cabin monoplane with seating for a pilot and four passengers.

In December 1926 Allan Loughead finally succeeded in obtaining capital to set up the new Lockheed Aircraft Company in a rented building in Hollywood, California. Northrop immediately set to work designing the aircraft, which was to be named Vega, starting the tradition of naming Lockheed aircraft after stars and other heavenly bodies.

Northrop employed the novel monocoque fuselage construction method which he had developed for the earlier Loughead S-1 biplane, using 3-mm thick half-shells created by laying up three laminations of spruce plywood in concrete female moulds. Rubber bags inflated to 137.9 KPa (20 lb/sq in) kept pressure on the casein-glued laminations for 24 hours, after which concentric wooden formers were glued in place and the two halves joined, model aeroplane-style, to form a lightweight, strong and extremely smooth structure. Northrop designed a cantilever high wing, a clean tripod-leg landing gear installation and selected a 149.2-kW (200-hp) Wright Whirlwind J5 radial engine to power the aircraft.

The Vega (2788) made its first flight from a hayfield at Inglewood, California (now the site of Los Angeles International Airport) on 4 July 1927, in the hands of test pilot Eddie Bellande. It had been sold before it had flown (for $12,500, less than it had cost to build), to San Francisco newspaper owner George Hearst Jr, who announced his intention to enter it in the $25,000 California-Hawaii Air race spon-

In airline service the Vega was an important stepping-stone for many carriers. The most notable was Varney Speed Lines, forerunner of today's Continental. This aircraft was restored to its former glory with help from the airline.

sored by James Dole of the Hawaiian Pineapple Company. Equipped with two 378.54-litre (100-US gal) tanks in its cabin in addition to the standard tankage of 606 litre (160 US gal), and fitted with survival gear, rapid-inflating flotation bags, cork lining to the lower fuselage and provision for dropping its undercarriage in the event of ditching, as well as a navigator's station with advanced navigation equipment and radio, the Vega, re-registered NX913, named *Golden Eagle* and flown by Hearst's pilot Jack Frost and navigator Gordon Scott, established a number of point-to-point records while preparing for the race. Frost and Scott left Oakland, California on 16 August 1927 – only six weeks after the Vega's first flight – as firm favourites to win the 3925-km (2,439-mile) race, but tragically *Golden Eagle* and its crew vanished over the Pacific in an incident- and death-ridden race that became known as the Doleful Derby.

Despite this tragedy, in a few weeks the 217-km/h (135-mph) Vega had established a reputation which brought orders flooding into Lockheed and resulted in production of another 128 aircraft in a plethora of standard models and numerous one-off 'specials' too varied to describe in detail here.

The first production version was the Vega 1 (28 built), powered by uncowled 169-kW (225-hp) Wright Whirlwind J5/A/AB/B/or C nine-cylinder engines. The Vega 2 (six built) differed principally in having 224-kW (300 hp) Whirlwind J6s. The major production model, of which 77 were built in four main variants, was the Vega 5, first flown in March 1928. This Vega had five/six passengers seats (one beside the pilot), and was powered, according to sub-type, by a 306-kW, 313-kW, 336-kW (410-, 420- or 450-hp) Pratt & Whitney Wasp engine. NACA cowlings were fitted to most Vega 5s, and streamlined wheel fairings were a popular option for this model, whose 298 km/h (185 mph) top speed earned Vegas the nickname 'Lockheed's Wooden Bullets'. Float landing gear could also be fitted, reducing top speed by just 13 km/h (8 mph).

Other variants of the Vega included those developed by the Detroit Aircraft Corporation after it acquired Lockheed in July 1929: the DL-1, DL-1B and DL-1 Special (so-designated to signify 'Detroit-Lockheed') featured Detroit-built all-metal duralumin fuselages and wooden wings made by Lockheed's Burbank plant, as did the unique 373-kW (500-hp) Pratt & Whitney R-1340-17 powered DL-1B Special/Y1C 17 delivered in December 1930 to the US Army Air Corps. With a top speed of 356 km/h (221 mph) the Y1C-17 was the USAAC's fastest aircraft. It was used by Captain Ira Eaker for an attempt on the eastbound US transcontinental speed record which ended in failure on 10 March 1931 when the aircraft crashed on take-off from Kentucky, though not before it had average 381 km/h (237 mph) on the 2800-km (1,740-mile) flight from Long Beach, California.

Racing and records

Widely used by airlines, businessmen and companies throughout the United States and abroad, the swift Vega was also the chosen mount for many pioneering flights, racing and speed record attempts. Too numerous to detail fully here, the Vega's multifarious achievements are the stuff of legend, beginning with explorer Captain Hubert Wilkins' first trans-Arctic flight from Alaska to Spitzbergen in a Vega 1 in April 1928, and Art Goebel and Harry Tuckers' first eastbound non-stop US transcontinental flight from Los Angeles to New York in August of that year, to Amelia Earhart's first solo transatlantic flight by a woman on 20/21 May 1932, when she flew her Vega 5B from Newfoundland to Ireland in 15 hours, 18 minutes, and three months later made the first women's transcontinental flight across the USA and set a new international distance record for women on her 3938-km (2,447-mile) journey between Los Angeles and

Lockheed's three main derivatives from the pioneering Vega are shown together. At left is the Orion, at right is the Altair and in the middle is the Air Express, produced by raising the wing of the Vega to a parasol position and moving the cockpit aft of the wing.

Left: *On account of its speed, the Vega proved exceptionally popular with a wide variety of customers, these including airlines, companies and wealthy private owners. Many were fitted with wheel spats for extra aerodynamic cleanliness.*

Considered as one of the most important aircraft in history, the prototype Vega is seen at Inglewood on the occasion of its first flight. The aircraft introduced a monocoque structure for the first time, the basic construction technique still applying to many of today's aircraft.

Newark, New Jersey.

But best known of all the Vega's many exploits were those achieved by the one-eyed Wiley Post, pilot for wealthy Oklahoma oilman F. C. Hall. Post flew Hall's first Vega, a model 5 named 'The Winnie Mae' after the owner's daughter, on business trips for his employer, but it was Hall's second Vega, a Model 5B NR-105W, also named 'The Winnie Mae', which was to make him a household name literally the world over.

After winning the 1930 National Air Races from Los Angeles to Chicago in 9 hours, 9 minutes, Post prepared for an attempt on the around-the-world record. With Harold Gatty as navigator, 'The Winnie Mae' took off from Roosevelt Field, New York, on 23 June 1931 and flew eastwards via Newfoundland and across the Atlantic to England, then to Germany, across the Soviet Union, and up through Alaska and Canada to land back at its starting point 8 days, 15 hours and 51 minutes after departure. Not content with that remarkable flight, Wiley Post set a record for circling the globe solo. Again flying 'The Winnie Mae', he took off from New York's Floyd Bennett Field on 15 July 1933, made the first-ever non-stop flight from New York to Berlin in 25 hours and 45 minutes, and continued his epic journey to reach New York again in 21 hours less than his flight with Gatty had taken, despite having damaged the Vega's undercarriage and propeller during a forced landing in Alaska.

Not resting from this achievement, Post began a co-operative effort with the B. F. Goodrich company to develop a pressure suit which would enable him to explore the possibilities of gaining speed increases through flying in the thin air of the sub-stratosphere, and also make an attempt on the world altitude record, then held by an Italian pilot at 14432 m (47,352 ft). In this he failed, despite the much-modified Vega reaching 16764 m (55,000 ft) (a recording barograph did not function, so the record could not receive official sanction). Three attempts were frustrated by technical difficulties. A fourth, for which 'The Winnie Maes's' Wasp engine had been fitted with a second supercharger to increase power output at high altitude and a jettisonable undercarriage to reduce drag (a central skid was provided for a dead-stick landing) ended in an emergency landing on Muroc Dry Lake in California's Mojave Desert (later the site of the now-famous test centre Edwards Air Force Base). A saboteur, never identified, had poured two pounds of metal filings into the Wasp's intake manifold. In 1935 Post made three attempts with the jettisonable-undercarriage Vega to establish a new transcontinental speed record by riding high-altitude tailwinds, but each was thwarted by technical problems despite 'The Winnie Mae' reaching a speed of 547 km/h (340 mph). Bought by the US government after Post's untimely death later that year, this most famous single-engined Lockheed is displayed in the National Air & Space Museum in Washington, DC.

Vega offspring

The Vega spawned a series of other designs based on the same basic airframe components. The performance of the early Whirlwind-powered Vegas attracted the interest of Western Air Express which operated Air Mail Route 4 between Los Angeles and Salt Lake City with Douglas M-2 biplanes. To meet the carrier's requirements for an aft cockpit location, Jack Northrop redesigned the Vega with a new cantilever wood wing of slightly greater area mounted parasol-fashion above the four-seat/9.29-m³ (100-cu ft) cabin/cargo area, and created a new open cockpit to the rear. Designated Model 3 Air Express, the first aircraft (4897) was completed and flown in April 1928 and was delivered to WAE two months later, but was badly damaged on landing at Las Vegas on its inaugural mail flight.

Though totally rebuilt by Lockheed, its Wright Whirlwind engine replaced by a 309-kW (410-hp) Pratt & Whitney Wasp with streamlined NACA cowling (the first such to be installed in a commercial aircraft), neither this

A small number of Lockheed 'singles' were used by the military, including this Altair (32-232). Designated Y1C-23, it was used for high-speed staff transport. It was one of the DL-2A aircraft with a Detroit-built metal fuselage.

The last of Lockheed's early 'singles' was the Model 9 Orion, which brought the design process full circle by uniting the capacious and sturdy fuselage of the Vega with the high performance wing of the Altair.

Specification
Lockheed Vega 5C
Wingspan: 12.5 m (41 ft)
Length: 8.38 m (27 ft 6 in)
Height: 2.6 m (8 ft 6 in)
Wing area: 25.5 m² (275 sq ft)
Powerplant: one 336-kW (450-hp) Pratt & Whitney Wasp radial engine
Passenger capacity: 6
Empty weight: 1163 kg (2,565 lb)
Maximum take-off weight: 2041 kg (4,500 lb)
Maximum speed: 298 km/h (185 mph)
Cruising speed: 266 km/h (165 mph)
Service ceiling: 5791 m (19,000 ft)
Maximum range: 1167 km (725 miles)

Specification
Lockheed Sirius DL-2
Wingspan: 13 m (42 ft 10 in)
Length: 8.5 m (27 ft 10 in)
Height: 2.8 m (9 ft 2 in)
Wing area: 27.3 m² (294 sq ft)
Powerplant: one 313-kW (420-hp) Pratt & Whitney Wasp C radial engine)
Passenger capacity: 1
Empty weight: 1342 kg (2,958 lb)
Maximum take-off weight: 2345 kg (5,170 lb)
Maximum speed: 282 km/h (175 mph)
Cruising speed: 233 km/h (145 mph)
Service ceiling: 5486 m (18,000 ft)
Maximum range: 1569 km (975 miles)

Specification
Lockheed Orion 9
Wingspan: 13 m (42 ft 9¼ in)
Length: 8.4 m (27 ft 8 in)
Height: 2.9 m (9 ft 8 in)
Wing area: 27.3 m² (294 sq ft)
Powerplant: one 306/313-kW (410/420-hp) Pratt & Whitney Wasp A/C radial engine
Passenger capacity: 6
Empty weight: 1551 kg (3,420 lb)
Maximum take-off weight: 2359 kg (5,200 lb)
Maximum speed: 354 km/h (220 mph)
Cruising speed: 282 km/h (175 mph)
Service ceiling: 6706 m (22,000 ft)
Maximum range: 1207 km (750 miles)

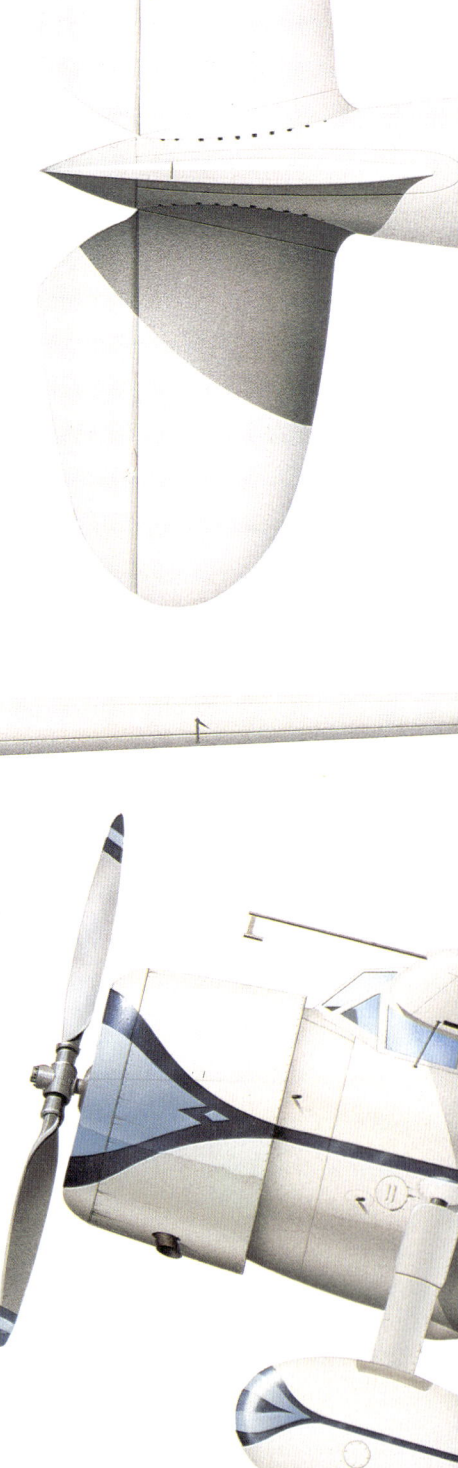

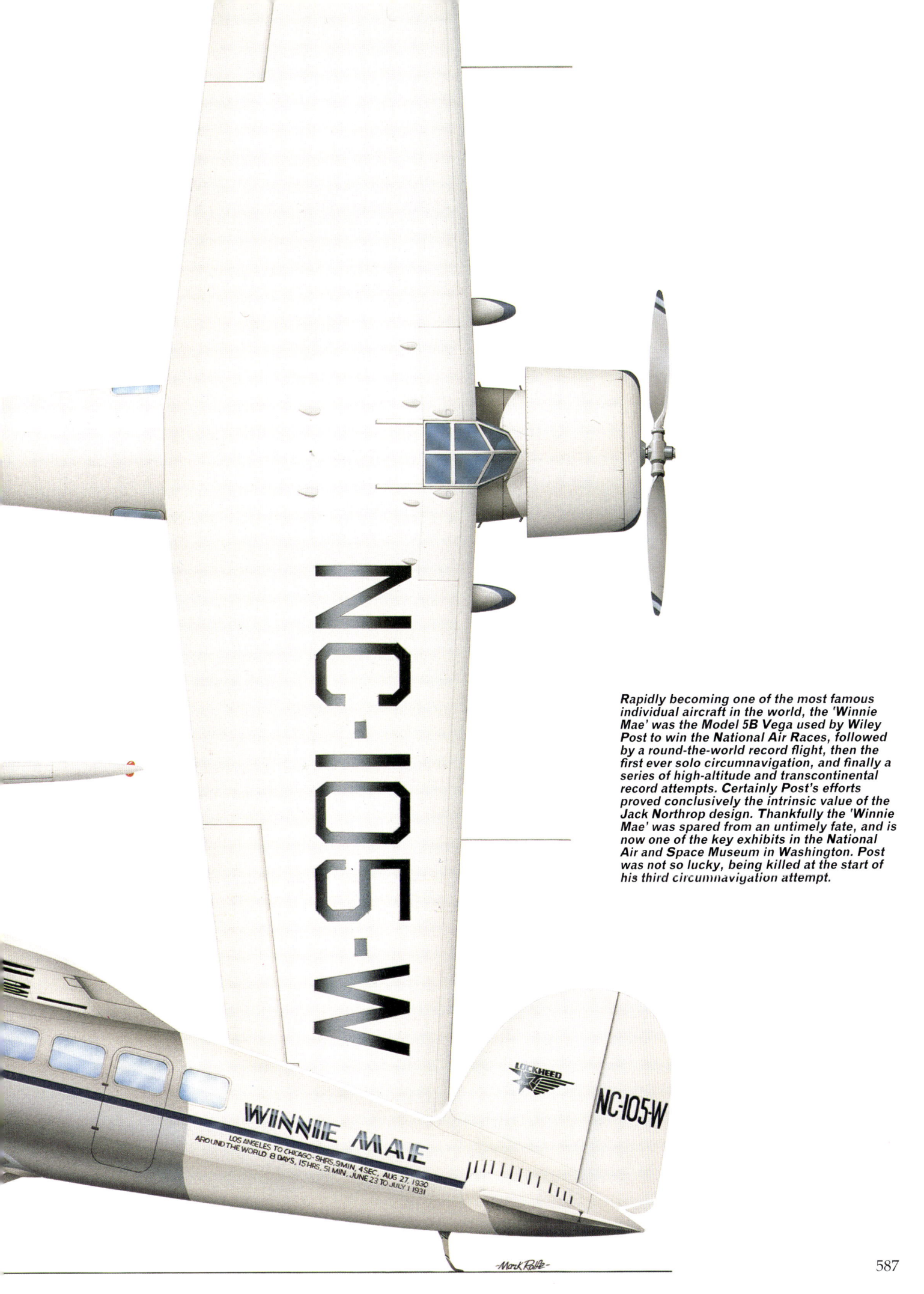

Rapidly becoming one of the most famous individual aircraft in the world, the 'Winnie Mae' was the Model 5B Vega used by Wiley Post to win the National Air Races, followed by a round-the-world record flight, then the first ever solo circumnavigation, and finally a series of high-altitude and transcontinental record attempts. Certainly Post's efforts proved conclusively the intrinsic value of the Jack Northrop design. Thankfully the 'Winnie Mae' was spared from an untimely fate, and is now one of the key exhibits in the National Air and Space Museum in Washington. Post was not so lucky, being killed at the start of his third circumnavigation attempt.

Two views show the famous Sirius 8 of Charles Lindbergh, taxiing towards and being maintained on HMS Hermes 'on the Yangtze River. The aircraft was damaged here and had to be returned to Lockheed for repair, during which time it was re-engined.

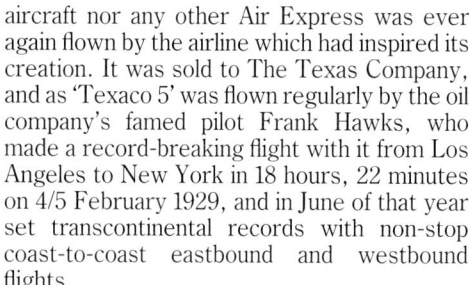

aircraft nor any other Air Express was ever again flown by the airline which had inspired its creation. It was sold to The Texas Company, and as 'Texaco 5' was flown regularly by the oil company's famed pilot Frank Hawks, who made a record-breaking flight with it from Los Angeles to New York in 18 hours, 22 minutes on 4/5 February 1929, and in June of that year set transcontinental records with non-stop coast-to-coast eastbound and westbound flights.

Seven other Air Expresses were built, including two for the New York, Rio and Buenos Aires airline (NYRBA). The most famous Air Express was the General Tire & Rubber Company's 392-kW (525-hp) P&W Hornet-powered NR3057, which won the 1929 National Air Race from Los Angeles to Cleveland at an average speed of 251 km/h (156 mph) before sale to the Gilmore Oil Company. Named 'Gilmore Lion' after the company's trade mark and the pet lion cub which was the constant companion of its flamboyant pilot 'Colonel' Roscoe Turner, this aircraft set many cross-country records, including an eastbound transcontinental trip in 18 hours 45 minutes. The last production Air Express, completed in May 1931, was used by aviatrix Laura Ingalls for an unsuccessful attempt to become the first woman to fly solo across the Atlantic, but three years later Miss Ingalls used the aircraft, then named 'Auto-da-Fé' (Act of Faith), for her 27192-km (16,897-mile) 23-country tour of Latin America, for which she was awarded the prestigious Harmon Trophy.

Least known and least successful of all the single-engined Lockheeds was the Explorer, which was also the first low-wing derivative of the Vega concept. Combining the single open-cockpit wooden fuselage of the Air Express with a new cantilever low wing of 14.6-m (48-ft) span and a 336-kW (450-hp) Wasp engine, the first Model 4 Explorer (NR856H) was completed in June 1929 after a protracted stop-go two-year building period, and delivered as 'City of Tacoma' to wealthy Washington State lumber dealer John Buffelen. Loaded with 3414 litres (902 US gal) of fuel bringing its maximum take-off weight to nearly three times its empty weight, the aircraft crashed on take-off from Pierce County Airport, Washington, on 28 July 1929 when attempting a non-stop trans-Pacific flight to Japan. Pilot Albert Bromley was unhurt, but the aircraft was destroyed. A second 'City of Tacoma' was built; it too crashed, during test flying. Undeterred, Buffelen ordered a third; it also crashed, fatally this time, during full fuel load take-off tests on Muroc Dry Lake. The single Model 7 Explorer was built for a planned attempt by Art Goebel on the New York-Paris speed record, but was never delivered to him, and, after being damaged twice in quick succession, was used to create the Orion-Explorer in which Wiley Post and Will Rogers were killed

Among the pilots who had flown and been impressed by the Air Express was Colonel Charles Lindbergh, who commissioned Lockheed's new chief engineer Gerry Vultee (Jack Northrop had left to form his own company) to design him a new low-wing monoplane. Known as the Sirius 8, this was based on the Explorer concept with two open cockpits, fixed spatted undercarriage and a 336-kW (450-hp) Wasp engine. Test flying of the aircraft (NR211) commenced in November 1929, and continued for some months in the hands of Lockheed's Marshall Headle and Lindbergh himself, leading to many modifications, including the addition of sliding cockpit canopies, before it was accepted by the pioneering aviator. On 20 April 1930 Colonel Lindbergh and his wife set off in the Sirius from Glendale, California and, after a brief refuelling stop at Wichita, set a new transcontinental record to New York, which they reached in 14 hours 45 minutes and 32 seconds.

To prepare for a number of survey flights which the Lindberghs were to undertake for Pan American Airways, the aircraft was then re-engined as a Sirius Special with a 429-kW (575-hp) Wright Cyclone and equipped with twin Edo floats, which also doubled as additional fuel tanks, raising the aircraft's total capacity from 1575 litres to 1953 litres (416 to 516 US gal). Between 27 July and 2 October 1931 the Lindberghs flew NR211 from New York via the Great Lakes to Alaska, and thence via the Bering Straits to the Aleutian Islands and down the Siberian coastline to Japan and China, where the Sirius Special came to grief at Hankow while being lowered from the British aircraft carrier HMS *Hermes* onto the Yangtze River. It was shipped back to Lockheed for rebuilding, again re-engined, with a 530-kW (710-hp) Cyclone driving a controllable pitch propeller, and – now named 'Tingmissartoq' (Eskimo for one who flies like a bird) – used for a 48279-km (30,000-mile) survey flight of the North and South Atlantic, after which it was donated to the American Museum of Natural History in New York. Like Post's

Early Vegas were powered by the Wright Whirlwind J5 or J6 engine, and featured uncowled cylinders. After export to Norway, this example demonstrates the type's versatility by operating on skis.

This Vega was originally built as a Vega 2, but is seen in later life after re-engining with a Wasp engine. The floats impaired performance only slightly, but did require a ventral fin for additional keel area.

Several Vegas (including this Model 5-C) kept active after World War II, although usually with different engines. The high wing and sturdy construction were suited to the rough treatment handed out by bush-fliers.

Orion numbers were low compared to the Vega, but the aircraft were prized by the airlines which bought them, being able to offer extraordinary speeds on mail-carrying sectors. TWA was one of the major operators.

Sir Charles Kingsford Smith used an Altair (a re-worked Sirius) for the first flight from Australia to the United States. The aircraft and crew were lost during an England-Australia record attempt.

The Altair/Sirius production run did not amount to many aircraft, but they became well-known as high-performance thoroughbreds. This Altair (note retractable undercarriage) was exported to the Japanese Mainichi newspaper.

globe-trotting 'The Winnie Mae', 'Tingmassartoq' is now part of the NASM collection.

The success of the Lindberghs' flights led to production of a further 14 Siriuses, one of which, named 'Anahuac', was used by Mexican Air Force Colonel Roberto Fierro to make the first non-stop flight between New York and Mexico City on 23 June 1930, and subsequently was modified for service as a fighter-bomber on the Republican side in the Spanish Civil War.

Charles Lindbergh's original specification for his Sirius had called for two sets of wings to be constructed, one of which was to incorporate a retractable undercarriage. Lockheed's engineers devised a manually-cranked retraction system in which the wheels retracted inwards to lie fully-flush with the lower surface of the wing. It was the world's first truly modern retractable landing gear, but in the event this set of wings was not needed by Lindbergh and in August 1932 was mated to a Lockheed-owned Sirius 8A to create the prototype Model 8 Altair. Four further Sirius airframes were converted to Altairs, and six more Altairs were manufactured from scratch. Among the total of 11 Altairs were two metal-fuselaged Model DL-2As supplied to the US Military, one as staff transport for the Assistant Secretary of War under the US Army designation Y1C-23, and the other as the US Navy's XRO-1, which served as command transport for the Assistant Secretary of the Navy; both were based at Washington, DC. Best known of the Altairs was 'Lady Southern Cross', purchased shortly after conversion from a Sirius Special by the Australian aviator Sir Charles Kingsford Smith for an abortive entry in the 1934 MacRobertson Air Race from London to Melbourne. Withdrawn because of technical difficulties, 'Lady Southern Cross' subsequently made the first eastbound flight from Australia to the United States between 20 October and 4 November 1934, but was lost with Sir Charles and his navigator Tommy Pethybridge when it is believed to have crashed into the Bay of Bengal on the Allahabad-Singapore leg of an attempted record flight from England to Australia.

The Sirius and Altair were specialised aircraft, unsuited to commercial use, but the performance advantages of low-wing monoplanes and retractable undercarriages were not lost on airline operators or aircraft manufacturers, and it was thus a natural progression for Lockheed to combine the Altair wing with the six-passenger cabin of the Vega to create the Model 9 Orion. First flown by Marshall Headle in March 1931, the prototype Orion (X960Y) was powered by a 313-kW (420-hp) Wasp A engine and received its Type Certificate on 6 May, subsequently entering service with Bowen Air Lines in Fort Worth, Texas.

Orion puchasers

Thirty-four wooden-fuselage Orions followed it, and a single metal-fuselage Model 9C Special variant converted from an Altair DL-2A. This latter aircraft was bought by the Shell Petroleum Corporation, named 'Shellightning' and flown by the celebrated Jimmy (later Brigadier-General) Doolittle. Later operated by Hollywood movie flier Paul Mantz, it was purchased in 1976 by the Swiss Transport Museum at Lucerne, where it is still displayed.

Swissair was the only foreign airline operator to buy Orions new, taking delivery of two 429-kW (575-hp) Wright Cyclone-powered Model 9Bs in 1932. They later found their way into Republican hands during the Spanish Civil War, along with 11 other examples of the breed, all of which were destroyed during the conflict. Among other airline operators of the Orion, which with a maximum speed of 354 km/h (220 mph) was faster than most contemporary pursuit aircraft in US service, were American Airways, Continental Airways, New York and Western Air Lines, Northwest Airways, Pan American, Transcontinental & Western Air (TWA) and Varney Speed Lines in the United States and Lineas Aéreas Occidentales SA of Mexico. Others were employed as executive aircraft, one was outfitted with broadcasting equipment and a wing-mounted serial camera pod as a flying newsroom for the *Detroit News*, and Laura Ingalls' specially-built Model 9D 'Auto-da-Fé' made the first non-stop transcontinental flight from Burbank to Newark, New Jersey, on 1 July 1935 in 13 hours, 34 minutes and five seconds.

Orion-Explorer

One of TWA's 336-kW (450-hp) Wasp-powered Model 9Es was acquired by Wiley Post in 1935 and rebuilt with the wing of the Model 7 Explorer to create the hybrid Orion-Explorer NR12283. Mounted on Edo floats, and flown by Post and his companion, Oklahoma-born comedian Will Rogers, who was a national figure in the USA, the unique aircraft suffered an engine failure shortly after take-off from a forced-landing at Walakpi, Alaska, at the start of a planned westbound around-the-world flight and crashed into a shallow river, killing both fliers.

In October 1934 legislation was passed in the United States forbidding the use of single-engined aircraft on scheduled passenger services at night or on routes over terrain unsuitable for forced-landings. The heyday of Lockheed's 'plywood bullets' was over.

A one-off was the Orion-Explorer, mating the fuselage of an ex-TWA Orion with the wings of the abortive Model 7. Wiley Post and Will Rogers were killed in this machine at the start of yet another round-the-world flight.

Lockheed TriStar

In the brash and brawling 1960s business deals surrounding the Lockheed TriStar and Douglas DC-10, only the shoot-out at the OK Corral was missing. Seldom have two aircraft been the focus of so much attention, each in its own unfortunate way. Nevertheless, the TriStar emerged as a good aircraft, despite the political and financial scandals that dogged its early career.

The mid-1960s witnessed the now famous operational request of American Airlines to the US aircraft industries for a 250-seat wide-bodied short- to medium-range commercial aircraft: it was a far-sighted specification for an aircraft tailored to the high-capacity Chicago to Los Angeles, and La Guardia (New York) to Chicago routes, with particular ability to operate from the confines of La Guardia, alongside Flushing Bay and within the shadows of Manhattan and the traffic patterns of nearby Kennedy International. Fuselage length was to be no longer than 36.39 m (185 ft), with the new generation of quiet and fuel-efficient turbofans providing the power. American made no choice, but laid before the aspiring companies the Pratt & Whitney JT9D, the General Electric CTF39 and the Rolls-Royce RB.178-51. Ease of maintenance, quick turn-around and automatic landing capabilities were also specified. The American Airlines' wish posed a demanding but possibly highly lucrative proposition, and both Douglas and Lockheed answered the call to action. Under the leadership of William M. Hannan, the design team of Lockheed-California started work on the project, initially the Model 193 but later changed to the Model 385 Lockheed L-1011 TriStar, in January 1966: some nine months later, on 11 September, Lockheed's president, Daniel J. Haughton, announced his company's ability to accept orders. The race was on.

Despite the heady business atmosphere of the time, unhindered by the nightmare of high energy costs, both McDonnell Douglas (merged in April 1967) and the Lockheed concern were under financial strain. The former had recently lost the competition for the USAF's super-heavy transport to Lockheed, which was destined to be crippled by the development costs of this project, resulting in the C-5A Galaxy. The airframe configurations now adopted by Lockheed and McDonnell Douglas were startlingly similar: both were wide-bodied aircraft with wings swept at 35° and with two pod-mounted turbofans under each wing and a third mounted in or on the rear fuselage. By now the McDonnell Douglas project was the DC-10, with either the General Electric CF6 or Pratt & Whitney JT9D engines as options; and Lockheed's intended first two customers (American Airlines and Delta Air Lines) had plumped for the Rolls-Royce RB.211, the British company receiving the order in March 1968, much to its delight. However, it was doubtful whether Lockheed had any insight into the technical problems that the development of the RB.211 would provide for Rolls-Royce, or of the rapidly deteriorating financial situation that this British company was experiencing largely as a result of the research and development costs of this powerplant.

Cut-throat competition

Desperate though the position was, the DC-10 design team of McDonnell Douglas excelled, and produced a product that featured great adaptability: through access to more powerful models of the CF6 and JT9D that were immediately on hand, the company was in a position to market the DC-10 to a far wider selection of buyers, while Lockheed's RB.211 had no long-haul capability. Both the DC-10 and the TriStar were entering the design stage when the auctioning commenced. On 19 February 1968 American Airlines, one of the 'Big Four', placed an order for 25 DC-10 Series 10s in a deal totalling $382 million at $15.3 million per unit. Haughton reacted with the usual Lockheed aggressiveness to offer to slash a million dollars off each L-1011 TriStar sold. Heady stuff – and it worked! March 1968 saw Lockheed announcing sales of no fewer than 118 TriStars to Eastern Airlines, TWA and Delta Air Lines, plus a substantial 50 to a British government/BUA/financial consortium, Air Holdings Ltd. The latter concern was formed by the current British government and Lock-

For many years Delta Air Lines has been a model of operating cost efficiency, and in this achievement the TriStar has featured with prominence. Seen in flight over the Rockies, N751DA is an L-1011-500 on a proving flight prior to delivery to Delta, and shows the shortened fuselage.

Lockheed L-1011 Model 385 TriStar of the Japanese internal operator, All Nippon Airways. Externally it is impossible to differentiate between the L-1011-1, -50, -100 and -200 models of the TriStar; the primary modifications include type of engine, tyre, wheel-rim and oleo strength, and internal fuel tankages.

Lockheed L-1011-1 Model 385 TriStar of the German carrier Lufttransport Unternehmen (LTU) which operates freight and charter work from Frankfurt, Bremen, Hanover and numerous airports in Western Germany. Near its peak, LTU's TriStar fleet numbered 11, including five L-1011s, one L-1011-100, one L-1011-200 and four extended-range L-1011-500s.

heed to push sales of the RB.211 and the TriStar to the mutual benefit of both. For McDonnell Douglas it appeared that the war was over: with orders for only 25 DC-10s compared with 168 TriStars, how could the situation have looked otherwise? But McDonnell Douglas fought back. Against offers of generous credit terms and at least $500,000 off each DC-10, the orders started to come in: 60 were ordered on 25 April 1968 by the giant United Air Lines, and this combined with the American Airlines order ensured the future production of the DC-10.

Foresight paid off handsomely for McDonnell Douglas: the KSSU complex (KLM, Swissair, SAS and UTA) ordered the long-range intercontinental DC-10 Series 30 version in June 1969: this was a magnificent aircraft, and appealed to the European and Far Eastern airlines, whereas the L-1011 (primary model of the TriStar) was still merely a short- to medium-range type. But this notwithstanding, it had its flair and glamour for the markets intended.

Resplendent in white and red livery, the first TriStar L-1011-1, registration N1011, was rolled out at Palmdale on 15 September 1970, to be crewed by H. B. Dees, R. C. Cokeley and G. E. Fisher on its first flight on 16 November of that year. It was powered by three derated RB.211-22F engines each giving a maximum of 16556-kg (36,500-lb) thrust. Price undercutting had forced the break-even sales figure to around 250, but the order books were filling, and few clouds were in sight.

The Rolls-Royce disaster

Unwise marketing policies and the development costs of the RB.211 engine, allied with the setback experienced over the use of revolutionary carbonfibre material in its design, were some of the factors involved in the voluntary receivership proceedings into Rolls-Royce Ltd in February 1971: the company was bankrupt. However, such was the strategic and economic importance of this concern that no sane government could see it go to the wall: on 23 February 1971 the company re-emerged from its crisis under government financing as Rolls-Royce (1971) Ltd with Lord Cole at its head. In connection with the RB.211 and the TriStar there now ensued frantic exchanges between the Heath and Nixon administrations, with Haughton as the running-boy. The immediate effect of the collapse was the halting of TriStar production, and the lay-off of some 9,000 Lockheed workers: the RB.211 was already running six months behind schedule in supplies. In short, it was a disaster.

The British government signed an underwriting contract for the RB.211 on 11 May 1971, having indemnified costs to the receiver during the intervening period – but there was a $240 million catch. The money for the RB.211 would be there providing that Washington guaranteed the future of the TriStar. Both Lockheed and Nixon cajoled the banks to provide $250 million on loan, which they did on the condition that Congress too lent its shoulder to the future survival of Lockheed, which was also in dire straits over the C-5A at this time. By just three votes the House of Representatives agreed to guarantee the US loans on 30 July 1971, while, three days later, the Senate cast 49 votes in favour and 48 against. The TriStar was allowed to survive, but only at massive costs and having suffered untold delay at the expense of its formidable rival.

On the line

Powered by RB.211-22Bs each of 19050-kg (42,000-lb) thrust, the first L-1011-1 was delivered to Eastern Airlines on 6 April 1972 for crew conversion. The line, based in Miami, flew its first schedules on 26 April: the aircraft proved reasonable for the domestic US trunk routes. The maximum take-off weight of the L-1011-1 is 195048 kg (430,000 lb), with a maximum payload of 37785 kg (83,300 lb); economical cruise speed is 890 km/h (553 mph) at 10670 m (35,000 ft), and the maximum range with full load is 5077 km (3,155 miles), which is good for internal US and Caribbean routes, but is marginal on longer distance legs such as Bahrain-London at high ambient temperatures for take-off. A total of 400 seats can be used in the all-economy (coach) class, whereas the norm is 256 in first and coach. Eight Class A-type access doors are installed, with catering prepared by an underfloor galley with two elevators. The aircraft has automatic pressurisation and air conditioning, three integrated drive generators, backed up by a Pratt & Whitney Canada ST6 auxiliary power unit (APU), four independent 207 bar (3,000 lb/sq in) hydraulic systems that are backed by air turbine motors, power transfer units, and a ram air turbine. The primary controls are power-operated with spring and Mach-feel.

Making its first flight on 16 November 1970, Lockheed's Model 385 TriStar, N1011, was the company's spearhead aimed at countering the threat of McDonnell Douglas DC-10, and at capturing the lucrative US internal network market. Ship One, as she was known, served with Lockheed as the Advanced TriStar variant.

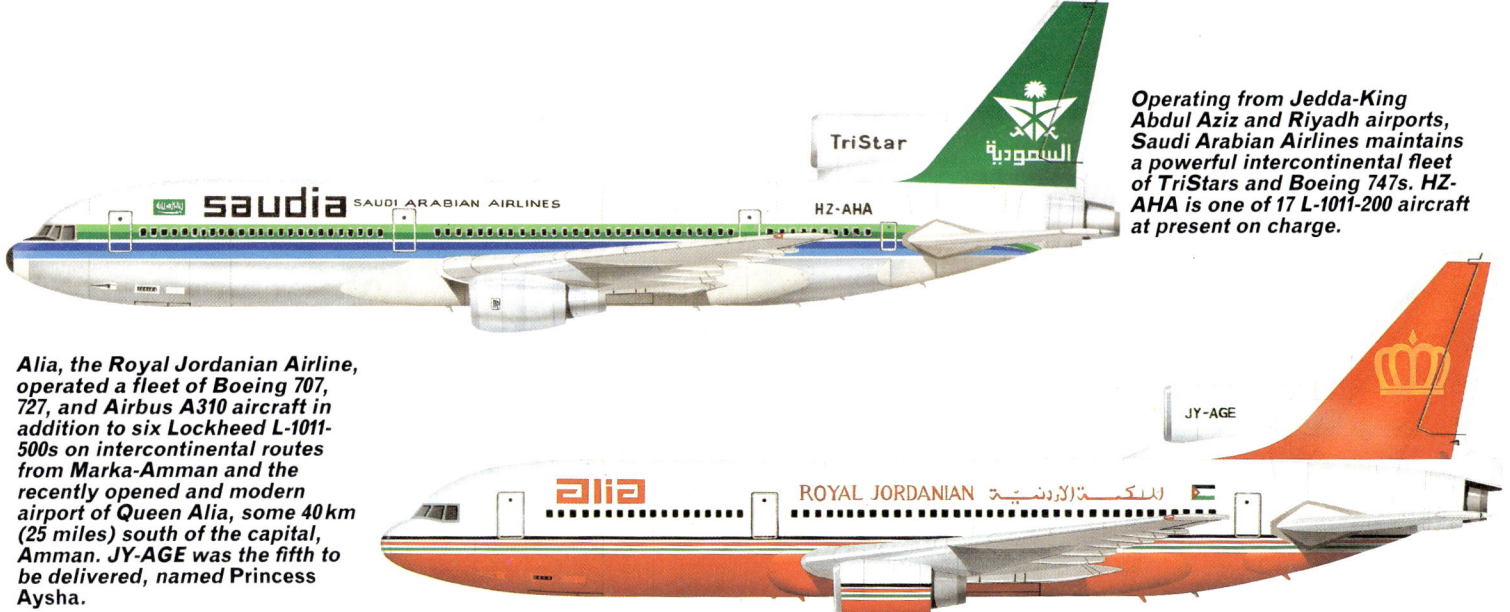

Operating from Jedda-King Abdul Aziz and Riyadh airports, Saudi Arabian Airlines maintains a powerful intercontinental fleet of TriStars and Boeing 747s. HZ-AHA is one of 17 L-1011-200 aircraft at present on charge.

Alia, the Royal Jordanian Airline, operated a fleet of Boeing 707, 727, and Airbus A310 aircraft in addition to six Lockheed L-1011-500s on intercontinental routes from Marka-Amman and the recently opened and modern airport of Queen Alia, some 40 km (25 miles) south of the capital, Amman. JY-AGE was the fifth to be delivered, named Princess Aysha.

In addition to Lockheed-Fowler flaps and leading-edge slats, the L-1011 has six computer-programmed spoilers on each surface of the wings for roll-assist, automatic ground spoiler (AGS) and direct lift control (DLC). The DLC offers a very stable approach with little or no change in pitch required to alter rates of descent: after the selection of flap beyond 30° the spoiler panels Nos 1-4 rise to a 7° null to move over the DLC range with control inputs. The Collins FCS-240 or Sperry autopilot systems are extremely good, and offer a wide range of facilities including altitude capture, auto-throttle, and autoland. In the field of low-visibility take-offs and landings the TriStar was, and still is, in the forefront. Take-offs can be made with the flight director with PVD (para-visual display) reducing airfield minima in some cases to as low as 75 m (246 ft). Approach-land mode enables the aircraft to make stunningly smooth landings automatically in limits down to Category 3B (4.6 m/15 ft with a runway visual range of 125 m/410 ft), while recently the British CAA authorised landings in RVRs down to 75 m (246 ft). Dual autopilots are used in Category 2 and Category 3 landings: at 456-m (1,500-ft) radio altitude the APFDS goes to approach-land mode with the Yaw SAS (stability augmentation system) in parallel with rudder movement. At 45 m (150 ft) Align mode occurs whereby any drift is kicked off by rudder and the aircraft assumes runway-localiser heading, and the into-wind wing is banked to compensate for drift. Flare takes place at 16 m (50 ft), throttles are progressively closed, and just after roll-out (1.5 m/5 ft radio altimeter) the aircraft lands, sometimes daintily on one main into-wind landing gear. Thereafter the aircraft remains on the centreline, with the pilot having recourse to the PVD after autopilot disconnect.

TriStars also carry up to three Carousel IV inertial navigation systems, with one or two tied to the flight management system (FMS): this digital computer is fed with software for two functions in the main. The first is a navigation programme that enables an entire route, say New York—Los Angeles, to be fed into the INS: the pilot merely punches in the code JFK-LAX, and all the waypoints are displayed. In the second, several options are available for the climb, cruise, engine-out, and descent speed and power schedules, two being the Min Cost and Max Fuel programmes. Such a system enables the pilot to make substantial savings in fuel costs.

Later variants

By late 1973 the L-1011-1 had been delivered, or was pending delivery, to Eastern, TWA, Air Canada, Court Line, Delta Air Lines, All Nippon Airways, and Lufttransport Unternehmen; British Airways had nine with options for nine more. Some 56 were in service, with orders and options for 199 more. But it was not really enough, and by this time the DC-10 Series 30 was making great inroads into the intercontinental market. There was the dire need, albeit belated, to increase the TriStar's range and payload. The first offering was the L-1011-100 (RB.211-22B turbofans) that had two fuselage tanks with 8165 kg (18,000 lb) of extra fuel, in addition to the four integral wing tanks: maximum take-off weight was also increased to

The first delivery of the Lockheed L-1011-500 to Pan American World Airways was N64911, Clipper Eagle, *which was powered by three RB.211-524B engines, each of 22680-kg (50,000-lb) thrust; the maximum take-off weight is increased to 228,610 kg (504,000 lb), making it the heaviest of the TriStar models.*

The TriStar 1 was tailored to meet the demands of transcontinental US carriers operating high density sectors. Pacific Southwest bought the type, this being their first aircraft. Note the smiling mouth applied to the airline's aircraft (now absorbed by American Airlines).

211375 kg (466,000 lb), with payload (maximum) range increasing to 6335 km (3,980 miles): several L-1011-100s were ordered by TWA, Air Canada, Gulf Air, Cathay Pacific and Saudi Arabian Airlines. Extra power came with the RB.211-524 (21773-kg/48,000-lb thrust) engine in the Lockheed L-1011-200, which received FAA certification on 26 April 1977: which became the most adaptable of the TriStar series, but its tankage (similar to the Dash 100) and relatively low zero fuel weight (ZFW) failed to match the range-payload capability of the DC-10 Series 30.

Long-range TriStar

In an attempt to alter the range properties of the TriStar radically, the L-1011-500 was produced, receiving FAA certification in December 1979: powered by RB-211-524Bs or B4s (22680-kg/50,000-lb thrust) the L-1011-500 had a shorter fuselage (reduced by 4.11 m/13 ft 6 in), six Class A doors, an engine no. 2 fairing, and no tail skid. A large fuselage fuel bay raised tankage to 96165 kg (212,000 lb) precluding an under-floor galley. The wing tips were lengthened by 1.37 m (4 ft 6 in), with load relief against gusts and turbulence catered for by the ACS (Active Control System): when the aircraft was clean, each outboard aileron was up-set to a null of 8° and was programmed to modulate with wing bending and load. By shortening the fuselage and lengthening the wings the L-1011-500 had lost some of its stability about the horizontal (pitch) axis at high Mach numbers: to guard against excessive accelerations, an RSB (Recovery Speed Brake) was fitted, and this deployed the speed brakes automatically at 0.85 Mach with applied *g* and in level flight in 0.885 Mach conditions. This facility is restricted to some operators. With a maximum take-off weight of 228615 kg (504,000 lb) the L-1011-500 is the heaviest of the TriStars, and has the best performance, with a range of 9697 km (6,025 miles). One British Airways L-1011-500 flew empty non-stop from Montevideo to London.

Few criticisms can be levelled at the TriStar, which continues to offer much passenger appeal. However, its always-threatened production came to a halt on 19 August 1983 when the 250th and last TriStar was rolled out at Palmdale. Several of the TriStar 1s were later modified to better standards (Series 250) with RB.211-524B engines, while a handful have been converted for RAF use as dual-role tanker/transports. The Series 50 was another updated model with increased take-off weight but retaining the RB.211-22B turbofans.

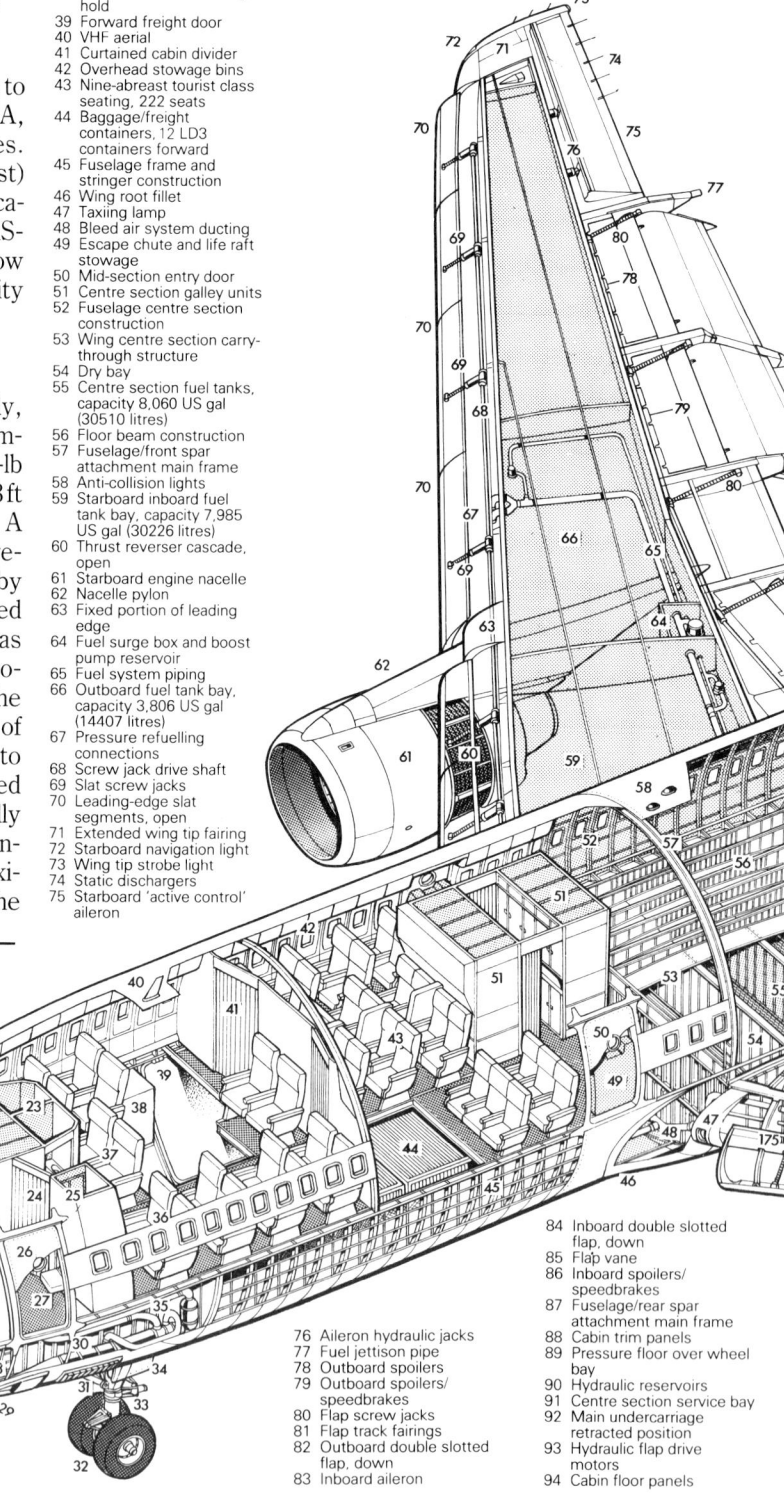

Lockheed TriStar 500 cutaway drawing key

1. Radome
2. VOR localiser aerial
3. Radar scanner dish
4. ILS glideslope aerial
5. Front pressure bulkhead
6. Curved windscreen panels
7. Windcreen wipers
8. Instrument panel shroud
9. Rudder pedals
10. Cockpit floor level
11. Ventral access door
12. Forward underfloor radio and electronics bay
13. Pitot tubes
14. Observer's seat
15. Captain's seat
16. First officer's seat
17. Overhead panel
18. Flight engineer's station
19. Cockpit roof escape hatch
20. Air conditioning ducting
21. Forward galley units
22. Starboard service door
23. Forward toilet compartments
24. Curtained cabin divider
25. Wardrobe
26. Forward passenger door
27. Cabin attendant's folding seat
28. Nose undercarriage wheel bay
29. Ram air intake
30. Heat exchanger
31. Nose undercarriage leg strut
32. Twin nosewheels
33. Steering jacks
34. Nosewheel doors
35. Air conditioning plant, port and starboard
36. Cabin window panel
37. Six-abreast first-class seating, 24 seats
38. Forward underfloor freight hold
39. Forward freight door
40. VHF aerial
41. Curtained cabin divider
42. Overhead stowage bins
43. Nine-abreast tourist class seating, 222 seats
44. Baggage/freight containers, 12 LD3 containers forward
45. Fuselage frame and stringer construction
46. Wing root fillet
47. Taxiing lamp
48. Bleed air system ducting
49. Escape chute and life raft stowage
50. Mid-section entry door
51. Centre section galley units
52. Fuselage centre section construction
53. Wing centre section carry-through structure
54. Dry bay
55. Centre section fuel tanks, capacity 8,060 US gal (30510 litres)
56. Floor beam construction
57. Fuselage/front spar attachment main frame
58. Anti-collision lights
59. Starboard inboard fuel tank bay, capacity 7,985 US gal (30226 litres)
60. Thrust reverser cascade, open
61. Starboard engine nacelle
62. Nacelle pylon
63. Fixed portion of leading edge
64. Fuel surge box and boost pump reservoir
65. Fuel system piping
66. Outboard fuel tank bay, capacity 3,806 US gal (14407 litres)
67. Pressure refuelling connections
68. Screw jack drive shaft
69. Slat screw jacks
70. Leading-edge slat segments, open
71. Extended wing tip fairing
72. Starboard navigation light
73. Wing tip strobe light
74. Static dischargers
75. Starboard 'active control' aileron
76. Aileron hydraulic jacks
77. Fuel jettison pipe
78. Outboard spoilers
79. Outboard spoilers/speedbrakes
80. Flap screw jacks
81. Flap track fairings
82. Outboard double slotted flap, down
83. Inboard aileron
84. Inboard double slotted flap, down
85. Flap vane
86. Inboard spoilers/speedbrakes
87. Fuselage/rear spar attachment main frame
88. Cabin trim panels
89. Pressure floor over wheel bay
90. Hydraulic reservoirs
91. Centre section service bay
92. Main undercarriage retracted position
93. Hydraulic flap drive motors
94. Cabin floor panels

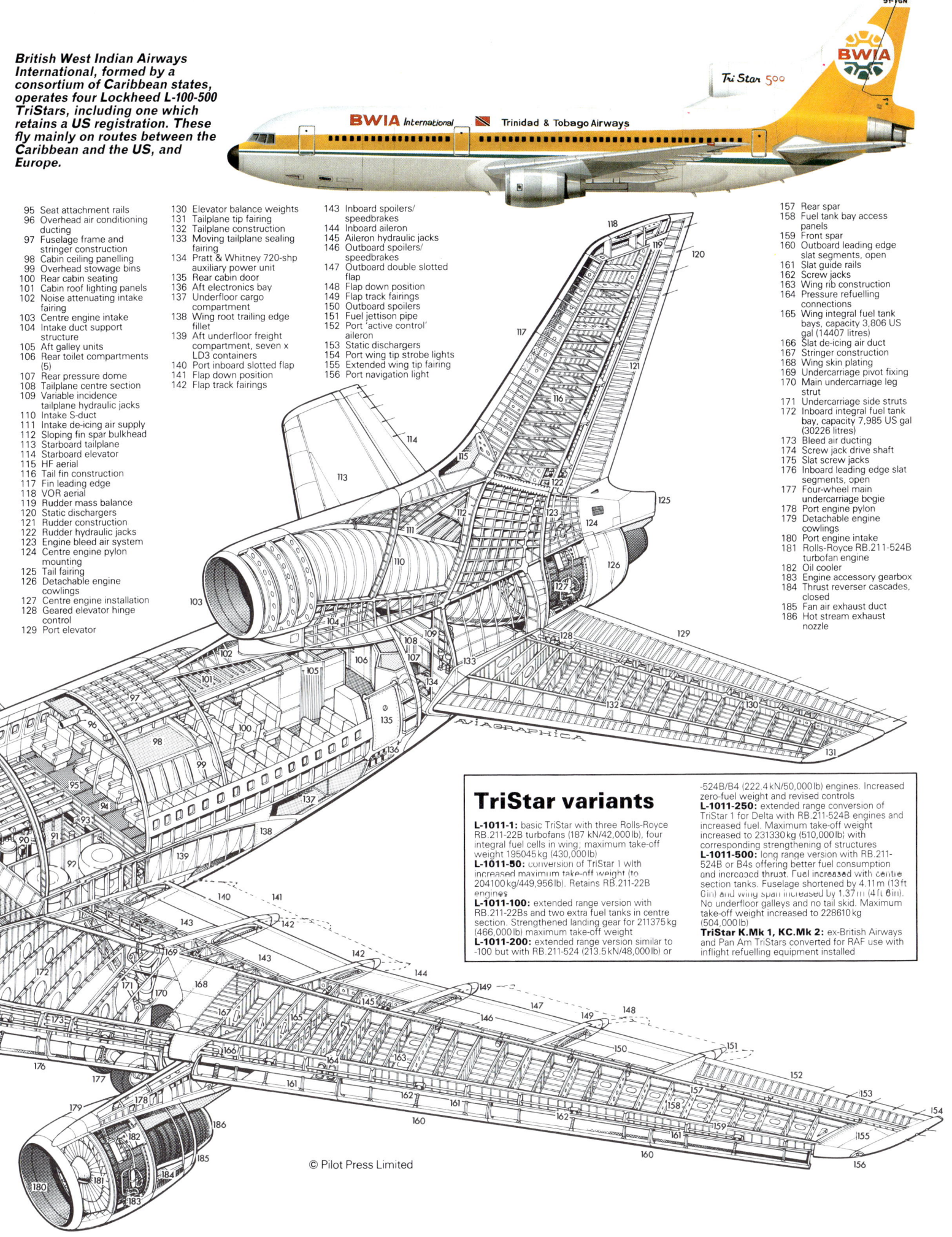

Lockheed Twins

Emerging from the stagnation of the great depression, an invigorated Lockheed Aircraft Company began to forge ahead with a new series of twin-engined aircraft. Its earlier landmark designs were succeeded by the new Electra family, encompassing the initial Model 10 and the later Electra Junior and Super Electra. Acting as everything from airliners to parachutist's jump-ships, they are among the most long-lived types ever built.

In June 1932 Lockheed Aircraft Corporation, which had gone into receivership during the Great Depression, was bought by a group of investors led by San Francisco broker Robert Gross and reorganised at Burbank, California.

Initial activity centred on completing some two dozen Vegas, Altairs and Orions which had been on the assembly line when the company closed, but a priority was the development of an entirely new aeroplane. At first the company planned to develop a single-engined 10-seat transport from a design which had first been conceived by Lockheed's assistant chief engineer Hall Hibbard and Lloyd Stearman when they worked for Stearman-Varney Inc., but Gross was able to persuade Lockheed's new management that a twin-engined aircraft was needed – a wise decision in the event, since two years later single-engined aircraft were banned from carrying passengers over the United States at night or over terrain unsuitable for emergency landings.

Design work on the new Lockheed Model 10 Electra was undertaken by Hibbard, Richard Von Hake and Lloyd Stearman, who proposed an all-metal low-wing monoplane powered by two NACA-cowled 336-kW (450-hp) Pratt & Whitney Wasp Junior radial engines driving two-bladed variable-pitch propellers, with split flaps to reduce landing speed, and a retractable tailwheel undercarriage. Wind-tunnel testing of a scale model of the Lockheed 10 was delegated to Professor Edward Stalker of the University of Michigan's Aerodynamics Department, where a young engineer named Clarence L. Johnson was assigned to the project.

Johnson reported that, in his opinion, changes should be made to the Lockheed 10's single fin configuration, forward-raked windscreen and large wing/fuselage fillets, which he believed would lead to handling difficulties. Although his superior Stalker did not fully endorse his findings in the report submitted to the manufacturer, Lockheed hired Johnson to redesign the aircraft.

Johnson designed a twin fin/rudder configuration for the Electra – creating a shape that was to become a Lockheed hallmark for many years – but the forward-raked windscreen and fillets were retained on the prototype (X233Y)

As with other models of the Lockheed Twin family, the Model 18 was not particularly popular with airlines in the United States, most of which were operating the slower but larger Douglas DC-3. One exception was that shown here, Alaska Star Airlines.

when it made its first flight in the hands of test pilot Marshall Headle on 23 February 1934. As test flying proceeded, Johnson's other recommendations soon proved correct, and the prototype Electra was given a V-shaped windscreen and had its wing root fairings modified. The Electra received its type certificate on 11 August 1934, by which time launch customer Northwest Airlines had already received several of its 13-fleet order and crashed one of them four days previously!

With a comfortable 10-seat cabin and a 306-km/h (190-mph) cruising speed which matched that of the bigger and much more powerful Douglas DC-2, the Electra proved popular with passengers, though limited capacity at a time of rapidly expanding demand for air transport made it less attractive to operators. A total of 148 production Electras was built between August 1934 and July 1941. There were four main commercial versions. The Model 10-A was the major production variant, 101 of which were built. It was powered by Wasp SB engines fitted with variable-pitch propellers, though constant-speed propellers became available during the production run and were retrofitted to some aircraft. The Model 10-B was the only Electra model not powered by Pratt & Whitney, having two 328-kW (440-hp) Wright R-975-E3 Whirlwinds; 18 were built between September 1935 and July 1937. Pan American Airways became the second customer for the Electra when it ordered eight Model 10-Cs with Wasp SC1 engines (of which it had a surplus stock) for its Alaska Division and Latin American subsidiaries Aerovias Centrales and Cubana. The Model 10-E was the most powerful model of all, with 447-kW (600-hp) Wasp 52H1 engines. Fifteen of those were built.

Apart from Northwest and Pan American, other US airlines which ordered Electras included Braniff, Chicago & Southern Airways, Continental Airlines, Delta, Eastern Air Lines, and National Airways. European operators included British Airways, which took delivery of seven in 1936/37; LOT of Poland, whose five Electras escaped to Bucharest after the German invasion of Poland and were bound for Imperial Airways' base at Cairo when they were seized by the Nazis before they could continue their journey; Aeroput of Yugoslavia with eight; and LARES of Romania with seven. British Airways' Model 10-A G-AEPR became famous when it was used by Prime Minister

Left: British Airways was a customer for the Model 10-A, taking seven in the late 1930s. These were used on European routes, including the trip to Scandinavia, which frequently encountered bad weather.

Below: This ungainly version of the Model 12 was the C-40B, delivered to the US Army Air Corps for tests of tricycle undercarriage and then for use as an airborne radio laboratory. A similar XJO-3 was flown by the US Navy for carrier trials with tricycle landing gear.

After having been satisfied with their Model 10s, British Airways looked to the Model 14 for increased capacity. This particular machine transported prime minister Neville Chamberlain to the Munich Conference in September 1938.

Neville Chamberlain for his September 1938 flight to Germany to appease Adolf Hitler in a fruitlessly optimistic quest for 'peace in our time'. Elsewhere Electras were operated by the Japanese trading company Okura & Co, which placed the first export order in 1935, by Ansett, Guinea Airways and MacRobertson-Miller in Australia, Union Airways in New Zealand, Trans-Canada Air Lines, LAV in Venezuela, and LAN-Chile, which received 10 Model 10-As, including the last production Electra on 18 July 1941.

The Electra also proved popular as a high-speed executive transport and for special missions both with civilian and military operators. The US Navy operated two Electras under the designations XR2O-1 and XR3O-1 as staff/command transports, while the US Army Air Corps ordered four new-build Y1C-36/37s and during World War II impressed into service 27 civilian Electras under the designation C-36A/B/C. Unique among the military Electras was the single XC-35, ordered in June 1936 by the US War Department as a research craft for development of pressurised cabins for high-altitude flight.

While selling the Electras to airlines, Lockheed's vice-president of sales Carl Squier discovered a potential market for a smaller aircraft of similar power to serve as a 'feeder liner'. The US Bureau of Air Commerce also identified this need and in late 1935 invited manufacturers to submit designs by 30 June 1936. Lockheed's Hall Hibbard and 'Kelly'

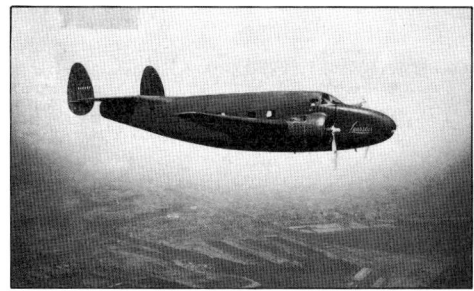

The most famous of the Model 18 modifications was the Learstar, the first example of which is seen here. Considerable aerodynamic refinement resulted in an aircraft which was much faster than the original, much to the chagrin of William Lear's erstwhile friend 'Kelly' Johnson, the designer of the Model 18.

Depicted in wartime camouflage (with rear fuselage red, white and blue flash), this is one of the Lodestars which performed sterling work in BOAC service during World War II. One of their important roles was a regular shuttle to Scandinavia.

Johnson immediately set to work scaling down the Electra to create the Model 12 Electra Junior, which was 1.3 m (4 ft 3 in) shorter than the Model 10, 862 kg (1,900 lbs) lighter at maximum take-off weight, and had a six-seat cabin with optional executive interior featuring club chairs or sofas, work tables, typewriters and other 'flying office' equipment. The Electra Junior retained the 336-kW (450-hp) Wasp SB engines of its bigger brother. An extraordinary effort by design, engineering and production staff saw the prototype Lockheed Model 12 (NX16052) lift off from Burbank on its first flight on 27 June (approximately at 1212 hours), just within the Bureau's deadline. The Electra Junior's excellent handling and top speed of 362 km/h (225 mph) won it both the Bureau of Air Commerce's approval and that of potential customers, so that by the time type certification had been achieved on 14 October the first two customer aircraft were already in the hands of their owners, railroad and oil drilling companies.

Diverse sales

Despite having won the feeder liner competition, the Lockheed 12 achieved little success among airlines, only six of the 130 aircraft built between 1936-42 going to US carriers, and one to Associated Airlines Pty of Australia. The remainder were dispersed among private and corporate operators and US and foreign government agencies and air arms. Among the former were many major oil corporations and the Indian Maharajahs of Jodphur, Jammu and Kashmir, and Jaipur, while overseas military operators included the Argentine army, which commissioned the only two 328-kW (440-hp) Wright R-975-E3D powered Model 12-Bs manufactured, the Brazilian air force, which acquired eight Model 12-As, and the Netherlands East Indies government, which became the largest single operator of the type with 36 aircraft. Sixteen of these were bomber crew training variants designated Model 212, and were equipped with a dorsal turret armed with a 7.7-mm (0.303-in) machine-gun, a similar gun in a fixed, forward-firing position, and under-fuselage bomb racks capable of carrying eight 100-lb bombs.

Two specially modified Electra Juniors with non-retractable tricycle undercarriages were delivered to the US Army Air Corps and US Navy under the respective service designations C-40B and XJO-3, the latter briefly undertaking deck landing trials aboard the aircraft-carrier USS *Lexington* in August 1939 to investigate the suitability of tricycle landing gear for shipboard operation.

About one quarter of the 130 Lockheed 12s built survived the war, and continued to give good service as executive transports before more modern business aircraft appeared. Several are still flying, including at least one which has had its radial engines replaced with 298-kW (400-hp) horizontally-opposed flat-eights.

Super Electra

In the autumn of 1935, while development of the Lockheed 12 was well under way, Lockheed's Hall Hibbard and Kelly Johnson, with project engineer Don Palmer, began design work on a larger transport to compete with the forthcoming Douglas Sleeper Transport (DC-3). Thus was born the third member of the Electra family, designated Model 14 Super Electra, the prototype of which (NX17382) first flew in the hands of Marshall Headle on 29 July 1937.

Though it shared the Electra name, the Model 14 was an entirely new design, featuring a deeper fuselage with accommodation for 14 passengers in single seats either side of a central aisle, and a highly loaded, relatively small and sharply tapered wing for high speed, with area-increasing Fowler flaps to reduce take-off distance and lower landing approach speeds. The distinctive twin-finned tail unit developed by Johnson for the Models 10 and 12 was retained. Other innovative features included under-floor baggage holds, and two-stage superchargers with feathering propellers for the aircraft's two 652-kW (875-hp) Pratt &

One of the best-known Electras was the 10-E Special for Amelia Earhart, fitted with much-increased tankage and additional navigation equipment. After success in the 1936 Bendix Trophy, Earhart and the Electra were lost during a 1937 round-the-world attempt.

The Model 12 Electra Junior was slightly smaller than the Model 10, and intended for the feederliner market. In the event the vast majority of these aircraft were delivered to corporate owners and governments, who appreciated the comfort and speed of the design.

Whitney Hornet S1E-G radial engines. After a brief and uneventful flight test programme the Super Electra received its Type Certificate on 15 November 1937.

Three principal versions of the aircraft were offered: Model 14-H/H2 with 652-kW (875-hp) Pratt & Whitney Hornet S1E-G or S1E2-G engines; export Model 14-WF62 with 671-kW (900-hp) Wright Cyclone SGR-1820-F62s; and Model 14-WG3B with 671-kW (900-hp) Cyclone GR-1820-G3Bs.

Success and setbacks

Launch customer for the Super Electra was Northwest Airlines, which had operated previous Lockheeds, the Orion and Model 10 Electra. It ordered nine Hornet-powered Model 14Hs before the prototype had flown, and introduced the aircraft into service on its Twin Cities-Chicago route in October 1937, before the aeroplane had been certificated. With a cruising speed in excess of 362 km/h (225 mph) and a more comfortable cabin than its predecessor, the Super Electra was well received by the travelling public, but suffered a serious setback when three of Northwest's aircraft crashed within the first 15 months of service.

The Model 14 achieved greater success abroad. The Dutch national carrier KLM, and its East Indies subsidiary KNILM, became the first export customer, taking delivery of 11 Model 14-WF62s between February and June 1938, operating them on the long haul from Amsterdam to Batavia, and in the West Indies. British Airways, a satisfied Model 10 operator, placed an order for four Model 14-WF62s.

Other European operators included LARES of Romania, with four Model 14-WG3s; Aer Lingus (two Model 14-Hs); Régie Air Afrique of France (five Model 14-Hs); and LOT of Poland, which took delivery of 10 Model 14-Hs, one of which became the first airliner ever to make a transoceanic delivery flight, when, in May 1938, it was ferried from Lockheed's plant at Burbank, California, via South and Central America, the South Atlantic and West Africa to Europe.

The Model 18 first flew in 1939, and proved to be a popular design. It had a stretched fuselage and raised tailplane position compared to the 14. This example was a corporate aircraft in the service of Stanley Dollar Industries, seen in 1952.

The biggest single order for the Super Electra came from Trans-Canada Air Lines, which bought 16 Model 14-Hs, later re-engined with 894-kW (1,200-hp) Pratt & Whitney Twin Wasp S1C3-Gs as Model 14-08s. But the largest operator of the type was Japan Air Transport, later Dai Nippon Koku KK, which bought 10 Wright Cyclone GR-1820-G3B-powered Model 14-WG3s, to which were added a further 20 acquired by the Japanese manufacturer Tachikawa Hikoki KK as part of a licence-production agreement. Tachikawa, in conjunction with Kawasaki Kokuki Kogyo KK, built a total of 119 Super Electras between 1940-42 as Army Type LO transports (Allied code name *Thelma*), which were powered by 671-kW (900-hp) Mitsubishi Ha-26-I engines.

Hughes' adventure

Lockheed built a total of 112 Super Electras, including four 820-kW (1,100-hp) Wright Cyclone GR-1820-powered Model 14-Ns supplied to private owners as executive aircraft. One of these, NX18973, was commissioned by Howard Hughes for a round-the-world flight. It was fitted with additional fuel tanks in its fuselage, raising the standard fuel capacity of 2438 litres (644 US gal) to 6980 litres (1,844 US gal). Flown by Hughes, with his four-man crew of co-pilot, navigator, radio operator and engineer, NX18973 left New York's Floyd Bennett Field on 10 July 1938 and circumnavigated the globe in a total elapsed time of 91 hours, 14 minutes, 10 seconds.

Because of the Super Electra's limited commercial success (its subsequent development as the Hudson bomber and maritime attack aircraft is beyond the scope of this article), Lockheed was forced to investigate ways of improving the aircraft's operating economics, high seat-mile costs being a major factor in the Model 14's lacklustre service entry, and thus broaden its customer appeal. Project engineer Jake Cowling headed the team assigned to the task. They decided to increase the aircraft's passenger capacity to 18 by increasing fuselage length by 1.67 m (5 ft 6 in) to enable two additional rows of seats to be installed, a bonus being enhanced directional stability. Three production Model 14s were modified to this new configuration, the first, ex-Northwest Airlines machine NX17385, flying again in its new guise as the prototype Model 18 on 21 September 1939 with Marshall Headle and Louis Upshaw in control.

Lockheed decided to suspend production of the Model 14 in favour of the new Model 18, and to overcome possible customer prejudice and emphasise that this was a new aeroplane, the Electra title was finally dropped and the name Lodestar chosen. The first new-production Lodestar (NX25604) made its maiden flight on 2 February 1940, but test flights of this and the three converted Model 14s revealed a problem with elevator buffeting which was not cured by the first chosen 'fix' of splitting the one-piece elevator into two parts and installing servo tabs. The solution was found in raising the position of the tailplane on a 'saddle' mount to remove it from the effects of turbulent airflow from the wing and extending the chord of the inboard wing trailing edge by 0.3 m (1 ft). Thus modified the Lodestar received its Type Certificate on 30 March 1940 and simultaneously entered commercial service with Mid-Continent Airlines.

As with the Super Electra, Lockheed offered customers a choice of powerplants on the Lodestar, and adopted a two-digit suffix to the model designation to indicate the sub-type.

Known as the B14L, this aircraft was a hasty modification of a Model 14 to Hudson standards for the Royal Air Force. The principal difference lay in the extensive nose glazing, necessary for its intended maritime reconnaissance role.

The major commercial variants were: Model 18-07 powered by 652-kW (875-hp) Pratt & Whitney Hornet S1E2-Gs; 18-08 and 18-10 powered by 895-kW (1,200-hp) P&W Twin Wasp S1C3-Gs; 18-14 powered by 895-kW (1,200-hp) Twin Wasp S4C4-Gs, 18-40 powered by 895-kW (1,200-hp) Wright Cyclone GR-1820-G104As; 18-50 powered by 895-kW (1,200-hp) Cyclone GR-1820-G202As; and 18-56 with similarly rated Cyclone GR-1820-G205As.

Beaten into second place

With similar power (and thus direct operating costs) to the Super Electra, but able to carry four more passengers (or three more, totalling 15, when a full galley was installed and a cabin attendant carried), the Lodestar's seat-mile costs were much better than the earlier aircraft's, and almost matched those of the slower Douglas DC-3. But, since major US carriers were mostly already committed to the Douglas design by the time the Lodestar appeared, its domestic popularity was again limited, Continental Airlines, Dixie Airways,

The Howard Aero company was heavily involved in Lodestar/Ventura modifications, producing some stunning examples such as this machine. Sleek fuselages, executive windows and interiors, huge radials driving four-bladed propellers and even tricyle undercarriages in some cases were hallmarks of the Howard range.

Inland Air Lines, Mid-Continent Airlines, National Airlines, Pan American Airways/Panair do Brasil and United Air Lines ordering a total of only 31 between them.

But, as with the Super Electra, foreign airlines were eager customers for the Lodestar, which found immediate favour with South African Airways, the major customer with a fleet order for 21; Trans-Canada Airlines (12); British Overseas Airways Corporation, which ordered nine for its trans-Africa routes; Air Afrique (five); Air France (three); and other carriers in Brazil and Venezuela.

The US military was to prove the major customer for the Lodestar, 324 of which were built as C-60A paratroop transports for the US Army Air Force, though many were diverted through the Lend-Lease programme to other Allied air arms. In addition, more than 100 civil Lodestars were impressed into USAAF use under the designations C-56 (Cyclone and Hornet engines), C-57 (Twin Wasp) and C-59 (Hornet), 36 Cyclone-engined Model 18-56s were impressed as C-60s/Lodestar IIs for the USAAF/Royal Air Force; and 18 were built for, and 76 impressed by, the US Navy and Marine Corps as R5O VIP/staff transports.

Lockheed built 625 Lodestars in civilian and military models, and, as the Super Electra evolved into the Hudson, so the Lodestar airframe spawned the PV-1 Ventura and PV-2 Harpoon, but these too are outside the scope of this article.

In peacetime Lodestars enjoyed a new commercial life, mostly with small 'third-level' airlines, although notable major operators of the type also included Trans-Australia Airlines, East African Airways and Linjeflyg of Sweden. Prior to World War II nine Lodestars had been delivered to private owners and corporations as executive transports, and it was in this role that the high-speed, comfortable twin was to find its forte in the early 1950s, long before the advent of purpose-built business aircraft. Most major US corporations had a Lodestar in their fleet. Best known, perhaps, was *The Liz*, owned by Hollywood produced Mike Todd and named after his wife actress Elizabeth Taylor; Todd was eventually killed in the crash of this aircraft.

Among operators of the Lodestar was an enterprising and innovative aeronautical engineer and inventor named William P. Lear, later to achieve lasting fame as the creator of the Learjet business jet. He had been given a surplus C-60A by the United States Air Force to act as a testbed for a new autopilot and blind landing system which he was developing for jet fighters. Lear enjoyed flying his Lodestar *Greenie Weenie*, but he believed he could extract much more performance from it by aerodynamic refinement.

Speed ship

Lear hired a team of aerodynamicists which included Gordon Israel and Benny Howard, creators of the pre-war Howard racing aircraft, and a young engineer called Ed Swearingen and set them to work at Santa Monica, California. The result, after two years' work, was the Learstar, first flown by test pilot Les Coan on 10 May 1954. It was the most meticulously refined conversion of a commercial airliner ever undertaken, effectively creating an entirely new aircraft, and the Lear team's attention to detail paid handsome dividends. With extended, squared-off wingtips, cropped tailplane tips, elongated nose and tailcone, a streamlined windscreen and flush cabin windows, smaller landing gear, smoothed, filled and contoured surface skinning and two 1063-kW (1,425-hp) Wright Cyclone R-1820-C9HD engines housed in close-fitting cowlings, the Learstar displayed a maximum speed of 516 km/h (321 mph) and a cruising speed of 450 km/h (280 mph) with a maximum range of 6115 km (3,800 miles) – far better than any contemporary commercial airliner. It could carry up to 12 passengers in elegant comfort, with cabin amenitites which could include a full-service galley, hot and cold running water, fold-away cocktail bar, divan beds and panoramic windows, making the Learstar a true limousine of the air, and at $650,000 the most expensive private aircraft then on the market. Lear sold the first Learstar conversion to the British American Oil Company.

Seen at Floyd Bennet Field, New York, this is Howard Hughes' Model 14-N being welcomed home following his epic round-the-world trip in 1938. He had stopped at Paris, Moscow, Omsk, Yakutsk, Fairbanks and Minneapolis along the way.

Specification
Lockheed Model 14 Super Electra
Type: medium-range airliner/executive transport
Powerplant: two Wright SGR-1820-F62 Cyclone radial engines rated at 671.4 kW (900 hp) each
Performance: maximum speed 402 km/h (250 mph); normal cruising speed 367 km/h (228 mph); service ceiling 7468 m (24,500 ft); maximum range 2558 km (1,590 miles); range with maximum payload 1368 km (850 miles)
Weights: empty 4876 kg (10,750 lb); maximum take-off 7938 kg (17,500 lb)
Dimensions: wing span 19.96 m (65 ft 6 in); length 13.51 m (44 ft 4 in); height 3.48 m (11 ft 5 in); wing area 51.19 m^2 (551 sq ft)
Accommodation: normal flight crew of two; seating for 10-14 passengers

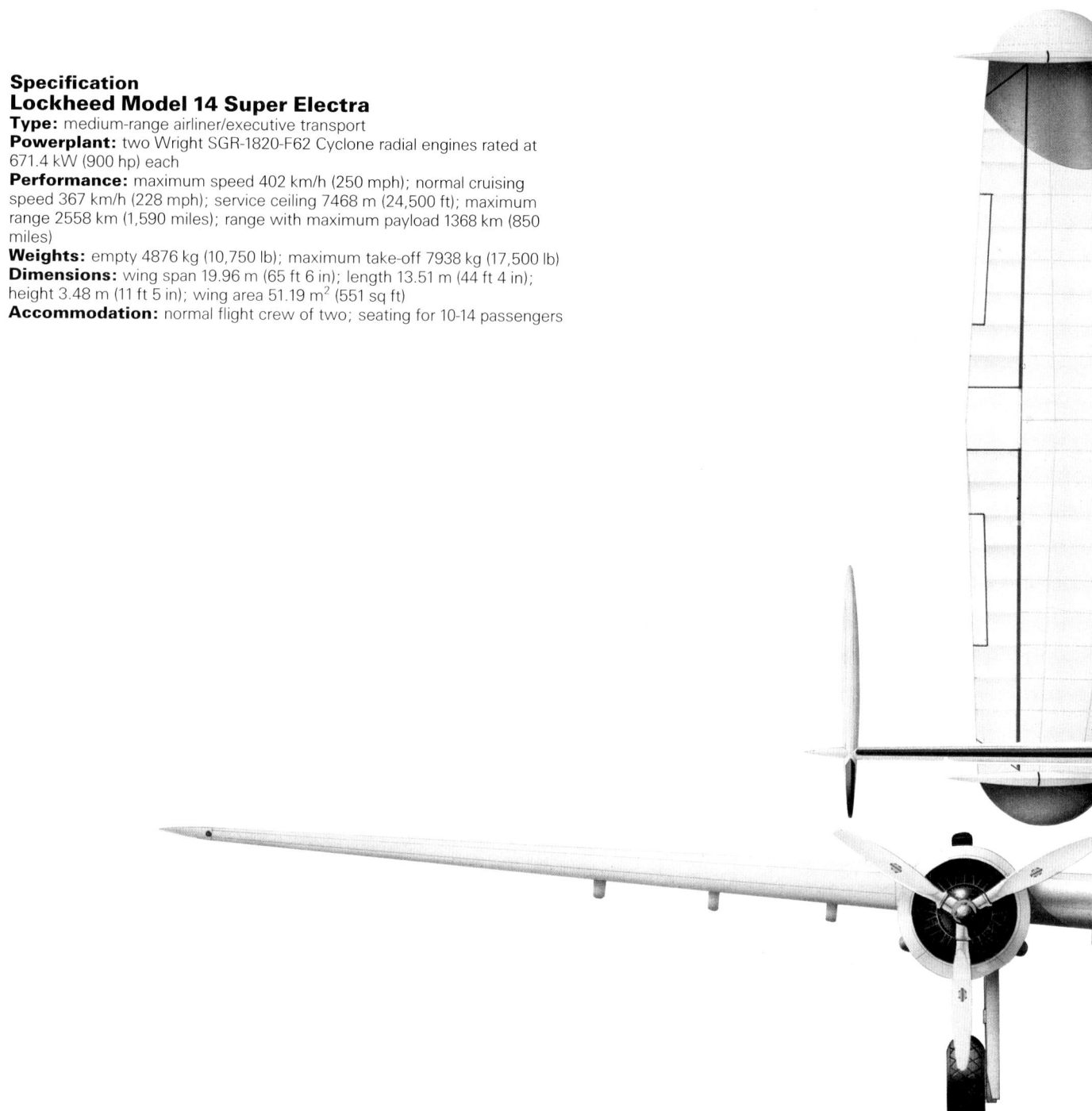

Left virtually unpainted to save weight, this is the Model 14-N Super Electra used by Howard Hughes for his round-the-world flight. By fitting extra tanks in the cabin (and in so doing blocking out most of the windows), fuel capacity was increased by 286 per cent to 6980 litres (1,844 US gal). Sophisticated navigation equipment was installed, as was survival gear and flotation bags. In addition to Hughes, the aircraft had a four-man crew comprising co-pilot, navigator, radio operator and flight engineer. The trip covered some 23611 km (14,672 miles) at an average speed of 331.6 km/h (206.1 mph)

Chapter 28

Martin 2-0-2 & 4-0-4

McDonnell Douglas DC9 & MD-80

McDonnell Douglas DC10

Martin 2-0-2 and 4-0-4

Though principally a manufacturer of military aircraft, Martin hit the mark with its new civil twins following World War II: the piston-engined 2-0-2 and later 4-0-4. The long service careers of these types, which continued until recently, saw them ranked alongside the greatest of the 'propliners'.

As the end of World War II approached, US domestic airlines began to look for replacements for the Douglas DC-3 for use on busy inter city routes. In early 1945 American Airlines issued its specification for such an aircraft, prompting the Consolidated Vultee (Convair) and Glenn L. Martin companies to begin work on suitable designs.

One of the United States' oldest established aircraft manufacturing companies, founder Glenn L. Martin having been a partner with Orville Wright in the Wright-Martin Aircraft Corporation before setting up his own company in 1917, Martin was principally a manufacturer of military aircraft – notably the B-26 Marauder bomber – but had established its reputation in the civilian market with the huge four-engined M-130 *Clippers* built for Pan American Airways in the mid-1930s.

To meet the American Airlines requirement it proposed a twin-engine, low-wing design powered by two Pratt & Whitney R-2800-CA18 Double Wasp 18-cylinder twin row radial engines. Designated Martin 2-0-2, the first of two prototype 'Martin-Liners', NX93001, made its first flight in the hands of company test pilot O. E. 'Pat' Tibbs from Martin's Baltimore factory on 22 November 1946, four months later than the rival Convair Model 110. The two aircraft were strikingly similar in configuration, but while the Convair Liner was almost totally redesigned as the Model 240 before it went into production, the Martin 2-0-2 was quickly granted U.S. Civil Aeronautics Board certification, achieved on 12 May 1947 – the first postwar twin-engined commercial aircraft to be certificated in the United States. Some changes had been incorporated as a result of inadequacies in roll and yaw stability and engine-out handling encountered in early flight testing, so the second prototype, NC93002, differed from the first in having increased dihedral on its outer wing panels and a large dorsal fillet added to its fin, the fairing sweeping up in a graceful curve almost to the fin tip.

A tricycle undercarriage design with cirular-section, an unpressurised cabin seating between 32-40 passengers and a crew of three, the Martin 2-0-2 introduced a number of novel features. These included a self-contained ventral entrance door/stairway or 'airstair' in the underside of the rear fuselage to facilitate rapid loading and unloading of passengers, who could board in safety even with the aircraft's engines running. The 15.4 m (50 ft 8 in) long, 2 m (6 ft 7 in) high cabin was comfortably appointed, usually configured for 36 passengers, with a 7.95 m^3 (281 cu ft) passenger-accessible baggage area so that travellers could hand-carry their bags aboard themselves if they wished; overhead luggage racks were provided for small items. A hot buffet was standard on the aircraft.

The Martin-Liner was of all-metal semi-monocoque construction with a tapered, two-

Systems International Airways (formerly Southern International Airways) was a long-term Martin operator. Its final fleet numbered two Martin 4-0-4s, but shown below is one of its earlier aircraft. Despite their age, the Martins were still a profitable type for such airlines until their engines reached the end of their supportable lives.

Above: The second Martin 2-0-2 returns after its first flight in January 1947. This aircraft was later converted to 2-0-2A standard before being totally overhauled to serve as the Martin 4-0-4 prototype.

Below: In 1952 Pioneer Air Lines boldly traded in its 11 DC-3s to buy nine Martin 2-0-2s from Northwest Airlines. This proved a bad move, as Pioneer lost its C.A.B. subsidy and was forced to revert to a DC-3 operation.

Above: Beating the Convair CV-240 into production, the Martin 2-0-2 gained large orders from airlines such as Northwest. Sadly, it was with Northwest that the type's wing problems occurred.

spar wing employing a Martin W-16 airfoil section. The wing spars enclosed rubber fuel cells of 3785 litres (1,000 US gal) capacity with provision for under-wing refuelling for rapid turnarounds. Double-slotted flaps were installed, and wing and tail leading edges had a hot air de-icing system. Like those of the Convair Liner, the Martin-Liner's Double Wasp engines had a thrust augmentor tube exhaust system to enhance engine cooling and provide a small measure of additional thrust, and drove three-blade Hydromatic fully-feathering propellers. The Double Wasps were rated at 1343.6 kW (1,800 hp) each at 2,600 rpm at an operating altitude of 2590 m (8,500 ft), but were boosted by water injection to 1790.4 kW (2,400 hp) for take-off, giving the aeroplane excellent performance from short airfields or in 'hot and high' conditions, and a cruising speed at 70 per cent power of 460.2 km/h (286 mph) at 3048 m (10,000 ft) – some 128 km/h (80 mph) better than the DC-3 it was designed to replace.

Airlines were eager to order the Martin-Liner on Martin's optimistic promise of early delivery. Eastern Air Lines and Northwest Airlines (NWA) each contracted for 50, while Pennsylvania Central Airlines (later renamed Capital Airlines) signed for 35, Colonial Airlines for 20, and Trans World Airlines for an initial batch of 12. Smaller numbers were ordered by Braniff, Chicago & Southern and Delta Airlines in the USA, and in South America by LAN-Chile and Linea Aeropostal Venezolana (LAV) of Venezuela.

Service and development

Public transport approval was granted by the C.A.B. on 13 August 1947. Initial deliveries were made to Northwest Airlines, although LAN-Chile was the first carrier to operate revenue services with the Martin 2-0-2, commencing on October 1947, while NWA began its operations the following month, and by September 1948 had replaced all its DC-3s with Martin-Liners.

Even before the Model 2-0-2 entered service, Martin was developing a pressurised version of the aircraft, designated Model 3-0-3, and a cargo variant, the Model 3-0-4. The prototype Model 3-0-3 was first flown by 'Pat' Tibbs on 3 July 1947, and attracted a launch order from United Airlines for 50 aircraft.

Meanwhile the Martin 2-0-2's initial success had been short-lived. Deliveries fell badly behind schedule, and operational problems disrupted airline schedules. A fatal crash in 1948 resulted in the aircraft being grounded for investigations which revealed structural weakness in the wing. As a result of this, and the production delays, airlines began cancelling orders. A slump in US domestic air traffic exacerbated the new aircraft's problems.

Above: This is the first of two very early production Martin 2-0-2s sold to Linea Aeropostal Venezolana in 1947. Named 'Rafael Urdaneta', it spent 13 years in South America. Its sister ship was used for spares.

Below: Between them TWA and Eastern Airlines bought 100 Martin 4-0-4s. Eastern's aircraft, the first shown here, was named 'Silver Falcon'. It carried a red eagle on the tail and entered service in January 1952.

Below: California Central Airlines was an inter-state carrier which made waves in the heavily regulated airline world of the 1950s by offering a fare of $9.99 on its Oakland to Burbank route, flown by 2-0-2s from 1951.

Northwest would accept only 25 of the 50 2-0-2s and 3-0-3s it had ordered; United Airlines cancelled its 50-aircraft order entirely. Martin quickly devised a modification programme for the wing structures of those Model 2-0-2s already in service, and developed an improved version of the unpressurised Martin-Liner, designated Model 2-0-2A for future production. This had water-injected 1940-kW (2,600-hp) Pratt & Whitney R-2800-CB16 engines, increased fuel capacity of 5110 litres (1,350 US gal), an increase in maximum take-off weight to 19391 kg (42,750 lb) and interior changes. The prototype Model 2-0-2 was modified as the first 2-0-2A, flying again in its new guise in July 1950. The Model 2-0-2A entered service with TWA on 1 September of that year. Only 12 were built, all for TWA, in addition to two prototypes and 31 production Model 2-0-2s.

Big brother

Following the problems encountered with the Model 2-0-2, Martin elected to abandon further development of the 3-0-3 and 3-0-4 in favour of the substantially redesigned Model 4-0-4. Again the original prototype was rebuilt to serve as the test and certification aircraft. The Model 4-0-4 had a fuselage lengthened by 1.0 m (3 ft 3 in) allowing an extra row of seats to be installed to provide standard accommodation for 40 passengers and three crew in a pressurised cabin. The R-2800-CB16 Double Wasps of the Model 2-0-2A were retained, while maximum take-off weight was increased to 20366 kg (44,900 lb). The prototype first flew on 21 October 1950. A much-improved aircraft, the Martin 4-0-4 quickly attracted an order for 35 aircraft from Eastern Air Lines, and another from Trans World Airlines for 30, later increased to 60 and 41 respectively. They were to be the only commercial customers for the aircraft, although the US Coast Guard ordered two commercial-standard Martin 4-0-4s in 1951 for use in logistical support and VIP transport duties. Outfitted with executive-style interiors, these aircraft entered service with the USCG in 1952 under the designation RM-1Z, and were redesignated VC-3A in 1962 under the new unified designation system adopted by the US military services. The two Martins were subsequently replaced by Grumman Gulfstream Is (VC-4As) and transferred to the US Navy.

TWA was the first carrier to introduce the new pressurised Martin-Liner on its routes, in October 1951, followed by Eastern in January 1952. Production ceased in early 1953 after 103 had been built. By that time the first Martin 2-0-2s were being disposed of, Northwest's fleet going to Pioneer Airlines of Texas, California Central Airlines and Transocean Airlines, and later to Allegheny Airlines (which

The last two Martin 4-0-4s to be built were delivered to the US Coast Guard as RM-1Zs (later VC-3As). They spent their service lives with the Coast Guard and the Navy before their withdrawal in 1969 and 1970.

In 1964 this former Eastern Airlines 4-0-4 entered service with Southern Airlines. In 1976 it was sold to Shawnee Airlines for its short-lived operations.

modified three for use as freighters) and Southwest Airways. TWA's Model 2-0-2As were sold to Allegheny, Modern Air Transport and Southeast Airlines.

The pressurised Model 4-0-4 established a fine reputation for reliability, safety and economic operation, and when the time came for Eastern and TWA to dispose of their fleets, there was no shortage of ready buyers among smaller US domestic carriers such as Mohawk Airlines, Pacific Air, Piedmont Airlines and Southern Airways, which was the last major user of the Martin-Liner, retiring the last of its fleet in 1978. Martins were then enjoying further careers with third level carriers and charter companies such as Florida Airlines, which had a fleet of seven Model 4-0-4s, Naples Airways with eight, and Marco Island Airways and Provincetown-Boston Airlines, each with three. The aircraft also proved popular as an executive transport before the advent of dedicated business aircraft such as Grumman Gulfstream I. A number were modified by specialist outfitters such as Remmert-Werner Inc. and Pacific Airmotive, with plush interiors, weather radar and other refinements. Among operators of executive Martin-Liners was the entertainer Frank Sinatra.

During the 1980s, the Martin fleet was depleted, most surviving examples ending their days hauling cargo in Latin America and none are now in service. Though it never enjoyed the market success, long production run, multifarious variants or aftermarket re-engining and upgrading programmes accorded to its rival the Convair Liner, the Martin-Liner, particularly the pressurised Model 4-0-4, was an advanced aircraft for its day, and in later life was the dependable, cheap-to-operate cornerstone on which many regional airlines and third level carriers were built.

'Skyliner Baltimore' was TWA's first Martin 4-0-4. One of the USA's 'big four' airlines, its order for significant numbers of aircraft, along with Eastern's, did much to save Martin after the Model 2-0-2 disaster.

McDonnell Douglas DC-9 and MD-80

If the capacity for development is the mark of a great design, the DC-9/MD-80 family deserves a place in that small and exclusive category. The latest model in the family carries very nearly twice as many passengers as the first, and weighs nearly twice as much.

The DC-9's capacity for development was inherited. Douglas, the then-independent aerospace and defence company which conceived it in the early 1960s, had based an entire line of commercial, military, medium-range and intercontinental transports on a single basic design, the DC-4, and had produced the line with great success over a 15-year period. The DC-4/6/7 family had been followed by a decision to move into the jet age with the DC-8, which entered service in 1959.

The appearance of the first jets on long-distance services contrasted sharply with the older aircraft used on shorter flights, particularly in the USA, but there was no basic agreement on the best way to replace them. Some argued that fast turboprop aircraft would not be much slower than the jets on the shorter routes, and would be more economical. Others felt that the travelling public would come to regard the jet as a standard, and would identify any propeller-driven aircraft as a product of an earlier era. The right size of such a new aircraft was also a matter of controversy. The market for new short-haul aircraft thus became a fierce and confusing battlefield. Lockheed and Vickers were pushing advanced turboprop aircraft. France had a short-haul jet already in production, in the shape of the Sud-Aviation Caravelle. Neither Boeing nor Douglas had any firm programme to offer, but each believed that jets were the right solution. Both started out by looking at scaled-down versions of their big jets, and by mid-1959 Douglas had made serious presentation of the first DC-9, with four Pratt & Whitney JT10 turbojets, to United Airlines and other major carriers. At the end of 1960, however, United and Eastern (the two biggest US domestic carriers) placed orders for the new and very advanced Boeing 727.

Douglas immediately saw that there was room for a smaller aircraft. Of its two US rivals, Boeing was preoccupied with the Model 727 and Lockheed was still salvaging the Electra programme; the only competition would come from abroad, most probably from an improved version of the Caravelle. United had ordered 20 Caravelles in 1959, and the US manufacturers were worried that lower European labour rates would give the French airliner a competitive edge.

Following the launch of the Model 727, Douglas, Sud-Aviation and General Electric began to hold very serious discussions about the joint development of an improved and Americanised Caravelle, powered by General Electric CJ805-23 aft-fan engines. But the Caravelle was, by that time, a design already eight years old, and it was a first-generation turbine airliner from the systems viewpoint. With the General Electric engines, too, it would be quite close in size to the Model 727. Then, in May 1961, the British Aircraft Corporation announced the go-ahead for its BAC One-Eleven, a somewhat smaller, but completely new aircraft. The Douglas people began to have doubts about the Caravelle.

In 1962, Douglas salesmen began to show airlines a completely new design, the D-2086. Like the Caravelle, One-Eleven and Model 727, it had rear-mounted engines, a clean wing and a short landing

The DC-9 family ended with the Series 80, this group of variants taking up the MD-80 designation. This family itself has extended further, now covering two basic fuselage lengths and a variety of powerplant/range performance and avionics. This is an MD-87 of SAS.

The Scandinavian Airlines System requirement for a 'hot-and-high' model of the DC-9 resulted in the Series 20, which retained the Series 10 short fuselage but added the long-span wings of the Series 30. Only ten were built, all for SAS.

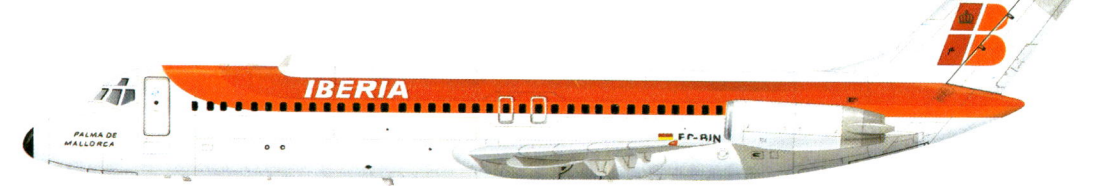

Right: For many years the DC-9 provided the backbone of several European airlines' short/medium haul routes. Typical of these is Iberia, which operates this Series 32. MD-87s were ordered to serve alongside these earlier aircraft but were then cancelled.

gear, the last being particularly important because the new jet was designed to operate in the absence of complex ground-handling facilities. Like the Model 727, the new Douglas aircraft would have to use runways shorter than those used by the big jets. The development of airports in the USA had lagged behind the expansion of the airlines, and the new jet would have to use the same runways as slower piston-engine types. The wing design philosophy, however, was closer to that of the Caravelle than the highly swept, extensively flapped design of the Model 727; Douglas selected a relatively large and moderately swept wing, with double-slotted trailing-edge flaps and no other high-lift devices. The powerplant was to be a pair of Pratt & Whitney JT8Ds, the very promising turbofan engines under development for the new Model 727. The JT8D was a little more powerful than necessary, but it provided room for growth and would be common to the airlines' Model 727 fleets.

This was the DC-9, launched with a 15-aircraft order from Delta Air Lines in April 1963. By this time, the BAC One-Eleven was only a month away from its first flight, and had secured major US orders from American and Mohawk Airlines; speed would be vital if the DC-9 was to catch up. The flight-test programme was intensive. The first aircraft flew in February 1965, and five were flying by June. The DC-9-10 initial production version was certificated in November, just over 30 months after the programme had been launched, and nine months from the first flight, setting an unbeaten speed record for development of a brand-new airliner.

Modified version

By the time Delta started operations with the new type, Douglas was well advanced with development of a new and considerably modified version, designed primarily for operations on the US east coast and in Europe, where runways in the 3050-m (10,000-ft) bracket were generally available, and where the near-transcontinental range of the Model 727 was not needed. The new version was to be stretched by some 4.57 m (15 ft), raising passenger capacity from 80 to 105 seats. The wing was slightly increased in span, and fitted with full-span leading-edge slats, and the JT8Ds were used at their full design rating. The first order for this new version, the DC-9-30, was received from Eastern Airlines in February 1965. The higher-rated engines were also made available on the basic aircraft, which was then designated DC-9-10 Series 15 or DC-9-15.

The DC-9-30 was substantially larger than any version of the One-Eleven, and was more economical, but it faced tough competition from the new Boeing 737, launched just two months later. Douglas held one decisive advantage; the DC-9-30 would enter service in early 1967, before the first Model 737 flew, and the airline industry was growing so fast that airlines were racing to be the first to bring jet service into competitive markets. To take advantage of its lead, Douglas decided to build up production as fast as possible, so that the maximum number of customers could get their DC-9s before Boeing could start deliveries.

Douglas was also prepared to meet any needs the customer might express, and even developed two versions of the DC-9 specifically for one airline, Scandinavian Airlines System. These were the DC-9-40, stretched by two seat rows compared with the DC-9-30 to match the seating capacity of the Model 737-200, and the DC-9-20, a 'hot-rod' version with the original short fuselage, the high-lift wing of the DC-9-30 and the same high-thrust engine as the DC-9-40. Douglas offered customers a huge variety of other options: different fuel capacities, different engine models and different weights, as well as a wide choice of finishes, internal configurations and other features.

The sales strategy was phenomenally successful. The DC-9 sold as no airliner had sold before, and Douglas had orders for more than 400 aircraft by the end of 1966. Douglas was also going broke, and doing so very quickly. The company was still spending money on the development of the DC-9 in all its versions, and new versions of the DC-8 were also on the point of certification. Moreover, Douglas was losing money on every DC-9 that it delivered; the company had sold

Wearing the colours of launch customer Delta Air Lines, this DC-9-14 illustrates to good effect the original short fuselage which offered a maximum passenger capacity of 90. Basically similar to the Series 10, it incorporated increases in fuel capacity and all-up weight.

Immaculately finished in the colours of the Special Air Missions Wing based at Andrews AFB, Maryland, this is one of three VC-9C Nightingales acquired for VIP transport for the USAF. Larger numbers are used in the aeromedical role worldwide by the service.

Designated C-9B Skytrain II in US Navy service, this militarised fleet logistics support transport has a 3.45 m (11 ft 4 in) × 2.06 m (6 ft 9 in) cargo door in the forward port fuselage, which permits the loading of standard military cargo pallets. A total of eight can be carried in the all-cargo configuration.

A foreign customer for the C-9B Skytrain II was the Kuwait air force, which purchased two aircraft in 1975. The aircraft, since redesignated C-9K, are operated in the dual passenger/cargo role. Another military operator is the Italian air force.

many aircraft at low 'introductory' prices, but they were proving more expensive to build than had been predicted. Because production had been built up so fast, a great many DC-9s were being produced 'off the top of the learning curve': assembly procedures were still being refined, and the workers were still learning their jobs, so that each aircraft was taking more man-hours (and costing more money) to build than would be the case later in the programme. That particular problem was compounded because Vietnam war production had already used up all Southern California's pool of trained aerospace workers, and because there were some 20 different airline configurations on the production line within a few months of the first deliveries. War production was also causing delays in the supply of components. The crisis came to a head when deliveries began to slip behind schedule, and some of the airline customers launched massive lawsuits to recover their estimated losses. Facing bankruptcy, Douglas was taken over by the McDonnell company of St Louis, Missouri, at the end of April 1967.

The new management brought DC-9 deliveries back on schedule, and the aircraft retained its hard-won status as the world's best-selling twin-jet airliner through the early 1970s. The DC-9 was selected by a number of large European carriers, outselling the Model 727 and Model 737 in that market; Swissair, KLM, SAS and Alitalia were among the biggest operators of the DC-9. The Douglas aircraft went on to equip affiliated charter airlines. Delta and Eastern were major US operators of the type; outside the trunk airlines, too, the DC-9 proved popular with the US regional airlines such as Allegheny, North Central and Ozark.

Nearly all the DC-9s sold in this period were DC-9-30s; Toa Domestic Airways (TDA) of Japan was the only customer for the DC-9-40 apart from SAS, and demand for the short-runway performance of the DC-9-10 diminished as airport development proceeded worldwide. The DC-9-20 remained an SAS special. An all-cargo version, the DC-9-30F, was delivered to Alitalia in 1968; a similar main-deck cargo door was fitted to the DC-9-30CF (convertible) and DC-9-30RC (rapid-change convertible) variants, both of which were delivered in some numbers. In the course of the 1970s, the DC-9-30 was made available with more powerful variants of the JT8D engine, higher gross weights and auxiliary fuel tanks in the lower fuselage; the last-named option proving attractive to European charter airlines, which needed an aircraft to fly nonstop from northern Europe to the Canaries.

Another DC-9-30 customer was the US Air Force, which placed an order for a specially-equipped aeromedical evacuation variant in August 1967. Designated C-9A Nightingale, these aircraft can carry 30 to 40 stretcher patients, and 21 were delivered between 1968 and 1971. There were two further military developments of the DC-9-30. One is the C-9B Skytrain II, ordered by the US Navy to carry priority freight to overseas naval bases; 15 are in service with the US Navy, and two were delivered to Kuwait. Finally, three C-9Cs are used for VIP transport by the USAF Special Air Missions wing at Andrews AFB, near Washington, DC.

Competition in the twin-jet market grew more intense in the early 1970s, as Boeing introduced its new Advanced 737 series. The DC-9, though, was an inherently easier aircraft to stretch than the Model 737; McDonnell Douglas took advantage of this attribute of the design in mid-1973 by launching the 135-seat DC-9-50. The second major stretch of the DC-9, the DC-9-50 compared with the DC-9-30 as the latter had compared with the DC-9-10. The fuselage of the new version was 4.34 m (14 ft 3 in) longer than that of the basic DC-9-30; higher-thrust engines, also to be offered on the DC-9-30, were standard, but the wing was externally unchanged and the maximum take-off weights were only slightly increased. The DC-9-50 was not intended as the replacement for the DC-9-30, but as a complement to it. It had better economics, but was less flexible in terms of range and runway performance.

Swissair (centrally located in Europe, with consequently short routes, and with no hot-and-high airport problems) was the first customer for the DC-9-50, and started operations in August 1975. The aircraft performed as advertised, but was, inevitably, noisier

The additional thrust available from the Pratt & Whiitney JT8D-15 and JT8D-17 turbofans resulted in the DC-9-50 being launched. Once again a fuselage stretch was incorporated, 1.87m (6 ft 4 in) over the Series 40. Illustrated is a DC-9-51 of the Venezuelan carrier Aeropostal, which had nine of this variant.

than the DC-9-30s to which the people around Swissair's base airports were accustomed. The community reaction caused Swissair to cut back its planned purchases of DC-9-50s, and the airline, a loyal Douglas customer, started to press for a new and quieter large-capacity DC-9.

In the early 1970s, when concern over the environment was at its peak, the US government had launched a number of programmes aimed at reducing aircraft noise. One of these was the development of a modified JT8D with a larger-diameter fan and other changes, specifically intended to make future versions of the Model 727, Model 737 and DC-9 significantly quieter. The new engine would also be more powerful and more efficient.

Meanwhile, McDonnell Douglas was engaged in a long drawn-out sales effort in Japan, where a number of airlines, of which TDA was the largest, were operating turboprop-powered aircraft out of 1200-m (4,000-ft) airfields, and where local opposition to extending runways was uncompromising. In early 1975, McDonnell Douglas proposed a DC-9-QSF (quiet, short-field) to the Japanese airlines, a DC-9-40-sized version with refanned engines and a highly modified wing. The main change was a new, wider centre-section, to which the existing outer wings were attached. Adding extra span in the middle, rather than at the tips, meant that the extra strength required could be built into the new centre-section, avoiding some redesign and retooling; the centre-section could also be made deeper to hold more fuel.

While the Japanese market never opened up, the new wing and engines formed the basis for a new DC-9 variant to supersede the DC-9-50. With a new wing, much increased weights and refanned engines, the revised aircraft would have better economics than the DC-9-50, the operating flexibility of the DC-9-30, and lower noise than either of its predecessor variants. First discussed in 1976, the type was initially known as the DC-9-RSS (refan, super-stretch) and then as the DC-9-55. It soon became clear that it would be the biggest redesign in the history of the DC-9. Its fuselage would be another 4.34 m (14 ft 3 in) longer than that of the DC-9-50. The bigger wing would mean new flaps and changes to the control system. The new centre-section would house a heavier landing gear, and the entire central structure (wing, centre fuselage and tail) would have to be 'beefed up' to handle higher weights. The greater weight and wing area meant a larger tailplane, which in turn would require modifications to the fin. To make the aircraft more attractive in the 1980s, the systems and electronics would also be thoroughly overhauled. The new version would take slightly longer to develop, from go-ahead to certification, than the original, brand-new DC-9-10.

Super 80 launch

Redesignated DC-9 Super 80, the new aircraft was launched in October 1977 with an order from Swissair. Development was not entirely smooth; a strike and production problems delayed the completion of the first aircraft, which flew in October 1979, and both prototypes were damaged in landing incidents during flight tests. The Super 80 was certificated and delivered some five months behind schedule, and entered service with Swissair in late 1980.

Three versions of the basic DC-9 Super 80 have been produced, with progressive increases in engine thrust, weight and range, and improvements in fuel economy. The basic model, the Super 81, was followed by the somewhat heavier Super 82, with JT8D-217 engines, which was launched in early 1981. The Super 82 largely superseded the earlier model, because the newer JT8D-217 engine proved more economical than the original JT8D-209. The next development of the series, the 72576-kg (160,000-lb) Super 83, was launched in early 1983 and entered service in mid-1985 with Alaska Airlines. It has the new JT8D-219 engines with more power and lower fuel consumption, a strengthened airframe and auxiliary fuel in the lower fuselage; its increased range allows it to carry a full load of passengers between Cincinnati and Los Angeles, or from London to Beirut. The MD-87 followed, with a shorter fuselage for a single-class accommodation of 130 passengers. Currently the last member in the family is the MD-88, powered by JT8D-219 engines and with an advanced cockpit. Both MD-83 and MD-87 are available in executive jet versions. All earlier versions of the DC-9 are now out of production, the last DC-9-30s having been delivered in 1982.

The designation of the type has changed twice in the past few years. McDonnell Douglas, concerned by image problems stemming from much publicised accidents to its DC-10, dropped the DC-9 designation in late 1982, and referred to the type simply as the Super 80. After the last commercial DC-10 was produced, in 1983, the company dropped the DC designation completely, and the Super 80 became the MD-80.

The MD-80, by any name, has proved highly successful in service. The combination of low noise and excellent economics (particularly in a high-density, 170-seat configuration) is unmatched by any current aircraft. The MD-80 now dominates the Californian intrastate market; PSA, which replaced many of its Model 727s with MD-80s in the course of one year, is probably the only airline in recent years to have experienced an actual decline in its fleet-wide direct operating costs. A number of other low-cost US carriers, such as New York and Frontier Airlines, also took delivery of MD-80s.

Studies of an advanced 110-seat replacement for the DC-9-30, known as the Super 80SF or Super 40, culminated in the announcement of a planned development known as the MD-90, intended to compete with the Model 737-300. However, this project was terminated by the company in November 1983. McDonnell Douglas considered other advanced versions of the basic design after this time,

McDonnell Douglas MD-80 cutaway drawing key

1. Radome
2. Weather radar scanner
3. Front pressure bulkhead
4. Pitot tube
5. Radio and electronics bay
6. Nosewheel well
7. Twin nosewheels
8. Rudder pedals
9. Instrument panel
10. Instrument panel shroud
11. Windscreen wipers
12. Windscreen panels
13. Cockpit eyebrow windows
14. First officer's seat
15. Overhead switch panel
16. Captain's seat
17. Nosewheel steering control
18. Underfloor electrical and electronics bay
19. Nose strake
20. Retractable airstairs
21. Door mounted escape chute
22. Forward passenger door, open
23. Entry lobby
24. Starboard service door
25. Forward galley
26. Toilet compartment
27. Wash hand basin
28. First-class seating compartment, 12 passengers four-abreast
29. D/F loop aerials
30. VHF aerial
31. Curtained cabin divider
32. Cabin window panel
33. Pressurization valves
34. Fuselage lower lobe frame construction
35. Wardrobe
36. Tourist class seating, 125 passengers five-abreast

Along with the Boeing 727, DC-9s and MD-80s are the mainstay of TWA's internal fleet. MD-82s are the main variant in use, originally leased from the manufacturer.

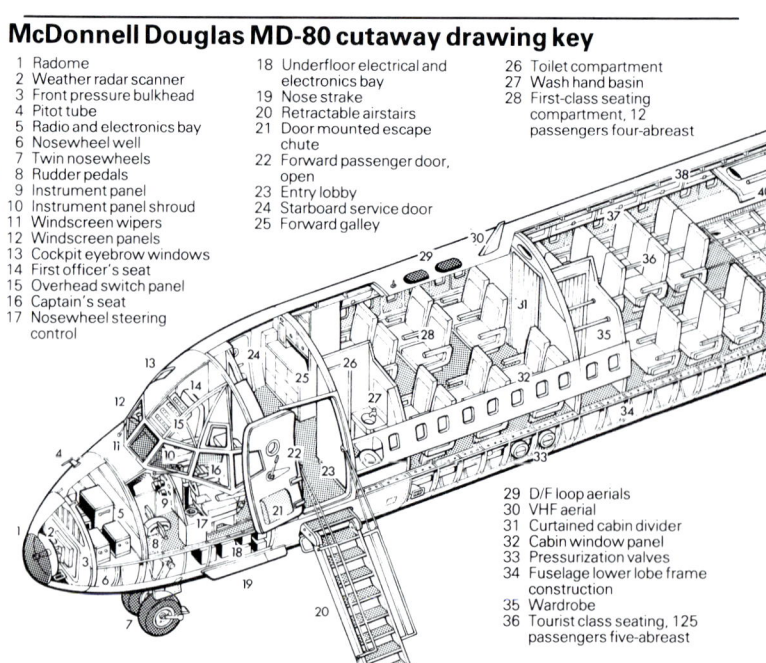

Another special variant for SAS was the Series 40, essentially a Series 30 with a modest fuselage stretch of 1.87m (6ft 4in). In addition to SAS, the variant was also ordered by Toa Domestic Airlines (TDA) of Japan.

Right: Swissair was the launch customer for the DC-9-50, followed by many others for this stretched DC-9-30 development. This aircraft flies with Ghana Airways alongside two Fokker F28s and a single DC-10 for long-range work.

including the possibility of a version engined with the internationally developed IAE V2500 for further reductions in noise and fuel consumption. In August 1997 McDonnell Douglas formally merged with Boeing under the latter's name, and the MD-80 series is now a Boeing product. Boeing decided that MD-80 production would cease in 2000 after the completion of the 1,191st aircraft.

37 Overhead stowage bins
38 Cabin roof frames
39 Air conditioning ducting
40 Cabin roof trim panels
41 Floor beam construction
42 Forward freight hold, capacity 849 cu ft (24.04 m³)
43 Forward freight hold rear door
44 Port overhead stowage bin rack
45 Fuselage frame and stringer construction
46 Leading edge slat central hydraulic jack control
47 Wing panel centreline joint
48 Floor beam construction
49 Centre fuselage construction
50 Cable drive to leading edge slats
51 Starboard wing integral fuel tank; total system capacity 5,779 US gal (21876 litres)
52 Fuel system piping
53 Ventral wing fence ('vortilon')
54 Pressure refuelling connections
55 Leading edge slat segments, open
56 Overwing fuel filler cap
57 Starboard navigation lights
58 Extended wing tip
59 Rear navigation and strobe lights
60 Static dischargers
61 Starboard aileron
62 Aileron tabs
63 Starboard outer double-slotted flap, down position
64 Flap hydraulic jacks
65 Flap hinge brackets
66 Outboard spoilers
67 Inner double-slotted flap, down position
68 Inboard spoiler
69 Starboard emergency exit windows
70 Pressure floor above wheel bay
71 Port emergency exit windows
72 Hydraulic reservoir
73 Main undercarriage wheel well
74 Rear cabin tourist class seats
75 Cabin attendant's folding seat
76 Rear service door/emergency exit
77 Rear underfloor freight hold door
78 Cabin wall trim panels
79 Overhead stowage bins
80 Starboard engine intake
81 Detachable engine cowlings
82 Cabin rear bulkhead
83 Rear galleys, port and starboard
84 Toilet compartments, port and starboard
85 Rear pressure bulkhead
86 Rear entry door
87 Engine thrust reverser, open position
88 Fin root fillet
89 Air conditioning ram air intake
90 Fin construction
91 VOR aerials
92 Rudder feel system pressure sensor
93 Tailplane trim jack
94 Starboard tailplane
95 Elevator horn balance
96 Starboard elevator
97 Elevator tabs
98 Tailplane bullet fairing
99 Elevator hinge controls
100 Tailplane pivot mounting
101 Port elevator
102 Tailplane construction
103 Rudder construction
104 Rudder tab
105 Static dischargers
106 Tailcone, jettisonable for emergency exit
107 Air conditioning louvres
108 Sloping fin attachment frames
109 Tailplane de-icing air duct
110 Rear entry airstairs tunnel
111 Air conditioning plant
112 Engine pylon
113 Port engine thrust reverser doors, closed
114 Radical lobe engine silencer
115 Nacelle strake
116 Bleed air piping
117 Pratt & Whitney JT8D-209 turbofan engine
118 Engine accessory gearbox
119 Port engine intake
120 Rear underfloor freight hold, capacity 445 cu ft (912.60 m³)
121 Wing root trailing edge fillet
122 Port inner double-slotted flap
123 Flap rib construction
124 Flap vane
125 Main undercarriage mounting
126 Main undercarriage leg strut
127 Inboard spoiler
128 Flap down position
129 Outer double-slotted flap
130 Outboard spoilers
131 Aileron tabs
132 Port aileron
133 Fixed portion of trailing edge
134 Static dischargers
135 Rear navigation and strobe lights
136 Retractable landing lamp
137 Port navigation lights
138 Leading edge slat segments (fully open position)
139 Slat guide rails
140 Front spar
141 Wing rib construction
142 Port wing integral fuel tank
143 Rear spar
144 Wing stringers
145 Ventral wing fence ('vortilon')
146 Wing skin plating
147 Twin mainwheels
148 Slat de-icing air duct
149 Air supply duct
150 Wing root fillet
151 Taxiing lamp

McDonnell Douglas DC-10

Despite the problems which dogged its early career, the Douglas DC-10 has become a stalwart of the world's airline fleets, used for long range wide-body transport where the capacity of the larger Boeing 747 is not required. Battling with the Lockheed TriStar to fill this all-important niche, it has overcome the bad publicity resulting from a series of crashes.

Last and largest in the DC (Douglas Commercial) series, the McDonnell Douglas DC-10 has experienced peaks of success and troughs of bitter tragedy such as have never previously been encountered by any aircraft programme. It is impossible to tell the DC-10 story without mentioning the succession of highly publicised accidents, some of them genuinely attributable to the aircraft, which made the DC-10 a household word in a very negative sense and generated an unprecedented spate of critical publicity including three hardback books. Yet far more design and engineering effort went into the DC-10 than into any of its predecessors, and dozens of customers and thousands of pilots in all parts of the world consider it one of the best commercial transports flying today.

Like DC-9, the designation DC-10 had been used before, in this case for a STOL transport in the class of the Breguet 941. The DC-10 actually built stemmed from diverse studies urgently undertaken after Douglas lost the CX-HLS heavy-airlift USAF competition to Lockheed's C-5A. These studies included enormous six-engined cargo aircraft and a 650-passenger version, and a big twin-engined aircraft looking like a Boeing 747SP but with two RB.207 engines. The studies were suddenly given a clear competitive direction by the decision in June 1966 of American Airlines to issue a requirement for a new transport that would offer the passenger appeal and the fuel economy of the Model 747 but be matched to domestic trunk routes.

In particular, American demanded the ability to use the small airport at La Guardia (New York), which meant a restricted body length and short take-off, while carrying 250 passengers to Chicago. From better airports longer range was demanded, and Douglas took it for granted that transcontinental sectors, such as New York to Los Angeles, would eventually be a requirement. Most of the early studies were twin-engined, because the CX-HLS competition had left a legacy of engine proposals in the 18144- to 21319-kg (40,000- to 47,000-lb) thrust class. Eventually a vague reaction by the US domestic market and the specific problem of engine-out performance at hot/high airports, such as Denver, prompted a three-engined layout, with engines of about 14515-kg (32,000-lb) rating. At all design stages two of the engines were hung in pods under the wing, but the third engine eventually was installed in a unique straight-through duct above the rear fuselage, with the fin above it. The solution was completed by the availability of computer-controlled machining of the four aluminium forgings, each initially weighing about 2041 kg (4,500 lb), which extend from the rear fuselage in the form of 'banjo rings' encircling the tail-engine duct and finally form the spars of the fin. After machining each forging weighs some 204 kg (450 lb). The main drawback of this tail-engine location is the difficulty of changing

Characterised by the centre engine's bulged cowl forward of the fin, the DC-10-40 was powered by the Pratt & Whitney JT9D. Japan Air Lines and Northwest Airlines were the only customers.

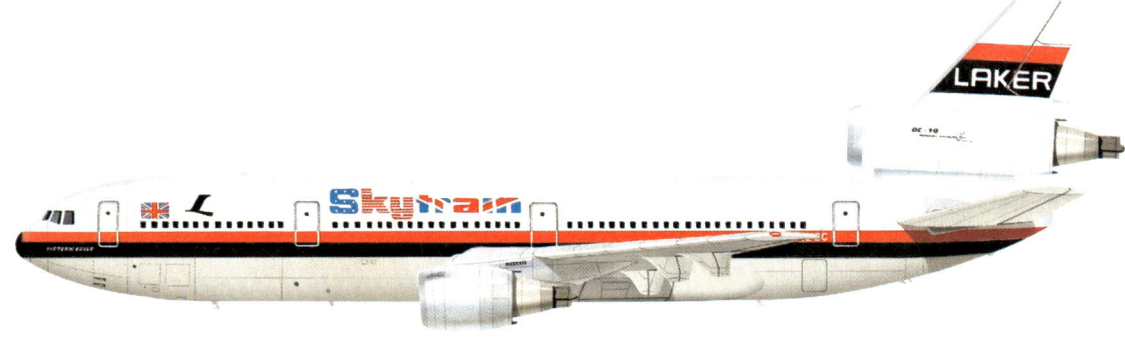

G-AZZC, Eastern Belle, was one of six Dash-10s bought at bargain prices by Laker Airways in its heyday, chiefly to operate the cut-price Skytrain routes, with 345-seat interiors. Freddie Laker made his licence application in 1971 but the USA held up the start of Skytrain until 27 September 1977. Sadly, the airline failed in 1982.

the engine.

Fuselage design was equally fluid, and involved various body diameters as well as figure-8 configurations with two DC-8 bodies superimposed or laid side-by-side. A double-deck body with a diameter of 6.71 m (22 ft) looked attractive, but eventually a diameter of 6.02 m (19 ft 9 in) was chosen, with an option of large cargo compartments or galleys (but not normally passengers) at the underfloor level. A DC-10A with a length of 51.82 m (170 ft) and a DC-10B with a length of 58.22 m (191 ft) were eventually superseded by a standard body length of 51.97 m (170 ft 6 in) with much increased floor area seating not the 230 mixed-class passengers of the DC-10A but up to 270, with 380 in an all-economy cabin configuration.

Swept wing

A crucial factor was the wing, which was to be swept at the high (perhaps too high) value of 35° and, to meet the demands of La Guardia and Denver with maximum payload, came out to a size of 329.8 m^2 (3,550 sq ft). A cruise Mach number of 0.86 was chosen, with a long-range cruise at 0.82, and for optimum efficiency the aerofoil shape, thickness and incidence were varied from root to tip. Three-position leading-edge slats were added, with a gap at the cutback engine pylon, and large-chord double-slotted trailing-edge flaps were chosen. Triplex hydraulics were provided to power the flight controls, full power being available from any one system, driving inboard ailerons and flight spoilers, augmented at low speeds by outboard ailerons, as well as four sections of elevator on a trimming tailplane (stabiliser) and upper and lower rudders.

Future development

Extremely comprehensive detail design and testing was needed to clear the structure to a lifetime of 60,000 hours and 42,000 landings. Passenger windows were made '25 to 35 per cent larger' than on other wide-body aircraft then in existence, and a unique feature for a wide-body transport was the provision of direct-vision windows on the flight deck which can be opened at indicated airspeeds up to 460 km/h (286 mph). Another very important factor was that from the very start of detail design in January 1968 the Douglas team at Long Beach planned for future development. Longer and shorter fuselages never reached the production line, but a long-range model overtook the original US domestic DC-10 in importance and far outsold it, as described later.

Early planning was beset by financial troubles, but these were resolved by the merger with McDonnell which took effect in April 1967, bringing Bob Hage from St Louis as the DC-10 Program Manager. By late 1967 the DC-10 had been selected over the rival L-1011 by the Board of American, but Douglas resisted the airline's recommendation of Rolls-Royce as supplier of the engines and instead sought acceptance of General Electric's CF6, derived from the same company's TF39 used in the C-5A but with a new single-stage fan of lower bypass ratio. The airline eventually accepted this US engine, and a contract for 25, with another 25 on option, was announced on 19 February 1968.

This was a valuable order, but not enough for a commitment to production. Intensive selling in competition with the almost identical L-1011 continued, but no further customer emerged until on 29 March 1968 Lockheed announced a massive deal for 144 TriStars soon followed up by 28 more, for a whole string of major carriers. It looked as if Lockheed had cleaned up the market, and the McDonnell Douglas Board met to consider cancelling the American order and putting the DC-10 on ice. Salvation came on 25 April 1968, when United signed for 30 plus another 30 on option, and Northwest followed later. The DC-10 was all systems go, under a new manager,

Federal Express, the dynamic air freight airline centred at Memphis, Tennessee, which has a huge fleet of Boeing 727s and Cessna Caravans got into the wide-body business in 1980 by buying four DC-10-10CF domestic-range cargo aircraft from Continental Airlines. Now 68 assorted DC-10s fly with heavy loads around the clock.

The last DC-10s on the Long Beach assembly line were the KC-10A Extenders for the US Air Force. Derived from the Dash-30CF, these extremely capable aircraft have extra fuel cells along the underside of the fuselage, a high-speed refuelling boom and a hosereel, as well as special provisions for carrying USAF cargo.

John C. Brizendine.

The DC-10 was clearly launched in head-on competition with Lockheed, and trailing the TriStar in both orders and number of customers. It had just one advantage, and that was a trump card. The TriStar was matched to the RB.211-22 and no growth version was possible at that time. By contrast General Electric planned a more powerful version of the CF6, the CF6-50, to be rated at 21455 kg (47,300 lb) to be available in 1974, while Northwest selected another powerful engine, the Pratt & Whitney JT9D-15. This enabled Douglas to offer its planned long-range version, for which provision had been made from the start.

Intensive testing

But it was the initial DC-10-10 that had to be certificated first, and five aircraft were used in an intensive flight programme with up to 11340 kg (25,000 lb) of instrumentation aboard each aircraft and direct link to recorders and computers on the ground. The first DC-10, registered N10DC and painted in company colours, was flown on 29 August 1970 from Long Beach by a test crew headed by Cliff Stout. Subsequent flying was based at Edwards and Yuma, and the FAA certificate was received on 29 July 1971. Powered by 18144-kg (40,000-lb) CF6-6 engines, and weighing 185976 kg (410,000 lb) fully loaded, the DC-10-10 could fly 255/270 passengers in mixed class from Los Angeles to Boston. American began operations on 5 August 1971. Nine days later United flew a DC-10 in line service from San Francisco to Washington.

On 28 February 1972 the first long-range model, a DC-10-20, flew on JT9D engines. Later this was redesignated DC-10-40, with 22408-kg (49,400-lb) JT9D-20 engines. Few were sold, the only other customer being Japan Air Lines with the 24041-kg (53,000-lb) JT9D-59A.

Aggressive sales, but then...

Almost all the long-range models were DC-10-30s, the first of which flew with 22226-kg (49,000-lb) CF6-50A engines on 21 June 1972. Like the Pratt & Whitney-engined models it had an extended wingspan, greatly increased fuel capacity and an extra twin-wheel landing gear on the centreline to spread the load over airport pavements. At first the DC-10-30 was certificated at 251748 kg (555,000 lb), but this was later raised to 259450 kg (572,000 lb) with an option at 263994 kg (582,000 lb), more than 75 per cent above the gross weight envisaged in 1967. This growth, not matched by Lockheed, opened the doors of airlines all over the world, and for a while the long-range DC-10 looked like fulfilling Brizendine's prophecy that 'The DC-10 will become the backbone of the commercial air fleet in the next decade.' Douglas sold the airliner aggressively on the basis that it could fly all routes from about 480 to over 8050 km (300 to over 5,000 miles) and thus could replace almost all previous aircraft, provided the route offered enough traffic for an aircraft in the 250/370-seat class. A major spur was the selection of the DC-10-30 by the European KSSU (KLM, SAS, Swissair and UTA) group in June 1969, and eventually the DC-10 was bought by 48 airlines, for a total of 386 aircraft. These included small numbers of DC-10-10CF and DC-10-30CF convertible freighters with large cargo doors, and cleared to a weight of 267624 kg (590,000 lb) with a cargo load of 53500 kg (117,947 lb), and the DC-10-15 for hot/high airports, which is essentially a DC-10-10 with CF6-50 engines.

During the first months of service there were a few major engine failures, including inflight break-up of a tail engine, but a much more serious failure was encountered by an American DC-10-10 over Michigan on 12 June 1972. The latches failed on the aft underfloor cargo door, allowing the door to burst open (ejecting a coffin incidentally) and causing the pressurised passenger cabin to collapse the main floor down through the control cables. This aircraft landed safely, but when a similar accident happened to a THY DC-10-10 over the Forêt d'Ermenonville soon after leaving Paris for London on 3 March 1974, all 346 on board were killed. Tremendous publicity followed, and many writers predicted the DC-10 would never fly

Specification
McDonnell Douglas DC-10-30
Type: high-capacity long-range commercial transport
Powerplant: three 23134-kg (51,000-lb) General Electric CF6-50C turbofans
Performance: maximum speed at 7620 m (25,000 ft) 982 km/h (610 mph); normal cruising speed 871 km/h (541 mph); maximum rate of climb (MTO weight) 884 m (2,900 ft) per minute; service ceiling 10180 m (33,400 ft); at 249475 kg (550,000 lb); range with maximum payload 7413 km (4,606 miles)
Weights: empty equipped 121198 kg (267,197 lb); maximum payload 48330 kg (106,550 lb); maximum take-off 259450 kg (572,000 lb)
Dimensions: span 50.41 m (165 ft 4 in); length 55.50 m (182 ft 1 in); height 17.70 m (58 ft 1 in)

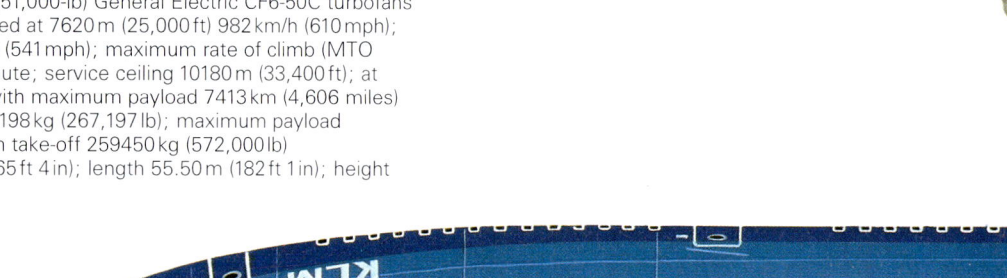

McDonnell Douglas DC-10 variants

DC-10-10: initial production version powered by 18144-kg (40,000-lb) thrust CF6-6 engines (later some fitted with 18598-kg/41,000-lb CF6-6D1); one fitted with drag-reducing winglets in 1981
DC-10-15: short-field version appearing at the end of the programme for Mexicana (5) and Aeromexico (2), a DC-10-10 airframe with 21092-kg (46,500-lb) thrust CF6-50C2F engines for use from hot/high airports
DC-10-30: standard long-range model, with various CF6 engines from 22226-kg (49,000-lb) thrust CF6-50A to 238140-kg (52,500-lb) thrust CF6-50C2; span increased by 3.05 m (10 ft), much greater fuel capacity, extra centreline landing gear
DC-10-30ER: extended-range model cleared to 263805 kg (580,000 lb) with additional fuel tank and 24494-kg (54,000-lb) thrust CF6-50C2B engines; used by Swissair and Finnair

DC-10-40: first long-range model, similar to later DC-10-30 but with Pratt & Whitney engines (see text for details); used by Northwest and Japan Air Lines
DC-10-10CF and **-30CF:** convertible passenger/freighter versions of DC-10-10 and DC-10-30 with increased maximum landing weight (and higher landing speed), heavy cargo floor with conveyors and loading system, large cargo door in left side, and provision for containers or pallets above floor and half-size containers or pallets at lower level
KC-10A Extender: US Air Force tanker/transport version of DC-10-30CF with greatly augmented fuel capacity, inflight-refuelling boom and hosereel for probe-equipped receiver aircraft. First flown 12 July 1980 and planned orders for 60 announced in 1982. All now in service with Strategic Air Command

The subject of this drawing is the fifth of KLM's fleet of ten DC-10-30s (four of which have been on long-term lease to other operators, leaving six in Royal Dutch Airlines service). All are named after famous composers. They operate in mixed class alongside Boeing 747-200Bs, 300s and 400s on long-haul trunk routes between Amsterdam and all parts of the world. Visible in this illustration are such features as the large external hinges for the flaps, the vortex-inducing strakes above the wing engine cowls to improve flow over the wing, the way the centre engine is hung at the extreme rear of its pod under a long beam cantilevered off the fin spars, the flush inlets under the nose to the air-conditioning system (with access doors above the nose gear bay), and the centreline main gear aligned with the rear bogie wheels – shown in grey in the front elevation.

Proudly wearing the logos of the first 15 customers, the very first DC-10, DC-10-10 with registration N10DC, lifts off the runway at Long Beach International on its first flight, on 29 August 1970. Vortex-inducing strakes above the wing engine pods have yet to be added. Later this DC-10 became American Airlines' N101AA.

The DC-10 proved immensely popular to airlines the world over, providing for many their first wide-body type. These still give excellent service, and many operators are following-on with the MD-11 derivative. JAT, the Yugoslav flag carrier, had four DC-10-30s and then ordered two MD-11s.

again. Mandatory changes were made to the latches, and DC-10s not only flew but kept selling.

Several other crashes were shown not to have been the fault of the aircraft, but a second shocking disaster occurred at the headquarters of the original customer on 25 May 1979, when the no. 1 engine fell off the left wing of an American DC-10-10 on a scheduled departure. The slat on that wing retracted, the aircraft rolled to the left and crashed inverted, killing all on board. The FAA certificate was withdrawn, and after a prolonged investigation it was concluded that the root cause was faulty maintenance procedures by American, which had been in the habit of offering up the complete pod and pylon on a fork-lift truck and causing local damage at the crucial highly-stressed pod mounting attachments.

McDonnell Douglas never succeeded in launching its various longer, shorter or otherwise modified DC-10s, but in December 1977 a modified DC-10-30CF won a USAF order for an off-the-shelf ATCA (advanced tanker cargo aircraft) with the designation KC-10A Extender. The ATCA was needed to support global deployment of aircraft, men and supplies, doing the work of several KC-135s and cargo aircraft with much better efficiency. Along the lower lobe of the fuselage are seven bladder fuel cells, and under the tail is a new type of long-reach computer-controlled refuelling boom. The KC-10A can transfer 90718 kg (200,000 lb) of fuel to fighter or other aircraft at a distance of 3540 km (2,200 miles) from its base, or it can carry a 45400-kg (100,000-lb) cargo load 11112 km (6,905 miles). Production of 60 enabled the line to be kept open until manufacture of the MD-11 began in March 1987, an externally similar but vastly upgraded successor which, McDonnell Douglas hoped, would continue the success story begun by the DC-10.

McDonnell Douglas DC-10 Series 30CF cutaway drawing key

1. Weather radar
2. Windshield
3. Instrument console
4. Flight deck
5. Captain's seat (Aircraft Mechanics Inc)
6. First Officer's seat (ditto)
7. Flight Engineer's position
8. Supernumary crew seat
9. Flight deck door
10. Forward starboard toilet
11. Forward port toilet
12. Crew and passenger forward entry door
13. Twin wheel nose gear (Abex or Dowty Rotol; Goodyear tyres)
14. Air conditioning access doors
15. Forward cargo bulkhead
16. Air conditioning bay (Garrett AiResearch equipment)
17. Forward lower galley area (used for containerised cargo)
18. Air conditioning trunking
19. Cargo deck lateral transfer area (omni-caster rollers)
20. Cargo deck pallet channels (rollers)
21. Main cargo door (fully open position)
22. VHF antenna
23. Frame-and-stringer fuselage construction
24. Main deck cargo (ten 2.23×3.17-m (pallets), capacity 140.4 m³ (4.958 cu ft)
25. Passenger door
26. Forward lower compartment (five 2.23×3.17-m pallets, capacity 53.5 m³ (1,890 cu ft)
27. Centre-section fuselage main frame
28. Centre-section front beam
29. Sheer-web floor support over centre-section fuel tank
30. Cargo/passenger compartment dividing bulkhead
31. Starboard engine pod (Rohr subcontract)
32. Engine intake
33. Nacelle pylon
34. Leading-edge slats
35. Integral wing fuel tank
36. Starboard navigation lights
37. Low-speed outboard aileron
38. Fuel ventpipe
39. Wing spoilers/lift dumpers
40. Double-slotted flaps
41. All-speed inboard drooping aileron
42. Passenger doors
43. Centre-section fuselage mainframe
44. Cabin air ducts
45. Centre undercarriage bay
46. Keel box structure
47. Fuselage/wing attachment points
48. Wing torsion-box construction
49. Leading-edge structure
50. Nacelle pylon
51. Engine intake
52. General Electric CF6-50 turbofan
53. Exhaust outlet
54. Four-wheel main undercarriage (Menasco Manufacturing; Goodyear tyres and brakes)
55. Leading-edge slats
56. Outboard slat extended
57. Port navigation lights
58. Low-speed outboard aileron
59. Fuel vent pipe
60. Outboard flap hinge fairings
61. Fuel pipes
62. All-speed inboard drooping aileron
63. Inboard flap hinge actuator and fairing
64. Undercarriage support structure
65. Flap construction

The last type of DC-10 for the commercial market was the Dash-15, certificated on 12 June 1981. Powered by General Electric CF6-50C2F engines giving constant 21092-kg (46,500-lb) thrust to no less than 43.3°C (110°F), they have a Dash-10 lightweight airframe and are tailored to extreme hot/high-altitude airports such as Mexico City.

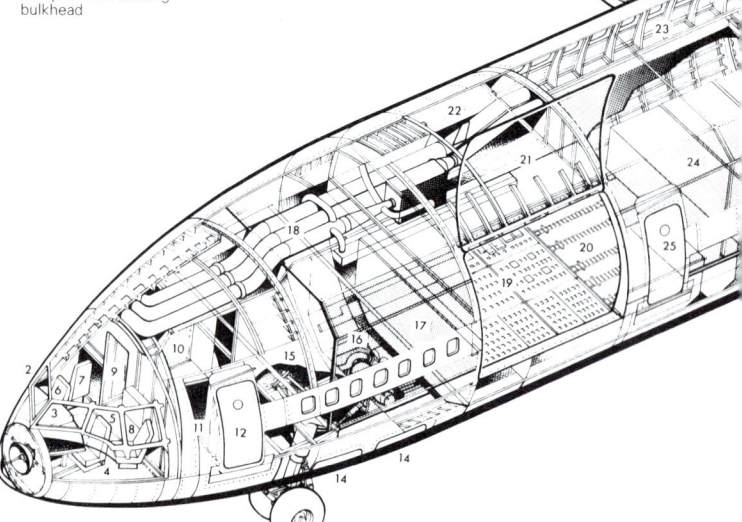

*Delivered in July 1972, **N116AA** was in fact the 17th **DC-10**, being one of the first batch of production Dash-10s numbered by Douglas 46501 to 46525, all for American Airlines. Powered by 18144-kg (40,000-lb) General Electric CF6-6 engines and with a maximum weight of 206384 kg (455,000 lb), they are tailored to US domestic trunk routes of up to 5794 km (3,600 miles).*

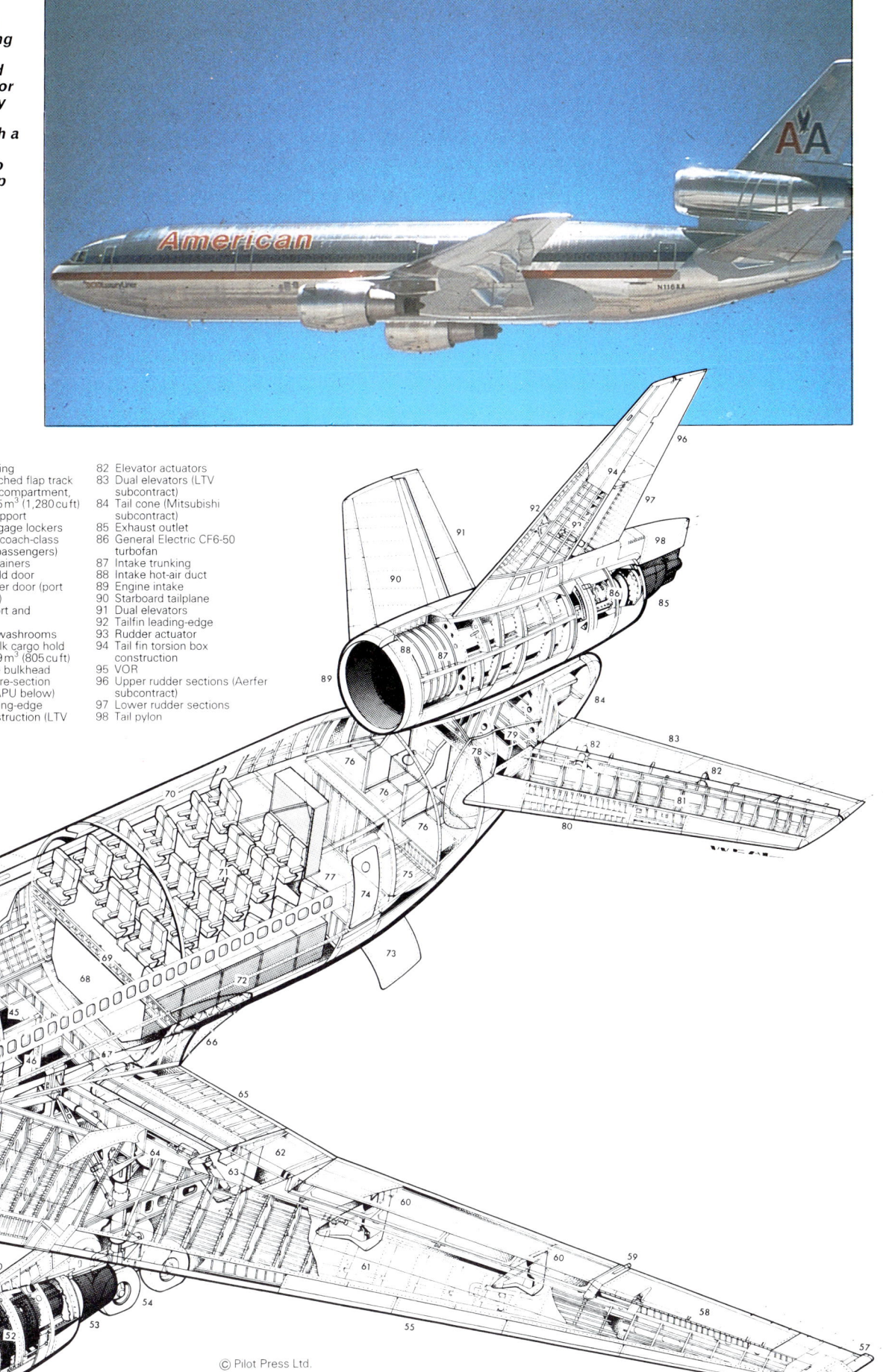

66 Wing root fairing
67 Fuselage-attached flap track
68 Centre cargo compartment, capacity 36.25 m³ (1,280 cu ft)
69 Cabin floor support
70 Overhead luggage lockers
71 Eight-abreast coach-class seating (147 passengers)
72 Baggage containers
73 Bulk cargo hold door
74 Rear passenger door (port and starboard)
75 Rear toilet (port and starboard)
76 Three toilets/washrooms
77 Underfloor bulk cargo hold capacity 22.79 m³ (805 cu ft)
78 Rear pressure bulkhead
79 Tailplane centre-section (AiResearch APU below)
80 Tailplane leading-edge
81 Tailplane construction (LTV subcontract)
82 Elevator actuators
83 Dual elevators (LTV subcontract)
84 Tail cone (Mitsubishi subcontract)
85 Exhaust outlet
86 General Electric CF6-50 turbofan
87 Intake trunking
88 Intake hot-air duct
89 Engine intake
90 Starboard tailplane
91 Dual elevators
92 Tailfin leading-edge
93 Rudder actuator
94 Tail fin torsion box construction
95 VOR
96 Upper rudder sections (Aerfer subcontract)
97 Lower rudder sections
98 Tail pylon

© Pilot Press Ltd.

Chapter 29

McDonnell Douglas MD-11

McDonnell Douglas MD-80 & MD-90

Metro

McDonnell Douglas MD-11

One of the most stylish of the current crop of airliners, the MD-11 has its roots in the DC-10 but is a very different animal. While McDonnell Douglas has had more than its fair share of problems with the programme, the MD-11 is gathering a respectable order book and looks set to fulfill its promise.

Although the Douglas Aircraft Company projected a number of 'stretched' versions of its DC-10 tri-jet transport, that aircraft was only built with one fuselage length throughout its 20-year, 386-airframe civilian production run.

Decline in airline activity throughout the early 1970s forced the manufacturer to shelve plans for variants with increased passenger capacity, but these were revived again in the summer of 1978 when Douglas announced design studies for the DC-10-30/40 Intercontinental Stretch, which was to have a fuselage 8.13 m (26.7 ft) longer than the DC-10-30, increasing passenger capacity to 353 in a mixed-class cabin configuration. With a choice of General Electric CF6-50C2 or Pratt & Whitney JT9D-59B engines and maximum fuel capacity of 112039 kg (247,000 lb), Douglas estimated that the DC-10-30/40 Intercontinental Stretch would have a range of 5,400 nm (10000 km/6214 miles), would offer seat-mile costs 14 per cent better than those of the DC-10-30 and burn 24 per cent less fuel per seat for equivalent range.

As design work progressed at Douglas' Long Beach, California plant alongside production of the DC-10, the designation 'DC-11' was unofficially adopted for the new aircraft. But in 1982 the parent McDonnell Douglas aircraft company decided to adopt a common 'MD' prefix for all future commercial aircraft lines, dropping the 'Douglas Commercial' appelation which had been applied to all civilian airliner Douglases since the DC-1 of 1933, and MD-100 was the chosen designation for the stretched DC-10, following on from the MD-80/90 twin-jet series.

Again development was postponed when, in late 1984, MDD ordered Douglas to cease all further work on future commercial aircraft projects pending reassessment of the airline market, but work on the 'MD-100' resumed during the autumn of 1984 in time for the company to formally announce the aircraft at the Paris air show in June 1985. By then the designation MD-11X had been chosen. As revealed in Paris, the aircraft was clearly as DC-10 derivative, but with a fuselage stretched by 6.82 m (22 ft 4 in) to accommodate 331 passengers in

Finnair was the initial airline to receive the MD-11, the first arriving in Helsinki on 29 November 1990. It was also the first carrier to put the aircraft into service. Its aircraft are configured to carry either 288 passengers in a two-class arrangement, or up to 403 in a single-class layout.

The prototype MD-11 flies unpainted over the California desert during flight trials from Edwards AFB. The most distinctive features of the MD-11 are its drag-reducing winglets.

two classes. Maximum take-off weight would be 267624 kg (590,000 lb) and range 6,070 nm (11242 km/6985 miles). A shorter range MD-11XR with maximum take-off weight of 226800 kg (500,000 lb) and range of 4,150 nm (7686 km/4776 miles) was also proposed. Drag-reducing winglets, then being flight-tested on a DC-10, were also a prospective feature of the new aircraft, for which Douglas began soliciting launch offers, primarily among existing DC-10 operators.

Design changes

By the time more detailed specifications had been published, the MD-11 had emerged with a more modest fuselage stretch of 5.70 m (18 ft 7 in) over that of the DC-10-30. Its wing was to have a 1.52-m (5-ft) extension in span at each tip and 2.7-m (8-ft 9-in) outward-canted winglets, and have a redesigned airfoil section with more trailing-edge camber, while the advanced design horizontal tail would be smaller in area than the DC-10's, with cambered airfoil section, reduced sweepback, integral 'trim tank' holding 7570 litres (2,000 US gal) of fuel, and an extended, drag-reducing tailcone. Other features were a restyled passenger interior making use of modern lightweight materials, and a state-of-the-art two-crew flight deck incorporating CRT displays, all-digital instrumentation and automatic flight control system, dual flight management systems and full-authority digital engine control (FADEC) for the 266.9/289.1-kg (60,000/65,000-lb st) General Electric CF6-80C2, Pratt & Whitney PW4000 series or Rolls-Royce RB211-524L (since renamed Trent 650) turbofan engines which were offered to customer choice.

Purchase plans

The first MD-11 customer was announced on 3 December 1986 when British Caledonian Airways ordered nine aircraft (later cancelled when BCAL was absorbed by British Airways), swiftly followed by the Japanese leasing organisation Mitsui with five, and Scandinavian Airline System which ordered 12. By the end of 1986 the order book had swelled to 92, of which 52 were firm orders and 40 were options, from 12 customers, and McDonnell Douglas was able to announce formal launch of the MD-11 programme on 30 December.

Other customers included Aero Lloyd, Air Europe, Air Outre Mer, Air Zaire, Alitalia, American Airlines, China Airlines, China Eastern Airlines, Delta Air Lines, Dragonair, Evergreen Airways, Federal Express Corporation, Finnair, Garuda Indonesian, Guinness Peat Aviation, International Lease Finance, Japan Air Lines, Korean Air, Luftransport Union (LTU), Minerve, Nigeria Airways, Singapore Airlines, Swissair, Thai Airways International, VARIG, VIASA, Zambia Airways and ZAS Airline of Egypt.

McDonnell Douglas used the first four MD-11s to complete the test and certification process, these flying from both Edwards and the company facility at Yuma, Arizona. This is the third aircraft.

Three versions of the aircraft were once offered. The basic MD-11 accommodates 293 passengers in three-class, 323 passengers in first class and economy, or 405 passengers in single-class cabin configurations, with a maximum range of 7,008 nm (12979 km/8065 miles). The MD-11F is an all-cargo aircraft with a payload of 81648 kg (180,000 lb) of freight. A cargo door, measuring 356 cm × 260 cm (140 in × 102 in), is situated in the port fuselage side ahead of the wing. The main cargo deck can accommodate up to 26 2.23/2.43-m × 3.13-m or 35 2.23-m by 2.74-m (88/96-in × 125-in or 35 88-in × 108-in) pallets, with additional freight carried in under-floor compartments. It has its cabin windows plugged and lacks cabin ceilings to facilitate carriage of bulky freight. The MD-11CB Combi freight/passenger version has a main deck cargo door on the port rear side of the cabin and a movable cabin bulkhead permitting four or six standard cargo pallets and between 168-240 two-class passengers to be carried, with a maximum range of 5,210-6,950 nm (9648-12871 km/5995-7998 miles) depending on configuration.

Delta Air Lines ordered 13 MD-11s with 29 on option, all powered by the PW4460. However, its first two aircraft were leased, and powered by the rival CF6 80CD1F. This is the first of those, temporarily registered N514MD for company trials.

Composite materials such as glass-fibre, carbon-fibre and Kevlar are employed in major elements of the MD-11's structure, including most moving control surfaces on the wings and horizontal tail surfaces, the wing/fuselage juncture fairings, and engine inlets and cowlings, while from the 14th production airframe onwards fuselage floor beams have been manufactured from aluminium-lithium. The remainder of the main airframe structure is of conventional metal construction. The fuselage is an aluminium alloy semi-monocoque failsafe structure with cabin and cargo holds pressurised to a maximum pressure differential of 0.6046 kg/cm^2 (8.6 lb/sq in). The MD-11's wings are formed around a two-spar structural box with chordwise ribs and bulkheads and have full-span two-position leading-edge slats, double-slotted trailing-edge flaps, five-segment upper surface spoilers/lift dumpers and

Swissair has purchased 15 MD-11s to replace its DC-10s on its extensive long-haul network. The machines are powered by PW4460s, and are each configured with 18 first-class, 72 business-class and 146 economy seats.

display system (CFDS); flight management system (FMS); automatic flight system (AFS) with Cat IIIb autoland, windshear detection system, and full-time longitudinal stability augmentation; laser inertial reference system (IRS); digital air data computer (DADC); and aircraft system controllers that provide automatic monitoring and control of the aircraft's electrical, hydraulic, environmental and fuel systems. Total fuel capacity is 117482 kg (259,100 lb).

Other manufacturers

Among Douglas' partners and sub-contractors in the MD-11 programme are Alenia of Italy, which supplies the vertical tail surfaces, winglets and some fuselage panels; CASA of Spain, which manufactures the horizontal tail surfaces; General Dynamics' (now Lockheed) Convair Division, which builds main fuselage sections; EMBRAER of Brazil, suppliers of the outboard flap sections; Rohr Industries, which

upper (metal) and lower (composite) winglets. The airfoil section is a proprietary Douglas design. The horizontal tail unit, which has electro-hydraulically actuated incidence adjustment, is of metal construction with carbon fibre trailing-edge and elevators. The main undercarriage comprises two four-wheel bogie units mounted inboard of the wing engine pylons and a twin-wheel unit mounted on the fuselage centreline.

On the flight deck, the advanced avionics fit, supplied by Honeywell, provides a high degree of automation for the two-pilot crew and includes an electronic flight instrument system (EFIS) comprising six 20.32-cm × 20.32-cm (8-in × 8-in) colour CRT tubes; a central fault

Success for airliners can often be assured by orders from the large US carriers. American Airlines has a large fleet of DC-10s, and was a natural early customer for the MD-11. With a current fleet of 14 such aircraft, the airline once considered the purchase of many more MD-11s.

manufactures the engine pylons; and Westland Aerospace of the United Kingdom, which builds flap vane and engine inlet duct components.

Construction of the first MD-11 began on 9 March 1987 with the rear engine/upper fin support 'banjo' fitting, and assembly of the airframe was started on 9 March 1988 when work began on rivetting the forward nose section. The first MD-11 – a Freighter destined eventually for Federal Express – was powered 273.7-kN (61,500-lb st) General Electric CF6-80C2D1F turbofans and made its maiden flight from Long Beach on 10 January 1990. It was followed by the second, another CF6-powered FedEx MD-11F freighter, on 1 March, and by the third aircraft, which was also the first equipped with 267-kN (60,000-lb st) Pratt & Whitney PW4460 engines, on 26 April. Four MD-11s (three GE-powered, one PW-powered) took part in the test programme, which was conducted at Yuma, Arizona and Edwards Air Force Base, California. Federal Aviation Administration certification was granted on 8 November 1990, with the first customer delivery (to Finnair) following on 29 November.

Route-master

The first revenue-earning flight by an MD-11 was made by Finnair on 20 December 1990 when its first MD-11 flew from Helsinki, Finland to Tenerife, Canary Islands with 360 passengers. Finnair's introduction of the aircraft to commercial service was followed by American Airlines, China Eastern, Delta Air Lines, Federal Express, Korean Air and Swissair. By early 1993 McDonnell Douglas held firm orders for nearly 200 MD-11s, plus options for a further 144.

The first delivery was made to Federal Express, which accepted its premier MD-11F, named 'Christy', on 27 June 1991. The second operator was Korean Air, followed by Delta, VASP, Thai Airways, Alitalia, VARIG, Swissair, China Eastern Airlines, China Southern Airlines, China Airlines, LTU, American Airlines, Aero Lloyd, ZAS of Egypt, Garuda (via GPA) and Finnair. Outstanding customers include EVA Air, KLM and Japan Air Lines. JAT Yugoslavian Airlines' order has fallen into abeyance.

Soon after its introduction into service the MD-11 was plagued by reports of airline dissatisfaction, as it fell well short of its range guarantees. McDonnell Douglas rapidly instigated a stepped fix program. Known as the 'A-1 Modification, PIP' (Performance Improvement Programme), these revisions will be finally implemented on all new-build aircraft by the end of 1993, and are available for retrofit. Changes have been made to the aircraft's aerodynamics, overall weight and fuel capacity, and the engine manufacturers involved are also have to improve their performance. The wing slatting has been changed, as has the rigging on the ailerons which are now drooped a further 3° in the cruise. McDonnell Douglas ultimately intends to offer a maximum take-off weight of 283729 kg (625,500 lb), as opposed to its current level. In a move similar to that made with the DC-10-30, an optional 7457-litre (1,970-US gal) fuel tank can be added to the cargo hold.

McDonnell Douglas had intended to produce an aircraft capable of flying 7000 nm (12964 km/8055 miles) with 293 passengers; by the end of this year it should be capable of 6900 nm (12778 km/7940 miles), with the advertised load. The Rolls-Royce Trent 650-powered version, which had been ordered only by the now-defunct British carrier Air Europe, was scheduled for certification during 1993 but this has been abandoned.

Future proposals

When the MD-11 was announced, McDonnell Douglas also revealed two future proposals for the extended-range MD-11ER with shorter DC-10 type fuselage accommodating 276 passengers in first and economy classes and a maximum range of 7,500 nm (13,890 km/8631 miles), and the MD-11 Advanced, with a 1.03-m (40-ft 7-in) fuselage stretch, which would accommodate 380 two-class passengers and be able to fly 7,380 nm (13667 km/8493 miles). Neither has been developed, but McDonnell Douglas at a later date undertook a large-scale study for development of a stretched version of the MD11, designated MD-12X

While this aircraft started life as a scaled-up MD-11, it was formally unveiled in its definitive

Like the DC-10 before it, the MD-11 has generated considerable interest overseas, notably among current DC-10 operators. Korean Air was an early recipient of the aircraft, taking five PW4460-powered aircraft.

form as a four-engined design, with a full-length double-deck fuselage, named the MD-12. McDonnell Douglas had hoped that the MD-12 would be their entrant in the market for a future ultra-long-range, high-capacity airliner, but in the light of uncertain demand, and rising losses in the civil aircraft market, the MD-12 has now been firmly shelved.

McDonnell Douglas also took under consideration several changes to the MD-11, the first being known as the MD-11 Super Stretch. Lengthened by a further 10.66 m (35 ft), passenger capacity increases to 372. A further option is the so-called 'Panorama Deck', installed in place of the forward cargo hold, providing up to 92 additional seats. A second planned version was the MD-11 Advanced Stretch, intended for ultra-long ranges with a load of 372 passengers. Following reassessment of its product lines in the aftermath of the merger of McDonnell Douglas into Boeing in 1997, however, Boeing decided that MD-11 production would end in 2000 with the completion of the 195th aircraft.

MD-80 and MD-90 Family

By the end of the 1980s the DC-9 had already been flying for 15 years, but McDonnell Douglas decided that it was time to open a new chapter in this story. The stretched MD-80 family upheld their forerunners' sales record and itself gave birth to a whole new family of airliners for the 1990s: the MD-90s.

In 1973 the Douglas Aircraft Company Division of the McDonnell Douglas Corporation, which was then finalising design of the DC-9 Series 50, began looking to the next major development of its versatile twin-jet airliner. A number of different proposals were considered, all based on quieter, more fuel-efficient re-fanned versions then becoming available of the Pratt & Whitney JT8D engine which had powered all DC-9s to date. Proposals under study included a DC-9-50RS (Re-fanned, Stretched), DC-9-50RSS (Re-fanned, Super Stretched), and the DC-9-55 and -60, all of which would have had a larger wing, longer cabin and more power than the Series 50, enabling them to carry greater passenger loads over the same range as the extended-range DC-9-54.

These, and several other options, were discussed with potential customers and eventually evolved into the DC-9 Super 80 which was launched in October 1977 when an order for 15 aircraft was received from Swissair.

While bearing a clear resemblance to its DC-9 forerunners, the Super 80 was to all intents and purposes an entirely new aircraft, a fact which McDonnell Douglas recognised by dropping the DC-9 title in favour of the designation MD-80, a generic term for the entire series of current-production MDD twin-jets, individual models of which are detailed below.

Compared with the DC-9 Series 50 which it succeeded, the Super 80/MD-80 has a fuselage lengthened to 45 m (147 ft 10 in) by means of an eight-frame (3.9 m/12 ft 8 in) plug forward of the wing and a single-frame (0.5 m/1 ft 7 in) plug in the rear fuselage. The cabin provides accommodation for 137 passengers in typical mixed-class configuration and up to 172 in high-density layout. Under-floor cargo volume is increased to 366 m³ (1,294 cu ft). The MD-80 series wing incorporates a new centre-section and a 0.6-m (2-ft) long extension at each wingtip, increasing wing area by 28 per cent and span by 4.4 m (14 ft 6 in). It also has new four-position full-span leading-edge slats and three-segment spoilers on the wing upper surfaces, the outer segments of which act as flight or ground spoilers, the inboard segments serving as ground spoilers/lift dumpers only. The enlarged wing provides space for an additional 5754-litre (1,520-US gal) fuel tank, taking standard fuel capacity to 21876 litres (5,779 US gal). Tailplane span is increased by 1.06 m (3 ft 6 in), and overall height by 0.43 m (1 ft 4 in).

The engine selected for the MD-80 was the 37.81-kN (18,500-lb st) Pratt & Whitney JT8D-209 high bypass ratio turbofan, with a further 3.33 kN (750 lb st) Automatic Power Reserve (APR) available in engine-out situations. Enlarged, acoustically-treated nacelles were designed to accommodate the engines, enabling the aircraft to meet FAA Federal Air

When McDonnell Douglas launched the first of their new generation twin-jets they retained the familiar DC-9 name. However, the clean lines of the Super 80's significantly stretched fuselage quickly set it apart from the earlier models.

The Super 80 was soon renamed the DC-9-81, and later MD-81, to become the basic version of the family. The first operator was Swissair, which received their initial delivery in late 1980. The aircraft serve on the carrier's European network.

Along with Finnair, Austrian Airlines was the launch customer for the MD-87, which saw the combination of MD-80 series technology with a much shorter fuselage equivalent to the DC-9-30.

Regulation Part 36 noise emission standards as well as the more stringent ICAO standards planned for introduction in the future.

Internally, the MD-80 was given a 'Wide Body Look' cabin with large overhead baggage lockers, acoustically-lined ceiling, and fluorescent lighting; entry is via a forward door with integral electrically-operated airstairs and a hydraulically-actuated ventral stairway, as on earlier DC-9 models. Systems improvements include AiResearch dual cycle air conditioning and pressurisation systems (maximum pressure differential 53.6 kPa/7.77 psi) using engine bleed air; dual hydraulic systems for operation of landing gear, flaps, spoilers, slats, rudder, brakes, thrust reversers and ventral stairs; a larger capacity auxiliary power unit, flow-through avionics cooling and a digital fuel quantity measuring system.

A key feature of the MD-80's advanced two-crew flight deck is its Sundstrand digital electronic integrated flight guidance and control system incorporating speed command with digital full-time autothrottles and Sperry CAT IIIA autoland. A Honeywell Electronic Flight Information System (EFIS) with CRT screens is available optionally in place of electro-mechanical flight instrument displays. An electronic Performance Management System (PMS) similar to that previously developed for the McDonnell Douglas DC-10 has been introduced as standard from 1983, coupling through the autopilot and autothrottle systems to control automatically the aircraft's pitch attitude and engine thrust to give optimum speed and fuel efficiency in climb, cruise and descent. A Honeywell-developed windshear detection and avoidance system was certificated for the MD-80 series in June 1989 and is now standard equipment on new-build airframes and retrofittable to earlier production aircraft.

First in line

The first aircraft in the Super 80/MD-80 series, an MD-81 N980DC, made its first flight from the Douglas Aircraft Company plant at Long Beach, California, on 18 October 1979, and was followed by the second and third prototypes (N1002G and N1002W) on 6 December and 29 February 1980 respectively. FAA certification was granted to the MD-81 on 26 August 1980. The first of launch customer Swissair's aircraft was delivered on 12 September, flying its inaugural revenue service from Zurich to Frankfurt on 5 October. Other customers for the MD-81 have included Air California, Hawaiian Airlines, Muse Air and Pacific Southwest Airlines (PSA) in the United States, Austral of Argentina, Japan Air System and Toa Domestic Airlines in Japan, and Austrian Airlines, Inex-Adria and Scandinavian Airlines Systems in Europe.

On 16 April 1979 McDonnell Douglas announced development of the MD-82. Powered by 88.97-kN (20,000-lb st) JT8D-217 turbofans with 3.78-kN (850-lb st) APR, this model is generally similar to the MD-81, but is optimised for operation from 'hot-and-high' airports and has a higher maximum take-off weight of 66679 kg (147,000 lb). The MD-82 first flew on 8 January 1981, was certificated by the FAA on 30 July and entered service with the US carrier Republic Airlines in the following month. An optional 67813-kg (149,500-lb) maximum take-off weight version with JT8D-217A engines was introduced in the autumn of 1982.

Finnair's association with Douglas began with the DC-3 and has steadfastly continued to this day. This is one of its early MD-82s delivered in 1983.

1987 marked the first time a propeller had been seen on a Douglas transport since the DC-7. Known first as the UDF (UnDucted Fan) programme, when fitted with a General Electric engine, and as UHB (Ultra High Bypass) when carrying an Allison powerplant, it involved the testing of one of these revolutionary new 'propfan' engines on a standard MD-80. A single JT8D turbofan was retained on the starboard side.

Right: Troubled American 'Major' TWA operates a large fleet of MD-82s and MD-83s alongside an equally sizeable number of DC-9-30s and -40s. This is its second MD-82.

On 12 April 1985 McDonnell Douglas signed an agreement with the China Aero-Technology Import/Export Corporation (CATIC) and the Shanghai Aviation Industrial Corporation of the People's Republic of China for the sale of 26 MD-82s for use by the Civil Aviation Administration of China, five aircraft already having been supplied for use by China Eastern Airlines and China Northern Airlines. This agreement provided for licence assembly of all but one of the 26 aircraft by the Shanghai Aircraft Manufacturing Factory (SAMF) using components and sub-assemblies shipped from Long Beach. A further 20 aircraft were added to the agreement in April 1990. MDD delivered the sole US-built CAAC MD-82 of this order on 30 September 1985; the first Shanghai-built aircraft flew on 2 July 1987, entering service with CAAC the following month. The first batch of 25 aircraft has now been completed in China. The 10 aircraft ordered in 1990 (five MD-82s and five MD-83s) will see the programme continuing throughout 1993.

Announced on 31 January 1983, the MD-83 is a long-range development of the MD-80 series, externally similar to the MD-81/2 but powered by 93.42-kN (21,000-lb st) JT8D-219 engines, with an additional 4391 litres (1,160 US gal) of fuel in two cargo bay tanks, and a maximum take-off weight of 72576 kg (160,000 lb). It was first flown on 17 December 1984, certificated at the end of 1985 and entered service early in 1986 with launch customers Alaska Airlines and Finnair. Finnair's first MD-83 made the longest non-stop flight ever recorded by an MD-80 series aircraft during its delivery flight on 14 November 1985, flying the 6308 km (3,920 miles) from Montreal, Canada to Finnair's base at Helsinki in 7 hours 26 minutes. An MD-83 of the Swedish charter airline Transwede made the first revenue-earning transatlantic flight by an MD-80 series aircraft in September 1987, flying from Stockholm to Fort Lauderdale, Florida, with en route stops in Norway and Newfoundland.

The only version of the basic MD-80 series to display a major airframe change is the MD-87, announced on 3 January 1985. This has a fuselage shortened by 5 m (16 ft 5 in), seating 109 passengers in mixed-class or 130 in single-class accommodation. Powered by two 88.97-kN (20,000-lb st) JT8D-217C turbofans, the MD-87 was the first aircraft in the series to offer EFIS, an attitude/heading referencing system (AHRS) and a head-up display as standard in its avionics/flight deck configuration, and also incorporates a cruise perform-

Above: Carrying the flags of Denmark, Norway and Sweden, Scandanavian Airline Service (SAS) has been steadily supplementing its DC-9s with MD-81s, -82s, -83s, and MD-87s.

Left: Rome-Fiumicino is home to Alitalia's vast fleet of MD-80s and DC-9s. They form the bulk of the airline's fleet and substantial numbers of MD-82s are still on order.

Below: Iberia is yet another DC-9 operator which has joined the MD-80 club. Since 1990 it has opted for the smaller MD-87. Note the type's revised tail design.

ance improvement package which includes a new fillet fairing between the rear fuselage and engine pod mounting pylons, an APU fairing, improved horizontal tail surface sealing, low-drag flap hinge fairings and an extended low-drag tailcone. The MD-87 was first flown on 4 December 1986; deliveries to launch customers Austrian Airlines and Finnair began in October 1987.

The fifth member of the MD-80 family, designated MD-88, was announced by McDonnell Douglas on 23 January 1986, following receipt of a launch order for 80 aircraft (subsequently increased to 110) from Delta Air Lines. Similar to the MD-82, this version is powered by 93.42-kN (21,000-lb st) JT8D-219 engines and features an EFIS flight deck, flight management system (FMS), inertial reference system (IRS), and windshear detection system as standard. Increased use has been made of composite materials in the airframe, and its 142-passenger, five-abreast seating cabin interior has been extensively redesigned with wider aisle and new overhead storage bins. The MD-88 first flew on 15 August 1987 and was certificated by the FAA on 9 December, entering service with Delta on 5 January 1988. Delta remains the principal customer of the type with small numbers having been ordered by Aerolineas Argentinas, Aeromexico, Aviaco of Spain, and American carriers Midway and Midwest Express.

In addition to airliner variants of the MD-80 series, McDonnell Douglas also offers MD-83 and MD-87 Executive Jets. Built to individual customer specifications, these typically are configured to carry 20/30 passengers and have maximum ranges of 4,100-4,500 nm (7593-8334 km; 4,718-5,179 miles).

By the end of 1991 McDonnell Douglas had delivered a total of 965 MD-80s of all models, and the aircraft was in service with more than 50 operators worldwide. The 1,000th was delivered on 23 March 1992, and orders had reached 1,458.

Jet for the 1990s

On 14 November 1989 McDonnell Douglas launched the MD-90 series of twin-jet airliners, to be powered by two International Aero Engines V2500 turbofans. The MD-90 employs the same fuselage cross-section, advanced high-lift wing and EFIS flight deck displays used on the MD-80 series, while the introduction of interchangeable, standardised modular fuselage components allows the aircraft to be assembled on a common production line as the MD-80s. According to McDonnell Douglas the MD-90 family can be referred to with a hyphen inserted between the 'MD' and '90'. When an individual model is being referred to, however, the hyphen appears only between the type and the variant, ie 'MD90-10'. Launch customer for the type is again Delta Airlines with an announced order for 26. So far 77 MD-90s are on order for seven customers, chief among those remaining being 20 aircraft for China. The new aircraft made its

Left: By far the biggest operator of the MD-80 is American Airlines, which has over 350 serving its extensive route network. The majority are MD-82s with the balance being made up by MD-83s.

Below: The MD-83 is fitted with uprated JT8D-219 engines and additional fuel tankage and has become a favourite with airlines specialising in lengthy holiday charters. One such operator is Spanair, which now possesses a total of 18 such aircraft.

maiden flight in August 1993 and entered service with Delta in April 1995.

Three versions of the MD-90 have been announced. The MD90-10 is the basic version, with a fuselage of the same length as the MD-87 accommodating 114 passengers (12 first class, 102 economy), and will be powered by two 97.87-kN (22,000-lb st) IAE V2500-D2 turbofans. The MD90-30 features an MD-80 fuselage, lengthened forward of the wing by 1.5 m (4 ft 9 in), and MD-87 tail surfaces. Accommodation is for 153 passengers (12 first class, 141 economy), and engines are two 111.21-kN (25,000-lb st) IAE V2500-D1s. Largest of the MD-90 variants is the MD90-40, with accommodation for 180 passengers in mixed-class configuration and powered by two 124.85-kN

Above: The MD-95 was planned as McDonnell Douglas's next medium-haul airliner, and its future was initially tied to the Chinese trunkliner programme.

Below: Newest development is the MD-88. With accommodation for 142 passengers, a full EFIS-equipped 'glass' cockpit and large-scale use of composite materials in its construction, the first MD-88 entered service with Delta in 1988.

Left: A change in company policy saw the abandonment of the familiar DC (Douglas Commercial) title for all airliners in favour of the more functional MD (McDonnell Douglas). Aircraft sales remained unchanged.

Below: The MD-89 was a projected but unflown MD-80 variant with a further stretch to the fuselage. It was intended to carry 173 passengers over a distance of almost 4830 km (3,000 miles).

(28,000-lb st) IAE V2500-D5s. A proposed version for the European market is designated MD90-40EC (European Configuration). All models have a maximum fuel capacity of 22108 litres (5,840 US gal).

McDonnell Douglas is modifying the now-abandoned MD-80 UHB propfan demonstrator to serve as the prototype for the MD90-30, with certification and first deliveries scheduled for the fourth quarter of 1994. MD90-10 certification is anticipated in the first quarter of 1995.

The final addition to the McDonnell Douglas family of twin-jet airliners was the MD-95 launched at the 1991 Paris air show. McDonnell Douglas, CATIC of China and Pratt & Whitney were collaborating on development of the aircraft, which was to have been powered by two Pratt & Whitney JT8D-218 or Rolls-Royce Tay 650 turbofans but now has BMW Rolls-Royce BR715-58 units, for high-frequency air routes. The smallest member of the current MDD twin-jet range, about 0.9 m (3 ft) longer than the DC-9-30, which it is expected to replace in many fleets, the MD-95 has accommodation for 106-130 passengers five abreast at a maximum take-off weight of 53706 kg (118,400 lb), and will have a maximum range of 3241 km (2014 miles). Many features of the MD-88's advanced two-crew flight deck are fully incorporated in the MD-95's cockpit, including Category III-a autoland capability and non-radio dependent vertical and lateral navigation systems. Development of the aircraft was conducted in conjunction with the People's Republic of China's Trunk Aircraft Programme, and a memorandum of understanding signed with CATIC provided for final assembly of the aircraft in China. Aerospace Technologies of Australia, Aerospatiale of France, Alenia of Italy and CASA of Spain are also manufacturing partners in the programme.

The launch of the MD-95 was initially conceived within the context of the success of the Sino-US trunkliner programme, in which both the MD-80 and MD-90 series were to have been

built in Shanghai as trunkliners for China's rapidly expanding internal route network, which was rightly assessed as urgently needing modern Western aircraft. The Shanghai-built aircraft were to have been built to a standard largely similar to that of the Long Beach-built aircraft, but with beefed-up landing gear to cope with the stresses of frequent operations to and from less well-equipped runways. There was considerable wrangling between the various parties, including IAE, for licensed production of its V2500 turbofan in China. The programme was then reduced from the planned 100 aircraft to an initial batch of 20 with more possibly to follow, and the scheme finally fell into virtual abeyance.

The MD-95 was finally launched in October 1995, the first aircraft flying in September 1998 after McDonnell Douglas had merged with Boeing. Boeing has decided to continue with the MD-95 programme as it offers a regional type with capacity below that of the 737-600, the smallest of Boeing's current types. In January 1998 Boeing renamed the MD-95 the Boeing 717.

1993 saw the arrival on the scene of the V2500-powered MD-90. A range of MD-87/MD-80-sized versions and larger are on offer. The first customer will be Delta Airlines.

METRO Fleet Feeder

The spindly fuselage of the Metro is a regular sight at most US airports, sizeable fleets of the aircraft hurrying between large fields and smaller ports-of-call to bring passengers into the main hub centres. This commuter traffic constitutes the bulk of the Metro's operations, although an increasing number are being used for cargo carriage, albeit operating in a similar fashion by feeding a central hub from outlying airfields.

Inside the United States one finds roughly half of the world's air transport movements. Many are big wide-bodies, many more are the vast numbers of narrow-body jets, but wending their way among all this traffic are hundreds of small turboprops, busy on local-service and so-called commuter services, often feeding in radially from small airports to a big central hub. Among the most numerous of the small turboprops are some with fuselages like long tubes, so constricted in diameter that one might wonder that grown-up passengers could squeeze their way inside.

Clearly, passsengers *can* squeeze inside, as shown by the fact that the number of these aircraft – Metros, designed by Swearingen and now made by Fairchild – together with the related family of cor-

The Metro III airliner and Merlin IVC executive aircraft offer rapid and economical transport of passengers and high-value cargo. The Garrett TPE331 turboprops produce 746 kW in dry thrust, but can be augmented with water/methanol boosting to give 820 kW.

porate aircraft already number well over 800 and may soon top the 1,000 mark. In the same way that the Learjets, in which it is impossible to stand up, have outsold all other business jets, so has the Metro outsold all other small local-service turboprops. To the criticism that it is impossible to stand up, the designers might say 'Who wants to? You don't stand up in a luxury car!'

Edward J. Swearingen started his little company at San Antonio, Texas, almost 30 years ago. His first major work was to build prototypes for others, and to fit more powerful engines to Beech Twin-Bonanza and Queen Air aircraft, but he yearned to produce aircraft of his (or his company's) own design. To start with, he took a modified Queen Air wing and Twin-Bonanza landing gear and added twin turboprop engines and a completely new fuselage and tail. Called the Merlin IIA, it first flew on 13 April 1965 and was a winner from the start. It needed few changes, and deliveries began soon after Certification in August 1966. This was the starting point for the entire Merlin/Metro family.

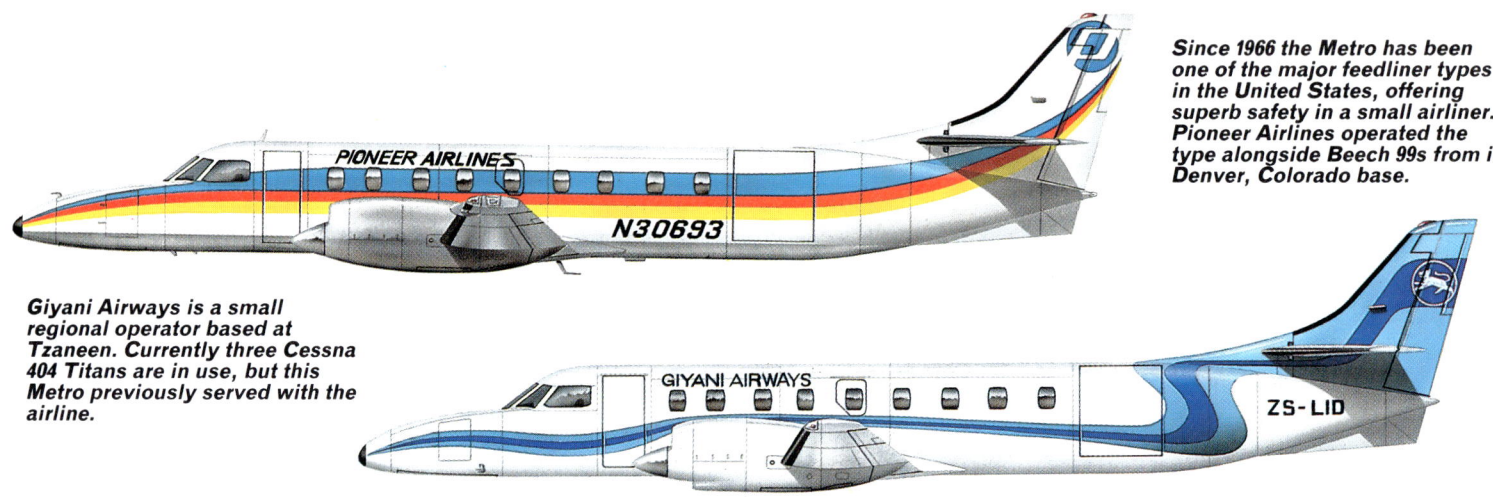

Since 1966 the Metro has been one of the major feedliner types in the United States, offering superb safety in a small airliner. Pioneer Airlines operated the type alongside Beech 99s from its Denver, Colorado base.

Giyani Airways is a small regional operator based at Tzaneen. Currently three Cessna 404 Titans are in use, but this Metro previously served with the airline.

Many European operators fly the Metro, its useful load and speed making it an ideal choice for short-range sectors between cities. NFD from Nürnberg flies seven of the type, alongside ATR.42-300s and a Dornier Do 228. This one is seen at the regional airport at Saarbrücken.

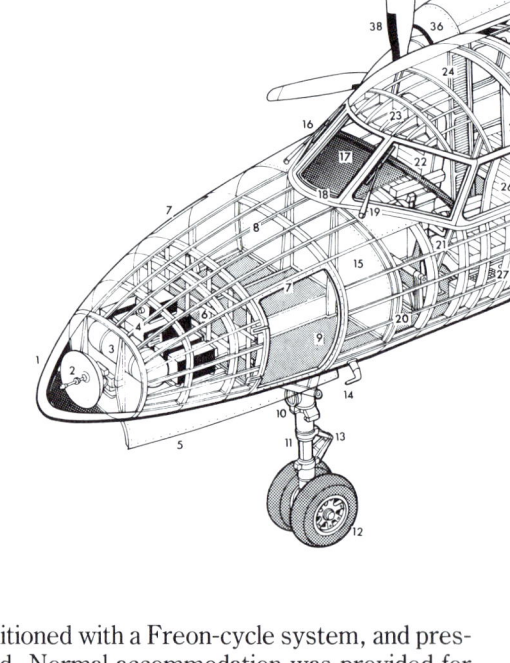

An expanding business, particularly in the United States, is the rapid movement of parcels by air, and United Parcels Service is the largest, operating a huge fleet of 727s, 747s, 757s and DC-8s. At the smaller end of the scale are a dozen Fairchild Expediters, the cargo version of the Metro.

Structurally, the original Merlin IIA was a completely traditional all-metal stressed skin machine, with a smooth flush-riveted exterior. Aerodynamically the only points worthy of comment were the efficient slotted flaps and the fact that both the vertical and horizontal tail surfaces were swept back, the horizontal surface quite sharply. Unlike most aircraft in its class the leading edges of the wings and tail surfaces were de-iced by Goodrich pneumatically inflated and deflated boots. The engines were 410-kW (550-hp) Pratt & Whitney Canada PT6A-20 turboprops, fed from integral tanks in the wing and with bleed-air anti-icing of the inlets. The retractable tricycle landing gear had a single wheel on each leg and, like the flaps, was operated electrically, the only hydraulic item being the wheel brakes. The cabin was fully air-conditioned with a Freon-cycle system, and pressurised by engine bleed. Normal accommodation was provided for two people on the flight deck and six more (three pairs) in the cabin, with a bulkhead with a sliding door behind the flight deck. Easy access was gained by an airstair door on the left side.

In 1968, when 36 aircraft had been produced, the engines were changed to the 496-kW (665-shp) Garrett TPE331-1-151G, with the inlets above the spinners of the Hartzell three-blade propellers, the latter being feathering and reversing and fitted with a synchrophasing system to reduce noise and 'beat' vibration. Though there have been projects to use PT6A engines, all subsequent production versions of the Merlin/Metro family have used various sub-types of Garrett TPE 331, a simple and robust single-shaft engine which has been greatly developed in power to meet the demands of successive aircraft.

Merlin IIB production continued to 1972, but back in 1968 work had begun on a completely redesigned aircraft owing nothing to the Twin-Bonanza or Queen Air. This aircraft, the Merlin III, had a fuselage 0.62 m longer, and totally new wings with a different aerofoil profile and fractionally greater span, and new landing gears with two small wheels on each leg. The new more-tapered wing had double-

Swearingen Metro II cutaway drawing key

1. Radome
2. Weather radar scanner
3. Oxygen bottle
4. Radio and electronics equipment
5. Nosewheel door
6. Baggage restraint net
7. Baggage doors, forward opening
8. Fuselage nose construction
9. Nose baggage hold
10. Landing and taxi lamp
11. Nosewheel leg
12. Twin nosewheels
13. Torque scissors
14. Pitot tube
15. Cockpit pressure bulkhead
16. Windscreen panels
17. Instrument panel shroud
18. Curved centre panel
19. Windscreen wipers
20. Rudder pedals
21. Control column
22. Co-pilot's seat
23. Cockpit roof construction
43. Starboard aileron
44. Static dischargers
45. Starboard flap
46. Tailpipe exhaust duct
47. Fuselage frames
48. Cabin interior trim panels
49. Passenger seats
50. Window side panel
51. Cabin floor construction
52. Seat rails
53. Air trunking
54. Cabin windows
55. Starboard emergency escape hatches
56. Main fuselage frames
57. Centre box construction
58. Port emergency escape hatch
59. Starboard seating
60. Port seating, nine passengers
61. Cabin rear bulkhead
84. Static discharger
85. Tailplane construction
86. Tail navigation light
87. Ventral fin
88. Rudder hinge control
89. Tailplane control cables
90. Fin attachment frame
91. Cargo hold rear bulkhead
92. Baggage/cargo hold floor
93. Rear fuselage frames
94. Seat fixing rails
95. Trailing edge root fillet
96. Port flap
97. Fuel pumps
98. Wing main spar
99. Wing spar attachment

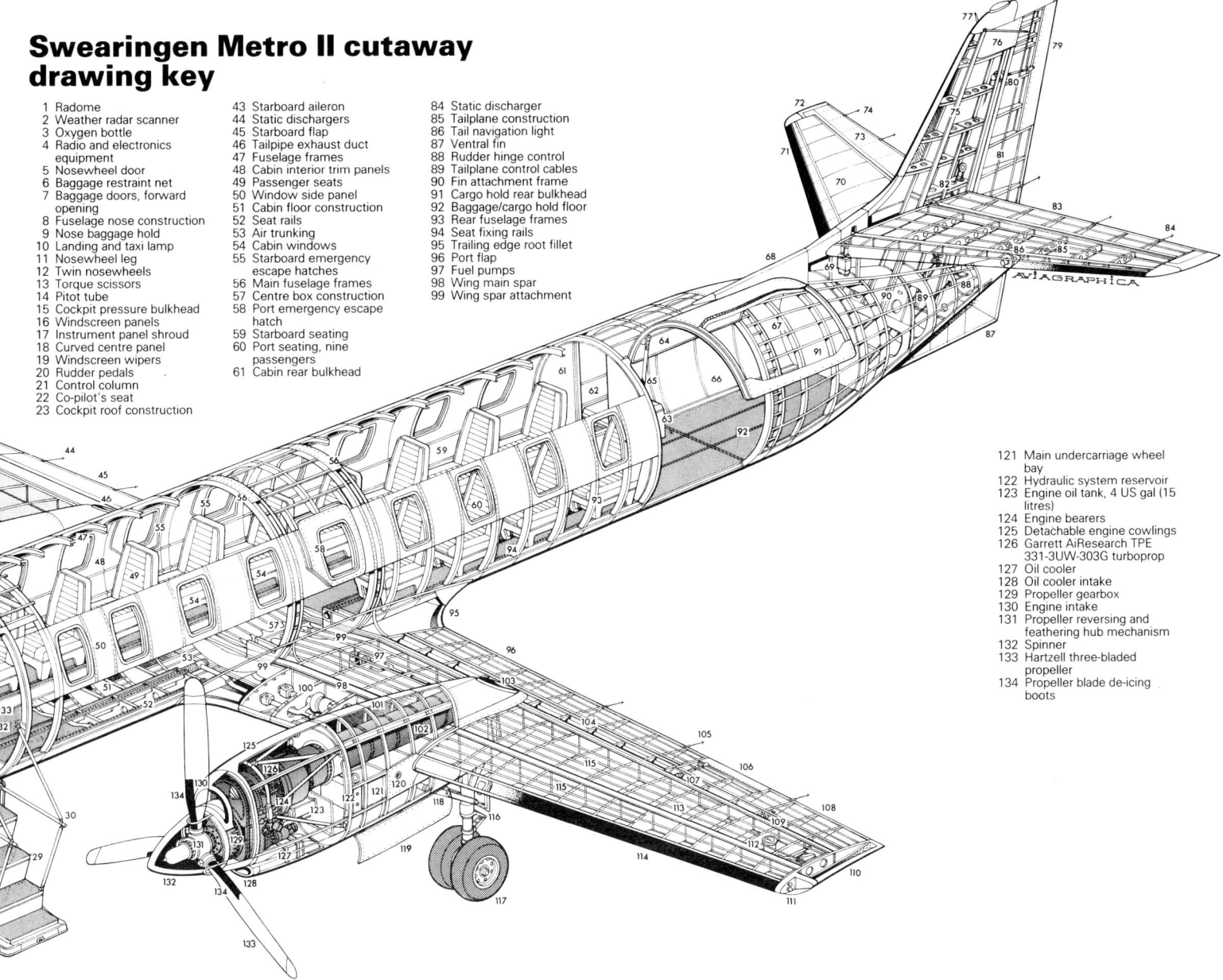

24. Cockpit bulkhead
25. Electrical panels
26. Pilot's seat
27. Pilot's side control panel
28. Passenger door
29. Airstairs
30. Handrails
31. Entry doorway
32. Cabin centre aisle floor
33. Air conditioning duct louvre
34. Forward fuselage frame construction
35. Starboard engine cowlings
36. Engine intake
37. Hartzell three-bladed constant-speed reversing and feathering propeller
38. Propeller de-icing boot
39. Leading-edge de-icing
40. Starboard wing fuel tank, capacity 324 US gal (1226 litres)
41. Starboard navigation light
42. Fuel filler cap
62. Toilet compartment door
63. Toilet
64. Rear cargo door
65. Door actuator
66. Rear cargo and baggage compartment
67. Fuselage frame and stringer construction
68. Fin root fillet
69. Tailplane electric trim jacks
70. Starboard tailplane
71. Leading edge de-icing
72. Elevator horn balance
73. Starboard elevator
74. Static dischargers
75. Fin construction
76. Rudder balance
77. Antenna
78. Anti-collision light
79. Rudder trim tab
80. Trim tab control jack
81. Rudder construction
82. Elevator hinge control
83. Port elevator
100. Air conditioning plant
101. Engine cowling construction
102. Tailpipe
103. Engine exhaust duct
104. Double slotted flap construction
105. Static dischargers
106. Aileron trim tab
107. Trim tab hinge control
108. Port aileron
109. Aileron hinge control
110. Port wing-tip
111. Port navigation light
112. Fuel tank filler cap
113. Wing rib construction
114. Leading-edge de-icing
115. Port wing fuel tank, 324 US gal (1226 litres)
116. Main undercarriage leg
117. Twin mainwheels
118. Retractable strut
119. Mainwheel door
120. Leading edge ice inspection light
121. Main undercarriage wheel bay
122. Hydraulic system reservoir
123. Engine oil tank, 4 US gal (15 litres)
124. Engine bearers
125. Detachable engine cowlings
126. Garrett AiResearch TPE 331-3UW-303G turboprop
127. Oil cooler
128. Oil cooler intake
129. Propeller gearbox
130. Engine intake
131. Propeller reversing and feathering hub mechanism
132. Spinner
133. Hartzell three-bladed propeller
134. Propeller blade de-icing boots

slotted flaps, and both these and the forward-retracted landing gears were operated hydraulically. Thanks to use of 626-kW (840-shp) TPE331-303G engines, the all-round flight performance was considerably enhanced. The Merlin III was certificated in July 1970, and was followed in production by the IIIA, IIIB and IIIC incorporating numerous mostly minor improvements, among the most important being 671-kW (900-shp) TPE331-10U-501G or -503G engines, driving four-blade propellers, giving cruising speeds up to 571 km/h (355 mph).

In parallel with the Merlin III, the growing Swearingen engineering staff designed the SA-226TC Metro. In most respects the two aircraft were identical, but the Metro was considerably longer, the passenger cabin being extended from 3.23 m (10 ft 7 in) to 7.75 m (25 ft 5 in) to seat 19 or 20 passengers. Internal cross section remained exactly as before, the maximum width near the seated passengers' elbow level being 1.57 m (5 ft 2 in) and the maximum height from the ceiling to the floor (which was lower than the floor level on each side, where the single seats were attached) being a mere 1.45 m (4 ft 9 in). Indeed, the claustrophobic effect was heightened by using small circular windows instead of big picture windows as in the Merlin III. One might have thought such cramped dimensions would have made the long, tube-like Metro unsaleable. Far from it! Airlines all over the world have queued to build up fleets of these aircraft, which are cheap to buy, relatively fast and, once you are in your seat, comfortable. Any feeling of claustrophobia, of course, depends on the individual.

The Metro, like the Merlin III, was a major effort for Swearingen, even though employment exceeded the 900 mark before the end of 1968. The development and marketing of the new aircraft was a joint venture with what was then Fairchild Hiller, and the new wings and various other parts were made at Fairchild's plant at Hagerstown, Maryland. The first Metro flew on 26 August 1969, and as with previous Swearingen designs it proved virtually 'right first time'.

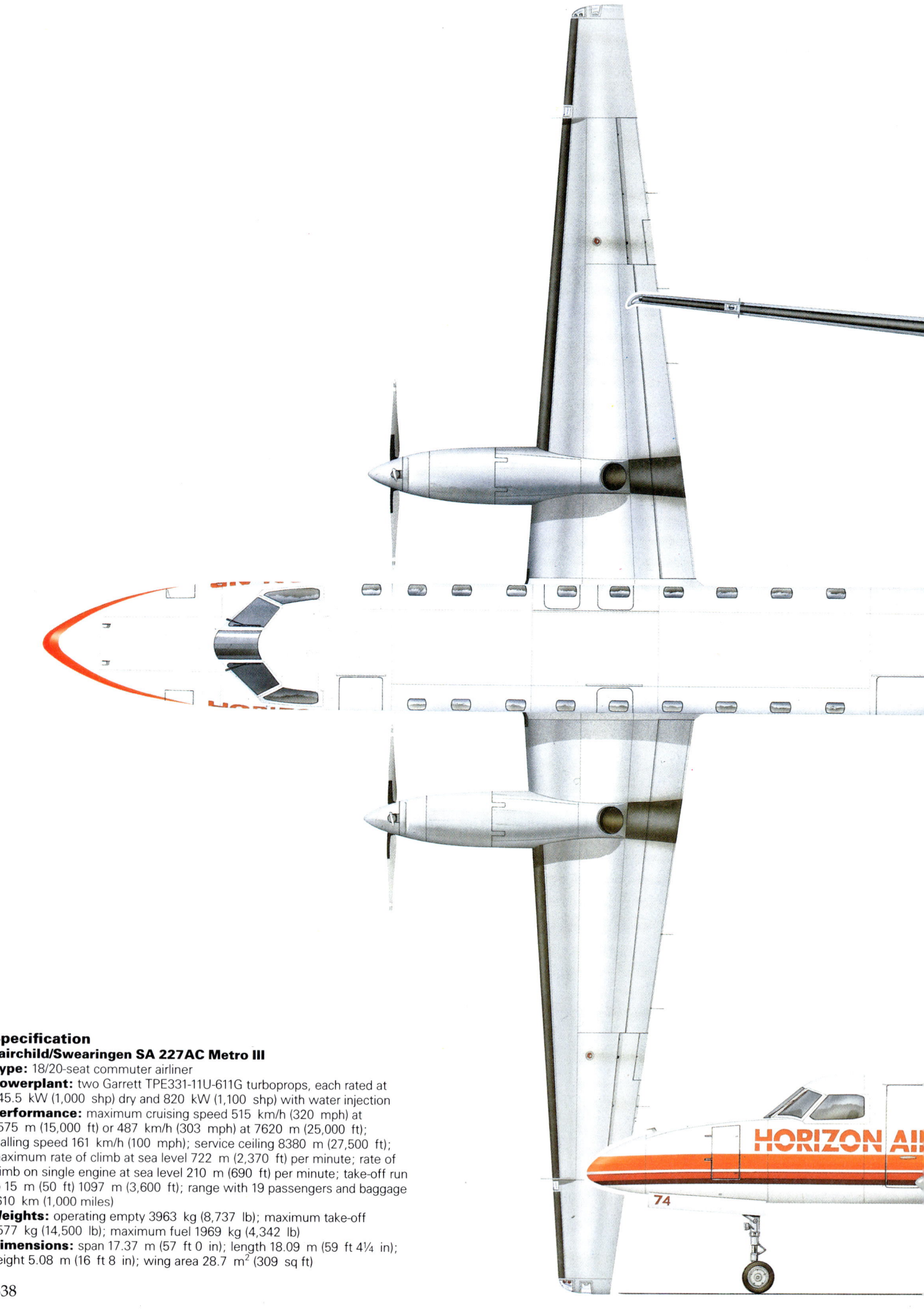

Specification
Fairchild/Swearingen SA 227AC Metro III
Type: 18/20-seat commuter airliner
Powerplant: two Garrett TPE331-11U-611G turboprops, each rated at 745.5 kW (1,000 shp) dry and 820 kW (1,100 shp) with water injection
Performance: maximum cruising speed 515 km/h (320 mph) at 4575 m (15,000 ft) or 487 km/h (303 mph) at 7620 m (25,000 ft); stalling speed 161 km/h (100 mph); service ceiling 8380 m (27,500 ft); maximum rate of climb at sea level 722 m (2,370 ft) per minute; rate of climb on single engine at sea level 210 m (690 ft) per minute; take-off run to 15 m (50 ft) 1097 m (3,600 ft); range with 19 passengers and baggage 1610 km (1,000 miles)
Weights: operating empty 3963 kg (8,737 lb); maximum take-off 6577 kg (14,500 lb); maximum fuel 1969 kg (4,342 lb)
Dimensions: span 17.37 m (57 ft 0 in); length 18.09 m (59 ft 4¼ in); height 5.08 m (16 ft 8 in); wing area 28.7 m² (309 sq ft)

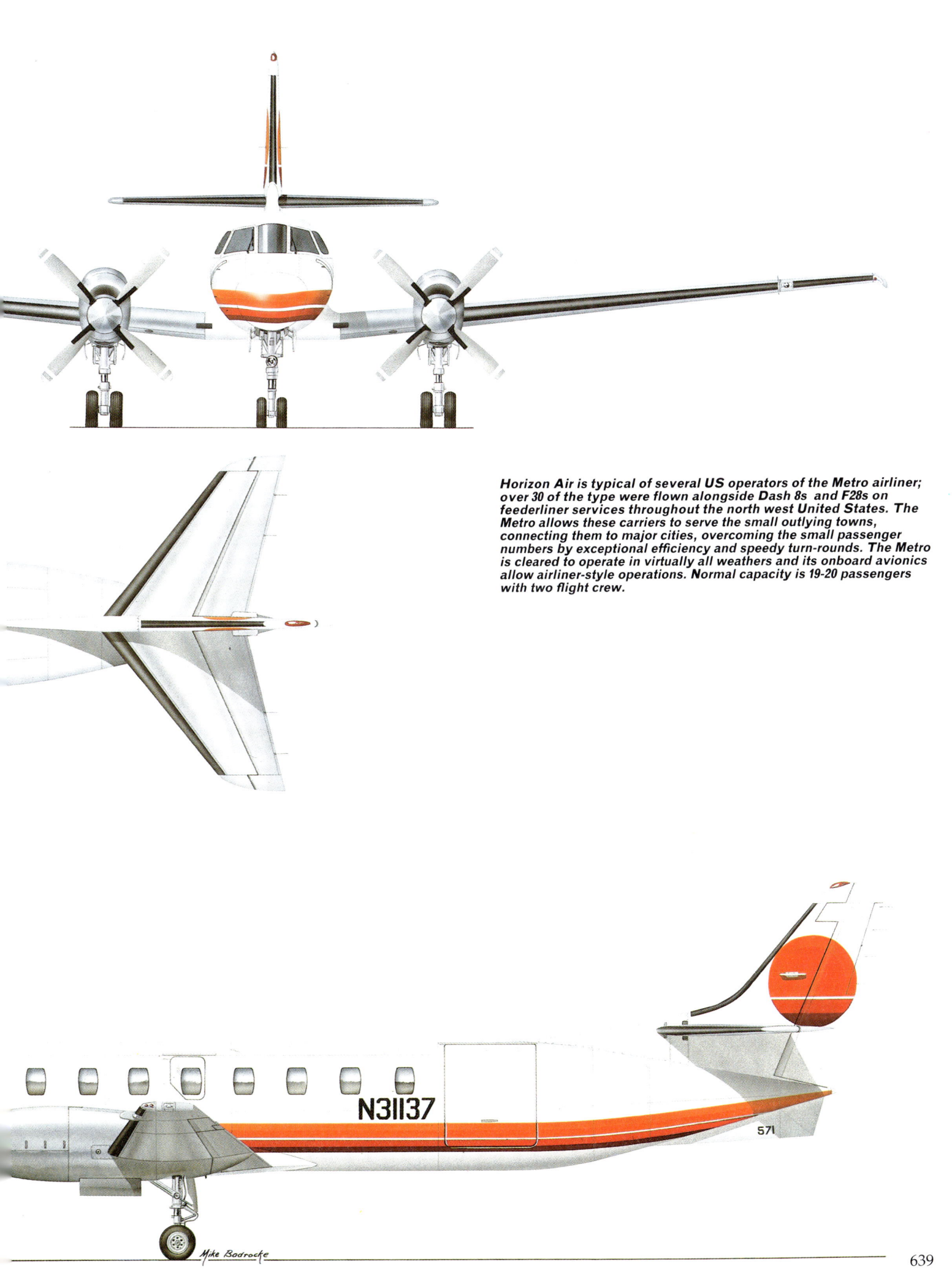

Horizon Air is typical of several US operators of the Metro airliner; over 30 of the type were flown alongside Dash 8s and F28s on feederliner services throughout the north west United States. The Metro allows these carriers to serve the small outlying towns, connecting them to major cities, overcoming the small passenger numbers by exceptional efficiency and speedy turn-rounds. The Metro is cleared to operate in virtually all weathers and its onboard avionics allow airliner-style operations. Normal capacity is 19-20 passengers with two flight crew.

Without doubt the most bizarre Metro is this Swedish air force aircraft, used for tests of an Ericsson airborne early warning radar carried in the over-fuselage canoe fairing. Extra equipment is used to down-link information from the radar to ground stations.

A/S Norving is a well-known Norwegian regional airline, flying Dornier Do 228s and Britten Norman Islanders on services throughout Scandinavia and northern Europe. Two Metros were on strength, supporting oil traffic operations over the North Sea.

In almost all major respects the rest of the Metro resembled the Merlin III, the engines being -303Gs driving three-blade Hartzell reversing propellers, and the fuel capacity being unchanged at 2452 litres (648 Imp gal). To increase the likely market Swearingen put the passenger door, with integral airstair built into the inner face, immediately aft of the flight deck and added a wide cargo door at the rear. The company offered a dedicated cargo version, a quick-change passenger/cargo Metro with stowing fold-up seats, a clip-in aisle filler and snap-in sections of carpet, plus a movable bulkhead between passenger and cargo compartments. Other versions offered included an executive version (later marketed separately as the Merlin IVA), an ambulance with 10 litters, a hospital casevac version equipped for emergency operations, and a surveillance model with single or dual camera installations.

Sales were certainly enhanced by the early switch of production to the Metro II, still called the SA-226TC. This greatly improved the psychological comfort of the cabin by replacing the tiny portholes by upright rectangular windows every bit as large as those found in typical jetliners. At the same time the flight deck was considerably improved, and there were various systems improvements. As already noted, Swearingen replaced the 'corporate Metro' by the Merlin IVA, first delivered in 1970. This introduced a modified fuselage with two fewer windows on each side, the interior being normally configured for 12 to 15 passengers, with a toilet and 4.05 m³ (143 cu ft) baggage compartment. Surprisingly, fuel capacity is reduced to 2096 litres (554 Imp gal). Another variant, offered from 1979, was the Merlin SM maritime surveillance aircraft, with extremely comprehensive navaids and numerous sensors including radars, IR scanners, cameras, searchlight and a wealth of other equipment. The SM would be operated in the Restricted category at weights up to 6350 kg (14,000 lb).

In 1979 Swearingen became a wholly-owned subsidiary of Fairchild Industries, but the work remained centred at San Antonio and the company operated as the Fairchild Aircraft Corporation.

It was partly the obvious desire to increase the weights of the Metro/Merlin family that led, in 1981, to a major upgrade. The chief change in this new Metro III version was a dramatic 3.05 m (10 ft) increase in span, to 17.37 m (57 ft). This allowed certification under FAA Special Regulations 41 and 41B at a normal weight of 6577 kg (14,500 lb), providing an increase in useful load of 672 kg (1,480 lb). Even greater capability results from selection of the high gross weight option at 7257 kg (16,000 lb). Another important change in the III was use of the TPE331-11U-612G engine, rated at 746 kW (1,000 hp), or 820 kW (1,100 hp) with water/alcohol injection, driving a McCauley propeller with four blades and diameter of 2.69 m (8 ft 10 in), turning at reduced rpm. The engines are in more streamlined

Metros have been sold to many European countries, where small operators find the type's economy attractive. Metro Air was based at Roskilde, Denmark, and flew two aircraft, regularly seen at Aberdeen airport in Scotland, centre for the North Sea oil industry.

France has many regional airlines connecting towns and cities within the region. Compagnie Aérienne du Languedoc was one, but is now part of Air Littoral, based at Montpelier with several types of feederliners in use. Six Metro IIIs were the backbone of the fleet.

nacelles, with improved all-round access cowlings. Rather remarkably, in view of the increase in weight from the original 5670 kg (12,500 lb), the tyre pressures were actually reduced, enabling the aircraft to use good unpaved surfaces if necessary, tyre pressure in the four main tyres being reduced from 7.03 to 5.6 kg/cm^2. The main landing-gear doors were also redesigned, to reduce drag on take-off and landing and provide better access for maintenance.

Options that did not get into production were the use of Dowty Rotol propellers (though Dowty Rotol propellers, in a larger size of course, were selected for the Saab/Fairchild SF 340) and Pratt & Whitney Canada engines. Fairchild designed an extremely efficient installation for the 820-kW (1,100-shp) PT6A-45R turboprop for the Metro IIIA, and flew a demonstrator on 31 December 1981. Compared with the Metro III the IIIA also had an environmental control system of considerably greater capacity, giving among other things improved cooling on the ground in heat ambient conditions. The engine cowlings were of the clamshell type, and McCauley four-blade propellers were fitted.

While the Metro III was entering production the design team at San Antonio continued the product-improvement process by upgrading the smaller Merlin III into the SA-227/TT41 Merlin 300. Production of the Merlin IIIC ended in late 1982. In 1983 development began of the Merlin 300, and deliveries of this have been in progress since the beginning of 1985. The most obvious change is the addition of 0.76-m (2-ft 5-in) wingtip fences, or winglets, which lean backwards and outwards from the ends of the wings. Other changes include improvements to the flight-control circuits (for example the pulleys are larger) and reshaped ailerons for greater roll power. Engines are 671-kW (900-shp) TPE331-10U-513Gs, driving four-blade Dowty Rotol propellers. Like the later Merlin III variants the rear pressure bulkhead is right back at the tail, giving a cabin long enough for up to 10 passengers, with 136 kg (300 lb) of baggage in the entrance vestibule at the rear.

The long-body Metro family continued to expand. United Parcel Service and DHL were two of the early fleet buyers of the Expediter I dedicated cargo version, with a payload of over 2268 kg (5,000 lb). Features included a reinforced floor, repositioned air conditioning ducts to give more cargo space, cargo restraint system and an empty weight reduced from 4164 kg (9,180 lb) to 3960 kg (8,730 lb).

In 1987 Fairchild offered a maritime patrol or anti-submarine version called the Air Sentry, but more recently these have been marketed as just one range of variants with special role fits. By mid-1990 35 such aircraft had been delivered, some able to fly a 10-hour mission at a radius of 1946 km (1,209 miles) with full reserves.

In 1986 the US Department of Defense allocated funds for the acquisition of standard Metro IIIs for use as US Air National Guard Operational Support Transport Aircraft and to be known as the C-26A. The original order, confirmed early in 1989, was for 10 aircraft but later increased to 20, with one example of the second batch designated C-26B. The first C-26A was delivered to the California ANG in April 1989 and despite the Company seeking protection under Chapter 11 of the US bankruptcy laws in February 1990, de-

One of the best-known European Metro operators was Crossair of Zurich, whose primary small-capacity types have become the Saab 340 and 2000.

liveries of 25 aircraft on order for the USAF and civil operators at the time of financial difficulties were expected to continue.

The corporate version of the Metro III is the Merlin IVC, which includes among other upgrades, certification of a Bendix electronic flight instrument system or 'glass cockpit'. Fitted with a luxurious interior which includes reclining seats, a couch, and sophisticated refreshment and entertainment centres, the Merlin IVC can accommodate between 11 and 14 persons. By January 1990 24 had been delivered.

In 1987, Fairchild announced the intention to expand the SA226/7 family by adding to the range the Metro V, Merlin V, Metro VI and Merlin 400SP, the Merlins being corporate versions. All were planned to feature T-Tails, a redesigned wing with Fowler area-increasing flaps over 90 per cent of the span and small ailerons inset from the tips. However, development of all the new variants was discontinued in 1989 in favour of the SA-227DC Metro 25.

Able to seat 25 passengers and powered by two 820-kW (1,100-shp) TPE331-12 engines, the Metro 25 higher gross weight modification of the Metro III has the cargo door deleted but features an increased number of windows in the rear of the cabin, where the extra passengers are accommodated in what was the baggage area. Luggage is housed in a large underfuselage pannier. A rebuilt Metro III incorporating Metro 25 features began flight tests on 25 September 1989 in order to assess design features prior to a full production decision, which went against the type.

In mid-1996 Fairchild acquired the Dornier company, Germany's main manufacturer of regional transport aircraft, to create Fairchild-Dornier, whose US arm is Fairchild-Dornier USA.

While the Metro is tailored for the small airliner role, the similar Merlin IVA has an executive interior with fewer seats. The Merlin/Metro family was designed and originally built by Swearingen, but the company was taken over by Fairchild in 1979, and the product renamed Fairchild in 1982.

Chapter 30

The Microlight Revolution

Mil's Civilian 'Helos'

Mitsubishi MU-2

NAMC YS11

The Microlight Revolution

Hang-gliders and microlights may seem to be a recent innovation, but the idea is the earliest in manned flight. It was not until the mid-1970s that the concept of a skeletal wing with loose covering took off again, sparking a massive increase in the number of these low-powered aircraft, and a line of development that would see the microlights becoming ever more sophisticated.

The first forays into the sport took the form of 'Ski-kites', which were diamond-shaped kites with a tubular framework along the two leading edge sides of the structure and fore and aft between the front and rear corners. The pilot sat on a suspended seat with a rudimentary control column and the whole contraption was towed behind a car for a running take-off. These ski-kites provided a lot of fun and they would rise to a height of up to 152 m (500 ft). They also established the weight shift method of control in which the pilot used the control column to move his body from side to side and thus bank the kite in the desired direction. Needless to say, there were exciting moments for most kiters, and it was not unusual to find an intrepid birdman hanging in a tree because of a gust of wind or some minor over-control.

The modern hang-glider revolution was actually prompted by work done by Francis Rogallo on a NASA project for a free-flying kite for use in rescuing spacecraft. In the early stages the gliders were built at home by amateurs from photographs of the Rogallo experimental models and it was not long before it became plain that power launches were unnecessary if the right hillside was available and if the pilot had sufficient courage to launch himself off the edge and trust his home-made creation. By 1978, however, it was estimated that over 65,000 hang-gliders were in use and the design of the wings was becoming more and more sophisticated. As time went by hang-glider pilots learned the techniques of soaring which were first explored by the rigid glider pilots of the 1930s, and very long flights became possible by using the updraughts coming off mountain ridges and the thermals which could be found if one persevered.

Some hang-glider enthusiasts started to work with rigid-wing structures. The Mitchell Wing, for instance, was much more akin to a sailplane layout with a rib structure, and its glide ratio was much better than that of the wind-filled canopy of the standard Rogallo. It was at this point of the development that it became practical for engines to be fitted to hang-glider-type structures, and for the microlight to come into its own. For many pilots this was a welcome development because hang gliding required a lot of luck to achieve satisfying performance and it was not without its casualties. Even the most experienced pilots could lose their life through structural failure, control error or weather conditions, and a fair amount of adverse comment was generated which sometimes resulted in hang gliding being banned from the more promising sites.

There was nothing really new about the idea of a powered hang-glider. In 1963 the US Army had worked with Ryan Aircraft on the XV-8A 'Fleep' project and this had been, basically, a Rogallo-style wing with a light open fuselage hung beneath it and a 156-kW (210-hp) Con-

Now marketed by Freedom Fliers in the USA, the Ascender II+ has an enlarged hang cage and larger diameter spars, plus other structural modifications allowing a substantial increase in payload. Overwing drag rudders and a stick-operated frontal elevator allow two-axis control, with spoilers available as an option.

While the vast majority of microlights are designed and flown for leisure, there are practical commercial applications which are being exploited by several companies. Mitchell Wing Inc. offers the AG-38 Falcon complete with 53-litre (14-US gal) spray tank, two spray heads and a spray pump. Over 500 have been sold.

Two-seat microlights are becoming increasingly popular, this example being a Maxair Drifter XP which features complete dual controls. Powered by a 36-kW (48-hp) Rotax 503 engine, the XP is a derivative of the single-seat DR.277 Drifter. The wings and tail surfaces have a double-surface Dacron covering.

tinental flat-six piston engine in a pusher installation acting as motive power. In England ML Aviation built the ML Utility inflatable wing aircraft which, again, had a crew pod with a pusher engine slung underneath. The first microlights of the new era became known as 'trikes' because, although they looked much like hang-gliders, they had a tricycle frame arrangement with a wheeled landing gear and an engine mounting, all slung underneath what looked like a hang-glider wing. They had much in common with the hang-gliders because control was by weight shift with a control bar hanging down in front of the pilot. On the other hand, the wing was a significantly stronger structure with heavier-gauge tube frames and much more wire bracing.

Introduced in 1984, the Eipper Quicksilver GT was a completely new design for the company. The initial standard model is the GT 280, with tapered wings embodying ailerons and flaps. The glass-fibre nose 'bullet' and windscreen are standard, with a wrap-around fuselage pod or cockpit enclosure as options.

Microlight variations

Three broad types of microlight have emerged. Firstly, there are basic trikes with their weight-shift control; secondly, there are the machines with rigid structure and control through movable wing and tail surfaces; and, thirdly, there are the advanced microlights which, but for the rules of the certification authorities, would be regarded as normal light aircraft. Because the sport has advanced so quickly and because the American and other airworthiness authorities did not feel they could cope with certification of these machines in the same category as 'proper' powered light aircraft, some rather arbitrary rules were set down to define microlights. Every country has its own definition (and its own rules on registration and airworthiness approval) but it is broadly correct to say that any powered aircraft of less than 150 kg (331 lb) is to be regarded as a microlight. This weight limit refers to the empty condition of the aircraft and in the United States the limit is somewhat lower at 115 kg (254 lb). And while they are called microlights in the UK, they are known as ultralights in the USA and as ULMs (Ultra Léger Motorisé) in France.

The trike type of aircraft are now normally two-seaters with the passenger seated in a nylon slung seat immediately behind the pilot. An increasing number of trikes are fitted with a fibreglass pod enclosure to give some protection to the crew in flight, but one will expect to pay around £3,000 for a flyable aircraft powered by a 37-kW (50-hp) pusher engine,

Microlights come in a bewildering array of airframe and powerplant configurations. This is the Zenair Zipper, featuring side-by-side twin engines.

but before fitting the pod or any other extras. Most trikes are purchased in two pieces. The wing comes from one manufacturer and the trike unit with engine comes from another. The most popular combination in the UK is a wing from Flexiform Sky Sails or Southdown Sailwings with an open tubular trike manufactured by Mainair Sports of Rochdale, Lancs. In the USA, of course, there are many manufacturers and combinations available. The aircraft is generally assembled for each flight by manhandling the wing up onto the head of the A-frame which forms a part of the trike structure and contains the engine mounting. Most aircraft now have electric engine starters, and throttle control is through a foot pedal which leaves the pilot's hands free to operate the suspended control bar.

Overcoming training troubles

For some time the great problem with microlights was that all the available types were single-seaters and it was very difficult to train new students. The anxious instructor had no alternative to watching his pupil aloft on his own and could only hope that the ground tuition had been absorbed sufficiently to ensure a safe landing. Two-seat trainers have now been developed, and aircraft like the Chargus Titan have side-by-side seating for instructor and pupil. With the aircraft controlled on the weight-shift principle, the two occupants have to be well co-ordinated. The throttle control and the control bar are combined on the Titan and, to get an even balance, it is necessary for the two occupants to cross their arms as they take hold of the bar, and this can lead to some complex aerobatics until the technique is fully understood. The Titan is powered by a 30-kW (40-hp) Fuji Robin engine driving a wooden propeller through a toothed belt drive system, and the whole engine installation is fitted behind the pilots' seats with a dual exhaust system exiting well to the rear of the microlight.

The three-axis type of microlight has become the most popular in recent years because it is much closer to a conventional light aircraft than its weight-shift cousins. The weight-shift microlight is physically demanding and the pilot of a three-axis control machine can have a more relaxed flight, although he pays more for his aircraft. The most successful models have come from the USA and include the American Aerolights Eagle, the Maxair Hummer, the Rotec Rally and the Eipper Quicksilver MX. The Quicksilver is probably

Several of the microlight designs have structures rigid enough for aerobatics, although these are limited due to the low power available. This machine is equipped with skis for snow landings.

the most popular of all microlights, and it is fitted with a stick controlling the rudder and elevators and foot pedals which are connected to 'spoilerons' on the wings which create rolling movement. The spoilerons can also be deployed on both wings at the same time to

One of the more unusual microlight designs, with its tandem-wing configuration, is the Belgian-designed Butterfly. Powered by a McCulloch MC-101 engine with direct drive to a pusher propeller, this small microlight has two-axis control, and is available in kit form for homebuilding.

Microlights are beginning to appear in the Soviet Union, with at least three designs identified. This is a LAK (Lithuanian Aircraft Construction) BROK-1M, a powered version of the Oshkinis BRO-23KR glider. A central skid or conventional tricycle landing gear is available.

The success of microlights in civil aviation has been followed by the evaluation of various designs for military purposes. These include the French Zenith Baroudeur M, four of which were delivered to the French army. They carry two or four rocket launchers for anti-tank operations.

increase or decrease the speed of descent of the machine on its landing approach. The Quicksilver MX is fitted with a tricycle landing gear and is normally sold as a home assembly kit at a price of $5,295.

In Europe a fair number of basic three-axis designs have been sold. The French Aéronautic Baroudeur has found many buyers and is available as either a single- or two-seater. It has a faired-in pilot's compartment on the single-seat version and uses a 30-kW (40-hp) Hirth HL2701 engine. In Belgium the Butterfly Company has produced a two-axis control aircraft designed on the tandem-wing principles propounded by the late Henri Mignet. The stick control on the Butterfly moves the forward wing to effect diving and climbing motion, and deflects the rudder to bank and turn the aircraft. A fixed tailwheel landing gear is fitted, and some versions have been equipped with a rudimentary pilot protection fairing in the nose.

Three-axis designs

The great challenge for the microlight designers, however, has been to produce a three-axis aircraft which fits into the rules for

From its initial design in 1972, the Eipper Quicksilver has spawned several successful derivatives, including the Quicksilver MX, the first of the three-axis control models. Illustrated is the twin-float-equipped MX variant, the Seaquick II. The MX has found use in many countries in roles such as SAR and border patrol.

microlight aircraft but is, in all important respects, a normal light aircraft. It is this class of microlight which has become predominant in recent years. One of the best known in the United Kingdom is the Dragon 150 which was built by a subsidiary company of British Air Ferries and first flew in July 1982. It has an empty weight of 149 kg (328 lb), which does not qualify it as an ultralight under the American certification rules, and is equipped with two seats in a fabric-covered forward section of its pod and boom fuselage. The engine, a 34-kW (45-hp) Robin, is fitted in the nose in conventional light aircraft fashion and the strut-braced high wing has normal half-span ailerons. Some 20 Dragons were built by Dragon Light Aircraft, but the company was subsequently dissolved and the machine was taken out of production.

Different looks

Another British design is the Euro Wing Goldwing which is a rather futuristic flying-wing aircraft. A single-seater, the Goldwing has a slim fuselage with an open cockpit and a canard foreplane which mounts the elevator. The high aspect ratio wings have fin/rudder assemblies mounted at the tips, and normal roll control is provided by ailerons which give upward movement only. Of rather more conventional appearance is the biplane Super Tiger Cub 440 built by Micro Biplane Aviation of Worksop, Notts. The Tiger Cub is virtually a miniature version of any normal sport biplane one might name and, indeed, it is not a great deal smaller than a Pitts Special or Christen Eagle. All the controls are normal light aircraft style units and the 37-kW (50-hp) Robin engine is mounted between the wings, driving a tractor propeller fitted in the nose ahead of the pilot's position. The Tiger Cub has folding wings and can be transported behind a motor car on a purpose-built trailer. More than 70 Tiger Cubs have been registered in the United Kingdom.

Speed machines

One of the highest-performance microlights in the world is the Nogrady AN-2 Avionette. Built by the French company, Avions Nogrady, this machine is a side-by-side two-seater with a glass-fibre fuselage and a wing constructed of Kevlar and aluminium with Dacron fabric covering. It has the external appearance of a motor-glider and is fitted with a single main wheel beneath the pilot's compartment, and has a butterfly tail. The cockpit is fully enclosed with a hinged bubble canopy and the Avionette can attain a maximum speed of 170 km/h (106 mph) on its 32.1-kW (43-hp) Hirth engine, which technically invalidates it as a microlight in certain countries where the certification basis involves a restriction on the maximum permitted speed.

Perhaps the most successful of the French high-performance 'ULM' aircraft is the Aviasud

Of British design, the Mainair Tri-Flyer 330 is one of a family of 'powered trike' units designed by the company to fit flexible hang-glider wings. This single-seat, weight-shift microlight has two-seat derivatives in the Striker Tri-Flyer and the Striker Gemini, which has a cockpit pod.

Sirocco. It has been available since January 1983 and has a glass-fibre fuselage with a streamlined cockpit pod in the front and a slim boom carrying the conventional fin and tailplane. A stout pylon is built-in behind the pilot and this carries the JPX-PUL425 pusher engine and the slightly swept high wing. The wing is designed to be folded down and stored in a special tube so the Sirocco can be taken to the flying field from home storage.

Scaled-down aircraft

In the USA Fisher Flying Products of South Webster, Ohio, has been selling kits for a series of sophisticated ultralights. The FP-101 and FP-202 Koala are high-wing aircraft with very much the appearance of scaled-down Piper Cubs. They are built of wood and fabric with fixed tailwheel landing gear, and have conventional wing and tail control surfaces. A kit for the FP-101 (including the engine) costs around $4,000, which makes this a very attractive home-built project. Fisher also sells the FP-303, which is a low-wing design of wooden geodetic construction and a layout similar to that of the Evans VP-1 homebuilt. This single-seater is equipped with a 21-kW (28-hp) Rotax 277 engine and the whole kit (including powerplant) is $2,798, which must make it one of the lowest-priced amateur-built aircraft on the market.

Several designers of normal amateur-built aircraft have become involved in the microlight

The old saying 'flying is for fun' seems to be particularly applicable to the microlight revolution of the last decade. Here a pair of Maxair Drifters fly along the Californian coastline powered by a 28-kW (38-hp) Kawasaki TA-440 two-cylinder two-stroke engine fuelled by a 19-litre (5-US gal) tank.

revolution. The Sorrell Hiperbipe has always been a popular homebuild and now Sorrell Aircraft has brought out a scaled-down version which is known as the Hiperlight. It has a completely enclosed cabin and is provided with a special transparent area just by the pilot's knee to ease the forward view for take-off and landing. The famous Claude Piel also designed a microlight just before he died, and several examples of this CP-150 Onyx are under construction. The Onyx is a Mignet-style aircraft with a large rear-mounted wing and a smaller forward wing on pylons ahead of the pilot's cockpit. This forward wing is used for pitch control and the wings are fitted with conventional ailerons. Large endplate fins are mounted at the tips of the main wings with normal rudder surfaces attached, and the 9-kW

Designs such as the Sorrell Hiperlight have much more in common with conventional light aircraft than with powered hang-gliders, featuring completely cowled engine and enclosed cockpit within a conventional fuselage.

(12-hp) Solo engine is mounted in pusher fashion in the tail.

Microlight aviation has come through a period of growing-up, and it is plain that new rules will have to be established which do not class an aircraft as a microlight merely on the basis of its empty weight and limited speed performance. What cannot be questioned, however, is the inventiveness which has been shown by microlight designers, and this must be a most encouraging development for the whole future of light aviation.

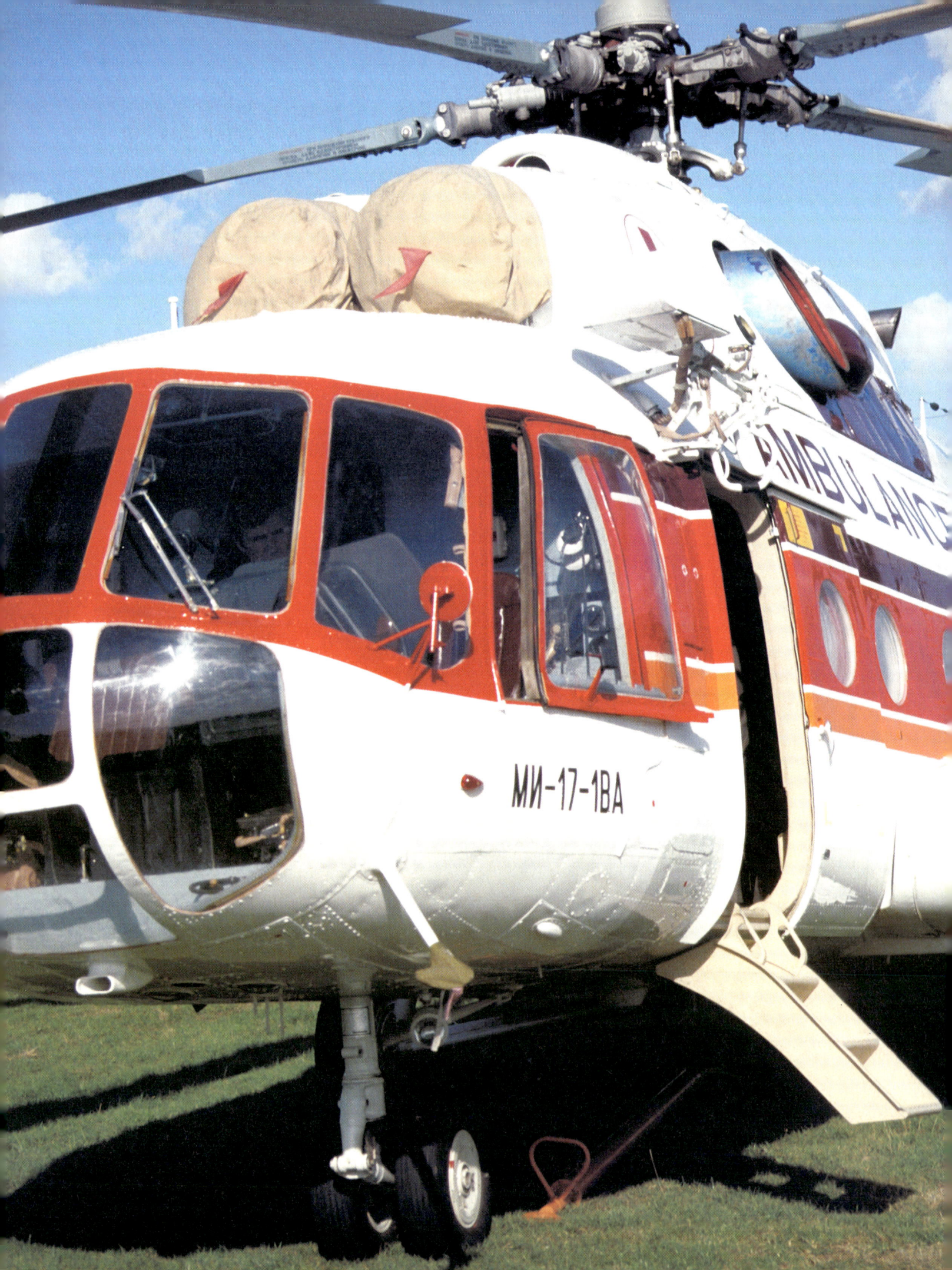

Mil's Civilian 'Helos'

The design bureau named for Mikhail Mil is one of the best-known names in the rotary-wing world. Most of its designs have been produced in sizeable quantities, and such has been their basic soundness and versatility that they have been employed by both civil and military users alike. Here we examine the civilian use of the Mil helicopter dynasty.

Mikhail L. Mil, who had been one of the Soviet Union's leading designers of autogyros, was permitted to form his own OKB (experimental construction bureau) in 1947. His first product was also the Soviet Union's first helicopter of conventional (so-called penny-farthing) design, with one main rotor and an anti-torque tail rotor. Its designation was GM-1, from Gelikopter Mil, the production designation being Mi-1.

The basic design was that of a simple helicopter with fixed nose-wheel landing gear, a four-seat cabin with a hinged door on each side, and a 429-kW (575-hp) Ivchyenko AI-26V seven-cylinder radial engine. The engine was mounted behind the cabin in the centre fuselage, which was made of welded steel tubing with detachable aluminium skin panels. To the rear was the metal fuel tank, and then the main fuselage ended abruptly, being joined to the slim semi-monocoque boom carrying the tail rotor and small adjustable stabilisers (tailplanes), and a long skid to protect the tail rotor from hitting the ground. The engine was mounted with the crankshaft horizontal, driving via a cooling fan and centrifugal clutch to a 90° angle box to the vertical rotor shaft with a rotor brake. The main rotor was based to some degree on Mil's previous autogyro designs, with three blades of mixed steel, plywood and fabric construction, mounted in a fully articulated hub.

First off the drawing board

Design of the Mi-1 began in September 1947, and the first flight was achieved exactly a year later. Not many modifications were needed, and over 1,700 were built. From 1954 the programme was transferred to WSK Swidnik in Poland, the designation becoming SM-1. The basic type was the Mi-1T, with a pilot and two or three passengers, plus radio. By 1952 alcohol de-icing of the rotors was standard. Other versions included the Mi-1NKh for crop spraying and the dual-control Mi-1U trainer. In 1960 a de luxe model (at first dubbed Moskvich) appeared with almost untapered all-metal blades, hydraulic control and improved equipment and soundproofing. Normal loaded weight was 2470 kg (5,445 lb), and range was typically 360 km (224 miles) at 140 km/h (87 mph).

The Mi-2 design was modified for turbine engines, and did not appear until much later. Accordingly the next Mil machine in production was the Mi-4, first flown in May 1952. This retained the pod/boom layout of the Mi-1, but it was strongly influenced by the American Sikorsky S-55, the engine being mounted diagonally in the nose to drive via a slanting shaft passing up between the two pilots seated above the front of the cabin. However, the Soviet helicopter was far more powerful, the engine being a Shvetsov ASh-82V two-row radial rated at 1268 kW (1,700 hp), matched to a gross weight which in the production versions was about 7800 kg (17,196 lb).

Almost the whole structure was light-alloy semi-monocoque, resting on the ground on four individual landing gears. The engine was installed at an angle of 25° in a fireproof compartment whose outer casing took the form of left and right side doors and upper/lower-hinged nose doors. The main rotor was virtually a scaled-up version of the Mi-1 rotor, but with a fourth blade. Diameter was originally 17.22 m (56 ft 6 in), but in the production Mi-4 this was considerably increased to 20.73 m (68 ft 11 in). Hydraulic control and liquid de-icing were fitted from the start. At the rear was the semi-monocoque boom carrying the stabilisers, tailskid and the tail rotor with three blades of bakelite-ply, a form of plastic-impregnated wood construction.

The main hold was 1.8 m (71 in) square in section and 4.14 m (13 ft 7 in) long. The main hinged door was on the left, and most versions also had left/right rear doors which, together with clip-on ramps, allowed heavy cargo and small vehicles to be loaded. Civil Mi-4P production began in 1954, with large rectangular windows and a comfortable and heated interior normally arranged with 10 passenger seats, stowage for 200 kg (441 lb) of coats and luggage, a toilet and often a seat for an attendant. A few Mi-4L helicopters were produced for VIP use, usually arranged luxuriously for six passengers. The Mi-4S agricultural version could undertake both dusting (with 1000-kg/2,205-lb hopper) and spraying (with 1600-litre/352-Imp gal tank and wide spraybars). In the course of production new all-metal blades were introduced, and various other improvements made. For over 15 years this was the standard local-service helicopter of Aero-

Unrivalled by any other helicopter in terms of power, the monstrous eight-bladed Mi-26 now equips many Aeroflot and Soviet air force heavy-lift units, and was notable for its exploits during the efforts to contain the Chernobyl nuclear disaster.

The Mi-1 was typical of early helicopters, but proved successful in operation. Many were built in Poland as the SM-1, an example of which is seen here. It is fitted with covered stretchers on the fuselage sides for the air ambulance role.

flot, which received about 1,000 of a Soviet total of over 3,500. Some 700 were exported, and about 1,000 were built under licence in China as the Z-5.

The leap from Mi-1 to Mi-4 was considerable, but from the Mi-4 to the Mi-6 was the greatest advance in the history of helicopter technology. This giant machine was designed to meet a joint civil/military requirement of 1954, calling for a transport helicopter of unprecedented size, power and capability. It was made possible only by the development of gas-turbine engines, which multiplied the power available from a given powerplant weight and bulk, and also enabled the engines to be mounted above the cabin adjacent to the rotor gearbox. The new engines also facilitated flight with a failed engine, burned the same fuel as jet aircraft, eliminated centrifugal clutches and eliminated engine-cooling problems (though of course with such high gearbox powers the need for continuous oil cooling became important).

The first Mi-6 made its first flight in September 1957. It caused gasps of astonishment, and incidentally showed the world the configuration that would be adopted by the majority of helicopters in future. The huge fuselage was entirely a semi-monocoque stressed-skin structure, with a length of 33.2 m (108 ft 11 in). The main section, forming a cargo hold in most versions but also occasionally equipped for up to 90 passengers, was 2.64 m (104 in) wide, 2.50 m (98.5 in) high (slightly lower at the front under the engines), and 12.00 m (39 ft 5 in) long. There were various side doors, and heavy cargo and vehicles could be loaded via the clamshell rear doors under the tail boom, the maximum cargo load being 12000 kg (26,455 lb). The nose was normally equipped for a flight crew of five, comprising a navigator in the glazed nose, two pilots side-by-side, a radio operator and flight engineer.

Sheer size and scale

The engines, specially developed for this helicopter, were Solovyev D-25Vs, each rated at 4103 kW (5,500 shp). Each drove directly into the massive R-7 rotor gearbox, an all-steel package 3.05 m (10 ft) high and 2.13 m (7 ft) square weighing 3.2 tonnes (more than the pair of engines). Apart from reducing the shaft speed through a ratio of 69:1 it also drove the oil cooler fan between the engines, a range of accessories and the shaft to the tail rotor. Normally turning at 120 rpm, the main rotor had a diameter of 35.00 m (114 ft 10 in), with five blades with steel spars and screwed-on aluminium trailing-edge pockets to give the correct aerofoil section. The tail boom carried powered stabilisers and a fin pylon on which was mounted the tail rotor with four bakelite-ply blades. The landing gear was of the fixed nosewheel type. A new

The Mi-4 became the stalwart of Aeroflot's helicopter services, based on the Mi-4P version equipped with 10 seats in a comfortable interior. This version even included a toilet. This example is seen at Moscow's Vnukovo airport.

feature was the addition of a wing, with a span of 15.29 m (50 ft 2 in), set at about 15° incidence and providing 20 per cent of the lift in cruising flight. This reduction in main-rotor lift helped increase cruising speed to 250 km/h (155 mph), and a specially prepared Mi-6 set a speed record around a 100-km (62.1-mile) circuit at 340 km/h (211.35 mph).

Over 800 of these monsters were delivered, split between Aeroflot and the military services. Most were of the Mi-6A type for cargo operations, with equipment for fitting 41 stretchers and seats for attendants in the emergency casevac role (for example, after earthquakes). A single luxurious Mi-6P was built, with 80 airline-style seats and large rectangular windows. For sustained operations in the crane role the wing could be removed. It was usual to assign a 74.6-kW (100-hp) APU (auxiliary power unit) to each helicopter. Mounted on a trolley, this was plugged in externally to start the main engines and furnish ground power, the unit then being wheeled on board.

In 1960 the Mil bureau flew the V-10, the prototype of a special crane helicopter using the engines, transmission and rotors of the Mi-6. It differed in having only a slim fuselage, carried high above the ground on four giant landing gears. The whole machine was designed to straddle its load, which could be a single-deck bus, or a container or a pallet measuring 8.53 m (28 ft) by 3.43 m (11 ft 3 in). Maximum load was 15000 kg (33,070 lb), and 28 passengers could be seated in a cabin in the fuselage. About 60 production crane helicopters were built, designated Mi-10, as well as a single Mi-10K designed purely for carrying loads slung on a cable, and thus fitted with shorter legs.

Success of the Mi-6 spurred development of two smaller twin-turbine machines, the Mi-2 and Mi-8. The Mi-8, in the form of the V-8 prototype, flew first, in early 1961. This was temporarily powered by a 2014-kW (2,700-hp) Ivchyenko AI-24V engine, the rotors being based on those of the Mi-4. It was then decided to design a more up-to-date rotor system, with main and tail blades having a spar of extruded aluminium alloy forming the complete nose, the profile being completed by 21 light honeycomb-stabilised rear pockets. The hubs were scaled down from those of the Mi-6, with hydraulic control. Mil copied Sikorsky's pressurised-gas method of indicating the presence of fatigue cracks in the blades, and added electrothermal de-icing.

The light-alloy semi-monocoque fuselage continued Mil's pod/boom preference, this again making possible left/right clamshell doors opening up the entire rear end of the cabin to bulky cargo or small vehicles. In the production Mi-8 the usable cabin is 6.40 m (21 ft) long, 2.36 m (7 ft 9 in) wide and 1.80 m (5 ft 11 in) high. Passenger versions can be equipped for up to 32 passengers with a toilet and large wardrobe for coats and baggage. Maximum internal cargo load is 4000 kg (8,820 lb). Above the cabin are the twin TV2-117 engines, especially developed by the S. P. Isotov bureau (today the Klimov/Leningrad KB). In the second prototype these were rated at 1044 kW (1,400 shp) each, but the production Mi-8 has TV2-117A engines of 1268 kW (1,700 shp). The VR-8 main gearbox has two stages of gearing reducing the 12,000 rpm input to the main-rotor speed of 192 rpm. At the front is a shaft driving the fan drawing air

Most Mi-2s have been built in Poland, where several advanced versions have since been developed. This agricultural version has large hoppers either side of the cabin and a chute for dusting crops.

through the oil cooler. A small 445-litre (98-Imp gal) fuel cell is under the floor, but the main tanks are external drums strapped on each side under the cabin windows, bringing total fuel to 1870 litres (411.5 Imp gal). Auxiliary tanks in the cabin bring the total for ferry purposes to 3700 litres (814 Imp gal). In nearly all Mi-8s the right external tank is faired at the forward end into a cabin heater with a ram inlet at the front. The cockpit in the nose is equipped for two pilots and an engineer. The fixed landing gear has twin nosewheels. Standard equipment includes autostabilisation and an autopilot.

At least 8,100 and probably more than 9,500 Mi-8 helicopters were built. Of this enormous total at least 2,200 were supplied to Aeroflot and civil export customers. The standard civil Mi-8P has airline-type seats for 32, or for 28 plus a toilet and wardrobe. The cabin has large square windows, and often spats are fitted to the main wheels. The Mi-8T utility transport has an austere cargo interior, round windows and no spats. Usually 24 tip-up seats are fitted, 12 along each side. Small numbers have been built of the luxurious Mi-8 Salon, with VIP accommodation for 11.

'Hip' replacement

First displayed in 1981, the Mi-17 replaced the Mi-8 in production. The airframe is virtually unchanged, but the engines are the Isotov (Klimov) TV3-117MT turboshafts each rated at 1455 kW (1,950 shp), and lighter, more compact and with reduced specific fuel consumption. The automatic engine control system senses any failure of drive torque and immediately increases power of the remaining engine to 1641 kW (2,200 shp). In 1989 Mil demonstrated a further uprated version, the Mi-17-1VA, powered by the 1678-kW (2,250-shp) TV3-117VM. All Mi-17 versions have engine intake baffles to prevent ingestion of sand or other foreign matter, as well as a gas-turbine APU to provide compressed air for starting the main engine. The Mi-17 can be distinguished from the Mi-8 by the shorter engine cowlings, and by the fact that the tail rotor is moved to the left side of the fin. Civil Mi-17 versions include the usual passenger, utility and Salon types, as well as rescue models with a 200-kg (440-lb) winch and provision for 12 stretchers and an attendant. The Mi-17-1VA displayed at the 1989 Paris air show was equipped as a mobile surgical hospital, while in 1991 Mil displayed the Mi-171, with airline interior.

Mil's Mi-2 was held back to await the development of the Isotov GTD-350 engine, a small turboshaft rated at 298 kW (400 shp). Two of these are mounted above the cabin of a helicopter similar to the Mi-1 in size, and with a rotor system which at first was identical with the Mi-1. But the greater power and new location of the engines enabled the cabin to be made much larger, matched to the greater payload. Thus, instead of a pilot and two or three passengers the Mi-2 can carry eight passengers, or four stretchers and an attendant. The landing gear has twin nosewheels, or skis can be fitted.

The V-2 prototype flew in September 1961, and many improvements were made before the whole programme was transferred to WSK-Swidnik in Poland in January 1964. Among the changes were new rotors with extruded light-alloy D-nose spars and honeycomb filled rear pockets, a new tail rotor with plastic bonded honeycomb blades, electrothermal de-icing of all blades, bleed-air de-icing of the inlets, an improved main hub with hydraulic dampers, and provision for auxiliary external tanks. Since 1965 the Swidnik (Lublin) factory has delivered over 5,500, including about 2,500 of six civil versions: passenger/cargo, casevac/rescue, agricultural, freighter, trainer and multisensor imagery (cameras, TV and IR). Swidnik has produced several developed versions.

Birth of the 'Halo'

Mikhail L. Mil died on 31 January 1970, his successor being Marat N. Tishchyenko. At that time the bureau was in the throes of testing a gigantic helicopter designated V-12, or Mi-12, with a fuselage similar in size to a Boeing 747 and a mass of struts and narrow wings carrying two complete Mi-10 twin-engine/rotor units side-by-side. Span across the two rotors was 67.00 m (219 ft 10 in), and the whole machine was ponderous. Development was abandoned in 1971, the heavy lift requirement being met by the far better Mi-26. This completely new helicopter remains by far the most powerful in service in the world. Amazingly, though engine power and lifting capability are more than doubled, the overall dimensions are more compact than those of the Mi-6.

A key to the Mi-26 is the new main rotor with no fewer than eight blades of superior profile, with a stainless steel spar and glass-fibre rear pockets with Nomex honeycomb cores. The hub is an unconventional titanium forging, and the tail rotor also has a titanium hub with five glass-fibre blades. The 33.73-m (110-ft 8-in) fuselage is a semi-monocoque structure with a cargo hold 15.00 m (49 ft 3 in) long, 3.25 m (128 in) wide and up to 3.17 m (125 in) high. Downward-hinged doors with integral stairs are fitted at the front on the left and on each side at the rear. For loading vehicles and bulky cargo the entire rear end can be opened by twin clamshell doors. Maximum payload is normally 20000 kg (44,090 lb), though 22000 kg (48,500 lb) is being cleared (late 1990) and one Mi-26 set records lifting 25000 kg (55,115 lb) to a height of 4100 m (13,451 ft). Comprehensive equipment is fitted for hoisting and positioning cargo. Military versions often have 85 tip-up seats round the walls, but no civil passenger versions had been disclosed by late 1990.

The twin engines are the specially developed Lotarev D-136, each rated at 8504 kW (11,400 shp). They are fed from 12000 litres (2,640 Imp gal) of fuel in underfloor rubber cells. The top of the rotor gearbox drives a fan drawing air through the oil cooler above and between the engines.

Large square windows identify the Mi-8P civil version of the 'Hip' helicopter, which often has spats fitted to the mainwheels. This version is the current backbone of Aeroflot's services, huge numbers being supplied for local transport duties with 28-32 passengers.

Specification
Mil Mi-26 'Halo'
Type: heavy transport helicopter
Powerplant: two Lotarev D-136 free-turbine turboshafts, mounted side-by-side and each developing 8380 kW (11,240 shp)
Performance: maximum level speed 295 km/h (183 mph); normal cruising speed 255 km/h (158 mph); service ceiling 4600 m (15,100 ft); hovering ceiling out of ground effect 1800 m (5,900 ft); range with maximum internal fuel at maximum take-off weight 800 km (497 miles)
Weights: empty 28200 kg (62,170 lb); maximum take-off 56000 kg (123,450 lb); maximum payload 20000 kg (44,090 lb)
Dimensions: main rotor diameter 32.00 m (105 ft 0 in); length overall, rotors turning 40.025 m (131 ft 3¾ in); length of fuselage excluding tail rotor 33.727 m (110 ft 8 in); height to top of rotor head 8.145 m (26 ft 8¾ in); width over mainwheels 8.15 m (26 ft 9 in)

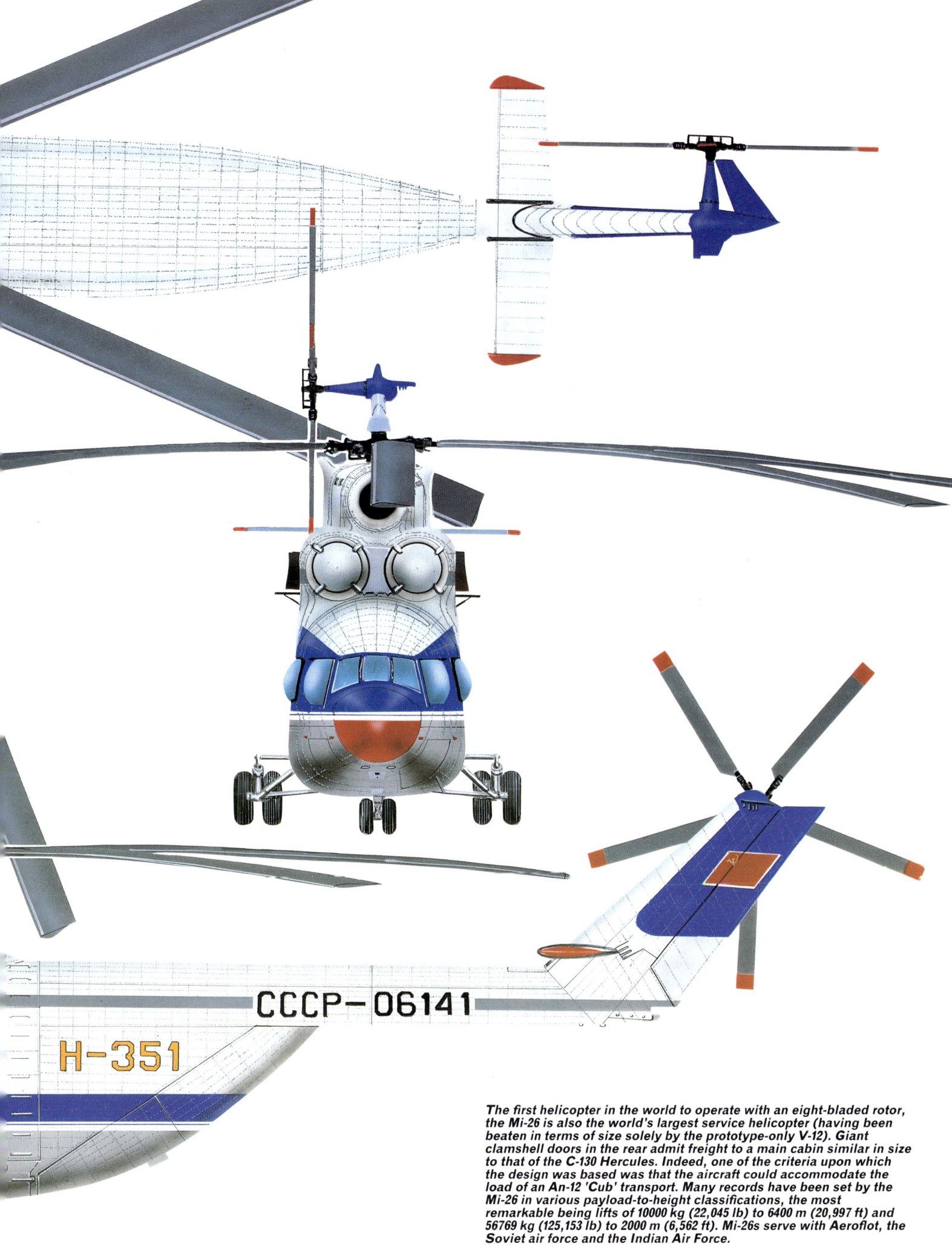

The first helicopter in the world to operate with an eight-bladed rotor, the Mi-26 is also the world's largest service helicopter (having been beaten in terms of size solely by the prototype-only V-12). Giant clamshell doors in the rear admit freight to a main cabin similar in size to that of the C-130 Hercules. Indeed, one of the criteria upon which the design was based was that the aircraft could accommodate the load of an An-12 'Cub' transport. Many records have been set by the Mi-26 in various payload-to-height classifications, the most remarkable being lifts of 10000 kg (22,045 lb) to 6400 m (20,997 ft) and 56769 kg (125,153 lb) to 2000 m (6,562 ft). Mi-26s serve with Aeroflot, the Soviet air force and the Indian Air Force.

The giant Mi-6 was fitted with large wings to unload the main rotor in cruising flight, allowing more power to be given to forward thrust and consequently increasing speed. The wings were removed for crane operations. Note the strap-on ferry tanks either side of the cabin.

The cockpit in the nose seats two pilots, a navigator, flight engineer and loadmaster, and immediately to the rear is a compartment for four passengers. The fixed landing gear is particularly neat, with small twin-wheel units under the nose and on each side of the fuselage. The main gears indicate total take-off weight on the engineer's panel, and can be adjusted in height to facilitate loading through the rear doors. Maximum take-off weight was in 1990 set at 56000 kg (123,450 lb), but is expected to be increased. Cruising speed is typically 254 km/h (158 mph), maximum speed being 295 km/h (183 mph). The first Mi-26 flew on 14 December 1977. Development was very successful, and by 1991 the number built was around 100, split roughly evenly between civil and military versions. Mr Tishchyenko has emphasised that this giant machine was designed to operate from unprepared sites in the harshest environments in the world.

Smaller designs

In 1981 the team began development of an attractive light helicopter seating a pilot and three passengers (or two pilots in a trainer version). The prototype Mi-34 flew in 1986, and one was displayed at the 1989 Paris air show. The 10.00-m (32-ft 10-in) rotor has a hub similar to that of the American MD 500, with no lead/lag hinges and four blades of composite construction. Flight controls are manual, and the engine is the well-tried 242-kW (325-hp) M-14V-26 radial piston engine. The streamlined pod/boom fuselage carries on its rear end a swept fin with a tailplane mounted symmetrically (with respect to the helicopter, not to the offset fin) on top. With a maximum weight of 1350 kg (2,976 lb), and maximum cruising speed of 180 km/h (112 mph), the Mi-34 is intended primarily for training and aerobatic competition, but could serve many other duties. A twin-turbine version is planned, international partners being sought.

In the late 1990s, the newest Mil helicopter was the Mi-38, which first flew in the first half of the decade. Similar in appearance and capability to the Anglo-Italian EH.101, the Mi-38 is totally new. The 21.23-m (69-ft 8-in) main rotor will have five blades and the tail rotor two blades set at 55°. All blades will be of composite construction, with electric de-icing, and mounted in a hub with elastomeric (flexible rubber) bearings. The engines will comprise two Klimov (formerly Isotov) TV7-117V turboshafts, each flat-rated at 1753 kW (2,350 shp) but with 2536 kW (3,400 shp) available in the event of failure of either engine. The engines will be housed at the rear of the streamlined rotor pylon. The fuselage contains a two-pilot cockpit in the nose, with provision for single-pilot operation, separated by the avionics racking from the 6.70-m (22-ft) main cabin, 1.83 m (6 ft) wide and 2.21 m (7 ft 3 in) high, which will accommodate 32 airline-type seats, or 5000 kg (11,020 lb) of cargo (loaded through clamshell rear doors) or various specialised loads. It is hoped that thousands of Mi-38s will be built from 1996 to replace the Mi-8 and Mi-17. Maximum weight will be 14500 kg (31,965 lb), cruising speed up to 275 km/h (171 mph) (made possible by fully retractable tricycle landing gear), and range with 3500-kg (7,715-lb) payload and full reserves 800 km (497 miles).

The Mil V-12 'Homer' was the largest helicopter ever built, powered by two complete Mi-10 rotor/engine assemblies on the ends of reverse-taper wings. The huge cabin had a rear loading ramp. The V-12 remained only in prototype form.

Harnessing the enormous power of the Mi-6, the Mi-10 was a dedicated flying-crane version for service in lifting giant structures in support of Soviet construction and mineral exploitation programmes. The Mi-10K (illustrated) was a one-off cable-lifter. Note the undernose gondola from where the helicopter was flown during lifting operations.

*Wearing the titles of **DOSAAF**, the state sport aviation organisation, this is an Mi-34, Mil's smallest helicopter to date. Intended primarily for training, the Mi-34 is being developed in a twin-turbine version, initially for export.*

Mitsubishi MU-2

A potent and purposeful twin turboprop transport, the MU-2 achieved excellent sales in the face of stiff competition from the established Beech, Cessna and Piper types. More than this, the type represented the rebirth of the Japanese aviation industry, bestowing confidence on its ability to produce indigenous designs.

Mitsubishi, which up to the end of World War II had been one of the most prolific of Japanese aircraft manufacturers with more than 80,000 aircraft to its credit, resumed manufacturing in 1956 when it started licence construction of 300 North American F-86F Sabre jet fighters for the Japanese Air Self Defence Force.

In September 1959 the company began design studies for a small twin turboprop utility transport aircraft aimed at both military and civilian markets, but it was not until 14 September 1963 that the first of four prototype Mitsubishi MU-2s made its maiden flight. The aircraft was an all-metal cantilever high-wing monoplane with conventional tail surfaces and a swept fin, pressurised 5/7 seat cabin, wingtip fuel tanks, a tricycle undercarriage with narrow track main landing gear units which retracted rearwards into the lower fuselage sides, and was powered by two French-manufactured Turboméca Astazou IIK turboprop engines.

Following extensive flight trials with the prototypes, three Astazou-engined production standard MU-2As were manufactured, by which time the decision had been taken to adopt as standard the US-manufactured Garrett AiResearch TPE331-25AA engine of 575 shp, driving three-bladed propellers. The first Garrett-engined aircraft flew on 11 March 1965 and subsequently entered production as the Mitsubishi MU-2B, customer deliveries of which began in 1966 following Japanese certification on 15 September 1965 and US Federal Aviation Administration approval on 4 November. Apart from the engine change and numerous system modifications, the MU-2B had its wingspan extended by one metre. The MU-2B had a maximum take-off weight of 8,930 pounds, a cruising speed of 240 knots, and a certificated ceiling of 25,000 feet.

Co-operation with Mooney

The United States was the principal market for executive versions of the MU-2, and there a marketing relationship was established in 1965 with Mooney Aircraft Inc. whereby Mooney was responsible for final assembly, outfitting, sales and support of MU-2s in North America, where 28 of the 34 MU-2Bs built were sold. In 1969 Mitsubishi established a wholly-owned US subsidiary, Mitsubishi Aircraft International, at San Angelo, Texas. The parent company in Japan shipped basic airframe components to San Angelo where MAI undertook final assembly, incorporating many American-manufactured components such as engines, propellers, avionics, brakes, tyres and interior furnishings prior to completion and sale.

The MU-2C was an unpressurised multi-role transport version,

A potent-looking machine, the MU-2 has a superb performance range. This is largely due to the adoption of overwing spoilers in place of ailerons for roll control, allowing the full span to be used for double-slotted Fowler flaps. This allows the use of very short fields.

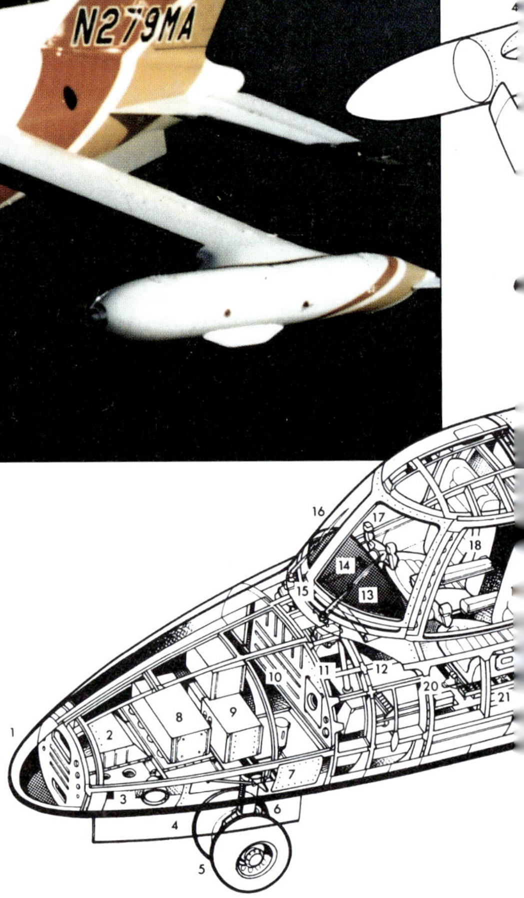

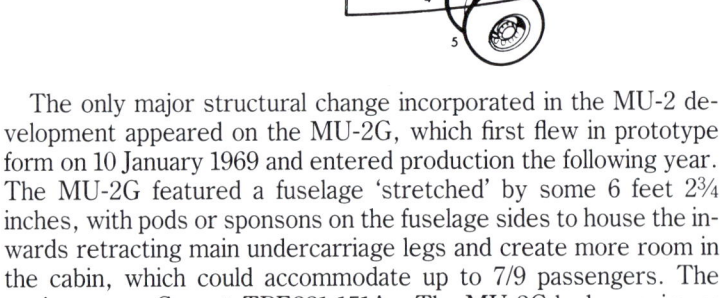

four of which were delivered to the Japanese Ground Self Defence Force. The civilian MU-2B was succeeded in 1968 by the MU-2D, which featured integral 'wet wing' fuel tanks in place of earlier bladder-type tanks, a higher cabin pressure differential, four-position trailing-edge flaps and an increased maximum take-off weight of 9,350 pounds. Cruising speed was increased to 250 knots.

The MU-2E, later designated MU-2S, was a dedicated search-and-rescue variant, 16 of which were manufactured for the Air Rescue Wing of the Japanese Air Self Defence Force. The MU-2S features a thimble-type nose radome housing Doppler search radar, bulged cabin side windows for downward vision, secure communications equipment, and a flight-openable door for dropping of rescue equipment and emergency supplies.

Uprated model

Development of the MU-2 proceeded rapidly, with another new model, the MU-2F, succeeding the MU-2D in 1968. This model, which remained in production until 1971, was powered by uprated 665 shp TPE331-151A engines. Wingtip tank fuel capacity was increased to meet the higher fuel flows of the new engines and maximum take-off weight increased to 9,920 pounds. A total of 95 was built.

The only major structural change incorporated in the MU-2 development appeared on the MU-2G, which first flew in prototype form on 10 January 1969 and entered production the following year. The MU-2G featured a fuselage 'stretched' by some 6 feet 2¾ inches, with pods or sponsons on the fuselage sides to house the inwards retracting main undercarriage legs and create more room in the cabin, which could accommodate up to 7/9 passengers. The engines were Garrett TPE331-151As. The MU-2G had a maximum

Introduced in 1977 were the companion MU-2N and MU-2P. The N was the long-body version (foreground) while the P stuck to the original cabin length (background).

The wing of the MU-2 has a distinctive plank-like appearance, further enhanced by the underslung engines and fuel tanks. This is a typical American executive transport, seen at Beef Island.

Its high transit speed, good field performance, inflight comfort and sporty looks made the MU-2 popular as an executive transport. Good sales were achieved in the United States, where the stretched MU-2J was marketed as the Marquise.

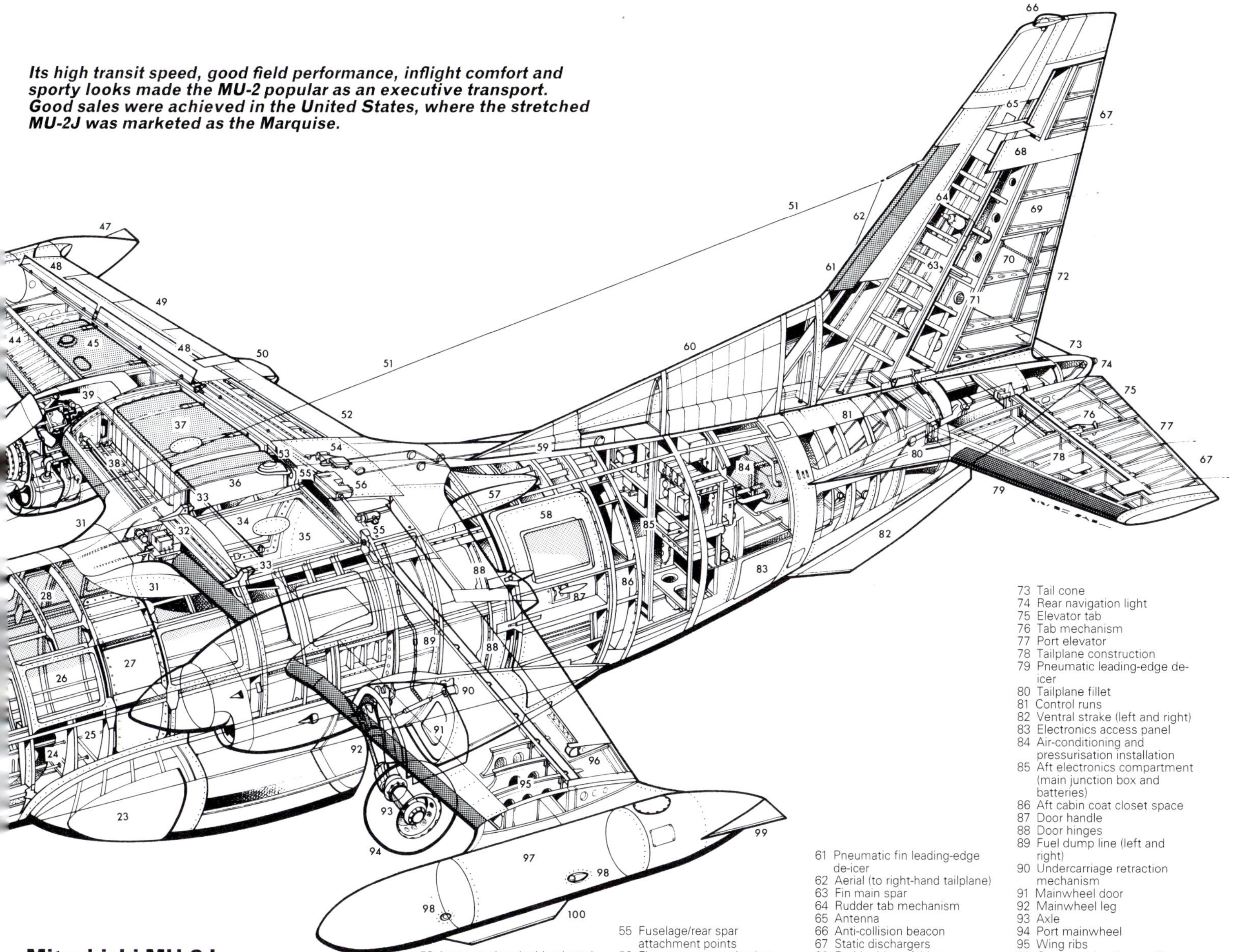

Mitsubishi MU-2J cutaway drawing key:

1. Nose cone
2. Hinged nose doors (left and right)
3. Hinged landing and taxi lamps (left and right)
4. Nosewheel doors
5. Forward-retracting twin nosewheels
6. Nosewheel leg
7. Landing gear access panel
8. Forward electronics compartment
9. Forward battery
10. Bulkhead
11. Control column
12. Rudder pedals
13. Windshield wiper
14. Instrument console shroud
15. Windshield de-icing installation
16. Two-piece curved windshield
17. Control yoke
18. Second pilot's seat
19. First pilot's seat
20. Seat adjustment mechanism
21. Circuit breaker panel
22. Floor support structure
23. Main undercarriage fairing
24. Underfloor control runs
25. Main passenger cabin floor
26. Three-a-side cabin windows
27. Strengthened anti-ice panel
28. Frame and longeron fuselage construction
29. Fuselage skinning
30. Aerial mast
31. Wingroot fairings
32. Leading-edge relay panel
33. Fuselage/front spar attachment points
34. Emergency escape window (right-hand rear)
35. Wing carry-through surface
36. Centre cootion fuel tank
37. No. one right-hand fuel tank
38. Fuel lines
39. Garrett AiResearch TPE331-6-251M turboprop
40. Intake
41. Airscrew spinner
42. Three-blade Hartzell propeller
43. Pneumatic leading-edge de-icer
44. Leading-edge ribs
45. No. two right-hand fuel tank
46. Auxiliary tip tank
47. Tip tank fin
48. Spoilers (extended)
49. Trim aileron section
50. Flap track fairing
51. Aerial
52. Inner section double-slotted flap
53. Centre-section anti-collision beacon
54. Spoiler mechanism
55. Fuselage/rear spar attachment points
56. Flap actuator mechanism
57. Wingroot fillet
58. Cabin entry door
59. Air-conditioning ducts
60. Dorsal fillet
61. Pneumatic fin leading-edge de-icer
62. Aerial (to right-hand tailplane)
63. Fin main spar
64. Rudder tab mechanism
65. Antenna
66. Anti-collision beacon
67. Static dischargers
68. Rudder hinge fairing
69. Rudder construction
70. Rudder tab control
71. Rudder post main beam
72. Rudder tab
73. Tail cone
74. Rear navigation light
75. Elevator tab
76. Tab mechanism
77. Port elevator
78. Tailplane construction
79. Pneumatic leading-edge de-icer
80. Tailplane fillet
81. Control runs
82. Ventral strake (left and right)
83. Electronics access panel
84. Air-conditioning and pressurisation installation
85. Aft electronics compartment (main junction box and batteries)
86. Aft cabin coat closet space
87. Door handle
88. Door hinges
89. Fuel dump line (left and right)
90. Undercarriage retraction mechanism
91. Mainwheel door
92. Mainwheel leg
93. Axle
94. Port mainwheel
95. Wing ribs
96. Outer-section flap profile
97. Port auxiliary tip tank
98. Wingtip lights (navigation and strobe)
99. Tip tank fin
100. Tip tank strake

The nose radar identifies the MU-2E rescue version. Other MU-2s in JASDF service are four lengthened MU-2Js that fly in the calibration and radio aid checking roles.

Mitsubishi MU-2C (army designation LR-1) of the Japan Ground Self Defence Force.

take-off weight of 10,800 lb and a cruising speed of 240 knots.

Forty-six MU-2Gs were built. It was succeeded in 1972 by the MU-2J, which featured 665-shp TPE331-251M engines increasing cruising speed to 280 knots, and a modified interior providing a further 11 inches of cabin room, with improved sound-proofing. The same -251M engines were installed in the MU-2K, which was otherwise similar to the short-fuselage MU-2F which it replaced on the production line but with a higher cruising speed of 300 knots. Production totals for the MU-2J and MU-2K were 36 and 83 respectively, MU-2J production including a number of military aircraft for the JASDF's airways and navigational aid Flight Check Squadron based at Iruma.

Production and development of short- and long-bodied MU-2s continued in parallel, the next model change taking place in 1975 when the MU-2M and MU-2L were introduced. The short-body MU-2M retained the 665-shp engines of the MU-2K, but had an increased maximum take-off weight of 10,470 lb and a higher cabin pressurisation differential permitting a maximum certificated altitude of 28,000 ft. The long-body MU-2L featured uprated 715-shp TPE331-251M engines and a gross weight increase to 11,575 lb but retained the lower certificated altitude of 25,000 ft. Production totals were 27 and 36 respectively.

In 1977 the MU-2N and MU-2P were introduced, succeeding the MU-2L and MU-2M respectively, from which they differed principally in having slower-turning versions of the Garrett TPE331-252M turboprops driving larger-diameter four-bladed propellers, reducing cabin noise levels by 10 dbA. Production totals were 36 and 31 respectively.

The final developments of the MU-2 series appeared on the market in late 1978 when the newly-named Solitaire and Marquise were introduced. The Solitaire, first flown on 28 October 1977 and certificated by the US Federal Aviation Administration on 2 March 1978, was developed from the short-bodied MU-2M. Modifications included installation of 665-shp Garrett TPE331-10-501M engines and maximum fuel capacity increased to 1,526 litres in five integral wing tanks and two tiptanks. Maximum certificated operating ceiling was increased to 31,000 ft, with a service ceiling of 33,500 ft and a single-engine ceiling of 16,900 ft. Maximum cruising speed was 321 knots at 20,000 ft, while normal cruising speed increased slightly over the MU-2P to 313 knots.

The long-bodied Marquise also adopted the -501M engines, each rated at 715 shp. Its service ceiling was increased to 29,750 ft and single-engine ceiling to 14,800 ft, while maximum cruising speed was 308 knots at 16,000 ft.

The Solitaire and Marquise remained in production until Mitsubishi Aircraft International ceased manufacturing operations in March 1986. The final production total for the Solitaire was 111, and for the Marquise 171.

Structurally all of the MU-2 series aircraft are similar. The wings, which have airfoil section NACA 64A415 at the root, NACA 63A212 (modified) at the tip and no dihedral, are constructed on two one-piece all-metal spars with chemically-milled aluminium alloy wings, and are of particularly small area (16.55 m^2) for an aircraft in this class. The MU-2 has no ailerons, roll control being provided by upper surface spoilers located between the rear spar, and full-span electrically-actuated double-slotted Fowler flaps. All primary control surfaces are manually operated. The fuselage is a semi-monocoque structure of circular section with a single entry door on the port side under the wing (short-bodied models) or in the rear fuselage (long bodied). The flight deck and cabin are pressurised to a differential of 6.0 psi in later models. Typical avionics fits include weather radar

The streamlined fuselage is pressurised, this being achieved by an automatic system which maintains a constant cabin pressure through descent and climb. When the wheels are on the ground a release valve carefully equalises internal and external pressure before the door can be opened.

The MU-2G and MU-2J introduced a 1.90-m increase in fuselage length to raise cabin accommodation. At the same time the undercarriage was altered, the main wheels retracting into fairings on the fuselage sides and the nosewheel moved back but retracting forwards.

Mitsubishi MU-2E patrol aircraft serve the Air Rescue Wing of the Japan Air Self Defence Force.

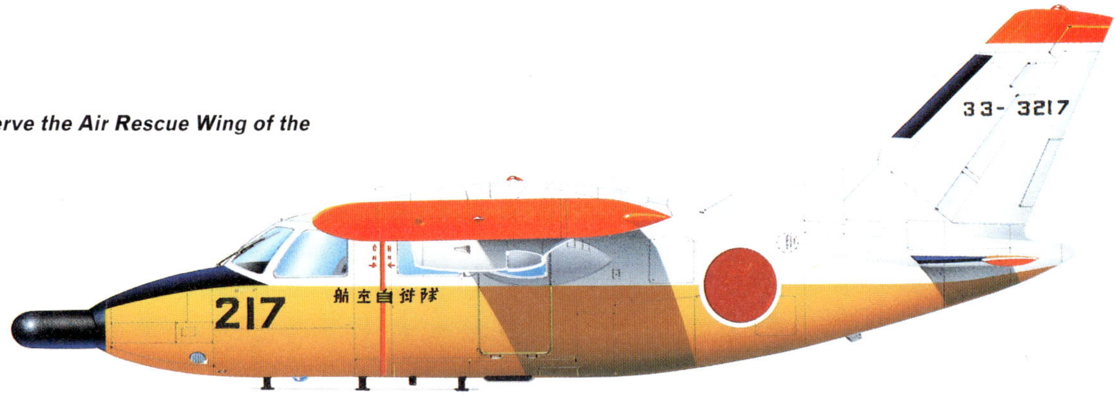

The Marquise (illustrated) and Solitaire were the last production variants. The long-body aircraft were distinguished by having noticeable fairings for the undercarriage.

and a Sperry SPZ-500 integrated autopilot/flight director system. Air conditioning and pneumatic de-icing boots on wing and tail surface leading edges are standard.

Cargo carrier

The high cruising speed of the MU-2 series has made them popular with overnight package airfreight operators, and several all-cargo versions have recently been developed in the United States. Among these is the Cavenaugh Cargoliner, which first flew in mid-1985. Developed by Cavenaugh Aviation Inc. of Conroe, Texas, this modification can be based on any of the long-bodied MU-2G, 'J, 'L, 'N or Marquise series. The conversion involves substitution of a crew door with integral step for the port side cockpit window, installation of a bulkhead between the cockpit and the cabin area, and provision in the port fuselage side for a one-piece, upward-opening cargo door. MU-2 Modifications Inc. of Dallas, Texas, has developed a similar cargo conversion of long-bodied MU-2s known as the MU-2 Express, and also offers a dedicated emergency medical evacuation conversion known as the MU-2 Medi-Vac Express with accommodation for three stretcher cases, three medical attendants and life-support equipment; and an executive/cargo quick change variant.

Local interest in the MU-2 generated an order for the Japan Ground Self Defence Force for the MU-2C/LR-1 variant. This is used for liaison and reconnaissance, having the ability to carry cameras or light guns. This is the first example, flying without the characteristic wing tanks.

NAMC YS-11

By design rather than accident, the YS-11 is a little-known aircraft outside its native Japan, even after more than 30 years of service. Closely tailored to the needs of domestic carriers in its homeland, decades of passenger and cargo hauling have underlined the type's reputation as a solid, hardworking airliner.

In 1956 the Japanese Ministry of International Trade and Industry issued a requirement for a short/medium-range airliner of indigenous design to meet the needs of the country's internal airlines and to serve as a replacement, both domestically and in hoped-for foreign sales, for the ubiquitous Douglas DC-3. With a subsidy from the Japanese government, which maintained a 54 per cent controlling interest in the project, the Transport Aircraft Development Association (TADA) was established in May 1957 to undertake design and development of the aircraft. The design team was headed by Dr Hidemasa Kimura.

As work proceeded TADA was superseded on 1 June 1959 by Nihon Kokuki Seizo Kabushiki Kaisha (Nihon Aeroplane Manufacturing Company, or NAMC) which was made up of a consortium of six established aircraft manufacturing companies: Fuji Heavy Industries, Kawasaki Aircraft, Mitsubishi Heavy Industries, Japan Aircraft Manufacturing (Nippi), Shin Meiwa Industry (formerly Kawanishi), and Showa Aircraft Industry which had built Douglas DC-3 transports during World War II. Each was to be responsible for building sub-assemblies for the aircraft, which by then had been given the designation NAMC YS-11.

The work-split among the consortium was as follows: Fuji (tail section); Kawasaki (wings and engine nacelles); Mitsubishi (main forward fuselage section, equipment installation and final assembly); Nippi (ailerons and flaps); Shin Meiwa (rear fuselage section); and Showa (light alloy honeycomb structures such as doors). NAMC was the overall controlling body responsible for co-ordinating their efforts, and for marketing the aircraft.

The finalised design was for a low-wing configuration with twin turboprop engines, a pressurised, circular-section fuselage, conventional tail unit and retractable tricycle undercarriage. Rapid growth in Japanese domestic airline traffic prompted the decision to design a larger aircraft than the otherwise comparable Avro 748, Fokker F27 Friendship and Handley Page Dart Herald then being developed in Europe. The YS-11 had a single-class cabin seating 60 passengers in four-abreast configuration with a central aisle and 34-in seat pitch.

After extensive evaluation of all available turboprop powerplants, engine choice was narrowed down to the British Napier Eland and Rolls-Royce Dart and US-manufactured Allison 501, with final choice falling to the R-R Dart RDa.10/1 driving four-bladed 4.45-m (14-ft 6-in) diameter Rotol propellers. Fuel, contained in integral 'wet wing' tanks, and bladders in the wing root area, totalled 7273 litres (1,600 Imp gal).

Two flying and two static test airframes were constructed, the first flying prototype taking to the air on 30 August 1962, followed by

All Nippon Airlines is a substantial YS-11 owner, though the aircraft are actually operated today by Tokoyo-Haneda-based subsidiary, Air Nippon. This aircraft is a YS-11A-523, delivered to All Nippon in 1970.

JA8611 was the first NAMC YS-11 to take to the air, making its maiden voyage on 30 August 1962. The initial variant was not as successful as had been hoped, largely on account of the low-power Rolls-Royce Dart engines originally fitted

Filipinas Orient Airways was one of three overseas customers for the original YS-11 variant. The airline had four aircraft, registered PI-C-962 to -965.

LANSA of Peru took three of the early YS-11s for service on its domestic network. Owing to the hot-and-high nature of its operations, capacity was restricted to 40 passengers.

the second on 28 December. The Japanese Civil Aviation Bureau (JCAB) awarded the YS-11 its type certificate on 25 August 1964; US Federal Aviation Administration approval was obtained on 7 September 1965.

Meanwhile the first production YS-11 had made its maiden flight on 23 October 1964, and was delivered to the JCAB in March 1965. The first airline operator of the type was Toa Airways, which flew its first revenue-earning YS-11 services between Osaka and Hiroshima on 1 April 1965, closely followed by Japan Domestic Airlines (JDA, subsequently to merge with Toa Domestic) a month later, and by All-Nippon Airways in July.

Production of the initial YS-11 version totalled 48 including prototypes, most of which were sold to Japanese carriers, plus six to the

Honolulu-based Mid Pacific Air has operated many YS-11s over the years, initially with a fleet of Series 600s. In 1998, as Mid Pacific Cargo, it operated one Series 100 and five cargo-configured Series 600 aircraft.

Japanese Self-Defence Forces. Four of the military aircraft were configured as 32/48-seat VIP transports and two as all-cargo aircraft.

Slow sales

Reflecting the aircraft's principal design philosophy to meet domestic needs, export sales of the YS-11 were sluggish, only Filipinas Orient Airlines (two, later increased to four), Hawaiian airlines (three leased) and LANSA of Peru (three) taking delivery of the aircraft. Hoped-for sales in the United States did not materialise initially, and The Charlotte Aircraft Corporation, which had sole responsibility for marketing the YS-11 in the USA, suggested improvements to the design to make it more attractive to US carriers.

These bore fruit in 1966 when NAMC announced development of the improved YS-11A in three variants, all featuring water-methanol injected 'hot-and-high' Rolls-Royce Dart Mk 542 turboprops rated at 2279 kW (3,060 shp) and increased maximum take-off

This 1966-vintage NAMC YS-11-117 was originally delivered to Hawaiian Airlines before returning to its native Japan two years later, in 1968.

weights. The YS-11A-200 proved to be the most successful variant of the range, with a 60-seat, five-abreast cabin and a 1270-kg (2,800-lb) increase in payload to 6949 kg (15,320 lb). It first flew on 27 November 1967, was approved by the Japanese authorities in January 1968 and by the FAA on 3 April, and entered service that summer with the US carrier Piedmont Airlines, which ultimately had a fleet of 22 aircraft. The Brazilian airline Cruzeiro do Sul was another export customer, ordering 12. The YS-11A made its European public debut at the 1968 Farnborough air show in England at the start of an extensive demonstration tour of the region.

The YS-11A-300CP was a mixed passenger/cargo derivative with a 2.48-m by 1.8-m (8-ft 1½-in by 6-ft) cargo door in the port side of the forward fuselage for loading the (10.2-m^3/

Brazilian domestic carrier Cruzeiro took a dozen YS-11A-202s for their network. Earlier a pair of YS-11-100s was leased from NAMC pending their first deliveries, which began in March 1968.

Versions of the YS-11A sold adequately in the Far East, including aircraft operated by Korean Airlines, Bouraq, Pelita and Merpati Nasantara of Indonesia, and Philippine Airlines. The Japanese domestic operators and defence forces accounted for the majority.

360-cu ft) forward freight hold, behind which was a 46-seat passenger cabin, with another 4.9 m³ (176 cu ft) of cargo space to the rear. The YS-11A-300CP first flew in 1968; 16 were manufactured.

The YS-11A-400 was an all-cargo version intended primarily for military use. First flown on 17 September 1969, it had a large 3-m by 1.8-m (10-ft by 6-ft) rear cargo door and offered a

An early customer was the Japan Air Self-Defence Force, which bought the YS-11 for general transport tasks. Some aircraft were later converted for electronic warfare and electronic intelligence gathering.

cargo capacity of 79 m² (2,790 cu ft). Nine were built, all delivered under the designation YS-11C to the Japanese Air Self-Defence Force for operation by Nos 401, 402 and 403 Squadrons of the JASDF Air Transport Wing based at Komaki, Iruma and Miho respectively, but now all serving with 402 Squadron. Three YS-11s delivered to the Japanese Air Self-Defence Force's Headquarters Squadron at Iruma were modified for electronic countermeasures (ECM) training and designed YS-11E, and six operated by the Japanese Maritime Self-Defence Force as crew trainers for Lockheed P-2J Neptune anti-submarine aircraft were designated YS-11T ASW. Other military aircraft were used by the JASDF's Flight Check Squadron at Iruma for calibrating and checking ground-based radio navigation aids.

The final versions of the YS-11 were the YS-11A-500, -600 and -700, all externally similar to the YS-11A-200/300/400 but with maximum take-off weight increased to 25000 kg (55,115 lb). Four YS-11A-500s and five -600s were built, bringing YS-11 production to a close in 1973 after a total of 182 had been manufactured. The manufacturing consortium NAMC was progressively dismantled, responsibility for supporting the aircraft falling to Mitsubishi.

9041 was the Japanese Maritime Self-Defence Force's first YS-11, assigned to 205 Kokutai at Shimofusa. Under the designation YS-11T it, and subsequent deliveries, were used for multi-engine training.

The YS-11 was Japan's first (and to date only) post-war civil airliner of original design, and the only large civil transport type ever to have been built in the country in large numbers, if pre- and post-war licence-manufactured versions of foreign designs are discounted. Apart from Japanese airlines and military services, which took delivery of more than half the total produced, other operators of new and used YS-11s included US carriers Hawaiian Airlines, Mid Pacific Air, Piedmont Airlines, Pinehurst Airlines, Pyramid Airlines and Reeve Aleutian, while operators elsewhere included Air Ivoire of the Ivory Coast, Austral of Argentina, Bouraq Indonesian Airlines, China Airlines, Korean Airlines, Merpati Nusantara of Indonesia, Olympic Airlines of Greece (the sole European customer for the aircraft), Pelita Air Service of Indonesia, Philippine Airlines, SGA of Zaire, Transair of Canada and VASP of Brazil. It is believed that 66 YS-11 aircraft were still operational in 1998.

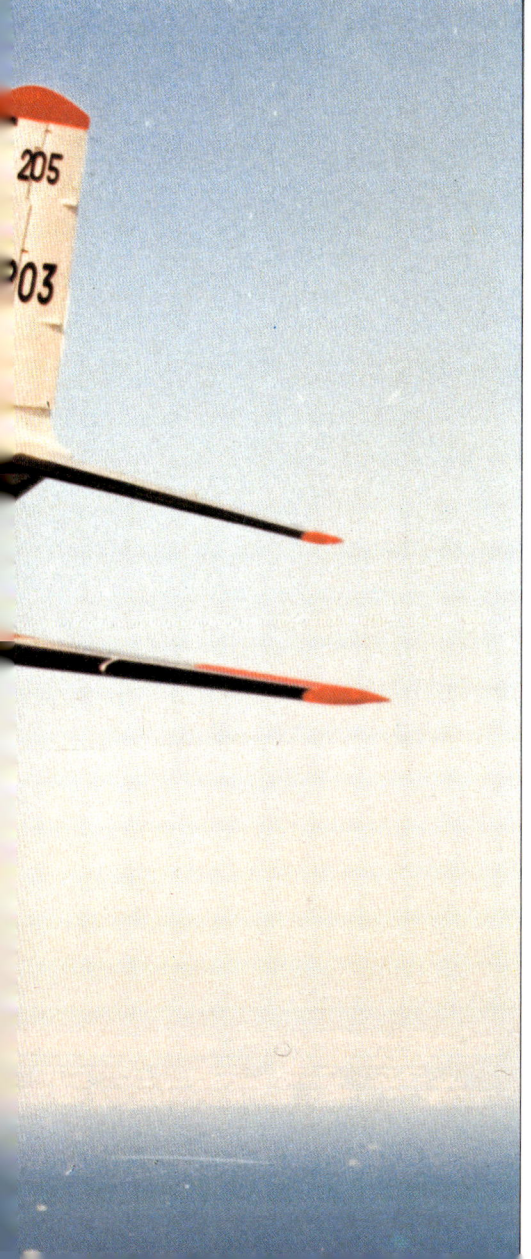

The YS-11A-300, 400, 600 and 700 all featured cargo doors in the port side, and several other aircraft have subsequently been modified for cargo carriage. This aircraft was originally the first YS-11 delivered to a US operator, a Series 205 for Piedmont.

After their service with major airlines was over, many YS-11s passed to small operators, notably in the United States. Texas-based Fort Worth Airways was one such operator.

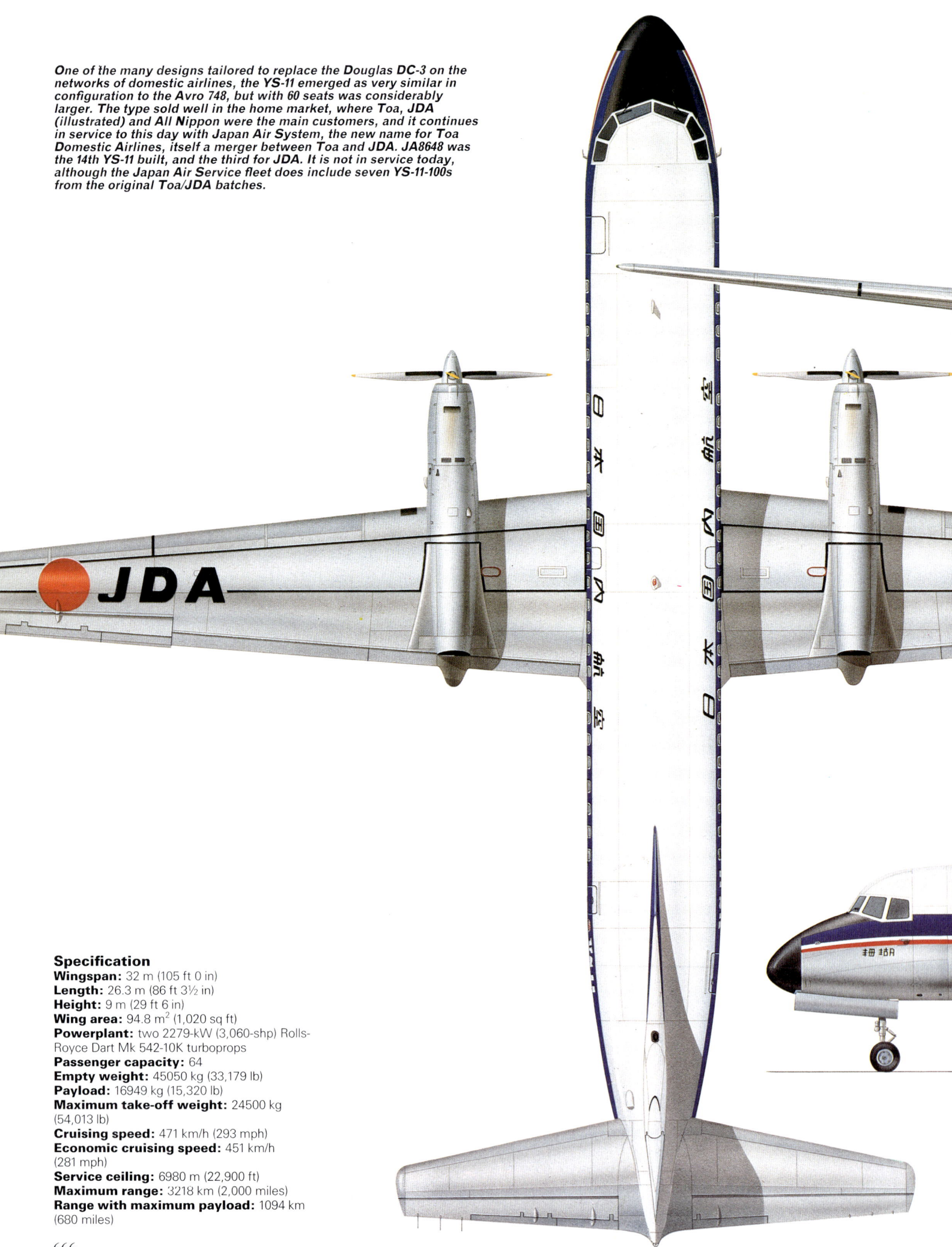

One of the many designs tailored to replace the Douglas DC-3 on the networks of domestic airlines, the YS-11 emerged as very similar in configuration to the Avro 748, but with 60 seats was considerably larger. The type sold well in the home market, where Toa, JDA (illustrated) and All Nippon were the main customers, and it continues in service to this day with Japan Air System, the new name for Toa Domestic Airlines, itself a merger between Toa and JDA. JA8648 was the 14th YS-11 built, and the third for JDA. It is not in service today, although the Japan Air Service fleet does include seven YS-11-100s from the original Toa/JDA batches.

Specification
Wingspan: 32 m (105 ft 0 in)
Length: 26.3 m (86 ft 3½ in)
Height: 9 m (29 ft 6 in)
Wing area: 94.8 m² (1,020 sq ft)
Powerplant: two 2279-kW (3,060-shp) Rolls-Royce Dart Mk 542-10K turboprops
Passenger capacity: 64
Empty weight: 45050 kg (33,179 lb)
Payload: 16949 kg (15,320 lb)
Maximum take-off weight: 24500 kg (54,013 lb)
Cruising speed: 471 km/h (293 mph)
Economic cruising speed: 451 km/h (281 mph)
Service ceiling: 6980 m (22,900 ft)
Maximum range: 3218 km (2,000 miles)
Range with maximum payload: 1094 km (680 miles)

Chapter 31

Piper Cherokee

Piper Navajo & Cheyenne

Pitts Special

RFB Fantrainer

Piper Cherokee family

The Cherokee is a classic of modern aircraft design. Since it first flew in 1960, a plethora of diverse versions has poured from production lines in the United States and abroad. Only the recent global recession has put a brake on the Cherokee's career.

Piper Aircraft Corporation began looking for an all-metal replacement for the high-wing, fabric-covered PA-22 Tri Pacer in the early 1950s, briefly investigating the possibility of buying existing designs such as the Mooney M.20, Erco Ercoupe and Thorp T-11 Sky Scooter. When the decision was taken in early 1957 to design an entirely new aircraft, designated PA-28, Sky Scooter and Ercoupe designers John Thorpe and Fred Weick were hired by the company, which established a new factory at Vero Beach, Florida where the aircraft would be built.

The PA-28 design philosophy was to create a simple low-wing, fixed-landing-gear, four-seat aeroplane whose structure would have the minimum number of component parts to facilitate economic high-volume production.

An experimental 111.75-kW (150-hp) development model PA-28 (N9315R) made its first flight from Vero Beach on 10 January 1960. With its constant chord, plank-like wing it revealed a family link with Thorp's earlier Sky Scooter design. Its creators had succeeded in minimising parts-count, the aircraft having only 1,200 components against 1,600 for the Tri Pacer and nearly 2,500 for the sophisticated retractable undercarriage PA-24 Comanche then in production at Piper's Lock Haven, Pennsylvannia headquarters. An innovation was the use of glassfibre components for non-structural items such as the tips of the wing, tailplane and fin, and the engine cowling nose-bowl, while other noteworthy features of the PA-28 included its one-piece all-moving 'stabilator' tailplane, wide-track landing gear and wing leading-edge fuel tanks, which greatly facilitated removal for maintenance.

A true prototype PA-28 (N2800W), powered by the 119.2-kW (160-hp) Lycoming O-320-D2A engine chosen for production aircraft, made its maiden flight in the spring of 1960. Certification of the aircraft, which had been dubbed Cherokee in keeping with the contemporary Piper practice of naming its aircraft after Indian tribes, was granted on 31 October 1960. Deliveries began shortly after the first production aircraft flew on 10 February 1961. Two versions of the PA-28 were offered initially: the Cherokee 150 with 111.75-kW (150-hp) Lycoming O-320-A2A engine, and the Cherokee 160 with the -D2A powerplant. Both were externally identical but available with a choice of four levels of cockpit equipment and cabin trim: Standard, Custom, Super Custom and Autoflite, the latter having a simple auto-pilot. The 134.1-kW (180-hp) Lycoming O-360-A2A-powered Cherokee 180 joined the range in August 1962, when improved Cherokee B versions of all three models were introduced, these having increased fuel capacity, streamlined wheel fairings and numerous internal improvements.

Production of the Cherokee had built rapidly: the 1,000th aircraft was delivered on 24 January 1963, less than two years after the

Success with the first Warrior led to the Warrior II, basically similar but able to fly with 100LL low-lead fuel. This became the basic Cherokee model (although the original name was dropped) and spawned a new club trainer in the shape of the Cadet.

The mainstream touring models for many years were the PA-28-180 and PA-28-235. In 1973 the Cherokee Challenger was produced, this being a PA-28-180 with longer wings, a fuselage stretch and numerous other refinements.

A logical step for the Cherokee development team was the adoption of a retractable undercarriage, this being the first such model. The PA-28-180R Cherokee Arrow proved extremely popular and the spur for considerable further development.

first aeroplane reached a customer. Later that year PA-28S floatplane versions of the Cherokee 160/180 were approved, and Piper introduced the six-cylinder Lycoming O-540-B2B5-engined Cherokee 235 – first flown on 9 May 1952 and certificated on 16 June 1963 – which had strengthened wings 0.6 m (2 ft) longer than the lower-powered variants containing tanks in the tips which raised usable fuel capacity from 182 to 310 litres (48 to 82 US gal). The Cherokee 235, with an empty weight of 640 kg (1,410 lb) and maximum take-off weight of 1316 kg (2,900 lb) had the best load-carrying capability in its class and was one of the few aircraft to have a useful load greater than its empty, equipped airframe weight. The 1966 Cherokee 235B had an improved fuel system and a constant-speed propeller in place of the 235's fixed-pitch unit.

In 1964 Piper introduced Cherokee C models in 111.75, 119.2 and 134.1-kW (150, 160 and 180-hp) variants, all featuring new streamlined engine cowlings similar to that fitted to the 175-kW (235-hp) model, better cabin and engine compartment soundproofing, restyled interiors and instrument panels, more baggage space and simplified fuel systems. Production of these, and the Cherokee 235B, continued until 1968 when the Cherokee 180D and 235C replaced the earlier models (these had a third cabin window on each side, elongated propeller spinners and a new instrument panel design), while the Cherokee 150/160 were dropped from the range. Production of the Cherokee 180 and 235 series continued with minor improvements and more or less annual model suffix changes until 1972, when Piper launched a new version of the Cherokee 180 which had a 12.7-cm (5-in) increase in fuselage length to improve rear seat room, a 0.6-m (2-ft) increase in wingspan, larger all-moving tailplane, streamlined fin/fuselage juncture fillet, wider entrance door, a 23-kg (50-lb) increase in maximum take-off weight, new seats and a padded instrument panel. It was named Cherokee Challenger, a title replaced a year later by Cherokee Archer. Similar modifications and a 45.36-kg (100-lb) increase in gross weight for the 175-kW (235-hp) model, which already had the longer wing, produced the Cherokee Charger, later Pathfinder.

Pending availability of the first Cherokees, Piper had closed the Tri Pacer production run with a stripped-down 'no-frills' two-seat trainer

The Piper PA-28-140 was the baseline model of the series, intended almost entirely for the club trainer market. This example shows the early house colours adopted for Piper products.

An outgrowth of the PA-28-140 trainer was the Cherokee 140-4, which could fit four seats, although the description 2+2 would be more accurate as the rear seats were best suited to children. Piper named this class the Cruiser.

version designated Colt 108. Carrying over this concept to the Cherokee, and as a means of competing with the best-selling Cessna 150 in the trainer market, in 1964 the company introduced the Cherokee 140, which was a simplified version of the original 111.75-kW (150-hp) model with just two seats, aimed principally at flying schools. A version convertible to four seats followed in 1966, and was known as the Cherokee 140-4. When Piper established a worldwide chain of more than 500 Piper Flight Centers between 1968-70, each teaching students to fly to a common company-developed programme using the latest audio-visual techniques for ground tuition, the basic Cherokee 140 became standard equipment at the schools as the Cherokee Flite Liner, while the two/four-seat Cherokee 140-4 was renamed Cruiser. Production of the various 111.75-kW (150-hp) Cherokee 140 series continued with only minor equipment and styling changes until 1977.

New wing

During June 1972 Piper had begun design work on a new four-seat 111.75-kW (150-hp) model which was eventually to introduce a significant change to the entire range. Publicly launched on 26 October 1973, the PA-28-151 Warrior combined the longer fuselage and larger tailplane of the more powerful Challenger/Charger with an entirely new long-span wing which foresook the so-called 'Hershey Bar' shape of all previous models (named after a popular brand of chocolate from Piper's home state of Pennsylvannia) for a semi-tapered planform, constant chord to mid-span point then tapering to the tip. The new wing was of the same area as the older design, but had lower drag thanks to its high aspect ratio, while its larger Frise-type ailerons enhanced roll response, particularly at low airspeeds. The enhanced lifting properties of the new wing also enabled the Cherokee Warrior to take off at a maximum weight 79 kg (175 lb) higher than the

In refined form the Cherokee 140 became the Flite Liner, and was used widely by Piper Flight Centers to train students to a common company-formulated syllabus. The 140 and its derivatives were the major challenge to the Cessna 150.

Above: Cherokees were offered STOL performance conversions by the Robertson Company. Among the aerodynamic improvements on offer were drooped wingtips, wing fences, additional fin area and more powerful flaps.

The Arrow II had a 200-hp engine and aerodynamic refinements, but the Arrow III took the major step of adopting the 'Warrior' wing as the PA-28-201R. Illustrated is the standard model, but the type was available with a turbo engine.

Perhaps the ultimate single-engined Cherokee development is the PA-32RT Lance II, which combined the six-seat fuselage of the Cherokee Six with all the advances of the Arrow IV, including T-tail and retractable undercarriage. The Saratoga was a fixed-gear counterpart.

similarly-powered 'Hershey Bar' Cherokee Cruiser.

This new wing design was introduced on the 134.1-kW (180-hp) Cherokee Archer in 1976 as the PA-28-181 Cherokee Archer II, and two years later on the 175-kW (235-hp) Cherokee Pathfinder, which became the PA-28-236 Dakota (the addition of one horsepower to the designation's suffix appears to have been for marketing purposes only, and was not matched by a corresponding increase in engine output, which remained 184.1 kW/180 hp and 175 kW/235 hp respectively). A turbocharged version of the latter aircraft, the PA-28-236T Turbo Dakota, was marketed briefly in 1979/80 but was not a sales success. Despite the suffix, it was not of 175-kW (235-hp) like its normally-aspirated stablemate, but was powered by a 149-kW (200-hp) Continental TSIO-360-FB.

On 27 August 1976 Piper first flew the Cherokee Warrior II, which had a 119.2-kW (160-hp) Lycoming O-320-D2G engine suitable for 100LL low-lead fuel, and this became the standard production version from 1977. In 1988 Piper, which had then been bought by California businessman M. Stuart Millar, announced a two/four-seat trainer version of the Warrior II called Cadet which, like its predecessor the Flite Liner/Cruiser, is a 'no frills' aeroplane aimed at flying schools and clubs. Available with a choice of Basic Training Group (for Visual Flight Rules training) or Advanced Training Group (for Instrument Flying training) avionics and instruments, it has a utility interior with two front seats only (a rear bench is

The ultimate Arrow version was the PA-28RT-201 Arrow IV, again available in either standard or turbo form. This had an all-moving tailplane mounted on top of the fin. Sales were good, despite some unpopularity concerning the T-tail configuration.

optional), and lacks the Warrior II's external baggage door and standard wheel fairings.

Total production of all fixed-undercarriage PA-28 models exceeds 30,000 aircraft. In addition to those manufactured at Vero Beach, Aeromercantile of Colombia, Chincul of Argentina, EMBRAER of Brazil and ENAER and Aero Salfa of Chile have also assembled from Piper-supplied kits or manufactured under licence various PA-28 models for their domestic markets in Latin America.

Retracting undercarriage

Piper announced the first retractable undercarriage derivative of the PA-28 series on 19 June 1967 when it released details of the PA-28R Cherokee Arrow, three prototypes of which had been flown at Vero Beach during 1966. Based on the then-current four-seat Cherokee 180D's airframe, but with a fuel-injected 134.1-kW (180-hp) Lycoming IO-360-A2A engine driving a constant-speed propeller, the Cherokee Arrow had an electro-hydraulically operated retractable tricycle landing gear which incorporated a self-lowering system activated automatically by a pressure sensor when airspeed reduced below 169 km/h (105 mph) with power reduced below 30 cm (12 in) of manifold pressure. Intended to prevent the not uncommon and always expensive occurrence of wheels-up landings, the auto-extension system was not universally acclaimed, since it would also lower the landing gear after engine failure, just when a pilot trying to stretch his glide for an emergency landing did *not* want the additional drag of extended undercarriage legs.

The Cherokee Arrow was certified on 8 June 1967, and with its maximum speed of 273 km/h (170 mph) immediately proved popular as a logistical 'step up' for pilots familiar with the fixed-gear Cherokees. A total of 1,161 was built before the PA-28-200R Cherokee Arrow II succeeded the initial model for the 1972 model year. Incorporating the longer fuselage, increased wingspan, larger tailplane and other refinements already described for the contemporary 134.1/175-kW (180/235-hp) fixed-gear Cherokee Challenger/Charger, this model was powered by a 149-kW (200-hp) Lycoming IO-360-CC, and benefitted from an 8-km/h (5-mph) increase in maximum speed, and a 68-kg (150-lb) increase in maximum take-off weight to 1202 kg (2,650 lb). A 134.1-kW (180-hp) version was also available in this airframe, but the majority of the 2,850 Cherokee Arrow IIs built between 1972-76 had 149-kW (200-hp) engines.

In common with the fixed-undercarriage models, the Arrow (the Cherokee prefix had been dropped) gained long-span semi-tapered wings, first flown in prototype form on 16 September 1975 and introduced from the 1977

A major change to the look of the Cherokee family was introduced by the PA-28-151 Warrior, which featured a new wing of revised planform. The wing was applied to other members of the family, the Challenger/Archer becoming the PA-28-181 Archer II.

An early PA-28 development was the PA-32 Cherokee Six, available in 260-hp (illustrated) or 300-hp form. This was very similar to the PA-28, but was enlarged to accommodate six or seven passengers.

model year on the PA-28R-201 Arrow III, which had a further 113-kg (250-lb) increase in maximum take-off weight and fuel capacity increased from 181.6 to 272.5 litres (48 to 72 US gal). Two versions were offered: a normally-aspirated model with 149-kW (200-hp) IO-360 engine as on the Arrow II, and the PA-28R-201T Turbo Arrow III with turbocharged 149-kW (200-hp) Continental TSIO 360 F engine. Both aircraft received FAA certification on 2 November 1976, deliveries beginning in the following January. A total of 492 aircraft of both models was produced between 1977-79.

In 1979 the final development of the Arrow series appeared as the PA-28RT-201/201T Arrow IV/Turbo Arrow IV, which differed externally from the previous model in having a new all-moving T-tail, as was the fashion on several other Piper models at the time. Prototypes of both variants were first flown during March 1978 and were certificated in November. Production of the normally-aspirated Arrow IV ceased in 1982 after some 500 had been delivered, but the Turbo Arrow IV was retained in the product line, with more than 900 built, until the Piper company was bought in 1987 by M. Stuart Millar. Responding to the less than enthusiastic reception accorded the T-tail, which some owners claimed had spoilt the Arrow's handling characteristics, Millar restored to production the earlier low-tail Arrow III/Turbo Arrow III models, until production of all Piper's single-engined types was suspended. To date nearly 7,000 Arrow variants of all models had been built.

Stuart Millar acquired the company in 1987, in the face of falling demand and rising costs, due in a large part of the US's product liability

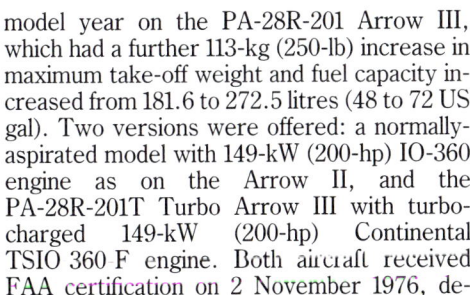

The PA-28-235 series continued with the new wing as the Dakota series. This example is one of the short-lived PA-28-236T Dakotas, a model which did not sell well.

laws which left aircraft manufacturers open to huge compensation claims in the event of an accident involving one of their products. By early 1990 Millar was looking for a buyer for Piper Aircraft Corporation, with France's SOCATA as the forerunner. Negotiations broke down, due chiefly to continuing uncertainty over the product liability rules. The Cherokee family continued in very low-volume production throughout this time, while its parent company filed for voluntary bankruptcy and Chapter 11 protection. This protection was extended while Piper continued its search for a buyer. Switzerland's Pilatus Aircraft has emerged as the most likely suitor, having had several legal restrictions on its bid lifted.

In addition to those already described, the basic PA-28 airframe has spawned a number of developments which are beyond the scope of this article, but which include the 'stretched' six/seven-seat 193.7/223.5-kw (260/300-hp) PA-32 SIX and Lance/Saratoga, the latter two built in fixed- and retractable-landing gear, normally-aspirated and turbocharged versions; the PA-34 Seneca twin six-seat twin; and the Arrow IV-derived PA-44 Seminole light twin. The Cherokee, in its many varied forms, is one of the most successful and adaptable light aircraft of all time, and with Piper having recovered from its bankruptcy of 1991-1995, the Cherokee series looks set to continue into the 21st century.

Piper's Twin Tribes: Navajo and Cheyenne

At airports and airfields around the world, wherever there are working aeroplanes earning a living, you will find Bill Piper's Navajos. Since its first flight in 1966, the PA-31 has become a byword for versatility and has served as the basis for the impressively sleek range of Cheyenne executive turbos.

In 1964 Piper Aircraft Corporation began design work on a new twin-engined business aircraft which, on the insistence of company founder William T. 'Bill' Piper, was to feature a cabin whose cross-section remained constant throughout its length, thus avoiding the poor headroom offered in the rear seats of contemporary twins whose tailcone taper started in the cabin area. At that time it was the largest aircraft ever built by Piper, and was intended to move the company up from being a manufacturer of 'light aircraft' into the 'cabin class' market then dominated by Beech and Cessna.

The designation PA-31 was assigned to the aircraft, for which the name Inca was chosen. The prototype, N3100E, made its first flight on 30 September 1964. A low-wing all-metal design, it was powered by two 230-kW (310-hp) Lycoming TIO-540 engines and featured a six/eight-seat passenger cabin with two large square windows on each side. During development two further cabin windows were added on each side. Two pre-production aircraft were flown from the company's headquarters at Lock Haven, Pennsylvania, during 1966.

Prior to its public launch in 1967 Piper renamed the PA-31 'Navajo'. Deliveries commenced on 30 March 1967. It was offered in two versions: the PA-31-300 Navajo with normally-aspirated 225-kW (300-hp) Lycoming IO-540Ms, and the PA-31-310 Turbo Navajo, with turbocharged 230-kW (310-hp) TIO-540-A1As, both driving two- or three-bladed propellers, and was available with a choice of six-

Following the development of the Chieftain, many features were applied to the short-fuselage Navajo family. This is the Navajo C/R, which introduced counter-rotating propellers to alleviate torque problems.

*Above: Under the designation **PA-31T-1**, Piper introduced in 1978 a low-cost version of the **Cheyenne** with less powerful engines (373-kW/500-shp) **PT6A-11**s). Named **Cheyenne I**, the type proved a good seller.*

*Like automobile manufacturers, the light aircraft industry favoured introducing yearly models to keep interest in their products alive. This is the 1981 **Chieftain** (the **Navajo** prefix having been dropped the previous year).*

An early development of the Navajo was the PA-31P Pressurised Navajo, and this in turn led to the turboprop-powered PA-31T Cheyenne family. The first version (illustrated) was redesignated Cheyenne II when a simpler version was introduced.

(430 km/h; 270 mph), while useful load was increased by more than 90 kg (200 lb); the higher fuel burn of the 315-kW (425-hp) engines reduced range to 925 nm (1710 km; 1065 miles) at 75 per cent power, however. Customer deliveries began in January 1970 and production continued with only minor changes to the aircraft's equipment until 26 August 1977, by which time a total of 259 Pressurized Navajos had been delivered.

First development

Meanwhile, in early 1970 Piper had converted two standard Navajos into prototypes for the PA-31-310B Navajo B. This version was powered by turbocharged 230-kW (310-hp) Lycoming TIO-540-A/Ss. Major changes incorporated in this model included new air conditioning and avionics systems, an optional pilot entrance door on the port forward side of the cabin, optional wide cargo/utility door and optional baggage lockers in extended compartments aft of the engine nacelles. Deliveries of the Navajo B (the 'Turbo' prefix was dropped to avoid the possibility of refuelling staff incorrectly fuelling the aircraft with turbine fuel) began in late 1970 and continued until 1974.

In keeping with its stated intention to develop a range of aircraft from the basic Navajo airframe, Piper commenced development in 1971 of the PA-31-350 Navajo II, intended to compete with the Cessna 402 in the commuter airliner market. Three prototypes were built, incorporating a 60-cm (24-in) fuselage stretch with an extra cabin window on each side, a

seat Standard or Executive or eight-seat Commuter seating arrangements. A horizontally-divided 'Dutch Door' with integral steps in the lower half was standard on the port side of the fuselage at the rear of the cabin. The PA-31-300 had a maximum speed of 196 kt (360 km/h; 225 mph), cruised at 184 kt (340 km/h; 210 mph) at 75 per cent power and had a with-reserves range of 965 nm (1790 km; 1110 miles), while the turbocharged PA-31-310's figures were 228 kt, 217 kt and 982 nm respectively. The 225-kW (300-hp) Navajo proved less popular than anticipated, only 14 being delivered, and in 1969 this version was dropped in favour of the PA-31-310.

Even before production of the PA-31-310 got under way Piper had begun preliminary development of several other versions of the basic airframe, which was readily adaptable to a variety of roles and configurations. First among these was the PA-31P Pressurized Navajo, on which the company began work in 1966. Basically a Navajo with a six-seat cabin pressurised to 40 kPa (5.5 lb/sq in) and powered by two 315-kW (425-hp) turbocharged and geared Lycoming TIGO 541 E1A6 engines, the prototype, N9200Y, first flew in March 1968, and was followed by a pre-production aircraft the following year. The powerful Pressurized Navajo was substantially faster than the standard model, with a maximum speed of 243 kt (450 km/h; 280 mph) and a 75 per cent power cruising speed of 233 kt

Left: The prototype PA-31 was characterised by three large windows in the fuselage side. It was later developed with two more. Originally named Inca, the PA-31 first flew in September 1964.

Right: The Navajo Chieftain was a lengthened Navajo accommodating up to 10 passengers. It became very successful, manufactured mostly at a purpose-built factory at Lakeland in Florida.

large cargo door, strengthened cabin floor to permit operation in all-cargo configuration, and 260-kW (350-hp) Lycoming L/TIO-540-J2BD engines with counter-rotating propellers to offset torque effects and avoid the 'critical engine' factor in asymmetric flight. With its longer fuselage the Navajo II could accommodate 10 passengers in Commuter configuration or six in luxurious club seating Executive layout, with space for up to 6 m^3 (217 cu ft) of cargo in the main cabin, 90 kg (200 lb) each in nose and rear cabin compartments and 70 kg (150 lb) in the rear of each engine nacelle, which could alternatively be used to house optional auxiliary fuel tanks, when baggage capacity was limited to 70 kg (50 lb) in each. The second prototype and first production aircraft were destroyed in a disastrous flood at Lock Haven in June 1972 when the Susquehanna river burst its banks, delaying introduction of the new model, which was renamed Navajo Chieftain before its public launch on 11 September 1972. First deliveries to an eager market began in early 1973, and by June of that year Piper had Navajo Chieftain production lines running at both Lock Haven and at a new purpose-built plant at Lakeland, Florida, to which all Navajo production was switching during 1974. From 1980 'Navajo' was dropped from the name and the aircraft was marketed as the Piper Chieftain. Production ceased in October 1984, by which time 1,825 Navajo Chieftains and Chieftains had been manufactured.

Better again

The 'standard' Navajo was further improved during 1974 with the introduction of the Navajo C with 230-kW (310-hp) Lycoming TIO-540-A2C engines, and the Navajo C/R, which was powered by 240-kW (325-hp) Lycoming L/TIO-540-F2BD powerplants with counter-rotating propellers. The 'C' replaced the earlier 'B' model; the C/R was externally distinguishable from it principally by engine nacelle luggage compartments/auxiliary fuel

A new designation, PA-42, was introduced for the Cheyenne III, reflecting the radical changes, which included lengthened fuselage and increased span, 537-kW (720-shp) PT6A-41 engines and a T-tail. The later Cheyenne IV switched to Garrett TPE331 power.

24-hours-a-day support centre at Tampa International Airport. Crew and engineer training were also offered. The first T-series to appear was the PA-31T-3 T-1020, an 11-seat commuter airliner based on the Chieftain airframe with engine time between overhaul (TBO) increased to 1,800 hours and featuring a harder wearing interior, lightened passenger seats, and strengthened landing gear with improved landing gear doors to cope with the high landing/flight hour rate experienced in short-leg commuter operations. Design of the T-1020 started at the end of April 1981. The prototype first flew on 25 September, with deliveries beginning after FAA certification in December. Among airlines taking delivery of T-1020s were Desert Sun Airlines and Air New Orleans. Production of the T-1020 ceased in 1985 after 21 had been delivered. A derivative known as the PA-31-353 was powered by 260-kW (350-hp) Lycoming L/TIO-540-X48 engines and combined the long-span wings of the Mojave with the fuselage, engine nacelles and vertical tail surfaces of the Chieftain and horizontal tail surfaces of the turboprop T-1040 (see below). A prototype and a pre-production aircraft were flown during 1985, but the PA-31-353 T-1020 never entered production.

Turbo-powered

Simultaneously with development of the T-1020, Piper had begun design work on the T-1040, which combined the Chieftain's fuselage with the nose, tail unit and 500-shp Pratt & Whitney Canada PT6A-11 turboprop engines of the PA-31T-1 Cheyenne executive transport, the wings and landing gear of the Cheyenne IIXL, and the engine nacelles and extended nacelle baggage lockers of the top-of-the-range Cheyenne IIIA, which proved an effective and economical short-cut to developing from scratch an unpressurised turboprop commuter airliner. The first of three pre-production T-1040s flew for the first time on 17 July 1981. FAA certification was achieved on 25 February 1982 and deliveries began the following May, when the aircraft was publicly launched at the Hanover Air Show in Germany. The T-1040 was also available in all-cargo configuration, offering a payload of 1310 kg (2,900 lb) and cargo capacity of 7 m^3 (246 cu ft), to which an optional ventral cargo pod added a further 0.9 m^3 (30 cu ft). Twenty-four T-1040s were built. A stretched 15-passenger T-1050 derivative with a fuselage lengthened by 3.5 m (11 ft 6 in) was projected to meet the requirements of overnight package carrier Federal Express, but was not proceeded with.

In November 1989 Piper announced the formation of a wholly-owned subsidiary, Piper North Inc., at its former birthplace at Lock Haven. It was proposed that this company would re-establish production of the Chieftain, but following financial difficulties encountered during 1990 this project was postponed, and no further production of any of the Navajo series is currently planned.

tank housings which extended aft of the wing trailing edges. The Navajo and Navajo C/R were available with a choice of five factory-installed avionics packages featuring an extensive range of navcom, autopilot and weather radar equipment from Bendix, Collins, King and Narco. Total deliveries of all 'short fuselage' Navajo variants, excluding pressurised versions, was 1,785.

In 1982 Piper began development of a successor to the Pressurized Navajo. Known as the PA-31P-350 Mojave it had the fuselage of the PA-31T Cheyenne (itself a turboprop development of the original PA-31P) mated to the 260-kW (350-hp) Lycoming L/TIO-540-V2AD engines and tail unit of the PA-31-350 Chieftain. The Chieftain's wings were also fitted, but with structural strengthening and span increased by 1.2 m (4 ft). Two prototype Mojaves were built during the summer of 1982 and certificated in June 1983. Mojave production was brief, terminating on 26 June 1984 after the 50th aircraft had been completed.

On 4 June 1981 Piper had announced the establishment of a new Airline Division to support commercial operators of the Chieftain and to develop and market a range of T-1000 (Transport) series aircraft, which were to be sold and supported factory-direct to air taxi, commuter airline and charter operators from a

Succeeding the PA-31P was the PA-31P-350 Mojave, a mixture of Cheyenne and Chieftain with long-span wings for better cruise performance. The type did not prove as successful as other variants, only 50 being built.

Following early successes with the standard Navajo in regional airline service, Piper developed the specialised T-1000 series tailored for airline work. The T-1020 (illustrated) was the piston-engined version, while the T-1040 had turboprops. Neither was a success, and production amounted to under 50 aircraft.

A number of modifications have been developed for the Navajo series. Perhaps best known among these are the 'Panther' conversions created by Colemill Enterprises Inc. of Nashville, Tennessee. Available for all short-fuselage Navajo or Navajo C/R models, the modifications involve replacing the standard engines with 260-kW (350-hp) turbocharged Textron Lycoming TIO-540-J2BDs driving Hartzell four-bladed 'Q-tip' propellers, and installing pressurised magnetos, continuous running fuel pumps, a digital fuel totaliser, redesigned engine nacelles, heavy duty brakes, and new wingtips which increase span to 13 m (42 ft 8 in), with optional winglets called 'Zip Tips' which increase span by a further six inches. The winglets improve low speed stability and reduce drag, resulting in cruising speed increases of 4-9 kt (7-16 km/h; 4-10 mph) at 45-65 per cent power settings. A similar conversion for the Chieftain series is known as the Panther II.

Schafer Aircraft Modifications Inc. of Waco, Texas, has developed turboprop conversions for the Pressurized Navajo and Chieftain. The Pressurized Navajo conversion is known as the Schafer Comanchero and involves installation of two 750-shp Pratt & Whitney Canada PT6A-135 turbine engines, flat-rated to 620-shp, with associated fuel and engine control systems. Useful load of the aircraft is increased by 345 kg (760 lb) and fuel capacity increased to 1360 litres (360 US gal) while cruising speed goes up to 282 kt and rate of climb to 1070 m (3,500 ft) per minute. The Comanchero was certificated by the Federal Aviation Administration in January 1981.

Schafer's Comanchero 500 is a turbine conversion of the Chieftain with either 715-shp PT6A-27s flat rated to 578-shp (Comanchero 500B) or 550-shp PT6A-20s (Comanchero 500A), the prototype of which first flew in August 1981. In addition to the engine installation, 130 or 340 litre (35 or 90 US gal) supplementary fuel tanks are installed in the rear section of each engine nacelle and structural strengthening incorporated to permit maximum take-off weight to be increased to 3630 kg (8,000 lb). Options include a 0.5 m^3 (15.5 cu ft) capacity underfuselage baggage pod. The Comanchero 500A/B has a cruising speed of 240 kt (440 km/h; 275 mph) and a maximum rate of climb of 730-850 m (2,400-2,800 ft) per minute. In 1984 Schafer licensed EMBRAER of Brazil to manufacture 50 Comanchero 500Bs. These aircraft have 550-shp PT6A-34 engines and are known as Neiva NE-821 Carajás.

Chincul S.A.C.I.F.I.A. of Argentina, EMBRAER of Brazil and Aero Mercantile of Colombia also locally assembled Navajos and/or Chieftains from Piper-supplied kits.

An early customer for the Navajo was West Coast Airlines, a regional operator in the Pacific Northwest. Navajos were used as eight-seat airliners on short sectors between towns in Washington, Oregon and Idaho. West Coast's buy represented the first use of 'executive' type aircraft for scheduled services.

Pitts Special

Few people would believe that the Pitts Special first flew as long ago as 1947. Despite being a biplane, its racey looks and the fact that it is still an omnipresent performer at air shows can mislead the onlooker as to the aircraft's lengthy pedigree. Once the unbeatable mount of all aerobatic champions, today the distinctive Pitts Special still lives in the fast lane.

Curtiss H. Pitts was born in Americus, Georgia, where he built his first aeroplane as a youth in 1932. It was a parasol-winged design powered by the engine of a Model T Ford automobile, and was wrecked when Pitts – who had no flying experience – lost control while taxiing it one windy day.

He left home for Ocala, Florida, where he became a carpenter on the railroad and began learning to fly in a Taylor E-2 Cub, then moved to Jacksonville where he built his second flying machine, assembled from parts of a Heath Parasol homebuilt and powered by a three-cylinder Szekely engine.

In 1940 Pitts became manager of the small airport at nearby St Augustine, and took a correspondence course in aircraft construction. Fascinated by aerobatics, and by the Great Lakes, Waco and Bucker biplanes favoured by the airshow performers of the day, but lacking the money to buy one, Pitts started to design an aerobatic aircraft of his own in 1942. It took him three years to complete the first 'Pitts Special', a tiny biplane powered by a 41-kW (55-hp) Lycoming engine that proved to be inadequate even though the aircraft weighed less than 226 kg (500 lb) empty. Within a short time Pitts had replaced it with a 67.3-kW (90-hp) unit, and developed a rudimentary inverted fuel system that proved to be its downfall. Shortly after it was sold to a hard-of-hearing crop duster pilot, the fuel system failed while the aircraft was in an inverted 45° climb. Not realising that the engine had stopped the pilot tried to 'pull through' to recover from the manoeuvre and mushed into the ground, destroying the first 'Special'.

"The biggest thrill for me was the first time I flew that little airplane. It was like a ride in a skyrocket after flying those other airplanes," Pitts recalled many years later. A replica of the aircraft was installed in the Experimental Aircraft Association's museum at Oshkosh, Wisconsin, during the EAA Convention in July 1991.

Moving again to Gainesville, Florida, Curtiss Pitts was commissioned to build 10 Pitts Specials for a local operator, who went bankrupt before the first was completed. This aeroplane, known as the Pitts '190' Special, was sold to Jess Bristow, owner of World Air Shows in Miami, and eventually passed on to a young lady pilot named Betty Skelton. She painted a cartoon skunk on it, christened the aeroplane *The Little Stinker*, and soon she and the tiny biplane became famous for their performances in all the major airshows and aerobatic contests. Betty Skelton took the US National Aerobatic Association's Women's title three years in a row between 1948-50. She brought the Pitts Special to Europe in 1949 for the *Daily Express* International Air Pageant at Gatwick Airport, shipping *The Litte Stinker* across the Atlantic as deck cargo aboard the liner *Queen Mary*. After many changes of ownership and re-engining which doubled its initial 63.4 kW (85 hp), Betty Skelton bought

The Pitts Special was for many years almost the universal mount of the world's aerobatic teams. Many, such as the Toyota team, flew the versatile two-seat S-2 model, as it allowed them to take passengers on orientation and demonstration rides.

back her beloved *Little Stinker* and donated it to the Sportsman's Hall of Fame Museum at Cypress Gardens in Florida, where it is displayed in a glass case.

Meanwhile Curtiss Pitts had gone into the crop-dusting business, but continued to build 'Specials' as time permitted, including Miss Caro Bayley's 93.2-kW (125-hp) *Black Magic*, and a one-off monster named *Samson* that was powered by a 335.7-kW (450-hp) Pratt & Whitney R-985 radial engine driving a 10-foot Hamilton Standard propeller from a Boeing 247D airliner. Jess Bristow commissioned the aircraft, which was bigger than the now familiar Pittses. *Samson* was destroyed by fire after a landing collision with Erco Ercoupe lightplane at an airshow in Fayetteville, North Carolina, but in 1985 American airshow pilot Steve Wolf completed a replica which is a popular performer with an abundance of power.

Early racers

Little-known aircraft built by Curtiss Pitts at this time were two midget racers. The low-wing *Pellett* was built in 1947, and was best placed coming third in the 1949 Continental Motors Race at 265.2 km/h (164.8 mph), flown by T. Bud Heisel, who was later killed in it. The second Pitts racer was a mid-wing design named *Li'l Monster*, which managed fourth in the Continental race in 1951 piloted by Bill Brennand, who had won the Goodyear Trophy in 1947 and 1949 flying Steve Wittman's famous *Buster*. Throughout its racing life *Li'l Monster* was plagued by engine troubles.

Following a number of spectacular and tragic accidents in the early 1950s, airshows and air

The British-based Rothmans team flew their blue-striped **S-2A S**pecials all around the world for many years. At first a four-ship disply team, it was later reduced to just a pair of aircraft.

Bob Herendeen, master of the never-ending flick-role, is seen in the **S-1 S**pecial in which he won the 1966 **US** national championship. Always in the top five of his day, he later flew Pitts with the American national team.

racing went out of fashion in the United States, and Curtiss Pitts devoted his energy to his cropdusting business. Then one day a visiting pilot from Tulsa, Oklahoma, asked if plans were available for the Pitts Special? Armed with Curtiss's rough-and-ready drawings he started to build his own aircraft, later finished by Dean Case of Wichita, Kansas, as *Joy's Toy* for his daughter Joyce, a promising aerobatic competitor. Several more homebuilt Pitts Specials appeared, and soon pressure on Pitts to release proper drawings persuaded him to hire a draughtsman to turn his own rough sketches into something an amateur builder might be able to work from. After a 'proof copy' homebuilt Pitts Special was displayed at the Experimental Aircraft Association Convention at Rockford, Illinois, large numbers of orders came in for plans, which went on sale in 1962.

The Pitts Special airframe was simple, "an old-fashioned airplane using the engineering knowledge that we've had since back in the mid-1920s," its designer admitted, comprising a welded steel tube fuselage with wooden stringers, and wooden wings with plywood-skinned leading edges, all fabric-covered. But it was immensely strong, stressed to $+9/-4.5g$; there is no record of any properly-built Pitts Special suffering catastrophic structural failure in the air.

By the time plans went on sale Pitts was completing a redesign to enable engines up to 134 kW (180 hp) to be installed, and in 1965 the open cockpit was lengthened 7.6 cm (3 in) – a concession, it was said, to the designer's broadening waistline. At that time the Pitt had ailerons on its lower wings only, and had a conventional M-6 section flat-bottomed asymmetrical lifting airfoil. As early as 1948 Curtiss had toyed with the idea of a symmetrical airfoil section to improve the aircraft's inverted performance, but had been discouraged by aeronautical experts at the University of Florida, in the early 1960s. A testbed set of symmetrical section wings was built, but proved to have unacceptable stall/spin characteristics. After the wings had been rebuilt twice the solution adopted was to give upper and lower wings the same incidence, but different airfoils. By this means the upper wing was made to stall first whether the aircraft was level or inverted, while the lower wing continued lifting and its ailerons continued to work.

In those days surplus military Stearman biplanes, Great Lakes, clip-winged Piper Cubs and Taylorcrafts, Bucker Jungmeisters and Jungmanns, Stampes and Zlins dominated aerobatics and airshow flying. Few people took the tiny Pitts Special seriously until Don Pittman won the 1962 US National Aerobatic Championships and astonished fans at the first Reno Air Races in 1964 with his spectacular flying in *Joy's Toy*.

Wild thing

But it was TWA captain Bob Herendeen who really put the aeroplane on the map when he twice won the US Nationals in it, in 1966 and 1970. After seeing Pittman's performance – "the wildest thing I ever saw" – Herendeen bought a 134-kW (180-hp) homebuilt Pitts Special, N66Y, fitted it with an inverted fuel system, and returned to Reno in 1965 to place third, thus qualifying for a place on the US team for the 1966 World Aerobatic Championships in Moscow where he placed 25th overall and gave Europe its first sight of the Pitts Special in competition.

Buoyed up by this modest success, Herendeen persuaded Curtiss Pitts to build him a new model S-1S with symmetrical, four-aileron wings. In this aeroplane, N266Y, he took third place in the 1968 World championships at Magdeburg, East Germany, and again took the US title in 1969. It was at RAF Hullavington in England at the 1970 world contest that the Pitts really exploded onto the international scene, flown by Herendeen, Gene Soucy, Bob Schnuerle and Mary Gaffaney. The Americans took team honours and the Nesterov Cup. Herendeen took second place in the Men's Championship – then the highest ever achieved by an American in world competition – just 125 points behind Russia's Igor Egorov. Except for an engine failure as he entered a spin during the Unknown Compulsory programme he might have been world champion. Mary Gaffaney took Bronze in the Women's competition. "Good aeroplane for fun. No good for contests," declared one disgruntled member of the Soviet team after flying a Pitts. The Pitts's astonishing roll rate, vertical and knife-edge flight capability and dazzling snap-rolls impressed even the most cynical observers, even if its tiny size gave it something of a 'wasp in a jamjar' appearance when performing at competition aerobatics height.

Aerobatic trainer

Meanwhile, by 1966 Curtiss Pitts' plane business, Pitts Aviation Enterprises, was going so well that he sold his crop-dusting operation to concentrate on the development of his designs. In 1964 he had begun design of an enlarged two-seat Pitts Special S-2, intended as an aerobatic trainer. He completed the 134-kW (180-hp) Lycoming-powered prototype N22Q, named *Big Stinker*, in time to take it to the 1967 EAA Fly-In at Rockford. Unlike the single-seater, this aircraft was intended from the outset for Federal Aviation Administration certification and series production, but at that time the FAA's rules, especially regarding stability and spin recovery, were such as to virtually make impossible the certification of any aeroplane actually *capable* of advanced aerobatics. It was not until 1971 that the S-2 was finally approved, and went into production as the 149.2-kW (200-hp) Lycoming AEIO-360-A1A engined Pitts S-2A at the former Callair agricultural aircraft plant at Afton, Wyoming, where Herb Anderson's Aerotek company built the aircraft.

It was the S-2A which turned the Pitts into a favourite formation aerobatic team mount. The late Manx Kelly, founder of the British Rothmans Aerobatic Team, was quick to see the

Gene Soucy flew this Pitts S-1S as part of the Red Devils Aerobatic Team. Even after re-equipping with Christen Eagles (and a name change) they continued as regular openers of the annual national championships at Reno.

683

One of the many teams to fly the Pitts Special was Jordan's 'Royal Falcons', whose bright red aircraft flew the flag throughout the 1970s and 1980s. They were sponsored by Alia, the national airline.

potential of the S-2A, five of which replaced the team's Stampes in 1973. Kelly also had a hand in forming the 'Carling Red Tops' (later 'Canadian Reds') in Canada. The 'Marlboro Aerobatic Team' and 'Royal Falcons' also adopted the two-seat Pitts, as did a Chilean Air Force team, 'Los Halcones' (Falcons), who have only recently replaced their Pitts with German Extra monoplanes.

On 13 February 1973 Curtiss Pitts also managed to obtain type certification for the single-seat S-1S with 134-kW (180-hp) Lycoming IO-360-B4A engine and symmetrical-section, four-aileron wings, and this too went into production at Afton. The S-1S's contest triumphs continued to mount. At the 1972 World Aerobatic Championships at Salon de Provence in southern France the American team were all Pitts-equipped, and triumphed totally in men's, women's and team championships.

Development of the design has continued, although Curtiss Pitts is no longer involved with the aircraft which made his name famous. He sold out his interests in 1977, and the type certificate and manufacturing and marketing rights to all Pitts Specials passed in 1981 to Frank Christensen, who had been US team manager at Salon. Christensen was a successful businessman who had developed advanced inverted fuel and oil systems for aerobatic aircraft, and had previously tried unsuccessfully to buy rights to the Pitts from Curtiss. Meeting flat refusal, he designed his own aeroplane, the two-seat 149.2-kW (200-hp) Christen Eagle II, which, though featuring spring-steel landing gear, a more streamlined cowl, reduced lower wing dihedral, new ailerons, a larger cockpit with huge bubble canopy, less rounded tail surfaces and a re-engineered structure, still looked like a Pitts S-2A.

Christen versions

By the time Christensen finally did get his hands on Pitts, three more models were in production at the Afton plant: the S-1T with 149.2-kW (200-hp) Lycoming AEIO-360-A1E engine driving a constant speed propeller, wings moved forward to compensate for additional engine weight and fully-enclosed cockpit; the S-2S, a 194-kW (260-hp) single-seat version of the S-2A which first flew in December 1977, entered production a year later and was FAA-certificated in June 1981; and the S-2B, a 194-kW (260-hp) AEIO-540-D4A5 powered two-seater with tandem cockpits enclosed by a large Eagle-style bubble canopy.

Several British Pitts Specials carry these G-III- series marks. This allows the registrations to be the regulation size and still fit on the tail, leaving the side of the aircraft free for the sponsor's logo.

This Pitts S-1 is the subject of our three-view artwork, but is seen here in the days following its radical modification with a three-bladed prop and uprated engine. The undercarriage legs have also been changed.

This is a good example of the plethora of modified Pitts Specials that have been built. The wings have been squared-off for more maneouvrability, while the tail area has been correspondingly increased for added control.

The S-1, S-2, and S-2S were also marketed in homebuilt kit form, amateur built versions having an 'E' suffix to their designations (S-1E, S-2E, S-2SE), while homebuilt aircraft constructed from drawings are the S-1C (two aileron) and S-1D (four ailerons). Total numbers of amateur-built Pittses is unknown, but probably exceeds 500 and continues to grow, although factory-supplied kits are no longer available. Production of the S-1T, S-2B and S-2S continues at Afton. According to the latest figures some 61 S-1Ss have been built, 64 S-1Ts, 259 S-2As, 196 S-2Bs and 17 S-2Ss.

Homebuilds and rebuilds

The most radical Pitts developments have come from private owners. One S-1S was modified with a nitrous oxide injection system which was said to boost power output to 270.7 kW (340 hp) and gave it a rocket-like vertical climb. Kermit Weeks, famed equally for his aerobatic prowess and his Miami-based historic aircraft collection, went even further, shortening his 149.2-kW (200-hp) S-1S's span and sweeping both wings (the standard aircraft has unswept lower wings) which were set at zero dihedral and incidence, adding gap-sealed symmetrical-section ailerons, a Hoffman CS prop, a new cowling and moving the cockpit 20.3 cm (8 in) aft and fitting it with a streamlined one-piece canopy. This 'S-1W' *Weeks Special* took him to second and third place respectively in the 1978 and 1980 WACs, and inspired his all-new 223.8-kW (300-hp) *Weeks Solution*, perhaps the ultimate aerobatic biplane.

While his little biplane may have been eclipsed at international competition aerobatics by modern monoplanes such as the French CAP 230, German Extra 300 and Russian Sukhoi Su-26M, Curtiss Pitt's old slogan holds true: A Pitts is Something Special.

RFB Fantrainer: High-tech tutor

One of the more unusual lightplanes to emerge in recent years is RFB's Fanliner/Fantrainer series. Powered by a ducted fan, these aircraft offer considerable improvements in fuel efficiency and noise output compared to conventionally powered machines. Most often seen in the civil colours of the company demonstrators, the Fantrainer is mostly tailored to the military light training market.

Watching a Fantrainer perform at a flying display is always a pleasure. Effortless agility, in both the horizontal and vertical planes, is combined with the grace of a ballerina, and with – instead of thunderous noise – a gentle hum like the proverbial sewing machine. But some of us feel uneasy, almost as if we are being cheated. Is it a propeller aircraft or a jet? Perhaps we can say it is a propeller aircraft that likes to pretend it's a jet.

The work was sparked off by the development in the 1970s of ducted-fan propulsors by the British Dowty company. These propulsors, which could be driven by any kind of shaft-driven engine, were very like the fans on the front of modern airline engines, but differed in that the blades were of the variable-pitch type. Alternatively, the propulsor could be regarded as a conventional variable-pitch and reverse-pitch propeller but of reduced diameter, with more than the usual number of blades and enclosed in a profiled surrounding duct. Though it adds weight the duct can increase thrust and efficiency, improve engine cooling, reduce dangers to ground staff, reduce vibration and, perhaps most important of all, dramatically reduce noise.

German interest

RFB, of Mönchengladbach, became attracted by the many advantages of ducted propellers, and quickly schemed a tandem-seat trainer called the AWI 2 Fantrainer. A model was exhibited at the 1970 Hanover air show. The most unusual feature was the overall configuration. Use of a small ducted propeller meant that the entire propulsion system could be in the rear fuselage, aft of the cockpits.

The Fantrainer was evaluated by the Luftwaffe in 1985, the service recommending various alterations. With its tandem cockpit and high performance, the Fantrainer offers the necessary performance for the advanced trainer role at a fraction of the cost of current types in service.

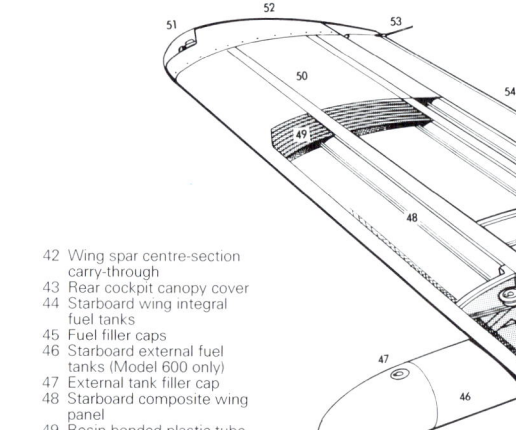

The ducted-fan principle evolved by RFB was first applied to the Fanliner, a light aircraft accommodating two side-by-side. First flight of this, the first prototype, occurred on 8 October 1973. RFB was collaborating with Grumman over development of the type.

RFB Fantrainer 400/600 cutaway drawing key

1. Pitot tube
2. Nose cone
3. Nosewheel bay
4. Nose section keel members, light alloy construction
5. Glassfibre nose skinning
6. Nosewheel doors
7. Nosewheel forks
8. Castoring nosewheel, forward and upward retracting
9. Ground power socket
10. Battery
11. Nosewheel hydraulic damper and retraction jack
12. Electrical system equipment
13. Nose compartment access door
14. Instrument panel shroud
15. Forward fuselage glassfibre skin/secondary structure
16. Rudder pedals
17. Cockpit fresh air intake
18. Fuselage keel unit, light alloy primary structure
19. Fold-out boarding step
20. Front cockpit floor level
21. Control column
22. Student pilot's instrument panel
23. Stand by compass
24. Frameless cockpit canopy, hinged to starboard
25. Side console panel
26. Student pilot's Stencel Ranger zero-zero rocket-assisted escape seat
27. Safety harness
28. Canopy handles, internal and external
29. Canopy latch
30. Engine throttle lever
31. Port side console panel
32. Lower VHF aerial
33. Rear cockpit floor level, 3 in (8 cm) above front
34. Keel unit bracing strut
35. Rear seat rudder pedals and control column
36. Rear instrument panel shroud
37. Canopy centre arch
38. Rear seat throttle lever
39. Cockpit aft sloping bulkhead
40. Rear canopy latch and handles
41. Instructor's Stencel Ranger seat
42. Wing spar centre-section carry-through
43. Rear cockpit canopy cover
44. Starboard wing integral fuel tanks
45. Fuel filler caps
46. Starboard external fuel tanks (Model 600 only)
47. External tank filler cap
48. Starboard composite wing panel
49. Resin-bonded plastic tube structure
50. Glassfibre skin panelling
51. Starboard navigation and strobe lights

So far the Royal Thai Air Force has been the only military customer for the aircraft. A 1982 contract covered 31 Fantrainer 400s and 16 Fantrainer 600s, all but two of which are assembled in Thailand. The GRP wings are being replaced by locally-built metal wings.

This had the important psychological effect of giving the trainer the aura of a jet, even though it was to be quite low-powered and be extremely economical to operate.

To gain experience with ducted-fan propulsion, RFB built the Sirius I and II powered sailplanes in 1971-74. These certainly proved the basic principle. In April 1974 RFB announced it would collaborate with Grumman American Aviation in a light two-seater called the Fanliner. This used various parts, including the wings and spatted tricycle landing gears, similar to those of the Grumman American Traveler. The fuselage, however, was quite new and comprised a neat nose cockpit with side-by-side seats, and a ducted fan aft of the wing driven by an NSU-Wankel rotary engine of 86 kW (114 hp). At the rear was a T-type tail. The prototype first flew on 8 October 1973, before the project was made public, and RFB said the results proved that, in addition to all the other benefits, a ducted fan used engine power more efficiently than a conventional propeller.

In March 1975 the Federal German government announced that it had awarded RFB a DM7.5-million contract to design and build two prototypes of the Fantrainer. These were intended each to be powered by twin 150-hp NSU-Wankel engines geared down to a single ducted fan, with airbrakes incorporated in the fan shroud. RFB had proposed that the Fantrainer could replace the Piaggio P.149D as the Luftwaffe's standard primary pilot trainer. It was expected that pupils would be able to transition directly from the Fantrainer to the Alpha Jet. The two prototypes were to be evaluated with this objective in mind.

Forward-swept wing

RFB duly built the two prototypes. Both had completely fresh airframes, owing nothing to the Fanliner. Structurally they were remarkable for the wing, made almost entirely of glassfibre and a plastics sandwich, and with 6° forward sweep to balance the long tandem cockpits ahead of the front spar. Electrically driven Fowler flaps were fitted, and the tricycle landing gear was also electrically driven, all units folding into the fuselage, the main legs folding straight upwards so that the wheels were actually housed in the roots of the wings. Most of the fuselage and T-type tail was aluminium alloy. All

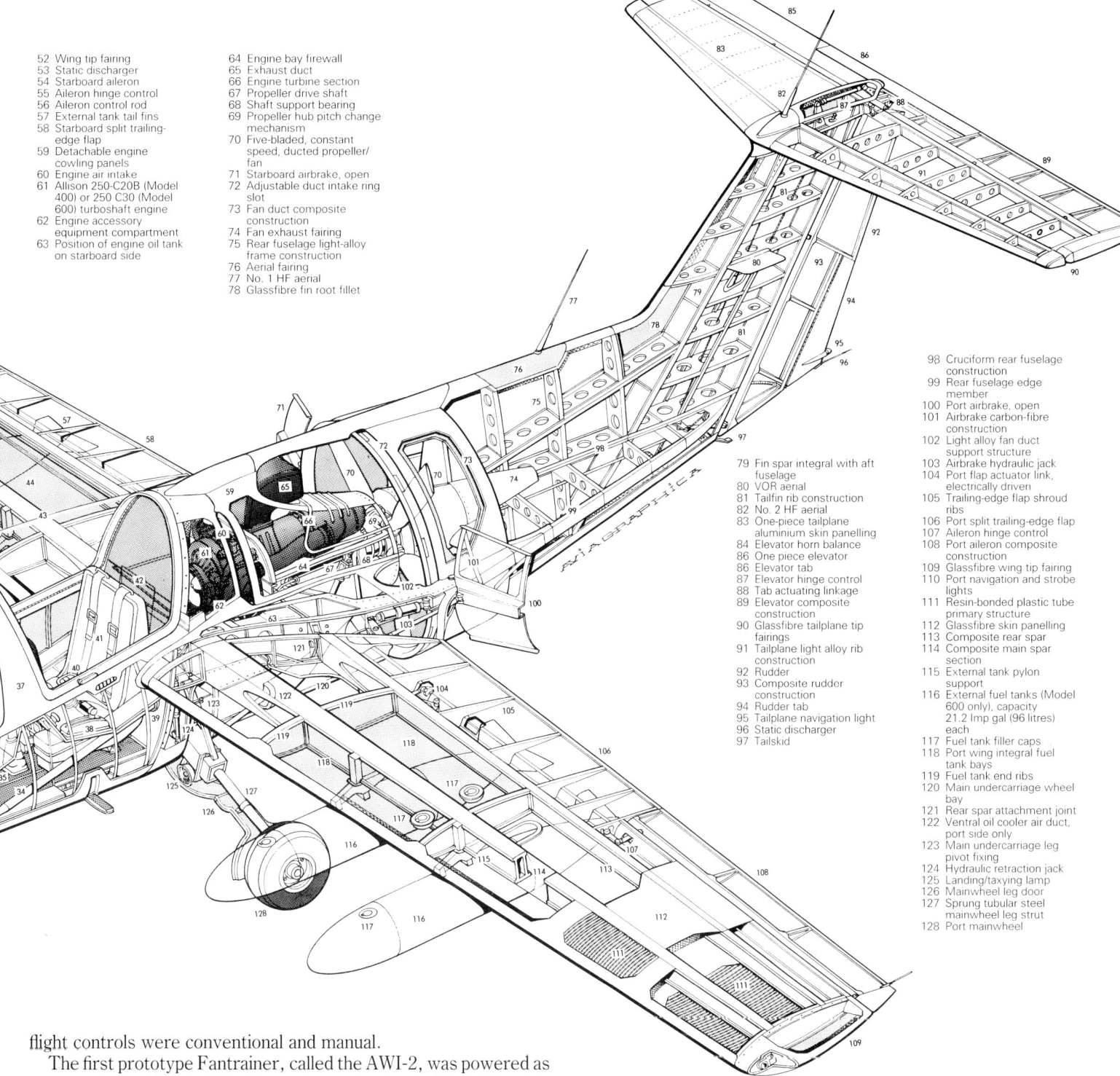

52 Wing tip fairing
53 Static discharger
54 Starboard aileron
55 Aileron hinge control
56 Aileron control rod
57 External tank tail fins
58 Starboard split trailing-edge flap
59 Detachable engine cowling panels
60 Engine air intake
61 Allison 250-C20B (Model 400) or 250 C30 (Model 600) turboshaft engine
62 Engine accessory equipment compartment
63 Position of engine oil tank on starboard side
64 Engine bay firewall
65 Exhaust duct
66 Engine turbine section
67 Propeller drive shaft
68 Shaft support bearing
69 Propeller hub pitch change mechanism
70 Five-bladed, constant speed, ducted propeller/fan
71 Starboard airbrake, open
72 Adjustable duct intake ring slot
73 Fan duct composite construction
74 Fan exhaust fairing
75 Rear fuselage light-alloy frame construction
76 Aerial fairing
77 No. 1 HF aerial
78 Glassfibre fin root fillet
79 Fin spar integral with aft fuselage
80 VOR aerial
81 Tailfin rib construction
82 No. 2 HF aerial
83 One-piece tailplane aluminium skin panelling
84 Elevator horn balance
86 One piece elevator
86 Elevator tab
87 Elevator hinge control
88 Tab actuating linkage
89 Elevator composite construction
90 Glassfibre tailplane tip fairings
91 Tailplane light alloy rib construction
92 Rudder
93 Composite rudder construction
94 Rudder tab
95 Tailplane navigation light
96 Static discharger
97 Tailskid
98 Cruciform rear fuselage construction
99 Rear fuselage edge member
100 Port airbrake, open
101 Airbrake carbon-fibre construction
102 Light alloy fan duct support structure
103 Airbrake hydraulic jack
104 Port flap actuator link, electrically driven
105 Trailing-edge flap shroud ribs
106 Port split trailing-edge flap
107 Aileron hinge control
108 Port aileron composite construction
109 Glassfibre wing tip fairing
110 Port navigation and strobe lights
111 Resin-bonded plastic tube primary structure
112 Glassfibre skin panelling
113 Composite rear spar
114 Composite main spar section
115 External tank pylon support
116 External fuel tanks (Model 600 only), capacity 21.2 Imp gal (96 litres) each
117 Fuel tank filler caps
118 Port wing integral fuel tank bays
119 Fuel tank end ribs
120 Main undercarriage wheel bay
121 Rear spar attachment joint
122 Ventral oil cooler air duct, port side only
123 Main undercarriage leg pivot fixing
124 Hydraulic retraction jack
125 Landing/taxying lamp
126 Mainwheel leg door
127 Sprung tubular steel mainwheel leg strut
128 Port mainwheel

flight controls were conventional and manual.

The first prototype Fantrainer, called the AWI-2, was powered as planned by two Wankel engines geared to a seven-bladed Dowty-Rotol constant-speed ducted fan. It made its first flight on 27 October 1977. It was followed by the slightly modified second prototype, designated as the ATI-2, which first flew on 31 May 1978. It differed mainly in that the Dowty-Rotol fan was driven by a 313-kW (420-shp) Allison 250-C20B turboshaft engine. Unfortunately this aircraft crashed in its 63rd flight hour on 7 September 1978.

By this time RFB had rebuilt the first Fantrainer as the prototype of the proposed Fantrainer 400 for the Luftwaffe, with an Allison C20B engine similar to that used in the No. 2 machine. In 1979 RFB teamed with Vought to propose a version to be known as the Vought 538 to meet the USAF's need for a T-37 replacement. This was to be powered by an Allison 250 C30 or, alternatively, a Textron Lycoming LTS101 engine, rated at 447 kW (600 shp). In the event the USAF picked the Fairchild T-46A, which was then itself cancelled.

This was a blow to RFB, and a much more severe blow was the consistent refusal of the Luftwaffe to adopt the Fantrainer. In June 1980 the second prototype was upgraded by being refitted with an Allison C30 engine rated at 650 shp. This made a major difference to the all-round performance, compared with the original prototype fitted with Wankel engines. For example the maximum sustained rate of climb at sea level was about 300 m/min (984 ft/min) originally; the Fantrainer 400 with the single C20B engine could climb at 472 m/min (1,550 ft/min), and the Fantrainer 600 with the C30 engine could reach 914 m/min (3,000 ft/min).

Despite the company's inability to sew up its own air force as a customer, the survival of the programme was assured by the award of a contract, in August 1982, for Fantrainers for Thailand. The Royal Thai Air Force decided to buy both types of Fantrainer, to the tune of 31 Fantrainer 400s and 16 Fantrainer 600s. Under the terms of the contract RFB was to build and test two of the Thai Fantrainers, subsequently supplying the remaining 45 aircraft in CKD (component knocked down) form; in other words, each aircraft should be shipped out as a kit for assembly and test in Thailand. This was a key

RFB's house colours adorn this Fantrainer 600, used as part of the test programme and to sell the product. The unusual design results from the desire to provide handling similar to tactical aircraft and incorporate the ducted-fan technology and tandem seating. While RFB figures show the Fantrainer to be a very efficient, economic, sweet-handling and environment-friendly aircraft, and despite spirited displays in front of major customers at such shows as Paris and Farnborough, Thailand is at present the only purchaser. However, the fact that Thai students progress straight from the Fantrainer to the Northrop F-5 is some measure of its potential as a trainer.

Specification
Rhein-Flugzeugbau Fantrainer 600
Type: two-seat basic trainer
Powerplant: one Allison 250-C30 turboshaft rated at 485 kW (650 hp) driving a five-bladed Hoffmann constant-speed ducted fan
Dimensions: wing span 9.7 m (31 ft 11 in); length 9.48 m (31 ft 1 in); height 3.0 m (10 ft 4 in); tailplane span 3.59 m (11 ft 9 in); wing area 14.0 m^2 (150.7 sq ft)
Weights: empty 1160 kg (2,557 lb); maximum take-off 2300 kg (5,070 lb) (non-aerobatic), 1600 kg (3,257 lb) (aerobatic); maximum fuel load 736 kg (1,623 lb) with drop tanks
Performance: never exceed speed 555 km/h (345 mph); maximum level speed 417 km/h (259 mph); cruising speed at 3050 m (10,000 ft) 370 km/h (230 mph); stalling speed 113 km/h (71 mph); maximum rate of climb 914 m (3,000 ft) per minute; take-off run 250 m (820 ft); landing run 250 m (824 ft); service ceiling 7620 m (25,000 ft); range 1037 km (645 miles) with maximum internal fuel; endurance 7 hours; g limits +6/-3

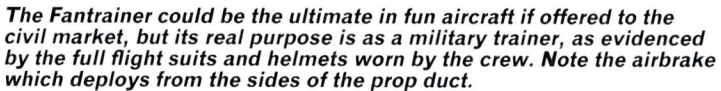

The Fantrainer could be the ultimate in fun aircraft if offered to the civil market, but its real purpose is as a military trainer, as evidenced by the full flight suits and helmets worn by the crew. Note the airbrake which deploys from the sides of the prop duct.

factor in the deal, because the Royal Thai Air Force was most eager to acquire the capability of initially assembling and later undertaking design work on modern aircraft. The Fantrainer was picked as being an aircraft with the right level of technology, besides having the qualities needed to serve as a low-cost trainer of RThAF pilots.

The first production Fantrainer, which was also the first Model 600, made its maiden flight at the Mönchengladbach flugplatz on 12 August 1984. This was also one of the first aircraft to fly with the uprated Allison C30 engine, which introduced an advanced single-stage centrifugal compressor which is lighter, more robust, more efficient and much simpler than the multi-stage axial compressor of earlier versions. This aircraft and the surviving prototype were allocated to the certification programme, which was completed successfully. FAA certification to FAR.Pt 23 in both utility and aerobatic categories was awarded in May 1985.

Compared with the prototypes the production Fantrainers have numerous, mostly minor, changes. The fan reduction gearbox is redesigned to permit the full shaft power of the C30 engine and the full thrust of the fan to be used. The original Dowty-Rotol fan is replaced by a five-bladed constant-speed Hoffmann product. The engine air inlets were moved above the wing, and the fuselage was lengthened by 150 mm (6 in). The tandem cockpits were upgraded in equipment detail, and the seats are Stencel Ranger rocket type, with car-

The second prototype Fanliner, flying in September 1976, featured a considerably redesigned fuselage offering even better visibility. The second prototype employed the wings and tailplane of the Grumman Cheetah, with power coming from a modified Audi/NSU KM 871 Wankel rotary engine.

A pair of Fantrainers poses for the camera, displaying the futuristic lines of the type. The ducted-fan principle offers great propeller efficiency while reducing noise to a minimum. The Fantrainer offers sparkling performance on relatively little power.

The Fantrainer prototypes were built to a German MoD contract, the first of which (AWI-2) first flew on 27 October 1977 powered by two Wankel engines. The second (ATI-2) flew in May 1978, this time powered by an Allison 250-C30 turboshaft.

tridge-fired ejection seats optional. The canopy was enlarged to improve the all-round view. In the Thai aircraft the avionics are of the King Gold Crown type. Though the flaps were left with electric actuation, a full hydraulic system was added to operate the revised landing gear and the large hinged airbrakes on each side of the fan duct. Structural commonality between the Fantrainer 400 and 600 is in the order of 92 per cent.

Armed Fantrainer

According to the contract the first two aircraft were duly delivered to Bangkok in October 1984. All 45 kits followed between the end of 1984 and early 1988. Meanwhile the Royal Thai AF SWDC (Science and Weapon Systems Development Centre) completed the design of metal wings, aerodynamically similar to the original German glass-fibre wings, but incorporating two hardpoints under each wing for the attachment of pylons. These wings went into production at the SWDC in 1987, and all 45 sets were scheduled to have been completed by early 1990 (19 pairs had been delivered by early 1988). Each wing has a two-spar structural box forming an integral fuel tank, with corrugated skin over the mainwheel bays near the wing roots. Compared with the original Fantrainer the metal-winged aircraft have slightly reduced internal fuel, and empty weight is marginally increased, to 1680 kg (3,700 lb).

In 1985 one Fantrainer 400 and two 600s underwent further evaluation by the Luftwaffe. As a result some modifications were recommended, including a revised instrument layout. The fan was modified to have blades with slightly swept-back and twisted tips, to reduce the already very low noise level still further. The first flight with the modified fan took place on 16 May 1986.

When, in October 1989, the Luftwaffe transferred primary training responsibility to the Bonanza-equipped Lufthansa training school in the USA, any further chance of selling the Fantrainer to the German armed forces finally evaporated. Nevertheless, in November 1980, MBB (responsible for RFB and now part of the Deutsche Aerospace consortium), together with Rockwell International, announced that they would jointly build and submit two prototypes of the turbofan-powered Fan Ranger pure jet variant for the USAF/USN JPATS requirement to replace the Cessna T-37. Powered by one Pratt & Whitney Canada JT15D-4 engine, but abandoning the ducted fan powerplant, the first flight of the Fan Ranger (later Ranger 2000) took place on 15 January 1993, but the JPATS winner was the Pilatus PC-9 Mk II, then ordered as the Raytheon T-6A Texan II.

Chapter 32

Reno-Formula One

•

Saab 340 & 200

•

Sabreliner

•

Shorts 330

•

Shorts 360

•

Shorts Civil Flying Boats

Reno: Formula One

While the Mustangs and Sea Furys battle it out in imposing fashion, a no less fierce and skillful competition is being waged in the Formula One class, albeit on a smaller scale. These racers are prone to the same vicissitudes of race day, the same trials and tribulations of their larger brethren, as they battle it out at Reno.

While few would deny that the Unlimited racers are the principal attraction at Reno, some of the most closely fought races occur in the Formula One or International Formula Midget (IFM) class. Unlike the Unlimiteds, all of which derive from military fighter aircraft, the IFMs are pure-bred racing aeroplanes, equating to Formula One racing cars. Their performance may be only half that of the 'big iron' warbirds but, remarkably, is achieved on less than five per cent of the horsepower produced by a typical Unlimited racer.

Formula One air racing had its origins in 1939 when a group of American racing pilots led by Art Chester, Benny Howard, Tony LeVier and Keith Rider devised a new class of racing aeroplane based on a standard powerplant of 190 cubic inches capacity. Their aim was to reduce the high cost of designing, building and operating racing aircraft, which in pre-war National Air Race days were often at the forefront of technology and not infrequently could out-perform their front-line military contemporaries.

America's entry into World War II prevented the new breed of racer from appearing until 1946, by which time technological progress spurred by wartime necessity had pushed the barrier of pure speed forever beyond the reach of amateur constructors. The new Formula One racers thus became a means whereby race pilots whose budgets could not stretch to surplus military aircraft could continue racing, and carry on the pre-war tradition of innovative design and 'backyard' building.

The first Midget races held in 1947 were sponsored by the Goodyear Tire & Rubber Company. Continental Motors Corporation later took up the sponsorship, though for many years midget racers were popularly known as 'Goodyears'. When air racing seemed certain

The Formula One hangar at Reno is a mass of activity as crew prepare their racers for competition. Meticulous care is taken with the aircraft, as their tremendous speed is gained not from giant engines, but from super-streamlined structures, with even the smallest rivet taped over to streamline it.

Jim Sharp was the 1982 champion and his Cassutt Racer AeroMagic was much fancied. Here Sharp rounds pylon 6 during the Bronze event, which he won at a speed of 230.513 mph.

The Formula One racers launch straight from the ground, entering a pace lap before starting the race proper at a 'scatter' pylon. Within around 7 or 8 minutes, the race will be over and the winner declared.

Formula One racers are very delicate pieces of machinery and require the utmost care on the ground for peak performance. This Owl Racer is being transported slowly around the Reno flightline before a race.

Heading for home after a long day, Gary Hubler's Cassutt Black Magic failed to start the Bronze race despite the best efforts of its crew. It was loaded back onto its trailer for the journey straight back to Caldwell, Ohio.

to disappear forever following the tragic crash of Bill Odom's Mustang at Cleveland in 1949 it was only the persistence of the Midget racer pilots which kept the sport alive and made possible its revival at Reno in 1964.

The 'Midget' formula devised in 1946 was almost identical to today's IFM rules which limit IFM racers to: a 200 cubic inch displacement engine, fixed pitch propeller, minimum wing area of 6.13 m^2 (66 sq ft), empty weight of 226.8 kg (500 lb), fuel tanks with a minimum capacity of 56.7 litres (15 US gal), and fixed landing gear, and also imposes standards for design, materials and workmanship.

Within the confines of the IFM formula designers have been encouraged to exploit contemporary approaches to materials and aerodynamics. Several unorthodox racers have appeared over the years, including tailless, twin boom and pusher designs, best known of which was Texan Jim Miller's *Texas Gem*, unique in its time among racers for its four-bladed pusher propeller, sharply swept T-tail and nosewheel landing gear.

However, the typical IFM racers look not so very different today than they did 45 years ago. Indeed, the most successful IFM

Kirk Hanna was clocking 224.386 mph in his Owl Racer Wise Owl in the qualifying Gold class races. However, on reaching the final he was disqualified for low flying.

racer in the sport's history was Ray Cote's famous midwing *Shoestring* which retired after taking the Formula One Championship at Reno in 1981, having first competed in the hands of Bob Downey at the 1949 National Air Races at Cleveland. Corporate pilot Ray Cote was the undisputed king of IFM racers, having taken the Reno title 11 times, including eight successive victories in *Shoestring* between 1968-1975. He held the Reno IFM class qualifying record in the aircraft at 396 km/h (246.06 mph). Another regular competitor, travelling much more slowly than in his past career, was former NASA astronaut 'Deke' Slayton, who flew the elegant elliptical-winged *Stinger*.

The IFM course at Reno is 5 km (3.1 miles) long, with six pylon turns. Unlike the T-6s and Unlimited events which feature air starts, IFM races begin with a 'race horse' start, eight racers taking off in groups of two for a pace lap before rounding the 'scatter' pylon for the start of the race proper. IFM races are short-lived, a typical eight-lap event taking just over seven minutes to complete. The essence of top class IFM racing – the only truly international air racing class, with regular competitions being held in Europe as well as the United States – therefore lies in careful airframe design, perfect engine/propeller matching, meticulous preparation and above all skilful piloting, for the short course has only two short straights and offers few opportunities to make up lost ground through pure speed.

Since the first Formula One event at Reno in 1964, won by Bob Porter in the veteran Miller *Little Gem* at 311.3 km/h (193.44 mph), race speeds have risen to close on 402.3 km/h (250 mph). In recent times the current record has stood at 400.8 km/h (249.01 mph) and was set by three-time Reno IFM champion John Parker in his svelte *American Special* in 1980 – a remarkable achievement for an aircraft powered by an engine designed to produce only 74.5 kW (100 hp).

Right: **Shoestring** *is one of the best-looking and most successful Formula One designs, seen here rounding a pylon. With six turns in a 5-km course, the turns are most important at Reno, where piloting technique comes to the fore. Only short straights are available to make up speed after a bad turn.*

Saab 340 and 2000

Before 1982, the last airliner that Saab had built was the Scandia of 1946 but, after nearly 40 years of successful military aircraft, the Swedish company felt that it had a world-beating civilian design to offer. Their faith in the Saab 340 has not been misplaced, as the aircraft has now far outsold all its rivals. The new-technology Saab 2000 should see this market leadership maintained well into the 21st century.

In the late 1970s the Swedish company Saab-Scania AB conducted a series of project and market studies for a commercial aircraft in the rapidly expanding commuter airliner sector. Burgeoning development and launch costs prompted the company to seek a risk-sharing partner in the project, and since North America represented at least half of the potential market for such aircraft it was naturally to the US industry that Saab looked.

Thus it was that, in June 1979, Saab and Fairchild Industries Inc. – manufacturer of the successful Merlin and Metro series of twin turboprop business and commuter aircraft – signed a preliminary agreement for a six-month joint study into the market potential and engineering feasibility of launching a 30-seat regional airliner, then known as Project 3000. Saab had already produced preliminary designs for a high-wing turboprop intended for both military and commercial roles, but by the end of 1979 a team of Saab and Fairchild engineers working at the Fairchild-Republic company's plant at Farmingdale, Long Island, had defined a new low-wing aeroplane exclusively aimed at airline use.

On 25 January 1980 first details of this aircraft were revealed in Stockholm: it was to be a 34-seat design which would employ extensive new technology in both airframe and engines to achieve significant gains in operating economics. On that day Saab and Fairchild signed a historic agreement jointly to develop and manufacture the aircraft – the first fully collaborative commercial aircraft venture between a European and an American company. The aircraft was to be known as the Saab-Fairchild SF-340, and would be the first to be designed according to the common worthiness standards (Joint Airworthiness Requirement 25) devised by the (then) nine signatories to the multi-national Joint Aviation Authorities agreement, while also meeting the requirements of the US Federal Aviation Administration's FAR 25, the highest certification standard in the US system.

A full project definition phase began immediately, with formal agreement drawn up between the two manufacturers for work-sharing: Fairchild would be responsible for the design and manufacture of the wings, tail unit and engine nacelles; Saab, which was to fund 75 per cent of development costs, would handle design and manufacture of the fuselage, systems integration, flight testing and certification, and would also undertake all final assembly of production aircraft at a new 25,000 m² plant at its Linköping factory. Those aircraft destined for the North American market would be ferried to Fairchild for completion and outfitting before customer delivery. A joint company was set up to handle worldwide marketing and sales.

At the Farnborough air show in September 1980 Saab-Fairchild announced formal go-ahead for the SF-340, and revealed further details of the design. The airframe would make extensive use of modern glassfibre/Kevlar/carbonfibre composites and sophisticated metal-to-metal bonding techniques to provide a longer fatigue life and enhance corrosion resistance. The configuration chosen was for a round-section fuselage with a cabin pressur-

Swiss regional operator Crossair was the first to put the 340A into service, and was again the launch customer for the 340B, the first example of which is seen here. In 1998 Crossair was operating 15 and 25 examples respectively of the Saab 340B and 2000.

Above: The SF-340QC was a quick-change passenger/freight version of the SF-340A. Five were ordered by Finnaviation, which uses them for commuter services with 34 seats by day, and as freighters by night.

Right: Then known as the Saab-Fairchild SF-340, the first of the breed flew on 25 January 1983. It wore Air Midwest colours on one side and Crossair colours on the other. Air Midwest currently flies three on its services from Wichita, Kansas, as part of the US Air Express network.

Below: In 1984 the prototype Saab-Fairchild SF-340 was fitted with winglets to test their effect on induced drag. These were not adopted for production aircraft. Note the test instrumentation nose boom, and the drogue sensor which deployed from the tailcone.

Below: KLM runs a sizeable domestic and regional subsidiary in the form of KLM Cityhopper, an amalgamation of NLM and Netherlines. Among the equipment are 12 Saab 340Bs and the fourth SF-340A leased from Swede

ised to maintain sea-level pressure up to 3685 m (12,000 ft) and seating 34/35 passengers three-abreast in a 1+2 arrangement. The high aspect ratio wing employed a modified NASA-developed low-drag airfoil. General Electric CT7 turboprops of 1214 kW (1,630 shp) driving Dowty four-bladed composite propellers were selected to power the SF-340, which was to have a King Gold Crown or Collins Pro-Line all-digital avionics system with autopilot and flight director as standard equipment – unique among small airliners at the time. A 12/19-seat executive/corporate transport version was to be developed, primarily for the US market. Launch customers for the SF-340 were Crossair of Switzerland, which placed an order for 24 aircraft two months after the programme was officially confirmed, and Comair of Cincinnati, which ordered 16.

New arrival

The prototype SF-340 (SE-ISF) was rolled out at Linköping on 27 October 1982 in front of 600 invited guests including King Carl XVI Gustaf of Sweden, and made its first flight on 25 January 1983, the third anniversary of the signing of the Saab-Fairchild agreement, in the hands of Saab test pilots Per Pellebergs and Erik Sjöberg. The second prototype (SE-1SA), and a pre-production aircraft (SE-ISB) made their first flights on 11 May and 25 August respectively. These, and two non-flying static and fatigue test airframes, were used for the test and certification programme, which had accumulated a total 1,730 flight hours by the time Swedish Board of Civil Aviation certification was granted on 30 May 1984. On 29 June the Joint Aviation Authorities and FAA approved the SF-340, followed by the Australian Civil Aviation Authority in October, making 12 approvals in all within the space of six months – an unique achievement.

Meanwhile, the first production SF-340 (N340CA for US carrier Comair) had flown on 5 March 1984, and the first delivery (HB-AHA) was made to Crossair on 6 June, enabling the Swiss airline to introduce the SF-340 into revenue service on its Base-Mulhouse to Paris route on 15 June. Comair's first SF-340 was delivered in July and entered service in the following month. The first corporate SF-340 entered service with the Mellon Bank of Pittsburgh, Pennsylvania in August 1985.

The SF-340's service entry was marred by technical problems, which resulted in three inflight engine shut-downs on Crossair's first two aircraft within weeks of them going into operation. SF-340s were briefly grounded in September 1984, but returned to service in the following month after modifications had been made to their General Electric CT7-5A engines. However, compressor problems resulted in a further grounding in November, again rapidly solved by the engine manufacturer.

At the Paris air show in June 1985 Saab-Fairchild launched a re-engined SF-340 powered by 1293-kW (1,735-shp) CT7-5A2 engines driving larger-diameter propellers and with maximum take-off weight increased from 11793 kg (26,000 lb) to 12872 kg (27,275 lb); aircraft already delivered were progressively modified to the new standard.

Fairchild falls out

Later that year, in October, Fairchild industries announced its withdrawal from the SF-340 programme, a culmination of its higher than anticipated costs and cancellation of the potentially lucrative Fairchild T-46A New Generation Trainer programme prompting the US company to abandon the civil aviation market altogether. On 1 November the SF-340 became a wholly Swedish-owned operation, with manufacture of those components previously supplied by Fairchild being progressively transferred to a new plant at Linköping. Once again the aircraft was beset with problems, when, following a number of inflight engine shut-downs while flying in icing conditions, the FAA grounded the aircraft for two weeks in December 1985 for modifications to provide

Right: Transportes Aereos Neuquen is a small Argentinian regional carrier based at Neuquen. Its fleet consists of two Turbo Commander 690Bs, two SA 227AC Metro IIIs and this solitary SF-340A. The Saab is leased from the manufacturer and configured for 35 seats.

Below: Saab 340s have been delivered to many small operators that wish to progress to a larger aircraft for their regional services. Typical is Linea Aérea Entre Rios of Parana, Argentina, which had this 340A on order.

continuous ignition for the CT7s. At the end of what had been a year of mixed fortunes, Saab had delivered a total of 40 SF-340s which had accumulated 50,000 hours of flight time in passenger service. Production at Linköping had then stabilised at 30 aircraft per year.

During 1986 Saab delivered the first SF-340QC quick-change cargo-passenger aircraft to the Finnish carrier Finnaviation. In the summer of 1987 – which marked the 50th anniversary of Saab's foundation – the company announced a further increase in maximum take-off weight to 12701 kg (28,000 lb), most aircraft again being retrofitted to the new standard, and on 4 September delivered the 100th production SF-340 to Swedish carrier Salair. That month also saw the final severing of Saab's links with Fairchild, and thereafter the aircraft was known simply as the Saab 340A. Concurrently Saab announced development of the Saab 340B.

This model, which replaced the Saab 340A on the production line from serial number 180 onwards in August 1988 and remains in current production, is optimised for hot-and-high operating conditions. Major changes include installation of General Electric CT7-9B turboprops with a take-off power rating of 1304 kW (1,750 shp) and automatic power reserve rating (APR) of 1393 kW (1,870 shp); a larger-span tailplane to permit a wider centre of gravity range, and maximum take-off weight increased to 12700 kg (28,500 lb). These changes offer enhanced payload/range and high altitude performance, better field performance, an increase in cruising speed from 272 kt (502 km/h; 312 mph) to 285 kt (526 km/h; 327 mph), and provide airlines with greater flexibility in interior configuration. In the latter case Saab, in conjunction with Metair Aircraft Ltd of the United Kingdom, which has been responsible for outfitting and painting all but the first six production aircraft, developed a new cabin interior known as Generation II which incorporates larger overhead baggage bins, improved ventilation, a redesigned aft lavatory, and cabin wall and ceiling panels made of materials which meet the latest ICAO smoke and heat release regulations. This interior was introduced on Saab 340As in the summer of 1988 and is standard on all 340Bs.

First for Crossair

Launch customer for the Saab 340B was again Crossair of Switzerland, which ordered five (later increased to 10) as soon as the improved aircraft was announced.

The second prototype Saab 340 was used as development aircraft for the new model, first flying with CT7-9B engines at the time of the 340B's announcement in September 1987. The first production aircraft flew in April 1989. Certification was obtained on 3 July 1989 and first delivery made to Crossair on 15 September.

1989 proved to be a significant year in the Saab 340 programme. Orders, which had previously been building very slowly, accelerated rapidly, passing 200 early in the year and including significant commitments from major carriers such as Air France and the Dutch national airline KLM, whose Saab 340Bs are operated on its Cityhopper services and were the first Saab 340s to carry provisions for full hot meals service outbound and return, plus duty free meals, thanks to the higher operating weights of the upgraded model. In May 1989 total orders rose to 300 when AMR Eagle, the regional carrier arm of American Airlines, placed a firm order for 50 Saab 340Bs plus options on a further 50 in what was at that time the largest single contract ever placed by a regional airline. This was followed in October by an order for 25 aircraft from Northwest Airlink.

American performance

The Saab 340B made its operational debut in the United States on 15 March 1990 with AMR's Nashville Eagle service. Two months later the total flight time recorded by the 185-strong Saab 340 fleet exceeded one million hours, in 1.3 million flights averaging seven each day with a despatch reliability of nearly 99 per cent. The 200th delivery was made in August 1990. By mid-1993 deliveries of the Saab 340A/B totalled over 320, more than half for US customers, with the remainder from operators in Europe, Australasia, Latin America and the Far East.

Production has been increased to 50 per year, and has been diversified between plants at Dagsberg (forward fuselage), Odeshog (rear fuselage), and Kramfors (sub-assemblies), with Westland in the UK supplying engine nacelles and Per Udsen of Denmark making cabin sections. Wings, tail units, final assembly

Left: The Saab 340B is outwardly similar to the 340A but has the CT7-9B engines with an additional power reserve for hot-and-high operations. A wide-span tailplane gives greater longitudinal stability, and thereby a wider range of gravity centre. This in turn allows greater flexibility of internal layout.

Right: A rapidly growing commuter market can be found in the Pacific Rim area. Saab's first sale in this region was to Taipei-based Fromosa Airlines, which took three SF-340As. Two are powered by the CT7-5A2 and one by the CT7-9B.

Right: Saab's most lucrative order has come from the American Eagle network, which bought 115 examples of the Saab 340B for operation by Flagship Airlines and Wings West. American Eagle's loyalty to Saab was further attested by an option to buy 50 Saab 2000s, but this option was not exercised.

Below: The Saab 2000 first prototype is seen during its roll-out ceremony. The 2000 is obviously an enlarged 340, employing the same fuselage cross-section but with extended fuselage and wings. Of note are the six-bladed propellers, designed to turn slowly and consequently produce less noise than standard airscrews.

Below: With military serial 100001, this Saab 340B was delivered to the Swedish air force under the designation Tp 100. It flies with the Royal Flight based at Stockholm-Tullinge on VIP transport duties. Note the additional communications aerials.

and flight testing are conducted at Linköping. After an inauspicious start, the Saab 340 has since captured more than 37 per cent of the worldwide market and 50 per cent of the European market for commuter airliners in its class.

In addition to the standard 37-passenger commuter configuration Saab also offers QC, Combi (19 passengers and 1500 kg (3307 lb of freight) and VIP versions of the 340B. The first VIP version was delivered to the Swedish air force on 23 February 1990. Designated Tp 100 in military service, it is used for transporting members of the Swedish royal family and senior government ministers.

Other special missions of the aircraft proposed, but not yet launched, include the 340T military transport, 340EZP economic zone protection/maritime surveillance aircraft, 340HQ airborne battlefield command post, 340FI flight inspection/navaid calibration aircraft and 340AEW airborne early warning platform which continues in development.

Known as the ERIEYE, and fitted with an Ericsson-developed radar, an order for a single aircraft with options on a further five has been received from the Swedish air force. First delivery will be in early 1995 and Saab forsees a small, but lucrative, market for the type around the world.

Orders for the Saab 340 stood at 394 by June 1993, thus giving Saab a convincing lead over its main rivals, the DHC-8-100 and EMB-120.

On 8 June 1993 the 100th aircraft for American Eagle (c/n 340) was handed over at Linköping. Saab is continuously improving the 340 with plans afoot to fit a 0.6-m (1-ft 10-in) wing extension to greatly improve performance from short runways or hot-and-high airfields. Another key programme is the SMOP (Saab 340 Maintenance Optimising Program) which seeks to increase the time between A-, B- and C- checks by a significant margin, ultimately doing away with B-checks altogether.

From the mid-1980s Saab had been studying possible developments of the 340 to meet the needs of regional carriers demanding greater speed and more seating capacity on high load factor or 'hub bypass' services. On 15 December 1988 the company announced its decision to develop a new 50-seat high-speed airliner, for which the designation Saab 2000 was chosen to reflect the 21st Century technology which was to be embodied in its design. Once again, Crossair became launch customer, confirming on the day of Saab's announcement an order for 25 Saab 2000s with options on a further 25.

The Saab 2000 retains the fuselage cross-section and many of the systems of the 340B, but has a longer fuselage with cabin accommodation for up to 58 passengers in three-abreast seating with a single aisle. The same advanced airfoil employed on the 340 series is retained, but in a wing increased in span by 15 per cent and in area by 33 per cent. Powerplant is two Allison GMA 2100A turboprops, flat-rated to 3096 kW (4,152 shp) for take-off, driving slow-turning Dowty propellers with six scimitar-shaped all-composite blades. The engines are mounted further outboard than on the 340, to reduce cabin noise levels by 10 dB below those of the smaller aircraft, and are equipped with full-authority digital engine control (FADEC). The aircraft features an advanced 'all glass' flight deck with a six-CRT EFIS display incorporating engine indicating and crew alerting system (EICAS), and an integrated suite of Rockwell-Collins Proline 4 avionics.

Though both Saab's new competitors in the 50-seat market, Canadair and EMBRAER, elected to develop jet-powered aircraft, Saab decided that it could provide equivalent performance over sectors anywhere between 160-1610 km (100-1,000 miles) with a turboprop. Maximum operating altitude is 9450 m (31,000 ft), and the Saab 2000 is capable of reaching its 6096-m (20,000-ft) cruising altitude in under 11 minutes, as compared with 18 minutes for medium twin-jets and 25 minutes for current turboprops.

The Saab 2000's airframe, which has been developed with extensive use of state-of-the-art CAD/CAM (Computer Aided Design/Manufacture) techniques, will have a service life of 60,000 flight hours or 750,000 landings. The forward fuselage is built at Saab's Dagsberg plant, centre-section at Malmo, and the rear assembly produced by Westland at Yeovil. Valmet in Finland provides the fin, rudder, horizontal tail and elevators. CASA builds the wings at their San Pablo factory, and hopes to use this same wing design in their own 72-seat CASA 3000. General Motors has taken a 50 per cent stake in Saab, and provides the GMA 2100A powerplant which is built around the T406 turboshaft and gearbox of the T56. These engines are primarily responsible for the aircraft's low cabin noise level of 76 dB, though their position has also been moved out further along the wing relative to those of the Saab 340.

The cockpit layout is arranged around the six CRTs of the Rockwell-Collins EFIS, with standard back-up instruments located at the base of the main panel. Saab resisted the temptation to fit side-stick controls, opting instead for more familiar control columns.

Saab 2000 on stage

Roll-out of the first aircraft (SE-001) took place at Linköping on 14 December 1991, with first flight on 26 March 1992. The test programme is now based at Skavsta Airport, some 100 km (62 miles) north-east of the company's main Linköping base, to avoid conflicting with the Gripen fighter project. The second aircraft (SE-002) flew on 3 July 1992, while the third followed swiftly on 28 August 1992. So far nearly 1,200 hours of flight trials have been chalked up. Aircraft no. 2 is involved in engine and performance tests, including adverse weather operations, while the third Saab 2000 is being used for accoustic trials, including Saab's new active noise-suppression system to reduce passenger cabin sound even further. This system, also planned for use with the Saab 340, uses a series of 24 loudspeakers and 47 microphones housed in the cabin which re-broadcast an inverse-wave nullifying intrusive vibration, airflow and engine noise. The effect of the new system has been listened to 'turning

Performance and new technology are two of the Saab 2000's greatest assets, as the manufacturer has graphically demonstrated in the sales effort by closely identifying it with Saab's new JAS 39 Gripen fighter.

the cabin volume on and off', and plans are well underway to apply the system to the Saab 340,

The fourth aircraft (flown March 1993) is being used for function and reliability tests, flying to destinations around Europe. Saab 2000 no. 5 had flown by June, and several other aircraft are on the Linköping line.

Saab has always promoted the aircraft's high speed and the smooth 1,000-hours-plus flight-test programme has vindicated the promise of a 360-kt (665-km/h; 413-mph) cruise speed. In fact, the Saab 2000 routinely flies several knots faster than this. On 19 May 1993 the type established a world climb record for its class, by reaching 9000 m (30,000 ft) in just seven minutes and 59 seconds. The previous holder was a Grumman E-2C Hawkeye. This was ahcieved in far from optimum conditions as part of a routine day's flying. During high-speed flutter testing a maximum speed of Mach 0.70 (794 km/h; 493 mph) has been reached in a dive, with no ill-effects.

The airliner made its public debut at the ILA show, Berlin, in 1992 and achieved a promising sales start despite a high purchase price of $12,800,000. Launch customer, for 25 aircraft, is Crossair of Switzerland, which was also first with the Saab 340 and 340B. Other announced

In 1995 Crossair continued their, by now, almost traditional role as SAAB's premier customer, when they took delivery of their first Saab 2000 wearing this, the airline's new colour scheme.

orders and options were held by Brit Air of France, Delta Air of Germany, Air Marshal Islands, plus AMR Eagle, Business Express, Comair, Northwest Airlink/Express Airlines, Skywest and Kendell Airlines in the United States. When Deutsche BA took over Delta Air they confirmed Delta's previous order for five aircraft, but Express Airlines deferred their 10 aircraft for one year while the Northwest Airlink system was restructured. In recent times things have looked slightly less rosy for the Saab 2000 as it gained no orders in 1992 and only five in 1991. Its potential orderbook (including options) once stood at over 150, but only 67 of these became firm orders. Even after the introduction of a 'lean' production system, enabling a rapid response to market conditions, matters did not improve and in December 1997 Saab announced its intention to forsake regional airliner production. Manufacture of both types was scheduled to end in 1999 after the completion of 455 and 67 examples respectively of the Saab 340 and 2000.

Sabreliner

Driven by a military requirement, which it was still fulfilling 45 years later, the North American (later Rockwell) Sabre formed the basis for the Sabreliner business jet. A range of aircraft were built over the next three decades, with most still in daily use.

In 1952 the United States was well and truly lost in its love affair with the jet. Across the nation, airlines were scrambling to sign up for the new designs taking shape at Seattle and Long Beach and were preparing to cast their classic 'propliners' into an early grave. The military, in particular, was fascinated by the new technology, pouring resources into new designs where speed was king. At that time the most glamorous cockpit in the USAF was probably the North American F-86 Sabre and, inspired by the fighter's qualities, North American Aviation (NAA) had decided it would attempt to translate its success into the civil field.

A new aircraft began to emerge from the designers' tables at Los Angeles, one with a swept wing and tail, a long rakishly-upswept nose, and stubby yet strangely sleek lines. Intended to accommodate up to eight passengers, the twin-jet would be comparable in size and weight to the F-86, with only a slightly lower maximum Mach number. The two aircraft even shared roughly the same wing design (along with the F-100 Super Sabre). Preliminary work on such an aircraft began in 1952, when the initial arguments of turboprop versus turbojet had been settled in favour of the General Electric J85 jet engine, and the aircraft's other essential qualities (engines mounted in and above the wingroots, coupled with a wing that passed under, rather than through, the fuselage) had been defined, at least on paper.

The proposal remained in this form until 1956, when the USAF announced its intention to acquire a small jet transport-training aircraft, on an 'off-the-shelf' basis. Regrettably, it was announced, the Air Force had no money to fund such a design itself, but should a manufacturer wish to respond to the requirement a significant order was in prospect. In August, Major General David H. Baker wrote to some 28 manufacturers expressing the Air Force's interest in 1,500 trainers and 300 larger transports, at a cost of approximately $200,000 to $400,000 each. No such American aircraft existed at that time, as the USAF well knew, but, with the lure of such a contract coupled with the later commercial potential, it would not be long before they sprang up.

The two General Design Specifications (GDS) were for a light twin-jet trainer, the UTX, and a medium four-engined utility transport, the UCX. NAA determined that it already had sufficient work completed to bid for the two-crew, four-passenger UTX, described by the GDS as "a twin-engined utility transport, the primary mission of which is combat readiness training." The aircraft's secondary mission was to be aerial target-towing, one of the few roles never actually fulfilled by the aircraft. North American lost little time in renaming their design the N.A.246 Sabreliner, identifying it closely with the superlative F-86, but found itself facing at least eight other eager manufacturers. Lockheed, Douglas and Fairchild had opted to chase the larger UCX goal, but Convair, Temco and Beechcraft were in direct competition with NAA for the UTX contract. Only NAA committed itself to actually proceeding with development, making an announcement to that effect on 27 August 1956. Both tunnel testing and extensive engineering mock-ups were soon embarked upon.

Both these investigations showed that a significant redesign was needed. This saw the Sabreliner's definitive faired-in wing and rear-engined layout emerge, but also pushed back the hoped-for first flight date of early 1958. Things were not helped by the unavailability of the flight-rated 11-kN (2,500-lb) J85-GE-X engines. The prototype N.A.246 (N4060K) was rolled out at the manufacturer's Inglewood plant on 8 May 1958, and made its maiden flight on 16 September at Palmdale, in the hands of pilots J. O. Roberts and Gage Mace. North American had presented the USAF with a *fait accompli*, and was notified that the Sabreliner had won the UTX competition. The aircraft was handed over to the USAF at Edwards AFB for flight testing, which proved very satisfactory, and a month after completing its Phase II evaluation the USAF placed an initial order for seven aircraft,

The final production Sabreliner was the Sabre 65, a TFE731-3-powered trans-continental range aircraft. It was also the first to feature the Raisbeck super-critical wing as standard.

With an F-100 Super Sabre acting as chase plane the first N.A.246 took to the air from Palmdale, on 16 September 1958. The winner of the UTX competition, it was soon in USAF hands.

A single Sabreliner 50 was built in 1964 to act as a company electronics/avionics testbed. Identifiable by its nose instrumentation boom, it is still in use in 1993, with Rockwell.

The Sabre 40 was followed by the stretched Sabreliner 60 (left) which could seat up to 10 passengers. Two extra windows were added on the starboard side.

The first of seven initial production T-39As for the USAF flew on 30 June 1960 and was delivered the following October. The project was funded as Weapons System SS452L.

designated T-39A (N.A.265). These aircraft would be fitted with uprated 13.35-kN (3,000-lb) Pratt & Whitney J60-P-3 engines.

While the initial batch of T-39As took shape, the USAF decided the type would make a suitable radar trainer for its new fleet of F-105D all-weather strike aircraft, the need for which was pressing. Consequently, the sixth T-39A was converted to T-39B (N.A.265-20) standard. On 15 January 1960 a follow-on order for 35 aircraft was placed, the first four of which would be built as T-39Bs. The first T-39A flew on 30 June 1960 and gained its provisional certification the following year, and full FAA certification was announced on 23 March 1962. Operational deliveries commenced on 4 June 1962, and the last of 143 T-39As was delivered in late 1963. In 1967 (on its XB-70 Valkyrie-headed paper) NAA announced the 'redelivery' of the first

One hundred and thirty-seven Sabre 40s were built, mostly for American customers. Recent years have seen a migration of older biz-jets such as these to Mexico and South America.

upgraded T-39A, fitted out for seven passengers with an increased maximum take-off weight. All the USAF's transport Sabreliners were so modified.

Fighter pilot trainer

While the bulk of aircraft were used for VIP and priority transport (initially with their unusual layout of inward-facing seats), a number were allocated to Air Training Command's Instrument Pilot Instructor School, at Randolph AFB. The T-39Bs were allocated to the 4520th Combat Training Wing at Nellis AFB. These aircraft had slightly larger noses to accommodate the F-105's R-14 radar and APN-131 Doppler navigation gear, and internally were equipped with three consoles for student F-105D pilots. The T-39B fleet moved about during its life and, when the F-105 was finally relegated to Air National Guard service in the 1970s, the T-39Bs flew with the Kansas ANG, the only Air Guard Sabreliners. A similar version to the T-39B, the T-39C, was proposed to train F-101B Voodoo interceptor pilots, but when F-101B acquisition was cut back the T-39C was soon cancelled. The final USAF version was the T-39F, the 'Teeny Weeny Weasel'. Three aircraft were specially modified by NAA to carry the ECM system of the F-105G SAM-suppression aircraft in use in Vietnam. 'Wild Weasel' crews were trained side-by side in the cabin, along with an instructor.

More military orders

The Navy (and Marine Corps) shared the Air Force's interest in North American's aircraft. In 1961 it began to order the T3J-1, a navalised T-39B equipped with the Magnavox APQ-94 intercept radar, to train pilots and RIOs for its F8U-2NE Crusaders and F4H Phantoms. In November 1962, before this version had actually flown, the USAF and USN designation system was brought into line and the T3J-1 became the T-39D (N.A.265-30). Forty-two were delivered to the Navy by November 1964, a projected buy of four for the Marine Corps having failed to materialise. North American had concentrated during these early years of the Sabreliner on fulfilling the initial flurry of military orders, but was well aware of the type's civilian potential, not least as its development had been paid for by military orders. Through its FAA/USAF certification process the Sabreliner already met CAR 4b jet transport requirements, and in 1962 the decision was made to proceed with a civil version. To the FAA, the military Sabreliner was certified as the N.A.265 and, sequentially, the first civil version was the N.A.265-40, so the name Sabreliner Series 40 was adopted. Essentially the same as the T-39A, the Sabreliner 40 was fitted with a civilianised-version of the J60 engine, the Pratt & Whitney JT12A-6A. Soon these were replaced by 14.7-kN (3,300-lb) JT12A-8 powerplants for improved hot-and-high performance, approved in November 1966. Along with the introduction of the new engine a third window was added to the cabin.

Distribution of the Sabreliner 40 was handled by Remmert-Werner, as NAA felt it was out of touch with the civil market, and aircraft were delivered 'green' to R-W for fitting out to customer specifications. The first aircraft to fly (N7820C) could be laid out for up to 11 passengers in a suitably well-appointed interior. At a price of $795,000 (not including avionics or interior), the aircraft sold well. By September 1965 NAA held 54 orders, upping the intended production rate to three per month, with customer deliveries beginning in October. At the same time, the military Sabreliner fleet passed the 800,000 flight hours mark. One hundred and thirty-seven Series 40s were built, and the aircraft set several records such as a non-stop distance achievement over the 3844 km (2389 miles) between Newfoundland and Lisbon, at an average speed of 909 km/h (565 mph).

A single Sabreliner 50 (N50CR later N287NA) was built in 1964 for avionics/ electronics research duties at the hands of North American's Autonetics Division. The

The T-39D began life as the T3J-1 and was the US Navy's equivalent to the T-39B radar trainer. Forty-two aircraft shared the reprofiled nose of the USAF version, housing (at first) APQ-94.

Re-engining the Sabre 70/75 resulted in the 'tall' fuselage Sabre 75A which offered stand-up accommodation for its passengers. Seventy-two were built until production ceased in 1982.

aircraft was a combination of Sabreliner 40 and T-39A/B/D, with a modified electrical system to cope with the aircraft's increased demand for internal power. Throughout its life it was fitted with various radomes and pods, wing tanks and racks, a smoke generator, and a permanent instrumentation boom above the nose. Equipment tested during its flight programme included low-light TV, infra-red sensors, laser rangefinders, and head-up and helmet-mounted displays. The Sabreliner 50 was heavily involved in development of the automatic terrain-following and Mk II navigation systems for the F-111D and F-111E.

Stretched variant

In 1967 the Sabreliner 40 was joined by the stretched Sabreliner 60. Some 96 cm (38 in) longer then its predecessor, the first Sabreliner 60 (N306NA) was FAA certified in May 1967 and sat 10 passengers, as opposed to the previous normal load of eight. Featuring five cabin windows in its lengthened fuselage, the aircraft's maximum operating speed was Mach 0.8 at a cruise altitude of 10670 m (35,000 ft). On 22 September 1967 North American Aviation was merged with the Rockwell-Standard Corporation to become the North American Rockwell Corp., this later evolving into the Sabreliner Division of Rockwell International. Production continued at El Segundo and, with the Sabreliner 40 rolling out alongside the Sabreliner 60, an improved yet simplified version of the initial model was unveiled in 1971. The seven-seat Sabreliner Commander was then priced at $995,000 to compete with aircraft such as the Learjet 23/24 in the smaller end of the market. In 1972 an increase in the aircraft's gross weight, from 8642 kg (19,052 lb) to 9045 kg (19,922 lb), saw it redesignated Sabreliner Commander 40A. The equivalent full-specification Sabreliner 40 also became the Sabreliner 40A.

The US Navy again became a Sabreliner customer when it ordered seven Sabreliner 40s in the form of CT-39Es (briefly VT-39E) in May 1967. Used in the fleet support role, necessitating world-wide deployment capability, the CT-39Es were fitted with military-standard TACAN navigation equipment and less-plush interiors. A second order, this time for the Sabreliner 60, was placed in September 1971. The first two were also designated CT-39E, until the CT-39G title was applied. Ten aircraft were acquired for both the Navy and Marine Corps.

By the early 1970s North American Rockwell was planning the next step forward for the Sabreliner family. Reflecting the year of its birth, the new aircraft (prototype N7572V) was named Sabreliner 70, and certified on 17 June 1970. Its deepened fuselage featured increased headroom for its passengers, but the most notable feature was the replacement of the Sabre's original rounded-triangular windows by five square ones. Sales were slow (only nine were produced) and the name was soon changed to Sabreliner 75; at the same time, the shortened 'Sabre' title was adopted by the marketing division, now a part of the parent company rather than Remmert-Werner. The slowing in sales was due to the increasing difficulty in selling a turbojet-powered aircraft in the face of newer, quieter and more-efficient turbofans. A trial fitting of a Garrett AiResearch ATF 3 engine was studied as the Sabre 80 but, instead, the General Electric CF 7002D-2 was chosen and tested on the Sabre 70 prototype.

Sabre 75A

Mated with the Sabre 70/75 fuselage, the new aircraft was still referred to as the Series 80 by Rockwell, but sold as the Sabre 75A from October 1972 onwards. Certification of the new model, with such refinements as cabin air conditioning, new Goodyear brake units and increased fuel capacity, was achieved in December 1973 and deliveries began soon after in the new year. A significant order for 15 aircraft came from the Federal Aviation Authority for use in the landing-aid role calibration and airways navaid checking world-wide, replacing existing DC-3s. In 1974 Sabre 40 production came to an end, leaving only the Sabre 60 and 75A available.

During the 1970s several developments of the Sabre were investigated but not pursued. The Sabre X was to be a three-engined development of the Sabre 60. Powered by Pratt & Whitney JT15D-4 turbofans, and with extended wings, the Sabre X was intended to cope with ranges of up to 3700 km (2,300 miles). The longer-range Sabre Y was a twin-engined development of the Sabre 60, powered by Garrett TFE731-4 engines, and would eventually form the basis of the Sabre 65. A boost for the Sabre 75A came in 1973 when the US Coast Guard announced that it would procure up to 65 HT-39H aircraft to replace its HU-16E Albatrosses in the maritime surveillance role. Unfortunately, Congress halted the buy, stating that "insufficient competitive procedures had been followed." In answer to a subsequent formal competition in 1975, Rockwell offered the S-75C Sea Sabre, powered by Avco Lycoming ALF-502R engines and with many changes relevant to its intended maritime role. The USCG looked favourably on the design, which was competing with the VFW-614 for the order, but the deal fell through after negotiations between the USCG and Rockwell broke down. A third attempt at fulfilling the Coast Guard's needs resulted in a purchase of the Dassault Falcon 20-derived HU-25.

The aircraft that would be the final pro-

The FAA used 15 modified Sabre 75As to calibrate airways navigational and airport landing aids. Later painted blue and white, they were replaced by BAe 125-800s (the ex-USAF C-29As).

Merging existing Sabres with the renowned Raisbeck wing resulted in a series of much improved 'Mark 5' aircraft. Note the tufts on this developmental Sabre 60 for airflow testing.

duction Sabre was unveiled in 1976, and made its public debut the following year. The Sabre 65 was a transcontinental-range aircraft, powered by a pair of Garrett TFE731-3 engines. The first production aircraft was rolled out in 1979, first flew on 8 April and was finally delivered in 1981. The most significant advance over previous aircraft was the Sabre 65's supercritical wing, developed for Rockwell by the Raisbeck Group, Seattle, achieving the first combination of such a wing with turbojet engines on a US business aircraft. The package of aerodynamic improvements for the Sabreliner, which included chord increases at root and wingtip, removal of the leading-edge slats, addition of wing stall fences and Fowler flaps, was referred to as the Mark Five system, and Raisbeck offered the new wing for retrofit to all existing Sabre models. In fact, Raisbeck became heavily involved in modifying earlier Sabreliners, and for the Sabre 60 offered a retrofit service with TFE731 turbofans. Other customers could buy new Sabre 60s with leased JT12A-8 engines pending their replacement by Garrett units. Sabre 75As fitted with the new wing were referred to as Sabre 80As.

With the success of these versions Rockwell planned to launch growth versions, the Sabre 80C and 85 stretching the Sabre 65 and 75 respectively. With a cabin extended to 7.62 m (25 ft), the Sabre 85 was to be Rockwell's first truly intercontinental biz-jet, and both the Rolls-Royce RB401 and Lycoming ALF502 were seen as potential powerplants. The company's financial position was far from secure and as production entered the 1980s only the Sabre 65 was still being built. Total Sabre 75A production had came to 72 and 76 Sabre 65s had been completed when, on 1 January 1982, production at El Segundo ceased, as the plant's lease expired on that day. There followed a hiatus of over a year until July 1983, when Rockwell finally sold the production rights for the aircraft to the St Louis-based Sabreliner Corporation, formed specifically by New York investment bankers Wolsey and Co.

Sabreliner Corp. attempted to take the preliminary designs for the Sabre 85 to Italy in the hope of forming a co-production agreement with Agusta. Aircraft would be built in Italy then ferried to Sabreliner's Perryville, Montana, facility for avionics installation, outfitting and painting. Several

Construction number 306-114 was the prototype Sabre 65 – a converted Sabre 60. A similar converted aircraft would also serve as the first production aircraft.

changes were incorporated to this 'new' design over its original Rockwell incarnation. Still combining the wing of a Sabre 65 with a stretched Sabre 75, the wing and fuselage stretches were further increased, winglets added and the flat windscreen replaced by a conical one. The existing 20-kN (4,500-lb) Garrett TFE731-5A powerplant was retained, and range with six passengers at Mach 0.725 was set at 3,060 nm (5667 km/3521 miles). Carrying a $7,000,000 price tag, first deliveries were pegged for 1988, but negotiations with Agusta proved fruitless and Sabreliner Corporation sought an alternative partner in a US manufacturer. None was forthcoming and the Sabre 85 was shelved as Sabreliner's attention turned elsewhere.

Soldiering on

Small numbers of the Sabre are still in service around the world, and the Sabreliner Corp. provides product support for all models. In 1987 it was awarded a five-year contract to support USAF and USN T-39s, and also completed a service life extension programme on 10 USAF aircraft (two T-39As and eight T-39Bs) which were delivered in 1988. The following year the company introduced its Excalibur programme for Sabre 40 and 60 aircraft, aimed at those examples approaching their 10,000 flight hours mark. Stripped of paint, demated, overhauled and reassembled, they are now known as Sabre 40EXs and 60EXs respectively, with airframe lives extended by a further 5,000 hours as a result. Sabreliner's most recent activities include the controversial Undergraduate Naval Flight Officer (UNFO) programme at NAS Pensacola. The UNFO contract covers the training of US Navy RIOs in 17 modified T-39N radar trainers, all converted by Sabreliner from civil Sabre 40s from 1991 onwards. Sabreliner lost the contract for a period on the grounds that their aircraft were unsuitable for the task, but were subsequently re-instated on appeal. As their rival's fleet of T-47s (modified Cessna 552 Citation IIs) was largely destroyed by recent flooding, Sabreliner looks set to see out the contract.

The Sabreliner is among several of the older generation of biz-jets which is being increasingly 'parted out' – that is, purchased and broken up for spares – and is becoming less and less attractive to operators in the light of rising costs and more modern alternatives. However, the vast majority of the aircraft built are still in service (including the first-built 40, 60, 65 and 75) chiefly in the United States and Mexico. The USAF and USN have steadily reduced their large fleets through retirement to the quieter acres of Davis-Monthan's MASDC storage facility. All the USAF's T-39s now sit under the Arizona sun, but Navy aircraft remain in service. The only other military operators are Ecuador and Argentina. They are all likely to accompany their civilian counterparts into the next century.

Now languishing in the sun at Davis-Monthan, this former MAC T-39 wears the 'European 1' wrap-around camouflage, an unusual scheme for a VIP aircraft.

The USAF may have relinquished its Sabres but the USN and USMC still operate immaculate late-model aircraft. This is one of the Marines' Sabre 60s designated CT-39G.

Shorts 330

A logical outgrowth from the successful Skyvan utility transport, the Shorts 330 was in the vanguard of the turboprop feederliner revolution. Although it has been overtaken by more recent and more advanced designs, is still in moderately widespread use, and has gained popularity as a light freighter, even adopting military colours for this role.

The history of the Shorts organisation dates from April 1901 when two brothers, Oswald and Eustace Short, first set up in business at Hove in Sussex as manufacturers of aerial balloons. In 1908 they were joined by their brother Horace, and at a new base at Shellbeach on the Isle of Sheppey near London commenced building, under licence, six of Orville and Wilbur Wright's biplanes.

During the late 1930s the main production facilities were moved to Belfast in Northern Ireland. Production began there in 1938 with a contract to build 50 Bristol Bombays, followed by 150 Handley Page Herefords. Then there came long production runs which included over 1,200 Stirlings and 130 Sunderland flying-boats. In 1964 the company built the Belfast freighter for the Royal Air Force, and the Skyvan followed in 1966. During the 1970s Shorts was producing many sub-assemblies including engine nacelles for the Rolls-Royce RB.211 and wings for the Fokker F28 Fellowship. At the Farnborough air show in September 1974, the sale of the 100th Skyvan was announced.

It was at the same air show that the company's next major airliner project was announced: the 30-seat SD3-30 (later renamed the 330). The prototype, registered G-BSBH, had first flown on 22 August, and eight days later had won its first order from Command Airways of Poughkeepsie, New York. The 330 was primarily intended for short-range regional and commuter air routes in the United States, although sales were quite good in other countries (e.g. DLT of West Germany). The choice of 30 seats for the Shorts 330 was dictated by the results of the company's market research survey, and especially since the US Civil Aeronautics Board had recently revised its rules to allow commuter and air taxi operators to use aircraft of up to 30 passengers with a maximum payload of 3400 kg. The Shorts 330 was also the logical growth vehicle to replace such aircraft as the Swearingen Metro, DHC-6 Twin Otter and Beech 99 and similar aircraft on developing networks. Briefly, the 330 combined unprecedented standards of luxury for the commuter passenger with design simplicity and reliability, these factors being very important for an up-and-coming commuter airline.

Airworthiness affirmed

Two prototypes (G-BSBH and G-BDBS) and the first production standard aircraft (G-BDMA) were used in the development programme. The UK Civil Aviation Authority awarded the first Certificate of Airworthiness in February 1976, quickly followed by approval from other countries, including West Germany and the USA. The

Avair operated this Shorts 330 on sectors between Irish cities on behalf of Aer Lingus Commuter, alongside a fleet of Beech King Air 200s and Beech 99 Airliners. ALC then operated a fleet of Shorts 360s, a move made by many former 330 operators.

The military application of the 330 seemed obvious from the start, yet only two nations operate it. One is Thailand, which has four 330-UTTs for the police and army.

One of the more exotic operators of the 330 was Coral Air, which operated this single example alongside Britten-Norman Islanders from its base at St Croix, Virgin Islands. The 330 was sold soon after its delivery in 1981 in favour of a Twin Otter.

first aircraft to be delivered (N51DD) was handed over to Command Airways in June 1976, and the very first Shorts 330 to enter service was a Time Air of Canada aircraft (C-GTAS) which commenced revenue flights on 24 August 1976.

The United Kingdom government agreed to commit £60 million to the company over a five-year period to enable a major programme of expansion to be undertaken. A large amount of this investment was earmarked for the development of the 330 airliner programme, which had orders in 1978 for 34 aircraft. Many new jobs were created by the Shorts 330, and by the end of 1978 Short Brothers employed 6,260.

Besides the basic passenger version, there are three other types of Shorts 330. A cargo conversion known as the 330C is achieved by the removal of all interior furnishings and the addition of a roller floor (optional). The commuter aircraft is then converted into a versatile and economic freighter, the cargo being loaded through the large forward door into a spacious 34.83 m^3 cargo area with additional space in the nose and rear compartments. In this configuration the Shorts 330C can carry a 3400 kg payload for up to 360 km with the normal IFR reserves.

Convertible freighter

The Shorts 330 Combi variant offers the full airline passenger comforts for 18 passengers in the rear cabin, but the front of the aircraft is sectioned off for the carrying of up to 1000 kg of cargo, or the aircraft can be converted to a full freighter with the interior wall furnishings retained so that the aircraft can be returned to an all-passenger or combi plane in a minimal time. A similar variant is the Shorts Sherpa (C-23A in USAF, C-23B in US Army service). This version is an all-freight type with rear ramp installed. The addition of this facility enables the shipment of four LD3 containers or a wide range of motor vehicles or aero engines. The 330-UTT is a Utility Tactical Transport with a strengthened floor and increased maximum take-off weight, operated by Thailand.

The United States provided the most lucrative market for Shorts, where commuter lines operated large fleets of small aircraft between cities and outlying airfields. Mississippi Valley Airlines was a major 330 operator, although it is now defunct.

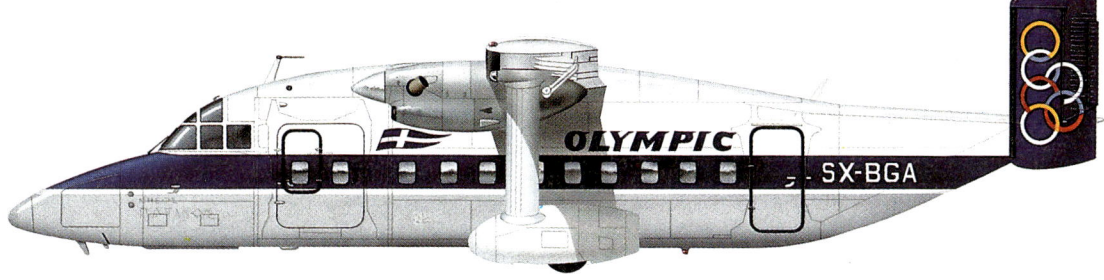

Five Shorts 330-100s served with the Greek national airline Olympic. A separate division from the international carrier, Olympic Aviation operated these, Skyvans and Dornier Do 228s on services around the myriad of islands in the Aegean.

The Shorts 330 was the first commuter aircraft tailor-made for the short-haul market. It retains the simplicity of design and operational characteristics of the first-generation commuter aircraft whilst offering all the passenger amenities of much larger and more complex turboprop and jet airliners, thereby allowing the operator to progress into the larger commuter airliner category with only modest expenditure. The main passenger features of the Shorts 330 include increased headroom and improved seating, washroom facilities, a galley, increased storage area and a lower noise level.

North American influence

The cabin interior was designed by the same US consultants who created the interiors for the Boeing 747; this design makes maximum use in flexibility of usage and layout. The landing gear, designed and manufactured by Menasco of Canada, is retractable and uses medium-pressure tyres with 5.3 kg/cm^2 for the main wheels and 4.1 kg/cm^2 for the nose tyre. The main tanks are located in the centre section wing fairing, which has a capacity of 2180 litres. There is single point pressure fuelling which is accomplished by one man standing on the ground. Power for the flap actuators, landing gear and brakes is supplied by engine-driven hydraulic pumps. A 28-volt DC system provides for general services, with a special AC source of 115V and 26V at 400 hz for certain instruments, avionics equipment and fuel booster pumps.

Two propjet UACL PT6A-45s free turbine engines drive five-bladed Hartzell 2.9-m diameter propellers. The powerplants are each flat-rated at 1156 shp for take-off and 1020 shp for maximum continuous operations. The cabin is fully air-conditioned by means of a Hamilton Standard air cycle refrigeration unit utilising engine bleed air. The engine intake ducts and lips, propellers and windshield are protected. Pneumatic aerofoil de-icing is also available as an optional extra.

Air Ecosse was a UK carrier, owned by Fairflight Ltd, for services in Scotland. The fleet was largely based on the EMBRAER Bandeirante, but some 360s and 330s were in service. Unfortunately, the airline ceased operations in 1987.

US Air Force C-23A

On Friday 2 March 1984 the Shorts Brothers aircraft company announced that the US Air Force had ordered 18 aircraft valued (at

Air UK is a major domestic carrier throughout Great Britain, connecting all the regional airports. Now centred on the Fokker F27, BAe 146 and Shorts 360, the Shorts 330 was an important early type in the regional network, complete with this colourful and patriotic scheme.

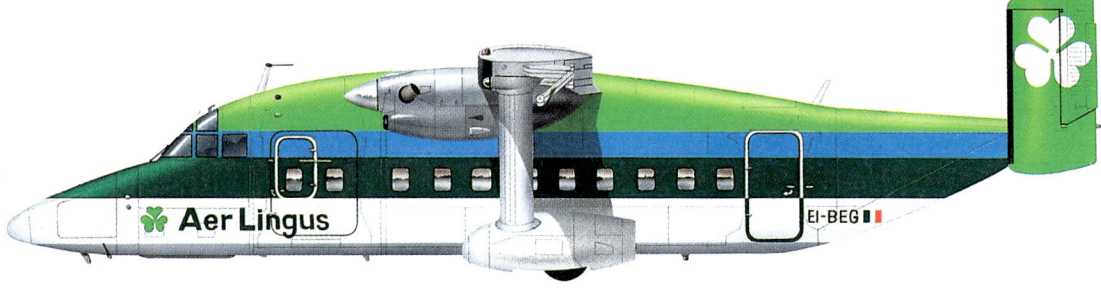

With the factory located in Belfast, Shorts is a major employer in Ulster, and brings considerable wealth and prestige to the troubled Province. South of the border, Aer Lingus Commuter has operated the type.

Although the first order came from Command Airways in California, the first aircraft to enter revenue-earning service was that delivered to the Canadian domestic operator Time Air, based at Lethbridge in Alberta. The first service was on 24 August 1976.

The Shorts 330 found favour with several overseas operators, who appreciate the load-carrying and short take-off ability of the type. Lineas Aereas Privadas Argentinas of Buenos Aires was one such airline, which purchased two. These have now been replaced by the Saab 340.

Shorts 330 cutaway drawing key

1 Glass-fibre nose cone
2 Weather radar installation
3 Nose skin panelling
4 Forward baggage compartment, 45 cu ft/400 lb (1.27 m³/181 kg) max
5 Upward-hinged baggage door, 30.5 in × 37.7 in (77.5 cm × 95.8 cm)
6 VHF 2 aerial
7 Hydraulically steerable rearward-retracting nosewheel
8 Nosewheel fork
9 Nosewheel oleo
10 Nosewheel pivot point
11 Nosewheel box
12 Nosewheel retraction mechanism and jack
13 Undercarriage emergency actuation accumulator
14 Hydraulics bay
15 Rudder circuit linkage
16 Avionics bay (port and starboard)
17 23 Amp/hr batteries (port and starboard)
18 Seat adjustment lever
19 Seat belt
20 Heated pitot head
21 Underfloor avionics equipment
22 Elevator circuit linkage
23 Control column
24 Pilot's seat
25 Rudder pedals
26 Windscreen wipers
27 Windscreen panels (electrically heated)
28 Instrument panel coaming
29 Central control console (trim wheels)

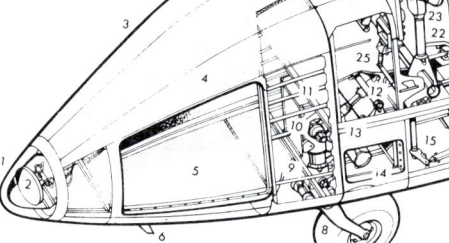

1984 prices) at US $165 million, plus options on a further 48; if the whole package were taken up it would be worth US $660 million. The order created an extra 600 jobs at the Belfast factory, and brought orders and options on the 330 up to 180 aircraft.

The first of the 18 aircraft, serial 83-0512, was delivered on 2 November 1984 and named *Zweibrücken*, and with the remaining 17 aircraft is employed on the USAF's European Distribution System Aircraft (EDSA) programme, serving approximately 20 bases in Europe. The hub for this operation is at Zweibrücken in Germany, with warehousing facilities at RAF Kemble in the United Kingdom and Torrejon in Spain.

The C-23A can accommodate four standard LD3-size air freight containers and can also accept aero engines that are used to power the F-15 and F-16 combat aircraft. The aircraft cruises at altitudes of around 30000 m with a cruise speed of around 320 to 380 km/h. With a maximum take-off weight of 10400 kg with a 3400 kg payload, the C-23A helped the 10th Military Air Squadron win the award for the 'Most Outstanding Military Air Command Support Squadron'

Air Puerto Rico ceased operations in 1987, having flown the Shorts 330 on services around the island and to the neighbouring Bahamas.

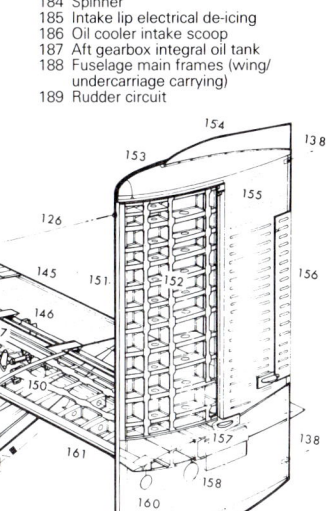

30 Co-pilot's seat
31 Overheat panel (AC/DC power supply)
32 Fuel cocks
33 Crew escape/ditching hatch
34 Flight deck/cabin sliding door
35 Aileron circuit linkage
36 Control cable conduit (rudder and elevator trim circuits)
37 Flight deck conditioned/ heating/de-misting air supply
38 Ambient-air intake
39 Combined VOR/Localiser/ILS guide-slope aerials
40 Blow-in door (ground running)
41 Turbine-blower intake
42 Heat exchanger
43 Air cycle installation
44 Engine bleed-air supply
45 Pre-cooler
46 Pre-cooler intake
47 Cabin conditioned/fresh air supply
48 Doorway-surround doubler plate

57 ADF sense aerials (port and starboard)
58 Rectangular fuselage section frames
59 Chemically-milled window panel
60 12-a-side cabin windows, 18.5 in × 14.4 in (74 cm × 36.6 cm)
61 Passenger accommodation: 30 seats, 3-abreast (single port/double starboard) arrangement
62 Engine bleed-air supply duct
63 Fuel tank mounting lugs
64 Forward multiple fuel tank (Cell 1)
65 Class II sealed tank dividing bulkhead
66 Fuel gravity filler
67 Forward multiple fuel tank (Cell 2)
68 Class 1 sealed tank dividing bulkhead
69 Forward multiple fuel tank (Cell 3)
70 Sealed containment area (tank seepage)
71 Tank/fuselage attachment
72 Wingroot fairing
73 Engine-propeller control cable runs
74 Hydraulics reservoir
75 Wing centre-section
76 Chemically-milled centre-section skinning
77 Dorsal anti-collision beacon

diameter
87 Propeller de-icing boots
88 Pratt & Whitney PT6A-45 turboprop engine
89 Oil filler cap
90 Outer/inner wing pin joints
91 Outer-section front spar
92 Outer wing support strut
93 Starboard landing/taxying lamp
94 Support strut pin joints
95 Strut attachment bracket
96 Fluid de-iced leading-edge (tank and pump unit mounted at rear of starboard mainwheel well)
97 Starboard navigation light
98 Glass-fibre wing-tip fairing
99 Starboard aileron
100 Aileron trim tab
101 Aileron hinge rib
102 Support strut box
103 Flap hinge ribs
104 Starboard outer flap section
105 Starboard centre flap section
106 Centre-section end rib
107 Starboard inner flap section
108 Flap actuating rod mechanism (mounted on spar rear face)
109 Water-methanol tank and pump
110 Gravity fuel filler
111 Aft fuel tank (Cell 4)

119 Buffet unit storage compartment (sandwiches/ biscuits etc)
120 Cabin furnishing profile
121 Coat closet
122 Toilet compartment
123 VHF 1 aerial
124 Skin outer panelling
125 Corrugated inner skin panelling
126 HF sense aerial
127 Rudder/elevator circuits
128 Emergency locator antenna

145 Three-section elevator
146 Elevator actuation quadrant
147 Rudder control linkage
148 Elevator spring strut
149 Trim cable pulleys
150 Port tailplane spar pin joints
151 Fluid de-iced leading-edge
152 Fin structure
153 Rudder aerodynamic balance
154 Rudder extension fairing
155 Port rudder

173 Contoured inner window surrounds
174 Cabin seating rearmost row (port seat omitted for clarity)
175 Rudder circuit linkage
176 Damper strut
177 Flap actuating rod
178 Centre-section ribs
179 Centre-section front spar
180 Firewall/bulkhead
181 Engine support structure
182 Engine mounting ring (with four dynafocal resilient mounts)
183 Exhaust duct
184 Spinner
185 Intake lip electrical de-icing
186 Oil cooler intake scoop
187 Aft gearbox integral oil tank
188 Fuselage main frames (wing/ undercarriage carrying)
189 Rudder circuit

156 Rudder trim tab
157 Rudder actuation lever fairing
158 Fin attachment access panels
159 Fin lower section
160 De-icing system access
161 Fluid de-iced leading-edge
162 Aft fuselage structure
163 Aft baggage door, 43 in × 57 in (109 cm × 145 cm)

190 Stub wing front and rear spars
191 Undercarriage mounting beam
192 Undercarriage retraction jack
193 Wing support strut attachment
194 Undercarriage pivot point
195 Undercarriage levered suspension leg
196 Port main landing-gear fairing
197 Retractable mainwheel
198 Shock-absorber strut
199 Port wing support strut
200 Port landing/taxying lamp
201 Hydraulic ground service panel (fairing hinged aft section)
202 Wing outer-section front spar
203 End ribs
204 Outer/inner wing pin joints
205 Port inner flap section
206 Outrigged flap hinge arms
207 Aileron trim tab cables
208 Port centre flap section
209 Hinged trailing-edge (controls) access panels
210 Port outer flap section
211 Aileron control rods
212 Support strut box
213 Multi-angle section diffusion members
214 Pressed ribs
215 Corrugated inner skin panels
216 Aileron actuating rod
217 Cable-operated trim tab jack
218 Trim tab actuating rod
219 Aileron trim tab
220 Port aileron
221 Outer-section rear spar
222 Aileron mass-balance weights
223 Wing skin outer panelling
224 Outer-section front spar
225 Outer-section leading-edge spar
226 End rib structure/tip attachments
227 Glass-fibre port wing-tip fairing
228 Port navigation light

49 Cabin forward emergency exits, port 37 in × 24.5 in (94 cm × 62 cm); starboard 42 in × 27 in (107 cm × 68.6 cm)
50 Forward freight door, 65.5 in × 55.6 in (167 cm × 141 cm)
51 Freight door hinges
52 Honeycomb-sandwich floor panels
53 Corrugated inner skin
54 Cabin air distribution duct
55 Seat mounting rails
56 Rudder circuit

78 Centre-section front spar
79 Leading-edge access panels
80 Oil cooler
81 Engine firewall
82 Engine mounting ring
83 Exhaust ducts
84 Air intake duct (with debris deflector)
85 Propeller pitch-change mechanism
86 Hartzell constant-speed five-bladed auto-feathering propeller, 9 ft (2.75 m)

112 Sealed containment area (tank seepage)
113 Tank/fuselage attachment
114 Elevator circuit
115 Cabin concealed ceiling lighting
116 Fuselage (detachable) top fairings
117 Overhead passenger hand-baggage lockers
118 Service door/emergency exit, 56.5 in × 28.4 in (143.5 cm × 72 cm)

129 Rectangular section aft frame
130 Tailplane spar pin joint strip
131 Tailplane structure
132 Rudder actuation lever
133 Rudder trim tab jack
134 Leading-edge de-icing fluid lines
135 Fin skin panels
136 Rudder aerodynamic balance
137 Rudder extension fairing
138 Static dischargers
139 Rudder trim tab
140 Starboard rudder
141 Trim tab actuating rod
142 Rear navigation light (starboard lower fin only)
143 Elevator trim tab
144 Trim tab actuating rod

164 Baggage door (open)
165 Baggage restraint net
166 Stepped aft baggage compartment, 100 cu ft/600 lb (2.83 m³/272 kg) max
167 Bulkhead
168 Doorway-surround doubler plate
169 Passenger entry door, 56.5 in × 28.4 in (143.5 cm × 72 cm)
170 Cabin electrics and communications panel
171 Buffet unit heated water container/cup stowage/trash bin
172 Cabin attendant's tip-up seat (lowered)

in 1985/6 and 1986/7. In October 1988 the C-23B was ordered, this flying with the US Army National Guard on logistic support duties.

Some 330 operators

Thai Airways announced, on 14 May 1982, an order for four 330-

713

A surprise order from the US Air Force led to the 330 adopting military camouflage as a light transport, designated C-23A Sherpa.

The C-23A Sherpas served with the 10th Military Airlift Squadron, 322nd Military Airlift Wing at Zweibrücken, but were regular sights at all USAFE bases. They shuttle spare parts between bases, providing a rapid distribution system throughout Europe.

200s, and on the same day took delivery of its first two aircraft, the first to be sold in South East Asia. They were bought to replace the Hawker Siddeley HS 748 on certain low-yield routes. The 330s are based at Bangkok, Chiang Mai and Hat Yai, and are used to feed the airline's Boeing 737 trunk routes. Command Airways is the oldest continuously-operating 330 carrier. It accepted its first 330 in June 1976 and has operated many of the type since that time. Currently 14 are operated under the American Eagle banner, feeding passengers to the American Airlines' network from points in New York State.

Suburban Airlines, the Pennsylvania-based carrier, has for a long time (since January 1979) operated two 330s. The carrier is a member of the Allegheny Commuter Group and has changed its name to adopt this title. Suburban has 10 aircraft operating in the northeastern states of the US, feeding passengers to the national US Air's network. Meanwhile in Europe, Olympic Aviation, a subsidiary of Olympic Airways, operates the 330 extensively from Athens to the Greek Islands in the Aegean Sea. The operator has five aircraft and the first, SX-BGA, was delivered on 22 May 1980. Several British operators adopted the type for regional transport.

Popular transport

One hundred and seventy-nine Shorts 330s had been ordered by the start of 1990, and have become increasingly popular as freighter/utility aircraft in the United States, where such carriers as Midnite Express and Mountain Air Cargo have used them for transportation on a small package service. Also in Great Britain, they were used as freighters by Talair and National Airways.

After more than 20 years of operations, the Shorts 330 is still flown in all parts of the world on varied types of operations, sometimes into remote airfields that would normally only accommodate small commuter airliners. Also, it fulfils the original intended functions of a short-range regional aircraft that gives passengers the comfort of larger turboprop and pure jet airliners in a 30-seat commuter airliner.

The Shorts 330-UTT version has a strengthened floor, parachuting doors and different avionics for its designated Utility Tactical Transport role, although it retains the cabin windows. Three are flown by the Royal Thai Police, and by the army.

Shorts 360

The success of the smaller Shorts 330, plus a rising demand from the market, encouraged Shorts to stretch and refine the design to produce the even more capable 360. Sturdy and straightforward, economical and reliable, the 'shed' was a benchmark commuter airliner on which many airlines cut their teeth.

Following the success of its twin-turboprop, 30-seat Shorts 330 commuter airliner among third-level airline operators, particularly in the United States, the Northern Ireland-based manufacturer Short Brothers began studying ways to improve the design. Usually, the development path of commercial airliners involves increasing passenger capacity, but in this respect Shorts was hampered by the US Civil Aeronautics Board Regulation 298 which limited American commuter airlines (the major market for the existing Shorts 330 and any planned derivative) to operating aircraft with seats for no more than 30 passengers. Although exemption was possible, compliance with the new requirements that would then become applicable increased operating costs to levels which frequently reduced load factors, thus making use of the larger aircraft self-defeating.

Initial studies for an improved Shorts 330 therefore focused on an increase in maximum weight from 10387 kg (22,900 lb) to 11113 kg (24,500 lb) to be achieved through substitution of 989-kW (1,327-shp) Pratt & Whitney PT6A-65R turboprops for the 893-kW (1,198-shp) PT6A-45Rs of the 330. However, in 1978 the US airline industry was deregulated, paving the way for third-level carriers to operate larger aircraft, and Shorts then began a redesign of the Shorts 330 to accommodate 36 passengers, optimised for stage lengths of 225 km (140 miles).

The new aircraft was initially given the internal company designation Shorts 336 (from the basic design number SD-3/36 seats). A prime aim of the design team was to increase cruising speed by reducing the drag of the rear fuselage and the 'H'-configured twin fin and rudder of the Shorts 330. A T-tail configuration was wind tunnel tested, producing a useful 18 per cent reduction in drag, though stalling qualities of the aircraft were unsatisfactory and a conventional single fin and rudder, sharply swept back, with unswept low-set tailplane, were adopted instead.

On 10 July 1980 Shorts first announced the new aircraft, now designated Shorts 360, revealing that it retained the unpressurised, square section, 1.9-m (6-ft 4-in) headroom 'wide-body' fuselage of the Shorts 330, but with a modest 'stretch' achieved by a 0.9-m (36-in) 'plug' forward of the wing and a new tapered rear fuselage, each permitting an additional three-seat row of seats to be added to

The Shorts 360 achieved fair sales outside its primary markets (the United Kingdom and United States), including an order from Malaysian Air Charter, which flew this aircraft under its Macair titles.

Left: The service between the Channel Islands and mainland Britain has provided a steady source of revenue for many carriers over the years. Jersey European has built up a sizeable fleet, including four Shorts 360s for use on the less important routes.

the cabin. Overall length of the new fuselage was 21.6 m (70 ft 10 in), an increase of 3.9 m (12 ft 9½ in) over the Shorts 330. The revised rear fuselage design not only reduced drag but also increased rear baggage compartment volume to 4.8 m^3 (170 cu ft), enabling Shorts to claim that the new aircraft would offer more baggage room per passenger (0.2 m^3/7 cu ft each, including overhead lockers in the cabin) than any other aircraft in its class.

The maximum take-off weight of the aircraft was increased to 11794 kg (26,000 lb), which necessitated some structural strengthening of the outer wing panels and lift struts and installation of a new undercarriage manufactured by Dowty rather than the Menasco unit employed on the Shorts 330. Composite materials were chosen for the main landing gear leg

Left: Danish carrier Maersk Air (using its subsidiary name Air Business) was an early customer for the 360, but has recently disposed of the last of its fleet. The carrier now flies the Fokker 50 and Boeing 737. This individual aircraft was the first to feature a fully-integrated flight control system.

Below left: Travel between the UK's Channel Islands is provided by Aurigny Air Services. The fleet is based on the Britten-Norman Trislander, which offers 16 seats, but a single Shorts 360 flies the 'trunk' sector between Jersey and Guernsey, and to Southampton.

The 360 showed its obvious parentage from the 330, but with the adoption of a conventional tail and refined lines the aircraft looked less purposeful and more elegant. The box-like fuselage was retained, giving operators great versatility in interior configuration.

'sponsons', nosewheel doors, wing- and tailplane tips, main cabin floor and other non-structural fairings, but the design relied heavily on utilising existing production jigs for the Shorts 330, enabling prototype development time and launch and production costs to be minimised.

Reliable worker

The Shorts 360 airframe is conventional, and, apart from those items already mentioned, of all-metal construction. The fuselage is manufactured in two sections: nose, incorporating flight deck, nosewheel bay and forward baggage compartment; and the centre/rear section which includes main wing spar attachment frames and transverse lower beams to carry the retractable main landing gear legs, rear baggage compartment and tail unit attachment structure. The wing uses a NACA 63A series (modified) airfoil section and is built on two spars in three sections. The wing centre-section, integral with the top of the fuselage, has taper on leading and trailing edges, while the strut-braced outer sections have parallel chord. Single-slotted, rod-actuated flaps and ailerons are installed. All fuel is contained in integral tanks in the wing centre-section and fuselage fairing, totalling 2182 litres (480 Imp gal).

The aircraft is configured for a flight deck crew of two, with main cabin accommodation for 36 passengers seated in 12 rows of three (two plus one with offset aisle) at a seat pitch of 76 cm (30 in). Overhead baggage lockers and externally accessible forward and rear baggage compartments provide a total of 7.5 m^3 (265 cu ft) of baggage volume. Integral passenger stairs are provided, and air conditioning is standard.

A launch order for four Shorts 360s was placed on 2 September 1980 by Suburban Airlines of Reading, Pennsylvania, an existing Shorts 330 operator. Shorts officially gave the 'go-ahead' for the new aircraft in January 1981, with the first flight of the prototype planned for a year later. This target was more than met, for the first Shorts 360, G-ROOM, was rolled out in May 1981 and made its first flight six months ahead of schedule on 1 June, temporarily powered by 861-kW (1,156-shp) Pratt & Whitney Canada PT6A-45 pending installation of production standard PT6A-65Rs when they became available in the following January. Three days after its maiden flight, with 10 hours of test flying logged, G-ROOM was flown to Le Bourget, Paris to make its public debut at the 1981 Paris air show. The second Shorts 360, G-WIDE, did not fly until 19 August 1982, by which time the British Civil Aviation Authority certification test programme was almost complete. Test flying revealed few shortcomings in the design, the only significant changes made consisting of the addition of vortex generators on the fin and inside the rudder shroud and deletion of one of the two trim tabs installed on each elevator. Certification was granted by the CAA on 3 September 1982, a month ahead of schedule. The US Federal Aviation Administration granted the Shorts 360 certification to FAR Parts 25 and 36 in November 1982.

First commuter days

Launch customer Suburban Airlines was first to receive the new aircraft, taking delivery of N360SA (formerly G-WIDE) on 11 November, and putting it into service on a 1127-km (700-mile), eight-city commuter route network on 1 December. On that day the first of 30 Shorts 360s ordered by Chicago-based Simmons Airlines was delivered, followed on 20 December by the first aircraft for a British customer, Genair.

Production gathered speed in 1983, with Shorts 360s coming off the Sydenham/Belfast Harbour Airport line at the rate of one every two weeks. Apart from Suburban and Simmons Airlines, carriers taking delivery during 1983/84 included Air Ecosse, Air UK, British

A useful export order for the 360 came from China, which bought seven in 1985. Four (including this aircraft) are assigned to China Eastern Airlines based at Shanghai, the remainder serving with China Southern Airlines at Guangzhou.

Left: Several UK carriers adopted the Shorts 360 for regional work and light freight carriage. Among them was Air Ecosse, which operated this specially-painted and registered aircraft on rapid parcel delivery services for the Royal Mail.

Right: Seen wearing UK registration (prior to taking up B-12277), this is the sole 360-300 delivered to Formosa Airlines in May 1989. The airline's mixed fleet is based at Taipei's Sung Shan airport, from where it flies to Taiwan's major cities.

Left: Simmons Airlines of Chicago was the second customer for the Shorts 360, receiving its first aircraft in December 1982 to begin commuter services in the Michigan-Illinois area. Today the carrier is part of the American Eagle network.

Right: The massive American Eagle commuter network had over 40 Shorts 360s, operated by Simmons Airlines and Flagship Airlines, the latter operating in the south-east United States. This aircraft is one of the original 360s delivered to Simmons in 1982.

With its exceptional reliability and excellent operating economics (allowing break-even with low load factors), the Shorts 360 was tailor-made for the US commuter market. It proved fairly popular, but in recent years has been overshadowed by more modern and economic designs.

Midland Airways, Manx Airlines and National Airways in the United Kingdom, Aer Lingus Commuter of Ireland, Maersk Air Business of Denmark, Murray Valley Airlines in Australia, Malaysia Air Charter Company, and Dash Air, Pennsylvania Airlines, Sunbelt Airlines and Time Air in North America.

In service the Shorts 360 quickly established a reputation for reliability, routinely recording a 99 per cent despatch rate thanks to trouble-free systems and easy maintenance which permitted such rapid turn-arounds that some hard-used aircraft in the United States made as many as 7,000 flights in a year. Economy was also a strong point, with airlines reporting break-even load factors as low as 39 per cent (14 seats filled). These factors led to the aircraft being given a British Design Council Award in 1984.

In 1985 Shorts introduced an upgraded version of the aircraft, known as the Shorts 360 Advanced (360 ADV). It featured more powerful and fuel efficient 1061-kW (1,424-shp) Pratt & Whitney Canada PT6A-65AR engines and associated systems changes, and became the standard production model from the 80th airframe, delivered to the Thai Airways Company of Bangkok in November 1985.

The Shorts 360 Advanced was superseded in 1987 by the final model Shorts 360-300. Improvements incorporated in this version included advanced technology six-blade Hartzell propellers with synchrophasing, new cambered lift struts, and a revised engine nacelle design with low-drag exhaust ports. Taken together, these enhancements provided significant improvements in cruising speed, take-off and climb performance at high temperatures/altitudes, and permitted an increase in maximum take-off weight to 12292 kg (27,100 lb). Optional equipment introduced on the Shorts 360-300 include a Rockwell Collins APS-65 Category II autopilot system, reclining passenger seats, a supplementary ground air-conditioning system, plumbed sink in the toilet and strengthened cabin floor with protective liners for freight carrying. The first two production Shorts 360-300s were delivered to Philippine Airlines on 18 March 1987, while the first aircraft certificated for the carriage of 39 passengers entered service with the now-defunct UK carrier Capital Airlines in October of that year.

Based at Hudiksvall, Sweden, Air Hudik operated a trio of Shorts 360s which were delivered between 1986 and 1991. This was its first aircraft, a Shorts 360 Advanced, which operates, along with its two sisterships, alongside a pair of Twin Otters.

At the same time, in what was now the twilight of the aircraft's career, Shorts announced a dedicated all-cargo version, hoping to prolong the aircraft's life in the face of failing airline orders. The 360-300F was initially designed with a rear loading ramp and a windowless fuselage, this arrangement having already been successful with the Shorts 330-derived Sherpa. In the event, it was similar to the 360-300, but with an optional large cargo door on the port side and no provision for seats. Its roller floor, with pallet locks and guidance rails, made it compatible with standard LD3 cargo containers, of which it could accommodate five. Aimed chiefly at the burgeoning overnight delivery market, the launch customer was Rheinland Air Service of Dusseldorf, which took delivery of two in March 1989.

These would be among the last of the Shorts 360s to be built at the Sydenham plant as the manufacturer was finding it increasingly difficult to sell its unsophisticated, unpressurised design in the face of more advanced competition. In the end the very qualities which had made the Shorts 360 such a success would be its downfall. One hundred and sixty-four (including the first production aircraft) were delivered between 1982 and 1991 when the line finally closed, still with several unsold aircraft on it. It marked the end of a family of aircraft which began with the Skyvan in 1963. Shorts' airliner activities are now confined to subcontract work, building airframe components for the Canadair RJ (now simply the Canadair Jet), Fokker 100, Boeing 737, 747, 757 and 777, and engine nacelles for Rolls-Royce, Textron-Lycoming and Rohr/IAE.

Time Air of Alberta is one of the network of Canadian Partner airlines operating feeder flights for Canadian International's services. Three Shorts 360s could at one time have been found among its fleet.

Short's Civil Flying Boats

Ask an aviation enthusiast about the golden age of British commercial aviation, and he will mention in the first breath the glorious Short Empire boats and their derivatives. Also spawning the immortal military Sunderland, the Short boats came to represent the utmost in luxury travel between Britannia and the far-flung corners of her empire. Here we present the story of these boats, including the amazing Mayo Composite and the post-war developments.

Short Brothers is one of the oldest company names in the history of aviation. Horace, Eustace, and Oswald Short, boat builders on the Isle of Sheppey, Kent, set up a balloon works at Queen's Circus, Battersea, at the turn of the century. By 1910 they had turned their attention to powered flight in time to produce the first torpedo-bombers for World War I. The Short Stirling was the first big four-motor bomber of World War II, and today the company's complex in Belfast continues to manufacture freighters, STOL craft and various military missile components.

But the Short products which for most enthusiasts characterise the company were the great four-engined monoplane flying-boats, built to serve the Empire in the 1930s; white giants with a style and grace which made them unique. Almost a dozen different types, both military and civil, came off the line between the mid-1930s and late 1940s, but all were variants of – and in some cases contemporary with – one master design: the S.23, designated 'C-class' but popularly nicknamed the 'Empire Boat'.

From the early 1920s onwards, Imperial Airways had been carrying passengers and a limited amount of mail out to India via Egypt and the Cape. Passengers travelled by landplane and seaplane, changing en route to seaplanes at points where landing strips were rudimentary or non-existent. The accent was on comfort and safety rather than speed. Along the Mediterranean and Egyptian routes service was provided, in the early days, by the three-engined Short

***G-AEUE** Cameronian was one of the second batch of S.23 'C-class' boats built for Imperial Airways, seen here flying along the Medway past Rochester Castle. It survived the war to be scrapped in 1947 with 15,562 hours on the clock.*

Originally ordered by QANTAS, three S.30 'C-class' boats were transferred to Tasman Empire Airways Ltd to extend the mail service from Australia to New Zealand. Before they were delivered to Auckland in March 1940, they were used for a short time by Imperial.

The Short-Mayo Composite was a remarkable attempt to provide a long range mailplane. The lightly-loaded lower section carried the mail-carrying upper section, releasing it in mid-air to continue its journey. The first mid-air separation took place on 6 February 1938.

Calcutta biplane, which was replaced in 1931 by the Short Kent, a 15-passenger four-engined biplane which was little more than an efficient stop-gap. In the next couple of years an increase in both postal and passenger traffic spurred Imperial to invite Shorts to submit a proposal for an improved Kent to carry 24 passengers and 1½ tons of mail at 150 mph, over a range of 800 miles. Almost simultaneously, the Air Ministry also invited the company to tender for Specification R 2/33 for a four-engined flying-boat.

The two specifications were almost identical and Oswald Short and designer Arthur Gouge began mulling over plans for a larger biplane. Then, in October 1934, they went to watch the start of the MacRobertson race from Mildenhall to Melbourne. The race was won by the D.H.88 Comet, but Short and Gouge were more impressed by the runner-up, a DC-2 entered by KLM airlines. They realised that the day of the biplane was over, and went back to the drawing board.

The two main problems with a monoplane boat were that a reliable wing juncture would have to be devised, and that the wing loading would be much higher than that of a biplane. Gouge solved this by means of a retractable electric trailing edge flap which increased the lift co-efficient by 30 per cent without penalty in trim or drag. To accommodate the simple wing attachment the hull was made taller

Although closely resembling the 'C-class' Empire Boat, the Maia lower section had larger wings, larger wing floats and numerous other differences. The tail was also raised to counteract the aerodynamic effects of the upper section.

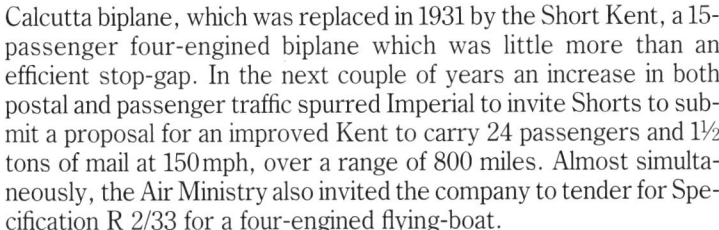

During World War II, BOAC received several Sunderland Mk IIIs to supplement the 'C-class' boats on vital empire links. These were stripped of military equipment and given austere seating. They went into service on the Poole-West Africa sector of the 'Horseshoe Route' in December 1942. With the war over, BOAC hastily fitted a luxury airline interior for 16 or 24 passengers, this aircraft (Hythe) being the first, also giving its name to the class. The Sunderland nose turret was retained for ease of mooring, and the blunt tailcone was also retained.

than was strictly necessary; this resulted in a two-deck layout which compensated for the narrowness of the cabin compared with that of the Kent. The hull itself was of classic Short Brothers design, with two massive built-up box-frames in line with the spar booms, forming double-skinned bulkheads with doorways. The stressed-skin construction was of 'Alclad' light alloy, apart from the tailplane and fin, which were fabric-covered between the spar booms.

Ingenious features

There were many ingenious design features, most of which were patented by Short Brothers. For instance, the engine nacelles were of monocoque construction and built into the leading edge, but adjacent sections of the latter were hinged to fold down and serve as maintenance platforms for the engineers while the aircraft was afloat – thus saving countless man-hours in dry-docking.

Four 985.6-kW (920-hp) Bristol Pegasus Xc radial engines were installed in long-chord cowls with exit gills adjustable in flight; the two-position propellers were DH Hamiltons. The two 326-gallon fuel tanks were located within the wing-spar truss between the engine nacelles.

The streamlined nose incorporated a large flight deck, with dual

Short S.23 C-class cutaway drawing key

1 Starboard elevator
2 Tailplane construction
3 Elevator tab
4 Tail navigation light
5 Tailcone
6 Rudder tabs
7 Fabric-covered rudder construction
8 Fin girder construction
9 Leading edge construction
10 Aerial cable
11 Port fabric-covered elevator
12 Port tailplane
13 Rudder and elevator control levers
14 Tailplane attachment double frames
15 Fuselage frame and stringer construction
16 Fuselage skin plating
17 Rear bulkhead
18 Baggage door
19 Aft main baggage bay
20 Cabin rear bulkhead
21 Window panels
22 Bilge keel construction
23 Aft cabin seating, six passengers

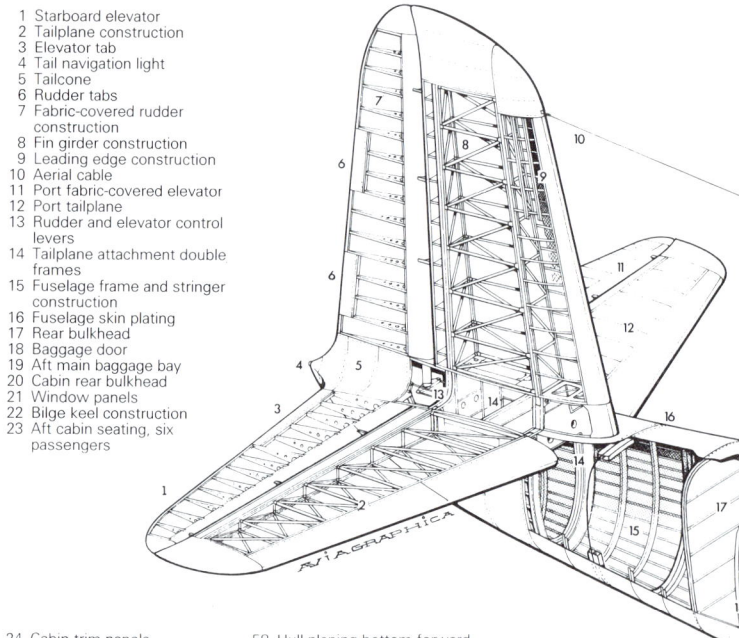

24 Cabin trim panels
25 Overhead luggage racks
26 Rear entry door
27 Bulkhead doorway
28 Wing root trailing edge fillet
29 Cabin roof bedding stowage
30 Window curtains
31 Wing root rib
32 Starboard aerial mast
33 Promenade cabin, eight passengers
34 Starboard flap shroud
35 Starboard Gouge-type flap
36 Girder construction rear spar
37 Trailing edge ribs
38 Starboard aileron
39 Fixed tab
40 Aileron control horns
41 Wing tip fairing
42 Starboard navigation light
43 Wire-braced wing rib construction
44 Front girder spar
45 Leading edge nose ribs
46 Float mounting struts
47 Diagonal wire bracing
48 Starboard wing-tip float construction
49 Landing/taxiing lamp
50 Wing stringers
51 Overwing exhaust outlet
52 Carburetter air intake
53 Starboard outer engine nacelle construction
54 Engine mounting ring
55 Exhaust collector ring
56 Detachable engine cowling
57 Oil cooler radiators

58 Hull planing bottom forward step
59 Midships cabin, three passengers
60 Midships window panel
61 Starboard inner engine nacelle
62 Cooling air flaps
63 Nacelle tail fairing
64 Heater intake duct
65 Cabin heater/exhaust heat exchanger
66 Wing/fuselage main spar attachments
67 Root rib cut-outs
68 Wing spar centre section carry-through
69 Port Gouge-type trailing-edge flap
70 Flap screw jack
71 Port aerial mast
72 Flap guide rails
73 Port aileron
74 Aileron control cables
75 Fixed tab
76 Port wing-tip fairing
77 Port navigation light
78 Landing/taxiing lamp
79 Port outer engine nacelle
80 Oil tank
81 Bristol Pegasus XC air-cooled 9-cylinder radial engine
82 De Havilland three-bladed propeller
83 Propeller hub pitch change mechanism
84 Port wing-tip float

Throughout their career, the Short boats evoked glamour and style. Imperial/BOAC fitted its aircraft with the most luxurious interiors. This is the buffet area of a BOAC Solent 3.

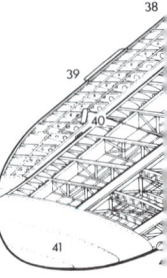

controls for the captain and first officer, a desk for the navigator-radio operator, and a compartment for the auxiliary electrical power unit. This deck also provided accommodation for the flight engineer – carried on long-haul routes – and the purser, along with stowage accommodation for bedding. Below the flight deck in the extreme bow was the mooring compartment with anchors, drogues, and tow ropes. Amidships was a steward's pantry, and in the normal configuration seats were arranged in front and rear cabins for 24 passengers; on long hauls, sleeping accommodation for 16 was provided. Amidships was the promenade cabin containing both seats and a walk-way, with an elbow rail under the windows on the port side. Right aft was the main freight hold, with a large loading hatch on the starboard side.

Even the adjustable tubular magnesium-alloy seats were of revolutionary design, developed by a firm which, by coincidence, occupied the premises in Battersea which had once housed the Short Brothers' balloon works.

Imperial Airways was delighted with the aeroplane's design, and made the bold decision to order 28; its need was so urgent that it took the technical risk of dispensing with prototype trials in advance of the contract signing, which was worth £1,750,000. The Air Ministry ordered a single prototype R.2/33, which eventually emerged as the Sunderland. The civil design was indexed as S.23; Imperial Air-

Three Sandringham 6s were ordered for DNL, the Norwegian national airline. These operated a coastal service from Oslo to Tromso during the summer months, although flying conditions were always dangerous, due to rough terrain. All three were lost in crashes, with two more being supplied as replacements.

85 Ram air intakes
86 Oil radiators
87 Outboard main fuel tanks
88 Port inner engine nacelle
89 Exhaust collector ring
90 Cooling air flaps
91 Overwing exhaust outlet
92 Exhaust pipe heat exchanger
93 Inboard main fuel tank; total tankage, 2727l (600 Imp gal)
94 Engine cowl flaps and fuel cock controls
95 Ship's clerk's station
96 Upper deck crew entry door
97 Access ladder between decks
98 Steward's galley
99 Port side toilet doors, two toilets
100 Upper deck level
101 Port mail and freight compartment
102 Sliding door
103 Forward entry door
104 Smoking lounge, seven passengers
105 Fuselage chine member
106 Forward fuselage porthole
107 Radio operator's seat
108 Radio racks
109 Aerial mast
110 Pitot tubes
111 Cockpit roof hatch
112 Chart table
113 Cockpit roof trim control cables
114 Pilot's seat
115 Sliding cockpit side windows
116 Co-pilot's seat
117 Control column
118 Rudder pedals
119 Instrument panel shroud
120 Curved windscreen panels
121 Mooring hatch
122 Marine equipment compartment
123 Mooring ladder
124 Anchor winch
125 Anchor stowage
126 Retractable mooring bollard
127 Towing cleat

Wartime successes with the Sunderland were the spur behind the Sandringham conversion, which featured a stylish low-drag nose and pointed tailcone. Five Sandringham 2s were supplied to the Argentine airline Dodero, this being the second delivered in December 1945.

ways designated it 'C-class' and eventually gave it the official name 'Imperial Flying-Boat', but to passengers and crew it remained the 'Empire Boat'.

Wing span was 34.75 m (114 ft 0 in) and length 26.84 m (88 ft 0 in). Maximum take-off weight was 19,732 kg (43,500 lb), with a maximum level speed of 311 km/h (200 mph) and a range of 1,245 km (760 miles).

The Empire takes to the air

The first Empire, named *Canopus* and registered as G-ADHL, made its successful maiden flight from the Isle of Sheppey on 4 July 1936, and was quickly followed by the *Caledonia* and the *Cambria* – all the Empires were subsequently were given names beginning with 'C'. All 28 S.23s were delivered, plus three for QANTAS in Australia. Subsequently 11 S.30s were built, eight for Imperial and three for Tasman Empire Airways, with 663-kW (890-hp) Perseus XIIc sleeve-valve engines and greater range – the first four being equipped for flight refuelling to greater weight. The final two Empire boats were S.33s, with increased weight and Pegasus engines.

During World War II, most of these huge aircraft were re-engined with the same 752.6-KW (1,010-hp) Pegasus 22 engines as their military version the Sunderlands, and flew on long-haul journeys, troop carrying and serving as RAF radar craft as well as maintaining the mail routes. After the war, most passed into the hands of BOAC, but

The ultimate Short flying boat was the Solent 4, ordered for TEAL and first delivered in November 1949. Aoteroa II was the first of the four acquired by the airline, launched by Princess Elizabeth in May 1949. TEAL's fleet was augmented by one ex-BOAC Solent 3.

were retired in 1947.

But this was in the future; even as the first S.23 was taking shape, Oswald Short and Arthur Gouge were grappling with an almost parallel project. Imperial Airways was eager to extend its airmail franchise westwards to Canada and the United States. Two routes were considered: the south Atlantic, with refuelling points in the Azores and Bermuda, and the 'Great Circle' via Ireland and Newfoundland. But the busy harbour at Horta in the Azores was too constricted to take large flying-boats, and weather conditions made the northerly route feasible only in mid-summer. Two solutions were at first considered; a floating barge in the Azores on which an amphibian could land and be re-launched by catapult, and a proposal by Sir Alan Cobham for mid-air refuelling. Later, this proposal was adopted with the S.30 version of the Empire, but meanwhile Major Robert Mayo, technical manager of Imperial Airways, had a third solution: an Empire flying-boat with a smaller aircraft on its back which would be launched in mid-Atlantic to complete the journey on its own.

As early as 1916 one such successful experiment had been carried out when a Porte Baby flying-boat had carried a Bristol Scout C. Accordingly, Mayo's idea went ahead, and in January 1936 he was granted a joint patent with Short Brothers for the 'Mayo-Short Composite'.

Controlling the Composite

When coupled, the two components of the Composite would behave like a single aircraft, with the upper controls locked until release. The bottom half was basically the same as an Empire boat but with a broader planing bottom and lowered wing floats to give more positive lateral stability on the water. The wing span was the same as the Empire's, but was 250 sq ft greater in area, and the engines were set further out to give clearance to the upper component. The tailplane was raised to suit both the 'biplane' composite and the solo lower half. The cabin was furnished for 18 passengers.

The upper component, designated S.20, was a clean twin-float mid-wing seaplane with four air-cooled Napier-Halford Rapier engines. Total fuel capacity was 1,200 Imperial gallons, giving a still-air cruise range of 3,500 miles, and a maximum solo weight of 16,000 lb. The pilot and navigator sat in tandem forward of the wing, while under and aft of the wing was a hold for 1,000 lb of mail. When ready for separation, lights in both cockpits of the Composite indicated trim and tension, and the upper pilot pulled a release lever.

The bottom half, registered G-ADHK and named *Maia* was first tested on 27 July 1937, while the upper component, registered G-ADHJ and named *Mercury*, followed on 5 September. The first Composite flight and separation was achieved on the Medway on 6 February 1938 while the second, with separation at only 700 feet,

Few flying boats remain from the golden age. This Solent 3 is currently undergoing restoration to flying condition at Oakland, California. It was originally part of the BOAC fleet, before passing to Trans-Oceanic Airways, South Pacific Airlines and Howard Hughes.

was performed on 23 February for press and newsreel cameras.

On 21 July, *Mercury* was launched over Foynes, Ireland, and arrived at Boucherville, Montreal, 20 hrs 20 minutes later, having flown 2,930 miles at a gross weight of 20,800 lb, including 600 lb of newspapers and newsreels. Despite headwinds of 25 mph and stronger, Captain D.T.C. Bennett still had 80 gallons of fuel in hand. The whole performance had been very much better than either Short Brothers or the Air Ministry had calculated.

On 21 September the Composite took off from Dundee and 42 hrs and 6,045 miles later, *Mercury* touched down in Orange River, South Africa, having broken the world distance record for a seaplane. Imperial Airways made plans for a small fleet of the Composites, but unfortunately war was looming, Short Brothers had more urgent work in hand, and the experiment was curtailed. *Maia* was destroyed at her moorings in Poole Harbour by enemy bombs on the night of 11 May 1941, and three months later *Mercury*, after a spell as a coastal reconnaissance trainer, was broken up at Rochester.

During the building of the Mayo-Short Composite, the Empire boats had begun their career to the great satisfaction of Imperial Airways. Accordingly, Imperial gave Shorts specifications for a larger version to carry mail on the North Atlantic run. They would be manned by a crew of five and be required to carry two tons of mail for 2,500 miles, against a 40-knot headwind. These 'G-boats' would be powered by four Bristol Hercules IV engines, serviced by six drum-type tanks containing 3,600 gallons of fuel within the deepened wing trusses. Three were built and registered as G-AFCI *Golden Hind*, G-AFCJ *Golden Fleece* and G-AFCK *Golden Horn*. With a hull depth of 19 ft and a wing span of 134 ft 4 inches, *Golden Hind* was one of the biggest flying-boats ever to be built when it emerged for its test flight on 21 July 1939.

G-class goes to war

By mid-September all three were handed over to Imperial Airways for crew training – rather optimistically, as it happened, for war had broken out and within days the giant trio was commandeered as X8275, 4, 3 respectively, along with their crews, and posted to Stranraer for long-range reconnaissance work. Their military service proved to be short.

On 20 June 1941, *Golden Fleece* suffered failure in two engines and made a forced landing off Cape Finisterre, breaking up almost immediately. On 6 December, the remaining two G-boats were handed back to BOAC for civilian duties on the Foynes to Lagos run via Lisbon, carrying mail and 40 passengers in 'austerity' seating. On 9 January 1943, *Golden Horn* crashed into the Tagus on a test flight after engine overhaul, and the remaining *Golden Hind* spent the rest of the war flying internal routes in South Africa. In 1945 it was returned to Short and Harland in Belfast for a re-fit and installation of Hercules XIV engines. Then, equipped to carry 24 first-class passengers, it began its first truly commercial peacetime career flying from Poole to Cairo, until it was sold to a private operator in 1947. On 17 April it made its last flight to moorings on the Medway, where it remained until scrapped in May 1954.

Civilian Sunderlands

Meanwhile, in 1943, the RAF had released several of its Sunderlands, re-equipped to carry 20 passengers, for civilian service with BOAC. Shorts used the basic Sunderland airframe, therefore, and with a few secondary structural changes and a complete interior re-arrangement issued the resulting aircraft to BOAC as the Sandringham. Powered by four 745.2-kW (1,000-hp) Bristol Pegasus 38 engines, it accommodated 24 passengers by day or 16 by night.

Re-equipped with Pratt & Whitney R-1830-92 Twin Wasps, the Sandringham also saw service with Argentinian, Tasmanian and Norwegian airlines.

And an even more successful offspring of the Sunderland, by way of the military Seaford variant, was the Short Solent. Twelve were operated by BOAC as 30-passenger Solent 2s, while a further six were 34-passenger Solent 3s. Power for these was provided by Bristol Hercules 637 engines, and all 18 were still in fine condition when, in 1950, BOAC abandoned the use of flying-boats, effectively bringing a memorable era of aviation history to an abrupt end.

The Solent was based on the Seaford military boat, the first aircraft (Solent 1) being converted from Seaford NJ201. The first order was from BOAC, for 12 Solent 2s with day interiors for 34 passengers. Sark was the fifth aircraft from the line.

Shorts S.45A Solent Mk 4
ZK-AMO 'Aranui'
Tasman Empire Airways Ltd

Chapter 33

Tupolev Tu-104

Tupolev Tu-114 'Rossiya'

Tupolev Tu-134

Tupolev Tu-144 'Charger'

Tupolev Tu-104

Following hot on the heels of Britain's Comet, the Tupolev Tu-104 was the second jet-powered airliner to enter service, and it was to suffer none of the immense and tragic difficulties of its rival. It was one of the many aircraft types to upset the widely-held belief that Soviet machines were inferior to those of the west.

Among the new aircraft which flew over Moscow's Red Square on May Day 1954 were nine enormous swept-wing twin-jets. They were the Tu-88 prototype and eight pre-production Tu-16s. Western observers, completely ignorant of such aircraft, at first called them Type 39 and then 'Badger'. The presence of so large a formation made it highly probable that this type was going into production (though one cannot take much for granted in analysing Soviet aircraft, and it might have indicated just the opposite), but not even Andrei N. Tupolev could have had the slightest idea of how important these aircraft would be, or how long a career they would have in so many versions.

Another thing not appreciated by outside observers was that the new jet was very much a thoroughbred. The great Tupolev OKB (experimental construction bureau) had previously built aircraft with almost the same fuselage, as well as big jet bombers and aircraft with swept wings and tail. In parallel it was building the even bigger turboprop Tu-95 ('Bear'). What it had never done before was to build an all-swept machine of this size, nor to use the very large Mikulin axial turbojet engine and bogie main landing gears.

It was in 1948 that the engine KB of Colonel-General A. A. Mikulin was given the task of creating a new turbojet more powerful than any previously attempted. There was no attempt at any extraordinary new technology: more thrust was to be gained by direct frontal attack, by making a bigger engine. Many German engineers, still effectively prisoners, were included in the design staff of over 150 headed by P.F. Zubets. As the M-209, this engine ran in 1950 and was qualified in January 1952 at a thrust of 6750 kg (14,881 lb), with plenty more to come. A classically simple engine, it had an eight-stage axial compressor with a pressure ratio of 6.4. When later cleared for production, it received the service designation of RD-3.

This big engine was the key to jet bombers of impressive capability, and it was picked for the Myasishchev M-4 four-jet bomber demanded by Stalin to have the capability of missions to the USA, a task in fact beyond the state of the art. Tupolev was more fortunate: at the same time in summer 1950 his OKB was given an order for three prototypes of an advanced bomber powered by two of the new engines. It did not have to reach the USA and, unlike Myasishchev, Tupolev had done most of it before, so the project moved very fast. The competition came from S. V. Ilyushin, whose Il-46 was smaller and less powerful, but even more of a known quantity, being a direct scale-up of the Il-28. The Il-46 was merely a less-capable insurance against any complete failure by Tupolev.

Round-the-clock development

At first the Tupolev was called Aircraft N, but it also received the OKB designation Tu-88. Naturally the smaller Tu-86, with a span of 25.5 m (83 ft 8 in) and TR-3 engines, was dropped; the Tu-88 was much more capable. Work went on, round the clock, on a three-shift

The first prototype Tu-104 (SSSR-L5400) displays its sleek and purposeful lines. Although it had a wider fuselage than the Tu-88 (Tu-16) bomber from which it was derived, the Tu-104 retained the swept wings and tail surfaces that gave it such good performance.

An early production Tu-104A displays the broad swept-wing, complete with trailing edge fairings that held the main undercarriage. The main units each had four-wheel bogies.

basis, and the Tu-88 prototype was rolled out in the first weeks of 1952. Tupolev later said it was the most beautiful aircraft he had ever built.

Laminar profile

The wing was scaled-up from that of the Tu-86, which in turn was a modified enlargement of that of the Tu-82, flown in February 1949. Like most fast Soviet jets of the day, the Tu-88 had a wing of SR-5S 'laminar' profile, with thickness tapering to 7.5 per cent at the tip, with leading-edge sweep 42° inboard to a structural break and thence 35° to the tip. Leading-edge de-icing was by hot engine bleed air, discharged through a grille at each tip. Machined skins made it easier to use integral tankage, fuel capacity in this bomber being five times as great as in the gigantic *Maksim Gorkii* of 20 years earlier, at 44900 litres (9,877 Imp gal), which increased in later versions. The wing was a considerable structural challenge, and the skins at the root were the thickest on any contemporary aircraft except the Boeing B-47. The big track-mounted slotted flaps were actuated electrically, but the ailerons were manual, with geared tabs.

This aircraft has of course become famous as the Tu-16 'Badger' bomber, which has proved its longevity and versatility time and again, and is still in widespread service in electronic, reconnaissance, tanking and missile-carrying roles. However, the superb wing and tail surfaces made the Tu-88 an obvious basis for a jet airliner. Tupolev took his drawings of the Tu-104 airliner derivative to Stalin a few weeks before the dictator's death, and received approval for work to begin.

Obviously the main new feature would be the fuselage, for the slim, pencil-like body of the Tu-88 bomber did not have sufficient volume for a passenger-carrying aircraft. Consequently the diameter was raised from 2.9m to 3.4m, retaining the pressurisation of the bomber. The main passenger cabin was 16.11m long, 3.2m wide and 1.95m high, consisting of a small forward area and a long rear area, the two divided by an awkward raised floor which passed over the wing centre section. This raised area was used on early aircraft to house a galley. Two toilets were located in the rear cabin, and another in the forward section. Up front in the cockpit was provision for up to five crew members, comprising two pilots, navigator, radio operator and flight engineer.

Smooth adaptation

Other differences between the airliner and its bomber predecessor were a wider wing centre section, and the engines mounted in complete nacelles with the intakes standing proud from the fuselage. The tailplane was moved down to be mounted on the fuselage, giving the airliner a neater look, while the nosewheel was moved forward. In fact the Tu-104 sat much higher off the ground, and with its enlarged cabin had an impressive air. Throughout its development, the Tu-104 had proved an astonishingly easy and smooth-running adaptation of the bomber design. All boded well for the future career of this major Soviet achievement.

The prototype, SSSR-L5400, was taken aloft by Y. I. Alasheyev on 17 June 1955, and reportedly took part in that year's Aviation Day at Tushino. Flight trials proceeded smoothly throughout the year and into the next, establishing the Tu-104 as an excellent design. These trials allowed the production aircraft to be finalised, this being a 50-seater. In the summer of 1955, crew training and route-proving for Aeroflot started, using a pair of demilitarised Tu-16s (SSSR-L5402 and -L5411). Known as *Krasnyi Shapochka* (Little Red Riding Hood), these interim trainers were designated Tu-104G.

On 22 March 1956, the prototype Tu-104 arrived at London's Heathrow Airport with a party of KGB officials, who had come to discuss a visit by Soviet leaders. The aircraft caused a great stir, for not only was it a rival to Britain's Comet, but its swept flying surfaces promised greater performance and technological superiority over

Two views of SSSR-42430, a Tu-104A. A notable difference between the bomber and airliner was the engine nacelle. On the Tu-104 it was part of the wing, a small section of leading edge joining it to the fuselage. The engines were mounted at the rear of the nacelle.

A Tu-104 takes off with tail bumper extended to avoid grazing the rear fuselage if the rotation was too great. The Tu-104 needed a lot of runway, and the climb-out was not impressive.

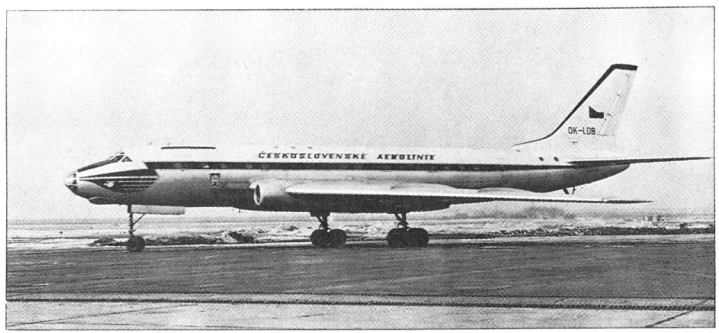

The Czech national carrier CSA was the only export customer for the Tu-104, somewhat surprising given the very low price tag. The six Czech Tu-104As had a different central window arrangement.

any Western airliner. In April, the aircraft returned to Heathrow with the first two production machines, SSSR-5412 and -1413, between them carrying Krushchev and his party. The Tu-104 was a force to be reckoned with.

Aeroflot service began on 15 September 1956, initially on the Moscow-Irkutsk route, followed by Moscow-Prague on 12 October. Every Aeroflot directorate was later to receive the aircraft, and throughout the late 1950s and early 1960s, the type became Aeroflot's most important. In early service, the Tu-104 had two small compartments ahead of the wing spars, each with tables and accommodating six and eight passengers. Behind the spars was another small eight-seat cabin, behind which was the main 28-seat cabin. The interiors were in mahogany and brass, lending a Victorian air to the accommodation.

Uprated model

Improvements to the Mikulin RD-3 engine (civil designation AM-3) allowed Tupolev to produce the Tu-104A in the winter of 1956-57. This had an increased gross weight, which in turn allowed the accommodation to be increased to 70 passengers, including five-abreast seating in the rear cabin. Production switched to this model at about the 11th aircraft, and many followed. On three days in September 1957, a Tu-104A set new world class records for altitude with 20000 kg load (11221 m), speed over 2000 km closed circuit with 2000 kg load (847.498 km/h) and speed over 1000 km circuit with 10000 kg load (970.821 km/h).

Further development of the airliner saw Tupolev add a small (1.21 m/3 ft 11½ in) stretch to the fuselage, producing the Tu-104B, and thereby increasing payload to take better advantage of the fully-rated AM-3M engines. The standard accommodation was 100, 30 passengers forward of the spars, 15 above them and the remainder behind, all seating being five-abreast with a single aisle. Numerous other improvements were made, including changes to the underfloor baggage holds and entry door, and increases in wing and flap chord to maintain take-off performance despite the extra weight.

This variant went into Aeroflot service on 15 April 1959 on the trunk Moscow-Leningrad route, and shortly after set some more class records, these being speed over 2000 km closed circuit with 15000 kg load (1015.86 km/h) and altitude with 25000 kg load (12799 m), signifying a considerable performance advance over the Tu-104A.

The Tu-104B was the definitive version of the aircraft, featuring a lengthened fuselage to even out payload/thrust ratios once the AM-3 engine was delivering its full potential. These initially seated 100.

Several variants were produced of the aircraft during its production run. Some Tu-104As were developed as the Tu-104V with new interiors to seat 100 passengers, these being housed 25 in the front cabin, 15 above the wing and 60 in the rear. The similar Tu-104D conversion raised accommodation to 85 only. The Tu-104E was a special Tu-104B variant used for further record-breaking, flying at 959.94 km/h over a 2000 km closed circuit carrying 15000 kg.

A considerable development of the basic design was the Tu-110, a four-engined derivative based on the Tu-104B. The four Lyul'ka AL-5 turbojets provided greater safety and better field length, while the aircraft possessed better handling. However, the Tu-110 remained in prototype form only, as the improvements over the Tu-104B were not considered great enough to warrant production. The single aircraft (SSSR-5600) was turned over to the air force.

Short-haul brother

Although a completely new aircraft, the Tu-124 short-haul airliner deserves mention, for it adhered very closely to the Tu-104 layout. Powered by Soloviev D-20 turbofans, the Tu-124 was roughly 25 per cent smaller in dimensions than the Tu-104, although at a distance the two were difficult to distinguish. First flying in June 1960, around the time Tu-104 production was closing down, the Tu-124 went on to compile an equally impressive service record, although only CSA, Interflug, and the Iraqi and Indian air forces bought the type.

Production of the Tu-104 terminated in 1960, with about 200 of all

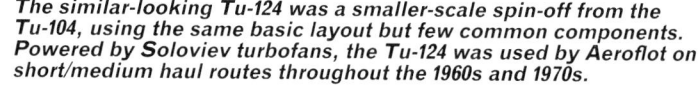

The similar-looking Tu-124 was a smaller-scale spin-off from the Tu-104, using the same basic layout but few common components. Powered by Soloviev turbofans, the Tu-124 was used by Aeroflot on short/medium haul routes throughout the 1960s and 1970s.

variants built. Despite a very low price tag of £425,000, the only export customer was the Czech airline CSA, which bought six Tu-104As from November 1957, each configured for 81 passengers. During the late 1960s, most of the Tu-104s were refurnished for 104 or 115 passengers. In 1981, the grand old lady was finally retired from Aeroflot service after having carried around 80 million passengers. Several passed into military hands for use as staff transports, while others performed zero-g training for cosmonauts. At least one was modified with a nose radar and cloud-seeding rockets on wing pylons for use as a meteorological research platform.

Now sadly gone, the Tu-104 achieved a notable place in aviation history. It was the second jet airliner into service, and considering the tragic difficulties which tarnished the early career of Britain's Comet, the Tu-104 may certainly be regarded as the first *successful* jetliner. In service it proved more reliable and safer than any other Soviet machine at the time, although it was heavy handling in some regimes and internally was very noisy. A lasting effect it had on Soviet civil aviation was to instigate a major lengthening programme of the nation's runways in order to accommodate its not inconsiderable take-off run, a fact which facilitated the enormous growth there in civil aviation during the 1960s.

In the later years of their service, Tu-104As were modified to raise their accommodation, as either Tu-104D (85 seats) or Tu-104V (100 seats). The badge carried on the nose of most aircraft was a stylised Cyrillic 'Tu'.

The Tu-124 fared only marginally better in the export market than its larger brother. CSA again bought the type, as did Interflug (East Germany) and two air forces (India and Iraq).

Tupolev Tu-114 'Rossiya'

The Tu-114 was Andrei Tupolev's response to Stalin's request for a Soviet civil transport which could span the globe. While it had its roots in Tupolev's military designs, it broke new ground in the history of air transportation. In Aeroflot service it became an airliner the like of which the world had never seen.

In 1953 the Soviet civil aviation organisation GVF issued requirements covering several completely new types of aircraft. Existing equipment was well matched to most small and ill-equipped airfields but, by international airline standards, was considered obsolescent. In particular, there was no aircraft available with either long range or large passenger capacity. So, before he died in February 1953, Stalin personally gave A. N. Tupolev permission to develop a transport aircraft derived from the Tu-88 bomber (service designation Tu-16), and this quickly took shape as the Tu-104 'Camel'. In 1954 Tupolev received a further instruction that he should attempt to meet Aeroflot's need for a large, long-range transport through a similar adaption of the Tu-95 bomber (service designation Tu-20).

At that time the Tu-95 'Bear' and its remarkable engines and propellers were still not fully developed. The prototype did not fly until the summer of 1954, and less than a year earlier a famous test pilot, A. D. Perelyet, had been killed by a propeller blade from a disintegrating engine on test in a Tu-4LL. Thus, while design of the Tu-114 transport went ahead, plans were quickly prepared to build three much simpler civilianised Tu-95s, designated Tu-116, to explore various expected compatibility problems with the GVF airfields and carry out intensive route testing of the unique propulsion system when flying to commercial type schedules. It was also expected that, in doing so, the Tu-116s would gain some impressive world records, even though this would give the West a reasonably accurate idea of the bomber's performance.

The main common elements in the Tu-95, -114 and -116 were the wing, propulsion system and landing gear. All were remarkable, especially in the context of the early 1950s. When the Tu-95 bomber was first seen publicly in 1955, Western observers were unable to comprehend how a propeller-driven aircraft could need sweptback wings and tail. It was not until the Tu-116, and later the Tu-114, began setting world records that it was understood that this particular engine/propeller combination drives all these aircraft at what is commonly called jet speed.

The engine was the NK-12. It bears the initials of N. D. Kuznetsov, who was then titular head of the engine experimental construction bureau which created it, even though the actual work was done mainly by a team of German prisoners headed by Austrian Ferdinand Brandner. A single-shaft turboprop, it was amazingly advanced for its day, besides being enormous. The initial production version was rated at 7994 kW (10,730 hp), but by 1957 the Tu-116 and the Tu-20 bomber had the NK-12M rated at 8936 kW (11,995 hp) and a year later the definitive NK-12MV became available with a maximum power of 11018 kW (14,790 hp). No other turboprop has ever even approached such power.

To convert this enormous shaft power into propulsive thrust the aircraft all use the AV-60N propeller. This was made up of two four-bladed units, rotating in opposite directions, each unit turning at a constant 750 rpm.

SSSR-76490 is believed to have been the last Tu-114 built, production totalling 31 aircraft plus the single prototype. The enormous height of the nosewheel strut is readily apparent.

Left: Another view of the last Tu-114. The massive airliners were used on international services to Havana, Conakry, Accra, Delhi, Paris, Copenhagen and Tokyo, in addition to one-off services often connected with government business.

Below: A Tu-114 displays typical Aeroflot colours. The motif on the nose was that of the design bureau, being a stylisation of the lower-case Cyrillic letters 'Tu'. The radar under the nose was the same as used by early 'Badger' and 'Bear' bombers.

Each four-bladed unit is independent, the engine's gearbox splitting the drive along two separate trains of gears. The diameter is 5.6 m (18 ft 4.5 in). The blades are thin solid duralumin, with leading edges de-iced by electric heater mats. In flight they are set to an extremely coarse pitch, matched to high airspeeds. After landing they can go into reverse pitch to shorten the run, especially important on icy runways.

Each engine is mounted on a rigid framework of welded steel tubes carried off the front spar of the wing. The air inlet is an annular ring immediately aft of the spinners. Nearly all the rest of the cowling is made up of hinged panels giving excellent access, though, like most of these aircraft, steps are needed. Underneath is the ducted oil cooler, and at the back the jetpipe is bifurcated (split) to exhaust on each side under the leading edge of the wing.

The wing, still one of the largest swept wings in the world, has a sweep angle of 37°, and root thickness/chord ratio of 12 per cent. Three fences run straight across the upper surface. Most of the trailing edge is occupied by double-slotted flaps, sometimes incorrectly described as of the Fowler type (unlike the Fowler, most of the retracted flap has no fixed structure above it). Outboard are long-span ailerons with hydraulic boost. Most of the inter-spar box is filled with fuel, total capacity being 71617 litres (15,754 Imp gal). This was slightly less than for the Tu-20, and not much more than half the capacity of later Tu-142 versions. The main landing gear in all early Tupolev bombers and transports was a bogie with four large wheels, with electric retraction to the rear. As the gear retracts, the bogie rotates until it lies inverted in a large streamlined fairing projecting aft of the wing.

Bomber lineage

The Tu-116 had exactly the same nose landing gear as the bomber, a tall forward-raked unit with twin hydraulically steered wheels, folding to the rear. The all-swept tail was also the same as that of the original bomber, with a fixed tailplane mounted part-way up the fin. The fuselage was structurally little altered, with the same 2.89-m (9-ft 6-in) diameter as the B-29 which was the parent of the whole range of Tupolev strategic aircraft. Of course, all armament and the weapon bay were removed, and the rear fuselage was arranged as a pressurised passenger cabin, seating 24 or 30 in double seats each side of the aisle.

Veteran Tupolev designer A. A. Arkhangyelskii had in effect been chief designer of all the large jet and turboprop bombers, and he remained in this position on the derived civil transports. Little difficulty was encountered with the three Tu-116 aircraft, the first of which flew in late 1956. It bore civil registration SSSR-76462, fourth in the number sequence allocated to the Tu-114 programme. The second and third aircraft bore military insignia and the large numbers 7801 and 7802. The civil aircraft was accepted by Aeroflot and called the Tu-114D (D for Dalnyi, long range). Surprisingly, it was never used for any impressive international or global flights, and with its low-powered engines it did not reflect the capability of the bomber or Tu-114. During 1958 no. 76462 did make a number of publicly announced flights, the most impressive of which was Moscow to Lake Baikal, 9600 km (5,965 miles), at an average speed of 740 km/h (460 mph). In another flight the 8500-km

Above: Because of the massive propellers, the aircraft had to stand high off the ground, creating numerous difficulties in providing the right size of stairs to reach the doors. One solution was to add another section of steps to existing truck-mounted airline steps.

Left: A JAL marshal is dwarfed by a Tu-114 as it arrives at Tokyo. The outer engines were usually stopped for taxiing.

(5,282-mile) round trip between Moscow and Irkutsk was flown at an average 800 km/h (497 mph), despite headwinds of up to 200 km/h (124 mph). The Tu-116/114D was said to have a gross weight of 121925 kg (268,796 lb), about 40825 kg (90,000 lb) less than might have been expected. Nothing was heard of these aircraft after 1958.

Aeronautical revolution

The Tu-114 prototype made its first flight on 3 November 1957, immediately prior to the 40th anniversary of the Revolution. It received the civil registration SSSR-L5611 and was occasionally referred to as the Rossiya (Russia). To produce it, the chief modification was exactly the same as in turning the Tu-88 bomber into the Tu-104 transport. A totally new fuselage was produced, mounted higher, so that the wing was in the low position. In turn, this meant that the tailplane could be moved down to be mounted centrally on the fuselage, and pivoted to serve as a powered trimming surface, and a new and much taller nose landing gear was provided. Other changes included increased wing anhedral and redesigned flaps with considerably greater chord, which in turn increased wing area and resulted in a kinked trailing edge.

Left: SSSR-76462 was the first of three aircraft designated Tu-116 or Tu-114D. These were, in effect, demilitarised Tu-95 'Bear' bombers, fitted with the low-powered NK-12M engines. The Tu-114Ds were used for route-proving and crew training.

There has been a vast amount of disinformation about these monster aircraft, and this is strange because they never had any military security classification. One standard Soviet reference book, *Aeroplanes of the Land of the Soviets*, gives the span at 54.00 m (177 ft 2 in) and the length as 47.2 m (154 ft 10.3 in), the latter figure being much less than the length of the bomber versions. The difficulty in discovering the truth is even stranger when it is recalled that, when it appeared, the Tu-114 was the largest and heaviest civil aeroplane in the world (indeed, apart from the B-52's wing span, it was the world's largest active aeroplane).

The enormous new fuselage had a circular cross-section, the shape giving minimum structural weight for a pressure cabin. Diameter was increased to 4 m (13 ft 1½ in), the internal cabin width being 3.92 m (154.3 in). The entire volume was pressurised back to the rear pressure bulkhead at the leading edge of the tailplane, at 57.84 KPa (8.39 lb/sq in). Height along the centreline above the floor was 2.18 m (86 in), and the main passenger cabin area measured 40.3 m (132 ft 2.6 in) long, with a volume of 332.2 m^3 (11,731.5 cu ft).

An extraordinary feature of the Tu-95/114 family was the swept wing, which allowed the propeller aircraft to cruise at 'jet' speeds. On retraction, the mainwheel bogies 'somersaulted' to lie flat in the extended nacelles.

The Tu-114 had the same wings as the 'Bear', but they were mounted low on the fuselage so that the airliner had a cabin unobstructed by the wing carry-through structure.

It is believed that 12 Tu-114s were converted to Tu-126 'Moss' standard, providing the Soviet Union with its first AEW aircraft. All have now been retired.

The nose resembled those of Soviet bombers, the navigator being seated in a compartment in the extreme nose with glazed front windows. Under his floor was the mapping radar, related to the mass-produced Argon set fitted to the Tu-16 and original Tu-20 bombers. Then, at the upper level, was the cockpit with two pilots side by side, with a central gangway and steps down to the nose. Further back were the stations for the engineer and radio operator. A pressure bulkhead was provided at the rear of the flight deck, the door being locked in flight. Next came the main cabin, which could be arranged in various ways. Unlike the Tu-104 the passenger floor was all at the same level, there being room beneath it for the whole of the wing centre-section. The floor was close with the mid-depth diameter of the fuselage, giving maximum floor width but at the cost of inward-sloping walls beside the window seats. The circular passenger windows were similar to those of the Tu-104.

Inside ideas

The standard cabin arrangement seated 170 passengers, which in the 1950s was a very exceptional number. The forward cabin seated 42 in seven rows of triple seats. All seats were rail-mounted, so that pitch could be varied. Next came large coat cupboards on each side, an essential feature in the Soviet Union in winter. At this time there was still fear of shedding propeller blades, and the coat cupboards neatly sterilised the danger region. In addition, large plates of steel armour were incorporated round this region of the fuselage, which was not done with any of the bomber versions. Next came the dining room, divided into four compartments each with facing triple seats separated by a table, a total of 48 seats. Next, marked by an absence of windows, came the galley, with electric lifts to bring the meals up from the kitchen at the lower level. Next came a stairway to the kitchen and a small compartment with seats for two of the cabin crew (normal complement was two galley/kitchen staff and three stewardesses). Then followed two small compartments on each side, each fitted with two triple seats and a folding bunk or two sleeping berths. Then came the main day cabin with nine rows of triple seats (total 54), followed by separate toilets and washrooms

Above and left: Two views of 'Moss' show the main features, including the dorsal rotodome, refuelling probe, extended tailcone and underfuselage blisters housing electronic warfare equipment, and the ventral fin.

and additional coat wardrobes. On the flight deck, ahead of the pressure bulkhead, were toilets and spare seats for the flight or cabin crew when off duty.

There were several other arrangements, the one which caused gasps of astonishment in the 1950s being the short-haul configuration. This used quad seat units in cabins arranged with rows of 3+4 or even 4+4 to accommodate a total of 220 passengers. The three main cabins seated 53, 66 and 76, the other 25 being in what in other arrangements were the four small cabins and galley. Conversely, there were long-haul versions seating only 100 or 104, with numerous bunks and a promenade bar. Surprisingly, the first-class section was invariably the aft cabin (which in a jet would be the noisiest), the seating being a luxurious 2+2.

Placing the floor across the fuselage diameter gave enormous underfloor volume, much of which was unused. The front and rear cargo/baggage holds had capacities of 240 m^3 and 46.0 m^3 (847.6 and 1,624.5 cu ft) respectively (smaller than the holds of a Sixty Series DC-8 with a much more constricted fuselage). At the leading edge of the wing was the enormous pressurisation and air-conditioning plant, with ram air fed in by a prominent projecting inlet under the fuselage. At the trailing edge was the kitchen, loaded via the staircase from the upper deck.

Apart from the prototype, L5611, it is believed that 31 production Tu-114s were built, numbered 76459-76490 (76462 being a Tu-116). From the start the service record of the Tu-114 was exemplary, though at first the engines were considerably derated. On 15 September 1959 the prototype was entrusted with conveying Prime Minister Khrushchev and other high officials non-stop to New York, 6698 km (4,162 miles) being covered with derated power in 11 hours. In March and April 1960 test pilot Ivan Soukhomlin used a production aircraft with fully rated NK-12MV engines to set an impressive number of world records. Most were closed-circuit records with payloads of up to 25 tonnes (55,116 lb). Several set speeds exceeding 850 km/h (530 mph), the flight of 9 April 1960 covering a 5000-km (3,107-mile) circuit at an average speed of 877.18 km/h (545.07 mph). No other propeller aircraft has yet even approached such speeds. In 1991, for example, Saab was preparing to build the Saab 2000 – 'the world's fastest turboprop' – with a maximum cruising speed of 677 km/h (421 mph) and normal cruising speed of 555 km/h (345 mph). In service, the Tu-114 normally cruised at 769-793 km/h (478-493 mph), and for 10 or more hours non-stop.

Throughout 1959-60 production aircraft made proving flights within the Soviet Union, while L5611 attended the Paris air show in June

Left: Thirty-two thrashing propeller blades driven by four of the world's most powerful turboprops gave the Tu-114 an impressive and distinctive noise signature. This is the prototype arriving at Le Bourget for the 1959 Paris air show.

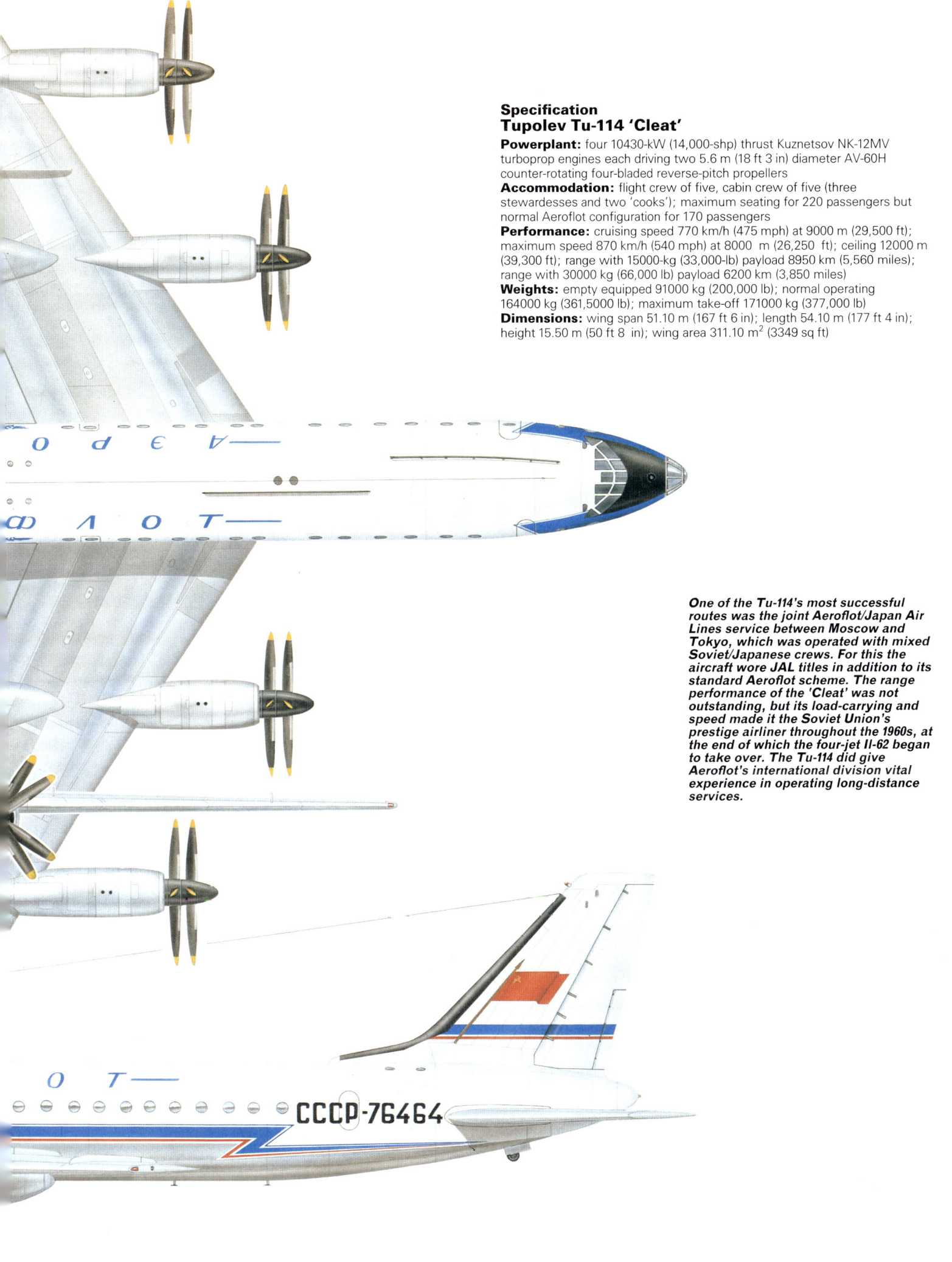

Specification
Tupolev Tu-114 'Cleat'

Powerplant: four 10430-kW (14,000-shp) thrust Kuznetsov NK-12MV turboprop engines each driving two 5.6 m (18 ft 3 in) diameter AV-60H counter-rotating four-bladed reverse-pitch propellers

Accommodation: flight crew of five, cabin crew of five (three stewardesses and two 'cooks'); maximum seating for 220 passengers but normal Aeroflot configuration for 170 passengers

Performance: cruising speed 770 km/h (475 mph) at 9000 m (29,500 ft); maximum speed 870 km/h (540 mph) at 8000 m (26,250 ft); ceiling 12000 m (39,300 ft); range with 15000-kg (33,000-lb) payload 8950 km (5,560 miles); range with 30000 kg (66,000 lb) payload 6200 km (3,850 miles)

Weights: empty equipped 91000 kg (200,000 lb); normal operating 164000 kg (361,5000 lb); maximum take-off 171000 kg (377,000 lb)

Dimensions: wing span 51.10 m (167 ft 6 in); length 54.10 m (177 ft 4 in); height 15.50 m (50 ft 8 in); wing area 311.10 m² (3349 sq ft)

One of the Tu-114's most successful routes was the joint Aeroflot/Japan Air Lines service between Moscow and Tokyo, which was operated with mixed Soviet/Japanese crews. For this the aircraft wore JAL titles in addition to its standard Aeroflot scheme. The range performance of the 'Cleat' was not outstanding, but its load-carrying and speed made it the Soviet Union's prestige airliner throughout the 1960s, at the end of which the four-jet Il-62 began to take over. The Tu-114 did give Aeroflot's international division vital experience in operating long-distance services.

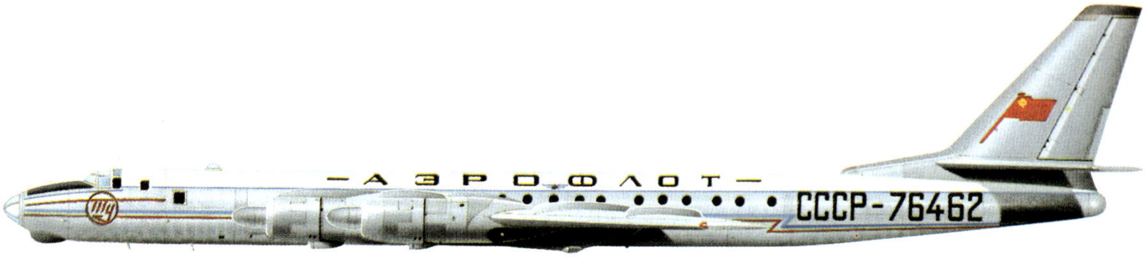

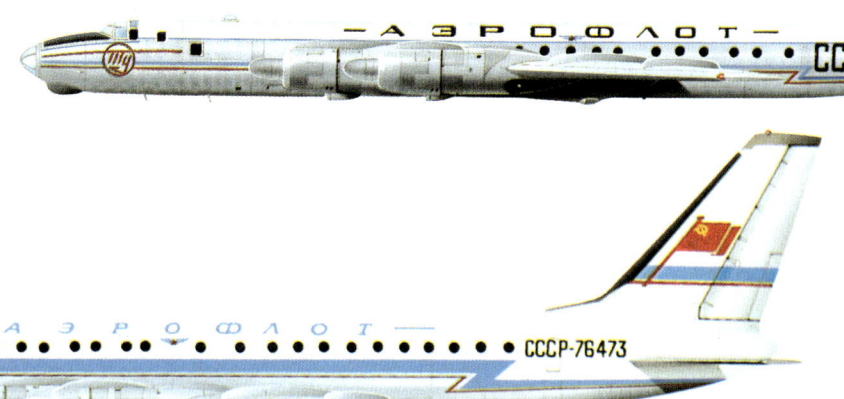

The first Tu-114D wore full Aeroflot colours and civil registration, while the other two had military markings. Windows were incorporated in the rear fuselage, and the tail gun turret was faired over. Otherwise the aircraft were externally identical to early 'Bears'.

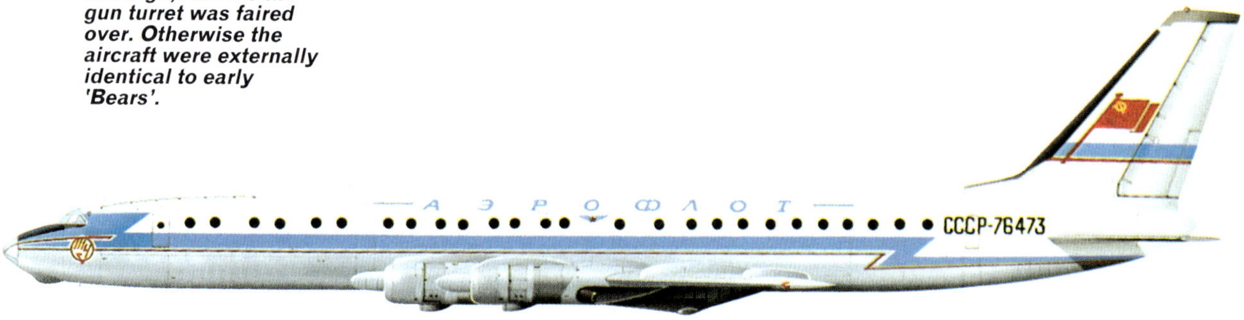

By comparison with the 'Bear' and Tu-114D, the standard airliner had a much fatter fuselage, able to take up to 220 passengers. One hundred and seventy was a more usual seating arrangement.

1959. Considerable difficulty was experienced with the strength and width of taxiways and runways, and with providing stairways able to reach even the underfloor holds 4.1 m (13 ft 5.5 in) above the ground. The passenger doors presented an even bigger problem, the sill of the forward door being 5.7 m (18 ft 8.5 in). At last, regular scheduled services began on 24 April 1961, on the Moscow-Khabarovsk route, the time for the 6800 km (4,226 miles) being 8 hr 15 min. By the end of 1961 this route was flown five times per week.

The Tu-114 gave Aeroflot an intercontinental capability it had never previously possessed. On 7 January 1963 services began from Moscow north to a refuelling stop at Murmansk, followed by a non-stop run to Havana, Cuba. On 25 March of that year the Tu-114 non-stop service began to Delhi, followed by a new route to Conakry and Accra on 19 August 1965, and another to Montreal on 4 November 1966. On 17 February 1966 a Tu-114 (not in line service) crashed taking off from Moscow-Sheremetyevo in a blizzard. This was the only major accident known to have been experienced by any aircraft of this type.

Mother of 'Moss'

In April 1967 services began to Tokyo. This route was operated in partnership with Japan Air Lines, whose titles were carried, using mixed Soviet/Japanese flight and cabin crews. Other cities served included Paris and (on the Montreal service) Copenhagen. From September 1967 the Il-62 began to replace the monster turboprop, but the process took until 1975 to complete. Meanwhile, from 1966, one Tu-114 was converted into an airborne early warning and control aircraft, with a large surveillance radar in a rotating dome carried on a pylon above the rear fuselage. This aircraft featured in a propaganda film in 1968, and led to the Tu-126, at least 12 of which were operated by the AV-MF (naval air force) in 1971-90. One was detached to the Indian Air Force in 1971 and used in the war against Pakistan.

The Tu-126 has a totally restructured fuselage, occupied by the main radar, consoles for a mission crew of 10 or 11 and a flight crew of four. Very comprehensive communications were carried, and an inflight-refuelling probe was added above the nose. These aircraft had an endurance of 25 hr at 10363 m (34,000 ft), compared with 11 hr at 8839 m (29,000 ft) for the Boeing E-3A Sentry.

Tu-114s flew their last airline services in 1975, by which time several had already been converted to Tu-126s. Details about the fate of the others are incomplete, but this example was put on display in new Aeroflot colours.

Tupolev Tu-134

Never able to challenge similar Western types in the areas of performance, reliability or economics, the Tu-134 has nevertheless been a most important type to the Soviet Union and its former client states, most of which took the type into the inventories of their national airlines. Most are still in service, although their days are numbered as Western aircraft and new Russian types become available.

Visitors to major airports throughout Western Europe have for nearly 25 years been pleased to see Tu-134s. Pleased because these aircraft were built in the Soviet Union, and thus are a welcome change from the routine diet of Boeing, McDonnell Douglas and Airbus. Moreover, the -134 is quite striking, with well-sprung bogie main gears that fold back into fairings behind the sharply swept wings, and one of the most pronounced 'T-tails' in the business.

It certainly looks much more modern than its predecessor, the Tu-124. And the -124 stemmed from the -104, which in turn was a passenger version of the Tu-88 (Tu-16 'Badger') bomber, so we can trace the lineage back to the 1940s! Indeed, it really began in the 1930s, with the project design at Seattle of what became the Boeing Model 135, produced as the B-29. Stolen by the Soviet Union and put into production in a modified form as the Tu-4, this was then stretched via the Tu-80 and Tu-85 into the Tu-88 twin-jet and Tu-95 four-turboprop, still with pretty much the same fuselage as the Tu-85 but with completely new swept-wings and tail. It was a remarkable short-cut that, in 1953, resulted in the simple and cheap design of a substantial and fast passenger transport merely by putting a completely new circular-section pressurised fuselage on the Tu-88. The result was the Tu-104, first flown on 17 June 1955. This was the first jet airliner in the world after the Comet 1 to enter service.

Aeroflot, the gigantic Soviet civil airline, knew from the start that the Tu-104 was somewhat uneconomic. In any case, for many routes it was too big. What was needed in the second half of the 1950s was a modern jet that could replace the old piston-engined Li-2s and Il-14s. The main replacement was to be the An-24 twin-turboprop, but there was also a need for a jet, and the simplest thing to do was to build a smaller edition of the Tu-104. This could have had two of some existing kind of fighter turbojet, such as the Tumanskii R-13. Such engines would have been tough, simple, cheap and already developed, to some degree matching the style of the Tu-124. Perhaps surprisingly, the decision was taken to develop a totally new engine of the turbofan type, the Soloviev D-20P of 5400 kg (11,905 lb) thrust. This reduced fuel burn by about 50 per cent and also reduced noise (though noise was not a significant consideration in the Soviet Union 30 years ago). When the Tu-124 entered service on 2 October 1962 it was the first turbofan short-haul transport in the world.

By this time, however, the giant OKB (experimental construction bureau) of Andrei N. Tupolev was well advanced with plans for a successor. By this time the original Tu-104 was distinctly dated, and there was a need for a new jet to replace it. In February 1961 the GVF (civil aviation organisation) and Aeroflot agreed to fund a new transport in the Tu-104 size class but, for the first time, the aircraft would inherit nothing from the Tu-88 bomber and was in all respects a modern and uncompromised aircraft. The task was assigned to Leonid Selyakov, with the older and more senior A. A. Arkhangelski as his deputy.

An option available for the Tu-134 series was a more streamlined nose, which housed the weather radar and dispensed with the undernose radome. This is a Tu-134A of CSA, seen at Frankfurt airport.

Following two prototypes, this aircraft (SSSR-65601) was the first of five pre-production aircraft, seen at Tokyo during a sales tour. Designated simply Tu-134, these demonstrated the original short fuselage of the type.

The prototype, L-45075, was simply slotted into the existing Tu-124 production line at Kharkov. Called a Tu-124A, it first flew in December 1963. Major parts of the fuselage and wings were similar to those of the Tu-124, including the engines, but these were moved to the sides of the rear fuselage. The centre-section of the wing was thus cleaned up, greatly reducing drag and improving the efficiency of the flaps which were increased in travel, all four sections being extended to 20° for take-off and 38° for landing. Span was increased by over 3.3 m (11 ft), and the ailerons were divided into two parts. The landing gears were strengthened for higher weights, and a distinctive feature was the addition of prominent sloping drag struts which allow the entire gear to pivot up to the rear, giving extra travel for rough-field operation. As before, all units retracted to the rear, the characteristic fairings aft of the wings being longer and more pointed. More powerful anti-skid brakes made a tail parachute no longer necessary.

New look takes shape

Of course, the new engine installation required redesign of the tail. The new tail was considerably larger than before, the tailplane being mounted on top of the fin with a pointed fairing over the junction. As in the -124 the tailplane was driven by an electric screwjack for trimming purposes. Again as before, all flight controls were manual, with geared tabs. The spoilers for lift dumping and augmented roll control were increased in power, and the powered airbrake under the fuselage was also retained to steepen the approach. De-icing was also unaltered, with electrothermal strips along the tailplane, engine-bled hot air for the wings, fin and engine inlets. No reversers were fitted.

Though the fuselage of the -124A was only 1.6 m (5 ft 3 in) longer, its capacity was increased by more than 50 per cent. A major improvement was that, unlike the Tu-104 and -124, the main spar box no longer projected into the cabin. Thus the floor could be level, with all seats and porthole-type windows at the same level. From the start the new fuselage was equipped for (typically) 64 passengers, with 16 first class at the front, 20 tourist in the middle cabin and 28 tourist in the rear cabin. Noise and vibration in the cabin were significantly better than in any of the previous Tupolev jets. The crew remained as before: two pilots, a navigator in the 'bomber-type' glazed nose and a steward or stewardess. There were no underfloor holds, baggage and cargo being stowed behind the flight deck and at the extreme rear.

In service delays

Nothing was released publicly, but flight testing was reasonably free from problems. The first details and pictures were released on 16 September 1964, and representatives of the press were carried on the 100th test flight on 29 September. By this time five pre-production aircraft had been completed, with consecutive numbers. These differed in having another completely new engine. Again from the bureau of P. A. Soloviev, it was designated D-30 and was both more powerful and more efficient than the D-20. Take-off rating was 6800 kg (14,990 lb), enabling the new transport to be developed to higher weights. With this engine fitted the designation was changed to Tu-134. The main production run followed straight on behind the Tu-124s, with registrations from 65000 onwards. Many of the production aircraft for Aeroflot had no first-class cabin, the seating being increased to 72 in consequence (44 in the front cabin, 28 in the unchanged rear cabin). In typical Soviet fashion, several years elapsed between first flight and Aeroflot service. No Tu-134s flew in line service until well into 1967, and even then cargo only was carried until, in September 1967, a -134 appeared on a service from Moscow to Stockholm. Subsequently, the Tu-134 became the most familiar Soviet transport throughout Europe.

By 1967 various modifications had been incorporated. One of these was to use inverters to supply most of the main electrical power loads as frequency-wild AC (alternating current). By 1969 negotiations had been completed for the supply from Britain of English Electric CSDs (constant-speed drives), as used on the One-

Seen in typical East European surroundings, this is one of Balkan Bulgarian's Tu-134s, the first of which was delivered in 1968. Note the glazed nose fitted to the early versions, with weather radar mounted underneath.

Several Tu-134s have been used by governments and military agencies as staff or VIP transports. This aircraft served the People's Republic of Mozambique, before it was downed in suspicious circumstances, killing all on board, including the country's president, Samora Machel.

LOT of Poland has been a long-time Tu-134 user, its aircraft being regular sights at European airports. This dramatic photograph of a Tu-134 shows the mainwheels as they 'somersault' backwards to lie horizontally within the wing nacelles, the latter a trademark of the post-war Tupolev aircraft.

Eleven, giving paralleled AC supplies. This was a 'first' in a Soviet aircraft. Just over a year later the auxiliary power systems were further enhanced by at last fitting an APU (auxiliary power unit). Installed in the rear fuselage, this subsequently provided electric power for starting the main engines, as well as power for air conditioning (heating or cooling) the cabin on the ground. Yet another major upgrade, in 1969, was the fitting of thrust reversers on the engines. These are of the twin-clamshell target type, and change the engine designation to Series II.

Introducing improvements

By 1970 production at Kharkov had switched to the Tu-134A. This had first flown prior to 1970, and it introduced a fuselage lengthened (mainly ahead of the wing) by 2.10 m (6 ft 10 in), overall length going up from 34.35 m (112 ft 6 in) to 37.10 m (122 ft 0 in). This retained the 28-seat rear cabin but made possible various front cabins seating up to 80 in all, or exceptionally up to 84. Seating throughout remained 2+2, so with 84 passengers 21 rows were needed. The longer fuselage also increased cargo/baggage space by 2 m³ (71 cu ft). All the aforementioned upgrades were incorporated, as well as the D-30 Series III (or D-32) engine, with an extra zero-stage on the LP compressor, enabling the existing ratings to be maintained with reduced turbine temperature or under adverse (hot and high) conditions, and maximum thrust to reach 7080 kg (15,608 lb). The main wheels and brakes were made as on the Il-18, but retaining the original tyre pressure, and the legs were strengthened. The wing was also strengthened to handle the increased weights, maximum take-off weight having gone up from 44500 kg (98,104 lb) in the -134 to 47000 kg (103,600 lb) in the -134A. The most obvious external change was that the fin/tailplane bullet was extended forwards as a long spike to form a VHF radio antenna.

Throughout the life of the Tu-134/134A the avionics fit was progressively increased, though it was always fairly basic and used 'state of the art' equipment. Arkhangelski once claimed the ability to make blind landings in ICAO Category III conditions, but this was certainly erroneous. The original -134 fit included the RO3-1 weather radar under the nose, the NAS-1A6 navigation system, based mainly on Doppler, and the BSU-3P ILS steering when coupled via the AP-6EM-3P autopilot. The latter was replaced in the -134A by the Course MP-1 navigation and landing system, with several Sperry items, including the SP-50 autopilot, VOR and ILS (later duplicated). Other additions included two SO-70 transponders, an RV-5 radio altimeter and twin ARK-15 radio compasses. The radar was changed for the ROZ-1, and the Doppler for the DISS-013.

On the whole the Tu-134, given the NATO name of 'Crusty', was certainly the best short-haul jet produced at that time in the Soviet Union. As early as 1968 export orders had been signed with Interflug of East Germany, Balkan Bulgarian, LOT of Poland, Malev of Hun-

Another Balkan Tu-134, this time seen at Vienna. It wears a later colour scheme, since replaced by the current style. The title 'Balkan' is worn in Cyrillic on the starboard side and in Roman script on the port side.

LOT's original fleet comprised the early short-fuselage Tu-134, but these were supplanted by the stretched Tu-134A. Eight of the latter remain on strength, each configured to carry 76 passengers in a single-class arrangement.

In Aeroflot service, the Tu-134 was largely used on domestic services, although some limited international services were flown in Europe and the Far East. Replacement of the type is now under way by the Yakovlev Yak-42.

gary and Aviogenex of Yugoslavia. All these operators later also ordered the Tu-134A, Interflug replacing the earlier type with 20 of the longer model. An additional customer for the 134A was CSA of Czechoslovakia, buying 13.

Aeroflot might not have required any change, but in 1971 Aviogenex suggested that the retention of a glazed nose had become archaic, and that radar performance, and possibly aerodynamic drag, might be improved if the Tu-134A were to be redesigned in line with Western transports, with a scanner antenna of different shape looking ahead from the nose. The first two aircraft for this operator were already on the production line, but it was found possible to incorporate the requested change on the No. 3 aircraft. Subsequently, the more modern nose became an option on all aircraft, and it was immediately adopted by CSA for its complete fleet. Several other aircraft were retroactively modified. Even the last batch for Aeroflot was fitted with radar in the nose.

The last batches were of two sub-variants incorporating further minor improvements. The Tu-134A-3 was fitted with lightweight seats of a less-bulky shape, enabling 24 rows to be accommodated in an all-tourist layout, a total of 96, giving a seat-mile cost 50 per cent lower than that of the original version. The Tu-134B had a revised flight deck, with no navigator but retaining an engineer at the side panel. The spoilers were also modified, so that the pilot could use them for DLC (direct lift control).

Malev's last six Tu-134s were of the A-3 variant with improved Soloviev D-30-III turbofans. These have the same power as the previous engine but offer far greater fuel efficiency.

Production of the Tu-134 family ended around 1978. The total was at least 300, about 200 having for many years carried a major part of the traffic on Aeroflot's shorter trunk-route sectors, up to about 1,000 miles. Roughly 100 were exported, including 56 to major airlines in Europe. Everyone who has used the Tu-134, especially in the later variants, has been well satisfied. It still offers an acceptable level of reliablility, adequate cruising speed of 750 km/h (466 mph), and a saleable interior. The only real problems are that, by modern standards, the fuel burn per seat mile is high and the noise does not meet current and future certification standards.

Rather than go to the cost of fitting noise suppressors, or new engines such as the Lotarev D-36, the GVF and Aeroflot are funding an entirely new aircraft which it was hoped would replace the Tu-124 from 1993. However, the CIS's financial problems meant that this Tu-334 first flew only in 1999.

Among later Tu-134 deliveries were the Tu-134B-3s delivered to Syrianair, which currently has six on strength, although three are used for government VIP duties. The B-3 has a revised interior with lightweight seats, and a revised forward-facing cockpit for two pilots.

Tupolev Tu-144 'Charger'

Often derided as a mere Soviet copy of the Anglo-French Concorde, the Tu-144 differed in a considerable number of ways, and actually preceded its rival into the air to become the first flying SST. However, the major difference lay in the fact that while Concorde has been an unqualified success, the Tu-144 was so beset by problems that it never achieved anything like its promise.

The production Tu-144 was the longest, heaviest, fastest and most powerful SST (supersonic transport) ever built. Moreover, the whole project began in the firmest possible way, with a planned production run of 75 for the national airline alone. But that was just one side of the story. The other side was a remarkable history of problems and failures, to such a degree that the entire aircraft was totally redesigned and eventually, after the expenditure of hundred of millions of roubles, and a second redesign, the whole project was abandoned. This does not alter the fact that all versions of the Tu-144 were among the most impressive aircraft of all time.

In the mid-1960s Andrei N. Tupolev explained the reason for the Tu-144. He pointed out that the Soviet Union was the largest country in the world. With such enormous distance, air travel could save a great deal of time, much more so than in smaller countries. He said that the civil transport operator, Aeroflot, had calculated that in 1962-63 its passenger operations saved an average of 24.9 hours per journey, compared with the best alternative surface travel. It had also calculated that widespread use of an SST cruising at over Mach 2 could increase this saving to over 36 hours per journey. Aeroflot and the GVF (the civil air fleet, of which Aeroflot is the operating branch) recognised that most long-haul passengers were important people – doctors, high-ranking officers, scientists and engineers, for example – whose time could be costed at a high rate per hour. The final outcome was that it was proved arithmetically that money could be saved by putting into service a fleet of 75 SSTs.

This is remarkable on several counts. First, international flights were of secondary importance, and almost ignored in the calculations. Second, there was no discussion of the sonic boom, which Tupolev – like other Russians – dismissed contemptuously with "Nobody is concerned with natural thunder, so what's the problem?" Third, no other airline in the world thought in terms of a fleet of 75 SSTs.

The task of designing and constructing prototypes was assigned to Tupolev's OKB (experimental construction bureau), with the General Constructor's son Dr Alexei Andreyevich Tupolev as chief designer. The engine bureau of N. D. Kuznetsov was instructed to develop a suitable engine from the existing NK-8 turbofan, as used in the Il-62 and Tu-154. Among other things the new engine, the NK-144, had to be fitted with an afterburner and variable nozzle, and be made of

Although in planform the second-generation Tu-144's wings were much squarer than those of the prototype, they introduced camber and droop, considerably improving the aerodynamics. The fuselage was considerably longer.

Like Concorde, the Tu-144 featured a drooping nose to allow adequate visibility during approach and landing. The similarity between the two designs led naturally to the Tu-144 being dubbed 'Concordski' by the popular media, while the NATO Standards Committee allocated the reporting name 'Charger'.

Unlike Concorde, which had the transatlantic service uppermost in its design, the Tu-144 was tailored to fast domestic services. Performance and load goals were similar, so it was no surprise that the two designs featured the same configuration. Where the Tu-144 scored was in having a larger fuselage cross-section.

different materials to withstand the high inlet temperatures when cruising at over Mach 2. Recognising the magnitude of the development task, the Kuznetsov bureau was assisted by the Central Institute of Aviation Motors, while the Tupolev bureau was supported by the largest team at the Central Aero- and Hydro-dynamics Institute ever assigned to a single aircraft project.

There was never much doubt that the cruising speed would be about Mach 2.2, which is the practical limit for aircraft constructed mainly of aluminium alloys. It was also decided early on that the configuration should be similar

The prototype Tu-144 seen on its maiden flight, on which it was accompanied by the A-144 'Analog', a MiG-21 with a similar double-delta planform wing. The main undercarriage featured 12 wheels in rows of four on each strut, while the nosewheel was positioned much further aft than on following machines.

to that adopted for Concorde, which had already been published. There was a strong body of opinion which insisted there should be either a tailplane or a foreplane, in order to maintain pitch control and enable flaps to be used to increase lift on take-off and landing. Mikoyan built the A-144, a MiG-21 with a scaled tailless delta wing as proposed for the Tu-144, and this confirmed that a horizontal tail could be dispensed with. It was proposed to use four engines grouped in a single box under the centreline, with a deep gash at the front separating the engine ducts just enough to admit the rearwards-retracting nose landing gear, which was very similar to that of the Tu-114. After studying nine other schemes it was decided to design each main landing gear with 12 small wheels, four on each of three axles, so that the whole unit could fit inside the very thin wing. Each main gear was designed to retract forwards, the reverse of usual Tupolev practice, the bogie rotating to lie inverted inside the wing just outboard of the engine box.

Shapely wing

The wing, manufacture of which was assigned to the O.K. Antonov bureau, was designed with a thickness/chord ratio as low as 2.8 per cent, with the leading edge at 76° inboard and 57° outboard, sweeping round to a large curved tip. The whole wing was relatively flat, the leading edge being fixed and the trailing edge comprising four almost square elevons on each wing, each driven by two power units (one in each hydraulic circuit) faired inside the wing. Likewise, the only other control sur-

The four engines of the first Tu-144 were grouped together along the central trailing edge, the intake trunks separated at the front to accommodate the long nosewheel strut.

face, the rudder, was divided into two units each driven by power units in separate circuits. In this case their fairings caused prominent blisters, those on the lower rudder being on the left and those on the upper rudder being on the right. The slender fuselage was given an almost circular cross-section, with a diameter of about 3.4 m (134 in), compared with 2.9 m for Concorde. Like the Anglo-French SST the nose was arranged to droop, pivoting down through 12° to give the pilots a view ahead on take-off and landing. Along each side were arranged 25 small windows, and it was planned to seat between 95 and 140 passengers depending on the interior layout and the choice of four- or five-abreast seating. A total of 87,500 litres (19, 247 Imp gal) of fuel was to be carried in integral tanks in the wings and lower fuselage. At quite a late stage it was decided to follow the practice adopted on Concorde and fit special trim tanks as far forward and aft as possible, to cancel the powerful shift in wing centre of pressure during transonic acceleration and subsequent deceleration back to subsonic flight. The pipe to the tail trim tank, being an afterthought, had to be added externally on the prototype.

Off the drawing board

In early 1964 Tupolev's bureau was awarded a contract for two flight articles and a static-test airframe. The first, No. 68001, was completed about nine months late at the Tupolev prototype factory at the Zhukovskii centre near Moscow. Accompanied by the 'Analog' it made its first flight on 31 December 1968, still comfortably ahead of the Concorde. The test pilots were Edward Elyan and co-pilot Mikhail Kozlov, assisted by engineer Yu Seliverstov and flight-test director V. N. Benderov. This flight was encouraging, and further testing proceeded according to plan, Mach 1 being exceeded on 5 June 1969 and Mach 2 on 26 May 1970, the latter at 16300 m (53,478 ft). The highest speed reached was Mach 2.4, design cruising speed having been set at Mach 2.35 (2500 km/h; 1,550 mph). Of course, No. 68001 was far removed from being an SST, as its interior was largely filled with test gear and the four crew-members had ejection seats fired through blow-off hatches in the fuselage.

No. 68001 was publicly displayed at Moscow Sheremetyevo on 21 May 1970. On that occasion it was announced that series production had already started at a GAZ (state aviation factory) at Voronezh. No announcement was made of the first flight of the second prototype, but clearly a vast amount of work had to be done to fit and develop the variable engine inlets with computer control, the Doppler/inertial

The third Tu-144 blasts off the le Bourget runway during the 1973 Paris air show. A tragic crash, never satisfactorily explained by the Soviets, killed the crew and nine people on the ground during a steeply climbing turn in the display routine on the last Sunday of the show.

Specification

Tupolev Tu-144 (early production aircraft)
Type: long-range supersonic transport
Powerplant: four Kuznetsov NK-144 turbofan, each rated at 127.5 kN (28,660 lb) thrust without afterburning and 171.6 km (38,580 lb) thrust with afterburning
Performance: maximum cruising speed Mach 2.35 (2500 km/h; 1,550 mph); normal cruising speed Mach 2.2 (2300 km/h; 1,430 mph); landing speed 280 km/h (174 mph); cruising altitude 16000-18000 m (52,500-59,000 ft); landing run 2500 m (8,530 ft); maximum range with 140 passengers at average speed of Mach 1.9 6500 km (4,030 miles)
Weights: operating empty 85000 kg (187,400 lb); maximum take-off 180,000 kg (396,830 lb); maximum fuel 9500 kg (209,440 lb); maximum payload 15000 kg (33,070 lb)
Dimensions: wing span 28.80 m (94 ft 6 in); length 65.70 m (215 ft 6½ in); height with wheels up 12.85 m (42 ft 2 in); wing area 438 m² (4,714.5 sq ft)

CCCP (USSR)-77144 was the production standard Tu-144 which visited the 1975 Paris air show, two years following the tragic crash of CCCP-77102. Noticeable are the tail-down sit of the aircraft, the uneven distribution of the actuating systems for the two rudder sections and the relatively-straight edges to the wing planform. The scrap view details the 12° movement of the drooping visor nose. The inscription on the nose of this Tupolev aircraft is Cyrillic for 'Tu-144' – surprisingly, no Russian name was applied.

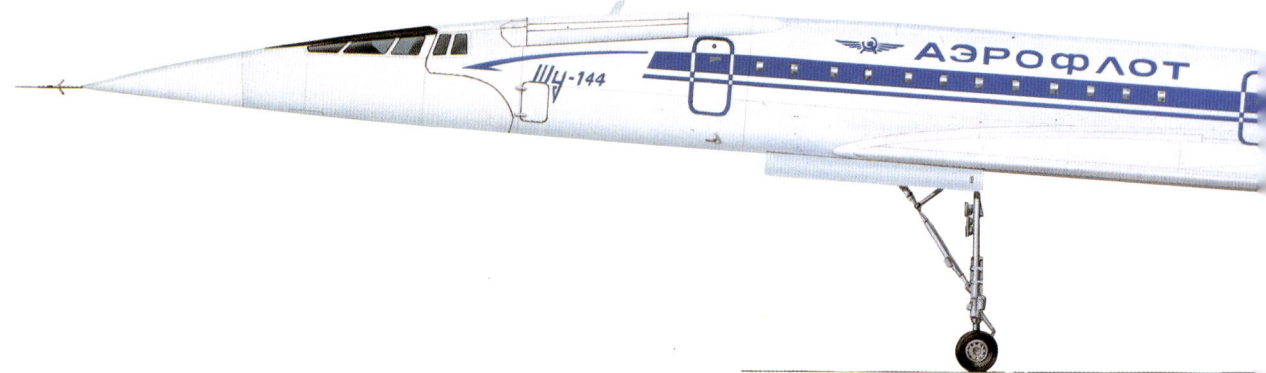

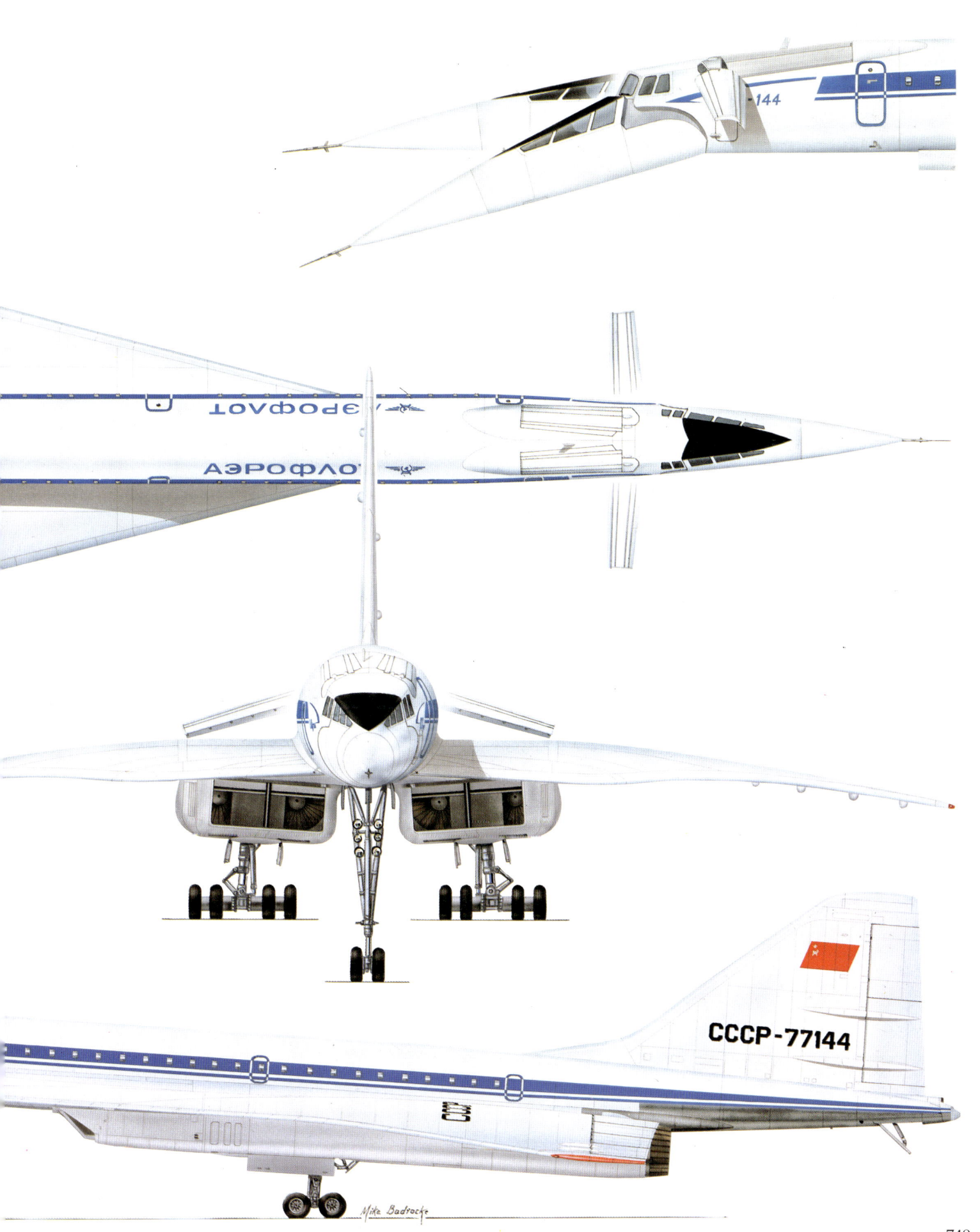

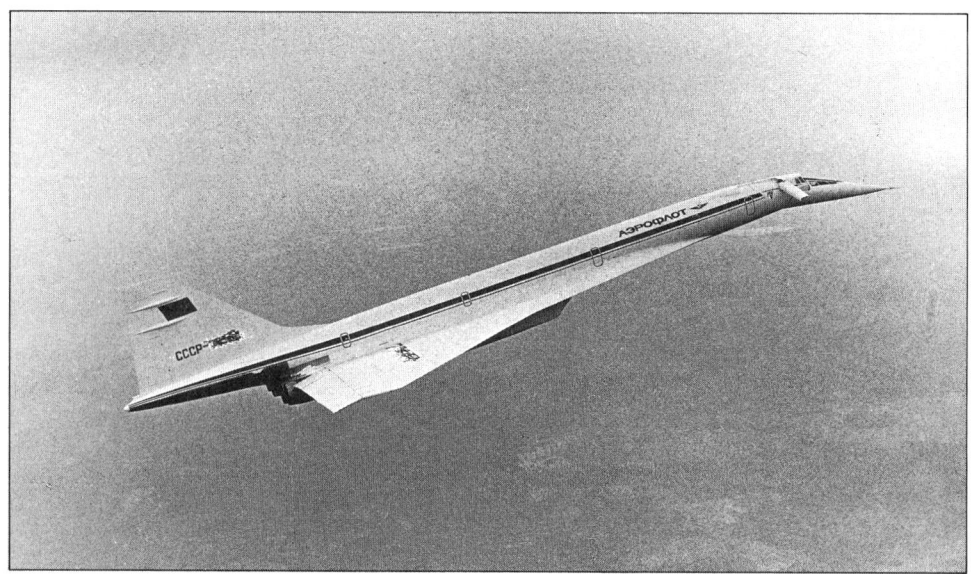

With serials crudely censored by the Soviets, this is one of the early Tu-144s seen during the late 1970s when the type was tentatively entering service. It demonstrates the use of the foreplane, which provided enough lift to enable the trailing edge elevons to be depressed, so gaining valuable lift.

The first Tu-144 had a more elegant, curved wing planform than later aircraft, more reminiscent of the Anglo-French Concorde. Note the ejector seat hatches above the cockpit and central fuselage, deemed necessary for testing such a radical machine.

navigation system, again with computer control, the production type environmental control system, the definitive flight controls and many other features. What nobody expected was that the Tu-144 would be largely redesigned; indeed, the second-generation Tu-144 had hardly any parts in common with the first. The changes extended to aerodynamic design, structure, materials and systems.

Revised lines

The wing was enlarged, area being increased from 421 to 438 m² (4,532 to 4715 sq ft). The new planform had a leading edge formed from two straight lines, with squared-off tips. Instead of being flat, the wing had pronounced camber (curvature), the leading edge being drooped and the trailing edge curving downwards, especially at the tips. The elevons were modified and extended right to the tip, with a revised control system. Wing structure was altered to include more titanium alloy, such as 6Al4V, more integrally stiffened members and more honeycomb skin, and with welding replacing many rivets. The fuselage was again modified in structure and materials, and lengthened by 6.3 m (20 ft 8 in), with 34 windows each side and Class 1 exits on each side at the rear. The engine nacelles were completely rethought, the left and right twin-engine nacelles being moved apart to be completely separate. The inlets were redesigned to be fully variable, very like those of Concorde, with variable roof and lower lips and a deep central splitter. The engines were moved aft, so that the nozzles projected behind the wing instead of stopping well short of the trailing edge. The nozzles themselves were redesigned, and instead of being all together were separated into two pairs by the width of the fuselage. The nose landing gear was redesigned to be even longer, and moved forward 9.6 m (31 ft 6 in), retracting forwards into an unpressurised compartment under the forward fuselage. The main gears were totally redesigned, with broad eight-wheel bogies (four tyres on each of two axles) attached to forged frames in the nacelles and retracting forwards and rotating 90° so that each unit could lie on its side in a narrow compartment occupying the full depth of the nacelle between the inlet air ducts.

A completely new feature, and the most externally obvious when in use, was the addition of retractractable high-lift foreplanes. These comprised miniature wings, pivoted to the top of the fuselage aft of the flight deck. When

The first Tu-144 thumps on to the runway, illustrating the closely-spaced wedge-type intakes of the first generation aircraft. No canard foreplanes were fitted, these being adopted by the 'production' machines.

Second-generation Tu-144 in clean configuration. The Tu-144 was marginally faster than Concorde, and entered limited revenue-earning service ahead. However, even in the improved and re-engined Tu-144D version, it never achieved any great success in operation.

extended, these canard surfaces had a span of 6.1 m (20 ft), with zero sweep but sharp anhedral. Each surface had a high-lift profile augmented by a double leading-edge slat and a double-slotted trailing-edge flap. They were extended for every take-off and landing, their extra lift far ahead of the centres of gravity and of wing pressure, enabling the elevons to be deflected downwards to enhance lift. Without the canards the elevons had to be deflected upwards to maintain the required angle of attack, in effect adding many tons to the weight. Under similar conditions that reduced take-off field length from 2600 to 1980 m (8,530 to 6,500 ft), despite the later aircraft being much heavier. This increase in weight, from 130 tonnes to 180, was almost entirely accounted for by a great increase in fuel capacity, from 87500 to 118750 litres (19,248 to 26,121 Imp gal). Whereas the prototypes had no hope of reaching even half the design range of 6500 km (4,039 miles), the production aircraft reached the full distance.

Catastrophe

The first of the second-generation aircraft, No. 77101, flew in spring 1973. It was quickly joined by 77102, but this aircraft was tragically lost at the Paris air show on 3 June 1973, killing all on board (including Kozlov and Benderov) and others on the ground. It was clear that catastrophic wing failure occurred during a violent pull-up manoeuvre, but no report was ever published and no modification was made to other Tu-144s, apart from (said Alexei Tupolev) "some updated equipment".

In December 1972 eight seemingly production aircraft had been seen on the assembly lines at Voronezh by French visitors. No. 77144 attended the 1975 Paris air show, being described by Tupolev as "the eighth aircraft, including prototypes but not including two static test airframes". They were said to have made 1,000 flights at that time. No. 77144 carried Aeroflot titles, and it was stated that initial route proving had been in progress since May 1974, including the longest sectors of all: Moscow-Tyumen-Vladivostok. On 26 December 1975, No. 77106 began regular cargo flights between Moscow and Alma Ata, routinely flying the 3260 km (2,026 miles) in under 2 hours total block time. By this time several production aircraft were available with passenger furnishings, normally arranged for 140: 11 first-class 2+1, 30 in the mid cabin 2+3, 75 in the rear cabin 2+3 and a final 24 at the back 2+2.

There were various problems, but on 22 February 1977 a series of 50 proving flights began between Moscow and Khabarovsk. At last, about five years late, passenger service was opened between Moscow and Alma Ata on 1 November 1977. Five of the next six flights were cancelled, but 102 revenue passenger flights had been made when the service was abruptly cancelled on 1 June 1978, following the crash of a Tu-144 not in Aeroflot service. At that time 13 Tu-144s had flown.

From 1970 there had been reports of a Tu-144D, with new engines. At last the first Tu-144D was announced as having made its first flight on 23 June 1979. The engines, the Rybinsk (Koliesov) RD-36-51A, were similar in power to the NK-144 but of totally new design, very similar in many ways to the Concorde Olympus apart from having variable-incidence stators throughout. It had been announced in 1973 that the new engines would have reversers, something absent from the Tu-144 and made up for by fitting twin braking parachutes to assist the wheel brakes on short or icy runways. The RD-36-51A remained an enigma until one was put on view in Moscow in 1990, when it was seen to have no reverser, but a shaft drive to a prominent accessory gearbox mounted above the engine inside the wing. Specific fuel consumption in cruising flight at Mach 2.2 was placarded as 0.88, roughly equivalent to the Olympus and certainly superior to the older Kuznetsov engine; but it was difficult to comprehend the comment made by the GVF in 1979 that the new engine made the Tu-144 "50 per cent more economical [than earlier Tu-144s] in operation". It was also stated that the revised Tu-144 "met international noise legislation" and was "ready for series production".

Disappearance

Since 1979 almost nothing has been heard of these great aircraft. In April 1982 it was announced that the Tu-144D was to start a regular cargo service between Moscow and Tashkent, but this has never figured in Aeroflot's subsequent annual reports. Certainly numerous Tu-144s have been 'seen' by USAF intelligence-gathering satellites, some of them at the Ramenskoye test airfield alongside prototypes of the Tu-160 strategic bomber. Some of the aircraft were clearly in the process of being re-engined, but the absence of further news makes it clear that Aeroflot decided against reopening revenue-earning service.

So convoluted and fascinating is the history of the various forms of Tu-144 that it has formed the subject of a full-length book, *Soviet SST*, by an American author, Howard Moon. This concentrates on the political and intelligence aspects, as well as the technical. It reads like a spy thriller.

Chapter 34

Tupolev Tu-154

Vickers Vanguard

Vickers Viking

Vickers Viscount

Vickers VC10

Tupolev Tu-154

When the Soviets came to replace their first-generation jets and turboprops on the medium-range, medium-density Aeroflot routes, Tupolev was a natural choice, as the design bureau was the nation's most experienced builder of large jet aircraft. The tri-jet Tu-154 was a typical Tupolev design, drawing heavily on previous airliners but also introducing new features inherited from successful Western designs.

The Soviet design bureau of A. N. Tupolev was one of the pioneers of passenger jetliners, with the Tu-104 of 1955. This led to the Tu-124 and -134, the latter built in very large numbers. This provided the strong background of experience needed to tackle a challenging GVF (civil air fleet) requirement of 1964 for a larger and much more powerful jetliner to replace the Tu-104 and the turboprop-powered An-10 and Il-18. Passenger capacity clearly had to exceed 120, and particularly difficult demands were that reduced payload sectors of up to 6000 km (3,725 miles) had to be flown, and that the new jet should be able to operate from airports with runways no longer than 2 km (6,500 ft) with a surface of gravel or packed earth.

Under General Constructor Andrei Tupolev, the Tu-154 project went ahead under S. M. Yeger, who was later replaced by Chief Designer Dmitri Markov. From the start there was never much doubt that the new jetliner would generally resemble an enlarged Tu-134, with three engines. The timing was perfect to adopt any good features from the British Trident and American Boeing 727. From the former it was decided to adopt triplexed (three independent) hydraulic systems, with one pump driven by each engine and triple actuators driving each powered flight control surface. From the 727 came two things never before attempted by the Tupolev bureau, leading edge slats and triple-slotted flaps.

Like its predecessors, the Tu-154 was given a fuselage of circular cross-section, with a diameter of 3.8 m (149.5 in), almost as large as that of the monster Tu-114. This was adequate for six-abreast seating and for the wing structure to be accommodated beneath the floor. Pressure differential was set higher than before at 0.63 kg/cm² (9 lb/sq in), and instead of large circular windows the passengers looked out through Boeing-style miniature rectangular windows. It was eventually decided not to fit a rear airstairs door under the tail. Instead, two main doors were fitted on the left, ahead of the wing, with service doors on the right-hand side.

The wing followed previous practice, with an aerofoil profile that could be traced back to the Tu-16 bomber prototype of 1952. Leading edge sweep was 40° and 38° outboard, but the angle at the quarter-chord line was taken as 35°, the same as before. Structurally the wing had three spars, the main torsion box being sealed to form integral tankage, typically for 41140 litres (9,050 Imp gal). The slats were arranged in five sections, driven by electric screwjacks, along the outer four-fifths of each wing. The large triple-slotted flaps were driven hydraulically, the inner and outer sections being separated by the main landing gear. Ahead of the inner sections were arranged hydraulically-powered airbrakes, used as lift dumpers after touchdown. Ahead of the outer flaps were three spoilers which were used to augment the ailerons for roll control, especially at low speeds, the two inboard sections also being usable symmetrically as airbrakes.

As in the Tu-134, a T-type tail was fitted, but with sweep increased to 45° on the leading edges. As with the ailerons and other wing movable surfaces, the tail skins were made of honeycomb sandwich. Triplexed power units drove not only the elevators but also the

Fourteen Tupolev Tu-154Ms served the Polish airline LOT. The Tu-154 and subsequent Soviet airliners are no longer assured of export sales to East Europe. Several carriers (and even Aeroflot) are beginning to adopt Western types, LOT already flying only Boeing airliners.

Tu-154s form the backbone of most East European airline fleets. Balkan Bulgarian Airlines had a fleet of 23, split between Tu-154B, B-1, B-2 (illustrated) and M variants, reduced by 1998 to only 15 Ms.

tailplane, which thus became a primary control surface. Hot high-pressure bleed air was piped to de-ice the leading edges of the wings and tail, and the centre engine inlet. The slats were electrically heated.

Rough field capability

To meet the severe field-length requirement considerable engine power was needed, so that though gross weight and seating capacity were actually both lower than for the 727-200, the engine chosen was much bigger and more powerful than the US counterpart. The engine selected was the NK-8-2, rated at 20,950 lb, a slightly lighter and less powerful version of the turbofan already being developed by N. D. Kuznetsov for the four-engined Il-62. The engines were grouped at the tail, the centre engine being fed via a curved S-duct. The outer engines were toed inwards, their nacelles being parallel to the taper of the rear fuselage, and were fitted with cascade-type reversers with exits above and below.

To meet the severe demand for operation from unpaved runways, the main landing gears were designed as six-wheel bogies. Each unit was basically similar in geometry to that of the Tu-134, the main legs leaning backwards and braced by a long drag strut from the front. The difference was that the bogie beam, which in the air with the gear extended slopes down to the front, was provided with three axles in a row, each carrying a pair of wheels with tyres inflated to 114 lb/sq in. Each unit was designed to retract hydraulically to the rear, the bogie beam somersaulting so that the bogie could lie inverted in the fairing projecting aft of the trailing edge. As in earlier Tupolev jetliners, the twin-wheel steerable nose-gear retracted to the rear also.

Wearing muted Aeroflot titles, the first prototype Tu-154 displays its angular yet attractive lines. First flying on 4 October 1968, the aircraft had been derived to replace the An-10, Il-18 and Tu-104.

The first of six development aircraft, SSSR-85000, was flown by test pilot N. Goryanov on 4 October 1968. No. 85006 was fully furnished and used for Press flights in August 1970, subsequently being delivered to Aeroflot in January 1971 for route proving. By this time Aeroflot already had No. 85701 and at least one other production aircraft. Initial Aeroflot services carried cargo only. Irregular passenger services were flown to Tbilisi from July 1971, scheduled services began to Mineralnye Vody via Simferopol on 9 February 1972 and the first international service was flown to Prague on 1 August 1972. Almost all early deliveries had 128 passenger seats, a few having a higher density configuration for 146, with reduced galley facilities.

From the outset all production was centred at Kuibyshyev, and strenuous efforts were made to clear the structure for a 30,000-hour

crack-free life (about half typical Western design targets). Meanwhile, the designers under Markov had by 1971 completed work on the Tu-154A, first delivered in April 1974 and in regular airline service from mid-1975. Almost indistinguishable from the Tu-154 externally, the 154A introduced the NK-8-2U engine, uprated to 23,150 lb, enabling maximum weight to be increased from 198,416 lb to 207,235 lb. Oddly, though most of the increase was taken up by an extra fuel tank in the centre section, the fuel in this tank could not be transferred elsewhere except on the ground. Thus, it was used merely to cut down purchases of fuel in foreign countries; it did nothing to extend the aircraft's range. Other new features in the 154A included an interconnect link between the slats, flaps and tailplane, the latter pivoting 3° down when flap was extended. The electrical systems were uprated, the baggage/cargo holds were strengthened and fitted with smoke detectors, two extra emergency exits at the rear enabled seats to be increased to 168, and an autolanding facility was added to the autopilot for eventual clearance to Cat. II conditions.

Tu-154B: Updated and uprated

In 1977 production switched to the Tu-154B, which was quickly refined to 154B-2 standard. This was cleared to 216,050 lb, and introduced two further emergency exits to enable passenger accommodation to be raised to a maximum of 180. The most important change in the B-2 was that the centre-section tank was made a normal part of the fuel system, usable in flight (strangely, though comparisons were difficult, the published figures for range were generally lower than for the Tu-154). Other changes in the B-2 included improved lateral-control spoilers and upgraded avionics, including French Thomson-CSF/SFIM autopilot and navigation equipment and a new weather radar. A curious feature was the addition of small actuators which, in cross-wind landings, can pivot the front axle (only) of each main landing gear to reduce tyre wear.

In 1982 the Tu-154C freighter was announced. So far as is known all of this version are conversions of As and Bs. The main differences include the provision of a large door in the left-hand side of the forward fuselage, stripping out of the passenger accommodation to give a clear interior with a volume of 72 m^3 (2,542 cu ft) and the fitting of special cargo flooring with ball mats inboard of the door and roller tracks along the main hold. The door measures 2.8 m by 1.87 m (110 × 73.5 in). Maximum cargo load, above and below the floor, is 20 tonnes (44,090 lb).

In 1981 a new Tupolev airliner was announced, the Tu-164, but this designation was subsequently changed to Tu-154M. By this time 600 of the earlier versions had been ordered, of which more than 500 had been delivered to Aeroflot. Thus, it may seem to have been a bit late to choose this time to carry out a major update of the 154, especially as the Tupolev bureau was already thinking about a new airliner in this class which has now materialised as the Tu-204. Be that as it may, a total revision of the entire aircraft was requested, and though there are obvious external differences – apart from the bigger forward-pointing fairing ahead of the tailplane – it is not possible to convert any earlier version to 154M standard.

The most important change is a switch to the Soloviev D-30KU-154-II engine, a newer and more efficient turbofan than the NK-8. Though the take-off rating of 23,380 lb is only slightly greater, the mass flow is significantly increased, necessitating enlarged inlets. The centre engine has a plain nozzle, instead of a scalloped shape, and the side engines have totally new pods arranged

Romania maintained a large national flag-carrier in the shape of Tarom. Ten Tu-154B-2s were on strength, each with 164 seats in a single-class arrangement.

As the standard Soviet airliner of the last two decades, the Tu-154 has been widely exported around the world. Perhaps the most unusual airline user was Guyana Airways, with this single Tu-154M.

The Tu-154M offers considerably better economy compared to earlier variants, thanks to the adoption of the Soloviev D-30KU turbofan. Czech carrier CSA's fleet contained four of the uprated model in 1998.

The Hungarian national carrier, Malev in 1998 had seven Tu-154B-2s among its fleet as the principal international type, serving most European capitals.

A handful of Tu-154s are in military service, mostly as staff or VIP transports. This single aircraft flies with the Czech air force's transport regiment at Mosnov.

Following a doctrine of buying from both East and West, China is steadily rebuilding its airline fleet with modern aircraft. From the Soviet Union comes the Tu-154M, the first being delivered in 1985.

A very early production machine (22nd aircraft) seen on a test flight accompanied by another Tu-154. From an early stage, the 'Careless' was designed to operate from semi-prepared airports with packed gravel strips.

almost axially and fitted with clamshell reversers. The gas-turbine APU (auxiliary power unit), previously above the centre engine, was moved to the centre fuselage. The slats have been made smaller yet more efficient, and the outboard spoilers have been considerably enlarged, from three sections on each side to four. As noted, the only obvious change is that the spike fairing projecting ahead of the tailplane is slightly modified, but what is less apparent is that the entire horizontal tail has been redesigned. The span remains unchanged at 13.4 m (43 ft 11.6 in) but the surface is cleaned up and increased in area from 40.55 m^2 (436.5 sq ft) to 42.2 m^2 (454.24 sq ft).

What is certainly remarkable is that over the years the basic operating weight (the equipped empty aircraft) has risen from 49,500 kg in the Tu-154 to 50,775 kg in the B and to 55,300 kg (121,915 lb) in the M version. This largely cancels out payload/range improvements gained by extra fuel and newer engines. Maximum take-off weight of the M is 100 tonnes (220,459 lb), yet, while the original Tu-154 could carry a payload of 6700 kg (14,770 lb) with full reserves a distance of 6900 km (4,287 miles), today's 154M can only carry 5450 kg (12,015 lb) payload for 6600 km (4,100 miles).

The Tu-154M was preceded by a Tu-154B-2 which was returned to Kuibyshyev and fitted with the D-30 engines. It began its flight test programme in 1982. Aeroflot took delivery of the first two production Tu-154Ms on 27 December 1984. Since then well over 100 of this version have been built, though output has been tapering off, and the new Tu-204 was to have completely replaced the older aircraft. Various Tu-154 versions were exported to Balkan Bulgarian, Cubana, Malev, Tarom, LOT, CSA, Syrianair, CAAC and Aeronica of Nicaragua.

A single Tu-154, SSSR-85035, has been modified to evaluate L/H$_2$ (liquid hydrogen) as a fuel, with the designation Tu-155. The obvious prospect of an eventual shortage of fossil petroleum could make L/H$_2$ very important, but it suffers from two major drawbacks. One is that its low density means that for equivalent heat energy the tankage has to be much bigger than previously (though the fuel is lighter). The other is that L/H$_2$ boils at $-253°C$ ($-423°F$), so that it has to be stored at this intensely cold temperature, which causes numerous problems (though most have long since been solved in developing L/H$_2$ rocket engines). Aircraft 85035 has had its centre engine replaced by a modified type designated NK-88. This is fed from a carefully lagged tank filling the rear of the cabin via a pipe in an insulated duct along the right-hand side of the rear fuselage. Flight testing began on 15 April 1988, and it was then expected that the test engine would later also be run on LNG (liquified natural gas).

Tu-154s are the backbone of Aeroflot's domestic and medium-range international network, serving in progressively advanced variants. These Tu-154B-2s were the most advanced available before switching to Soloviev power.

Vickers Vanguard

The Vanguard was one of the 'second generation' of propeller-driven airliners to be developed during the 1960s. Like the Lockheed Electra, it was born into a time when jet aircraft numbers were steadily rising, and airlines and passengers alike were somewhat averse to this 'old' technology. Correspondingly the Vanguard's production run was low, and its time spent in top-of-the-line service brief. Yet the Vanguard has only relatively recently left the scene, and in the hands of smaller operators was a capable freighter.

In late 1953 Vickers Aircraft Limited began design studies for a four-engined turboprop airliner intended to be a faster, longer range 'big brother' to the Vickers Viscount, which was destined to become the most commercially successful British civil aircraft. Initially it was intended to 'stretch' the Viscount to enable it to carry up to 100 passengers, but this proved impractical and Vickers decided on an entirely new design.

The new aircraft was designated Vickers Type V870, and was prompted by a detailed specification issued by British European Airways on 15 April 1953 for an aircraft capable of operating economically on short-to-medium haul routes with stage lengths of 320-4000 km (200-2,500 miles). By coincidence, on the same day that BEA issued its requirement, Trans Canada Airlines (which was a customer for the Viscount) also issued a request for proposals for a larger turboprop airliner capable of carrying 60 passengers over transcontinental ranges.

Some 60 separate studies were undertaken, including several with swept wings and others with high wing layouts, which BEA considered desirable to provide a good downward view for passengers. As the design evolved it became apparent that the requirements for a high (640 km/h, 400 mph) cruising speed combined with a low landing speed dictated a straight wing, and BEA was persuaded to forgo its high wing preference in favour of the low wing layout stipulated by TCA. BEA's seasonal traffic patterns led to under utilisation of many of its aircraft in the winter months, and inspired Vickers' to suggest a 'double bubble' fuselage cross section for the new aircraft with two capacious lower deck cargo compartments in the lower half of the 'bubble' totalling 38 m³ (1,350 cu ft) capacity and capable of carrying up to 7711 kg (17,000 lb) of freight. With full cargo capacity utilised it would be capable of carrying its maximum payload of 9525 kg (21,000 lb) with only thirty per cent of seats occupied, enhancing utility and profitably.

BEA decides on the Vanguard

The final design was redesignated Vickers Type 900 and was to be the first aircraft powered by the new Rolls-Royce RB.109 turboprop engine. The name Vanguard was chosen by BEA, which ordered six Type 901 aircraft in 1955 and a further 14 in 1956. A contract for these 20 Vanguards was signed on 20 July 1956, calling for service entry of the aircraft during 1960. An increase in maximum take-off weight to 61236 kg (135,000 lb) and selection of 3714-kW (4,985-shp) Rolls-

Eight Vanguards graced the Indonesian register for 10 years in the hands of Merpati Nusantara, beginning in 1971. This aircraft was the very last Vanguard to be built, and was bought by EAS in 1981. It ended its days at Perpignon.

Along with BEA, Trans-Canada Air Lines were the prime mover behind the Vanguard. The Canadian airline favoured Vickers' 870 design, with its low wing rather than BEA's preferred option of a high-winged aircraft. The Vanguard's projected freighter role swung the argument in favour of the final configuration.

Royce Tyne R.Ty1 Mk.506 turboprop engines led subsequently to the designation being changed to Type 951. BEA's Vanguards were to be configured for five crew and up to a maximum of 126 passengers in all-economy seating. In typical mixed-class configuration the Vanguard could seat 42 first class passengers in four-abreast seating with 99-cm (39-in) pitch, and 55 economy class in six-abreast configuration with 86-cm (34-in) seat pitch.

BEA's specification did not then meet Trans-Canada's requirements, so Vickers developed the Type 952 Vanguard for the Canadian carrier. It had increased fuel tankage to meet the transcontinental range requirement, six-abreast seating for 139 passengers, 4131-kW (5,545-shp) Tyne R.Ty11 Mk.512 engines, an increased payload of 10886 kg (24,000 lb) and a maximum take-off weight of 63957 kg (141,000 lb) (later increased to 66452 kg/146,500 lb). TCA (later Air Canada) ordered 20 Vanguards in January 1957, subsequently adding a further three aircraft to its requirement.

Improvements on the line

Construction of the prototype Type 950 Vanguard was under way at Vickers' factory at Weybridge, Surrey when BEA opted for the increased fuel capacity and higher 63957 kg (141,000 lb) maximum take-off weight of the TCA variant, since the additional useful load it afforded could be used for increased payload when maximum range was not required. BEA's second batch of 14 aircraft thus became Type 953s, but retained the lower-powered Tyne engines of the six Type 951s.

The prototype Vanguard, G-AOYW, was built on production tooling. It was rolled out at Weybridge on 4 December 1958 and made its first flight on 20 January 1959, flown by Vickers' chief test pilot G.R. 'Jock'

G-APEA was the second Vanguard built and the first to be delivered to British European Airways. The Vanguard 951 entered service with BEA on 1 March 1961, and won considerable passenger favour in the face of the new jets.

Bryce. Production models followed quickly, BEA's first aircraft, G-APEA, named *Vanguard,* flying on 22 April. The second BEA Vanguard, G-APEB *Bellerophon,* made the type's public debut at the Farnborough Air Show in September 1959. The remainder of the BEA fleet, named after warships of the Royal Navy were: *Sirius, Defiance, Indefatigable, Ajax, Euryalus, Victory, Arethusa, Audacious, Dreadnought,*

Above: Nine Vanguards found their way to British operator Invicta International, who used the aircraft as freighters. This aircraft was leased and then bought outright from Air International Holdings in 1973.

Below: Originally delivered to TCA in January 1961, this V.952 was later sold by Air Canada to Air Holdings, who regularly leased it out to other operators. While flying for its new French owner, Inter Jet, in 1989 it crashed at Marseilles.

Right: For a period during its flight tests, the prototype Vanguard carried a rounded dorsal fin. Fitted to cure the tendency of the rudder to 'hunt' at small angles of deflection, the problem was cured on the production aircraft.

Leander, Agamemnon, Valiant, Orion, Superb, Amethyst, Swiftsure, Temeraire and *Undaunted*. The first Type 952 for TCA made its maiden flight on 21 May 1960.

First deliveries

Although the aircraft's performance met design parameters, a compressor fault in the Tyne engine which had been discovered during bench tests by Rolls-Royce delayed the certification programme and caused the aircraft to be grounded while modifications were made to the engine, and it was not until 2 December 1960 that the Airworthiness Requirements Board issued a full public transport certificate of airworthiness to the Vanguard. BEA took delivery of its first aircraft on the next day, and operated its first route-proving passenger service with the new aircraft on 17 December on the London-Paris route,

The last operator of passenger Vanguards was Merpati Nusantara of Indonesia, which was also a committed Viscount user. As late as 1983 Merpati had a pair of Vanguards flying alongside a substantial Viscount fleet.

although full Vanguard services did not commence until 1 March 1961. Trans Canada operated its first Vanguard service on 1 April 1961, to the Caribbean. A month later BEA introduced the Type 953 on its European routes. The last BEA Vanguard was delivered on 30 March 1962.

Despite Vickers' sales efforts, no further orders were received for the Vanguard, total production totalling just 44 aircraft,

including the prototype, which was broken up at the company airfield at Wisley in 1962. Like its American counterpart the Lockheed L.188 Electra, the Vanguard suffered from making its appearance at the time when the first short/medium range jet airliners such as the One-Eleven, Caravelle and DC-9 were also emerging, and these proved far more attractive to airlines and their passengers than turboprops.

A short life

By 1967 Trans Canada began converting some of its Vanguards to freight configuration, and started retiring and disposing of its fleet from 1969 onwards, the last aeroplane leaving TCA in 1974. British European Airways also began removing Vanguards from its passenger services from

Between 1972 and 1973 Sweden's Air Trader leased a trio of Vanguard 952s from Air Holdings. Other Nordic airlines to fly the type included Air Viking and Thor F/S of Iceland.

G-APEE 'Euryalus' was the penultimate Vanguard 951 built for British European Airways. Vickers introduced the V.953 from the seventh production aircraft, certifying it for operations at a higher gross weight over a slightly shorter range than the V.951. When BEA learned of these plans it changed its order to six V.951s and 14 V.953s, production being too far advanced to revise the initial half dozen. Vanguards entered service on route-proving trials in February 1960, but in the final week of trials service entry was formally delayed by a hitch with the power supply unit. BEA had planned a summer service with the Vanguard and was extremely unhappy with the delay. Problems were resolved by December, however, and G-APEE was used extensively for crew-training at Stansted from the second of the month onwards. Sadly Euryalus was to have a short career with BEA. Delivered in early December 1960, 'Euryalus' was lost when it overshot the runway and crashed, in fog, at Heathrow on 27 October 1965.

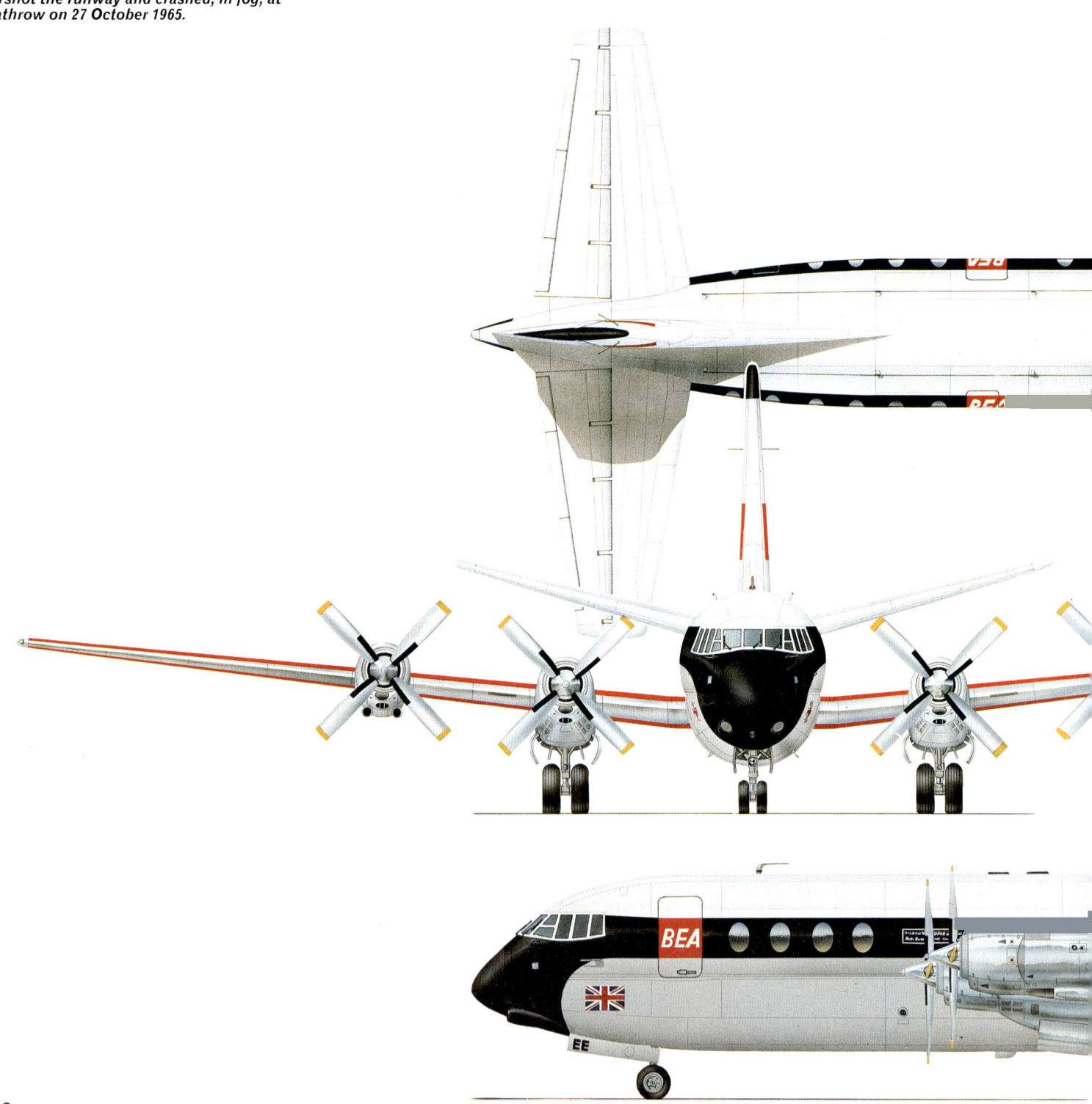

Specification
Vickers Vanguard Type 952
Wingspan: 36.17 m (118 ft 7 in)
Length: 37.21 m (122 ft 10 ½ in)
Height: 10.39 m (34 ft 11 in)
Wing area: 142 m² (1,529 sq ft)
Powerplant: 4 x 4131-kW (5,545-shp) Rolls-Royce Tyne R.Ty11 Mk 512 turboprops, driving de Havilland four-bladed fully-reversing and feathering propellers.
Passenger capacity: 139
Empty weight: 37422 kg (82,500 lb)
Fuel weight: 18597 kg (41,000 lb)
Maximum payload: 16783 kg (37,000 lb)
Maximum take-off weight: 66452 kg (146,500 lb)
Maximum speed: 684 km/h (425 mph)
Cruising speed: 676 km/h (420 mph)
Service ceiling: 9144 m (30,000 ft)
Maximum range: 4988 km (3,100 miles)
Range with maximum payload: 2945 km (1,830 miles)

Above: Despite the introduction of jet equipment by BEA, the Vanguards continued to make good economic sense, especially over sectors of less than 800 km (500 miles). Typical of these was the regular (full) run to Malta with passengers and freight.

Left: Two BEA Vanguards were lost in accidents, leaving 17 in service by April 1972 when British Airways was created. The last BA Merchantman freighter was disposed of only in 1980.

October 1968. Aviation Traders Limited of Southend Airport engineered an all-cargo conversion for BEA which involved strengthening of the aircraft's cabin floor, installation of a large upward-opening side-loading freight door in the port side of the forward fuselage, and provision of a roller conveyor system for freight handling. Designated Type 953C Merchantman, the first converted aircraft made its maiden flight on 10 October 1969, and was followed by a further eight of BEA's Vanguards which became Merchantmen with the carrier's cargo division. The Vanguard found its niche in cargo operations with BEA, and gave sterling service until 1980, when the last of its Merchantmen was retired. During service with BEA two Vanguards were destroyed in fatal accidents. G-APEE crashed and burned on the runway at London-Heathrow Airport on 27 October 1965 while attempting to land in fog, and G-APEC crashed near Arselle, Belgium on 2 October 1971 after an inflight structural failure which was attributed to corrosion of the aft pressure bulkhead and rear fuselage area caused by leakage from the rear toilet.

Vanguard homecoming

Many of Trans Canada's Vanguards returned to Europe, and with surplus BEA aircraft served with Europe Aero Service of France which became the major operator of the type with a fleet of 11, Air Trader of Sweden, Invicta Air Transport of Great Britain, Air Viking and Thor Cargo of Iceland and, further afield, with PT Merpati Nusantara Airlines and Angkasa Civil Air Transport of Indonesia and Lebanese Air Transport. Only Air Bridge Carriers of East Midlands, remained faithful to the type maintaining, and even increasing, its Merchantman fleet by returning several withdrawn aircraft to service during the 1980s. As the numbers in service fell once more, Air Brigade (renamed Hunting Air Cargo and then Hunting Cargo Airlines) flew three V953Cs until replaced by the expanding Electra fleet on overnight delivery services around Europe. The aircraft retained their former-BEA registrations and names (*Agamemnon, Superb,* and *Swiftsure*) keeping alive the Vanguard's line, and the whistle of its Tyne engines.

Elan couriers (courtesy of Air Bridge Carriers) resurrected this former EAS aircraft in 1987, a period when the Merchantman fleet enjoyed a small renaissance owing to its desirability for the growing overnight-delivery market.

Vickers Viking

Developed as an interim airliner from the Wellington bomber, the perky Viking remained in service for considerably longer than was planned. Although not the world's best or most successful aircraft, it was instrumental in resuming services as normal following the cessation of hostilities in Europe.

In October 1944 the British Ministry of Supply and Aircraft Production ordered from the Vickers Aircraft Company three prototype 'Wellington Transport Aircraft', to specification 17/44, to meet post-war needs for a short/medium range airliner.

To save time and development costs Vickers designer Rex Pierson proposed to use as much as possible of the airframe of the Wellington bomber, including its 1,675-hp Bristol Hercules sleeve valve radial engines, nacelles, landing gear, and fabric-covered geodetic structure outer wing panels, mated to a new all-metal stressed skin fuselage with accommodation for up to 21 passengers and a crew of four. The portly fuselage was unpressurised, and featured large square cabin windows, with a single passenger entry door to the rear just ahead of the tailplane.

Little can be seen of the Wellington in the outward appearance of the Viking, yet there is an unmistakable air of the old 'Wimpey' about the airliner. It saw its most valuable service in BEA colours, replacing Douglas DC-3s and revitalising the pre-war routes to Europe after a devastating war.

The first prototype, known as the Vickers Type 491, made its maiden flight from Wisley aerodrome in Surrey on 22 June 1945 in the hands of Vickers' famous test pilot J. 'Mutt' Summers, and was the first post-war British commercial aircraft to take to the air. The second and third aircraft, designated respectively Type 495 and 496, flew later in the year and by the end of 1945 the three aircraft had jointly amassed 160 hours of test and development flying, by which time the name Viking had been selected. Although it was known generically as the Vickers Commercial 1, it was Vickers' practice to allot each customer their own type number.

An initial order for 19 Type 498 Viking IAs, powered by 1,690-hp Bristol Hercules 630 engines, was received from British European Airways. The first of these made its maiden flight on 23 March 1946. Three aircraft were used for crew training at BEA's base at Northolt Airport near London, five were modified internally and delivered to British West Indian Airways for operations out of Trinidad, and the remainder became BEA 'V Class' airliners, replacing Douglas DC-3s. The first passenger service by a Viking was flown by G-AHOP *Valerie* between Northolt-Copenhagen on 1 September 1946, following a proving flight by *Vagrant* to Stavanger and Oslo 10 days previously.

BEA subsequently ordered a further 13 Type 614 Viking Is, these aircraft having stressed skin rather than geodetic wings and Hercules 634 engines. Five of these were diverted to overseas operators – three to BWIA and one each to Central African Airways and South African Airways. By the end of 1947 BEA Vikings had replaced DC-3s on all the airlines' principal overseas routes to Belgium, France, Germany, Holland, Ireland, Norway, Spain, Sweden and Switzerland, and on domestic services between Northolt-Glasgow. Many fabric-winged Viking IAs were later upgraded to Viking I standard.

One of BEA's fleet, *Vimy*, made headline news on 5 April 1948 when it crashed, killing 10 passengers, following a mid-air collision with a Soviet Yak fighter whilst approaching to land at Gatow, Berlin.

Total production of the Viking I/IA was 48 aircraft, of which 12 were delivered to the Ministry of Supply as Viking C.2s. Four of these served The King's Flight which had been re-formed at RAF Benson, Oxfordshire on 1 May 1946. The King and Queen each had a personal Viking, the two others serving as support crew transport and mobile workshop. In February 1947 all four flew to South Africa for a major Royal Tour by the King, Queen and Princesses Margaret and Elizabeth. A fifth King's Flight Viking was acquired during 1947.

On 6 August 1946 the first Viking IB was flown. This model featured a 28-inch extension to the forward fuselage section permitting three more passengers to be carried on short range routes. BEA took delivery of an initial batch of eight, followed by 39 more, while other orders came from the Argentine Air Force, Royal Australian Air Force, Central African Airways, DDL of Denmark, Misrair of Egypt, Aer Lingus of Eire, Air India, Iraq Airways, the Arab Legion Air Force of Jordan, the Pakistani Air Force, and Southern Rhodesia Airways, bringing total Viking IB production to 113 aircraft by the time final deliveries – to South African Airways – took place at the end of 1947.

In January 1951 British European Airways introduced a modification which enabled its Vikings to seat up to 36 passengers with an increased maximum take-off weight of 36,712 lb. At the same time a new and soon to be familiar BEA livery of dark red and white was introduced and the Viking fleet was progressively renamed as 'Admiral class' aircraft, the first 36-seater, G-AIVL, *Vigilant*, becoming *Lord Hawke*. As a result of the increased seating capacity BEA was able to offer the first 'cheap return' fare of £9. 15s. 0d (£9.75) on the London-Paris route which was inaugurated on 1 October 1952 by *Lord Fisher*, one of eight bought back from South African Airways to supplement BEA's hard-working fleet.

British European airways operated Vikings until 1954 when the venerable aircraft were replaced by turboprop Vickers Viscounts. In eight years of service BEA Vikings had carried 2,748,000 passengers over 65 million miles in 414,000 airborne hours, earning the carrier £35 million in revenue.

After retirement by BEA, Vikings began second careers with independent carriers such as Airwork, Eagle Aviation and Hunting Air Transport on trooping, freight and charter flights, and later with a host of British inclusive tour operators such as Air Ferry, Air Safaris, Autair, BKS Air Transport, Channel Airways, Continental Air Services, Eros Airline, Falcon Airways, Independent Air Travel, Invicta, Orion Airways, Pegasus Airlines and Tradair, some continuing in service until the end of the 1960s. A substantial number were also operated in Germany by charter carrier Luftransport Union (LTU).

One Viking was to play a significant part in the post-war development of British commercial aviation. The 107th airframe off the production line was completed to Ministry of Supply order with a pair of 5,000-lb st thrust Rolls-Roce Nene turbojet engines mounted in underslung wing pods in place of the usual Hercules piston engines.

Other modifications included heavier skinning on the wings, metal-skinned elevators in place of the standard fabric-covered control surfaces, and entirely different undercarriage with four separate main legs, each carrying a single wheel, and changes to the fuel system and cockpit. When 'Mutt' Summers took off in the Type 618 Nene Viking from Wisley on 6 April 1948, it became the first British pure jet transport aircraft to fly.

The Nene Viking had a maximum speed of 468 mph and could cruise at 393 mph over a range of 345 miles. On 25 July 1948, the 39th anniversary of Louis Blériot's first Channel crossing, it carried a full complement of 24 passengers between London and Villacoublay, near Paris in 34 minutes, seven seconds – less than half the scheduled flight time of BEA's standard piston-engined Vikings and faster than the previous record between the capital cities set by a Supermarine Spitfire. After years of successful trials the Nene Viking was converted back to piston-engined Mark 1B configuration in 1954 and operated as freighter *Lord Dundonald* by Eagle Aviation, eventually going to the scrapheap in 1962.

In addition to its wide-ranging role as a post-war airliner and cargo aircraft, the Viking was also developed for service with the Royal Air Force as the Vickers Valetta troop transport and freighter. The prototype Valetta made its maiden flight on 30 June 1947. It differed from the contemporary Viking IB in having a strengthened fuselage floor with large double loading doors on the port side, new landing gear with long-stroke oleo legs, a modified fuel system, and more powerful 1,975-hp Bristol Hercules 230 engines. The Valetta C.1 could carry up to 34 fully equipped troops, 20 paratroops, or freight which could be air-dropped. The Valetta C.2 was structurally similar but configured for VIP transport with accommodation for 9-15 passengers, and had increased fuel capacity for longer range. The

BEA aircraft operated regularly on routes in Europe, and compiled a good safety record for the day. Certainly the most unusual loss was that of a G-AIVP, which suffered a mid-air collision with a Soviet fighter along the air corridor to West Berlin.

The majority of Vikings were the Mk 1B variant, which featured a lengthened forward fuselage. This increased seating from 21 to 24 initially, although later aircraft were configured to carry a maximum of 38 passengers. Note the early BEA scheme.

The first of the Mk 1Bs went not to BEA but to Indian National Airways, this being the first aircraft. INA bought six of the type, and were followed by Air India with nine. These were later pooled in the Indian Airlines Corporation.

After service with BEA, many of the early Vikings were sold to small British airlines. G-AHOY was a typical example, flying successively with Hunting, Eagle, Pegasus, Autair (illustrated), Claydon Aviation and Invicta.

Valetta C.1 entered service with No. 204 Squadron, RAF, at Kabrit, Egypt in May 1949, and progressively replaced the Douglas Dakota in RAF Transport Command, giving a good account of itself during the Malayan crisis when aircraft from the Far East Transport Wing dropped more than 2,000 tons of supplies in support of British forces in the jungle.

A 'flying classroom' version, designated Valetta T.3, entered service with the RAF in 1950 for training of navigators. This aircraft had a row of roof-mounted astrodomes above the cabin to enable pupil navigators to take star and sun shots. Forty Valetta T.3s were built. The Valetta retired from RAF transport service in 1966.

The final development of the Viking airframe was also a military aircraft. The Vickers Varsity T.1, first flown on 17 July 1949, was designed as a replacement crew trainer for the Wellington T.10, to be used for training multi-engine pilots, navigators and bomb-aimers. Although of similar appearance to the Viking and Valetta, the Varsity was essentially a new aircraft, having a wing span six feet greater, a longer fuselage with a ventral 'panier' accommodating a bomb-aiming position and bomb bay for 25-lb practice bombs, tricycle undercarriage, and 1,950-hp Bristol Hercules 264 engines. One hundred and sixty Varsities were built for the RAF, the last retiring in 1976, although one is still on the strength of the Royal Aircraft Establishment. Two civilianised Varsities were used for research purposes by Smith's Industries during the mid-1960s.

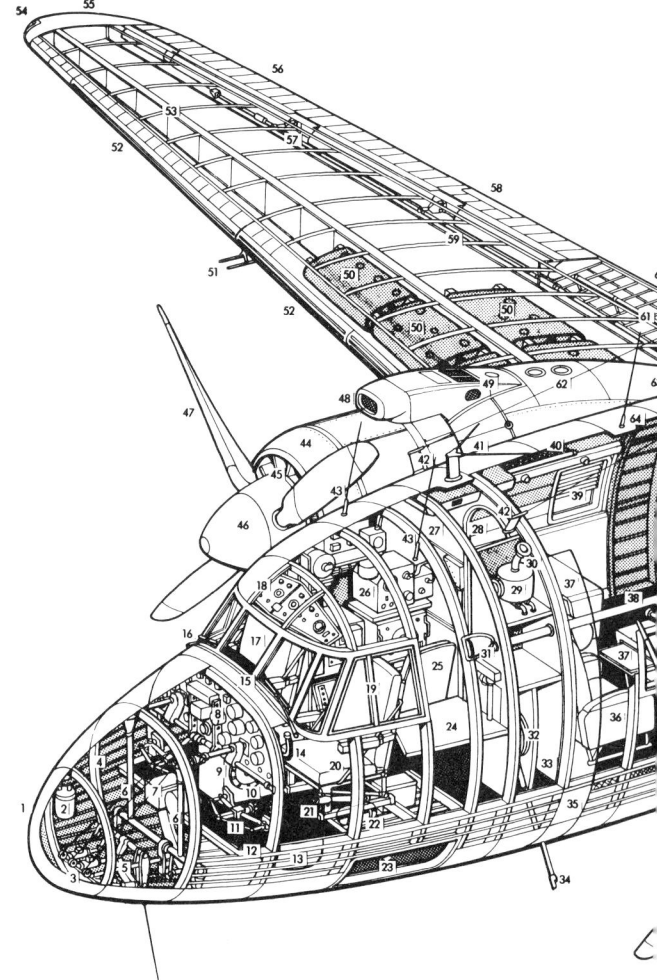

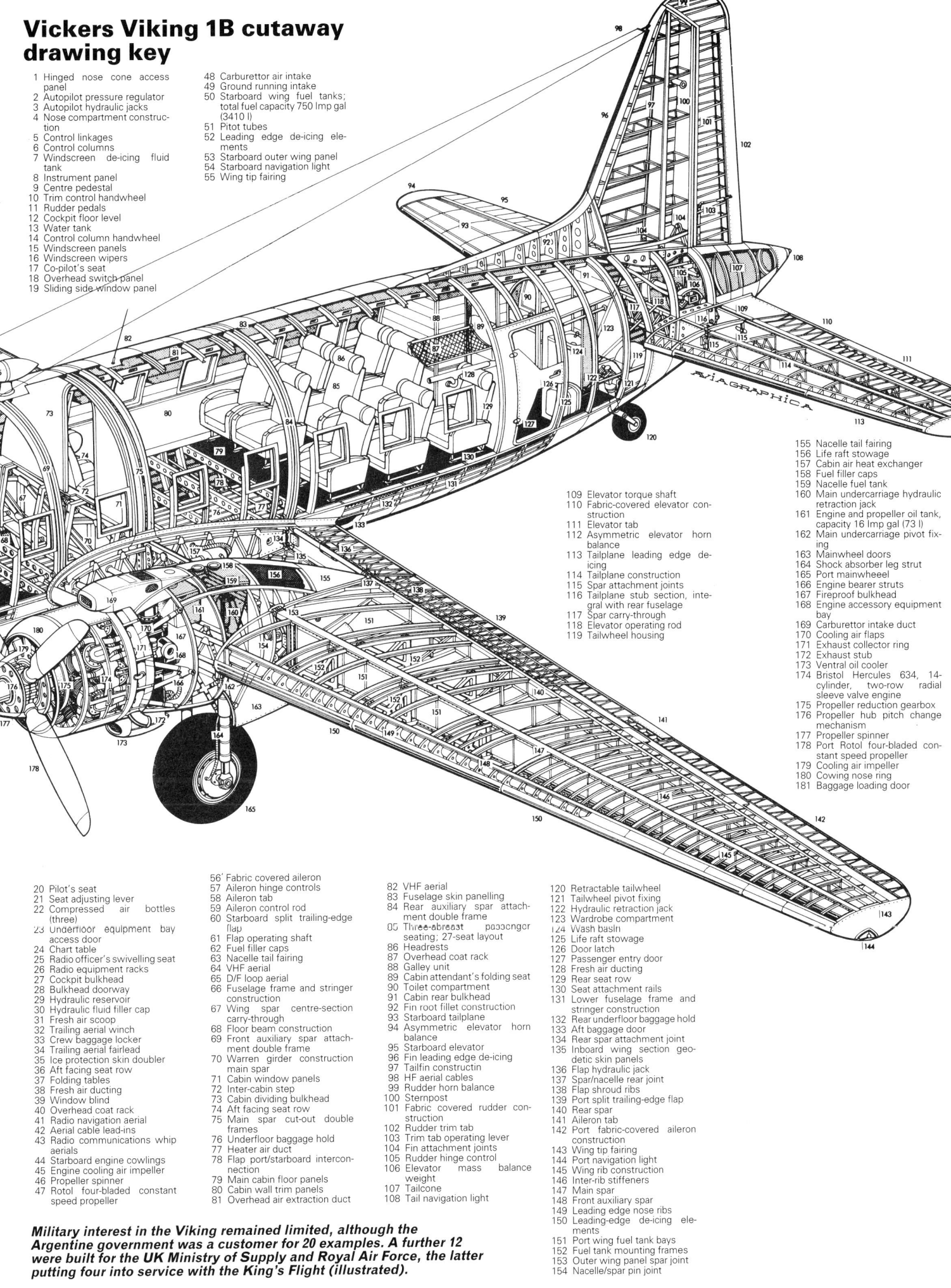

Vickers Viscount

For 30 years between 1935 and 1965, the American dominance of the world commercial aircraft market encountered only one formidable competitor: the silky-smooth Viscount. The world's first true turboprop aircraft, the Viscount emerged as a superb design, offering comfort, reliability and efficiency far in advance of that available with reciprocating engines, and consequently sold in sizeable numbers on both sides of the Atlantic.

With most aircraft, and no less so with engines, it takes a touch of genius to get the design right – too clever, and it takes too long to achieve reliability and the costs go through the roof; too crude, and it will be uncompetitive. With the Dart engine Rolls-Royce appeared to have erred on the side of crudity: in 1966 Rolls-Royce chief executive Sir Denning Pearson called the engine 'agricultural machinery', but he also commented that at that time nobody had ever junked a Dart-engined aeroplane. In fact mating four of these engines with the Viscount airframe proved so successful that in 1955 Convair, the immediate rival, tried to compete by offering the airlines a look-alike called the Convair Dart. It was never built, because the customers kept buying Viscounts.

Towards the end of World War II the UK tried hard to make up for lost ground by reconvening the Brabazon Committee to make specific proposals on which types of civil transport should be built when peace returned. It was rightly believed that the nation had a world lead in gas-turbine engines, and though for the immediate future the industry had to make do with 'interim' airliners converted from bombers, for the longer term it was planned to design 'clean sheet of paper' aircraft which, where appropriate, would have turboprop or turbojet engines.

One of the most important requirements was the Brabazon Type II, intended to produce a major short-haul airliner for European routes, and clearly with global sales prospects. It was split into Type IIA (piston) and Type IIB (turboprop) concepts. The former category was filled by Arthur Hagg's aerodynamically outstanding Airspeed Ambassador, with two Centaurus engines. BEA bought 20, but for reasons not unconnected with de Havilland's takeover of the Airspeed company, this remained the only order. The Type IIB clearly posed a higher technical risk, so the Ministry of Supply ordered prototypes of two rival types, the Armstrong Whitworth A.W.55 Apollo and the Vickers-Armstrong VC2. The latter appeared to be the result of careful studies at Weybridge started in mid-1944 to design a successor to the VC1 Viking, an interim DC-3 type machine derived from the Wellington bomber. First ideas centred on a stretched Viking with tricycle landing gear and stressed-skin wings carrying four 745.7-kW (1,000-hp) turboprops, but by January 1945 double-bubble pressurised drawings began to appear. In April 1946 the Type IIB specification (8/46) was published, calling for 24 seats, a 3402-kg (7,500-lb) payload and 1609-km (1,000-mile) range.

In September Rex Pierson was promoted to be chief engineer, so

Stretching the V.630 into the V.700 did wonders for the Viscount's appearance. Here the first V.700, G-AMAV, is seen early in its career in 1950. It never belonged to BEA, although in 1953 it was named **Endeavour** *as a member of the airline's 'Discovery' class, and with race number 23 on the tail took part in the race to Christchurch, New Zealand.*

Originally the second Viscount 701 to be delivered to British European Airways in February 1953, G-ALWF flew as part of the airline's 'Discovery' class until transfer to Channel Airways. Subsequent operators were British Eagle and Cambrian Airways, the latter's colour scheme being shown here.

TCA's order in November 1952 was most significant: for the first time a British company realised that an operator outside Britain might be able to improve the product and make it more acceptable in the world market. CF-TGK was No. 3 for TCA, delivered on 4 February 1955, and finally scrapped by Air Canada in July 1970.

his place as chief designer (Weybridge) was filled by G.R. Edwards. Sir George Edwards, as he became, was probably the most important leader the British aircraft industry has ever had, and he was to direct the Viscount, Valiant, Vanguard, VC10, TSR.2 and Concorde with the skill and confidence lacked by most other British companies but to be found in abundance in their American rivals. Without him the Viscount would almost certainly never have happened, because most British managements would have reacted differently at several crucial points in the programme.

For a start he decided to design for not 24 but 32 seats. He decided against a double-bubble and went for an almost circular fuselage, but with the cockpit roof added as an extra section built separately and attached on top. The wing was derived from that of the Wellington, via the Viking, with a single main spar but all stressed-skin structure, and with thermal de-icing and double-slotted flaps. The four engines were to be in pencil-slim nacelles, in turn calling for main gears each with two small wheels (which even then would be bulkier than the engines). Unlike de Havilland with the Comet, Vickers took great care to avoid any small-radius corners in the fuselage cut-outs, and both doors and windows were elliptical neutral holes. In later years the giant windows were to be highly popular with passengers, and Fokker adopted them for the F.27 Friendship along with many other Viscount parts, including the engines and propellers.

At first the best engine seemed to be the Armstrong Siddeley Mamba, an advanced turboprop with an axial compressor. This was naturally picked for the rival Apollo, because both were Hawker Siddeley companies. When the Ministry placed its order on 9 March 1946 it specified two Mamba-powered VC2s (now the Vickers Type 609 and named Viceroy, G-AHRF/RG), with Vickers to build a third at its own expense. The third was possibly to have the Rolls-Royce Dart engine, a seemingly more primitive turboprop with two centrifugal compressors derived from the Griffon supercharger and seven can-type combustion chambers around the outside. Edwards talked with Lionel Howorth, the Dart designer, and with Lord Hives, who assured him of Rolls-Royce's commitment to the engine. Because of the reliability of the centrifugal compressor and the fact that at that time the whole turbine experience had been with this type of engine, Edwards decided in March 1947 to pick the Dart. The designation with this engine was changed to V.630, and manufacture began at the secure experimental centre at Foxwarren of the two V.630s plus, to a later schedule, G-AJZW funded by Vickers, the V.640 with four 1118.6-kW (1,500-hp) Napier Naiads. In August 1947 India gained independence and the name Viceroy was changed to Viscount (names that hardly assist marketing, and in 1954 Capital Airlines was to print thousands of very necessary stickers saying 'Pronounced Vi-Count').

Success and failure

G-AHRF was flown by 'Mutt' Summers and 'Jock' Bryce from Wisley on 16 July 1948. The results could hardly have been better. The radical new engines ran like sewing machines, and the aircraft proved beautiful to fly. Edwards was just daring to believe that he might have a world-beater when, on 22 September 1948, BEA ordered its fleet of 20 Ambassadors. To say this was a body-blow is to understate the case; with loss of the prime customer there seemed little point in continuing. The third prototype was cancelled and work on the second slowed down. The existing machine was repainted in Ministry markings as VX211, but continued in low key to fly outstandingly well to gain a restricted certificate of airworthiness on 19 August 1949. Meanwhile Edwards had been informed of the RDa.3 rating of 1044 kW (1,400 hp) compared with 738.2 kW (990 hp), and he then obtained from Peter Masefield and Bob Morgan of BEA an indication of renewed interest in the Viscount provided it continued to prove reliable and could be stretched to take advantage of the greater power. By this time the 32-passenger specification could be seen to be still too small, and in late 1948 Edwards cleared drawings for the RDa.3-powered V.700 version with a 47-

The V.630 prototype appeared in numerous civil and Ministry colour schemes, and is seen here over Poole Harbour in January 1949 in Vickers house livery, with civil registration. A few months later it was VX211, but still in Vickers livery; then it carried a subdued BEA title, and by early 1950 had been repainted in full BEA airline livery.

One of the numerous ministerial nonsenses of the immediate post-war period was Britain's failure to use the world-beating Nene and Tay turbojets. The only British Tay-powered aircraft was the second Viscount, the V.663. Like its predecessor the Nene-Viking, it had four main landing gears, retracting on either side of the jetpipes.

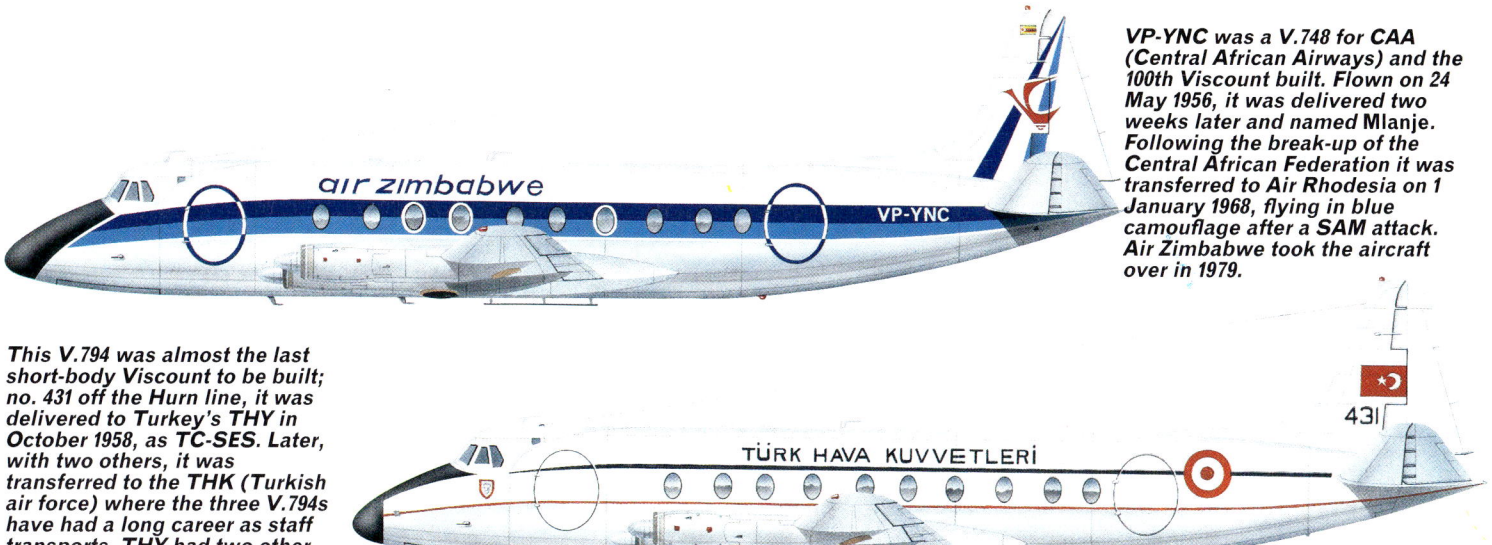

VP-YNC was a V.748 for CAA (Central African Airways) and the 100th Viscount built. Flown on 24 May 1956, it was delivered two weeks later and named Mlanje. Following the break-up of the Central African Federation it was transferred to Air Rhodesia on 1 January 1968, flying in blue camouflage after a SAM attack. Air Zimbabwe took the aircraft over in 1979.

This V.794 was almost the last short-body Viscount to be built; no. 431 off the Hurn line, it was delivered to Turkey's THY in October 1958, as TC-SES. Later, with two others, it was transferred to the THK (Turkish air force) where the three V.794s have had a long career as staff transports. THY had two other V.794s, which crashed at Ankara and Gatwick.

seat fuselage lengthened by 2.23 m (88 in), and with 762 mm (30 in) added to the root of each wing, not only to increase lift but also to give larger propeller/fuselage clearance to reduce noise.

The second airframe had been diverted as a Ministry research aircraft powered by two Rolls-Royce Tay turbojets, each of 2835-kg (6,250-lb) thrust (it was the only British aircraft to use this engine). Serialled VX217, it was put on trials to support the Vickers Valiant bomber, being leased to Boulton Paul Aircraft for work on powered flight controls and then to Louis Newmark and Decca Navigator – overall spending eight years pioneering FBW (fly-by-wire) control systems. Its type number was V.663. As for the V.630, this received clearance on 27 July 1950 for one month of scheduled airline operation, and on 29 July it took off from London Northolt in the hands of Captain H.R. 'Dickie' Rymer on a regular service to Paris-le Bourget. This was the first-ever airline service with any form of turbine power, and it was a revelation. G-AHRF made 35 further Paris trips, followed by eight to Edinburgh. The passenger reaction, and total reliability, confirmed BEA in its decision to buy the stretched version.

The prototype of the latter, the V.700 G-AMAV, was Ministry-funded and greatly speeded by using parts from the cancelled G-AJZW and bringing in other Vickers-Armstrong factories such as Itchen (Southampton) for the wings and South Marston (Swindon) for the fuselage. Thus G-AMAV was able to fly from Weybridge on 28 August 1950, and it proved every bit as good as had been hoped. Despite the increase in size, and in gross weight from 18144 to 22680 kg (40,000 to 50,000 lb), performance was considerably enhanced and operating economics were much better. The turning point came on 3 August 1950 when at last BEA confirmed an order for 20 (later increased to 26) of the V.701 type with one-class seating for 47 or 53 and augmented freight capacity both below and (at the rear) above the floor. The first V.701 (G-ALWE) flew on 20 August 1952. On 11 February 1953 Lady Douglas named it *Discovery*, all subsequent members of the class being named for famous explorers. By this time gross weight had been cleared at 25402 kg (56,000 lb), a full certificate of airworthiness was granted on 17 April 1953, and regular service started on the following day on the route from London Heathrow, via Rome and Athens, to Nicosia.

This was effectively the true start of the turbine age on the world's airlines (the jet Comet suffering a major hiccup), and orders began to come in as never before for a British airliner. First to sign were Air France and Aer Lingus, followed by TAA of Australia (which already used the immediate rival, the Convair-Liner). During 1952 Canada's TCA (now Air Canada) had begun discussions at Weybridge. The airline's list of engineering change orders soon exceeded 100 (one, for an on-board potable water dispenser, appearing in Weybridge documents as 'portable water'). A few years earlier a self-centred British industry might have declined to bother, but Edwards knew that to do the impossible and crack the North American market would demand a much more 'North American' aircraft, in many ways a better aircraft. It had much greater electric power, two-pilot cockpit, new fuel system, provision for cold-weather operation and a revised interior, the specification being agreed as the V.724, and 15 were ordered in November 1952. This was just the start; TCA came back for V.757s of the V.700D family with 1193-kW (1,600-hp) Dart 510 engines and a gross weight of 27216 kg (60,000 lb), and altogether bought 51 Viscounts. In Washington, Capital Airlines staked its whole operation on the British aircraft and took delivery of 60.

In October 1953 a race was flown from London to Christchurch, New Zealand. BEA considered that, though the Viscount was a short-hauler, it could get favourable publicity from entering. G-AMAV, still on Ministry charge, was fitted with a big rectangular fuel tank in the cabin, given the 'Discovery' name *Endeavour* and, under Captain W. Baillie and with chief executive Peter Masefield in the crew, averaged 467 km/h (290 mph) over the 18982 km (11,795 miles), which did nothing to harm either BEA or the Viscount. On the other hand Hughes Tool Company was one of the customers who flocked to Weybridge in 1953, Howard Hughes himself having a personal interest. The specification for the single V.763 was the longest ever drawn up, and a resident team of Hughes men inspected every part and almost every rivet. Eventually the V.763 had to be shunted to one side, because it was getting in the way of the production line, and three years later it was re-registered YS-09C and sold to TACA of El Salvador, Hughes by this time being interested in Bristol Britannias and Boeing 707s.

The Dart 510 (RDa.6 rating) clearly made possible further increases in capability. Some V.700D aircraft had slipper tanks outboard of the outer engines for increased range; the first being the fifth V.720 for TAA. More significant was the decision to stretch the fuselage, in partnership with BEA. At first the V.801 was studied with a massive stretch, but it was then found that, by moving back

G-AOYV was the prototype for the ultimate V.810 series, and was painted in the markings of the first customer (for the V.812). G-AOYV flew in December 1957 and spent over a year on company research from Wisley, in the course of which it was fitted with massive water spray rigs and a dummy Vanguard fin for de-icing certification trials.

Typical of the smaller airlines which recognised the value offered by the Viscount is Manx Airlines, an operator of relatively recent origin that leased this particular V.813 aircraft from British Midland Airways; the aircraft had seen over 25 years of operational flying.

the rear pressure bulkhead, a 2.82-m (111-in) internal stretch could be achieved with a fuselage only 1.17 m (46 in) longer. On 14 April 1954 BEA ordered 12 (later 24) of these longer V.802 aircraft, still sticking to names of discoverers. As before, many customers followed suit, but Rolls-Royce was by this time ardently developing the trusty Dart and kept offering increased power. In 1954 the 1267.7-kW (1,700-shp) Dart 520 (RDa.7) led to the V.806, with unchanged 515-km/h (320-mph) economical cruising speed but gross weight raised to 29257 kg (64,500 lb). In 1955 the Dart 525 (RDa.7/1) of 1342.3 kW (1,800 shp) led to the considerably revised V.810 series with structure restressed for greater weights and higher indicated airspeeds. The only visible change was a simpler rudder aerodynamic balance, but in fact gross weight was cleared to 32885 kg (72,500 lb) and cruising speed to 579 km/h (360 mph).

Healthy sales and production

The V.810 prototype flew on 23 December 1957, by which time Viscounts were flooding off the assembly lines, the V.800 series all being assembled at a newly-established factory at Hurn. The V.700 was by this time in Vickers hands and it was re-engined with RDa.7s to speed Viscount V.810 series certification. At the end of this programme in 1963 it was unceremoniously given to the Fire School at Stansted, which was sad because the original V.630 prototype had been written off at Khartoum in 1952.

The plan in 1956 had been to offer a V.840 series with RDa.8 engines of 1864.3 kW (2,500 ehp), for a cruising speed of 644 km/h (400 mph). When Robert F. Six of Continental signed for 14 V.812s he intended to re-engine them later, and several other customers were interested in the '400-mph Viscount', but it was never built. This was partly because with Boeing 707s in service 644 km/h (400 mph) was less impressive and partly because the engine-out certification would have required a substantial redesign of the rear fuselage and tail. As it was, the V.810 family kept selling very healthily, until the very last order was received from CAAC of the People's Republic of China (that country's first purchase from a Western source). The 444th and last Viscount was first flown on 2 January 1964, 438 of these being regular aircraft sold to customers. Several were bought new by executive owners (the first being the Canadian Department of Transport in 1954) and air forces (the first being the Indian Air Force, again in 1954). In 1955 United States Steel bought three, specially equipped for luxurious inflight conferences. When ex-airline Viscounts came on to the second-hand market every single example was snapped up by other customers, many of them being scheduled operators but including an increasing proportion of corporate customers.

Four aircraft with long lives in the UK included two ex-Capital V.745s used by the Empire Test Pilots School and two stretched examples (an ex-Ghana V.838 and ex-Austrian V.837) which did sterling work for the Royal Radar (now Royal Signals and Radar) Establishment. Several countries have used Viscounts as flying testbeds for radars and other systems, and in February 1982 a Canadian aircraft began flight testing the new PW100 turboprop (ironically a Dart replacement) with the new engine mounted far ahead of the original nose. In 1998 eleven civil Viscounts were still active, the largest surviving fleet being the three of Bouraq Indonesia. To the passenger the Viscount is still a very pleasant way to travel, in no way inferior to the short-haulers now being built. It will be some time before its distinctive sound disappears from the scene.

Vickers Viscount 810/840 Series cutaway drawing key

1 Radome
2 Weather radar scanner
3 Radar tracking mechanism
4 Front pressure bulkhead
5 Radome hinge panel
6 Nose section construction
7 Pressurization relief valve
8 Control system linkage
9 Rudder pedals
10 Pneumatic system air bottle
11 Nosewheel doors
12 Forward retracting twin nosewheels
13 Nosewheel steering jack
14 Pitot head
15 Cockpit floor level
16 Seat mounting rails
17 Conditioned air delivery duct
18 Nosewheel steering control
19 Control column handwheel
20 Instrument panel
21 Weather radar display
22 Instrument panel shroud
23 Windscreen wipers
24 Windscreen panels
25 Overhead systems switch panels
26 Co-pilot's seat
27 Direct vision opening side window panel
28 Pilot's seat
29 Cockpit rear bulkhead
30 Cockpit pressure dome
31 Folding observer's seat
32 Cockpit doorway
33 Windscreen de-icing fluid reservoir
34 Wing and engine inspection floodlight
35 Radio and electronics equipment racks
36 Underfloor autopilot controllers
37 Cockpit section frame construction
38 Folding airstairs
39 Folding handrail
40 Entry lobby
41 Forward entry doorway
42 Starboard side baggage compartment
43 HF aerial mast
44 Hydraulic equipment compartment
45 Cabin front bulkhead
46 Cockpit pressure dome aft fairing
47 Forward passenger compartment
48 Four-abreast passenger seating
49 Forward entry door, open
50 Door latch
51 Door hinge link
52 Forward 'pull-out' emergency exit window
53 Underfloor cargo hold, 250 cu ft (7.08 m³) capacity
54 Toilet compartments, port and starboard
55 Ventral cargo hold door, starboard side
56 Toilet compartment doors
57 Magazine rack
58 VHF aerial
59 Starboard engine nacelles
60 Engine cowling panels
61 Propeller spinner
62 Rotol four-bladed constant speed propeller
63 Propeller blade root de-icing
64 Starboard wing central fuel cells; total fuel system capacity 1,916 Imp gal (8710 litres)
65 Fuel system piping
66 Outboard fuel cells
67 Overwing fuel filler cap

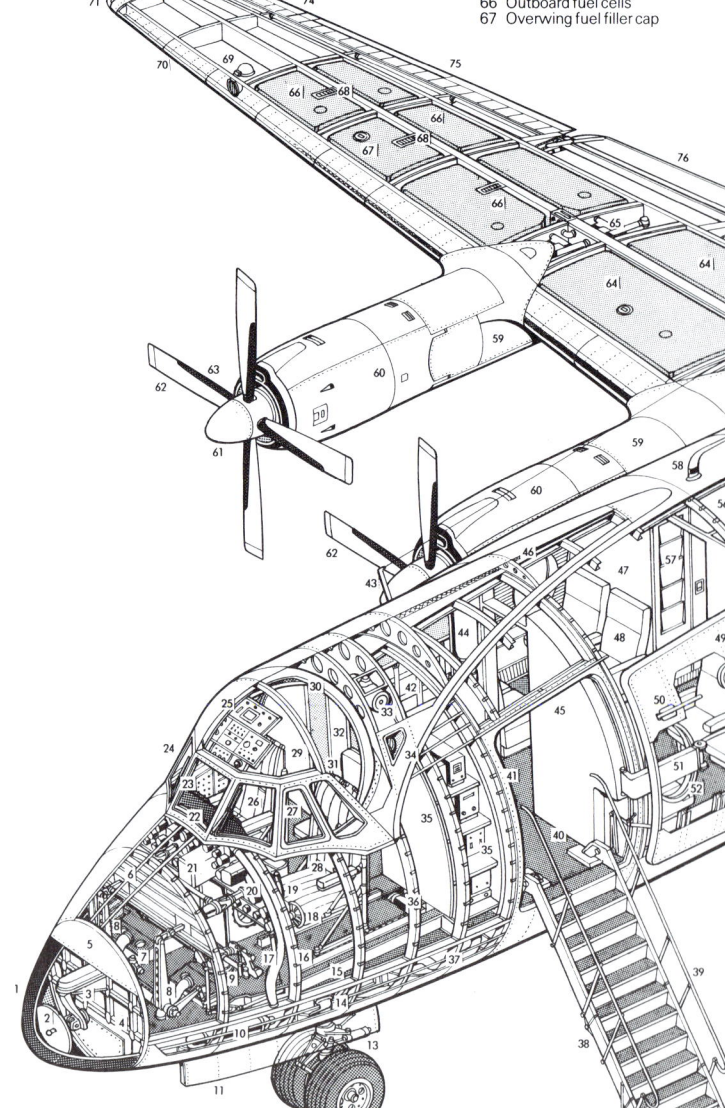

9G-AAU came very near the end of the line as no. 446, being delivered to Ghana Airways on 26 November 1961. Note the rectangular doors swinging aft on parallel arms which on the stretched 800-series replaced the original elliptical type to improve freight loading. Subsequently no. 446 was G-BCZR with Field, BMA, Southern International and Dan-Air. Viscount construction numbers included undelivered or unfinished aircraft, no. 446 being in practice no. 431.

- 68 De-icing air outlet louvres
- 69 Retractable landing taxiing lamp
- 70 Leading edge hot air de-icing
- 71 Starboard navigation light
- 72 Wing tip fairing with hot-air exhaust
- 73 Static dischargers
- 74 Starboard aileron
- 75 Aileron trim tab
- 76 Starboard double-slotted Fowler-type flap (down position)
- 77 Flap guide rails
- 78 Flap operating torque shaft and chain drive
- 79 D/F loop aerials
- 80 Fuselage frame and stringer construction
- 81 Floor beam construction
- 82 Main cabin passenger seating
- 83 Control rod runs
- 84 Front auxiliary spar attachment main frame
- 85 Main cabin 'pull-out' emergency exit windows
- 86 Main spar centre-section carry-through
- 87 Spar boom bolted joints
- 88 Cabin wall soundproof lining
- 89 Electrical system static inverters
- 90 Central flap drive motor
- 91 Main spar attachment double frame
- 92 Cabin trim panelling
- 93 VHF aerial
- 94 Overhead air conditioning ducting
- 95 Rear auxiliary spar attachment main frame
- 96 Cabin window panels
- 97 Main cabin four-abreast passenger seating, 52-seat layout (alternative 65-seat tourist or 75-seat coach class layouts)
- 98 Fresh air ducting
- 99 Individual passenger service units
- 100 Overhead coat/light luggage racks
- 101 Cabin rear bulkhead
- 102 Fin root fillet construction
- 103 Starboard tailplane
- 104 HF aerial cable
- 105 Starboard elevator
- 106 Elevator trim tab
- 107 Fin rib construction
- 108 Leading edge double skin de-icing air duct
- 109 VOR/ILS aerial
- 110 Fin tip fairing with hot-air exhaust
- 111 Rudder tab
- 112 Rudder construction
- 113 Tailcone
- 114 Tail navigation light
- 115 Spring tab
- 116 Anti-balance tab
- 117 Static dischargers
- 118 Port elevator construction
- 119 Tailplane rib construction
- 120 Leading edge de-icing air duct
- 121 Tailplane attachment main frame
- 122 Rudder and elevator control linkages
- 123 Rear pressure bulkhead
- 124 Wardrobe
- 125 Lounge compartment
- 126 Buffet/galley units, port and starboard
- 127 Rear entry/service door, open
- 128 Cabin attendant's folding seat
- 129 Rear entry/service doorway, port and starboard
- 130 Fuselage skin panelling
- 131 Cabin floor panelling
- 132 Underfloor air conditioning plant
- 133 Wing trailing edge root fillet
- 134 Flap rib construction
- 135 Flap shroud ribs
- 136 Port double-slotted Fowler-type flap
- 137 Flap down position
- 138 Optional fuel jettison pipe
- 139 Port aileron rib construction
- 140 Aileron hinge control
- 141 Aileron tab
- 142 Static dischargers
- 143 Port wing tip fairing with hot-air exhaust
- 144 Port navigation light
- 145 Outer wing panel rib construction
- 146 Retractable landing/taxiing lamp
- 147 Leading edge de-icing hot air duct
- 148 Leading edge double skin panelling
- 149 Outer wing panel fuel tank bays
- 150 Rear auxiliary spar
- 151 Main spar
- 152 Outer wing panel spar joints
- 153 Petal-type engine cowlings, open position
- 154 Engine bearer struts
- 155 Main engine mounting ring frame
- 156 Engine annular air intake
- 157 Spinner
- 158 Intake lip de-icing
- 159 Oil cooler air intake
- 160 Forward retracting twin mainwheels
- 161 Main undercarriage leg strut
- 162 Rear drag strut
- 163 Central fuel tank bays
- 164 Inner wing panel rib construction
- 165 Main undercarriage hydraulic retraction jack
- 166 Mainwheel leg breaker joint
- 167 Engine nacelle mounting rib
- 168 Wheel bay door operating link
- 169 Engine fire extinguisher bottle
- 170 Mainwheel bay
- 171 Air ducting to conditioning system
- 172 Inboard fuel cells
- 173 Water/methanol tank, port and starboard, total capacity 75 Imp gal (340 litres)
- 174 Leading edge air ducting
- 175 Engine-driven cabin air compressor (three)
- 176 Engine accessory equipment gearbox
- 177 Jet pipe
- 178 Engine bearer struts
- 179 Fireproof bulkhead
- 180 Accessory gearbox drive shaft
- 181 Engine flame tubes
- 182 Oil cooler
- 183 Rolls-Royce Dart RDa. 7/1 (Mk 525) turboprop engine
- 184 Propeller hub pitch change mechanism
- 185 Rotol or DH four-bladed constant speed propeller

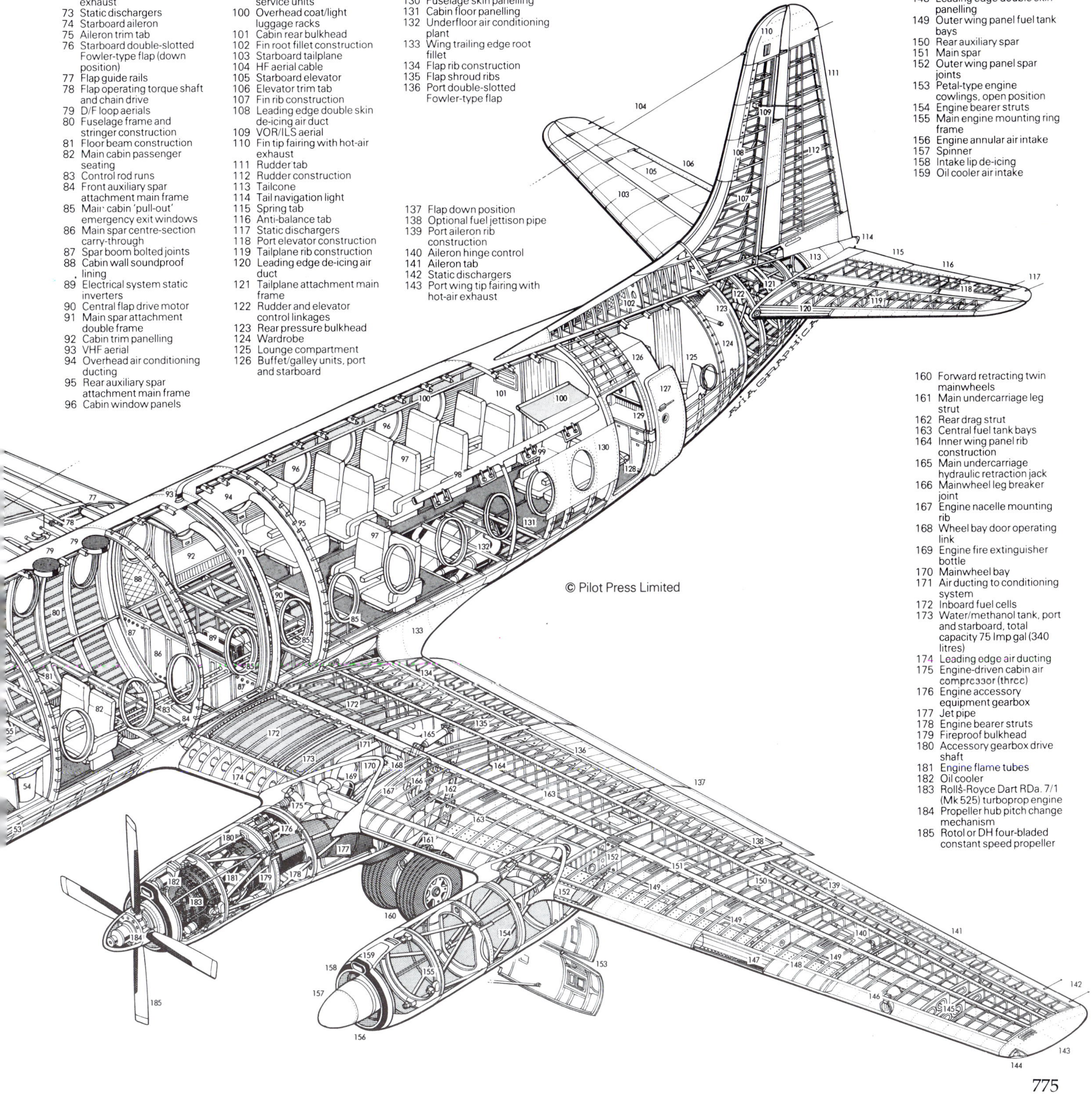

Vickers VC10

Although not a commercial success, the VC10 was nevertheless a superb airliner, blessed with sparkling performance and exquisite handling. Praised above all others by crews and passengers alike, the VC10 was the flagship of the British airliner fleet, and its longevity has allowed it to adopt a new career as a tanker.

In 1956 Vickers Aviation Limited based at Weybridge, Surrey, began initial design studies for an intercontinental jet airliner intended to meet a British Overseas Airways Corporation requirement for an aircraft to operate on its Commonwealth routes. The specification called for an aircraft capable of carrying a payload of 35,000 lb over a range of 4,000 miles, cruising at speeds up to Mach 0.8, and being able to operate with full payload from short-runway or 'hot and high' airfields such as those encountered at Kano, Nairobi and Johannesburg on BOAC's services to Africa, and others on the Far East and Australasian services.

Vickers elected to use a clean wing, rear-engined T-tail configuration, the first time such a layout had been chosen for a long-range jet airliner. It offered the prospect of a quiet passenger cabin and afforded the opportunity to create a wing of very clean design, with no breaks in the full-span leading-edge slats and large Fowler flap high-lift devices for engine pylons.

On 22 May 1957 BOAC issued a letter of intent for 35 aircraft, known as the VC10, for service entry in 1963, at a total cost of some £68 million. A firm contract was signed on 14 January 1958, with an option to purchase a further 20 aircraft. At the time BOAC's order for VC10s was the largest single airliner order ever placed with a British aircraft manufacturer.

The VC10 was to seat 135-152 passengers and to be powered by four 90.63 kN (20,370 lb) thrust Rolls-Royce Conway RCo42 Mk 540 turbofan engines, mounted in podded pairs. Parallel to the standard VC10, Vickers developed the Super VC10, initially envisaged with a 21-foot fuselage stretch to accommodate 212 economy-class passengers for BOAC's transatlantic routes, where

The remaining Super VC10s of British Airways were bought by the RAF for conversion to tankers as the K.Mk 4. Some were scrapped or burned, but the others then rested at Abingdon airfield while awaiting reconfiguration to triple-point tanker status.

Inset: Five Gulf Air Standards and four East African Supers were converted to tanker configuration for the RAF's No. 101 Squadron at Brize Norton. Commendably safe and reliable, they now form the backbone of the UK's inflight-refuelling assets.

airfield performance was not a major factor and runway performance could be exchanged for greater passenger carrying capability.

On 23 June 1960 BOAC placed an order for 10 Super VC10s with options on a similar number. However, the airline subsequently revised its future traffic forecasts and asked the manufacturer to reduce the passenger capacity of the Super VC10, and in its final form, announced in 1961, the aircraft featured a fuselage stretch of 13 feet over the standard VC10, accommodating 163-187 economy-class passengers. Additional modifications included increased fuel capacity in an integral fin fuel tank to give the aircraft range comparable to that of the standard VC10, a 26300-kg (57,980-lb) payload, higher operating weight, uprated 96.94-kN (21,800-lb) thrust Conway RCo43 Mk 550 turbofans, and a four per cent chord wing leading-edge extension. BOAC subsequently revised its order for VC10s to 15 (later 12) standard aircraft and 30 Super VC10s. In 1963 the airline asked for eight of the Supers to be convertible passenger/cargo aircraft, but this was later cancelled when BOAC further reduced its order to just 17 Supers, all configured for passenger use.

The prototype VC10, known by the Vickers designation Type 1100, made its first flight from Brooklands Airfield, Weybridge, to the Vickers test airfield at nearby Wisley on 29 June 1962, flown by G. R. 'Jock' Bryce and Brian Trubshaw, and was followed by the first production aircraft, Type 1101, on 8 November. Certification was granted by the British Airworthiness Requirements Board on 23 April 1964 after protracted flight testing to cure excessive cruise drag. Six days after certification the standard VC10 entered BOAC service on the London-Lagos, Nigeria, route. The prototype was later modified to full airline standard and sold to Laker Airways.

Inset below: Despite its Nigerian Airways titles, this aircraft was delivered to BOAC. However, another BOAC aircraft (G-ARVA) was later sold to the African carrier. Other operators in the continent were Ghana Airways, Air Malawi and East African Airways (Kenya-Tanzania-Uganda joint operation).

The first Super VC10, Type 1151, flew for the first time on 7 May 1964 and inaugurated BOAC's services with the type on the London-New York route on 1 April 1965, following acrimonious disputes between the Ministry of Aviation and BOAC, which had wanted to cancel its Super VC10 orders altogether in favour of buying US-built Boeing 707-320Bs, and finally accepted Super VC-10s "against our commercial judgement". BOAC Super VC10 deliveries were not completed until May 1969.

Passenger comfort

While BOAC management may have had severe reservations about the aircraft, both VC10 models quickly won praise from the travelling public for the comfort and quiet of their cabins, which many seasoned travellers claim has never been bettered (or even equalled) by modern wide-body long-haul aircraft. Throughout their service BOAC (later British Airways) recorded better passenger load factors for its VC10s than on any other aircraft of its fleets.

Other VC10 customers included Ghana Airways, which ordered two Type 1102s; British United Airways, with three Type 1103s, (both these and the Ghana aircraft having freight doors and the extended leading-edges of the Super); and East African Airways, which operated five Type 1154 Super VC10s, also with strengthened floors and large freight doors on the port side. Additionally, BOAC aircraft were leased to Nigeria Airways and Air Ceylon, and the former prototype and one of the Ghana Airways aircraft were leased to Middle East Airlines, the latter being destroyed by Israeli commandos in an attack on Beirut Airport on 28 December 1968. One BOAC Super VC10 was also destroyed by explosives planted by Palestinian guerrillas at Dawson Field in the Jordanian Desert on 17 September 1970, after being hijacked from Beirut along with a Trans World Airways Boeing 707 and a Swissair Douglas DC-8. A former British United/British Caledonian VC10 was sold to Air Malawi.

The Type 1100 prototype first flew on 29 June 1962, and after an extensive flight trials programme, was reconfigured for airline use with Laker Airways as the Type 1109. It later served with Middle East Airlines and British Caledonian before being scrapped in 1974.

1106), followed by another six in 1962 and three more in July 1964 to serve the strategic long-range force of Transport (now Strike) Command. The RAF aircraft was broadly similar to the standard VC10, but incorporated the Super VC10's uprated Conway engines and fin fuel tank, with other modifications including a side-loading freight door, strengthened floor, rearward-facing seats for 150 passengers or up to 78 stretcher cases, and a tailcone-mounted auxiliary power unit. The aircraft was to have a range of 3,670 miles on internal fuel, but with provision for modification for inflight-refuelling which would extend endurance to 17 hours and double unrefuelled range.

The first RAF VC10 C. Mk 1 flew from Weybridge on 26 November 1965, with deliveries to No. 10 Squadron at RAF Fairford commencing in the following July. Regular route flying began on 4 April 1967, and eventually No. 10 Squadron's aircraft operated 27 scheduled services each month to the Far East, offering the RAF high-speed, high-capacity, global mobility.

When the renamed British Airways Overseas Division withdrew its standard VC10s from service in 1975/76, five were sold to Gulf Air, while others went to serve as private transports in the Middle East. British Airways' Super VC10s continued in service until 1979.

Apart from commercial customers for the VC10, in September 1961 the Royal Air Force placed an initial order for five VC10s (Type

Thirteen VC10 C.1s, five of which were modified for inflight refuelling capability in 1986, continue to serve with No. 10 Squadron from RAF Brize Norton, Oxfordshire, and are also used regularly for flights by government ministers and members of the royal family. All RAF VC10 C.1s carry individual aircraft names of holders of the Victoria Cross. The other aircraft, from the total RAF order of 14,

Throughout the 1960s and 1970s, the VC10 was the standard-bearer of British commercial aviation overseas, serving on Commonwealth and North Atlantic routes. Twelve of the Type 1101 Standard VC10 were delivered to British Overseas Airways Corporation, one of which is seen here at Dubai.

One of the Royal Airforce transports was sold to Rolls-Royce to test the RB.211 high bypass ratio turbofan intended for the TriStar. With its civil registration, the aircraft had its port Conways replaced with a single RB.211, and undertook a successful test programme.

British United Airways was the second home operator to order the type, using three Type 1103 aircraft. When the Gatwick-based company merged with Caledonian Airways, the VC10s adopted the recently-disappeared British Caledonian titles.

Oman provided an unusual retreat for retiring BOAC/BA aircraft, no fewer than five joining Gulf Air, the national airline. Another flew with the Qatar Royal Flight, one with the Abu Dhabi Royal Flight, and this ex-BCal aircraft which flew with the Sultan of Oman Royal Flight.

was diverted to Rolls-Royce Limited in 1969 to serve as a flying test-bed for the RB.211 engine destined for the Lockheed L1011 TriStar. This VC10 first flew in modified form on 6 March 1970 with an RB.211 replacing both Conways on the port side.

The next chapter in the VC10 story began in April 1978 when the UK Ministry of Defence awarded British Aerospace a design study contract for the conversion of surplus civil VC10s into flight refuelling tanker aircraft, primarily for air defence fighters such as the Phantom and Tornado. A contract for conversion of nine aircraft was awarded three months later. The Royal Air Force acquired five

Seventeen Type 1151 Super VC10s were bought by BOAC for long-distance sectors, particularly transatlantic routes. Delivered between 1965 and 1969, they were withdrawn in the early 1980s, after having adopted British Airways colours.

The first Type 1103 for BUA was the 14th VC10 built. All three were sold by BCal, this example becoming 7Q-YKH, the sole Air Malawi machine. The others went to the Omani Royal Flight and the Royal Aircraft Establishment at Bedford, the latter for equipment trials.

standard VC10s from Gulf Air (originally BOAC aircraft) and four Super VC10s from East African Airways, to be converted by British Aerospace at Filton, Bristol, into VC10 K.Mk 2 and K.Mk 3 tanker variants respectively, while a further 14 Super VC10s were bought from British Airways and placed into store for future conversion (see below) and spares use.

The tanker conversion involved installation of underwing pylons to carry Flight Refuelling Mk 32/2800 refuelling pods, each capable of transferring fuel at a rate of 1270 kilograms per minute, and provision in the lower rear fuselage of a Flight Refuelling 17B HDU (hose-drum unit) with remotely-operated closed-circuit television for monitoring receiving aircraft, and floodlighting for night refuelling missions. Five additional 3182-litre fuel tanks were installed in the fuselage, interconnected to the standard fuel system, giving a total capacity of 94,272 litres in the K.2 and 102,782 litres in the K.3. In addition the tanker VC10s were equipped with their own flight refuelling probes and tailcone-mounted Rolls-Royce Turboméca Artouste Mk 520 auxiliary power units, and had many systems modifications to offer commonality with the RAF VC10 C.1 transports.

The first converted VC10 K.2 flew from Filton on 22 June 1982 and was subsequently used for flight trials and for further testing at A&AEE Boscombe Down. Operational deployment with No. 101 Squadron, RAF Strike Command at RAF Brize Norton, Oxfordshire, took place on 1 May 1984. The first VC10 K.3 flew on 3 July 1984 and entered service on 20 February 1985. Delivery of all nine VC10 tankers to No. 101 Squadron was completed on 24 September 1985. In 1990, work began on converting five more ex-BA Super VC10s to tanker standard, to be designated K.Mk 4. At the same time it was announced that the C.Mk 1 fleet would be given underwing refuelling pods for a secondary tanking role.

Total production of the VC10 was 37 aircraft, the last, for the RAF, being delivered in 1968, while 27 Super VC10s were built, culminating in the final delivery to East African Airways in 1970.

CHRONOLOGY OF CIVIL AVIATION

1783

25 April 1783
The brothers Joseph and Etienne Montgolfier achieve the world's first successful flight with a hot-air balloon at Annonay near Lyons in France.

21 November 1783
Piltre de Rœier and the Marquis d'Arlandes make the world's first successful manned flight in Paris using a hot-air balloon.

1 December 1783
Professor Jacques Charles and Noel Roberts make the world's first manned flight in a hydrogen-filled balloon in Paris.

1796

26 June 1796
The French Army makes the world's first practical use of a manned balloon as an observation platform in the Battle of Fleurus.

1852

24 September 1852
Henri Giffard of France makes the world's first semi-controlled and powered flight in a steam-powered cigar-shaped dirigible airship.

1853

Summer 1853
Sir George Cayley achieves the world's first flight in a manned, heavier-than-air craft with the despatch of a servant across a valley in a glider.

1896

10 August 1896
Otto Lilienthal, a German pioneer of manned, fully controlled, heavier-than-air flight, is killed when of one of his hang-gliders crashes.

1900

2 July 1900
General Ferdinand Graf von Zeppelin, late of the German Army, reveals his LZ1 airship, the starting point for practical airship transport.

1903

17 December 1903
The American brothers Orville and Wilbur Wright succeed in making the world's first powered, sustained and controlled flight in their Flyer.

1906

12 November 1906
Brazilian Alberto Santos-Dumont makes the first successful, heavier-than-air flight in Europe, with his No.14-bis, but its design is far behind the Wrights'.

1907

29 September 1907
The Breguet brothers achieve the first take-off with a rotary-winged aircraft at Douai in northern France with four men stabilising the machine.

13 November 1907
The first free-flight of a rotary-winged aircraft is made by Paul Cornu near Lisieux in France. Unfortunately, this machine lacks any real means of control in the air.

1908

9 August 1908
Wilbur Wright stuns Europe with highly impressive flights at Le Mans, proving the overall superiority of the Americans' achievements.

1909

25 July 1909
Louis Blériot creates a huge surge in European air-mindedness with the first flight across the English Channel to Dover in his Type XI monoplane.

22–29 August 1909
The Aviation Week at Rheims in France is a huge success in terms of flights and records, and boosts the image of flight to a marked degree.

1910

28 March 1910
Henri Fabre becomes the first man in the world to take off from water when he flies his Hydravion on the Mediterranean coast.

1911

18 February 1911
The world's first official delivery of mail by air takes place in India.

4 July 1911
The first recorded air freight is a box of Osram light bulbs, carried from Shoreham to Hove in southern England by Horatio Barber in a Valkyrie monoplane.

1914

January 1914
The world's first scheduled, heavier-than-air passenger service starts with a Benoist XIV flying boat between St Petersburg and Tampa in Florida.

1915

14–15 June 1915
John Alcock and Arthur Whitten-Brown achieve the first non-stop flight over the Atlantic between Newfoundland and Ireland in a Vickers Vimy.

1916

5 October 1916
The world's first airline to be formally registered is the Aircraft Transport and Travel Ltd, created by George Holt Thomas in the UK.

1918

11 March 1918
The world's first scheduled, international air mail service starts between Vienna in Austria-Hungary and Kiev in Ukraine using Hansa-Brandenburg C I biplanes.

28 July–12 December 1918
A Handley Page O/400 bomber makes the first air transit from the UK to India, with a long stopover in Egypt, to indicate the type of long-range route-proving flights that characterize the period after World War I.

1919

5 February 1919
Deutsche Luft-Reederei starts the world's first sustained commercial daily passenger service by heavier-than-air craft.

3 March 1919
The first US international air mail service is flown between Seattle, Washington and Victoria, British Columbia, by a Boeing Type C floatplane of Hubbard Air Service.

8–31 May 1919
A portent for the future of air travel is found in the first successful transatlantic flight achieved by one of three Curtiss NC flying boats of the US Navy.

Mid-1919
The first aircraft to receive a British civil registration is a de Havilland D.H.9 operated as a mailplane between London and Paris by Aircraft Transport and Travel.

25 August 1919
The first scheduled, international air services for passengers begin with a flight between London and Paris by a de Havilland D.H.16 of Air Transport and Travel.

7 October 1919
KLM is established as the Dutch national airline, the oldest such operator still in existence.

1 November 1919
The first US scheduled, international air service begins with a flight between Key West, Florida and Havana, Cuba by Aeromarine West Indies Airways.

12 November–10 December 1919
The brothers Ross and Keith Smith complete the first flight between the UK and Australia in a converted Vickers Vimy bomber.

1920

4 February–20 March 1920
Pierre van Ryneveld and Christopher Quintin Brand complete the first flight between the UK and South Africa in a converted Vickers Vimy bomber.

16 November 1920
Australia's first airline, the Queensland and Northern Territory Aerial Service (QANTAS) is established, with services following six days later.

14 December 1920
The first fatal accident to a British airliner occurs with the crash of a Handley Page O/400, in which four of the six occupants are killed.

1921

22–23 February 1921
The USA's first coast-to-coast air mail service operates between California and New York with a DH-4 biplane.

1922

7 April 1922
The first mid-air collision takes place over northern France when a Farman Goliath of Grands Express Aériens and a de Havilland D.H.18 of Daimler Airway collide, killing all seven on board the two aircraft.

1923

9 January 1923
Juan de la Cierva makes the first flight of a practical rotary-winged aircraft, his C.4 Autogiro, at Getafe in Spain.

10 April 1923
Daimler Airway starts the first scheduled air service between London and Berlin with halts at Bremen and Hamburg.

1924

1 April 1924
The UK's first national airline is created as Imperial Airways with 13 aircraft as a result of the forced merger of British Marine Air Navigation, Daimler Airway, Handley Page Transport and Instone Air Lines.

6 April–28 September 1924
Two of four Douglas DWC aircraft of the US Army complete the first aerial circumnavigation of the world, covering 26,345 miles in just over 363 flying hours.

1 July 1924
The first American transcontinental air mail service is started between New York and San Francisco, California, with 14 intermediate stops.

1925

15 May 1925
The Junkers G.23, the world's first three-engined transport of all-metal construction, enters service with A. B. Aerotransport of Sweden.

1927

20–21 May 1927
Charles Lindbergh completes the first solo non-stop crossing of the north Atlantic with a flight in his Ryan NYP between New York and Paris.

14–15 October 1927
Dieudonné Costes and Joseph Le Brix complete the first non-stop crossing of the South Atlantic between Senegal and Brazil in a Breguet Bre.19.

19 October 1927
Pan American Airways, one of the most celebrated names in aviation history, begins operations with a service between Key West, Florida and Havana, Cuba.

1928

31 May–9 June 1928
The Fokker F.VIIB/3m, captained by Charles Kingsford-Smith, completes the first aerial crossing of the Pacific between California and Australia with three intermediate stops.

23 December 1928–25 February 1929
The world's first major airlift evacuates 586 persons from Kabul in Afghanistan to India using eight Vickers Victoria transports.

1929

30 March 1929
Imperial Airways starts the world's first air service, from the UK to India with Armstrong Whitworth Argosy, Short Calcutta and de Havilland D.H.66 aircraft, in a time of seven days, incorporating train travel between Basle in Switzerland and Genoa in Italy.

1930

15 May 1930
Ellen Church becomes the first air stewardess in the world as a member of the Boeing Air Transport service between San Francisco, California and Cheyenne, Wyoming.

8–29 August 1930
Commanded by Hugo Eckener, the Zeppelin airship Graf Zeppelin completes the first circumnavigation of the world by a lighter-than-air craft.

25 October 1930
Transcontinental and Western Air launches the first all-air passenger service between the US east and west coasts with a service linking New York and Los Angeles, California.

14 November 1930
First flight of the Handley Page H.P.42, last of the classic four-engined biplane airliners offering low performance but excellent comfort.

1932

25 March 1932
The USSR revises its Dobroflot airline into Aeroflot, which rapidly expands into the world's single largest civil aviation enterprise.

27 April 1932
Imperial Airways launches the first regular air service between the UK and South Africa using de Havilland D.H.66 airliners for the 11-day route.

1933

8 February 1933
First flight of the Boeing Model 247, the world's first 'modern' airliner that presages the advent of the larger and altogether more successful Douglas DC series.

18 February 1933
Imperial Airways completes its first 10 million miles of flying. Its low accident rate also sees a reduction in insurance charges from 12 shillings to one shilling per £1,000, a rate comparable for the first time with surface transport insurance.

1 July 1933
First flight of the Douglas DC-1, real progenitor of the 'modern' airliner with a low-wing, cantilever configuration, all-metal, stressed-skin construction, twin-engined powerplant with variable-pitch propellers, trailing-edge flaps, tailwheel landing gear with retractable main units and fully enclosed accommodation.

22 July 1933
Wiley Post completes the first solo round-the-world flight in the Lockheed Vega, landing in New York after a journey of 7 days 18 hours 49 minutes.

31 October 1933
Air France is created as the French national airline as a merger of four existing operators: Air Orient, Air Union, CIDNA and SGTA.

1934

14 May 1934
The Douglas DC-2 first enters service with TWA with what is, in effect, the DC-1 prototype.

28 July 1934
Swissair introduces the first air stewardess, Nelly Diener, on European services, working on a Curtiss Condor biplane transport.

8 December 1934
Launch of the first weekly air mail service between the UK and Australia by Imperial Airways, Indian Trans-Continental Airways and Qantas Empire Airways.

1935

13 April 1935
Inauguration of the first through passenger air service between the UK and Australia by Imperial Airways and Qantas Empire Airways.

16–17 April 1935
First airline-proving flight between California and Hawaii by a Sikorsky S-42 four-engined flying boat of Pan American Airways.

29 October 1935
British Airways is created by the merger of three existing operators, Hillman, Spartan and United Airways.

22 November 1935
The first scheduled air mail flight across the Pacific is flown by a Martin Model 130, a four-engined flying boat of Pan American Airways, between California and the Philippines with four intermediate stops.

17 December 1935
Douglas flies the first example of its DST (Douglas Sleeper Transport), which paves the way to the classic DC-3 and C-47/Dakota series.

1936

26 June 1936
The Focke-Achgelis Fa 61 makes its first free flight as the world's first fully practical helicopter, the problem of torque reaction being overcome by the use of a side-by-side pair of counter-rotating rotors on the ends of long outrigger arms.

4 July 1936
The Short 'C' or 'Empire' class four-engined flying boat makes its first flight, and the type later enters service for luxury services linking the UK with Africa, the Far East and Australasia.

5–6 July 1936
Simultaneous east–west and west–east proving flights are flown across the North Atlantic between Ireland and Newfoundland by a Short 'C' class flying boat of Imperial Airways and a Sikorsky S-42 flying boat of Pan American Airways.

27 October 1936
A Martin Model 130 flying boat of Pan American World Airways completes the first scheduled passenger service across the Pacific, arriving at Manila in the Philippine Islands after a four-halt journey from San Francisco, California.

1937

6 May 1937
The era of the passenger airship comes to an end with the loss of the Hindenburg as it catches fire as it tries to land at Lakehurst, New Jersey, resulting in the deaths of 36 persons.

1938

25 September 1938
First flight of the Douglas DC-4E, the world's first 'modern' airliner of the four-engined type but really too advanced for its time and soon abandoned as a subject of further development.

31 December 1938
Boeing flies the first example of its Model 307 Stratoliner, a four-engined airliner based on the B-17 Flying Fortress bomber but with pressurized 33-seat accommodation.

1939

4 August 1939
British Overseas Airways Corporation is created as the UK's national airline through the merger of Imperial Airways and British Airways.

11 August 1939
Pan American World Airways completes the first revenue-earning passenger service across the north Atlantic with the Boeing Model 314 flying boat.

1940

13 May 1940
Igor Sikorsky achieves the initial free flight with his Vought-Sikorsky VS-300 helicopter, the world's first fully practical helicopter of the single-rotor type with cyclic-pitch control for full controlability.

1945

24 October 1945
American Overseas Airlines resumes transatlantic passenger services, after the end of World War II, with a Douglas DC-4 service between New York and Hurn, near Bournemouth.

1946

15 February 1946
Douglas flies the first example of its XC-112 prototype, which leads to the DC-6 four-engined airliner, a much improved development of the DC-4 with pressurised 86-seat accommodation.

8 March 1946
The Bell Model 47 becomes the first helicopter in the world to achieve full civil certification.

1947

16 March 1947
Convair makes the first flight of its CV-240 twin-engined airliner, designed to succeed the DC-3 and also the antecedent of the excellent CV-340 and CV-440 airliners.

1949

1 April 1949
Pan American World Airways introduces the new Boeing Model 377 Stratocruiser to revenue-earning service as a pressurised airliner with two-deck accommodation.

1950

29 July 1950
The Vickers Viscount becomes the first turboprop-powered airliner in the world to make a revenue-earning flight with a journey between London and Paris.

13 October 1950
Lockheed flies the first example of its L1049 Super Constellation, a 'stretched' and more powerfully-engined development of its L-649 Constellation designed to rival the Douglas DC-6.

1952

2 May 1952
The de Havilland Comet 1 becomes the first turbojet-powered airliner in the world to make a revenue-earning flight with a British Overseas Airways Corporation service between London and Johannesburg.

1953

3 March 1953
A de Havilland Comet 1 becomes the first turbojet-powered airliner in the world to suffer a fatal accident when a machine of Canadian Pacific Airlines crashes in a take-off accident at Karachi, killing all 11 on board.

18 May 1953
Douglas flies the first example of its DC-7 airliner, a 'stretched' and more powerfully-engined development of the DC-6 designed to outmatch the Lockheed Super Constellation.

1954

15 July 1954
Boeing flies its Model 367-80 prototype as the world's first transport with swept flying surfaces and a four-turbojet powerplant. The machine is the prototype for the Model 707 airliner and KC-135 flight-refuelling tanker, and as such the real precursor of the world's current, mass, air-transport market.

1955

27 May 1955
The Sud-Est Caravelle makes its first flight as France's first turbojet-powered airliner, and the world's first airliner with its two engines attached to the sides of the rear fuselage.

1957

1 July 1957
TWA inaugurates the Lockheed L-1649 Starliner, rival to the Douglas DC-7, on the prestigious North Atlantic service.

19 December 1957
British Overseas Airways Corporation (BOAC) flies the first transatlantic service with a turbine-powered airliner, a Bristol Britannia turboprop type, between London and New York.

21 December 1957
Boeing flies the first production example of its epoch-making Model 707 four-jet airliner.

1958

30 May 1958
Douglas flies the first example of its DC-8, created specifically to rival the Boeing Model 707 but, ultimately, not as successful in commercial terms.

4 October 1958
British Overseas Airways Corporation flies the first transatlantic services with turbojet-powered airliners, both de Havilland Comet 4 aircraft, in simultaneous east–west and west–east flights between London and New York.

1959

27 January 1959
Convair flies the first example of its CV-880, intended as a rival to the Boeing Model 707 and Douglas DC-8, but in fact a commercial failure.

10 October 1959
Pan American World Airways launches the world's first round-the-world scheduled service with the Boeing Model 707 four-engined airliner.

1962

9 January 1962
The Hawker Siddeley (originally de Havilland) Trident flies as the world's first medium-range jet airliner with a three-engined powerplant.

29 June 1962
Vickers flies the prototype of its VC10 long-range airliner, with its four turbofan engines grouped on the sides of the rear fuselage, designed to provide the British with a rival to the Boeing Model 707 and Douglas DC-8.

1963

2 February 1963
Boeing flies the first example of its Model 727, a three-turbofan airliner designed for the rapidly expanding medium-range segment of the airline market.

1965

25 February 1965
Douglas flies the first example of its DC-9, a new twin-turbofan airliner designed to capture the major slice of the short-haul airline market.

1967

6 June 1967
The Boeing company deliver their 1,000th jet airline, a 707-320.

1968

31 December 1968
The USSR achieves the maiden flight of the world's first supersonic airliner, the Tupolev Tu-144.

1969

9 February 1969
First flight of the Boeing 747, the world's first wide-body jumbo airliner.

2 March 1969
First flight of the Anglo-French Concorde supersonic airliner.

1970

16 November 1970
First flight of the Lockheed TriStar three-turbofan transport.

1971

21 January 1971
A Boeing 747 named 'Clipper Young America' enters service with Pan Am, heralding a new era for commercial aviation.

1972

28 October 1972
The first Airbus A300 makes its maiden flight.

1974

1 September 1974
Air France and British Airways begin using Concorde airliners for scheduled passenger services.

1975

27 March 1975
The de Havilland Canada DHC-7 Dash 7 four-turboprop airliner flies for the first time.

1976

21 January 1976
The era of supersonic air transport begins with the inauguration of Air France's Concorde service between Paris and Rio de Janeiro, Brazil.

22 December 1976
First flight of the Ilyushin Il-86, the USSR's first jumbo jet.

1979

18 October 1979
First flight of the McDonnell Douglas MD-80 (originally Super 80) as a modernized version of the DC-9.

1981

28 March 1981
Dornier flies the first example of its Do 228 commuter-liner.

3 September 1981
McDonnell Douglas delivers the 1,000th example of the DC-9.

3 September 1981
British Aerospace's maiden flight of BAe146 four-turbofan airliner designed for short-range operations and very quiet field performance.

1981

26 September 1981
Boeing flies the first example of its 767 twin-turbofan airliner for medium-range operations.

1982

19 February 1982
Boeing flies the first 757, a twin-turbofan type designed as successor to its 727 three-turbofan type.

3 April 1982
First flight of the Airbus A310 twin-turbofan airliner for medium-range operations.

1983

15 June 1983
First flight of the Saab-Fairchild SF 340 twin-turboprop commuter-liner, later known as the Saab 340.

20 June 1983
First flight of the de Havilland Canada DHC-8 Dash 8 twin-turboprop transport for the short-range market.

27 June 1983
First flight of the EMBRAER EMB-120 Brasilia twin-turboprop regional airliner.

1984

16 August 1984
First flight of the ATR 42 twin-turboprop for the regional airliner market.

1985

1 February 1985
The Boeing 767 becomes the first twin-turbofan airliner to start scheduled services across the North Atlantic, in this instance in the hands of TWA.

1986

23 December 1986
Dick Rutan and Jeanna Yeager complete the first non-stop unrefuelled round-the-world flight in the Rutan Voyager after 9 days, 3 minutes and 44 seconds in the air.

1987

22 February 1987
First flight of the Airbus A320, the first airliner in the world designed largely with CAD (Computer-Aided Design) and CAM (Computer-Aided Manufacture) techniques.

1988

27 October 1988
First flight of the ATR 72 as an enlarged development of the ATR 42 for the regional airliner market.

1991

25 October 1991
First flight of the Airbus A340 long-range airliner, the first four-turbofan type designed and built by the European consortium.

6 December 1991
First flight of the Dornier Do 328 twin-turboprop for the regional airliner market.

1992

20 March 1992
First flight of the Saab 2000, designed as a larger and faster version of the successful Saab 340.

2 November 1992
First flight of the Airbus A330, the twin-turbofan counterpart of the four-turbofan A340 for the medium/long-range market.

1993

22 February 1993
First flight of the McDonnell Douglas MD-90, a stretched development of the MD-80 series with an updated powerplant.

1994

12 June 1994
First flight of the Boeing 777 as the company's new twin-turbofan contender in the market for long-range airliners.

1999

21 March 1999
End of the first successful flight round the world by a hot-air balloon, the Breitling Orbiter III, which covers more than 26,000 miles in a time of 19 days, 21 hours and 55 minutes.

SPECIFICATIONS

The following pages list specification details from a selection of aircraft in this book. Entries are listed alphabetically by manufacturer.

CARAVELLE
Model/variant: Caravelle III
Manufacturer: Aérospatiale (Sud-Est/Sud Aviation)
Country of Origin: France
Type: Short/medium-range passenger/cargo transport
Accommodation (typical): 96 (one-class)
Flight deck crew: 2
Prototype maiden flight: 27 May 1955
POWERPLANT
Manufacturer: Rolls-Royce
Model: Avon RA.29/3 Mk 527
Type: Turbojet
Number: 2
Output: 51.5kN (5200kg; 11,450lb) thrust
PERFORMANCE
Maximum cruising speed: 805km/h (500mph) at 7,620m (25,000ft)
Service ceiling: 12,000m (39,370ft)
Maximum range: 1845km (1140 miles)
WEIGHTS
Empty: 24,185kg (53,207lb)
Maximum take-off: 46,000kg (101,200lb)
DIMENSIONS
Span: 34.3m (112.5ft)
Length: 32m (195ft)
Height: 8.7m (28.5ft)
Wing area: 146.5m² (1577.4sq ft)

CONCORDE
Model/variant: -
Manufacturer: Aérospatiale/British Aerospace
Country of Origin: France/Great Britain
Type: Supersonic long-range passenger transport
Accommodation: up to 144
Flight deck crew: 3
Prototype maiden flight: 2 March 1969
POWERPLANT
Manufacturer: Rolls-Royce/SNECMA
Model: Olympus 593 Mk 610
Type: Turbojet
Number: 4
Output: 169kN (17260kg; 38,050lb) thrust with afterburning
PERFORMANCE
Maximum cruising speed: Mach 2.04 (equivalent to 2180km/h; 1355mph) at 15,635m (51,300ft)
Service ceiling: 18,290m (61,000ft)
Maximum range: 6230km (3870 miles) (with max. payload at Mach 2.02 cruise)
WEIGHTS
Empty: 78,700kg (173,500lb)
Maximum take-off: 185,100kg (408,000lb)
DIMENSIONS
Span: 25.55m (83.8ft)
Length: 62.1m (203.75ft)
Height: 11.4m (37.5ft)
Wing area: 358m² (3855sq ft)

AIRBUS A310
Model/variant: A310-300
Manufacturer: Airbus Industrie
Country of Origin: France/Germany/Great Britain
Type: Medium-range passenger/cargo transport
Accommodation (typical): 220 (two-class)
Flight deck crew: 2
Prototype maiden flight: 3 April 1982
POWERPLANT
Manufacturer: General Electric or Pratt & Whitney
Model: CF6-80C2A2 or PW 4152
Type: Turbofan
Number: 2
Output: 222.5kN (22,720kg; 50,000lb) thrust
PERFORMANCE
Maximum cruising speed: 965km/h (595mph) at 10,000m (32,800ft)
Service ceiling: 12,000m (39,370ft)
Maximum range: 9175km (5700 miles)
WEIGHTS
Empty: 70,275kg (154,930lb)
Maximum take-off: 157,000kg (346,125lb)
DIMENSIONS
Span: 43.9m (144ft)
Length: 46.7m (153ft)
Height: 15.8m (51.9ft)
Wing area: 219m² (2357sq ft)

AIRBUS A320
Model/variant: A320-200
Manufacturer: Airbus Industrie
Country of Origin: France/Germany/Great Britain
Type: Short/medium-range passenger/cargo transport
Accommodation (typical): 154 (two-class)
Flight deck crew: 2
Prototype maiden flight: 22 February 1987
POWERPLANT
Manufacturer: CFM International or IAE
Model: CFM56-5-A1 or V2500-A1
Type: Turbofan
Number: 2
Output: 111.25kN (11,360kg; 25,000lb) thrust
PERFORMANCE
Maximum cruising speed: 965km/h (595mph) at 10,000m (32,800ft)
Service ceiling: 12,000m (39,370ft)
Maximum range: c5300km (3270 miles)
WEIGHTS
Empty: 40,150kg (88,515lb)
Maximum take-off: 73,500kg (161,700lb)
DIMENSIONS
Span: 33.9m (111.2ft)
Length: 37.6m (123ft)
Height: 11.8m (38.8ft)
Wing area: 112.5m² (1210sq ft)

AIRBUS A340
Model/variant: A340-300
Manufacturer: Airbus Industrie
Country of Origin: France
Type: Long-range passenger/cargo transport
Accommodation (typical): 335 (two-class)
Flight deck crew: 2
Prototype maiden flight: n/a
POWERPLANT
Manufacturer: CFM International
Model: CFM56-5-C2
Type: Turbofan
Number: 4
Output: 139kN (14,205kg; 31,250lb) thrust
PERFORMANCE
Maximum cruising speed: 1035km/h (640mph) at 10,000m (32,800ft)
Service ceiling: 12,000m (39,370ft)
Maximum range: c12,500km (7700 miles)
WEIGHTS
Empty: 125,500kg (276,680lb)
Maximum take-off: 251,000kg (553,360lb)
DIMENSIONS
Span: 58.7m (192.6ft)
Length: 63.7m (209ft)
Height: 16.75m (55ft)
Wing area: 362m² (3892sq ft)

AN-2 'COLT'
(NATO REPORTING NAME)
Model/variant: An-2P (An-2R)
Manufacturer: Antonov [WSK-PZL, Poland]
Country of Origin: Soviet Union
Type: Short-range STOL passenger/cargo transport (crop-sprayer)
Accommodation (typical): 12 (-)
Flight deck crew: 2
Prototype maiden flight: 31 August 1947
POWERPLANT
Manufacturer: Shevestov
Model: ASh-621R
Type: 9-cylinder radial
Number: 1
Output: 746kW (1000bhp)
PERFORMANCE
Maximum speed: 260km/h (160mph) at 1750m (5750ft)
Service ceiling: 4400m (14,425ft)
Maximum range: 900km (560 miles) (with 500kg (1100lb) payload)
WEIGHTS
Empty: 3450kg (7605lb)
Maximum take-off: 5500kg (12,125lb)
DIMENSIONS
Span: 18.2m (59.7ft) (upper); 14.25m (46.8ft) (lower)
Length: 12.75m (41.9ft)
Height: 4m (13.1ft)
Wing area: 71.5m² (301.5sq ft)

AN-12 'CUB'
(NATO REPORTING NAME)
Model/variant: -
Manufacturer: Antonov
Country of Origin: Soviet Union
Type: Medium-range mixed passenger and cargo transport
Accommodation (typical): 14 plus cargo
Flight deck crew: 2/3
Prototype maiden flight: n/a
POWERPLANT
Manufacturer: Ivchenko
Model: AI-20K
Type: Turboprop
Number: 4
Output: 2985kW (4000ehp)
PERFORMANCE
Maximum cruising speed: 600km/h (375mph) at 7620m (25,000ft)
Service ceiling: 10,200m (33,465ft)
Maximum range: 3400km (2100 miles) (with 10,000kg (22,000lb) payload)
WEIGHTS
Empty: 28,000kg (61730lb)
Maximum take-off: 61,000kg (134,480lb)
DIMENSIONS
Span: 38m (124.7ft)
Length: 33.1m (108.7ft)
Height: 10.5m (34.5ft)
Wing area: 121.7m² (1310sq ft)

AN-225 MRYIA
Model/variant: -
Manufacturer: Antonov
Country of Origin: Soviet Union
Type: Very heavy lift cargo transport
Accommodation (typical): -
Flight deck crew: 3/4 plus relief
Prototype maiden flight: 21 December 1988
POWERPLANT
Manufacturer: Lotarev
Model: D18-T
Type: Turbofan

Number: 6
Output: 230kN (23,500kg; 51,600lb) thrust
PERFORMANCE
Maximum cruising speed: 805km/h (500mph) at 9750m (32,000ft)
Service ceiling: 12,000m (39,370ft)
Maximum range: 4500km (2795 miles)
WEIGHTS
Empty: n/a
Maximum take-off: 600,000kg (1,320,000lb)
DIMENSIONS
Span: 88.4m (290ft)
Length: 84m (275.5ft)
Height: 18.1m (59.4ft)
Wing area: n/a

AN-24 'COKE'
(NATO REPORTING NAME) [Y-7]
Model/variant: An-24RV
Manufacturer: Antonov [Xian, People's Republic of China]
Country of Origin: Soviet Union
Type: Short-range passenger/cargo transport
Accommodation (typical): 50
Flight deck crew: 2/3
Prototype maiden flight: April 1960
POWERPLANT**Manufacturer:** Ivchenko
Model: A1-24A
Type: Turboprop
Number: 2
Output: 1900kW (2550ehp)
PERFORMANCE
Maximum cruising speed: 450km/h (280mph) at 3500m (11,500ft)
Service ceiling: 8500m (27,600ft)
Maximum range: 550km (340 miles) (with max. payload)
WEIGHTS
Empty: 13,300kg (29,320lb)
Maximum take-off: 21,000kg (49,300lb)
DIMENSIONS
Span: 29.2m (95.8ft)
Length: 23.5m (77.2ft)
Height: 8.32m (27.3ft)
Wing area: 75m^2 (807sq ft)

AW.650 ARGOSY
Model/variant: Series 100
Manufacturer: Armstrong-Whitworth
Country of Origin: Great Britain
Type: Medium-range cargo transport
Accommodation (typical): -
Flight deck crew: 2/3
Prototype maiden flight: 8 January 1959
POWERPLANT
Manufacturer: Rolls-Royce
Model: Dart 526
Type: Turboprop
Number: 4
Output: 1506kW (2020shp)
PERFORMANCE
Maximum cruising speed: 475km/h (295mph) at 3500m (11,500ft)
Service ceiling: 6100m (20,000ft)
Maximum range: 3220km (2000 miles)
WEIGHTS
Empty: 20,865kg (46,000lb)
Maximum take-off: 39,920kg (88,000lb)
DIMENSIONS
Span: 35m (115ft)
Length: 26.45m (86.75ft)
Height: 8.25m (27ft)
Wing area: 135.5m^2 (1460sq ft)

AW.650 ARGOSY
Model/variant: Series 220
Manufacturer: Armstrong-Whitworth
Country of Origin: Great Britain
Type: Medium-range cargo transport
Accommodation (typical): -
Flight deck crew: 2/3
Prototype maiden flight: 8 January 1959
POWERPLANT
Manufacturer: Rolls-Royce
Model: Dart 532
Type: Turboprop
Number: 4
Output: 1664kW (2330shp)
PERFORMANCE
Maximum cruising speed: 460km/h (285mph) at 3500m (11,500ft)
Service ceiling: 6100m (20,000ft)
Maximum range: 3500km (2150 miles)
WEIGHTS
Empty: 22,135kg (48,800lb)
Maximum take-off: 42,185kg (93,000lb)
DIMENSIONS
Span: 35m (115ft)
Length: 26.45m (86.75ft)
Height: 8.25m (27ft)
Wing area: 135.5m^2 (1460sq ft)

AW.660 ARGOSY
Model/variant: C. Mk 1 (RAF designation)
Manufacturer: Armstrong-Whitworth
Country of Origin: Great Britain
Type: Medium-range cargo transport
Accommodation (typical): -
Flight deck crew: 2/3
Prototype maiden flight: 8 January 1959
POWERPLANT
Manufacturer: Rolls-Royce
Model: Dart 536
Type: Turboprop
Number: 4
Output: 1843kW (2470shp)
PERFORMANCE
Maximum cruising speed: 440km/h (275mph) at 3500m (11,500ft)
Service ceiling: 6100m (20,000ft)
Maximum range: 5600km (3450 miles)
WEIGHTS
Empty: 25,400kg (56,000lb)
Maximum take-off: 44,000kg (97,000lb)
DIMENSIONS
Span: 35m (115ft)
Length: 26.45m (86.75ft)
Height: 8.25m (27ft)
Wing area: 135.5m^2 (1460sq ft)

ATR 42
Model/variant: -500
Manufacturer: ATR
Country of Origin: France/Italy
Type: Short/medium-range passenger transport
Accommodation: 42-50
Flight deck crew: 2
Prototype maiden flight: n/a
POWERPLANT
Manufacturer: Pratt & Whitney Canada
Model: PW127E
Type: Turboprop
Number: 2
Output: 1790kW (2400shp)
PERFORMANCE
Maximum cruising speed: 560km/h (345mph) at 5500m (18,000ft)
Service ceiling: 7620m (25,000ft)
Maximum range: 1950km (1200 miles)
WEIGHTS
Empty: 11,250kg (24,750lb)
Maximum take-off: 18,600kg (40,920lb)
DIMENSIONS
Span: 24.5m (80.5ft)
Length: 22.6m (74.3ft)
Height: 7.3m (24.9ft)
Wing area: 54.5m^2 (585sq ft)

ATR 72
Model/variant: -500
Manufacturer: ATR
Country of Origin: France/Italy
Type: Short/medium-range passenger transport
Accommodation: 64-74
Flight deck crew: 2
Prototype maiden flight: n/a
POWERPLANT
Manufacturer: Pratt & Whitney Canada
Model: PW127F
Type: Turboprop
Number: 2
Output: 2050kW (2750shp)
PERFORMANCE
Maximum cruising speed: 425km/h (262mph) at 5500m (18,000ft)
Service ceiling: 7620m (25,000ft)
Maximum range: 2700km (1700 miles)
WEIGHTS
Empty: 14,600kg (32,120lb)
Maximum take-off: 22,500kg (49,500lbs)
DIMENSIONS
Span: 27m (88.75ft)
Length: 27.1m (89ft)
Height: 7.6m (25ft)
Wing area: 61m^2 (656sq ft)

BEECH MODEL 200 'SUPER KING AIR'
(C-12 - US MILITARY DESIGNATION)
Model/variant: -
Country of Origin: United States of America
Manufacturer: Beech Aircraft Corp.
Type: Short/medium-range light passenger/cargo transport
Accommodation (typical): 13 (commuter model)
Flight deck crew: 2
Prototype maiden flight: 27 October 1972
POWERPLANT
Manufacturer: Pratt & Whitney Canada
Model: PT6A-2
Type: Turboprop
Number: 2
Output: 634kW (850shp)
PERFORMANCE
Maximum cruising speed: 525km/h (325mph) at 7620m (25,000ft)
Service ceiling: 10,670m (35,000ft)
Maximum range: 3760km (2335 miles)
WEIGHTS
Empty: 3420kg (7540lb)
Maximum take-off: 5670kg (12,500lb)
DIMENSIONS
Span: 16.6m (54.5ft)
Length: 13.35m (43.75ft)
Height: 4.6m (15ft)
Wing area: 28.15m^2 (303sq ft)

BEECH MODEL E18S 'SUPER 18'
(C-45 - US MILITARY DESIGNATION)
Model/variant: -
Manufacturer: Beech Aircraft Corp.
Country of Origin: United States of America
Type: Short/medium-range light passenger transport
Accommodation: 8-15
Flight deck crew: 2
Prototype maiden flight: 10 December 1953 (original model, 15 January 1937)
POWERPLANT
Manufacturer: Pratt & Whitney
Model: R-985-AN-14B Wasp Junior
Type: 9-cylinder radial
Number: 2
Output: 336kW (450bhp)
PERFORMANCE
Maximum cruising speed: 354km/h (220mph) at 3050m (10,000ft)
Service ceiling: 6525m (21,400ft)
Maximum range: 3060km (1530 miles)
WEIGHTS
Empty: 2650kg (5840lb)
Maximum take-off: 4490kg (9900lb)
DIMENSIONS
Span: 15.2m (49.7ft)
Length: 10.75m (35.25ft)
Height: 2.9m (9.5ft)
Wing area: 33.5m^2 (360sq ft)

BOEING 307 'STRATOLINER'
(C-75 - US MILITARY DESIGNATION)
Model/variant: Model S-307
Manufacturer: Boeing
Country of Origin: United States of America

Type: Medium-range passenger/cargo transport
Accommodation (typical): 33
Flight deck crew: 2/3
Prototype maiden flight: 1936
POWERPLANT
Manufacturer: Wright
Model: GR-1820 Cyclone
Type: 9-cylinder radial
Number: 4
Output: 670kW (900bhp)
PERFORMANCE
Maximum cruising speed: 355km/h (220mph) at 6100m (20,000ft)
Service ceiling: 7985m (26,200ft)
Maximum range: 3850km (2390 miles)
WEIGHTS
Empty: 13,610kg (30,000lb)
Maximum take-off: 19,050kg (42,000lb)
DIMENSIONS
Span: 32.6m (107ft)
Length: 22.7m (74.3ft)
Height: 6.3m (20.75ft)
Wing area: 138m² (1486sq ft)

BOEING 314 'CLIPPER'
(C-98 - US MILITARY DESIGNATION)
Model/variant: Model 314A
Manufacturer: Boeing
Country of Origin: United States of America
Type: Long-range passenger transport flying-boat
Accommodation: up to 74
Flight deck crew: 4/5
Prototype maiden flight: 7 June 1939
POWERPLANT
Manufacturer: Wright
Model: GR-2600 Double Cyclone
Type: 18-cylinder radial
Number: 4
Output: 1193kW (1600bhp)
PERFORMANCE
Maximum cruising speed: 295km/h (185mph) at 3050m (10,000ft)
Service ceiling: 4085m (13,400ft)
Maximum range: 5635km (3500 miles)
WEIGHTS
Empty: 22,800kg (50,270lb)
Maximum take-off: 37,420kg (82,500lb)
DIMENSIONS
Span: 46.3m (152ft)
Length: 32.3m (106ft)
Height: 8.4m (27.5ft)
Wing area: 266m² (2860sq ft)

BOEING 377 'STRATOCRUISER'
Model/variant: Model 377-10-28
Manufacturer: Boeing
Country of Origin: United States of America
Type: Long-range passenger/cargo transport
Accommodation (typical): 81
Flight deck crew: 4
Prototype maiden flight: 8 July 1947
POWERPLANT
Manufacturer: Pratt & Whitney
Model: R-4360 Wasp Major
Type: 28-cylinder radial
Number: 4
Output: 2610kW (3500bhp)
PERFORMANCE
Maximum cruising speed: 550km/h (340mph) at 7620m (25,000ft)
Service ceiling: 9755m (32,000ft)
Maximum range: 6760km (4200 miles)
WEIGHTS
Empty: 37,875kg (83,500lb)
Maximum take-off: 66,135kg (145,800lb)
DIMENSIONS
Span: 43m (141ft)
Length: 33.6m (110ft)
Height: 11.65m (38ft)
Wing area: 164.5m² (1770sq ft)

BOEING 707
Model/variant: Model 707-320C (VC-137 - US military designation)
Manufacturer: Boeing
Country of Origin: United States of America
Type: Long-range passenger/cargo or mixed passenger and cargo transport
Accommodation: 147 (two-class) or 219 (high density) (49)
Flight deck crew: 3
Prototype maiden flight: 15 July 1954 (as Model 367-80)
POWERPLANT
Manufacturer: Pratt & Whitney
Model: JT3D-7
Type: Turbofan
Number: 4
Output: 84.5kN (8620kG; 19,000lb) thrust
PERFORMANCE
Maximum cruising speed: 974km/h (605mph) at 9750m (32,000ft)
Service ceiling: 11,900m (39,000ft)
Maximum range: 9260km (5750 miles) (with 147 passengers)
WEIGHTS
Empty: 66,400kg (146,400lb) passenger, 64,000kg (141,100lb) cargo
Maximum take-off: 151,320kg (333,600lb)
DIMENSIONS
Span: 44.4m (145.75ft)
Length: 46.6m (153ft)
Height: 12.9m (42.5ft)
Wing area: 283.35m² (3050sq ft)

BOEING 727
Model/variant: Model 727-200
Manufacturer: Boeing
Country of Origin: United States of America
Type: Medium-range passenger/cargo transport
Accommodation (typical): 163 (two-class) 189 (high density)
Flight deck crew: 3
Prototype maiden flight: 9 February 1963
POWERPLANT
Manufacturer: Pratt & Whitney
Model: JT8D-9A
Type: Turbofan
Number: 3
Output: 64.4kN (6575kg; 14,500lb) thrust
PERFORMANCE
Maximum cruising speed: 920km/h (570mph) at 7620m (25,000ft)
Service ceiling: 11,900m (39,000ft)
Maximum range: 4000km (2485 miles) (with maximum payload)
WEIGHTS
Empty: 46,700kg (103,000lb)
Maximum take-off: 95,025kg (209,500lb)
DIMENSIONS
Span: 32.9m (108ft)
Length: 46.7m (153.25ft)
Height: 10.35m (34ft)
Wing area: 158m² (1700sq ft)

BOEING 737
Model/variant: Model 737-200
Manufacturer: Boeing
Country of Origin: United States of America
Type: Medium-range passenger/cargo transport
Accommodation: 130 (high density)
Flight deck crew: 3
Prototype maiden flight: 8 August 1967
POWERPLANT
Manufacturer: Pratt & Whitney
Model: JT8D-15
Type: Turbofan
Number: 2
Output: 68.8kN (7030kg; 15,500lb) thrust
PERFORMANCE
Maximum cruising speed: 925km/h (565mph) at 9145m (30,000ft)
Service ceiling: 11,900m (39,000ft)
Maximum range: 4265km (2650 miles) (with max. payload)
WEIGHTS
Empty: 27,700kg (60,940lb)
Maximum take-off: 53,070kg (117,000lb)
DIMENSIONS
Span: 28.35m (93ft)
Length: 30.5m (100ft)
Height: 11.3m (37ft)
Wing area: 91m² (980sq ft)

BOEING 747
Model/variant: Model 747-200B
Manufacturer: Boeing
Country of Origin: United States of America
Type: Long-range passenger/cargo transport
Accommodation (typical): 440 (two-class)
Flight deck crew: 3
Prototype maiden flight: 9 February 1969 (as Model 747)
POWERPLANT
Manufacturer: Pratt & Whitney
Model: JT9D-7FW
Type: Turbofan
Number: 4
Output: 222kN (22,680kg; 50,000lb) thrust
PERFORMANCE
Maximum cruising speed: 940km/h (585mph) at 9150m (30,000ft)
Service ceiling: 13,715m (45,000ft)
Maximum range: 9625km (5980 miles)
WEIGHTS
Empty: 171,000kg (377,000lb)
Maximum take-off: 365,150kg (805,000lb)
DIMENSIONS
Span: 59.65m (195.7ft)
Length: 70.5m (231.3ft)
Height: 19.35m (63.5ft)
Wing area: 511m² (5500sq ft)

BOEING 747
Model/variant: Model 747-400
Manufacturer: Boeing
Country of Origin: United States of America
Type: Long-range passenger/cargo transport
Accommodation (typical): 416 (three-class)
Flight deck crew: 2
Prototype maiden flight: 1989
POWERPLANT
Manufacturer: General Electric, Pratt & Whitney or Rolls-Royce
Model: CF6-80C2, PW4000 or RB211-524
Type: Turbofan
Number: 4
Output: c270kN (27,500kg; 60,500lb) thrust
PERFORMANCE
Maximum cruising speed: 910km/h (550mph) at 9150m (30,000ft)
Service ceiling: 10,580m (34,700ft)
Maximum range: 13,590km (8440 miles)
WEIGHTS
Empty: 180,890kg (398,800lbs)
Maximum take-off: 397,000kg (875,000lb)
DIMENSIONS
Span: 64.3m (211ft)
Length: 70.7m (232ft)
Height: 19.4m (63.7ft)
Wing area: 525m² (5600sq ft)

BOEING 757
Model/variant: Model 757-200
Manufacturer: Boeing
Country of Origin: United States of America
Type: Medium-range passenger/cargo transport
Accommodation (typical): 201 (two-class)
Flight deck crew: 2
Prototype maiden flight: February 1982
POWERPLANT
Manufacturer: Pratt & Whitney or Rolls-Royce
Model: PW2040 or RB211-535E
Type: Turbofan
Number: 2
Output: 185.6kN (18,950kg; 41,700lb) or 178.5kN (18,230kg; 40,100lb) thrust
PERFORMANCE
Maximum cruising speed: 965km/h (595mph) at 9150m (30,000ft)
Service ceiling: 11,900m (39,000ft)
Maximum range: 7315km (4550 miles)

WEIGHTS
Empty: 57,040kg (125,750lb)
Maximum take-off: 100,000kg (220,000lb)
DIMENSIONS
Span: 38m (125ft)
Length: 47.5m (155.5ft)
Height: 13.6m (44.5ft)
Wing area: 181.25m^2 (1950sq ft)

BOEING 767
Model/variant: Model 767-300ER
Manufacturer: Boeing
Country of Origin: United States of America
Type: Medium-range passenger/cargo transport
Accommodation (typical): 245 (three-class)
Flight deck crew: 2/3
Prototype maiden flight: 26 September 1981 (as 767-200)
POWERPLANT
Manufacturer: General Electric, Pratt & Whitney or Rolls-Royce
Model: CF6-80C2B, PW4000 or RB211-524
Type: Turbofan
Number: 2
Output: c213.5kN (21,800kg; 47,950lb) thrust
PERFORMANCE
Maximum cruising speed: 965km/h (595mph) at 9150m (30,000ft)
Service ceiling: 10,725m (35,200ft)
Maximum range: 11,390km (7080 miles)
WEIGHTS
Empty: 79,380kg (175,000lb)
Maximum take-off: 172,365kg (380,000lb)
DIMENSIONS
Span: 47.6m (156ft)
Length: 54.9m (180.25ft)
Height: 15.8m (52ft)
Wing area: 283.35m^2 (3050sq ft)

BOEING 777
Model/variant: Model 777-300
Manufacturer: Boeing
Country of Origin: United States of America
Type: Long-range passenger transport
Accommodation (typical): 386 (three-class)
Flight deck crew: 3
Prototype maiden flight: n/a
POWERPLANT
Manufacturer: General Electric, Pratt & Whitney or Rolls-Royce
Model: GE.90-75B, PW 4077 or Trent 875
Type: Turbofan
Number: 2
Output: 339.5kN (34,655kg; 76,400lb), 342kN (34,930kg; 77,000lb) or 346kN (35,335kg; 77,900lb) thrust
PERFORMANCE
Maximum cruising speed: 900km/h (560mph) at 9150m (30,000ft)
Service ceiling: 13,135m (43,100ft)
Maximum range: 10,800km (6720 miles)
WEIGHTS
Empty: 159,760kg (352,200lb)
Maximum take-off: 299,375kg (660,000lb)
DIMENSIONS
Span: 60.9m (200ft)
Length: 63.75m (209ft)
Height: 18.5m (60.75ft)
Wing area: 427.8m^2 (4605sq ft)

BRISTOL FREIGHTER
Model/variant: Type 170 Freighter Mk 21
Manufacturer: Bristol Aeroplane Co
Country of Origin: Great Britain
Type: Short-range mixed passenger and cargo transport
Accommodation: up to 32 plus cargo
Flight deck crew: 2/3
Prototype maiden flight: 2 December 1945 (as Type 170 Mk I)
POWERPLANT
Manufacturer: Bristol Aeroplane Co
Model: Hercules 672
Type: 14-cylinder radial
Number: 2
Output: 1260kW (1690bhp)
PERFORMANCE
Maximum cruising speed: 265km/h (165mph) at 3500m (11,500ft)
Service ceiling: 6400m (21,000ft)
Maximum range: 800km (500 miles)
WEIGHTS
Empty: 12,015kg (26,490lb)
Maximum take-off: 18,150kg (40,000lb)
DIMENSIONS
Span: 33m (108ft)
Length: 21m (68.5ft)
Height: 7.62m (25ft)
Wing area: 138m^2 (1489sq ft)

BRISTOL FREIGHTER
Model/variant: Type 170 Freighter Mk 32
Manufacturer: Bristol Aeroplane Co
Country of Origin: Great Britain
Type: Short-range mixed passenger and cargo transport
Accommodation: up to 32 plus cargo
Flight deck crew: 2/3
Prototype maiden flight: 2 December 1945 (as Type 170 Mk I)
POWERPLANT
Manufacturer: Bristol Aeroplane Co
Model: Hercules 734
Type: 14-cylinder radial
Number: 2
Output: 1475kW (1980bhp)
PERFORMANCE
Maximum cruising speed: 265km/h (165mph) at 3500m (11,500ft)
Service ceiling: 6400m (21,000ft)
Maximum range: 1320km (820 miles)
WEIGHTS
Empty: 13,405kg (29,950lb)
Maximum take-off: 19,960kg (44,000lb)
DIMENSIONS
Span: 33m (108ft)
Length: 22.4m (73.5ft)
Height: 7.62m (25ft)
Wing area: 138m^2 (1489sq ft)

BAE 146
Model/variant: -200
Manufacturer: British Aerospace
Country of Origin: Great Britain
Type: Short-range passenger transport
Accommodation (typical): 109
Flight deck crew: 2
Prototype maiden flight: 1 August 1982
POWERPLANT
Manufacturer: Textron
Model: Lycoming ALF 502R-5
Type: Turbofan
Number: 4
Output: 31kN (3170kg; 6970lb) thrust
PERFORMANCE
Maximum cruising speed: 776km/h (480mph) at 7925m (26,000ft)
Service ceiling: 11,280m (37,000ft)
Maximum range: 2180km (1355 miles) (with max. payload)
WEIGHTS
Empty: 23,270kg (51,300lb)
Maximum take-off: 42,185kg (93,000lb)
DIMENSIONS
Span: 26.2m (86ft)
Length: 28.6m (93.3ft)
Height: 8.6m (28.25ft)
Wing area: 77.3m^2 (831sq ft)

JETSTREAM
Model/variant: Jetstream 31
Manufacturer: British Aerospace
Country of Origin: Great Britain
Type: Short-range light passenger transport
Accommodation (typical): 18
Flight deck crew: 2
Prototype maiden flight: 28 March 1980
POWERPLANT
Manufacturer: Garrett (AlliedSignal)
Model: TPE331-10
Type: Turboprop
Number: 2
Output: 671kW (900shp)
PERFORMANCE
Maximum cruising speed: 490km/h (305mph) at 7620m (25,000ft)
Service ceiling: 9630m (31,600ft)
Maximum range: 1250km (780 miles) (with max. payload)
WEIGHTS
Empty: 4580kg (10,100lb)
Maximum take-off: 7350kg (16,200lb)
DIMENSIONS
Span: 15.85m (52ft)
Length: 14.35m (47.1ft)
Height: 5.3m (17.5ft)
Wing area: 25.1m^2 270sq ft

BAC ONE-ELEVEN
Model/variant: Series 500 (Series 560)
Manufacturer: British Aircraft Corporation/British Aerospace [ROMBAC, Romania]
Country of Origin: Great Britain
Type: Medium-range passenger transport
Accommodation (typical): 119 (one-class)
Flight deck crew: 2/3
Prototype maiden flight: 20 August 1963
POWERPLANT
Manufacturer: Rolls-Royce
Model: Spey Mk 512DW
Type: Turbofan
Number: 2
Output: 55.8kN (5695kg; 12,550lb) thrust
PERFORMANCE
Maximum cruising speed: 740km/h (470mph) at 8535m (28,000ft)
Service ceiling: 10,670m (35,000ft)
Maximum range: 2750km (1700 miles)
WEIGHTS
Empty: 24,455kg (53,910lb)
Maximum take-off: 47,400kg (104,500lb)
DIMENSIONS
Span: 28.5m (93.5ft)
Length: 32.6m (107ft)
Height: 7.5m (24.5ft)
Wing area: 95.8m^2 (1030sq ft)

CANADAIR CL-215
Model/variant: -
Manufacturer: Canadair (later Canadair Division of Bombardier Aerospace)
Country of Origin: Canada
Type: Firefighting amphibian
Payload: 5455kg (12,000lb)
Flight deck crew: 2
Prototype maiden flight: 23 October 1967
POWERPLANT
Manufacturer: Pratt & Whitney
Model: R-2800-CA3
Type: 18-cylinder radial
Number: 2
Output: 1566kW (2100bhp)
PERFORMANCE
Maximum cruising speed: 290km/h (180mph) at 3050m (10,000ft)
Service ceiling: 6100m (20,000ft)
Maximum range: 2095km (1300 miles) (with 1600kg (3600lb) payload)
WEIGHTS
Empty: 12,160kg (26,810lb)
Maximum take-off: 19,730kg (43,500lb) (land), 17,100kg (37,700lb) (water)
DIMENSIONS
Span: 28.6m (93.8ft)
Length: 19.8m (65ft)
Height: 8.9m (29.25ft)
Wing area: 100m^2 (1080sq ft)

CESSNA 152 [F-152]
Model/variant: Aerobat
Manufacturer: Cessna Aircraft [Reims Aviation, France]
Country of Origin: United States of America
Type: Two-seat cabin monoplane with dual controls and aerobatic capability

Accommodation (typical): -
Flight deck crew: 2
Prototype maiden flight: 1976
POWERPLANT
Manufacturer: Avco
Model: Lycoming O-235-N2C
Type: 4-cylinder horizontally-opposed
Number: 1
Output: 80kW (180hp)
PERFORMANCE
Maximum speed: 200km/h (125mph) at sea level
Service ceiling: 4480m (14,700ft)
Maximum range: 1160km (720 miles)
WEIGHTS
Empty: 500kg (1100lb)
Maximum take-off: 760kg (1680lb)
DIMENSIONS
Span: 10m (32.8ft)
Length: 7.35m (24ft)
Height: 2.6m (8.5ft)
Wing area: 14.6m² (157sq ft)

CESSNA CITATION
Model/variant: Model 501 Citation I
Manufacturer: Cessna Aircraft
Country of Origin: United States of America
Type: Short-range light passenger transport
Accommodation (typical): 6
Flight deck crew: 2
Prototype maiden flight: 15 September 1969
POWERPLANT
Manufacturer: Pratt & Whitney Canada
Model: JT15D-1
Type: Turbofan
Number: 2
Output: 9.8kN (1000kg; 2200lb) thrust
PERFORMANCE
Maximum cruising speed: 665km/h (410mph) at 10,700m (35,000ft)
Service ceiling: 12,500m (41,000ft)
Maximum range: 2500km (1550 miles)
WEIGHTS
Empty: 3000kg (6600lb)
Maximum take-off: 5375kg (11,850lb)
DIMENSIONS
Span: 14.35m (47ft)
Length: 13.35m (43.7ft)
Height: 4.4m (14.3ft)
Wing area: 25.8m² (278.5sq ft)

CESSNA CITATION
Model/variant: Model 650 Citation III
Manufacturer: Cessna Aircraft
Country of Origin: United States of America
Type: Short/medium-range light passenger transport
Accommodation (typical): 10
Flight deck crew: 2
Prototype maiden flight: 30 May 1979
POWERPLANT
Manufacturer: Garrett (AlliedSignal)
Model: TFE731-3-100
Type: Turbofan
Number: 2
Output: 16.25kN (1660kg; 3650lb) thrust
PERFORMANCE
Maximum cruising speed: 875km/h (545mph) at 10,700m (35,000ft)
Service ceiling: 15,550m (51,000ft)
Maximum range: 4350km (2700 miles)
WEIGHTS
Empty: 5350kg (11,800lb)
Maximum take-off: 10,000kg (22,000lb)
DIMENSIONS
Span: 16.35m (53.5ft)
Length: 16.9m (55.5ft)
Height: 5.15m (16.8ft)
Wing area: 29m² (313sq ft)

CONVAIR 240
Model/variant: - (C-131A - US military designation)
Manufacturer: Consolidated-Vultee (Convair)
Country of Origin: United States of America
Type: Short/medium-range passenger/cargo transport
Accommodation: 40
Flight deck crew: 2/3
Prototype maiden flight: 16 March 1947
POWERPLANT
Manufacturer: Pratt & Whitney
Model: R-2800-S1C3-G
Type: 18-cylinder radial
Number: 2
Output: 1566kW (2100bhp)
PERFORMANCE
Maximum cruising speed: 450km/h (275mph) at 3660m (12,000ft)
Service ceiling: 6100m (20,000ft)
Maximum range: 1930km (1200 miles)
WEIGHTS
Empty: 13,400kg (29,500lb)
Maximum take-off: 19,300kg (42,500lb)
DIMENSIONS
Span: 28m (91.75ft)
Length: 22.7m (74.7ft)
Height: 8.2m (27ft)
Wing area: 76m² (817sq ft)

CONVAIR 440
Model/variant: - (C-131D - US military designation)
Manufacturer: Consolidated-Vultee (Convair)
Country of Origin: United States of America
Type: Short/medium-range passenger/cargo transport
Accommodation: 52
Flight deck crew: 2/3
Prototype maiden flight: 6 October 1955
POWERPLANT
Manufacturer: Pratt & Whitney
Model: R-2800-CB16
Type: 18-cylinder radial
Number: 2
Output: 1865kW (2500bhp)
PERFORMANCE
Maximum cruising speed: 485km/h (300mph) at 3960m (13,000ft)
Service ceiling: 7590m (24,900ft)
Maximum range: 2100km (1300 miles)
WEIGHTS
Empty: 15,110kg (33,310lb)
Maximum take-off: 22,225kg (49,000lb)
DIMENSIONS
Span: 32.1m (105.3ft)
Length: 24.1m (79.2ft)
Height: 8.6m (28.2ft)
Wing area: 85.5m² (920sq ft)

CONVAIR 580
Model/variant: -
Manufacturer: Consolidated-Vultee (Convair)
Country of Origin: United States of America
Type: Short/medium-range passenger/cargo transport
Accommodation: 52 (high density)
Flight deck crew: 2/3
Prototype maiden flight: 19 January 1960 (with Allison 501 engines)
POWERPLANT
Manufacturer: Rolls-Royce
Model: Dart 542
Type: Turboprop
Number: 2
Output: 2256kW (3025shp)
PERFORMANCE
Maximum cruising speed: 485km/h (300mph) at 6100m (20,000ft)
Service ceiling: 7590m (24,900ft)
Maximum range: 3140km (1950 miles)
WEIGHTS
Empty: 13,735kg (30,275lb)
Maximum take-off: 25,855kg (57,000lb)
DIMENSIONS
Span: 32.1m (105.3ft)
Length: 24.1m (79.2ft)
Height: 8.6m (28.2ft)
Wing area: 85.5m² (920sq ft)

CONVAIR 640
Model/variant: -
Manufacturer: Consolidated-Vultee (Convair)
Country of Origin: United States of America
Type: Short/medium-range passenger/cargo transport
Accommodation: 56 (high density)
Flight deck crew: 2/3
Prototype maiden flight: 19 January 1960 (as Model 580 with Allison 501 engines)
POWERPLANT
Manufacturer: Rolls-Royce
Model: Dart 542
Type: Turboprop
Number: 2
Output: 2256kW (3025shp)
PERFORMANCE
Maximum cruising speed: 485km/h (300mph) at 6100m (20,000ft)
Service ceiling: 7590m (24,900ft)
Maximum range: 3140km (1950 miles)
WEIGHTS
Empty: 13,735kg (30,275lb)
Maximum take-off: 26,400kg (58,165lb)
DIMENSIONS
Span: 32.1m (105.3ft)
Length: 24.1m (79.2ft)
Height: 8.6m (28.2ft)
Wing area: 85.5m² (920sq ft)

CURTISS CONDOR
Model/variant: AT-32B Condor II
Manufacturer: Curtiss
Country of Origin: United States of America
Type: Short-range light passenger/cargo transport
Accommodation (typical): 12 (sleeper berths) or 15 (seats)
Flight deck crew: 2
Prototype maiden flight: 30 January 1933
POWERPLANT
Manufacturer: Wright
Model: SCR-1820-F2 Cyclone
Type: 9-cylinder radial
Number: 2
Output: 537kW (720bhp)
PERFORMANCE
Maximum cruising speed: 270km/h (165mph) at 3050m (10,000ft)
Service ceiling: 6700m (22,000ft)
Maximum range: 1150km (715 miles)
WEIGHTS
Empty: 5550kg (12,235lb)
Maximum take-off: 7950kg (17,500lb)
DIMENSIONS
Span: 25m (82ft)
Length: 15.1m (49.5ft)
Height: 5m (16.3ft)
Wing area: 118.5m² 1275sq ft) (biplane)

MYSTÈRE/FALCON 10
Model/variant: Model 100
Manufacturer: Dassault Aviation
Country of Origin: France
Type: Short/medium-range light passenger transport
Accommodation (typical): 6
Flight deck crew: 2
Prototype maiden flight: 1 December 1970
POWERPLANT
Manufacturer: Garrett (AlliedSignal)
Model: TFE731-2
Type: Turbofan
Number: 2
Output: 14.4kN (1465kg; 3230lb) thrust

PERFORMANCE
Maximum cruising speed: 912km/h (566mph) at 7620m (25,000ft)
Service ceiling: 13,715m (45,000ft)
Maximum range: 3550km (2200 miles)
WEIGHTS
Empty: 4880kg (10,760lb)
Maximum take-off: 8500kg (18,740lb)
DIMENSIONS
Span: 13.1m (43ft)
Length: 13.9m (45.5ft)
Height: 4.6m (15.1ft)
Wing area: 24m^2 (259.5sq ft)

MYSTÈRE/FALCON 20
Model/variant: Model 200 (Falcon 20H)
Manufacturer: Dassault Aviation
Country of Origin: France
Type: Short/medium-range passenger/cargo transport
Accommodation (typical): 10
Flight deck crew: 2
Prototype maiden flight: 30 April 1980
POWERPLANT
Manufacturer: Garrett (AlliedSignal) or General Electric
Model: ATF3-6-2C or CF700-2-D2
Type: Turbofan
Number: 2
Output: 24.5kN (2520kg; 5540lb) or 20kN (2040kg; (4500lb) thrust
PERFORMANCE
Maximum cruising speed: 850km/h (530mph) at 7620m (25,000ft)
Service ceiling: 12,800m (42,000ft)
Maximum range: 3300km (2050 miles)
WEIGHTS
Empty: 7530kg (16,600lb)
Maximum take-off: 13,000kg (28,600lb)
DIMENSIONS
Span: 16.3m (53.5ft)
Length: 17.15m (56.25ft)
Height: 5.3m (17.3ft)
Wing area: 41m^2 (441.5sq ft)

MYSTÈRE/FALCON 50
Model/variant: -
Manufacturer: Dassault Aviation
Country of Origin: France
Type: Medium/long-range light passenger transport
Accommodation (typical): 8
Flight deck crew: 2
Prototype maiden flight: 7 November 1976
POWERPLANT
Manufacturer: Garrett (AlliedSignal)
Model: TFE731-3-1C
Type: Turbofan
Number: 3
Output: 16.5kN (1680kg; 3700lb) thrust
PERFORMANCE
Maximum cruising speed: 880km/h (545mph) at 7620m (25,000ft)
Service ceiling: 13,800m (24,275ft)
Maximum range: 6300km (3915 miles)
WEIGHTS
Empty: 9150kg (21,170lb)
Maximum take-off: 17,600kg (38,800lb)
DIMENSIONS
Span: 18.9m (62ft)
Length: 18.5 (60.75ft)
Height: 7m (22.9ft)
Wing area: 46.8m^2 (504sq ft)

FALCON 900
Model/variant: -
Manufacturer: Dassault Aviation
Country of Origin: France
Type: Medium/long-range light passenger transport
Accommodation (typical): 12
Flight deck crew: 2
Prototype maiden flight: 21 September 1984
POWERPLANT
Manufacturer: Garrett (AlliedSignal)
Model: TFE731-5AR-1C
Type: Turbofan
Number: 3
Output: 18.4kN (2040kg; 4500lb) thrust
PERFORMANCE
Maximum cruising speed: 925km/h (575mph) at 7620m (25,000ft)
Service ceiling: 15,550m (51,000ft)
Maximum range: 7840km (4235 miles)
WEIGHTS
Empty: 10,240kg (22,575lb)
Maximum take-off: 20,640kg (45,500lb)
DIMENSIONS
Span: 19.3m (63.5ft)
Length: 20.2m (66.3ft)
Height: 7.55m (24.75ft)
Wing area: 49m^2 (528sq ft)

D.H. 89 DRAGON RAPIDE
Model/variant: D.H. 89A
Manufacturer: de Havilland Aircraft
Country of Origin: Great Britain
Type: Short-range light passenger transport
Accommodation (typical): 10
Flight deck crew: 1/2
Prototype maiden flight: 17 April 1934
POWERPLANT
Manufacturer: de Havilland
Model: Gypsy Queen 2
Type: 6-cylinder in-line
Number: 2
Output: 150kW (200bhp)
PERFORMANCE
Maximum cruising speed: 225km/h (140mph) at 3050m (10,000ft)
Service ceiling: 4875m (16,000ft)
Maximum range: 850km (525 miles
WEIGHTS
Empty: 1465kg (3230lb)
Maximum take-off: 2720kg (6000lb)
DIMENSIONS
Span: 14.65m (48ft)
Length: 10.5m (34.5ft)
Height: 3.1m (10.25ft)
Wing area: 31.2m^2 (336sq ft) (biplane)

D.H. 104 DOVE
Model/variant: Dove 1
Manufacturer: de Havilland Aircraft
Country of Origin: Great Britain
Type: Short/medium-range light passenger/cargo transport
Accommodation (typical): 11
Flight deck crew: 2
Prototype maiden flight: 25 September 1945
POWERPLANT
Manufacturer: de Havilland
Model: Gypsy Queen 70-3
Type: 6-cylinder in-line
Number: 2
Output: 246kW (330bhp)
PERFORMANCE
Maximum cruising speed: 265km/h (165mph) at 2440m (8000ft)
Service ceiling: 6100m (20,000ft)
Maximum range: 1245km (805 miles)
WEIGHTS
Empty: 2570kg (5655lb)
Maximum take-off: 3855kg (8500lb)
DIMENSIONS
Span: 17.4m (57ft)
Length: 12m (39.3ft)
Height: 4.1m (13.3ft)
Wing area: 31m^2 (335sq ft)

D.H. 106 COMET
Model/variant: Comet 4B
Manufacturer: de Havilland Aircraft
Country of Origin: Great Britain
Type: Medium-range passenger/cargo transport
Accommodation (typical): 99
Flight deck crew: 3/4
Prototype maiden flight: 27 June 1959

POWERPLANT
Manufacturer: Rolls-Royce
Model: Avon 524
Type: Turbojet
Number: 4
Output: 51kN (5215kg; 11,500lb) thrust
PERFORMANCE
Maximum cruising speed: 855km/h (530mph) at 12,800m (42,000ft)
Service ceiling: 14,325m (47,000ft)
Maximum range: 5190km (3225 miles)
WEIGHTS
Empty: 33,480kg (73,810lb)
Maximum take-off: 70,760kg (156,000lb)
DIMENSIONS
Span: 32.9m (108ft)
Length: 36m (118ft)
Height: 8.7m (28.5m)
Wing area: 191.3m^2 (2060sq ft)

D.H. 114 HERON
Model/variant: Heron 2B
Manufacturer: de Havilland Aircraft
Country of Origin: Great Britain
Type: Short-range passenger/cargo transport
Accommodation (typical): 14
Flight deck crew: 2
Prototype maiden flight: 10 May 1950
POWERPLANT
Manufacturer: de Havilland
Model: Gypsy Queen 30-2
Type: 6-cylinder in-line
Number: 4
Output: 186kW (250bhp)
PERFORMANCE
Maximum cruising speed: 295km/h (185mph) at 2440m (8000ft)
Service ceiling: 5640m (18,500ft)
Maximum range: 1475km (915 miles)
WEIGHTS
Empty: 3610kg (7960lb)
Maximum take-off: 5900kg (13,000lb)
DIMENSIONS
Span: 21.8m (71.5ft)
Length: 14.8m (48.5ft)
Height: 4.75m (15.5ft)
Wing area: 46.4m^2 (500sq ft)

SAUNDERS ST-27
Model/variant: -
Manufacturer: de Havilland Aircraft (modified by Saunders Aircraft Corp.)
Country of Origin: Great Britain/Canada
Type: Short-range passenger/cargo transport
Accommodation (typical): 23
Flight deck crew: 2
Prototype maiden flight: n/a
POWERPLANT
Manufacturer: Pratt & Whitney Canada
Model: PT6A-34
Type: Turboprop
Number: 2
Output: 560kW (750shp)
PERFORMANCE
Maximum cruising speed: 370km/h (230mph) at 3050m (10,000ft)
Service ceiling: 5640m (18,500ft)
Maximum range: 1320km (820 miles)
WEIGHTS
Empty: 3175kg (7000lb)
Maximum take-off: 6590kg (14,500
DIMENSIONS
Span: 21.8m (71.5ft)
Length: 14.8m (48.5ft)
Height: 4.75m (15.5ft)
Wing area: 46.4m^2 (500sq ft)

DHC-7 (DASH 7)
Model/variant: Series 100
Manufacturer: de Havilland Canada (later de Havilland Division of Bombardier Aerospace)
Country of Origin: Canada
Type: Short-range passenger/cargo transport
Accommodation (typical): 50
Flight deck crew: 2

Prototype maiden flight: 27 March 1975
POWERPLANT
Manufacturer: Pratt & Whitney Canada
Model: PT6A-50
Type: Turboprop
Number: 4
Output: 835kW (1120bhp)
PERFORMANCE
Maximum cruising speed: 435km/h (270mph) at 2440m (8000ft)
Service ceiling: 6400m (21,000ft)
Maximum range: 1280km (795 miles)
WEIGHTS
Empty: 12,540kg (27,650lb)
Maximum take-off: 19.960kg (44,000lb)
DIMENSIONS
Span: 28.35m (93ft)
Length: 24.6m (80.7ft)
Height: 8m (26.25ft)
Wing area: 79.9m^2 (860sq ft)

DHC-8 (DASH 8)
Model/variant: Series 100
Manufacturer: de Havilland Canada (de Havilland Division of Bombardier Aerospace)
Country of Origin: Canada
Type: Short-range passenger transport
Accommodation (typical): 36
Flight deck crew: 2
Prototype maiden flight: 20 June 1983
POWERPLANT
Manufacturer: Pratt & Whitney Canada
Model: PW120A
Type: Turboprop
Number: 2
Output: 1490kW (2000shp)
PERFORMANCE
Maximum cruising speed: 500km/h (310mph) at 4575m (15,000ft)
Service ceiling: 7620m (25,000ft)
Maximum range: 2010km (1250 miles)
WEIGHTS
Empty: 10,000kg (22,000lb)
Maximum take-off: 15,650kg (34,500lb)
DIMENSIONS
Span: 25.9m (85ft)
Length: 22.25m (73ft)
Height: 7.5m (24.5ft)
Wing area: 54.35m^2 (585sq ft)

DHC-2 BEAVER
Model/variant: Beaver I (L-20/U-6 - US military designation)
Manufacturer: de Havilland Canada
Country of Origin: Canada
Type: Short-range light utility STOL transport
Accommodation (typical): 7
Flight deck crew: 1
Prototype maiden flight: 16 August 1947
POWERPLANT
Manufacturer: Pratt & Whitney
Model: R-965 Wasp Junior
Type: 9-cylinder radial
Number: 1
Output: 336kW (450hp)
PERFORMANCE
Maximum cruising speed: 230km/h (145mph) at 1525m (5000ft)
Service ceiling: 5485m (18,000ft)
Maximum range: 1180km (735 miles)
WEIGHTS
Empty: 1295kg (2850lb)
Maximum take-off: 2315kg (5100lb)
DIMENSIONS
Span: 14.6m (47.75ft)
Length: 9.25m (30.25ft)
Height: 2.75m (8.75ft)
Wing area: 23.2m^2 (250sq ft)

DHC-6 TWIN OTTER
Model/variant: Model 300
Manufacturer: de Havilland Canada
Country of Origin: Canada
Type: Short-range light utility STOL transport
Accommodation (typical): 20 (high density)
Flight deck crew: 2
Prototype maiden flight: 20 May 1965
POWERPLANT
Manufacturer: Pratt & Whitney Canada
Model: PT6A-27
Type: Turboprop
Number: 2
Output: 485kW (650shp)
PERFORMANCE
Maximum cruising speed: 340km/h (210mph) at 3050m (10,000ft)
Service ceiling: 8140m (26,700ft)
Maximum range: 1300km (810 miles) (with 1135kg (2500lb) payload)
WEIGHTS
Empty: 3365kg (7415lb)
Maximum take-off: 5670kg (12,500lb)
DIMENSIONS
Span: 19.8m (65ft)
Length: 15.8m (51.75ft)
Height: 5.95m (19.5ft)
Wing area: 39m^2 (420sq ft)

DO R SUPER WAL (DO 15)
Model/variant: Do R4
Manufacturer: Dornier
Country of Origin: Germany
Type: Long-range passenger transport flying-boat
Accommodation: 19
Flight deck crew: 4
Prototype maiden flight: September 1936
POWERPLANT
Manufacturer: Bristol/Siemens
Model: Jupiter
Type: 9-cylinder radial
Number: 4
Output: 392kW (525bhp)
PERFORMANCE
Maximum cruising speed: 180km/h (112mph) at 1000m (3300ft)
Service ceiling: 2000m (6600ft)
Maximum range: 2000km (1235 miles)
WEIGHTS
Empty: 9850kg (21,720lb)
Maximum take-off: 14,000kg (30,800lb)
DIMENSIONS
Span: 28.6m (93.9ft)
Length: 24.6m (80.8ft)
Height: 6m (19.7ft)
Wing area: 137m^2 (1475sq ft)

DO X
Model/variant: -
Manufacturer: Dornier
Country of Origin: Germany/Switzerland
Type: Long-range passenger transport flying-boat
Accommodation (typical): 100
Flight deck crew: 5
Prototype maiden flight: 12 July 1929
POWERPLANT
Manufacturer: Bristol/Siemens or Curtiss or FIAT
Model: Jupiter, GV-1570 Conqueror or A-22R
Type: 9-cylinder radial, V-12 or V-12
Number: 12
Output: 392kW (525bhp) or 477kW (640bhp) or 432kW (580bhp)
PERFORMANCE
Maximum cruising speed: 175km/h (110mph) at 1000m (3300ft)
Service ceiling: 1250ft (4100ft)
Maximum range: 2200km (1370 miles)
WEIGHTS
Empty: 32,675kg (71,885lb)
Maximum take-off: 56,000kg (123,200lb)
DIMENSIONS
Span: 48m (157.5ft)
Length: 40m (131.3ft)
Height: 10m (33ft)
Wing area: 450m^2 (4844sq ft)

DC-2
Model/variant: - (C-33/C-34 - US military designation)
Manufacturer: Douglas Aircraft
Country of Origin: United States of America
Type: Short/medium-range passenger/cargo transport
Accommodation (typical): 14
Flight deck crew: 2
Prototype maiden flight: 1 July 1933
POWERPLANT
Manufacturer: Wright
Model: SGR-1820-F52 Cyclone
Type: 9-cylinder radial
Number: 2
Output: 652kW (875bhp)
PERFORMANCE
Maximum cruising speed: 305km/h (190mph) at 2240m (8000ft)
Service ceiling: 6845m (22,450ft)
Maximum range: 1620km (1000 miles)
WEIGHTS
Empty: 5630kg (12,410lb)
Maximum take-off: 8420kg (18,560lb)
DIMENSIONS
Span: 25.9m (85ft)
Length: 18.9m (62ft)
Height: 5m (16.3ft)
Wing area: 87.25m^2 (940sq ft)

DC-4 SKYMASTER
Model/variant: C-54G (US military designation)
Manufacturer: Douglas Aircraft
Country of Origin: United States of America
Type: Long-range troop transport
Accommodation (typical): 50
Flight deck crew: 4
Prototype maiden flight: 14 February 1942
POWERPLANT
Manufacturer: Pratt & Whitney
Model: R-2000-9
Type: 14-cylinder radial
Number: 4
Output: 1081kW (1450bhp)
PERFORMANCE
Maximum cruising speed: 350km/h (220mph) at 3050m (10,000ft)
Service ceiling: 6800m (22,300ft)
Maximum range: 4025km (2500 miles) (with 5200kg payload)
WEIGHTS
Empty: 19,640kg (43,200lb)
Maximum take-off: 33,100kg (72,800lb)
DIMENSIONS
Span: 35.8m (117.5ft)
Length: 28.6m (93.9ft)
Height: 8.4m (27.5ft)
Wing area: 136m^2 (1460sq ft)

EMB-110 BANDEIRANTE
Model/variant: EMB-110P1
Manufacturer: EMBRAER (Empresa Brasileira de Aeronáutica)
Country of Origin: Brazil
Type: Short/medium-range light passenger transport
Accommodation (typical): 18
Flight deck crew: 2
Prototype maiden flight: 26 October 1968
POWERPLANT
Manufacturer: Pratt & Whitney of Canada
Model: PT6A-34
Type: Turboprop
Number: 2
Output: 560kW (750shp)
PERFORMANCE
Maximum cruising speed: 410km/h (255mph) at 3050m (10,000ft)
Service ceiling: 6550m (21,500ft)
Maximum range: 2000km (1240 miles)
WEIGHTS
Empty: 3555kg (7840lb)
Maximum take-off: 5680kg (12,500lb)
DIMENSIONS
Span: 15.35m (50.25ft)
Length: 15.1m (49.5ft)

Height: 4.9m (16.1ft)
Wing area: 29m² (313.5sq ft)

METRO
Model/variant: SA 227 Metro III
Manufacturer: Fairchild Swearingen
Country of Origin: United States of America
Type: Short-range light passenger transport
Accommodation (typical): 20
Flight deck crew: 2
Prototype maiden flight: 13 April 1965 (as Merlin IIA)
POWERPLANT
Manufacturer: Garrett (AlliedSignal)
Model: TPE331-11U-601G
Type: Turboprop
Number: 2
Output: 820kW (1100shp)
PERFORMANCE
Maximum cruising speed: 515km/h (320mph) at 4575m (15,000ft)
Service ceiling: 8380m (27,500ft)
Maximum range: 1600km (1000m)
WEIGHTS
Empty: 3970kg (8740lb)
Maximum take-off: 6580kg (14,500lb)
DIMENSIONS
Span: 17.4m (57ft)
Length: 18.1m (59.3ft)
Height: 5.1m (16.7ft)
Wing area: 28.75m² (310sq ft)

F27 FRIENDSHIP
Model/variant: Mk 200
Manufacturer: Fokker [Fairchild, United States of America]
Country of Origin: Netherlands
Type: Short/medium-range passenger/cargo transport
Accommodation (typical): 40
Flight deck crew: 2
Prototype maiden flight: 24 November 1955
POWERPLANT
Manufacturer: Rolls-Royce
Model: Dart 536-7R
Type: Turboprop
Number: 2
Output: 1730kW (2320shp)
PERFORMANCE
Maximum cruising speed: 480km/h (300mph) at 6100m (20,000ft)
Service ceiling: 8990m (29,500ft)
Maximum range: 1925km (1200 miles)
WEIGHTS
Empty: 12,100kg (26,620lb)
Maximum take-off: 20,450kg (45,000lb)
DIMENSIONS
Span: 29m (95.1ft)
Length: 23.6m (77.3ft)
Height: 8.5m (27.9ft)
Wing area: 70m² 753.5sq ft)

F28 FELLOWSHIP
Model/variant: Mk 4000
Manufacturer: Fokker
Country of Origin: Netherlands
Type: Short/medium-range passenger/cargo transport
Accommodation (typical): 85
Flight deck crew: 2/3
Prototype maiden flight: 9 May 1967
POWERPLANT
Manufacturer: Rolls-Royce
Model: RB183-2 Mk 555-15P
Type: Turbofan
Number: 2
Output: 44kN (4490kg; 990lb) thrust
PERFORMANCE
Maximum cruising speed: 845km/h (525mph) at 7000m (22,965ft)
Service ceiling: 10,670m (35,000ft)
Maximum range: 2100km (1300 miles)
WEIGHTS
Empty: 17,645kg (38,820lb)
Maximum take-off: 34,200kg (75,240lb)

DIMENSIONS
Span: 25.1m (82.3ft)
Length: 29.6m (97.1ft)
Height: 8.5m (17.9ft)
Wing area: 79m² (850sq ft)

TRI-MOTOR
Model/variant: 5-AT-B
Manufacturer: Ford
Country of Origin: United States of America
Type: Short-range passenger transport
Accommodation (typical): 15
Flight deck crew: 2
Prototype maiden flight: 11 June 1926
POWERPLANT
Manufacturer: Pratt & Whitney
Model: Wasp C-1
Type: 9-cylinder radial
Number: 3
Output: 313kW (420bhp)
PERFORMANCE
Maximum cruising speed: 195km/h (120mph) at 3050m (10,000ft)
Service ceiling: 5640m (18,000ft)
Maximum range: 885km (550 miles)
WEIGHTS
Empty: 3450kg (7600lb)
Maximum take-off: 5740kg (12,650lb)
DIMENSIONS
Span: 23.7m (77.9ft)
Length: 15.3m (50.3ft)
Height: 3.9m (12.8ft)
Wing area: 77.6m² (835sq ft)

H.P.42E
Model/variant: -
Manufacturer: Handley Page Aircraft
Country of Origin: Great Britain
Type: Short-range passenger transport
Accommodation: 18 or 24
Flight deck crew: 4
Prototype maiden flight: 17 November 1930
POWERPLANT
Manufacturer: Bristol
Model: Jupiter XIF
Type: 9-cylinder radial
Number: 4
Output: 365kW (490bhp)
PERFORMANCE
Maximum cruising speed: 160km/h (100mph) at 2000m (6560ft)
Service ceiling: 3800m (12,450ft)
Maximum range: 800km (500 miles)
WEIGHTS
Empty: 8050kg (17,750lb)
Maximum take-off: 12,700kg (28,000lb)
DIMENSIONS
Span: 39.6m (130ft)
Length: 28.1m (92ft)
Height: 8.25m (27ft)
Wing area: 278m² (2992sq ft) (biplane)

H.P.42W (H.P.45)
Model/variant: -
Manufacturer: Handley Page Aircraft
Country of Origin: Great Britain
Type: Short-range passenger transport
Accommodation: 38
Flight deck crew: 4
Prototype maiden flight: 17 November 1930
POWERPLANT
Manufacturer: Bristol
Model: Jupiter XFBM
Type: 9-cylinder radial
Number: 4
Output: 414kW (555bhp)
PERFORMANCE
Maximum cruising speed: 180km/h (110mph) at 2000m (6560ft)
Service ceiling: 3800m (12,450ft)
Maximum range: 480km (300 miles)
WEIGHTS
Empty: 8050kg (17,750lb)
Maximum take-off: 13,760kg (30,335lb)

DIMENSIONS
Span: 39.6m (130ft)
Length: 28.1m (92ft)
Height: 8.25m (27ft)
Wing area: 278m² (2992sq ft) (biplane)

H.P.R.7 HERALD
Model/variant: Series 200
Manufacturer: Handley Page Aircraft
Country of Origin: Great Britain
Type: Short/medium-range passenger/cargo transport
Accommodation (typical): 56
Flight deck crew: 2/3
Prototype maiden flight: 25 August 1955
POWERPLANT
Manufacturer: Rolls-Royce
Model: Dart 527
Type: Turboprop
Number: 2
Output: 1570kW (2105shp)
PERFORMANCE
Maximum cruising speed: 445km/h (275mph) at 4570m (15,000ft)
Service ceiling: 8505m (27,900ft)
Maximum range: 1785km (1110 miles)
WEIGHTS
Empty: 11,700kg (25,800lb)
Maximum take-off: 19,500kg (43,000lb)
DIMENSIONS
Span: 28.9m (94.75ft)
Length: 23m (75.5ft)
Height: 7.3m (24ft)
Wing area: 82.3m² (886sq ft)

TRIDENT
Model/variant: 1C
Manufacturer: Hawker Siddeley
Country of Origin: Great Britain
Type: Medium-range passenger/cargo transport
Accommodation (typical): 103
Flight deck crew: 2/3
Prototype maiden flight: 9 January 1962
POWERPLANT
Manufacturer: Rolls-Royce
Model: Spey RB163-25 Mk 505-5F
Type: Turbofan
Number: 3
Output: 44kN (4480kg; 9850lb) thrust
PERFORMANCE
Maximum cruising speed: 935km/h (585mph) at 7620m (25,000ft)
Service ceiling: 11,580m (38,000ft)
Maximum range: 2170km (1350 miles) (with max. payload)
WEIGHTS
Empty: 30,850kg (67,870lb)
Maximum take-off: 53,200kg (117,000lb)
DIMENSIONS
Span: 27.1m (89.9ft)
Length: 35m (114.75ft)
Height: 8.2m (27ft)
Wing area: 126m² (1358sq ft)

TRIDENT
Model/variant: 1E
Manufacturer: Hawker Siddeley
Country of Origin: Great Britain
Type: Medium-range passenger/cargo transport
Accommodation: 139 (high density)
Flight deck crew: 2/3
Prototype maiden flight: 9 January 1962
POWERPLANT
Manufacturer: Rolls-Royce
Model: Spey RB163-25 Mk 511-5F
Type: Turbofan
Number: 3
Output: 50.7kN (5180kg; 11,400lb) thrust
PERFORMANCE
Maximum cruising speed: 935km/h (585mph) at 7620m (25,000ft)
Service ceiling: 11,580m (38,000ft)
Maximum range: 4345km (2700 miles)

WEIGHTS
Empty: 31,650kg (69,630lb)
Maximum take-off: 64,640kg (142,200lb)
DIMENSIONS
Span: 28.9m (95ft)
Length: 35m (114.75ft)
Height: 8.2m (27ft)
Wing area: 134.5m² (1446sq ft)

TRIDENT
Model/variant: 2E
Manufacturer: Hawker Siddeley
Country of Origin: Great Britain
Type: Medium-range passenger/cargo transport
Accommodation (typical): 115 (two-class)
Flight deck crew: 2/3
Prototype maiden flight: 9 January 1962 (as Trident 1)
POWERPLANT
Manufacturer: Rolls-Royce
Model: Spey RB163-25 Mk 512-5W
Type: Turbofan
Number: 3
Output: 53kN (5425kg; 11,960lb) thrust
PERFORMANCE
Maximum cruising speed: 955km/h (595mph) at 9150m (30,000ft)
Service ceiling: 11,580m (38,000ft)
Maximum range: 3965km (2465 miles)
WEIGHTS
Empty: 33,200kg (73,000lb)
Maximum take-off: 65,320kg (143,700lb)
DIMENSIONS
Span: 29.8m (98ft)
Length: 35m (114.75ft)
Height: 8.2m (27ft)
Wing area: 135m² (1456sq ft)

TRIDENT
Model/variant: 3B
Manufacturer: Hawker Siddeley
Country of Origin: Great Britain
Type: Medium-range passenger/cargo transport
Accommodation (typical): 180 (high density)
Flight deck crew: 2/3
Prototype maiden flight: 22 March 1970
POWERPLANT
Manufacturer: Rolls-Royce
Model: Spey RB163-25 Mk 511-5F (RB162-86)
Type: Turbofan
Number: 3 (plus 1)
Output: 51kN (5180kg; 11,400lb) (plus 23kN (2380kg; 5250lb)) thrust
PERFORMANCE
Maximum cruising speed: 870km/h (540mph) at 9150m (30,000ft)
Service ceiling: 11,580m (38,000ft)
Maximum range: 3600m (2235 miles)
WEIGHTS
Empty: 37,085kg (81,590lb)
Maximum take-off: 68,040kg (150,000lb)
DIMENSIONS
Span: 29.8m (98ft)
Length: 39.9m (131.2ft)
Height: 8.2m (27ft)
Wing area: 135m² (1456sq ft)

HS 125
Model/variant: Series 800
Manufacturer: Hawker Siddeley (later British Aerospace)
Country of Origin: Great Britain
Type: Medium-range light passenger transport
Accommodation: up to 14
Flight deck crew: 2
Prototype maiden flight: 26 May 1983
POWERPLANT
Manufacturer: Garrett (AlliedSignal)
Model: TFE 731-5-R
Type: Turbofan

Number: 2
Output: 19.1kN (1955kg; 4300lb) thrust
PERFORMANCE
Maximum cruising speed: 740km/h (455mph) at 11,280m (37,000ft)
Service ceiling: 13,100m (43,000ft)
Maximum range: 5560km (3000 miles)
WEIGHTS
Empty: 6850kg (15,100lb)
Maximum take-off: 12,425kg (27,400lb)
DIMENSIONS
Span: 15.7m (51.5ft)
Length: 15.5m (51ft)
Height: 5.4m (17.75ft)
Wing area: 34.75m² (374sq ft)

HS 748
(ANDOVER - RAF DESIGNATION)
Model/variant: Series 2A
Manufacturer: Hawker Siddeley/British Aerospace [Hindustan Aircraft, India]
Country of Origin: Great Britain
Type: Short/medium-range passenger/cargo transport
Accommodation: 48
Flight deck crew: 2
Prototype maiden flight: 24 June 1960
POWERPLANT
Manufacturer: Rolls-Royce
Model: Dart Mk 535-2
Type: Turboprop
Number: 2
Output: 1700kW (2280shp)
PERFORMANCE
Maximum cruising speed: 425km/h (280mph) at 4570m (15,000ft)
Service ceiling: 7620m (25,000ft)
Maximum range: 1300km (810 miles) (with max. payload)
WEIGHTS
Empty: 11,645kg (25,670lb)
Maximum take-off: 21,100kg (46,500lb)
DIMENSIONS
Span: 31.25m (102.5ft)
Length: 20.4m (67ft)
Height: 7.6m (24.8ft)
Wing area: 77m² (828sq ft)

HE 70 'BLITZ'
Model/variant: He 70G
Manufacturer: Heinkel
Country of Origin: Germany
Type: Short-range light passenger transport and mail carrier
Accommodation: 4
Flight deck crew: 1
Prototype maiden flight: 1 December 1932
POWERPLANT
Manufacturer: BMW
Model: VI
Type: V-12 in-line
Number: 1
Output: 560kW (750bhp)
PERFORMANCE
Maximum cruising speed: 360km/h (225mph) at 3050m (10,000ft)
Service ceiling: 5485m (18,000ft)
Maximum range: 1250km (775 miles)
WEIGHTS
Empty: 2530kg (5580lb)
Maximum take-off: 3640kg (7630lb)
DIMENSIONS
Span: 14.8m (48.5ft)
Length: 12.2m (40ft)
Height: 3.25m (10.7ft)
Wing area: 36.5m² (393sq ft)

ILYUSHIN IL-14 'CRATE'
(NATO REPORTING NAME)
MModel/variant: Il-14M [-; Avia 14-32]
Manufacturer: Ilyushin [DDR; Czechoslovakia]
Country of Origin: Soviet Union
Type: Short-range passenger/cargo transport
Accommodation (typical): 28

Flight deck crew: 2
Prototype maiden flight: 1952
POWERPLANT
Manufacturer: Shvetsov
Model: ASh-82T
Type: 9-cylinder radial
Number: 2
Output: 1417kW (1900bhp)
PERFORMANCE
Maximum cruising speed: 320km/h (200mph) at 3050m (10,000ft)
Service ceiling: 7400m (24,250ft)
Maximum range: 1305km (810 miles) (with max. payload)
WEIGHTS
Empty: 12,600kg (27,700lb)
Maximum take-off: 18,000kg (39,600lb)
DIMENSIONS
Span: 31.7m (104ft)
Length: 22.3m (73.2ft)
Height: 7.9m (26ft)
Wing area: 99.7m² (1075sq ft)

ILYUSHIN IL-18 'COOT'
(NATO REPORTING NAME)
Model/variant: Il-18D
Manufacturer: Ilyushin
Country of Origin: Soviet Union
Type: Medium-range passenger/cargo transport
Accommodation (typical): 122
Flight deck crew: 2/3
Prototype maiden flight: 4 July 1957
POWERPLANT
Manufacturer: Ivchenko
Model: AI-20M
Type: Turboprop
Number: 4
Output: 3170kW (4250shp)
PERFORMANCE
Maximum cruising speed: 675km/h (420mph) at 8000m (26,250ft)
Service ceiling: 12,200m (40,000ft)
Maximum range: 3700km (2300 miles) (with max. payload)
WEIGHTS
Empty: 35,000kg (77,000lb)
Maximum take-off: 64,000kg (141,000lb)
DIMENSIONS
Span: 37.4m (122.75ft)
Length: 35.9m (117.8ft)
Height: 10.2m (33.4ft)
Wing area: 140m² (1507sq ft)

IL-62 'CLASSIC'
(NATO REPORTING NAME)
Model/variant: Il-62M
Manufacturer: Ilyushin
Country of Origin: Soviet Union
Type: Long-range passenger/cargo transport
Accommodation (typical): 174
Flight deck crew: 5
Prototype maiden flight: January 1963
POWERPLANT
Manufacturer: Soloviev
Model: D-30KU
Type: Turbofan
Number: 4
Output: 107kN (11,000kg; 24,200lb) thrust
PERFORMANCE
Maximum cruising speed: 900km/h (560mph) at 10,000m (32,800ft)
Service ceiling: 12,800m (42,000ft)
Maximum range: 7800km (4850 miles) (with max. payload)
WEIGHTS
Empty: 69,400kg (152,700lb)
Maximum take-off: 165,000kg (363,000lb)
DIMENSIONS
Span: 43.2m (141.7ft)
Length: 53.1m (174.25ft)
Height: 12.35m (40.5ft)
Wing area: 282.2m² (3040sq ft)

IL-76 'CANDID'
(NATO REPORTING NAME)
Model/variant: Il-76T (Il-76DMP)
Manufacturer: Ilyushin
Country of Origin: Soviet Union
Type: Medium/long-range cargo transport (fire-fighter)
Payload: 40,000kg (88,000lb)
Flight deck crew: 3/5
Prototype maiden flight: 25 March 1971
POWERPLANT
Manufacturer: Soloviev
Model: D-30KP
Type: Turbofan
Number: 4
Output: 117.5kN (12,000kg; 26,400lb) thrust
PERFORMANCE
Maximum cruising speed: 800km/h (500mph) at 10,000m (32,800ft)
Service ceiling: 15,500m (50,850ft)
Maximum range: 3650km (2265 miles) (witn max. payload)
WEIGHTS
Empty: 92,000kg (202,400lb)
Maximum take-off: 170,000kg (375,000lb)
DIMENSIONS
Span: 50.5m (165.7ft)
Length: 46.5m (152.9ft)
Height: 14.7m (48.5ft)
Wing area: 300m^2 (3230sq ft)

IL-96
Model/variant: Il-96-300
Manufacturer: Ilyushin
Country of Origin: Soviet Union
Type: Medium/long-range passenger transport
Accommodation (typical): 300 (one-class)
Flight deck crew: 3
Prototype maiden flight: 28 September 1988
POWERPLANT
Manufacturer: Soloviev
Model: PS-90A
Type: Turbofan
Number: 4
Output: 156.7kN (16,000kg; 35,200lb) thrust
PERFORMANCE
Maximum cruising speed: 900km/h (560mph) at 12,100m (39,700ft)
Service ceiling: 14,600m (47,900ft)
Maximum range: 7500km (4660 miles)
WEIGHTS
Empty: 117,000kg (257,500lb)
Maximum take-off: 216,000kg (475,500lb)
DIMENSIONS
Span: 60m (197ft)
Length: 55.3m (181.5ft)
Height: 17.5m (57.5ft)
Wing area: 391.5m^2 (4215sq ft)

IAI 1124 WESTWIND
Model/variant: Westwind 1
Manufacturer: Israeli Aircraft Industries
Country of Origin: Israel
Type: Short/medium-range light passenger transport
Accommodation (typical): 12
Flight deck crew: 2
Prototype maiden flight: n/a
POWERPLANT
Manufacturer: Garrett (AlliedSignal)
Model: TFE731-3 1G
Type: Turbofan
Number: 2
Output: 16.5kN (1680kg; 3700lb) thrust
PERFORMANCE
Maximum cruising speed: 750km/h (465mph) at 9150m (30,000ft)
Service ceiling: 13,715m (45,000ft)
Maximum range: 3985km (2475 miles)
WEIGHTS
Empty: 5580kg (12,275kg)
Maximum take-off: 10,600kg (23,500lb)
DIMENSIONS
Span: 13.65m (44.8ft)
Length: 15.95m (52.25ft)
Height: 4.8m (15.8ft)
Wing area: 28.65m^2 (308.5sq ft)

F 13
Model/variant: -
Manufacturer: Junkers
Country of Origin: Germany
Type: Short-range light passenger transport
Accommodation: 4
Flight deck crew: 2
Prototype maiden flight: 25 June 1919
POWERPLANT
Manufacturer: Junkers
Model: L-5
Type: 6-cylinder in-line
Number: 1
Output: 156kW (220hp)
PERFORMANCE
Maximum cruising speed: 140km/h (87mph) at 2500m (8200ft)
Service ceiling: 4000m (13,100ft)
Maximum range: 900km (560 miles)
WEIGHTS
Empty: 1150kg (2530lb)
Maximum take-off: 1950kg (4290lb)
DIMENSIONS
Span: 17.75m (58.25ft)
Length: 9.6m (34.5ft)
Height: n/a
Wing area: 44m^2 (474sq ft)

G 38
Model/variant: -
Manufacturer: Junkers
Country of Origin: Germany
Type: Short-range passenger transport
Accommodation: 34
Flight deck crew: 6
Prototype maiden flight: 6 November 1929
POWERPLANT
Manufacturer: Junkers
Model: Jumo 204
Type: V-12 in-line
Number: 4
Output: 560kW (750bhp)
PERFORMANCE
Maximum cruising speed: 185km/h (115mph) at 2000m (6560ft)
Service ceiling: 3100m (10,200ft)
Maximum range: 1850km (1150 miles)
WEIGHTS
Empty: 16,400kg (36,100lb)
Maximum take-off: 23,000kg (50,600lb)
DIMENSIONS
Span: 44m (144.4in)
Length: 23.2m (76ft)
Height: 7.2m (23.75ft)
Wing area: 305m^2 (3283sq ft)

LONGHORN 55
Model/variant: -
Manufacturer: Learjet (Learjet Division of Bombardier Aerospace)
Country of Origin: United States of America
Type: Short/medium-range light passenger transport
Accommodation (typical): 10
Flight deck crew: 2
Prototype maiden flight: 19 April 1979
POWERPLANT
Manufacturer: Garrett (AlliedSignal)
Model: TFE731-3A-2B
Type: Turbofan
Number: 2
Output: 16.4kN (1675kg; 3700lb) thrust
PERFORMANCE
Maximum cruising speed: 740km/h (460mph) at 14,235m (47,000ft)
Service ceiling: 15,545m (51,000ft)
Maximum range: 4150km (2580 miles)
WEIGHTS
Empty: 5500kg (12,100lb)
Maximum take-off: 9525kg (21,000lb)
DIMENSIONS
Span: 13.35m (43.75ft)
Length: 16.8m (55.1ft)
Height: 4.45m (14.7ft)
Wing area: 24.6m^2 (265sq ft)

LET L-410 TURBOLET
Model/variant: L-410UVP-E
Manufacturer: Let
Country of Origin: Czechoslovakia
Type: Short-range light passenger transport/air ambulance
Accommodation (typical): 19
Flight deck crew: 2
Prototype maiden flight: 16 April 1969
POWERPLANT
Manufacturer: Walter
Model: M-601B
Type: Turboprop
Number: 2
Output: 545kW (730shp)
PERFORMANCE
Maximum cruising speed: 365km/h (225mph) at 3050m (10,000ft)
Service ceiling: 6100m (20,000ft)
Maximum range: 550km (340 miles)
WEIGHTS
Empty: 3985kg (8765lb)
Maximum take-off: 6400kg (14,100lb)
DIMENSIONS
Span: 19.5m (63.9ft)
Length: 14.5m (47.5ft)
Height: 5.85m (19.1ft)
Wing area: 35.2m^2 379sq ft)

L-188 ELECTRA
Model/variant: L-188A
Manufacturer: Lockheed
Country of Origin: United States of America
Type: Short/medium-range passenger/cargo transport
Accommodation (typical): 80
Flight deck crew: 2/3
Prototype maiden flight: 6 December 1957
POWERPLANT
Manufacturer: Allison
Model: 501-D13
Type: Turboprop
Number: 4
Output: 2795kW (3750shp)
PERFORMANCE
Maximum cruising speed: 650km/h (405mph) at 7620m (25,000ft)
Service ceiling: 8655m (28,400ft)
Maximum range: 3540km (2200 miles)
WEIGHTS
Empty: 26,000kg (57,200lb)
Maximum take-off: 51,250kg (112,750lb)
DIMENSIONS
Span: 30m (99ft)
Length: 32.15m (104.5ft)
Height: 10.25m (32.75ft)
Wing area: 121m^2 (1300sq ft)

L-1011 TRISTAR
Model/variant: L-1011-500
Manufacturer: Lockheed
Country of Origin: United States of America
Type: Long-range passenger/cargo transport
Accommodation (typical): 400 (high density)
Flight deck crew: 2/4
Prototype maiden flight: 17 November 1970
POWERPLANT
Manufacturer: Rolls-Royce
Model: RB211-524B
Type: Turbofan
Number: 3
Output: 222kN (22,680kg; 50,000lb)
PERFORMANCE
Maximum cruising speed: 975km/h (605mph) at 9150m (30,000ft)
Service ceiling: 12,800m (42,000ft)

Maximum range: 9655km (6000 miles) (with max. payload)
WEIGHTS
Empty: 109,300kg (241,000lb)
Maximum take-off: 225,000kg (496,000lb)
DIMENSIONS
Span: 47.35m (155.3ft)
Length: 50m (164ft)
Height: 16.9m (55.3ft)
Wing area: 321m² (3456sq ft)

L-1049 SUPER CONSTELLATION
Model/variant: L-1049G
Manufacturer: Lockheed
Country of Origin: United States of America
Type: Long-range passenger/cargo transport
Accommodation (typical): 109
Flight deck crew: 4/5
Prototype maiden flight: 9 January 1943 (as L-49)
POWERPLANT
Manufacturer: Wright
Model: 972TC-18DA-3
Type: 18-cylinder radial
Number: 4
Output: 2535kW (3400bhp)
PERFORMANCE
Maximum cruising speed: 570km/h (355mph) at 6890m (22,600ft)
Service ceiling: 7225m (23,700ft)
Maximum range: 8200km (5100 miles) (with max. payload)
WEIGHTS
Empty: 36,125kg (79,475lb)
Maximum take-off: 65,775kg (144,700lb)
DIMENSIONS
Span: 38.5m (126.25ft) (including wingtip fuel tanks)
Length: 32.15m (104.5ft)
Height: 10.25m (32.75ft)
Wing area: 103m² (1300sq ft)

ORION 9
Model/variant: Orion 9
Manufacturer: Lockheed
Country of Origin: United States of America
Type: Short-range light passenger transport
Accommodation (typical): 6
Flight deck crew: 1
Prototype maiden flight: February 1931
POWERPLANT
Manufacturer: Pratt & Whitney
Model: Wasp C
Type: 9-cylinder radial
Number: 1
Output: 315kW (420bhp)
PERFORMANCE
Maximum cruising speed: 280km/h (175mph) at 3050m (10,000ft)
Service ceiling: 6700m (22,000ft)
Maximum range: 1160km (720 miles)
WEIGHTS
Empty: 1650kg (3630lb)
Maximum take-off: 2360kg (5190lb)
DIMENSIONS
Span: 13m (42.9ft)
Length: 8.65m (28.3ft)
Height: 2.95m (9.7ft)
Wing area: 27.5m² (295sq ft)

SIRIUS 8
Model/variant: DL-2
Manufacturer: Lockheed (Detroit Aircraft Corp.)
Country of Origin: United States of America
Type: Aerial survey
Accommodation: 1
Flight deck crew: 1
Prototype maiden flight: November 1929
POWERPLANT
Manufacturer: Pratt & Whitney
Model: Wasp
Type: 9-cylinder radial
Number: 1
Output: 335kW (450bhp)

PERFORMANCE
Maximum cruising speed: 235km/h (145mph) at 3050m (10,000ft)
Service ceiling: 5470m (18,000ft)
Maximum range: 1570km (970 miles)
WEIGHTS
Empty: 1345kg (2960lb)
Maximum take-off: 2345kg (5150lb)
DIMENSIONS
Span: 13m (42.9ft)
Length: 8.5m (27.9ft)
Height: 2.8m (9.2ft)
Wing area: 27.5m² (295sq ft)

SUPER ELECTRA 14
Model/variant: Super Electra 14-WF62
Manufacturer: Lockheed
Country of Origin: United States of America
Type: Short-range passenger/cargo transport
Accommodation (typical): 10
Flight deck crew: 2
Prototype maiden flight: 29 July 1937
POWERPLANT
Manufacturer: Wright
Model: SGR-1820-F62 Cyclone
Type: 9-cylinder radial
Number: 2
Output: 670kW (900bhp)
PERFORMANCE
Maximum cruising speed: 370km/h (230mph) at 3050m (10,000ft)
Service ceiling: 7500m (24,600ft)
Maximum range: 1370km (850 miles) (with max. payload)
WEIGHTS
Empty: 4875kg (10,750lb)
Maximum take-off: 7940kg (17,500lb)
DIMENSIONS
Span: 20m (65.5ft)
Length: 13.5m (44ft)
Height: 3.5m (11.5ft)
Wing area: 51.2m² (551sq ft)

VEGA
Model/variant: Vega 5C
Manufacturer: Lockheed
Country of Origin: United States of America
Type: Short-range light passenger transport
Accommodation (typical): 7
Flight deck crew: 1
Prototype maiden flight: 4 July 1927 (as Vega 1)
POWERPLANT
Manufacturer: Pratt & Whitney
Model: Wasp
Type: 9-cylinder radial
Number: 1
Output: 335kW (450bhp)
PERFORMANCE
Maximum cruising speed: 265km/h (165mph) at 3050m (10,000ft)
Service ceiling: 5,500m (16,400ft)
Maximum range: 885km (550 miles)
WEIGHTS
Empty: 1165kg (2565lb)
Maximum take-off: 2155kg (4750lb)
DIMENSIONS
Span: 12.5m (41ft)
Length: 8.4m (27.5ft)
Height: 2.6m (8.5ft)
Wing area: 25.6m² (275sq ft)

DC-8
Model/variant: Super 71
Manufacturer: McDonnell-Douglas
Country of Origin: United States of America
Type: Long-range passenger/cargo transport
Accommodation (typical): 259 (one-class)
Flight deck crew: 2/3
Prototype maiden flight: 30 May 1958 (as DC-8)
POWERPLANT
Manufacturer: CFM International
Model: CFM56-2-1C
Type: Turbofan

Number: 4
Output: 106kN (10,890kg; 24,000lb) thrust
PERFORMANCE
Maximum cruising speed: 855km/h (530mph) at 10,650m (35,000ft)
Service ceiling: 13,800m (42,000ft)
Maximum range: 7500km (4660 miles)
WEIGHTS
Empty: 73,800kg (162,700lb)
Maximum take-off: 147,415kg (325,000lb)
DIMENSIONS
Span: 45.3m (148.5ft)
Length: 57.1m (187.5ft)
Height: 12.9m (42.5ft)
Wing area: 272m² (2927sq ft)

DC-10
Model/variant: Series 30
Manufacturer: McDonnell-Douglas
Country of Origin: United States of America
Type: Long-range passenger/cargo transport
Accommodation (typical): 380
Flight deck crew: 2/3
Prototype maiden flight: 29 August 1970 (as DC-10-10)
POWERPLANT
Manufacturer: General Electric
Model: CF6-50C
Type: Turbofan
Number: 3
Output: 227kN (23,135kg; 51,000lb) thrust
PERFORMANCE
Maximum cruising speed: 908km/h (564mph) at 9150m (30,000ft)
Service ceiling: 10,180m (33,400ft)
Maximum range: 7410km (4600 miles) (with max. payload)
WEIGHTS
Empty: 121,200kg (267,200lb)
Maximum take-off: 263,100kg (580,000lb)
DIMENSIONS
Span: 50.4m (165.3ft)
Length: 55.5m (182ft)
Height: 17.7m (58ft)
Wing area: 368m² (3960sq ft)

MD-11
Model/variant: MD-11P
Manufacturer: McDonnell-Douglas
Country of Origin: United States of America
Type: Long-range passenger transport
Accommodation (typical): 405 (one-class)
Flight deck crew: 2
Prototype maiden flight: 10 January 1990
POWERPLANT
Manufacturer: General Electric or Pratt & Whitney
Model: CF6-80C2D1F or PW4460
Type: Turbofan
Number: 3
Output: 273.5kN (27,955kg; 61,500lb) or 257kN (27,270kg; 60,000lb) thrust
PERFORMANCE
Maximum cruising speed: 930km/h (580mph) at 9150m (30,000ft)
Service ceiling: 10,200m (33,500ft)
Maximum range: 9270km (5260 miles)
WEIGHTS
Empty: 125,870kg (277,500lb)
Maximum take-off: 273,300kg (602,500lb)
DIMENSIONS
Span: 51.6m (169.5ft)
Length: 61.2m (200.9ft)
Height: 17.6m (57.75ft)
Wing area: 339m² (3650sq ft)

MD-80
Model/variant: MD-81
Manufacturer: McDonnell-Douglas [Shanghai Aviation, China]
Country of Origin: United States of America
Type: Medium-range passenger/cargo transport
Accommodation (typical): 137
Flight deck crew: 2/3
Prototype maiden flight: 19 October 1979

POWERPLANT
Manufacturer: Pratt & Whitney
Model: JT8D-209A
Type: Turbofan
Number: 2
Output: 82kN (8410kg; 18,500lb) thrust
PERFORMANCE
Maximum cruising speed: 850km/h (530mph) at 10,650m (35,000ft)
Service ceiling: 13,900m (42,000ft)
Maximum range: 2900km (1800 miles)
WEIGHTS
Empty: 35,630kg (78,550lb)
Maximum take-off: 63,000kg (138,600lb)
DIMENSIONS
Span: 32.8m (107.9ft)
Length: 45m (147.9ft)
Height: 9m (29.7ft)
Wing area: 118m² (1270ft)

MD-80
Model/variant: MD-82
Manufacturer: McDonnell-Douglas [Shanghai Aviation, China]
Country of Origin: United States of America
Type: Medium-range passenger/cargo transport
Accommodation (typical): 137
Flight deck crew: 2/3
Prototype maiden flight: 8 January 1981
POWERPLANT
Manufacturer: Pratt & Whitney
Model: JT8D-217A
Type: Turbofan
Number: 2
Output: 89kN (9070kg; 20,000lb) thrust
PERFORMANCE
Maximum cruising speed: 850km/h (530mph) at 10,650m (35,000ft)
Service ceiling: 13,900m (42,000ft)
Maximum range: 3800km (2360 miles)
WEIGHTS
Empty: 35,630kg (78,550lb)
Maximum take-off: 67,900kg (149,500lb)
DIMENSIONS
Span: 32.8m (107.9ft)
Length: 45m (147.9ft)
Height: 9m (29.7ft)
Wing area: 118m² (1270ft)

MD-80
Model/variant: MD-87
Manufacturer: McDonnell-Douglas [Shanghai Aviation, China]
Country of Origin: United States of America
Type: Medium-range passenger/cargo transport
Accommodation (typical): 112 (two-class)
Flight deck crew: 2/3
Prototype maiden flight: 4 December 1986
POWERPLANT
Manufacturer: Pratt & Whitney
Model: JT8D-200
Type: Turbofan
Number: 2
Output: 89kN (9070kg; 20,000lb) thrust
PERFORMANCE
Maximum cruising speed: 850km/h (530mph) at 10,650m (35,000ft)
Service ceiling: 13,900m (42,000ft)
Maximum range: 3800km (2360 miles)
WEIGHTS
Empty: 33,630kg (74,140lb)
Maximum take-off: 63,500kg (140,000lb)
DIMENSIONS
Span: 32.8m (107.9ft)
Length: 39.75m (130.5ft)
Height: 9m (29.7ft)
Wing area: 118m² (1270ft)

MARTIN 2-0-2
Model/variant: -
Manufacturer: Martin
Country of Origin: United States of America
Type: Short/medium-range passenger/cargo transport
Accommodation: 36
Flight deck crew: 2
Prototype maiden flight: 22 November 1946
POWERPLANT
Manufacturer: Pratt & Whitney
Model: R-2800-CB16 Double Wasp
Type: 18-cylinder radial
Number: 3
Output: 1790kW (2400bhp)
PERFORMANCE
Maximum cruising speed: 445km/h (275mph) at 6100m (20,000ft)
Service ceiling: 6900m (22,600ft)
Maximum range: 2510km (1550 miles)
WEIGHTS
Empty: 11,380kg (25,100lb)
Maximum take-off: 18,100kg (39,900lb)
DIMENSIONS
Span: 28.4m (93.25ft)
Length: 21.7m (71.3ft)
Height: 8.6m (28.5ft)
Wing area: 80.2m² (864sq ft)

MARTIN 4-0-4
Model/variant: -
Manufacturer: Martin
Country of Origin: United States of America
Type: Short/medium-range passenger/cargo transport
Accommodation: 40
Flight deck crew: 2
Prototype maiden flight: 20 June 1947 (as Martin 3-0-3)
POWERPLANT
Manufacturer: Pratt & Whitney
Model: R-2800-CB16 Double Wasp
Type: 18-cylinder radial
Number: 3
Output: 1790kW (2400bhp)
PERFORMANCE
Maximum cruising speed: 450km/h (280mph) at 6100m (20,000ft)
Service ceiling: 8900m (29,200ft)
Maximum range: 4185km (2600 miles)
WEIGHTS
Empty: 13,210kg (29,125lb)
Maximum take-off: 20,400kg (44,900lb)
DIMENSIONS
Span: 28.4m (93.25ft)
Length: 22.7m (74.5ft)
Height: 8.6m (28.5ft)
Wing area: 80.2m² (864sq ft)

MARQUISE
Model/variant: -
Manufacturer: Mitsubishi
Country of Origin: Japan
Type: Short/medium-range light passenger transport
Accommodation (typical): 7
Flight deck crew: 2
Prototype maiden flight: 14 September 1963
POWERPLANT
Manufacturer: Garrett (AlliedSignal)
Model: TPE331-10-501M
Type: Turboprop
Number: 2
Output: 533kW (715shp)
PERFORMANCE
Maximum cruising speed: 570km/h (355mph) at 4575m (15,000ft)
Service ceiling: 9070m (29,750ft)
Maximum range: 2580km (1600 miles)
WEIGHTS
Empty: 3470kg (7650lb)
Maximum take-off: 5250kg (11,550lb)
DIMENSIONS
Span: 11.95m (39.2ft)
Length: 12m (39.5ft)
Height: 4.2m (13.7ft)
Wing area: 16.5m² (178sq ft)

YS-11
Model/variant: YS-11A-200
Manufacturer: NAMC
Country of Origin: Japan
Type: Short/medium-range passenger/cargo transport
Accommodation (typical): 64
Flight deck crew: 2/3
Prototype maiden flight: 30 August 1962
POWERPLANT
Manufacturer: Rolls-Royce
Model: Dart Mk 542-10K
Type: Turboprop
Number: 2
Output: 2457kW (3060shp)
PERFORMANCE
Maximum cruising speed: 470km/h (290mph) at 4575m (15,000ft)
Service ceiling: 7000m (23,000ft)
Maximum range: 1100km (680 miles) (with max. payload)
WEIGHTS
Empty: 15,500kg (34,000lb)
Maximum take-off: 24,500kg (54,000lb)
DIMENSIONS
Span: 32m (105ft)
Length: 26.3m (86.3ft)
Height: 9m (29.5ft)
Wing area: 95m² (1021sq ft)

SABRELINER
Model/variant: Sabreliner 65
Manufacturer: North American/Rockwell
Country of Origin: United States of America
Type: Short/medium-range passenger transport
Accommodation (typical): 10
Flight deck crew: 2
Prototype maiden flight: 29 June 1977
POWERPLANT
Manufacturer: Garrett (AlliedSignal)
Model: TFE731-3-1D
Type: Turbofan
Number: 2
Output: 16.5kN (1680kg; 3700lb) thrust
PERFORMANCE
Maximum cruising speed: 900km/h (560mph) at 9150m (30,000ft)
Service ceiling: 13,715m (45,000ft)
Maximum range: 4450km (2765 miles)
WEIGHTS
Empty: 6420kg (14,155lb)
Maximum take-off: 10,900kg (24,000lb)
DIMENSIONS
Span: 15.4m (50.5ft)
Length: 14.3m (47ft)
Height: 4.9m (16ft)
Wing area: 35.3m² (380sq ft)

ARCHER II
Model/variant: PA-28-181
Manufacturer: Piper
Country of Origin: United States of America
Type: Light cabin monoplane
Accommodation: 3
Flight deck crew: 1
Prototype maiden flight: 14 January 1960 (as PA-28-150)
POWERPLANT
Manufacturer: Avco Lycoming (Textron)
Model: O-360-A4M
Type: 4-cylinder horizontally-opposed
Number: 1
Output: 134kW (180bhp)
PERFORMANCE
Maximum cruising speed: 220km/h (137mph) at 3050m (10,000ft)
Service ceiling: 4575m (15,000ft)
Maximum range: 1460km (900 miles)
WEIGHTS
Empty: 630kg (1390lb)

Maximum take-off: 1160kg (2550lb)
DIMENSIONS
Span: 10.7m (35ft)
Length: 7.3m (24ft)
Height: 2.2m (7.25ft)
Wing area: 15.8m² (170sq ft)

CHEROKEE 160
Model/variant: PA-28-160
Manufacturer: Piper
Country of Origin: United States of America
Type: Light cabin monoplane
Accommodation: 3
Flight deck crew: 1
Prototype maiden flight: 14 January 1960 (as PA-28-150)
POWERPLANT
Manufacturer: Avco Lycoming (Textron)
Model: O-320-B2B
Type: 4-cylinder horizontally-opposed
Number: 1
Output: 119kW (160bhp)
PERFORMANCE
Maximum cruising speed: 200km/h (125mph) at 3050m (10,000ft)
Service ceiling: 4575m (15,000ft)
Maximum range: 1300km (810 miles)
WEIGHTS
Empty: 550kg (1210lb)
Maximum take-off: 1000kg (2200lb)
DIMENSIONS
Span: 9.15m (30ft)
Length: 7.1m (23.3ft)
Height: 2.2m (7.25ft)
Wing area: 14.9m² (160sq ft)

CHEYENNE IIIA
Model/variant: PA-42
Manufacturer: Piper
Country of Origin: United States of America
Type: Short/medium-range light passenger transport
Accommodation (typical): 11 (high density)
Flight deck crew: 1/2
Prototype maiden flight: 1978
POWERPLANT
Manufacturer: Pratt & Whitney
Model: PT6A-61
Type: Turboprop
Number: 2
Output: 537kW (720bhp
PERFORMANCE
Maximum cruising speed: 565km/h (450mph) at 7620m (25,000ft)
Service ceiling: 9755m (32,000ft)
Maximum range: 4150km (2580 miles)
WEIGHTS
Empty: 2900kg (6390lb)
Maximum take-off: 5080kg (11,200lb)
DIMENSIONS
Span: 14.5m (47.7ft) (over wingtip tanks)
Length: 13.25m (43.4ft)
Height: 4.5m (14.75ft)
Wing area: 27.2m² (293sq ft)

DAKOTA
Model/variant: PA-28-236
Manufacturer: Piper
Country of Origin: United States of America
Type: Light cabin monoplane
Accommodation: 3
Flight deck crew: 1
Prototype maiden flight: 14 January 1960 (as PA-28-150)
POWERPLANT
Manufacturer: Avco Lycoming (Textron)
Model: O-540-J3A5D
Type: 6-cylinder horizontally-opposed
Number: 1
Output: 175kW (235bhp)
PERFORMANCE
Maximum cruising speed: 260km/h (160mph) at 3050m (10,000ft)
Service ceiling: 5350m (17,500ft)
Maximum range: 1500km (935 miles)

WEIGHTS
Empty: 730kg (1610lb)
Maximum take-off: 1360kg (3000lb)
DIMENSIONS
Span: 10.7m (35ft)
Length: 7.5m (24.5ft)
Height: 2.2m (7.25ft)
Wing area: 15.8m² (160sq ft)

TURBO ARROW IV
Model/variant: PA-28RT-201
Manufacturer: Piper
Country of Origin: United States of America
Type: Light cabin monoplane
Accommodation: 3
Flight deck crew: 1
Prototype maiden flight: 14 January 1960 (as PA-28-150)
POWERPLANT
Manufacturer: Teledyne-Continental
Model: TSIO-360-FB
Type: 6-cylinder horizontally-opposed
Number: 1
Output: 149kW (200bhp)
PERFORMANCE
Maximum cruising speed: 320km/h (200mph) at 3050m (10,000ft)
Service ceiling: 4575m (15,000ft)
Maximum range: 1300km (810 miles)
WEIGHTS
Empty: 785kg (1730lb)
Maximum take-off: 1315kg (2900lb)
DIMENSIONS
Span: 10.8m (35.5ft)
Length: 8.35m (27.5ft)
Height: 2.5m (8.25ft)
Wing area: 15.8m² (170sq ft)

PITTS SPECIAL
Model/variant: S-1S
Manufacturer: Pitts
Country of Origin: United States of America
Type: Specialist aerobatics and display
Accommodation (typical): -
Flight deck crew: 1
Prototype maiden flight: n/a
POWERPLANT
Manufacturer: Avco Lycoming (Textron)
Model: IO-360-B4A
Type: 4-cylinder horizontally-opposed
Number: 1
Output: 134kW (180bhp)
PERFORMANCE
Maximum speed: 255km/h (160mph)
Service ceiling: 6125m (20,100ft)
Maximum range: 550km (340 miles)
WEIGHTS
Empty: 455kg (1000lb)
Maximum take-off: 680kg (1500lb)
DIMENSIONS
Span: 5.3m (17.3ft)
Length: 4.7m (15.5ft)
Height: 1.94m (6.4ft)
Wing area: 9.15m² (98.5sq ft) (biplane)

PITTS SPECIAL
Model/variant: S-2B
Manufacturer: Pitts
Country of Origin: United States of America
Type: Specialist aerobatics and display
Accommodation (typical): 1
Flight deck crew: 1
Prototype maiden flight: n/a
POWERPLANT
Manufacturer: Avco Lycoming (Textron)
Model: AEIO-540-B4A
Type: 6-cylinder horizontally-opposed
Number: 1
Output: 194kW (260bhp)
PERFORMANCE
Maximum speed: 280km/h (175mph)
Service ceiling: 6125m (20,100ft)
Maximum range: 520km (320 miles)
WEIGHTS
Empty: 522kg (1150lb)
Maximum take-off: 740kg (1625lb)

DIMENSIONS
Span: 6.1m (20ft)
Length: 5.8m (19ft)
Height: 1.94m (6.4ft)
Wing area: 11.6m² (125sq ft) (biplane)

SAAB 340
Model/variant: S340B
Manufacturer: Saab
Country of Origin: Sweden
Type: Short/medium-range passenger transport
Accommodation (typical): 35
Flight deck crew: 2
Prototype maiden flight: 25 January 1983
POWERPLANT
Manufacturer: General Electric
Model: CT7-9B
Type: Turboprop
Number: 2
Output: 1495kW (1870shp)
PERFORMANCE
Maximum cruising speed: 522km/h (325mph) at 4575m (15,000ft)
Service ceiling: 7620m (25,000ft)
Maximum range: 1810km (1125 miles)
WEIGHTS
Empty: 8035kg (17,715lb)
Maximum take-off: 13,050kg (28,800lb)
DIMENSIONS
Span: 21.45m (70.3ft)
Length: 19.75m (64.7ft)
Height: 6.9m (22.5ft)
Wing area: 41.8m² (450sq ft)

SHORT S.20 'MERCURY'
Model/variant: -
Manufacturer: Short Bros
Country of Origin: Great Britain
Type: Long-range cargo transport floatplane
Accommodation (typical): -
Flight deck crew: 2
Prototype maiden flight: 6 February 1938 (first airborne separation)
POWERPLANT
Manufacturer: Napier
Model: Rapier VI
Type: in-line
Number: 4
Output: 255kW (340bhp)
PERFORMANCE
Maximum cruising speed: 340km/h (210mph) at 3660m (12,000ft)
Service ceiling: n/a
Maximum range: 9820km (6100 miles) (airborne 'launch')
WEIGHTS
Empty: 4615kg (10,165lb)
Maximum launch: 12,160 (26,800lb)
DIMENSIONS
Span: 22.2m (73ft)
Length: 15.5m (51ft)
Height: n/a
Wing area: 56.8m² (611sq ft)

SHORT S.21 'MAIA'
Model/variant: -
Manufacturer: Short Bros
Country of Origin: Great Britain
Type: Launch vehicle for S.20 'Mercury'
Accommodation (typical): -
Flight deck crew: 3
Prototype maiden flight: July 1936 (as S.23)
POWERPLANT
Manufacturer: Bristol
Model: Pegasus XC
Type: 9-cylinder radial
Number: 4
Output: 685kW (920bhp)
PERFORMANCE
Maximum cruising speed: 320km/h (200mph) at 3660m (12,000ft)
Service ceiling: 6100m (20,000ft)
Maximum range: 1225km (760 miles)

WEIGHTS
Empty: 11,235kg (24,745lb)
Maximum take-off: 24,740kg (54,430lb) (including S.20)
DIMENSIONS
Span: 34.7m (114ft)
Length: 25.9m (85ft)
Height: 9.7m (31.9ft) (solo)
Wing area: 162.5m² (1750sq ft)

SHORT 330

Model/variant: 330-200 (Sherpa - UK military designation; C-23 - US military designation)
Manufacturer: Short Bros
Country of Origin: Great Britain
Type: Short-range passenger/cargo utility transport
Accommodation (typical): 30
Flight deck crew: 2
Prototype maiden flight: 22 August 1974
POWERPLANT
Manufacturer: Pratt & Whitney Canada
Model: PT6A-45R
Type: Turboprop
Number: 2
Output: 895kW (1200shp)
PERFORMANCE
Maximum cruising speed: 350km/h (220mph) at 3050m (10,000ft)
Service ceiling: n/a
Maximum range: 875km (545 miles) (with max. payload)
WEIGHTS
Empty: 6680kg (15,730lb)
Maximum take-off: 10,400kg (22,900lb)
DIMENSIONS
Span: 22.8m (74.7ft)
Length: 17.7m (58ft)
Height: 4.95m (16.25ft)
Wing area: 42.1m² (453sq ft)

SHORT 360

Model/variant: 360-100
Manufacturer: Short Bros
Country of Origin: Great Britain
Type: Short-range passenger/cargo transport
Accommodation (typical): 36
Flight deck crew: 2
Prototype maiden flight: 1 June 1981
POWERPLANT
Manufacturer: Pratt & Whitney Canada
Model: PT6A-65R
Type: Turboprop
Number: 2
Output: 895kW (1200shp)
PERFORMANCE
Maximum cruising speed: 355 km/h (220mph) at 3050m (10,000ft)
Service ceiling: n/a
Maximum range: 1700km (1060 miles)
WEIGHTS
Empty: 7700kg (16,950lb)
Maximum take-off: 11,800kg (26,000lb)
DIMENSIONS
Span: 22.8m (74.9ft)
Length: 21.6m (70.9ft)
Height: 7.2m (23.7ft)
Wing area: 42m² (138sq ft)

SHORT 360

Model/variant: 360-300
Manufacturer: Short Bros
Country of Origin: Great Britain
Type: Short-range passenger/cargo transport
Accommodation (typical): 36
Flight deck crew: 2
Prototype maiden flight: February 1987
POWERPLANT
Manufacturer: Pratt & Whitney Canada
Model: PT6A-67R
Type: Turboprop
Number: 2
Output: 1060kW (1425shp)

PERFORMANCE
Maximum cruising speed: 400 km/h (250mph) at 3050m (10,000ft)
Service ceiling: n/a
Maximum range: 1600km (1000 miles)
WEIGHTS
Empty: 7900kg (17,380lb)
Maximum take-off: 12,300kg (27,000lb)
DIMENSIONS
Span: 22.8m (74.9ft)
Length: 21.6m (70.9ft)
Height: 7.2m (23.7ft)
Wing area: 42m² (138sq ft)

TU-104 'CAMEL'
(NATO REPORTING NAME)

Model/variant: Tu-104B
Manufacturer: Tupolev
Country of Origin: Soviet Union
Type: Medium-range passenger/cargo transport
Accommodation (typical): 100
Flight deck crew: 3/4
Prototype maiden flight: 17 June 1955
POWERPLANT
Manufacturer: Mikulin
Model: AM-3M-500
Type: Turbojet
Number: 2
Output: 95kN (9700kg; 21,385lb) thrust
PERFORMANCE
Maximum cruising speed: 820km/h (510mph) at 9150m (30,000ft)
Service ceiling: 11,500m (37,750ft)
Maximum range: 2650km (1650 miles) (with max. payload)
WEIGHTS
Empty: 41600kg (91,710lb)
Maximum take-off: 76,000kg (167,550lb)
DIMENSIONS
Span: 34.5m (113.3ft)
Length: 40m (131.3ft)
Height: 11.9m (39ft)
Wing area: 183.5m (1975sq ft)

TU-114 'CLEAT'
(NATO REPORTING NAME)

Model/variant: -
Manufacturer: Tupolev
Country of Origin: Soviet Union
Type: Long-range passenger/cargo transport
Accommodation (typical): 170
Flight deck crew: 5
Prototype maiden flight: 3 October 1957
POWERPLANT
Manufacturer: Kuznetsov
Model: NK-12MV
Type: Turboprop (counter-rotating propellers)
Number: 4
Output: 11,035kW (14,795shp)
PERFORMANCE
Maximum cruising speed: 770km/h (480mph) at 9000m (29,500ft)
Service ceiling: 12,000m (39,370ft)
Maximum range: 8950km (5560 miles) (with 15,000kg (33,000lb) payload)
WEIGHTS
Empty: 91,000kg (200,000lb)
Maximum take-off: 171,000kg (377,000lb)
DIMENSIONS
Span: 51.1m (167.5ft)
Length: 54.1m (177.3ft)
Height: 15.5m (50.9ft)
Wing area: 311m² (3349sq ft)

TU-134 'CRUSTY'
(NATO REPORTING NAME)

Model/variant: Tu-134A
Manufacturer: Tupolev
Country of Origin: Soviet Union
Type: Short/medium-range passenger/cargo transport
Accommodation (typical): 72 (two-class)
Flight deck crew: 3/4
Prototype maiden flight: 1962

POWERPLANT
Manufacturer: Soloviev
Model: Soloviev
Type: D-30-II
Number: Turbofan
Output: 30.25kN (3085kg; 6800lb) thrust
PERFORMANCE
Maximum cruising speed: 885km/h (550mph) at 9150m (30,000ft)
Service ceiling: 11,900m (39,050ft)
Maximum range: 1890km (1175 miles) (with max. payload)
WEIGHTS
Empty: 29,050kg (64,050lb)
Maximum take-off: 47,000kg (103,620lb)
DIMENSIONS
Span: 29m (95.1ft)
Length: 37m (121.5ft)
Height: 9.15m (30ft)
Wing area: 127m² (1370sq ft)

TU-144 'CHARGER'
(NATO REPORTING NAME)

Model/variant: -
Manufacturer: Tupolev
Country of Origin: Soviet Union
Type: Long-range supersonic passenger transport
Accommodation (typical): 140
Flight deck crew: 3
Prototype maiden flight: 31 December 1968
POWERPLANT
Manufacturer: Kuznetsov
Model: NK-144
Type: Turbofan
Number: 4
Output: 196kN (20,000kg; 44,000lb) with afterburning
PERFORMANCE
Maximum cruising speed: 2500km/h (1555mph) at 18,000m (59,000ft)
Service ceiling: 18,000m (59,000ft)
Maximum range: 6500km (4040 miles) (with max. payload)
WEIGHTS
Empty: 85,000kg (187,400lb)
Maximum take-off: 180,000kg (396,830lb)
DIMENSIONS
Span: 28.8m (94.5ft)
Length: 65.7m (215.5ft)
Height: 12.85m (42.1ft)
Wing area: 438m² (4715sq ft)

TU-154 'CARELESS'
(NATO REPORTING NAME)

Model/variant: Tu-154M
Manufacturer: Tupolev
Country of Origin: Soviet Union
Type: Medium-range passenger/cargo transport
Accommodation (typical): 180 (high density)
Flight deck crew: 3
Prototype maiden flight: 1971
POWERPLANT
Manufacturer: Kuznetsov
Model: NK-8-2U
Type: Turbofan
Number: 3
Output: 102.8kN (10,500kg; 23,150lb) thrust
PERFORMANCE
Maximum cruising speed: 900km/h (560mph) at 12,000m (39,370ft)
Service ceiling: 13,000m (42,650ft)
Maximum range: 2750km (1700 miles)
WEIGHTS
Empty: 50,775kg (111,950lb)
Maximum take-off: 94,000kg (207,235lb)
DIMENSIONS
Span: 37.5m (123.2ft)
Length: 47.9m (157.1ft)
Height: 11.4m (37.5ft)
Wing area: 201.5m² (2169sq ft)

SUPER VC10
Model/variant: Type 1151
Manufacturer: Vickers
Country of Origin: Great Britain
Type: Long-range passenger/cargo transport
Accommodation: 139 (two-class)
Flight deck crew: 3
Prototype maiden flight: 29 June 1962 (as Type 1100 VC10)
POWERPLANT
Manufacturer: Rolls Royce
Model: Conway Mk 550
Type: Turbofan
Number: 4
Output: 97kN (9905kg; 21,800lb) thrust
PERFORMANCE
Maximum cruising speed: 935km/h (580mph) at 9450m (31,000ft)
Service ceiling: 11,600m (38,000ft)
Maximum range: 7600km (4725 miles) (with max. payload)
WEIGHTS
Empty: 71,940kg (158,600lb)
Maximum take-off: 152,000kg (335,000lb)
DIMENSIONS
Span: 55.6m (146.25ft)
Length: 52.3m (171.7ft)
Height: 12m (39.5ft)
Wing area: n/a

VANGUARD
Model/variant: Type 952
Manufacturer: Vickers
Country of Origin: Great Britain
Type: Short/medium-range passenger/cargo transport
Accommodation (typical): 139
Flight deck crew: 2/3
Prototype maiden flight: 20 January 1959 (as Type 950)
POWERPLANT
Manufacturer: Rolls-Royce
Model: Tyne Mk 512
Type: Turboprop
Number: 4
Output: 4135kW (5545shp)
PERFORMANCE
Maximum cruising speed: 685km/h (425mph) at 6100m (20,000ft)
Service ceiling: 9150m (30,000ft)
Maximum range: 2945km (1850 miles) (with max. payload)
WEIGHTS
Empty: 37,420kg (82,500lb)
Maximum take-off: 64,000kg (141,000lb)
DIMENSIONS
Span: 36.1m (118.5ft)
Length: 37.45m (121.9ft)
Height: 10.65m (35ft)
Wing area: 142m^2 (1527sq ft)

VIKING
Model/variant: Type 491 Viking 1B
Manufacturer: Vickers
Country of Origin: Great Britain
Type: Short-range passenger/cargo transport
Accommodation: 36 (high density)
Flight deck crew: 2/3
Prototype maiden flight: 22 June 1945 (as Viking VC1)
POWERPLANT
Manufacturer: Bristol
Model: Hercules
Type: 14-cylinder radial
Number: 2
Output: 1260kW (1690bhp)
PERFORMANCE
Maximum cruising speed: 340km/h (210mph) at 1850m (6000ft)
Service ceiling: 7250m (23,750ft)
Maximum range: 840km (520 miles)
WEIGHTS
Empty: 10,550kg (23,250lb)
Maximum take-off: 15,350kg (34,000lb)
DIMENSIONS
Span: 27.2m (89.25ft)
Length: 19.9m (65.2ft)
Height: 5.95m (19.5ft)
Wing area: 82m^2 (882sq ft)

VISCOUNT
Model/variant: Type 700
Manufacturer: Vickers
Country of Origin: Great Britain
Type: Short/medium-range passenger/cargo transport
Accommodation (typical): 56
Flight deck crew: 2/3
Prototype maiden flight: 16 July 1948 (as Type 630)
POWERPLANT
Manufacturer: Rolls-Royce
Model: Dart Mk 506
Type: Turboprop
Number: 4
Output: 1045kW (1400shp)
PERFORMANCE
Maximum cruising speed: 500km/h (310mph) at 6100m (20,000ft)
Service ceiling: 7250m (23,750ft)
Maximum range: 1510km (940 miles) (with max. payload)
WEIGHTS
Empty: 14,665kg (32,330lb)
Maximum take-off: 23,810kg (52,500lb)
DIMENSIONS
Span: 28.6m (93.75ft)
Length: 25m (82ft)
Height: 8.1m (26.75ft)
Wing area: 89.5m^2 (963sq ft)

INDEX

A-144 'Analog' 746
A.D.C. Cirrus I *see* de Havilland D.H.60
Aeritalia AIT 320 89
  see also ATR
Aero Commander
  see also IAI Jet Commander
  Ag Commander S2D 105, 106
  Jet Commander 264, 537–8
Aero Spacelines/Conroy Guppy 229, 305–8
  Guppy 101 229, 307
  Guppy 201 229, 305, 307, 308
  Mini Guppy 229, 306, 307
  Pregnant Guppy 229, 306, 307
  Super Guppy 229, 306
AeroMagic 694
Aéronautic Baroudeur 646
Aérospatiale
  see also ATR; British Aerospace; Sud Aviation; Airbus Industrie
  AS-35 89
  Caravelle
    11 17, 761
    10A 12, 14, 17
    10B1.R 14, 17
    10B3 13, 17
    10B 'Super Caravelle' 141–5, 17, 300
    11R 14, 17
    12 14, 17
    12/58T 15
    I 12, 13, 17
    IA 13, 17
    III 12, 13, 15, 161–7
    SE 210 11–12
    VI 14
    VII 14, 17
    VIN 14, 17
    VIR 12, 13, 14, 17
Air Express *see* Lockheed Air Express
Air Tractor
  AT-301 107
  AT-502 105
Airbus Industrie
  see also Aérospatiale; British Aerospace; Sud Aviation
  A300 19–24, 20–9
  A300-600 21, 23–4, 28, 31
  A300-600C 21
  A300-600ER 21
  A300-600F 21
  A300-600R 20–1, 23, 24
  A300-605R 24
  A300-622R 24
  A300B1 20, 21
  A300B2 22, 23
  A300B2-100 21
  A300B2-200 21
  A300B2-201 21
  A300B2-220 21
  A300B2-300 21
  A300B2-400 19
  A300B2-600 21
  A300B4 19, 22
  A300B4-100 21, 22, 23

A300B4-101 21
A300B4-120 21
A300B4-200 19, 20, 21, 22, 23
A300B4-600 21
A300B9 43
A300B11 43
A300B 20, 22–3, 26–7, 43, 201
A300C4 21, 22
A300F4 21
A310 26–31, 34, 202, 211, 424, 520, 535
A310-200 29–30
A310-200C 30
A310-200F 30
A310-203 31
A310-221 30
A310-222 27, 31
A310-300 30–1
A310-304 27, 28, 30, 31
A310-322 30
A310-324 27
A320 35–41, 171, 202, 205, 210
A320-100 36, 41
A320-110 38–9
A320-200 40, 41
A321 36, 41, 176
A330 43–6, 216
A340 43–6, 216
A340-200 43, 45
A340-300 45–6
SA-1 35
SA-2 35
Skylink Guppies 307–8
Super Airbus Transporter (SAT) 24
Airco D.H.121 489
Aircraft Manufacturing Company *see* Airco
Airspeed Ambassador 771
Alenia *see* Aeritalia
Allison 'Prop Jet Super Convair' 312, 313
Altair *see* Lockheed Altair
American Aerolights Eagle 645
American Special 697
Andover
  CC.Mk 2 97, 98, 101
  C.Mk 1 97, 98, 101, 484
  E.Mk 3 97, 101
Annushka *see* Antonov An-2 'Colt'
Antonov
  An-2 'Colt' 47–51
  An-2 LALA-1 49
  An-2F Fedya 49–50
  An-2K 49–50
  An-2L 49
  An-2M 49, 50
  An-2NRK 49–50
  An-2P 48, 50
  An-2PP 49
  An-2PT 49
  An-2SKh 48–9
  An-2T 48, 49
  An-2TD 48, 49
  An-2V 49
  An-2ZA 48, 49
  An-3 50, 51

An-4 49
An-6 49
An-8 53
An-10 53, 61
An-12 'Cub' 53–60, 515, 528, 533
An-12 'Cub-A' 54, 56
An-12 'Cub-B' 56, 60
An-12 'Cub-C' 56, 60
An-12 'Cub-D' 56, 60
An-12B 'Cub' 54
An-12BP 'Cub' 53–4, 56–7, 60, 523
An-24 61–5, 73–6, 80, 741
An-24P 74, 76
An-24RT 74, 76
An-24RV 61, 62, 63, 74, 76
An-24T 63, 74, 76
An-24TV 63
An-24V Series I 62, 63
An-24V Series II 63, 64–5, 74–5, 76
An-26B 'Curl-A' 64, 74, 77
An-26 'Curl' 63–4, 73, 74, 76–7, 80
An-28 50
An-30 'Clank' 63, 64, 74, 77, 80
An-30M 'Sky Cleaner' 77
An-32 'Cline' 64, 65, 74, 77
An-124 67–72
An-225 67–72
An-235 194
An-322 64
Apollo *see* Armstrong Whitworth A.W.55
Araguaya *see* EMBRAER EMB-120
Argosy *see* Armstrong Whitworth Argosy
Armstrong Whitworth Argosy 81–7
  Argosy 100 82, 84
  Argosy 101 87
  Argosy 102 86, 87
  Argosy C.Mk 1 83, 85, 86
  Argosy E.Mk 1 86
  Argosy Mk 200 87
  Argosy Mk 220 81
  Argosy Mk 222 81, 84, 87
  Argosy T.Mk 2 86, 87
A.W.55 Apollo 771, 772
Army Type 92 Super Heavy Bomber 558
Ascender II+ 643
Astra *see* Jet Commander 1125
AT-7 Navigator 127, 130
AT-11 Kansan 126, 127, 130
ATM 42L 901
ATP *see* British Aerospace Advanced Turboprop
ATR
  42 89–94, 336, 339
  42-100 89
  42-200 89, 90
  42-300 90, 92, 94, 636
  42F 90
  52C 94
  72 91–4
  72-201 92
  72-210 94
  XX *see* ATR 72
ATR/Dassault-Breguet Petrel 42/72 93–4

Avia
   14 508, 509
   14 Salon 509
   14-32A 509
   32 509
   F.IX 443
Aviasud Sirocco 646–7
Aviation Traders Carvair 412, 413, 481
Aviolanda
   Do 24K-2 400
   Wal 398
Avionette *see* Avions Nogrady AN-2
Avions de Transport Régional *see* ATR
Avions Marcel Dassault *see* Dassault
Avions Nogrady AN-2 Avionette 646
Avro
   642 443
   740 489
   748 96–7, 666
   748 Series 2 97–8
   780 *see* Avro/Hawker Siddeley H.S.748 MF
   ATP 545
   Lancastrian VM703 367
   Ten 445
Avro/Hawker Siddeley
   H.S.748 96–101, 714
   H.S.748 Coastguarder 99
   H.S.748 Series 2A 98, 101
   H.S.748 Series 2B 99
   H.S.748 Super 2B 96
   H.S.748-234 101
   H.S.748-256 101
   H.S.748MF 98, 484
Ayres
   Leo Thrush 110
   S2R-600 110
   S2R-R3S 107–10
   S2R-R1820/510 Bull Thrush 110
   S2R-T65/400 Turbo Thrush NEDS 110
   S2R-T Turbo Thrush 106, 107, 110
   Thrush 105–10
   Turbo Sea Thrush 110
   V-1-A Vigilante 110

B-17 *see* Boeing B-17 Flying Fortress
BAC One-Eleven 111–15, 163, 453–4, 541, 580, 609–10, 742–3
   see also British Aerospace; Vickers
   Freighter 114, 115
   Series 200 112, 113
   Series 300 112, 113
   Series 400 112–13, 115
   Series 401 112–13, 261
   Series 402 115
   Series 475 114, 115
   Series 479 FU 115
   Series 487 115
   Series 500 111, 113, 114, 115
   Series 525 115
   Series 537 112
   Series 670 112–13
   'Super One-Elevens' 113–14
Bandeirante *see* EMBRAER Bandeirante
Bartini T-117 507
Beaver *see* de Havilland DHC-2
Beech
   99 543, 709
   AT-7 Navigator 127, 130
   AT-11 Kansan 126, 127, 130
   C-45 Expeditor 119, 127, 130
   JRB-1 Voyager 126
   JRB-3 127
   JRB-4 127
   King Air 119–23
   King Air 90 119, 122
   King Air 100 120
   King Air 200 512, 709
   King Air 200T 121–2
   King Air A90 119
   King Air A100 120–1
   King Air B90 119
   King Air B100 121
   King Air C90 120
   King Air C90-1 120
   King Air C90A 119, 120
   King Air E90 120, 121
   King Air F90 120, 121
   King Air F90-1 120
   King Air Maritime Patrol 200T 122
   King Air Model 1300 Commuter 122
   Model 17 'Staggerwing' 125
   Model 18 125–30
   Model 18B 125–6
   Model 18D 126
   Model 18S 126
   Model 50 Twin Bonanza 119
   Model 87 119
   Model 120 119
   Model 400A Beechjet 264
   Model B18S 126–7
   Model C18C 126
   Model C18S 127
   Model D18C/CT 130
   Model D18S 127–30
   Model E18S Super 18 125, 128–9, 130, 432
   Model G18S 130
   Model S-18A 125
   Queen Air 80 119
   SNBs 127, 130
   Super H18 130
   Super King Air 200 120, 121, 122
   Super King Air 300 122
   Super King Air 350 122, 123
   Super King Air 350C 123
   Super King Air B200 121, 122
   U-21 355
Beechcraft-Hawker BH 125 250
Bird Dog *see* Cessna Model 305A
'Bison' *see* Myasishchyev M-4
Black Magic 695
BN-2 Islander *see* Britten-Norman BN-2 Islander
BN-2A Mk III Trislander *see* Britten-Norman BN-2A Mk III Trislander
BN-2B Islander *see* Pilatus Britten-Norman BN-2B Islander
BN-2T Turbine Islander *see* Pilatus Britten-Norman BN-2T Turbine Islander
Boeing
   7J7 41, 44
   7N7 201, 209, 210
   7X7 209, 210
   247 131–4, 155, 220–1, 324, 404, 477
   247A 132
   247D 132–4
   247E 132, 133
   247Y 133, 134
   307 Stratoliner 135–9, 414, 478, 571, 572
   367 Stratocruiser 155, 224
   367-80 149, 152, 155–6, 423
   377 Stratocruiser 155, 209, 224–31, 270, 305–6, 511
   707 13, 148–54, 156, 199, 209, 261, 315, 423–4, 774
   707-3J9C 154
   707-020 156
   707-120 149, 152, 156, 423, 424
   707-120B 152, 157
   707-121 149, 152, 154
   707-123 149, 154
   707-138 149, 152, 156
   707-153 153
   707-220 149, 152, 155, 156
   707-320 Intercontinental 149, 152, 154, 156, 423, 424
   707-320B 152, 154, 156, 430
   707-320C 152–3, 156
   707-338C 153
   707-351C 152
   707-353B 154
   707-358C 149
   707-420 152, 154, 157, 370
   707-465 148
   717 156, 711
   720 152, 155–8, 317–18, 423
   720-022 158
   720-023B 158
   720-040B 158
   720-047B 157, 158
   720-051B 157, 158
   720-058B 157
   720-059B 157
   720-060B 157
   720-061 157, 158
   720B 152, 157
   727 157, 159–62, 163, 201, 209, 489, 580, 609–10, 753
   727-2B6 161
   727-100 160, 162
   727-100 Business Jet 162
   727-100C 162
   727-100QC 162
   727-113C 162
   727-200 160, 162
   727-200F 162
   727-212 161
   727-224 161
   727-243 161
   727-256 159, 162
   727-277 159
   727-300 201
   737 35, 157, 163–75, 209, 261, 610, 719
   737 'Next Generation' 176–81
   737 Surveiller 165
   737-100 163, 164, 168
   737-130 173
   737-200 40, 163, 164, 165, 166–7, 168–9, 610
   737-230 173
   737-298C Advanced 164
   737-300 165–6, 169–71, 172, 173
   737-330 173
   737-400 166, 171–2, 173
   737-500 166, 172–3
   737-530 173
   737-600 176–7, 180–1
   737-700 176, 177, 178, 179
   737-800 176, 177, 178–9
   737-900 176, 177–8
   737X 176
   747 19, 164, 183–9, 191–9, 209, 531, 719
   747-100 184, 186, 187
   747-100B 184, 186
   747-200 184, 186–7
   747-200B 184, 185, 186, 619
   747-200B Combi 184, 186
   747-200C Convertible 184, 186
   747-200F 184, 185, 186
   747-300 (SUD) 184, 185, 186, 191, 619
   747-400 184, 185, 186, 191–9, 619
   747-400F 195

747-400M Combi 195
747SP 184–5, 186, 187, 261, 615
747SR 184, 186
747SUD see Boeing 747-300
757 201–8, 719
757 Corporate 77-52 205
757-2M6 202
757-200 202–3, 204, 210
757-200 Combi 205
757-200PF 205
767 192, 201, 204, 209–15
767 AOA 211–12
767-200 210
767-200ER 210
767-238ER 209
767-300 211
767-300ER 211, 214–15
767-336 210
767-338ER 209
767-400 214, 223
767-X 214
767ER 214
777 176, 216–23, 719
777-200 218–19, 222
777-200 A Market 219, 222
777-200 B Market 219, 222
777-200ER 222
777-200X 222–3
777-300 222
777-300X 222–3
Advanced 727-200 160–116-2
Advanced 737 164, 611
B-9 bomber 131
B-17 Flying Fortress 135, 136, 141, 155
B-29 Superfortress 155, 224, 289, 734, 741
B-47 148, 155, 163, 228, 730
B-52 bomber 148, 155, 228
Business Jet 178
C-73 134
C-75 Stratoliner 137, 138, 139
C-97 Stratofreighter 155, 225, 226, 228, 305–6
C-97A Stratofreighter 225
C-98 Clipper 143
C-135 149
Clipper 141–6
E-3A Sentry 576, 740
KC-97 Stratofreighter 148, 270
KC-97A Stratofreighter 228
KC-97E Stratofreighter 228–9
KC-97G Stratofreighter 226, 227, 228, 229
KC-97L Stratofreighter 224, 226, 229
KC-130 Stratofreighter 226
KC-135 Stratofreighter 149, 156, 158, 209
KC-135A Stratofreighter 224, 229
KE-3A 152
Model 40 155
Model 80 131
Model 200 Monomail 131, 155
Model 299 see Boeing B-17 Flying Fortress
Model 314 see Boeing Clipper
SA-307B 137
SA-307B-1 138
T-43A 164, 166
VC-137A 152, 153
VC-137B 152, 153
VC-137C 152, 154
XC-97 Stratofreighter 224, 225
YC-97 Stratofreighter 224
YC-97A Stratofreighter 224–5
YC-97B Stratofreighter 224, 225, 226
YC-97J Stratofreighter 226, 229
Bristol

see also BAC
200 489
Bombay 709
Britannia 233–7, 511, 541
101 234
102 234, 235
200 234
250 234
300 234
301 235
302 235
305 235
310 234
312 234, 235, 236–7
313 234, 235
314 234
318 235
C.Mk 1 233
C.Mk 2 233
Type 170 239–40, 241, 242, 244
Type 175 233–4, 265
Type 187 235–6
Freighter 239–45
Freighter Mk 21 240–1, 245
Freighter Mk 21A 240–1
Freighter Mk 21E 241
Freighter Mk 21P 241
Freighter Mk 31 240, 242, 243
Freighter Mk 31M 239, 242
Freighter Mk 32 242–3, 245
Freighter XI 240, 241
Type 198 300
Wayfarer 239, 240, 242
British Aerospace (BAe)
see also BAC
125 247–51, 376
125 Series 1B 248
125 Series 700 247, 251
125 Series 800 248–9, 250–1, 263–4, 707
146 101, 202, 253–9, 336, 339, 454, 711
146 CC.Mk 1 254
146 CC.Mk 2 254
146 Statesman 257
146-100 254–5, 256–7
146-200 257
146-300 257
146-350 257
146-QT 253, 257
146M 257
146MSL 257
146MT 257
146STA 257
748 Coastguarder 99, 544
1000 251
Advanced Turboprop (ATP) 99, 452
Corporate Jetstream 544
Jetstream 31 541, 542, 543, 544, 545
Jetstream 31EZ 544
Jetstream 41 544–5
Jetstream 61(ATP) 545
Jetstream Super 31 544
Jetstream T.Mk 3 544
Super 748 99, 100
British Aerospace (BAe)/Aérospatiale Concorde 297–303
British Aerospace Corporation see BAC
Britten-Norman
see also Fairey-Britten-Norman; Pilatus Britten-Norman
BN-2 Islander 365, 501–6, 640
BN-2A Islander II 502, 503
BN-2A Mk III Trislander 503–6, 716

BN-2A Mk III-2 Trislander 505
BN-2A Mk III-3 Trislander 505
BN-2A Mk III-4 Trislander 505
Islander Super 503
Broussard 432
Bull Thrush see Ayres S2RR1820/510
business jets 261–4
Butterfly 646

C-2A Fokker VII 443
C-3A Ford Tri-Motor 465
C-4A Ford Tri-Motor 465
C-9 Nightingale 178
C-9A Douglas DC-9 (Nightingale) 611
C-9B Douglas DC-9 (Skytrain II) 611
C-9C Douglas DC-9 611
C-10A Jetstream 543
C-12 Super King Air 121
C-20 Gulfstream 470–1
C-21A Learjet 562, 563
C-23A/B Shorts Sherpa 710, 711–13, 714
C-26A Metro III 641
C-29A BAe 125 257
C-32A Douglas DC-2 407
C-33 Douglas DC-2 407
C-34 Cessna 290
C-36 Lockheed Electra 598
C-39 Douglas DC-2 407
C-40B Lockheed Electra 598, 599
C-41 Douglas DC-2 407
C-42 Douglas DC-2 407
C-45 Expeditor 119, 127, 130
C-45F Beech Model 1 8 127
C-54 Skymaster 409–12, 414
C-54A Skymaster 409, 410
C-54B Skymaster 409
C-54C Skymaster 409
C-54D Skymaster 410, 411
C-54E Skymaster 410
C-54G Skymaster 410, 412
C-54P Skymaster 412
C-54Q Skymaster 411
C-56 Lockheed Lodestar 601
C-57 Lockheed Lodestar 601
C-59 Lockheed Lodestar 601
C-60A Lockheed Lodestar 601
C-69 Lockheed Constellation 571–2, 576
C-73 Boeing 247 134
C-75 Boeing Stratoliner 137, 138, 139
C-95 Bandeirante 432–3, 434, 440
C-95A Bandeirante 433, 434, 439
C-95B Bandeirante 434
C-97 Stratofreighter 155, 225, 226, 228, 305–6
C-97A Stratofreighter 225
C-98 Boeing Clipper 143
C-118 Douglas DC-6 Liftmaster 414
C-118A Douglas DC-6 415, 416, 419
C-118B Douglas DC-6 415, 417, 419
C-121 Lockheed Constellation 572, 573
 C-121A Lockheed Constellation 571, 576
 C-121C Lockheed Constellation 573, 576
 C-121G Lockheed Constellation 573, 576
C-121J Lockheed Constellation 576
C-131A Convair (Samaritan) 310
C-131B Convair 311, 313
C-131F Convair 313
C-135 Boeing 707 149
Cammacorp/McDonnell Douglas DC-8 Super 71 424–5, 427, 430, 431

807

Canadair
  C-5 412
  Challenger 261, 262
  CL-2 DC-4M North Star 411, 412
  CL-28 Argus 236, 265
  CL-44 265–70
  CL-44-6 26–5, 266
  CL-44-O 268, 270, 307
  CL-44D4 266–70
  CL-44D 235, 236
  CL-44G 270
  CL-44J 235, 236, 267, 270
  CL-44O 236, 237
  CL-66B 312
  CL-66C 312
  CL-215 271–6
  CL-215T 275, 276
  CL-600 Challenger 262, 277–81
  CL-601 Challenger 262, 278, 279–81
  CL-601-1A Challenger 280, 281
  CL-601-3A Challenger 279, 281
  CL-601-3A/ER Challenger 281
  CL-601-S Challenger 281
  CL-610 Challenger E 279
  DC-4M-2 412
  Global Express 471
  Regional Jet (RJ) 262, 281, 339, 719
Canadian Commuter *see* de Havilland Dash-8
  Canadian Commuter
CAP 230 684
Carstedt Jet Liner 600 377, 378
Carvair 412, 413, 481
CASA
  3000 703
  Wal 398, 399
  Cassutt Racer 694, 695
  Cavenaugh Cargoliner 659
  CC-106 Yukon 236, 237, 265–6
  CC-109 Convair Cosmopolitan 312, 336
  CC-132 Dash-7 331, 332
  CC-138 Twin Otter 363–4
  CC-142 Dash-8 336, 337
  CC-144 Canadair Challenger 279, 280, 281
Centurion *see* Cessna 210
Cessna
  402 677
  402 Utililiner 438
  404 Titan 636
  552 Citation II 708
  Citation 283–7
  FA.152 Aerobat 294, 295
  Fanjet 500 283, 284
  120 289, 290, 295
  140 289, 295
  140A 290
  142 291
  150 Trainer 290, 291, 295
  150D 291
  150F 291
  150K 291
  150L 291
  152 290, 291, 295
  165 Airmaster 289
  170 289–90, 295
  170A 290
  170B 290
  172 Cutlass RG 292, 295
  172A Skyhawk 290–1
  172B Skyhawk 291
  172J 291
  172Q Cutlass 295
  175 290, 291, 295
  175A Skylark 291
  177 291, 295
  177RG Cardinal RG 291, 292, 295
  180 290, 291, 295
  182 Skylane 290, 291, 292, 295
  182A Skylane 290
  182C Skylane 291
  185 Skywagon 290, 291
  188 Agwagon 295
  190 289, 295
  195 295
  205 *see* Cessna Model 210-5
  206 Super Skywagon 292
  207 Stationair 292–3, 295
  208 Caravan 1 294
  208A Cargomaster 294
  208B Super Cargomaster 294
  210 Centurion 292, 293, 295
  210-5 292, 295
  210A 292
  210B 292
  210C 292
  210D 292
  305A Bird Dog 290, 295
  305B 290
  305C 290
  306 Bird Dog 291
  321 295
  500 Citation 283, 284
  500 Citation I 263,284
  501 Citation I/SP 284
  525 Citation Jet 264, 286–7
  550 Citation II 284, 285
  551 Citation II/SP 283, 284
  552 Citation II 263, 284–5
  560 Citation V 264, 286
  650 Citation III 263, 285, 286, 287
  650 Citation VI 285, 287
  660 Citation VII 264, 285–6
  670 Citation IV 285
  750 Citation X 264, 287
  A150K Aerobat 291, 295
  A150L 291
  A185F Ag Carryall 295
  A188 Ag Truck 295
  P172D Skyhawk Powermatic 291
  P206 Super Skylane 292
  R172 Reims Rocket 295
  R182 Skylane RG 295
  S550 Citation S/II 284, 285
  Skyhawk 172 290, 291, 292–3, 295
  T188C Ag Husky 295
  U206 Stationair 289
  U206E Stationair 292
  U206G Stationair 295
  P210N Pressurized Centurion 293, 295
  single-engined family 289–95
  Titan 438, 542
Chargus Titan 645
Charlotte Aircraft Corps DC-4 conversion 412
Cherokee *see* Piper PA-28
Cheyenne *see* Piper PA-31T
Cheyenne I *see* Piper PA-31T-1
Cheyenne III *see* Piper PA-42
Chieftain *see* Piper PA-31-350 Chieftain
'China Bomber' *see* Curtiss Condor BT-32
Christen
  Eagle 646
  Eagle II 684
Clipper *see* Boeing Clipper
CMASA Wal 398
CN-235 336
Coastguarder *see* British Aerospace 748
  Coastguarder

Comanchero *see* Schafer Comanchero
Comet *see* de Havilland Comet
Commodore Jet *see* IAI 1123 Commodore Jet
Concorde 297–303, 745, 746, 747, 750, 751, 772
Conroy CL-44-O conversion 270, 307
Conroy/Aero Spacelines
  Guppy 229, 305–8
  Guppy 101 229, 307
  Guppy 201 229, 305, 307, 308
  Mini Guppy 229, 306, 307
  Pregnant Guppy 229, 306, 307
  Super Guppy 229, 306
Constellation *see* Lockheed Constellation
Convair
  880 157, 159, 315–19
  900 157
  990 317, 318–19
  990A Coronado 317, 318
  B-36 Peacemaker 315
  B-58 Hustler 315, 316
  CV-240 Liner 261, 309–12, 507, 509, 605–6
  F-102 Delta Dagger 315
  F-106 Delta Dart 315
  18 316
  22 316
  30 *see* Convair 990
  110 309, 605
  240-21 Turbo-Liner 312
  340 310–11, 312, 313
  340B 311
  440 Metropolitan 311, 312
  540 Cosmopolitan 312
  580 310, 312, 313
  600 309, 313
  640 312, 313
  Super 580 313
  Twins 309–13
Convair-Fort Worth SC-54D Skymaster 413
Corporate Jetstream 544
Cox Air DHC-3-T Turbo Otter 358–9
CP-107 Canadair 265
CP-150 Onyx 647
CSR-123 Otter 354
CT-39E Sabreliner 707
CT-39G Sabre 707, 708
CT-142 Dash-8 337
Curtiss
  AT-32 Condor II 323, 325–7
  AT-32A Condor II 326
  AT-32B Condor II 326
  AT-32C Condor II 326
  AT-32D Condor II 326
  AT-32E Condor II 326
  B-2 Condor 323, 326
  BT-32 Condor II ('China Bomber') 324, 327
Condor 18 323, 324, 325, 326
Condor II 323–7
  CT-32 Condor II 327
  T-32 Condor II 324–5, 326, 327

Dash-7 329–33, 335, 336
  7R Ranger 330–1
  Series 100 329
  Series 101 330, 332
  Series 150 332
  Series 300 332
Dash-8 Canadian Commuter 332, 335–9
Dash-8-102 336
Dash-8M-100 337
  Series 100 336, 339, 545, 702
  Series 200 338
  Series 300 335, 338–9
  Series 400 335, 338–9

Dash-600 see Airbus Industrie A300-600
Dassault
   9000 471
   Falcon 341–4, 468
   Falcon 10 (Mini-Falcon) 264, 342, 343
   Falcon 10MER 342, 343
   Falcon 20 250, 261, 342–3, 344, 707
   Falcon 20 Cargo Jet 343
   Falcon 20-5 343, 344
   Falcon 20C 342
   Falcon 20D 342
   Falcon 20E 342
   Falcon 20F 342
   Falcon 20G 342, 343
   Falcon 20H see Dassault Falcon 200
   Falcon 20T 342, 343
   Falcon 30 342, 343
   Falcon 40 343
   Falcon 50 261, 341, 342, 343, 344
   Falcon 100 341, 342, 343
   Falcon 200 263, 341, 342, 343
   Falcon 900 262, 263, 341, 343, 344
   Falcon 900B 344
   Falcon 2000 261, 344
   Falcon V10F 343
   Falcon X see Dassault Falcon 2000
   Mercure 35
   Mystere 20 341–2, 560
   Mystere 20 Fan Jet Falcon 261
Dassault-Breguet/ATR Petrel 42/72 93–4
   DC-1 see Douglas DC-1
   DC-2 see Douglas DC-2
   DC-3 see Douglas DC-3
   DC-4 see Douglas DC-4
   DC-6 see Douglas DC-6
   DC-7 see Douglas DC-7
   DC-8 see Douglas DC-8
   DC-9 see McDonnell Douglas DC-9
   DC-10 see McDonnell Douglas DC-10
   DC-118B (DC-7) 417
de Havilland
   see also Hawker-Siddeley
   119 489
   Comet 148, 261, 297, 315, 367–73, 573
      1A 368, 369, 370, 371
      1XB 371
      2 369, 370, 371
      2E 370, 371
      2(RAF) 371
      2X 369, 371
      3 369–70, 371
      3B 370, 371
      4 369, 3701, 371, 423
      4A 370, 371
      4B 369, 370, 371, 372–3
      4C 369, 370, 371
      C.Mk 2 367, 371
      R.Mk 2 371
      T.Mk 2 371
   Dash-7 329–33, 335, 336
      Series 100 329
      Series 101 330, 332
      Series 150 332
      Series 300 332
      7R Ranger 330–1
   Dash-8 Canadian Commuter 332, 335–9
      Series 100 336, 339, 545, 702
      Series 200 338
      Series 300 335, 338–9
      Series 400 335, 338–9
      -102 336
      -8M-100 337
   D.H.51 385

D.H.60 Moth 379, 385
D.H.60G Gipsy Moth 385–6, 387
D.H.60G-III Gipsy Moth 386
D.H.60M Metal Moth 386, 387
D.H.60T Moth Trainer 386, 387
D.H.60X Moth 385
D.H.61 Giant Moth 379, 386
D.H.66 Hercules 379
D.H.71 Tiger Moth 386–7
D.H.75 Hawk Moth 387
D.H.80 Puss Moth 345, 387
D.H.81 Swallow Moth 387
D.H.82 Tiger Moth 387–8
D.H.82 Tiger Moth Mk I 387
D.H.82 Tiger Moth Mk II 387–8
D.H.82B Queen Bee 388
D.H.83 Fox Moth 379, 388
D.H.84 Dragon 379–80, 381, 382–3
D.H.84 Dragon 1 379
D.H.84 Dragon II 380
D.H.84M Dragon 380
D.H.85 Leopard Moth 388, 389
D.H.86 Express Air Liner 377, 380, 381
D.H.86B 381
D.H.87 Hornet Moth 388
D.H.87B Hornet Moth 389
D.H.88 Comet 406, 721
D.H.89 Dragon Rapide 375, 380–2
D.H.89 Dragon Rapide Mk 2 382
D.H.89 Dragon Rapide Mk 3 382
D.H.89 Dragon Rapide Mk 4 382
D.H.89 Dragon Rapide Mk 5 382
D.H.89 Dragon Rapide Mk 6 382
D.H.89A Dragon Rapide 380, 381, 383
D.H.89B Dragon Rapide (Dominie Mk I/II) 381–2
D.H.89M Dragon Rapide 381
D.H.90 Dragonfly 381
D.H.94 Moth Minor 388
D.H.104 Dove 247, 375–8, 505, 541, 559
D.H.104 Dove 1 375
D.H.104 Dove 1B 376
D.H.104 Dove 2 376
D.H.104 Dove 2B 376
D.H.104 Dove 4 376, 378
D.H.104 Dove 5 376
D.H.104 Dove 6 376
D.H.104 Dove 7 376
D.H.104 Dove 8 376–7
D.H.106 see de Havilland Comet
D.H.108 367
D.H.114 Heron 375, 376, 377–8
D.H.114 Heron Mk 2 378
D.H.114 Heron Mk 2C 378
D.H.114 Heron Mk 2D 378
D.H.120 489
D.H.121 489, 490
D.H.121 Trident 159
D.H.123 253
D.H.125 Jet Dragon 247–8, 376, 541
D.H.126 253
D.H.131 253
DHC.1 Chipmunk 345
DHC-2 Beaver 345–51, 353
DHC-2 Beaver AL.Mk 1 347
DHC-2 Beaver II 347, 349, 350
DHC-2 Turbo Beaver 346, 350
DHC-2 Turbo Beaver II 350
DHC-2 Turbo Beaver III 349, 350–1
DHC-3 Otter 332, 353–9
DHC-3-T Turbo Otter 359
DHC-3/1000 Otter 358

DHC-6 Twin Otter 329–30, 332, 335, 336, 358, 360–5, 709
DHC-6 Twin Otter Series 100 361–2
DHC-6 Twin Otter Series 200 362, 363
DHC-6 Twin Otter Series 300 362–3, 364
DHC-6 Twin Otter Series 300M 365
DHC-6 Twin Otter Series 300MR 365
DHC-6 Twin Otter Series 300S 363
Moths 385–91
Sea Devon 376, 378
Type IV 367
Vampire TG278 367
De Schelde Do 24K-2 400
Defender (BN-2 Islander) 502, 503, 505
Detroit-Lockheed Vega 584
DHC-2 see de Havilland DHC-2 Beaver
DHC-3 see de Havilland DHC-3 Otter
DHC-6 see de Havilland DHC-6 Twin Otter
DHC-7 see Dash-7
DHC-8 see Dash-8 Canadian Commuter
Dominie Mk I 381–2, 383
Dominie Mk II 381–2
Dominie T.Mk 1 (HS 125) 248, 250
Dornier
   Do 15 (Militär-Wal) 399
   Do 17F 494, 495
   Do 18 399–400
   Do 18D 398, 399, 400
   Do 18E 399
   Do 18G 398
   Do 18H 398
   Do 18N 398
   Do 24 397, 400
   Do 24K-1 400, 401
   Do 24K-2 400
   Do 24N-1 400
   Do 24T 400–1
   Do 24T-1 399
   Do 24T-2 399, 400, 401
   Do 24T-3 400
   Do 24TT 397, 400
   Do 228 501, 636, 640, 711
   Do 328 545
   Do X1A 393
   Do X2 395–6
   Do X3 395–6
   Do X 391–6, 398, 399
   Flying Boats 397–401
   Gs I 397
   Gs II 397
   Militär-Wal 33 399
   RS I 397, 399
   Type J IId 398
   Type J Wal 397–9
   Type R Super Wal 399
Douglas
   see also McDonnell Douglas
   A-26 289
   C-32A 407
   C-33 407
   C-39 407
   C-41 407
   C-42 407
   C-54 Skymaster 409–12, 414
   C-54A Skymaster 409, 410
   C-54B Skymaster 409
   C-54C Skymaster 409
   C-54D Skymaster 410, 411
   C-54E Skymaster 410
   C-54G Skymaster 410, 412
   C-54P Skymaster 412
   C-54Q Skymaster 411
   C-118 Liftmaster 414

809

C-118A 415, 416, 419
C-118B 415, 417, 419
Dakota 261, 768
DC-1 324, 403–4
DC-2 324, 403–7, 598, 721
DC-2-120 406
DC-2A 407
DC-2B 407
DC-2K 407
DC-3 47, 253, 407, 447, 597, 599, 601, 765
see also Lisunov Li-2
DC-4 Skymaster 409–13, 414, 571, 609
DC-4A 409
DC-4E 409, 410
DC-6 233, 311, 411, 414–22, 572
DC-6A 414–15, 416
DC-6B 415, 416, 418, 419, 422
DC-6B 'Super Six' models 418–19
DC-6BF 416
DC-6C 415, 416
DC-7 417, 419–22
DC-7B 417, 422
DC-7C 209, 417, 418, 422, 574
DC-7D 417, 422
DC-7F 'Speedfreighter' 417
DC-8 149, 159, 315, 422, 423–31, 609
DC-8 Series 32 423
DC-8 Series 33 431
DC-8 Super 61 426, 430
DC-8 Super 62 426, 430, 431
DC-8 Super 63 426, 430, 431
DC-8 Super 63CF 426, 430, 431
DC-8 Super 71 424–5, 426, 427, 430, 431
DC-8 Super 72 426, 431
DC-8 Super 73 426, 431
DC-8 Super Sixty 183
DC-8-10 423, 424, 426
DC-8-20 424, 426
DC-8-30 424, 426
DC-8-40 424, 426
DC-8-42 431
DC-8-50 424, 426, 430
DC-8-53 431
DC-8-54 424, 426
DC-8-55 424, 426
DC-8-55CF 425
DC-8-55F 424, 426, 430
DC-8-60 169
DC-8-70 169
DC-8F Jet Trader 424, 425, 426
DC-118B 417
M-2 585
MC-118A 417, 419
R2D-1 Skymaster 407
R5D-1 Skymaster 410
R5D-2 Skymaster 411
R5D-4 Skymaster 412
R6D-1 414, 416, 419
R6D-1Z 414, 416, 419
SC-54D Skymaster 413
VC-118 414, 417
VC-118A 417
VC-118B 415, 417, 419
XC-54F Skymaster 410
XC-54K Skymaster 410
XC-112A 414, 416
XC-114 Skymaster 410
XC-116 Skymaster 410
YC-34 DC-2 407
YC-112A 416
Dove see de Havilland D.H.104
Dragon 150 646
Dragon see de Havilland D.H.84

Dragon Rapide see de Havilland D.H.89
Dragon Six see de Havilland D.H.89
Dragonfly see de Havilland D.H.90
Dumod Liner 127

E-4B 747 185–6
E-9A Dash-8 337–8
EC-35A Learjet 562
EC-95 Bandeirante 433, 434
EC-121C Lockheed Constellation 576
EC-121H Lockheed Constellation 576
EC-121J Lockheed Constellation 576
EC-121K Lockheed Constellation 576
EC-121L Lockheed Constellation 576
EC-121M Lockheed Constellation 576
EC-121P Lockheed Constellation 576
EC-121Q Lockheed Constellation 576
EC-121R Lockheed Constellation 574, 576
EC-121S Lockheed Constellation 576
EC-121T Lockheed Constellation 576
Eipper
    Quicksilver GT 644
    Quicksilver MX 645–6
Electra see Lockheed Electra
Electra Junior see Lockheed Electra Junior
EMBRAER
    Bandeirante 432–9, 711
    Comanchero 500B 680
    EMB-100 Bandeirante 432, 434
    EMB-110 Bandeirante 432, 434–5
    EMB-110A Bandeirante 434
    EMB-110B1 Bandeirante 434
    EMB-110B Bandeirante 434
    EMB-110C Bandeirante 433, 434, 439
    EMB-110E Bandeirante 433, 434
    EMB-110J Bandeirante 433
    EMB-110K1 Bandeirante 434
    EMB-110P1 Bandeirante 433, 434, 438, 440
    EMB-110P1A Bandeirante 434, 439
    EMB-110P1K Bandeirante 434, 439
    EMB-110P1SAR Bandeirante 434
    EMB-110P2 433, 434, 438, 39
    EMB-110P Bandeirante 433, 434
    EMB-110S1 Bandeirante 434
    EMB-111 Bandeirante 433, 434, 438
    EMB-111A Bandeirante 433, 434
    EMB-111A(N) 434, 439, 440
    EMB-120 545, 702
    EMB-120 Araguaya 434
    EMB-121 Xingu 434, 439
    EMB-123 Tapajos 434
    EMB-145 Amazon 339
    Ipanema 432
    Tucano 432
'Empire Boats' see Shorts S.23 'C-class'
Empresa Brasileira de Aeronautica S.A. see EMBRAER
Euro Wing Goldwing 646
Evans VP-1 647
Explorer see Lockheed Explorer
Express Air Liner see de Havilland D.H.86
Extra 300 684

F-2 Beech Model 18 126
F-101B Voodoo 706
Fairchild
    see also Saab-Fairchild
    82 345
    Air Sentry 641
    Expediter 636
    F.27 447
    Merlin 400SP 641
    Merlin IVC 641

Merlin V 641
Metro V 641
Metro VI 641
SA 227/TT41 Merlin 300 641
SA-227DC Metro 25 641
T-46A 687, 700
Fairchild-Hiller
    FH-227 449, 450, 454
    FH-227B 450
    FH-228 454
Fairchild/Swearingen SA 227AC Metro III 638–9, 640–1, 701
Fairey-Britten-Norman Turbo Islander 505–6
Falcon see Dassault Falcon
Fan Jet Falcon
    see Dassault Falcon
FFA P-16 559
Fisher
    FP-101 647
    FP-102 Koala 647
    FP-303 647
Flugschiff see Dornier Do X
'Flying Banana' see Handley Page H.P.42
Flying Fortress see Boeing B-17 Flying Fortress
Fokker
    50 452, 456, 716
    50 Maritime Enforcer Mk 2 452
    50 Maritime Mk 2 452
    70 339
    100 256, 452, 453, 455,456–7, 719
    Do 24T-1 399
    F.24 447
    F.27 Friendship 62, 73, 329, 447–53, 479, 481–2, 772
    F.27 Maritime 452
    F.27 Mk 100 Friendship 447, 448, 451
    F.27 Mk 200 Friendship 449–50
    F.27 Mk 300 Friendship 449, 450
    F.27 Mk 400M Friendship 449
    F.27 Mk 500 Friendship 449, 450–1, 452
    F.27 Mk 525CRF Friendship 450
    F.27 Mk 600 Friendship 449
    F.28 Fellowship 253, 256, 450, 453–7, 613, 709
    F.28 Mk 1000 Fellowship 453, 454, 455, 457
    F.28 Mk 2000 Fellowship 455
    F.28 Mk 3000 Fellowship 455
    F.28 Mk 4000 Fellowship 453, 454, 455, 456–7
    F.28 Mk 5000 Fellowship 455
    F.28 Mk 6000 Fellowship 455
    F.28 Mk 6600 Fellowship 455
    F.29 450, 455–6
    F.32 443–4
    F-10A 403
    F.I 441
    F.II 441
    F.III 441
    F.IV 441–2
    F.IX 443, 444
    F.VII 441–5, 459
    F.VII-3m 442
    F.VIIA 442, 443
    F.VIIA-3m 442–3, 444
    F.VIIB-3m 443, 444–5
    F.VIII 443
    F.XII 444
    F.XVIII 444
    F.XX 444
    F.XXII 444
    F.XXXVI 444
    P.275 447
    P.335 450
    Super F.28 455
Fokker/McDonnell Douglas MDF-100 450, 456

Ford Tri-Motor 458–65
  Model 3-AT 459, 460
  Model 4-AT 459–62
  Model 4-AT-A 462
  Model 4-AT-B 462
  Model 5-AT 460, 462, 465, 474, 475, 478
  Model 5-AT-B 460
  Model 5-AT-C 465
  Model 5-AT-CS 459
  Model 5-AT-D 460–1, 465
  Model 6-AT-1 465
Formula One races 693–7
Fox Moth *see* de Havilland D.H.83

G-2 *see* Grumman Gulfstream II
G-boats *see* Shorts G-boats
Gajaraj *see* Ilyushin Il-76MD
Gates Learjet 277, 561–2
  EC-35A 562
  Longhorn 28 562, 563
  Longhorn 29 562
  Longhorn 50 562–3
  Longhorn 55 263, 561, 562
  Model 23 264
  Model 24 264, 560, 562
  Model 24B 560, 561
  Model 24C 561
  Model 24D 561
  Model 25 264
  Model 25B 559, 561
  Model 25C 561
  Model 25D 562
  Model 25G 562
  Model 31 562
  Model 35 264, 561
  Model 35A 561, 562, 563
  Model 36 264, 561
  PC-35A 562
  RC-35A 562
  RC-36A 562
  UC-35A 562
  UC-36A ECM 562
General Dynamics *see* Canadair
Giant Moth *see* de Havilland D.H.61
Gipsy Moth *see* de Havilland D.H.60G
Gloster Meteor 447
Golden Arrow *see* Convair 880
Goldwing *see* Euro Wing Goldwing
Grumman
  see also Grumman American; Gulfstream American
  E-2 576
  E-2C Hawkeye 703
  F4F Wildcat 467
  F6F Hellcat 467
  G.73 Mallard 467
  G.159 *see* Gulfstream
  G.1159 *see* Gulfstream II
  G.1159A *see* Gulfstream III
Gulfstream 467–72
  Gulfstream I 607, 608
  Gulfstream I-C 468
  Gulfstream II 261, 262, 454, 468–9
  Gulfstream IIB 262
  Gulfstream III 262
  Gulfstream II(TT) 469
  Gulfstream IV 262–3, 456
    HU-16 Albatross 468
    TBM-3 Avenger 572
    TC-4C Academe 468
    TF-1 Trader 467
    VC-4A 468
    Grumman American Fanliner 686

Gulfstream III 469
  Traveler 686
Gulfstream 467–72
  C-20 470–1
  I 607, 608
  I-C 468
  II 261, 262, 454, 468–9
  IIB 262, 469, 470
  III 262, 469–71, 512
  II(TT) 469
  IV 262–3, 456, 470, 471
  IV-SP 467, 471
  SMA-3 470
  SRA-1 470
  SRA-4 471
  V 471
Gulfstream American Corporation
  C-20 Gulfstream 470–1
  Gulfstream IIB 469, 470
  Gulfstream III 262, 469–71
  Gulfstream IV 470, 471
  SMA-3 Gulfstream 470
  SRA-1 Gulfstream 470
  SRA-4 Gulstream 471
Gulfstream SA-30 Gulfjet *see* Swearingen Jaffe SJ30 Fanjet
Guppy *see* Aero Spacelines Guppy

Hamilton
  H-18 473, 474
  H-45 Metalplane 'Silver Streak' 473–4, 475, 476, 477, 478
  H-47 Metalplane 'Silver Eagle' 474–5, 476, 477, 478
  Metalplane 473–8
  VC-89 Metalplane 477
Handley Page
  Herald 61, 438, 448, 541
  Hereford 709
  H.P.42 485–8
  H.P.42E 485, 486, 487, 488
  H.P.42W 485, 486, 487, 488
  H.P.45 485
  H.P.124 481, 484
  H.P.125 484
  H.P.127 Jet Herald 484
  H.P.129 Mini Herald 484
  H.P.137 Jetstream 542–3
  H.P.R.3 479, 480
  H.P.R.4 479, 480–1
  H.P.R.7 Herald 479–84
  H.P.R.7 Herald Series 100 481
  H.P.R.7 Herald Series 101 484
  H.P.R.7 Herald Series 200 480, 481, 482–3, 484
  H.P.R.7 Herald Series 206 479
  H.P.R.7 Herald Series 207 480, 481
  H.P.R.7 Herald Series 209 481
  H.P.R.7 Herald Series 210/11 481
  H.P.R.7 Herald Series 214 481
  H.P.R.7 Herald Series 215 481
  H.P.R.7 Herald Series 300 484
  H.P.R.7 Herald Series 401 481
  H.P.R.7 Herald Series 500 484
  H.P.R.7 Herald Series 700 484
  H.P.R.8 Car Ferry 481
Harpoon *see* Lockheed PV-2
Hawk Moth *see* de Havilland D.H.75
Hawk XP *see* Cessna Model R172
Hawker Sea Fury 447
Hawker Siddeley
  see also British Aerospace; de Havilland
  HS.125 264, 468
  HS.125 Series 2 248, 250

HS.125 Series 3 248
HS.125 Series 3A 248
HS.125 Series 3A-RA 248
HS.125 Series 3B 248
HS.125 Series 3B-RA 248
HS.125 Series 400 250, 251
HS.125 Series 600 250, 251
HS.125 Series 700 250, 251
HS.136 253
HS.144 253
HS.146 256
Trident 489–92, 541
Trident 1 491, 492
Trident 1C 160, 491
Trident 1E 160, 489, 490, 491–2
Trident 1E-140 492
Trident 1F 160, 162
Trident 2E 490, 491, 492
Trident 3B 492
Trident III 490
Trident Super 3B 492
Hawker Siddeley/Avro
  H.S.748 96–101, 714
  H.S.748 Coastguarder 99
  H.S.748 Series 2A 98, 101
  H.S.748 Series 2B 99
  H.S.748 Super 2B 96
  H.S.748-234 101
  H.S.748-256 101
  H.S.748MF 98
Heinkel
  He 65 493
  He 70 493–9, 558
  He 70A 494, 496, 497
  He 70B 493, 494
  He 70C 493, 494, 498–9
  He 70D 494
  He 70D-0 495
  He 70E-1 bomber 494
  He 70F-1 494
  He 70F-2 494–5, 496–7
  He 70G 493, 494
  He 70G-1 494, 499
  He 70V-1 494
  He 170 495
  He 270 495
  Helio Courier 346
Herald *see* Handley Page H.P.R.7
Hercules *see* de Havilland D.H.66
Heron *see* de Havilland D.H.114
HFB320 Hansa 264
Hornet Moth *see* de Havilland D.H.87
HU-16E Albatross 707
HU-25 Falcon 707
HU-25A Falcon (Guardian) 342, 343
HU-25B Falcon 342
HU-25C Falcon (Interceptor) 342, 343
HU-39H Sabre 707
Hudson *see* Lockheed Hudson
Hunting H-107 111
Hurel Dubois HD-31 11

IAI Jet Commander 537–40
  112-1 538
  1123 Commodore Jet 537, 538, 540
  1124 Westwind I 263, 264, 538–40
  1124A Westwind 2 538–9, 540
  1124N Sea Scan 538, 540
  1125 Astra 540
IFM *see* International Formula Midget races
Ilyushin
  Il-12 507–9, 512
  Il-12D 509

Il-12T 508, 509
Il-14 61, 62, 77, 507–9
Il-14D 509
Il-14F 509
Il-14G 509
Il-14M 509
Il-14P 509
Il-14T 509
Il-18 'Coot' 61, 511–15, 531, 743
Il-18B 512
Il-18D 512, 513
Il-18DORR 515
Il-18I 512–13
Il-18T 515
Il-18V 512, 513
Il-18Ye 513
Il-20 'Coot-A' 512, 513, 515
Il-22 515
Il-28 515, 729
Il-38 513–14, 515
Il-46 729
Il-62 517–21, 531, 739, 740, 745, 754
Il-62M 517, 519, 520–1
Il-62MK 520, 521
Il-76 'Candid' 523–9
Il-76 'Mainstay' 528, 529
Il-76 'Midas' 528, 529
Il-76M 524, 525, 528, 529
Il-76MD 'Candid-B' 524–5, 529
Il-76T 'Candid-A' 524, 526–7, 528, 529
Il-76TD 529
Il-78 576
Il-86 'Camber' 520, 531–5
Il-90 535
Il-96 534–5
Il-96-300 535
l-96M 535
International Formula Midget (IFM) races 693–7
Ipanema see EMBRAERIpanema
IPD-6504 432
Islander see Britten-Norman BN-2 Islander
Israel Aircraft Industries see IAI

JC-121C Lockheed Constellation 576
JC-121K Lockheed Constellation 576
JEC-121P Lockheed Constellation 576
Jet Commander 537–40
　112–1 538
　1123 Commodore Jet 537, 538, 540
　1124 (Westwind I) 263, 264, 538–40
　1124A (Westwind 2) 538–9, 540
　1124N Sea Scan 538, 540
　1125 Astra 540
Jetstream 541–5
　3M 543
　31 541, 542, 543, 544, 545
　31EZ 544
　41 544–5
　61 (ATP) 545
　200 543
　Super 31 544
　T.Mk 1 543
　T.Mk 2 542, 544
　T.Mk 3 544
JRB-1 Voyager 126
Jumbo Jet see Boeing 747
Junkers
　F 13 547–52
　F 13bi 548–9
　F 13da 550–1
　F 13ke 552
　F 13W 549
　G 23 442, 552

G 24 552
G 38 553–8
J 10 547
J 13 547–8
Ju 52 552
Ju 52/3m 552, 558
Ju 86 558
Ju 160 558
W 33 552
W 34 552
Junkers-Larsen JL 6 552

Kawasaki Wal 398
KC-10 tanker 19
KC-10A Extender (DC-10) 617, 618, 620
KC-97 Stratofreighter 148, 270
KC-97A Stratofreighter 228
KC-97E Stratofreighter 228–9
KC-97G Stratofreighter 226, 228, 229
KC-97L Stratofreighter 224, 226, 229
KC-130 Stratofreighter 226
KC-135 Stratofreighter 149, 156, 158, 209
KC-135A Stratofreighter 224, 229
KC-135R Stratofreighter 169
KE-3A Boeing 707 152
Ki-20 558
King Air see Beech King Air
'King Beaver' see de Havilland DHC-3 Otter
K.Mk 2 VC10 782
K.Mk 3 VC10 782
K.Mk 4 VC10 778, 782

L-19 Cessna Bird Dog 290, 291, 295
L-19A-1T Cessna 290
L-19E Cessna 290
L-20A Beaver 347
L-23/U-8 Seminole 119
LAK BROK-1M 646
Learjet 250, 277, 467, 559–63, 635
　EC-35A 562
　Longhorn 28 562, 563
　Longhorn 29 562
　Longhorn 50 562–3
　Longhorn 55 263, 561, 562
　Model 23 264, 560
　Model 24 264, 560, 562
　Model 24B 560, 561
　Model 24C 561
　Model 24D 561
　Model 25 264
　Model 25B 559, 561
　Model 25C 561
　Model 25D 562
　Model 25G 562
　Model 31 562
　Model 35 264, 561
　Model 35A 561, 562, 563
　Model 36 264, 561
　PC-35A 562
　RC-35A 562
　RC-36A 562
　UC-35A 562
　UC-36A ECM 562
Learstar 277–8, 559, 599, 601
Leo Thrush see Ayres Leo Thrush
Leopard Moth see de Havilland D.H.85
LET
　L-410 565–9
　L-410A 566, 568
　L-410AB 566, 568
　L-410AF 565, 567
　L-410M 567
　L-410MA 567, 568

L-410MU 566, 567
L-410UVP 566, 567–8
L-410UVP-E 568, 569
L-410UVP-K2 569
L-610 565, 568–9
Lisunov Li-2 61, 73, 507, 508, 512
Lithuanian Aircraft Construction see LAK
Little Gem 697
Lockheed
　1649A Starliner 574, 576
　Advanced Tristar 592
　Air Express Model 3 584, 585, 588
　Altair 584
　Altair 32-232 585
　Altair DL-2A 589
　Altair Model 8 589
　C-5 Galaxy 67, 68, 69
　C-5A Galaxy 183, 591, 615, 616
　C-130 53, 160
　C-141 Starlifter 523, 527, 528
　CL-303 577
　Constellation 209, 233, 267, 571–6
　Constellation Model 049 571–2, 576
　Constellation Model 649 572, 576
　Constellation Model 749 414, 572, 576
　Constellation Model 749A 572, 576
　Electra Junior Model 12 598, 599, 600
　Electra Junior Model 12-A 599
　Electra Junior Model 12-B 599
　Electra L-188 159, 513, 577–81, 759, 761
　Electra L-188A 577, 579, 580, 581
　Electra L-188AF 580
　Electra L-188C 577–9, 580, 581
　Electra L-188CF 580
　Electra Model 10 597–8
　Electra Model 10-A 598
　Electra Model 10-B 598
　Electra Model 10-C 598
　Electra Model 10-E 598, 599
　Electra Model 212 599
　Explorer 588
　Hudson 571, 601
　JetStar 261, 264, 283, 468
　L.1049C Super Constellation 419
　L-749 159
　L-1649 Starliner 209
　Lodestar Model 18 597, 599, 600–1
　Lodestar Model 18-07 601
　Lodestar Model 18-08/10 601
　Lodestar Model 18-14 601
　Lodestar Model 18-40 601
　Lodestar Model 18-50 601
　Lodestar Model 18-56 601
　Model 14 571
　Orion Model 9 584, 585, 586, 589
　Orion P3V-1 581
　Orion P-3 513, 581
　Orion-Explorer 588, 589
　P-38 571
　PV-1 Ventura 601
　PV-2 Harpoon 601
　'singles' 583–9
　Sirius 8 588–9
　Sirius DL-2 586
　Super Constellation Model 1049 572–3, 576
　Super Constellation Model 1049B 573, 576
　Super Constellation Model 1049C 573, 574–5, 576
　Super Constellation Model 1049D 573, 576
　Super Constellation Model 1049E 573–4, 576
　Super Constellation Model 1049G 573–4, 576
　Super Constellation Model 1049H 572, 574, 576
　Super Constellation Model 1249 574, 576

Super Electra Model 14   599–600, 602
Super Electra Model 14-N   600, 601, 602–3
Super Electra Model 14-F62   600
Super Electra Model 14-G3B   600
Super Electra Model 14H/H2   600
Tristar K.Mk 1/2   595
Tristar L-1011 1920,   591–5, 615, 616–17, 782
Tristar L-1011-1   592, 595
Tristar L-1011-50   592, 594, 595
Tristar L-1011-100   592, 593–4, 595
Tristar L-1011-200   592, 593, 594, 595
Tristar L-1011-250   595
Tristar L-1011-500   591, 592, 593, 5945
Twins   597–603
Vega   583–5, 586–7, 588
Vega 2   584, 588
Vega 5   584–5
Vega 5B   584, 585, 586–7
Vega 5C   587, 588
XC-35   571
Lodestar *see* Lockheed Lodestar
Longhorn *see* Learjet Longhorn
Loughhead S-1   583
LR-1 Mitsubishi   658, 659

McDonnell Douglas *see also* Douglas
   DC-8 Super 63CF   426, 430, 431
   DC-9   14, 159, 163–4, 256, 609–13, 632, 761
   DC-9 Super 80 *see* McDonnell Douglas MD-80
   DC-9 Super 81   612
   DC-9 Super 82   612
   DC-9 Super 83   612
   DC-9-10   610, 611, 612
   DC-9-14   610
   DC-9-15   610
   DC-9-20   610, 611
   DC-9-30   163, 610, 611
   DC-9-30CF   611
   DC-9-30F   611
   DC-9-30RC   611
   DC-9-32   610
   DC-9-40   610, 611, 613
   DC-9-50   611–12, 613, 629
   DC-9-51   611
   DC-9-55   612
   DC-9-RSS   612
   DC-9QSF   612
   DC-10   19–20, 430, 591–2, 612, 615–21, 626, 630
   DC-10-10   616, 617, 618, 620, 621
   DC-10-10CF   616, 617, 618
   DC-10-15   617, 618, 620
   DC-10-20   617
   DC-10-30   592, 593, 594, 617, 618–19, 623, 627
   DC-10-30/40 Intercontinental Stretch   623
   DC-10-30CF   617, 618, 6201
   DC-10-30ER   618
   DC-10-40   615, 617, 618
   DC-10A   616
   MD90-10   633
   MD90-30   633
   MD90-40   633–4
   MD-11   216, 620, 623–7
   MD-11CB   625
   MD-11F   625, 627
   MD-11X   624
   MD-11XR   625
   MD-12   627
   MD-80   35, 40, 165, 171, 202, 256, 612–13, 629–34
   MD-81   630, 632
   MD-82   612, 630–1, 632, 633
   MD-83   612, 631, 632, 633

MD-83 Executive Jet   632
MD-87   609, 610, 612, 630, 631–2
MD-87 Executive Jet   632
MD-88   612, 632, 634
MD-90   171, 613, 632–4
MD-91   41
MD-92   41
MD-95   634
MD-100 *see* McDonnell Douglas MD-11
McDonnell Douglas/Cammacorp DC-8 Super 71   424–5, 427, 430, 431
McDonnell Douglas/FokkerMDF-100   450, 456
Mainair Tri-Flyer 330   646
'Mainstay' (Il-76)   528, 529
Marquise *see* Mitsubishi MU-2J
Marsh S2R-T Turbo Thrush   107, 110
Martin
   2-0-2   605–8
   2-0-2A   606, 607, 608
   3-0-3   606
   4-0-4   605, 607–8
   B-26 Marauder   605
   M-130 Clipper   141, 605
Maxair
   DR.277 Drifter   644, 647
   Drifter XP   644
   Hummer   645
Mayo-Short Composite   721, 724–7
   MBB Fan Ranger   691
   MC-118A   417, 419
Merchantman *see* Vickers Merchantman
Metal Moth *see* de Havilland D.H.60M
Metalplane *see* Hamilton Metalplane
Metro *see* Swearingen MetroMH.52   432
Micro Biplane Aviation Super Tiger Cub   440 646
microlights   643–7
'Midas' (Il-76)   528, 529
MiG-21   746
Mil
   'Helos'   649–54
   Mi-1   649
   Mi-1NKh   649
   Mi-1T   649
   Mi-1U   649
   Mi-2   649, 650, 651
   Mi-4   649, 650
   Mi-4P   649, 650
   Mi-4S   649
   Mi-6   650, 654
   Mi-6P   650
   Mi-8   650–1
   Mi-8P   651
   Mi-8T   651
   Mi-10   650, 654
   Mi-12   651
   Mi-17   651
   Mi-17-1VA   651
   Mi-26   649, 651, 654
   Mi-34   654
   Mi-38   654
   SM-1   649, 650
   V-12   654
Militär-Wal *see* Dornier Do 15
Mini Guppy *see* Conroy/Aero Spacelines Mini Guppy
Mini-Falcon *see* Dassault Falcon 10
Mitchell Wing   643, 644
Mitsubishi
   Army Type 92 Super Heavy Bomber   558
   MU-2   655–9
   MU-2 Solitaire   658
   MU-2A   655

MU-2B   655
MU-2C   655–6, 658, 659
MU-2D   656
MU-2E   656, 657, 659
MU-2F   656
MU-2G   656–8
MU-2J Marquise   656–7, 658, 659
MU-2K   658
MU-2L   658
MU-2M   658
MU-2N   656, 658
MU-2P   656, 658
MU-2S   656
MU-300 Diamond   264
ML Aviation ML Utility   644
Model 53 Condor CO Transport *see* Curtiss Condor 18
Moth *see* de Havilland D.H.60
Moth Major *see* de Havilland D.H.60G-III
Moth Minor *see* de Havilland D.H.94
Moth Trainer *see* de Havilland D.H.60T
MU-2 *see* Mitsubishi MU-2
MU-2 Modifications Inc.
MU-2 Express   659
MU-2 Medi-Vac Express   659
Myasishchyev M-4 'Bison'   528, 729
Mystere 10 MiniFalcon *see* Dassault Falcon 10
Mystere 20 *see* Dassault Mystere
NAA *see* North American Aviation
NAMC
   YS-11   661–7
   YS-11-100   664, 666
   YS-11-117   663
   YS-11-205   665
   YS-11A-200   663
   YS-11A-202   664
   YS-11A-300CP   663–4
   YS-11A-400   664, 665
   YS-11A-500   664
   YS-11A-523   661
   YS-11A-600   664, 665
   YS-11A-700   664, 665
Navajo *see* Piper PA-31
NC-121K Lockheed Constellation 576
NC-131H Convair   311, 313
Neiva NE-821 Carajs   680
Nene Viking *see* Vickers Nene Viking
Nihon Aeroplane Manufacturing Company *see* NAMC
Nogrady AN-2 Avionette   646
North American Aviation (NAA)
   see also North American Rockwell
   F-86 Sabre   468, 655, 705
   F-100 Super Sabre   560, 705, 706
   N.A.246 Sabreliner   705, 706
   Sabreliner   538
   Sabreliner 40   706, 707
   Sabreliner 50   706–7
North American Rockwell
   see also North American Aviation
   S2R Thrush Commander   105–6, 107
   Sabre 60   707, 708
   Sabre 65   705, 708
   Sabre 70/75   707
   Sabre 75A   707, 708
   Sabre 85   708
   Sabreliner   261, 263, 264, 705–8
   Sabreliner 60   706, 707
   Sabreliner 70   707
   Sabreliner Commander 40A   707
   T-39 Sabreliner   285
North Star *see* Canadair CL-2 DC-4M

Northrop
   F-5 688
   T-38A Talon 559–60

Onyx *see* CP-150 Onyx
Orion *see* Lockheed Orion
Orion-Explorer *see* Lockheed Orion-Explorer
Otter *see* de Havilland DHC-3
Owl Racer 695, 696

P-95 Bandeirante Patrulha 433, 434, 438
PacAero 'Super Convair' 312
PC-35A Learjet 562
Pilatus Britten-Norman
   BN-2B Islander 506
   BN-2T Turbine Islander 506
Piper
   Apache 501
   Aztec 501, 542
   PA-22 Tri-Pacer 669
   PA-24 Comanche 669
   PA-28 Cherokee 669–73
   PA-28 Cherokee Warrior II (Cadet) 669, 672
   PA-28-140 Cherokee 140-4 670–1
   PA-28-150 Cherokee 669
   PA-28-151 Cherokee Warrior 671–2, 673
   PA-28-160 Cherokee 669
   PA-28-180 Cherokee 669, 670
   PA-28-180 Cherokee Archer 670
   PA-28-180R Cherokee Arrow 670, 672
   PA-28-181 Cherokee Archer II 672, 673
   PA-28-200R Cherokee Arrow II 672
   PA-28-235 Cherokee 670
   PA-28-235B Cherokee 670
   PA-28-236 Cherokee Dakota 672, 673
   PA-28R-201 Cherokee Arrow III 672–3
   PA-28RT-201 Cherokee Arrow IV 672, 673
   PA-28RT-201T Cherokee Turbo Arrow IV 673
   PA-31 Navajo 675–80
   PA-31 Navajo C/R 675, 678–9
   PA-31 Navajo Panther 680
   PA-31-300 Navajo 675, 677
   PA-31-310 Turbo Navajo 675, 677
   PA-31-310B Navajo B 677
   PA-31-350 Chieftain 676, 677–8, 679
   PA-31-350 Chieftain Comanchero 500 conversion 680
   PA-31-350 Navajo II *see* Piper PA-31-350 Chieftain
   PA-31P Pressurized Navajo 677
   PA-31P-350 Mojave 679, 680
   PA-31T Cheyenne 677, 679
   PA-31T-1 Cheyenne I 676, 679
   PA-32 Cherokee Six 673
   PA-32RT Lance II 672, 673
   PA-34 Seneca 673
   PA-42 Cheyenne III 679
   PA-44 Seminole 673
   T-1020 679, 680
   T-1040 679, 680
Pitts
   '190' Special 681–2
   S-1 Special 682
   S-1S Special 683, 684
   S-1T Special 684
   S-1W Special 684
   S-2 Special 681, 683
   S-2A Special 682, 683–4
   S-2B Special 684
   S-2S Special 684
   Special 646, 681–4
PO-1W Lockheed Constellation 572, 576

Pregnant Guppy *see* Conroy/Aero Spacelines Pregnant Guppy
Puss Moth *see* de Havilland D.H.80

Queen Bee *see* de Havilland D.H.82B
Quicksilver *see* Eipper Quicksilver

R2D-1 Douglas DC-2 407
R4C-1 Condor II 325, 327
R4Y-1 Convair 311, 313
R5D-1 Skymaster 410
R5D-2 Skymaster 411
R5D-4 Skymaster 412
R6D-1 DC-6 414, 416, 419
R6D-1Z DC-6 414, 416, 419
R7V-1 Lockheed Constellation 573, 576
R7V-2 Lockheed Constellation 574, 576
R50 Lockheed Lodestar 601
R70-1 Lockheed Constellation 572, 576
R-95 Bandeirante 433, 434, 440
RC-12 Super King Air 121
RC-35A Learjet 562
RC-36A Learjet 562
RC-121C Lockheed Constellation 576
RC-121D Warning Star 576
Reims Cessna
   see also Cessna
   172 Skyhawk 292–3
   FA.152 Aerobat 294, 295
Reims Rocket *see* Cessna Model R172
Reno Formula One 693–7
Republic Rainbow 572
RFB
   Fanliner 686
   Fantrainer 685–91
   Fantrainer 400 686–7, 691
   Fantrainer 600 686–7, 688–9, 690, 691
Riley
   see also Jetstream
   Turbo Executive 400 377
   Turbo Skyliner 378
   RM-1Z Martin 4-0-4 607, 608
Rockwell (North American)
   see also Aero Commander; North American Aviation
   CT-39 562
   S2R Thrush Commander 105–6, 107, 108–9
   S-75C Sea Sabre 707
   Sabre 60 707, 708
   Sabre 65 705, 708
   Sabre 70/75 707
   Sabre 75A 707, 708
   Sabre 85 708
   Sabreliner 261, 263, 264, 705–8
   Sabreliner 60 706, 707
   Sabreliner 70 707
   Sabreliner Commander 40A 707
Rombac 1-11 115
Rotec Rally 645
RR-4 Ford Tri-Motor 465
Ryan Aircraft XV-8A 643–4

Saab
   see also Saab-Fairchild
   340 699–703, 712
   340A 699, 700, 701
   340B 545, 699, 700, 701, 702
   340B Combi 702
   340B VIP 702
   340BQC 702
   2000 699, 702–3, 737
   JAS 39 Gripen 703

Scandia 447, 507, 699
Saab-Fairchild
   SF-340 432, 641, 699–700
   SF-340QC 700, 701
SAAC-23 559–60
   see also Learjet
SABENA DC-6B conversion 419
Sabre *see* North American Aviation F-86; North American Rockwell Sabre
Sabreliner *see* North American Aviation Sabreliner; North American Rockwell Sabreliner
Sabreliner Corp.
   Sabre 40EX 708
   Sabre 60EX 708
   Sabre 85 708
Sandringham *see* Shorts Sandringham
SAT *see* Airbus Industrie Super Airbus Transporter
Saunders
   ST-27 378
   ST-28 378
   SC-54D Skymaster 413
   SC-95B Bandeirante 433, 434
Schafer
   Comanchero 680
   Comanchero 500 680
   SCMP Wal 398
Sea Thrush *see* Ayres Turbo Sea Thrush
Seaford *see* Shorts Seaford
'Seven Seas' *see* Douglas DC-7C
Shaanxi Yunshuji-8 56, 60
Shin Meiwa Tawron 378
Shoestring 697
Short-Mayo Composite 721, 724–7
Shorts
   330 439, 709–14, 715, 717
   330 Combi 710
   330-100 711
   330-200 713–14
   330-UTT 710, 714
   330C 710
   360 438, 709, 711, 715–19
   360 Advanced 718, 719
   360-300 718
   360-300F 719
   Calcutta 720–1
   Civil Flying Boats 228, 720–7
   G-boats 727
   Kent 721
   S.23 'C-class' 720, 722–4
   S.30 'C-class' 721, 724
   S.33 'C-class' 724
   Sandringham 723, 724, 727
   SD3-30 *see* Shorts, 330
   Seaford 727
   Sherpa (C-23A/B) 710, 711–13, 714
   Skyvan 709, 711
   Solent 722, 724, 727
   Stirling 720
   Sunderland 722, 723–4, 727
Sikorsky S-42 141
'Silver Eagle' *see* Hamilton H-47
'Silver Streak' *see* Hamilton H-45
Sino-German MPC.75 339
Sirius *see* Lockheed Sirius
Sirocco *see* Aviasud Sirocco
Skyhawk *see* Cessna Model 172
Skylane *see* Cessna Model 182
Skylark *see* Cessna Model 175A
Skymaster *see* Douglas DC-4
Skywagon *see* Cessna Model 185
SMA-3 Gulfstream 470

SNB-1 Beech Model 18 127
SNB-4/5/5C Beech Model 18 130
SNCASE
   see also Sud Aviation
   Model 200 11
   X-210 11
Snow
   S1 105
   S2 105
   S2A 105
   S2B 105
   S2C 105, 106
   S2D 105, 106
   S2R Thrush Commander 105–6, 107
Société Nationale de Construction Aéronautique Sud-Est *see* SNCASE
Solent *see* Shorts Solent
Solitaire *see* Mitsubishi MU-2 Solitaire
Sorrell
   Hiperbipe 647
   Hiperlight 647
Speedfreighter *see* Douglas DC-7F
SRA-1 Gulfstream 470
SRA-4 Gulfstream 471
Starlifter *see* Lockheed C-141
Stationair 7 *see* Cessna Model 207
Stinger 697
Stout 2-AT Tri-Motor 475
Stratocruiser *see* Boeing 367; Boeing 377
Stratoliner *see* Boeing 307
Striker
   Gemini 646
   Tri-Flyer 646
Sud Aviation
   see also Aérospatiale; Airbus Industrie; British Aerospace
   Caravelle 11–17, 159, 580, 609, 761
   Caravelle 10A 12, 14, 17
   Caravelle 10B1.R 14, 17
   Caravelle 10B3 13, 17
   Caravelle 10B 'Super Caravelle' 14–15, 17, 300
   Caravelle 11R 14, 17
   Caravelle 12 14, 17
   Caravelle 12/58T 15
   Caravelle I 12, 13, 17
   Caravelle IA 13, 17
   Caravelle III 12, 13, 15, 16–17
   Caravelle SE 210 11–12
   Caravelle VI 14
   Caravelle VII 14, 17
   Caravelle VIN 14, 17
   Caravelle VIR 12, 13, 14, 17
Sukhoi
   S-21 471
   Su-26M 684
   Su-27 528
Sunderland *see* Shorts Sunderland
Super B *see* Aérospatiale Caravelle 10B
Super Broussard 432Super Guppy *see* Conroy/Aero Spacelines Super Guppy
'Super Six' *see* Douglas DC-6B 'Super Six' models
Super Skylane *see* Cessna Model P206
Super Skywagon *see* Cessna Model 206
Super Wal *see* Dornier Type R
Swallow Moth *see* de Havilland D.H.81
Swearingen
   Merlin IIA 635–6
   Merlin IIB 636
   Merlin III 636–7, 641
   Merlin IVA 640, 641
   Merlin IVC 635
   Merlin SM 640

Metro 635–41, 709
SA-226TC Metro II 636–7, 640
Swearingen Jaffe SJ30 Fanjet 264
Swearingen/Fairchild SA 227AC Metro III 635, 638–9, 640–1, 701

T3J-1 Sabreliner 706, 707
T-1A Jayhawk 264
T-2 Fokker F.IV 441–2
T-29A Convair 310, 313
T-39A Sabreliner 705–6, 708
T-39B Sabreliner 706, 708
T-39F Sabreliner 706
T-41A Mescalero 291
T-41B Cessna 291
T-43A Boeing 737 164, 166
T-47A Citation II 263, 284–5
T-50 Cessna 289
TA9 *see* Airbus Industrie A330
TA11 *see* Airbus Industrie A340
Tapajos *see* EMBRAER EMB-123
Tawron 378
   TC-4C Academe 468
   TC-121C Lockheed Constellation 576
   TC-121G Lockheed Constellation 576
Terr-Mar Aviation Turbo Sea Thrush 110
Texas Gem 695
Thrush *see* Ayres Thrush
Tiger Moth *see* de Havilland D.H.82
'Tin Goose' *see* Ford Tri-Motor
TL-19D Cessna 290
Tp102 Gulfstream 471
Tp 100 Saab 340B 702
Trident *see* Hawker Siddeley Trident
Trislander *see* Britten-Norman BN-2A Mk III Trislander
Tristar *see* Lockheed Tristar
TSR.2 772
Tucano *see* EMBRAER Tucano
Tumanskii R-13 741
Tupolev
   Tu-4 741
   Tu-16 'Badger' 729, 730, 733, 736, 753
   Tu-20 733, 734, 736
   Tu-70 531
   Tu-82 730
   Tu-86 729, 730
   Tu-88 729–30, 733, 741
   Tu-95 'Bear' 60, 531, 729, 733, 735, 741
   Tu-104 'Camel' 61, 729–32, 733, 741
   Tu-104A 730, 731, 732
   Tu-104B 731, 732
   Tu-104D 732
   Tu-104E 732
   Tu-104V 732
   Tu-110 732
   Tu 114 'Rossiya' 517, 519, 531, 733–40
   Tu-114D 734–5, 740
   Tu-116 733, 734–5, 740
   Tu-124 732, 741, 753
   Tu-124A 742
   Tu-126 'Moss' 60, 528, 736, 740
   Tu-134 'Crusty' 741–4, 753, 754
   Tu-134A 741, 743–4
   Tu-134A-3 744
   Tu-134B 744
   Tu-134B-3 744
   Tu-142 734
   Tu-144 'Charger' 745–51
   Tu-144D 751
   Tu-154 745, 753–7
   Tu-154A 755

Tu-154B 754, 755, 757
Tu-154B-1 754
Tu-154B-2 754, 755, 756, 757
Tu-154C Freighter 755
Tu-154M 753, 755, 757
Tu-155 757
Tu-204 755, 757
Tu-334 744
Turbine Defender (BN-2 Islander) 504, 506
Turbo Commander 690B 701
Turbo Islander *see* Fairey-Britten-Norman Turbo Islander
Turbo Stationair *see* Cessna Model 207
Turbo Thrush *see* Ayres S2R-T
Turbolet *see* LET L-410
'Twin Beech' *see* Beech Model 18
Twin Otter *see* de Havilland DHC-6

U-1A Otter 354, 355
U-1B Otter 354, 355
U-2/F King Air 121
U-6 Beaver 347
U-17 Skywagon 290
UC-1 Otter 354, 355, 358
UC-12 Super King Air 121
UC-35A Learjet 562
UC-36A ECM Learjet 562
UC-78 Bobcat 289
UC-880 Convair 317
ULMs *see* microlights
ultralights *see* microlights
UV-10B Twin Otter 362
UV-18A Twin Otter 362, 365

Valetta *see* Vickers Valetta
Vanguard *see* Vickers Vanguard
Vardax Vazar Dash 3 357, 359
Varsity *see* Vickers Varsity T.1
VC-3A Martin 4-0-4 607, 608
VC-4A Gulfstream 468, 607
VC-9C Nightingale (DC-9) 610
VC-10 *see* Vickers VC-10
VC-89 Metalplane 477
VC-118 DC-6 414, 417
VC-118A DC-6 417
VC-118B DC-6 415, 417, 419
VC-121A Lockheed Constellation 576
VC-121B Lockheed Constellation 576
VC-121C Lockheed Constellation 576
VC-121E Lockheed Constellation 576
VC-121G Lockheed Constellation 576
VC-131A Convair VIP Transport 310
VC-131D Convair 311, 312–13
VC-137A Boeing 707 152, 153
VC-137B Boeing 707 152, 153
VC-137C Boeing 707 152, 154
Vega *see* Lockheed Vega
Ventura *see* Lockheed PV-1
Vickers
   see also BAC
   Merchantman 515, 764
   Nene Viking 767, 772
   Super VC10 778–9, 780, 781–2
   Valetta C.1 767–8
   Valetta C.2 767
   Valetta T.3 768
   Valiant 772
   Vanguard 20, 515, 541, 579, 758, 759–64, 772
   Vanguard Type 900 759
   Vanguard Type 951 760, 762–3
   Vanguard Type 952 760, 761, 763
   Vanguard Type 953 760, 761, 762